KV-370-741

RESERVE STOCK
AVAILABLE
ON REQUEST

COLLINS HF: Introducing the HF-9000 series. Slim, trim and only 23 lbs. Built for the new generation of narrower, lighter tactical aircraft, Collins' HF-9100 system weighs one-third less than similar military performance equipment, and is sized down to about 800 cu. inches. Yet it delivers 175 watts PEP over the 2.0 to 29.9999 MHz frequency range. ■ Pilots will find the Collins HF-9100 easy to operate with 20 programmable preset channels. All available at the flick of a single switch. ■ Two microprocessors make possible automatic operation plus Built-In Test (BITE). Also, the direct digital frequency synthesizer provides the frequency agility required in tactical situations. ■ The HF-9000's use of fiber optics not only saves weight, it eliminates the effects of electro-magnetic interference and can handle large amounts of data in a very short time. This is essential to cost-effective system growth, part of Collins' Pre-Planned Product Improvement Program. ■ For information on the Collins HF-9000 series, contact Collins Defense Communications, Cedar Rapids, Iowa 52498, U.S.A. (319) 395-2690, Telex 464-435. ■ **COLLINS HF says it all.**

Rockwell International

...where science gets down to business

HF FOR TOMORROW'S LEAN MACHINES.

07106 08306 0879 DC

CAMBRIDGESHIRE
LIBRARIES

**REFERENCE
AND
INFORMATION
SERVICE**

CAMBRIDGE
REFERENCE LIBRARY

LIBRARY

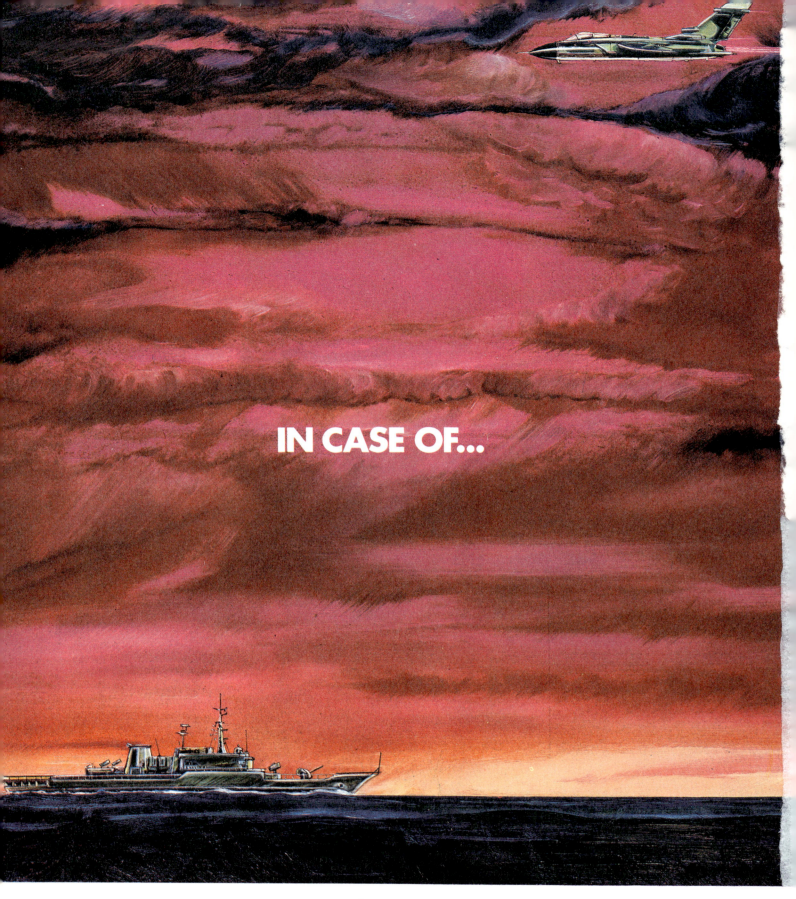

IN CASE OF...

If you need to identify...
Italtel's IFF system capability: fixed and
mobile surface interrogators, airborne
interrogators, transponders, antennas
and decoder groups developed and
manufactured in the L'Aquila facility.

IRI-STET GROUP

IF YOU NEED TO IDENTIFY.

ITALTEL DEFENSE TELECOMMUNICATION DIVISION - 00187 ROMA (ITALY) 66, VIA DUE MACELLI - PHONE (+39.6) 672121

JANE'S AVIONICS
1986-87

Jane's Publishing Company Limited, 238 City Road, London EC1V 2PU, England
Jane's Publishing Inc, 4th Floor, 115 5th Avenue, New York, NY 10003, USA

07106 08306
629.1355 REF

Alphabetical list of advertisers

WORKING FOR TOMORROW.

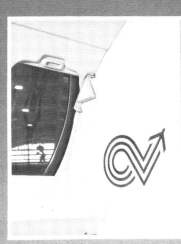

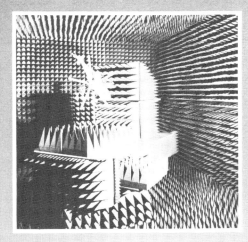

At Aeritalia – Avionic Systems and Equipment Group – we are working hard to develop advanced systems with new technologies. We want to improve aerospace products and make them suitable to various roles for both civil and military applications. Our work mostly means study, development, testing and at last manufacturing of avionic systems, airborne and ground-based electronic equipment, aerial reconnaisance and defence systems. It means to use new technologies making flight safer and more confortable, respecting the environment and man, and strenghtening national defence. It means research and innovation to be competitive in the international market. Aeritalia is working today for tomorrow.
Avionic Systems and Equipment Group - 10072 Caselle Turin (Italy)

AERITALIA IRI finmeccanica
società aerospaziale italiana

Classified list of advertisers

The companies advertising in this year's publication have informed us that they are involved in the fields of manufacture indicated below:-

Accelerometers
Crouzet Aérospatial
Ferranti
Rockwell International
SFIM
Smiths Industries

Accelerometers for Inert   Navigation
Crouzet Aérospatial
Ferranti
SFIM
Smiths Industries

Accident Recorders
SFIM

Actuators
Lucas Aerospace
Rockwell International
SECA
SFIM

Advanced Airborne HF Communications
Elbit Computers
Rockwell International

Aerials/Antennas
Elbit Computers
MEL
SECA
SMA

AFCS
Hughes Aircraft
SFIM

AIDS (Aircraft Integrated Data System)
SFIM

Airborne ASW Detection Equipment
Crouzet Aérospatial
Ferranti
MEL
SMA
Thomson-CSF

Airborne Communication Systems
Control Data
Elbit Computers
Electronique Aérospatiale
Hughes Aircraft
Rockwell International
TRT

Airborne Data Handling Systems
Control Data
Crouzet Aérospatial
Elbit Computers
Ferranti
Hughes Aircraft
MEL
SFIM
Thomson-CSF

Airborne Digital Computer Systems
Control Data
Crouzet Aérospatial
Elbit Computers
Ferranti
MEL
Smiths Industries
Thomson-CSF

Airborne Weapon Systems Simulation
Crouzet Aérospatial
Elbit Computers
Ferranti
Hughes Aircraft
MEL

Aircraft Instruments
Crouzet Aérospatial
Electronique Aérospatiale
Electro-Optics Industries (EL-OP)
Ferranti
SECA
SFIM
Smiths Industries
Thomson-CSF
TRT

Aircraft Radar
Ferranti
Hughes Aircraft
MEL
SECA
SMA
Thomson-CSF
TRT

Air Data Computers
Control Data
Crouzet Aérospatial
Ferranti
Hughes Aircraft
Rockwell International
Smiths Industries

Air Data Flight Instruments
Crouzet Aérospatial
Smiths Industries

Air Defence Data Handling and Display
Crouzet Aérospatial
Hughes Aircraft
MEL

Air Traffic Control Systems, Civilian and Military
Ferranti
Hughes Aircraft
TRT

Alignment Equipment
SMA

Altimeters
Crouzet Aérospatial
MEL
Rockwell International
SECA
Smiths Industries
Thomson-CSF
TRT

Amplifiers
Rockwell International
SECA

Analysers
Rockwell International

Antennas
Crouzet Aérospatial
Elbit Computers
Hughes Aircraft
Italtel
MEL
Rockwell International
SECA

Artificial Horizons
Elbit Computers
Ferranti
SECA
SFIM

ASW
Control Data
Crouzet Aérospatial
Hughes Aircraft
MEL
SFIM

ATC Airborne Equipment
Rockwell International
SECA
TRT

ATC Transponders
Electronique Aérospatiale
Hughes Aircraft
SECA
TRT

Attack Radar
Control Data
Ferranti
FIAR
Hughes Aircraft
MEL
SMA
Thomson-CSF

Attenuators, Avionic
Control Data
Rockwell International

Attitude/Heading Flight Instruments
Control Data
Elbit Computers
Electronique Aérospatiale
Ferranti
Litton Italia
SECA
SFIM
Smiths Industries
Thomson-CSF

Autopilot Approach Monitors
Smiths Industries

Autopilots
Lucas Aerospace
Rockwell International
SFIM
Smiths Industries

Autothrottle
Smiths Industries

Avionics
Control Data
Crouzet Aérospatial
Elbit Computers
Electronique Aérospatiale
Electro-Optics Industries (EL-OP)
Ferranti
Hughes Aircraft
Italtel
MEL
Rockwell International
SATT Communications
SECA
SFIM
Smiths Industries
Thomson-CSF
TRT

CAD/CAM Systems
Control Data
Elbit Computers
Ferranti
Italtel
Rockwell International

Calibration Equipment & Services
MEL
Rockwell International
SECA
SFIM

Cockpit/Flight Deck Displays
Control Data
Elbit Computers
Ferranti
Hughes Aircraft
Lucas Aerospace
MEL
SMA
Smiths Industries
Thomson-CSF

Cockpit Management Systems
Control Data
Elbit Computers
Electronique Aérospatiale
MEL
Rockwell International

Combined Map & Electronic Displays
Control Data
Elbit Computers
Ferranti
Hughes Aircraft
MEL
SMA
Smiths Industries
Thomson-CSF

Command Systems & Equipment
Crouzet Aérospatial
Hughes Aircraft
SMA

SUPER SEARCHER ASW Radar
– created by MEL as a complete airborne command and control centre

Already ordered by the Royal Australian Navy, SUPER SEARCHER is now the preferred fit for the Sikorsky S-70B Seahawk; the only British ASW radar to ever achieve such a distinction.

SUPER SEARCHER fitted to Westland Sea King Mark 42B helicopters has been ordered by a major Far Eastern country.

SUPER SEARCHER is an advanced ASW radar with the capability to counter a variety of threats in difficult maritime and ECM environments.

SUPER SEARCHER clearly presents a total operational scenario.

A single operator can integrate information from own and external sensors to locate, display and track numerous sub-sea, surface and airborne threats. From this compact command and control centre an operator can coordinate ASW forces, vector fighters and guide on-board active or semi-active missiles such as SEA SKUA, or provide over-the-horizon guidance for third party missiles, including EXOCET, HARPOON and PENGUIN.

Superb radar coverage and detection performance, role adaptability and stretch potential based on modular hardware and software provide a system that can be tailored in single or multiple screen configurations to match existing and future mission requirements.

For further information on SUPER SEARCHER, microwave landing systems, NAVAIDS etc...contact MEL...the creative force in Avionics.

A division of Philips Electronics and Associated Industries Limited. MEL, Manor Royal, Crawley, West Sussex, England RH10 2PZ. Telephone 0293 28787. Telex 87267.

CLASSIFIED LIST OF ADVERTISERS

Computers
Crouzet Aérospatial
Elbit Computers
Ferranti
Hughes Aircraft
Lucas Aerospace
MEL
Thomson-CSF

Consoles
Ferranti
Hughes Aircraft
MEL
SMA
Thomson-CSF

Contrast Enhancement Filters
Electro-Optics Industries (EL-OP)
Pilkington PE

Data & Information Processing Systems
Crouzet Aérospatial
Hughes Aircraft
SATT Communications
SMA
Thomson-CSF

Data Links
Elbit Computers
Ferranti
Hughes Aircraft
MEL
Thomson-CSF

Defensive Electronic Countermeasure (DECM) System
MEL
SATT Communications
Thomson-CSF

Designator Lasers
Elbit Computers
Ferranti
Hughes Aircraft
Thomson-CSF

Display, Alphanumerical Large Scale
Crouzet Aérospatial
Elbit Computers
Hughes Aircraft
MEL
SMA

Display, Cathode Ray Tube
Electro-Optics Industries (EL-OP)
Ferranti
Hughes Aircraft
MEL
SMA
Thomson-CSF

Electronic Displays
Control Data
Elbit Computers
Electro-Optics Industries (EL-OP)
Ferranti
Hughes Aircraft
Lucas Aerospace
MEL
Rockwell International
SFIM
SMA
Smiths Industries
Thomson-CSF

Electronic Flight Displays
Control Data
Crouzet Aérospatial
Elbit Computers
Electro-Optics Industries (EL-OP)
Ferranti
Hughes Aircraft
MEL
Rockwell International
Smiths Industries
Thomson-CSF

Electronic Warfare (COMINT)
Elbit Computers
Hughes Aircraft
MEL
Thomson-CSF

Electronic Warfare (ECM)
Elbit Computers
Electronique Aérospatiale
Hughes Aircraft
MEL
SATT Communications
Thomson-CSF
TRT

Electronic Warfare (ELINT)
Elbit Computers
Hughes Aircraft
MEL
SATT Communications
Thomson-CSF

Electronic Warfare (ESM)
Elbit Computers
Ferranti
Hughes Aircraft
MEL
SATT Communications
Thomson-CSF

Electronic Warfare (SIGINT)
Hughes Aircraft
MEL
Thomson-CSF

Encryption & Scrambling Equipment
Control Data
MEL
Rockwell International
TRT

Engine/Thrust Management—Health Monitoring
SATT Communications
Smiths Industries

Environmental Testing
Control Data
Hughes Aircraft
SMA

Flight Data Recorders
Control Data
SFIM

Flight Directors
Control Data
Rockwell International
SECA
SFIM

Flight Management Computer Systems
Smiths Industries

Flight Performance Management
Control Data
Rockwell International
Smiths Industries

Flight Simulation
Ferranti
Flytsim
Hughes Aircraft

Flight Simulation Components eg Motion System/Control Force
Flytsim

Flight Simulation for Helicopters & Fixed-wing Aircraft
Ferranti
Flytsim

Flight Testing Systems
Control Data
Rockwell International
SFIM

FLIR
Elbit Computers
Electro-Optics Industries (EL-OP)
Ferranti
FIAR
Hughes Aircraft
Pilkington PE
TRT

Fuel Quantity Gauging System
SECA
TRT

Guidance & Control Systems
Crouzet Aérospatial
Litton Italia
SMA
TRT

Gyroscopes
Ferranti
SECA
SFIM
Smiths Industries

Head Down Display Optics
Elbit Computers
Electro-Optics Industries (EL-OP)
Ferranti
MEL
Pilkington PE

Head Up Display Optics
Electro-Optics Industries (EL-OP)
Ferranti
Hughes Aircraft
Pilkington PE

Height Radar
Hughes Aircraft

Helicopter Radars
Ferranti
FIAR
Hughes Aircraft
MEL
SMA
Thomson-CSF
TRT

Helicopter Stabilisation Systems
Ferranti
SFIM
Smiths Industries

Helicopter Weapon Sights
Crouzet Aérospatial
Elbit Computers
Electro-Optics Industries (EL-OP)
Ferranti
Hughes Aircraft
Pilkington PE
SFIM
Thomson-CSF

Helmet Displays
Elbit Computers
Electro-Optics Industries (EL-OP)
Ferranti
Hughes Aircraft
Thomson-CSF

High Performance HF Radio
Hughes Aircraft
Rockwell International

Holographic Helmet Display
Electro-Optics Industries (EL-OP)

Horizon Gyro Units
SECA
SFIM

Horizontal Situation Indicator (HSI)
Electronique Aérospatiale
Ferranti
SECA
Thomson-CSF

HUD
Electro-Optics Industries (EL-OP)
Ferranti
FIAR
Hughes Aircraft
Smiths Industries
Thomson-CSF

Hybrid Microcircuits
Control Data
Crouzet Aérospatial
Ferranti
Hughes Aircraft
Lucas Aerospace

Inertial Navigation, Digital
Elbit Computers
Ferranti
Litton Italia
SFIM

SCIENCE/SCOPE ®

An innovative digital receiver is being developed to alert military aircraft when they are approaching enemy radars and electronic warfare systems, thereby putting them at less risk while on a mission. The device, designed for electronic support measures (ESM), will be approximately 1/20 the weight and substantially smaller than current receivers. It will search for, intercept, record, analyze, and locate sources of radiated electromagnetic energy. The receiver can store this information. Or, if an enemy signal poses a threat, it can pass this information along to another type of electronic warfare system, such as a jamming device. Hughes Aircraft Company is developing the receiver with independent research and development funds.

Some of the most efficient and productive results in high-technology design and manufacturing have been achieved during the development of complex antennas for Intelsat VI communications satellites. The antennas will concentrate transmissions to four heavily populated areas. Hughes pushed computer-aided design/computer-aided manufacturing (CAD/CAM) into new realms in meeting the tough engineering requirements. CAD proved four times faster than conventional methods of design. CAM boosted productivity, cutting equivalent manufacturing time from 2,400 hours to 600 hours.

Programmable software formats within a night vision system for helicopters allow new features to be added as needed to meet new threats. The Hughes Night Vision System (HNVS) is a low-cost, forward-looking infrared system that provides excellent imagery and object detection day or night in all weather. It has extensive built-in test and fault isolation test capabilities. Among the features that may be modified to meet specific requirements are flight symbology, navigational data, automatic set-up mode, system status data, and push-buttons around the display face.

NATO will upgrade its air defense network with eight long-range radars for four of its member nations. The new HR-3000 radars are a new generation derivative of the Hughes Air Defense Radar (HADR) operating in West Germany, Malaysia, and Norway. The radar is fully transportable and can be set up and torn down in hours. It also has better electronic counter-countermeasures, improved capability for rejecting clutter, and a faster rotating antenna to accommodate NATO's requirement for a higher data rate. The radars will be installed in Turkey, Greece, and Italy. They will be integrated into the Hughes-developed NATO Air Defense Ground Environment (NADGE) system. In addition, another radar will be installed in Portugal.

U.S. Marine Corps A-4M Skyhawk attack aircraft are as much as three times more effective when equipped with the Angle Rate Bombing Set (ARBS), according to pilots who participated in the first overseas operational deployment of such aircraft. ARBS is a computerized bombing system that can be used with bombs, gunfire, and rockets. Its dual-mode TV and laser tracker locks onto and tracks targets either identified by the pilot on a cockpit display or designated by a laser device. Marine Attack Squadron 311 pilots using ARBS in bombing exercises concentrated bomb hits 50% to 75% closer to targets than did pilots using previous manual techniques. These tests demonstrated how the Hughes-built ARBS can cut the number of missions over enemy territory and improve the A-4M's primary close air support mission of attacking targets located near friendly troops. The Marines are also employing ARBS on the AV-8B.

For more information write to: P.O. Box 45068, Los Angeles, CA 90045-0068 USA

© 1986 Hughes Aircraft Company

Subsidiary of GM Hughes Electronics

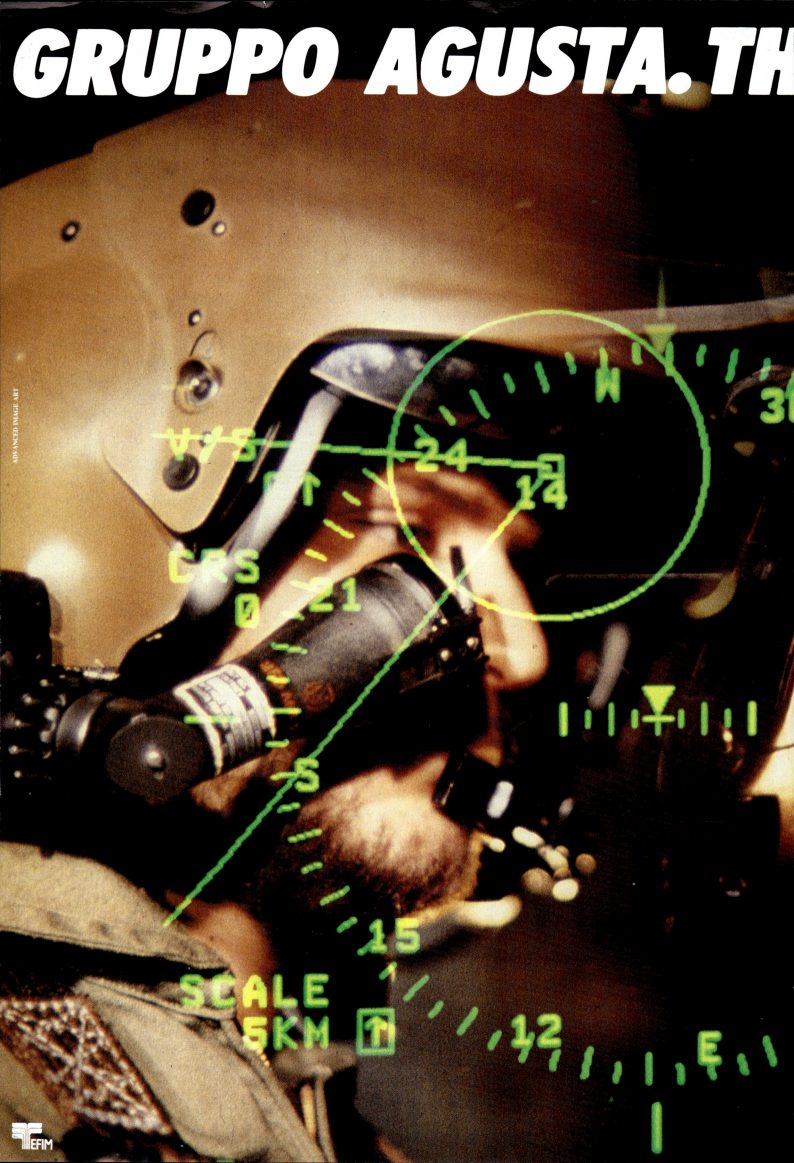

GRUPPO AGUSTA. TH

E MISSION GOES ON.

The Agusta Group, an Italian enterprise involved in major international projects, is at the forefront of the most advanced technological achievements in the field of helicopters, airplanes and avionic/aerospace systems. The continuous challenge of high technology requires the development of ever more complex and specialized systems to accomplish increasingly difficult tasks. To meet this challenge, the Agusta Group has organized its 22 companies into three divisions — helicopters, airplanes and systems — and invests over 20% of total turnover in research. The Agusta Group combines a forward-looking philosophy with enormous production potential and technical know-how. That's why every flight of the Agusta Group is a mission towards the future.

GRUPPO

AGUSTA

Research and Aerospace Technology

CLASSIFIED LIST OF ADVERTISERS

Infra Red Telescopes
Electro-Optics Industries (EL-OP)
Pilkington PE

Laser Ranging
Electro-Optics Industries (EL-OP)
Ferranti
Hughes Aircraft
Thomson-CSF

Lightweight Airborne HF Transceivers
Hughes Aircraft
MEL
Rockwell International
TRT

Low Altitude Airfield Attack Weapons Equipment
Ferranti
Hughes Aircraft

Low-Light TV
Electro-Optics Industries (EL-OP)
FIAR
Hughes Aircraft
Thomson-CSF

Marked-Target Seeker Lasers
Ferranti
Hughes Aircraft

Marker Beacons
Ferranti
Rockwell International
TRT

Meteorological Equipment Meters
Hughes Aircraft
SMA

Microwave Communication Equipment
Ferranti
Hughes Aircraft
MEL
TRT

Microwave Landing Systems
MEL
SMA
TRT

Military Standard 1553 Data-Bus Equipment
Control Data
Ferranti
MEL
SATT Communications
SFIM
Smiths Industries

Missile-Guidance—Thermal Imagery
Elbit Computers
Hughes Aircraft
TRT

Missile-Guidance TV
Hughes Aircraft

Mission Planning Equipment
Ferranti
MEL
SFIM

Mounts, Antenna
Hughes Aircraft
MEL
SMA

Moving Map Displays
Control Data
Ferranti
Hughes Aircraft
MEL
Pilkington PE
Smiths Industries
Thomson-CSF

Nav/Attack Systems
Control Data
Elbit Computers
Ferranti
MEL
SFIM
Smiths Industries
Thomson-CSF

Naval Combat Systems
Ferranti
Hughes Aircraft
MEL
SMA

Naval Data Processing
Ferranti
Hughes Aircraft
MEL
SMA

Navigation/Nav-Com (DME)
Crouzet Aérospatial
Electronique Aérospatiale
MEL
SECA
Thomson-CSF
TRT

Navigation/Nav-Com (GNAV)
FIAR

Navigation/Nav-Com (INS)
Ferranti
SFIM

Navigation/Nav-Com (Magnetic Compass)
Crouzet Aérospatial
SECA
SFIM

Navigation/Nav-Com (MLS)
FIAR
TRT

Navigation/Nav-Com (TACAN)
Electronique Aérospatiale
Rockwell International
SECA
TRT

Navigation/Nav-Com (VOR/ILS)
Crouzet Aérospatial
Electronique Aérospatiale
SECA
TRT

Panoramic Reconnaissance Camera
Electronique Aérospatiale
TRT

Pilots Night Vision Goggles
Electro-Optics Industries (EL-OP)
Ferranti
Hughes Aircraft
MEL
Pilkington PE
TRT

Platforms, Stable
Electro-Optics Industries (EL-OP)
Ferranti
SMA

Plot Extractors
MEL
SMA

Pre-planned Product Improvement (P3I)
Control Data
Hughes Aircraft

Radar, Airborne
Ferranti
Hughes Aircraft
MEL
SECA
SMA
Thomson-CSF
TRT

Radar, Ground
Hughes Aircraft
SMA

Radar, Ground-Ship-Based
Hughes Aircraft
SMA

Radar Modelling
Control Data
Hughes Aircraft
MEL

Radar Processing
Ferranti
Hughes Aircraft
MEL
SMA
Thomson-CSF

Radar Systems
Control Data
FIAR
Hughes Aircraft
Italtel
SMA
Thomson-CSF

Radio Communications (HF)
Hughes Aircraft
Rockwell International
SECA

Radio Communications (Integrated Management)
TRT

Radio Communication (UHF)
Electronique Aérospatiale
Italtel
SECA
TRT

Radio Communication (VHF)
Electronique Aérospatiale
Italtel
SECA
TRT

Radomes
Lucas Aerospace
SMA

Reconnaissance Equipment
Electro-Optics Industries (EL-OP)
Ferranti
SMA
Thomson-CSF

Reconnaissance Radar
Hughes Aircraft
SMA
Thomson-CSF
TRT

Satellite Navigation
Control Data
Crouzet Aérospatial
Hughes Aircraft
Rockwell International
Smiths Industries
TRT

Secondary Surveillance Radar
Hughes Aircraft
SMA

Secure Data Links (Microwave)
Ferranti
Hughes Aircraft
Italtel

Selcal
Rockwell International

Simulation Systems
Elbit Computers
Ferranti
Flytsim
Hughes Aircraft
Rockwell International

Stall Warning Systems
Ferranti
SECA

Stores Management
Control Data
Elbit Computers
Rockwell International

Surveillance Radar
Control Data
Ferranti
FIAR
Hughes Aircraft
SMA
TRT

MLS
Microwave Landing System

Absolute safety in any environmental condition

MLS is an accurate approach and guidance system designed to operate in the most severe weather conditions.
Compact, flexible and modular, it drastically reduces installation and maintenance costs and problems.

- It is adopted by ICAO and is available on the market.

- It can be rapidly installed and commissioned with out site excavation at any airport.

- Ideal for all types of aircraft including STOL, VSTOL and helicopters allowing for different approaching profiles.

- Indispensable at small airports and heliports sited in difficult terrain or bad landing conditions.

- MLS eliminates present limitations of ILS systems and may even be co-located with them.

Philips S.p.A. - Defence & Control Systems Division - P.zza Monte Grappa, 4 - 00195 Roma - Tel. 06/3302.1 - Telex 610042 Phirom I

 Defence & Control Systems

PHILIPS

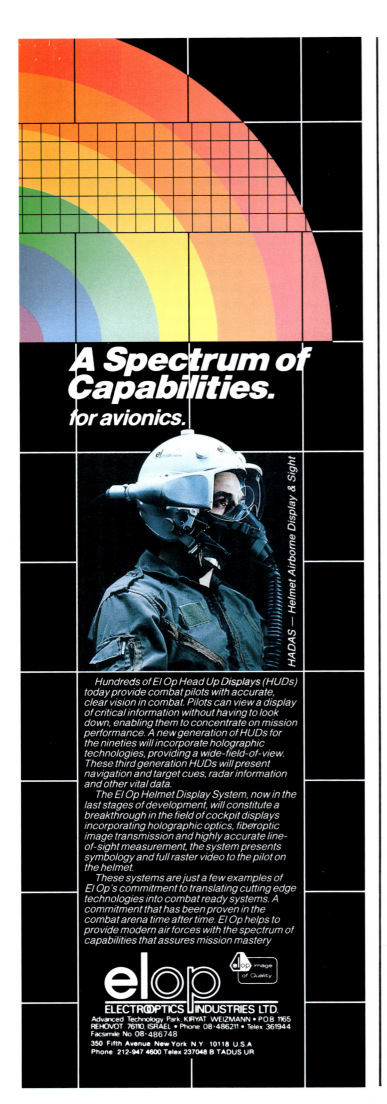

A Spectrum of Capabilities.
for avionics.

HADAS — Helmet Airborne Display & Sight

Hundreds of El Op Head Up Displays (HUDs) today provide combat pilots with accurate, clear vision in combat. Pilots can view a display of critical information without having to look down, enabling them to concentrate on mission performance. A new generation of HUDs for the nineties will incorporate holographic technologies, providing a wide-field-of-view. These third generation HUDs will present navigation and target cues, radar information and other vital data.

The El Op Helmet Display System, now in the last stages of development, will constitute a breakthrough in the field of cockpit displays incorporating holographic optics, fiberoptic image transmission and highly accurate line-of-sight measurement, the system presents symbology and full raster video to the pilot on the helmet.

These systems are just a few examples of El Op's commitment to translating cutting edge technologies into combat ready systems. A commitment that has been proven in the combat arena time after time. El Op helps to provide modern air forces with the spectrum of capabilities that assures mission mastery

elop
ELECTROOPTICS INDUSTRIES LTD.
Advanced Technology Park, KIRYAT WEIZMANN • P.O.B. 1165
REHOVOT 76110, ISRAEL • Phone 08·486211 • Telex 361944
Facsimile No 08·486748
350 Fifth Avenue New York N.Y. 10118 U.S.A.
Phone 212-947 4600 Telex 237048 B TADUS UR

elop Image of Quality

Tactical Command Systems
Control Data
Ferranti
Hughes Aircraft
SMA

Tactical Control Systems
Control Data
Crouzet Aérospatial
Hughes Aircraft

Tactical Guidance & Landing Systems
Control Data
Hughes Aircraft
TRT

Target Designation & Acquisition
Crouzet Aérospatial
Electro-Optics Industries (EL-OP)
Ferranti
Hughes Aircraft
SFIM
SMA
TRT

Telemetry
Electro-Optics Industries (EL-OP)
Hughes Aircraft
SFIM
TRT

Test Equipment
Ferranti

Thermal Imagery
Control Data
Elbit Computers
Electro-Optics Industries (EL-OP)
Hughes Aircraft
SFIM

Trajectography Systems
SFIM

UHF Airborne Transceivers
Electronique Aérospatiale
Rockwell International
SECA
TRT

UHF Line-of-Sight/Satellite Transceivers
Hughes Aircraft
Rockwell International

UHF/VHF Radio
MEL

Ultra Compact VHF/UHF Multimode Systems
Control Data
Rockwell International
TRT

VHF
Rockwell International
SECA
TRT

Video-Cameras
Ferranti

Video-Playback Facilities
Ferranti

Video Recording Systems
Ferranti

Video-Recorders
Ferranti

Visual Simulation Technology
Ferranti
Flytsim
General Electric

VLF
Rockwell International
SECA

Weapon Delivery & Navigation
Elbit Computers

JANE'S
MICROFORM
SERVICE

JANE'S MICROFICHE PUBLISHING PROGRAMME

This service, offered as a result of continued requests from industry, the Services and government, will ultimately provide a comprehensive collection of Jane's publications previously only available in bound book and magazine formats.

- ☐ JANE'S FIGHTING SHIPS
- ☐ JANE'S ALL THE WORLD'S AIRCRAFT
- ☐ JANE'S INFANTRY WEAPONS
- ☐ JANE'S WEAPON SYSTEMS
- ☐ JANE'S DEFENCE REVIEW
- ☐ JANE'S DEFENCE WEEKLY

Please send orders and requests for further information to:

JANE'S PUBLISHING CO. LTD.
238 City Road, LONDON EC1V 2PU. Tel: 01-251 9281.
or
Jane's Publishing Inc.,
115 Fifth Avenue,
New York, New York 10003, USA
Tel: 212-254 9097

The only one...

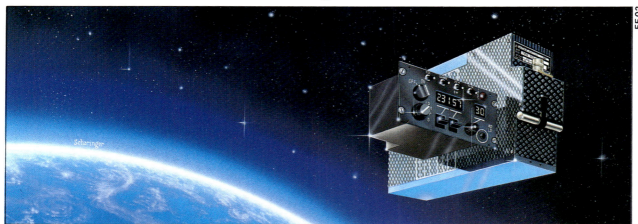

ECMRITS* is the only air force voice and data radio-communication system ensuring real protection against jamming and interception.
- The only French system proven in actual airborne operating conditions.
- The only one with truly modular design and upgrading capability.
That's why it has been selected by the most discerning operators!

Ensuring secure communications, that's special, that's Aerospatiale.

eas
groupe aerospatiale

B.P. 51 - 93350 Le Bourget - France
Tel. : (33/1) 48 62 51 95 - Telex : 220 809 F.

* (Electronic Counter-Measures Resistant Information Transmission System)

that's special. that's aerospatiale.

JANE'S DEFENCE WEEKLY

BE INFORMED...
No weekly Defence Magazine brings you the news faster

JANE'S DEFENCE WEEKLY is the first defence journal with both the authority and ability to provide an objective assessment of developments in world defence and the frequency of publication to keep you constantly informed of the latest changes in the defence scene worldwide. Building on the international recognition won by its predecessor, Jane's Defence Review, **JANE'S DEFENCE WEEKLY** is available **only** on subscription.

Full subscription details on application.

Write to:

Jane's Publishing Co. Ltd.,
238 City Road,
London EC1V 2PU.

JANE'S
where the facts are found

For a complete list of all

JANE'S YEARBOOKS

please write to:
The Information Department
Jane's Publishing Company Limited
238 City Road, London EC1V 2PU

At Seca, exceptional service is the rule.

Airplane and helicopter maintenance. All engine O/H and repairs.

Avionics: authorized service center
for all major American manufacturers.

Airframe: – maintenance, repairs, and paint jobs.
– A/C modification for special assignments.
– Customized interiors.
– Engine refits.

Engines: Pistons - Turboprops - Turbojets - Turboshafts.
Specialists: LYCOMING - CONTINENTAL -
ALLISON - GENERAL ELECTRIC - GARRETT -
PRATT & WHITNEY.

aerospatiale

seca
groupe aerospatiale

FB 01

Aéroport du Bourget - BP 32 - 93350 Le Bourget - France
Tel: (33/1) 48.35.99.77 - Telex: 220 218 F SECBG

25 YEARS EXPERIENCE

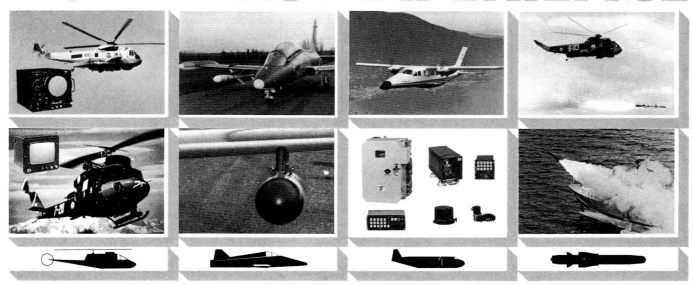

SMA is active in development, design and construction of radar systems for military ground, ship, applications since 1948. The size of the Company gives SMA flexibility and versatility. SMA since 1960 designs and manufactures a large series of **avionic radar products for searching, tracking, mapping, positioning and missile assignment in S.A.R., A.S.W., A.W.W. missions.** The **MM/APS-705** search radar is a large production equipment already in use in several navy's helicopters employing frequency diversity. Three antenna sizes and suitable units to interface weapon systems are available. The **APS-707** is a state of the art search radar designed for helos and

SMA
SEGNALAMENTO MARITTIMO ED AEREO

patrolling aircraft applications. Frequency agility, high power transmission, raster scan presentation, signal and data processing features, including track while scan, I.F.F. integration, different antenna sizes, reduced weight and power consumption are the main characteristics. The **SM-1** is a homing head employable in fire and forget control A.S.M. The **INTRA system** (MM/UPX-718 interrogator, MM/UPX-719 transponder) is an x band codified beacon suitable for long range, low coverage applications. System definition regarding a **nose multirole radar** and radiometer for combat aircraft has been already completed and the development phase is about to start.

P.O. BOX 200-FIRENZE(ITALIA)-TELEPHONE:055/27501-TELEX:SMARAD 570622-CABLE:SMA FIRENZE

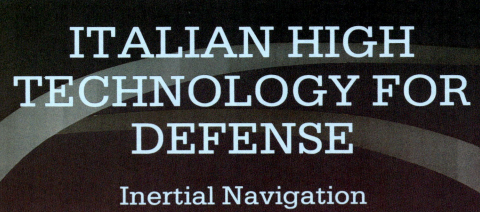

ITALIAN HIGH TECHNOLOGY FOR DEFENSE

Inertial Navigation
Guidance & Control
Attitude/Heading Reference

Command & Control

Litton
Italia

Via Pontina Km. 27,800 - 00040 Pomezia, Rome (Italy)
Telephone: 91.19.21 - Telex: 610391 LITAL I

Stile Regina Editrice srl

SHOWING THE WAY IN DEFENCE TECHNOLOGY

For active service.

For over forty years, Ferranti has been a supplier of advanced systems and components to the armed forces of the world.

By harnessing the resources of over 22,000 people employed in more than 75 locations worldwide, Ferranti is able to develop and deliver effective defence technology. From advanced multi-function avionics for aircraft to sonar and computer-based systems for submarines and surface ships.

Ferranti is at the forefront in the development of C^3I, taking you into the next generation of battlefield technology.

Ferranti has a track record of success, an enviable reputation for solving complex problems and for delivering workable combat-proven solutions – both real-time and simulated.

If you are seeking defence technology of the highest quality with a competitive edge, look no further than Ferranti for active service.

Ferranti plc Bridge House Gatley Cheadle Cheshire SK8 4HZ

FERRANTI

DTC/C/07/046/B1

JANE'S ALL THE WORLD'S AIRCRAFT 1986-87

ALL THE WORLD'S LATEST AIRCRAFT LAND HERE FIRST

This is where they all come in. Jane's All the World's Aircraft describes the latest developments in civil and military aircraft, airships and balloons, microlights, sailplanes, sports and racing aircraft of all kinds.

Now in its 77th edition, Jane's All the World's Aircraft has climbed to its pre-eminent position as the leading annual aircraft review because of its comprehensive scale, constant updating, and a meticulous attention to accuracy.

With sections on official records, first flight details, aero engines and air-launched missiles, it is the only truly all-embracing work in its field.

Not least in this essential manufacturer-by-manufacturer datasource of 1000 pages and 2000 illustrations is the highly-regarded annual review of the commercial, technical and political worlds of the air.

Those in the aviation business have come to rely on Jane's constantly. If you need to land the latest aircraft information, from anywhere in the world, your approach is clear.

First flights • Official records • Aircraft • Sport aircraft • Hang gliders • Sailplanes • Lighter-than-air • RPVs and targets • Air-launched missiles • Aero engines

Jane's Publishing Co. Ltd.,
238 City Road, London EC1V 2PU
Tel: 01-251 9281 Tlx: 894689

Jane's Publishing Inc.,
115 5th Avenue, 4th Floor, New York, NY10003
Tel: (212) 254 9097 Tlx: 272562 VNRC UR

Reach for
JANE'S

ALL THE WORLD'S AIRCRAFT 1986-87

[18]

Computer-generated image of McDonnell Douglas F-15 STOL Aircraft by COMPU-SCENE IV. Courtesy of McDonnell Douglas.

DETAILS COUNT

COMPU-SCENE® IV
Visual Simulation Technology from GE

In combat, where survival depends on how fast you react to where you are and what you see—details count. That's why scene detail is a critical measure of effectiveness in combat mission training. COMPU-SCENE IV* systems deliver detail. They deliver detail where it counts and when it counts.

Scene detail is traditionally expressed by the number of separate faces (or polygons) viewable at any one time. But with the advent of photographic cell texture, face count is no longer an adequate measure—it is just another number.

COMPU-SCENE IV visual systems incorporate photographic cell texture, a process that allows real-world photographs to be included in visual data bases without using additional faces. Not only does this technology breakthrough

achieve extraordinary realism, it also frees up faces to be used where they count most.

But authentic visual simulation is more than just a pretty face. COMPU-SCENE IV data bases are built with automated modeling tools that can instantaneously transform DMA source data into large area, real-world environments. And to support complex NOE mission training scenarios, COMPU-SCENE IV incorporates dynamic occulting features which allow multiple moving models in rolling terrain and pinpoint accuracy of collision detection throughout the environment.

Since its introduction in 1984, COMPU-SCENE IV has been an unprecedented success with more than a dozen systems already ordered and many more forecast in the near term.

International Marketing
General Electric Company (U.S.A.)
Simulation and
Control Systems Department
P.O. Box 2500
Daytona Beach, Florida (U.S.A.) 32015
Tel: (904) 258-2331

U.S.A.

*COMPU-SCENE and ⓖⓔ are registered trademarks of General Electric Company

JANE'S AVIONICS

FIFTH EDITION

EDITED BY
STEPHEN R BROADBENT MRAeS

1986-87

ISBN 0 7106 08306
JANE'S YEARBOOKS
"Jane's" is a registered trade mark

Copyright © 1986 by Jane's Publishing Company Limited, 238 City Road, London EC1V 2PU, England

All rights reserved. No part of this publication may be reproduced, stored in retrieval systems or transmitted,
in any form or by any means, electronic, mechanical, photocopying, recording or otherwise,
without the prior written permission of the publishers.

In the USA and its dependencies by
Jane's Publishing Inc, 4th Floor, 115 5th Avenue, New York, NY 10003, USA

MEL
the Creative Force
in Avionics

Customised Avionic Systems, shaped from modular hardware and software, are created by MEL to match operational requirements.

In airborne ASW radars like SUPER SEARCHER, new, state-of-the-art microelectronic components, sophisticated digital signal processing and the very latest colour display technology are cost-effectively tailored to meet both existing and future needs.

Role adaptability, stretch potential and sensor integration capability are fundamental aspects of MEL's radar systems. SUPER SEARCHER can integrate information from own and external sensors to track numerous targets, vector fighters and guide on-board active or semi-active missiles, or third party missiles.

MEL has the professional people, the technology and in-depth resources of the Philips Group to create reliable new Avionics Systems...quickly...and to give them full support throughout their life.

Excellent design and quality manufacturing combine with product training, technical publications, post design services and refurbishment to ensure low through-life costs and up-to-date performance in all MEL Avionics Systems.

For new horizons in airborne radars and microwave landing systems, contact MEL...
the creative force in Avionics

A division of Philips Electronics and Associated Industries Limited.

MEL, Manor Royal, Crawley, West Sussex, England RH10 2PZ.

Telephone 0293 28787. Telex 87267.

Contents

BATTLE BRED

Elbit's Weapon Delivery and Navigation Systems are ready for action.
Developed in interaction with and flown by the Israel Air Force,
Elbit's WDNS, the choice for retrofit programs.
System 81. An IMU-based, digital computerized system featuring smart
weapons handling, head-up displays (HUD), accurate navigation and
advanced A/G and A/A modes.

System 82. Offering the same state-of-the-art features. Enhanced with a
Stores Management System.
Low cost WDNS. Enhanced A/G and full A/A modes. Digital computation
combines with head-up display (HUD), increasing aircraft operational
capability. A moderately priced system developed as the intermediate stage
between expensive, full performance WDNS and very low cost, low
performance gunsights. Particularly compatible with light attack aircraft,
advanced trainers and interceptors.
Ally your forces with experience. With systems bred in battle. Forged in
action. Elbit Weapon Delivery and Nagivation Systems, on the front line.

 **BATTLE BRED**

ELBIT COMPUTERS LTD • P.O.Box 5390 • Haifa 31053 • Israel • Tel: (04)556677 • Telex: 46774

FOREWORD

Spurred by the perceived need to counter an ever more sophisticated threat, the technology of military avionics is pushing forward at an unceasing pace. The desire for better detection of and retaliation against 'The Threat' means that avionics is at the very forefront of electronic technology, and is the driving force behind each succeeding advance in the state of the art.

In the commercial world of the airlines, the requirement to present the crew with better information about their flight, and to automate yet further their task, so that commercial flying is made safer and more economic is also driving the art of the black box into new areas of capability.

In both areas, civil and military, the rate of change is fast, so fast, indeed, that in key areas much has changed since the last edition of *Jane's Avionics* was published. And yet, paradoxically, the gestation period for a new aircraft gets ever longer — the European Fighter Aircraft, which was given the go-ahead this year and which is the focus of attention of most of Europe's leading avionics companies, is (together with the broadly-similar French Rafale) the only such aircraft starting life on the drawing boards of Europe's aerospace industry and it can be expected that it will be at least the year 2000 before the boards are full with its replacement.

The same goes, to a certain extent, for the American Advanced Tactical Fighter, and even when the other 1990s projects such as the Saab JAS 39, IAI Lavi, US Advanced Tactical Bomber and designs such as Osprey and LHX are accounted for, there are precious few targets at which the avionics industry can aim its new designs.

And the same also applies in the commercial world — the 'next generation' of airliners, the Airbus A330 and 340, the MD-11 and the Boeing 7J7 are set for service in between five and ten year's time, and their cockpits will be a step ahead, in technological terms from their predecessors such as the Airbus A320 and Boeing 767. These cockpits are being designed today, but the following generation of

transports will not be at the same stage until well after the year 2000. True, the evolution is a continuous one, since the competing airliners are not being developed in parallel, older types are being updated with new avionics and yet other designs will emerge, but the technological jump from 'clockwork' to 'glass' cockpit is taking place now. Who knows what will come next?

The paradox — slow rate of airframe development, fast avionics, is solved by considering the stakes. The company that finds its technology on, say, the EFA, has secured its own well-being, at least in that area, into the next generation. The company that loses will, by and large, lose the impetus and will find staying in that market area very difficult. There are very few ball games in town, to use an Americanism, and companies are making every endeavour to get their admission tickets. The drive is being fuelled by governments, anxious that their latest flying technological piece of wizardry is better, first of all than the equipment deployed by the 'opposition' and also better than the rival design from a neighbouring country, ostensibly an ally.

A key trend has unfolded during the past few years. When I started in the British aerospace industry, over 20 years ago, the word 'avionics' had not even been coined, and electronics was in its infancy. Airframe companies made airframes, honed to the last per cent of performance by boffins with slide rules, reams of graphs and much knowledge in the ways of the air. Those companies would never have dreamt of poaching on the traditional skills of the engine manufacturer, the forger of undercarriages or the designers of ejection seats, all of whom sent 'reps' to the airframe shrine, to be told when and where their product was to be fitted. Design the aerodynamics, grudgingly give structures a look in, let engines and other ancilliaries fit into the space available, and, down the corridor, out of sight, the radar company rep sat, like a hungry dog at his master's table, waiting for the last pickings. No wonder radios

It has been suggested that when the Fokker 100 enters service it will be equipped with the world's most advanced flight deck. Extensive use of crts and computers for monitoring of all phases of flight and aircraft systems will emphasise the emerging role of the captain as a flight manager. These extensive facilities are provided to ensure that any system or instrument malfunction shall be presented to the crew only if it requires any action on their part. Consequently they are able to concentrate, undisturbed, on the correct priorities. The number of aural warning signals currently in use on airliner flight decks will be effectively halved.

SMITHS INDUSTRIES INSTRUMENTS AND SYSTEMS –THE LOGICAL CHOICE

Head Up Displays and Weapon Aiming Systems.

Civil Electronic Flight Information Systems.

Military Multi-Purpose Colour Displays.

Data Transmission Systems.

Flight Management Systems.

Flight Deck Displays.

Smiths Industries has been providing instruments and systems for aviation since 1910. Today the company supplies over 450 airlines and air forces flying 150 different types of aircraft operating in 150 countries.

SMITHS INDUSTRIES
AEROSPACE & DEFENCE
SYSTEMS COMPANY
Childs Hill, 765 Finchley Road, London NW11 8DS
Tel: 01-458 3232 Telex: 928761

Flight Control Systems · Weapon Systems · Micro Circuit Products · Flight Deck Displays
Head Up and Head Down Electronic Displays · Flight Management Systems · Engine and Fuel Management Systems.

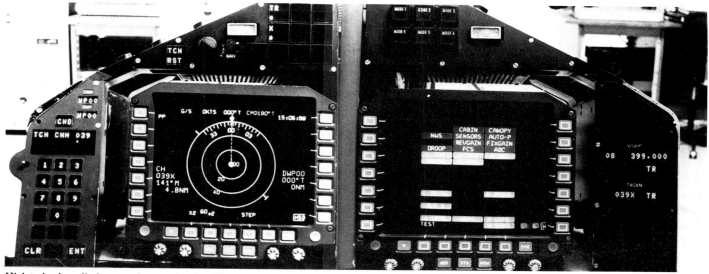

High technology displays are also entering the military cockpit – Smiths Industries has developed these two multi-function cathode ray tube displays for the British Experimental Aircraft Programme (EAP) demonstrator

failed first, radars were of doubtful usefulness and other sensors just did not sense − the seat of the pants pilot and the aerodynamicist ruled!!

Even in combat aircraft in service today, the number of avionics systems which are hung on, as an after thought, are many. Only now is the idea that the radar, thermal imager and laser tracker are the keys to mission success gaining positive acceptance, and in the next generation of combat aircraft, the avionics companies' desires for prime location, adequate supply of power and cooling and proper account of cockpit real estate for display is taking precedance.

Avionics now rules the design office and the technology has developed from poor cousin to prime mover in the space of 20 years. Aircraft noses are profiled to give proper radar antenna size for the detection job demanded by the user, not to suit the aesthetic desires of the Chief Aerodynamist: pods and bulges sprout everywhere to house the sensors where once artistic merit would have denied inclusion.

The trend is evidenced most strongly by the airframe companies themselves. Twenty years ago, as I said, they would never have dreamed of entering engine design, for example, but, in this edition of *Janes' Avionics,* almost every airframe manufacturer is represented, either by having a design leadership on a major avionics project, or by having a dedicated avionics division. Those companies have well realised the vital part avionics technology is to play in their future business fortunes.

The trend will continue. Twenty years ago the electrical system was last in the queue when space was allocated: ten years ago it got a fair shout; today it gets prime position and the argument is not which aircraft system gets first shout, but which avionics system − radar, thermal imager or whatever.

But avionics still comes in 'black boxes', generally rectangular, with wires between and space around where blended aerodynamic shape and cubic box do not quite meet. In the next generation, avionics will fill the space 'wall to wall', not in boxes, but constrained by the aircraft structure, with one volume of electronics, covering a multitude of functions. There will not be a 'Company X' thermal imager in one box, linked by cables to 'Company Y's' radar processor and 'Company Z's' threat analyser: it will all be one avionics 'set', with expertise from several sources, but all co-ordinated by one body − which is why the airframe companies have expanded into this area so quickly.

Rationalisation is a natural outcome of this process. One of the headaches of editing *Janes Avionics* is to keep track of who owns who. Bloggs Avionics (in America: John Doe Electronics Inc) suddenly becomes 'A Division of Mega-Electronics' and this trend, too, will gather pace. Europe, in particular, has far too many avionics companies chasing far too few projects, and once Tornado production tails off and the battle over participation in EFA is won and lost, there will be a severe pruning.

Britain alone has at least five companies with airborne radar capability, four with combat nav/attack system expertise (the most aggressive of which is, in parallel with the above related trends, British Aerospace itself). Yet there is no 'all British' military aircraft project in design, so even when the very few European combat aircraft projects have been carved up and the British taken a sub-system here, a design leadership there, it is improbable that there is the business to keep even one company at the forefront of, say, airborne radar technology, into the 21st century.

There is the same problem in the USA, but it is not as severe − there are far more projects, with far bigger production runs, and while US technology can find a home in European projects, and the export potential of US aircraft, thanks to the scale effects of size and of 'Foreign Military Sales', makes for even longer production runs for US radars and processors. With very few and highly noteworthy exceptions, the success of Europe in America (and, it must be said, of France in Britain and Britain in France, for the same nationalistic reasons) is rare indeed.

The rationalisation also affects this book: this edition of *Janes Avionics* could well have more individual company names in it than any succeeding issue and by the 1990 edition the number of names of companies with leading edge of technology capability will be severely reduced. The trend will also be evidenced by it becoming increasingly difficult to place each product in a specific section, a task already hard enough. There will not be a 'Type X Avionic System' listed in a relevant section, with a note saying it is being designed for such and such aircraft, since the Type X will have become embedded in the electronics of an all-embracing avionic system around which the aircraft is designed.

The technologies
What, then, are the key technologies for the 1990s?

The main driving force in avionics is not even a part of this book. Software, the invisible wizardry, rules the performance of every

Artificial Intelligence made the transition from laboratory to flight-test in 1986 with the McDonnell Douglas AIMES (Avionics Integrated Maintenance System) system installed on an F/A-18. Artificial Intelligence will become a vital part of an aircraft's avionics suite within the next few years.

ELECTRONIC WARFARE

SATT Communications AB, formerly the Communications Division of SATT Electronics AB, is one of the oldest companies in Sweden supplying defence electronics.

We specialize in custom designed ESM and ECM systems. Our range of products includes:

○ Radar Warning Systems
○ Surveillance Receiver Systems
○ Radar Jamming Systems
○ EW Training Systems
○ Jet Engine Monitoring Systems

SATT Communications AB

Box 32701, S-126 11 STOCKHOLM. SWEDEN,
Phone Int. +46 8 81 01 00. TELEX 15325 SATCOM S

SATT

avionic system which includes an element of computing, and software expertise is a key factor in deciding the place a company might have in the market place. It is very difficult to think of a book 'Janes Airborne Software' but if there were one it would point out the future.

A key element of software, however, is mentioned within these pages, and the proportion of space devoted to the topic will rise dramatically in the next few editions. Artificial Intelligence is firmly on the road from academia to battlefield. The amount of data which an airborne platform can now sense and hold within it, or relay to the ground, is so many times what can be sensibly and efficiently dealt with by even the brightest crew. Artificial Intelligence aims to bring into the computer the combined heuristic knowledge of teams of experts in the specific task, so that the avionics can sense and process all the data and deduce a best course of action, presenting the pilot with a single decision or simple options in order for him to effect his mission fully. The first military Artificial Intelligence systems flew in 1986; by 1996 they will be everywhere.

The second key technology is the ubiquitous microchip. Avionics is the driving force behind the rapidly growing capability of the 'chip' as more and more data has to be processed in a small volume, reliably, with no heat dissipation and while being subjected to a hostile nuclear environment. The USA's very high-speed integrated circuit programme (vhsic) is the engine which keeps the USA's military avionics industry dominant in the world, and what the military does today becomes available for the commercial world tomorrow.

The military fighter scene

At the time of writing two demonstrators, anticipating two European fighters are about to fly for the first time (though eventually there may be just the one version). Two others, in Sweden and Israel, will follow shortly, and there is considerable activity along similar lines under wraps in the USA. It is therefore appropriate to review the avionics of one, the British EAP, as being representative of the trends which will take the fighter pilot into the next century.

The British Experimental Aircraft Programme (EAP) demonstrator is different things to different people: it is certainly the most politically important British aerospace project since Concord or TSR-2, even though it is not a prototype for any future combat aircraft.

Why then, has British industry invested over £50 million of its own money in the 'one off' demonstrator? — because it is just that, a demonstrator for British military avionics. Without the EAP, certain areas of the British avionics industry would have had difficulty in making the ever-widening jump from Tornado technology of 10 to 15 years ago to the European Fighter Aircraft (EFA) project, whose avionics are about to be defined.

Without EAP, the avionics companies would not have had the technology to fight for a large share of the EFA systems, and back in 1982 it was largely the avionics companies, led by GEC Avionics, Ferranti and Smiths Industries (but including many others), which fought hard to have the project launched. EAP brings together for the first time in a flying aircraft many new technologies vital to EFA's proposed capabilities.

The most physically evident of these new avionics are the displays. The three multi-function head-down displays, developed by Smiths Industries revolutionise the presentation of flight management information (navigation, weapon selection, attitude reference and systems data) by eliminating dial and pointer gauges in favour of versatile colour cathode ray tube displays having both cursive (for presentation of instruments) and raster (for sensor data display) capability. Even the standby instrument panel (also by Smiths, with GEC Avionics providing the standby engine instrument panel) uses light-emitting diode (led) technology rather than 'analogue' dials.

Over recent years, Smiths has also developed a new concept in avionics, the integrated utilities management system, the technology of which is being demonstrated in the EAP. This system brings into four linked line-replaceable units the management of power, hydraulics, engines, fuel, environment, oxygen etc, which were previously managed individually by many separate units. The result is, as with many advanced avionic systems, a reduction in cost, weight and system complexity and an increase in reliability and overall performance.

GEC Avionics has kept at the forefront of head-up display technology by achieving notable sales successes in the US military market, and has thus been able to bring a 'ready made' wide field of view holographic head-up display to the EAP which represents both the most advanced technology available and a system which has already been exported. Among many other systems being supplied by this company are the radios (the radio management system is from Racal), Tacan and air data computer, as well as the maintenance panel which presents the ground crew with an led display of faults on the avionics data-bus.

However, perhaps the most critical part of the EAP avionics is the flight control system, also developed by GEC Avionics from the system which has been so successfully tested in the Jaguar FBW aircraft. The EAP is Britain's first naturally unstable aircraft and the fly-by-wire system thus has no mechanical back-up. The Jaguar FBW programme gave confidence that this technology was practical, technology which has been taken one step further with EAP, on the way, GEC Avionics would hope, to the EFA system.

It was a significant decision to equip the EAP with a Ferranti FIN 1070 inertial navigation system, similar to the units being supplied for the Jaguar and Buccaneer refit programmes, rather than selecting the technologically more advanced ring-laser gyro based navigation systems. Both Ferranti and British Aerospace are developing RLGs, development work at the latter company being seemingly more advanced: EFA will almost certainly have an RLG navigation system and the battle between the two British companies for that order, if the navigation system is indeed British, will be intense.

Tying all the avionics systems together (of which the above is only a selected review, British Aerospace lists almost 20 contributing avionics companies, including those in Germany and Italy) is the MIL-STD-1553B data-bus and again the EAP is the first new British military aircraft to incorporate this technology. Much relevant research work has been done on the '1553 bus' rig at the British Aerospace facility at Brough, with the executive and control functions in the EAP being provided by Smiths. Without the proper functioning of the data-bus the whole avionics suite fails, and the integrity of the system is almost as critical as that of the flight control system.

The EAP has provided the stimulus for advances in a wide range of British avionics, the flight-test programme will prove the actual capabilities and the battle will then be on to turn them into firm order once the EFA programme is underway. Without EAP the UK avionics industry would have greatly declined and would have been significantly weaker when facing the sales battles ahead.

Postscript — problems in the commercial world

As already noted, the glass cockpit revolution is now with the airlines, whether they like it or not. The pilot is not a pilot, but an avionics systems manager, watching an array of TV screens for signs of a discrepancy from the normal indications — indeed, one school of thought says the pilot should have nothing displayed unless it is a deviation from the norm, if all is well, and engine, systems and flight parameters are within limits and the aircraft on track, the cockpit remains 'dark'. At that stage the argument will start as to whether two crew is really 100 per cent too many.

Yet in some respects the airlines still remain obstinately embedded in the era when the captain was king, a stout and sturdy hero who not only flew the machine with beads of sweat and white knuckles, but also landed when appropriate to mend the machine or forage for accommodation or fuel. The airlines certainly have to come to terms with a different kind of avionics — those brought aboard by the passengers.

Most people, except those rather selfish folk who use them recklessly, detest the squeaks and thumps which leak from the personal stereo tape players which are seemingly on every train and bus. They can annoy when used improperly, but there are a host of electronic items which are both harmless to the fellow passenger and which might be considered quite essential to the user — computer games, to keep the children amused, or personal computers so that the businessman can work while travelling. Such devices, however, are an anathema to cabin staff, who generally continue to demand the stowage of the offending items, 'as the captain says they interfere with the aircraft's navigation instruments'.

A more amazing nonsense in a high technology environment it is hard to imagine and it is to be hoped that the survey conducted recently by Airbus Industrie, which indicated there to be no such interference, will be taken to heart by the operators.

HOLOGRAPHIC HEAD UP DISPLAY OPTICS
CONTRAST ENHANCEMENT FILTERS
PILOT'S NIGHT VISION GOGGLES
HEAD DOWN DISPLAY OPTICS
INFRA RED TELESCOPES

Pilkington PE is a world leader in the design and manufacture of Head Up display optics – over 4,000 are in service with the free world's airforces. Investment in Holographic optics enables Pilkington PE to supply optical elements for the GEC Avionics Lantirn HUD and retrofit programmes for many other aircraft, providing dramatic visual improvements for the pilot. Pilkington PE leads in Head Down displays, too. Pilots of all Tornado and F/A 18 Hornet squadrons maintain navigational accuracy at low altitude and high speed with our moving map projection optics. Pilkington PE is breaking barriers in thermal imaging technology with a range of compact zoom telescopes which are proving ideal for FLIR and surveillance applications.

Our latest developments, including contrast enhancement filters and pilot's night vision goggles, demonstrate our strength in avionic optics and illustrate our commitment to remain at the leading edge of electro-optics.

PILKINGTON
◄ **Electro-optical Systems** ►
Pilkington PE Limited
Glascoed Road St Asaph Clwyd LL17 0LL UK
Telephone St Asaph (0745) 583301 Telex 61291
Fax 0745 584258

Leading edge avionic optics

. . .and the military

If self-expression by passengers is the airlines' blind spot, the equivalent on the military is security. As a Press Officer, I once gave a journalist two completely unclassified brochures on the same avionics subject. He published the words from both brochures in the same article and promptly the sirens went off when someone decided that the result was classified as secret. Visits were made, everyone in sight interviewed, brochures shredded, all because that person's view of the world was different from the person, presumably his predecessor, who had passed the brochures for publication in the first place.

Any intelligence agent worth his salt can deduce, for example, the frequency of an airborne radar. In the UK there are companies with libraries of tapes giving the signatures of Warsaw Pact radars, such data coming, it is obvious, from intelligence sources (how else could electronic countermeasures systems be designed?) and doubtless

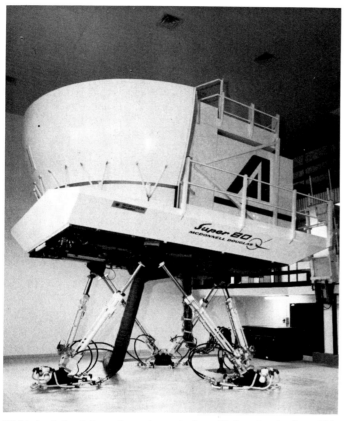

Flight simulation is becoming an increasingly sophisticated technology. This Rediffusion MD-80 simulator for Alitalia features a full motion system and the latest in computed generated visuals.

there are similar libraries in Warsaw Pact offices. And yet, I, as a journalist, or you, as an 'uncleared' reader, are not allowed even to see the front face, or a picture of an airborne radar antenna, in case you can deduce from even a casual glance the frequency characteristics of the radar. Just look at the number of antennas depicted in this book, all pointing coyly away from the camera, or covered with a modesty board.

Of course, it's not this book that matters, it's the principle that people can be so myopic, and thrive on it — this myopia affects this book in another way: too many times the phrase 'in production and in service with foreign air forces' appears. Which foreign air forces I cannot say, not even for some relatively innocuous item like a uhf radio. Why? Because the foreign purchaser has said he wants no one to know what is in his aircraft. Everyone knows what aircraft he has, and its standard equipment and capability, but the list of systems must be kept secret for fear of giving offence. At the very least the manufacturers are denied an important piece of publicity about a sales success, but it is, I fear, an attitude we shall have to endure.

Proper security yes, but hiding behind the apparent mystique of advanced avionics to create a false impression of grandeur, be it in civil airliner or military hardware, is pure waste.

Acknowledgements

To paraphrase a well known quotation, 'Some propose Yearbooks, some achieve Yearbooks, and some have Yearbooks thrust upon them'. For your new editor, the last category applies, and accomplishing the herculean task in the very short time available has only been possible thanks to the tremendous help offered by one or two individuals.

Appreciation is expressed to the many people around the industry who have contributed to these pages, but certain individuals' efforts are worthy of specific note. To Bernard Blake, who once again compiled the Radio Communications section, and to Don Parry, who took on the basically civil sections in the middle of the book at very short notice, go my grateful thanks: without their friendly efforts, large parts of the book would look very much like last year's edition. I am indeed thankful for their efforts under pressure, and their continuing helpful advice.

The production staff at Jane's London Office have coped with the vague whims of this tyro with amazing good humour. The bulk of the burden fell on Veronica Knapman, who has been totally thorough and meticulous. Sadly, Veronica has chosen to move on to better pastures, the final dotting of I's and crossing of T's passing to Anne Horgan and Christine Moss. Grateful thanks to all at City Road who have helped this book on its way.

Finally, without the total support and highly-valued friendship of ex-Jane's production girl Viv Harper, who gave me inspiration through some very dark days, there simply would not have been a book at all. This book is dedicated to Viv.

Stephen R Broadbent

INSTANT REPLAY

Reconnaissance data must be available in near-to-real time to be of value to command decision makers. Conventional photographic data can be several hours old—too stale to be of real value.

Control Data helps overcome this problem with their new Reconnaissance Management System (RMS). RMS provides instant onboard replay of sensor data while continuing to record additional imagery. In real time. Day or night. And those important instant replays can be transmitted by digital data links for immediate interpretation and processing at a ground station.

RMS is one way Control Data uses proven technol-

ogies to help you solve problems. For more information and a free, full-color "Instant Replay" poster, call 612/853-5000. Or write Government Systems Resource Center, P.O. Box 0, Minneapolis, MN 55440. In the United Kingdom call Computing Devices Co. Ltd., (0424) 53481.

PRIME MOVER IN DEFENSE SYSTEMS

⊖⊖ CONTROL DATA

Career opportunities available.

[32]

INTRODUCTION

About the contents of this book

Avionics, unlike ships or aircraft, for example, do not fit neatly into categories; selecting into which section some products should fall is a very finely-balanced decision. In editing this edition of *Jane's Avionics* I have tried to reduce sharply the number of sections, in the hope that readers may find their required texts more easily, and I have also re-defined the contents of some sections.

At the same time, avionics is a very rapidly changing and expanding technology. Companies enter and companies leave the market-place, and there are frequent changes of name as mergers and take-overs occur. This is primarily a reference book of avionics in development, production and service, so items may well appear long after they have ceased to be marketed or developed by a particular company. Nevertheless, some entries have been deleted, and historic data from others removed to make room for up to date information – the deleted entries can be identified in the index, which now gives the last year of appearance for deleted topics.

In modifying the running order of the book, I have followed the path of data entering an aircraft, breaking the subject into two main sections: Sensors and Data Processing, Management and Displays.

Sensors

This section includes anti-submarine warfare sensors (ie sonar), radar and electro-optics. This last includes 'non visual' sights such as thermal imaging and reconnaissance systems and also lasers. The radar section includes all forms of active radar and secondary radar transponders, but Doppler radar used for navigation is in the navigation sub-section.

Passive sensors, in other words ECM, ECCM and ESM commonly grouped as 'electronic warefare' form the fourth part of this section.

Data processing, management and displays

Where to put computers is the biggest problem in sectionalising avionics, and in particular computers with specific functions, eg navigation, radar signal processing, etc, will be found in the appropriate section. For what can only be admitted to be editorial convenience, air data computers are to be found with their related air data instruments (see below). In this second main section, therefore will be found general-purpose computers, specifically those designed to MIL-STD-1750, their related languages and data links, including data highways (MIL-STD-1553B etc).

An adjacent sub-section describes all forms of data recording, including crash recorders, display recorders and engine health monitors.

All forms of navigation system are together as the second part of this section. Next comes flight, performance and engine management systems and flight control systems, again all grouped together: high technology flight control (fly-by-wire, -light and -voice) are separate, since the technologies involved are distinct and fast-developing.

Having sensed, processed and managed the data, the next stage in the chain is to display it, and this is the next part of the book. Head-down displays (ie symbology viewed other than through the cockpit canopy or windscreen) and including warning displays, are in one grouping, but head-up displays are separated, partly because of the largely military applications, and partly because this topic is closely related to electro-optics, and separation groups a specific technology. With head-up displays are helmet-mounted sighting and display systems and roof-mounted sights; the dividing line between this and electro-optics is that the latter is concerned with invisible light.

Stores management, as the final link in the chain in a military aircraft, comes next.

Radio communications

This is a section all to itself, but the previous entries for headset, microphones and in-flight entertainment have been deleted from this years book.

Training systems

Simulation provides a real headache, since, by definition, it is not avionics, remaining firmly anchored to the ground. Nevertheless, it is a very rapidly expanding market, involving many advanced technologies, and well within Jane's area of business. The section appears last in the book, has been considerably expanded, and re-titled more accurately to reflect the market.

Finally, comes an unresolved problem. The trend in avionics is to integrated systems and complete aircraft up-dates. Several of these are described in this volume, their position in the book being largely related to the main entry of the major contractor, but it is hoped that this will be the subject of a major expansion in the scope of the 1987-88 edition of *Jane's Avionics*.

TOTAL CAPABILITY IN AVIONICS

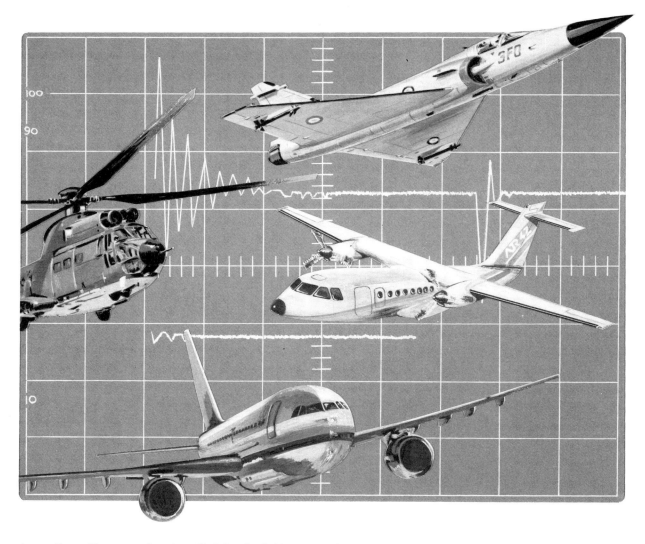

Leading Europe in the field of civil and military radio altimeters, the TRT Group also ranks high in aerial reconnaissance, thermal imagery, airborne radars, radio communications and radio navigation.

Over 90 international airlines, almost every aircraft builder in the world, and the armed forces of many countries, specify and use the products and systems designed and manufactured by the TRT Group.

TELECOMMUNICATIONS RADIOELECTRIQUES ET TELEPHONIQUES

88, rue Brillat-Savarin -
75640 PARIS CEDEX 13 - FRANCE
Phone : 33 (1) 45 81 11 12 - Telex : 250838 F

Glossary

ac	Alternating current	DFDAU	Digital flight data acquisition unit	HSI	Horizontal situation indicator
ACARS	ARINC communications addressing and reporting system	DG	Directional gyro	HUD	Head-up display
		Dicass	Directional command active sonobouy system	IAS	Indicated airspeed
a-d	Analogue-to-digital (conversion)			ic	Integrated circuit
ADAU	Auxiliary data acquisition unit	Difar	Directional acoustic frequency analysis and recording	ICAO	International Civil Aviation Organisation
ADC	Air-data computer				
ADF	Automatic direction finder	DITS	Digital information transfer system	Icap	Increased capability
ADI	Attitude director indicator	DME	Distance measuring equipment	icu	Indicator control panel
Advcap	Advanced capability	DMET	Distance measuring equipment with respect to time	icw	Intermittent continuous wave
AEW	Airborne early warning			if	Intermediate frequency
AFB	Air Force Base (USA)	DMS	Domestic military sales	IFF	Identification friend or foe
afc	Automatic flight control	dmu	Data management unit	IFR	Instrument Flight Rules
afcs	Automatic flight control system	DoD	Department of Defense (USA)	ILS	Instrument landing system
afds	Autopilot/flight-director system	Dolram	Detection of laser, radar and millimetric waves	IMC	Instrument meteorological conditions
AFSAT	Air Force satellite				
AFTI	Advanced fighter technology integration	DPSK	Differentially coherent pulse shift keying	IN	Inertial navigation
				INS	Inertial navigation system
AGC	Automatic gain control	DRLMS	Digital radar landmass simulation	INU	Inertial navigation unit
AHRS	Attitude and heading reference system	drts	Detecting, ranging and tracking system	I/O	Input/output
				IR	Infra-red
AI	Airborne interception	dsb	Double sideband	IRCM	Infra-red countermeasures
AIDS	Aircraft integrated data system	DSU	Data storage unit	irls	Infra-red laser system
ALCM	Air-launched cruise missile	DTM	Data transfer module	IRS	Inertial reference system
AM	Amplitude modulated	DTS	Data transfer system	ITU	International Telecommunications Union
AMRAAM	Advanced medium-range air-to-air missile				
		EADI	Electronic attitude director indicator		
amti	Airborne moving target indicator	earom	Electronically alterable read-only memory	Jato	Jet-assisted take-off
APR	Automatic power reserve			JTIDS	Joint Tactical Information Distribution System
APU	Auxiliary power unit	ECAM	Electronic centralised aircraft monitor		
ARBS	Angle rate bombing system				
ARINC	Aeronautic Radio Incorporated	ECCM	Electronic counter-countermeasures	Kips	Kilos (thousands) of instructions per second
ARM	Anti-radiation missile	ECM	Electronic countermeasures		
ASD	Aeronautical Systems Division (US Air Force)	eeprom	electrically erasable programmable read-only memory	Kops	Kilos (thousands) of operations per second
ASI	Airspeed indicator	EFIS	Electronic flight instrument system	K words	Kilo-words (1024 words of computer storage)
ASPJ	Airborne self-protection jammer	egt	Exhaust-gas temperature		
ASV	Air-to-surface vessel	ehf	Extremely high frequency	Lamps	Light airborne multi-purpose system
ASW	Anti-submarine warfare	EHSI	Electronic horizontal situation indicator	Lantirn	Low altitude navigation targeting infra-red night
ATC	Air traffic control				
Ategg	Advanced technology experimental gas generator	EICAS	Engine indicator and crew-alerting system		
				Lapads	Lightweight acoustic processing and display system
Atlis	Automatic tracking and laser illumination system	elf	Extremely low frequency		
		Elint	Electronic intelligence	lcd	Liquid-crystal display
atu	Antenna tuning unit	EMC	Electromagnetic compatibility	led	Light-emitting diode
auw	All-up weight	EMI	Electromagnetic interference	lf	Low frequency
AWACS	Airborne warning and control system	epr	Engine pressure ratio	LLTV	Low-light television
		eprom	Electrically programmable read-only memory	LOC	Localiser
BFDAS	Basic flight data acquisition system			Lofar	Low frequency omni-directional acoustic frequency analysis and recording
bisjet	Business-jet aircraft	ESM	Electronic support measures		
BIT	Built-in test	ETA	Estimated time of arrival		
bit	Single pulse in digital system	EW	Electronic warfare	LQA	Link quality analysis
BITE	Built-in test equipment	Excap	Expanded capability	LRMTS	Laser ranger marked target seeker
BPSK	Bi-phase shift keyed			lru	Line-replaceable unit
BTH	Beyond the horizon	FAA	Federal Aviation Administration (USA)	lsb	Lower sideband
byte	8 bits of digital data			lsi	Large-scale integration
		FAC	Forward air control	lst/scam	Laser spot tracker/strike camera
CAA	Civil Aviation Authority (UK)	Fadec	Full authority digital engine control	ltds	Laser target designator system
CAMS	Command active multi-beam sonobuoy	fbw	Fly-by-wire	LVDT	Linear variable differential transformer
		fcs	Flight control system		
ccd	Charge-coupled device	FDAU	Flight data acquisition unit		
ccip	Continuously computed impact point	FDEP	Flight data entry panel	Mach	Mach number (1 equals local speed of sound)
CDI	Compass director indicator	FDS	Flight director system		
cdu	Control and display unit	FET	Field effect transistor	MAD	Magnetic anomaly detector
cep	Circular error probability	Flir	Forward-looking infra-red	MCU	Management control unit
CGI	Computer-generated image	FM	Frequency modulated	MDGT	Mission data ground terminal
cmos	Complementary metal-oxide semiconductor	fmc	Flight management computer	MDT	Mission data terminal
		fmcs	Flight management computer system	MIL-STD	Military Standard
Comint	Communications intelligence	fmicw	Frequency modulated intermittent continuous wave	Mips	Millions of instructions per second
Comsat	Communications satellite			MLS	Microwave landing system
Comsec	Office of Communications Security (USA)	FMS	Flight management system Foreign military sales	M_{MO}	Maximum permitted operating Mach number
CNI	Communications, navigation and identification	fov	Field of view	Mops	Mega (million) operations per second
		fsk	Frequency shift keying	mos	Metal-oxide semiconductor
cpt	Cockpit procedures trainer			mosfet	Metal-oxide semiconductor/field-effect transistor
cpu	Central processing unit	GA	General aviation		
crt	Cathode ray tube	GaAs	Gallium arsenide	MPA	Monolithic phased array
css	Cockpit system simulator	gci	Ground control interception	MR	Maritime reconnaissance
cvr	Cockpit voice recorder	gmti	Ground moving target indicator	MRAAM	Medium-range air-to-air missile
cw	Continuous wave	GPS	Global positioning system	mrad	Milliradian
		GPWS	Ground proximity warning system	MRASM	Medium range air-to-surface missile
DAIS	Digital avionics information system	GRE	Ground readout equipment	MRTU	Multiple remote terminal
dB	Decibel	GS	Glideslope	msi	Medium-scale integration
dc	Direct current			mtbf	Mean time between failures
DF	Direction finder	hf	High frequency		

HIGHWAY TO TECHNOLOGY

SPACE
DEFENSE
ROBOTICS & INDUSTRIAL AUTOMATION
LOGISTICS SUPPORT

Headquarters: Via Montefeltro, 8 - 20156 Milan - Italy
Tel.: (02) 35790.1 - Telex: 331140 FIARMO I
Telefax: INFOTEC 6002 Tel.: (02) 342030

FIAR LEADS THE WAY

mtbo	Mean time between overhauls
mtbr	Mean time between removals
MTI	Moving-target indicator
mttr	Mean time to repair
N_1, N_2	Fan, compressor rotational speeds
NADGE	NATO Air Defence Ground Environment
NASA	National Aeronautics and Space Administration (USA)
NATO	North Atlantic Treaty Organisation
NBAA	National Business Aircraft Association (USA)
ndb	Non-directional beacon
NiCd	Nickel cadmium
nmos	N-type metal-oxide semiconductor
novram	Non-volatile ram
NVG	Night-vision goggles
OEM	Original equipment manufacturer
ON	Omega navigation
OTH	Over the horizon
pas	Performance advisory system
pcb	Printed circuit board
pcm	Pulse-coded modulation
pdcs	Performance data computer system
pds	Passive detection system
pep	Peak envelope power
PLSS	Precision location strike system
pmc	Performance management computer
pmcs	Performance and management computer system
pmos	P-type metal-oxide-semiconductor
PNVS	Pilot night vision system
ppi	Plan-position indicator
pps	Pulses per second Pilot's performance system
prf	Pulse repetition frequency
pri	Pulse repetition interval
prom	Programmable read-only memory
PVS	Pilot's vision system
QPSK	Offset quadraphase shift keyed
R&D	Research and development
RAE	Royal Aircraft Establishment (UK)
RAF	Royal Air Force (UK)
ram	Random access memory
rcs	Radar cross-section
rf	Radio frequency
RFP	Request for proposals
RLG	Ring-laser gyro
RMI	Radio magnetic indicator

RNav	Area navigation
rom	Read-only memory
RPV	Remote-piloted vehicle
RSRE	Royal Signals and Radar Establishment (UK)
RT	Radio telephone
RTCA	Radio Technical Commission for Aeronautics
rvr	Runway visual range
rwr	Radar warning receiver
Sadang	Système Acoustique d'Atlantique Nouvelle Generation
SAR	Search and rescue Synthetic aperture radar
selcal	Selective calling
SFDR	Standard flight data recorder
shf	Super high frequency
sid	Standard instrument departure
Sigint	Signal intelligence
Sincgars-V	Single channel ground and airborne radio system-vhf
SIT	Silicon intensified target
SLAR	Sideways-looking airborne radar
sms	Stores management system
SOJ	Stand-off jamming
sos	Silicon-on-sapphire
SRAAM	short-range air-to-air missile
SRS	Survival radio set
sru	Shop-replaceable unit
ssb	Single sideband
ssi	Small-scale integration
ssr	Secondary surveillance radar
STANAG	Standard NATO Agreement
star	Standard terminal-arrival route
STS	Support and test system
swr	Standing wave ratio
TACAMO	Take charge and move out
Tacan	Tactical air navigation
TADS	Tactical airborne designator system
TAS	True airspeed
TDMA	Time-division multiple access
TDS	Tactical data store
TEMS	Turbine engine monitoring system
Tercom	Terrain contour-matching
Terec	Tactical electronic reconnaissance
TICM	Thermal imaging common module
Tiseo	Target identification system, electro-optical
TRAM	Target recognition and attack multi-sensor
TSO	Technical Service Order
TTL	Transistor-transistor logic

TTM	Tape transport magazine
TVOR	Terminal vhf omni-directional radio
'twin'	Twin-engined light aircraft
TWS	Track-while-scan
twt	Travelling-wave tube
tx/rx	Transmitter/receiver
uhf	Ultra-high frequency
ULA	Uncommitted logic array
Ulaids	Universal locator airborne integrated data system
UMA	Unmanned aircraft
URR	Ultra-reliable radar
USAF	United States Air Force
usb	Upper sideband
USMC	US Marine Corps
USN	US Navy
UTM	Universal transverse Mercator
uverom	Ultra-violet erasable read-only memory
VASI	Visual approach slope indicator
VCR	Video cassette recorder
vdu	Visual display unit
VFR	Visual Flight Rules
VG	Vertical gyro
vhf	Very high frequency
vhpic	Very high-power integrated circuit
vhsic	Very high-speed integrated circuit
vlad	Vertical line-array Difar
vlf	Very low frequency
vlsi	Very large-scale integration
V_{MO}	Maximum permitted operating speed
VNav	Vertical navigation
VOGAD	Voice-operated gain adjustment device
VOR	Very high frequency omni-directional radio
VOR/LOC	Very high frequency omni-directional radio and ILS localiser
Vortac	Combined VOR and Tacan
V_{REF}	Reference speed (eg for approach)
VSI	Vertical speed indicator
V/Stol	Vertical/short take-off and landing
vswr	Voltage standing-wave ratio
VTM	Voltage-tuned magnetron
Vtol	Vertical take-off and landing
WBS	Weight and balance system
WRA	Weapon replaceable assembly
YAG	Yttrium aluminium garnet
YIG	Yttrium indium garnet

FLYTSIM LTD.
UNIT 9
VALLEY CENTRE, SLATER ST.
HIGH WYCOMBE, BUCKS.
HP13 6EQ (0494) 459545
Telex: 849462 Telefac G

FOR FIXED WING AND
HELICOPTER SIMULATORS
Agents for IVEX Visual Systems

SIMTECH LTD.
UNIT 14
SOUTHERN RD.
AYLESBURY, BUCKS.
Tel: (0296) 87255
Telex: 837520 ADTRAV G

For simulated instruments and
surface vehicle simulation

FLYTSIM TRAINING LTD.
CAA APPROVED FOR
I/R RENEWALS ETC.
(0494) 459545

sfim

- **AIDS** and **flight test systems.**
- **Automatic Flight Control Systems.**
- **Navigation systems.**
- **Guidance and/or stabilisation systems.**
- **Observation, Target Acquisition and Detection System.**

for aircraft, missiles, helicopters, ships, land vehicles...

VENDRE 3026

Société de **F**abrication d'**I**nstruments de **M**esure
F 91344 MASSY CEDEX
Tél.: 33-(1) 69 20 88 90 - Télex: SFIM 692.164 F

JANE'S AVIATION YEARBOOKS

Regarded by service and industry leaders throughout the world as the most definitive and authoritative sources of reference available, Jane's Yearbooks put the facts that matter at your fingertips.

Written and updated by professionals for professionals, each specialist yearbook details the latest available data in depth and comprehensively evaluates the most recent and significant developments country by country worldwide.

Jane's Yearbooks.

Subscribed to and used by political, military and civilian organisations throughout the world—they are where the facts are found.

For details of priority despatch, standing order schemes, contact the Marketing Department:

Jane's Publishing Co. Ltd.,
238 City Road, London EC1V 2PU
Tel: 01-251 9281 Tlx: 894689

ORDER NOW

Jane's Publishing Inc.,
115 5th Avenue, 4th Floor, New York, NY 10003.
Tel: (212) 254 9097 Tlx: 272562 VNRC UR

Reach for
JANE'S
AVIATION YEARBOOKS

Electromagnetic spectrum

Many of avionics sensors described in this book use electromagnetic radiation, and are referred to by category names. These do not always form self-evident classifications, and to provide some guidance to the total range of categories in everyday use a chart of the electromagnetic spectrum is given below.

For example, hf (high frequency) radio is a historical name; in fact it now occupies almost the lowest frequency band of the electromagnetic spectrum that is commonly used by avionic equipment. The chart emphasises that lasers and visible-light sensors also use electromagnetic radiation and are thus related to radio and radar equipment.

Electromagnetic radiation travels at the speed of light and the relationship between the frequency and wavelength of any electromagnetic radiation is given by $c = fl$, where f is the frequency (hertz), l is the wavelength (metres), and c is the speed of light ($2 \cdot 99793 \times 10^8$ metres/s).

Frequency is quoted in preference to wavelength in most applications, and is usually expressed in Hertz by international agreement. Thus:-

1 Hz	1 hertz	1 cycle/s
1 kHz	1 kilohertz	1000 cycle/s
1 MHz	1 megahertz	1000 000 cycle/s
1 GHz	1 gigahertz	1000 million cycle/s

US and NATO frequency bands

Many military systems operate between approximately 1 and 100 GHz, and it has long been a recognised practice to refer to waveband classifications in this range. Not a few companies still use the official US military band classifications, but there is a trend towards international adoption of a set of NATO standard categories. A chart showing the relative spread of each of these categories is shown below.

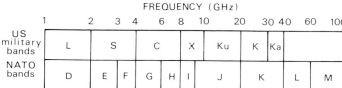

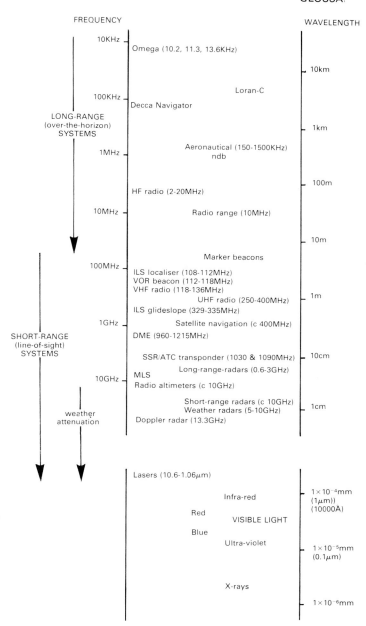

Standard case sizes

To achieve the most efficient use of the always limited avionics bay space available, the US organisation Aeronautical Radio Inc (ARINC) has devised a set of standard avionics box sizes. These are used internationally by commercial equipment manufacturers and are now found increasingly in military specifications too. A table of ARINC ATR (Air Transport Radio) case sizes is presented below. Both ATR Short and ATR Long descriptions are provided. ATR Dwarf is used occasionally and is based on a 85·8 mm (3·38 inches) height dimension. There is also an ATR Tall specification, based on a 269·88 mm (10·625 inches) height dimension, but this is rarely, if ever, used.

A recently introduced standard, which was adopted to bring a more readily usable classification for the smaller equipment sizes, is that of MCU (modular concept unit) case sizes. These are related to ATR cases, such that ⅛ ATR Short = 1 MCU. Hence a 1 ATR Short box is the same as an 8 MCU box.

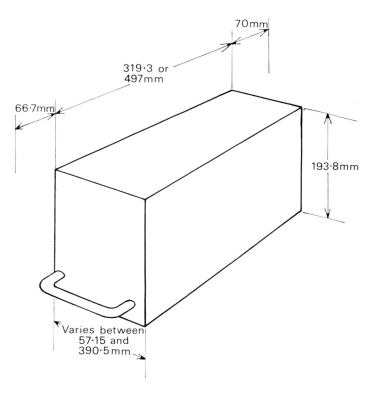

ATR size	Volume	Width	Length
¼ Short	215 in³/0·0035 m³	2·25 in/57·15 mm	12·6 in/319·3 mm
¼ Long	335 in³/0·0055 m³	2·25 in/57·15 mm	19·6 in/497 mm
⅜ Short	340 in³/0·0056 m³	3·56 in/90·43 mm	12·6 in/319·3 mm
⅜ Long	530 in³/0·0086 m³	3·56 in/90·43 mm	19·6 in/49 mm
½ Short	470 in³/0·0076 m³	4·88 in/123·95 mm	12·6 in/319·3 mm
½ Long	725 in³/0·0119 m³	4·88 in/123·95 mm	19·6 in/497 mm
¾ Short	720 in³/0·0118 m³	7·5 in/190·5 mm	12·6 in/319·3 mm
¾ Long	1120 in³/0·0184 m³	7·5 in/190·5 mm	19·6 in/497 mm
1 Short	970 in³/0·0159 m³	10·13 in/256·8 mm	12·6 in/319·3 mm
1 Long	1510 in³/0·248 m³	10·13 in/256·8 mm	19·6 in/497 mm

AN Numbered Communications Equipment

The Joint Electronics Designation System (JETDS)

```
                                      AN/   G  R  C  -  26 A   (X, Y or Z)   (V)
COMPLETE SET
INDICATES JETDS SYSTEM _____|  |  |     |  |      |         |
INSTALLATION _____|  |     |  |      |         |
TYPE OF EQUIPMENT _____|     |  |      |         |
PURPOSE _____|  |      |         |
MODEL NUMBER _____|      |         |
MODIFICATION LETTER _____|         |
CHANGES IN VOLTAGE. PHASE OR FREQUENCY _____|
VARIABLE GROUPINGS _____|
```

SAMPLE OF COMPONENT
USED WITH PARTICULAR SET   C-808/GRC-26A

SAMPLE OF COMPONENT
NOT USED WITH PARTICULAR SET   S-69/GRC

Set or equipment indicator letters

Installation
A Airborne
B Underwater
C Air transportable
D Pilotless carrier
F Fixed
G Ground, general
K Amphibious
M Ground, mobile
P Pack, portable
S Water surface craft
T Ground, transportable
U General utility
V Ground, vehicular
W Water, surface and underwater

Type of equipment
A Invisible light, heat radiation
B Pigeon
C Carrier

D Radiac
E Nupac
F Photographic
G Telegraph or teletype
I Interphone and pa
J Electromechanical
K Telemetering
L Countermeasures
M Meteorological
N Sound in air
P Radar
Q Sonar
R Radio
S Special types
T Telephone (wire)
V Visual
W Armament
X Facsimile or television
Y Data processing

Purpose
A Auxiliary assemblies
B Bombing
C Communications
D Direction finding
E Ejection release
G Fire control
H Recording
L Searchlight control
M Maintenance and test assemblies
N Navigational aids
P Reproducing
Q Special or combination of purposes
R Receiving
S Detecting range bearing
T Transmitting
W Control
X Identification and recognition

Type of service and operation explained

Type of service

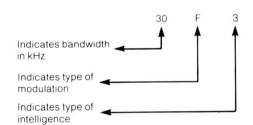

```
         30    F    3
```

Indicates bandwidth in kHz
Indicates type of modulation
Indicates type of intelligence

Type of service indicators

MODULATION
A Amplitude
F Frequency or phase
P Pulse

Type of operation
a. Duplex (radio), full duplex or FDX (cable): Simultaneous operation in opposite directions. Transmission and reception over 2 frequencies.
b. One-half duplex, half duplex or HDX: System arranged to permit operation in either direction but not simultaneously, with break-in capability.
c. One-way reversible: Operation in 1 direction at a time without break-in capability. Utilises 1 radio frequency.

INTELLIGENCE (PARTIAL LISTING)
0   None
1   Telegraphy (cw, fsk, psk)
2   Modulated cw (mcw)
3   Telephone (voice)
3a Ssb, reduced carrier
3b 2 isb, reduced carrier
3h Ssb, full carrier
3j Ssb, suppressed carrier
4   Facsimile
5   Television
9   Composite, or not otherwise covered

Sensors

Royal Air Force/British Aerospace Nimrod MR2 maritime patrol aircraft are equipped with most advanced sensor suite including systems of British, Canadian and American manufacture

Anti-submarine warfare

CANADA

CAE Electronics

CAE Electronics Limited, PO Box 1800, Saint-Laurent, Quebec H4L 4X4
TELEPHONE: (514) 341 6780
TELEX: 05-824856

AN/ASA-64 magnetic anomaly detector

CAE has completed a product improvement programme in support of the AN/ASA-64 submarine anomaly detector originally built for the US Navy/Lockheed P-3C Orion maritime patrol aircraft. The improved version has increased processing power. A variant for helicopters is proposed.

The AN/ASA-64 identifies and marks local distortions in the earth's magnetic field induced by the presence of submarines. The operator is alerted by visual and aural alarms, thereby reducing the level of experience needed to operate the system. As the system does not require constant monitoring the operator can devote more time to other sensors.

Dimensions: (control unit) 4 × 5.75 × 3.5 inches (102 × 146 × 90 mm)
(ID-1559 processor) 9 × 5.9 × 6 inches (229 × 150 × 153 mm)
Weight: (control unit) 1.5 lb (0.68 kg)
(ID-1559 processor) 6.5 lb (3 kg)
Power: 20 W at 115 V
Specification: MIL-E-5400

STATUS: in service but no longer in production.

CAE AN/ASA-64 magnetic anomaly detector

AN/ASA-65(V) nine-term compensator

CAE developed this semi-automatic magnetic anomaly detector (MAD) compensator to improve the effectiveness of MAD on aircraft with only manual compensation for aircraft interference on the earth's magnetic field. Previously, fixed permalloy strips and copper coils were mounted in the MAD boom to create induced and eddy-current fields equal and opposite to those caused by the aircraft. These compensators had to be custom designed for each individual aircraft, took a long time to adjust on flight-test and did not cater for the changes which take place during the aircraft's life. CAE also says that new, more sensitive MAD equipment needs greater precision than fixed compensators can provide.

The AN/ASA-65(V) compensates for permanent interference after only 5 minutes' flying, compared with about an hour needed for manual compensators. The system allows for manoeuvre interferences after 30 to 45 minutes, improving MAD detection range, especially when frequent manoeuvres are performed and conditions are turbulent.

CAE AN/ASA-65(V) compensator group adapter

This compares favourably to manual compensation procedures which traditionally took 90 minutes.

The all-solid-state AN/ASA-65(V) is compatible with all current MAD systems. Internal patch connectors are used to adjust the system for the aircraft concerned. The nine-term compensator is used by the Royal Air Force/British Aerospace Nimrods and the US Navy/Lockheed P-3C Orion and S-3A Viking ASW aircraft.

Figure of merit: less than 1 gamma
Max compensation field: 50 gamma on each side of aircraft
Mtbf: >1800 h
Dimensions: (control indicator) 9 × 5 × 6.5 inches (229 × 146 × 165 mm)
(electronic control amplifier) 7.75 × 5.875 × 13.625 inches (197 × 149 × 346 mm)
(magnetometer assembly) 6 inches (152 mm) cube
(coil assembly) 3.5 inches (89 mm) cube
Total weight: 29.5 lb (13.4 kg)
Power: 100 W at 115 V plus 10 W at 28 V dc or ac for panel lamps

STATUS: in service but no longer in production.

AN/ASA-65(V) compensator group adapter

CAE has devised this add-on device for the AN/ASA-65(V) MAD compensator to automate the data-gathering and thus reduce the time required for the compensation exercise. The compensator group adapter requires no modification to the original compensator and is operational with Royal Air Force/British Aerospace Nimrod MR.2s and US Navy/Lockheed

S-3A Vikings and P-3C Orions. The compensator group adapter comprises an indicator and a micro-computer which calculates the changes required in each of the nine terms to provide optimum compensation. The compensation update requires only 6 minutes' flying, during which time low amplitude manoeuvres are conducted on four different headings. The operator then has to adjust the compensation terms manually, as shown on the indicator. The compensator group adapter also compensates for four terms after weapon drop. This takes 2 minutes and requires small random manoeuvres to be conducted on two or four headings. As well as automatic data acquisition, the device includes built-in test equipment, which reduces the time taken in pre-flight checks and trouble-shooting. Aircraft availability is thereby improved considerably.

Dimensions: (indicator) 3.75 × 5.75 × 5 inches (95 × 146 × 127 mm)
(computer) 8.7 × 9 × 16.5 inches (221 × 229 × 418 mm)
Weight: (indicator) 3 lb (1.35 kg)
(computer) 22 lb (10 kg)
Power: 65 W at 115 V 400 Hz plus 10 W at 5 V ac/dc for panel lighting and 5 W at 28 V dc for annunciator lamps

STATUS: in service but no longer in production.

AN/ASQ-502 magnetic anomaly detector

CAE claims that the AN/ASQ-502 is the most sensitive airborne MAD available. The quoted sensitivity is 0.01 gamma, or a change in the earth's magnetic field of one part in five million. The AN/ASQ-502 is in service on the CP-140

CAE AN/ASA-65(V) nine-term compensator

Auroras of the Canadian Armed Forces, and CAE has received enquiries from several other countries, including Britain, where it has been evaluated by the Royal Air Force.

CAE has abandoned the normal multiple-cell detection method in favour of a single caesium cell mounted on gimbals to allow for dead zones and heading errors, which are normally taken care of by the use of multiple cells. The device also features a stable self-oscillator caesium sensor loop, and a low-noise frequency-to-voltage converter which is claimed to have a virtually infinite dynamic range.

Sensitivity: 0.01 gamma (1 × 10⁻⁷ Oersted)
Detecting head figure-of-merit: (uncompensated) 1.5 gamma
(compensated) 0.3 gamma
Larmor frequency of detector: 3.5 Hz/gamma
Operating range: 0.2–0.75 Oersted
Larmor output (total field): 14 Hz/gamma
Specification: MIL-E-5400
Mtbf: 1500 h
Power: 120 W at 115 V, plus 10 W at 5 V ac or dc for panel lighting

STATUS: in service but no longer in production.

AN/ASQ-504(V) AIMS advanced integrated MAD system

The AN/ASQ-504(V) advanced integrated MAD system (AIMS) is an inboard system for helicopters, fixed-wing aircraft and lighter-than-air platforms. This fully automatic system improves detection efficiency while reducing significantly the operator's work-load. For helicopter installation, the detecting head is fixed to the aircraft body, thus providing 'on-top' contact when over a target by eliminating the time delay inherent in a towed detecting-head system.

The AN/ASQ-504(V) system combines sensitivity and accuracy with ease of operation, eliminates aircraft-generated interference, reduces geological and solar interference and provides automatic contact alert both visually and audibly. Detection data, via the control indicator or an avionics bus interface, allows the operator to determine if the aircraft is within target acquisition range.

AIMS eliminates the hazard associated with towed systems. It also allows surveillance and manoeuvrability at higher speed, thereby increasing patrol range, detectability and reducing the incidence of false alarms. When used with dipping sonar, transition between the systems can be performed quickly and effectively.

AIMS weighs less than 50 lb (23 kg) and comprises a 0.005 nanotesla optically pumped magnetometer, vector magnetometer, amplifier computer and control indicator. It can operate independently or it can accept and execute commands from common control/display units via a MIL-STD-1553 digital data-bus.

Dimensions: (control indicator) 7.5 × 5.7 × 5.7 inches (190 × 145 × 145 mm)
(amplifier computer) 7.6 × 10.1 × 22 inches (193 × 257 × 559 mm)
(vector magnetometer) 6 × 6 × 6 inches (152 × 152 × 152 mm)
(detecting head) 7 × 32 inches (178 × 813 mm)
Weight: 50 lb (23 kg)
Power: 200 VA at 108/118 V 380/420 Hz single-phase
Specification: MIL-E-5400
Sensitivity: 0.01 gamma (in-flight)
Feature recognition: automatic target detection; operator alert, visual and audible; estimated slant range, up to 2800 feet (860 metres)

STATUS: in production.

OA-5154/ASQ automatic compensation system

The 16-term fully-automatic compensation system (FACS) is the next step after the nine-term semi-automatic AN/ASA-65(V). It is in service with the Canadian Armed Forces/Lockheed

CAE AN/ASQ-502 magnetic anomaly detector

Canadian Armed Force/Lockheed CP-140 Aurora, with CAE AN/ASQ-502 magnetic anomaly detection system in rear-fuselage

CAE AN/ASQ-504(V) magnetic anomaly detector

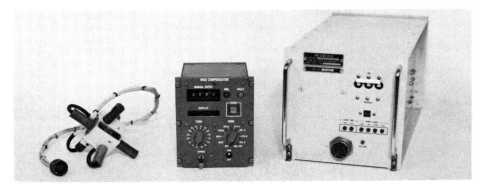
CAE OA-5154/ASQ fully automatic compensation system

CP-140 Auroras, and has been delivered to the West German and Netherlands navies for their Atlantic ASW aircraft. FACS conditions the raw magnetometer data, using its mini-computer for displays and other aircraft systems. The other necessary input is the orthogonal vector magnetometer signals which provide heading, manoeuvre, and total earth-field data. The use of completely electronic compensation obviates the need for output coils to generate opposing fields and no operator input is needed.

The compensation flight-programme lasts for no more than 6 minutes and comprises four 1-minute low-amplitude manoeuvres on headings approximately at right angles to each other. Additional trapping circles and clover-leaf manoeuvres are optional. As well as automating the recompensation exercise, the FACS allows the operator to update the system with minor magnetic variations at the touch of a button. Built-in test equipment is included.

Figure of merit: <0.4 gamma
Residual interference signals: 0.03 gamma average, or magnetometer internal noise if that is greater
Input: Minimally filtered analogue magnetometer detector signal
Power: 150 W at 115 V plus 5 V ac/dc for panel lighting, plus 10 W at 28 V dc for annunciator lights
Mtbf: 1900 h

Specification: MIL-E-5400
Dimensions: (control indicator) 7.2 × 5.8 × 6.5 inches (182 × 147 × 166 mm) (electronic amplifier) 9 × 9 × 21.5 inches (229 × 229 × 546 mm) (vector magnetometer) 6 inches (152 mm) cube
Weight: 31.5 lb (14.2 kg)

STATUS: in service but no longer in production.

Computing Devices

Computing Devices Company, A Division of Control Data Canada Limited, PO Box 8508, Ottawa, Ontario K1G 3M9
TELEPHONE: (613) 596 3810
TELEX: 05-34139

Fast Fourier transform analyser for AQS-901 acoustic processing system

Computing Devices provides the fast Fourier transform analyser of the AQS-901 acoustic processing system, which is built by GEC Avionics. The analyser can perform a 2048 complex transform in 11.25 milliseconds and forms an important part of the system. Computing Devices is also involved in the GEC Avionics advanced signal processing system, probably with a similar role.

STATUS: in production.

AN/UYS-503 sonobuoy processor

The UYS-503 sonobuoy processor, also known as the SBP 1-1, has been designed around a new computer architecture which exploits modern microprocessor and dense memory techniques; it interfaces easily with digital data highways and the platform's navigation systems. The SBP 1-1 can also function independently, driving a display and/or hard copy units.

Any existing American sonobuoy, working through any standard receiver, can be used in conjunction with the UYS-503. A Computing Devices patented de-multiplexer enables any buoy, even a digital one, to be de-multiplexed without the need for a digital receiver.

In processing information from passive buoys, the UYS-503 continuously stores the previous 20 minutes of Lofar data for all buoys and for all bands. Four such processing

AN/UYS-503 sonobuoy processor with programmable electro-luminescent control panel

modules, together with the requisite input/output circuitry, are housed in a single 1 ATR box, giving the system the capability to process up to four Difar or eight omni buoys in any combination. Each of the four modules acts independently, so that if one section fails, the remaining three processors are unaffected. Each of these modules, or slices, can process seven bands when working with Difar buoys, or 14 bands with omni units. The operator thus positions each of the seven bands from each slice anywhere in the 0 to 2560 Hz frequency range, providing rapid and simple initialisation.

For active processing, each processing slice of the UYS-503 can be used with one range-only or one Dicass buoy, the latter in both continuous wave and FM modes, the former in continuous wave only. A single bathythermal buoy can also be processed, the data being spectrum analysed and the temperature dependent tone is detected and converted to temperature.

The data from a dipping sonar can also be processed by a UYS-503 processor. Each of the four 'slices' is capable of processing a 90° sector.

Format: 1 ATR short 9 × 7.5 × 12 inches (229 × 190 × 305 mm)
Weight: 44 lb (20 kg)
Power: 350 W at 115 V 3-phase 400 Hz
Input channels: 8 standard sonobuoy receivers
Frequency band: 0-2560 Hz

STATUS: in production for Canadian Armed Forces and for Swedish and Royal Australian navies.

AN/AQS-801 Barra side processor

To enable standard sonobuoy processors such as the UYS-503, UYS-1 and 2, AQA-7 or AQA-5 to process Barra sonobuoys, Computing Devices has developed a Barra side processor (BSP) as a simple add-on to current systems.

The Barra side processor is a stand-alone device with its own demultiplexer and operator controls. A unique interface technique eliminates the need to 'invade' the existing processor circuitry in order to connect and operate the BSP. In addition to exploiting the broadband characteristics of the Barra sonobuoy, the BSP also retains a directed capability (three beams per buoy) to spectrum analyse a broadband detection for confirmation purposes. The BSP is a lightweight, single (1 ATR short) unit that contains four individual Barra processors for simultaneous processing of four Barra buoys.

Format: 1 ATR short 9 × 7.5 × 12 inches (229 × 190 × 305 mm)
Weight: 44 lb (20 kg)
Power: 350 W at 115 V 3-phase 400 Hz
Input channels: 4 sonobuoy receivers
Frequency range: (broadband) 500-1000 Hz and 1000-2000 Hz (narrowband) 10-2000 Hz

STATUS: in production for Royal Australian Navy

FRANCE

Crouzet

Crouzet SA, Division Aérospatiale, 25 rue Jules Védrines, 26027 Valence Cedex
TELEPHONE: (75) 42 91 44
TELEX: 345807

Magnetic anomaly detector for ATL2

Crouzet, a leading French magnetometer company, has been involved in the field since 1974. It is currently building the improved magnetic anomaly detector (MAD) for the Dassault-Breguet ATL2 for the French Aéronavale. This comprises twin magnetic sensors mounted longitudinally in the aircraft's MAD boom. Since the two sensors are in different positions with respect to the mass of the aircraft, any change in the aircraft's magnetic field due to dropping stores or manoeuvring will be detected by the combination. The use of sum and difference calculations from the two sensors

obviates the need for separate compensation equipment.

Flight trials took place in autumn 1983 aboard an Atlantique aircraft of Aéronavale. The flights were arranged to include those areas of the world with a wide variation of the earth's magnetic field, notably close to the magnetic poles and the equator.

Subsequently, this system was selected for the ATL2 and production of the first MAD systems began early in 1985, with aircraft deliveries due in 1988.

STATUS: in production.

Towed magnetometer Mk 3 for helicopters

This version of Crouzet's MAD equipment is intended for ASW helicopters. It detects the presence of a submersible by measuring the resulting disturbance to the earth's magnetic field. In order to eliminate disturbances created by the carrier helicopter, the detector probe is placed in a streamlined 'bird' which is towed at the end of a 70-metre cable. The digital computer measures the signals from the sensor and transforms them into suitable formats for the graphic recorder at the operator's station.

The measurement probe operates according to the nuclear magnetic resonance principle with electronic pumping. It offers the following advantages: it is ready to use as soon as it is switched on, even down to –40° C; it is easy to use; the sensor lifetime is virtually infinite; weight, size and electrical power consumption are low; and sensitivity is very high. The Mk 3 MAD comprises a detection element, computer, graphic recorder, and the winch and cradle assembly. Maintainability is assisted by in-flight checking of the 'bird' (the towed magnetometer), computer, control unit and recorder. The ground-test points are easily accessible and the sub-assemblies are plug-in units.

The 'bird' assembly includes the detection probe and an electronic unit, combining to produce a nuclear oscillator whose Larmor frequency is a function of the magnetic field exerted on the probe. The geometry of the probe is chosen so that its position with respect to the magnetic-field vector can be ignored.

The computer processes the signals from the probe and delivers the measurement to the recorder. Two processors are included in the computer. The first is a high-precision frequency meter which performs the frequency/voltage conversion of the signal from the probe. The second performs signal formatting, filtering, and amplification so that the operator sees pre-processed data. The computer also controls the power supply to the 'bird' and control unit, performs the cyclic and triggered tests, and interfaces with the control unit. The latter allows the operator to switch on and off, choose the sensitivity, and operate the built-in test equipment. A third, digital processor can be accommodated in the computer if required. It improves the signal filtering by adapting the filter characteristics to the ambient noise spectrum, and analyses mathematically the signals received as a further aid to target detection. This computed data is displayed to the operator on one of the two graphic recorder traces, the other remaining in its original state.

The graphic recorder sub-assembly is a potentiometric device which simultaneously displays two signals, recorded by red and black stylii on paper which unrolls at a constant speed. A light-emitting plate gives a permanent recording display. The operating controls are on/off, speed selection, test, and lighting.

The winch and cradle assembly comprises a winch, gear system, drum, cradle with locking device, power unit and control unit. The probe is connected to the computer by three coaxial cables, and a cutter is provided for emergency jettisoning.

Background noise: typical deviation 0.006 gamma
Measurement range: 25 000–70 000 gamma
Sensitivity for relative field output: 1, 2, 5, or 10 gamma for 100 mm stylus deviation
Paper speed: 6, 75, or 300 mm/minute

Streamlined body
Length: 1300 mm
Weight: 16 kg
Diameter: 160 mm

Computer
Dimensions: 346 × 124 × 194 mm
Weight: 7.5 kg
Power: 100 W at 200 V ac 400 Hz, 3-phase, plus 5 W at 28 V dc

Control unit
Dimensions: 165 × 150 × 57 mm
Weight: 1 kg

Recorder
Dimensions: 190 × 150 × 190 mm
Weight: 4 kg
Power: 50 W at 115 V ac 400 Hz, plus 20 A at 28 V dc

Winch and cradle
Dimensions: 1330 × 350 × 755 mm
Weight: 44 kg
Power: 40 A at 27 V dc

STATUS: helicopter version in production and service.

Mk 3 inboard-mounted magnetic anomaly detector for helicopters

Crouzet is developing a new MAD in conjunction with the French Navy. First trials were undertaken on a Dassault-Breguet ATL2 aircraft, with the system mounted in the rear MAD boom, and then in February 1985 on a Lynx helicopter of the French Navy to allow definition of the equipment for helicopter use. The system has been demonstrated to Aéronavale and the Royal Navy and Crouzet is teamed with

Crouzet MAD probe on ATL2

*Crouzet started flight trials of its Mk 3 inboard mounted helicopter MAD system in February 1985.
Actual positioning of boom is subject to alteration during development work*

MAD system operator at Crouzet graphic recorder

Dowty Electronics in proposing the Mk 3 system to the latter.

The MAD sensor is installed in a boom fixed to the side of the helicopter, eliminating the 'flying bird' MAD probe which previously had to be employed on helicopters in order to get the detectors as far away from the helicopter's magnetic influence as possible.

The measurement probe operates in accordance with the nuclear magnetic resonance principle with electronic pumping, a technique said by Crouzet to offer significant advantages in terms of reliability, weight, size, power consumption and sensitivity.

The sensor and its electronic unit form a nuclear oscillator with a Larmor frequency proportional to the magnetic field excited on the probe. The geometry of the probe is designed so that its position with respect to the magnetic field can be ignored.

A 32-bit Crouzet Alpha 732 computer analyses and processes the data from the probe before passing the results to the operator. It ensures that the effects of the host vehicle are precisely compensated for and a sensitive frequency meter ensures accurate conversion of the frequency signal from the nuclear oscillator.

The Mk 3 MAD can be connected into a MIL-STD-1553B data-bus, and also to a graphic

Crouzet MAD probe on Lynx helicopter

recorder. The system detects signals in the 25 000 to 70 000 gamma range with a discrimination in the order of 0.006 gamma.

Dimensions: (detection unit) 1900 mm dia × 1400 mm
(computer) 320 × 124 × 194 mm
(cdu) 145 × 146 × 190 mm
(recorder) 232 × 146 × 187 mm

Weight: (detection unit) 10 kg
(computer) 8 kg
(cdu) 3.5 kg
(recorder) 4.8 kg
Power: 60 VA at 115 V, 40 Hz and 20 W at 28 V dc

STATUS: in production.

Thomson-Sintra

Thomson-Sintra, 1 avenue Aristide Briand, 94117 Arcueil Cedex
TELEPHONE: (1) 42 53 1580
TELEX: 250675
FAX: (1) 42 53 5262

HS/DUAV-4 sonar for helicopters

The HS/DUAV-4 is a directive lightweight active/passive sonar system designed for light shipborne helicopters of the Westland/Aérospatiale Lynx class, although it is also applicable to small vessels as a variable-depth or dipping system. Signal processing of the HS/DUAV system is designed to improve the system's effectiveness in shallow and particularly noisy waters. The display may be operated in a general surveillance format for initial target detection, or in a plotting format which provides precision bearing information. In its active mode, the system gives range and bearing information, together with target radial velocity with respect to the sensor, but in passive, 'listening' mode, only bearing information is provided.

STATUS: in production and in service with French Navy and other maritime defence forces.

HS-12 sonar for helicopters

The HS-12 has been designed for installation aboard light helicopters such as the Westland/Aérospatiale Lynx, Aérospatiale Dauphin and Agusta-Bell 212 as well as naval helicopters such as Aérospatiale Super Puma, Westland Sea King and Kawasaki KV107II. A hydraulic winch system has been developed for raising and lowering the transducer to and from the aircraft. The cable, which may be up to 300 metres long, can be winched at rates up to 5 metres a second.

The HS-12 is a panoramic active-passive sonar and is particularly suitable for operation in shallow and/or noisy waters. The system has three operating frequencies around 13 kHz, can operate in both continuous wave and FM modes, and can automatically track two simultaneous target echoes. The HS-12 is microprocessor-controlled and digital signal processing is used. Display is of 12 pre-formed beams, each of which is subject to the system's adaptive signal-processing capability. Digital output to other systems aboard the aircraft,

HS-12 sonar console in Aérospatiale Dauphin helicopter

such as plotting units or weapon control equipment, is available.

Weight: < 240 kg

STATUS: in series production for French and foreign navies.

TSM 8200 & TSM 8210 acoustic processors for ATL2

Thomson-CSF is the project leader for the acoustic systems on board the Dassault-Breguet ATL2, for which it developed the TSM 8200 acoustic processing system, formerly called Sadang (Système Acoustique d'Atlantique Nouvelle Génération). The ATL2 is fitted with this system, comprising two identical modules designated TSM 8210.

The TSM 8210 system is available in single units for smaller aircraft or helicopters. It can process simultaneously eight omni-directional passive sonobuoys or two omni-directional active sonobuoys and can also process directional buoys. It can therefore handle most types of sonobuoy now available or under development.

Vhf receiver The TSM 8210 has a 99-channel vhf receiver capable of receiving the sonobuoy

Display and controls for Thomson-Sintra TSM 8210 sonobuoy processing system

transmissions from eight buoys simultaneously. A pre-amplifier is located close to the antenna, and a remote control unit is integrated into the operator's console. For testing the system, an acoustic signal source generator is incorporated. This simulates the transmissions from sonobuoys, and can be used at any time during a mission.

Signal processing The demodulated signals are processed successively in analogue and then digital form, depending on the types of sonobuoys being monitored and the types of processing selected by the operator. For position-fixing with passive sonobuoys, the processor uses various methods based on launching several waves of buoys from different positions and comparing their returns. Discrepancies in amplitude, frequency, time or phase between the signals received from two or more sonobuoys are used to calculate the position of the target submarine. Closer to the attack, active sonobuoys are used to provide range-only or range-plus-azimuth (as with the Dicass buoy). The Sadang processor can deal with these types as well.

Intelligibility of the noises sensed by the hydrophones in the sonobuoys is improved by the possibility of hearing selected frequency bands or listening in selected directions.

Sadang can arrange for this, and the operator can also listen simultaneously to transmissions from two different sonobuoys, one in each ear.

Displays A special processing unit is devoted to the man-machine interface for maximum efficiency and flexibility of communication with the system. Inputs are made via two physically identical keyboards, one alpha-numeric and one functional, plus a rolling ball to control the marker on the cathode ray tube display. The output is shown on a cathode ray tube in raster-scan mode, and on a graphic recorder. The cathode ray tube is used to display a wide variety of pictures selected by the operator according to the tactical situation and the types of sonobuoy in use. Several pages are updated simultaneously in the mass memory, each page being available for display at any time. Each page displays a spectral analysis of the acoustic signals detected from the sonobuoys, within various frequency bands and integration times. Alphanumeric infor-mation is also available. A graphic recorder keeps a permanent trace of the processed data, printing out four simultaneous channels each with a resolution of 750 points.

Successive analyses of the frequency lines are memorised and displayed on the cathode ray tube or on paper, thus benefiting from visual correlation by enhanced sensitivity and resolution. The various detectable frequency lines can be gathered into families with the help of the electronic facilities at the operator's disposal. Fundamental frequencies can be compared with signatures stored in a library, and the noise emitter can be identified or at least classified as belonging to a category of vessels. The frequency lines in the lower part of the spectrum (under 100 Hz) are especially typical, and modulate the amplitude of the higher part of the frequency spectrum. In order to detect these lines, a processing method called Demon (demodulation of noise) is available.

Active sonobuoy remote control The operation of active sonobuoys must be as quick and efficient as possible, since they can be detected by the target submarine. For this reason, some active sonobuoys are equipped with a remote-control capability by which a single pulse, or a few pulses, may be triggered where strictly necessary with pre-selected transmission characteristics (pulse duration, continuous wave or modulated frequency, etc). This function is performed by a sonobuoy remote control unit connected to a uhf transmitter tuned to the receiving frequency of the sonobuoy. The transmitter may be either a dedicated transmitter or the centralised com-munication system's transmitter.

On-top position indicator Target fixes are given relative to the sonobuoy positions, but the buoys drift with winds and sea currents, so their positions must be periodically updated by flying over the various buoys in succession. Flying towards these buoys is made possible by a radio direction-finder connected to a detection and display system, the on-top position indicator.

Navtac The sonobuoy data is transmitted via a digital data-bus to the tactical computer, and then to the tactical co-ordinator. The tactical computer, called Navtac, correlates the data from sonobuoys, radar, forward-looking infra-red, electronic support measures, magneto-meter, and navigation systems. The tactical co-ordinator sees all this information on his two-colour display, and can thus make de-cisions such as choice of sonobuoy launch patterns, and order the update of sonobuoy positions, leading to the attack.

Options The TSM 8210 has optional sub-units for recording acoustic signals for later exam-ination, and a sonobuoy launcher remote control system.

The TSM 8210 system has recently been developed from a 31 to a 99 channel system, giving it greatly increased capability, with several new models of high-performance

Thomson-Sintra TSM 8200 acoustic processing system for Dassault-Breguet ATL2 (two stations nearest camera)

Second prototype of ATL2 long-range maritime patrol aircraft

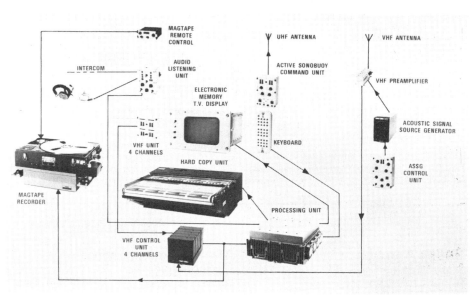

Thomson-Sintra TSM-8220 Lamparo acoustic detection system

acoustic processing computers being tested. It is understood that the company is offering this system to both French and overseas services.

STATUS: in production for Dassault-Breguet ATL2 and under consideration for other types, such as Franco-German Transall C-160S, British Aerospace HS748 Coastguarder, and Fokker F27 Maritime.

TSM 8220 Lamparo acoustic processor

Thomson-Sintra derived the Lamparo as a light-weight, compact version of the TSM 8210, suitable for smaller aircraft and light heli-copters. The aim was to provide the perform-ance level of the basic system, but with fewer functions. For example, the processing of directional sonobuoys of the Difar and Dicass type has been deleted. The Lamparo cannot handle as many sonobuoys, but it can still process data from four omni-directional passive or two omni-directional active buoys simultaneously.

Lamparo has all the analogue, digital and management processing functions gathered within a single unit and the audio processing is simplified. One keyboard and the rolling-ball marker have been dropped, leaving a single keyboard with a cathode ray tube display. The hard-copy recorder is optional, its deletion reducing the space requirement considerably. The vhf receiver can handle only four channels. Lamparo is fully compatible with Navtac (described in the TSM 8200 entry above).

Despite the reduction in size and functions, Lamparo can be upgraded if the user has the space and requires extra capability. More extensive audio processing can be added, as can the remote control and processing of active

Aérospatiale Dauphin 365 helicopter, candidate for Thomson-Sintra Lamparo system

sonobuoys. Likewise, a graphics recorder and a magnetic-tape recorder can be added.

STATUS: in production.

ITALY

Elettronica

Elettronica SpA, Via Tiburtina km 13 700, 00131 Rome
TELEPHONE: (6) 43641
TELEX: 611024

ELT 810 sonar performance prediction system

Elettronica has developed the ELT 810 sonar performance and propagation prediction sys-

tem in conjunction with its Elettronica Ingegneria Sistemi (EIS) subsidiary. It enables the sonar propagation characteristics to be fully exploited for both defensive and offensive purposes. The system comprises a single compact unit, which houses the computer, its memory, a graphic cathode ray tube display, and a dedicated command and control panel. Installation is foreseen in both fixed-wing aircraft and helicopters, as well as on submarines and surface vessels.

STATUS: in production and in service with Italian Navy.

Selenia

Selenia Industrie Elettroniche Associate SpA, Via Tiburtina km 12 400, 00131 Rome
TELEPHONE: (6) 43601
TELEX: 613513

Falco submarine locating system

Few details have been revealed of Selenia's anti-submarine warfare activities, but the Falco

system is known to be an airborne equipment designed to increase the capability for helicopters and fixed-wing aircraft to detect, classify and locate submarines. The system operates on the low-frequency noise radiated by targets and gathered by directional low-frequency passive sonobuoys.

Noise spectral analysis, target data processing and display are performed by a real-time digital computer. It is claimed that the Falco

system can determine target position with an accuracy sufficient to carry out an attack with automatic homing torpedoes, and with errors due to sonobuoys drifting automatically cancelled out. The equipment is produced in several versions.

STATUS: in production and service, probably on Italian Navy Atlantic aircraft and Agusta-Sikorsky SH-3 helicopters.

NETHERLANDS

Van der Heem

Van der Heem Electronics, Regulusweg 15, PO Box 16060, 2500 AB The Hague

SP1-04 sound-ray path analyser

The SP1-04 sound-ray path analyser is designed for service aboard fixed-wing aircraft

and helicopters, surface vessels and submarines. Microprocessor-based, it provides real-time predictions of underwater sound-ray paths over short, medium and long range. The data is based on sound velocity in water data derived from expendable bathythermograph and sound velocimeter information. Paths are displayed at spacing of 0.5° on an 8-inch (203 mm) cathode ray tube presentation.

The system features seven adjustable layers with sound velocity settings, reflective variable bottom depth selection and a numerical light-emitting diode profile setting display.

STATUS: in service with Royal Netherlands Navy and other navies.

UNITED KINGDOM

British Aerospace

British Aerospace plc, Naval Weapons Division, PO Box 5, Filton, Bristol BS12 7QW
TELEPHONE: (0272) 693 831
TELEX: 449452

Helras long-range sonar for helicopters

British Aerospace has teamed with Bendix Oceanics to develop the Helras long-range sonar for helicopters. Initial trials were carried out in 1984 in the Mediterranean and North Seas and development of a prototype is under way; entry into service is scheduled for 1989, a

prime sales target being the Anglo-Italian EH 101 helicopter now in development.

A large low-frequency volumetric array is used, which folds away when not operational. British Aerospace claims that this type of array will produce a performance significantly better than existing designs.

STATUS: in development.

GEC Avionics

GEC Avionics Limited, Maritime Aircraft Systems Division, Airport Works, Rochester, Kent ME1 2XX
TELEPHONE: (0634) 44400
TELEX: 96333

AD 130 sonar homing and direction-finding receiver

The AD 130 sonar homing and direction-finding receiver is designed for both fixed-wing aircraft and helicopters. It provides sonobuoy-homing and direction-finding facilities on 99 channels over the frequency range 136 to 173.5 MHz with a channel spacing of 375 kHz.

The system comprises a receiver with computer interface and a remote control unit together with associated antenna switching controls. Additional items can include a direction-finding antenna and a left/right and fore/aft antenna for 'sonobuoy on top' indication, plus cross-pointer and radio magnetic indicators.

Either computer or manual channel-selection may be employed. In computer mode, selection is accomplished by decoding of the serial bit stream and in manual, by means of pairs of thumbwheel switches. A light-emitting diode readout indicates the channel selected. Dual receivers, with independent channel selection, may be operated from a single controller. The homing system provides left/right and fore/aft deflections on a meter and the direction-finding section detects and processes AM signals from a direction-finding antenna for an azimuth display. Built-in self-test circuitry continuously monitors all modules, with interruptive test-switching for homing and direction-finding facilities, a failure mode indication for both homing and direction-finding, and slew switching for checking the direction-finding operation.

The system is said to be compatible with all known and planned NATO and allied services sonobuoys.

Dimensions: (receiver) 7.7 × 4.9 × 7.5 inches (196 × 124 × 190 mm)
(control) 5.9 × 3.9 × 7.9 inches (150 × 99 × 201 mm)
(switched antenna, main) 1.9 × 3.1 × 7.5 inches (50 × 80 × 193 mm)
(antenna switch fore and aft) 1.9 × 3.1 × 7.5 inches (48 × 79 × 191 mm)
Weight: (receiver) 17.46 lb (7.94 kg)
(control) 5.06 lb (2.3 kg)
(switched antenna, main) 1.76 lb (0.8 kg)
(antenna switch fore and aft) 1.76 lb (0.8 kg)

STATUS: in production and service.

AQS-901 acoustic processing system

The GEC Avionics AQS-901 sonobuoy processing system is installed in both the Royal Air Force Nimrod MR2 and the Royal Australian Air Force Lockheed P-3C Orion MPA. The AQS-901 is the only processing system in service able to handle data from all types of sonobuoy in the NATO inventory. This includes advanced sonobuoys, such as Barra, Cambs and Vlad which are intended to combat the movement of quieter submarines in deeper water. AQS-901 fully exploits the advantages of digital processing techniques. It displays processed data to the acoustic operators on a choice of cathode ray tubes and chart recorders. The processing mode can be varied to provide configurations suitable for different operational environments. Ease of processing and display mode changes increase effectiveness and the probability of mission success.

The AQS-901 receives sonobuoy data via an advanced eight-channel receiver, with each channel tunable to any of the 99 NATO sonobuoy frequencies. The command system allows control of the pulse length, type and repetition frequency of deployed active buoys.

Royal Navy Sea King HAS5 fitted with AQS-902

AQS-901 operators' station in Royal Air Force Nimrod MR2

GEC Avionics AQS-901 is installed in the Royal Air Force Nimrod MR2s

Both the receiver and command systems meet the latest international standards.

Beam-forming, spectral analysis and broad-band power analysis are used by AQS-901 to process data received from multi-hydrophone array buoys. Three passive location processing techniques are employed: for omni-directional buoys - hyperbolic and Doppler fix processing, for Barra buoys - direct-bearing processing. When extremely accurate data is required, as in

the attack phase, bearing, range and relative velocity of the submarine is provided by directional-array active buoys. Processing is achieved using frequency domain beam-forming and correlation analysis.

The AQS-901 incorporates a target-oriented auto-alert feature, which informs the operator of target detection. The relevant thresholds and parameters are set by the operator in order to distinguish real targets from spurious ones, such as surface shipping and friendly submarines. The AQS-901 can continue with other processing tasks whilst scanning for these features which may be altered at will. Detection and tracking of the target are automatic with displays optimised to assist long-range detection and to separate slow-moving targets from reverberation and shipwrecks.

STATUS: in production and service. AQS-901 installed in Royal Air Force Nimrod MR2s and Royal Australian Air Force P-3C Orions. In 1983 a £25 million order was placed to equip a second Australian P-3C squadron with AQS-901.

AQS-902 Lapads lightweight acoustic processing and display system

To complement the AQS-901, GEC Avionics has developed the AQS-902 range of acoustic processing systems for helicopters and smaller maritime patrol aircraft. The modular design allows configuration to suit a wide range of ASW applications. The AQS-902 is in service on the Royal Navy's Sea King HAS5 ASW helicopter, and in production for the Indian Navy Sea King Mk 42B. It has also been supplied to the Swedish Navy and has been selected for installation aboard the Fokker Enforcer MPA.

Suitable for aircraft without a current sonobuoy capability, AQS-902 is also an ideal candidate for installation during mid-life updates of existing ASW aircraft. Its modular construction allows cost-effective configurations to be specified for both passive MPA and active ASW roles.

The AQS-902 has been designed to process and display data from both dipping sonar and sonobuoys, effectively removing the dipping sonar constraint for the ASW helicopter. This combined processing has been achieved to a standard only previously attainable when using separate systems and will substantially improve the helicopter's flexibility, endurance, search area, mission range and success probability. With its autonomous passive sonobuoy processing AQS-902 now provides the helicopter with a covert search capability.

Processed data is presented on hard copy or cathode ray tube displays, or both, in fully annotated form. Interrogation is easily achieved using a simple control panel, and roll ball or stiff stick. Electronic frequency dividers and markers are also available. The operator is provided with aids to correlate, detect and record target data. Using these facilities, manual, Doppler or hyperbolic fixing and Lloyds mirror depth analysis may be carried out.

The AQS-902 system is compatible with all existing sonobuoy receivers and can process all of the following sonobuoy ranges: AN/SSQ-36, 41B, 47B, 53, 57, 62, 77, 801, 904, 905, 937 and 963.

Some examples of AQS-902 configurations are:

AQS-902A Passive omni-directional buoy processing with hard-copy display

AQS-902B Passive omni-directional buoy processing with cathode ray tube display

AQS-902C Passive omni-directional and directional buoy processing with hard-copy display

AQS-902D Passive omni-directional and directional buoy processing with cathode ray tube display

AQS-902D-DS AQS-902D with dipping sonar and display.

Capabilities:
AQS-902C
Buoys processed: 4 Jezebel, 1 Bathy, or 4 Difar
Weight: 112 lb (51 kg)
Volume: 5.675 ft³ (0.16 m³)
Power: 555 W
AQS-902F
Buoys processed: 4 Jezebel, 1 Bathy, 4 Ranger, 4 Difar, 2 Dicass or 4 Vlad
Weight: 130 lb (59 kg)
Volume: 3.5 ft³ (0.1 m³)
Power: 567 W

Fokker Maritime Enforcer has mission avionics suite integrated by GEC Avionics

Mission avionics suite in Fokker Enforcer, incorporating AQS-902 (foreground) and Tattix (centre)

Tattix in Fokker Enforcer

Other configurations are available to meet customer's requirements.

With a choice of buoy processing modes, a versatile data display, extensive detection, classification and fixing aids and, for the helicopter improved dipping sonar processing, the AQS-902 offers a high standard of performance. Built-in test equipment incorporates interruptive and non-interruptive confidence testing. This indicates system readiness, and provides go/no go testing and failure isolation to a high level of confidence. The built-in test circuitry also accomplishes first-line fault diagnosis. Testing to module level is achieved in second-line or hot-rig test equipment.

STATUS: in service with the Royal Navy, and Royal Swedish Navy. In production for the Indian Navy and on trial with the US Navy. AQS-902 has also been selected for the Fokker Enforcer MPA as part of the GEC Avionics integrated mission system.

ASN-902 TATTIX tactical processing system

The ASN-902 Tattix tactical processing system provides a means of correlating and processing data for display from all the aircraft's sensors and navigation systems. Based on proven AQS-902 technology, it presents the Tacco with a display which enables him to solve complex navigation, intercept and attack problems.

STATUS: in production.

AQS-903 signal processor

GEC Avionics is developing the AQS-903 signal processor aimed at the next generation

Westland-Agusta EH 101 to be fitted with AQS-903

of ASW helicopters, fixed-wing aircraft and major helicopter and fixed-wing avionic updates. The AQS-903 is planned for use in the projected Westland-Agusta EH 101, which will replace the Royal Navy's present Sea King HAS5 ASW helicopter.

The AQS-903 will process and display all current and projected sonobuoy types, dipping sonar and MAD. The use of distributed processing architecture with multiple processing pipelines will ensure that software system development and enhancements may be readily achieved. The system will employ a number of 'pipelines', each able to handle a certain quantity of passive and active sonobuoys. Four of these 'pipelines' will provide a processing power eight times that of an AQS-901, at one-quarter the weight.

STATUS: under development for the Westland-Agusta EH 101.

Systems integration

GEC Avionics is also able to offer various levels of systems integration. This improves the efficiency and flexibility of the mission avionics suite, minimises the weight of combinations of multiple sensors and simplifies logistics and training problems with common control units and multi purpose displays. The Maritime Aircraft Systems Division at GEC Avionics is at present completing systems integration for the Fokker Enforcer MPA. This programme incorporates AQS-902 and Tattix into the total avionics suite.

Plessey
Plessey Marine Limited, Wilkinthroop House, Templecombe, Somerset BA8 0DH
TELEPHONE: (0903) 70551
TELEX: 46268

Submersible unit of Plessey Cormorant sonar showing arms extended

Plessey Marine/GEC Avionics
Plessey Marine Limited, Wilkinthroop House, Templecombe, Somerset BA8 0DH
TELEPHONE: (0903) 70551
TELEX: 46268

GEC Avionics Limited, Military Aircraft Systems Division, Airport Works, Rochester, Kent ME1 2XX
TELEPHONE: (0634) 44400
TELEX: 96333

Type 195 sonar for helicopters

The Plessey Type 195 sonar is a helicopter dunking system designed principally for Royal Navy/Westland Sea King helicopters. It provides full 360° coverage and is said to be effective at ranges up to 8000 yards (7300 metres). Full azimuth coverage is provided in stepped fashion over 90° arcs progressively, but manual control permits the operator to concentrate on any particular sector. Alternatively the system may be programmed to undertake an automatic search. The operator is provided with audio, visual Doppler and visual sector sonar information and a close contact maintenance facility is provided for the tracking of nearby targets and those at greater depths.

The Type 195 may be employed for either surveillance or attack control, or for both simultaneously. Pulse length and detection range settings are operator-selectable to opti-

mise the system according to sea conditions and the tactical situation.

The system uses transistorised circuitry and modular techniques permit rapid replacement of faulty sections.

The current model is the Sonar 195M, which incorporates many improvements over the original equipment. Future developments to enhance further the overall performance and to meet naval requirements, include a longer cable to increase the operating depth of the submersible unit, and integration of the sonar with modern high performance processing systems.

Weight: 269.2 kg

STATUS: in production with over 250 systems in service with Royal Navy and several overseas navies.

Hisos helicopter integrated sonar system

Hisos combines the Plessey Cormorant lightweight dipping sonar with GEC Avionics Lapads AQS-902D-DS processor and display system to provide a comprehensive anti-submarine detection capability, for ASW helicopters having an all-up weight of 4100 kg or more, such as the Royal Navy/Westland Lynx.

The Hisos sonics suite provides the maximum tactical flexibility by exploiting Cormorant's high performance active and passive dipping sensor achieving long-range active detection against quiet targets. It is particularly effective in noisy environments and can be employed in a complementary role to sonobuoys as well as being used in autonomous covert surveillance.

The advanced design of Cormorant is said to overcome the size and weight constraints that have limited the operational performance of previous generations of dipping sonar. This new design is based on an expanding array

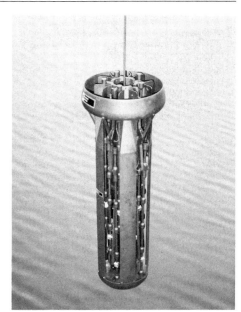

Plessey Cormorant lightweight dipping sonar (arms retracted)

technique which achieves a large acoustic receive-aperture. The sonar transducer itself comprises a central body to which are attached five folding arms, each arm carrying three staves containing the hydrophones. Once the submersible unit has reached its operating depth, the hydrophone array is powered out into position.

Cormorant transmits at approximately half the frequency of current dipping sonars. Reception is over a broad bandwidth and the

extended arm array gives a wide acoustic aperture allowing high bearing accuracy in both active and passive modes. Received signals from each of the 15 staves auxiliary sensor data and command information is transmitted up the cable on a digital data link for processing in the helicopter.

Mosaic integration techniques developed by GEC Avionics are claimed to permit more efficient exploitation of the Plessey sensor array.

STATUS: dipping trials of Cormorant started in 1985, with production starting in 1986. Some elements of the sonar are already being installed in the Royal Navy's Sea King HAS5 helicopters as part of their current update programme, which includes substantially new avionics equipment.

UNITED STATES OF AMERICA

Bendix

Allied Signal Inc, Bendix Oceanics Division, 15825 Roxford Street, Sylmar, California 91342
TELEPHONE: (818) 367 0111
TELEX: 662900
TWX: 910 496 1506

AN/AQS-13 sonar for helicopters

The Bendix AN/AQS-13 is a development of the company's earlier AN/AQS-10 system which has undergone considerable technical refinement and performance enhancement since its introduction in the late 1950s. Two variants of the AQS-13 are currently available, the A and B models. They are similar in terms of performance, although the B version is some 200 lb (91 kg) lighter and is said to possess improved servicing and maintenance characteristics.

Both are active/passive, long-range systems with a detection range in excess of 20 000 yards (18 300 metres). The systems can detect, locate and track submerged vessels and provide high-accuracy course and speed data and some degree of target classification. In the case of the B version, the introduction of advanced circuit and packaging techniques has not only reduced size and weight but has also permitted the incorporation of a high level of automation. Likewise, built-in test functions have been included, leading to a reduced maintenance requirement.

A digital adaptive processor system (APS) has been developed for the AQS-13, providing fast Fourier transform (FFT) narrow-band analysis. This facility provides improved shallow water detection capability, improved Doppler measurement with consequently more precise velocity measurement and a reduced false-alarm rate. The introduction of APS also gives the operator a digital readout of the target's Doppler radial velocity component, in addition to the usual plan-position indicator range and bearing presentation. APS also provides a longer, shaped, continuous wave pulse of higher energy which, together with the FFT narrow-band analysis, improves performance over a wide range of water depths.

Operational modes include active, passive, moving-target-indicator, APS and voice or key communications. The 13-A variant has a visual display of range and bearing data only, whereas the 13-B system has in addition range-gate and operator-verification facilities.

The AQS-13 is in service with the US and a number of other maritime defence forces throughout the world. An advanced version of the system, featuring further size and weight reduction together with improved performance, is believed to be in development.

Weight: (13-A) 820 lb (373 kg)
(13-B) 620.5 lb (282 kg)

STATUS: in production and service.

AN/AQS-18 sonar for helicopters

The Bendix AN/AQS-18 is a long-range active/passive sonar system which is based largely on earlier systems produced by the company. It is, however, considerably more advanced than the earlier models, employing digital electronic techniques, improved signal processing and updated displays. It is also lighter than its predecessors rendering it more suitable for service in the lighter classes of helicopter.

The system features rapid use in operation, attributable to the sensor's high sink rate and equally rapid speed of retrieval. A built-in multiplexer permits use of a single 333.5-foot (101-metre) cable for both suspension and power signal transmission to the transducer package. An adaptive processor system, similar to that of the AN/AQS-13 sonar system (see preceding entry) is optionally available for use with the AN/AQS-18 system.

Operating frequencies are selectable at 9.23, 10 and 10.77 kHz and operating modes include moving-target indicator and voice communication.

Weight: (main system) 554.4 lb (252 kg)
(adaptive processor) 29.25 lb (13.3 kg)

STATUS: in production and service.

Cubic

Cubic Corporation, Defense Systems Division, 9333 Balboa Avenue, San Diego, California 92123
TELEPHONE: (619) 277 6780
TWX: 910 335 2010

Sonobuoy Reference System

Cubic has developed a full line of Sonobuoy Reference System (SRS) units to fix the location of sonobuoys and compute the actual position of a target submarine, including its location relative to the ASW aircraft.

The SRS measures the position of a sonobuoy whenever the buoy is within line-of-sight of the aircraft. When operating in a passive mode, the SRS utilises angle measuring equipment (AME). In this mode, the SRS receives the sonobuoy's radiated vhf signal. By comparing the phase of the signal received by each antenna in the receiving array, the SRS computes the signal's angle of arrival. This gives the relative bearing of the transmitting sonobuoy. By receiving multiple lines of bear-

ing as the aircraft moves, SRS can compute the position fix for any type of sonobuoy.

Some Cubic SRS units also have the capability to operate in an active mode, using distance measuring equipment (DME). This technique enables the SRS simultaneously to measure both the bearing and range of certain types of active sonobuoys for an instant position fix. Range is measured using the round-trip phase delay on a modulated signal sent from the aircraft to the buoy via the uhf command link and returned to the aircraft on the sonobuoy's normal vhf signal. All active sonobuoys can be modified to function in this ranging mode.

AN/ARS-2 sonobuoy reference system

Designed and built for the S-3A ASW aircraft, the AN/ARS-2 provides passive AME and active DME location for sonobuoys operating on up to 31 standard sonobuoy channels. The system

uses the aircraft's on-board Univac 1832 computer for processing and display.

Ten fuselage and wing-mounted antennas receive radio signals from the sonobuoy, which the SRS computer uses to determine the angles and slant ranges to the detector.

A variant of the AN/ARS-2, the ARS-501, is built for the special requirements of the Canadian Armed Forces' CP-140 Aurora ASW aircraft. The ARS-501 also uses the aircraft's onboard computer to process and display up to 31 channels of sonobuoy location data.

Dimensions: 8.75 × 9 × 19 inches (222 × 229 × 483 mm)
Weight: (receiver-converter only) 35 lb (15.9 kg)
Power: 150 W at 115 V
Temperature range: −54° to 71° C
Cooling air: 0.54 lb/m (0.24 kg/m)
Vibration specifications: MIL-T-5422E
Mtbf: 2650 h
Specification: MIL-E-5400

STATUS: in service on US Navy/Lockheed S-3A Vikings.

AN/ARS-3 sonobuoy reference system

The AN/ARS-3 is a major part of the Update II improvements to the US Navy's P-3C ASW aircraft. Comprising 10 blade antennas and a receiver converter, the system provides for the passive detection and location of sonobuoys transmitting on 31 standard channels. The system uses the P-3C's Univac CP-901 computer for processing and display. It uses the aircraft's inertial navigation system to determine the geographical position of the buoys. In conjunction with the CP-901, the AN/ARS-3 can locate a a faulty module 95 per cent of the time. Unlike the AN/ARS-2, the system needs no cooling air.

Dimensions: 11 × 21.5 × 12.25 inches (279 × 546 × 311 mm)
Weight: (including antennas) 79 lb (35.8 kg) (receiver-converter only) 54 lb (24.5 kg)
Power: 150 W at 115 V
Temperature range: –54° to 71°C
Cooling air: none
Vibration specifications: MIL-T-5422E
Mtbf: 1000 h
Specification: MIL-E-5400

STATUS: in service on Lockheed P-3C Orions flown by the US Navy, Australia, Japan and the Netherlands.

AN/ARS-4 sonobuoy reference system

The AN/ARS-4 is a modified version of the AN/ARS-2 designed for the S-3B ASW aircraft. It is being retrofitted into all S-3 aircraft in the US Navy inventory. The unit's updated electronics provide improved performance, reliability and maintainability, as well as the ability to locate sonobuoys operating on the 99 sonobuoy channels.

Dimensions: 8.75 × 9 × 19 inches (222 × 229 × 483 mm)
Weight: (receiver-converter only) 35 lb (15.9 kg)
Power: 150 W at 115 V
Temperature range: –54° to 71°C
Cooling air: 0.54 lb/m (0.24 kg/m)
Vibration specifications: MIL-T-5422E
Mtbf: (SBR only) 2862 h
Specification: MIL-E-5400

US Navy/Lockheed S-3B Viking ASW aircraft are equipped with Cubic AN/ARS-4 sonobuoy reference systems

STATUS: currently being retrofitted into all US Navy/Lockheed S-3 aircraft.

AN/ARS-5 sonobuoy reference system

Designed for the P-3C Update III, Cubic's AN/ARS-5 is an improved version of the AN/ARS-3. It has 99 channel capability as well as state-of-the-art design and built-in expansibility for such additional capabilities as anti-jamming, communications intelligence (Comint), and search and rescue. The AN/ARS-5 incorporates the latest advancements in self-test antenna radiation and built-in test equipment.

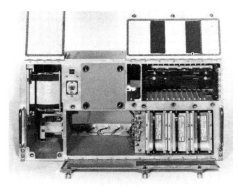

Cubic AN/ARS-5 sonobuoy reference system

Dimensions: 11 × 21.5 × 12.25 inches (279 × 546 × 311 mm)
Weight: (including antennas) 79 lb (35.8 kg) (receiver-converter only) 54 lb (24.5 kg)
Power: 150 W at 115 V
Temperature range: –54° to 71°C
Cooling air: none
Vibration specifications: MIL-T-5422E
Mtbf: 1000 h
Specification: MIL-E-5400

STATUS: in production for the US Navy's Lockheed P-3C Update III programme.

Stand-alone sonobuoy reference system

Under an independent research and development programme, Cubic is developing a stand-alone SRS that can allow virtually any fixed-wing aircraft or helicopter to have full SRS capabilities. This unit will incorporate such additional features as anti-jammng, DME, Comint, and search and rescue. Other features include self-calibration and built-in fault isolation for easier maintenance.

This stand-alone capability can be incorporated into either the AN/ARS-4 or the AN/ARS-5 chassis and can interface with any aircraft data-bus. Using its own imbedded computer system, the system can operate in both active and passive modes with sonobuoys operating on up to 99 channels.

Performance specifications for the stand-alone SRS are identical to the AN/ARS-5.

Emerson

Emerson Electric Company, Electronics and Space Division, 8100 West Florissant Avenue, St Louis, Missouri 63136
TELEPHONE: (314) 553 3232
TELEX: 209903
TWX: 910 761 1126

Emerson's anti-submarine warfare activities include acoustic processing and magnetic-anomaly detection. The company is responsible for the sonobuoy processing equipment in the West German Navy Atlantics under the project leadership of Dornier. Emerson is supplying a modification kit for the sonobuoy processing system, as part of the Atlantic improvement programme. The new equipment can handle a larger frequency spectrum and analyses eight buoys simultaneously. Better directional accuracy and an improved signal-to-noise ratio extends detection to quieter submarines, and

they can be located with more accuracy. Digital technology and the use of electronic rather than mechanical switching improves maintainability and reliability.

Emerson also produces the AN/ASQ-10A magnetic anomaly detector, which is used by the Nimrod MR2s of the Royal Air Force.

Calypso acoustic processing system

Calypso is an advanced, lightweight ASW system designed for the almost simultaneous processing of acoustic data from several sonobuoys to secure accurate position 'fixes'. The system is suitable for service in both fixed-wing aircraft and helicopters.

Calypso has been flight tested aboard a Sikorsky CH-124A Sea King of the Canadian Armed Forces, which is understood to be considering procurement of a number of these systems.

STATUS: under development.

Emerson Electric modification kit for sonobuoy processing system in West German Navy Atlantic ASW aircraft

Fairchild

Fairchild Communications and Electronics Company, 20301 Century Boulevard, Germantown, Maryland 20874
TELEPHONE: (301) 428 6000
TELEX: 892468

AN/ASQ-164 control indicator set

Fairchild's AN/ASQ-164 control indicator set was developed for the US Navy/Sikorsky SH-60B Lamps III helicopter. A microprocessor-based system which provides avionics control for the aircraft's tactical operator and sensor operator, it consists of a C-10487/ASQ-164 control indicator for use by the tactical operator and a C-10486/ASQ-164 control indicator for the other operator's use. Each of these units, which are known as 'keysets', allows control of the functions assigned to both operator stations.

Both units have switch/indicators used to select programming, through a MIL-STD-1553 data-bus, for the computer control of the helicopter's radar, navigation, acoustic detection and tactical activities. Analogue, two-axis force-stick controllers are also provided for use with the tactical status display screen.

The switch/indicators are used to display avionics function status. Black indicates that the function is not available, green depicts availability and amber indicates that the function is activated. The tactical operator's keyset has 74 main switch/indicators and that of the sensor operator has 69. Additional switch/indicators are provided for lamp test and for main microprocessor-controlled built-in test equipment fault indication. The self-test circuitry provides a 98.3 per cent fault indication capability. System panels are edge-lit and the lamp intensity can be varied by the operator.

The AN/ASQ-164 uses low-power Schottky 54LS-series logic circuitry. The mean time between failure is 3778 hours for the tactical operator's keyset and 3816 hours for the other keyset, at an ambient temperature of 71°C. Redundant over-temperature sensing devices are employed and the normal operating temperature range is from –40 to 71°C with an upper limit extension to 95°C for periods up to 30 minutes. The system is compatible with automatic test equipment at both unit and module level and 12 out of 13 sub-assembly modules are interchangeable between keysets.

Dimensions: (tactical and sensor units) 16.5 × 5.75 × 9.16 inches (419 × 146 × 233 mm)
Weight: (tactical unit) 22.8 lb (10.36 kg) (sensor unit) 22.6 lb (10.27 kg)

STATUS: in production and service.

General Electric

General Electric Company, Aerospace Electronic Systems Department, French Road, Utica, New York 13503
TELEPHONE: (315) 797 1000

AN/AYA-8C data-processing system

General Electric has been producing the data processing system for the US Navy/Lockheed P-3C Orion anti-submarine warfare aircraft since 1968. It provides the interface between the CP-901 central computer and the aircraft systems and so constitutes a major part of the P-3C's mission avionics.

The system is connected to all the crew stations on the aircraft: tactical co-ordinator (Tacco), non-acoustic sensor, navigation/communications, acoustic sensor, and flight-deck. In addition, it communicates directly with the radar interface unit, armament/ordnance

US Navy/Lockheed P-3C Orion has General Electric AN/AYA-8B ASW data processing system

system, navigation systems, sonar receiver, submarine anomaly detector, infra-red detection system, electronic support measures, sonobuoy reference system, and omega. A few systems, such as a digital magnetic tape system, and the data link, go directly into the CP-901.

The P-3C's data processing activities are divided into four logic units, which form separate boxes in which all electronic operations are conducted in a combination of digital and analogue formats. Various keysets and panels complete the system hardware.
Logic unit 1 interfaces between the central computer and four types of peripheral information system: manual data entry, system status, sonobuoy receiver, and auxiliary readout display.

The manual entry sub-system provides the communication between the various operator stations and the central computer. Each operator has a panel of illuminated switches and indicators by which he communicates with the central computer. System status for the navigation and submarine anomaly detector is received and stored by part of logic unit 1, and transmitted to the central computer. The computer receives the status information on demand, or when any status changes. Finally, the auxiliary readout display logic interfaces between the central computer and the auxiliary displays at the Tacco and nav/com stations. The radar, sonar antenna, infra-red detection system and electronic support measures interfaces are also achieved by logic unit 1.
Logic unit 2 is concerned with communicating between the central computer and the navigation, and armament/ordnance. This logic unit transmits to the computer Doppler and inertial navigation data, and instructions to launch search-and-kill stores.
Logic unit 3 controls the cathode ray tube displays provided for the Tacco, sensor

operators, and pilot. The Tacco and pilot displays can generate characters, vectors and conics, while the sensor displays use characters and vectors only.
Logic unit 4 is mainly an expansion unit, comprising two items: the data multiplexer sub-units (dms) and the drum auxiliary memory sub-unit (dams). These provide extra input/output capacity and memory capacity respectively. The dms can service four input and output peripherals, as selected by the central computer. One output channel presents characters, vectors, and conics for the auxiliary sensor display, and one input/output channel is used for the aircraft's omega navigation system, a command launch system for the Harpoon anti-ship missile, and a sonobuoy reference system. The dams was incorporated to give an additional 393 216 words of memory to the computer, so that the operational program could be expanded to accommodate extra functions and equipment.

Various keysets and control panels allow access to the central computer via the data processing system. A universal keyset allows the transfer of information between the computer and the nav/com operator. The pilot uses his own keyset for controlling the information presented on his cathode ray tube display, entering navigation stabilisation data, dropping weapons and flares, and entering information on visual contacts. The ordnance panel displays the commands which the computer has given to the ordnance operator concerning status and position of the search stores, such as sonobuoys, which are available for deployment. Finally, there is an armament/ordnance test panel, which monitors the output from the data processing system logic unit 2 to those systems.

STATUS: in production for Lockheed P-3C Orion.

Hazeltine

Hazeltine Corporation, Greenlawn, New York 11740
TELEPHONE: (516) 462 5100
TELEX: 967800

AN/ARR-78 & AN/ALQ-158 advanced sonobuoy communication link

US Navy/Lockheed P-3C Orions are receiving the Hazeltine advanced sonobuoy communication link (ascl) as part of their Update III improvement, and the system has been selected for the up-date of 160 of the US Navy/Lockheed S-3B Viking IIs. Hazeltine describes the ascl as a versatile and effective high-performance system designed to satisfy the most demanding

current and future ASW requirements. The standard configuration provides 20 receiver channels, each of which can operate on any of the 99 standard sonobuoy vhf frequencies. In the P-3C, a single AN/ALQ-158 antenna processor feeds two AN/ARR-78 receivers, which send 40 channels of data into the aircraft's Proteus advanced signal processor.

The ALQ-158 comprises a vhf blade antenna, radio frequency pre-amplifier, processor unit, and operator's control box. The system is computer controlled, with a microprocessor-based control unit. Components, including radio frequency amplifiers and mixers, have a high dynamic range to avoid third-order intermodulation effects, and surface-acoustic-wave filters and delay lines are used because of their linear phase characteristics, low signal distortion, and lack of field alignment requirement.

Hazeltine says that the ascl receiver, the AN/ARR-78, is the first such sonobuoy device to be computer controlled, and the first to have 99 channels. It has five units: the AM-6875 pre-amplifier, R-2033 receiver assembly unit, C-10126 indicator control unit, ID-2086 receiver status indicator, and C-10127 ADF receiver control. The incorporation of an ADF function is another 'first' claimed for the ascl. The ADF transmitter is used as a sonobuoy on-top position indicator, telling the crew when the aircraft is directly over a sonobuoy. This means that a separate unit is not required to monitor sonobuoy positions, which change due to current and wind conditions.

The receiver unit has two separate front-ends. One is used as an amplifier/filter for the ADF signal and drives one of the 20 receiver modules, which is dedicated to the ADF function and which provides the ADF antenna drive error signal. The second receiver front-end amplifies and filters the sonobuoy signals before splitting them 19 ways for the remaining receiver modules. Sixteen of these modules are identical and provide the FM demodulated acoustic outputs to the Proteus processor. The other three are auxiliary receiver modules, which provide miscellaneous functions and contain AM and FM demodulators. Two of the auxiliary modules provide audio outputs to the operator's headphones, and the other monitors radio frequency signal strength, sending an indication to the Proteus.

The common receiver modules each measure 10 × 7.5 × 0.67 inches (254 × 191 × 17 mm), and make extensive use of custom hybrid circuits in the synthesiser, intermediate frequency and demodulation areas. Only slight differences in the latter distinguish between the normal and auxiliary acoustic modules. Surface-acoustic-wave filters are used to allow a low-profile mechanical design and confer high selectivity. Each module is divided into five compartments: dc regulator, high-level mixer, synthesiser, amplifier, and FM demodulator.

The receiver assembly unit contains an input/output module which acts as the interface with the three control and status units. This input/output unit contains an 8-bit microprocessor that communicates with the 32-bit Proteus via a module which slows down the 32-bit signals into sets of four 8-bit words. The Proteus can command all operational and test modes and request detailed receiver status information such as sonobuoy type, radio frequency channel, and power level. All commands that can be sent from the Proteus can also be initiated by the operator on his panel,

Five modules make up Hazeltine AN/ARR-78 advanced sonobuoy communication link

using switches and a 12-digit keyboard on the indicator control unit (icu). This unit contains lighted status and test message indicators and a row of light-emitting diodes to indicate which commands are being processed. The icu has its own microprocessor to format commands and decode messages from the receiver assembly.

Hazeltine has built extensive self-test equipment into the AN/ARR-78 which can be initiated automatically by the Proteus, or manually by the operator. This system includes a radio frequency-modulated signal-generator capable of 99-channel operation, special baseband test validation circuitry, and test-related microprocessor routines. A number of tests are run automatically by the microprocessor before each flight. There are also loop tests of the digital circuits, to verify that all the data transfer lines are operational. A check

sum routine in the two microprocessor memories makes sure that they are both working. Hazeltine claims that built-in test equipment can detect 96 per cent of faults, and isolate 99 per cent of these to the defective module.

In the P-3C Update III, the 20 receiver modules are inserted vertically into the box, along with five other modules such as the computer input/output unit. The power supply and reference oscillator modules are placed above this part of the receiver. Hazeltine has also repackaged the system into a smaller, uniform box which places all the modules alongside each other in boxes with the same form factor. Fewer receiver channels (about ten) are provided, but this is suitable for applications in smaller aircraft such as the S-3B Viking, or in helicopters.

Dimensions: (AM-6875) 3 × 5.75 × 4.25 inches (76 × 146 × 108 mm)
(R-2033) 12 × 21.3 × 15.3 inches (309 × 541 × 389 mm)
(C-10126) 9 × 5.75 × 6.5 inches (229 × 146 × 165 mm)
(ID-2086) 10.5 × 5.75 × 5 inches (267 × 146 × 127 mm)
(C-10127) 2.25 × 5.75 × 3.25 inches (57 × 146 × 83 mm)
Weight: (AM-6875) 2.1 lb (0.95 kg)
(R-2033) 101 lb (45.9 kg)
(C-10126) 6.6 lb (3 kg)
(ID-2086) 4.1 lb (1.9 kg)
(C-10127) 1.4 lb (0.6 kg)
Power: 500 W at 115 V ac, 380–440 Hz, 3-phase, plus 7 W at 18–32 V dc plus 50 W at 26.5 V ac, 400 Hz 3-phase
Electro-magnetic interference: MIL-STD-461A

STATUS: in production.

IBM
IBM, Federal Systems Division, 9500 Godwin Drive, Manassas, Virginia 22110
TELEPHONE: (703) 367 2476

AN/UYS-1 signal processor
The IBM UYS-1 advanced signal processor has been fitted to the first of the US Navy's updated Lockheed S-3A Viking ASW aircraft and flight

trials began early in 1984. A total of 160 aircraft will be so modified, beginning in 1987. A similar processor also equips the Lockheed P-3C Update III standard aircraft, deliveries of which began to the US Navy in 1984.

Dimensions: 57.3 × 23.4 × 11.2 inches (1455 × 594 × 284 mm)
Weight: 240 lb (109 kg)
Power: 1530 W

STATUS: in production. AN/UYS-1 has also been selected for the US Navy's SH-60B Lamps III helicopter and for a number of other, non-airborne, applications.

Rospatch
Rospatch Electronic Systems Division, 7500 Main Street, PO Box 750, Fishers, New York 14453-0750
TELEPHONE: (716) 924 4000
TWX: 510 254 2896
FAX: (716) 924 5732

AN/AKT-22(V)4 telemetry data transmitting set
This system relays up to eight channels of sonobuoy data from an ASW helicopter to a ship. It comprises the T-1220B transmitter-multiplexer, the C-8988A control indicator, an AS-3033 antenna, and a TG-229 actuator. The sonobuoy signals are received by dual

Rospatch AN/AKT-22(V)4 telemetry data transmitting set

AN/ARR-75 radio receiving sets and passed into the transmitter-multiplexer. The control indicator has four trigger switches, each of which disables two data channels. Composite trigger tones are brought into the multiplexer separately via the control indicator, and combined with the sonic data channels and the single voice channel. The resulting FM signal is used to modulate the transmitter.

The data-transmitting set has a Difar operating mode, in which the extra voice channel is inoperative, and the composite FM modulating signal is disconnected from the transmitter input. Two Difar sonobuoy transmissions enter the transmitter-multiplexer on dedicated channels. After conditioning in an amplifier-adapter, the Difar signal is split into two components, Difar A and Difar B. The A signal is conditioned in a low-pass filter, while the B signal drives a variable-cycle oscillator centred at 70 kHz, and then passes through a bandpass filter. The two filter outputs are combined linearly, and the resulting composite modulation signal drives the transmitter. A switch on the controller-indicator controls whether the normal sonic or composite Difar signals are transmitted.

The AS-3033 antenna has two sections: a vhf element for receiving the sonobuoy signals, and a uhf part which sends the multiplexed data down to the ship. The ship receives the information on an AN/SKR telemetric data receiving set.

Primary power input: 115 V ac, 3-phase, at 0.85 A, 1.5 A, and 1 A respectively

Warm-up time: <1 m in standby mode; <15 m under environmental extremes
Operating stability: >100 h for continuous or intermittent operation
Output frequency: 2200–2290 MHz, 1 of 20 switch-selectable S-band channels
Multiplexer inputs: (a) 8 sonar data channels (7 with 10–2000 Hz bandwidth; 1 with 10–2800 Hz), at 0.16–16 V
(b) 4 sonar trigger channels, 26–38 kHz, at 1–3 V
(c) 1 voice channel, 300–2000 Hz bandwidth, at 0–0.25 V
(d) 2 composite Difar channels, 10–2000 Hz bandwidth at 3–6 V or 10.6 V
Channel phase correlation: difference in phase delay between any 2 passive data channels <1° (10–500 Hz)

STATUS: in service with US Navy.

AN/ARR-72 sonobuoy receiving system
This system is used on the Lockheed P-3C Orions of the US Navy and Japanese Maritime Self-Defence Force to receive, amplify and demodulate the vhf/FM signals from sonobuoys deployed by those aircraft. The AN/ARR-72 is included in the P-3C Updates I and III, and in the US Navy/Sikorsky SH-60 Lamps III helicopter system. It is compatible with the AN/SSQ-36, 41, 47, 50, 53, and 62 sonobuoys and receives on 31 channels. The receiver system is in five parts: AM-4966 pre-amplifier, CH-619 receiver, SA-1605 audio assembly,

C-7617 control indicator, and SG-791 acoustic sensor signal generator (assg), which performs diagnostic functions.

The radio signals from the sonobuoys contain data on the frequency and directional pattern of the sound pattern received by the buoy's hydrophone array. These signals are amplified by the AM-4966 before passing into the radio receiver unit, where a multi-coupler distributes them to 31 fixed-tune receivers according to their frequency. The receivers further amplify and demodulate the signals to provide base-band audio and radio level signals. Each receiver channel contains two plug-in converters, a discriminator/amplifier, and a filter; all parts are identical for each channel except for the crystal oscillator. The receiver power supply, designated PP-5000, uses integrated circuits, as does the SA-1605 audio assembly. This unit contains 19 audio channels, each of which has a 31 by 1 switching matrix to select the output of a given receiver. The signal is amplified a final time before transmission to the data processing equipment. The desired receiver channel may be selected by the mission computer, or manually on the C-7617 dual-channel control indicator.

Built-in test equipment is contained within the system, which generates test radio signals for the pre-amplifier, the multicoupler, or a radiating antenna. Internal circuits generate simulated signals to test processing equipment for sonobuoys such as Lofar, extended Lofar, and range-only. End-to-end checks of sophisticated equipment such as the AN/AQA-7(V) Difar sonobuoy indicator are performed by accepting modulation from their target generators. The 31 radio frequencies are generated by a frequency synthesiser comprising a multi-frequency oscillator which reduces the channelling time to less than a second.

Weight: (pre-amplifier) 2 lb (0.9 kg)
(receiver) 5.5 lb (2.5 kg)
(audio) 39 lb (17.55 kg)
(assg) 19.2 lb (8.64 kg)
(8 control indicators) 32 lb (14.4 kg)
Total weight: 97.7 lb (44 kg)
Power: 300 W max at 115 V ac, 400 Hz single-phase, plus 280 mA at 28 V dc for panel lighting of each control indicator and assg, plus 250 mA at 18 V dc for assg annunciators
Frequency range: 31 channels between 162.25 and 173.5 kHz
Noise figure: 5 dB max (3.5 dB is typical)
If rejection: 66 dB minimum (more than 100 dB is typical)
Image rejection: 66 dB minimum (more than 100 dB is typical)
High audio level: 16 V at ±75 Hz deviation
Standard audio level: 2 V at ±75 Hz deviation
Crosstalk: 54 dB minimum
Output isolation: 60 dB minimum
Audio frequency response: ±1 dB from 20 Hz–20 kHz; ±6 dB from 5 Hz–40 kHz
Specifications: MIL-E-5400, MIL-STD-781, AR-5, AR-10, AR-34, MIL-I-6181
Warm-up time: 30 s max
Mtbf: 500 h including assg and 10 dual-channel control-indicators
Operational stability: 500 h
Operating life: 20 000 h

STATUS: in production.

US Navy's SH-60 Lamps III helicopters have Rospatch AN/ARR-72 sonobuoy receivers

AN/ARR-75 sonobuoy receiving set

The AN/ARR-75 is a 31-channel sonobuoy receiving set intended mainly for helicopters. It is currently specified for the Kaman SH-2 Lamps I, Sikorsky SH-3, and Sikorsky SH-60 Lamps III helicopters. Independent receiver modules provide four simultaneous demodulated audio outputs each capable of selecting one of the 31 standard frequencies. Its volume is about a third of its predecessor, the AN/ARR-52A, which also weighs twice as much. The reliability of the new system is ten times that of the earlier one, through the use of solid-state electronics and fixed-tuned circuits which remove tuning adjustments.

The AN/ARR-75 comprises two units, the OR-69 receiver group assembly and the C-8658 (or C-10429) radio set control. The former unit contains most of the electronics: the power supply, four receiver modules and an electrical equipment chassis. The power supply contains the transformer and rectifier assembly and four switching regulator boards which each supply a very pure (low-noise) voltage. Each receiver module contains a suppression filter, local oscillator, converter, amplifier-discriminator, and the audio frequency output circuit-board. The electrical equipment chassis contains the antenna relay and filter, pre-amp-multicoupler, reference oscillator, built-in test reference divider, suppression filter cavity and interface connections.

The control unit provides independent selection and meter monitoring of any one of the 31

Rospatch Mark 1634, active remote module used for ground test of AN/ARR-75

Rospatch AN/ARR-75 sonobuoy receiver

channels for each of the four modules. A push-button built-in test function is included. Maintainability is improved by the modularity of the electronics. Each circuit board can be replaced without soldering, and built-in test functions down to plug-in module level are provided by the optional Mark 1634 Active Remote Module support equipment.

Rospatch says that the main features of the AN/ARR-75 are a large dynamic range and high sensitivity over a broad base-wavelength. A wide, linear-phase audio output is provided, with close uniformity between the four audio channels.

Construction: modular, with 36 quick-replacement assemblies
Dimensions: (receiver) 8 × 7.3 × 12 inches (203 × 186 × 305 mm)
(controller) 4.8 × 5.7 × 2.8 inches (122 × 145 × 71 mm)
Weight: (receiver) 21.5 lb (9.8 kg)
(controller) 2.5 lb (1.1 kg)
Input power: 75 W at 115 V ac, plus 5.6 W for lighting at either 27 V (C-8658) or 5 V (C-10429)
Frequency range: 162.25–173.5 MHz (31 channels)
Noise figure: 5 dB at 50 ohms
Specifications: MIL-STD-461, 462, 463, 781, MIL-E-5400, MIL-R-81681, AR-5, 8, 10, 34
Mtbf: 1500 h
Life: 20 000 h

STATUS: in production.

AN/ARR-75( ) sonobuoy receiving set

Rospatch has designed this replacement for the AN/ARR-75, but it has not yet been purchased and its designation remains AN/ARR-75( ). In physical size and audio characteristics, the system is the same as its predecessor, but channel capability goes up from 31 to 99. Four AN/ARR-75( ) sets may be used to replace the AN/ARR-72 in a large ASW aircraft. This multiple installation improves reliability through redundancy, as well as by modernising the electronics. No changes are necessary to the aircraft's electrical harness for such a retrofit.

This improved receiver uses the same electronic design as the AN/ARR-75, with a few improvements such as higher data output capability, digital output at radio frequency

Modules of Rospatch AN/ARR-72 sonobuoy receiver (from left): pre-amplifier, acoustic sensor signal generator, receiver, audio assembly and control indicator

Rospatch AN/ARR-75( )

level for remote indication, better components, and computer controlled built-in test. The same active remote module is used for diagnostic testing.

In October 1985 Rospatch announced that a development of the ARR-75 was to form part of the GEC Avionics AQS-903 acoustic processor for the Royal Navy's EH 101 helicopters. The units will be produced under licence by Dowty: programme value is put by Rospatch at in excess of $2 million. Dowty designates this processor R609.

Specifications: as AN/ARR-75 except:
Weight: (receiver) 22.5 lb (10.2 kg)
(controller) 2.5 lb (1.1 kg)
Power: 70 W

STATUS: available.

R-1651/ARA on-top position indicator

The US Navy/Lockheed P-3C Orion Updates I and III and the Sikorsky SH-60 Lamps III helicopter all carry the R-1651. This unit is a sonobuoy on-top position indicator. It is used in locating sonobuoys, and as a receiving converter when used with the aircraft's automatic direction finder equipment. The AN/ARA-25 automatic direction finder control box contains a switch which commands the R-1651 to switch the direction finder from a uhf automatic direction finder to a vhf sonobuoy. To establish the position of a sonobuoy, which moves with wind and sea currents, the R-1651 uses the vhf signals from the buoy. The system has the standard 31 channels, the correct one being chosen by binary code.

Dimensions: 3 × 4.9 × 5.3 inches (76 × 124 × 135 mm)
Weight: 3.5 lb (1.6 kg)
Input power: 15 W at 115 V ac, plus 20 W at 28 V dc for lighting

Rospatch R-1651/ARA sonobuoy on-top position indicator

Frequency range: 162.25–173.5 MHz (31 channels)
Sensitivity: 10 dB minimum
Noise: 6 dB nominal
Specifications: MIL-R-81680, MIL-E-5400, MIL-T-5422, MIL-STD-461, 704, 781, 785, MIL-I-6181
Temperature: –54 to 55° C
Mtbf: 1000 h
Operational stability: 1000 h
Operating life: 5000 h

STATUS: in production.

SG-1196/S acoustic test signal generator

Rospatch developed the SG-1196 acoustic test signal generator for the US Navy/Lockheed P-3C Orion Update III programme. The unit provides test patterns representing the formats of most active and passive sonobuoys for in-flight testing of AN/ARR-78 sonobuoy receivers and Proteus acoustic signal processors. It acts on all 99 standard sonobuoy vhf channels, and it has been designed for calibration duties as well as functional tests. The SG-1196 is interchangeable mechanically and electrically with the SG-791, the acoustic sensor signal generator used in conjunction with the

AN/ARR-72 on the P-3C. The acoustic test signal generator can thus be retrofitted easily, and will then provide a greater number of available test signals, which are more accurate and more stable. A single 23 MHz crystal oscillator synthesises all the radio frequencies in the SG-1196, while coherent baseband signals are generated by a 7 to 68 MHz crystal oscillator. A pseudo-random digital source provides simulated sea noise.

Weight: 21 lb (9.5 kg)
Frequency accuracy: ±5 kHz
Output level accuracy: ±0.5 dB
Channelling time for an on-channel signal: 100 ms typical, 250 ms max
Distortion and noise: 1% for deviations to 75 kHz and modulations to 50 kHz
1.5% for deviations to 150 kHz and modulations to 100 kHz

Difar and Lofar modes
Number of targets: 8
Target frequency accuracy: ±0.01%
Signal-to-noise ratio accuracy: ±0.5 dB

Range-only mode
Number of targets: 84, in 2 patterns
Number of Doppler frequencies: 11 per pattern; each accurate to ±0.1 Hz
Number of signal levels: 4
Signal-to-noise accuracy: ±0.5 dB
Timing accuracy: ±1 ms
Repeat time: 12.8 s

Cass mode
Number of targets: 8, in 2 patterns
Number of Doppler frequencies: 4 per pattern, each accurate to ±0.1 Hz
Number of signal levels: 4
Signal-to-noise ratio accuracy: 0.5 dB
Timing accuracy: ±1 ms
Repeat time: 12.8 s

Dicass mode
Number of targets: 8, in 2 patterns
Number of bearings: 4, accurate to ±1°
Doppler frequencies: as in cass mode

B/T mode
B/T frequency: 1700 Hz
Number of levels: 4, accurate to ±3%

STATUS: in production.

99-channel sonobuoy receiver

Rospatch has developed a 99-channel sonobuoy receiver and on-top position indicator which has been selected for the Sikorsky SH-60F CV-Helo carrier inner-zone helicopter. A $2.3 million contract, with options totalling a further $8.7 million was received in mid-1985.

Sparton

Sparton Corporation, 2400 East Ganson Street, Jackson, Michigan 49202
TELEPHONE: (517) 787 8600
TWX: 810 253 1925

TD-1135 directional sonar processor

The TD-1135 directional sonar processor accepts a composite Difar signal from any sonobuoy received and demultiplexes it to provide north-south, east-west and omni-directional outputs. This equipment features precise target bearing resolution, immunity to phase pilot noise sideboards, built-in test, outputs for computer analysis, selectable lock retention-time constants, and manual or remote gain control.

Frequency range: broadband Difar audio
Variable gain: (manual or remote) 48 dB
Time constants: 0.1, 1.0 and 10 s
Phase pilot bandwidth: 1 Hz max
Processor analysis bandwidth: 1 Hz max
Mtbf: 1500 hours min
Demultiplexer outputs: N-S, E-W and omni-directional
Processor outputs: N-S, E-W and omni-directional

Texas Instruments

Texas Instruments Inc, PO Box 226015, MS 3127, Dallas, Texas 75266
TELEPHONE: (214) 480 1417
TELEX: 470900
TWX: 910 867 4702

AN/ASQ-81(V) magnetic anomaly detector system

The AN/ASQ-81(V) MAD system was developed by Texas Instruments for the US Navy for

use in the detection of submarines from an airborne platform. The system operates on the atomic properties of optically pumped metastable helium atoms to detect variations of intensity in the local magnetic field. The Larmor frequency of the sensing elements is converted to an analogue voltage which is processed by bandpass filters before it is displayed to the operator.

Four configurations of the AN/ASQ-81(V) are available, two for use within an airframe and two for towing behind an aircraft. The US Navy uses the AN/ASQ-81(V)1 in the land-based

Lockheed P-3C Orion, where it is housed in a tail 'sting'. The AN/ASQ-81(V)3 is installed in the carrier-based Lockheed S-3A Viking aircraft, where it is extended on a boom.

The AN/ASQ-81(V)2 is a towed version employed by the US Navy on Sikorsky SH-3H and Kaman SH-2D helicopters It is also in service with other countries including the Netherlands, for use on the Westland Lynx, Japan, for use on the Mitsubishi HSS-2, and with forces employing the Hughes 500D helicopter.

The second towed version, the

AN/ASQ-81(V)4, is used by the US Navy on the Sikorsky SH-60B Lamps III helicopter.

All versions of the AN/ASQ-81(V) have the same C-6983 detecting set control, AM-4535 amplifier and power supply unit. The AN/ASQ-81(V)1 and 3 use a DT-323 magnetic detector, while the AN/ASQ-81(V)2 and 4 have a TB-623 magnetic detecting towed body. The towed version is controlled by the C-6984 reel control, which works the RL-305 magnetic-detector launching and reeling machine.

STATUS: in production and service.

AN/ASQ-81( ) magnetic anomaly detector system

The ASQ-81( ) is the latest derivative of the ASQ-81(V) (see previous entry) which uses digital electronics and advanced microprocessor technology to achieve aircraft compensation. Two sub-variants are available; one for inboard use and the other for towed installations. The new version is compatible with MIL-STD-1553B data-bus standards.

The ASQ-81( ) version uses the same wiring as the (V) model, requiring only one cable change for vector sensing. Compared with earlier MAD systems, the new equipment, which is suitable for both new aircraft and for retrofitting, offers enhanced reliability and performance as well as considerably reducing the number of units required to make up the total system.

Royal Netherlands Navy/Lockheed P-3C Orion uses Texas Instruments AN/ASQ-81(V)1 MAD system

STATUS: in development.

Radar

CANADA

Litton Canada

Litton Systems Canada Limited, 25 City View Drive, Rexdale, Ontario M9W 5A7
TELEPHONE: (416) 249 1231
TELEX: 06-989406
TWX: 610 492 2110

AN/APS-140 radar

The AN/APS-140(V) is an advanced derivative of Litton's APS-504 family of airborne search radars. The system employs a fully coherent travelling-wave tube transmitter with wideband frequency agility, high-ratio pulse compression, scan-to-scan integration and complementary digital signal (CFAR) processing. Selected combinations of these attributes achieve the primary performance objective of high-detection probability for small targets in high sea states. As a supplementary benefit, the wide-band frequency agility endows the system with relative immunity to ECM and to unintentional interference from other emitters.

STATUS: in production.

Units of Litton APS-504(V) forerunner of AN/APS-140(V)

MacDonald Dettwiler

MacDonald Dettwiler Technologies Limited, Airborne Radar Division, 3751 Shell Road, Richmond, British Columbia V6X 2Z9
TELEPHONE: (604) 278 3411
TELEX: 04-355599

IRIS synthetic aperture radar

MacDonald Dettwiler introduced the military version of its airborne synthetic aperture radar (SAR) reconnaissance system, IRIS, in 1985.

The IRIS represents the state of the art in SAR reconnaissance systems. At a stand-off distance of 100 km and a height of 15 km, it will produce in real-time and on board the aircraft strip imagery on dry paper (or film if desired) with resolutions of approximately 16 m over a 64 km swath.

Targets 1 metre across are detectable in this mode. The system can be switched instantly to a narrow swath mode where resolutions of 6 metres or less are possible over a 12 km swath.

The IRIS can be entirely mounted in an executive-sized jet or turboprop to provide a most economical system. Four configurations are available:

IRIS 1000 consists of radar antenna, inertial guidance system, SAR processor and a dry paper imaging device.
IRIS 1010 adds a downlink capability using a vhf transmitter. The receivers are small and rugged, thereby adding significant operational flexibility and responsiveness. Ground-based commanders are able to receive and analyse the SAR imagery as it is produced on the aircraft. In addition, the low cost and simplicity of these receivers allows for the set up of arrays along a border providing the most economical deployment of a SAR border surveillance system. This deployment also permits a very rapid area coverage capability.

IRIS 1020 includes a high density tape recorder onto which the unprocessed SAR data can be written. This tape is transferred to a ground-based precision SAR image processor which will produce SAR imagery with 1 metre resolution.
IRIS 1030 offers the basic system with the downlink and precision processing hardware in a single package. Current systems can include the following options: moving target indications, on-board switching from maritime patrol radar to SAR mode, and a sophisticated video display and image enhancement package.

In its civilian application, four IRIS (and IRIS prototypes) are flying, and the prototype IRIS has flown over 2000 hours in a difficult environment (the Canadian Arctic) without failure.

Further information on the IRIS system is given in the Addenda.

STATUS: in development.

CHINA, PEOPLE'S REPUBLIC

Airborne early warning/control system

Unconfirmed reports suggest that China may be investigating the possibility of producing its own airborne early warning system. The aircraft to carry this could be a version of China's Y-10 airliner, a four-engined transport resembling a Boeing 707. As well as American or Franco-American engines, it would have a surveillance radar almost certainly of US or UK origin (Westinghouse, General Electric or GEC Avionics). This system would assist in protecting China's very long frontier with the USSR by providing a 'force multiplier' to enhance the effectiveness of the 50 divisions deployed there.

FRANCE

Electronique Aérospatiale

Electronique Aérospatiale, BP 51, 93350 Le Bourget
TELEPHONE: (1) 48 62 51 51
TELEX: 220809

AT880R Migrator series ATC transponder

The AT880R is a fully modular ATC transponder, with full 'Mode C' capability and suitable for business and training aircraft and helicopters.

Weight: 1.7 kg
Dimensions: 62 × 123 × 276 mm
Frequency: (receive) 1030 MHz (transmit) 1090 MHz
Power: 0.8A at 28 Vdc
Power output: 330 W nominal
Altitude: up to 30 000 ft
Prf: 1200 replies/s

19

BCT 2535 digital transponder control unit

The BCT 2535 is a digital dual transponder control panel for use in commercial aircraft. The system allows two ARINC 718 ATC transponders to be remotely-controlled with ATC codes being inserted via a keyboard.

Dimensions: 146 mm × 66.6 mm × 140 mm
Weight: 1.5 kg max
Display: non-emissive dichroic lcd, white on black background, contrast ratio 15:1
Interface: ARINC 429
Power: two inputs 115 V ac/400 Hz < 100 mA each

Electronique Aérospatiale BCT 2535 transponder control unit

Electronique Serge Dassault

Electronique Serge Dassault, 55 quai Carnot, 92214 St Cloud
TELEPHONE: (1) 46 02 50 00
TELEX: 250 787

ESD 3000 IFF transponders

A range of transponders, with the generic designation ESD 3000, has been developed by ESD for fixed-wing aircraft and helicopters. They can be used, without modification, in conjunction with an automatic code-switching system. There are two versions, employing the same modules: the ESD 3300, in a single box, and the ESD 3400, in a two-box configuration.

Dimensions: (ESD 3300) 133 × 146 × 145 mm
(ESD 3400 control unit) 133 × 146 × 75 mm
(ESD 3400 transmitter-receiver) 194 × 90.5 × 319 mm
Weight: (ESD 3300) 4 kg
(ESD 3400) 7.8 kg
Frequency: (transmit) 1090 MHz
(receive) 1030 MHz
Power output: 500 W peak
Sensitivity: –77 dBm
Power: 50 W at 28 V dc
Reliability: 1500 h
Qualification: STANAG 5017 (Mk XA/IFF 3300; Mk XII/IFF 3400), Air 7304, ICAO Appendix 10 and ARINC 572

STATUS: in production.

Antilope V radar

The Antilope V radar has been designed for the Dassault Mirage 2000N. Its basic functions are terrain-following, air-to-air, air-to-sea, air-to-ground and navigation with ground mapping and navigation updating. It employs a travelling-wave tube J band coherent transmitter and a flat slotted-plate antenna. Radar information is displayed on a head-up display and on a three colour multi-mode cathode ray tube head-down display. The system can provide terrain-following commands at 300 feet (91 metres) and 600 knots computing a preset obstacle-clearance height and with a pre-selected g level.

ESD Antilope V terrain following and navigation radar

ESD Aida II automatic fire-control radar for guns and infra-red missiles

STATUS: Electronique Serge Dassault is design leader on this programme which is shared with Thomson-CSF. Mirage 2000N radar is planned for service in 1986.

Aida II interception radar

This miniature lightweight system is designed for light interceptors and aircraft with restricted accommodation in the nose. It can also be installed in pods for under-wing mounting as seen in some versions of the Dassault Mirage V. The system automatically searches for, acquires and provides ranges of air or sea targets within a cone of 18°. The antenna is fixed and the pilot points the aircraft in the direction of the target. Used in conjunction with a gyroscopic gunsight it supplies all the information for interception and attack with guns, rockets, bombs or missiles. The operating frequency lies in the I to J band and the transmitter power is between 80 and 100 kW.

RDN 80B navigation Doppler radar for helicopters

The RDN 80B represents a new generation of navigation Doppler radars featuring small size and low cost and intended for installation in lightweight helicopters, for which cost, size, ease of installation and weight are of major importance. It comprises a single unit, containing both antennas and the transmitter/receiver and produces the three orthogonal components of helicopter velocity. These components are used in association with a processor for providing dead reckoning navigation and also for pilot assistance or autopilot control in hovering mode.

Radar operation is checked in flight and on the ground by monitoring its main parameters by means of a built-in test system. This equipment has been developed under licence from Racal-Decca.

Dimensions: (with radome) 416 × 390 × 82 mm
Weight: (with radome) 9 kg
Transmission frequency: 13.325 MHz ± 10 MHz
Transmitter operating mode: cw
Transmitted power: 30 mW
Velocity range: –50 to +350 kt
Operational envelope: 0-20 000 ft altitude
Accuracy: (incremental data) ±0.5%
(numerical data) ±0.4%

Electronique Serge Dassault ESD 3300 single-box IFF transponder

Reliability: >2000 h mtbf
Transmit and receive antennas: 2 arrays of slots photo-etched on pcb
Transmitter: Gunn-diode oscillator
Power consumption: 45 W at 28 V dc

ESD RDN 80B navigation radar

LCT

Laboratoire Central de Telecommunications, 18-20 rue Grange-Dame-Rose, 78140 Vélizy-Villacoublay Cedex
TELEPHONE: (3) 946 96 15
TELEX: 698892

Orchidée battlefield radar

Flight trials of Orchidée (Observations Radar Coherent Heliporté d'Investigations des Element Ennemis) began in 1985. The radar is intended for use on board Aérospatiale SA 332 Super Puma helicopters of the French Army; about 20 helicopters will be equipped with the X band pulse-Doppler radar, becoming operational in 1992. Electronique Serge Dassault is responsible for the air-to-ground data link and that company is also developing an expert system to help decide tactics and strategy based on radar-detected data. Aérospatiale is responsible for system integration.

It is intended that the helicopter will fly at an altitude of about 3000 metres (10 000 feet) some 50 km (30 nautical miles) behind the battle lines and from there be able to detect enemy troop movements and dispositions, up to 100 km (60 nautical miles), the other side of the battle line.

Super Puma trials helicopter with prototype Orchidée antenna installation. Antenna rotates backwards through 90° for take-off, transit and landing

A data link will pass data into mobile ground terminals up to 60 km (37 nautical miles) away, the data being transferred into the French Army's Rita communications network.

Following the current phase of development, flight trials of a full prototype system are scheduled for 1988-91.

STATUS: flight trials of the antenna began in late 1985.

LMT

LMT Radio Professionelle, 46 quai Alphonse le Gallo, 92103, Boulogne-Billancourt Cedex
TELEPHONE: (1) 608 60 00
TELEX: 202900

LMT is a subsidiary of Thomson-CSF

NRAI-4A IFF transponder

The NRAI-4A transponder is the airborne element of the Mk X IFF unit, compatible with the Mk XII IFF system. It includes Mode C altitude coding and an ISLS (interrogator side-lobe suppression) function that inhibits replies to sidelobe signals from the interrogator. Friendly aircraft can be identified without manoeuvring. An IP (identification of position) capability permits specific friendly aircraft to be identified on the radar screen. The single-box system contains an automatic self-test function.

Dimensions: 127 × 129 × 145 mm
Weight: 2.6 kg
Frequency: (transmission) 1090 MHz (reception) 1030 MHz
Power output: 500 W
Sensitivity: -74 dBm
Modes available: 1,2,3/A,C
Codes: 32 on Mode 1, 4096 on Modes 2 and 3/A, 2048 on Mode C

STATUS: in production for Aérospatiale SA 330 helicopters and Sepecat Jaguar and Dassault Mirage F1 aircraft.

NRAI-7A IFF transponder

The NRAI-7A IFF solid-state, diversity transponder inhibits replies to interrogator side-lobe transmissions, and automatically codes special replies, for example, emergency, and assists in the position identification of particular aircraft. The pilot can also insert codes such as radio failure alert and warning of hijackers aboard. The system incorporates a diversity function, that is, it uses two antennas at different positions on the aircraft and transmits a reply via the antenna that receives the strongest interrogation signal. This permits more accurate identification, particularly during combat, where manoeuvres can blanket or interrupt signals. A one or two-box format is offered. The system is contained in a single box.

Dimensions: 130 × 127 × 145 mm
Weight: 3 kg
Frequency: (transmit) 1090 MHz (receive) 1030 MHz
Power output: 500 W peak
Sensitivity: -77 dBm
Modes available: 1,2,3/A, C, and Mode 4 compatibility
Power: 30 W at 28 V dc

LMT NRAI-7A IFF transponder for Mirage 2000

Number of codes: 32 in Mode 1, 4096 in Modes 2 and 3/A, 2048 in Mode C

STATUS: in production for Dassault Mirage 2000.

NRAI-9A IFF transponder

Essentially a two-box version of the NRAI-7A, the NRAI-9A incorporates several improvements. They include the elimination of side-

LMT NRAI-10A IFF transponder for Dassault ATL2

LMT NRAI-9A IFF transponder

lobe response, and automatic special-code referral, with positive identification permitting a ground operator to localise a particular aircraft. Special emergency codes may also be employed, chosen by the pilots, such as radio failure. Dual receiver channels connected to upper and lower antennas along with an antenna switch and comparison circuits, provide a diversity function.

Dimensions: (transmitter-receiver) 58 × 193 × 361 mm, (control unit): 127 × 130 × 80 mm
Weights: (transmitter-receiver) 2.5 kg, (control unit) 1.4 kg
Power output (peak) 500 W
Frequency: (transmit) 1090 MHz
Modes available: 1, 2, 3/A, C, and Mode 4 compatibility
Codes: 32 on Mode 1, 4096 on Modes 2 and 3/A, 2048 on Mode C

STATUS: pre-series production.

NRAI-10A IFF interrogator

This device permits an aircraft to interrogate another aircraft or ship. Coded interrogation signals are transmitted synchronously with pulses from the aircraft's radar, the coded reply supplying the identity or altitude of the responding aircraft. The single-box NRAI-10A incorporates a coder/decoder, transmitter and dual receiver, and signal returns are passed to a radar screen on the aircraft. The system is used in conjunction with a monopulse antenna having two different radiation patterns (one with a maximum, the other with a minimum) on the antenna axis. Unwanted returns due to sidelobe interrogation are suppressed and valid returns sharpened on the display.

Format: 3/8 ATR Short
Weight: 10 lb (4.2 kg)
Frequency: (transmit) 1090 MHz (receive) 1030 MHz
Sensitivity: –81 dBm
Power output: 1 kW
Modes available: 1,2,3/A and C. Mode 4 can be obtained with suitable coder/decoder

STATUS: in production for Dassault ATL2 maritime patrol aircraft.

Rasit battlefield surveillance radar

LMT and US company TCOM Corporation (a subsidiary of Westinghouse) have jointly produced a tactical surveillance radar called Rasit for installation in balloons, so overcoming the traditional disadvantage of ground forces of visibility being limited by natural obstacles such as hills or woods. The French radar weighs 55 kg, and is suspended from TCOM's Stars balloon as a two-axis, gyro stabilised package so that its antenna remains pointing in the same direction irrespective of the motion of the balloon. The radar was originally developed for surface use. The balloon is 25 metres long, has a volume of 700 cubic metres, and is tethered to a truck for mobility and ease of deployment. The maximum operating altitude is 750 metres, from where the system can under the best conditions detect individual persons at a range of 20 km, light vehicle at 30 km, and heavy vehicles at its maximum range of 38 km. Electrical power is fed to the radar via a cable connecting it to a generation system in the truck, and information is fed from the radar to an operator's console in the truck via an optic-fibre. The truck contains all processing, control and display equipment.

STATUS: in production. The system has been supplied to some 12 countries.

Omera

Société d'Optique, de Mécanique, d'Electricité et de Radio Omera-Segid, rue Ferdinand Berthoud, 95101 Argenteuil Cedex
TELEPHONE: (1) 39 47 09 42
TELEX: 696797

ORB 37 radar

This radar for fixed-wing aircraft and helicopters is the company's first venture into weather warning systems of this type but is based on a substantial avionics and radar background, including the ORB 32 surveillance radar. The monochromatic ORB 37 can also be used in a ground-mapping mode for navigation and there is an interrogation facility for beacon-homing.

The system comprises seven units: a slotted-array flat-plate antenna, separate transmitter and receiver, a plan-position indicator high-definition circular display for ground-mapping at a navigator's position, a rectangular weather display on the flight deck, and two control units, one for each station.

For maximum efficiency the antenna scans at a low rate for the weather mode and at a high rate for the ground-mapping mode. The corresponding pulse widths are 2.5 and 0.4 microseconds.

Principal components of Omera ORB 37 weather radar

STATUS: production deliveries for updated version of Franco-German C-160 Transall transport aircraft began in 1981, and derivative has been ordered by French Army for its Aérospatiale Puma and Super Puma helicopters.

ORB 32 radar

Omera's Heracles II/ORB 32 family of radars was developed from the earlier Heracles I/ORB 31.

The ORB 32 is a modular system whose sub-units can be tailored to optimise specific tasks.

The radar fulfils a wide range of missions under the headings of maritime patrol, anti-submarine warfare, active missile fire control, search and rescue, radar navigation and weather warning. Units common to all versions of the system are the antenna drive mechanism, junction box and control unit, while other parts of it, notably the antenna itself, transmitter-receivers and indicators and control/display consoles, are specified by the user. There are essentially two types of display: a 5-inch (127 mm) high-brightness digital system for mounting on the flight decks and a 9 or 16-inch (229 or 406 mm) tactical unit for an observer or weapons-system operator. The antenna drive mechanism permits continuous 360° scanning or 60°, 120°, or 180° sector scan, 12 and 24 rpm rotation speeds, ±15° beam elevation adjustment, antenna line-of-sight stabilisation up to ±20° and roll stabilisation up to ±35°. The system can drive wide-band antennas of various sizes which consist of a parabolic reflector with rear illumination.

Detection ranges claimed are 50 nautical miles for a snorkel, 75 nautical miles for a trawler, 100 nautical miles for a destroyer and 200 nautical miles in the weather mapping mode.

STATUS: in production and installed on Nord 262, Beech Super King Air, Aérospatiale SA 321 Super Frelon, SA 332 Super Puma, SA 365N Dauphin, and Boeing Vertol 107 helicopters.

Radar display of Omera ORB 32 aboard Aérospatiale SA 321 Super Frelon helicopter

Thomson-CSF

Thomson-CSF, Division des Equipements Avioniques, 178 boulevard Gabriel Péri, 92240 Malakoff
TELEPHONE: (1) 46 55 44 22

Cyrano IV radar family

Introduced in 1972, the Cyrano IV was the first of a family of air-to-air and air-to-ground radars. The following models are available or planned: Cyrano IV, IV-1, IV-2, IV-3, IV-M, IV-MR and IV-M3. The French Air Force operates three Dassault Falcon 20 aircraft modified to carry Cyrano radar for the training of fighter pilots. The right-hand pilot's panel is fitted with instruments and displays appropriate to the particular fighter being represented, eg Dassault Mirage III, IV, F1, or 2000.

Cyrano IV radar

The first Cyrano radar, a monopulse system, was designed for air-to-air interception, searching for and tracking hostile aircraft and providing flight, firing, and break-off information to the pilot via a weapons sight. In the search mode the radar scans ±60° in azimuth and ±30° in elevation. When the pilot puts a marker on his display to designate a particular target, the radar moves into the track mode, measuring the

range and relative velocity. In the final interception mode, the system signals the earliest and optimum times to fire a designated weapon, and then indicates when the firing 'window' has ended. The system, which operates in the 8 to 10 GHz band (the generator being a coaxial magnetron), was originally designed for the French Air Force/Dassault Mirage F1s.

STATUS: remains in production for export versions of Mirage F1.

Cyrano IV-1 radar

By adding a fixed-target rejection filter the Cyrano IV radar can be upgraded to include moving-target indicator capability so that hostile aircraft can be tracked amid ground clutter in the 'look-down' mode.

STATUS: in production.

Cyrano IV-2 radar

With the addition of real-beam-sharpening circuits and other modifications, the Cyrano IV converts into the first multi-mode member of the family. Besides air interception, the radar can be used for ground-mapping and low-altitude navigation. The latter capability includes contour mapping, terrain avoidance and blind penetration. For ground-attack the system

can provide range-to-target with an accuracy which matches the weapons' accuracy.

STATUS: in production.

Cyrano IV-3 radar

This progressive development of the series brings together all the improvements incorporated in the -0, -1, and -2, radars.

STATUS: in production.

Cyrano IV-M radar

The first Cyrano system to be designed specifically for multi-function operation (though still aimed basically at air interception), the -M can accomplish a wide variety of air-to-air, air-to-ground and air-to-sea missions and has a high degree of resistance to electronic countermeasures. Radar video signals are put up on a head-down cathode ray tube with a B-type display for air-to-air operation and a plan-position indicator presentation for air-to-surface operation. An interception and firing computer optimises the radar at medium and short-range combat. Interception guidance information is displayed on a large field of view head-up display. The 57 cm diameter inverted Cassegrain antenna handles 200 kW of power from a coaxial magnetron. Built-in test circuits permit rapid fault diagnosis down to individual line-replaceable unit level.

STATUS: French Air Force has upgraded Cyrano IV equipment to IV-M standard, and it is installed on Dassault Mirage F1s.

Cyrano IV-MR radar

This version of the Cyrano IV-M has been optimised for multi-sensor reconnaissance aircraft such as the Dassault Mirage F1CR. In combat or interception it provides the same information as the Cyrano IV-M, but ground-mapping (with optional Doppler beam-sharpening), contour-mapping and 'blind' penetration are additional capabilities to permit low-altitude operation in any weather. Information is presented on a television-type (raster-scan) head-down display, and can be integrated via a digital data link with other aircraft systems.

Thomson-CSF Cyrano IV airborne interception radar

Thomson-CSF Cyrano IV-M multi-mode radar

Thomson-CSF Cyrano IV-M3 radar

Two French Air Force F1CR squadrons are operational with Cyrano IV-MR.

STATUS: in production.

Cyrano IV-M3 radar

A development of the -M, this radar is designed for multi-role versions of the Dassault Mirage 50 and for retrofitting on Mirage III or V with air-to-air, air-to-ground and air-to-sea missions. Information is presented to the pilot on head-up and head-down displays similar to those employed with the Cyrano IV-M. Intended for export, or for aircraft with an earlier generation of electronics, this system employs technology developed for the new RDM and RDI radars (for details see separate entries), in for example resistance to electronic countermeasures. Modular design and built-in test make for easy maintenance.

STATUS: development has been completed.

RDI (radar Doppler à impulsions)

This is one of two pulse Doppler radars (the other being the RDM) developed for France's Dassault Mirage 2000 and stems from a French Ministry of Defence initiative in 1976. The RDI is intended for the all-altitude, air superiority and interception version, and is based on a travelling-wave tube, I to J band transmitter radiating from a flat, slotted plate antenna. The range is said to be around 90 km, and the radar is designed to work in conjunction with the 40 km range Matra Super 530D semi-active homing air-to-air missile. The performance of the radar, and of other systems on the aircraft, benefits from the digital signal handling and transmission of information by data-bus. Considerable electronic countermeasures resistance is built into the equipment, which can operate in air-to-

air search, long-range tracking and missile guidance, and automatic short-range tracking and identification modes. Although designed for air-to-air operation, the system incorporates ground-mapping, contour-mapping, and air-to-ground ranging modes.

A two-year delay in development has obliged a rescheduling of equipment among the first 50 Mirage 2000s, which accordingly have been fitted with the Thomson-CSF RDM radar. The fighter's airframe was too small to accommodate the original radar design and subsequent delays were the results of substantial design alterations, incorporating hybrid-circuit technology. It is expected, however, that the remaining 150 aircraft will be fitted with the RDI radar. The system is being developed by Thomson-CSF in collaboration with Electronique Serge Dassault, which is undertaking 30 per cent of the work.

STATUS: in production. Initial deliveries for flight testing and qualification were made in the summer of 1986. About 14 radars, including three prototypes, are involved with the flight test programme. Thomson-CSF and French Defence Ministry are already discussing an improved version for subsequent Mirage 2000s.

RDM (radar Doppler multifonction)

This monopulse Doppler X band radar is in production for the French Dassault Mirage 2000. Whereas the RDI is designed for interception and air combat, the coherent, multi-mode, all-digital, frequency-agile RDM is intended largely for the multi-role export version. It operates in air defence/air superiority, strike, and air/sea modes.

In the air-to-air role, the system can look up or down, range while searching, track while scanning, provide continuous tracking, illuminate targets for medium-range air-to-air missiles, generate aiming signals for air combat, and compute attack and firing envelopes. For the strike role it provides real-beam ground-mapping, navigation updating, contour-mapping, terrain-avoidance, blind let-down, air-to-ground ranging, and GMTI (ground moving target indications). In the maritime role it provides long-range search, track while scan and continuous tracking, and can designate targets for active missiles.

Options include a continuous wave illuminator and Doppler beam-sharpening, IFF identification and raid assessment.

Verification of the RDM design has been conducted with five prototype and three pre-production radars, and the system was first flown in January 1980. By January 1983 the company had delivered the first production

system and a further three were undergoing acceptance tests, some of them aboard Mirage 2000s. The planned eventual production rate is eight a month. Development is understood to have cost about Fr350 million (about US $48 million) and to have been partly, if not wholly, company funded.

Performance estimates show that 90 per cent of typical fighter-type targets, with radar cross-sections of 5 square metres, will be detected out to distances of 85 km with a four-bar search pattern scanning 120° in azimuth. The range increases to 100 km if single-bar pattern is used, and the scan is reduced to 30°. In the air-to-ground mode the system can ground-map a track 60° on each side of the centreline. For terrain-avoidance the system computes two clearance planes, shown on the head-down display. In a maritime application the radar can detect vessels down to patrol-boat size at ranges of up to 100 km.

The RDM system comprises eight line-replaceable units: antenna, antenna drive, X band transmitter, master oscillator, programmable data processor, two processing units, and power supply. Optional beam-sharpening and continuous-wave illumination (for use with the Matra S530D air-to-air missile for look-up/look-down interception) is also available. The inverted Cassegrain antenna is roll-stabilised and has a diameter of 655 mm with a corresponding beam-width of 3.6°. Self-test circuits localise faults to line-replaceable unit level and significant improvements in reliability are the results of microprocessor technology, hybrid circuits, and microwave integrated circuits. Comprehensive electronic counter-counter measures are incorporated.

Thomson-CSF, in conjunction with its subsidiary LMT is developing an IFF transponder for the RDM radar. Radar and IFF outputs are shown on a VE 130 head-up display and three-colour VMC 180 head-down display, which can also be used to present television pictures.

STATUS: RDM radar equips Mirage 2000 fighters of the Egyptian, Indian, Greek, Peruvian and United Arab Emirates air forces, as well as the first Mirage 2000 squadrons in the French Air Force.

RACASS/RDX next-generation radar for combat aircraft

The company has initialised a technology demonstrator programme for a future French or European multi-function fighter radar. It has been working on the project for five years, the technology being crystallised around a project sometimes known as RACASS (combat aircraft

Thomson-CSF RDI radar equips Dassault Mirage 2000

Thomson-CSF RDM multi-function radar in Dassault Mirage 2000

air-to-air and air-to-ground), the operational version of which has been designated RDX.

RDX development is scheduled to meet the in-service date of the production version of the Dassault Rafale experimental aircraft.

While the system has yet to be defined in detail, the company is investing heavily in the technology needed to optimise the performance of the new aircraft and its systems; vlsi, vhsic for programmable signal processors, new travelling-wave tubes, new phased-array antennas, and electronic scanning in vertical and horizontal planes. Some 40 functions have been defined for a next-generation multi-role radar, many of them dependent on infra-red and laser sensors.

In the long term, some of the technology developed for the RACAAS system would be appropriate to the current RDI and RDM radars, and could be fed back as retrospective improvements, perhaps in a Mirage 2000 mid-term refit programme. The term RDY is already being applied to proposed upgraded versions of the RDI and RDM radars.

STATUS: under development. Pre-production systems were delivered during 1985, with flight trials of definitive system beginning in 1986.

Thomson-CSF Varan surveillance radar

Iguane/Varan/Agrion surveillance/strike radars

Under French government contract, Thomson-CSF has been developing a family of airborne radars for sea surface surveillance and maritime warfare applications. Key factors are X band operation, pulse compression over several pulse widths and frequency agility. Three members of the family have been named: Iguane, Varan, and Agrion 15.

Iguane

This system is in production to replace the DRAA2A sea surveillance radar fitted in the Breguet Alizé and for the Dassault Atlantique 2 long-range maritime patrol aircraft.

STATUS: in production.

Varan

This is essentially an Iguane radar with a smaller antenna which makes it suitable for virtually all the present and planned lightweight maritime patrol aircraft and helicopters. It has been fitted to the Dassault Falcon Gardian of the French Navy and selected for the naval version of the ATR 42 transport and Aérospatiale SA 365F Dauphin 2 helicopter.

The system provides real-time pollution- and ice-detection. Key factors are I band operation, pulse compression over several pulse widths, frequency agility for electronic counter-countermeasures, and beacon detection. The unspecified but low peak power level, associated with high receiver sensitivity, increases the difficulty of detection by hostile radars. Typical

detection ranges, in sea state 3/4, are: snorkel 30 nautical miles, fast patrol boat 60 nautical miles, and freighter 130 nautical miles. Total weight of the system's six units is 244 lb (111 kg).

STATUS: in production.

Agrion

A variant of the Iguane and Varan, with the same pulse compression and frequency agility characteristics, the Agrion also offers over the horizon targeting and provides guidance for the Aérospatiale AS 15TT air-to-surface missile. Several types of antenna are proposed in order to meet the requirements of all appropriate aircraft.

STATUS: in production.

Agave attack radar

This lightweight, multi-role radar, initially designed for the Super Etendards of the French Navy and more recently chosen for the Jaguar International strike aircraft, is designed for naval use. However, the system has considerable air interception/ground attack capabilities.

Basic functions are search (air-to-surface and air-to-air), target designation for the homing head of long-range active missiles or a head-up display, automatic tracking (air-to-surface and air-to-air), ranging (air-to-air and air-to-surface) and mapping.

As the system was designed for operation by a single crew member it is automatic as far as

Thomson-CSF Iguane maritime surveillance radar

possible; for example, the system has an instantaneous automatic gang control whereby ground clutter or other unwanted echoes are considerably reduced in comparison with the

Thomson-CSF Agrion 15 lightweight search and attack radar in circular radome gives distinctive look to Aérospatiale SA 365F Dauphin 2 helicopter

Thomson-CSF Varan surveillance radar in nose of Dassault Gardian light maritime patrol aircraft

pin-point reflections of target aircraft and surface vessels. In another case, when the air-to-ground mode for ground-mapping is chosen, the best possible tracking elevation for maximum ground coverage ahead of the aircraft is automatically set up. The pilot can still override the computed elevation and have the option of final adjustment when short-range, high power echoes are present.

In the system developed for the Jaguar International all radar information can be displayed on a raster head-up display of a scan-converter fitted to the combined map/reader head-down display.

STATUS: in production.

Raphael and Arcana
Raphael is a pod-mounted cartographic radar, designed to give very fine detail of the earth in real time.

From it has been developed Arcana, which is fitted to the French Air Force's Dassault Mirage IV-P aircraft as an aid for up-dating the navigation system. Arcana is a pulse-Doppler radar with advanced pulse-Doppler techniques.

TRT
Télécommunications Radioélectriques etTéléphoniques (TRT), Defence and Avionics Commercial Division, 88 rue Brillat Savarin, 75460 Paris, Cedex 13
TELEPHONE: (1) 581 11 12
TELEX: 250838

TSR-718 ATC transponder
TRT's TSR-718 air traffic control transponder complies with ARINC 718 primarily for air-transport aircraft and provides identification on Mode A with altitude-reporting on Mode C. Transmitted power is 400 watts typical on 1090 MHz.

This digital system employs fully solid-state construction, is microprocessor controlled, and uses software analysis of the input data to generate the uhf output. Internal desensitisation techniques provide echo protection. Other features include pulse-width verification, side-lobe suppression, and a suppression pulse which inhibits other pulse equipment in the aircraft each time a reply group is transmitted.

Self-test circuits monitor all TSR-718 functions, including uhf power amplifier output and frequency stability. Automatic fault identification is standard, and a non-volatile fault-memorisation facility is an available option. The unit's case is hinged at each side to provide direct access to all components for maintenance and repair. A mean time between failures of more than 7000 hours is claimed. Antenna mismatch protection is incorporated and the transponder is not damaged by antenna short-circuit or open-circuit operation.

TRT TSR-718 transponder

Format: 4 MCU (ARINC 600)
Weight: 10 lb (4.2 kg)
Power: 30 VA at 28 V dc
Power outputs: 400 W

STATUS: in production and service.

TSR-718S Mode S ATC transponder
TRT is developing a Mode S transponder derived from the TSR-718. In addition to Modes A and C, it will have a data link capability for selective addressing and uplink and downlink message transmission.

Format: 6 MCU per ARINC 600
Weight: 5 kg

STATUS: in development.

Over the horizon target designation system
TRT, in conjunction with Omera, has produced a system called DOTH (Désignation d'Objectif Trans Horizon) to guide Exocet and Otomat missiles from launch to targets below the horizon and therefore unable to be tracked by surface radars. The system comprises an Omera-Segid ORB-32 radar carried aboard a ship-borne helicopter, and a guidance system on the ship itself. The ORB-32 can operate in two modes: tactical illumination of several targets or tracking of two of them. The helicopter carries a 360° scanning antenna with IFF, a tactical display console, and a jam-resistant voice and data transmission system incorporating a frequency-hopping unit, a Type ERA-8700 uhf transceiver and antenna.

The ORB-32 radar has a frequency of 8500 to 9600 GHz, frequency agility, high electronic countermeasures immunity, a peak power of 80 kW, and a pulse-repetition frequency of 300, 600 or 1200 Hz. For long-range attack (beyond 100 km) the system tracks, computes and commands mid-course correction via the TRT remote-control transmitter. The data link between aircraft and ship operates in the uhf frequency range 225 to 400 MHz, with a data-rate of 10 K bits a second at 15 W and in A9, F1 and F9 Modes.

Format: (frequency-hopping unit) ¼ ATR Short (ERA-8700 uhf transceiver) ¼ ATR
Weight: (ERA-8700 uhf transceiver) 8 kg

STATUS: system has been approved for operation in conjunction Exocet and Otomat missile systems.

GERMANY, FEDERAL REPUBLIC

Becker
Becker Flugfunkwerk GmbH, Postfach 1980, Niederwaldstrasse 20, D-7550 Rastatt
TELEPHONE: (7222) 121
TELEX: 781271

ATC 2000 transponder
This single-box panel-mounted air-traffic control transponder is designed for general aviation and permits automatic aircraft identification and position reporting. The ident button can for convenience be remotely located.

Dimensions: 146 × 47.5 × 196 mm
Weight: 1.2 kg

Frequency: (receiver) 1030 MHz (transmitter) 1090 MHz
Power: 0.5 A at 28 V dc, 1.1 A at 14 V dc
Power output: 250 W
Replies: 1200 replies/s
Modes: A, A/B and C (B after conversion). Mode C coding in 100 ft increments from −1000 to 62 700 ft

---

ISRAEL

Elta
Elta Electronics Industries Limited, PO Box 330, Ashdod 77102
TELEPHONE: (55) 30333
TELEX: 371807

Elta is a subsidiary of Israel Aircraft Industries Limited.

EL/M-2001B radar
This is a range-only radar for single-seat tactical aircraft operating in air-to-air and air-to-ground modes. The target is detected visually while acquisition and tracking is accomplished automatically by the radar. The system can operate in heavy ground clutter. Information from the radar can be displayed on the head-up display or fed into the weapon control computer for weapon delivery

computation. The six line-replaceable units are based on solid-state technology, with the exception of the travelling-wave tube, and have considerable reserves for future growth.

Base diameter: 450 mm
Length: 790 mm
Antenna diameter: 195 mm
Power: 115 V 3-phase 400 Hz

STATUS: in service with IAI Kfir fighters of Israeli Air Force.

EL/M-2021B radar
The EL/M-2021B is a second generation coherent radar operating in air-to-air and air-to-surface modes, designed specially for single-seat aircraft, with multi-mode operation and integrated control. The system operates either in the long-range mode with a built-in track-

Production of EL/M-2001B radar sensing heads

while-scan function or in the short-range mode with an automatic lock-on feature. After lock-on to the target, a track-while-scan mode provides the pilot with additional information on other enemy aircraft nearby. In the air-to-surface mode the radar provides range-to-target as part of the aircraft weapon delivery system and it can also be operated in a ground mapping mode. The performance can be improved in this mode by selecting Doppler beam sharpening. It is also possible to operate the radar in the air-to-sea search mode, when boats can be detected in both high and low sea states. Navigation, especially at low altitudes and in poor weather, is facilitated by using terrain following and terrain avoidance features. There is a beacon display in both air-to-air and air-to-ground modes which further improves navigation accuracy.

The system can operate independently or as an integral part of a full fire-control system. As part of a full weapons system, aircraft information and pilot commands to the radar are controlled by a central computer. The EL/M-2021B data is fed to the computer and information is displayed on the system's head-up and head-down displays. Communication with the system is mainly through a multiplexed bus. In the stand-alone configuration the radar communicates directly with the appropriate aircraft sub-systems. Information is presented to the pilot on a head-down display.

A version of the EL/M-2021B is being designed for the IAI Lavi fighter. It will have a coherent transmitter and stable multi-channel receiver for reliable look-down performance over a broad band of frequencies and for high-resolution mapping. A programmable signal processor, backed by a distributed computer network, will provide optimum allocation of computing power and great flexibility for growth and the updating of algorithms.

Weight: 120 kg
Power: 115 V 400 Hz 2.5 VA
Maintainability: bite system locates fault to specific lru (of any of 6), simultaneously alerting pilot
Reliability: designed for 100 h mtbf, complying with MIL-E-5400N

STATUS: under development.

Elta Industries EL/M-2001B radar

Radome swung aside on Israeli F-4E Phantom revealing Elta Industries EL/M-2021B attack radar

ITALY

FIAR

Fabrica Italiana, Apparecchiature Radio-elltriche, via Montefeltro, Milan
TELEPHONE: (2) 30 65 91
TELEX: 331140

RDR-1500 surveillance radar

Bendix and the Italian company, FIAR, collaborated in 1981 to produce a new, lightweight search, rescue and weather radar for the Agusta A109 utility helicopter. The system is derived from Bendix's RDR-1300C radar, chosen by the US Coast Guard for its Sikorsky HH-65A search and rescue helicopter.

Altair radar

In conjunction with the US company Westinghouse, FIAR is launching a lightweight multi-mode fire-control radar called Altair. Potential applications include the version of the Anglo-American AV-8B/GR5 Harrier 2, and the AM-X aircraft in development by Aeritalia and Aermacchi in Italy and Brazil's Embraer. The radar is based on FIAR's company-funded 90 kg I/J band Grifo radar, in development since 1982. It will also incorporate elements of the Westinghouse AN/APG-66 radar, the principal combat sensor in the General Dynamics F-16 fighter.

STATUS: in development.

SMA

SMA-Segnalamento Marittimo Ed Aereo, PO Box 200, via del Feronne-Soffiano, 50100 Florence
TELEPHONE: (55) 27501
TELEX: 570622

MM/APS-705 radar

This search and navigation radar was designed for naval helicopters, in particular the Agusta AB 212 and SH-3D types, built under licence in Italy from Bell and Sikorsky respectively. The antenna system is tailored to the space and location on the aircraft. For example, on the SH-3D the antenna is placed in the dorsal position, on top of the fuselage. Line-of-sight stabilisation is provided and there are alternative selectable antenna rotation rates, 20 or 40 rpm. Manually controlled antenna tilt provides ±20° of movement.

The display unit incorporates a 9-inch (230-mm) diameter cathode ray tube plan-position indicator presentation with electronic and mechanical cursors and markers, and complemented by a separate digital readout x-y reference display.

The system provides outputs for other displays and extractor units. Other facilities include sector transmissions and blanking, interfaces for beacon receiver, identification friend or foe, anti-submarine warfare and electronic countermeasure systems, built-in test, data link, track-while-scan and dense environment tracker. The radar can be integrated with the SMA UPX-710 beacon system.

Transmitter-receiver: two identical transceivers for frequency diversity operation, I-band, transmitting 25 kW, with option for 75 kW transmitter-receiver with frequency agility.
Pulse width/prf: 0.05 μs/1600 Hz – 1.5 μs/650 Hz
Range settings: 0.5, 1, 2, 5, 10, 20, 40, 80 n miles
System weight: 80 kg

STATUS: in service with AB 212 and SH-3D helicopters of Italian Navy and several other navies.

MM/APS-707 radar

This system exploits the basic circuits of the -705 radar described above but uses a single 20 kW transmitter-receiver for fixed frequency operation in the I band. It is intended for application where low weight, power consumption and cost is required. It is a military-qualified search radar featuring 360° scan for surface surveillance, small target detection for ASW operations, target designation for ASV attacks and radar mapping. The APS-707 can also display IFF, ESM, beacon and sonar data and the radar operates in the X band with frequency agility.

SWEDEN

Ericsson

Ericsson Radio Systems, Airborne Electronics Division, PO Box 1001, 543126 Mölndal
TELEPHONE: (31) 67 10 00
TELEX: 20905

PS-37/A radar

This multi-mode X band, monopulse radar was designed for the AJ 37 attack version of the Saab Viggen. It stems from mid-1960s designs and is largely integrated with the navigation, display and digital computer-based data processing sub-systems. It comprises two units: a scanner and a transmitter-receiver package, the latter being made up of 13 line-replaceable units.

Except for some of the high-frequency components, the PS-37/A is of solid-state design. Elaborate signal-processing provides a high degree of immunity from both natural interference and electronic countermeasures, and accuracy is improved by lobe-shaping, whereby the aperture (effectively the scanner diameter) is artificially increased to provide better resolution and reduce side lobe effects. The radar is semi-automatic in operation to reduce the work-load of the single crew-member, and information is presented on both head-up and head-down displays.

Operating modes are search, target acquisition, air-to-target ranging, obstacle warning, beacon-homing and terrain-mapping. By adding a further unit a terrain-following capability can be provided, the pilot flying the aircraft in response to head-up display demands.

STATUS: in service.

PS-46/A radar

This software-controlled, multi-mode, pulse-Doppler AI radar has been developed for the JA 37 fighter versions of the Swedish Air Force's Viggen. In view of the numerically small size of Sweden's defence force, great emphasis has been placed on operational availability and readiness, and all-weather capability and effectiveness in an electronic countermeasures environment are also important requirements. Designed to cope with high-performance aircraft, transports, and helicopters, the system

has wide-angle coverage, look-down capability, and can operate at all altitudes.

The multi-mode requirements of the PS-46/A are air-to-air and air-to-ground. The latter are met by using conventional non-coherent pulse waveforms, but the former call for more sophisticated waveforms. The standard radar functions are controlled by a data processor that extracts information from the raw radar and transfers it to other aircraft systems. A digital bus distributes all signals within the radar itself with minimum wiring. For the guidance of semi-active homing missiles an illuminator transmits a continuous wave radio-frequency signal through the radar antenna.

Control of the system through suitable software, says Ericsson, enables parameters to be changed or optimised according to the needs of flight development programmes without time-consuming equipment changes; similar changes can be introduced during service according to changing military requirements, and radar signatures adopted for peace-time training and exercises can be easily changed during conflict to thwart enemy intelligence and countermeasures.

Type: medium prf pulse-Doppler
Frequency: X band, bandwidth >8% of spectrum

Ericsson-designed PS-46/A multi-mode pulse-Doppler radar for Saab JA 37

Performance: detection range >50 km in look-down mode
Weight: 300 kg
Power: (coherent transmitter) 50 kW (continuous wave illuminator) 200 W
Antenna: 700 mm dia
Processor: high-speed 16-bit word-length system with 32 K word program memory
Modes: search, acquisition (automatic via HUD, semi-automatic via HDD), tracking (track-while-scan, continuous track), target illumination, air-to-ground ranging
High-resolution ground-mapping is optional
Reliability: mtbf 100 h

STATUS: in service.

Multi-mode radar for JAS 39

As a member of the five-company consortium responsible for designing and building Sweden's Saab JAS 39 Gripen, Ericsson is to provide the multi-mode pulse-Doppler radar, which forms the primary sensor for the new weapons system. The JAS 39 radar is claimed to be the first such European-built equipment that combines air combat, ground attack and reconnaissance in a single system. As with the Ericsson radar for the JA 37 Viggen fighter, most functions are controlled by software in a special processor which can be programmed for optimum performance in a dense ECM environment.

The antenna is a lightweight planar-array type in carbon-fibre composite for quick response in high-*g* conditions. The radar employs what Ericsson calls a 'flexible waveform' for long-range detection and all-hemisphere coverage of all targets. Frequency modulation, pulse compression, and synthetic-aperture techniques give high resolution in range and angle-measurement.

Ericsson announced in March 1983 that it had chosen Ferranti as a partner on the radar system, and had signed a contract relating to its development and production. The British company (which had already been involved with project definition work on the radar) is responsible for the antenna pedestal and participates in system work and the design and development of the signal processor.

The Gripen will have three principal functions: air combat, ground attack and reconnaissance. In the fighter role the radar will provide

Ericsson PS-37/A radar for strike version of Viggen

Ericsson PS-46/A radar for Viggen fighter

search and track-while-scan for target detection at long range, together with wide-angle rapid scanning and lock-on at short range, and control of gun and missiles. For air-to-ground and reconnaissance missions the radar will provide detection and ranging in a dense ECM environment, designating targets for the Gripen's RB15F missile. The system will have obstacle avoidance and navigation capabilities, and will generate ground maps at both normal and high resolution. Ericsson notes that the Gripen radar has three times as many functions as the radars in the Viggen, while occupying only 60 per cent of the space.

STATUS: under development. Gripen programme, launched to replace Viggen, embraces potentially 140 aircraft, with first flight set for 1987 and production deliveries to begin in 1992.

Side-looking airborne radar

This inexpensive, real aperture SLAR is designed exclusively for maritime patrol and surveillance (for fishery protection, oil pollution monitoring, and search and rescue), with emphasis on simplicity and ease of operation. It comprises five line-replaceable units: antenna, transceiver, digital signal processor, control unit and television display. The radar images can be recorded on standard video recorders. Digital radar video can also be recorded for subsequent computer processing. With a video link between aircraft and ground station or ship, images can be transmitted with no degradation of quality.

The digital signal processor is designed around a 1.6 M bit television memory display. Up to 2000 range cells can be processed to an accuracy of 6 bits (giving 64 grey-tone levels). The presentation and performance can be varied to suit customers' needs, and other features include a reference grey scale, level mapping, positive or negative picture representation and automatic positioning of targets. Monochromatic television images recorded in the air can be converted to colour on the ground.

Installation: for double-sided coverage, glass-fibre pod containing two antennas carried under fuselage, or antennas carried in individual pods on each side of aircraft
Beamwidth: (horizontal) 0.5°
(vertical) 33°
Frequency: 9.4 GHz
Peak output power: 10 kW
Prf: 1 kHz
Power: 15 A at 28 V dc
System weight: 70 kg

STATUS: in production.

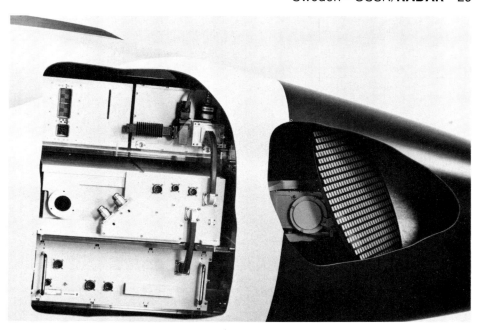

Engineering model of Ericsson multi-mode radar for JAS 39 fighter

Artist's impression of JAS 39 Gripen

Swedish Coastguard version of Cessna O-2 Skymaster has Ericsson side-looking airborne radar with ventrally located antenna

UNION OF SOVIET SOCIALIST REPUBLICS

Big Nose radar

This radar was the principal sensor in the Tupolev Tu-28 Fiddler, the large two-man long-range all-weather intercepter that entered service in 1961. The radar operates in conjunction with AA-5 Ash air-to-air missiles. It is thought that about 120 of these aircraft remain in service out of some 2000 built for home defence.

Izumrud radar

This interception radar was fitted, as interim equipment, to a number of MiG-15s and MiG-19s during the late 1950s.

Downbeat radar

This NATO designation has been given to the nose-mounted bombing and navigation radar in the Tu-22M Blackjack variable-geometry

medium bomber that entered service in the mid-1970s. The aircraft has tail-mounted defensive armament in the form of guns, controlled by a Bee-Hind radar.

SRD-5M High Fix radar

This I band radar is installed in the intake cone of the Sukhoi Su-20/22 family of swing-wing attack aircraft known as Fitter.

MiG-21 Fishbed fighter carries airborne interception radar in centre-body shock-cone

Jay Bird radar

Jay Bird is the NATO designation given to the air-to-air interception radar installed in the shock-cone intakes of some versions of the USSR's still widely used MiG-21 Fishbed fighter. It is reportedly also installed in export versions of the MiG-23, designated Flogger E. The system operates in the J band, between 1280 and 1320 MHz, according to one source and 10 to 20 GHz according to another. Antenna size is limited probably to about 400 millimetres and three pulse-repetition frequencies have been quoted: 2042 to 2048, 1592 to 1792 and 2716 to 2724 pulses per second. Search and tracking ranges for this 'short-legged' fighter are perhaps 30 km and 20 km respectively with lobe-switching for tracking targets. Transmitter peak power may be around 100 kW. Acquisition range against typical targets is said to be less than 20 nautical miles and there is almost certainly no look-down capability. Fishbed-J may carry a radar-homing version of the Atoll air-to-air missile, in which case a target illumination capability would be necessary.

GEC Avionics has proposed a version of its Skyranger lightweight radar for one MiG-21 operator (thought to be Egypt) as a replacement in view of the reported poor serviceability of the Jay Bird radar, the Soviet Union's withdrawal of spares and maintenance to that operator and the limited performance of the Soviet system.

The MiG-21 is likely to remain a numerically strong element of USSR and Warsaw Pact nations for some years, and emphasis on a continued effort to improve the radar's major areas of weakness can be expected; increased range and a look-down capability would be the two fundamental areas of improvement.

Look Two radar

This is the NATO designation for an I band weapon-delivery and navigation radar. Western electronic intelligence gives the operating band as 9245 to 9508 MHz in conjunction with frequency agility and four pulse-repetition frequencies: 320 to 336, 619 to 623, 1247 to 1253 and 1871 to 1879 pulses per second. This radar may be fitted in some versions of the Yakovlev Yak-28 Brewer, a 1950s-vintage swept-wing tactical bomber equivalent to the BAC Canberra which is still operational but no longer in front-line service with the Soviet Air Force.

Scan Fix radar

This is the NATO code for the interception radar fitted in some versions of the MiG-17 and -19 interceptors. The intake configuration of these aircraft would not have permitted either a large antenna or a wide scan angle. Both I and E/F band versions are thought to have been produced. The former may have equipped MiG-19 Farmers, the latter going to MiG-17 Frescos. These fighters have long since been withdrawn from front-line service with the Soviet Air Force, though they serve elsewhere, eg in Warsaw Pact countries, and a version of the MiG-19 known as the Shenyang F6 has been built in China and equips more than 40 air regiments there. The radar, like the aircraft themselves, is obsolescent.

Scan Odd radar

This NATO code-name is for an interception radar fitted to some versions of the MiG-19 Farmer interceptor. An I band system operating at 9300 to 9400 MHz, it is believed to be a later system than Scan Fix. If, as reported, it has an unusually complicated scan pattern, this is probably due to the limitations imposed by the installation in the pitot-type nose intake. Since the system is geared to the MiG-19, it must be considered obsolescent.

Scan Three radar

An I band radar operating in the 9300 to 9400 MHz range of frequencies, the interception radar, NATO code-named Scan Three, is fitted to the Yak-25 Flashlight two-seat all-weather fighter, a late 1940s design that came into service in 1955. It was based on the SCR-720 radar developed in the USA during the Second World War.

Short Horn radar

This weapons delivery and navigation radar appears from electronic intelligence information to be an example of relatively recent Soviet technology. It operates in the J band around 1400 to 1500 MHz with frequency agility. Four pulse-repetition frequency/pulse-width combinations have been identified: 313-316/1-1.8, 496-504/0.5-1.4, 624-626/0.4-1.3 and 1249-1253/0.01-0.09 (the second set of figures in each case being in microseconds). Circular and sector scans have been recorded and the system may have anti-surface vessel and maritime applications. Aircraft reported to have this equipment include the B, C, D and E versions of the 1960s-designed Yak-28 Brewer, Tu-105 Blinder (A, C and G models), and Tu-16 Badger H. About 150 maritime patrol Blinder Cs are still in service, together with some 800 of the Badgers.

Skip Spin radar

This X band interception radar (for which the name Uragan 5B has also been mentioned) was introduced in the early-1960s and is fitted to Su-11 Fishpot C, Su-15 Flagon A and Yak-28P Firebar fighters. Estimated output power is 100 kW with a range of 25 nautical miles operating at 8690 to 8995 MHz, and pulse-widths of about 0.5 microsecond and pulse-repetition frequencies of 2700 to 3000. The radar presumably provides the searching and tracking modes for the Anab air-to-air missiles which arm these fighters, as well as illumination for the radar-homing version of the Anab.

R1L/R2L Spin Scan radar

This is the NATO designation for a family of S-band search and track radars that commonly equip the USSR's most well-known fighter, the MiG-21 Fishbed. The system was also fitted to initial production versions of the Su-9 (Fishpot-B) fighter. Soviet designations are believed to be R1L and R2L, corresponding to Spin Scan A and Spin Scan B. The latter is an export version, for example to India. Specific models of the Fishbed so equipped are the D and F. The radome of the Fishbed D is larger, and presumably accommodates a bigger antenna than that of the Fishbed F. The system has a range of about 19 km, but is believed to be ineffective below 1000 feet owing to the presence of ground-clutter. Performance and reliability deficiencies have caused some MiG-21 operators (notably Egypt) to investigate the possibility of fitting Western radars.

Radome of this MiG-21 Fishbed D may house Spin Scan interception radar

Fox Fire radar

This radar has been developed for intercepter of the MiG-25 Foxbat, and has a reported output power of 600 kW. The range has been put at 85 km.

High Lark radar

This radar is intended to equip the MiG-23S Flogger B variable-geometry fighter that appeared in 1971. According to US intelligence, it is comparable with the Westinghouse APG-59/AWG-10 air combat radar in the US Navy/McDonnell Douglas F-4J Phantom.

Radars for new combat aircraft

Advanced, look-down, track-while-scan radars are (according to US sources) incorporated in the three new Soviet fighters whose characteristics were summarised in the US President's 1985 budget request to Congress in January 1984. The aircraft are the MiG-31 Foxhound interceptor, a development of the MiG-25 Foxbat that first appeared in 1967; the Su-27 Flanker (the Soviet equivalent of the McDonnell Douglas F-15 Eagle; and the MiG-29 Fulcrum (the equivalent of the General Dynamics F-16).

The MiG-31, deployment of which begun in 1983, is the first Soviet aircraft, according to US intelligence, to have a look-down radar capability. With a speed of Mach 2.4 and able to reach 80 000 feet, the MiG-31 is designed to intercept low-flying aircraft and cruise missiles, using a long-range pulse-Doppler radar and the new AA-9 radar-homing missiles. The radar may be equivalent to the Hughes AN/AWG-9 in the US Navy/Grumman F-14 fighter. This long-range radar is said to have been tested against a variety of targets, including drones, to simulate the size and performance of cruise missiles, at Vladimirovka, a research and testing base near the Caspian Sea.

The single-seat, twin-engined Su-27 became operational in 1984. The radar for this and the MiG-29 are thought by US Intelligence to be comparable in performance to the AN/APG-63 and AN/APG-66 radars fitted to the F-15A and F-16A fighters.

The Su-27 and MiG-29 are both said to have infra-red search and tracking equipment, digital data links and head-up displays.

MiG-25 Foxbat has Fox Fire interception radar

This version of Tu-16 Badger is one application of Short Horn nav/attack radar

Puff Ball radar

This NATO designation has been applied to a radar which has been stated to equip the Myasishchev M-4 Bison that still operates in small numbers with the Soviet Long-Range Aviation Force. It is probably also in some versions of the Tu-16 Badger and the Tu-142 Bear, both of which have broadly similar strategic roles.

Puff Ball is an I band surveillance radar for large-area surface-vessel detection and it may perhaps be employed in the guidance of such air-to-surface missiles as Kangaroo, Kennel, Kelt and Kipper. American reports state that the system can provide friendly surface-to-surface missile batteries with target co-ordinates for missile guidance, the information being transmitted to the missile site by means of a data link.

Tupolev Tu-142 Bear-D, thought to carry Puff Ball surveillance radar

UNITED KINGDOM

Cossor

Cossor Electronics Limited, The Pinnacles, Elizabeth Way, Harlow, Essex CM19 5BB
TELEPHONE: (0279) 26862
TELEX: 81228

IFF 2720 transponder

This is a complete micro-miniature IFF Mk 10A identification friend or foe/secondary surveillance radar transponder for use on all types of military aircraft and helicopters. On Modes 1, 2, 3/A and B the full 4096 codes are available, and 2048 codes are available on Mode C for altitude reporting. In addition there are circuits for the identification facility (SPI or I/P) and military

Indonesia's British Aerospace Hawks have Cossor IFF 2720 transponder

Cossor IFF 2720 transponder (left) and IFF 2743 control unit

Cossor IFF 3500 interrogator

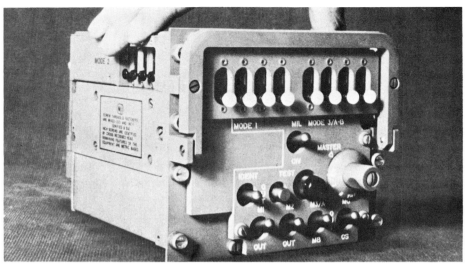

Cossor IFF 3100 transponder

emergency. The system comes in two units: a transmitter-receiver and controller, each of which features easily accessible circuit boards.

Electronic warfare provisions include resistance to continuous wave, modulated continuous wave and pulse jamming, sidelobe rate limiting, short pulse and spurious interference protection, single pulse rejection, and long pulse discrimination.

A new microprocessor-based control unit for use with this system is designated IFF 2743. It has a fully dimmable light-emitting diode display.

Dimensions: (IFF 2720 transponder) ⅜ ATR Short case, 3.52 × 7.625 × 12.562 inches (90 × 194 × 314 mm)
(IFF 2743 full facility control unit) 5.75 × 2.25 × 3.86 inches (146 × 57 × 98 mm)
Weight: (IFF 2720 transponder) 10 lb (4.6 kg)
(IFF 2743 full facility control unit) 1.54 lb (0.7 kg)
Frequency: (transmitter) 1090 MHz
(receiver) 1030 MHz
Power output: 27 dBW (500 W) minimum under all service conditions
Receiver sensitivity: –76 dBm
Dynamic range: ⩾ 50 dB
Sidelobe suppression: 3-pulse
Qualification: STANAG 5017 Edition 2, ICAO Annex 10

STATUS: IFF 2720 system in current production is Cossor's standard export IFF, and equips many strike, transport, combat and trainer aircraft; it is fitted to all British Aerospace export Hawk strike/trainers. More than 900 systems have been delivered to over 20 countries.

IFF 3100 transponder
The IFF 3100 is a single-package transponder tailored to the Royal Air Force/Panavia Tornado aircraft. Claimed advantages over previous systems are small size, lower weight, and simpler installation. Although the component density is high, reliability is ensured by the use of high-grade, close-tolerance circuits and a four-port circulator protects the output stages from the effects of any antenna mismatch. Open or short circuit conditions at the antenna do not damage the transponder.

Extensive integrity monitoring is incorporated during operation and when the test button is pressed. Checks cover receiver sensitivity, receiver centre frequency, mode decoding, aircraft reply coding and transmitter power level.

Interrogation Modes are 1, 2, 3/A, B and C and the reply capability 4096 codes for Modes 1, 2, 3/A and B. Provision for the use of the X pulse is included and there are 2048 codes for Mode C.

Dimensions: 5.75 × 5.2 × 6.5 inches (146 × 132 × 165 mm)
Qualification: ICAO Annex 10, STANAG 5017 Edition 2
Weight: 11.7 lb (5.3 kg)

STATUS: in production for Panavia Tornado with UK equipment fit.

IFF 3500 interrogator
The IFF 3500 airborne interrogator employs monopulse techniques to achieve high accuracy in the measurement of target bearing and incorporates an automatic code-changing system (IFF 3502) to enhance security and eliminate the possibility of incorrect code setting.

The transmitter employs P2 emphasis to provide antenna beam sharpening. P1 and P3 are transmitted on the antenna sum channel and P2 on the difference channel. Selectable 3 or 6 decibels of P2 emphasis is available. Advanced video-processing circuits for degarbling, defruiting, decoding and for echo- and multi-path suppression are contained within a single unit. Passive and active decoding are provided and two channels of passive decoding enable comparison during the overlap period between code changes. Active decoding provides serial readout of the 4096 reply codes.

Manual and continuous automatic built-in test circuitry checks transmitter power, interrogation coding, receiver sensitivity, defruiting/decoding, bearing accuracy and integrity of the transmission feeders.

Format: 1 ATR Short case to ARINC 404A
Weight: 45.5 lb (20.7 kg)
Frequency: 1030 MHz
Power output: (P1, P3) 30.5 dBW
(P2) 0, +3 or +6 dBW above P1 power
Spurious outputs ⩾76 dB below 1 W
Receiver frequency: 1090 MHz
Sensitivity (decoding): –80.5 dBm
Dynamic range: 60 dB
Spurious responses: 60 dB down outside pass-band
Bearing resolution: dependent on antenna configuration, but around 5% of angle between intersection points of control and interrogate patterns
Qualification: compatible with NATO STANAG 5017 Edition 3

STATUS: in production for Royal Air Force/Panavia Tornado F2, Nimrod AEW3 and F-4K/M Phantoms, and Royal Navy Sea King AEW helicopters.

Ferranti
Ferranti Defence Systems Limited, Ferry Road, Edinburgh EH5 2XS
TELEPHONE: (031) 332 2411
TELEX: 72141

Radar for European fighter aircraft
In June 1985 Ferranti was selected as the British lead contractor in the development of the radar for the European fighter aircraft (EFA) in collaboration with the specialist radar companies of the other participating European countries.

STATUS: initial development.

Blue Fox interception radar
Blue Fox is a lightweight (less than 190 lb, 86 kg), high-performance, pulse-modulated non-coherent radar that combines airborne

Sea Harrier FRS1 carries Ferranti Blue Fox radar

interception and air-to-surface search and strike for the Sea Harriers of the British Royal Navy and Indian Navy. The specific roles are:

Search, in which there is a choice of scan patterns in azimuth and elevation for use against air-to-air and air-to-surface targets.

Detection, in which the radar embodies special features to facilitate detection of small air and surface targets.

Lock-on track, to provide the input to the weapon aiming computer for weapon release.

Radar lock-on visual acquisition, where the pilot flies the aiming mark on the head-up display onto either air or surface targets and then locks the radar onto the target by pressing the 'accept' button.

Air-to-surface ranging, where during an attack the radar boresight can be slaved continuously to the aiming mark to provide ranging irrespective of whether or not the surface target is radar discrete.

Identification, in which transponder returns from friendly ships and aircraft are displayed for identification and location.

Visual identification, where the radar incorporates a short-range scale to enable the Sea Harrier to close to within a short distance so that the pilot can identify visually any intruder aircraft.

Navigation, in which the radar can be used in a ground-mapping role as well as interfacing with the navigation computer.

The main radar unit is housed within the folding nose of the aircraft and comprises five line-replaceable units: the transmitter, receiver, processor, amplifier electronic control, and scanner. The remaining four units within the cockpit are also first-line replaceable. The display which makes up two of them is divided between the actual presentation and its associated drive unit. Of the other two units, the radar control set contains those controls which can be preset before an attack and the hand controller contains those that may need to be operated during an attack.

The radar has a flat aperture, slotted array antenna which provides better detection ranges than more conventional forms; it also reduces sidelobes, another benefit in an electronic countermeasures (ECM) environment. It is roll-stabilised to provide increased accuracy during air-to-surface attacks. Source of the rf energy is a magnetron. Within the processor unit is a digital scan-converter which increases the brightness of the display in all conditions of ambient lighting, even when the sun is shining directly onto the face of the display; its storage and variable persistence can be used to advantage elsewhere. By selecting a 'freeze' facility the radar picture can be retained indefinitely after the transmitter has been

Ferranti Blue Fox attack radar assemblies

switched off, a useful capability in an ECM environment when approaching a potential surface target for subsequent attack. Alternatively, a 'tails' selection can be made which will cause moving targets to produce a comet-tail effect on the display and thus indicate relative track histories in an air-to-air encounter. The system is frequency-agile, ie within a wide frequency band each pulse is transmitted on a different frequency from the previous one, the rate of change of frequency being very rapid.

This technique results in several advantages:

Effects of sea and rain clutter are reduced and the system's detection capability under these conditions increases.

It combats the effect of ECM since the enemy is forced to jam over a wider frequency band, decreasing its effectiveness.

It counters the effect of glint. When the radar target with a complex reflecting surface such as a naval vessel carrying radar and radio antennas changes aspect relative to a radar transmitter, the effective radar centre of the target changes; it can even move to a point outside the physical outline of the target. The effect on the radar tracking target is to introduce low frequency noise into the tracking loop as the radar attempts to follow the rapidly shifting centre of reflection. Using the frequency-agile mode, this glinting action takes place much more quickly so that in a short time it is possible to determine a sharply defined average position of the target.

The elimination of glint is important for missile guidance systems.

Frequency agility eliminates the confusing effects of second trace returns; these are return echoes from large targets outside the range scale that appear on the display in random fashion. This advantage is particularly evident when operating near coast lines or other features having large radar echoes.

STATUS: in production for British Aerospace Sea Harrier.

Red Fox radar

Electronically identical to the Blue Fox radar in the Sea Harriers (see previous entry), Red Fox has been repackaged to suit other aircraft types with a different nose profile. This lightweight, I band pulse-modulated radar is designed for airborne interception and air-to-surface search and attack, with an overland capability. Potential applications include the retrofit of Soviet-designed radars in MiG-21 fighters.

Blue Vixen radar for Sea Harrier FRS2

At the 1984 Farnborough Air Show Ferranti unveiled a mock-up of the Blue Vixen radar that will form a major element in the Royal Navy's Sea Harrier mid-life update programme. Approval to proceed with the new radar was given in November of that year. Replacing Blue Fox, Blue Vixen is a multi-mode coherent pulse-Doppler radar that will provide the Sea Harrier with a look-up/look-down, beyond-visual-range all-weather air defence capability against targets over land or sea, and a much greater detection range.

The system is designed to operate in conjunction with infra-red and radar-guided weapons, specifically the new American advanced medium-range air-to-air missile (AMRAAM). Novel signal-processing methods and technology are said to provide a significant performance advance over equipment currently used in Europe or the USA. This feature provides, among other things, high-resolution land- and sea-search modes for navigation and surface target detection. Air-to-air modes include long-range and intermediate-range air interception, and a head-up combat mode for automatic target acquisition and tracking in high-*g* manoeuvring flight. Air-to-surface modes include high-resolution land and sea search and track for navigation and surface-target detection, and air-to-surface ranging for weapon delivery. The system is derived from the company-funded radar development programme known as Blue Falcon and aimed at

Ferranti Blue Fox radar showing flat-plate antenna

Ferranti Red Fox airborne interception radar

Blue Kestrel radar

Ferranti's Radar Systems Department is developing the Blue Kestrel radar for the UK Ministry of Defence version of the EH 101 helicopter. This radar has a flat aperture antenna of advanced design which, together with a travelling-wave tube transmitter, pulse compression, sophisticated digital data and signal processing, will provide a high performance radar for the EH 101 in its ASW and ASV roles. The first model for flight trials was delivered to the Royal Aircraft Establishment, Bedford, in late-1985 for fitment in a modified Westland Sea King helicopter. Flight trials began in Spring 1986.

Blue Falcon radar

Blue Falcon is essentially a core radar to provide data for the next generation of fighter radars, and has been under study for some seven years. It has provided some of the basis for the Blue Vixen radar which has been ordered for the Royal Navy/British Aerospace Sea Harrier mid-life update, and also for the radar being developed jointly with Ericsson for Sweden's new JAS 39 fighter.

Blue Parrot Radar

Blue Parrot is the radar used in Royal Air Force/British Aerospace Buccaneer aircraft.

Ferranti Blue Vixen radar with associated interface unit, display and processor

Impression of Anglo-Italian EH-101 ASW helicopter, which will have Ferranti Blue Kestrel radar

generating the technology foreseen as the basis of the next-generation radar systems. The pulsed rf energy is generated by a travelling-wave tube. The solid-state system contains very little hybrid circuitry, and both the processing speed and capacity have been improved by comparison with Blue Fox.

To accommodate the new radar the Sea Harrier's nose has been re-profiled and the signal processor located in the rear fuselage, linked to the radar sensor by a MIL-STD-1553B digital data-bus. Some elements of the system are being designed and built by Ericsson in Sweden as part of an agreement by which the UK company participates in the development of the attack radar for the Saab JAS 39 Gripen.

Blue Vixen is to be test-flown on a British Aerospace One-Eleven; the first prototype unit was delivered to the Royal Aircraft Establishment, Bedford, in early 1986 and flight trials were due to begin a few months later. Further flight trials will be conducted in Ferranti's own BAe 125 and a similar aircraft being adapted as a Sea Harrier avionics test-bed by British Aerospace. Production deliveries are to begin in 1988, and the upgraded Sea Harrier is due to enter service in 1989.

STATUS: in development. The mid-life update programme will convert the Sea Harrier FRS1 into FRS2 standard (FRS = fighter/reconnaissance/strike).

Flight trials have begun of Ferranti Blue Kestrel radar mounted in nose of this RAE/Westland Sea King helicopter

Project definition began in May 1983 on an update programme for this aircraft's avionics equipment, largely involving this radar. In February 1985 British Aerospace was appointed prime contractor for the Buccaneer S2B Update involving 32 aircraft. Blue Parrot was originally designed for optimum performance over water, but was later tuned for their new land-based mission. Now it will be re-optimised for over-water missions, in conjunction with the British Aerospace Sea Eagle anti-ship missiles.

Seaspray radar

Seaspray is a lightweight, high performance I band radar specifically developed and produced for the Royal Navy to counter fast patrol boats, its initial application being the Westland Sea Lynx helicopter.

The Mk 1 system, designed to detect small surface targets and provide guidance for the Sea Skua missile, comprises five line-replaceable units: a scanning antenna, transmitter, receiver, control/display unit and processor. With these is associated a heat exchanger. The control/display unit is a bright, flicker-free television display. A Ferranti scan converter displays the radar pictures in a television raster format which would allow the display to be used for other sensors such as forward-looking infra-red and low-light television. Target range and bearing is provided on the screen in alpha-numeric form and the display persistence may be varied according to the operator's requirements.

There are eight operational modes:
Search: the targets are large and small surface vessels in bad weather and high seas, low flying aircraft, and periscope detection in the anti-submarine warfare role
Weapon system integration (employing mono-pulse tracking techniques): guidance for the Sea Skua air-to-surface missile and accurate target range and bearing inputs (through a data link) for surface-to-surface missiles
Identification: the system can be used in an interrogate mode for target identification

Royal Navy/Westland Lynx HAS2 uses Ferranti Seaspray attack radar

Navigation: in normal operation, and by integration with a Decca tactical air navigation system
Search and rescue
Station keeping
Weather warning
Self let-down and recovery to base in bad visibility.

The adoption of frequency-agility affords protection from ECM, rejection of clutter and glint, and the elimination of second trace returns.

In August 1982 Ferranti announced two technology improvements, resulting in the Mk 2 and Mk 3 versions of the Seaspray radar. Seaspray Mk 2 brings a new processor to upgrade Mk 1 equipment. Seaspray Mk 3 is the result of technology improvements, and includes track-while-scan (TWS), permitting a number of targets to be tracked on the observer's display, with position, speed and heading. The system also permits ground-stabilised selectable-overlay picture expansion. Greater processing capability and substitution of a new control panel provides the additional facilities required for TWS. This modification benefits over-the-horizon targeting (where the target is invisible to the missile-armed vessel acting in concert with the Seaspray-equipped helicopter, but which can now receive information from the aircraft to provide the launch data for the missiles long-range). It can be applied retrospectively. The Seaspray Mk 3 TWS uses advanced processing techniques, Kalman filtering, read-only memories and digital-scan conversion methods, and there is sufficient spare memory to accommodate software developments.

Seaspray Mk 3 also features full 360° scan, and the system remains compatible with the Lynx helicopter, and it is also suitable for fixed-wing aircraft, surface vessels and shore installations.

Weight: 140 lb (64 kg)
Antenna dimensions and scan: 27 inches wide (686 mm), 10 inches deep (254 mm), ±90° azimuth about fore and aft axes.

Development model of modified Ferranti Blue Parrot radar, with digital interface unit, undergoing tests

Seaspray radar control and display panel in Lynx cockpit

Ferranti Seaspray radar in Lynx helicopter

STATUS: Seaspray Mk 1 is in production and in service with Sea Lynx helicopters of British Royal Navy and seven export customers, including Royal Netherlands Navy, Royal Danish Navy and Brazilian and Argentinian navies. By early 1985 more than 250 sets had been ordered.

In November 1984 the Italian company Agusta chose Seaspray 3 and British Aerospace Sea Skua missile combination for its licence-built Bell 212 ASW helicopters ordered by the Turkish Navy.

In April 1983 the West German Navy ordered two Mk 3 Seasprays for evaluation on West-land-built SH-3D Sea King helicopters. The first of these radars was delivered in March 1986 and at the same time Ferranti confirmed that 20 radars were to be supplied to Messerschmitt-Bölkow-Blohm for fitment to West German Navy Sea Kings.

CASTOR corps airborne stand-off radar

Ferranti was selected in April 1984 by the UK Ministry of Defence to conduct an engineering definition study for a stand-off battlefield surveillance system to meet a General and Air Staff requirement. It is one of two companies competing to build the radar, and handling trials have been completed on a Norman Islander modified to carry an advanced radar.

Pilatus Britten-Norman Islander light twin with Ferranti CASTOR radar in nose-mounted radome

Purpose of the system is to provide to corps headquarters primary intelligence information in the immediate battle zone and beyond. One application of the radar data will be to assist with the mission designation of remotely piloted vehicles, for example the UK's Phoenix project.

STATUS: competitive flight trials completed. UK MoD decision awaited.

GEC Avionics

GEC Avionics Limited, Airport Works, Rochester, Kent ME1 2XX
TELEPHONE: (0234) 44400
TELEX: 96333/4

AI 24 Foxhunter interception radar

This pulsed-Doppler coherent airborne interception radar has been developed for the Panavia Tornado F2, and it is claimed to be the first pulse-Doppler air-combat radar of UK design to enter production. Ferranti is a major sub-contractor to GEC Avionics in the programme and provides the scanner mechanism and transmitter. The total value of the radar systems for the 165 aircraft is put at about £100 million.

The primary role of the AI 24 system is the search for, and detection and tracking of airborne targets, but it also provides target illumination for the UK-designed Skyflash medium range air-to-air missile. Significant size and weight reductions have resulted from the adoption of microprocessors and hybrid and thick film electronics.

The system employs a travelling-wave tube for high efficiency at large power outputs, and with low noise. The antenna has a hydraulic scanning mechanism. The sensor, or 'front end' of the system, is analogue, but all-digital processing is employed. The entire system (12 line-replaceable units) is contained within the nose cone, and the standard of built-in test and accessibility is such that the mean time to replace a unit is less than 15 minutes.

Designed to operate in and around Europe and the Atlantic approaches, potentially the world's most dense electronic counter-measures environment, the system relies heavily on sophisticated counter-counter-measures and is designed for two-man operation. The navigator has two multi-function television tabular displays on which he can portray navigation, search, tactical evaluation, and approach-course information. The pilot has a head-down display that can repeat any of the information available to the navigator, and a head-up display for approach and attack guidance.

During the search phase the I band (3 cm) radar, operating in pulse-Doppler mode and with high pulse repetition frequency for maximum range and performance against low-flying aircraft, picks up targets at a range of about 100 nautical miles. The navigator can then build up a picture of the tactical situation by designating those that constitute potential threats and instructing the radar to continue automatic surveillance of the others. This ability to track targets while scanning for others is made possible by the radar's fmicw (frequency-modulated interrupted continuous wave) mode of operation. All selected targets are automatically tracked and information on their positions, heights, speeds and flight paths is continuously updated for display to the crew. Interrogation for target identification is provided by a Mk 10 IFF system. Several targets can be designated for simultaneous attack, the radar computing the firing 'windows' for each one. The radar also is required to operate at short range so that the pilot can visually identify potential targets, especially in peacetime.

In fast-moving situations, such as close combat, the pilot can direct the scanner as he wishes. When targets appear in the designated area lock-on follows immediately and is maintained during periods of high angular rates of motion, and medium or short-range missiles or guns can be used as appropriate.

The Foxhunter system is designed to provide the Tornado with an autonomous combat capability, since it was reasoned that battle-damage could cancel assistance from ground or airborne early-warning radar. However, in line with more recent thinking, there is now a requirement for a JTIDS (joint tactical information distribution system) data link so that information can be pooled with other land, sea and air vehicles and units.

STATUS: in production and in service with RAF. Will equip export versions of Tornado air defence variant. Technical problems which arose at a late stage in the radar's development caused the RAF's Tornado F2s to be without radars for a period in mid-1985. The problems have since been rectified.

GEC Avionics Foxhunter radar is installed in RAF Tornado F2 aircraft

Skyranger radar

This X band frequency agile lightweight air-combat radar is in full production (for export, it is believed, to Egypt and China). The radar has been designed as part of an integrated avionics suite, the other sub-systems being an air data computer, radar altimeter, head-up display, weapon-aiming computer and secure communications system. The radar is designed for air-to-air ranging and has no air-to-ground modes; it will be used in conjunction with infrared homing air-to-air missiles and guns. The system measures range and closing speed for presentation on the head-up display and is generally suitable for light attack aircraft such as the single-seat version of the British Aerospace Hawk, and other light fighters.

The radar comprises three line-replaceable units: a transmitter-receiver, combined signal processor/power supply, and a fixed, forward-view antenna. The system is modular at printed circuit board level, so individual boxes can be constructed to fit the available space. The total weight is 40 kg.

Typical ranges are 5 km in a gun mode, and 15 km in missile mode. The system has a design mean time between failures of more than 200 hours.

STATUS: in production for export.

Display model of GEC Avionics Skyranger radar

Airborne early warning radar for Nimrod AEW3 and export

Development of the airborne early warning version of the British Aerospace Nimrod goes back into the 1970s, and a production order for 11 aircraft, converted from Nimrod MR1s was announced in March 1977.

GEC Avionics is responsible for the mission system avionics, which include the pulse-Doppler radar and the data handling and display system made by that company. The integral IFF is by Cossor and the electronic support measures by the US company Loral. Unlike the Boeing E-3A Sentry, the Nimrod AEW3 has two antennas, fed sequentially from a single transmitter and placed in the nose and the tail of the aircraft where they scan from side to side, in the horizontal plane, scanning the forward and rear sectors.

The radar has basically two modes of operation. A high pulse-repetition frequency tracks fast-moving low-flying aircraft while a low pulse-repetition frequency is used for periodic updating of ship positions. The two modes are interleaved so that both fast and slow moving targets can be surveyed continuously.

The radar operates at E/F band frequencies. Pulsed transmissions provide the range measurement, while Doppler filtering is used for MTI and clutter rejection. The receiver analyses groups of returned pulses to detect

any Doppler shift. The antennas have been specially designed to minimise sidelobes, and both are of identical shape. Their 180° scans are synchronised and they are automatically roll- and pitch-stabilised by a pair of gyro platforms which compensate for structural flexing and overcome the cyclic errors present in other systems.

STATUS: first aircraft was originally scheduled to have been handed over to the RAF in April 1984, but service acceptance has been continually delayed by technical problems. One aircraft was delivered to the RAF base at Waddington in December 1984, followed by a second a year later. In 1985 the British company CAP Scientific was commissioned by the UK Ministry of Defence to investigate the reported technical problems and propose solutions.

By spring 1986 three Nimrod AEW3s had been delivered to the RAF/GEC Avionics Joint Trials Unit at Waddington and all 11 Nimrods had been converted and all 11 MSA's completed.

During spring and summer 1986 the UK Ministry of Defence evaluated proposals from American companies (Lockheed, Boeing and Grumman) to replace the Nimrod with other aircraft, while GEC Avionics was finalising its plans for solving the outstanding technical problems. A decision whether or not to continue with the Nimrod AEW project is expected in September 1986.

Sky Guardian early warning radar

GEC Avionics unveiled its Sky Guardian family of F band coherent pulse-Doppler airborne early warning radar systems in January 1984. Sky Guardian is based on the early warning system designed for the Royal Air Force/British Aerospace Nimrod AEW3 (see separate entry), using modules from that equipment wherever possible. Versions of the new radar proposed range from simple combinations of radar and air-to-ground data link to systems with IFF, electronic support measures, multi-frequency communications (probably with JTIDS) and operator display consoles.

Fixed-wing applications could include aircraft such as the Canadair Challenger (ie top-of-the-market corporate jets) and the Airbus A300, all using twin, fore and aft synchronised antennas like those of the Nimrod but sized to new airframe and operational requirements. For smaller fixed-wing aircraft and helicopters, GEC Avionics has patented a deployable antenna. A notable feature, claimed to be unique to Sky Guardian, is an automatic track-initiation function and associated software which reduces operator workload.

STATUS: number of versions have been proposed in varying degrees of detail, many of which have proceeded beyond the initial study stages. The most comprehensive scheme includes a derivative of the Lockheed-Georgia

Part of GEC Avionics mission system avionics suite in Nimrod AEW3

RAF Nimrod AEW3 airborne early warning aircraft. Distinctive bulges fore and aft house synchronised scanners for GEC Avionics surveillance radar

C-130, the existence of which programme was formally announced jointly by Lockheed and GEC at the 1985 Paris Air Show. The main feature of this combination is that it provides AWACS mission system effectiveness and patrol endurance at much less than half the cost of the Boeing E-3A. The improved volume and payload available in the C-130 also enables this version to carry a significantly enhanced sensor suite to match threat developments and a large tactical communications suite for the multi-role overland/oversea capability. Wind tunnel tests on a Lockheed Hercules AEW were carried out in 1985. A slightly smaller derivative on which engineering studies are nearing completion involves the G-222 of Aeritalia.

Cutaway model of proposed Lockheed C-130 Hercules with GEC Avionics Sky Guardian airborne early warning radar system

MEL

MEL, Defence and Avionics Systems, Manor Royal, Crawley, Sussex RH10 2PZ
TELEPHONE: (0293) 28787
TELEX: 87267

Super Searcher radar

The latest in MEL's range of airborne radars is Super Searcher, a development of Sea Searcher, first shown publicly at the 1982 Farnborough Air Show. Designed for multi-threat maritime operations, the new addition is a lightweight I band command and control radar with a unique 42-inch (1067 mm) horizontal aperture antenna providing a high definition display on a 14-inch (356 mm) colour cathode ray tube that can show true motion or a centred plan position indicator with variable sector scan mode.

In comparison with Sea Searcher it has a greater detection, target-tracking and guidance performance. The colour display replaces the earlier manned plan-position indicator plot. The system has an in-built guidance capability for the British Sea Skua anti-surface-skimming missile.

The system incorporates three selectable pulse-widths, including an ultra-short one to give high definition of small targets in bad weather. Contact recognition is also improved by the use of microprocessor techniques and signal processing algorithms. The colour cathode ray tube includes freeze-frame and memory storage facilities as well as a moving target indicator. It can be set to give either true motion/offset modes or relative motion latitude/longitude lines, grids, previous-mission intelligence or other reference frames, and provides accurate and permanent plotting on more than 32 individually vectored point markers. The system also has a multiple track-while-scan capability.

The system operates in primary and secondary modes, either separately or in combination. A comprehensive library of symbols, that can be vectored as desired, eases the operator's task, particularly in intelligence storage and extraction. Super Searcher is compatible with IFF/SSR interrogators and can be supplied with an optional electronic surveillance measures antenna covering the C-D and E-J bands

STATUS: in production. Super Searcher is the principal sensor aboard the 20 Westland Sea King Mk 42B ASW helicopters ordered by the Indian Navy.

In October 1985 MEL announced that the Super Searcher had been selected to equip the Royal Australian Navy's Sikorsky SH-70B Seahawk helicopters. Eight helicopters are to be

MEL Super Searcher radar system

delivered in 1988. The order for the radars is put at US$7.5 million.

Marec II radar

The Marec II maritime reconnaissance radar, based on the company's helicopter radar system, was developed to meet the need for a small but efficient and low-cost means of conducting sea-patrol duties, including search and rescue, fishery and oil-rig protection, pollution control and coastguard surveillance. Suitable for fixed-wing aircraft and helicopters, it has a true-motion plotting table display covering 360° in azimuth and a range of 250 nautical miles, together with a pilot's display.

The system has flown in the British Aerospace Coastguarder demonstrator, the prototype Airship Industries Skyship 500 airship, and a Dornier Do228 Skyservant assigned to maritime patrol work in Cameroon. It has been selected for the Indian Coast Guard/Dornier Do 228 light twins.

MEL Marec II radar control/display system

Plotting table display ranges: 17–219 n miles
Plotting table size: 430 × 430 mm
Pilot's indicator ranges: 10–250 n miles
Frequency: I band 9345 MHz
Power output: 80 kW
Pulse width: 0.4 and 2.5 μs
Prf: 200 and 400 Hz

STATUS: in production for Indian Coast Guard. Three sets being supplied by MEL, remaining 33 sets for the 36-aircraft fleet of Dornier Do 228s to be built under licence by Hindustan Aeronautics.

Sea Searcher radar

A lightweight high-performance 360° radar for ASW and ASV roles with over the horizon target tracking, Sea Searcher is suitable for both helicopter and fixed wing aircraft and is in production for Royal Navy/Westland Sea King helicopters. The system employs digital processing techniques permitting easy connection with other avionic equipment, and was specifically designed to interface with sonar equipment and Racal TANS navigational computers.

The north-orientated ground stabilised (true motion) radar display is formed on a 430 mm flat plotting-table, which allows transparent navigation charts to be overlaid. This provides the operator with extra tactical information and allows navigation data to be plotted without the need to transfer radar data to a separate chart or vice versa. This facility is enhanced by the inclusion of an I band (30 mm) transponder channel for identification of support craft (airborne as well as surface), coded transponder replies being displayed on the clutter-free plotting-table surface. The dual pulse-width transceiver includes built-in test circuits which together with the radar display enables in-flight or safe ground testing of the complete system.

A first level of electronic counter-counter-measures protection is achieved through manual adjustment, in flight, of the transmitter frequency, whose pulse repetition frequency is subject to a degree of random jitter, though its primary function is to eliminate unintentional jamming by radars operating in the vicinity.

A dedicated computer permits the automatic tracking of two targets simultaneously, allowing over-the-horizon targeting to be performed. The high accuracy of this digital tracking system enables it to be used for pre-launch guidance of missile systems such as Otomat and Exocet, or via a telemetary link to a surface weapon system.

To meet the installation needs of different aircraft, a family of antenna sizes are available ranging from 610 to 1070 mm horizontal aperture, the largest offering optimum performance.

The antenna normally operates in a continuous 360° scan mode, but sector scanning is available whereby the antenna scans about any bearing selected by the operator. This facility has the two-fold effect of maintaining radar silence in the non-sectored area and increasing the data-rate over the illuminated sector.

STATUS: in production and service.

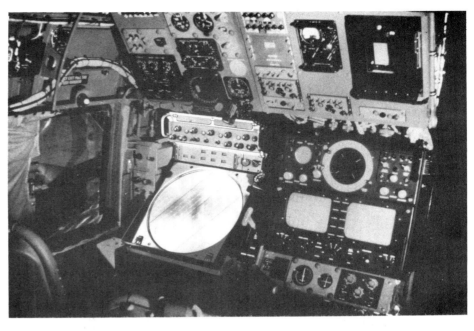

MEL Sea Searcher radar display in Westland Sea King helicopter

ARI 5983 I band transponder

The ARI 5983 was developed for use in naval helicopters and fixed-wing aircraft but can also be used on ship or shore. It comprises a transponder set, a remote control unit and one or two antennas as required.

When interrogated by an I band radar the transponder replies in code on another, dedicated reply frequency. The receiver design allows many diverse types of radars on aircraft, ships or shore to interrogate the transponder.

Interrogation bands: 9190-9290 and 9360-9460 MHz
Reply frequency: 9310 MHz
Reply power: Hi Mode 135 W
Lo Mode 10 W
Reply format: single pulse and 16 codes
Power: 28 V dc 45 W
Built-in test: operator initiated/in-flight
Dimensions: (transponder) 160 × 217 × 82 mm
(controller) 147 × 117 × 48 mm
(antenna) type depends on craft
Weight: (transponder) 2.7 kg
(controller) 0.5 kg
(antenna) 0.5 kg

ARI 5955/5954 radar

This radar for ASV, ASW and search and rescue roles was designed specifically for installation on helicopters. ARI 5955 designates the radar sensor and processing system, ARI 5954 I band transponder being the corresponding IFF

Royal Navy/Westland Sea King ASW helicopter uses MEL ARI 5955/5954 radar

system providing identification of friendly aircraft and surface craft.

The system operates in the I band and has an antenna that can be gyro-stabilised to compensate for aircraft motion. Antenna tilt is adjustable, both above and below a horizontal datum, from the radar operator's position. The receiver gain is adjustable and variable swept-gain facilities are provided. A selection of range presentations up to 50 nautical miles is available and information is displayed on a projection cathode ray tube in conjunction with a Schmidt optical system. This provides unusually high

brightness over a 430 mm square plotting surface. An illuminated parallel-line protractor is incorporated. The operator has the choice of three modes of presentation: conventional plan-position indicator, ground stabilised or ground stabilised with offset.

STATUS: no longer in production, but in service on Royal Navy Sea King and Wessex helicopters, and Royal Air Force Sea Kings. Several other countries have also bought this equipment.

Microwave Associates

Microwave Associates Limited, Dunstable, Bedfordshire LU5 4SX
TELEPHONE: (0582) 601 441

ARI 5983 IFF transponder

The ARI 5983 transponder is used in Royal Navy helicopters operating beyond visual range of their parent or co-operating ships to provide a means of enabling interrogation by normal primary radars aboard the ships. This provides navigational assistance, enhances ship's radar range performance and gives an identification facility in the course of patrol, ASW, and sonar operations involving ships and helicopters.

The equipment operates at frequencies in the I band and consists of two units, transponder and control unit. These operate in conjunction with two antennas mounted on the exterior of the helicopter.

The transponder receives interrogating signals, via the antenna, from pulse radars at any

Microwave Associates IFF transponder

frequency in two bands, 100 MHz wide. On the receipt of such interrogation the transponder will transmit a reply at a closely controlled fixed frequency in the same region. This will be either a single rf pulse, which provides enhancement of the radar return, or a coded group of up to six pulses which provide identification. Sixteen different reply codes are available. These facilities may be selected by the operator at the transponder control unit.

Reception and transmission can take place via either of two antennas, the required antenna being selected by means of an rf switch in the transponder which in turn is controlled from the control unit. Another switch causes a reduction of the transmitted power of approximately 11 decibels. The equipment is arranged so that it will be suppressed during the operation of other 3 cm equipment in the aircraft. Similarly, a suppression pulse is fed to other 3 cm equipments when the transponder is replying.

Each transponder can transmit a choice of differently coded replies, selected by the operator simply setting a switch on the operator's control unit. A self-test generates an interrogate signal which is fed into the system input and if the transponder is functioning correctly a green acceptance lamp is illuminated.

The same basic transponder design can have several forms. For example, the system can be designed so that it requires double or multiple pulse interrogation to provide a reply selectively where there are many interrogating radars (such as in a harbour or other busy area). Where there are few radar-bearing vessels, a single pulse is usually sufficient.

Transponder: I band
Receiver frequency: 2 bands, centre F_1 ± 85 MHz
Bandwidth: ±50 MHz
Sensitivity: –93 dBW sufficient to trigger reply
Transmitter frequency: F_2 ± 7 MHz

Output power: 135 W minimum to 300 W max peak
Pulse duration: 0.4 μs ±0.1 μs
Reply code: 6-pulse code, 16 settings, single-pulse reply capability
Pulse spacing: 2.9 μs nominal
Duty cycle: 0.005 max
Weight: 2.5 kg

STATUS: in production for the Royal Navy. Fittings include current types such as the Lynx and the Sea Harrier. Ten sets have been supplied to the Royal Marine helicopter squadron, the system has also been fitted in Royal Air Force Harriers.

Plessey Avionics

Plessey Avionics, Martin Road, West Leigh, Havant, Hampshire P09 5DH
TELEPHONE: (0705) 486 391
TELEX: 86227

PTR446A IFF/ATC transponder

The PTR446A lightweight transponder identifies aircraft in response to secondary radar interrogation and covers civil and military modes. Emphasis was placed on reliability combined with small size and low weight and these qualities have been achieved by the use of specially designed microelectronic circuits. A digital shift register replaces conventional delay lines in the decoder/encoder circuits, so providing time delays independent of temperature. Integrated circuits are used for the logic and video processing circuits and the logarithmic response intermediate frequency amplifier. These techniques have reduced weight and size by a factor of six in comparison with other equipment, according to Plessey. Decoder, encoder and associated switches in the control unit reduce the number of interconnecting wires to five and substantially cutting down the installation weight. The transmitter-receiver houses the pulse-selection and power-supply

modules. Three-pulse side-lobe suppression is incorporated.

Two control units are available for use with the transmitter-receiver. The smaller of the two is the PV447, of which there are six versions with the following capabilities:-
PV447 — Mode 1 or 3A/B and Mode C or off
PV447A — Mode 1 or 2 and Mode C or off
PV447B — Mode 2 or 3A/B and Mode C or off
PV447C — Mode 1 or 3A and Mode 2 or off
PV447D — Mode A or B and Mode C or off
PV447E — Mode A/B or off and Mode C or off
An alternative to the PV447 is the PV1447 control unit, which meets the requirements of NATO STANAG 5017 for IFF Mk XA and provides Modes 1, 2, 3/A and C. Automatic code changing (ACC) is provided on Modes 1 and 3A with storage capacity for 48 codes in each mode. Manual code entry for these modes is via a front-panel keypad. Mode 2 codes are entered through screwdriver-adjusted switches accessed through the top cover of the unit.

The transponder can be used with either control unit without modification to the transmitter-receiver. Comprehensive self-test is incorporated in all units.

Power output: 24.7 dBW
Pulse rate: 1200 replies/s, each containing up to 14 reply pulses
Triggering sensitivity: –72 to –80 dBm

Plessey PV1447 control unit

Dimensions: (transponder) 2.25 × 5 × 10 inches (57 × 127 × 254 mm)
(PV447 control unit) 5.75 × 2.25 × 4 inches (146 × 57 × 102 mm)
(PV1447 control unit) 5.75 × 3.75 × 6.5 inches (146 × 95 × 165.1 mm)
Weight: (transponder) 3.7 lb (1.7 kg)
(PV447) 1 lb (0.48 kg)
(PV1447) 3.5 lb (1.6 kg)

STATUS: in production.

Racal Avionics

Racal Avionics, Burlington House, 118 Burlington Road, New Malden, Surrey KT3 4NR
TELEPHONE: (01) 942 2464
TELEX: 22891/28588

ASR 360 maritime patrol radar

The ASR 360, Racal Avionics' first venture into airborne maritime surveillance, is a low-cost X band system with 360° coverage, suitable for general aviation aircraft and helicopters. Its principal applications are seen to be the policing of long and sparsely inhabited shorelines, fisheries and ice patrol, and search and rescue. The system contains the facilities of more expensive equipment, including true motion display and Clearscan video processing, providing the high resolution required to detect small targets and isolate the signals from rain or sea clutter.

The ASR 360 is derived from the solid-state marine radars produced by the former Decca Navigator Company and comprises a 12-inch (305 mm) true-motion display, 25 kW transceiver/processor and antenna and drive unit. The display has eight range scales covering 0.5 to 95 nautical miles with variable range markers. It was designed originally to meet a Royal Norwegian Navy requirement in 1979 for maritime surveillance and submarine detection equipment on its single- and twin-engined Cessna aircraft. In conjunction with this Navy, Racal-Decca Norge A/S produced a prototype system that was test-flown for a year on a Cessna 206 Skywagon. The Royal Norwegian Navy subsequently ordered equipment for the Cessna 337 Skymaster, which became the first application of the new radar.

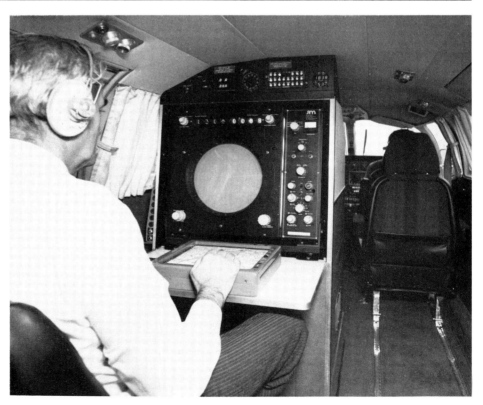

Racal Avionics ASR 360 surveillance radar installation in Cessna Titan fishery-protection aircraft

STATUS: in production. In January 1984 the system was ordered by the Sultanate of Oman Air Force for its 15 Shorts Skyvan aircraft used in policing that country's territorial waters. ASR 360 is also used aboard Royal Navy/British

Aerospace Jetstream for training ASW helicopter observers. Another system is employed for fishery protection on a Cessna Titan operated by the Department of Agriculture and Fisheries in Scotland.

Thorn EMI

Thorn EMI Electronics Limited, 135 Blyth Road,
Hayes, Middlesex UB3 1BP
TELEPHONE: (01) 573 3888
TELEX: 22417

Searchwater radar

Searchwater is standard equipment on the
Royal Air Force's Nimrod MR2 maritime
reconnaissance aircraft: it was designed to
replace the ASV Mark 21 (AN/APS-20) radar.
The system forms part of a major mid-life refit
and 35 Nimrod MR1s have been converted to
MR2s by the addition of Searchwater and other
improvements.

Searchwater is designed for all-weather, day
or night operation outside the defensive range
of potential targets. Systems of this kind pose
severe design challenges: high sea states tend
to obscure targets and falsify radar returns;
high target densities, and the complexity of the
radar system itself, lead to operator fatigue and
error, which is compounded by the presence of
electronic countermeasures; and an apparently
cost-effective design may prove to be too
inflexible to meet a constantly evolving threat or
new requirements.

The system comprises a frequency-agile
radar which uses pulse compression techniques
and a pitch and roll stabilised scanning antenna
with controllable tilt and automatic sector scan.
IFF equipment is included to interrogate surface
vessels and helicopters.

A signal processor enhances the detection of
surface targets (including submarine periscopes) in high seas and an integrating digital
scan converter permits plan-corrected presentation and classification of target and transponder returns. The single radar observer in
this aircraft is presented with a selection of
bright, flicker-free television-type plan-position
indicator B- and A-scope displays in a variety of
interactive operating modes. Weather radar
and navigation facilities are provided within the
system. A real-time dedicated digital computer
relieves the radar operator of many routine
tasks, while continuously and automatically
tracking, storing, and analysing data on a
number of targets at the same time.

The facilities offered by Searchwater reduce
the vulnerability of the host aircraft by permitting them to operate effectively in an entirely
'stand-off' mode, obviating the need to fly over
the target to confirm its type by eye. The system
is entirely modular, with the interfaces and
mechanical construction designed for ease of
fault location and replacement. Major units are
functionally self-contained as far as possible
with a minimum of interconnections with other
parts of the system. Extensive use is made of
hybrid and integrated circuit techniques. The
transmitter uses solid-state frequency-gener-
ators and mixers, followed by two cascaded
travelling-wave tubes. A fluorocarbon liquid
cooling system is employed. The scanner,
which both transmits and receives radar and
IFF signals, uses a reflector of lightweight
construction based on resin-bonded carbon
fibre.

As a result of demands from the Royal Navy
for more effective local defensive surveillance
following operations in the Falklands, Searchwater has now been modified for use aboard
converted Westland Sea King HAS5 anti-
submarine helicopters, re-designated Sea King
AEW2. The combination represents a highly
effective airborne early warning platform that
may be operated from CVs, RFAs, or other
standard naval surface vessels. Searchwater
systems have been acquired to permit 24-hour
AEW coverage for the Royal Navy. The system
has also been supplied to the Spanish Navy for
its Sikorsky SH-3D ASW helicopters under a
£13 million deal announced in September 1984.

STATUS: in production. Searchwater entered
service aboard Royal Air Force Nimrod MR2s in
1979. System has been evaluated on Lockheed
P-3B anti-submarine aircraft by US Navy. The
first deliveries of Sea King AEW2s fitted with

*Thorn EMI Searchwater antenna on Royal Navy/Westland Sea King AEW helicopter. System
rotates clockwise through 90° to give clearance for landing*

Thorn EMI Searchwater radar display in Royal Air Force Nimrod MR2

Pilatus Britten-Norman Defender with Thorn EMI CASTOR radar sensor in nose radome

modified Searchwater radars was in April 1985
and were initially assigned to No 849 Naval Air
Squadron.

CASTOR corps airborne stand-off radar

This UK Ministry of Defence sponsored project

is intended to provide detailed over-the-border
surveillance of the land battle, and of major
hostile ground forces. Denial of surprise and
defensive force multiplication are the primary
objectives, and the system will operate in
conjunction with the UK's Phoenix remote-
piloted vehicles and mortar-locating radars

from the late 1980s onwards. The system proposed by Thorn EMI (one of the two companies competing to build the radar) comprises a long-range airborne stand-off multi-mode radar sensor adaptable to most fixed-wing aircraft; a secure data link to corps-based mobile ground-stations; and computer-assisted facilities for the interpretation of moving-target and synthetic aperture data. Extensive theoretical studies and flight trials in conjunction with Royal Signals and Radar Establishment, have demonstrated the value of the proposed techniques in scanning 24-hour real-time targeting intelligence.

The system has been flying in a specially modified Pilatus Britten-Norman Defender. This aircraft carries the Thorn EMI software-driven synthetic-aperture I band radar and its attitude-stabilised narrow-beam antenna that can scan 360°, or over a smaller sector. The system provides over the horizon targeting, with accurate range measurement, and is managed by an operator sitting at a control/display console. The system has also been flying in a trials Canberra aircraft operated by the Royal Aircraft Establishment, Farnborough. The advantage of this aircraft is its ability to remain on station for long periods at 50 000 feet.

STATUS: in competitive evaluation with Ferranti CASTOR system.

Skymaster multi-mode early warning/maritime reconnaissance radar

From its experience with the Searchwater long-range maritime radar Thorn EMI is developing for the export market a multi-mode, software-driven airborne early warning and maritime reconnaissance radar. It will have optimum performance in the AEW role, with facilities for the automatic detection and tracking of fast-moving targets at high or low level, over land and sea. Choice of a high-powered I band (30 mm) transmitter and antenna, with optimised signal-processing techniques, will enable the system to detect airborne and surface targets in high levels of clutter and high sea states at ranges of more than 100 nautical miles.

Thorn EMI Skymaster airborne early warning radar trials aircraft

Operator's console of Thorn EMI Skymaster AEW radar suite in Defender aircraft

A choice of operating methods includes a look-up mode employing frequency agile and within-beam integration technique to enhance the detection and tracking of high-altitude targets; a look-down mode using medium pulse repetition frequency pulse-Doppler techniques for the automatic acquisition and tracking of low-altitude, fast-moving targets over land and sea; a frequency-agile maritime surveillance mode for the detection of surface vessels; and coastal navigation and adverse-weather warning modes.

Radar information is presented to an operator on a bright, flicker-free 15-inch (381 mm) monochromatic or colour television raster display. This composite display provide grid-referenced indication of confirmed targets, with 'raw' information presented in ground-stabilised format or rolling map, with supplementary alphanumeric annotation. The system is controlled, via a touch-sensitive plasma panel, permitting the software to configure the system according to the mode selected by the operator. Skymaster is being offered in a Pilatus Britten-Norman Islander light utility twin, and the cost of this aircraft with a Skymaster system is reported to be £5 million to £7 million.

STATUS: in development. Although primarily intended for installation in the Pilatus Britten-Norman Defender, it is known that Thorn EMI has proposed several other aircraft as platforms for the Skymaster system, the Shorts Skyvan being a notable example.

UNITED STATES OF AMERICA

Aerojet

Aerojet Electro Systems Company, PO Box 296, Azusa, California 91702
TELEPHONE: (818) 334 6211
TELEX: 670431

Aereye maritime surveillance system

Aereye is a sensor and data processing system being developed by Aerojet for the US Coast Guard. It will be fitted to about 35 of the 41 Dassault Falcon Jet HU-25A maritime patrol aircraft ordered by the US Coast Guard for off-shore environmental protection, law enforcement, search and rescue, and detection of hazards such as wrecks or icebergs.

Aereye stems from joint studies by Aerojet and the US Coast Guard during the 1970s which led to the testing of two versions of an oil-pollution detection system. This background culminated in a contract awarded to Aerojet in August 1980 for a definitive equipment.

The system consists of several sets of equipment carried in two wing-mounted pods and a single ventral pod. The equipment incorporates a Motorola AN/APS-131 sideways-looking radar, a Texas Instruments RS-18C infra-red ultra-violet scanner (for oil-slick verification), a low-light television system and a Chicago Industries Aerial KS-87B reconnaissance camera. The AN/APS-131 is a 200 kW peak power, X band radar with vertically polarised transmissions for optimum oil-slick detection on a calm sea. An image-enhancement computer system, developed by Aerojet for this application, 'freezes' the best video frame and moves the television camera in to record the situation in more detail (eg the ship's name). Although not part of the Aereye system, a Texas Instruments AN/APS-127 forward-looking radar will provide an initial search function.

STATUS: prototype Aereye system began flight trials on Falcon Jet during early 1984.

Bendix

Allied Signal Inc, Bendix Avionics Division, 2100 NW 62nd Street, Fort Lauderdale, Florida 33310
TELEPHONE: (305) 776 4100
TELEX: 514417
TWX: 510 955 9884

TPR 2060 ATC transponder

The Bendix TPR 2060 is a lightweight, compact air traffic control transponder designed for light aircraft and general aviation. It responds automatically to Mode A and Mode C interrogations and, with a suitable encoding altimeter input, will transmit aircraft altitude information with the normal reply pulses. A Mode B capability is optionally available for use in areas employing B Mode interrogation.

The TPR 2060 features special DME (distance measuring equipment) suppression circuitry to prevent interference between the transponder and DME installations when the antennas for the two systems are sited in close proximity. The system also permits transmission of a special identification pulse for a 20-second period by an ident button on the front panel. A reply lamp remains lit during this time to reassure the user that the transponder is identing.

Self-test facilities are incorporated. During self-test operation, the unit's coding and de-coding circuits are exercised in the same manner as they would be during actual radar interrogation. The unit, which may be panel, console or roof mounted, is in a single case and is of large-scale integrated circuit-type construction.

Dimensions: 1.75 × 6.31 × 8.46 inches (45 × 160 × 215 mm)
Weight: 2.6 lb (1.18 kg)

STATUS: in production and service.

RDR-150 weather/multi-function radar

One of the simplest models in the Bendix range of weather radars, the RDR-150 is designed for light and medium twins and the smaller helicopters. Two versions are available: the less expensive black-and-white Weathervision system, or the three-colour Colorvision type, both employing the same transmitter-receiver and antenna. The former may be upgraded to the latter, if space permits, by substituting a slightly larger control/display unit. There is a choice of antenna: normal parabolic, or slotted planar array for greater range with the same diameter.

Storm intensity is indicated by three grey tones in the Weathervision and by three colours in the Colorvision version, measured as a function of rainfall. Red indicates severe weather, with rainfall over 12 mm an hour; yellow shows more moderate conditions, with rainfall in the 4 to 12 mm an hour range, and green depicts rainfall up to 4 mm an hour. A weather-alert mode causes the red storm centre to blink on and off, drawing the crew's attention to the possibility of severe weather. The standard control/display unit has a track line that plots an accurate diversion around severe weather.

The system also provides a ground-mapping mode that shows up prominent surface features such as lakes, bays, rivers, channel markers and offshore oil rigs. This can be a valuable navigational cross-check when flying in poor visibility or above cloud.

The addition of a checklist control panel and a modified Hewlett-Packard HP-67 pocket calculator permits the system to store and display in alphanumeric form up to 16 pages of normal and emergency flight or aircraft procedures such as standard instrument departures or approaches or engine fire drills. Another optional feature is a navigation mode. By the addition of information from Bendix BX-2000 communications, navigation, identification and horizontal situation indication systems, entire flight profiles can be displayed with waypoints, course deviation and track shown in different colours. In a third option, weather information can be overlaid with navigation information to form a moving-map display.

Characteristics for both Weathervision and Colorvision are given below, except where indicated.

Frequency: X band (9375 MHz)
Rf power output: 8 kW/3.5 μs
Antenna scan angle and rate: 90°, 16 looks/min
Antenna tilt: manually selectable to any angle between ±15°
Display storage: digital memory
Range: 160 n miles
Antenna stabilisation: none
Pressurisation: none needed
Antenna size and weight: 12-inch (305 mm) parabolic, or 10- or 12-inch (254 or 305 mm) flat-plate; parabolic 2.7 lb (1.2 kg), flat-plate 5.5 lb (2.5 kg)
Transmitter-receiver size and weight: 1/2 ATR Short, 10.5 lb (4.8 kg)
Control/display unit size and weight
(Weathervision) 6.25 × 4 × 9.875 inches (159 × 102 × 251 mm), 5.5 lb (2.5 kg)
(Colorvision) 6.25 × 4.7 × 12.06 inches (159 × 119 × 306 mm), 10 lb (4.55 kg)
Qualification: TSO C-63b

STATUS: in production and service since 1976.

RDR-160 radar

This low-cost weather radar was designed to suit the majority of general aviation light twins and some singles. Recognising that space and economy are at a premium in this class of aircraft, Bendix has combined the transmitter-receiver and antenna into a single unit, obviating the need for waveguides and cable looms between the two, and resulting in a system weight of only 15.5 lb (6.9 kg), perhaps the lightest weather radar available anywhere. As

Bendix RDR-160 transmitter-receiver and antenna set

with the RDR-150, from which it was developed, the -160 is available either as the monochromatic Weathervision or the three-colour Colorvision and provides essentially the same facilities, namely weather and terrain-mapping and weather alert. The main differences from the -150 are a reduction in transmitter power, from 8 to 6 kW, and the non-availability of a flat-plate, slotted-array antenna. The control/display unit is identical to the corresponding unit for the RDR-150 Colorvision.

STATUS: in production and service since 1978.

RDR-230HP radar

The RDR-230HP is essentially an upgraded version of the X band RDR-160 with improved performance, mounting and packaging. Featuring a non-stabilised 12-inch (305 mm) flat-plate antenna and backed by a 5 kW peak-power transmitter, the all-colour system has a 240 nautical mile display range and a weather-avoidance range of 200 nautical miles. A special feature of the RDR-230HP is what Bendix calls Extended STC, which permits significant storm regions to show up as bright red even at great distances. Without this feature such weather developments might not appear in red until they grew stronger or the range was closed.

The system continues the Bendix-initiated ART idea which combines the antenna and transmitter-receiver into a single unit (again, a

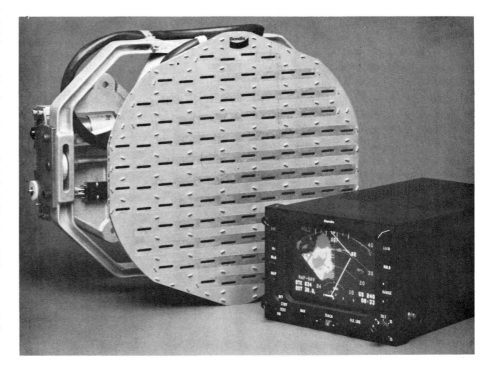

Bendix RDR-230HP radar

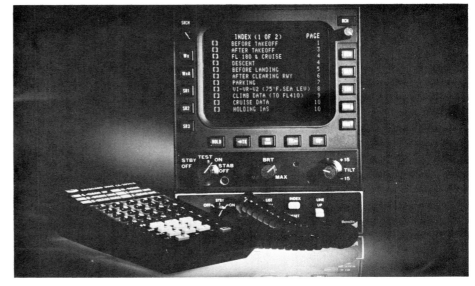

Bendix RDR-230HP radar indicator with data-entry keyboard

feature of the RDR-160). The 12-inch (305 mm) flat-plate antenna makes more effective use of the radar energy than does the parabolic dish by concentrating the beam into a narrower cone, providing better defined weather patterns and ground mapping. The antenna scans ±45° about the aircraft axis and tilts ±15° on pilot command.

Another feature is the track-line cursor. This is a line on the display that starts at the aircraft and can be rotated left or right as the pilot directs, the angular deviation from the aircraft axis being displayed in alphanumerics on the radar screen so that the pilot can choose a convenient course to avoid the worst weather. The system also accommodates all the most widely used radar options including a 32-page programmable checklist with automatic high-priority page call-up for alerts, and the moving-map display option so that information from the most popular area navigation, omega, inertial navigation, and Loran systems can be displayed.

STATUS: in production and service since 1981.

RDR-1100 radar

This is a lightweight, pitch and roll stabilised weather radar system with a small-face control/display unit for situations where space is limited, for medium turboprop and turbine twins. The RDR-1100 is available in monochrome Weathervision or three-colour Colorvision alternatives, both using the same transmitter-receiver, antenna, and drive units. A version designated the RDR-1150 has a fourth colour (magenta) for weather. The systems also use the same aircraft wiring and are plug-to-plug compatible so that upgrading to colour from monochrome entails simply the substitution of a slightly larger control/display unit.

Standard features for the colour system, apart from the usual three-colour presentation of storm intensity, are a movable azimuth track line permitting the pilot to choose and read off for air traffic control purposes a new course to avoid bad weather; weather alert, whereby the red storm centres blink on and off to attract the crew's attention; and a ground-mapping mode.

Optional features available in the Colorvision version are checklist, navigation and weather/navigation overlay modes. In the first of these, an additional unit, the pilot-programmable CC-2024B, C, or D unit provides 16 or 32 pages for the display of flight checklists or other information in alphanumeric form. In the navigation mode, an appropriate interface unit permits the pilot to display waypoints, course deviation and planned course changes. Bendix provides interface units for its own BX 2000 navigation system and for the most widely used area navigation, omega and inertial navigation systems.

Characteristics are for both monochrome and colour except where stated.

Frequency: X band (9375 MHz)
Rf power output: 8 kW
Antenna scan angle and rate: 120° (120°/60° colour), 24°/s (24°/s, 48°/s colour)
Display range: 200 n miles (240 n miles colour)
Antenna size and weight: 12-inch (305 mm) flat-plate, 8 lb (3.6 kg)
Transmitter-receiver size and weight: ½ ATR Short, 6 lb (2.7 kg), colour 10.5 lb (4.77 kg)
Control/display unit size: 6.25 × 4 × 9.875 inches (159 × 102 × 251 mm)
Colour has same face dimensions but is 12.06 inches long (306 mm)
Stabilisation: ±30° combined pitch, roll and tilt
Pressurisation: none required
Power: 3.5 A at 28 V dc
Qualification: TSO C63b

STATUS: in production.

RDR-1200 radar

This weather radar pioneered digital memory display weather radars for general aviation and is now recognised as one of the industry standards; it came on the market in 1974. The system is aimed at the heavy turboprops and jets of the corporate aircraft market, and is available in Weathervision (monochrome) or Colorvision versions. Special storage circuitry permits a continuous read-out of radar video and the system can be frozen for extended periods by a hold mode so that growth and movement in storm cells can be seen by switching back to the normal scan mode.

The system, which weighs 34 lb (15.5 kg) in its colour version, feeds 10 kW output power into a pitch and roll stabilised 12- or 18-inch (305 or 447 mm) flat-plate antenna.

STATUS: in production and service.

RDR-1400 weather/multi-function radar

This radar is designed for commercial helicopters, particularly those associated with the large international offshore oil and gas industry. It differs from almost all other Bendix weather radars by having a beacon interrogator that exploits the increasing use of portable radar beacons in these industries.

The original monochromatic RDR-1400 has now been joined by a colour-radar version with greater performance. The following operational modes are available:

Beacon navigation. The growing popularity of portable beacons is supported by several special RDR-1400 capabilities. Beacon signatures are denoted by short lines or obliques on the display, the actual location of the device being determined by the middle of the line and the pilot can overlay beacon returns on the weather map. The beacon's discrete code can be displayed for positive identification, an important factor when the pilot is trying to locate a specific rig in a drilling farm where numerous rigs may be transponder-equipped. Beacon-detection range is up to 160 nautical miles depending on altitude.

Beacon Trac. This mode, peculiar to Bendix, generates and displays on the weather radar screen an inbound course to the discrete beacon. This course line can be rotated 360° about the beacon by rotating the horizontal situation indicator course selector, thus allowing the pilot to choose a convenient course to the beacon. A number of programmes are under way to assess the role of these beacons in published approaches, for example as a final fix or for establishing a course.

OBS Trac. This mode provides another course-following option. When in a weather or search mode, a track line or course-bearing cursor can be generated from the aircraft position and controlled by the horizontal situation indicator course selector to provide a course line to the chosen target. The OBS Trac heading is displayed digitally in the lower right-hand corner of the indicator. Left/right deviations can be determined by comparing heading information to this number and by observing the movement of the track line in relation to background targets.

Search. Three search modes are available. Search 1 has special sea-clutter rejection circuitry to detect small boats or buoys, for example, down to the minimum range. Search 2 is for precision ground-mapping in situations where high target resolution is important. Search 3 includes maximum return clutter and can detect and track oil slicks.

With these capabilities the system is suitable for search and rescue, surveillance, aerial survey work and law enforcement applications, in addition to rig-servicing.

Frequency: X band
Rf power output: 10 kW
Antenna size and scan angle: 12- or 18-inch (305 or 457 mm) flat-plate, 120° or 60°
Display size: 4.34 × 3.33 inches (110 × 85 mm)
Dimensions: (transmitter-receiver) ½ ATR Short, 5 × 6.25 × 13.875 inches (127 × 159 × 352 mm)
(control/display unit) 6.25 × 6.25 × 10.875 inches (159 × 159 × 276 mm)
System weight: 34.1 lb (15.47 kg) (12-inch antenna)
Qualification: TSO 63b

STATUS: in production and service.

RDR-4A radar

Chosen by Boeing as standard equipment for the 767 and 757 transports, the Bendix RDR-4A is claimed to be the most advanced multi-function colour radar currently available.

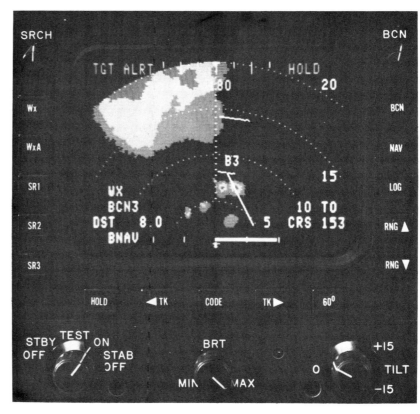

Bendix RDR-1400 control/display unit with weather plot

Bendix RDR-4A radar

Boeing 767 (in foreground) and 757 (behind) have Bendix RDR-4A radars

Designed to meet the new ARINC 708 requirements, the X band system features a solid-state transmitter and line-of-sight antenna with split-axis performance, and is compatible with the EFIS flightdecks of the Boeing 767, 757, Airbus A310, McDonnell Douglas MD-80, DC-10 and Lockheed L-1011 transports and other designs. The range is 320 nautical miles.

STATUS: in production and in service with, among others, Airbus A300s of TOA Domestic Airlines of Japan. In October 1982 it was chosen for Southwest Airline's 737-300s, and in November 1982 by Singapore Airlines for its eight Boeing 747-300s. Also in service with Delta, American, Pakistan International and Northwest Airlines and Saudia and Varig. In March 1986 Bendix announced orders for the RDR-4A from Finnair and Austrian Airlines. In both cases the radar will equip McDonnell Douglas MD-87 airliners.

RDR-1FB (AN/APS-133) radar

This digital colour radar is a high-performance weather, beacon-homing and terrain-mapping system designed for large commercial and military transports; it is said to represent about half the division's revenue from the military sector. The three-colour display can be used in conjunction with other optional equipment to show programmable checklists or to super-impose navigation or other information on the weather map. The system employs digital processing and microcomputer techniques, radio frequency semiconductors, and a solid-state modulator, three-colour plan-position indicator, and control unit.

Significant landmarks and continental shore-lines up to 300 nautical miles away can be portrayed in the ground-mapping mode by using the high power output concentrated into a pencil beam. At the same time discrete details such as lakes, rivers, bridges, runways and runway approach reflectors, readily show up on the colour display. To improve range resolution at short ranges the system operates with 0.5-microsecond pulses in contrast to the 5-microsecond pulses used for long-range ground and weather mapping.

In the air-to-air mapping mode, the RDR-1FB detects and tracks other aircraft during rendezvous, formation and air-refuelling. Aircraft of C-130/C-141 size can be tracked to 12 and 20 nautical miles, but may still be resolvable at ranges as little as 600 yards (550 metres) depending on relative bearing, aspect and altitude.

To provide long-range homing to remote ground destinations or tanker aircraft, the RDR-1FB operates at X band frequencies (9375 MHz) for beacon interrogation and reception. The identification of closely spaced pulse reply codes at long ranges is made possible by the marker and delay modes of the radar indicator. In the marker mode a variable marker is positioned on the screen just in front of the beacon reply. When switched to the delay mode the display presentation starts at the marker range. The range switch can then be moved to select a shorter range scale yielding an expanded view of the area containing the beacon reply. A version without the beacon facility is designated RDR-1F.

Derived from the Bendix RDR-1F used on many hundreds of airliners, the AN/APS-133 comprises four line-replaceable units: a 30-inch (762 mm) fully stabilised split-axis parabolic antenna that provides specially shaped search or fan beams for terrain-mapping, a transmitter-receiver, a colour display and a control unit.

In November 1984 the company delivered to the US Marine Corps the first units of the Improved Land-Mapping version for its fleet of Boeing KC-135 tankers and C-135 transports. It was specially tailored to US Marine Corps requirements, with a high pulse repetition frequency, short pulse-width, enhanced

computing and selectable scan-sectors to improve radar navigation at low level. The US Marine Corps is to acquire 62 sets of this version.

Frequency: X band (9375 MHz)
Power output: 65 kW
Prf: 200 pulses/s
Pulse widths: (weather) 5 μs
(beacon) 2.35 μs
(mapping) 0.5 μs
Weight: (antenna) 37 lb (15.8 kg)
(transmitter-receiver) 55 lb (24.9 kg)
(control/display unit) 13.8 lb (6.3 kg)
(control unit) 2 lb (0.9 kg)

STATUS: in service with US Air Force transport aircraft, notably Lockheed C-141 and C-5A Galaxy and McDonnell Douglas KC-10 Extenders, also on some US Coast Guard Lockheed C-130 transports.

In early 1986 Presidental Airways of the USA selected the RDR-4A to equip its 12 Boeing 737-200s.

RDS-82 radar

Launched in 1983, the digital, X band RDS-82 represents the first of a new line of Bendix weather radars and will eventually replace several of the more inexpensive models such as the RDR-160, RDR-150 and RDR-230HP. The change of designation, from RDR- to RDS-, indicates that a sufficiently large technological advance has been achieved to justify the description S (for 'system'), and a further indication is that this and future radars with an RDS- designation will be EFIS-compatible.

The RDS-82 has a four-colour display. Magenta indicates rainfall in excess of 2 inches (51 mm) an hour (Bendix pioneered the use of magenta for this purpose) and attenuation circuits automatically adjust the radar's sensitivity. These features were found earlier only in the more expensive RDR-1150 system. The stabilisation/voltage ratio can be adjusted to match any attitude gyro. The system is based on ARINC 429 digital data-handling, so that all information and commands between the radar head and the control/display unit are carried on a twisted cable-pair. The antenna can be a 10- or 12-inch (254 or 305 mm) flat-plate, scans 90° at the rate of 30° a second and radiates 1 kW. The system operates up to 55 000 feet without pressurisation.

STATUS: in production.

AN/APN-215 radar

The AN/APN-215 colour radar is a weather, surface search and precision terrain-mapping system derived from the successful and widely used RDR-1300 commercial system. It is designed for heavy twins, turboprops and transport helicopters. Low weight and a 240 nautical mile range suit it to utility and reconnaissance aircraft, and it was chosen for

Bendix AN/APS-133 four-unit weather radar

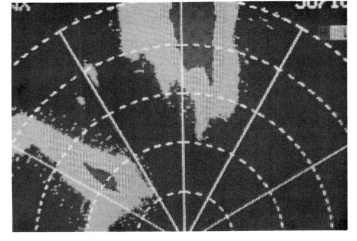

Moisture content of cloud ahead represented as three colours on AN/APS-133 system

the US Army's U-21s and RU-21s, military versions of the Beech King Air.

In conjunction with other equipment the APN-215 can display navigation pictorial information overlaid on the weather map, together with pilot-programmable pages of checklist information such as en route navigation data and emergency procedures.

The system comprises three units: a 12-inch (305 mm) pitch and roll stabilised antenna, transmitter-receiver and colour control/display unit.

STATUS: in production and service.

Series 3 RDS-84 Quadra radar

One of two digital colour radars announced by Bendix in April 1984, the RDS-84 is an addition to the company's range of Series 3 (ie third-generation) digital avionics for general aviation. It is intended for heavy pistol twins and light jet types. Providing a 120° antenna scan, and weighing 20 lb (9 kg) like the RDS-82 introduced in 1983, the RDS-84 features a combined transmitter-receiver and antenna. The designation Quadra indicates the use of four colours.

STATUS: in production.

Series 3 RDS-86 Quadra radar

This addition to the Bendix range of Series 3 digital avionics for business aircraft was announced in April 1984. It is intended for heavy, corporate jets such as the Dassault Falcon 900, Gulfstream IV and Canadair CL-600/601 Challenger. Like the RDS-84, the new radar has a combined transmitter-receiver and antenna, and displays four colours (as indicated by the description Quadra). The system has three features new to the Bendix colour-radar family: automatic range limitation, whereby area giving signal returns of an unreliably low level are painted blue as a warning, antenna stabilisation, to maintain a steady picture during climb and descent; and a long-range navigation mode in which data up to 1000 nautical miles ahead can be displayed (this is in addition to eight selectable distance scales ranging from 5 to 320 nautical miles).

The system weighs 22 lb (10 kg).

STATUS: in production.

Boeing

Boeing Military Airplane Company, Box 3707, Seattle, Washington 98124-2207
TELEPHONE: (206) 655 1198
TELEX: 329 430

Offensive avionics system for B-1B

In October 1981, after a long evaluation of possible aircraft to carry cruise missiles, President Reagan reinstated the Rockwell B-1A in an improved form as the follow-on to the huge but ageing fleet of Boeing B-52 bombers. This step reversed the June 1977 decision by President Carter to cancel the B-1 in favour of continued cruise missile development. The new programme calls for the acquisition by US Air Force Strategic Air Command of 100 aircraft, now designated B-1Bs, by June 1988.

During the suspension of the programme, however, refinements with the B-1A's avionics equipment continued, and the resulting system for the B-1B shows many improvements in accuracy, reliability and performance. The original B-1A equipment enjoyed considerable commonality with that planned for the B-52 electronic update (see separate entry) and this policy has carried over to the B-1B so that life cycle costs can be reduced as far as possible. The generic name for the avionics suite on the B-52 and B-1B that enables the aircraft to navigate to their targets and align and launch their weapons is the offensive avionics system, or OAS. The suite of 89 line-replaceable units, weighing some 5000 lb (2270 kg) is provided by Boeing Military Airplane Company, which also provided similar equipment for the B-1A. Boeing is also responsible for supporting elements of the defensive avionics and tail-warning systems.

The OAS will use an improved multiplex data-bus interface between the majority of equipment boxes. It minimises the need for special adapters or interface units in the aircraft and allows future growth to be catered for in a much simpler way. The sub-systems are controlled through the data-bus by computer software programs.

The navigation and weapons delivery system includes dual high-accuracy Singer-Kearfott SKN-2440 inertial navigation units (a modified version of the system used in the General Dynamics F-16), an AN/APN-218 Doppler velocity sensor by Teledyne Ryan, dual AN/APN-224 radar altimeter systems by Honeywell, and dual IBM AP101F terrain-following computers. The Westinghouse AN/APQ-164 dual-channel multi-mode offensive radar system, or ORS, incorporates a low-observable, phased-array antenna (the first such device to be developed for airborne use) and generates the terrain-clearance data which is processed by the terrain-following computers, producing command signals to the flight control system in the terrain-following mode (see also under Westinghouse). The system has a number of advantages over earlier

OFFENSIVE & SUPPORTING ELEMENTS OF DEFENSIVE AVIONICS SYSTEM

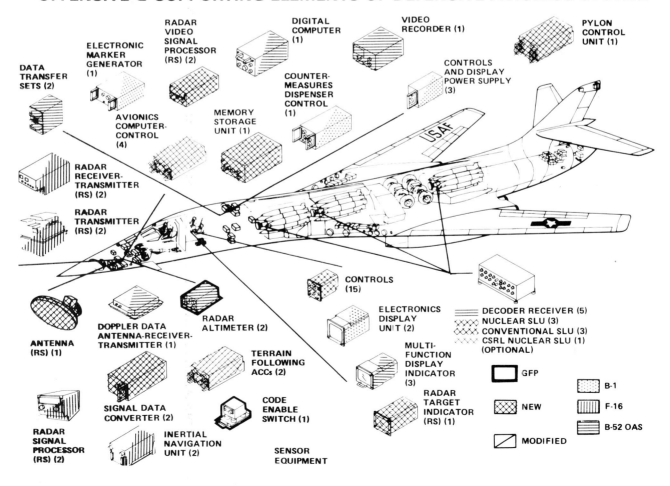

Offensive and defensive avionics for Rockwell B-1B swing-wing bomber

equipment, including low radar cross-section, improved reliability, full redundancy, and commonality with the AN/APG-66 radar equipping the US Air Force's General Dynamics F-16.

As part of the stringent contractual requirements for the B-1B, it was stipulated that the OAS must out-perform the system proposed for the original B-1A by factors of two or three to one in key functional areas, and be readily adaptable to accommodate future development. Two such growth areas currently being incorporated are the facilities to carry and launch cruise missiles and the use of what Boeing calls the Common Strategic Rotary Launcher for internally carried weapons.

Avionics control computers, together with a memory storage device and data transfer units, provide the instructions to navigation, weapon delivery, control and display, defensive, terrain-following and central test functions.

Three multi-function indicators, an associated display electronics unit, a video recorder (all by Sperry) and a Westinghouse radar display unit permit appropriate information to be shown and recorded at both offensive and defensive operators' stations. Threat analysis, countermeasures capability and performance are displayed graphically to the defensive system operator on two Sanders electronic display units. Data transfer units to load and store mission and flight information are provided by Sundstrand Data Corporation.

STATUS: in flight-test and production. Boeing received initial contract for full-scale OAS development in October 1981, followed by definitive contract in May 1982. B-1A prototype No 4, modified to incorporate B-1B flight characteristics and fitted with B-1B avionics, made its first flight as avionics test-bed for the new aircraft in July 1984. First production B-1B made its initial flight in October 1984. First ship-set of avionics hardware was also delivered in that month. First operational B-1B was handed to the US Air Force in June 1985 and all 100 aircraft should be operational by mid-1988.

Offensive avionics systems update for Boeing B-52

In parallel with the B-1B avionics development a huge OAS update programme is under way at Boeing Wichita for selected versions of the US Air Force's fleet of Boeing B-52s, the first major electronics refurbishment activity since the aircraft entered service in the mid-1950s. In July 1979 the US Air Force nominated Boeing as prime contractor to produce five OAS update systems for flight-test installations aboard three B-52Gs and two B-52Hs, together with the modifications needed to integrate the medium-range US Air Force/Boeing AGM-86B air-launched cruise missiles. Power was applied to the first test aircraft in mid-May 1980.

As a result of the trials, follow-on contracts for the B-52 OAS update were awarded, covering the production of 264 kits for B-52G and B52-H versions between November 1979 and March 1986. The first production kits were delivered in June 1981.

Principal items of equipment comprising the OAS update on the B-52 are: redundant 64 K general-purpose IBM computers for navigation and weapons delivery, with either computer capable of shouldering all the mission-essential functions; Teledyne Ryan AN/APN-218 common strategic Doppler radar; Lear Siegler attitude/heading reference system as used on the Lockheed C-141 StarLifter transport; Honeywell standard precision navigator gimballed electrostatic aircraft navigation system (SPN/GEANS), a higher accuracy version of the system developed for an earlier B-52D avionics update; Honeywell radio altimeters with terrain-correlation capability; modifications by Norden to the terrain-mapping and avoidance radar with the aim of improving its reliability (the radar has been a major maintenance item on the B-52); and new multi-function controls and displays that will reduce the space occupied and maintenance requirements.

Although reliability and lower maintenance, rather than better performance, were the principal aims of the OAS update, the change from heavy, power-consuming vacuum-tube technology to solid-state digital electronics will provide crews with more flexibility and a smaller workload. Reliability is expected to improve by a factor of about ten. The intention is that data on each mission will be programmed into the OAS via the IBM computers, using Sundstrand data-transfer units. Four such units will be used· on each aircraft: one for data relating to the basic computer programs, one for data associated with cruise missile terrain-correlation and targeting, one to record faults as they develop within the system for subsequent analysis and one for aircraft mission information.

STATUS: first B-52 to have OAS update made its initial flight in September 1980 and, after flight test programme, was handed over to US Air Force in August 1981. Improvement programme continues.

Strategic radar update for B-52G and H

In late-1985 Boeing received a $88 million contract from the US Air Force for the production of strategic radar kits to update Boeing B-52G and H aircraft. The kits will be installed, beginning in June 1987. Norden is the prime sub-contractor to Boeing for the production of these kits (for further details see entry under Norden).

Collins

Collins Divisions, Rockwell International, 400 Collins Road NE, Cedar Rapids, Iowa 52498
TELEPHONE: (319) 395 1000
TELEX: 464421
TWX: 910 525 1321

TDR-950 ATC transponder

Collins TDR-950 is a panel-mounted air traffic control transponder for light aircraft. It responds to interrogation on 4096 codes, covering Modes A and C, but can be converted to Mode B or to B/C coverage. When used with a suitable encoding altimeter, it has an altitude reporting capability of up to 62 000 feet. The system is interrogated on a frequency of 1030 MHz, responding with a nominal output of 250 watts on the 1090 MHz frequency.

The TDR-950 employs lsi-mos integrated circuitry, with all encoding and decoding functions being carried out by a chip. Self-test facilities are incorporated.

Dimensions: 6.25 × 1.625 × 8.15 inches (159 × 41 × 207 mm)
Weight: 2 lb (0.91 kg)

STATUS: in production and service.

TDR-90 ATC transponder

The TDR-90 is an air traffic control mode A/C transponder with 4096 codes and an altitude-reporting capability of up to 126 000 feet when used with an encoding altimeter. It is a remotely controlled system designed primarily for general aviation.

The system has a transmitter output power of 325 watts nominal (250 watts minimum) on a response frequency of 1090 MHz. Positive sidelobe suppression facilities are incorporated in order to provide a cleaner 'paint' on the interrogator's trace. Two-way mutual suppression avoids interference with distance-measuring equipment. Another feature is a strip-line duplexer to control receiver front-end noise while retaining high sensitivity and frequency stability, regardless of antenna matching.

A built-in test facility for both the transmitter and receiver functions is included. Test signals are injected at just above the minimum sensitivity level to ensure that receiver, decoder, encoder and transmitter are functioning correctly.

The system's CTL-90 control unit has two-knob code selection, ident, self-test, standby and altitude-reporting on/off controls. An optional system selection switch can also be incorporated for use in dual installations. The display is of the gas-discharge type. The TDR-90 electronic unit can, however, interface with most conventional transponder controllers as well as the CTL-90 unit.

Format: 1/4 ATR Short
Weight: 3.5 lb (1.59 kg)

STATUS: in production and service.

621A-6 ATC transponder

The Collins 621A-6 transponder is an air-transport system operating in Modes A, B and C on 4096 codes. Space is also available for Mode D operation and it has a provision for X-pulse. Mode C altitude-reporting extends to 126 700 feet when the transponder is used with an altitude digitiser or an air data computer. The unit operates on the normal receiver and transmitter frequencies of 1030 and 1090 MHz respectively and the transmitter output power is nominally 700 watts. The system is remotely controlled.

Solid-state digital microcircuitry construction is used throughout the 621A-6. Delay-lines are replaced by digital delay-circuits and all timing functions are performed by digital shift-register techniques. Front-end noise has been eliminated by using a strip-line duplexer which also permits both receiver and transmitter to operate through a single antenna. A lightweight quarter-wave cavity is employed in the transmitter section. A ferrite isolater between this and the antenna ensures constant loading regardless of antenna matching and contributes to frequency stability.

Self-testing is carried out by an internally generated signal which is injected into the receiver at just over the minimum sensitivity level. This checks receiver sensitivity and frequency, decoder tolerance, monitor circuits and the selected mode together with Mode C. The test can be performed from either the controller or at the transponder itself.

During normal operation, the monitor circuits maintain a continuous check on transmitter frequency, transmitted power, transmitted code, reply pulse spacing and internal timing circuits. If a fault is detected, a visual warning is given. A facility for continuous automatic self-test when the aircraft is beyond the range of normal ground interrogation is optionally available.

Maintenance monitors isolate any malfunctions to the antenna and its cables or to the transponder unit itself. Latching maintenance monitors on the front panel change from black to yellow if a malfunction occurs and remain yellow until reset. A separate test connector on the back panel facilitates ground maintenance and is used to interface the unit with automatic test equipment.

Format: 3/8 ATR Short
Weight: 13.6 lb (6.16 kg)

STATUS: in service.

621A-6A ATC transponder

Based largely on the design of the earlier 621A-6 model transponder (for which the 621A-6A is intended as a replacement), the later model exploits advances made in microcircuitry techniques to improve performance and reliability.

Compared with its predecessor, the 621A-6A has 190 fewer components and the functions of five cards in the earlier unit are now performed by a single card. Furthermore, technological improvements have resulted in reduction of 15 per cent in the initial purchase cost, and an increase of 100 per cent in mean time between failures is also expected.

The 621A-6A shares a large number of components with the 621A-6 and the two systems are directly interchangeable so that retrofits are simple. There is, apart from the reduced number of parts, one significant design change, namely the addition of pulse-width adjustment facilities. This permits adjustment to offset the effects of ageing in the main transmitter valve, extending this item's working life with, consequently, a lower replacement burden.

Format: ⅜ ATR Short
Weight: 13.6 lb (6.16 kg)

STATUS: in production and service.

Simple presentation of Collins weather radar indicator showing cloud 20 and 40 nautical miles ahead

Micro Line WXR-200A radar

Designed for light twins and some singles, the monochromatic WXR-200A is the smallest system in the Collins range of weather radars. A high-resolution picture, the result of a memory-enhancement technique to provide the equivalent of 128 000 bits of memory, smooths out block-edged storm outlines and minimises target shift and smearing. The result is that fine detail in the picture, such as hooks, anvils and scallops, can be more clearly seen. For typical weather-mapping the system operates at 5.5-microsecond pulse widths, but for close-in weather and for ground-mapping, it automatically shifts to 1 microsecond for better picture clarity and definition.

The system comprises three units: a 12-inch (305 mm) slotted-array flat-plate antenna (optional alternatives are a 10-inch (254 mm) flat-plate of 12-inch (305 mm) parabolic dish) with pitch stabilisation, transmitter-receiver and control/display unit. The bright display is readable under high ambient lighting conditions and is presented in four levels. Maximum cell activity is shown as a black patch or hole surrounded by areas of lesser rainfall. Pilot-selectable receiver gain has four levels to help in some ground-mapping situations and to assist in weather analysis.

Frequency: X band (9345 MHz)
Power output: 5 kW, 5.5 μs for long range, 1 μs for short range
Range: 180 n miles
Antenna scan angle: ±45° about fore-and-aft axis
Dimensions: (control/display unit) 6.25 × 4 × 10.1 inches (159 × 102 × 257 mm)
(transmitter-receiver) 5 × 5 × 12.5 inches (127 × 127 × 315.5 mm)
Weight: (antenna) 6.1 lb (2.77 kg)
(transmitter-receiver) 10.7 lb (4.86 kg)
(control/display unit) 5.9 lb (2.67 kg)
Qualification: FAA C63b

STATUS: in production and service.

Pro Line WXR-250A radar

For the larger twins, corporate jets and small/medium helicopters, the monochromatic WXR-250A incorporates many of the features of the smaller Micro Line WXR-200A described above. For long range mapping the system automatically generates 5.5-microsecond pulses, providing the power needed to scan the weather within a range of 240 nautical miles. When the pilot selects the mapping mode the system switches itself to 1 microsecond pulse-width for better resolution.

Seven modes are provided:
WX gives cyclic contouring on alternate antenna scans
WX HOLD freezes the display

WX ID (weather ident) displays only the contoured areas
NORM displays all levels (no contour or cyclic contour)
MAP gives maximum gain without cyclic contouring (there are four lower gain levels)
TEST energises a test pattern to verify operation of the control/display unit
STBY energises the system but without transmission.

At the time of its introduction in the mid-1970s the WRX-250A was perhaps the most economically packaged radar in its category. The transmitter-receiver package is still one of the smallest, enabling it to be mounted close to the antenna so that the wave-guide can be as short as possible, thus increasing performance.

A development of this radar, the WXR-270, is planned as its replacement.

Power output: 5 kW from magnetron, 5.5 μs long range, 1 μs short range
Frequency: X band (9345 MHz)
Range: 240 n miles
Antenna: 12-inch (305 mm) flat-plate fully stabilised, scan angle ±60°

STATUS: in production and service.

Pro Line WXR-300 radar

Designed for the larger twins and turboprop and jet corporate aircraft, the WXR-300 was the first of Collins' second-generation colour radars and the first such system to incorporate push-button control of mode and range. Good picture resolution is obtained by using 250 000 bits of display. One of the advantages of the radar is its capability as an alphanumeric and graphics information display when connected to an appropriate computer. Information such as normal and emergency operating procedures or navigation data can be entered through the pilot through an optional keyboard entry device.

Frequency: X band (9345 MHz)
Power output: 5 kW, 5.5 μs long range, 1 μs short range
Range: (10-inch (254 mm) antenna) 217 n miles (12-inch (305 mm) antenna) 256 n miles (18-inch (457 mm) antenna) 353 n miles
Weight: (system with 18-inch (457 mm) antenna) 28.6 lb (13 kg)
Qualification: FAA TSO C63b

STATUS: in production and service.

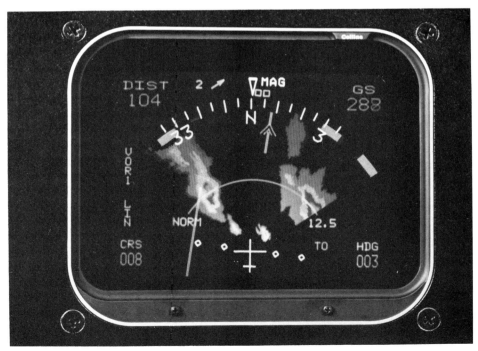

Presentation for Collins WXR-350 weather radar

Centre-panel presentation of Collins RNS-300 combined weather radar and navigation display

Collins WXR-700 weather radar display showing area of turbulence at right-hand edge of pattern

WXR-350 weather radar

Collins introduced the WXR-350A weather radar in 1984. It is intended to work specifically in conjunction with electronic flight instrument systems and features four levels of precipitation, the fourth being displayed in magenta. There is also a path-attenuation-correction alert function which warns pilots of possible weather cells which are hidden, or have apparently reduced intensities, behind heavy rainfall areas.

STATUS: in production.

RNS-300 weather/navigation radar system

The RNS-300 is essentially the Collins WXR-300 weather radar with an additional processing unit that makes it possible to superimpose, on the weather map, aircraft position with respect to ground stations, waypoints and course and heading lines. En route waypoints or navigation stations appear to move from the top of the screen to the bottom, a major advantage being that, with course and heading lines being displayed simultaneously on the same cathode ray tube, any drift shows up immediately. This indication is particularly helpful when flying in the vicinity of storms, when considerable and rapid changes in wind speed and direction may occur due to local and random pressure changes.

The radar/navigation functions of the system make it possible to plot course lines to area navigation waypoints, Vortacs, or the intersections defined by heading and course lines, or omega tracks. Two course lines and a heading line can be plotted simultaneously to form several legs in a navigation segment such as a SID or a STAR. Clear presentation of such information facilitates the choice time and fuel-saving routes.

Collins claimed that, at the time of its introduction, the RNS-300 was the only system on the market capable of plotting intersecting course lines using navigation information from dual navigation radios or area navigation systems.

STATUS: system was introduced during 1980 and has been adopted by Gates Learjet, Beech, and Sabreliner Division of Rockwell International. It has also been fitted to British Aerospace HS125-700s.

WXR-700C/WXR-700X radar

The Collins WXR-700 series digital colour radar is one of the most advanced systems to appear and is the first airline-category digital colour weather radar to fly. Geared to ARINC 708, the system replaces the magnetron of earlier radars with solid-state power generation as the basis for alternative C or X band radars. Along with electronic advances, the design team worked closely with the US National Severe Storms Laboratory (NSSL) to obtain the most up to date understanding of how cells form and grow in storms.

The WXR-700 comprises four units: a slotted-array flat-plate antenna (for good sidelobe reduction), a microprocessor-controlled transmitter-receiver, a display unit, and a control unit. The microprocessor control system in the transmitter-receiver supervises all control and data transfers, programmes and controls the rf processes such as pulse-width, bandwidth and pulse repetition frequency selection, and directs antenna scan and stabilisation. The unit also contains circuits to reduce ground-clutter suppression when operating in the weather mode. An optional feature is described by Collins as pulse pair Doppler processing, whereby, with the addition of a single circuit board to the transmitter-receiver, the horizontal velocity of rainfall can be sampled. This technique is recommended by the NSSL as being particularly suitable for the analysis of storm cells.

The cathode ray tube indicator uses a high-resolution shadow-mask tube with multi-colour display scheme. The cathode ray tube provides alphanumeric identification of radar modes and incorporates annunciators and controls. The receiver has a sufficiently wide dynamic range to 'see' the Z-5 and Z-6 levels of rainfall that indicate a high probability of hail.

Apart from the traditional single-channel configurations based on one set of equipment with perhaps a second indicator, the flight-deck standards of such aircraft as the Airbus A310, Boeing 757 and 767 offer alternative possibilities. For example, dual transmitter-receivers fed from a single antenna could feed multi-purpose EFIS indicators instead of dedicated weather cathode ray tubes (though these themselves can already display operating procedures or flight information).

In December 1982 Collins Air Transport Division announced TSO approval for a new version of the WXR-700 series radar incorporating a facility to detect atmospheric turbulence. The system entered production during 1983 and the first units were due for delivery at the end of the year. Much of the impetus for the development came from United Air Lines and British Caledonian, and the system was flown in tests by the NSSL.

The new facility permits a considerable advance in weather interpretation, going beyond the conventional method of assessing turbulence by measuring areas of high rainfall. By means of a Collins-patented technique involving Doppler processing, the system can measure changes in rainfall velocity rather than absolute rainfall rates, which is now concluded to be a more reliable guide to the presence of turbulence. Weather radars currently in service depend on operators' abilities to interpret radar echoes by their shape, intensity and gradient. The new method shows, in magenta on the radar screen, areas of turbulence by a more direct (and therefore more reliable) method than has previously been possible.

Two factors are responsible for this capability. First, the perfecting by Collins of a solid-state coherent transmitter-receiver that generates stable frequencies with very small dispersion, as opposed to the earlier magnetrons that produced a relatively broad band of frequencies. Secondly, Collins' modification of

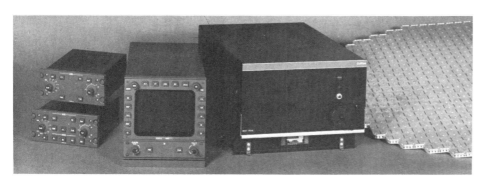

Four-unit Collins WXR-700 weather radar showing two versions of control unit

Collins WXR-850 weather radar antenna, 12, 14 and 18-inch diameter flat-plate antenna are available as options

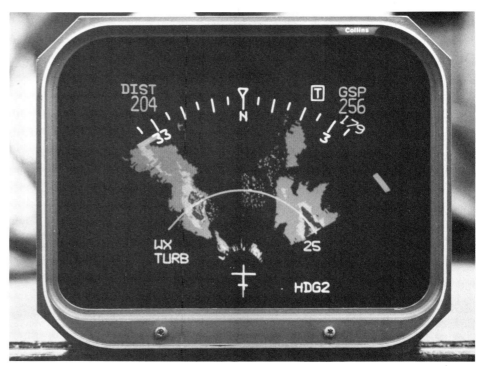

Collins WXR-850 cockpit display

the pulse-pair detection technique for measuring velocity changes permits much greater accuracy than with current radars. By receiving the in-phase and quadrature components of signals simultaneously, errors due to aircraft speed and frequently poor antenna angles with respect to the weather under surveillance are effectively cancelled.

Frequency: (C band) 5440 MHz
(X band) 9330 MHz
Power output: (C band) 200 W, prf 180 – 1440, pulse-widths 2 – 20 μs
(X band) 100 W, prf 180 – 1440, pulse-widths 1 – 20 μs
Range: (C band) 240 n miles
(X band) 320 n miles
Dimensions: (identical for C and X band)
(antenna) ARINC 708
(transmitter-receiver) 8 MCU
(indicator) ARINC 708 Mk II
(control unit) 5.75 × 2.625 × 6 inches (146 × 67 × 152 mm)

Weight: (antenna) 27 lb (12.25 kg)
(transmitter-receiver) 27 lb (12.25 kg)
(indicator) 18 lb (8.16 kg)
(control unit) 2.3 lb (1.04 kg)

STATUS: system in production and service, and is standard on Airbus A310. Equipment incorporating turbulence facility became available in late 1983.

WXR-850 weather radar
The WXR-850, introduced in September 1985, is claimed to be the first weather radar suitable

for business aircraft which can detect, and display, turbulence using the Doppler principle. The system uses solid state electronics technology, has an integrated transmitter-receiver, antenna, weighing less than 20 lb (9 kg) and a separate colour display. Maximum weather detection range is 300 nautical miles (550 km), turbulence can be detected out to 50 nautical miles (85 km).

STATUS: in production.

Eaton
Eaton Corporation, Command Systems Division, 815 Broad Hollow Road, Farmingdale, New York 11735
TELEPHONE: (516) 595 5000

AN/APX-103 IFF interrogator
This IFF interrogation system was developed for the US Air Force/Boeing E-3 Sentry AWACS operational with the US Air Force and with NATO and has been ordered by Saudi Arabia. The AN/APX-103 tags IFF replies received by the surveillance radar aboard the E-3 as representing either friendly or hostile aircraft. The interrogator section of the equipment can query other aircraft selectively in either conventional air traffic or military modes. The receiver/processor section automatically decodes the identities and locations of co-operative aircraft and feeds digital data to the E-3's central command and control computer, which also stores primary radar data. Both types of data may be called up subsequently by operators aboard the early warning aircraft.

STATUS: in production and service.

AN/APS-128 surveillance radar
The AN/APS-128 is one of the most widely used maritime patrol radars. It has come into its own largely as a result of the international 200 nautical mile fishing boundaries which need constant patrolling, but by low-cost aircraft to be economically effective.

The system comprises a rectangular flat-plate antenna and pedestal, transmitter-receiver, radar control module, range and bearing control module, and azimuth/range indicator.

Boeing E-3 Sentry AWACS early warning aircraft is equipped with Eaton AN/APX-103 IFF interrogator mounted back-to-back with primary radar antenna in rotating radome

The AN/APS-128 is installed aboard all the Beech 200T maritime-patrol aircraft operated by the Japanese Maritime Safety Agency, and has become operational with the Uruguayan Navy. The Brazilian Air Force bought 16 sets of equipment for its Embraer EMB-111 Bandeirante MR aircraft, and Gabon has also chosen the system for its EMB-111s. The Royal Malaysian Air Force has the AN/APS-128 in its Lockheed C-130 Hercules, and it is also offered as standard equipment for all MR versions of the Rockwell Sabreliner and Cessna's Conquest.

Target detection ranges are:
Fishing vessel (assumed 6 metres long, 10 square metres cross-section, in sea state 3), 25 nautical miles
Trawler (160 metres long, 150 square metres cross-section, sea state 5), 50 nautical miles
Freighter (360 metres long, 500 square metres cross-section, sea state 5), 80 nautical miles
Tanker (600 metres long, 1000 square metres cross-section, sea state 5), 100 nautical miles.
The system also functions as a weather radar with a range of 200 nautical miles.

Maritime-patrol Beech Super King Air with radome for Eaton AN/APS-128 surveillance radar on underside of fuselage

Component modules of Eaton AN/APS-128 Model D surveillance radar, with flat-plate rectangular scanning array, operator's display, control units and processor

Weight: 174 lb (79.1 kg)
Frequency: 9375 MHz
Frequency agility: 85 MHz peak/peak
Power output, pulse width, and prf: 100 kW peak, 2.4 and 0.5 μs, and 267, 400, 1200, and 1600 Hz
Antenna rotation rate: 15 and 60 rpm
Antenna stabilisation: automatic compensation for pitch and roll up to ±20°, with tilt to ±15°
Azimuth/range indicator: ppi, P-7 phosphor, 178 mm crt, north or aircraft heading orientated, range scales 25, 50, 125 n miles

STATUS: in production and service.

AN/APS-128 Model D surveillance radar

An upgraded version of the AN/APS-128, the Model D is an all-digital radar with a scan converter to present information in television raster format. It contains target enhancement and clutter reduction circuits for frequency agility, sensitivity time control, constant false-alarm rate and scan-to-scan integration.

The system comprises a flat-plate antenna and pedestal, transmitter-receiver, radar control unit, digital scan converter, a trackball or joystick cursor and bright display. A dual display for cockpit weather presentation is available, as well as cabin displays. A pro-

grammable microprocessor provides the alphanumerics and graphics to meet various mission requirements, while the antenna size can be altered to suit aircraft and operational needs. An alternative parabolic antenna with dual polarisation provides a pencil beam for sea search and a shaped beam for ground-mapping.

Claimed target detection ranges are:
Snorkel or fishing vessel (assumed 6 metres long, 10 square metres cross-section, sea state 3), 30 nautical miles
Trawler (assumed 160 metres long, 150 square metres cross-section, sea state 5), 60 nautical miles
Freighter (600 metres long, 500 square metres cross-section, sea state 5), 100 nautical miles
Tanker (600 metres long, 1000 square metres cross-section, sea state 5), 120 nautical miles.

When functioning as a weather radar the range is 200 nautical miles.

Versions of the APS-128D in service have a 30-target track-while-scan facility and narrow/wide-band data links.

Weight: 202 lb (91.8 kg)
Frequency agility: 85 MHz peak-to-peak
Power output: 100 kW peak
Pulse width and prf: 2.4 and 0.5 μs (0.1 μs pulse-compression optional), 400, 1200 and 1600 Hz

STATUS: in production.

AS-128 DITACS surveillance radar

A development of the AN/APS-128 Model D (see separate entry), DITACS (digital tactical system) works with other sensors as well as the basic radar for better multi-mission applications. Tactical navigation and radar information can be shown alternately or collectively on the existing bright display. Control and management functions can be stored, recalled and displayed in formats determined by the operator. The radar and display are microprocessor controlled, providing bright presentation with alphanumeric designations, tailored graphics and multi-sensor overlays.

Keyset functional firmware is based on specialised mission operational sequences. DITACS requires only the addition of a keyboard, the remaining features being represented by random access and read-only memories. The DITACS facility adds about 15 lb (6.5 kg) to the basic weight of an APS-128.

STATUS: installed on Beech 200T multi-mission aircraft and under development for US Navy/Grumman S-2E Tracker and several helicopter applications. Specialised versions of DITACS are installed in an RCA/US Coast Guard Aerostat system.

Emerson

Emerson Electric Company, Electronics and Space Division, 8100 West Florissant Avenue, St Louis, Missouri 63136
TELEPHONE: (314) 553 3232
TELEX: 44869

AN/APQ-159 radar

In 1973-74 Emerson developed the transistorised AN/APQ-159 for retrofit application, under government authorisation, with increased range and angle-tracking over the earlier AN/APQ-153/157, and providing data for a television display and scan converter. The system has a flat-plate antenna, but is non-coherent. It is still in production, though down to five or six a month from the original rate of 30 a month.

Emerson forsees a retrofit market among the many non-Warsaw Pact nations which operate MiG-21 fighters. The thermionic valve-based Soviet radar in this aircraft is widely recognised as a major weakness, with low reliability and (by current standards) poor performance, with short-range and no look-down capability. Egypt is a likely candidate for a radar retrofit; the system would operate in conjunction with the Sparrow missile system also proposed for the MiG-21. Integration would be done by US

company Vought, and installation of between 75 and 100 aircraft conducted in Egypt.

AN/APG-69(V) radar for F-5

This radar, first shown at the 1984 Farnborough Air Show, is a pulse-Doppler, digital coherent, air-to-air and air-to-surface system. Flight trials began in 1984, 59 flights being conducted in a Cessna 421B based in St Louis, followed by 151 sorties fitted to Northrop F-5s operating from Williams Air Force Base Arizona (the US F-5 training base). The application for this medium-pulse repetition frequency, travelling-wave tube based company-funded radar continues to be the Northrop F-5, in conjunction with the AIM-9L air-to-air missiles. The company foresees about a dozen countries as the potential retrofit market, along with the US Air Force Aggressor unit, whose aircraft simulate Soviet fighters for the purpose of realistic simulation.

Whereas the AN/APQ-153 (Emerson's first forward-mounted fire-control radar) had a range of about 10 nautical miles against a 5 square metre target, the new radar can read out to 25 to 30 nautical miles for the same target. The radar also introduces a look-down capability lacking in previous Emerson radars. The company claims that the improvements

Dual Emerson AN/ASG-21 fire-control radar in tail of Boeing B-52H

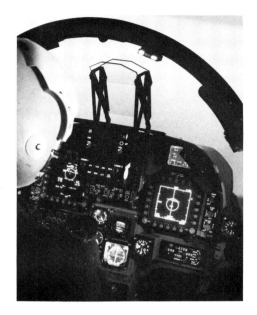

Display of Emerson AN/APG-69 radar is high-lighted in this photograph taken during evaluation in McDonnell Douglas AV-8B simulator

Emerson AN/APG-69(V) radar in nose of Northrop F-5 fighter

equate to a force-multiplier of between 6 and 12 times. The system comprises the radar transmitter receiver, and processor, with balanced monopulse antenna, signal data converter, video indicator (radar scope) and control unit.

Weight: 161 lb (73 kg)
Volume: 2.15 ft^3 (0.05 m^3)
Power: 1485 W at 115 V 400 Hz + 250 W at 28 V dc
Frequency: X band (9.3 GHz)
Output power level: 8 kW
Antenna: 19-inch dia flat-plate
Detection range:
(look-up, 5 m^2 target) 20 n miles
(look-down, 5 m^2 target) 14 n miles
(air combat, 5 m^2 target) 8 n miles
(surface vessel, 50 m^2 target) 33 n miles
Reliability: 200 h mtbf at the 150th unit

STATUS: in production as retrofit for Northrop F-5 also for US Navy/Cessna T-47A UNFO (Undergraduate Naval Flight Officer) trainer, a modified Cessna Citation II.

Radar for AV-8B

For some time there has been a campaign by radar manufacturers and some elements of the military establishments in both the USA and the UK to fit a radar to the McDonnell Douglas/British Aerospace AV-8B (US) and Harrier GR5 (RAF) aircraft.

Emerson took a practical step down this path in 1985 with the simulation of a version of its AN/APG-69 radar on the AV-8B simulator at McDonnell Douglas. Several pilots, from the US and Spanish forces flew the simulation, which was developed under a $4.5 million US Navy contract.

While (at the beginning of 1986) there was no firm US Department of Defense requirement for a radar in the AV-8B, other radars are also being proposed, including the Hughes AN/APG-65, the Westinghouse/FIAR Altair radar, the General Electric AN/APG-67 fitted in the Northrop F-20 and the Ferranti Blue Vixen selected for the Royal Navy's Sea Harrier mid-life update programme (see separate entries).

STATUS: conceptual development.

General Electric

General Electric Company, Aerospace Electronic Systems Department, French Road, Utica, New York 13503
TELEPHONE: (315) 793 7000

AN/APQ-113 attack radar

This system was designed in the early 1960s for the US Air Force/General Dynamics F-111A and F-111E and the Australian version, the F-111C.

STATUS: in service.

AN/APQ-114 attack radar

A member of the AN/APQ-113 family, the APQ-114 was developed in the mid-1960s specifically for the General Dynamics FB-111. The APQ-114 is similar to other members of the family, but features a north-orientated display, beacon-homing mode, and automatic photo-recording.

STATUS: in service with US Air Force/General Dynamics FB-111 aircraft.

AN/APQ-144 attack radar

This is another member of the APQ-113 family of attack radars, and was developed in the late

General Electric AN/APQ-114 attack radar in nose of US Air Force/General Dynamics FB-111

1960s for the General Dynamics F-111F tactical fighter. In addition to the basic features of that family, the radar contains a number of improvements. Notable among these are the addition of a 0.2 microsecond pulse-width capability and a 2.5 nautical mile display range.

Schemes for adding digital moving target indication and K band transmission capabilities were successfully flight-tested, but have not been incorporated into the service aircraft. A modified APQ-144 radar was built in small numbers for the US Air Force/Rockwell B-1A strategic bomber programme; this system was designated AN/APQ-163.

STATUS: in service with US Air Force/General Dynamics F-111F tactical fighters.

AN/APQ-161 attack radar

Developed for the US Air Force/General Dynamics F-111F, the APQ-161 is a modification of the AN/APQ-144 radar that provides a digital scan converted television raster display to operate in conjunction the Ford Aerospace Pave Tack laser designator and ranger.

STATUS: no longer in production, but continues in service.

AN/APQ-165 attack radar

This system is a modification of the APQ-161 radar to suit the requirements of the General Dynamics F-111C operated by the Royal Australian Air Force. In addition to changes needed for compatibility with the Ford Aerospace Pave Tack laser designator and ranger, the system has been upgraded to operate with the McDonnell Douglas AGM-84A Harpoon anti-ship missile.

STATUS: no longer in production, but continues in service.

AN/APQ-169 attack radar

The family of radars developed for the US Air Force/General Dynamics F-111 tactical fighter and coming under the designations AN/APQ-113, -114 and -161 is undergoing modifications to improve reliability and ease of maintenance. In addition to a guaranteed mean time between failures, the system incorporates a television raster display with scales ranging from 2.5 to 200 nautical miles, the addition of a pulse-compression mode, and the ability to generate and transmit energy pulses of only 0.25 microsecond width.

STATUS: in development.

AN/APG-67 attack radar for F-20

The multi-mode radar for the Northrop F-20 Tigershark, was unveiled in November 1982.

In June 1981 General Electric's Aircraft Equipment Division announced a $38 million contract from Northrop to develop a very advanced, lightweight, multi-purpose attack radar (AN/APG-67) for its F-5G fighter, unveiled that month at the Paris Air Show. GE had been in competition with Norden/Elta, Hughes Aircraft and Westinghouse. A contributory factor to GE's success was its background of radar research, notably involving the MSR (modular survivable radar) demonstrator programme, from which the F-20 proposal was a direct development. An additional $19 million was earmarked for ground test equipment. Like the aircraft, the radar was a risk-taking, private venture.

With a design mean time between failures of 200 hours (a contractual commitment imposed by Northrop at a time when the F-15 system was specified as 60 hours, the F-16 75 hours, and the F-18 125 hours) and weighing only 270 lb (123 kg), the APG-67 comprises a radar data computer, radar target data processor, transmitter, and steerable, phased-array antenna assembly, packaged into a total volume of 3 cubic feet (0.085 cubic metre) and operating in X band. It is the principal sensor for the F-20's six main missions, which comprise three air-to-air roles (supersonic intercept, combat air patrol and air superiority) and three air-to-ground (interdiction, close air support and sea detection). A significant characteristic of the APG-67, and stemming from its MSR ancestry, is the choice and layout of circuitry so that the system can be updated to incorporate new technologies, such as very large-scale integration, as they appear, without having to redesign the line-replacement units. Growth features can also be added in this way. Minimum feature size in the GE custom-designed and -built integrated circuits is about 5 microns.

In air-to-air missions the system searches for and tracks targets in both look-up and look-down situations, automatically acquiring and tracking them in combat. In its air-to-surface role the radar provides real-beam mapping, high-resolution Doppler beam-sharpened ground-mapping, and air-to-ground ranging.

The radar data computer is the 'brain' of the system, and contains two MIL-STD-1750A central processing units. One handles the radar mode control via the internal data-bus, permitting interface with F-20's mission computer, a Teledyne system. It also conducts built-in test functions. The other central processing unit carries out radar processing associated with

motion compensation and target tracking, and provides a raster scan output to the cockpit display.

The target data processor integrates 'raw' target data from the antenna so that reliable reports can be established. The unit incorporates custom large-scale integrated circuits developed by GE, and their use is considered to be the key to what is claimed to be this unit's exceptional performance. The processor performs a variety of functions, including fast Fourier transformation, moving-target indication, pulse compression, motion compensation, and can operate with variable waveforms and band-width. The transmitter produces coherent X band radiation from a travelling-wave tube, and can operate at low, medium and high pulse repetition frequencies, with variable power outputs and pulse-widths. The antenna is a flat-plate slotted array with low sidelobe sensitivity and can scan ±60° in azimuth and elevation. Scan width is selectable, and elevation search can be accomplished in 1, 2 or 4-bar modes. The receiver has a low-noise front-end to maximise detection range.

Weight: 270 lb (122 kg)
Volume: 3.3 ft³ (0.09 m³)
Operating frequency: X band
Power: 2400 VA
Transmit power: 200 W
Reliability: design 200 h mtbf
Antenna size: 16.7 × 26.4 inches (424 × 671 mm)
Detection range (fighter-size targets):
(look-down) 24 n miles
(look-up) 35 n miles
(sea targets, 50 m²) 35 n miles sea state 1, 30 n miles sea state 2
(max range, all modes) 80 n miles
Accuracy (air-to-ground): 50 ft or 0.5% of range

GE has built seven AN/APG-67(V) radars. One has been flying in a Douglas C-54 flight-test aircraft based near GE's Aerospace Electronics Systems Department at Utica, New York. Two others have been flying in F-20 prototypes at

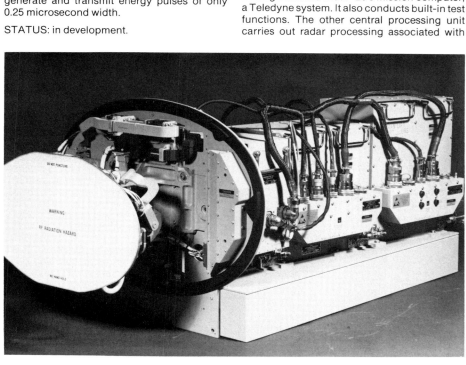

General Electric APG-67 radar on test

Principal sensor on Northrop F-20 Tigershark is General Electric AN/APG-67 radar

Northrop's facility at Edwards Air Force Base, California. By early 1985 the APG-67(V) had flown more than 600 flights and had demonstrated all operating modes including track-while-scan, air-to-ground weapons delivery and continuous wave illumination in association with the AIM-7 Sparrow missile. In over 15,000 hours of radar operation a mean time between failures of 200 hours had been established, claimed to be the highest achieved by any airborne radar.

STATUS: in pilot production to support Northrop F-20 Tigershark.

Multi-mode radar for F-5E

General Electric has developed a lightweight advanced multi-mode radar suitable for retrofitting into the Northrop F-5E. Operating in the X band, it uses digital pulse compression and coherent signal integration techniques and has high antenna gain and low sidelobes.

The radar has two main modes, with the antenna scanning over ±45°:

Air-to-ground with Doppler beam-sharpening (to 20 nautical miles range), ground track moving target indication (to 40 nautical miles range), air-to-ground ranging, ground-mapping (to 80 nautical miles range), freeze frame (which is radar silent) and scale expansion facilities. Similar capability is possible over the sea.

Air-to-air with look-up and look-down search and track, and air combat automatic acquisition in boresight, super search and vertical scan sub-modes. In super-search the radar scans the 20° × 30° field of view of the head-up display: vertical scan is 10° × 40°. In these sub-modes, and in boresight, the radar automatically locks on to a target and switches to single target track. In the look-up air-to-air mode range is 27 nautical miles with a ±45° antenna scan and 31 nautical miles with a ±10° scan. For look-down, these figures are 17 and 21 nautical miles respectively.

In air-to-air search, accuracy is put at 3 milliradians, 15 metres and 15 metres a second. In air-to-ground the range resolution is 45 metres and accuracy is 15 metres.

The radar interfaces with the standard MIL-STD-1553B data-bus and uses the MIL-STD-1750 computer architecture.

Antenna: 11 × 18.5 inches (279 × 470 mm)
Weight: 230 lb (104 kg)
Reliability: better than 150 h mtbf

STATUS: in development.

General Electric AN/APS-138 surveillance radar is being retrofitted to US Navy/Grumman E-2C Hawkeye AEW aircraft

MSR modular survivable radar demonstrator programme

GE's MSR (modular survivable radar) is a bench-top demonstrator programme to investigate the modular technology that the company believes will be needed in future airborne radars. The aim is to demonstrate a radar whose modules are so planned that changes or improvements can be incorporated simply by substituting new line-replaceable units with the minimum impact on weight and volume.

This modularity, says GE, is not just a sophisticated method of system partitioning, but a design discipline that extends down to shop-replaceable level. MSR is termed 'survivable' because its features permit it to operate in a severe electronic countermeasures environment.

The term 'technological transparency' was applied to the system by GE to indicate that it is broken down for design purposes into a number of functionally specified modular elements that can be upgraded with new technology. Equivalent functions can therefore be performed by succeeding generations of MSR at a fraction of the cost of replacing the entire system as a result of newer technology in just a few areas.

This functional building-block approach to the programmable signal processor, which is based on large scale integrated circuits, permits it to be matched to whatever level of processing required. The company has made a major commitment to very large scale integration technology and this is compatible with MSR in handling foreseen improvements during the 1980s.

STATUS: research tool, from which AN/APG-67 combat radar for F-20 design has been partly derived.

AN/APS-125 surveillance radar

In order to improve the overland performance of the AN/APS-120 radar to include an automatic detection capability, the US Navy in 1972 placed contracts with both Grumman and General Electric to develop for the E-2C Hawkeye early warning aircraft an advanced radar processing system that would combine greater sensitivity in noise and clutter with a greater resistance to false alarms. It would also have major electronic counter-counter-measures features new to airborne applications. Another improvement was the substitution of a digital airborne moving target indicator system for the analogue unit of the earlier APS-120.

Initial deliveries for production aircraft began in mid-1976 and were scheduled to continue into the mid-1980s. A refurbishment programme to upgrade APS-120s to -125 standards began in 1977. Four E-2Cs were ordered by Israel in 1977.

In 1981 Lockheed announced that it was studying a projected version of the Hercules designated EC-130ARE (the last letters denote airborne radar extension) that would mount the 24-foot (7.3-metre) diameter APS-125 radar dish on the transport's fin. In this position, says Lockheed, the uhf system and its associated IFF interrogation system could simultaneously track up to 300 targets, information being processed aboard the aircraft for action by the mission operators or transmitted by data link to an air-defence centre on the ground.

STATUS: in production for Grumman E-2C Hawkeye. AN/APS-125 radar has also been proposed for other aircraft, notably Lockheed P-3 Orion and S-3 Viking ASW aircraft. A P-3C Orion is being modified as a demonstrator aircraft, to carry APS-125 or APS-138 radar, and was being operationally tested during 1985.

The E-2C continues in production at a rate of about six per year. Egypt took delivery of its first machine, the 100th E-2C to be built, in October 1985, and will have all five of its order in service by May 1987.

Lockheed and General Electric have proposed version of P-3C Orion with AN/APS-125/ AN/APS-138 radar as relatively inexpensive early warning aircraft

AN/APS-138 surveillance radar

This is the designation for an improved version of the AN/APS-125. General Electric was awarded a $38 million contract in May 1983 to begin updating the APS-125 to APS-138 standards. By that date 84 radars had been built for the Grumman E-2C programme. A further change is in process, with the designation APS-138X scheduled for testing in 1986. The system has also been proposed for export to countries already operating the Lockheed P-3 Orion maritime patrol and ASW aircraft. A modified P-3B has been built as a Lockheed demonstrator, and first flew in June 1984.

Howls target-detection radar

One of the major new programmes initiated by General Electric following its re-entry into the airborne radar field in the mid-1970s is the experimental Howls (hostile weapons location systems) radar, sponsored by the US Defense Advanced Research Projects Agency and managed by MIT's Lincoln Laboratory.

With an advanced coherent phased-array antenna operating in the J band, the system is designed to investigate the technology needed to detect and locate enemy ground-based weapons, especially artillery, in the presence of heavy clutter. The principal element in the system is the antenna, designed and built by GE, that combines the radiating elements and pin diode phase-shifter into a single structure to reduce weight and space.

The system can operate in six basic modes: contextual ground-mapping, fixed-target detection, moving target detection, low-level Doppler signal analysis, Doppler beam-sharpening, and projectile or missile detection and tracking. Through microprocessor control the radar can be operated in up to four simultaneous modes by interleaving waveforms.

The technology now in development would be appropriate to a battlefield surveillance system for fixed-wing aircraft, helicopters, and particularly for remotely piloted vehicles, where low weight, space, and power consumption are important attributes.

STATUS: research tool since late 1970s; Howls has been evaluated in comprehensive flight-test programme.

Multibeam modular survivable battlefield surveillance radar

During 1982 General Electric began flight trials of a battlefield surveillance radar containing a number of the operational features expected to be included in the new US Air Force/US Army Joint Surveillance Target Attack Radar System, known as Joint-Stars (although GE never entered the competition for this contract). The equipment, known by GE as Multibeam modular survivable radar, comprises an electronically scanned phased-array antenna and a real-time all-digital large-scale integrated (lsi) radar

General Electric Howls experimental phased-array radar under test

Douglas DC-4/C-54 trials aircraft with General Electric battlefield radar carried externally in pannier

signal processor. It is based on the company's MSR modular survivable radar work (see under GE modular survivable radar programme), and may be considered as a version of that equipment.

The antenna has a low sidelobe to minimise susceptibility to jamming, and has been developed from a contract awarded in 1979 by the US Air Force to demonstrate enhanced electronic counter-countermeasures and multi-mode radar techniques, while the signal processor used in the trials has been provided by GE.

In setting out to develop the system, GE had three objectives: to demonstrate the company's multi-mode radar capability and technology; to develop modular building blocks appropriate to a variety of potential new applications; to acquire a flexible, advanced technology radar and suitable test vehicle to serve as a low-cost test-bed that can support new US Defense Department initiatives and projects.

The system operates in the Ku band, using linear and non-linear frequency modulation and binary-coded waveforms with bandwidths of 5 and 20 MHz. The pod-mounted, sideways-looking antenna has an effective radiating aperture of 2.5 metres and can scan ±60° in

both single and multi-beam modes. Real-time signal processing, and in particular pulse compression and spectral analysis, is accomplished with fast Fourier transform building blocks implemented in custom GE lsi devices. More than 600 such devices, of eight different types, are incorporated in the system. With such extensive lsi implementation, reductions of about eight to one have been achieved in weight, volume and power consumption by comparison with a similar system based on medium-scale integration.

After proof-of-concept flight trials on the multi-beam antenna in 1982, GE began a follow-on series of flights in the same Douglas C-54 aircraft to demonstrate the system's wide-area surveillance capabilities and the detection of slow-moving targets and the distinction between wheeled and tracked vehicles.

In one such demonstration, conducted in September 1982, the C-54 and its radar monitored a series of US Army exercises and manoeuvres around Fort Dix, New Jersey for 5 hours. Flying at around 5000 feet and at ranges of generally not less than about 15 nautical miles, the radar was able to detect and display moving vehicles either singly or in clusters, and the appropriate topographical details, such as roads, rivers and railways.

These trials served to emphasise that an effective airborne wide-area/moving target indication capability is a valuable adjunct to the command and control functions needed to engage large numbers of targets in the field. Constant surveillance over wide areas must be maintained while at the same time already identified targets of high priority must be continuously tracked until they can be dealt with.

STATUS: under development. No specific application for system has been disclosed.

Goodyear

Goodyear Aerospace Corporation, 1210 Massillon Road, Akron, Ohio 44315-0001
TELEPHONE: (216) 796 2121
TELEX: 986439
TWX: 810 431 2080

AN/UPD-4 side-looking radar

This radar scans a swathe of ground on either side of the host aircraft out to a distance of some 30 nautical miles. Radar signals are converted within the system into optical data, what is then recorded on photographic film on the aircraft, which is subsequently processed on the ground. Alternatively the raw data can be

transmitted down to a ground receiving station, so reducing the time delay in getting the intelligence to the ground forces.

STATUS: in production for US Marine Corps/McDonnell Douglas RF-4B, some US Air Force RF-4C, and German Air Force RF-4E reconnaissance aircraft.

Hazeltine

Hazeltine Corporation, Building 10, Greenlawn,
New York 11740
TELEPHONE: (516) 261 7000
TELEX: 967800

AN/APX-72 IFF transponder

The RT-859A/APX-72 system, which Hazeltine claims to be the most widely used IFF transponder in the world, provides military and civil air traffic controllers with aircraft identification data, traffic information, and automatic altitude reporting with a suitable altitude digitiser. The system receives, decodes and replies to the characteristic interrogations of operational Modes 1, 2, 3/A, C, and 4, the last of which uses a KIT-1A/TSEC computer. Pressurisation is provided for high-flying aircraft.

Transmitter
Power output: 27 dBw nominal
Duty cycle: 1% max
Frequency: 1090 MHz

Receiver
Frequency: 1030 MHz
Dynamic range: > 50 dB
Sensitivity: –90 dBv nominal

Decoder/coder
Mode C (altitude): decodes interrogations
Reply modes: (Mode 1) 32 codes
(Mode 2) 4096 codes
(Mode 3/A) 4096 codes
(Mode C) according to ICAO standards
(Mode 4) compatibility and full capability via a plug-in card

Dimensions: 5.76 × 6.39 × 13.37 inches (146 × 162 × 340 mm)

Hazeltine AN/APX-72 IFF transponder

Weight: 15 lb (6.8 kg)
Reliability: 300 h mtbf (a fully solid-state transmitter, now available, improves this figure)
Qualification: Mk 12, STANAG 5017, ICAO Annex 10, DOD AIMS 65-1000

STATUS: in production and service, more than 40 000 having been built. Users include all US Services and many foreign customers.

AN/APX-76A/B IFF interrogator

The Hazeltine AN/APX-76A/B IFF interrogator is an L band interrogator for all-weather interceptors and other tactical aircraft, with full AIMS capability in Modes 1, 2, 3/A, and 4. Narrow antenna beam-width and reduction of 'fruit' are achieved via interrogation and receiver sidelobe suppression circuits in conjunction with special antennas having sum (mainlobe) and difference (sidelobe) suppression patterns.

Bracket-decoded video and discrete-coded video are displayed on the radar scope to provide unambiguous correlation between IFF and radar targets. The system comprises four units: transmitter-receiver, switch amplifier, interrogator control and electrical synchroniser. In addition, dipoles are installed on the main radar antenna.

Transmitter
Frequency: 1090 MHz
Duty cycle: 1% max
High-power output: 2 kW

Receiver
Frequency: 1030 MHz
Sensitivity: –83 dBm

System weight: 37 lb (16.8 kg)
Mtbf: (AN/APX-76A) 225 h
(AN/APX-76B) 400 h
Qualification: DoD AIMS 66-252 Modes 1, 2, 3/A, and 4

STATUS: in production and service with aircraft of US Air Force and Navy, and other countries. Types so fitted range from McDonnell Douglas F-4, F-15 and Grumman F-14 to airborne early warning and anti-submarine warfare aircraft, and more than 5000 sets have been built.

AN/APX-76B, in current production, incorporates solid-state transmitter which is directly interchangeable with valve transmitter in AN/APX-76A.

AN/APX-104(V) IFF interrogator

Hazeltine claims that its AN/APX-104(V) air-to-air and air-to-ship interrogator set is the most technologically advanced, lightweight IFF interrogator yet produced. It provides positive identification and range and bearing measure-

Solid-state IFF transceiver for Hazeltine AN/APX-104(V) IFF system

ments of other friendly aircraft and ships in an all-weather environment, and is a useful aid to setting up a rendezvous with a refuelling tanker, and for maintaining contact with other aircraft on a common mission.

The AN/APX-104 is intended as a replacement for earlier IFF offering substantially better performance in terms of reliability, space and weight. It has full AIMS operation in Mode 1, 2, and 3/A, and also in Mode 4 when a KIR-1A/TSEC computer is fitted. The transmitter is a fully-solid-state device employing high peak power transistor amplifier sub-modules paralleled to give the required output power. This provides better reliability and a significantly longer time between maintenance than electron tube/cavity amplifier transmitters. The elimination of high-voltage power supplies and the need for pressurisation permits significant volume and weight savings and also contributes towards reliability. The local oscillator employs surface acoustic wave technology for simplicity, low production costs, good stability, and lack of field alignment requirements.

Frequency: (receive) 1090 MHz
(transmit) 1030 MHz
Decoding sensitivity: –83 dBm
Power output: 1.2 kW, switchable to half-power
Mtbf: 1000 h

STATUS: private venture. Qualification tests have been completed.

MM/UPX-709 multiplex diversity IFF transponder

The Hazeltine multiplex diversity control box transponder provides military and civil air traffic controllers with a vehicle identification, air traffic control and automatic altitude reporting in aircraft furnished with a suitable altitude digitiser. Data is provided independent of weather or radar environment; antenna blind spots are virtually eliminated by the automatic diversity feature using IFF antennas mounted

Remote-mounted version of Hazeltine MM/UPX-709 IFF transponder

Components of Hazeltine AN/APX-76A IFF system, including one short dipole antenna. Such dipoles are usually mounted on scanning radar antenna

on both upper and lower surfaces of the aircraft. The system was developed in conjunction with Italtel in Italy.

Reliability and low logistics and life cycle costs derive from the use of standard components and a highly ruggedised solid-state transmitter. The unit mounts remotely or in the panel space of a C6280A(P)APX control box, and can be provided with the MIL-STD-1553A multiplex interface to minimise aircraft wiring.

An extensive built-in test facility shows up faults in the transponder, Mode 4 computer, altitude digitiser, or either of the antennas.

Dimensions: (remote version) 5.375 × 5.375 × 8.375 inches (136 × 136 × 213 mm)
Weight: (remote version) 9 lb (4.1 kg)
Transmitter frequency: 1090 MHz
Receiver frequency: 1030 MHz
Frequency stability: ±3 MHz

Interrogation modes: 1, 2, test, 3/A, C, AIMS Mode 4
Sidelobe suppression: decoding suppressed on all modes
Codes: in accordance with DoD AIMS 65-1000
Reliability: 1000 h mtbf

STATUS: in production by Italtel in Italy for national and international customers.

Hughes Aircraft

Radar Systems Group, Hughes Aircraft Company, PO Box 92426, Los Angeles, California 90009
TELEPHONE: (213) 648 2345
TWX: 910 348 6681

AN/AWG-9 weapons control system for F-14A

In the mid-1950s airborne radars still had one major weakness: they could not look down without being blinded by clutter. Target aircraft could therefore escape detection by flying close to the ground where their echoes were masked by the much stronger ground echoes. This was eliminated by the advent of coherent pulse-Doppler radar. One of the first such systems was the Hughes AN/ASG-18 designed for the North American F-108 interceptor. When that project was cancelled, work continued on the radar and the technology was incorporated in the Mach 3 US Air Force/Lockheed YF-12 interceptor, although it never entered service.

Further development of the pulse-Doppler radar continued in the 1960s with the Hughes AIM-54 Phoenix weapon system for the US Navy/General Dynamics F-111B and was continued after that project was cancelled, ultimately resulting in its first volume-production application, the Hughes AN/AWG-9 weapon control system for the Grumman F-14 Tomcat. The F-14's radar is unique in several respects. To take advantage of the long-range detection capability conferred by the high pulse-repetition frequency waveform, it was designed in conjunction with Phoenix missile, capable of launch ranges of more than 100 nautical miles (185 km). The radar incorporates velocity-search, range-while-search and track-while-scan modes for target detection and tracking.

Hughes Aircraft AN/AWG-9 weapon control system for F-14 fighter 'opened up' for maintenance. Note IFF array on front face of flat-plate antenna

Range-while-search is a technique for coding pulses in the high pulse-repetition frequency Doppler waveform so that target distance information can be recovered. Range obtained in this way is not as accurate as that achieved using a low pulse-repetition frequency waveform, but it can be obtained at longer distances,

which is a valuable capability for fleet defence. This feature, coupled with what Hughes claims is the most powerful general-purpose airborne computer ever to be used with a fighter radar, made it possible to implement a track-while-scan mode in which the system can track up to 24 targets simultaneously. Prior to this, tracking could be accomplished only by pointing the antenna at the target, restricting the fighter's capability to single-target engagement. The AWG-9 system can engage up to six targets simultaneously with six Phoenix missiles while maintaining the complete tactical picture.

Some years ago the US Navy and Hughes proposed an enhancement programme, a key element of which is the programmable signal processor. Apart from the much greater flexibility in adding or changing modes, the new system will improve the countermeasures capability, widen the missile launch zone, add non-cooperative target recognition, and provide improved low pulse-repetition frequency and mapping quality. This is now being implemented (see separate entry).

Weight: 1300 lb (591 kg)
Volume: 25 ft³ (0.7 m³)
Dish diameter: 36 inches (914 mm)

STATUS: the AWG-9 radar remains in production and service with F-14As of both the US Navy and Iran.

AN/APG-63 fire control radar for F-15

This radar is the principal sensor for the McDonnell Douglas F-15 Eagle, which has already been proved in combat by the Israeli Air Force. McDonnell Douglas officials were quoted in March 1984 as claiming that Israel's F-15 force had destroyed 56 Soviet-built aircraft

Electronics of Hughes AN/AWG-9 radar

US Navy/Grumman F-14A fighter has Hughes AWG-9 weapons control system

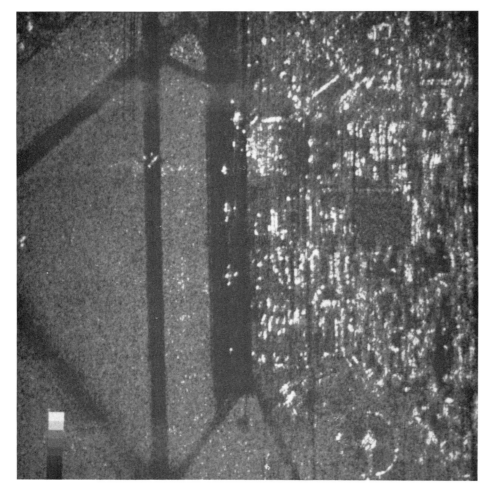

High-resolution radar mapping (top) compared with conventional air photography shows performance of AN/APG-63 radar modified to include synthetic-aperture capability. Picture resolution is said to be 10 times greater than previously obtainable with tactical radars

(including three MiG-25 Foxbats) without losing a single aircraft to combat damage. The APG-63 is claimed to be the first all-digital radar to be built in the USA.

The system was designed around three main objectives: capability, reliability, and maintainability. Capability includes one-man operation that can locate and track hostile aircraft at long range; close-in and look-down capability in situations that would blind earlier-generation radars; a clutter-free display with all appropriate target information; tracking and steering guidance information on the head-up display so that the pilot can keep his eyes continuously on the target; simplified controls and weapons coordination; and a certain air-to-ground capability.

The AN/APG-63 was the subject of reliability requirements that called for a mean time between failures of 60 hours, a level never before approached for this class of equipment. As a corollary, and despite its superior capabilities, maintenance time is reduced to about a quarter of that of previous systems. High reliability and relatively simple maintenance schedules reduce the turn-round time, and this is reflected in the smaller numbers of flight-line and maintenance personnel needed. This is largely due to the use of hybrid and integrated circuits to reduce the number of components and to the use of only four basic module sizes in the entire system.

Primary controls for the multi-mode, pulse-Doppler X band radar are located on the control column, allowing the pilot to keep his head 'out of the cockpit' during fast-moving situations. Three special modes are provided for close-in combat, supersearch, vertical scan and boresight, enabling automatic acquisition of, and lock-on to, targets within 10 nautical miles. In the supersearch mode the radar locks on to the first target entering the head-up display field of view and displays its position. This alerts the pilot and greatly increases the range of visual detection. In the vertical scan mode the radar locks on to the first target that enters an elevation scan pattern at right-angles to the aircraft lateral axis. In the boresight mode the antenna is directed straight ahead and the radar locks on to the nearest target within the beam.

The system comprises nine line-replaceable units: exciter, transmitter, antenna, receiver, analogue processor, digital processor, power supply, radar data processor, and control units. It has a wide look-angle and its antenna is gimballed in all three axes to hold target lock-on during roll manoeuvres. The clutter-free head-down radar display gives a clear look-down view of target aircraft silhouetted against the ground, even in the presence of heavy 'noise' from ground returns. This look-down, shoot-down ability is achieved by using both high and medium pulse repetition frequencies (a technology perhaps unique at the time of the radar's appearance), by the digital data processing and by using Kalman filtering in the tracking loops. The system's gridded travelling-wave tube permits variation of the waveform to suit the tactical situation.

False alarms are eliminated regardless of aircraft altitude and antenna look-angle by a low-sidelobe antenna, a guard receiver, and frequency rejection of both ground clutter and vehicles moving on the ground so that only real targets are displayed.

Although designed specifically for air-to-air roles, the F-15 has emerged as a potent ground-attack fighter, and the APG-63 does have target-ranging for automatic bomb release for visual attacking, a mapping mode for navigation, and a velocity update for the inertial navigation system.

In March 1984 the F-15 was selected as the US Air Force's new dual-role (air control and interdiction) fighter. The upgraded radars for the two-seat F-15 version will incorporate very high-speed integrated circuit technology to increase computational speed and improve electronic countermeasures performance. It

Hughes AN/APG-63 multi-mode radar in McDonnell Douglas F-15 Eagle

F-15 Eagle fighter with nose radome swung aside to give access to Hughes AN/APG-63 radar

has been designated AN/APG-70 (see separate entry).

The 486 lb APG-63 is compatible with both the AIM-7F Sparrow and AIM-9L Sidewinder air-to-air missiles. Antenna search patterns and radar display presentations are selected automatically for the type of weapon (medium-range radar-homing, or short-range infra-red missiles or M61 cannon) to be used; the pilot chooses the weapon by means of a three-position switch on the throttle.

The ability to handle equally effectively both air-to-air and air-to-ground situations is the result of combining high and medium pulse repetition frequencies and the APG-63 claimed to be the first radar to do this. High pulse repetition frequencies are necessary to detect targets at long range but are not suitable for measuring range because there is insufficient time for pulses to return and be correlated before the next ones are being transmitted. However, medium pulse repetition frequencies that enable accurate range measurement and elimination of ground clutter do not have the power to give detection echoes at long range. But the APG-63 interleaves, for the first time, both high and medium pulse-repetition frequency waveforms. The key to this technology was the substitution of heavy and bulky Doppler filters with a digital signal processor. By this means, incoming signals are sampled and their frequency content analysed by performing Fourier transforms on individual samples.

While the eventual reliability goal specified by MIL-STD-1781 was 60 hours, a figure met consistently on bench tests, but in the field, service equipment is currently giving about 30 to 35 hours mean time between failures.

All APG-63 radars produced since mid-1980 incorporate a programmable signal processor, a high-speed digital computer that enables the system to respond quickly to new tactics or weapons by changes to software rather than extensive hardware modifications. This change was introduced with the improved F-15C and F-15D models. Similar aircraft produced before the processor was introduced were scheduled to receive it as a retrofit item.

In December 1983 the US Air Force awarded McDonnell Douglas a $274.4 million contract to upgrade the control computer and armament control system as part of the MSIP (multi-staged improvement programme) launched in February of that year. MSIP arose out of a McDonnell Douglas study, beginning in June 1982, which showed the need to improve the avionics system, and specifically the radar, in order to maintain reasonable combat superiority over likely Soviet adversaries. Radar improvements constitute a memory increase to 1 M words and a processing speed tripled to

1.4 million operations a second. The system will also incorporate a programmable signal generator, a facility not available until the late 1970s. The outcome of this will be the AN/APG-70 (see separate entry).

A total of six Lockheed P-3 Orions, on loan from the US Navy, are being modified for anti-drug smuggling operations with the US Navy to include a Hughes AN/APG-63 radar mounted in the aircraft's nose.

Number of units: 9
Total weight: 486 lb (221 kg)
Total power: 12.975 kW
Reliability: (design) 60 h mtbf
(current) 30-35 h mtbf

STATUS: in production and service with McDonnell Douglas F-15 Eagle, by early 1986 some 650 sets had been built, including co-production in Japan.

AN/APG-71 fire control radar for F-14D

In July 1984 the US Navy contracted Grumman to proceed with a major improvement programme with its F-14 Tomcats. Among many airframe, engine and equipment changes will be an enhancement for the radar section of the AWG-9, the new system being designated AN/APG-71. These changes are the result of US Navy concern to maintain the F-14's air-superiority performance in the first rank well into the 1990s. The upgraded aircraft will be known as the F-14D.

The APG-71 is essentially a digital version of the radar section of the AWG-9, with greatly improved electronic countermeasures performance acknowledging the new and vastly more sophisticated jamming technologies that have appeared since design of the F-14 was frozen. New modes include monopulse angle tracking, digital scan control, target identification and raid assessment, but the number of boxes will come down from 26 to 14; the system will also employ some of the elements developed by Hughes for the new F-15 radar, the AN/APG-70, APG-71 will incorporate non-cooperative target identification, by which radar contacts may be identified as friendly or hostile through close examination, at high resolution, of the returns; the technique obviates deficiencies and ambiguities in IFF equipment.

STATUS: in development. First engineering development model due for delivery in the last quarter of 1986. First F-14D will enter service in the spring of 1990.

AN/APG-65 multi-mission radar for F/A-18

Escalating procurement costs have underlined the need for multi-mission tactical aircraft, but until recently the necessary compromises in

airframe and equipment caused by inadequate technology have prevented initial promises from being fully fulfilled. The problem has been that air-to-air systems need extremely fast data processing rates (closing speeds can be up to 1 mile a second) and variable waveform flexibility to achieve all-aspect, all-altitude target-detection capability, while air-to-surface systems need very large data storage and processing capabilities for high-resolution mapping. These requirements have hitherto been incompatible with one another.

The digital breakthrough stemming from the early 1970s has largely changed the situation and the McDonnell Douglas F/A-18 Hornet is perhaps the first combat aircraft to embody in one airframe and fire-control system an optimised air superiority and ground attack capability. For this reason the Hornet is the first US aircraft to have the dual designation F/A, indicating fighter/attack.

The key to combining the two into a single flexible unit is the programmable signal processor. While the AN/APG-63 in earlier models of the McDonnell Douglas F-15 fighter introduced digital signal processing to tactical fighters in the form of 'hard-wired' logic with a fixed repertoire of modes, the Doppler filter and range gate configurations that make up the programmable signal processor's capability in the APG-65 (and in the APG-63 of the improved, F-15C/D models) are defined by program coding; existing modes can be modified or new codes added by changes to software.

In the air-to-air role the APG-65 radar incorporates the complete range of search, track and combat mode variations, including several previously unavailable in an operational radar. Specifically these modes are head-up display acquisition (the system scans the head-up display field of view and locks on to the first target it sees within a given range); vertical acquisition (in which the radar scans a vertical 'slot' and automatically acquires the first target it sees, again in a given range); and boresight (the radar acquires the target after the pilot has pointed the aircraft at it).

In the raid assessment mode the pilot can expand the region around a single target that is being tracked, giving increased resolution around it and permitting separation of closely spaced targets.

As a gunsight the radar operates as a short-range tracking and lead-computation device using frequency-agility to reduce errors due to target 'scintillation'. All the pilot has to do is to put the gunsight 'pipper' on the target and press the gun-button.

Other modes are long-range velocity search (using high pulse-repetition frequency waveform to detect oncoming aircraft at high relative velocities); range-while-search (high and medium pulse-repetition frequency waveforms being interleaved to detect all-aspect targets, ie not only oncoming ones but those at any line-of-sight crossing angle); and track-while-scan, which can track simultaneously up to 10 targets and display eight. When combined with future autonomous missiles such as AMRAAM, this mode will confer a 'launch and leave' capability as well as the simultaneous engagement of multiple targets.

In the F/A-18's air-to-ground role, the APG-65 has six modes: terrain-avoidance, for low-level penetration of hostile airspace; precision velocity update, when the radar provides Doppler speed signals to update the aircraft's inertial navigation system and if necessary to align it; the tracking of fixed and moving ground targets; surface-vessel detection, in which the system suppresses sea clutter by a sampling technique; air-to-surface ranging on designated targets; and ground mapping. Two Doppler beam-sharpening modes are provided for these air-to-ground modes. During tests in early 1983 the system demonstrated (a year ahead of schedule) the 106 laboratory test hours mean time between failures required by contract. Two systems chosen at random for a reliability demonstration test operated for a total of 149 hours without requiring maintenance.

All trouble-shooting is conducted with BIT (built-in test); according to Hughes, there is an order-of-magnitude improvement between the radars for the F-14, F-15 and F/A-18. In its first months of operation, the initial Marine Air Group II at El Toro in California had only six radar engineers to support a 12-aircraft unit flying 30 hours a month. The BIT facility currently operates at module line-replaceable unit level. The biggest improvement in BIT technology with the AN/APG-65 is the reduction in false-alarm rate. The US Navy requirement is to be able to detect and locate 98 per cent of all radar faults by BIT. The US Navy also specifies 12 minutes mean time to repair, which involves running a BIT test, locating fault, removing the faulty line-replaceable unit and replacing it, and running a BIT check on the new unit.

In addition to the US Navy and Marine Corps, the AN/APG-65 equips the F/A-18 chosen by Canada, Australia and Spain.

Under the terms of an AIP (Australian industrial participation) programme signed in December 1981, the APG-65 will be built in Australia by Philips Electronics Systems. Philips is initially engaged in final assembly and test, and co-produces the data-processor. In February 1984 it was announced that Marconi Española SA had been licensed to build low-voltage power supply modules for the radars

Hughes Aircraft AN/APG-65 in F/A-18 Hornet. Radome swings aside for easy access

being built for Spain's F-18s. Production of the system for all applications was in early 1985 running at about 20 a month.

In May 1985 the West German Ministry of Defence chose the AN/APG-65 as a major part of its ICE improved combat efficiency effort to enhance the effectiveness of the West German Air Force's McDonnell Douglas F-4F Phantoms; it had been in competition in this application with Westinghouse's AN/APG-68. The new radar will displace the 1960s-design Westinghouse AN/APQ-120 in the F-4F, the limitations of which have become evident in the past decade; most seriously, it cannot detect air targets in ground clutter because it does not have the Doppler velocity information needed for MTI, and it cannot track multiple targets. Reliability also will benefit. The new system will be used in conjunction with the Hughes AIM-20 AMRAAM, which can knock out targets in the presence of ground clutter. The F-4F radar-refit programme will involve some 75 aircraft at a

cost of $300 million. MBB is the responsible company, and initial operational capability is set at 1989/90. The West German Air Force also has an option to retrofit its fleet of F-4Es with the APG-65.

Number of lrus: 5
Antenna: low sidelobe planar array with fully balanced direct electric drive replacing hydraulics and mechanical locks of previous systems
Transmitter: liquid-cooled, contains software-programmable gridded travelling-wave tube amplifier
Receiver: contains A-D converter
Radar-data processor: general-purpose with 250 K 16-bit word bulk-storage disc memory
Signal processor: fully software programmable, runs at 7.2 million operations/s
Reliability: 106 h mtbf. Built-in test equipment detects 98% of faults and isolates them to single replaceable assemblies that can be changed in 12 minutes without adjustments or setting up
Weight: 340 lb (154.6 kg)
Volume: < 24.5 ft³ (0.42 m³) excluding antenna

STATUS: in production and service with the US Navy, US Marine Corps, the Canadian Armed Forces, the Royal Australian Air Force and the Spanish Air Force. Total production rate is about 20 sets a month.

AN/APG-70 radar for F-15E

In March 1984 the US Air Force chose the McDonnell Douglas F-15E as its new dual-role fighter to back up its General Dynamics F-111s. The radar for this two-seat combat aircraft is a substantially improved version of the APG-63, designated AN/APG-70. It has also been chosen to upgrade fighter versions of the F-15 under the US Air Force's MSIP multi-stage improvement programme.

By comparison with the earlier system, the APG-70 will have a 75 per cent greater rf bandwidth, a larger look-down target detection range, a one-third increase in meantime between failures, and be packaged in six units instead of seven. Four of the line-replaceable units (radar data processor, programmable signal processor, analogue signal converter and receiver/exciter) are completely new, while the transmitter and control unit have been modified; only the power supply remains unchanged. Expanded built-in test will be

USN/McDonnell Douglas F/A-18 Hornet uses Hughes Aircraft's AN/APG-65 radar

McDonnell Douglas F-15E demonstrator. Operational aircraft will have Hughes AN/APG-70 radar

provided, giving unambiguous fault-detection and isolation, and the electronic counter-countermeasures capability will be improved to combat new and more advanced threats. The system will be compatible with existing and new missiles such as AIM-7F/M Sparrow,

AIM-9L Sidewinder and AMRAAM, and 20 mm cannon.

STATUS: in development both for follow-on F-15 fighters and for new F-15E.

ASARS-2 advanced synthetic aperture radar system

The high-resolution side-looking radar was designed for the US Air Force/Lockheed TR-1 high-altitude battlefield surveillance aircraft, and satisfies a US Army requirement for a stand-off intelligence-gathering system. ASARS-2 was launched in 1977 under the designation AN/UPD-X (AN/UPD- designations denote a family of synthetic aperture radars) and first test-flown aboard a Lockheed U-2R reconnaissance aircraft in 1981. It is likely that the resolution approaches limits set only by altitude, residual airframe vibration, and atmospheric turbulence. Guidance for the ASARS-2 antennas will be provided in part through several other sensors, and information will be transmitted via a data link to a special ASARS-2 deployable processing station on the ground. Here the signals are converted into strip-maps and spotlights, and are available within minutes to army commanders. Selection of Hughes system puts that company, for the first time, into the reconnaissance radar field.

STATUS: in production for US Air Force/ Lockheed TR-1 high-altitude reconnaissance aircraft. Details of the radar remain classified.

Motorola

Motorola Inc, Government Electronics Group, 8201 E McDowell Road, PO Box 1417, Scottsdale, Arizona 85252
TELEPHONE: (602) 949 4176

AN/APS-94 SLAR side-looking airborne radar

AN/APS-94 side-looking X band radar was produced for the US Army/Grumman OV-1B through to OV-1D Mohawk aircraft. The primary sensor in the OV-1D is the APS-94F which has a slotted-waveguide planar-array antenna housed in a 5.5-metre long pod slung under the aircraft. The transmission frequency is X band and can be varied in flight for optimum performance. Detection range against typical targets is approximately 60 nautical miles. The system is teamed with a film-recorder and data link to form the AN/UPD-7 radar surveillance system.

STATUS: the AN/APS-94 is out of production but remains in service.

Maritime surveillance radars

This family of X band, airborne side-looking radars was originally designed for the US Coast Guard. The AN/APS-131 is installed on the

US Coast Guard HU-25 Falcon 20 aircraft, and its primary mission is oil slick detection. A second version is the AN/APS-135 mounted on the US Coast Guard HC-130 aircraft. The primary mission for this system is ice mapping and iceberg detection as a part of the US Coast Guard international ice patrol treaty obligation in the North Atlantic. The APS-131 and APS-135 also are used in search and rescue applications.

A version of this radar, trade marked SLAMMR (side-looking airborne modular multi-mission radar), is offered internationally for a wide variety of maritime surveillance and border patrol applications. SLAMMR has been delivered in a maritime surveillance configuration on the Boeing 737-200 aircraft for Indonesia and to another country in a border patrol configuration on a C-130H aircraft.

SLAMMR can be installed on a wide variety of aircraft ranging from the Beech 1900 or Fairchild Merlin IV through such larger aircraft as the Fokker F-27 Sentinel, Gulfstream III, C-130 or Boeing 737.

The radar is designed to accommodate a variety of applications, such as ice patrol, fishery protection, anti-drug smuggling, mapping, and search and rescue. In particular, its performance over the sea is sufficiently high to reveal details of oil-spills from tankers, and provide maritime reconnaissance data to a

maximum range of 100 nautical miles on either or both sides of the aircraft. The sensor uses a planner-array 2.5 or 5 metres long mounted below the aircraft or along the side of the fuselage feeding an operator's control/display console. At the 1984 Farnborough Air Show a SLAMMR installation was displayed on a Gulfstream III. The two antennas are arranged to point perpendicular to the aircraft longitudinal axis. From this position the two antennas can survey the sea or ground out to either side of the aircraft out to the horizon.

A notable application of SLAMMR is the patrol maintained by the US Coast Guard to provide iceberg warning to shipping. Before the advent of side-looking radar, iceberg observations were conducted visually, and therefore could only be performed in good visibility. Now, with SLAMMR, the US Coast Guard HC-130 Hercules can maintain a much more consistent monitoring. Gulfstream Aerospace at the 1984 Farnborough Air Show exhibited a military version of its Gulfstream III in Danish Air Force markings, designated SRA-1, with a SLAMMR pod below the fuselage. Another system was shown mounted on a Fokker Sentinel, a military version of the F-27 transport.

STATUS: in production.

Narco

Narco Avionics, 270 Commerce Drive, Fort Washington, Pennsylvania 190483
TELEPHONE: (215) 643 2900
TELEX: 846395

Narco took over production of the King KWX 56 and KWX 58 weather radars in 1985.

KWX 56 digital colour radar

Introduced at the National Business Aircraft Association's 1981 Convention, this system is the result of a two-year development programme to build an inexpensive pitch/roll stabilised digital colour radar.

The system comprises two units: a five-inch (127 mm) diagonal KI 244 or KI 248 panel-mounted high-contrast black matrix display and a KA 126 or KI 128 combined antenna/ transmitter-receiver. The latter can be stabilised using a flight-director or vertical gyro; the addition of roll-stabilisation eliminates screen blanking caused by ground returns during

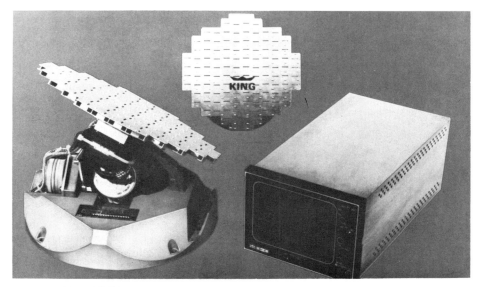

Narco KWX 56 two-unit colour radar system including front-view of antenna

medium or steep turns. The flat-plate antenna has a diameter of either 10 or 12 inches (254 or 305 mm), and can be supplied either pressurised or unpressurised for altitudes of up to 20 000 and 50 000 feet (6100 and 15 200 metres) respectively.

In August 1983 King began deliveries of its KGR 356 radar graphics unit, announced at the 1982 NBAA exhibition. The KGR 356 combines with the KWX 56 and a new version of the KNS 81 Silver Crown integrated navigation system to give a self-contained panel-mounted area navigation system. With the KGR 356 in the Nav mode, the radar can display a weather plot with the superimposed location of the active Vortac and waypoints stored in the memory of the KNS 81.

Power: nominal peak pulse 7.5 kW, 3.5 μs pulse width

Ranges: 10, 20, 40, 80, and 160 n miles by rotary switch
Display: Conventional colours. In mapping mode: red becomes magenta, green becomes blue, and yellow remains unchanged
Stabilisation: From KI 256 flight director and ARINC-standard vertical gyros associated with most autopilot/flight-director systems, the KWX 56 can be used with Century 41, Century IV, Cessna ARC 400, 800 and 1000 series autopilots

STATUS: in production and service.

KWX 58 digital colour radar
Based on the earlier KWX 56, this weather radar, announced early in 1984, uses two of its elements, the KI 128 combined antenna/

transmitter/receiver and the KI 248 5-inch (127 mm) colour display. Weather plots are shown in green, yellow, red and magenta, according to rainfall density, on a high-contrast, non-fading black matrix screen. The system uses a 10 or 12-inch (254 or 305 mm) diameter planar-array roll/pitch stabilised antenna radiating 7.5 kW. Narco claims that the penetration compensation feature gives a more accurate detection of storms behind closer areas of rainfall. Display ranges are 5, 10, 20, 40, 80, 160 and 320 nautical miles. The KWX 58 system weighs 18.7 lb (8.5 kg).

STATUS: production deliveries began in spring 1984.

Norden
Norden Systems, United Technologies Corporation, Box 5300 Norwalk, Connecticut 06856
TELEPHONE: (203) 852 5000
TWX: 710 468 0788

AN/APQ-148/156 radar for A-6E
The principal sensor for the US Navy's Grumman A-6E Intruder all-weather attack aircraft, the Norden AN/APQ-148 has been continuously upgraded throughout its 20-year service history. It provides guidance for terrain-avoidance, and target acquisition and tracking. The most recent development, designated AN/APQ-156, interfaces with the latest version of the A-6E's TRAM (target recognition attack multi-sensor system). As part of the US Navy's A-6E development programme, Norden is applying synthetic-aperture technology to the APQ-156 for accurate weapon delivery at stand-off ranges.

A variation of the APQ-156 is being built for the B-52 OAS (offensive avionics system) update. It incorporates solid-state electronics, with microprocessor-based digital computers, to improve the bomber's navigation and weapon delivery and to integrate the Boeing AGM-86B ALCM air-launched cruise missile. Development began in 1978 under contract to Boeing, with the objective of improving reliability by replacing the 1950s' thermionic valves with solid-state circuitry.

STATUS: in service.

Radar for A-6E update
In early 1985 Norden won a contract to develop the radar for the Grumman A-6E update programme. The new radar, over 300 of which will be required, will have longer range and higher reliability than the APQ-156 and will have better jam-resistance. A ship classification (target identification for stand-off weapons)

and air-to-air mode for Sidewinder and AMRAAM missiles will be incorporated.

STATUS: in development. Flight trials will start in late 1986.

Strategic-radar update for B-52
Norden is the subcontractor to Boeing Military Airplane company for the forward radar update that forms part of the Boeing OAS (offensive avionics system) improvement programme for the B-52 bomber. The US Air Force has established that the structural integrity of the B-52G and -H versions is sufficient to support operational service out to the year 2010. The priority in the OAS update is better reliability, with the change from valve-technology to solid-state, and sometimes from analogue to digital; in many cases—such as the radar—systems had become logistically unsupportable.

Two major changes to the original strategic radar involve improvements to terrain-avoidance and ground-mapping performance. A new terrain-avoidance feature is the provision of a high-resolution computer sketch of the ground directly ahead of the aircraft (the B-52 was originally designed in the 1950s for high-altitude missions, but is now obliged to operate at low altitude to avoid detection). An improved ground-mapping system will cover areas up to 100 nautical miles across with improved clarity. The number of boxes has been reduced from 50 to 8. A $100 million flight-development programme was completed in February 1984, and a $109 million initial contract to support upgrading the radars on 264 B-52Gs and -Hs was awarded to Boeing in July 1984. An $80 million follow-on contract to cover production of 59 sets of equipment was negotiated in November 1984.

Boeing awarded Norden a $30 million contract at the end of 1985 to provide receiver-transmitter modulators, radar processors and display generators. This was as a result of a larger main contract, awarded to Boeing by the

US Air Force for radar kits to be delivered beginning in June 1987.

Millimetre wave radar
Norden claims that flight trials of a millimetre wave airborne radar, developed under a joint US Air Force/Navy contract let in March 1981, have demonstrated an improved performance in poor visibility compared with current infrared systems. The 95 GHz Talons (tactical avionics for low-level navigation and strike) was built to demonstrate the feasibility of millimetre-wave radiation for improved ground-mapping, terrain-avoidance and terrain-following, and weapons guidance in poor weather and where smoke obscures battlefield targets.

During flight trials the Talons system was carried in a pod under the fuselage of a T-39 Sabreliner light military communications jet. The equipment comprised a gimballed antenna, microwave receiver, transmitter-modulator, intermediate-frequency processor and power supplies. Controls and displays for measuring and assessing performance were carried in the aircraft's cabin.

STATUS: feasibility study for US Air Force Aeronautical Systems Division, Wright-Patterson Air Force Base; 18-month trial was concluded in late 1982.

AN/APS-130 mapping radar
The AN/APS-130 radar, another development of the APQ-156, is standard equipment in the US Navy/Grumman EA-6B electronic warfare aircraft and the KA-3B tanker. It provides real-time mapping and is essentially a navigation radar. Position updates are obtained from prominent radar check-points. The APS-130 operates in the Ku band, and has display ranges of 15, 30, 75 and 150 nautical miles.

STATUS: in production and service.

Sperry
Sperry Corporation, Electronic Systems, Marcus Avenue, Great Neck, New York 11020
TELEPHONE: (516) 574 2100
TELEX: 960167

AN/APN-59E(V) search radar
The APN-59E(V) was developed in the late 1970s to replace the earlier AN/APN-59B with improved performance and reliability in a variety of retrofit situations. Sperry considers the new X band radar a very mature design; it was tailored to a closely defined mission profile with unusually stringent quality assurance demands. For example, component selection was made on the basis of a number of engineering and data bank recommendations

such as the Government/Industry Data Exchange Program, and the resulting system has been verified by rigorous testing to AGREE (Advisory Group on Reliability of Electronic Equipment) type testing. Mean time between failures is given as 219 hours.

The principal modes are search, navigation, weather mapping and beacon homing, and to accommodate all these the operator can choose pencil or fan beam with a variety of pulse lengths and repetition rates, and the system can be set up for angle sector or 360° scan.

All line-replaceable units are interchangeable with those of the earlier system so that separate stocks of spares are not needed for flight line support, and gradual upgrading of a system can be accomplished over a period of time and without standing aircraft down.

US Air Force/Lockheed C-130 Hercules tactical transport will have APN-59E(V) radar

Alternative configurations range from single azimuth/range displays driven by the radar as a single and independent system, to more complex installations with up to three displays. Where requirements are particularly critical the system can be connected to a compass and dead-reckoning computer for the most accurate navigation fixes. The weight of a typical configuraton is about 185 lb (84 kg).

STATUS: in production under US Air Force contract to retrofit Lockheed C-130, C/KC-135 and RC-135 fleets, and C-130 fleets of certain other air forces.

Sperry Corporation, Aerospace and Marine Group, Avionics Division, PO Box 29000, Phoenix, Arizona 85038
TELEPHONE: (602) 863 8000
TELEX: 668419

WeatherScout I weather radar

Introduced in June 1978 the WeatherScout I has a combined antenna and transmitter-receiver designed for installation in wing leading-edges of light singles. The system's weight is 15.5 lb (7 kg).

STATUS: no longer in production, but remains in service.

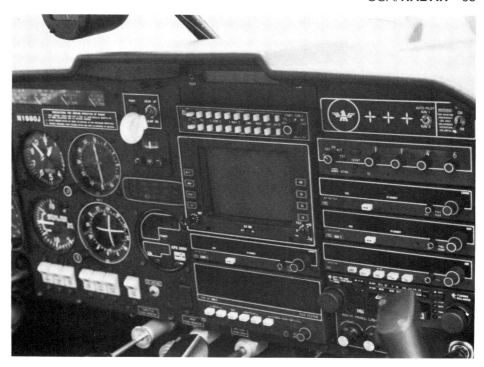

WeatherScout II radar scope in Mooney 231 flight panel

WeatherScout II weather radar

Introduced in 1978, this two-unit system is intended for single-engined aircraft (pod-mounted under a wing), and light twins. Weather up to 120 nautical miles away can be displayed at three levels of monochrome, using x-y scanning so that the screen area is completely filled.

Qualification: TSO C63b (DO–160) Class 6
Power: 2 A at 28 V dc
Frequency: X band, 9.345 GHz

Control/display unit
Dimensions: 6.25 × 4 × 9.8 inches (159 × 102 × 249 mm)
Display size and type: 5 inches (127 mm), 90° sector scan
Tilt control: ± 12°
Weight: 4.8 lb (2.18 kg)

Antenna-transmitter-receiver
Dish size: 10 or 12 inches (254 or 305 mm)
Average power: 2.28 W
Prf: 228 pps
Beam size: 10° at 10 inches (254 mm), 8° at 12 inches (305 mm)
Scan rate: 17.8 looks/minute
Pulse-width: 4 μs for 12 and 30 n mile ranges 10 μs for 60 and 120 n mile ranges

STATUS: in production.

Primus 100 ColoRadar

This is a colour version of the WeatherScout II having a 200 nautical mile range, target-alert and freeze mode. Weather is displayed in three colours: green (light rainfall), yellow (medium),

and red (heavy precipitation). The radar can also be used for mapping as a navigational aid and to avoid confusion maps are shown in magenta, yellow and blue according to the strength of returns. Least expensive of the company's colour radars, the system can be installed in a pod for under-wing mounting on single-engined types or, conventionally, in the nose for twins.

Characteristics are similar to those of the Primus 100 monochromatic system, except that the control/display unit is housed in a 12.38-inch (314 mm) deep case.

STATUS: in production.

Primus 200 ColoRadar

This system represented a step forward in design by combining a compact, 4.375 × 6.25 inch (111 by 159 mm) control/display unit with a lightweight transmitter-receiver and flat-plate antenna for use with light twins. Weather is displayed in red, yellow, and green, with maps in magenta, yellow and blue.

Primus 300SL ColoRadar

This second-generation colour radar is packaged into three units: 10-, 12- or 18-inch (254, 305 or 457 mm) antenna, transmitter-receiver, and control/display unit, and with a 300 nautical mile range appropriate to the faster turboprop twins or jets.

Qualification: TSO C63b
Power: 4.5 A at 28 V dc
Frequency: X band, 9.375 GHz

Control/display unit
Dimensions: 6.25 × 4.37 × 12.38 inches (159 × 111 × 315 mm)
Display size and type: 5 inches (127 mm), x-y scan
Tilt control: ± 12°
Weight: 10.9 lb (4.95 kg)

Transmitter-receiver
Dimensions: ½ ATR Short, 5.06 × 7.68 × 12.65 inches (129 × 195 × 321 mm)
Weight: 13 lb (5.9 kg)

Antenna
Size: 10-, 12-, or 18-inch flat-plate (254, 305 or 457 mm)
Weight: 8.4 lb (3.8 kg) with 12-inch (305 mm) antenna
12.8 lb (5.8 kg) with 18-inch (457 mm) antenna
Stabilisation: line of sight ± 30° pitch and roll
Scan rate: 12.5 looks/minute

STATUS: in production.

Primus 500 ColoRadar

This is claimed to be the first colour radar to combine simultaneously beacon position and weather mapping. Intended for helicopters and fixed-wing aircraft, the three-unit system can interrogate a radar beacon and show its position on the control/display unit in relation

Sperry Primus 200 antenna and display

Sperry Primus 300SL antenna, transmitter and control/display unit

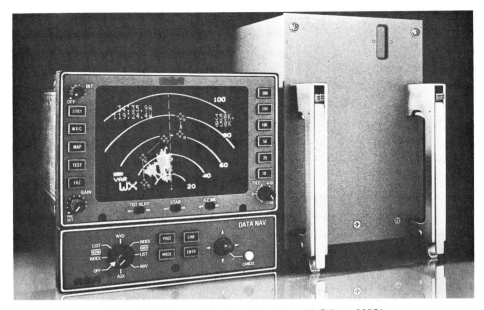

Sperry Data Nav system in conjunction with Primus 300SL

to weather and topography; the beacon-homing mode is particularly useful in parts of the world off normal routes, or for special-purpose operations such as servicing the off-shore oil and gas industry.

When combined with the RCA-designed Sperry Data Nav display systems, the Primus 500 can show 120 pages of normal and emergency checks, such as take-off drills and engine-fire procedures, performance tables, and other information that can be entered by operators. The system also interfaces with a number of long-range navigation systems so that flight plans can be displayed on the Primus control/display unit, the waypoints pictorially associated with the range markers and weather.

Qualification: TSO C63b (DO–138) Class 7
Power: 5.6 A at 28 V dc
Range: 200 n miles
Frequency: 9.375 GHz

Control/display unit
Dimensions: 6.36 × 6.36 × 12.5 inches (162 × 162 × 318 mm)
Weight: 12.7 lb (5.77 kg)
Scan angle: 60° and 120°
Tilt control: ± 15°
Display modes: radar only, beacon and radar simultaneously, beacon only

Transmitter-receiver
Format: 3/4 ATR Short
Weight: 17.5 lb (8 kg)
Peak power: 10 kW
Prf: 120 MHz

Antenna
Dish size: 12 or 18 inches (305 or 457 mm)

Weight: 8.8 lb (4 kg)
Scan rate: 14 looks/minute at 120° or 28 at 60°
Stabilisation: ± 30° pitch and roll

Primus 450 and 650 weather radars
The Primus 450 and 650 weather radars were introduced by Sperry in 1985 as small, advanced and highly-reliable systems primarily designed for installation in turboprop business aircraft. Both systems are compatible with Sperry's electronic flight instrument systems (EFIS), with Sperry's Data Nav displays and with the Stormscope WX-12 weather mapping system.

The Primus 650 has a flight-plan facility to enable the aircraft's intended route to be super-imposed on the display.

STATUS: in production.

Primus 800 radar
Sperry's Primus 800, introduced in 1983, is the first EFIS-compatible ColoRadar with what Sperry calls REACT (rain echo attenuation compensation technology), a new safety feature that displays in blue areas in which distant storms may be hidden. The system is aimed at the new generation of long-range business and corporate jets and turboprops, such as the Gulfstream IV, Canadair CL-600 Challenger and Dassault Falcon 900.

When coupled with an EFIS system, and with a 'smart' microprocessor-controlled antenna, the Primus 800 can function in two modes and at two different ranges on alternate sweeps of the antenna, updating the radar screen and

EFIS display. This feature allows the flight crew to view simultaneously low-altitude weather characterising departure conditions (the display being enhanced by GCR, a ground-clutter reduction feature) and scan en route weather.

The system can use 12, 18 or 24 inch (305, 457 or 610 mm) diameter antennas, each with a digital tilt-readout and giving a 300 nautical mile range. The transmitter can operate at four pulse-widths, selectable from the cockpit, for optimum efficiency at a given range. Primus 800 is compatible with Sperry's family of Data Nav electronic check-list and real-time navigation mapping display systems (see below). The system can be used with the 5 × 5 inch (127 × 127 mm) EDZ-600 and the 5 × 6 inch (127 × 152 mm) EDZ-800 EFIS displays.

STATUS: in production.

Primus 708 radar
This third-generation weather radar is designed around ARINC 708 digital interface and installation requirements for flight-decks with the new EFIS displays. In its primary mode the system provides a seven-colour display of weather within a range of 300 to 320 nautical miles. Colours are shown with equal intensity against a black background for visibility in bright ambient light. Red, yellow and green show heavy, medium and light precipitation and the other colours can be used for ground-mapping, turbulence detection or for other multi-function display modes.

The Primus 708 has a solid-state impact diode transmitter which, according to Sperry, has less circuitry and is more reliable and easier to maintain in the field than varactor multiple chains. Two sets of pulse-widths are employed, one for weather mapping, the other for long and short range ground-mapping. Receiver band-widths are matched to the transmitter pulse widths and are automatically designated by microprocessor circuits when range and mode are selected. Antenna stabilisation is also microprocessor controlled. Optimum ground clutter removal is obtained by dedicated active circuitry used in conjunction with a flat-plate, slotted-array antenna with minimum sidelobe performance. Another feature is what Sperry calls REACT (rain echo attenuation compensation technology), which maintains the radar signal at the correct level in the presence of intervening precipitation.

STATUS: in production and service.

Data Nav display systems
During the early 1970s, when the aviation industry began to investigate the benefits of electronic displays for centralised aircraft management, RCA (whose radar business is now a part of Sperry) recognised the potential of cathode ray tubes to show more than just radar

Sperry Primus 500 weather radar system

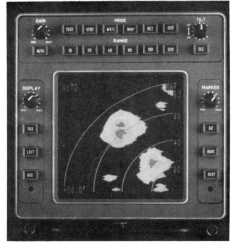

Display unit for Sperry Primus 708 radar

images of weather, and development of a more general graphic display capability began in early 1974. Working in conjunction with consultants in the industry, and with leading suppliers of navigation equipment, RCA produced a range of interface units (essentially signal converters) that can process information into a form suitable for display on its weather radar control/display units.

The information displayed falls into two categories: check-list data (for take-off and descent and emergency procedures) in alphanumeric form, and pictorial data displayed as conventional weather plots on which is superimposed alphanumeric or symbolic information such as flight-plan waypoints. In the former category the systems can store for display up to 120 pages of check-list, procedures or other information. For check purposes each function can be 'ticked off' by a cursor which changes the colour of the list: green (unchecked), yellow (being checked) and blue (check completed).

Three systems are currently offered: Data Nav I electronic checklist, Data Nav II en route navigation display for JET DAC-7000 navigation system, and Data Nav III en route navigation display for use with Delco Carousel IVa IN, Garrett AiRNAV 300, Global GNS 500A omega/vlf, Litton LTN 211 omega/vlf, and Litton LTN 72R IN. These Data Nav systems are compatible with all the RCA (now Sperry) ColoRadar systems.

Apart from the Primus radar, the system comprises four units: an interface computer, control unit, control panel and pilot-entry keyboard (effectively a pocket calculator whose display is the radar screen).

STATUS: in production.

Stormscope

Stormscope Weather Mapping Systems, Building 223-3N-01 3M Center, St Paul, Minnesota 55144-1000
TELEPHONE: (612) 736 2943

WX-8, WX-10A, WX-11 and WX-12 Stormscope thunderstorm detection systems

These units work on an entirely different principle to radar systems but are included in this section because their purpose and presentations are similar to those of weather radars.

These patented systems indicate the presence and location of thunderstorm activity. They detect the location relative to the aircraft of the electrical discharge activity generated by convective shear motion between vertical air currents in thunderstorms, and present the information in the form of a plan-position indicator display. Electrical discharge activity generated by atmospheric convection directly corresponds to the same factor that produces gust loads on an aircraft; that is, convective shear. The system detects the presence and intensity of thunderstorms by picking up radio frequency energy from the lightning discharges that accompany intense convection, processing it and displaying it on a 360° azimuth indicator.

The system basically comprises three units: a receiving antenna, computer/processor and display. The 3 ATI display is mounted in the panel and contains all the controls necessary to operate the system.

Each electrical discharge is analysed to provide azimuth and range and is plotted as a bright green dot in the appropriate position on the display. Azimuth is determined in the same way as for a radio beacon, and range is determined by finger-printing the discharges in accordance with an understanding of the behaviour of lightning discharges. As repetitive discharges occur, clusters of such dots form, indicating the extent of thunderstorm activity and the location of the thunderstorm. Since the discharges are momentary, the image is temporarily held in the system's memory to allow these clusters of dots to form. The rapidity with which the dots occur, indicates the severity of the thunderstorm. Range settings can be varied to show activity within 25, 50, 100 and 200 nautical miles.

The system can display 256 discharges on the screen at any one time. When the 257th discharge occurs, the earliest recorded image is erased to make way for the new information. Any discharge that has been on the screen for 4 minutes (2 minutes for jets) is cleared automatically. In addition, a clear button is provided to erase the display manually at any time. The data received from the thunderstorm is low frequency; consequently, the Stormscope system can be used on the ground as well as in the air. In addition, the Stormscope system is omni-directional (unlike a conventional weather radar). This capability is useful in giving the crew the big picture view of thunderstorm activity in the air, and also giving real time enroute and terminal area thunderstorm information on the ramp and while taxiing, even before departure. 3M has recently developed a system that interfaces with Sperry ColoRadar. The system will overlay electrical discharge data on the precipitation information in the radar mode and give the user a 360° view with radar in standby mode. This system is compatible with virtually all Sperry ColoRadar systems, including current and previous production models.

WX-8 thunderstorm detection system

This is a two-unit system comprising an antenna and three-colour, liquid-crystal display that also contains a micro-computer to analyse and process thunderstorm activity.

STATUS: in production and service.

WX-10A thunderstorm detection system

A 360° crt display is provided. Maximum range is 220 nautical miles. Display size is 3 ATI. Operating ranges are switch selectable 25, 50, 100 and 200 nautical miles.

Weight: (computer processor) 4.3 lb (1.9 kg)
(antenna) 2 lb (0.9 kg)
(control/display unit) 3.4 lb (1.5 kg)
Qualification: FAR Part 135 Section 173 approved for thunderstorm and severe weather avoidance

STATUS: in production and service.

WX-11 thunderstorm detection system

This system is similar in appearance to the WX-10A. The WX-11 allows for heading inputs from the aircraft to allow the dot patterns to rotate as the aircraft turns. The heading inputs are compatible with most synchro sources found onboard aircraft.

Weight: (computer processor) 4.3 lb (1.9 kg)
(antenna) 2 lb (0.9 kg)
(control/display) 3.4 lb (1.5 kg)

Qualification: FAR Part 135 Section 173 approved for thunderstorm and severe weather avoidance

STATUS: in production and service.

WX-12 thunderstorm detection system

This system interfaces with Sperry ColoRadar systems and it is retrofittable on nearly all Primus Radar systems. This system consists of a receiving antenna and a remote-mounted computer processor, and a panel or pedestal mounted controller. Range is selected on the radar display. The electrical discharge data can be superimposed over the precipitation information, and the 360° display can be shown when the radar is in the standby mode.

STATUS: in production and service.

Stormscope WX-8 display

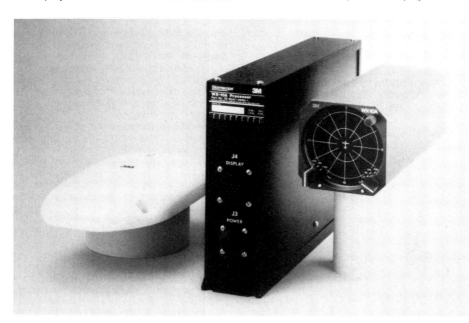

Stormscope WX-10A weather sensor and display

Texas Instruments

Texas Instruments, Equipment Group, PO Box 660246, MS 3127 Dallas, Texas 75266
TELEPHONE: (214) 480 1417

Terrain-following radar for Tornado

As one of the principal specialist producers of terrain-following radars (TFRs) in the West, Texas Instruments was chosen to design and develop the system for the Panavia Tornado IDS variant; the radar has no official designation, but is often referred to as the Tornado Nose Radar (TNR).

The all-weather radar comprises two essentially separate systems that share a common mounting, power supply and computer/processor. They are TFR and the GMR (ground-mapping radar). The first is used for automatic high-speed, low-level approach to the target and escape after an attack. The second is the primary attack sensor for the Tornado and operates in air-to-ground and air-to-air modes to provide high-resolution mapping for navigation updating, target identification and fire control.

The radar enables the crew to fix the aircraft's position by updating the Doppler-monitored inertial navigation system, provides range and tracking information for offensive or defensive weapon delivery, and commands, via the autopilot, a contour-hugging flight profile that shelters the aircraft as far as possible from detection by hostile air-defence radars. It makes extensive use of ECCM to provide relative immunity from interference in severe ECM environments. The three units comprising the system are the radar sensor (transmitter-receiver package for TFR and GMR), a digital scan converter and a radar display unit in which a moving map image can be superimposed on to a radar 'picture' for navigation updating and target identification.

The GMR operates in the Ku frequency band with nine modes: readiness, test, ground-mapping, bore-sight contour mapping, height-finding, air-to-ground ranging, air-to-air tracking, land/sea target lock-on, and beacon homing.

The TFR operates at Ku frequencies and has three modes: readiness, test, and terrain-following. In the latter mode the pilot can have the aircraft flown automatically or fly it himself through head-up display steering commands. He can also select ride comfort (for a given speed, the closer the allowable ground clearance, the less comfortable the ride owing to the greater g-levels needed to stay on the commanded flight profile).

Both systems have extensive built-in test features to ensure a high degree of fault isolation and comprehensive reversionary modes.

STATUS: in production by consortium of companies under licence to Texas Instruments.

Texas Instruments terrain-following radar equips the Tornado IDS variant

Texas Instruments advanced terrain-following/avoidance radar is pod-mounted under General Dynamics F-16 for flight trials

Companies are Ferranti/GEC Avionics (UK), AEG-Telefunken/Siemens (West Germany), and FIAR/Elletronica Aster (Italy). In 1985 Texas Instruments began development of the Phase 1 improvement for the Tornado TFR radar. Production of this improved version, which increases the speed and power of the radar's computing, begins in 1987. Further phases of enhancement are expected to follow.

Terrain-following radar for Lantirn

Fourth in the series of development of terrain-following radars developed by Texas Instruments (the first three equip the General Dynamics F-111, McDonnell Douglas F-4 and Panavia Tornado respectively) the TFR for the US Air Force's Lantirn (low altitude navigation and targeting, infra-red, by night) programme is claimed to offer orders of magnitude improvements in ECCM, weather detection, and performance in turning flight by comparison with previous equipment. Such improvements enhance the survivability of combat aircraft during penetration of hostile airspace.

A new level of operational flexibility, said to be unobtainable with earlier systems, derives from the use of a high-throughput digital processor, an advanced adaptive algorithm, and terrain-data storage techniques.

The Lantirn TFR retains the basic facilities available in the Tornado including the ability to operate under enemy jamming in bad weather and at very low level but changes in system performance enable the aircraft to turn at a maximum rate to 5.5° per second compared with the Tornado's 2° per second, and to bank of up to 60° whilst still able to maintain terrain-following capability. Whereas the Tornado and previous terrain-following radars employed magnetrons, the Lantirn TFR has a travelling-wave tube giving considerably better performance, particularly in an ECM environment and higher reliability. The pilot can control the aircraft's flight-path by selecting terrain-clearance heights of 100, 200, 300, 400, 500 or 1000 feet.

The Lantirn radar began test flying in a General Dynamics F-16 during 1985 and quickly demonstrated the ability to fly down to 100 feet above the terrain. The system is scheduled to equip the Fairchild A-10,

Two of Texas Instruments attack radar units: digital scan converter (left) and radar display for interdictor version of Panavia Tornado

Texas Instruments attack radar for Panavia Tornado GR1 showing attack and terrain-following antennas

McDonnell Douglas F-15E and F-16 aircraft in US Air Force service. Texas Instruments has received a production contract for some 700 systems for delivery between 1990 and 1991 with a contract value of $407 million.

STATUS: entering production.

AN/APQ-168 multi-mode radar for HH-60 helicopter

The Sikorsky HH-60 Night Hawk helicopter has been developed for the US Air Force to replace the HH-3 and HH-53 combat rescue helicopters that have been in service since the early 1960s. It is intended to penetrate hostile territory during darkness to rescue downed aircrews or deliver and retrieve special-operations teams. The task calls for long-distance nap-of-the-earth flying and accurate nagivation.

To meet this requirement, Texas Instruments is building a version of the Lantirn terrain-following radar (see previous entry) that retains commonality with five of the six line-replaceable units of the equipment now under development for the US Air Force. The radar can operate in terrain-clearance, terrain-avoidance, air-to-air ranging and cross-scan modes, the latter combining ground-mapping or terrain-avoidance with terrain-following.

A terrain-storage facility permits the radar to have a reduced duty cycle thereby reducing the probability of detection by enemy ESM equipment.

The HH-60 multi-mode radar entered full scale development in December 1982 and two systems have been built: flight testing beginning in April 1986. In this configuration the radar

weighs 250 lb (114 kg) and occupies 6.98 cubic feet.

The same radar has also been selected for the V-22 Osprey (JVX) aircraft, again mounted alongside the fuselage but with a reduced length pod and with the interface unit mounted remotely from the rest of the radar. This radar, which also includes beacon and weather modes of operation, will equip the US Air Force and Navy V-22's and some of those destined for the Army but not the bulk of the machines which are going to the US Marine Corps.

Texas Instruments have studied several possible applications for the AN/APQ-168 radar, including mounting it in an Alpha Jet, where all the line-replaceable units will be remote from the scanner, and also in the US Navy's F/A-18's where the antenna diameter has been reduced to 8 inches (203 mm) and the system is located under the aircraft's nose to give the F/A-18 a complete attack capability.

The system has increased electronic counter-measures resistance, improved weather penetration, better guidance in turning flight, a power management function for semi-covert operation and low beam reflectivity. Extensive built-in test equipment provides a high degree of fault isolation and detection. The system is carried in a pod alongside the forward fuselage. The system can be readily fitted or removed according to the mission being flown.

STATUS: under development.

ETMP enhanced terrain masked penetration programme

While terrain-following radars provide a unique way of permitting ground attack aircraft to fly safely at low altitude, they have the severe drawback that powerful radars can easily be detected by enemy sensors making it improbable that an attack can be made covertly. Texas Instruments is now working in conjunction with the US Air Force's Wright Aeronautical Laboratories to reduce this problem with the Enhanced Terrain Masked Penetration programme (ETMP). The first phase of flight-testing in the company's Convair 580 was completed in 1985 and a second phase followed in 1986.

The ETMP takes the AN/APQ-168 multi-mode radar processor and adds to it a digital map from the Defense Mapping Agency. This provides the aircraft with a very accurate map of the route to be followed and the radar is only required to confirm that the aircraft is in fact on course and that the data being presented for navigation is accurate. The digital mapping information would be co-ordinated with inertial and global positioning navigation information and correlated with intelligence information provided to the aircraft either prior to or during the flight. Intelligence data can be relayed to the system in real time via the Joint Tactical Information Distribution System (JTIDS) or the aircraft's own sensors, which are detecting the enemy radar transmissions for example, can be used to amend the flight path in the optimum

fashion. The crew therefore only need to give the start point of the flight and the proposed target, together with any pre-known intelligence information and the ETMP equipment will work out the best possible route in three dimensions and, if the aircraft is so equipped, feed the navigation information direct to the autopilot.

In the first phase of flight-testing the Convair 580 demonstrated the navigation and radar integration in both vertical and lateral manoeuvres in severe terrain, using existing military qualified microprocessor-based electronics and the Jovial standard language. In the second phase threat-avoidance and fuel and flight management were added for real-time route planning.

In the Advanced ETMP programme, which will be the next stage in Texas Instruments' terrain-following radar technology development, an expert system will be added to the radar processing for real-time route planning. Threat avoidance functions will also be added.

STATUS: ETMP in flight test.

AN/APQ-99 terrain-following radar for RF-4C

This sensor is the forward radar for the reconnaissance version of the US Air Force/McDonnell Douglas F-4C Phantom II. Initial work is under way under a $29.8 million contract from the US Air Force Warner Robins Air Logistics Center awarded in February 1985 to improve the reliability and ease of maintenance of this radar. The programme, affecting some 340 aircraft, began with a $9.3 million award in March 1984 for the design and development stage. The total programme will be covered by a $96 million multi-year contract that will cover modification kits for the principal line-replaceable units and a new display system. The US Air Force says that the effort now under way will result in a three to one improvement ratio in reliability, with better mapping and terrain-following performance, and extend the useful life of the radar beyond the year 2000. Flight trials began in spring 1986 with deliveries expected to begin before the end of the year. The improvements will be adopted by the West German Air Force and probably by Japan.

AN/APQ-126 terrain-following radar for A-7D/E

This radar equips the LTV A-7D/E Corsair II aircraft. The equipment for this mid-1960s design continues in production, most recently for the A-7P and TA-7P aircraft and ordered by Portugal in 1984. The cost of 30 sets of APQ-126 for these aircraft was given as $10.4 million.

AN/APS-124 search radar

The AN/APS-124 search radar was specially designed to be part of the comprehensive avionics suite for the US Navy/Sikorsky SH-60B Seahawk ASW helicopter built to satisfy the Lamps (light airborne multi-purpose system) Mk III requirement. One of the problems associated with the operation of these medium-size helicopters from the *Spruance*-class destroyers on which they serve is that of stowage, particularly in height limitation. The APS-124 is therefore designed around a low-profile antenna and radome.

Optimum detection of surface targets in rough sea is accomplished by several unique features, including a fast-scan antenna and an interface with the companion OU-103/A digital scan converter to achieve scan-to-scan integration. The system is associated with a multipurpose display and with the Lamps data link so that radar video signals generated aboard the aircraft can be displayed on Lamps-equipped ships.

The system operates in three modes covering long and medium range search and navigation and fast scan surveillance. Mode 1, long-range

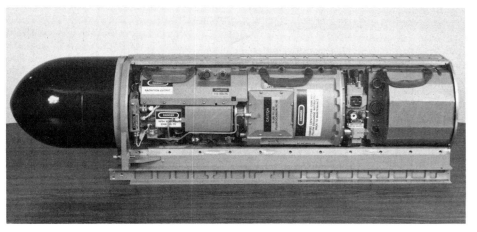

Texas Instruments AN/APQ-168 radar for Sikorsky HH-60 Night Hawk helicopter

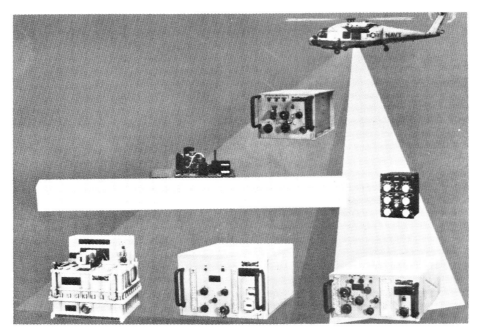

Principal detection system of US Navy/Sikorsky SH-60B Seahawk ASW helicopter is Texas Instruments AN/APS-124 radar

search, is characterised by long pulse length, low pulse repetition frequency, and slow scan, actual values being 2 microseconds, 470 pulses a second, and 6 rpm. Display ranges are selectable out to 160 nautical miles. In the medium range Mode 2, these values change to 1 microsecond, 940 pulses a second, and 12 rpm respectively. For Mode 3 they become 0.5 microsecond, 1880 pulses a second, and 120 rpm. The display ranges are selectable up to 40 nautical miles and the false-alarm rate is adjustable to suit conditions.

The system is designed around the MIL-STD-1553 digital data-bus to communicate with other aircraft equipment, and the modular design facilitates installation on other aircraft. The under-fuselage antenna provides 360° coverage, and the entire six-unit APS-124 weighs 210 lb (95 kg).

STATUS: in production; service deployment of Seahawk began in 1983.

AN/APS-134 (V) periscope detection radar

The APS-134 (V) anti-submarine warfare and maritime surveillance radar is what Texas Instruments calls the 'international successor' to the US Navy's AN/APS-116 periscope-detection radar. Texas Instruments says that the APS-116 is the world's only radar specifically designed to detect submarine periscopes under high sea state conditions and that its performance makes it the main-stay of the US Navy's Lockheed S-3A Viking ASW aircraft fleet. The APS-134 (V) incorporates all the features of the former system while improving performance and adding capabilities, including a new surveillance mode.

The heart of the new radar is a fast-scan antenna and associated digital signal processing which, says Texas Instruments, form the only proven and effective means of eliminating sea clutter. This technique is used in two of the three operating modes, the third being a conventional slow scan for long-range mapping and navigation. The transmitter puts out 500 kW.

In Mode 1, periscope detection in sea clutter, high resolution pulse compression is employed with a high pulse repetition frequency and a fast-scan antenna, actual values being 1.5 feet (0.46 metre), 2000 pulses a second and 150 rpm. Display ranges are selectable to 32 nautical miles. There is an adjustable false-alarm rate to set the prevailing sea conditions and scan-to-scan processing is employed.

Mode 2, long range search and navigation, operates at medium resolution and with a low pulse repetition frequency, low scan and display ranges selectable to 150 nautical miles. Actual values are 500 pulses a second and 6 rpm.

Mode 3 operates, again at high resolution, for maritime surveillance. A low pulse repetition frequency (500 pulses a second) is used in conjunction with an intermediate scan speed of 40 rpm. Display ranges are selectable to 150 nautical miles and an adjustable false-alarm rate is used together with scan-to-scan processing.

The system is also available in an off-line configuration, with its own 10 × 10 inch (254 × 254 mm) cathode ray tube control/display unit. On-line operation (ie in linked with other aircraft systems) is accomplished via a MIL-STD-1553 digital data-bus, with the digital scan converter providing faster-scan video for other aircraft displays. The weight of the entire APS-134(V), including the wave-guide pressurisation unit, is 527 lb (237 kg). The equipment is compatible with the inverse synthetic aperture radar techniques developed by Texas Instruments for long-range ship classification.

STATUS: in service with US Navy/Lockheed S-3A Viking and with Dornier/Breguet Atlantic ASW aircraft of West German Navy. Initial deliveries are under way for Royal New Zealand Air Force/Lockheed P-3B Orion update programme. The US Coast Guard has also chosen the APS-134(V) for its Lockheed HC-130 Hercules.

AN/APQ-122 radar

A modification programme is under way to enhance the performance of several hundred AN/APQ-122 high resolution radars installed in Lockheed C-130 transports and special-mission aircraft. The modification comprises the addition of a scan converter, made by Systems Research Laboratories (SRL) that accepts range and bearing radar returns and converts it to high-resolution raster-scan cathode ray tube in accordance with standard display formats such as American Industries RS-343. The APG-122 comprises about a dozen line-replaceable units, and the scan converter will replace several of them, including the plan-position indicator, which will give way to a raster-scan video display. The significance of this enhancement lies in the resolution of the unit: scan converters for use with fire-control system are generally of low resolution, typically 400 × 400 pixels. The SRL unit will provide a 1024 × 810 pixel format, matching and exploiting the radar imagery itself, with eight grey shades.

STATUS: in production.

US Navy/Lockheed S-3A Viking-based sub-hunter carries AN/APS-134(V)1 surveillance radar

United States Air Force
Electronic Systems Division, Air Force Systems Command, Hanscom AFB, Massachusetts

Joint-Stars
Joint-Stars (Joint Surveillance Target Attack Radar System) is a joint US Air Force/US Army venture to develop a stand-off battlefield surveillance system. The objective of this system is to control the flow of second echelon enemy forces to allow friendly ground and air elements to effectively engage the enemy at close range without the danger of an overwhelming enemy breakthrough in the friendly lines of combat. To accomplish this critical need to observe manoeuvring enemy forces across the forward line of troops and into second echelons, Joint-Stars will provide all-weather, day or night, battlefield surveillance and attack control for the airland battle.

Joint-Stars combines the operational requirements of the US Air Force Pave Mover target acquisition and guidance system that was under development for its Assault Breaker family of weapons and the US Army's Stand-Off Target Acquisition System (Sotas). In May 1982, the US Army was instructed to combine its requirements for the targeting radar with that of the US Air Force. At the same time, the US Air Force was designated the executive service responsible for joint programme management. Initially, it was envisioned that the Joint-Stars airborne equipment would be carried on the US Army's Rockwell OV-1D, and/or the US Air Force's Boeing C-18 (a derivative of the commercial 707) or the Lockheed TR-1 derivative of the U-2 high-altitude reconnaissance aircraft.

By November 1983, a number of industry teams had lined up to study requirements for

Joint-Stars. They were Grumman/Norden/ TRW, Hughes Aircraft/E-Systems, Boeing/ General Electric and Lockheed/Westinghouse. In the spring of 1984 Boeing/General Electric decided not to bid (though General Electric was heavily involved in battlefield radar research and demonstration, see above). Requests for proposals (RFPs) due in May 1984 were revised in November 1984 so that the three remaining teams could optimise the designs in light of a September 1984 US Air Force/Army agreement to use the Boeing C-18 as the host aircraft.

Defense Systems Acquisition Review Committee (DSARC) approval for Joint-Stars was granted in August 1985 and shortly thereafter (in September) the US Government awarded a 60-month full-scale development contract to the prime contractor, Grumman, valued at $657 million. Of this, $74 million is allocated to Boeing Military Airplane Company for the modification of the first two 707-320s, designated C-18 by the US military. Modification work on the first of these used aircraft began in spring 1986. There is an option to procure a further two airframes at a later date, and it is anticipated that there will eventually be 10 aircraft in service.

The C-18s will be equipped with a 25-foot (8-metre) steerable, conformal radar antenna mounted under the fuselage, developed by the Nordern Division of United Technologies. This will be operated as either a conventional sideways-looking radar for the detection of fixed or stationary targets, or in a Doppler mode to track slow moving targets such as tanks or troop formations. First flight of a prototype system is set for late 1988.

Inside the aircraft, radar data will be processed and displayed at 10 operators' consoles, with full colour displays, the display processing

Boeing Military Airplane Company is to modify at least two used 707-320 aircraft beginning in spring 1986 as development aircraft for US Air Force and Army's Joint-Stars programme

being developed by Aydin. Information required by ground-based commanders will be passed by the Joint Tactical Information Distribution System (JTIDS), while the US Army commanders will receive raw radar data via a secure data link.

The C-18s will be equipped with the Have Quick frequency-hopping secure radios and elements of the US Army's Sincgars (single channel ground/air radio system).

In late 1985 Control Data received a $32 million contract to develop the programmable signal processor for Joint-Stars, and Cubic received a $25 million contract, also from Grumman, to provide data links between the C-18s and the ground bases.

STATUS: in full scale development.

Westinghouse
Avionics Division, Westinghouse Electric Corporation, PO Box 746 Friendship Site, Baltimore, Maryland 21203
TELEPHONE: (301) 765 1000

AN/AWG-10 radar for F-4
In 1963 Westinghouse was awarded a US Navy contract for the AN/AWG-10 to provide fire control for guns and Sparrow and Sidewinder air-to-air missiles on the McDonnell Douglas F-4 Phantom. The first radars were delivered in 1966 and became standard equipment. The AWG-10 was claimed to be the first interceptor radar to feature transistorised circuitry.

During 1966 the US Navy announced a look-down requirement for its F-4s, and the AWG-10 became the first multi-mode radar with pulse-Doppler look-down capabilities and a comprehensive built-in test system. The UK bought the AWG-11 and -12, slightly modified versions of the AWG-10, for its F-4K and F-4M Phantoms.

AWG-10A radar for F-4
During 1973 Westinghouse was contracted by the US Navy to improve the performance and reliability of the AWG-10. The most significant change affecting reliability was the substitution of a solid-state transmitter using a klystron power amplifier. A digital computer was added

for the more effective solution of launch equations, for example target manoeuvring information could now be used. A servoed optical sight display was also added and these modifications also permitted the full range of air-to-ground ordnance to be exploited.

Only three of the AWG-10's 29 lrus remain unmodified; there are six new lrus and seven units of the AWG-10 system are deleted.

The success of the AWG-10A computer led to its procurement by the US Air Force for the APQ-120 radar in the F-4E to provide computer-aided target acquisition. The West German Air Force, in a programme known as Peace Rhine, also procured the computer to upgrade its F-4F Phantoms.

STATUS: AWG-10 family has long been out of production, but is still in widespread service. In early 1984 US Defense Department announced possibility of flight demonstrating McDonnell Douglas F-4s with new engines and avionics suite so as to create new market for F-4s currently in US inventory. Fundamental change in avionics would be substitution of Westinghouse AN/APG-66 radar, as used in General Dynamics F-16 fighter.

AN/AWG-11/12
AN/AWG-11 and -12 are the designations given to the AWG-10 radars fitted to the F-4K Phantoms of the Royal Air Force and the F-4Ms of the Royal Navy in the 1970s. Ferranti is the UK Design Authority. The updates which gave rise to the AWG-10A designation were implemented in the UK to give the current AN/AWG-11A and -12A standards in RAF service.

AN/APG-66 radar for F-16
Westinghouse was chosen in October 1975 to build the X band coherent pulse-Doppler radar for the F-16 after a competitive fly-off with a Hughes system. The Westinghouse radar was a direct development of an in-house development called the WX-200, claimed by the company to

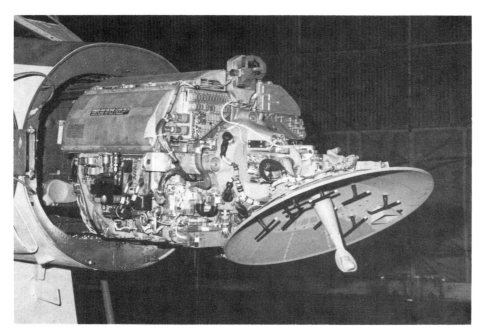

Westinghouse AWG-10A installed in McDonnell Douglas F-4 Phantom

Westinghouse AN/APG-66 radar in nose of General Dynamics F-16

Westinghouse AN/APG-68 radar for General Dynamics F-16 fighter

General Dynamics F-16, prime vehicle for Westinghouse AN/APG-66 and APG-68

be the first airborne radar with a programmable signal processor. Designed to operate in conjunction with Sparrow medium-range and Sidewinder short range missiles, the AN/APG-66 has a look-down range, in ground clutter, of 20 to 30 nautical miles and a look-up range of 25 to 40 nautical miles. The production system has no associated hydraulics, rate gyros or roll gyro, and there are only 9500 components. This reflects in a formally demonstrated mean time between failures of 97 hours, and a mean time to repair of 5 minutes.

There are ten operating modes, several of them associated with frequency agility to resist jamming. All functions, including self-test, are controlled by a computer via a Digibus serial digital data-bus. The company claims that the WX-200 radar was the first airborne radar to incorporate programmable signal processor, and this technology was absorbed into the APG-66.

The AN/APG-66 is of modular design, with six functional line-replaceable units, each with its own power supply, so that foreseen improvements can be easily incorporated. One such improved version has been chosen by the Japanese Air Self Defence Force to replace the current Westinghouse APQ-120 fire-control system on its McDonnell Douglas/Mitsubishi F-4EJ Phantom fighters.

Line-replaceable units are co-produced under licence by the European F-16 operators,

the Netherlands, Belgium, Norway and Denmark. The system entered service in 1978.

The radar has also been installed in six Cessna Citation and eight modified Piper Cheyenne III aircraft operated by the US Customs for anti-smuggling and illegal immigration surveillance duties.

Weight: 296 lb (134.5 kg)
Volume: 3.6 ft³ (0.102 m³)
Frequency range: X band
Operating modes: (air-to-air) look-up, look-down, search and track; air combat search; automatic tracking
(air-to-ground) real-beam mapping; 8:1 Doppler beam-sharpening; and scan freeze; ranging; beacon-homing; sea search (sea states up to 4 and above)
Transmitter: travelling-wave tube
Antenna: planar array, 29.1 × 18.9 inches (740 × 480 mm)
Search angle and range: 120° in azimuth and elevation, 80 n miles (148 km)
Electronic protection: frequency agility in certain modes
Reliability: 97 h demonstrated mtbf
Maintenance: 5 minutes mttr

STATUS: in production for US Air Force and air forces of Belgium, Netherlands, Denmark, Norway, Israel, Egypt, Pakistan and Venezuela. By early 1985 more than 1000 systems had been built. In mid-1985 Westinghouse announced

that a development was in hand which would permit the radar in the General Dynamics F-16A/B to track a specific target while continuing to monitor and display other targets. Retrofitting of this modification was completed in 1985.

AN/APG-68 radar for F-16

In August 1980 the US Air Force authorised Westinghouse to develop a substantially improved version of the AN/APG-66 fire control radar for the General Dynamics F-16 as part of the multinational staged improvement programme (known as MSIP). It will improve the fighters' ability to deliver air-to-air and air-to-ground weapons in all weather conditions; in particular, it will permit F-16 pilots to launch from beyond visual range the Hughes AIM-20 AMRAAM advanced medium range air-to-air missile, and will improve the ground-mapping performance. Formerly designated the Improved AN/APG-66, the system has been redesignated AN/APG-68 to emphasise the difference between it and the earlier system. The major change is the incorporation into one unit of all digital processing activities; the new processor weighs 100 lb (45 kg), occupies about 1 cubic foot (0.03 cubic metre), and dissipates 1 kW. The new radar has more than 22 air-to-air and air-to-ground modes. All program instructions are stored in a 384 K word, non-volatile, block-orientated random access memory. The radar computer is programmed in Jovial J73, the US Air Force-approved high-order language. Initial operational capability for the F-16C and -D versions, the first to have the APG-68, was in March 1985.

These important new capabilities will be gained by the addition of a programmable signal-processor and a dual-mode transmitter. The former incorporates advanced modular processing architecture and a reliable high-density, solid-state memory. The latter will select the best waveform for each mode of operation, ranging from low pulse repetition frequency (prf) in air-to-surface engagement to medium and high prf for long-range air interception.

In the air-to-air mode, the range at which on-coming targets are detected is increased by using a high-prf, velocity-search mode (ie the radar is searching for targets with high relative velocities). Once detected, a medium-prf range-while-search mode can be employed against targets with any aspect to gain additional range and angle information. In the track-while-scan mode, the radar can track simultaneously a number of targets, assess the degree of threat from each, and launch missiles as appropriate. By using high-resolution Doppler techniques, closely spaced targets can be distinguished and tracked separately.

In the air-combat mode, the radar scans selected airspace and automatically acquires the nearest target. In 'look-down' situations (where targets are seen against ground), land-based targets such as moving vehicles or vessels are ignored because the radar rejects returns with less than a specified threshold. This ground-mapping terrain radar technique ensures that only airborne targets are portrayed on the radar scope.

The air-to-ground performance of the new system has been improved by an eight-fold increase in the ground-mapping resolution and substantial spare computing power is available for future growth, which could include, for example, automatic terrain-following or terrain-avoidance or very high resolution synthetic aperture mapping.

Prototype systems were delivered in 1981, and by January 1983 three of them had completed 23 test flights aboard a Sabreliner trials aircraft to complete Phase 1 of the Improved AN/APG-66 (as it was then still known) development programme.

Westinghouse had originally rescheduled 60 Phase 1 flights, but the benefits of real-time processing of airborne data enabled problems to be located and solved during flight. A software development test bench, using simulated data, was used to refine and debug the software, so that the system operated acceptably even on its first flight.

This phase also demonstrated integration of the programmable signal processor and dual-mode transmitter. By that time Phase 2 had begun, with equipment installed on a test F-16.

More than 1500 US Air Force F-16s are to have the improved radar. A version of the APG-68, designated APQ-164, will form part of the offensive avionics system in the US Air Force/Rockwell B-1B (see following entry).

Weight: 336 lb (153 kg)
Frequency range: X band
Operating modes: (air-to-air) range-while search; track-while-scan (10 targets); velocity search and cued track-while-scan (high/medium prf); look-up search (medium prf); raid cluster assessment; multiple anti-acquisition air combat
(air-to-ground) real-beam mapping with Doppler beam-sharpening and scan freeze; sea surface search; beacon-homing; ground moving-target indication and tracking; fixed-target tracking; terrain-following and -avoidance
Special capabilities: ability to be extensively reprogrammed for growth and weapons' changes; multiple target awareness; rapid-response ECCM; high-resolution ground-mapping; target identification beyond visual range
Transmitter: gridded, multiple peak power travelling-wave tube
Antenna: planar array, 29 × 19 inches (740 × 480 mm)
Search range: 160 n miles
Number of units: 5 lrus

STATUS: in production, with orders for US Air Force, Israel, Korea, Turkey, Greece and Egypt.

AN/APQ-164 radar for B-1B

Westinghouse is a sub-contractor to Boeing respecting the APQ-164 ORS offensive radar system for the Rockwell B-1B which replaces the separate terrain-following and ground-mapping radars in the B-1A. This system is an outgrowth of two US Air Force programmes that have been in existence for several years. The first of these was the EAR (electronically agile radar), that started in 1974 to develop the next generation strategic radar. The B-1 took in from EAR the modal technologies, that is, the fundamental capabilities and functions that were needed for a strategic system, together with the phased-array technology which enabled those capabilities to be implemented in the EAR system.

The second major contributor to the B-1 is the radar for the F-16C and -D, the AN/APG-68. Key hardware technology developed or utilised on the APG-68 programme consists of a dual-mode transmitter, allowing extensive multi-mode operation for air-to-ground and air-to-air

use; a programmable signal processor, a direct outgrowth of the processor development of the EAR programme, which allows versatility in the modes that may be installed in any given system; and the modular low-power rf (MLPRF) system, which provides all the receiver and stable local oscillator functions.

The B-1 radar generates data for navigation, penetration, weapon delivery, and for certain other functions, such as air refuelling. There are really four modes in the APQ-164 system that provide the navigation capability. The primary mode is a high-resolution synthetic aperture radar mapping mode, backed-up by a real-beam ground-mapping mode. The system also detects weather ahead and can display ground-beacon returns over any radar image. The penetration functions of the radar include automatic terrain-following and terrain-avoidance. For weapon delivery the radar provides three different functions. The first is a velocity update mode, similar to a Doppler navigator, which generates velocity information for the inertial navigation system. Second, there is a ground moving-target detection and tracking capability. There is also a high-altitude altimeter function that provides a very accurate measure of height above ground-level.

The synthetic aperture mode provides the operator a high-resolution image of an area of ground that can be chosen by the inertial navigation system or the operator. Long-range maps can be made and five different map scales displayed: 0.625 nautical miles on a side; 1.25, 2.5, 5 and 10 nautical miles. The synthetic-aperture mapping-mode accepts the coordinates of a waypoint from the inertial navigation system and makes a map centred on that point. To make an image the antenna is electronically scanned out in the direction the waypoint should be located. The radar transmits a burst of energy, gathers data for the image, and then switches itself off. At the same time, the image is stored in the radar and presented on the display unit in a rectangular, ground coordinate display.

The radar provides the basic data required for automatic terrain-following. It scans the ground in front of the aircraft and measures the terrain in a range versus altitude profile out to 10 nautical miles and stores that data in the radar computer. That profile data is sent across the multiplex bus to the terrain-following control unit, where the data is used to generate the climb/dive commands. This flight profile is then automatically fed into the pilot's flight control system. Since the radar is not continuously scanning in terrain-following, a very low update rate is utilised helping to reduce the risk of detection. This rate is variable and depends upon flight altitude, terrain roughness, and aircraft ground speed. Under normal conditions updates are made at 3 to 4 second intervals. However, if the terrain demands it, data can be gathered continuously.

The AN/APQ-64 in the B-1B is a dual-redundant system, with two complete and independent sets of lrus, except for the phased array antenna, the first airborne application of this technology for combat aircraft. Only one channel is used at a time, the other being maintained at standby.

The phased-array antenna is an outgrowth of the antenna developed on the EAR programme. It contains 1526 phase control modules and allows virtually instantaneous beam movement to any point in the antenna field of regard. When the radar computer commands the desired pointing angle, the antenna is physically movable to three different positions on a roll detent mount. The APQ-164 can therefore look off to either side of the aircraft or forward by rolling the antenna about an axis. The normal antenna position is looking forward. However, when the antenna is rolled off to the side, the field of view extends from the aircraft nose back to about 105°, permitting a look off to the side at an area of interest without having to change aircraft heading. Once physically moved to one of the three available positions, the antenna is

locked into a detent. From that fixed spot, the antenna can be scanned electronically ±60° in azimuth and elevation by means of a box on the antenna called the beam steering control, which controls all 1526 phase modules.

Weight: 1256 lb (570 kg) including both channels
Frequency range: X band
Operating modes: (air-to-ground) high resolution mapping, real beam mapping, automatic terrain-following, manual terrain-avoidance, velocity update, ground moving target detection and track, high altitude calibrate, ground beacon
(air-to-air) weather mapping, air-to-air beacon, rendezvous mode. System provides for growth to full conventional standoff capability and a full air-to-air mode complement
Transmitter: gridded, multiple, peak-power travelling-wave tube, (identical with that on the APG-68 radar)
Antenna: First airborne phased-array electronically scanned antenna, 44 × 22 inches (1118 × 559 mm)

STATUS: in production for US Air Force/Rockwell B-1B bomber; 100 of these aircraft are to be acquired, the first of which was delivered in June 1985.

Flight trials, involving a converted British Aerospace One-Eleven airliner were carried out from July 1984 to December 1985. The One-Eleven is still being used to evaluate operational improvements to the AN/APQ-164 system.

USN night/adverse weather attack radar

This compact, lightweight, millimetre wave NAW attack radar was designed for close-support aircraft operating at night or in bad weather. As a non-coherent sensor operating in the Ka band it offers four simultaneous air-to-surface modes:
Terrain-following/terrain-avoidance: in which the system displays on a head-up display contour lines indicating ground profiles up to 3 nautical miles ahead of the aircraft. A steering 'box' on the display permits the pilot to fly as low as 200 feet above ground level.
Emitter locator: in which, by the use of the broad-band characteristics of the reflector antenna, the radar can detect and isolate high-priority targets in the Ku/Ka bands (such as anti-aircraft missile radars) and provide azimuth steering information for fire-control.
Ground moving target indication: in which the radar can discriminate vehicles, vessels, or other mobile targets moving at speeds as low as 3 nautical miles an hour using clutter-referenced rejection techniques. Helicopter- or tank-size targets can be detected at ranges up to 10 nautical miles.
Ground-mapping: in which picture resolution is sufficient to permit navigation and the identification of tactical targets up to 15 nautical miles distant.

Future modes envisaged for the system, which is based on the Westinghouse WX series that (claims the company) pioneered digital and modular technologies, include air-to-ground, ranging and beacon-homing.

US Air Force/Fairchild A-10 has carried experimental Westinghouse NAW attack radar

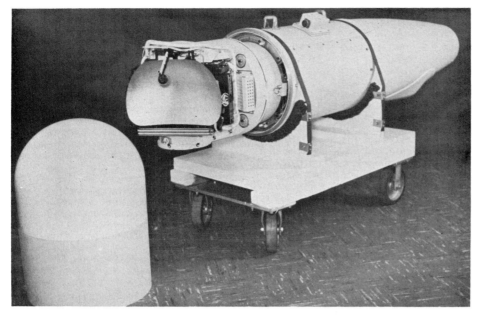

Prototype of Westinghouse night/adverse weather attack radar

An earlier version of the radar, designated WX-50, was flight-tested by the US Navy on Douglas TA-4J Skyhawk and Grumman OV-10 Mohawk aircraft, and on a Bell UH-1N Huey helicopter. An evaluation set was later supplied to Fairchild to demonstrate the A-10's ability to operate as an all-weather battlefield bomber. Subsequent A-10 trials have explored the potential for single-seat operation at night and in bad weather, using this radar and the standard A10 equipment.

Volume: 3.4 ft³ (0.1 m³)
Weight: (nose-mounted) 145 lb (65.9 kg)
(in pod) 290 lb (131.8 kg)
Installation: external pod is alternative to nose-mounting

STATUS: private venture research programme.

URR ultra-reliable radar technology

A major factor in the day-to-day availability of front-line combat aircraft is the current unreliability of radar equipment. Unreliability is a 'force divider', reducing the effectiveness of a given fleet of aircraft, and pushing up maintenance costs. To combat this inherent weakness, the Avionics Laboratory of the US Air Force Aeronautical Systems Division in February 1985 awarded Westinghouse a $28.6 million contract to develop what is known as URR (Ultra-Reliable Radar) technology that can be used to upgrade aircraft such as the Rockwell B-1B, McDonnell Douglas F-15 and General Dynamics F-16. The technology can also be incorporated into the radar for the Air Force's ATF advanced tactical fighter.

By contrast with the 40 to 50 hours between failures of current radars, URR technology will permit 400 to 500 flying hours before unscheduled removal of a box. Radar elements to be tested will include an active-element phased-array antenna incorporating some 2000 individual power-handling modules. Up to 5 per cent of these modules can cease functioning without seriously degrading the system; more 'graceful' degradation will permit greater flexibility in scheduling removals, and better mission effectiveness. To meet reliability criteria established by the US Air Force for the rest of the system, very high-speed integrated circuit (vhsic) technology will be employed. The principal purpose of vhsic in this application is to permit a 50 or 100-fold increase in data throughput, at the same time conferring a 10-fold reduction in size, weight, power required, failure-rate and life-cycle cost. A major contribution to reliability will be the much smaller number of interconnections needed.

The URR system will have an interface with other aircraft systems in accordance with the specification for advanced avionics architecture being developed by the Avionics Laboratory in its Pave Pillar programme.

STATUS: US Air Force will begin testing URR radar at Westinghouse's Baltimore facility in April 1988. Current development contract runs to January 1990.

AN/APY-2 surveillance radar for E-3A

The principal sensor for the Boeing E-3A AWACS airborne warning and control system is the Westinghouse AN/APY-2 developed specifically as a surveillance and early warning radar. The scanner, which has a range of several hundred miles from the Sentry's operating altitude, is mounted back-to-back with a complementary identification friend or foe, secondary surveillance radar antenna and both are contained in a radome carried over the rear fuselage. The system works at X band frequencies (10 cm) and has seven modes of operation. These modes are:

Pulse-Doppler non-elevation scan (PDNES): enabling aircraft to be detected and attacked down to ground level, though their height is not measured.

Pulse-Doppler elevation scan (PDES): in which target elevation is measured by electronic scanning of the beam in the vertical plane.

Beyond-the-horizon (BTH): using pulse radar without Doppler for extended range surveillance where ground clutter is in the horizon shadow.

Passive: in which the radar transmitter can be shut down in selected sectors of the scan while receivers continue to process electronic countermeasures information coming from that direction. A single strobe line passing through the position of each jamming source is generated on the display console.

Maritime surveillance: involving the use of very short pulses to reduce the effect of sea clutter and enhance the detection capability of moving or stationary surface vessels.

Test maintenance: in which control of the system is delegated to the radar technician for maintenance.

Standby: in which the radar is kept in an operational condition ready for immediate use, but the receivers are shut down.

Total weight of the radar system is 8250 lb (3742 kg).

Full-scale development on the E-3A did not begin until the radar was proved to be capable of conducting the operational mission. Accordingly, in July 1970, Boeing received a US Air Force contract to develop and flight test two competing radar designs, one built by Hughes and the other by Westinghouse. Flight tests on both systems were conducted in 1972, as a result of which the Westinghouse radar was selected by Boeing with the approval of the US Air Force. In January 1973 the US Air Force authorised Boeing to proceed with full-scale development for a fleet of 34 aircraft. The first AWACS was delivered to the US Air Force in March 1977, and the 552 AWACS Wing at Tinker Air Force Base became operational in 1978.

The type began service with Norad (North American Air Defence Command) in January 1979. Deployment in Europe, the Far East and Alaska was completed to test the interoperability

Boeing E-3A Sentry with rotordome housing Westinghouse AN/APY-2 surveillance scanner

of the system with existing air defence installations.

In December 1978 the NATO Defence Planning Council, made up of defence ministers of 13 member nations, gave approval for the purchase of a fleet of 18 AWACS for the NATO airborne early warning requirement, at a reported cost of $3487 million. An international team was established with Boeing as leader and Dornier in charge of the mission-systems installation and checkout. ESG, also in Germany provided 30 engineers at Boeing's Seattle plant to assist in developng the software. AEG Telefunken worked with Westinghouse on radar production, while SEL with IBM provided the data-processors. Displays were built by Siemens with Hazeltine. Substantial improvements over the US Air Force System permit the tracking of ships, while the IBM CC-2 computer provides twice the speed and four times the memory of the earlier CC-1 system.

The first NATO aircraft was delivered in January 1982 and the last in April 1985.

US Air Force Sentries have been deployed on a number of occasions in trouble spots throughout the world, including the monitoring of the Iran/Iraq war. The sale of five E-3As to Saudi Arabia was approved in 1981, and delivery of these will begin in 1986, though they will not have the JTIDS communication system or the latest-standard electronic counter-countermeasures. Discussions concerning possible E-3A sales have also been conducted with several countries including Japan, France and China.

STATUS: in production and service.

Balloon-borne radar programmes

Under contract to the US Defense Advanced Research Projects Agency (DARPA), Westinghouse has provided radars for use with advanced aerostat or balloon platforms. In the mid-1960s DARPA initiated an R & D programme to develop a stable balloon platform and to exploit the use of such platforms for sensor and communications applications. Early work was carried out using modified British barrage balloons and these were later superseded by purpose-designed aerostats, of various sizes. Typical dimensions range between 8.5 and 35 metres in length; and 9 and 30 metres in diameter.

This family of tethered balloons is stated to be capable of operations at altitudes of 3000 and 4500 metres. A powered tether system and low helium leakage allow the aerostat to operate at altitude for extended periods, providing greatly increased radar surveillance or communications coverage due to the increased line-of-sight ranges obtainable. Typical radar applications proposed include surveillance and AEW, location, and tracking of targets, fire control and weapons direction. Communications, EW, and target location/designation by other means, eg electro-optical sensors, are other possibilities.

Radars employed at first were all of the non-coherent variety but Westinghouse later received a contract from the US Air Force Rome Air Development Center to build a coherent system to detect airborne moving targets.

Airship Industries has signed an agreement with Westinghouse to co-operate in plans for a battle surveillance airship system for the US Navy.

The following brief details of this programme refer to the Westinghouse balloon-borne radar projects which have been identified.

LASS low altitude surveillance system

This is a static tethered balloon-borne system which is able to carry a radar such as the AN/TPS-63. It can carry 1800 kg to a height of several thousand metres for periods of up to 20 to 30 days. Westinghouse has received a $12.6 million contract from Saudi Arabia for a demonstration of a LASS system to supplement the E-3A airborne early warning aircraft. The system consists of an AN/TPS-63 radar installed in a 10 340 cubic metre aerostat, having a nominal altitude of approximately 3000 metres. The demonstration and trials will take place in Saudi Arabia in 1986.

STARS small tethered aerostat relocatable systems

A tactical, transportable system built by TCOM Corporation, a subsidiary of Westinghouse. Normal use is for coastal surveillance and at 750 metres altitude a small balloon-borne radar can detect a target with a radar cross-section of 100 square metres at ranges of up to 110 km. TCOM has linked with LMT Radio Professionelle (a Thomson-CSF subsidiary) to offer the RASIT battlefield surveillance radar for tactical use.

Electro-optics

DENMARK

Jørgen Andersen Ingeniørfirma

Jørgen Andersen Ingeniørfirma a-s, 1 Produktionsvej, DK-2600 Glostrup, Copenhagen
TELEPHONE: (1) 475 939
TELEX: 848906

SIT/ISIT low-light level television system

Jørgen Andersen's SIT/ISIT system is essentially a man-portable low-light level television system designed for use as an electronic news gathering equipment by broadcasting organisations. It has, however, been applied to a number of airborne military, paramilitary and civil roles, particularly aboard helicopters. These include reconnaissance, fisheries patrol and inspection, police night surveillance activities, mountain rescue and night electronic news gathering. In such applications, the camera may be shoulder-mounted for use from a helicopter cabin footstep or used with a Ronford 15S mounting head combined with a helicopter shock mount to counteract the effects of rotor-induced vibration.

The system operates over a wide range of lighting conditions down to moonlight and starlight illumination levels. For helicopter work the system is supplied with a 4.5-inch (114 mm) viewfinder and a double pan bar for use with the 15S head. A full range of accessories is available and special equipment for outside broadcast applications includes a video transmitter-receiver operating in the 2 GHz band, a real-time video image processing system and a man-portable video tape recording unit. The company also supplies a range of motorised zoom lens assemblies which meet the special optical requirements of low-light level television systems. With these lenses, says the manufacturer, low-light level television systems can operate over a range of high or low ambient lighting conditions.

Weight: variable, according to lens system fitted, but typically totalling approximately 26.5 lb (12 kg) including camera, lens system and nickel cadmium belt-mounted battery power supply.

STATUS: in production and service.

Type 771 low-light level television system

Developed originally for naval applications in conjunction with director radar weapon control systems, the Jørgen Andersen Type 771 camera is a ruggedised low-light level television sensor particularly suited to harsh operational environments. It operates over an ambient lighting range from sunlight to starlight at extremes of temperature and is well-adapted to aircraft applications. Possible applications include target identification, weapon-aiming, and use with other sensors such as laser rangefinders. Of primary importance to aerospace users is the Type 771's reliability, which includes electronic circuit-protection measures. A mean time between failures of over 8000 hours is claimed for this system.

Dimensions: 29.5 × 13.75 × 12.9 inches (750 × 350 × 330 mm)

STATUS: in production and service.

FRANCE

Cilas

Compagnie Industrielle des Lasers, Route de Nozay, 91460 Marcoussis
TELEPHONE: (6) 449 12 95
TELEX: 692415

TCV 115 laser rangefinder

This is a low repetition-rate, short-wavelength laser system, suitable for use in ground vehicles as well as helicopters, and using an avalanche-photodiode detector unit. The airborne version is integrated with the gyrostabilised APX M334-04 helicopter sight. In addition to the basic rangefinder, the system also has an optical assembly which matches the rangefinder and sight, so ensuring that the cross hairs are projected into the eyepiece of the sight. Operator eye-protection is ensured by an optical filter that provides 70 dB attenuation at the operating wavelength.

Dimensions: 390 × 190 × 180 mm
Weight: 8 kg with optical assembly
Operating wavelength: 1.06 microns
Measurement range: 450-19 900 m

Accuracy: ±10 m
Pulse duration: 25 × 10 ns
Energy: 100 mJ
Peak power: 4 MW
Repetition rate: 1 measurement every 2s for 3 successive shots
12 measurements/minute
Power: (rangefinder) 19-28 V dc
(graticule projector) 115 V 400 Hz

STATUS: in production.

Electronique Aérospatiale

Electronique Aérospatiale, BP 51, Le Bourget 93350
TELEPHONE: (1) 48 62 51 51
TELEX: 220809

ATAL television surveillance system

Electronique Aérospatiale's ATAL (appareilage de télévision sur aeronef léger) is a simplified airborne surveillance system designed for civil and paramilitary applications on all classes of light aircraft and helicopters. Typical applications include crowd control, traffic survey and surveillance. It is comprised principally of commercially available components and subsystems contained in a pod which is mounted externally on the aircraft.

The pod contains a television camera and an FM transmitter which relays imagery to a ground station over visual ranges. The

Electronique Aérospatiale ATAL system

transmitter, which has a power output of 15 watts, operates in the uhf band over the frequencies 1350 to 1540 MHz and 2000 to

2400 MHz. A monochrome television camera is standard but a colour system is an available option. A video monitor is installed in the aircraft and tilt, pan and focus control are exercised through a unit strapped to the operator's leg. The ground station is equipped with a monitor for real-time operation, a video cassette recorder and a steerable antenna system for tracking the aircraft. The FM data link is also used to transmit the operator's commentary to the ground control station.

The system has been in service on fixed and rotary wing aircraft operated mainly by French and foreign police forces. It has also been fitted aboard the Airship Industries Skyship 500 during demonstrations.

Dimensions: (pod) 400 mm dia × 1120 mm length
Weight: 14 kg

STATUS: in production and service.

Omera

Société d'Optique, de Mécanique, d'Électricité et de Radio Omera-Segid, 49, rue Ferdinand Berthoud, BP 68, 95101 Argenteuil
TELEPHONE: (3) 39 47 09 42
TELEX: 696797

Irold reconnaissance camera

Irold is a long-range slant reconnaissance camera system carried in an under fuselage pod. A video unit relays the camera's picture to the pilot's radar screen and a symbology generator superimposes some flight data on the picture. The photograph can be taken with either left or right slant, there being a 90° mirror within the optical train. The slant angle can be varied between 65° and 86° from the vertical, giving a resolution of better than 0.3 metres at 65° down to 1.5 metres at 86° when the aircraft is at 40,000 feet altitude.

Dimensions: 0.3 m long × 0.53 m diameter
Weight: 340 kg

Image format: 114 mm × 111 mm
Film capacity: up to 1200 frames (according to film type)
Filming rate: up to 0.6 s/frame
Lens focal length: 1700 mm
Field of view: 3.84° longitudinal × 3.74° lateral
Power requirements: 417 W at 28 V dc, 90 W at 115 V, 400 Hz single-phase
Resolution: 1m at 100km

STATUS: in production and service.

Sopelem

Société d'Optique, Précision, Électronique et
Mécanique, 102 rue du Chaptal, BP 223, 92306
Levallois-Perret Cedex
TELEPHONE: (1) 47 57 31 05
TELEX: 620111

CN₂H night-vision binoculars

The CN₂H (conduit nuit, second generation,
helicopters) night-vision binoculars are part of
a range of night-vision systems produced by
Sopelem for military applications; the binoculars are particularly designed for use in helicopters and fixed-wing aircraft for night observation, instrument or map reading. They are
fixed to the helmet, with a power-pack on the

back, and a specially-designed support bracket
enables them to be immediately discarded in an
emergency: focussing and positional adjustments are available to suit the wearer. The
goggles are compatible with US and NATO
helmets and they incorporate second or third
generation light intensifiers according to requirements.

Weight: 0.93 kg including battery pack
Operating life: 10 or 20 hours according to
battery type and light levels
Field of view: 40°
Magnification: × I

STATUS: in production. Adopted by the French
Army, under the designation OB56, and exported.

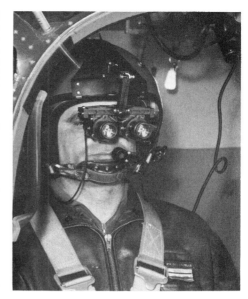

Sopelem CN₂H night-vision binoculars

Thomson-CSF

Thomson-CSF, Division Equipements Avioniques, 178 boulevard Gabriel-Péri, 92240 Malakoff
TELEPHONE: (1) 46 65 44 22
TELEX: 204780

Atlis laser designator/ranger pod

This is a high-precision air-to-ground designation system which comprises television and
laser sensors mounted in a single pod. It is
suitable for single-seat aircraft, and has been
demonstrated on Sepecat Jaguar, Dassault
Mirage 2000 and General Dynamics F-16
fighters. Development of the system was
undertaken jointly by Thomson-CSF, which
developed the monitoring/control system and
displays, and Martin Marietta Orlando Aerospace, which developed the sensor pod, using a
laser illuminator by the French company Cilas.
It is designed for low-level, high-speed attack
by single-seat aircraft operating autonomously
(that is, without guidance or control from
ground stations or other aircraft).

Initial flight tests took place in France during
late 1976 and early 1977, the equipment being
installed in a single-seat Jaguar. These tests
assessed the pod's ability to mark targets for
laser-guided missile attack at ranges of up to
10 km. The Mk 1 Atlis pod used in these trials
has been superseded by the shorter and lighter
Atlis 2 pod. It is more suitable for use on Jaguar
and F-16 fighters and flight-testing was successfully completed in late 1980, using laser-guided
bombs and AS.30 laser missiles.

The television sensor provides a × 20 magnified picture to the pilot and tracks targets
using area-correlation or contrast techniques,
performed at long-ranges and aided slightly as
a high-contrast scene is produced by the
particular electro-optic spectrum used. The
laser shares a common optical port with the
television sensor and can be directed either for
designation or range information. Sightline

Atlis pod installed on Jaguar

prediction capability built into the system
allows dead-reckoning tracking of targets
obscured temporarily by high-ground or
patches of cloud.

The most common application of the system
is as a designator for autonomous operation by
a single-seat aircraft using laser-guided bombs,
such as the Texas Instruments Paveway and
the Matra Arcole, or the Aérospatiale AS.30L
missile.

The Atlis 2 pod equips French Air Force
Jaguars and has been ordered for export to
equip Mirage 2000 and F-16 aircraft. Qualification for use on the F-16 was achieved late in
1985 following a six-month flight trials programme in the USA. The qualification was in
relation to the order for F-16s placed by
Pakistan.

Atlis 2
Pod length: 2.52 m
Max dia: 0.305 m
Weight: 170 kg
Operating wavelength: 1.06 microns
Max operating range: over 10 km
Attachments: standard 760 mm NATO bomb
rack
Angle of regard
(roll) unlimited
(pitch) –160° to +15°
Power output: 2.3 kW 115 V 400 Hz

STATUS: in production.

TAV 38 laser rangefinder

This system has been installed in the French
export versions of Sepecat Jaguar and Dassault
Mirage F1 aircraft.

The system comprises two units, the power
supply and the laser head. The former is largely
a power-supply transformer and distribution
centre, providing auxiliary supplies and a high
energy direct current voltage output for the
flash exciter. The laser head, developed by
Cilas, houses the laser cavity, which comprises
a neodymium rod, flash exciter and triggering
device, plus receiver elements such as the
avalanche photodiode, digital range-counter
and optics. The range counter uses a clock
frequency of 29.98 MHz, corresponding to
distance increments of 5 metres. The optics
steer the laser beam using a beam-deflection

Jaguar with Atlis pod, releasing Aérospatiale AS.30 laser-guided weapon

Thomson-CSF TAV 38 laser ranger

Thomson-CSF TMV 630 laser ranger

system developed by Marcoussis Laboratories' CGE research centre.

Dimensions: (laser head) 415 × 120–150 mm dia (power supply) ¾ ATR Short
Weight: (laser head) 7.5 kg
(power supply) 13 kg
Operating wavelength: 1.06 microns
Max range: 10 km in clear air
Accuracy: ± 27 m (fixed error ± 22 m)
Angular resolution: better than 1 milliradian
Angular movement: ±10° in elevation and azimuth
Pulse rate: 12 shots a minute; 1 s between shots
Operating temperature: –30 to +70°C
Power: 200 VA at 115 V 400 Hz 3-phase, plus 1A at 28 V dc

STATUS: in production.

TMV 630 airborne rangefinder

This equipment has been designed to meet single-unit, small installation requirements and can be easily fitted to a wide range of aircraft. It provides high-precision aircraft-to-target range measurement and is claimed to increase considerably the performance of conventional weapon-aiming systems. The large field-of-view provided is compatible with all head-up displays and the electrical interfaces are compatible with almost all aircraft types. The high speed and accurate laser beam steering is specifically adapted for continuously computed impact point attacks irrespective of terrain or the nature of the weapons, or aircraft altitude.

Dimensions: 190 × 190 × 520 mm
Weight: 15 kg
Operating wavelength: 1.06 microns
Max range: 10 n miles (19 km)
Accuracy: better than 1 milliradian
Power: 12 A at 28 V dc

Thomson-CSF Raphael TH pod fitted to Dassault Mirage

STATUS: in production as part of nav/attack system on Dassault Mirage and Dassault-Breguet/Dornier Alpha Jet aircraft.

Raphael reconnaissance pod

Thomson-CSF is producing the Raphael TH system for long-range stand-off reconnaissance. In series production, it equips the Dassault Mirage F1 and is being considered for the Mirage 2000.

It is composed of a pod-mounted radar which combines pulse compression and synthetic aperture processing. Radar information is transmitted to a ground station, several hundreds of kilometres away via a high-speed data link. The radar image is presented in real-time on display systems housed in the ground stations and also recorded in flight and on the ground.

TRT

Télécommunications Radioélectriques et Téléphoniques (TRT), Defence and Avionics Commercial Division, 88 rue Brillat Savarin, 75460 Cedex 13, Paris
TELEPHONE: (1) 45 81 11 12
TELEX: 250838

TRT has been concerned with thermal imaging technology since 1964 and has worked under a number of French Government contracts in developing a range of experimental equipment designed to check the validity of technical approaches, and to explore their operational possibilities. A number of operational systems have been produced under this development programme. The aim of further work, in collaboration with Société Anonyme de Télécommunications (SAT), has been to develop a range of common modules for the French Ministry of Defence.

Hector thermal imager

TRT's Hector system is a thermal imager designed for Aérospatiale's SA 360 Dauphin helicopter for night and poor-visibility firing of the Franco/German HOT anti-tank missile, operating in the 8 to 13 micron band. The sensor has an array of fewer than 60 detector elements and a series-parallel scanning method is employed, providing an output suitable for television displays. In the Dauphin application, information is presented on a micro-monitor of the missile operator's day-firing eyepiece. The Hector camera has two fields of view and is integrated into a gyro-stabilised platform.

STATUS: in production and service.

Modular thermal imaging system

Under a common modules development programme for the French Defence Ministry, TRT, in collaboration with SAT, has developed a

TRT Hector thermal imaging camera installation on Aérospatiale SA 360 helicopter

Elements of TRT common modules

TRT Tango Flir system mounted on nose of Dassault ATL2

range of thermal imaging modules to meet the requirements of the three armed services. Under an inter-company agreement, each is responsible for the manufacture of specific modules and either one may be the prime contractor, depending on particular project requirements and current workload, for any system application. Design co-ordination authority, however, is vested in TRT.

Initiation of the common modules programme followed a study, conducted by TRT and SAT on behalf of the French Ministry of Armed Forces, to define a thermal imaging system to meet national specifications and to establish the technical design. The resulting parameters were 8 to 12 micron spectral bandwidth with series-parallel scanning; full television display compatibility on a standard CCIR 625-line cathode ray tube; a single, double or triple field of view optical system requiring a minimum number of lenses; a limited number of detector elements; possibility of application to light-emitting diode displays; ease of coupling with image-processing systems; open or closed-circuit cryogenic cooling; highly stabilised optic axis; and reliable behaviour in the presence of heat sources such as missile traces or fires.

The design included a number of requirements which are similar to those of the common modules programmes of other nations. These include easy adaptation to a variety of applications; an increase in the number of systems produced due to standardisation of the camera (that is, sensor) units; easy incorporation of new developments; and facilitation of fault-finding and qualification testing of camera

systems through the use of already approved modules. Additional advantages from the programme include a reduction in study, development and production costs for a given application, simplification of maintenance, enhanced reliability and reductions in production time-scales and training.

The basic module range comprises a scanner module, designed to explore the field of view at the appropriate frame and line frequencies in order to feed the detectors with a suitable signal, using series-parallel scanning techniques; a detector module comprising a cadmium mercury telluride, two-dimensional mosaic detector cooled to 77 K and associated pre-amplifiers; a linear electronics module designed to amplify, rephase for each line, and shape the signals from the various detector module channels; a signal processing module for producing a television-compatible, standard CCIR 625-line video output from simultaneous signals produced by the linear electronics unit; display modules for direct viewing of the field of view by the operator through a light-emitting diode viewing lens, or for indirect viewing on a remote cathode ray tube monitor screen, depending on the application; a cryogenic module designed to cool the detector array; and a test and diagnosis module which runs internal self-tests for fault-detection and isolation.

The modules may be assembled in various configurations depending on particular requirements. They fulfil the basic functions of a thermal imager while additional elements tailor the unit to particular applications; for example, a larger or smaller field of view would require a

new optical train, and the television monitor display would be added when remote viewing of the image is desired.

Typical performance and characteristics quoted for a long-range system are diagonal fields of view of 2.6° and 7.7° with a detection range of 27 km against a small maritime target (such as fast patrol boat), and a reconnaissance range of 18 km against a land target such as a tank. For missile-guidance a system would offer fields of view of 3.6° and 10.6° with detection and reconnaissance ranges of 5 km and 3.5 km respectively, against land targets such as tanks.

TRT modular systems have been applied to both fixed-wing aircraft and helicopters for a number of surveillance roles including target detection, identification and designation, and for reconnaissance and navigational purposes.

Current aeronautical applications for these common modules include the Tango thermal camera for the Dassault ATL2 maritime reconnaissance aircraft for the French Aéronavale. In this application the system is used as a maritime surveillance aid with two fields of view and is integrated into a gimballed platform. An experimental system in which the thermal camera is also in use is being tested at the Centre d'Essais en Vol, at Bretigny. Here, in conjunction with a head-up display, the equipment is used for military aircraft developmental work as a pilot's aid. This system has two fields of view, 20° and 10°.

Tango specification
Turret weight: 85 kg
Diameter: 600 mm
Field of view: azimuth ±110°, elevation +15° to –60°
Rate of rotation: up to 60°/s

STATUS: in production, service and further development.

GERMANY, FEDERAL REPUBLIC

Eltro

Eltro GmbH, Postfach 102120, D-6900 Heidelberg 1
TELEPHONE: (6221) 705 214
TELEX: 461811

CE 626 laser ranger

Eltro has developed a laser transmitter for rangefinding in a wide variety of ships, aircraft, helicopters, tracked- and wheeled-vehicle applications.

It is a nitrogen-cooled neodymium-YAG laser (1.06 micron wavelength) which normally

operates at 9 to 11 Hz pulse-repetition frequency, but with a 40 Hz maximum capability. Initial application is in German-operated Panavia Tornado aircraft.

STATUS: under development.

MBB

Messerschmitt-Bölkow-Blohm GmbH, Postfach 80 11 60, 8000 Munich 80
TELEPHONE: (89) 6000 0
TELEX: 5287975

Night vision systems for helicopters

Using a BO 105 helicopter as a test vehicle, MBB has run, since 1981, a flight-test programme aimed at the definition of advanced cockpits, visual aids and navigation systems which will enable future helicopters to operate at night and in adverse weather conditions beyond present limitations. MBB is conducting flight tests with (mainly) prototype systems of various national and international equipment manufacturers.

One example of such a system is the Ophelia mast-mounted sight designed for observation in the following helicopter roles: search and rescue, surveillance (eg police missions, border patrol), military reconnaissance/combat. The system provides an unobstructed 360° view without extensive structural modifications to the fuselage (minimal modifications are required to the BO 105 rotor head) and creates no centre of gravity problems. Installation of a sensor package with a line-of-sight approximately 110 cm above the rotor plane allows the helicopter to maintain observation whilst under maximum cover; a valuable advantage in anti-terrorist or military engagements. Inclusion of a thermal imager in the payload ensures extended operations at night and during inclement weather conditions. Sensor images (TV and infra-red) are displayed on either a head-up or head-down display installed in the cockpit. Images are superimposed with symbology created by a computer symbol generator. The sensor platform, Ophelia, is produced by the French company Sfim.

Trials with nose-mounted sensor systems have also been carried out, including the PVS (pilot's vision system), a dual-sensor, visually coupled system where the sensor line-of-sight follows exactly the pilot's head movements.

MBB BO 105 helicopter with PISA system

Images of the outside world are relayed to the pilot from electro-optical sensors mounted on a platform installed in the nose of the BO 105. They are displayed on a miniature crt fitted to the pilot's helmet. These sensor images are superimposed with computer-generated symbology allowing the pilot to fly the aircraft without recourse to normal flight instrumentation. The system comprises a stabilised platform (by Sfim) and two electro-optical sensors, a Flir and a low-level light TV camera. The pilot is free to choose images from either sensor, depending on the flying conditions.

Another nose-mounted system flight tested on the BO 105 is the PISA (pilot's infra-red sighting ability) system developed by the MBB Dynamics Division. PISA is a wide-angle sensor providing night-vision capability for orientation and observation purposes. The system consists of an infra-red thermal imager with Sprite detectors, a single-axis steering mechanism (azimuth), control panel and symbol generator. It provides an extremely large field of view of 30° × 60° (elevation × azimuth), high-image resolution, homogeneous over the total field of view, low infra-red signature on account of specially integrated cooling and can be operated either fully autonomously or integrated in a gunner's sighting system.

In addition to these aircraft-mounted systems, MBB has also carried out tests with night-vision goggles, an obstacle warning system (radar), various types of displays and symbology, a combined Doppler/strapdown navigation system and control and display units.

STATUS: under development.

ISRAEL

Tadiran

Tadiran Limited, 11 Ben Gurion Street, Givat Shmuel, PO Box 648, Tel Aviv 61006
TELEPHONE: (3) 713 111
TELEX: 341692

Reconnaissance payload for Mastiff RPV

Tadiran has developed a reconnaissance payload for the Mastiff Mk 3 remotely piloted vehicle. Intended for day-time target detection, identification and range-finding, the payload includes a stabilised television camera which can sweep 360° in azimuth and from +5° to –88°

Tadiran Mastiff Mk 3 mini-RPV

Tadiran RPV reconnaissance payload

in elevation. Typically an armoured fighting vehicle can be identified at a 4 km range and detected at 5 km. The picture is transmitted to a ground station, from which the RPV also receives its commands. An Ashai television camera is employed with a 525-line resolution and a focal length of from 11 to 110 mm (22 to 220 mm with a beam expander). This corresponds to a field of view of from 34° × 48° (wide) to 3.4° × 4.8° (narrow) or 1.7° × 2.3° with the beam-expander. The full zoom range can be traversed in 7 seconds.

Mastiff was one of the two Israeli RPVs used to provide intelligence during the Lebanon conflict of 1982.

Dimensions: 440 mm long × 330 mm dia
Weight: 16 kg
Power: 300 W at 28 V dc

STATUS: in production and service.

ITALY

OMI

OMI-Ottico Meccanica Italiana SpA, via della Vasca Navale 79, Rome
TELEPHONE: (6) 547 881
TELEX: 610137

CIRTEVS infra-red television system

OMI's CIRTEVS (compact infra-red television system) is a recently-developed, low-cost system which operates in the 8 to 14 micron band and requires no cooling system or rotating or scanning mirrors. The system requires an 80-watt primary power supply which may be either ac or dc. The system is CCIR television compatible, is available in various 'packages' according to role requirements, and is supplied with a choice of optical systems giving focal lengths of between 50 and 500 mm.

STATUS: in production.

SWEDEN

Ericsson

Ericsson Radio Systems, Defence and Space Systems, PO Box 1001, S-43126 Mölndal
TELEPHONE: (31) 671 000
TELEX: 20905

Airborne laser ranger

The Ericsson airborne laser ranger is a compact neodymium-YAG unit which can be integrated into sighting and weapon delivery systems, where it provides higher accuracy. By giving continuous range data it also enhances safety in dive attacks.

The unit is of modular construction, consisting of transmitter, receiver, range-counter and deflection unit. Laser aiming can be slaved to the aircraft sighting system via the deflection unit. There are also built-in safety features to improve personnel safety during operation and maintenance. Typical operating range is 5 nautical miles (9 km) at an optical visibility better than 10 nautical miles (18 km). Range is measured to an accuracy of 16 feet (5 metres).

Weight: 14 kg
Operating wavelength: 1.06 microns
Pulse repetition frequency: 1–10 Hz

Ericsson airborne laser ranger with training filter

Field of view: 0.5 mrad
Pulse length: approx 15 ns
Output: ARINC 429
Power: 280 W 28 V dc

STATUS: in production.

UNITED KINGDOM

British Aerospace

British Aerospace, Air Weapons Division, Manor Road, Hatfield, Hertfordshire AL10 9LL
TELEPHONE: (07072) 62300
TELEX: 22324

Linescan 214 infra-red reconnaissance/surveillance system

The 214 Linescan is a member of British Aerospace's range of 200 Series equipment, which were initially developed for the UK Government for military application and several hundred 201 systems have been delivered and are in service in a number of countries. The 201 system is, for example, the standard equipment aboard the Canadair CL-89 surveillance remotely piloted vehicle. Another member of this series is the Linescan 212 system, designed for use in light aircraft and helicopters. Likewise, a number of these systems are also in service throughout the world.

The 201 and 212 systems record their imagery on standard 70 mm photographic film, but the 214 Linescan has been developed to meet a requirement for real-time imaging, although facilities for recording the output have been incorporated.

The 214 system operates in the 8 to 14 micron band. It has an instantaneous field of view of 1.5 mrad with an across-track scan of 120°. Thermal resolution (rms noise equivalent temperature) is 0.25° C at 22° C. The system is cooled by liquid nitrogen and will operate up to 5 hours from a single reservoir charge. It incorporates an integral film recorder, using standard aerial photographic film, and the 6-metre capacity cassette provides a typical coverage of 65 km by 1 km at 300 metres altitude (40 miles by 0.6 miles at 1000 feet). Velocity/height range is within the band 0.1 to 1 radian per second depending on the recording method in use. Selection of velocity/height ratio is made at the cockpit control unit in a series of eight steps. The unit also contains an automatic gain-control switch which provides overland/oversea selection in four steps. A video inversion control is also incorporated

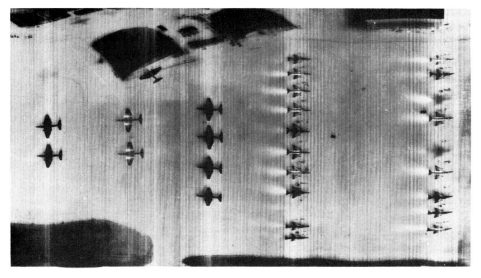

Combat airfield pictures by British Aerospace Linescan system. White areas denote heat; note hot engine casings on two British Aerospace (BAC) Canberra light bombers

British Aerospace Linescan 214

British Aerospace Linescan 204

allowing hot targets to be displayed in either black or white polarity.

The mixed video and synchronisation pulse output is compatible with a wide range of recording and data transmission systems and equipment. Real-time oscillographic recorders suitable for use with Linescan 214 are available from a number of manufacturers, including Honeywell and Medelec, and there is a choice of recording paper and image development processes according to the particular application. Tape-recording systems for recording and playback, either in the air or on the ground, are also available from manufacturers such as Racal or Sony. Suitable data transmission equipment is manufactured by Meteor and Pacific Aero Systems of Italy and the United States respectively.

Dimensions: (linescan sensor) 10.16 × 17.2 × 12.5 inches (258 × 437 × 318 mm)
(cooling pack) 10.75 × 5.5 × 4.125 inches (273 × 140 × 105 mm)
(control unit) 4.4 × 5.75 × 4.4 inches (113 x 146 × 113 mm)
Weight: (linescan sensor) 26.4 lb (12 kg)
(cooling pack) 3.74 lb (1.7 kg)
(control unit) 2.42 lb (1.1 kg)

STATUS: in service.

Linescan 401 infra-red reconnaissance/surveillance system

The British Aerospace 401 infra-red linescan system is designed for low-level, high-speed air reconnaissance. The system operates in the 8 to 14 micron band and employs a five-element mercury-cadmium-telluride detector array which is cooled to liquid-air temperature by a Joule-Thompson high-pressure air minicooler. Typical endurance times from a standard air cylinder are 120 minutes at 20° C and 45 minutes at 70° C soak temperatures. Detector cool-down time is less than 1 minute.

Imagery is recorded on standard 70 mm photographic film which is contained in a large-capacity, quick-change magazine. Automatic velocity/height ratio control is provided and the flight navigation data is recorded on the film edge, reducing both aircrew and photo interpreter workloads.

British Aerospace Linescan 401

British Aerospace Linescan 401 day (top) and night pictures of yacht marina and oil storage tanks

The transverse field of view covers 120° and can be offset left or right to provide horizon-to-horizon coverage. Roll stabilisation to ±55° is provided and allows high angles of aircraft bank without loss of continuity of the scanned area or the resulting imagery.

Cockpit control items include an event-marker and a film-footage indicator. Built-in self test circuitry activates a cockpit warning lamp in the event of equipment unserviceability.

Details of spatial resolution, thermal sensitivity and velocity/height ratio are classified. The equipment is said to be compatible with a number of high performance strike aircraft.

Dimensions: (linescan) 23.75 × 12.6 × 11 inches (604 × 320 × 280 mm)
(roll swept dia) 14.4 inches (366 mm)
(cooling pack) 15.3 × 9 × 4.75 inches (390 × 230 × 120 mm)
Weight: (linescan) 74.8 lb (34 kg)
(cooling pack) 19.8 lb (9 kg)

STATUS: in production and service. Seven countries have Linescan 401s in service and it is fitted to the Royal Air Force's Sepecat Jaguar reconnaissance aircraft. Over 400 Linescan systems, of all variants, have been sold to a total of 20 countries. Total sales value exceeds £40 million.

Rigel infra-red linescan/television reconnaissance system

The Rigel system comprises a British Aerospace Linescan 214 sensor, a low-light level television camera, and the Italian Meteor air-to-ground data link. This assembly is normally pod-contained for either underwing or fuselage centre-line mounting but may be installed inside the fuselage if space permits. The system's low-light level television camera is mounted in the nose of the pod and has a motorised zoom/focus facility for ground control. The Linescan 214 infra-red sensor is centrally mounted, together with its cooling-pack and control unit, and views the ground

through an open window in the underside of the pod with an unobstructed arc of 120° across track. All Linescan functions are controllable from the ground, including on/off selection of photographic-film recording in addition to real-time air-to-ground imagery transmission.

Also contained within the pod are electronic units including the command receiver/decoder and power amplifier, the television/telemetry transmitter, altitude and airspeed transducers, antennas for television/telemetry transmission and command signal reception and the electrical supplies. A small battery for ground-checkout is mounted at the rear of the pod.

In normal use, the ground-based operator can select remotely all major functions without reference to the pilot in the aircraft, but facilities to display sensor information can be installed aboard the aircraft if required. During surveillance the ground operator can scan the terrain with the low-light level television camera and select individual targets for the infra-red Linescan unit. Typically, infra-red imagery is available at the ground station within 10 seconds of the target being overflown. The infra-red sensor may also be operated in a continuous-scan mode.

The pod itself has NATO standard (STANAG 3726) attachments at 14-inch (356 mm) centres.

The Rigel ground station consists of three mobile units mounted on military-pattern, four-wheel-drive truck chassis each carrying a 250-volt 50 Hz 250 kVA power supply. One of the trucks contains a fully air-conditioned ground station on which is mounted the main antenna. In the station is a television monitor screen to display the low-light level television camera picture, and a trace recorder for infra-red imagery. The station also contains a map-plotter display which continuously records the position of the aircraft. A control panel is provided whereby the operator can select remotely the various sensor parameters, including high-quality photographic film recording for subsequent processing on the ground. The second truck contains the film processing and interpretation facilities, while the third vehicle accommodates calibration and maintenance equipment for the aircraft pod and the ground station.

The aircraft infra-red sensor specification is the same as that of the standard Linescan 214 equipment (see separate entry). The low-light level television camera can be either a 625-line, 50 Hz or 525-line 60 Hz unit with automatic gain control. Signals from both the thermal imaging and television sections are transmitted to the ground over L band links with a choice of five FM frequencies. The command link likewise is a five-frequency unit but can be either L band or uhf. The airborne vehicle can be tracked at ranges of up to 80 nautical miles.

In the ground station, the aircraft position is displayed on a 30 × 30 inch (760 × 760 mm) plotting board with map presentation. Altitude and airspeed are displayed graphically and numerically, range and azimuth position in numeric form only. Television images are shown on a 12-inch (304 mm) monitor. Infra-red linescan recording is oscillographic with a paper development time of between 5 and 15 seconds. Grey scales are from 6 to 16 and resolution of the system is eight lines a mm. A magnetic tape signal recording system is optionally available.

A simplified variant of Rigel, known as Rigel Sim, is similar to the major system but dispenses with the low-light television element. In the Rigel Sim airborne pod, the low-light television camera and its gimbal mountings are replaced by ballast weights and the altitude and airspeed sensors are also deleted. The Rigel Sim ground station is correspondingly simplified and contains no television monitor or aircraft plotting facilities.

Dimensions: (pod) 57.1 × 15 inches (1450 × 380 mm dia)
(infra-red sensor) 10.16 × 17.2 × 13.5 inches (258 × 437 × 343 mm)
(cooling pack) 10.75 × 5.5 × 4.01 inches (273 × 140 × 105 mm)
Weight: (pod) 132 lb (60 kg)
(infra-red sensor) 26.4 lb (12 kg)
(cooling pack) 4.4 lb (2 kg)

STATUS: in service.

Linescan 4000 Tornado reconnaissance system

Flight trials began in mid-1985 in a Tornado aircraft of the infra-red sensors for a new generation reconnaissance system being developed by British Aerospace and Computing Devices Company for the Royal Air Force.

The Tornado system incorporates a British Aerospace sideways looking infra-red (SLIR) system and Linescan 4000 airborne infra-red surveillance system. Computing Devices is responsible for the signal processing and video recording system. For the first time in such a tactical aircraft, cameras have been completely replaced by an all-video system integrated with an advanced infra-red sensor system. When fully operational the Tornado reconnaissance system will provide a very high quality picture of the battlefield and offers tremendous flexibility.

The RAF requirement for the new system stipulates the need to identify targets close to the horizon while the aircraft is flying fast and low by day or night and in all weather. Mounted internally in the fuselage, the system provides horizon-to-horizon across-track coverage with roll stabilisation, and gives a real-time TV display in the cockpit.

In the Tornado, the output from the infra-red sensors is instantly video tape recorded and the operator in the rear cockpit can also monitor the scene while the sortie is under way, both in real-time, directly from the sensors or in near real-time by replaying video tape recorders. The system offers a high definition thermal picture which can be magnified or enhanced at the touch of a button.

Linescan 4000 system

This small lightweight sensor equipment is for high-speed, low-level tactical aerial reconnaissance with high-performance aircraft. It can be installed in an external pod underwing or underfuselage or in the fuselage of the aircraft. The sensor design is such that, although it has a very wide field of view, it only requires a very small aperture in the pod or fuselage.

Operating in the 8 to 14 micron waveband infra-red region of the electro-magnetic spectrum, the Linescan sensor collects the IR radiation from the terrain being overflown and by means of highly advanced conversion and processing electronics, provides a video output signal. Thus the collection of the infra-red radiation and its instantaneous conversion into a video signal enables the IR imagery of the scene below the aircraft to be displayed in real-time on a TV monitor, recorded on a video tape recorder and data transmitted to a ground receiving station. The ground station would contain similar display and recording equipment as above and could also have a moving roll photographic paper recorder for the production of hard copy.

The Mini-Linescan 4000 equipment consists of two units, the Infra-Red sensor unit and the Infra-Red electronics unit.

The sensor unit is constructed with an aluminium alloy main frame in the form of a platform on top of which are the detector assembly, the cooling engine and the housing for the electronic circuitry for the detector outputs, scanner pick-offs and speed control and the associated BITE (built in test equipment). The IR optical components, that is the scanner with its drive motor, the parabolic mirrors and the ridge mirror are on the underside of the main frame platform.

The electronics unit is a metal box designed to fit into a standard rack in the aircraft. It contains most of the electronic circuitry for the system for video amplification, processing and roll correction cooling control, data-bus and video tape recorder interfaces with appropriate power supplies and BITE.

The sensor unit infra-red optics are similar in principle to that of earlier Linescans, where the scanner with its rotational axis parallel to the direction of flight collects radiation from the ground and transmits it to a pair of parabolic

mirrors which focus via a combining ridge mirror on to the detector.

While earlier models have a square section scanner, Linescan 4000's scanner is of triangular section with three mirror facets. This improved design not only permits a horizon-to-horizon scan coverage but with the advanced technology multi-element detector used, a much smaller infra-red collecting area is adequate. The infra-red collecting area for Linescan 4000 is one-quarter that of earlier Linescans with the consequent reduction in size of the required viewing aperture in the aircraft or pod structure. The Linescan 4000 infra-red optical system, scanner, parabolic and ridge mirrors are of aluminium alloy with diamond machined faces. These high quality optical surfaces are protected by an evaporated coating. The scanner rotates at 12 000 rpm to give 600 scans per second. An optical encoder on the scanner shaft provides continuous reference of scanner position and speed to the processing electronics.

The multi-element cadmium-mercury-telluride detector encapsulation accommodates a 'cold finger' to cool the detector to the required low operating temperature. The cold finger is itself cooled by a split cycle Stirling engine which is part of the sensor unit. The cooling engine is electrically powered and has electronic control circuitry to maintain the precise cryogenic temperature level at the detector.

The Linescan 4000 uses the UK developed Sprite (signal processing in the element) detector technology which can be operated at higher data rates than earlier generation infra-red detectors, with increased performance. A special Split-Stirling Cycle Cryogenic cooling engine, developed by British Aerospace, supplies continuous cryogenic temperature cooling to the detector. The design features of the cooling system have led to a very robust system capable of long life in the operational environment aboard the aircraft.

SLIR system

The SLIR system developed by British Aerospace embodies elements of the UK Thermal Imaging Common Module System and is the only airborne application of this system currently in full scale development for intended production. This system also embodies the Sprite detector and the British Aerospace cooling engine.

This technology and the unique development and integration experience of the UK common modules is available for a number of other applications on UK, US and West German high-performance aircraft.

Dimensions: (sensor unit) 11.25 × 8.25 × 9.75 inches (285 × 212 × 247 mm)
(electronics unit) 12.5 × 11.75 × 7.75 inches (320 × 300 × 200 mm)
Weight: (sensor unit) 21.5 lb (9.7 kg)
(electronics unit) 22 lb (10 kg)
Power requirement: 56 W at 28 V dc, 110 VA at 115 V 1-phase, 90 VA at 200 V 3-phase

STATUS: in development.

Computing Devices

Computing Devices Company Limited, Castleham Road, St Leonards-on-Sea, East Sussex TN38 9NJ
TELEPHONE: (0424) 53481
TELEX: 95568

RMS 1000 reconnaissance management systems

Computing Devices has for many years produced a range of management systems for control of all aspects of airborne reconnaissance equipment. Current systems are microprocessor-based to permit maximum flexibility

with adaptation to a wide range of parameters and formats under software control, with built-in spare processing capacity. This permits complex calculations to be made and provides such services as conversion of latitude and longitude to grid coordinates (national grid over the UK automatically changing to UTM coordinates elsewhere); image motion correction for sensor systems according to aircraft height/velocity ratio; and automatic roll stabilisation for infra-red linescan equipment. Other parameters controlled include heading, speed and camera-framing rate, with control and synchronisation of each forward-, side- and downward-looking camera to allow the

construction of composite stereoscopic pictures of the target area. A feature of the data processing system is the alphanumeric annotation achieved by strobing a simple seven-dot matrix, available to each frame, and providing a record of the complete data package over three consecutive frames.

A system of this type, developed for the Sepecat Jaguar fighter-bomber, has been in Royal Air Force service for some years and versions of the system have been exported to unspecified overseas customers.

STATUS: in production and service.

RMS 3000 reconnaissance management system

The development of a system is in hand for the Royal Air Force's Tornado reconnaissance aircraft, which provides a number of additional facilities compared with the RMS 1000. These include a video recorder to capture high-resolution video from all infra-red sensors with associated synchronisation pulses, digital data and event-markers, and it may be removed from the aircraft for ground replay and analysis. The system provides in-flight replay of required sections of the recorded imagery, can magnify both along-track and across-track directions, and can also freeze-frame and replay at several different rates. It can distribute control and annotation data within the total reconnaissance sub-system.

The incorporation of special purpose cockpit displays or the use of existing television tabulator systems aboard the aircraft permit real-time imagery from all electro-optic sensor types, including LLTV, infra-red and radar, to be put up for the pilot or crew. A finger-position overlay is used as a high integrity interface between the crew and the system for control purposes. Multi-channel recording of all processed sensor imagery is standard for in-flight replay and post-mission ground analysis.

STATUS: in development.

Ferranti

Ferranti Defence Systems Limited, Electro-Optics Department, Robertson Avenue, Edinburgh EH11 1PX
TELEPHONE: (031) 337 2442
TELEX: 72529

Type 221 thermal imaging surveillance system

The Type 221 thermal imaging surveillance system is designed for service with military helicopters. Developed by Ferranti in conjunction with Barr & Stroud Limited, the system incorporates an IR18 thermal imager and telescope by the latter company, with sightline stabilisation steering provided by a Ferranti stabilised mirror. The assembly is contained in a pod which can be mounted either beneath the nose of a helicopter or can project through an aperture in the aircraft's floor.

The Barr & Stroud IR18 imager unit provides a normal field of view of 38° in azimuth and 25.5° in elevation. In the Type 221 application, users can choose a telescope magnification of either × 2.5 or × 9, with corresponding wide or narrow fields of view. The wider field of view (15.2° azimuth by 10.2° elevation) would be used for general surveillance, target acquisition or navigation. The narrow field of view permits detailed observation for target identification or engagement of targets detected in the wide field of view mode. In the narrow field of view, the angles are 4.2° in azimuth and 2.6° in elevation. With the Ferranti sighting mirror in the pod installation the system has fields of regard of +15° to -30° in elevation and ±178° in azimuth. The entire sensor system is vertically mounted above the mirror which is angled, periscope fashion, at 45° to the horizontal, to provide views in the horizontal plane.

The sensor system employs Mullard Sprite detector units cooled by a Joule Thompson minicooler supplied with high-pressure compressed air. The air source is a bottle, mounted on the equipment and charged immediately before flying. This has a capacity of 1 litre and provides a system operation time of approximately 2.5 hours. If greater endurance is required other cooling options, involving the use of mini-compressors permanently connected to the equipment, are available. The system operates in the 8 to 13 micron band and has a sensitivity of between 0.17 and 0.35° C to target background and surroundings.

The Ferranti-produced mirror has an aluminium reflective element which is diamond machined to a flatness of two fringes at 550 nanometres. This mirror has a reflectivity of greater than 97.5 per cent at 45° incidence, averaged over 8 to 12 microns. Its stabilisation system comprises a two-axis device with integrating rate gyros as rate sensors. The mechanism is driven in each axis by a direct-drive dc torque motor while steering is obtained by torquing the integrating rate gyro. Angular information is derived from a resolver fitted to each axis. The mirror sightline is controlled by signals from an electronics unit which also provides power and signals to the main turret azimuth drive on the pod. System control may be exercised through a digital computer or by a hand controller unit. According to Ferranti, the use of the mirror system provides a higher

Ferranti/Barr & Stroud Type 221 thermal surveillance system

degree of stabilisation than is normally attainable by other methods and claims that the resultant image is blur-free under typical aircraft vibration conditions. Infra-red spectrum vision is obtained via a germanium window on the front of the mirror turret assembly.

The display may be presented on either 525- or 625-line television monitors. The output is either in CCIR or EIA composite video formats, as required. This television-compatible output may be displayed on one or more monitors situated at various locations throughout the aircraft, and Barr & Stroud emphasises the desirability of having monitors at both the pilot and winch operator positions of a helicopter in order to co-ordinate crew-members for better hover control during night and bad visibility winching operations.

The Type 221 system has been designed to UK Ministry of Defence standards and is in service with the Royal Air Force. It is intended for use in medium-to-large helicopters and roles envisaged include maritime reconnaissance, search and rescue, and integration with on-board weapon systems to improve their all-weather capability.

The Ferranti Type 221 Thermal imager has been fitted to a number of Aérospatiale/Westland Puma helicopters operated by the Royal Air Force in Northern Ireland under the project name Pleasant 3.

Dimensions: (pod unit) 34 inches (865 mm) × 16.5 inches (420 mm) max dia
(electronics unit) 5 × 17 × 13 inches (127 × 432 × 330 mm) max

Weight: (scan head unit) 8.36 lb (3.8 kg)
(electronic processor unit) 5.28 lb (2.4 kg)
(power supply unit) 2.64 lb (1.2 kg)
(total IR18 thermal imager) 16.28 lb (7.4 kg)
(× 2.5/ × 9 telescope) 18.48 lb (8.4 kg)
(pod complete) 165 lb (75 kg)
(electronics unit) 17.5 lb (8 kg)
(total system) 183 lb (83 kg)

STATUS: in service.

Type 234 thermal imaging/laser system

Designed originally as a sensor payload for an unmanned aircraft, the Type 234 can be adapted easily to manned aircraft platforms. Developed by Ferranti, in conjunction with Rank Pullin Controls (RPC) Limited, the system incorporates an RPC high performance imager operating in the 8 to 14 micron band. There is space provision in the sensor ball for a Ferranti laser designator and rangefinder. The imager has a dual field of view telescope which can be switched to either 3° × 4° or 12° × 16°. The ball allows full lower hemisphere coverage and can be tilted through +20° to -120°, and panned through ±220°. A Joule Thompson expansion valve provides cooling, air being supplied by compressor or gas bottle. Pointing accuracy is 3 milliradians, with stabilisation of better than 50 microradians. There is a choice of video outputs from the imager: 625 lines at 25 Hz frame rate or 525 lines at 30 Hz frame rate.

Dimensions: (ball diameter) 330 mm
(height of ball installation) 470 mm
(electronics unit) dependent on facilities required
Weight: (sensor head with laser) 24 kg
(electronics unit) dependent on facilities required

STATUS: under further development to offer multi-sensor payloads.

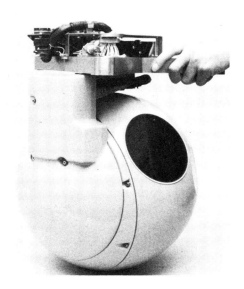

Ferranti Type 234 thermal imaging/laser system was originally developed as a contender for the UK Ministry of Defence's Phoenix RPV project

LRMTS laser ranger and marked-target seeker

Development of the LRMTS began in 1968 under a government contract, and prototype units were first flown in 1974. Deliveries to the Royal Air Force, for installation in nose housings of the British Aerospace Harrier and Sepecat Jaguar aircraft and accounting for over 200 units, was completed in 1984. Deliveries for the Royal Air Force's Tornados continue, and by early 1986 over 500 LRMTS had been delivered to the Tornado, Jaguar and Harrier programmes.

The LRMTS is a dual-purpose unit which can be used as a self-contained laser ranger or as a target-seeker with simultaneous rangefinding. In the target-seeking role it can be used to detect and attack any target designated by ground troops with a comsat laser, enhancing the effectiveness of battlefield close air support.

The LRMTS is a neodymium-YAG laser mounted in a stabilised cage, which allows beam-pointing and stabilisation against aircraft movement. The seeker can detect marked targets outside the head-movement limits. It operates at a relatively high pulse-repetition frequency, thus allowing continuous updating of range information during ground-attack. As range is a crucially important parameter for accurate weapon-delivery, and yet virtually unobtainable on a non-laser equipped aircraft, the LRMTS is a vital additional sensor.

In a typical operation, the ground operator with pulsed-laser target-designation equipment, who is called the forward air controller, directs the aircraft to a location within laser-detection range before switching on the ground-marker equipment. Radio communications between the forward air controller and aircraft crew are minimised and positive identification of even small, hidden or camouflaged targets is assured.

Once the unit has recognised the target from reflected laser energy, it provides steering commands to the pilot on the head-up display. Ranging data is also shown and fed directly into weapon-aiming computations for the accurate and automatic release of weapons.

The LRMTS is easy to install and harmonise with other aircraft systems, and is said to be more effective than any alternative sensor during operations at the grazing angles used in low-level ground-attack. By improving weapon-delivery accuracy, the sensor ensures a high probability of success in single-pass, high-speed attacks.

Associated with the LRMTS head is an electronics unit which contains power supplies,

Ferranti LRMTS optics in nose of British Aerospace Harrier GR3

and ranging and seeker processing. The laser needs a transparent window, and in the Jaguar is mounted behind a chisel-shaped nose, with two sloping panels. For the Harrier, where there is more chance of debris accumulation during Vtol operations, the optics are protected by retractable eye-lid shutters.

A specially-designed installation for the Royal Air Force/Panavia Tornado has incorporated the LRMTS into an underbelly blister. In addition to the systems in service with the Royal Air Force, equipment is also used by three overseas air forces.

LRMTS head
Dimensions: 300 × 269 × 607 mm
Weight: 21.5 kg
Operating wavelength: 1.06 microns
Pulse repetition frequency: 10 pulses/s
Angular coverage: (elevation) +3° to –20° (azimuth) –12° to +12°
Roll stabilisation: ±90°
Detection angle: ±18° from aircraft heading
Max range: >9 km

Electronic unit
Dimensions: 330 × 127 × 432 mm
Weight: 14.5 kg
Power: 700 VA at 200 V 400 Hz 3-phase plus 1 A at 28 V dc

STATUS: in service.

Type 105 laser ranger

The Type 105 equipment is a series of high repetition rate, neodymium-YAG, steerable laser rangefinders, developed privately by Ferranti to provide compact, low-cost, accurate

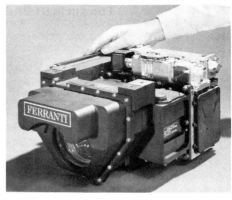

Ferranti Type 105 laser ranger

target-ranging sensors for ground attack aircraft.

A prototype flew in the Fairchild Republic Night Attack A-10 in 1979. The D version is in production for the Royal Danish Air Force Draken retrofit and has contributed to the more recent Fairchild single-seat night attack programme.

The 105 series includes the D, E and S variants. The D and S versions are single packages combining the Ferranti Type 629 transceiver, a beam steering unit and beam steering electronics.

The E version, functionally identical, has the beam steering unit packaged separately.

Target range is measured to 3.5 metres standard deviation to a range of 10 km, effectively removing the largest source of error in air-to-surface weapon delivery.

Low power consumption, ease of integration with aircraft avionic systems, flexible configuration, small size and frontal area are seen to be important factors in Type 105 potential for fit, or retrofit, in ground attack aircraft.

Dimensions: 198 × 261 × 357 mm
Weight: 13 kg
Operating wavelength: 1.06 microns
Pulse repetition frequency: 10 pps
Angular coverage: within 10° semi-apex angle cone
Range: ≤10 km
Accuracy: 3.5 m standard deviation
Power: 200 W 28 V dc
Reliability: >1000 hours mtbf

STATUS: in service.

Type 105D laser ranger

This is a private development of the Type 105 laser ranger system and it uses a smaller transceiver. In July 1980 the Type 105D was ordered by the Royal Danish Air Force for inclusion in the Weapon Delivery and Navigation System refit programme associated with that service's Saab Draken fighters.

Operating characteristics are virtually identical to the Type 105, although a mean time between failures of 1000 hours is quoted and the basic pulse-repetition frequency is 10 pulses a second.

Dimensions: 210 × 264 × 308 mm
Weight: 8.7 kg
Operating wavelength: 1.06 microns
Pulse repetition frequency: 10 pps
Angular coverage: within 10° semi-apex angle cone
Roll stabilisation: ±360°
Operating range: ≤5 km
Accuracy: ±5 m standard deviation
Power: 9 A at 28 V dc

STATUS: in production.

Type 106 laser seeker

This unit essentially comprises the passive elements of the LRMTS used in Royal Air Force Jaguar and Harrier close-support aircraft. The

Ferranti LRMTS mounted in nose of Royal Air Force/Sepecat Jaguar GR1

Ferranti Type 105D laser ranger is fitted to Saab Draken

Ferranti Type 122 helicopter laser ranger

system permits a crew to identify a laser-designated ground target and can present appropriate director information on the head-up display or gunsight. It is claimed to increase significantly the effectiveness of close air support against difficult targets, and to reduce the briefing time required between the forward air controller on the ground and the pilot.

The laser receiver optics are contained in an approximately cylindrical pressurised unit which requires a forward field of view equivalent to the optic angular slewing limits. A small electronic unit is also installed containing power supplies and target-seeking electronics.

To find a marked target, the optical sensor is set at a pre-determined depression angle and scanned horizontally until the designation is detected. The laser receiver then locks onto the target and produces steering commands to the pilot, ensuring accurate target-recognition at the earliest opportunity. Blind attacks can be carried out.

Ferranti has proposed the Type 106 laser seeker with the ISIS D-209 RM gyro gunsight as part of a low-cost target acquisition system for ground-attack aircraft.

Dimensions: 275 × 247 × 312 mm
Weight: 21 kg
Operating wavelength: 1.06 microns
Angular coverage: (elevation) –25° to +20° (relative to roll axis)
(azimuth) –18° to +18° (relative to heading)

Ferranti Type 116 CO$_2$ laser ranger

Roll stabilisation: ±90°
Power: 300 VA at 200 V 400 Hz 3-phase, plus 3 A at 28 V dc

STATUS: under development.

Type 116 CO$_2$ laser ranger

This high pulse-repetition frequency carbon-dioxide laser transceiver operates at an eye-safe wavelength, thus overcoming one of the training disadvantages of the neodymium-YAG laser rangers now in service. The unit is being

developed for the Royal Aircraft Establishment, Farnborough.

Operating wavelength: 10.6 microns

STATUS: under development.

Type 117 laser designator/ranger

Developed for the US Navy/McDonnell Douglas F/A-18 Hornet attack aircraft, this unit is supplied to the Ford Aerospace and Communications Corporation for integration into the forward-looking infra-red pod produced by Ford for the US Navy. The laser is a high-repetition rate neodymium-YAG device. An initial $2 million production contract was awarded in 1983, and further production worth an additional $20 million is anticipated.

Operating wavelength: 1.06 microns

STATUS: in production.

Type 118 lightweight laser designator/ranger

This laser transceiver is suitable for integration with the visual optics of future or existing helicopter sights to designate targets for spot tracker or laser guided weapon applications. It has been selected for the mast-mounted sight in the US Army's OH-58D AHIP (army helicopter improvement programme) development.

Operating wavelength: 1.06 microns

STATUS: in production.

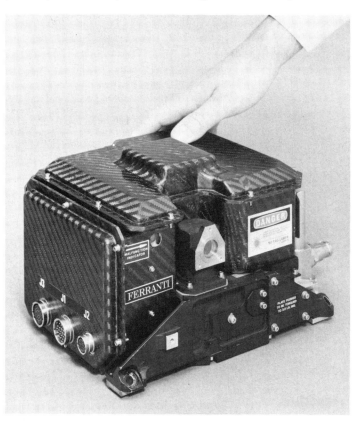

Ferranti Type 118 lightweight laser designator

Ferranti Type 118 laser being used in conjunction with helicopter mast-mounted sight

Type 122 laser ranger for helicopters

The Type 122 is a low-cost neodymium-YAG laser ranger suitable for integration with the visual optics of existing or future helicopter sights. It can provide accurate range measure- ments from a scout or attack helicopter in many tactical situations. A Type 520 laser transceiver is used and the 122 series ranger is installed in the company's AF542 roof-mounted sight.

Operating wavelength: 1.06 microns

STATUS: under development.

GEC Avionics

GEC Avionics Limited, Christopher Martin Road, Basildon, Essex SS14 3EL
TELEPHONE: (0298) 22822
TELEX: 99225

Heli-Tele television system for helicopters

The GEC Avionics Heli-Tele is an airborne television system combined with a microwave link and designed for helicopter applications. The system provides long-range, real-time ground surveillance to meet the requirements of police and military forces and other security authorities.

Heli-Tele comprises a colour camera, with a 20:1 zoom lens giving a narrow, 1° field of view, mounted on a gyro-stabilised platform aboard the helicopter. The operator can control the pointing angle of the camera and adjust the field of view to display any selected area on the ground. Video information thereby obtained is transmitted to any number of ground stations via an omni-directional microwave link. Receiving stations may be located at distances up to approximately 50 nautical miles from the helicopter.

While the standard Heli-Tele System employs a daylight colour camera, alternative sensors can be installed for night operation. A TICM II thermal imager, developed by GEC Avionics and Rank Taylor Hobson under a UK Ministry of Defence programme, provides this additional capability.

The airborne equipment, which has been certificated by the UK's Civil Aviation Authority, has been fitted to nine types of helicopter and is in use with a number of police forces.

STATUS: in service.

Thermal imaging common modules

GEC Avionics and Rank Taylor Hobson are producing a range of thermal imaging common modules under contract from the UK Ministry of Defence (Procurement Executive). These modules, which provide indirect view, tele- vision-compatible thermal images, can be used as the building blocks to assemble Flir systems that will meet a variety of requirements from the three armed services.

The imagers operate in the 8 to 13 micron band, permitting smoke and haze penetration. They operate from 28 volts dc and are designed

Prototype GEC Avionics Phoenix sensor payload

to meet a comprehensive range of applications, including a number of airborne roles.

The effectiveness of the modular design and its operational success can be judged by the fact that using the same set of common modules, many different imager configurations are being produced for customers worldwide. Applications include pilot night-vision and reconnaissance systems for fixed wing aircraft, helicopters and RPVs, vehicles and ships.

STATUS: in production.

Sensor payload for Phoenix RPV

GEC Avionics is the prime contractor for the British Army's new Phoenix remotely piloted targeting and battlefield surveillance system. The company was awarded a fixed price development and production contract for Phoenix by the UK Ministry of Defence.

The system comprises a small air vehicle with advanced avionics and thermal imaging sensor, an air-to-ground data link, a mobile ground station and logistics vehicles for launch and recovery.

The Electro-Optical Surveillance Division is responsible for the development of the air vehicles thermal imaging sensor payload, the

Prototype RPV thermal imaging sensor

GEC Avionics common module units

Thermal imager assembled from GEC Avionics common modules

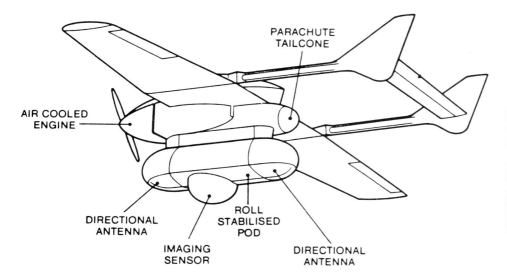

Flir image on pilot's head-up display

Drawing of GEC Avionics Phoenix air vehicle

data link and for various related parts of the mobile ground control station.

The sensor payload employs a thermal imager based on Class II thermal imaging common modules (TICM II). The imager with a zoom telescope is configured with the electronics processing and cryogenics in a gyro-stabilised turret resulting in a lightweight maintainable package. The sensor turret, with 360° look-around capability, will provide real time surveillance coverage day or night or in poor visibility by display in the ground control station where the image can either be automatically or manually controlled.

STATUS: in development.

Forward-looking infra-red systems
GEC Avionics Electro Optical Surveillance Division is developing Flir night-vision systems for UK Tornado IDS and Harrier GR5 aircraft under a £48 million fixed price contract from the Ministry of Defence. The same Flir system will also be installed in the US Marine Corps AV-8B aircraft by McDonnell Douglas.

The Flir system, which is based on the modular UK Class II thermal imager, is installed inside the aircraft's fuselage with the optics looking forward through a small blister near the front of the aircraft.

The Flir provides an automatically optimised high resolution image of the terrain ahead, with automatic cueing of potential targets, for viewing on the pilot's head-up display.

Regarded as a major 'force multiplier', this passive system will enable close support missions to be undertaken at night and in poor weather and will significantly increase frontline aircraft utilisation.

This new system is the first pilot-aid Flir to be used by the Royal Air Force.

STATUS: in development.

GEC Avionics Atlantic Flir on General Dynamics F-16

Atlantic podded Flir systems
Electro Optical Surveillance Division's Atlantic Flir is a pod-mounted system designed to give ground attack aircraft night and poor weather capability on high speed low-level missions.

Atlantic employs a Flir configured from UK Class II (TICM II) modules and incorporates an advanced thermal cuer for early target detection and maximum reaction time by alerting the pilot to potential targets in the head-up display field of view. The system, with multi-aircraft compatibility, includes a MIL-STD-1553 data-bus interface and can be integrated with existing weapon and avionic systems.

Atlantic is involved in continuing flight trials and demonstrations in the USA and Europe fitted to a General Dynamics F-16 aircraft to assess suitability for export F-16s. It is expected that the system will be demonstrated to potential US and European customers on other aircraft during 1986.

STATUS: in development.

Navigation Flir pod
The GEC Avionics navigation Flir pod is a cost effective and flexible solution to the pilot night-vision and target-detection requirements of a variety of aircraft.

The navigation Flir uses full production standard UK Class II thermal imaging common modules which are housed in a Vinten Vicon 70 pod together with a VCR for in-flight recording.

The high sensitivity image can be viewed on either a head-down display or a head-up display with a 1:1 overlay on an outside world for increased pilot confidence.

The navigation Flir pod system is designed for role flexibility and easy retrofit on aircraft

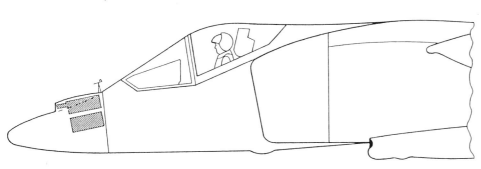

Location of Flir components in US Marine Corps AV-8B

Artist's impression of GEC Avionics passive identification device

The head-up display is a modified version of the GEC Avionics dual-mode display, more than 200 of which have been produced for night attack A-7E aircraft (out of a total A-7D/E head-up display production of more than 2000). It was modified from an 875-line to a 625-line video standard, with the display symbology interlaced with the Flir raster display of the night scene.

The rear-seat monitor is an existing electronic head-down display from current production, but modified with a tactile, or touch-sensitive, screen. Targets indicated by the Flir can be designated to the mission system by merely touching the display surface.

Cats Eyes night-vision goggles

Night-vision goggles were originally developed for use by ground forces, and were only later employed by helicopter aircrews. NVGs were successful in this role, and so their use was further extended to embrace trials with high-performance fixed-wing aircraft. Some limitations were encountered in these trials, however, and accordingly GEC Avionics has developed a new type of goggles called Cats Eyes, which has an optical combiner for each eye. Conventional NVGs of the binocular type do not permit a direct view of the outside world, so that the pilot has to monitor cockpit instruments by looking round and under the goggles, preventing the use of any peripheral visual cues which are outside the instrument's field of view.

With the Cats Eyes system the images from the two visual input paths, one direct and the other from the intensifier, are registered in a 1:1 relationship and so complement one another. The advantages of this arragement have been established during extensive low-level night-flying trials (as part of the Royal Aircraft Establishment's Night Bird programme) involving a fully integrated night vision cockpit, with compatible lighting, conventional raster-scan head-up display (portraying infra-red images of the outside world produced by a fixed forward-looking infra-red pod), and a head-down multi-function display.

GEC Avionics tabulates the following confirmed advantages.
1. The resolution and dynamic range of the head-up display seen through Cats Eyes is not impaired since it is viewed directly through the combiners as opposed to the image intensifiers;
2. Monitoring of all cockpit instruments is much easier using see-through combiners;
3. Pilot awareness of the outside world is improved because of the direct vision and ability to scan with eyes on either side of the combiners;
4. The pilot's sense of orientation is improved from the combination of direct vision and better peripheral vision;

fuselage or wing store pylons on NATO standard 14 or 30 inch (356 or 762 mm) attachments.

STATUS: in production.

PID passive identification device sensor turret

GEC Avionics Electro Optical Surveillance Division is developing a system to be installed on Royal Navy Lynx helicopters which will provide a day or night long-range target detection and identification capability.

The modular system uses a GEC Avionics thermal imaging sensor mounted in a highly stabilised turret platform sited on the nose of the helicopter.

The high resolution thermal image, automatically optimised for maximum picture quality, is displayed in the cockpit. The employment of advanced signal processing provides potential target cueing automatically, further reducing cockpit workload.

The new system will be completely integrated with the aircraft avionics, including auto operation and manual control via a joy-stick if required.

Night attack system

Between April and October 1984 trials were conducted on a LTV TA-7C aircraft of the US Marines Corps of a GEC Avionics integrated night-vision and attack system. In all, 78 flights, totalling 150 hours of trials, were accomplished at the US Naval Weapons Center at China Lake, California. The programme demonstrated the feasibility of night-time low-level close support missions using a low-cost Flir pod, raster head-up display, night-vision goggles and a touch-sensitive high-resolution head-down display. The system was developed and integrated in fewer than six months from go-ahead. Much of the original development work was done by the Royal Aircraft Establishment, Farnborough (see separate entry).

The podded Flir system is built from the UK's Class II (indirect view) TICM II thermal imaging common modules currently in service with British and overseas forces. The high-resolution thermal imagery gives good long-range target detection and is provided by electronic processing of the signal from the detector head. The system owes much of its success to the automatic gain control techniques employed, helping to minimise pilot workload.

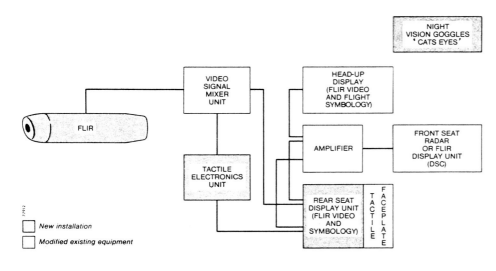

GEC Avionics night-attack system

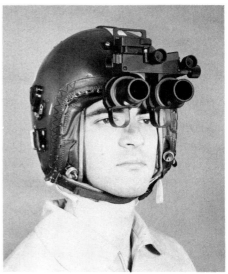

GEC Avionics Cats Eyes night-vision goggles

5. Dusk to dark transitions are less problematical because the image-intensifier display becomes more noticeable as the direct, outside-world view becomes fainter;

6. The shorter length of the Cats Eyes system permits the pilot a greater degree of head mobility than possible with conventional NVGs.

Weight: (including mount and helmet bracket) 1.7 lb (0.78 kg)
Field of view: 30° circular
Magnification: unity
Eye relief: 1 inch (25 mm)
Exit pupil: 0.4 inch (10 mm)

STATUS: research and assessment.

Lasergage

Lasergage Limited, Newtown Road, Hove, East Sussex BN3 7DL
TELEPHONE: (0273) 770 341
TELEX: 877062

LT 1065 hand-held thermal imager

The Lasergage hand-held thermal imaging system was originally developed by the Philips Electronics MEL Division. However, MEL disposed of its opto-electronics business in early 1982, withdrawing from this sector of the industry and further development, manufacturing and marketing of the thermal imager has since been undertaken by Lasergage, a company well-known for lightweight laser ranging and target-designation equipment.

Lasergage has been acquiring certain marketing rights for the Flir Systems Inc 100A thermal imaging system. The opportunity to obtain the assets of MEL's thermal imaging development and production facilities, together with the services of key scientists and technicians, enabled the company to form the basis of development, manufacturing and support capability to promote the sales of both systems.

The hand-held system is compact, lightweight and self-contained, being powered from a small battery pack which may be either belt-mounted or carried over the operator's shoulder by a sling strap. Operating in the 3 to 5 micron band, the imager is unusual in that it requires no external cooling supplies, the detector cooling being achieved by thermo-electric methods. This factor makes the system particularly suitable for use in aircraft or in any confined space where the presence of pressurised gas bottles could constitute a hazard. Its independence of a cooling gas supply also extends the operational duration of the system, a single battery charge being capable of maintaining operation for periods of over 8 hours. While the battery powered method of operation is ideal for portable use, the system may, however, derive its power supply from standard 12 or 24 volt vehicle batteries.

In standard form, the unit has a field of view of 10° in azimuth and 7° in elevation with a lens providing a ×1 magnification. A ×0.44 magnification wide-angle lens is an available option offering a field of view of 22.5° azimuth and 16° elevation. Other lenses, offering longer focal lengths for long ranges and greater angular resolutions, and shorter focal lengths for wider fields of view, are understood to be under development.

The Lasergage imager employs a mechanical system for scanning an array of 12 detectors placed at the focal point of the objective lens. Sequential scanning is carried out by an eight-faceted rotating mirror and the detector output, following processing, is fed to a bank of light-emitting diodes which are again scanned by the mirror at a point diametrically opposite the detector bank. Light output from the light-emitting diodes is fed to a small eyepiece screen and appears to the operator as a picture

Lasergage hand-held thermal imager aboard JetRanger helicopter

Lasergage/Barr and Stroud IR18 helicopter infra-red systems

similar to that produced by a 96-line television screen which, despite the relatively small number of lines, produces a high-quality image of good resolution.

The system is easy to operate, possessing only four controls, an on/off switch and knobs which control focus, brightness and contrast. It is normally held to the user's eye by grip straps on the side of the unit's casing but a removable pistol grip is also provided. For prolonged operations from the footstep of a helicopter, a chest harness may be used to relieve the operator of the need to support the imager but it is in fact small and light enough to be supported by bungee straps suspended from the top of a cabin or cockpit door aperture. This removes the need to support the unit and also eliminates the problem of aircraft vibration, which tends to affect the use of some portable, direct viewing eyepiece systems.

The thermal imager's eyepiece may be fitted with adaptors to permit attachment of still or cinematography cameras to obtain permanent records. A television camera interface may also be used for remote display on a monitor or for videotape recording. When not in use the entire system, including the nickel-cadmium rechargeable battery pack, can be stored or transported in a briefcase-sized carrying case.

The system is operable over a temperature range varying from –20° to +40° C and can detect targets exhibiting a temperature differential of less than 1° C from their background or surroundings. This degree of sensitivity led to the system, aboard a JetRanger helicopter, being instrumental in the discovery, after less than one hour's flying, of a dead body which had remained undetected despite searches by 100 police officials, aided by dog teams, over a two-week period.

Dimensions: (hand-held imager (excluding pistol grip)) 12.2 × 6.3 × 4.75 inches (310 × 160 × 120 mm)
(shoulder battery pack) 5.25 × 2.9 × 2.75 inches (133 × 75 × 70 mm)
(carrying case) 18 × 13.6 × 6.5 inches (457 x 345 × 165 mm)
Weight: (hand-held imager) 6.6 lb (3 kg)
(battery pack) 4.4 lb (2 kg)
(carrying case (empty)) 2.2 lb (1 kg)

STATUS: in production and service.

HIRS helicopter infra-red system

HIRS is a high-resolution IR18 thermal-imaging system, developed in conjunction with Barr and Stroud and mounted in a steerable pod which provides real-time pictures on a crew's display. The system is modular, making it adaptable to a number of applications, and is normally gimbal-mounted, although a fixed mount can be provided. Objects as close as 1.5 metres can be viewed and a × 6 infra-red telescope is used for long-range detection. Pictures are presented on a 525/625-line monochrome display, which is also provided with the associated controls. More than one display can be linked to the sensors, and the picture can be video recorded or data-linked to a ground station.

Field of view: (normal) 38 × 25.5
(× 6 telescope) 6.3 × 4.25
Angle of view: (elevation) +10° to –100°
(azimuth) ±100°
Power consumption: 32 W at 24 V dc
Weight: (total system including monitor and air bottle) 35 kg
(× 6 telescope) 3.9 kg extra

STATUS: in production and service.

Royal Aircraft Establishment

Royal Aircraft Establishment, Farnborough, Hampshire
TELEPHONE: (0252) 24461
TELEX: 858134

Obstacle-warning device for helicopters

The Royal Aircraft Establishment is developing a device that warns helicopter pilots of obstacles, particularly power cables, in the aircraft's flightpath and while hovering at low level behind cover.

The system is being developed, in collaboration with British industry, to meet the nap-of-the-earth flight regime requirement for tactical battlefield use, especially in northern Europe where electricity grid power cables proliferate. It consists of a laser fitted in a rotating mount, which acts as a scanning mirror. When operating in the obstacle warning mode, during low hovering flight, there is coverage of a full 360° hemisphere below the aircraft. Indication of obstacles within 50 metres of the aircraft is provided. In forward flight, the laser is locked into a forward scanning mode and covers a discrete area ahead of the

helicopter, giving warning of 'live' and 'dead' cables, crossing the flight path. The laser operates at a wavelength outside the normal detection band of battlefield thermal imaging systems and is said to be eye-safe at a distance of less than 1.5 metres. The system may be used as an autonomous unit or in conjunction with a mission management system.

STATUS: under development.

Nightbird programme

The Nightbird research programme has been undertaken by the Flight Systems Department, RAE Farnborough, to investigate the use of passive electro-optic sensors on fixed-wing, ground-attack aircraft for use as flight and weapon aiming aids during poor weather by day and at night. The major achievement of this programme has been the recognition of the value of the combined use of night-vision goggles (NVGs) and forward-looking infra-red (Flir).

Development and demonstration of the system has been performed using a two-seat Hunter aircraft. The NVGs, mounted on the pilot's helmet, permit him to view the terrain at night, hence he can use daylight flying techniques whilst manoeuvring at high speed and

low level. NVG compatible cockpit lighting has also been developed. The UK Class II common module thermal imager mounted in the nose of the aircraft produces a high resolution Flir image of the scene ahead of the aircraft which is displayed on a head-up display and a head-down display. This combined system enables flight at night through a wider range of conditions than aircraft equipped with either system alone.

The high resolution thermal image assists in locating targets by night and day through mist, fog, smoke and camouflage. Target detection is further enhanced by use of a thermal cueing aid which automatically processes the infra-red signal to locate potential targets and indicate their position to the pilot. Current research work is investigating the integration of this wide fov system with a steerable narrow fov thermal imager and carbon dioxide laser rangefinder to give a target acquisition and weapon aiming system. This work is being performed on the RAE Buccaneer avionics flight test facility.

Early in 1986 a Ferranti cockpit display, with colour for the map presentation and monochrome (green/black) for the thermal imaging picture was flown in the Buccaneer. This was said to be the first airborne trial involving shadow mask cathode ray tube technology.

Thorn EMI

Thorn EMI Electronics Limited, 120 Blyth Road, Hayes, Middlesex UB3 1DL
TELEPHONE: (01) 573 3888
TELEX: 22417

Thermal imagers

Thorn EMI has developed a number of thermal imaging systems as part of the company's involvement in the UK Ministry of Defence thermal imaging common modules programme.

The MRTI (multi-role thermal imager), which uses Class I modules from the TICM programme, has been developed as a direct and indirect view system and both systems are now in quantity production for the British Army and the Royal Navy. MRTI has a wide range of applications on the battlefield, for offshore patrol, airborne surveillance, weapon-aiming and mortar and artillery fire observation. Successful flight trials with MRTI have been conducted on an Army Air Corps Scout helicopter.

Thermal radiation is detected by a cadmium mercury telluride detector array which is scanned across the scene in a series-parallel mode. The detector array operates at 80K and can be cooled to its working temperature by a Joule-Thompson cooler using compressed air or by a closed-cycle cooling engine. MRTI is configured to use a 0.6-litre air bottle which, when charged to a pressure of 300 atmospheres, will cool the detector for 4.5 hours. Alternatively, the closed-cycle cooler has a power consumption of 40 watts.

The system operates in the 8 to 13 micron band and has a temperature sensitivity of less than 0.1K. Two fields of view are provided, with the standard MRTI telescope, setting 4.9° × 3.2° and 12.9° × 8° in the narrow and wide settings respectively. Telescopes are also available giving a wider field of view. The equipment maintains focus when it is switched from one field of view to the other. Controls for temperature offset and temperature window enable the operator to emphasise different temperature ranges within the thermal scene under surveillance. Image polarity, either hot/white or cold/white, is selected by a switch.

When used in the direct-view mode, the thermal scene is reconstructed using a scanned light-emitting diode array. By the addition of a further module to the direct-view system, a television-compatible (CCIR) output to a remote

Thorn EMI hand-held thermal imager being used for airborne observation tasks

display can be obtained to form the indirect-view system. The indirect-view image monitor can be mounted on the imager or removed to a required position.

Dimensions: (thermal imaging unit) 19 × 11 × 7.5 inches (480 × 280 × 190 mm)
(indirect viewing module) 7.5 × 5.5 × 2 inches (190 × 140 × 60 mm)
Weight: (thermal imaging unit (including compressed-air supply bottle and battery)) 25.3 lb (11.5 kg)
(indirect viewing module) 2.86 lb (1.3 kg)

STATUS: in quantity production.

Hand-held thermal imagers

Thorn EMI produces lightweight hand-held thermal imagers, suitable for night/bad weather observations in a number of roles, including airborne reconnaissance.

STATUS: in production and service.

Obstacle-warning device for helicopters

This system is designed to warn helicopter pilots of an obstacle in their flight-path, and it is claimed that even 'dead' electricity power-lines can be detected over a 400-metre range; detection ranges of over 1 km have been demonstrated for larger objects. The device, which started flight trials in summer 1983, uses a laser-ranging technique to give accurate distance and bearing to an obstacle. The beam is rotated at 60 rpm with a hemispherical motion beneath the helicopter to derive a laser map of the terrain which is then processed and displayed to the pilot. The picture is updated every second.

STATUS: under development.

UNITED STATES OF AMERICA

Boeing

Boeing Military Airplane Company, PO Box 7730, Wichita, Kansas 67277
TELEPHONE: (316) 526 3153
TELEX: 417484
TWX: 910 741 6900

Digital scan converter

Early in 1985 Boeing was awarded a $14 million US Air Force contract to produce a digital scan converter which will replace the existing forward-looking infra-red signal processor in its B-52G and -H bombers. A total of 291 units will be delivered between February 1986 and July 1987.

STATUS: in development.

Airborne sensor package

In mid-1984 Boeing received a $289.4 million, five-year research and development contract from the US Army to demonstrate the use of airborne infra-red sensors and data processors in an anti-ballistic missile defence system. The programme will involve a modified Boeing 767 equipped with long-wave infra-red sensors developed by Aerojet Electro Systems and Hughes Aircraft. A series of flights on the Kwajalein Missile Range, using missiles fired from the Vandenburg launch site, is envisaged and the project is the airborne optical adjunct of the US Army's layered anti-ballistic missile system.

STATUS: in development.

Fairchild

Fairchild Communications and Electronics Company, 20301 Century Boulevard, Germantown, Maryland 20874
TELEPHONE: (301) 428 6000
TELEX: 892468

Integrated sensor control system

Fairchild's ISCS integrated sensor control system has been developed as the primary control equipment for all reconnaissance subsystems on the Northrop RF-5E Tigereye reconnaissance aircraft. The system is of all-solid-state construction and utilises microprocessor control and low-power technology. It consists of a control panel unit – side (CPU-S), a control panel unit – front (CPU-F), a control interface unit (CIU) and four annotation display units (ADUs).

The CPUs provide manual control of the reconnaissance system including sensor selection, overlap, scan-angle and direction, cloud compensation, time clock synchronisation, data-selection and self-test. The ADUs provide seven-segment alphanumeric annotation within the sensors. Interface within the ISCS with the sensors, the associated inertial navigation system and the radar altimeter is accomplished by the CIU. The system's capability of operation with the optional 'pallet' configurations of sensors is claimed to be unique. The CPU also provides pilot display of real-time, elapsed time, groundspeed and track information during aircraft operation.

Features include sunlight-readable switch/indicators, microprocessor control of functional status, interfaces and built-in test and compatibility with a variety of sensors. The system has four sensor-control channels and light-emitting diodes are used for alphanumeric annotation, some through the film backing. The system meets a large number of MIL-STD specifications and operates over a temperature range of –54°C to +71°C at pressure altitudes up to 50 000 feet.

Dimensions: (CPU-S) 5.25 × 5.75 × 5.25 inches (133.4 × 146.1 × 133.4 mm)
(CPU-F) 3.05 × 7.50 × 6.55 inches (77.5 × 190.5 × 166.4 mm)
(CIU) 11 × 9 × 4.62 inches (279.4 × 228.6 × 117.4 mm)
(ADU) 1.25 inches (31.8 mm) dia × 4.5 inches (114.3 mm) length

STATUS: in production.

Flir Systems

Flir Systems Inc, 16505 SW 72nd Avenue, Portland, Oregon 97224
TELEPHONE: (503) 684 3731
TWX: 510 101 0797

Flir Systems Inc is a manufacturer of both standard and specialised infra-red imaging equipment; primarily for airborne platforms. The earlier models 100A and 1000A have been superseded by the 2000 series which has expanded to five models in 1985. In addition to these standard systems, the company has developed special OEM packages and is currently developing a land based, surveillance-oriented system.

2000 Series thermal imaging systems

The 2000 series of airborne thermal imaging systems consists of five models varying in resolution and field of view and price. The simplest system is the basic model 2000A which has two fields of view (28° × 15° and 7° × 3.5°) and excellent resolution at 1.9 mrad (wide) and 0.47 mrad (narrow). The system is mounted in an unstabilised, unaided positioning system and is primarily used for surveillance, law enforcement and search and rescue. The most sophisticated system is the model 2000G, also with two fields of view (28° × 15°, 5° × 2.6°) but with higher resolution, 1.4 mrad in wide, 0.25 mrad in narrow field of view. The 2000G is also equipped with an inertial pointing to aid tracking and electronic caging to lock the system in a fixed orientation relative to the axis of the aircraft. This allows the 2000G to be readily adapted for low-cost fire control systems.

All models are available in both the US and European video standards (EIA RS 170, CIR) and are compatible with conventional video accessories, including recorders, downlinks, and digital video processors.

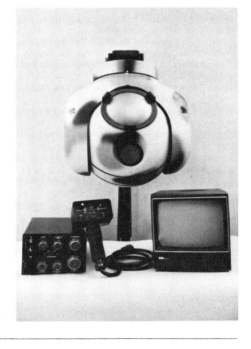

Flir Systems Model 2000 infra-red sensor system

Ford Aerospace

Ford Aerospace and Communications Corporation, Ford Road, Newport Beach, California 92660
TELEPHONE: (714) 720 4512
TELEX: 678470
TWX: 910 596 1354

AN/AAS-38 forward-looking infra-red system for F/A-18

The Ford F/A-18 Flir system has been designed to provide the US Navy's McDonnell Douglas F/A-18 Hornet with a day, night and adverse-weather attack capability by presenting the pilot with real-time thermal imagery in a television formatted display for the location, identification and attack of tactical targets. The system also enhances the air-to-air capability of the F/A-18 by passive detection and identification of threat aircraft.

The system's sensor and electronic section is contained in a pod-mounted on the lower left side of the aircraft's fuselage at a Sparrow guided-missile station. In the cockpit, the pilot can select a displayed field of view of 3° × 3° or 12° × 12° and is able to control the system's stabilised line-of-sight over a field spanning +30° to –150° in elevation with ±540° of roll freedom. Targets may be automatically tracked in either field of view throughout the field of regard.

The F/A-18 Flir interfaces with the aircraft and its avionics systems through the aircraft mission computer, receiving commands directly over a MIL-STD-1553A multiplexed digital data-bus. The system provides the mission computer with accurate target line-of-sight pointing angles, angle-rates and a complete pilot-initiated and automatic periodic built-in test evaluation of readiness. The system is the first US Navy Flir attack sensor designed for operation in a supersonic environment, and the pod is subject to dynamic flexure at such speeds. To compensate for this, boresight compensation is employed to ensure appropriate pointing accuracies throughout the flight envelope.

Special emphasis has been placed on re-liability and maintainability of this system which is expected to exceed a 100-hour mean time between failures requirement. The system comprises ten weapon replaceable assemblies which can be readily accessed and replaced without the need for calibration, alignment, special tools or handling equipment. The mean time to repair is predicted to be significantly less than the 12 minutes required for replacement of these assemblies on the flight line.

Full-scale development of the F/A-18 Flir

commenced in March 1978 and progressed through completion of one development unit and seven pre-production models. Initial flight testing on a Rockwell T-39D/Sabreliner, amounting to 50 hours of airborne operation, was conducted between November 1980 and mid-January 1981 and demonstrated satisfactory results with regard to stabilisation, automatic tracking and video quality. Further flight tests on F/A-18 aircraft commenced at the US Navy's Patuxent River Air Test Center in June 1981 and were completed in December that year, the system having successfully shown compliance with the specification. Additional flight-testing on a US Navy/McDonnell Douglas F/A-18 was conducted at China Lake, California, Fallon Air Force Base, Nevada and Yuma, Arizona during August and September 1982. Full-scale development, including formal qualification, and a Navy Technical Evaluation, was completed in early 1984. Over 175 F/A-18 Flir test flights were flown during the FSD phase of the programme. In 1984, production pods were flown an additional 175 test flights during a production verification flight test programme. Production orders totalling over 100 pods have been placed by McDonnell Douglas on behalf of the US Navy. Over fifty of these systems have been delivered and are operational with US Navy and Marine Corps F/A-18 squadrons. Total production potential exceeds 700 systems.

Ford Aerospace is developing a laser target designator/ranger (LTD/R) under a full-scale development contract from McDonnell Douglas and the US Navy. Two additional weapon replaceable assemblies will be added that provide the pod with a laser designation capability for the delivery of laser guided weapons and a laser ranging capability for improved ranging accuracy. The LTD/R subsystem is scheduled to be incorporated with the FY-86 Flir pod procurement.

Dimensions: 13 inches dia × 72 inches long (330 mm dia × 1829 mm long)
Weight: 340 lb (154.5 kg)

STATUS: first production pods delivered at the end of 1984. Over 50 pods had been delivered up to November 1985 and are operationally deployed with US Navy and Marine Corps F/A-18 squadrons.

Nite Owl forward-looking infra-red system

Ford Aerospace, together with Texas Instruments, has developed a Flir system called Nite Owl. This small lightweight pod features exceptionally good performance while achieving 100 per cent commonality with all performance related components of the F/A-18 Flir which is now operational with US Navy and Marine Corps F/A-18 squadrons. The Nite Owl system is compatible with a number of high-to-medium performance single or two-seat aircraft. It provides enhanced mission effectiveness 24 hours a day in air-to-air and air-to-ground applications. The pod-mounted system has an improved infra-red sensor which provides a real time high resolution television format picture for the pilot. Nite Owl has a self-contained pod cooling unit and incorporates a laser target designation/ranger (LTD/R) sub-system that

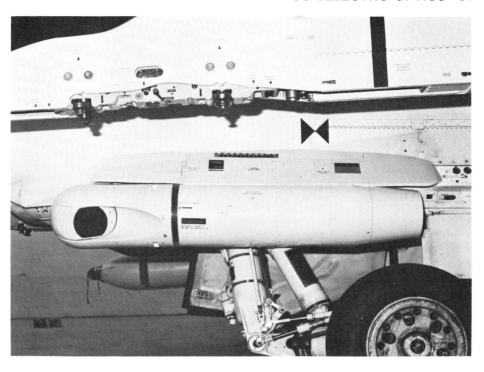

Ford Flir pod installed on US Navy/McDonnell Douglas F/A-18

provides a laser designation capability for the delivery of laser guided weapons, and a laser ranging capability for extremely accurate weapon delivery. This system provides high-performance aircraft with adverse weather day/night precision strike capability using conventional or laser guided weapons. The Nite Owl system has been tested to MIL-STD levels of burn-in, random vibration and acceptance testing. Nite Owl has been successfully integrated and flight demonstrated on both the McDonnell Douglas F-15 and General Dynamics F-16 aircraft.

AN/AVQ-26 Pave Tack laser designator/ranger

Pave Tack employs Flir target-acquisition sensors and laser designation/ranging. Ford Aerospace provides the pod structure, electronic control, environmental-control system and stabilisation equipment and is responsible for integration with the sensors. The Flir unit is manufactured by Texas Instruments, and the laser is produced by Litton Laser Systems.

The project was initiated by a request for competitive tenders, issued by the US Air Force Aeronautical Systems Laboratories, which led to an initial $15 million award to Ford Aerospace in 1974. Flight-testing began two years later, and in 1977 evaluation was stated to be 95 per cent complete for US Air Force/McDonnell Douglas RF-4C and F-4E Phantom operations by day or night and in various weather conditions. A US Air Force/General Dynamics F-111F installation was at the same stage by August 1978 and the first Pave Tack production contract, worth $48.5 million, was let in the same year. Production deliveries began in

August 1980 with Phantom initial operational clearance achieved in late 1980, and F-111F in early 1981.

Total US Air Force production of 149 pods is completed. Of these 25 are used on RF-4C Phantom, 45 on F-4E Phantom and 75 on F-111Fs. The number of aircraft converted to accept the Pave Tack pod is expected to comprise 60 RF-4Cs, 180 F-4Es and 100 F-111Fs. Pods, with support equipment, have been stated to cost $153.5 million, and aircraft modifications an additional $84.3 million. The first production order in 1978 embraced 23 pod sets, a further 48 sets were ordered in February 1979 and 78 sets were contracted in February 1980. Additional production was completed in 1983 to deliver Pave Tack equipment (10 sets) worth $26.8 million for use on Royal Australian Air Force F-111Cs. Another 10 sets are being produced for equipping F-4E aircraft operated by an unspecified export customer.

The aim of the Pave Tack system is to provide unvignetted full lower-hemisphere coverage for the sensors installed, and thus to permit the crew almost total freedom in its choice of flight path, both approaching and leaving the target area. Delivery techniques can be varied to suit the weapon and the stand-off range desired, the crew having the choice of dive, loft, glide, toss or level-release of weapons, with virtually no manoeuvrability or flight-envelope constraints.

When not in use the system has the optical port at the aft end of the pod retracted into its most streamlined position to minimise cruise drag. The pod is in two sections: the base-section assembly (the forward portion), and the head-section assembly (the rear portion).

The base section contains a digital aircraft interface unit, which permits direct integration with either the Phantom AN/ARN-101 digital avionics system or F-111 avionics. There is also a cathode ray tube interface unit. The electronics provide precise target-location data to the weapon-system computer aboard either aircraft. Special modes, incorporated as software modifications, permit Pave Tack to be used as a terrain-following sensor. In this mode of operation the pod line-of-sight is pointed along the aircraft velocity vector. There is also a tape recorder which records the crew's video display and provides bomb-assessment data.

The head section contains an optical bench and turret structure. Flir, laser, range-receiver and stabilised-sight equipment is mounted on the optical bench, which is linked to the turret

Ford Aerospace Pave Tack installation under fuselage, aft of nose gear, on General Dynamics F-111F

structure and base section in such a way that 180° of pitch movement is possible by using turret motion, and roll motion is provided by rotating relative to the base section.

The Flir and laser are boresighted and stabilised. Flir-imagery provides a wide field of view display for target acquisition, but also has a narrow field of view, high magnification, target identification and target-tracking mode. The AN/AVQ-25 laser designator/ranger is compatible with existing 1.06-micron laser-guided weapons.

Cockpit equipment includes a cathode ray tube which displays the Flir imagery and operational data in alphanumeric format. The system constitutes a multi-function display surface in the cockpit where the crew can view radar, Flir, television and weapons data. Initial design was based on taking the space occupied by the AN/APQ-144 indicator/recorder system in the F-111F, and the configuration has now been adapted to replace the AN/APQ-120 radar-plot indicator used in the F-4 Phantom rear cockpit.

Pave Tack is a day or night, clear and adverse weather sensor which enhances target acquisition and permits accurate release of laser-guided or conventional weapons onto any target visible to the electro-optics system.

STATUS: in service. System has also been evaluated for McDonnell Douglas F-15E dual-role version of the standard US Air Force fighter.

General Electric

General Electric Company, Aerospace Electronic Systems Department, French Road, Utica, New York 13503
TELEPHONE: (315) 793 7000

Infra-red detector research

In 1985 General Electric developed and successfully tested linear infra-red detector arrays using a new epitaxial technique to fabricate indium antimony wafers.

The implementation of these wafers in future airborne systems will have a marked effect on system performance since they can operate at 110K at which temperature no bulky cooling system is required. Currently, most infra-red systems use cadmium mercury telluride detectors which need to be cooled to 80K or below.

AN/ASQ-145 low-light level television system

The AN/ASQ-145 is a low-light level television system designed to provide fire-control facilities for the US Air Force/Lockheed AC-130 Hercules gunship aircraft, on which it is a primary sensor. The system is unusual in that it uses a dual camera installation in order to provide both wide and narrow fields of view simultaneously. General Electric was the contractor responsible for system integration, test and alignment as well as production.

STATUS: in production and service, currently being updated.

GE LLLTV low-light level television system

General Electric has introduced what it terms GE LLLTV (General Electric low-light level television). A general-purpose, multi-role system, it is designed for maritime surveillance missions such as search and rescue, monitoring of territorial waters and policing offshore environmental laws. Principal features are high resolution and sensitivity at low light levels and a small (16 mm diagonal) format. It is claimed to be capable of resolution densities in excess of 30 lines a millimetre and to have a wide dynamic range, which provides useful imagery around brightly lit areas of the area surveyed.

The camera head can be mounted in any attitude and is designed for 'hands-off' operation. It uses a 16 mm vidicon tube and an 18 mm second-generation hybrid-type photocathode intensifier, which has extended sensitivity at the red end of the spectrum.

The system's electronic unit has a switchable line rate of either 525- or 875-lines a frame as standard but other line/frame rates are optionally available. Normal aspect ratio is 4:3 but a version with an aspect ratio of 1:1 is also available. Frame rate is 30 Hz. Resolution is 600 television lines horizontal, and the dynamic range permits operation at face illumination levels from 0.1 foot candle down to starlight conditions. The system operates from a 28-volt dc nominal power source and requires a power of 30 watts.

The GE LLLTV has been chosen by the Spanish Navy for a shipboard fire-control application. The US Coastguard has also selected an active gated version for possible use with its Dassault-Breguet HU-25 Guardian

aircraft, the application being the monitoring and policing of territorial waters. In this configuration the equipment has been designated AN/ASQ-174 active gated television (AGTV) and was flight-tested during 1984/85.

Dimensions: (camera head) 3 inches dia × 9 inches (76 × 228 mm)
(electronics unit) 9.5 × 7.26 × 6.5 inches (241 × 184 × 165 mm)
Weight: (camera head) 3 lb (1.36 kg)
(electronics unit) 10 lb (4.54 kg)

STATUS: in production and service.

Infra-red search and track system

General Electric is developing an airborne infra-red search and track system (IRSTS) under what is described as 'multi-million dollar funding' from the US Air Force's Wright Aeronautical Laboratories at Wright-Patterson Air Force Base, Ohio. Under this programme, General Electric is to build and conduct flight evaluation of an IRSTS which employs modern focal-plane array technology and sophisticated lsi-circuitry signal/clutter processing techniques. The system is intended to permit the multiple tracking of thermal energy emitting targets at extremely long ranges to augment information supplied by conventional tactical radars.

The General Electric programme entered the Phase 2B final design, fabrication and flight demonstration stage during the latter half of 1982 following a 16-month Phase 1 and 2A early design effort.

STATUS: under development.

GTE Sylvania

GTE Sylvania Systems Group, Western Division, 100 Ferguson Drive, Mountain View, California 94042
TELEPHONE: (415) 966 9111

Have Lace programme

Early in 1984 the US Air Force's Aeronautical Systems Division awarded GTE Sylvania a $419 000 contract to produce two laser communications terminals under the Have Lace (laser airborne communications experiment) programme. Each terminal will consist of a

laser transceiver and an acquisition and tracking system, with a terminal in each of two co-operating Boeing C-135 aircraft. Communications of voice and data (up to 20 000 bits a second) at ranges up to 100 nautical miles will be tested.

Honeywell

Honeywell Inc, Military Avionics Division, Box 31, 2600 Ridgeway Parkway, Minneapolis, Minnesota 55413
TELEPHONE: (612) 378 4141
TELEX: 290631

ATR automatic target recogniser for helicopters

Under direction from the US Army's Night Vision and Electro-Optics Laboratory, Honeywell is devising a computer-based forward-looking infra-red/television discrimination system for battlefield helicopters that detects and recognises ground targets during short 'pop-up' manoeuvres from natural cover. The purpose is to reduce greatly the time spent by the aircraft above cover, and thus its vulnerability to ground-fire, while the crew searches for targets.

The system in its prototype form incorporates four functions, built into a single box. The four elements are a video multiplexer, segmentation

processor, recognition processor, and executive controller.

As the helicopter moves up from cover into a search position the ATR drives the forward-looking infra-red and television sensors through a pre-determined scan pattern. Analogue image signals are digitised in the video processor and passed to the segmentation processor, where 'objects' are separated from 'background' by identifying their temperatures. The edges of objects are also identified, and so the system can discern shapes. Objects are then sorted automatically into man-made devices or clutter. The recognition processor further sorts images in the first category into classes of targets (tanks, trucks, armoured personnel carriers and others) by comparing them with target algorithms contained within the system's memory. Finally, the target images are assigned priorities according to categories determined at the pre-flight briefing. Thus for a particular mission, armoured personnel carriers may be the most important targets, and the system is set up to display these to the crew, with priority targets

Hughes AH-64 Apache attack helicopter would be prime application for an ATR system

according to range for example. The ATR uses statistical classification to detect and recognise objects.

The high-speed algorithms are written in microcode for efficiency. The system necessarily operates in real-time, and very rapidly so as to keep down exposure time. Image data necessarily requires a large bandwidth, typically 10 MHz, and processing information in the short time permitted entails extremely high processing speeds; operational ATR systems will require around 5000 million to 10 000 million computations a second, and will therefore oblige the use of very high-speed integrated circuit technology, both for speed and to keep down the space needed.

Honeywell began privately sponsored research into ATR technology in the 1960s, leading under US Air Force sponsorship to a statistical classifier in the early Seventies. By the mid-70s algorithms had improved and microprocessors and large-scale integrated devices had developed sufficiently so that flight-test demonstrations became feasible. The first such demonstration was conducted in 1980, based on the PATS I (prototype automatic target screener) of 1977 built in conjunction with the US Army. A second-generation PATS II is now under test, with improved performance and for the first time compatible with operational electro-optical sensors. A flight-standard system was begun in 1984 leading to trials aboard an attack helicopter in early 1985.

STATUS: research, with flight trials beginning in 1985.

Honeywell Inc, Electro-Optics Division, 2 Forbes Road, Lexington, Maryland 02173
TELEPHONE: (617) 863 6222

AN/AAD-5 infra-red reconnaissance set

The Honeywell AN/AAD-5 is a dual-field, high-performance, infra-red reconnaissance set which scans the ground area beneath an aircraft's flightpath. Received infra-red energy is reflected through an optical scanner system and focused on two detector arrays enclosed in a vacuum sealed Dewar chamber. These arrays are maintained at a temperature of approximately 80K by a cryostat, which is part of a closed-cycle nitrogen cooling system, inserted into the Dewar. These different-sized,

Honeywell D-500 (left) and AN/AAD-5 infra-red reconnaissance sets

12-element detector arrays convert received infra-red energy into electrical signals. The analogue voltage representing the velocity/height (V/H) ratio determines the number and the type of detectors to be used. At low V/H ratio the small area array yields better resolution information, which improves high-altitude sensor performance and also provides a better scale for image interpretation. At high V/H ratio, the large area array is used to meet the higher data rate. The V/H ratio also determines the number of detectors used in each array, from a minimum of one to a maximum of 12, and establishes the optimum number of overlaps.

Detector signals are processed into a video format, which is displayed as an intensity modulated trace on a 5-inch (127 mm) multi-beam cathode ray tube or crt in the system's recorder. The crt image is optically coupled to the film magazine and recorded on moving film, the speed of which is also controlled by the V/H ratio analogue voltage. The system compensates for aircraft roll by using inputs from the aircraft's inertial navigation system.

In aircraft or pod installation, the AN/AAD-5 line-replaceable units may be separated from each other by a distance of up to 20 feet (6 metres) but the film magazine must remain coupled to the recorder.

The seven major line-replaceable units of the AN/AAD-5 comprise the receiver, the recorder, the film magazine, a control indicator, an infra-red performance analyser, a cooler, and a power supply.

The receiver, which scans the traversed terrain and converts radiation received into electrical signals, includes the scanning optics, cooler cryostat, two 12-element detector arrays, 24 pre-amplifiers and the associated buffer electronics. The detector arrays are mercury-cadmium-telluride photoconductors which are sensitive to infra-red radiation in the 3 to 14 micron band. One array is used for the wide field of view and the other for a narrower field of view. The receiver also contains the spin-motor regulator and an encoder to provide system timing pulses to the recorder.

The recorder converts the video signals and time pulses from the receiver into the information required to produce a film record. It consists of the crt, recording optical components, video and sweep electronic cards, digital timing circuits, a high-voltage power supply, film-speed control circuits and an ADAS (auxiliary data annotation set) film annotating crt.

The high-resolution multibeam crt provides spatial correspondence between the readout and the detector array. It also affords flexibility in the selection of field of view options for the infra-red reconnaissance set. The crt phosphor is of a fine grain type with aluminized backing, optimised for photo-recording. A spot size of 0.001 inch (0.025 mm) is achieved. The tube is

Honeywell night navigation pod for General Dynamics F-16B

Honeywell Mini-Flir is based on standard modules

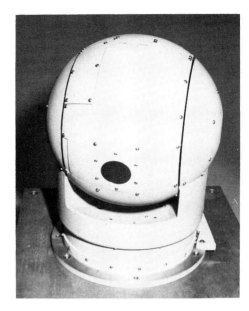

Honeywell helicopter night-vision system housed in lightweight two-axis gimballed dome

of a ruggedised type suitable for airborne applications.

An Autofocus opto-electronic sub-system is used to maintain optimum focus in the electronic optics of the multibeam crt. It also maintains correct spot brilliance on the faceplate and provides a sensor for the built-in checkout system which monitors performance of the display. A recorder head assembly for the ADAS is also provided. According to Honeywell, an advantage of using a crt as the recorder is that the recorder system can be mounted remotely from the receiver section.

The cryogenic refrigerator compressor receives gaseous nitrogen from the receiver cryostat, compresses it to approximately 2200 psi and returns it to the cryostat for expansion to a liquid state. This compressor contains a motor, turn-on relays, fans, absorbers and safety thermostats.

The AN/AAD-5 was initially designed for service with US Air Force/McDonnell Douglas RF-4C aircraft but has also been procured by the US Navy and Marine Corps for use with McDonnell Douglas RF-4B and Grumman F-14 aircraft respectively. Following extensive testing, the first production systems were delivered in December 1976. The system is in service on

various types of aircraft with the armed forces of several nations including Australia, Greece, Iran and Turkey. The system is suitable for high performance reconnaissance aircraft and a smaller version of the equipment, designated D-500, is available for smaller aircraft and remote-piloted vehicles.

Dimensions: (power supply) 10 × 17 × 7 inches (250 × 430 × 180 mm)
(receiver) 18 × 15 × 13 inches (460 × 380 × 330 mm)
(recorder) 16 × 23 × 9 inches (410 × 580 × 230 mm)
(magazine) 15 × 7 × 10 inches (380 × 180 × 250 mm)
(control indicator) 6 × 4 × 3 inches (150 × 100 × 80 mm)
(cryogenic refrigerator compressor) 16 × 8 × 12 inches (410 × 200 × 300 mm)
(IR performance analyser) 6 × 2 × 12 inches (150 × 50 × 300 mm)
Weight: (power supply) 43 lb (20 kg)
(receiver) 82 lb (37 kg)
(recorder) 97 lb (44 kg)
(loaded magazine) 34 lb (15 kg)
(control indicator) 2 lb (1 kg)
(cryogenic refrigerator compressor) 25 lb (11 kg)
(IR performance analyser) 4 lb (2 kg)
(total system weight) 287 lb (130 kg)

STATUS: in production and service.

Night navigation pod for F-16B

The Honeywell night navigation Flir forward-looking infra-red pod, primarily designed for the two-seat General Dynamics F-16B attack aircraft, can be mounted on the fuselage or wing pylons. It is also suitable for the Northrop F-5 and Fairchild Republic A-10 aircraft.

Honeywell Mini-Flir installation US Army Aquila RPV

The equipment generates information for a raster-scan cockpit display of the detected imagery. Two 14-element mini-Flir monolithic liquid-crystal displays give fields of view of 19° × 28° and 7.5° × 11.5°. Focus, gain and level controls are automatic and the pod is compatible with MIL-STD-1553B data-bus controllers, or it can be hardwired.

Dimensions: 150 mm dia × 1470 mm
Weight: 36 kg

STATUS: under development.

Night-vision system for helicopters

Honeywell's helicopter night-vision system is a miniature, forward looking infra-red system designed principally to provide a night time navigation capability. It may also be used to assist helicopter aircrews to locate targets in battlefield smoke and dust conditions.

The system's sensors are mechanically cooled and comprise a single 24° vertical × 37° horizontal field of view optical unit mounted in a two-axis gimbal and accommodated in a sphere suitable for 'chin' site mounting on a helicopter. The sensor is of modular construction and employs monolithic circuitry. The sensor elements are serially-scanned. Images produced are claimed to be bloom-resistant and of uniform quality.

The system is television-compatible and may be readily integrated with helmet and remote displays, automatic trackers and target cuers, video processors and recorders.

Dimensions: 10 inches (254 mm) height
7.99 inches (203 mm) sphere dia
7 × 7 inches (178 × 178 mm) base
Weight: 24.2 lb (11 kg)

STATUS: in production.

Flir payload for Aquila RPV

In mid-1984 Honeywell teamed with Ford Aerospace and began full-scale development of an enhanced payload for night use on the US Army/Lockheed Aquila RPV. Under a $40.8 million contract, Honeywell is providing its Mini-Flir system and dual-mode auto-tracker. During a programme scheduled to last 41 months, nine prototype payloads will be produced for trials with the Aquila RPV.

STATUS: in development.

Hughes

Hughes Aircraft Company, Missile Systems Group, 8433 Fallbrook Avenue, Canoga Park, California 91304
TELEPHONE: (818) 702 1000
TWX: 910 494 4997

AN/ASB-19(V) ARBS angle rate bombing set

The Hughes AN/ASB-19(V) ARBS angle rate bombing set was designed for US Marine Corps aircraft to improve day and night bombing accuracy when operating in the close-support role using unguided weapons. The system provides accurate delivery, irrespective of target velocity, wind velocity, target elevation or dive angle. It is also compatible with guided ordnance and can also be used to direct gunfire and air-to-ground rockets. ARBS was originally designed for application to the US Marine Corps/McDonnell Douglas A-4M and is also the primary weapons delivery system for the US Marine Corps and Spanish Navy AV-8Bs and the Royal Air Force's Harrier GR5 aircraft.

ARBS comprises three main sub-systems: a dual-mode tracker, weapon-delivery computer, and control units. The tracker comprises laser

and pilot-controlled television tracking equipment, both using a common optical system. The dual-mode tracker automatically tracks targets designated by a laser either ground-based or in a partner aircraft, or targets which are television-designated by the pilot of the ARBS-equipped aircraft itself. The tracking system's common optics enable transition from laser to television tracking mode to be accomplished without losing the target.

Tracking information is passed from the dual mode tracker to the weapon delivery computer. The weapon delivery computer performs computations for weapon trajectory and fire control, position control of the dual-mode tracker during target acquisition, digital filtering of the dual-mode tracker angular rate signal outputs and automatic fire or weapon-release signals to the armament and other systems.

The weapon delivery computer receives aircraft-target line-of-sight angle and angle rate data from the tracker when that unit has achieved target lock-on. This data, combined with true airspeed and altitude information from the air data computer, and processed by the weapon delivery computer, yields the weapon delivery solution. Target position, weapon release and azimuth steering information are generated and presented to the pilot via a head-

up display. The function of the ARBS control units is to interface with the weapon delivery computer and to provide the pilot with control of the tracker modes, entry of display and target information, navigation and maintenance information.

When operating in the laser mode, the sensor automatically acquires the target, which is illuminated from an external laser source. The sensor, via the weapon delivery computer, presents steering signals to the pilot on the head-up display. In television mode, visible light from the common optical system is directed to form an image on the television display. Tracking of small, low contrast, poorly defined moving targets is said to be possible even with changes in aspect ratio, in the presence of competing clutter or when the target is partially lost to view behind ground obstructions.

Following head-up acquisition in the television mode, the pilot is shown a × 7 magnified image of the target on the cockpit television monitor display. He may at this stage use a hand control to slew the tracker gate onto a new track point or onto a nearby alternative target.

Weapon release in either laser or television tracking mode may be made automatically or manually, with a weapon release time and a

Hughes ARBS in nose of US Marine Corps AV-8B

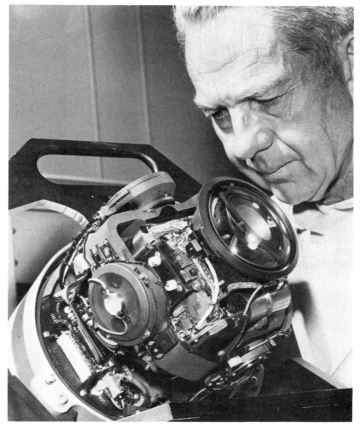

Hughes ARBS being adjusted before delivery

continuously computed impact point being generated by the weapon delivery computer. A weapons data insert panel provides entry into the weapon delivery computer of weapon characteristics and rack type for those stores being carried for a specific mission.

ARBS permits delivery of any type of weapon at any altitude and airspeed combination at any dive angle or in level flight. The system accuracy is said to be sufficient for first-pass precision delivery in close air support or interdiction missions against land or sea-borne targets, but navigation steering commands are also generated to provide second passes against hardened targets. This reattack facility provides the pilot with head-up steering information for return to a designated target, the location of which is retained in the system memory until a new target is designated by the pilot.

The system's television tracking element also provides a limited air-to-air capability for daylight operations, its × 7 magnification being particularly useful for visual identification and tracking of airborne targets.

Angular coverage of ARBS extends from ±35° in azimuth and from +10° to −70° in elevation at roll angles of ±450°. These limits are sufficiently wide to allow extensive flexibility in weapon selection and delivery profiles and permit multiple-roll manoeuvres to take place on the attack approach run without loss of target lock-on. The tracking equipment's laser is NATO-coded, but is compatible with other current laser designators such as Hughes' TRAM (target recognition and attack multisensor) at ranges adequate to ensure the success of first-pass strikes.

The ARBS weapon delivery computer processor on the A-4M is a 16 K word capacity system extensible to 32 K words. With 16 K words, the system has far more capability than is required for weapon release and steering command computation and may be used for additional purposes such as the pilot initiated self-test of the ARBS system which performs operational readiness testing and fault-isolation functions. By modification of the computer input/output scaling factor, the ARBS can be

FACTS Flir system on test bench at Hughes' factory

configured to interface with aircraft avionics other than those of the A-4M and the Harrier AV-8B. Three basic aircraft interfaces are required: a vertical reference to provide pitch and roll data, true airspeed and a servoed optical sight or head-up display on which to present azimuth steering commands, continuously computed impact point and bomb-release information.

ARBS systems, therefore, can be configured, in either internally- or pod-mounted versions, to suit the requirements of other aircraft. Several pod configurations have been formulated for candidate aircraft with attachment hooks at 14 inches (356 mm) and 30 inches (762 mm) standard centres, compatible with standard pylon or fuselage centreline mountings. Where existing cockpit controls cannot be used, two small ARBS control units may be fitted at any accessible position. The A-4M

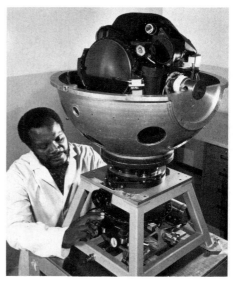

Hughes detecting and ranging set for US Navy/Grumman A-6E attack aircraft

control units are configured with standard 5.75-inch (146 mm) widths adaptable to most cockpit consoles.

Exploratory development of ARBS began in 1965 at the US Navy's Weapons Test Center, China Lake, California, although concept feasibility was not demonstrated until 1970. In 1974 Hughes received an engineering development contract for the construction and test of six production-standard prototypes following a competitive flight test programme with other systems. A full production contract ($48 million for 92 aircraft sets) was awarded in 1979 following extensive flight testing at both China Lake and the Naval Air Test Center at Patuxent River. Production commenced in late 1979 and deliveries began in July 1982 for retrofit to US Marine Corps A-4M Skyhawks and, in 1984 to US Marine Corps Harrier AV-8B aircraft. The Royal Air Force has also specified the system

Nose-mounted FACTS sensor in Bell AH-1S Cobra light attack helicopter

for its Harrier GR5s and conducted an evaluation of the system during a series of special test flights from the US Marine Corps Air Station at Cherry Point, North Carolina.

Dimensions: 2.5 ft³ (0.07 m³)
Weight: 128 lb (58.18 kg)

STATUS: in production and service.

Hughes Aircraft Company, Electro-Optical and Data Systems Group, PO Box 902, El Segundo, California 90254
TELEPHONE: (213) 616 1375
TWX: 910 348 6666

AN/AAS-33 TRAM target recognition and attack multi-sensor

The Hughes AN/AAS-33 target recognition and attack multi-sensor (TRAM) detecting and ranging set is an integrated day/night weapon delivery system developed for the US Navy/Grumman A-6E Intruder aircraft, under a contract from that service's Air System's Command. The equipment comprises a forward looking infra-red sensor, a laser designator-ranger and a laser receiver, all housed in a precision-stabilised turret mounted under the nose of the aircraft. The TRAM system is designed to provide sighting and guidance facilities for a wide range of laser guided and conventional weapons. TRAM also fulfils a navigation and target location function, when aligned and operated with the aircraft's radar system.

The Flir system is said to be sensitive enough to detect the quantities of oil in storage-tank depots on the ground, solely through temperature differentials caused by varying tank levels. In operation, the weapon-operator/

navigator acquires the target by using the aircraft radar, and then engages the bore-sighted Flir sensor. The latter is fitted with a zoom lens which brings the target into close-up view for final recognition purposes. Full travel of the optical system from minimum to maximum field of view and vice versa takes less than three seconds and during zoom the scene image remains on the display so that no time is lost re-acquiring a target.

Once a target is satisfactorily acquired and recognised by the Flir element, the weapon-operator uses the laser (which is also bore-sighted) for designation, marking the target with a laser spot and so providing a radiant source for a laser-guided bomb to home on. The TRAM system provides either autonomous attack facilities for the aircraft on which it is fitted, or permits the aircraft to attack a target which is laser illuminated by another aircraft or by a ground laser source. Gyro-stabilisation of the TRAM turret ensures that both forward looking infra-red and laser sensors remain accurately aligned on the target during high-*g* attack manoeuvres.

STATUS: in production and service.

Cobra-Nite airborne TOW system upgrade

Cobra-Nite is the designation given to FACTS, a Flir-augmented weapon sighting system, which was developed in 1979 by the Hughes Aircraft Company's Electro-Optical and Data Systems Group to provide an enhanced night and adverse visibility capability for Cobra helicopters equipped with TOW missiles. FACTS employs US Army thermal imaging common modules which have been manufactured by Hughes under contract from the Army's Night Vision and Electro-Optics Laboratories, Fort Belvoir, Virginia. Modules designed and produced under this programme have been applied to a number of night-vision and thermal detection and sighting systems for service with armoured fighting vehicles, including the XM-1 tank, and those armoured fighting vehicles equipped with the fighting vehicle system turret armed with TOW/Bushmaster weapon systems.

The FACTS system has been successfully tested aboard Bell AH-1S helicopters during trials conducted at Fort Hood, Texas and Fort Polk, Louisiana. According to Hughes, the

system performed well and during live firing tests five hits were achieved in the same number of firings.

FACTS also provides the Cobra with a target sight for use with unguided rockets and during conventional gun attacks. The system has also been used effectively during exercises to monitor 'hostile' forces and to direct ground troops from the helicopter, resulting in envelopment of the 'enemy' by 'friendly' forces. One modified airborne TOW system with FACTS has been delivered to the US Army.

In the autumn of 1984 Hughes received a production contract to commence retrofitting over 500 of the US Army's Bell AH-1S Cobra helicopters with a production version of FACTS. The night-time Flir system will be used with the TOW 2 missile and the helicopter's cannons and rockets and is designated Cobra-Nite.

Hughes also received a $21 million contract from the US Army in late 1984 to furnish 17 Flir systems for fitment to Hughes 500E helicopters.

STATUS: Cobra-Nite in production.

AN/AAQ-16 helmet night-vision system

Hughes is developing a helmet-mounted night-vision system for use in attack helicopters and other vehicles. The pilot's head movement is transmitted through a helmet servo linkage to the Flir turret and the resulting Flir image is presented to the pilot on his visor, superimposed with flight information. In this mode, the helmet's field of regard is ±115° in azimuth and from +60° to –30° in elevation.

The Flir turret can also be controlled by a joystick for even greater angular range: ±210° in azimuth and from +85° to –180° in elevation.

Dimensions: (Flir turret) 17 inches (43 cm)
Reliability: > 300h mtbf
Total system weight: 99 lb (45 kg)

STATUS: in development.

Detecting and ranging set

The Hughes detecting and ranging set is contained within an improved turret for the US Navy's Grumman A-6E Intruder. It incorporates a thermal imager and laser devices to enable the crew to attack surface targets in darkness and poor weather.

Hughes TRAM sensor package installed in turret

Hughes night-vision system; pilot is wearing head-up display visor, sensor package is seen below helicopter's nose

Compared with the previous infra-red turret installed on the A-6E, the new device has a clamshell design which reduces the length of cabling by 35 feet (11 metres) and gives easy access to all the sensors, thereby significantly reducing maintenance. The new turret entered service at the end of 1983 and more than 150 of the 199 sets ordered had been delivered by early 1986.

STATUS: in production. In addition to an earlier contract for 199 sets, the system is now being bought under the US Department of Defense multi-year acquisition policy, which in this case will save about $72 million (or about 20 per cent) on the cost of four separate annual contracts. The new contract calls for construction of a further 125 sets.

LAAT laser-augmented airborne TOW sight

The standard M65 TOW-missile sight produced by Hughes Aircraft for use on US Army/Bell AH-1S Cobra light-attack helicopters contains a miniature laser transmitter, designated the LAAT sight.

The laser is enclosed in a box measuring 130 × 130 × 40 mm. It produces four pulses a second for up to 5 seconds, then switches off for a 5-second cooling period. The unit provides ranging data to objects in the LAAT sight, information being fed directly to the fire-control computer for accurate target sighting, irrespective of helicopter or target motion.

A novel cooling technique is employed to reduce size and weight: the laser flash-tube is embedded in a heat-conductive and highly-reflective layer of material, and this obviates the need for more conventional cooling equipment which is bulky and heavy.

STATUS: in production.

IRST infra-red search and track system

In the 1970s, Hughes equipped several hundred US Air Force F-101, F102 and F-106 aircraft with infra-red search and track (IRST) systems.

In mid-1985 Hughes proposed transferring 300 of these sets to equip F-4s of the US Air National Guard, at a total cost in the order of $30 million which, according to Hughes would be one-tenth the price of a new system.

The IRST system was flight-tested by the 119th Fighter Interceptor Group in 1985.

Hughes laser-augmented airborne TOW sight

International Laser Systems

International Laser Systems, Litton Industries, 3404 North Orange Blossom Trail, Orlando, Florida 32804
TELEPHONE: (305) 295 4010
TWX: 810 850 0195

Laser transmitter assembly for Aquila RPV

A sub-system of the US Army/Lockheed Aquila RPV programme, the laser transmitter assembly is supplied to Westinghouse as a component of that company's mission payload sub-system. It operates in conjunction with a television camera to provide intelligence gathering, navigation update, range-to-target information and target illumination without endangering human life. The system comprises a heat exchanger, cooling system, power supply, pulse-forming network, optics and associated electronics.

Weight: 3.2 kg
Environmental specification: MIL-STD-810C

STATUS: under development.

Lantirn designator/ranger laser

Lantirn (low-altitude navigation and targeting infra-red system for night) is under development for use on US Air Force/General Dynamics F-16 and Fairchild A-10 attack aircraft. It is intended to provide first-pass attack capability at a greater stand-off range than with conventional systems.

The laser designator/ranger is fitted into the targeting pod and provides the laser energy for illuminating/ranging targets for the delivery of conventional and laser-guided ordnance. The unit is claimed to be one of the most advanced systems of its kind under development. Of modular construction, the lightweight device is microprocessor controlled and uses digital interfaces to provide serial or parallel commands directly to the Lantirn navigation pod. The device has a self-contained, automatic or manually initiated built-in test capability, and is fitted with connectors for automatic test equipment fault-isolation checks.

The system comprises four line-replaceable units: a laser transmitter-receiver, an electronic unit, a high-voltage power supply, and a synchroniser/ranger computer.

STATUS: under development.

Pave Tack designator/ranger laser

The only laser designator/ranger system currently deployed by the US Air Force, the AN/AVQ-25 Pave Tack designator and ranger (see Ford Aerospace entry) uses a laser system developed by ILS. It comprises a laser transmitter, power supply, laser receiver and a range-counter computer. The transmitter features an ILS-patented cross-porro prism interferometer for maximum alignment stability over the range of shock, vibration and temperature conditions specified by MIL-STD-810C.

STATUS: in service. Similar systems have been developed for the TADS target acquisition and designator sight on the US Army/Hughes AH-64A Apache helicopter and also as part of the Northrop mast-mounted sight in the Army's OH-58D AHIP Army helicopter improvement programme.

In 1984 eight AN/AVQ-26 Pave Tack systems, total value $50 million, were ordered by South Korea for equipping McDonnell Douglas F-4E aircraft.

Martin Marietta

Martin Marietta, Orlando Aerospace, PO Box 5837, Orlando, Florida 32855
TELEPHONE: (305) 352 2211
TELEX: 564414
TWX: 851 850 4125

Lantirn low-altitude navigation and targeting infra-red for night system

One of the most important avionic system projects in the US, with a currently-estimated programme cost of $3.16 billion, Lantirn will provide the means by which single-seat close-support aircraft will be able to acquire their targets and deliver guided and unguided weapons around-the-clock. The initial application will be the two-seat McDonnell Douglas F-15E chosen by the US Air Force early in 1984. Lantirn will also equip General Dynamics F-16C/D, and possibly the Fairchild A-10. Compatibility of the navigation portion of the Lantirn system with single-seat operation has been demonstrated and targeting pod capability will be verified during operational system testing.

The system comprises two separate sets of equipment each contained in its own pod suitable for underwing or under-fuselage attachment. Either or both pods can be carried, depending on the particular mission requirement. This option enhances flexibility and diminishes support demands. The equipment within the pods is supplied by a number of manufacturers, but overall responsibility rests with Martin Marietta as the prime contractor.

The first, navigation, pod contains a wide field of view forward looking infra-red unit and a terrain-following radar installation, together with the associated power supply, pod control computer and environmental system. Flir imagery from the pod is displayed on a new, wide field of view holographic head-up display developed by GEC Avionics for the Lantirn system. This provides the pilot with night vision for safe flight at low-level. The Texas Instruments Ku band terrain-following radar permits operation at very low altitudes with en route weather penetration and blind let-down capability. Terrain-following is accomplished manually by means of symbology presented to the pilot on the head-up display. The A-10 does not have an automatic flight control system designed for automatic terrain-following. However, the F-15E and F-16 will have fully automatic terrain following with the production navigation pods. Automatic implementation can be accomplished, says Martin Marietta, but is not included as part of the present programme.

The second, targeting, pod contains a stabilisation system, wide and narrow-field Flir, laser designator/ranger, automatic multi-mode tracker, automatic infra-red Maverick missile hand-off system, environmental control unit, pod control computer, power supply and provision for an automatic target recogniser. The targeting pod interfaces with the aircraft controls and displays as well as the fire control system to permit low-level day/night manual target acquisition and semi-automatic weapon delivery of guided and unguided weapons. It may be arranged as a laser designator-only pod, for use with laser-guided munitions and conventional weapons, by deleting Maverick hand-off sub-system.

Both pods have environmental control units to ensure that their systems will function satisfactorily over a wide range of temperatures. Aircraft interfaces include a MIL-STD-1553 multiplex data-bus, video channels, and power supplies. Service ground support employs typical three-level maintenance: organisational, where no special test equipment is required;

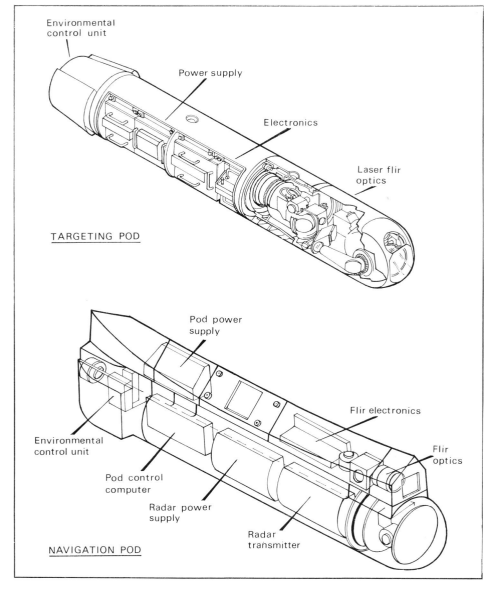

TARGETING POD

- Environmental control unit
- Power supply
- Electronics
- Laser flir optics

NAVIGATION POD

- Pod power supply
- Environmental control unit
- Pod control computer
- Radar power supply
- Radar transmitter
- Flir electronics
- Flir optics

Internal details of Martin Marietta Lantirn targeting and navigation pods

Hughes AH-64A Apache with Martin Marietta TADS/PNVS system mounted on nose

intermediate, which employs automatic test equipment; and depot servicing.

Development of Lantirn commenced in 1980, under a $94 million contract awarded to Martin Marietta by the US Air Force, and completion was initially scheduled to take three years. The programme is directed by the US Air Force Aeronautical Systems Division, at Wright-Patterson Air Force Base, Ohio. Under the development contract, six prototype units were designed, developed and tested and initial production began. Following the award to Martin Marietta of a $87 million production contract for two systems in April 1985 production of the navigation pod began in 1985 but, because of the need for further trials, production of the targeting pod did not start until 1986. The procurement of 720 pod sets and associated support has been authorised under a $3190 million firm fixed price contract.

As well as Martin Marietta, GEC Avionics and Texas Instruments, a number of other companies are also involved in the Lantirn development programme. These include Hughes Aircraft, responsible for the target recogniser and boresight correlator for Maverick hand-off; General Motors Delco Division, which supplies computers; Sundstrand Aerospace, which provides the environmental control; and International Laser Systems, suppliers of the laser designator.

The first Lantirn navigation pod was accepted by the US Air Force Aeronautical Systems Division in late February 1983 and delivered to General Dynamics, Fort Worth, Texas for integration with an F-16 fighter for engineering ground tests. Delivery of the first targeting pods took place in 1983 and they have been installed and flight tested, over some 40 sorties aboard an F-16 at Edwards Air Force Base, California. This phase ended in December 1985 and initial operation evaluation and testing began in January 1986 at McChord Air Force Base, Washington.

In October and November 1984 a series of 55 initial operational tests were carried out on a Canadian Air Force range using three US Air Force F-16s. The trials were described by the US Air Force as "highly successful".

STATUS: navigation pod in production: targeting pod began operational flight testing in January 1986. System is scheduled to become operational in 1989, prime application being the US Air Force F-15E to be followed by the F-16C and A-10.

TADS/PNVS target sight and night-vision sensor for AH-64A helicopter

Martin Marietta's TADS/PNVS is designed to provide day, night and limited adverse weather target information and navigation capability for the US Army/Hughes AH-64A Apache battlefield helicopter. This aircraft, after a practical development programme, was scheduled to become operational in 1985.

TADS/PNVS comprises two independently functioning systems known as the target acquisition designation sight and the pilot night vision sensor. The first of these provides the co-pilot/gunner with search, detection and recognition capability by means of direct-view optics, television or forward looking infra-red sighting systems which may be used singly or in combinations according to tactical, weather or visibility conditions. The second system provides the pilot with flight guidance symbology which permits nap-of-the-earth flight to,from and in the combat area at altitudes low enough to prevent or delay detection by enemy forces.

TADS consists of a rotating turret, mounted on the nose of the helicopter and containing the sensor sub-systems, an optical relay tube located at the co-pilot/gunner station, four electronic units in the avionics bay, and cockpit controls and displays. TADS turret sensors have a field-of-regard covering ±120° in azimuth and from +30° to -60° in elevation. By day,

This flight-trials General Dynamics F-16A carries under its fuselage both navigation and targeting pods of Martin Marietta Lantirn system

Martin Marietta TADS/PNVS installation on nose of Hughes AH-64A Apache attack helicopter. PNVS night-vision sensor is uppermost of two units. TADS system is below and contains the Flir (lower left) and TV (lower right)

either direct-vision or television viewing may be used. The direct-vision system has a narrow field of view, 4° at a × 16 magnification, and a wide field of view, 18° at ×3.5 magnification. The television system provides a narrow field of view of 0.9° and a wide field of view of 4°. For night operations the TADS sensors have three fields of view: narrow of 3.1°, medium of 10.1°, and wide of 50°.

Once acquired, targets can be tracked manually or automatically for autonomous attack with guns, rockets or Hellfire anti-tank missiles. A laser may also be used to designate targets for attack by other helicopters or by artillery units firing the laser-guided anti-armour Copperhead weapon.

PNVS consists of a forward looking infra-red sensor system packaged in a rotating turret mounted above the TADS, an electronics unit located in the avionics bay, and the pilot's display and controls. The system covers a field of ±90° in azimuth and from +20° to –45° in elevation. Field of view is 30° × 40°.

TADS is designed to provide a back-up PNVS capability for the pilot in the event of the latter system failing. The pilot or the co-pilot/gunner can view, on his own display, the video output from either TADS or PNVS, raising the probability of mission success. Although designed primarily for combat helicopters flying nap-of-the-earth missions, PNVS may also be used as a single entity in tactical transport and cargo helicopters.

Special attention has been paid to reliability, as has the ability to remove and replace units easily and rapidly on the flight-line. TADS/PNVS underwent final evaluation during the latter part of 1981 and a US Army production contract for $130 million was awarded to the manufacturer in mid-1982. Initial production called for 13 systems with support equipment. Follow-on contracts worth $300 million were also expected.

In January 1983 Northrop was awarded an $8.5 million contract by the US Army Aviation Research and Development Command, St

Louis, Missouri, to modify four Bell AH-IS Cobra helicopters to accommodate the Martin Marietta PNVS and other vision equipment. Northrop is also to furnish an electronics controller unit, enabling PNVS to work with software previously developed for the US Army. The modified helicopters are to be operated by the US Army Training Command, Fort Rucker, Alabama for use in the US Army's/Hughes AH-64 Apache training programme. A production model TADS/PNVS system carried out successful trials fitted to a AH-64 attack helicopter at the Yuma Proving Ground, Arizona, in the summer of 1984.

STATUS: in production. Programme is expected to run for ten years, with production of 700 to 800 systems. Ultimately, all 515 Apaches planned for US Army will have system. US Marine Corps and West German and Italian armies with their respective Hughes AH-64A, Eurocopter PAH-2 and Agusta A129 helicopters are also potential users. Saudi Arabia has been mentioned in connection with this system.

AN/ASQ-173 LDT/SCAM laser detector tracker/strike camera

Developed specifically for the US Navy/McDonnell Douglas F/A-18 Hornet, the laser-spot tracker permits crews to identify laser designated targets illuminated by either ground forces or co-operative aircraft, and to achieve more accurate tracking and weapon-release performance than a non-laser equipped aircraft. Target-position data from the laser detector tracker is fed directly to the F/A-18 mission computer and used to provide weapon-aiming and ordnance-release information.

The laser-spot tracker optics are stabilised using attitude data passed to the unit from the aircraft inertial navigator. The detector is a four-quadrant photodiode type, mounted behind a hemispherical dome. Laser-pulse decoder electronics are also contained in the detector.

Located in the same pod, for which Martin Marietta is prime contractor, is a strike-camera. This unit is in the aft section and has a wide field of view in the lower hemisphere. It can be slaved to the laser-spot tracker or independently controlled by the mission computer. Photographic data from this unit permits rapid strike-damage assessment after aircraft attacks.

Units have been flown for test and evaluation since 1980, and production deliveries to the US Navy and Marines commenced in 1983. For strike operations with laser-spot tracker/strike-camera on the F/A-18, the laser-spot tracker/strike-camera pod will be mounted on the starboard fuselage side, and an infra-red target acquisition/tracker pod will be carried on the port fuselage side.

Ldt/scam pod length: 2.29 m
Body diameter: 0.2 m
Weight: 73 kg
Operating wavelength: 1.06 microns
Scan-patterns: pre-programmable/pilot-selectable.

STATUS: in production.

AN/AAS-35(V) Pave Penny laser-tracker

This is an advanced miniaturised day/night laser-based target identification set. Used in conjunction with a laser-designation system, either ground-based or in a co-operating aircraft, targets can be recognised and identified rapidly, and accurate steering-data provided to assure quick pilot reaction and accurate delivery of weapons. The equipment was installed initially on US Air Force/Fairchild A-10 close-support aircraft and later on LTV A-7D Corsair IIs. Pave Penny is also to be used on the General Dynamics F-16, and is suitable for the McDonnell Douglas F-4 Phantom, Northrop F-5 and Dassault-Breguet/Dornier Alpha Jet.

A silicon pin diode detector head is used, with full lower-forward hemisphere coverage plus some lock-up capability. Pilot's controls permit selection of several seeker scanning patterns to improve early designator recognition. It can be used to improve the accuracy of conventional weapon-delivery, or to lock-up laser-guided munitions. The system is contained in a relatively small pod, which is usually fuselage-mounted to allow easy harmonisation with other on-board sensors. Pre-flight bore-sighting of A-10 units is claimed to allow pod attachment in a matter of minutes. An aircraft adapter module is used to integrate the sensor data with on-board processors and a pilot's control panel provides for easy use and built-in test operations.

The first operational Pave Penny/A-10 unit was delivered in March 1977 and the US Air Force has been equipping its entire fleet of A-10s, originally numbering 733 aircraft. Operations began with the 354th Tactical Fighter Wing in January 1978. Up to 380 US Air Force A-7D Corsairs are also to have Pave Penny installed, and Pave Penny/F-16 development is complete.

Pod length: 0.833 m
Max diameter: 0.2 m
Weight: 14.5 kg
Operating wavelength: 1.06 microns
Scan coverage: (elevation) –90° to +15° (azimuth) –90° to +90°
Selectable scan patterns: wide, narrow, depressed, offset
Output: direction cosines of line-of-sight
Power: <18 A at 28 V dc

STATUS: in service.

Martin Marietta AN/ASQ-173 LDT/SCAM fitted to US Navy F/A-18 Hornet

Martin Marietta Pave Penny laser tracker installation on Fairchild A-10 Thunderbolt II uses special pylon on front starboard fuselage

Northrop

Northrop Corporation, Electro-Mechanical Division, 500 East Orangethorpe Avenue, Anaheim, California 92801
TELEPHONE: (714) 871 5000
TELEX: 655417
TWX: 910 592 1268

Seehawk forward-looking infra-red system

Seehawk is a thermal imaging system using tri-service common-module Flir sensors to provide high-resolution imagery. Designed for service with aircraft or surface vessels, the system's principal application has been aboard a US Coast Guard/Sikorsky HH-52A helicopter on which a prototype unit has been installed for trials. Its primary role in service would be search and rescue although subsidiary tasks include law-enforcement, maritime environmental control and reconnaissance, fisheries control, disaster relief, marine traffic control, and navigational assistance.

As applied to the HH-52A, the Seehawk Flir comprises a nose-mounted, sealed turret containing the sensors and an electronic processing

and display system carried in the cabin. The internally-cooled detector array has two selectable fields of view, a wide field of 30° × 40° and a narrow field of 10° × 13°. The field of regard covered by the turret is ±90° in azimuth and from +30° to –80° in elevation. The sensor system, caged in a gyro-stabilised gimbal mounting, is protected by a window when not in operation. Two display systems are used in this trial installation, a 5-inch (127 mm) unit on the cockpit centre instrument panel and a 10-inch (254 mm) system in the helicopter's main cabin. The displays presently used are standard television-type cathode ray tubes and facilities for video-tape recording the field of view are also incorporated into the current system.

Control of the system can be exercised from the cockpit or cabin positions. A control stick with top-mounted thumb switches is used to control sensor scan in azimuth and elevation, focus, scan centring, and field of view selection. An automatic scan facility which provides constant search coverage in elevation and azimuth is also controlled from switches on the control stick. The auto-search mode is enhanced by inclusion of automatic lock-on which reacts to either large or small targets, as selected by the operator. Small dots appear

Northrop infra-red monitor in TH-1S Cobra helicopter

around the display's crosswire centre-reticle when lock-on is made, and these commence flashing if break-lock occurs. Controls for the auto-search, scan rate, scan angle and continuous or step scan are above the cathode ray tube display. As an aid to target recognition, the Seehawk display system has selectable white-hot or dark-hot display polarity.

The Seehawk is also installed in the Beech 200T Special Mission Aircraft which is suitable for maritime patrol and surveillance.

Weight: (Flir turret) 70 lb (31.8 kg)
(Flir electronics and power supply unit) 35 lb (15.9 kg)

STATUS: ready for production.

Northrop Tiseo system mounted on US Air Force/McDonnell Douglas F-4 fighter

Tiseo electro-optical target-identification system

Northrop's Tiseo (target identification system electro-optical), claimed by the company to be the first operational electro-optical system, is a television-based, passive, daytime automatic target acquisition and tracking system. Comprising a high resolution closed-circuit television sensor combined with a two fields of view telescope, Tiseo is mounted on the port wing leading-edge of many US Air Force/McDonnell Douglas F-4 fighters. More than 500 Tiseo systems are known to have been delivered to the US Air Force.

STATUS: in service.

TCS television camera system

The Northrop TCS (television camera system) is similar to the Tiseo (see preceding entry) but offers enhanced capabilities. It is a passive, daytime, automatic search and acquisition system which can be either manually operated or slaved to an air-interception radar. The system not only acquires targets automatically, but presents multiple fields of view and is operational on US Navy/Grumman F-14 Tomcat fighters.

STATUS: in production. Northrop has orders for 305 units.

VATS video-augmented tracking system

Under a contract from Ford Aerospace Communications Corporation, Northrop is working on the engineering development of a system called VATS (video augmented tracking system), designed for retrofit to the Pave Tack system used by US Air Force/McDonnell Douglas F-4E and General Dynamics F-111 fighters. Pave Tack is an air-to-ground, laser-designated weapons delivery system mounted in a pod containing a Flir sensor. VATS would automatically track ground targets, eliminating or reducing the weapons operator's need to follow the monitor by eye. It is micro-processor controlled, with a self-contained power supply and a digital interface to the aircraft's flight computer.

STATUS: ready for production.

Infra-red search and track systems

Northrop is currently involved in the development of infra-red search and track systems,

designed for passive long-range target acquisition at night and in adverse weather. This equipment is known to have a multi-target capability. Other roles envisaged for these systems include weapon delivery, surveillance and fire control. One such system is called the focal plane array. It is an infra-red sensor that can direct optical target acquisitions at night or in poor visibility, and was tested on a US Army helicopter in 1984. The unit comprises 16 384 individual sensors, each of only 0.001 inch in diameter, feeding signals into a microprocessor. The resulting information is displayed on the television monitor of the host optical system.

STATUS: under development.

Infra-red display

Northrop has developed a cockpit-mounted infra-red monitor and associated 1-inch (25 mm) helmet-mounted display to facilitate training of pilots on the US Army's TH-1S Cobra helicopters. This training leads to the pilots flying the PNVS-equipped AH-64 Apache helicopters.

IRVAT infra-red video automatic tracking

In May 1985 Northrop announced a $8 million contract from Grumman to develop an infra-red video automatic tracking (IRVAT) system for the Grumman A-6 aircraft operated by the US Navy.

IRVAT will operate in conjunction with the A-6's target recognition and attack multi-sensor (TRAM) weapons system: it will computerise and automate the tracking portion of TRAM.

STATUS: in development. Pre-production systems due for delivery in late-1986.

AN/AVQ-27 laser target designator set

This designator has been developed by Northrop for installation aboard two-seat F-5B and F-5F light fighters and can also be fitted in other two-seat aircraft. Several customers have selected this system, which permits target-designation during operations using laser-guided munitions.

The system is installed in the rear cockpit, and can be removed when not required. The installation consists of a high-power laser, stabilised direct-view optics with two fields of view, and a 16 mm recording camera. The designator set package is only 100 mm wide, and fits on the lower canopy rail, maintaining ejection-seat clearance and not interfering with other cockpit functions. A viewfinder/sight swings across into the pilot's field of view, so that the crewman can track targets. This can be conducted manually, using a two-axis hand controller, and with rate or rate-aided tracking modes. The front cockpit is fitted with a sight and canopy markings which allow the pilot to assist the crewman in initial target acquisition and then maintain the target within the set's field of view. One-hand operation is claimed.

Northrop TCS system installed in US Navy F-14

Normal operations would call for designation by one aircraft, with weapon-delivery conducted by an accompanying aircraft.

STATUS: in production.

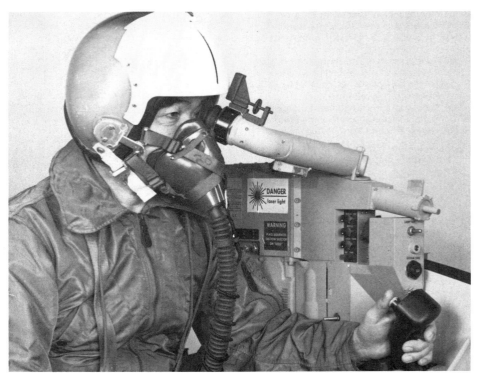

Northrop AN/AVQ-27 laser target designator set

Rockwell

Rockwell International Corporation, Strategic Defense and Electro-Optical Systems Division, Defense Electronics Operations, 3370 Miraloma Avenue, Anaheim, California 92803
TELEPHONE: (714) 762 7551
TELEX: 678437
TWX: 910 591 1654

Common module Flir digital scan converter

Rockwell International's digital scan converter performs sophisticated video processing of thermal imaging systems for displays in fixed- and rotary-wing aircraft, armoured ground vehicles and surface ships. It is the connection between the target-sensing common module Flir system and the video display presented to the operator.

The device provides improved target acquisition and identification. It performs high-capacity signal processing in real-time, obviating the need for the operator to adjust controls when approaching a target. It also eliminates the need for light-emitting diodes and drivers, visible-collimator imaging optics and the camera required by conventional electro-optical multiplexer Flir systems. These systems are replaced by the converter's high-performance electronic multiplexer, signal-conditioner and re-formatting circuits to achieve improved performance and volume reduction.

The scan converter improves common module flir performance through automatic detector equalisation and image-enhancement, wide-dynamic-range image processing interfaces and automatic (hands-off) operation. It

increases tactical flexibility through its frame freeze, a 2:1 and 4:1 electronic zoom and frame integration features.

Potential growth applications include automatic target recognition, fire control and navigation. For these purposes the system permits the Flir to present fine details in the presence of large variations in background temperature without operator intervention. The digital scan converter provides wide dynamic range capability permitting a variation of up to 50°C in background temperature without loss of detail.

In addition to acting as a format converter the system is claimed to provide a number of unspecified features which enhance sensor-derived information. The system is compatible with the MIL-STD-1553B digital data-bus and incorporates built-in self-test circuits.

Output formats cover frame rates of 25 or 30 Hz, and either 525, 625, 875, 925 or 1024 lines per frame on RS-330 and RS-343 standards. Input formats include 60 to 180 channels at aspect ratios of 2:1, 4:3 or 1:1. The dynamic range compression is 30 dB non-linear and the AGC range is 34 dB. Power consumption is less than 120 W. The system is claimed to have a mean time between failures of 2000 hours and a mean time to repair of less than 1 hour.

The system was launched in 1980, when the US Army's Night Vision and Electro-Optics Laboratory chose Rockwell to develop a digital scan converter to support a common-module Flir system for a variety of day/night applications.

Dimensions: 90-350 inches³ (1500-6000 cm³)
Weight: 3.5-12 lb (7.7-26 kg) according to type

STATUS: in production.

Rockwell International Corporation, Defense Electronics Operations, 1800 Satellite Boulevard, Duluth, Georgia 30136
TELEPHONE: (404) 497 5269
TWX: 810 766 4917

Airborne laser-tracker

This is a laser-seeker head, with associated components, which Rockwell produces for the Bell AH-1S Cobra light attack helicopter and, in derivative form for the target acquisition designation system/pilot's night viewing system (TADS/PNVS) in the US Army/Hughes AH-64 armed helicopter.

Production of AH-1S equipment was initiated in September 1981, and first delivery was in May 1984. TADS/PNVS production began in June 1981 and deliveries began in January 1983.

The seeker is a wide field of view unit, sensitive to 1.06 microns radiation, which can detect target designations from ground troops or co-operative aircraft. Coded-pulse data is used to minimise jamming, and a four-quadrant silicon detector head is used.

Dimensions and weight
(receiver unit) 214 mm × 188 mm dia; 9 kg
(electronics unit) 152 × 152 × 188 mm; 3.4 kg
(control panel) 144 × 66 mm area; 0.6 kg
Scanning field of view: (elevation) -60° to +30° (azimuth) -90° to +90°
Instantaneous field of view: (elevation) 10° (azimuth) 20°
Optical port diameter: 5 inches (127 mm)
Focal length ratio: 0.3:1
Power: 400 Hz and 28 V dc

STATUS: in production.

Texas Instruments

Texas Instruments Equipment Group, PO Box 660246, Dallas, Texas 75266
TELEPHONE: (214) 995 2011
TELEX: 73324
TWX: 910 867 4702

Texas Instruments has for many years been involved in the development of infra-red detection and recognition technology, having developed its first infra-red linescan mapping system as early as 1958. The company also claims to

have developed the first forward-looking infra-red system in 1964 and by 1967 Texas Instruments' Flir forward-looking infra-red systems were serving aboard US Air Force gunship aircraft in South-East Asia. In 1972, at the request of the US Department of Defense, the company conducted studies to examine ways of reducing maintenance costs associated with thermal sensor systems. The results of these studies emerged in the form of a family of common modules that can be packaged to meet custom requirements. A wide range of common module-based equipment is now in production.

Three recent developments have brought step changes to the performance of airborne thermal imagers. It has become possible to make the cyrogenic cooler far smaller and far cheaper, while the advent of focal plane arrays, is beginning to produce results which will give highly detailed pictures with a far less complex system than was previously the case with serial and parallel detectors. Also coming up is possible replacement of the traditional germanium lenses with gallium arsenide, the new material being particularly appropriate to installation in high speed aircraft since germanium becomes opaque above 80°C.

Studies by Texas Instruments show that in focal plane array technology performance is optimised at around 1000 elements and the company is very close to achieving the required standards of detection having made steps forward during 1985. The company says that it believes that focal plane technology will be as mature in 1988 as common module technology was in 1978.

Texas Instruments is also involved in laser technology and claims that its carbon dioxide lasers have made significant steps forward recently, particularly in the area of reliability. The company's expertise in CO_2 lasers is linked closely with that of the cyrogenic cooler of the thermal imager, the two techniques being common and the company believes that CO_2 laser rangefinders will become standard in the military, displacing the traditional neodymium-YAG in this application where eye safety is paramount. A CO_2 operates at 10.6 microns, exactly in the middle of the traditional range employed by thermal imaging sensors and thus the two units can be integrated together to provide accurate range information and picture sensing in a single system.

OR-89/AA forward-looking infra-red system

The Texas Instruments OR-89/AA Flir system sensor is a direct development of the company's earlier AN/AAD-4 and AN/AAD-7 Flir sensors which were used extensively for the US Air Force gunship programmes. Development of the sensor, for use on the US Navy/Lockheed S-3A Viking aircraft, began under a contract from Lockheed-California in August 1969 and the first prototype was delivered in March 1971. The system was subjected to extensive qualification testing, in addition to US Navy testing which continued over 22 months. The first production equipment was delivered to Lockheed in August 1972 and the system was introduced into the fleet early the following year. Deployment continued for at least the next five years, with the delivery of well over 400 S-3A or derivative systems, together with support equipment and spares.

The OR-89/AA Flir system uses mercury cadmium telluride detector arrays packaged in a gimbal system with self-contained attitude stabilisation. This provides an azimuth coverage of ±200° and 0° to –84° in elevation. Output is shown on an 875-line RS-343 composite television display. The system comprises three

basic modules or weapon replaceable assemblies, the infra-red viewer, power supply video converter, and servo control converter.

The system is controlled by a general-purpose digital computer which sends outputs in serial word form to the forward-looking infra-red system control converter, which translates them into set-control commands or analogue servo commands. Set-control commands provide such control functions as standby/off, activate, servo on/off, polarity, gain level and field of view. Servo commands control azimuth and elevation drive and brake functions. Position data is supplied by the control converter, which converts it into digital format and then transmits it to the general purpose digital computer for status purposes. On certain configurations, off-line ancilliary units produced by Texas Instruments are used for control and display. These include set and slew control weapon replaceable assemblies, a position indicator, cathode ray tube displays, and a forced-air shock-mounting tray.

The OR-89/AA's primary function aboard the S-3A is that of detecting and classifying surface vessels at night, but the obvious use of the system led to the development of a family of derivative systems to fulfil a wide range of applications. These derivatives have been successfully installed and operated in 12 other types of aircraft ranging from light twins to large transports and including helicopters and high-performance jets. Examples include the incorporation of the system in the US Navy/ Lockheed P-3B Orion for ASW missions and for non-military use with the US Customs Service in Linebacker operations and other special missions.

The US Navy is upgrading the OR-89 by incorporating a common module receiver. The upgraded Flir system is designated OR-263 and the programme is scheduled for completion in the late 1980s.

Weight: 266 lb (120.9 kg)

STATUS: in service.

OR-5008/AA forward-looking infra-red system

The Texas Instruments OR-5008 Flir system is a derivative of the company's OR-89/AA equipment (see preceding entry), adapted for use on the Lockheed CP-140 Aurora ASW aircraft operated by the Canadian Armed Forces. It was designed to support a variety of mission

requirements in the maritime patrol field including search and rescue, shipping and fisheries surveillance, mapping, ice reconnaissance and defence surveillance.

The system is mounted in the lower part of the CP-140 radome in a similar manner to that of the AN/AAS-36 Flir system used in the US Navy's Lockheed P-3 aircraft. The specification is similar to that of the OR-89/AA except that it uses the US Navy-type P-3 interface casting, provides for a 5° up-look capability, uses off-line control for the extend-retract function, and has additional composite video outputs. The use of this system, has also improved mean time between failures and mean time to repair.

STATUS: in service.

AN/AAQ-9 infra-red detecting set

The Texas Instruments AN/AAQ-9 infra-red detecting set is a Flir system developed for the US Air Force Pave Tack programme to provide autonomous target acquisition for delivery of standard and laser-guided (Paveway) bombs. Besides this target-designation capability the system allows the weapon systems operator to locate, acquire and track ground targets in day, night or adverse weather, and forms an integral component of the AN/AVQ-26 Pave Tack target designator system.

The system is designed primarily for use on US Air Force/McDonnell Douglas F-4E, RF-4C and General Dynamics F-111 fighters and strike aircraft. The Royal Australian Air Force has also purchased Pave Tack for F-111C application, and a number of other international customers are understood to be considering it for use with other airborne weapon designation and delivery systems.

Development of the AN/AAQ-9 commenced during the mid-1970s, with Texas Instruments being selected as the sole-source supplier following a competitive design and flight test programme. The system entered production in 1977 and the first production unit was flown at Eglin Air Force Base in Florida in April 1979.

Design is based on the Texas Instruments-produced US Department of Defense common modules. Features include high resolution, automatic thermal and range optical compensation and a two field of view optical system. The optical system is designed for pod installation and derotation compensation is provided with a coverage of ±190° at 100° a second. Video output is provided to either 525-line RS-170 composite video standards or 875 lines to RS-343. In addition to the 4:1 infra-red optical field of view switching provided, a 2:1 image magnification is selectable.

The AN/AAQ-9 system has undergone particularly stringent reliability, qualification and flight testing with a view to achieving high reliability and a current mean time between failures rate of over 730 hours has been attained. The system is of modular construction for improved maintainability and is supplied with AN/AAM-59 test equipment to permit field maintenance and alignment down to intermeciate servicing level. For rapid in-the-field harmonisation, the equipment is mounted on an adjustable boresight platform.

Weight: (receiver) 107 lb (48.63 kg)
(control electronics unit) 28 lb (12.72 kg)

STATUS: in service.

AN/AAQ-10 forward-looking infra-red system

Another derivative of the Texas Instruments OR-89/AA system (see preceding two entries), the AN/AAQ-10 Flir system has been designed for use in helicopter search and rescue applications. A single-window turret has been provided to supply the increased field of coverage necessary to optimise the helicopter's capability in this role and special attention has been given to the design of the stabilisation

Texas Instruments OR-89/AA Flir system

Texas Instruments infra-red detectors (centre) surrounded by common modules

servo elements to counter the much greater vibration levels in helicopters. Video output is shown on a RS-343A 875-line television display.

Development of the AN/AAQ-10 sensor began in late 1974 with the prototype system being delivered to the US Air Force in mid-1975. Flight test and familiarisation missions with this system continued for a 32-month period and in March 1978 a production contract was awarded for 10 systems. The first production unit was delivered in December that year.

Additional production of the system was contracted by the US Air Force in May 1980. The delivery of 21 sets started in March 1981, and the batch was completed by July 1982. These systems were contracted on an FMS (foreign military sales) basis, and are installed on Israeli Air Force/Sikorsky CH-53 helicopters and Israeli Navy/IAI Westwind light reconnaissance aircraft. The AN/AAQ-10 has been adapted to the Westwind's performance capabilities by the installation of a specially-designed high-speed turret.

The AN/AAQ-10, as configured for use on the US Navy/Sikorsky HH-53 helicopter, consists of an infra-red receiver, power supply, electronic control amplifier, control indicator and mounting base. All controls and monitoring functions are off-line and contained in the control indicator, allowing the operator complete one-hand authority over sensor operation and servo slew commands.

The system is a self-contained, slewable infra-red sensor requiring only external power, a gimbal position indicator and a video display to become fully operational. Accurate gimbal position indication is a critical factor in defining a target location to the helicopter's navigation computer and this may be accomplished either by three-wire synchro outputs or a buffered voltage. A Texas Instruments' position indicator, using the synchro outputs and a video display, is an available option.

STATUS: in service.

AN/AAS-36 infra-red detection set

The Texas Instruments AN/AAS-36 infra-red detection set is a Flir system designed for US Navy/Lockheed P-3C maritime patrol aircraft to detect surface vessels, surfaced or snorkeling submarines and drifting survivors in darkness and limited visibility. The system was initially designed and developed to meet a P-3C update programme requirement, but the equipment has also been fitted to earlier P-3C and P-3B aircraft.

Production of the system commenced in 1977 following a testing, evaluation and demonstration programme which used 10 pre-production systems to assure the US Navy that design specifications were either met or exceeded. The service's Initial Operational Capability (IOC) was realised in 1979.

Based on Texas Instruments' US Department of Defense common modules, which employ mercury cadmium telluride detectors, the AN/AAS-36 is a stand-alone system requiring only electrical power for operation. The common module infra-red receiver is mounted in an azimuth-over-elevation stabilised gimbal and provides lower hemisphere coverage of ±200° in azimuth and from +15° to –82° in elevation. Additional weapon replaceable assemblies provide system power, servo control, forward-looking infra-red system control, slew commands and a real-time video display. The display presentation is on a 875-line RS-343 composite television monitor which permits the operator to identify, as well as observe, vessels.

Features include automatic optical temperature compensation, gimbal-pointing outputs for servo platform slaving, self-contained stabilisation, and a two field of view optical system (15° × 20° or 5° × 6.7°). A digital computer interface is available for on-line gimbal control. The system contains built-in self test facilities which permit checkout down to weapon replaceable assembly level, and these themselves are compatible with automatic test equipment. Mean time between failures is 300 hours.

The system is also being supplied to many non-US operators of the P-3 for upgrading to US Navy standards. The receiver-converter weapon replaceable assembly of AN/AAS-36 has also been fitted to Cessna Citation light twinjets, Sikorsky CH-53 helicopters, Beechcraft E-90, King Air 200 and other, unspecified aircraft.

By the end of 1985 six production lots, totalling 279 systems had been delivered, and in September 1985 Texas Instruments announced the Lot 7 order for 39 systems, value at $17.4 million.

Weight: 300 lb (136.36 kg)
Power: between 950 VA and 2500 VA at 115 V 400 Hz 3-phase and between 30 and 100 W at 28 V dc

STATUS: in production and service.

AN/AAS-37 infra-red detection set

The Texas Instruments AN/AAS-37 equipment is a version of the company's AN/AAS-36 detection set but is a more sophisticated system, being combined with a laser designation and ranging capability. Developed for the US Marine Corps/Rockwell OV-10D forward air control aircraft, the AN/AAS-37 provides infra-red vision for day or night operations under degraded environmental conditions, automatic target-tracking and laser target-designation. The laser provides ranging and illumination of ground targets for laser-guided weapons such as the Paveway bomb or the Hellfire missile, either on an autonomous basis for aircraft equipped with the AN/AAS-37 or for a co-operating aircraft armed with appropriate weapons.

Specifications of the Flir sensor and associated equipment are virtually identical to those of the AN/AAS-36 system from which it was developed. It does, however, possess a number

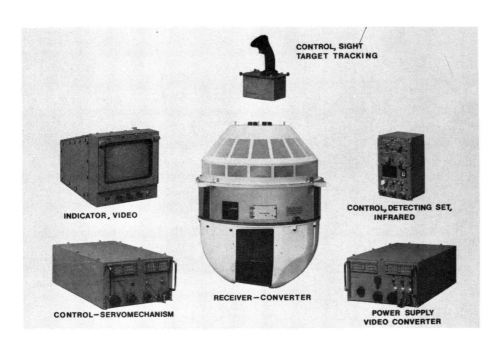

CONTROL, SIGHT TARGET TRACKING

INDICATOR, VIDEO

CONTROL, DETECTING SET, INFRARED

CONTROL-SERVOMECHANISM

RECEIVER-CONVERTER

POWER SUPPLY VIDEO CONVERTER

Texas Instruments AN/AAS-36 Flir system

of additional features, derived mainly from incorporation of the laser section. These include direct-readout laser-ranging and designation capability. The system can be used as a target sight aligned with the aircraft boresight by means of electronic adjustment, and line-of-sight depression may be set by the operator for precision air-to-ground delivery of weapons. There are interfaces with aircraft systems for a radar altimeter and remote gyroscope, and an accelerometer provides the system with rate-aided automatic target tracking capability using an adaptive gate centroid tracker. Offset tracking from a target or another landmark is also possible. The system's display is daylight visible.

Over 3000 hours of reliability testing were completed on the laser designator/ranger before the award of a production contract. Production deliveries to the US Marine Corps began in late 1979 and the system is installed in Rockwell OV-10D forward air control aircraft. According to Texas Instruments, numerous successful missions have been flown using both conventional and laser guided bombs with the laser designator/ranger.

Between 1978 and 1981 23 systems were delivered and a further 60 systems are scheduled for delivery between 1985 and 1989.

Laser designator/ranger
Azimuth coverage: ±200°
Elevation coverage: –82° to +16°
Weight: 417 lb (189 kg)
Power requirements: up to 3 kVA at 115 V, 400 Hz, 3-phase and up to 1800 W, 28 V dc

STATUS: in production. Texas Instruments was awarded a contract in mid-1985 for the production of a further 10 tracking computers for use in the AAS-37 systems.

AN/AAS-38 forward-looking infra-red system

The AN/AAS-38 is the standard Flir pod in US Navy F/A-18 fighters. The pod, integrated by Ford Aerospace, is equipped with a Texas Instruments Flir sensor (see separate entry on Iris).

STATUS: in production.

AN/AAR-42 infra-red detection set

The Texas Instruments AN/AAR-42 is a Flir system designed for use on the US Navy/LTV A-7E strike aircraft. It is designed to provide a night window or bombsight which permits the pilot to perform single-seat night attack, close air support and reconnaissance missions by day or night and during poor weather con-

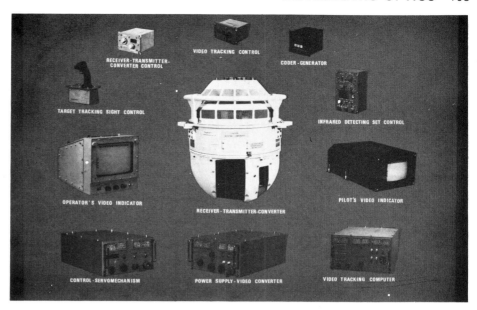

Texas Instruments AN/AAS-37 Flir system

Texas Instruments AN/AAR-42 Flir system

ditions. The system is said to allow the night delivery of conventional bombs with accuracy better than or equal to that demonstrated in day bombing using more conventional sighting methods.

The AN/AAR-42 is installed in a pod, with a gimballed Flir unit which provides stabilised imagery on the pilot's head-up display system (specially developed by GEC Avionics for the A-7). The Flir system uses components of the US Department of Defense common modules developed by Texas Instruments. It provides an azimuth coverage of ±20° and an elevation coverage from +5° to –35°. Both wide and narrow selectable fields of view are provided. The wide field of view, giving a × 1 magnification, is employed for pilot orientation, navigation update and target acquisition, while the narrow field of view, with a × 4 magnification, is used for target identification and weapon delivery. Features include automatic thermal focus compensation and sensor window de-icing.

Deliveries of AN/AAR-42 flir systems commenced in late 1977 following US Navy operational evaluation earlier that year. The system is currently operational on the A-7E and is said to exceed the specifications in both performance and in reliability, with a mean time between failures of 390 hours.

Weight: (canister assembly) 210 lb (95.45 kg) (servo electronics) 40 lb (18.18 kg)

STATUS: in production and service.

Demonstrator Flir pod

Texas Instruments' demonstrator Flir pod comprises a production-standard, military-qualified Flir system designed to meet acquisition and attack requirements of standard weapon-equipped aircraft. The system is housed in an 18-inch (450 mm) diameter pod and when mounted on existing high-performance aircraft provides thermal imagery for night navigation, target acquisition and weapon delivery. The basic Flir system has been deployed in the US military inventory since 1977 and features a two-axis common module Flir receiver and servo electronics. A video tracker is optionally available to increase accuracy and reduce operator workload.

The receiver has two fields of view, either narrow at 3° × 3° nominal or wide-angle at 12° × 12°, and is provided with automatic thermal focus compensation. The two-axis servo provides manual control over a total slew angle of ±20° in azimuth and from +5° to –35° in elevation at a slew rate of 30° a second in both

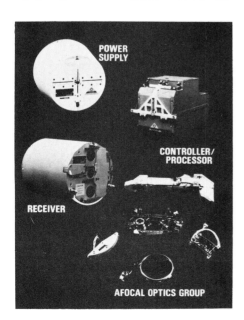

Texas Instruments AN/AAS-38 Flir system

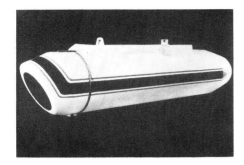

Texas Instruments demonstrator Flir pod

axes. The system is line-of-sight stabilised. The optional tracker control provides centroid tracking geometry and has a tracking accuracy better than ±0.5 pixel. The tracker has automatic gain and level control, automatic coast mode and built-in test facilities. It interfaces with a MIL-STD-1553 data-bus.

The system's video output may be supplied in either 875, 525 or other line standard specified by a customer, and complies with RS-343, RS-170 or other, adapted, standards. The pod is normally mounted at a standard weapon station. An environmental control unit is an available option for those installations which may require cooling air.

Dimensions: 18 inches (450 mm) dia × 88 inches (2235 mm) length
Weight: 500 lb (227 kg)

STATUS: basic Flir in production.

Iris infra-red imaging sub-system

This Texas Instruments system is part of AN/AAS-38 Flir imaging system, a self-contained pod designed for use on the US Navy/McDonnell Douglas F/A-18 fighter. Major sub-assemblies of Iris manufactured by this company include the infra-red receiver, controller-processor, power supply, and the infra-red afocal optics group. The pod and gimbal systems are manufactured by Ford Aerospace and Communications Corporation. The AN/AAS-38 Flir system is designed for target acquisition and recognition, weapons delivery, and reconnaissance under day and night and adverse weather conditions.

The Iris includes dual fields of view of 12° × 12° and 3° × 3° with automatic thermal focus compensation, image derotation for natural horizon display, and microprocessor-controlled built-in-test circuits affording 98 per cent fault detection. A key feature of Iris is the automatic video tracker contained in the controller-processor, which provides automatic target acquisition, line-of-sight control, and offset designation for accurate weapons delivery. This capability is embodied in the two SBP-9989 microprocessors and the 32 K eprom memory within the controller-processor.

The controller-processor provides the video processing necessary to output the 875-line modified RS343 television video with required track and field of view reticles for cockpit displays. Communication with the aircraft AN/AYK-14 mission computer is via a remote MIL-STD-1553 digital data-bus terminal contained in the controller-processor.

The infra-red receiver design is based on the US Department of Defense common modules produced by Texas Instruments and is similar in packaging to the AN/AAQ-9 infra-red receiver.

The Iris power supply utilises advanced switching regulator designs and state-of-the-art packaging to provide highly efficient primary-to-secondary power distribution, control and regulation for the Iris and pod system.

Full-scale engineering development of Iris began in 1978 and was followed by the first production contract award in 1981.

STATUS: in production.

Automatic video tracker

The automatic video tracker equipment has been developed by Texas Instruments as an optional 'add-on' to forward-looking infra-red systems to improve performance, especially weapon delivery accuracy. The microprocessor controlled system processes forward-looking infra-red video signals and provides servo control commands, allowing hands-off operation. It also provides display symbology and offers a number of operating modes.

Typical symbology options offered by the automatic video tracker include a target cursor, automatic tracking gate and coast mode indication (all with adjustable display brilliance, and the latter also with size adjustment), narrow field of view indicators, and range readout.

Four different tracking modes are provided. In manual mode, the operator maintains track via a slew stick control but he can select the computer-aided mode to assist in target acquisition and maintenance of manual track by removing the effects of aircraft motion. In automatic track mode, the track is accurately maintained under microprocessor control while the operator is free to deal with other tasks. The offset automatic track mode is similar to that of normal automatic track but permits the operator to hold a target at the centre of the reference crosshairs by locking the tracker onto another, more stable reference target in the system's field of view

The automatic video tracker is claimed to have sufficient tracking accuracy for laser designation and other fire control operations, and the sensitivity is adequate for the tracking of either large or small targets with high or low contrast against their background. In the latter case, the system's use is enhanced by automatic target polarity selection. The system's so-called 'coast mode' is also a useful operational feature. If a target is screened momentarily, or if its apparent size or shape changes rapidly, then the automatic video tracker estimates its probable position to allow rapid, automatic re-acquisition. Such a facility is of value in high-speed manoeuvres, particularly during tactical combat at low altitudes.

Control of the automatic video tracker system is exercised by a Texas Instruments' SBP-9989 microprocessor, a MIL-qualified system, which provides a high degree of flexibility in that parameters may be optimised for

particular variations of the system by simply changing the software. Use of the microprocessor has also permitted significant improvements and additions to the traditional adaptive gate, centroid tracking algorithms.

Texas Instruments has built automatic video trackers for Rockwell OV-10D, McDonnell Douglas F/A-18, Lockheed P-3C and LTV A-7 aircraft.

Two new types of automatic video trackers are currently in development. The first type is an increased performance, adaptive gate, centroid tracker designed for a lower cost than previous tracking systems. This device has been used on the P-3C aircraft and is now being built for the Sikorsky HH-60D Night Hawk helicopter, AN/AAQ-15 system. The second type is a combined adaptive-gate centroid correlation signal-processing algorithm-based system. This tracker has been designed for high-performance tracking requirements, such as those calling for precision weapons delivery. The combined algorithm tracker is built up from a recently developed set of common digital module electronics developed to provide a standardised set of electronics for different automatic video tracking system requirements.

STATUS: in development and production.

RS-700 series infra-red line-scanner

The Texas Instruments' airborne infra-red line-scanner type RS-700 operates in the 8 to 14 micron band, where absorption by carbon dioxide and water vapour is at a minimum. The detectors are of the mercury-cadmium-telluride type and use the common module closed-cycle cooling subsystem. The optical system focuses radiation on to the detectors which produce video electrical signals that correspond with the picture formed by the radiation pattern scanned. After processing, the video signals are converted to visible wavelengths for recording by light-emitting diodes.

The RS-700 is composed of three sub-assemblies which are mechanically mounted together to form a single assembly for aircraft installation. Among the operational features of the equipment are: manual or automatic gain selection; manual or automatic level control; video compression of unusually hot or cold objects; continuous scanning over whole

Typical imaging from Texas Instruments RS-700 series infra-red line scanner shows 'hot' (white) details of area surrounding port

Texas Instruments RS-710 infra-red line-scan equipment for Panavia Tornado

velocity/height range; adjustable hot-spot marker; event marker; and built-in test equipment.

Data annotation is an optional feature of the system. Numerical data annotation in the film margin can be used for mission identification, date, time, heading and other information. MIL-STD-782 C code matrix data annotation is also available. Another option which can be

provided is roll stabilisation, which allows the RS-700 to be attitude stabilised up to angles of ±30° from the nadir.

Oblique slew, when added to the roll-stabilised version, permits offsets of up to ±15° from the nadir, enabling hostile targets to be overflown while providing profile data from special-interest targets. The RS-700 also can be supplied in a reconnaissance pod.

Since the introduction of the original RS-700 linescan, Texas Instruments has developed a variety of other models. The current model is typified by the RS-710 which was developed for the West German Navy/Panavia Tornado aircraft and which is also being retrofitted to that service's Lockheed F-104Gs. Other variants have been produced for US and foreign aircraft.

Scan mirror facets: 4
Optical aperture: 38.4 cm²
Detector cooling: closed cycle, 77 K
Detector type: mercury-cadmium-telluride
Recording light source: gallium-arsenide-phosphor diodes
Film width: 70 mm
Film capacity: 46 m, 70 m
Velocity/height range: 0.2-5
Thermal resolution: 0.2°C
Spatial resolution: 0.5-1.5 mrad

Total field of view: 120°
Weight: 32 kg (without roll stabilisation); 42 kg (with roll stabilisation)

STATUS: in production. RS-700 series linescan systems have been supplied to air forces of Denmark, West Germany, Italy, Singapore, Sweden, Switzerland, Saudi Arabia, Malaysia, and USA.

Lightweight Flir system

Texas Instruments has developed a lightweight Flir system for airborne and other applications, with three fields of view. The wide field is 15.3° × 30.6° and is used for surveillance or navigation, while the medium field is one-third as wide, and is for target tracking and acquisition. The narrow field, one-third the size again, is for target identification. The minimum resolvable temperature is 0.2°C. The system is gyro-stabilised and can be developed to provide auto-tracking and other facilities.

Dimensions: (turret) 14 inches dia × 17 inches high (356 mm dia × 432 mm high)
Weight: (turret) 40 lb (18 kg)
(total system, typical) 66 lb (30 kg)

STATUS: in development.

Westinghouse

Westinghouse Electric Corporation, Aerospace and Electronic Systems Division, PO Box 746, Friendship Site, Baltimore, Maryland 21203
TELEPHONE: (301) 765 1000
TELEX: 087828

AN/ASQ-153 Pave Spike laser designator/ranger

The Pave Spike development programme was initiated in 1971 and delivery of 156 pod sets to the US Air Force was completed by August 1977, by which time 327 McDonnell Douglas F-4D Phantoms had been converted to accept the system. A further 82 sets were delivered up to September 1979, for foreign use, including a substantial number for the Royal Air Force and some to the Turkish Air Force.

The system is contained within an externally mounted pod, the nose section of which revolves about the pod axis to provide roll stabilisation, and a cylindrical forward portion which rotates in pitch to provide elevation stabilisation. Virtually complete lower-hemisphere coverage is thus provided in a relatively compact and light arrangement. The nose section is sealed and pressurised with nitrogen, maintained at a constant temperature for optimum sensor performance. The centre section provides umbilical connections between the nose and rotating sections, between the nose and rotating sections, between the aircraft and the aft electronics system. In the aft section is a cold plate onto which are mounted the electronic line-replaceable units. These comprise a low-voltage power-supply and pod control, servo drivers, laser control, laser power-supply and interfaces. The pod contains a television-tracking sensor and laser designator/ranger. The television sensor can be used for target acquisition and the designator permits accurate delivery of laser-guided munitions. Laser-ranging can be used to improve the delivery accuracy of conventional weapons.

Initially the unit was procured for use only on US Air Force/McDonnell Douglas F-4D and F-4E Phantoms, but it is now employed on several other types, including British Aerospace

Royal Air Force Buccaneer equipped with Pave Spike laser-designator/ranger and Paveway II laser-guided bomb mounted respectively on inboard and outboard pylons

Pave Spike target-designator/ranger under port engine air intake of McDonnell Douglas F-4 Phantom

Buccaneers in the UK. Since August 1978 Ferranti has held a technical support contract for systems operated by the Royal Air Force.

A shorter model, Pave Spike-B, was offered for the General Dynamics F-16 fighter, but has not entered production, and Westinghouse also proposes a '24-hour' version of the system, Pave Spike-C.

The overall AN/ASQ-153 system comprises the AN/AVQ-23 pod and several system components in the aircraft. These include a line-of-sight indicator, control panel, range indicator, modified radar-control handle and weapon-release computer. The system can be used with Paveway laser-guided bombs and several other laser-guided munitions.

Pod length: 3.66 m
Pod diameter: 0.25 m
Weight: 193 kg
Operating wavelength: 1.06 microns

STATUS: in service.

AFTI/F-16 electro-optics targeting set

Flight testing began early in 1985 of the Westinghouse electro-optic sensor/tracker set, a part of the AFTI/F-16 advanced fighter technology integration programme. The sensor/tracker allows the single-seat General Dynamics F-16 to deliver ordnance very accurately while maintaining high manoeuvrability during low-level attacks or high-altitude intercepts.

The system incorporates a coded laser for ranging and designation as well as Flir and television trackers. The gimballed sensor head, which can be cued onto the target by the radar, is conformally mounted in the aircraft's starboard wing strake. Two additional units, a power supply and a MIL-STD-1750A processor are mounted in the fuselage.

STATUS: in development.

Electronic warfare

CHILE

ENAER

Empresa Nacional de Aeronautica – Chile,
Gran Avda J M Carrera 11087, Santiago
TELEPHONE: (2) 58 85 82
TELEX: 645115

Chile has built up a considerable electronic
warfare capability, in response to the various
threats which surround it. Details of the suite
developed for airborne applications are given
below.

Caiquen II radar warning receiver

This system equips the Chilean Air Force's
British Aerospace Hunter FGA.71 aircraft and is
understood to be under development for
installation in the country's Mirage 50, among
others.

The frequency band between 2 and 18 GHz is
split into four sub-bands, each with its own
wide-band crystal video receiver and antenna,
offering coverage through a full 360° in azimuth
and ±40° in elevation.

The cockpit control panel has 12 displays,
indicating the type and bearing of the received
signals.

The Caiquen operates up to 40 000 feet, and
weighs 8 kg.

Eclipse chaff/flare dispenser

Operating in conjunction with the Caiquen, the
Eclipse incorporates a control unit, from which
the various countermeasures can be selected
from up to four dispensers: total system weight
is 70 kg. Each launcher can carry 40 RR-170
chaff and 20 MJU-7B flare cartridges, and
operation can be manual, at the command of

the pilot (single or multiple launches), automatic (under the command of the rwr). The
Chilean Hunters are understood to carry two
launchers, carried under the rear of the
fuselage.

Itata elint system

Suitable for land, air or sea-borne use, the Itata
is tuned to the electronic emissions from the
radars located in the regions around Chile. It
uses a superheterodyne receiver covering the
30 MHz to 18 GHz band, operating over either
the whole band, or specified sub-bands. The
Itata system is used on the Chilean Air Force's
Beech 99A aircraft.

STATUS: all three systems are in operation in
Chile, and are available for export.

FRANCE

Alkan

R Alkan et Cie, rue du 8 Mai 1945, 99460
Valenton
TELEPHONE: (1) 43 89 39 90
TELEX: 203876

Chaff/infra-red countermeasure dispensers

Alkan has developed a new generation of
countermeasure dispensers that can accommodate either chaff or infra-red flare cartridges
or a combination of the two. The cartridges are
arranged in interchangeable, easily-handled
magazines loaded in the modular dispenser.
Typically, a single dispenser will accommodate

five to seven modules. Each module houses a
magazine containing, for instance, either eight
60 mm diameter infra-red cartridges, or 18
40 mm diameter chaff cartridges.

The electronic management system of the
dispenser whether or not connected to a rwr
performs the firing sequences created by
software. It permanently manages the inventory
of available cartridges and provides the necessary information to the cockpit control unit
which displays the status of the complete
equipment.

The system is in production for the French
Air Force's Jaguar aircraft. Two dispensers are
fitted under the wing, in a conformal installation
near to the aircraft fuselage. Each dispenser
contains seven modules (Alkan Type 5020).

The same system is used for the Dassault
Mirage F1 and the Mirage 2000 aircraft as the
core of the complete Matra self-protection
system called respectively Sycomor and
Spirale.

The Alkan Type 5080 pod is designed to fit
either to the Jato point of the MiG-21 or to any
14-inch (356 mm) standard armament hard
point. It is in series production.

Various other applications of the same
concept are under development for the Mirage
III/5/50 as well as for transport aircraft such as
the Transall C-160 in its role of supporting army
combat units and for armed maritime patrol
aircraft.

Electronique Serge Dassault

Electronique Serge Dassault, 55 quai Carnot,
92214 Saint-Cloud
TELEPHONE: (1) 46 02 50 00
TELEX: 250787

Electronique Serge Dassault is a leading
supplier of electronic warfare equipment for all
models of Dassault Mirage. The largest recent
programme is believed to have been an
internally-mounted electronic countermeasures system for Mirage IV strategic
bombers operated by the French Air Force. At
least four variants of this system, Agiric,
Agosol, Become and Agacette, have been

reported. Continuous development has provided noise-jamming, wide-band barrage
jamming and deception jamming capability.
Production of this system is probably complete.

A new generation of internally-mounted
electronic countermeasures equipment is
believed to be under development for the
Mirage 2000/4000 and may enter production
soon.

Lacroix

Société E Lacroix, BP 213 Route de Toulouse,
31601 Muret
TELEPHONE: (61) 51 03 37
TELEX: 531478

ECM equipment

Lacroix is a European specialist in the
development and production of infra-red flares,
electromagnetic chaff and electro-optic cartridges, and collaborates with countermeasure-
system manufacturers in the design of
expendable devices. It has been involved with
this activity for some 16 years.

The company is associated with Philips

Elektronikindustrier AB of Sweden in the
production of infra-red and electromagnetic
cartridges for the BOZ and BOP families of
countermeasure pods. These are used on
Swedish Air Force/Saab Viggens and Luftwaffe
Tornados. These companies also collaborate in
the production of BOH pods for helicopters for
which Lacroix supplies electromagnetic and
infra-red countermeasure cartridges.

Lacroix, together with Matra, produces infra-
red and electromagnetic cartridges for the
Sycomor and Spirale countermeasure systems
used on the Dassault Mirage F1 and Mirage
2000 fighters. Similar equipment is also supplied
for the Saphir helicopter protection system.

The company has agreements with Alkan to
supply cartridges for various countermeasure

systems. These include 40 mm devices for
various versions of the Mirage III and F1 fighters
and for the LC530 system used on the Sepecat
Jaguar and Dassault Super Etendard, 55 mm
cartridges for transport and maritime patrol
aircraft such as the Transall and Atlantic
aircraft and smaller units for light aircraft and
drones. Cartridges are also produced for US-
designed expendable launchers, current commitments including supplies to the Belgian Air
Force (AN/ALE-40 on F-16 and AN/ALE-39 on
Mirage V fighters) and West German Air-Force
(Lambert system on McDonnell Douglas F-4
Phantom). Lacroix also produces chaff war-
heads for 68 mm and 2.75-inch rockets intended
to jam radars between 1 and 10 km ahead of the
launch aircraft.

Matra

Matra SA, 37 avenue Louis Bréguet, Velizy 78140
TELEPHONE: (3) 946 96 00
TELEX: 698077

ECM equipment

Matra has been active in the electronic countermeasures (ECM) field for some 15 years and has developed four main types of countermeasures equipment. They are: Phimat, Sycomor, Spirale and Saphir, all of them chaff or chaff/flare dispensers. Phimat comprises a dispenser and control unit. The dispenser is a 3.6-metre long 180 mm diameter tube containing the chaff packs, ejection mechanism and drive electronics. The system weighs 105 kg

and is operational with the French Air Force and Navy and the Royal Air Force. The Sycomor chaff/flare system is intended for the various versions of the Dassault Mirage F1 fighter, and can be packaged either in an externally-mounted pod or in a conformal pack. Apart from the French Air Force, Sycomor is operational with Mirage F1s flown by other air forces. Spirale (chaff/flare) is currently under development for the Dassault Mirage 2000 fighter and will be operational in 1987. Saphir (chaff/flare) is optimised for use with helicopters.

STATUS: all in production.

Pod-configured Matra Sycomor ECM system on Mirage F1 fighter

Thomson-CSF

Thomson-CSF, Equipements Avioniques et Spatiaux, 178 boulevard Gabriel-Péri, 92240 Malakoff
TELEPHONE: (1) 46 55 44 22
TELEX: 204780

Serval radar warning receiver

Fitted to French Air Force and export versions of the new Dassault Mirage 2000 fighter, Serval warns the pilot when his aircraft is being illuminated by surface or airborne threat radars of a hostile nature; friendly/hostile discrimination is done by comparing the characteristics of the illumination energy with those of their emitters held in a reprogrammable threat library held in the system.

Serval uses four detection antennas mounted on the wing-tips and fin, and feeding a hybrid analogue/digital processor. The cathode ray tube display unit shows the strength and direction of the threat emitter, and its nature (ground-based or airborne). Details of several emitters can be shown simultaneously. At the same time an audio alarm sounds in the pilot's head-set.

STATUS: in production.

BF radar warning receiver

Type BF radar warning receivers are used in Dassault Mirage F1-A, F1-C and IIIZ, Super Etendard and Netherlands Air Force/Northrop F-5 fighters. A version for helicopters, designated TMV 008H, is also produced. The system provides the crew with warning of most categories of airborne and surface radar threats, and with an indication of their direction.

Four wide-band antennas are used, linked to a video receiver and, when necessary, a synchronisation unit. In the Mirage the system control and display unit is integrated into other cockpit equipment, but it can be provided separately for other applications.

The receivers comprise photographically-

etched spiral antennas and microwave circuits for an rf test oscillator, limiter modulator diodes, high-pass filter, detector circuit, video modulation and pre-amplification. The two side-mounted antennas lie flush with the fin structure, while fore and aft antennas on the fin have conical radomes. An audio alarm is generated when threats are detected and approximate threat direction is indicated by one or more of four signal lamps. The threat is also categorised by one of three lamps which indicate: conventional pulse radar, continuous wave or intermittent continuous wave radar, track-while-scan ground-threat radar.

Format: (video receiver) ¼ ATR Short
Dimensions: (flat antennas) 148 mm dia × 53 mm deep
(conical antennas) 82 mm dia × 360 mm deep
(synchronisation unit) 209 × 110 × 45 mm
(control box) 146 × 95 × 40 mm
(indicator unit) 68 × 61 × 61 mm
Weight: (video receiver) 3.3 kg
(synchronisation unit) 0.6 kg
Total system weight: 9.2 kg
Power: <500 VA at 200 V 400 Hz

STATUS: in production.

Sherloc radar warning receiver

The Sherloc rwr is a low-cost system designed for fixed-wing aircraft or helicopters. It incorporates a high speed digital processor and a radar signals library, which is easily reprogrammable on the flight line.

Radar data is presented to the pilot in the form of alphanumeric symbology on a high brightness cathode ray tube; the symbols denoting the type of threat detected and their position on the crt being relative to the threat's bearing and range. Alternatively, a simple light-emitting diode display can be used, indicating threat classification and relative strength. Sherloc operates in the E to J bands (2 to 20 GHz). System weight is 10 kg.

STATUS: in production.

DB-3141 noise jamming pod

The DB-3141 H to J band (8 to 20 GHz) noise jamming pod has a single receiver, a travelling-wave tube jammer, and fore and aft transmitter antennas. It provides a simple active electronic warfare capability for Dassault Mirage fighters and possibly also for that company's Super Etendard. The system has a 'look-through' capability, enabling it to discontinue jamming as soon as threat reception ceases.

Pod length: 3.5 m
Max diameter: 0.25 m
Weight: 175 kg
Power: 1.7 kVA at 200 V 400 Hz

STATUS: in production.

DB-3163 noise jamming pod

Developed primarily to equip the French Air Force's Dassault Mirage 2000, this pod provides I to J band coverage and is compatible with other Mirage variants. It is designed to provide self-protection against both air and ground radar threats. Pulse and continuous wave emitters can be detected, identified and countered. A superheterodyne receiver performs a frequency-scan search on emissions received by antennas at both ends of the pod. During pre-flight preparation, up to three bands, from a choice of six, can be selected for simultaneous use: up to three threats can be jammed simultaneously. An internal bootstrap air-cooling system is employed and the system is energised from the aircraft's power supplies.

Pod length: 3.52 m
Diameter: 0.25 m
Weight: 175 kg
Max power: 1.7 kVA at 200 V 400 Hz

STATUS: in production.

Caiman noise/deception jamming pod

Probably developed from the earlier Alligator pod, Caiman is designed for dedicated electronic warfare aircraft which are protecting groups of similar types in ground-attack roles. It can be installed underwing or on a fuselage pylon. The pod is self-contained, ram-air entering the unit through an annular intake to drive a power turbine and to provide cooling. Within the pod are fore and aft receiver antennas, and two independent jammers, each weighing 130 kg. Caiman is suitable for the Dassault Mirage F1 fighter and has been supplied to export customers for several aircraft types.

Pod length: 5.95 m
Diameter: 0.41 m
Weight: approx 500 kg

STATUS: in production.

Caiman jammer on Dassault Mirage F1 fighter

Barem jamming pod

The Barem self-protection jammer is designed to protect tactical aircraft from radar-directed missiles and similar threats. Housed in a pod cleared for flight at speeds above Mach 2, the system has receive and transmit antennas coupled to the receiver and transmitter which work under automatic micro-processor control over the 6 to 20 GHz range (H, I and J bands). Threats detected by the antennas are analysed and compared with known signals in the computerised radar library; received signals are recorded in flight for subsequent analysis and storage in the library.

Length: 3.45 m
Diameter: 0.16 m
Weight: 85 kg
Power required: 700 VA

STATUS: in production.

Thomson-CSF Barem jamming pod can be flown on tactical aircraft beyond Mach 2

Elisa elint receiver

This highly-sensitive Elint system has been designed for rapid acquisition of radar transmissions. It can be fitted on many aircraft types, and on ships and ground-based systems. The superheterodyne receiver covers C to J band (0.5 to 18 GHz) and can programme search parameters on-line (called 'smart-scan') enabling the operator to adapt the frequency and direction of search. Sensitivity is such that there is continuous detection of radar sidelobes and scattered radiation.

Elisa capacity can be varied to suit different installations. Received data is analysed automatically during the search phase and this function can be controlled by operator demands. Results are shown in graphical and numeric form. A display shows pulse repetition frequency, pulse width and direction histograms of sensed radiation. This presentation technique is designed to eliminate spurious and uncertain measurements. Elisa uses an antenna array which will have an application-dependent configuration, a frequency-transposition unit, a reception and data processing unit, and control and display console. The latter uses a plasma-plate display screen with keyboard controls and a magnetic cassette recorder.

STATUS: under development.

TMV 018 Syrel elint pod

This is a fully-automatic electronic reconnaissance system attached by a special centreline pylon on Dassault Mirage F1 aircraft. It can be used during tactical penetration missions at medium or low altitudes to acquire and record automatically data relating to the identification and location of ground-based electronic

systems. It is intended to provide reliable information on radars for early-warning systems, search and acquisition, ground control interception and fire-control for anti-aircraft artillery or missiles.

The pod has two antennas at both front and rear, and receiver units, an amplifier and recorders in its centre-section. The pylon houses a cooling system which has a ram-air intake in the pylon leading-edge. High-speed operation is assisted by thick-film and microwave circuit assemblies on ceramic substrates. Thomson-CSF also produces first- and second-line maintenance equipment for use with the pod.

Pod length: 3.75 m

STATUS: in production.

TMV 026 ESM system

Suitable for patrol aircraft and helicopters, the system uses up to six DR2000 antennas and can

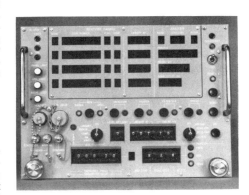

Thomson-CSF Dalia processor for TMV 026 system

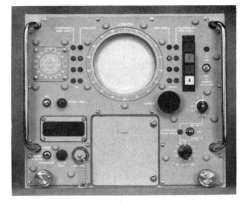

DR2000 display and control unit is part of Thomson-CSF TMV 026 ESM system

distinguish pulse and continuous wave threats. On-board equipment comprises a DR2000 receiver/display unit and either a Dalia 500 or Dalia 1000 (according to the number of threat types held in the library) or Arial 15 (15 threat types in library) analyser. The system provides pulse repetition frequencies, pulsewidth, radio frequency level, antenna rotation rate and jitter information on identified threats. Instantaneous audio and visual indications of threats are provided. System weight, excluding antennas, is 72 kg and power requirement is 700 VA, 200 V ac.

STATUS: in production.

TMV 202/DR 4000(A) ESM system

A new-generation electronic surveillance measures (ESM) system, the DR 4000(A) ESM suite is intended for surveillance aircraft and helicopters, and intercepts all signals between the D and J bands. With reprogrammable logic and a three-colour display, the probability of interception, in both direction-finding and frequency discrimination, is claimed to be 100 per cent with only a single pulse, as a result of the crystal video amplified techniques used. The sensitivity is sufficient to intercept pulse-compression signals. The system can be interfaced with any data-handling system and ECM suite (chaff-launched or jammer) through suitable data-buses or point-to-point links.

Weight: (including antennas) 169 kg
Power: 1800 VA

STATUS: in development.

ARAR/AXAX ESM receivers

These electronic surveillance receivers are used in European-operated Dassault-Breguet Atlantic and the British Aerospace Nimrod long-range ASW aircraft. In both aircraft the equipment is housed in fin-tip pods.

STATUS: in production.

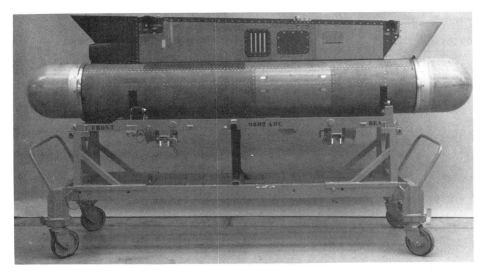

Thomson-CSF TMV 018 Syrel Elint pod

GERMANY, FEDERAL REPUBLIC

AEG-Telefunken

AEG-Telefunken, Elisabethenstrasse 3, D-7900 Ulm
TELEPHONE: (731) 1921
TELEX: 712723

AEG-Telefunken has collaborated on the development of the Elettronica EL/70 deception repeater jammer for German and Italian NATO/Lockheed F-104G Starfighters. This is a combined receiver/jammer system which operates in the 2.6 to 5.1 GHz and 8 to 10 GHz bands.

AEG-Telefunken was involved primarily with design of the lower band components.

The company also has a co-production agreement with Itek Corporation for the manufacture of enhanced radar warning equipment.

MBB

Messerschmitt-Bölkow-Blohm GmbH, Postfach 80 11 60, D-8000 Munich 80
TELEPHONE: (89) 6000 6312
TELEX: 5287-960

Tornado ECR

MBB, in conjunction with the Tornado prime contractor Panavia, began design work on the Tornado Electronic Combat and Reconnaissance (ECR) version in 1985, US $50 million having been budgeted by the German government for this work in 1986. It is anticipated that the West German Air Force will order 40 ECR aircraft, with deliveries starting in 1989 and ending in 1991/92. Total project cost for the 40-aircraft fleet is estimated at over US $1000 million. Improved computing, thermal imaging and electronic sensors will be installed and pod-mounted line-scanning and counter-measures equipment will be carried as standard. Although no firm equipment has been specified at the time of writing, it is known that 'off the shelf' rather than all new systems will be installed, to minimise costs.

Artist's impression of proposed Tornado ECR carrying electronic warfare and reconnaissance pods and missiles

---

ISRAEL

Elisra

Elisra Electronic Systems Limited, 48 Mivtza Kadesh Street, Bene Berak 51203
TELEPHONE: (3) 754 5111
TELEX: 33553

SPS-200 self-protection jammer

Designed for retrofit into existing jammer installations, the new system comprises a digital signal analyser that can process information and interface with other equipment. In combination with an optional MPMN-36(V) flare/chaff pack it can act as a complete self-protection system, or can function with radar and other jamming equipment on the aircraft. The system operates in the 1 to 18 GHz band, continuous wave radar threat information being

presented on a 3-inch (76 mm) cathode ray tube as alphanumeric data on type, angle-of-arrival, relative lethality and status. The system incorporates an extensive library of radar files that can be updated by means of a portable field-loader unit. The system is qualified to MIL-E-5400 Class 2.

STATUS: available.

SPS-20 warning system

Low-cost, low-volume and lightweight, this radar warning system is designed to fit existing helicopters and aircraft. It detects and displays pulsed radar threats operating within the 0.7 to 18 GHz frequency range. A 3-inch (76 mm) display unit provides an alphanumeric representation on the type, course, angle-of-arrival,

relative lethality and status of the analysed radar threats. It is equipped with a fast microprocessor which executes data processing and interfacing tasks.

The system features wide-band acquisition including C to D band, optional high sensitivity cw detection, optional recording of flight events and emitter parameters for play-back and evaluation.

The system is fully programmable and there is an optional provision for tie-in with the flare/chaff dispensing system.

Power: 28 V dc (MIL-STD-704)
Environmental specification: MIL-E-5400

STATUS: in production.

Elta

Elta Electronics Industries Limited, A Subsidiary of Israel Aircraft Industries Limited, PO Box 330, Ashdod 77102
TELEPHONE: (55) 30333
TELEX: 31807

EL/K-1250 vhf/uhf comint receiver

This compact, synthesised receiver operating in the 20 to 510 MHz band, is used as a building block for larger Comint or EW systems such as Arava EW. The unit has four selectable intermediate-frequency filters which demodulate AM, FM, continuous wave and single sideband signals. Intermodulation protection is claimed from rf pre-selection by voltage-tracking filters, and the fast-tuning synthesiser settles within 500 microseconds between channel changes. Remote digital control operation is possible and the compact dimensions, low weight and power consumption are achieved by extensive use of advanced microcircuit technology. The

unit is in service with Israeli and other armed forces.
Dimensions: 190 × 57 × 496 mm
Weight: 11 lb (5kg)
Frequency range: 20-510 MHz
Frequency accuracy/stability: ±1 ppm
Synthesiser settling time: 500 µs
If bandwidth: select 4 of 10, 20, 50, 100, 300, 600, 1000 KHz
Noise figure: 12dB 20-180 MHz
11dB 180-510 MHz

STATUS: in production.

EL/K 7010 tactical communications jammer system

This is a modular family of communication-jamming systems designed for stand-alone operation and providing for signal-search and acquisition, preset channel monitoring, and automatic jamming modes. It can also be integrated within a larger electronic warfare system. Computer control of power

Elta EL/K-1250 compact vhf/uhf Comint receiver

management, fast-reaction jamming and multi-kilowatt effective radiated power provides simultaneous multiple target capability. New interception tasks are accomplished rapidly by a fast scanning receiver and automatic signal sorting in the system computer. In addition to airborne installations, ground-based systems suitable for armoured personnel carriers and air-conditioned shelters are available.

STATUS: in production.

EL/K-7032 airborne comint system

The K-7032 is designed for the surveillance and interception of radio signals in the 20 to 500 MHz frequency range, operation being largely computer controlled.

A typical airborne system would include a supervisor's station, having a computer controller, two vhf/uhf radios, a display and a data recorder. This station can work with up to four traffic collection stations, each having up to four radios and data recorders.

The K-7032 has a frequency resolution of 1 kHz (with 10 Hz an option), and can intercept AM, FM, cw and ssb transmission, as required.

STATUS: in service and production.

CR-2800 airborne elint/ESM set

The CR-2800 is a large sophisticated airborne Elint/ESM system suitable for use in transport aircraft for Elint gathering and maritime surveillance. Simultaneous 360° coverage of the 2 to 18 GHz band is offered, together with a very accurate determination of signal bearing, due to the use of four sets of antenna/receivers: low noise rf amplifiers are located adjacent to the aerials to give high sensitivity. Frequencies are determined by a digital IFM receiver.

The real-time analyser uses advanced hardware and software technologies to process all received signals on a real-time basis. Each signal undergoes file processing and is instantly compared with an on-board library.

Communications between operator and system is via a colour display, keyboard and tracker ball, with a soft menu to aid the operator. Received information is recorded for post-flight analysis.

Dimensions: (console and equipment rack) both 1150 × 566 × 610 mm
Weight: (both units) each 125 kg
(df receivers) 2 at 25 kg each
(antennas) typical configuration about 80 kg

STATUS: in service.

Elta El/L-8202 advanced self-protection jammer, pod-mounted on Israel Air Force Kfir

EL/L-8202 advanced self-protection jamming pod

This combined receiver/jammer system is compatible with the IAI Kfir fighter and any similar aircraft type able to accommodate the pod, which resembles that of the US Air Force AN/ALQ-131. F to J band (3 to 20 GHz) coverage is provided with high-power broadband jammer outputs through either fore or aft antennas. Beam-shaping features are incorporated, and the threat library plus jamming techniques and stored mission data can be reprogrammed on the flight-line. There is an integral ram-air/liquid-cooling system. The logic unit within the pod interfaces the jammer with remote radar-warning receivers and cockpit displays.

Pod dimensions: 2900 × 260 × 390 mm
Power: 2.3 kVA

STATUS: in production.

EL/L-8230 internal self-protection jammer

Suitable for IAI Kfir/General Dynamics F-16-size aircraft, this unit is designed to combat both surface- and air-radar threats. It operates across the G to J (4 to 20 GHz) bands with separate receiver and transmitter antenna groups, and can generate noise or repeater jamming signals. It is designed to use standard avionics-bay cooling air and, being fully contained within the aircraft, does not penalise aircraft drag. The radio frequency unit, radio frequency power amplifier and logic unit are combined in one box, measuring 250 × 240 × 600 mm. Waveguide installation can dictate location in aircraft. Weight, excluding antennas, is 42 kg and maximum power consumption is 1.8 kVA.

STATUS: in production.

EL/L-8231 internal self-protection system

The L-8231 is an internally-mounted ECM set, operating in the H, I and J bands, designed to protect combat aircraft or helicopters from attack by missiles with cw radar guidance. Incoming signals are automatically analysed and the appropriate jamming signal transmitted. The set also interfaces with the aircraft's rwr.

Power consumption: (transmit) 600 VA
Weight: 18 kg
Volume: 15 litres (0.53 ft³)

STATUS: in production.

EL/L-8300 airborne sigint system

In operation in large aircraft such as the Israel Air Force's converted Boeing 707s, the L-8300 is a long-range, highly sophisticated Sigint system, said to be capable of detecting communications and other electronic signals at ranges up to 450 km. Received and processed data can be transmitted to a ground command and control centre (EL/L-8353) for further processing, evaluation and dissemination of data.

The L-8300 incorporates the L-8312A Elint and K-7032 Comint systems (see separate entry) and the L-8350 command and analysis station on the aircraft. Training for operators of the L-8300 can be undertaken on the L-8351 simulator (see Training Systems section), and post-mission analysis is done on the L-8352 system.

STATUS: in service.

EL/L-8303 ESM system

This is a computerised ESM system for shipborne, airborne and ground-based applications. It is designed to intercept, digitise, identify, display and record radar signals. The processed data can be examined by an

Elta EL/L-8231 self-protection jammer being installed in Israel Air Force F-4 Phantom

Elta EL/L-8300 Sigint system being operated in Israel Air Force Boeing 707 EW aircraft

operator or used to activate automatic ECM or chaff systems. The system incorporates computer control and processing features. It has an omni-directional instantaneous direction-finding capability which is claimed to be highly sensitive and accurate with a high intercept probability across a wide bandwidth. Alphanumeric keyboard inputs and graphical display output of data are standard features, and the operator can access part of the emitter library. The system also includes built-in test circuits.

Intercepted radar signals are measured and classified according to pulsewidth, frequency, azimuth location, power and time of arrival and are then fed to the processor. This conducts signal sorting, recognition functions and emitter identification, and produces graphical data outputs. A modular system configuration permits optimisation of the system to meet customer cost and performance requirements.

Frequency range: 2-18 GHz (optional down to 0.5 GHz)
Azimuth accuracy: 5° rms (3° option)
Max pulse rate: 500 000/s
Sensitivity: –60 dBm wide-band; –80 dBm narrow-band
Dynamic range: 60 dB
Reaction time: 1 s (typical)
Installation: two 19-inch (487 mm) racks
Power: 4 kVA at 115 V, 400 Hz
Environmental specification: MIL-E-5400

STATUS: in production.

EL/L-8312A elint ESM system
The L-8312A system forms a part of the L-8300 Sigint set (see separate entry), or can operate

Elta EL/L-8310 wide-band Elint/ESM receiver system in Arava light transport/EW aircraft

alone, covering the 0.5 to 18 GHz frequency band and able to detect, analyse and identify signals out to the radio horizon. Direction-finding is fast and accurate and bearings are stored in the associated computer and correlated with the aircraft's navigational data to give a display of the actual location of selected transmissions on a colour graphic console. The

L-8312A incorporates the L-8312R receiver, L-8320 signal parameters measurement equipment and the L-8610 computer.

STATUS: in production and service.

Comint/elint system for Arava
IAI Airborne Systems Division has developed a Comint/Elint system for the Arava twin-engined light transport. The equipment can be rolled on/off a multi-purpose aircraft or be permanently installed. The Comint operator uses sensors such as the Elta EL/K-1250 with omni-directional antennas, direction-finding and signal recording. The Elint operator has 0.5 to 18 GHz bandwidth coverage with omni-directional antennas, signal analyser, data processor and recorder facilities. Jamming can be accomplished using antenna arrays in the aircraft tail, giving approximately 240° azimuth coverage in the lower hemisphere. Spot and band jammers of 20 to 400 watts power are used.

STATUS: in production.

EW system for Lavi aircraft
Elta is the prime contractor for many of the Lavi's avionics systems, including the electronic warfare suite. An integrated ECM/ESM computer allows the incoming signal to be detected and analysed rapidly, providing information to the crew and automatically triggering automatic response using jamming and deception techniques.

STATUS: in development.

Tadiran
Tadiran Limited, 11 Ben-Gurion Street, Givat-Shmuel, PO Box 648, Tel Aviv 61006
TELEPHONE: (3) 713 111
TELEX: 341692

Automatic vhf/uhf jammer
The airborne system consists of four transmitters (covering the 100 to 200 and 200 to 400 MHz frequency bands), a vhf/uhf receiver controller with diskette and terminal, broadband antennas, an hf transceiver and a power supply unit. It is capable of long-distance tactical ground and ground-to-air communications jamming. Up to 16 simultaneous missions can be jammed, and all equipment is self-contained once installed on board an aircraft.

The system operator feeds frequencies to be jammed into the controller and the computer

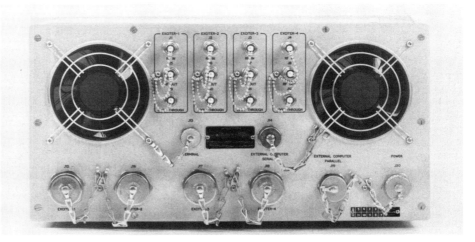

Tadiran automatic vhf/uhf jammer control

initiates time-shared jamming. Three main functions are performed: spectrum scanning from 100 to 400 MHz, with reporting of any activity either unknown or unrecognised; scanning lists of known frequencies to update activity status on operators' screens; and enabling the operator to monitor, analyse and record any active station features.

Frequency range: 100-400 MHz
Look-through: optional
Jammed missions: 16
Coverage: 360°

STATUS: under development.

Automatic vhf/uhf comint and direction-finding system

This is an airborne, self-contained system designed for acquisition and monitoring of tactical radio nets which operate in the 200 to 500 MHz frequency band. The system features spectrum-scanning in programmable frequency sectors, and provides active display of up to 100 pre-programmed frequencies with identification, last direction-finder, signal level and time-of-reception data, and fully automatic direction-finding. It has an operator-oriented 'hands off' monitoring capability and built-in test equipment. Operation has been simplified by adopting a question-and-answer operation mode. All possible exits from current status are continuously displayed and most functions require only a single keystroke. Main functions are to provide spectrum scanning, from 200 to 500 MHz, with reporting of new active, previously unknown or unrecognised stations, scanning of known frequencies and updating operators' activity displays, and providing a monitor, analysis and recording facility for any active stations.

Frequency range: 200-500 MHz
DDA accuracy: 2.5° rms typical
Pre-programmed frequencies: 100

STATUS: in production and service.

RAS 1B elint and ESM system

This airborne interferometric Elint system, installed on large transport aircraft, covers the 0.7 to 18 GHz frequency band (0.5 to 40 GHz optional) to obtain an enemy electronic order of battle (EOB). It can measure frequency, pulse repetition frequency (prf), amplitude and direction-of-arrival (DOA) data on an average of two emitters per second, with exceptionally high accuracy. Operation has been optimised for weak-emitter recognition in high-density electromagnetic environments, and an activity file is maintained which contains identification, classification and updated data on recognised threats. A combination of wide-bandwidth-acquisition ifm receiver and a narrow-band superheterodyne analysis receiver is used to ensure 100 per cent probability of detection and high system throughput. Operation can be fully automatic, or man-machine oriented in simple manual mode which can accommodate up to two operators.

Frequency range: 0.7-18 GHz (0.5-40 GHz optional)
DOA accuracy: 1° typical
De-interleaving capability: Up to 8 pulse trains in the same frequency-DOA cell

STATUS: in production and service.

TDF-500 vhf/uhf automatic direction-finding system

This is an airborne automatic direction-finding system using a powerful central computer, high-performance twin-channel receiver and antenna dipoles and monopoles mounted on the aircraft wings and under surfaces. It covers the 20 to 500 MHz band and can determine direction of arrival to 1.5°. The antenna ports connect to an rf switch matrix which selects dipole pairs and feeds them to the twin-channel receivers. In each sampling period a phase measurement is performed between one dipole-pair, and successive phase measurements are processed to obtain direction-of-arrival information. Response time is claimed to be less than 300 milliseconds.

Frequency range: 20-500 MHz
Accuracy: 2.5° rms between 30 and 500 MHz
4° rms between 20 and 30 MHz
Coverage: (azimuth) 360°
(elevation) –2° to –7°
Polarisation: vertical

STATUS: in production and service.

Tadiran automatic Comint and direction-finding system

Array of Tadiran TDF-500 aerials on Boeing 707 model

Tadiran RAS-2A layout and console

RAS-2A ESM and elint system

This airborne, wingtip-mounted, interferometric tactical ESM system is deployed on small and medium-size transport aircraft. In the 0.5 to 18 GHz frequency band, it can measure frequency, pri, amplitude and direction-of-arrival data with excellent accuracy, 360° around the aircraft. Operation has been optimised for a single operator who can both display an EOB scene and passively target an ARM or anti-ship missile. The high sensitivity of the system ensures weak-signal processing, even at long ranges. Automatic file activity is maintained for quick identification and classification of threats.

Frequency coverage: 0.5–18 GHz
Azimuth coverage: 360°
Sensitivity: –80 dBm
Instantaneous bandwidth: 1, 5, 15, 40 MHz

STATUS: under development.

TACDES

TACDES is an airborne Sigint system consisting of the RAS-1B Elint and the automatic vhf/uhf Comint and DF units. The system is installed on large-size transport aircraft such as Boeing 707s.

STATUS: in production.

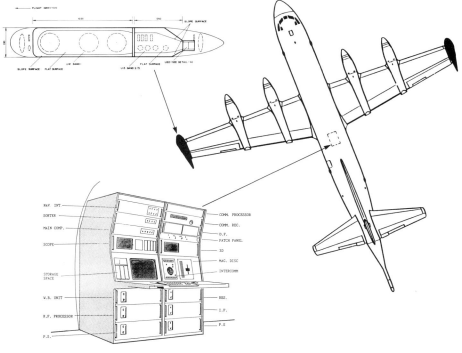

Tadiran RAS-1B operator's console

ITALY

Elettronica

Elettronica SpA, Via Tiburtina Km 13 700, 001 31 Rome
TELEPHONE: (6) 43641
TELEX: 611024

ELT/156 radar warning receiver

Suitable for fighter, light-strike/attack aircraft and helicopters, this lightweight passive detection system provides broadband 360° azimuth coverage and can distinguish anti-aircraft artillery, surface-to-air missile and air-to-air

missile illuminations in bearing and range. Miniature crystal video receiver technology and advanced signal processing, including built-in test equipment, contribute to minimise system weight. A dual-mode cathode ray tube display, on which synthetic or raw video data can be shown, is standard. An audio output is available to complement the display. Latest versions of the equipment are lighter than earlier production systems.

System weight breakdown:
(4 antennas) 0.33 lb (0.15 kg) each
(2 rf heads) 2.2 lb (1 kg) each
(signal processor) 11.5 lb (5.2 kg)
(display unit) 4.3 lb (1.9 kg)
(control panel) 1.3 lb (0.6 kg)
Total weight: 22.66 lb (10.3 kg)

STATUS: in production.

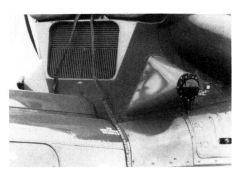

Above photographs show typical Colibri system installation on Agusta-Bell 212 helicopter, and cockpit display and control panel

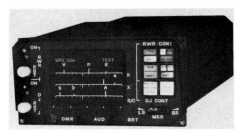

ELT/261 radar warning receiver display unit for Elettronica Colibri system

Colibri integrated ESM/ECM system

Developed for autonomous helicopter operations, especially for carrier-borne forces, the Colibri system includes receiver and jammer elements. The basic electronic support measures facilities are the ELT/161 system, plus ELT/261 displays and instantaneous frequency measurement (ifm) receiver units. Jamming is based on the ELT/361 noise- and ELT/562 deception-jammers. Basic receiver equipment can comprise up to 10 antennas, a warning processor, df/rf band receiver, junction box, radar warning receiver display, blanking unit and preset unit. An electronic surveillance measures display and control console and ifm receiver can be added. Noise jammer components are an antenna and transmitter and driver units. Deception jamming requires an extra receiver and transmitter unit. The system is capable of the following instantaneous surveillance coverage:

Altitude	Effective radius against radar type (n miles)	
	I band (8-10 GHz)	F band (3-4 GHz)
300 ft (91 m)	26	34
1000 ft (305 m)	41	50
2000 ft (610 m)	47	63

STATUS: in production.

ELT/263 ESM system

This is an integrated ESM system designed and manufactured to meet the electronic surveillance requirements of patrol aircraft. The

Elettronica ELT/156 radar warning receiver

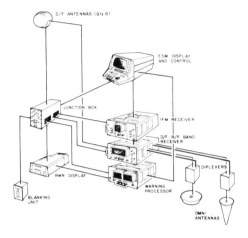

Schematic illustration of Elettronica ELT/263 ESM receiver

Gates Learjet 35A maritime patrol variant with Elettronica ELT/263 ESM system

Mechanical check-out of Elettronica ELT/460 ECM pod ram-air turbine

Elettronica ELT/555 deception jammer pod on Gates Learjet 35A multi-mission trainer

equipment detects, analyses and identifies all electromagnetic emissions in the E to J band (2 to 20 GHz). Additionally, the system performs bearing measurements to permit source location by triangulation. Information can be transferred to external users, such as ground stations, naval units and other operators aboard the aircraft. Data can be recorded by a printer. Built-in test facilities are provided and the system is designed to MIL-STD environmental specifications. The equipment is in current production for several customers.

STATUS: in service.

Altitude	Effective radius against radar type (n miles)	
	I band (8-10 GHz)	F band (3-4 GHz)
2000 ft (610 m)	42	65
6000 ft (1830 m)	53	100

STATUS: in service.

ELT/457-460 supersonic noise jammer pod

A set of four related noise jammer pods is available for use on any high-performance strike/fighter or light-attack aircraft, mounted on an underwing pylon. All units are self-contained with a ram-air turbine-driven generator in the nose section and each is dedicated to a particular waveband. A heat-exchanger is

situated behind the turbine, and there are fore and aft antennas in all pods. The ELT/459 and ELT/460 versions have additional antennas beneath the body of the pod. Processing within the system allocates threat-jamming power on proportional bases against any type of pulsed-radar threat, and includes built-in test equipment and control of blanking with other aircraft systems. All units are claimed to be highly resistant to ECCM, including any type of frequency-agility, and incorporate a large threat library. Maximum operating speeds are Mach 1.1 at sea level and Mach 1.5 at 40 000 feet (12 200 metres).

Pod length: 3120 mm
Diameter: 340 mm
Weight: 145 kg
Interfacing: STANAG 3726 and MIL-A-8591D

STATUS: in service.

ELT/555 supersonic deceptive jammer pod

This has similar configuration to ELT/457 system, with the same ram-air turbine on the nose of the pod. Internal equipment includes fore and aft facing antennas which receive pulsed and continuous wave signals and

transmit pulsed responses, plus separate fore and aft facing continuous wave transmitter antennas on the undersurface. The system is sensitive to H to J band (6 to 20 GHz) threats, and is designed to operate in dense electromagnetic environments. It has multiple-target contrast capability and features built-in test equipment. Operating envelope and mechanical interface specifications are the same as those of the ELT/457. The cockpit display can be tailored to customers' requirements.

Pod length: 3000 mm
Diameter: 340 mm
Weight: 140 kg

STATUS: in service.

ELT/562 and ELT/566 deception jammers

These are repeater jammer systems for internal installation on aircraft and helicopters. ELT/562 is designed to counter pulse threats, while ELT/566 combats continuous wave threats. One or both jammers may be associated with an airborne ESM system such as the Elettronica Colibri. The jammers are effective against H to J band (6 to 20 GHz) threats and are intended to protect aircraft and helicopters in battlefield environments.

STATUS: in production.

Fast jam ECM system

This is a communications ECM system developed for helicopter installation. It provides surveillance and automatic-jamming of emissions and can differentiate between friendly and unfriendly emissions. In normal operations the operator specifies a set of instructions and the system then operates in automatic mode. Operator-initiated updates can be introduced at any time during a mission, the mix of automatic or manual control being specified by him. In automatic mode the system achieves extremely short reaction times, both in reception and transmission. Sophisticated jamming techniques permit simultaneous jamming and look-through without loss of jamming efficiency. The system has an analysis and monitoring receiver with associated display, a computer for system management, one or more all-solid-state jammers and a plasma-panel display.

STATUS: in production.

Smart Guard ESM system

The Smart Guard ESM system has been developed for helicopter operations to provide continuous and automatic surveillance of the vhf/uhf radio communications bands. The system can intercept, analyse and measure the direction of any signal in the waveband, even on short-duration transmissions. The processor will also automatically correlate successive intercepts of the same emitter for location. Automatic surveillance is conducted simultaneously in one or more frequency ranges, there is automatic control of pre-determined channels (this includes technical analysis, direction-finding, fixing and network analysis)

Elettronica ELT/460 supersonic ECM pod

and automatic demodulation of intercepted emissions for operator audio monitoring or recording. The system uses a high-scan velocity receiver, an analysis and monitoring receiver with associated display, a high-precision direction-finder, a computer to manage the sensors and process gathered data into display formats, a dual-track audio recorder and a digital mission recorder. All data acquired is displayed to the operator for interpretation, evaluation and immediate use.

STATUS: in production.

ELT/999 comint system

The ELT/999 is a communications intelligence gathering system designed specifically for airborne applications. It can survey rapidly the uhf and vhf bands, picking up even the shortest transmissions, locating the emitters and analysing the signals for monitoring or recording purposes. Information processed by a computer is presented to operators on alphanumeric or panoramic displays. Intended for medium-size aircraft, this strategic intelligence-gathering system can survey large areas.

Of modular construction, it can be reconfigured rapidly to meet different Comint requirements. Advanced signal-processing is employed and the different functions are controlled by software, so that the system is flexible in operation. With a high degree of automation, it can be controlled by a single supervisor, while up to three operators perform monitoring tasks. Weight of the system is around 900 kg depending on the configuration chosen.

STATUS: in production.

Selenia

Selenia, Special Equipment and Systems Division, Via dei Castelli Romani 2, 00040 Pomezia, Rome
TELEPHONE: (6) 43601
TELEX: 613513

IHS-6 ESM/ECM system

Designed for tactical electronic surveillance measures missions such as stand-off jamming, air-strike support or fleet protection, the system has two major components: RQH-5/2 ESM system and TQN-2 jamming system. The ESM component can detect, analyse and identify threats in the 1 to 18 GHz band and has a wide-open receiver with instantaneous frequency measuring capability. It provides monopulse direction finding and can accommodate a

library of 2000 threats. At least 50 targets can be tracked simultaneously. Data is displayed on a cathode ray tube with alphanumeric or graphic presentation.

The TQN-2 jammer can operate on up to four bands, using steerable I to J band (8 to 20 GHz) antennas and fixed installations for lower frequencies. Spot, barrage or hybrid jamming is possible, with automatic release of chaff countermeasures.

Systems using two antennas and operating in I to J band have been shown for the Agusta 109A helicopter, and a full four-band system has been installed in Egyptian-operated Westland Commando helicopters.

STATUS: in production.

SL/ALQ-34 ECM pod

Available as a self-defence aid for high-

performance aircraft, although no customer has been clearly identified with the system. The SL/ALQ-34 provides a radar warning facility and can jam anti-aircraft artillery or surface-to-air missile radars. Current versions operate across a 6 GHz bandwidth in the H to J band (6 to 20 GHz) but lower frequency systems (C, D, E and F bands), effective against air-defence search radars, are in prospect.

The receiver processor conducts threat assessment and ranking, using a stored library of threat data, and performs jamming power-management tasks. Travelling-wave tube transmitters (probably two) are used, and cooled by a closed-loop liquid cooling system. Supersonic, high-altitude clearance has been obtained.

STATUS: in production.

JAPAN

Mitsubishi

Mitsubishi Electric Corporation, Mitsubishi Denki Building, 2-3 Marunouchi 2-chome, Chiyoda-Ku, Tokyo 100
TELEPHONE: (3) 218 2111
TELEX: 24532

ALQ-5 ESM system

Used on ESM variants of the Kawasaki C-1 medium transport and in development since 1978, the system receives and jams surface-to-air missile radars and is complementary to the ALQ-3 system. (See *Jane's Avionics 1986-87*).

ALQ-6 jamming system

In development to provide an airborne radar

jamming capability for McDonnell Douglas F-4EJ and Mitsubishi F-1 fighters.

ALQ-8 jamming system

In development for F-15J, and may be a variant of the ALQ-6 set. It is understood to provide countermeasures in the 1 to 4, 4 to 8 and 7.5 to 18 GHz bands. ALQ-8 jammers integrate with the APR-4 radar-warning system.

Tokyo-Keiki

Tokyo-Keiki Company Limited, 16 Minami-Kamata 2-chome, Ohta-Ku, Tokyo
TELEPHONE: (31) 732 2111
TELEX: 2466194

APR-3 radar warning system

Used on Mitsubishi F-1 support aircraft, but origins and capability unknown. May still be in limited production.

APR-4 radar warning system

Used in McDonnell Douglas F-15J/DJ Eagle fighters. Its specification calls for the ability to process multiple inputs simultaneously in a dense electromagnetic environment. The

system incorporates a digital processor with reprogrammable software which permits reconfiguration to meet developing threats. A tactical situation cathode ray tube display presents multiple-threat data in alphanumeric and graphic form. Interfaces with other on-board electronic countermeasures systems, such as ALQ-8 jammers, can be accommodated.

STATUS: in production.

SWEDEN

Ericsson

Ericsson Radio Systems AB, Airborne Electronics Division, S-163 80 Stockholm
TELEPHONE: (8) 757 0000
TELEX: 13545

Loke radar jammer family

Loke radar jammers constitute a family of such devices, effective against both continuous wave and pulsed transmissions from airborne or ground-based fire-control systems. They are available for both tactical and training purposes, and differ from one another mainly in their software control, cooling arrangements, frequency range and jamming techniques.

STATUS: in production for and operation with Swedish Air Force/Saab Viggen multi-mission fighters. Other versions are in development.

Loke 220 jammer

A member of the Loke family, the 220 is a fully automatic system effective against pulse and cw radar transmissions. Velocity, range and angular deception modes are all available, and the equipment can communicate with the aircraft's avionics via a standard data-bus. The Loke 220 is carried in a pod, 4250 mm long and 425 mm in diameter.

Frequency coverage: H, I and J bands
Output power: 100-150 W

STATUS: in service.

Ericsson Loke 220 automatic jammer

Type A ECM system for JAS 39

Ericsson has been chosen to conduct the integration of the electronic warfare suite on the JAS 39 Saab Gripen currently in development for the Swedish Air Force. The system will employ power-management and self-adaptive techniques in multi-mode operation to meet likely threats up to the end of the century. In addition to its role as integration contractor, Ericsson is also designing the active ECM element of the system.

STATUS: in development.

Philips

Philips Elektronikindustrier AB, Defence Electronics, S-175 88 Jarafalla
TELEPHONE: (758) 100 00
TELEX: 11505

BOH 300 ECM dispenser

This is a chaff/flare dispenser in development for helicopter applications, designed to provide protection in hovering and forward flight. It has processor control, reprogrammable operation features and data-bus interface capability. It is suitable for radar, optical and infra-red decoy deployment from attack, anti-tank or scout helicopters. The dispensers can be mounted on landing gear, fuselage structure or weapon hard points.

Installation: (dispenser) 210 × 154 × 500 mm
(control unit) 76 × 147 × 165 mm
Weight: (empty) 33 lb (15 kg)
(loaded) 55 lb (25 kg)
Power: 100 VA at 115/220 V 400 Hz

STATUS: under development.

BOP 300 ECM dispenser

This dispenser system is suitable for combat aircraft, and shares many common items with the BOH 300 helicopter system. The processor can control up to four different dispensers simultaneously and will have data-bus interface capability. Installation can be internal or external flush-mounted to the aircraft structure.

Installation: (dispenser) 154 × 210 × 715 mm
(control unit) 76 × 147 × 165 mm
Weight: (4 dispensers, empty) 32 kg
(4 dispensers, loaded) up to 66 kg
Power: 100 VA at 115/220 V 400 Hz

STATUS: under development.

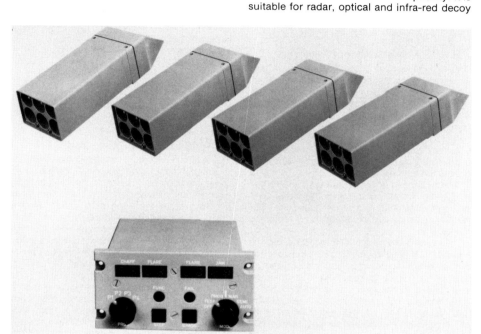

Philips BOP 300 ECM dispenser system

BOZ 3 training chaff dispenser

This is a high-capacity unit originally developed to a Swedish Air Force requirement, and since adapted for training use in view of its electro-mechanical design. It can be manually or automatically initiated and has both break-lock and corridor chaff modes.

Length: 2.94 m
Diameter: 0.37 m
Weight: 270 kg
Attachments: 14 or 30-inch (356 or 762 mm) bomb lock

STATUS: in production.

BOZ 10 chaff pod

This chaff pod, developed jointly by France, Germany and Sweden, is believed to be in use on light-attack versions of Dassault/Dornier Alpha Jet and Saab 105.

STATUS: in production.

BOZ 100 ECM dispenser

This is an advanced ECM dispenser originally developed for the Swedish Air Force, but also supplied to overseas customers. It is capable of sophisticated break-lock and corridor operation at subsonic and supersonic speeds. The unit has a comprehensive EW system interface and is suitable for use in deep-penetration, strike, reconnaissance and electronic warfare roles. It is microprocessor controlled and has a reprogrammable program memory.

Length: 4 m
Diameter: 0.38 m
Weight: 325 kg
Attachments: 14 or 30-inch (356 or 762 mm) bomb locks

STATUS: in production.

Philips BOZ 3 training chaff dispenser

Philips BOZ 100 chaff/flare dispenser for high-performance aircraft

SATT

SATT Communications AB, PO Box 32701,
S-126 11 Stockholm
TELEPHONE: (8) 810 100
TELEX: 15325

Radar warning system

No designation appears to have been allocated, but the system is used on the Saab 105G and Viggen ground-attack aircraft. The system has four detector heads aligned to provide 360° azimuth coverage, and can determine quadrant of I to J band pulse and continuous wave radar threats. E to F and G to H band options are available.

The Saab 105G has detectors in the tail-plane/fin bullet fairing, while the Viggen uses two forward-facing antennas in wing-tip installations and two rearward-facing antennas in the fuselage. A circular display together with aural warning is provided in the cockpit. Equipment facilities include non-threat data rejection, provision for radar blanking and built-in test equipment.

Detector heads: 430 × 105 × 105 mm
Weight: 3 kg each
Indicator: 79 mm dia × 69 mm
Frequency band: 8-20 GHz (4-8, 2-4 GHz options)
Power: 300 mA at 28 V dc (excluding indicator lamps)

AR-765 radar warning receiver

This is a lightweight, low-cost system developed for helicopters, and particularly suitable for the Agusta-Bell JetRanger, where the installation replaces the battery access panel and the radar warning receiver takes the place of the battery cover plate. Dual forward-facing circularly-polarised spiral antennas provide full forward hemisphere coverage with ±45° elevation. A visual indication of the threat quadrant is provided, together with aural warning.

STATUS: in production.

AQ-31 deception jamming pod

The AQ-31 is a continuous wave radar-jamming system suitable for aircraft pod mounting and able to provide full coverage in the E to J band (2 to 20 GHz). A high power output is claimed and AM, FM noise/sawtooth jamming modes are incorporated. The pod is ram-air cooled and has mainly digital operation, with interfaces suitable for connection to onboard digital signal processors. Two pods can be used simultaneously, connected to a single control panel in the cockpit. The company also produces a companion receiver pod, the AR-753.

STATUS: in production.

AR-753 jammer/set-on receiver pod

Although operating details have not been disclosed, this appears to be a home-on-jam pod, and to have no active elements within it. It is claimed to offer high resolution over a wide operating band and it has its own cockpit control panel, containing the display, antenna selection switch and gain control.

Pod dimensions: 140 mm dia × 1600 mm length
Weight: 25 kg

STATUS: in production.

AR-777 airborne threat receiver

The AR-777 is designed to provide rapid and unambiguous detection and identification of emitters between 1 and 8 GHz in a dense signal environment. A computer-controlled airborne unit which uses a heterodyne receiver with a high-performance YIG-tuned rf head, the AR-777 is claimed to have excellent frequency accuracy and resolution on account of a special if processor and a microprocessor-controlled calibration routine.

Three rf heads cover the full bandwidth (one octave per head), each including YIG-filters, YIG-tuned oscillators and drive circuits. The frequency range can be modified by replacing rf heads.

Two sweep widths are selected, covering 100 and 300 MHz ranges, and the YIG-filters overlap so that switching between filters is avoided, so optimising frequency tracking and linearity. YIG-filter specifications provide 25 MHz bandwidth to the 3 dB points, spurious depression outside band 70 dB and skirt selectivity of 24 dB per octave. A switch selects any one of six receiver inputs, which include double-balanced mixer and pre-amplifiers generating an intermediate frequency of 160 MHz. Log amplifiers and bandpass filters are used with a video processor to achieve a high level of selectivity and unambiguous readouts.

The receiver and display units communicate through a 32-bit serial digital highway which conveys information for the tuning, control functions and parity checks. The display unit has a keyboard on which the centre frequency can be selected. Received signals are displayed on a spectrum-display comprising two rows each containing 21 lamps. The upper row has a memory and stores displayed data for a preset time while the lower row gives momentary indication of receiver perceived activity. A fixed-tuning mode, or one of several search patterns, can be initiated by the operator. In fixed-tuning mode the received-signal prf (pulse repetition frequency) is evaluated and the result displayed on two digital displays. These can simultaneously display two prfs, but if more than two prfs are present at one radio frequency they are displayed sequentially. A separate indication of garbled or jittered prfs is provided, and all spectrum analysis output is available from a co-operating jammer system. Stored data can be analysed in ground-based evaluation equipment.

Dimensions: (receiver) 497 × 257 × 194 mm
(display) 580 × 355 × 173 mm
Frequency range: 1–8 GHz
Power: 300 VA at 200 V 3-phase 400 Hz
Sweep width: 100-300 MHz
Sweep rate: 10 sweeps/s
Frequency accuracy: 5 MHz
Frequency resolution: 5 MHz
Prf measurement: 200-2000 Hz
Prf accuracy: 1 Hz
Sensitivity: -65 dBm
Dynamic range: 65 dB

STATUS: in development, production expected.

AQ-800 noise jamming pod

The AQ-800 jammer is intended for use in the training of radar operators and for evaluation of radar-station performance in a jamming environment. It is pod-mounted and hung from a pylon under the wing or fuselage; several pods can be carried by a large aircraft.

The pod contains a travelling-wave tube-type transmitter, with three signal generators, and a sweeping superheterodyne set-on receiver. A single pod can cover one band, either S (2 to 4 GHz), C (4 to 8 GHz), X (8 to 12 GHz) or Ku (12 to 20 GHz) band. There is a standard cockpit unit which will handle two pods simultaneously with any set of frequency combinations. A single transmitter can generate up to three simultaneous jamming signals, the modulation, centre-frequency and bandwidth of each signal being independently controllable. Modulation can be AM-noise, FM-noise or FM-sawtooth and noise. Both fore and aft antennas are available for the transmitter and receiver and direction of activity is selectable from the cockpit. Ram-air cooling is a standard provision.

Pod length: 3.9 m
Max width: 0.43 m
Max depth: 0.52 m
Weight: 220 kg
Power: 200/115 V 3-phase 400 Hz 3 kVA
Signal generator bandwidths:
S band: 700 MHz
C band: 1200 MHz
X band: 2000 MHz
Ku band: 2000 MHz
Modulation types:
(AM) double sideband, suppressed carrier, noise with selectable bandwidth up to 200 MHz.
(FM) sawtooth (20 Hz to 8 Hz) and noise (10 MHz bandwidth). Selectable bandwidth up to 200 MHz.
Output power:
(AM) 200 W (S, C and X bands)
100 W (Ku band)
(FM) 120 W (S, C and X bands)
50 W (Ku band)
Antenna coverage: 70° horiz × 30° vert
Receiver bandwidth: –70 dBm typical
If bandwidth: 2.5-6 MHz
Gain control: 50 dB

STATUS: in service.

AQ 900 airborne self-protection system

The AQ 900 is a pod-mounted automatic jamming system which is available in different wideband versions covering the frequency range from S to Ku band. The complete system consists of two pods, operated from one

SATT AQ-800 jamming pod under wing of Lansen fighter

cockpit control unit, one pod for jamming search radars in S band and one for jamming fire control radars in X or Ku band. The pods are systemwise interchangeable and are designed for pylon-mounting on the aircraft.

The S band and X/Ku band version are essentially identical in operation and consist of a receiver, a threat analysis unit, a decision and control unit and a transmitter. Each pod has its own ram-air turbine generator and cooling system.

The receiver contains all the rf and signal processing circuitry for reception and threat identification. The signal is then passed to the analysis unit where the signals are analysed and compared with data stored in the system library. The decision and control unit includes a microcomputer that gives the system tactical flexibility and allows threat priority, technique selection, power management, signal generation and control, and cockpit control and indicator interface.

The transmitter contains all the signal generating circuitry and the high power twt amplifier. The X/Ku band version also contains a unit for various deception techniques to counter fire control and tracking radars.

STATUS: in production.

UNITED KINGDOM

British Aerospace

British Aerospace plc, Electronic Systems and Equipment Division, Downshire Way, Bracknell, Berkshire RG12 1QL
TELEPHONE: (0344) 483 222
TELEX: 848129

Infra-red jammer for helicopters

Designed to be mounted externally on a helicopter, this system operates by radiating modulated infra-red energy from a unique optical system. The infra-red source is a graphite-radiating element, hermetically sealed within a sapphire envelope. An optical assembly surrounds the lamp and rotates at high speed, the unit being powered by a brushless dc motor contained in the base of the jammer assembly. The outer casing has 16 longitudinal slots which are covered by windows to provide environmental protection and to mask visible emissions so that the system is covert at night. Radiated power is claimed to be sufficient to protect most helicopter types against infra-red

missiles. An experimental unit has completed trials, and production units have been delivered.

Dimensions: 200 mm dia × 405 mm high
Weight: 30 lb (13 kg)
Coverage:
(azimuth) 360°
(elevation) +30°
Power: 24-29 V dc 1.2-2 kW

STATUS: in production.

British Aerospace helicopter infra-red jammer

Ferranti

Ferranti Instrumentation Limited, Aircraft Equipment Department, Lily Hill House, Lily Hill Road, Bracknell, Berkshire RG12 2SJ
TELEPHONE: (0344) 424 001
TELEX: 848117

Datar radar warning system

Datar (detection and tactical alert of radar) has been developed jointly with E-Systems for use in helicopters, fixed-wing aircraft and maritime patrol vehicles. In addition to providing a warning of hostile radar emissions Datar analyses the radar signal to characterise, classify and establish the priority of threats.

A main feature of the system is its ability to be programmed in the field to update the threat

library held in the computer's database and to determine priorities on a mission by mission basis.

Datar operates in both pulse and continuous wave environments covering all common radar bands. It analyses received signals to ascertain frequency and pulse repetition interval. This information is then compared with data held in the threat library. The threats are displayed to the aircrew on a small 3-inch (76 mm) crt indicator in plan position format. The display is heading-compensated and shows the threat's relative range and bearing with symbology depicting the type of threat. An imminent threat triggers an audio alarm. The system can also be used to initiate ECM sub-systems.

Datar is packaged into four line-replaceable units, a receiver set, central processor, indicator

and control unit. A typical installation uses five antennas. Facilities are also included to interface with other tactical sensors such as laser warning which can be integrated into the same display.

STATUS: in development.

Ferranti Defence Systems Limited, Electro-optics Department, Robertson Avenue, Edinburgh EH11 1PX
TELEPHONE: (031) 337 2442
TELEX: 72529

Laser warning receiver

Ferranti is developing a laser warning receiver, which will provide the pilot with an audible and visible warning of designation by a laser, in much the same way as he already gets warning of illumination by radar.

To counter the problems of detecting the very narrow laser beam, which at any instant would only be illuminating a small part of the airframe, a number of dispersed sensors are positioned over the aircraft, so as jointly to provide total coverage. (The receiver can, of course, be applied to any required vehicle or

Ferranti Datar system units: (left) receiver, (centre) antennas and display, (right) central processor

Ferranti Datar cockpit display

object.) The laser light detected by the sensors is analysed to provide a bearing from which the beam is coming. The sensors protrude 25 mm above, and to 50 mm below the aircraft's skin, and are connected to the processor with fibre-optic cables to reduce the false warnings which might be triggered by electromagnetic signals or interference.

STATUS: in development.

Marconi Defence Systems

Marconi Defence Systems Limited, Chobham Road, Frimley, Camberley, Surrey GU16 5PE
TELEPHONE: (0276) 63311
TELEX: 858289

ARI 18223 radar warning receiver

This is a self-contained radar warning receiver system suitable for single-seat aircraft, and it is installed in Royal Air Force Jaguar and Harrier aircraft. Systems have also been supplied in some overseas deliveries of the two types. Frequency coverage is 2 to 20 GHz (E to J band), and a lamp-type display is used to show the quadrant in which the highest priority threat is determined. Antennas are on the vertical fins of the aircraft and an audio-alarm is also triggered when a threat is displayed.

STATUS: in service.

ARI 18228 radar warning receiver

This system uses the same antennas and has the same frequency-band coverage as the ARI 18223 system. It is designed for two-seat aircraft such as the British Aerospace Buccaneer and McDonnell Phantom. It is believed that the British Aerospace Vulcan also used this system. The display is a small circular cathode ray tube on which the direction of several threats from the aircraft can be shown simultaneously.

STATUS: in service.

Radar homing and warning receiver for Tornado F2

Based on previous experience Marconi Defence Systems has developed an advanced radar warning receiver for the Panavia Tornado F2. This new technology system will be able to handle threats likely to be met in Europe. Based on a modular design, the system has its own processor, thus allowing a quick and accurate analysis of threats and the flexibility inherent in a software-based system.

STATUS: under development.

Sky Shadow ECM pod

Marconi Defence Systems is the prime contractor in the production of the Sky Shadow ECM pod, developed for use with the Royal Air Force/Panavia Tornado. Other components are provided by Racal, British Aerospace, GEC Avionics and Plessey. Marconi Defence Systems is responsible for overall design, development and subsequent testing. Sky Shadow flight-tests were completed in 1980.

A high-power travelling-wave tube amplifier is used with a dual-mode capability for deceptive and continuous-wave jamming. A voltage-controlled oscillator uses a varactor-tuned Gunn diode to cover the full frequency band. The set-on receiver is of Decca design, and signal processing uses GEC Avionics hardware.

The pod incorporates both active and passive electronic warfare systems, and includes an integral transmitter-receiver, processor and cooling system. It has radomes at both ends and is stated to be capable of countering multiple ground and air threats, including surveillance, missile and airborne radars. Some automatic power management can be assumed, and modular construction probably allows the system to be 'tuned' to differing operational missions.

Approximate pod dimensions: 380 mm dia × 3350 mm long

STATUS: in production.

Hermes ESM system

Hermes is a software-controlled superheterodyne receiver that automatically searches for, acquires and identifies radar energy impinging on the host aircraft, compares its frequencies with an emitter library in a computer, displays processed information in the cockpit, and records it for subsequent analysis. The system is modular in nature and can be tailored to suit the particular configuration required in ground-stations and ships and land-based vehicles as well as aircraft. It can also provide the appropriate signal and power management to drive jammers.

STATUS: in development. The system is reported to be under study by Northrop (which has an agreement with Marconi Defence Systems to build part of the Zeus ECM system, see separate entry) for its RF-5E reconnaissance fighter. A Hermes system, with antennas mounted in two under-wing pods, has been fitted to an AEW demonstrator version of the Pilatus Britten-Norman Defender.

Zeus internal ECM system

In May 1984 Marconi Defence Systems was chosen by the Royal Air Force to provide a new defensive electronic warfare system called Zeus for its 60 McDonnell Douglas/British Aerospace Harrier GR5s. The contract, worth around £100 million, will be shared with Northrop (whose Defense Systems Division will build the rf transmitter and will jointly market the system with Marconi in the US and elsewhere) to the extent of about $42 million.

Zeus is intended both for new combat aircraft, such as the Harrier GR5 and its US-built version, the AV-8B, and for retrofit on older types. Modular in nature, the system can be configured in many ways, from simple warning

Royal Air Force Jaguar equipped with Marconi Defence Systems ARI 18223 radar warning receiver units near fin tip

Royal Air Force Tornado carrying Marconi Defence Systems ARI 23246/1 ECM pod on each outer pylon

Marconi Defence Systems Zeus internal ECM system

equipment to a complete radar countermeasure suite that can drive chaff dispensers, infra-red flares and decoys. It can therefore meet threats such as radar-controlled anti-aircraft guns and surface-to-air and air-to-air missiles.

A typical installation could include a power supply, receiver, techniques generator, data processor, video signal demodulator, control unit, display, two transmitters, and antennas. The latter could comprise two pairs of high and low frequency detectors located on or near the nose and scanning in the forward direction, and two similar pairs on the tailplane or rear fuselage looking aft. In addition there would be probably at least two transmit antennas, one looking forward, the other aft. A typical system weighs around 240 lb (109 kg).

The jamming system is activated by the radar warning receiver digital processor under software control. The transmitters can jam both pulse and continuous wave radars, at the same time preventing weapons employing home-on-jam guidance equipment from acquiring the system and its host aircraft.

STATUS: in development. Zeus systems have been 'paper optimised' for British Aerospace Hawk, Sepecat Jaguar, Dassault Mirage, Northrop F-5, General Dynamics F-16, and McDonnell Douglas F/A-18 and A-4 aircraft in addition to its first application. Phase-7 upgrade of the existing Royal Air Force Harrier GR3 fleet has also been under discussion.

Apollo ESM/ECM system

This is a lightweight, modular package providing comprehensive protection for ground-attack and air-defence aircraft, and continuous platform monitoring in the warning role. The intercept element of Apollo is represented by the Guardian series of radar warnings receivers.

STATUS: entering service.

Guardian series radar warning receivers

These exploit the most recent large-scale

integration technology to provide a family of small, lightweight receivers. They are microprocessor-based for operational flexibility, and data specific so particular missions can be entered as part of the pre-flight check. Guardian receivers represent the passive intercept and power-management element for the Apollo ESM/ECM system (see separate entry).

STATUS: under development.

Jammers for RPVs

Marconi Defence Systems is developing miniaturised active and passive EW equipment suitable for RPVs. It is also working on a range of miniature unattended jammers for use against hostile radar and communication equipment.

STATUS: under development.

MEL

MEL, Manor Royal, Crawley, West Sussex RH10 2PZ
TELEPHONE: (0293) 28787
TELEX: 87267

Airborne Matilda threat-warning system

Matilda is a radar-intercept device capable of providing warning of a locked-on fire-control radar or the target-seeking radar beam of a guided weapon. The Matilda unit for small ships is also being applied to the development of a miniaturised, lightweight system for helicopters and light aircraft.

The surface-vessel version of Matilda is designed to react to either pulsed or continuous wave illuminating radar signals. The criteria used to initiate an alarm are either more than 250 pulses of similar bearing, pulse width and amplitude in a 500-millisecond period (in the case of a pulsed radar), or continuous wave emission detection for more than 100 milliseconds.

Outputs take the form of an aural warning and a visual display and, if desired, an automatic countermeasures initiating signal which activates chaff dispensing equipment. In the surface-vessel case, a voice module is used for verbal indication of the sector bearing of the threat. Simultaneously, the visual display provides a pictorial presentation of the same data. The airborne unit will probably dispense with the voice alarm and will use a warning tone in its place, although sector bearing will be displayed on a panel instrument.

The current Matilda unit uses four wide-beam antennas which cover 360° and are deployed around the vessel's masthead and MEL is presently investigating the substitution of suitable replacements for airborne uses. Direction sensing of the received signal is carried out by amplitude comparison of the signal received through different antennas.

A detector diode in the receiver section produces a dc-to-10 MHz video signal which is amplified for processing and for simplification of bearing measurement. A video processor is used to compare channel outputs and the signal output from this is digitised for transmission to a data processor. Digitised values of bearing, amplitude and pulse width are assembled to form an address to an 8 K word

times 8-bit random access memory. Each time a particular location is accessed by pulse data, its contents are increased and threat-warning is initiated when a preset level is reached within any one integration period of 0.5 second. A test facility which simulates a locked-on radar is incorporated for confidence checking.

The present display system takes the form of an eight sector lamp indicator presentation which is likely to be reduced from its present size to fit a standard aircraft display for panel mounting.

Format: (processor unit) ¾ ATR

STATUS: under development.

Katie radar warning receiver

Katie (killer alarm threat indication and evasion) is an airborne radar warning receiver to provide warning of illumination by pulsed or continuous wave radar-directed weapon systems. Four antennas and associated receivers provide all-round azimuth coverage using amplitude comparison techniques to obtain accurate threat-bearing information. The system has a digital processor in which radar parametric data is stored and used to identify radars by type, bearing and threat mode. Threat radars are identified on a heading-orientated cathode ray

tube display which uses alphanumeric symbols and is readable in bright sunlight or through pilot's night-vision goggles. A voice synthesiser output is also available to give speech identification and warnings, or warning tones.

Frequency detection is available over a wide bandwidth and a version covering a limited bandwidth is also available. Suitable for helicopters or light strike aircraft, the system comprises a processor, two receiver units, a control unit, display and four separate antennas. The threat-identification library is held in eeprom storage for rapid field reprogramming. The display can also accept data from other sensors such as laser warning receivers, while the processor can provide hand-off data for automatic operation of chaff, flare, smoke and expendable decoy stores, or an integrated jammer system.

Frequency range: 2-18 GHz, (option) 0.5 to 2 GHz
Weight: 12 lb (6 kg) approx
Installation: processor, two receiver units, control unit, display, four antennas

STATUS: under joint development by MEL and Dalmo Victor. MEL will design a new receiver for a more advanced version, while Dalmo Victor will further develop the processor and display system of the present Mk 3 system.

MEL/Dalmo Victor Katie radar warning system, with four antennas, receivers, processor, control unit, and warning indicator

MS Instruments

MS Instruments Limited, Rowden Road, Beckenham, Kent BR3 4NA
TELEPHONE: (01) 650 7233

Hofin hostile-fire indicator

The hostile-fire indicator (Hofin) uses shock-wave sensors to detect gun-fire aimed at the host aircraft, and is a passive-warning system suitable for helicopter use. It uses a five-armed sensor (four arranged in one plane, at 90° to each other, and one perpendicularly) which is mounted beneath the helicopter. Shock-waves generated by projectiles are converted to electrical signals, and then processed in an electronic unit which occupies a ³/₈ ATR Short box. This produces outputs which can be used to generate audio or visual warning of nearby arms fire. An audio warning lasts for about one second after a shock-wave has been detected, and a visual warning is presented on a 4 ATI size unit which has eight 45° wide segments. Four segments which indicate the approximate direction from which the shock is detected will illuminate for about 5 seconds.

Dimensions and weight: (sensor array) 305 × 305 × 195 mm; 4.25 lb (1.93 kg) (computer) 94 × 418 × 228 mm; 6 lb (2.72 kg) (indicator) 106 × 106 × 125 mm; 2.2 lb (1 kg)
Power: 60 W at 22-28.5 V dc
Sensitivity: responsive to supersonic projectiles at 20 m
Operating temperature range: –20 to +50° C
Humidity limit: 95% non-condensing

STATUS: in service.

Racal-Decca

Racal-Decca Defence Systems Limited, Davis Road, Chessington, Surrey KT9 1TB
TELEPHONE: (01) 397 5281
TELEX: 894015

HWR-2 radar warning receiver

This is a miniature hand-held radar warning receiver. It is suitable for use in light helicopters and other small craft, where weight and volume restrictions might preclude a permanent panel-mounted installation. However, the system is self-contained and can be hard-mounted if necessary.

The unit is sensitive to radar frequencies from 2 to 11 GHz and is scanned manually, providing a bearing indication to within approximately ±10°. Antenna polarisation, when the unit is held with the handle vertical, is 45°, so vertical, horizontal and circularly-polarised signals are all received. An audio tone indicates reception of a signal, and by rotating the receiver about the axis pointing towards the indicated target, polarisation is shown by an illuminated arrow on the back of the unit. Knowledge of polarisation characteristics is a valuable guide to deducing the type of radar illuminating the aircraft. The audio signal-frequency is representative of the radar pulse repetition frequency detected, and to determine frequency more accurately either E/F or I/J band suppression can be selected. Audio signal intensity is representative of range and power, and if required a 15 dB audio reduction can be obtained by operating a sensitivity switch.

Production deliveries have been made to unspecified customers.

STATUS: in service.

MIR-2 ESM system

Production of over 100 MIR-2 sets has been reported for Royal Navy/Westland Sea King

Racal-Decca MIR-2 ESM system

and Lynx helicopters, and total production since 1978, including exports, is over 200. The system was originally designed for Royal Navy helicopters, but has also proved suitable for light naval vessels such as patrol boats. The system was installed aboard Sea King helicopters in early 1982, just before the Falklands conflict, so that its airborne trials in this aircraft were conducted largely under operational circumstances. Racal has also fitted on MIR-2 suite on its own British Aerospace Jetstream for fixed-wing trials.

The airborne ESM system combines airborne-warning system and search receiver functions using an advanced digital receiver served by a fully-solid-state wide-band antenna system covering the C to J band radar frequencies (0.6 to 18 GHz). Six antennas monitor full 360° in azimuth and provide a very high intercept probability, improving the crew's location and identification of friendly and enemy radars. The control-indicator unit, in the cockpit coaming, is a compact light-emitting diode display which indicates frequency band, amplitude and relative bearing on the main display. On an associated fine-bearing unit the true bearing and radar pulse-repetition frequency are indicated to a high degree of accuracy. The complete system of eight units comprises six antennas, a processor and the control-indicator unit.

Westland Sea King with Racal-Decca Kestrel ESM system

Frequency range: 0.6-18 GHz

STATUS: in production.

Kestrel ESM system

An ESM system under development to meet requirements for an airborne Elint capability, Kestrel will gather information about potential hostile threats during any operation and assess their deployment and movement during active operations. Long range information can be obtained without the hazards associated with other reconnaissance methods.

The system receives and processes radar emissions between 0.6 to 18 GHz. A six-port amplitude-comparison system measures the bearing of the emissions, and provides instantaneous digital information over 360° in azimuth. At the same time, a frequency-measurement receiver, with omni-directional azimuth coverage, provides an instantaneous frequency indication.

Digitised information on all threats, as accumulated pulse-by-pulse, is passed to a pre-processor. In this equipment overlapping pulse trains from different radars are derived and their pulse repetition frequency and frequency-agility characteristics determined. The regularised data is then passed into the main processor where long-term information is extracted. Radar identification is by comparing measured and derived radiation parameters with those stored in a library of known emitters. Full information about the radar-signal environment is then presented to the system operator on an ordered tabular or tactical display or, alternatively, the data can be transferred to remote displays using a standard data-highway.

STATUS: under development.

Prophet radar warning receiver

Prophet is a private-venture system is designed to reduce helicopter and light aircraft vulnerability to radar-associated threats on the battlefield. It provides the pilot/observer with a timely and unambiguous threat warning, permitting appropriate countermeasures, such as manoeuvres or chaff discharges, to be employed. System design has taken account of the likely need to integrate laser warning, hostile-fire indicator and frequency extensions in order

Royal Navy/Westland Lynx equipped with Racal-Decca MIR-2 warning receiver

to increase the survivability of the battlefield helicopter.

The receiver detects signals in H to J bands (6 to 20 GHz) and, by means of a processor with a programmable threat library, alerts the pilot to imminent hostile action. The system is designed to operate in a dense rf environment. A four-port antenna provides octantal direction-finding, using signal amplitude comparison to derive bearing. The system is claimed to have a low false-alarm rate and threats are recognisable with a high degree of confidence. An audible warning is provided by a tone-generator and fed to the aircraft internal communications system.

The Prophet display uses light-emitting diodes and has three lines, each dedicated to a threat. Each line can show an arrow indicating threat direction, and a three-character alpha-numeric identifier. Display colour and brightness is such that it can be viewed in bright sunlight and through night-vision goggles.

Frequency range: H to J bands (6-20 GHz)
Azimuth coverage: 360°
Elevation coverage: 30°
Power: 28 V dc 80 W
Weight: 15 lb (7 kg)
DF accuracy: within 45° sector
Detection time: <1 s

STATUS: under development. The Prophet system was selected in mid-1985 for the MBB BK 117A-3M multi-role helicopter project.

Full-back radar warning receiver
A version of Prophet for tactical fighter and advanced trainer and known as Full-back is currently under investigation by Racal. As well as providing audible and visual threat-warning, Full-back will also activate chaff and flare decoy systems.

UNITED STATES OF AMERICA

AEL
American Electronic Laboratories Inc, 305, Richardson Road, Lansdale, Pennsylvania 19446
TELEPHONE: (215) 822 2929
TWX: 510 661 4976

AN/APR-44(V) radar warning receiver
With the company designation CMR-500B, this low-cost unit is based on the AN/APR-42, with which it is electronically interchangeable and compatible. It has been adopted by the US Army and Navy.

The system consists of three units. The AEL Model AS-3266 antenna is a monopole unit which provides omni-directional, vertically-polarised, coverage within a 50° elevation sector and is designed to be mounted on a flat horizontal surface. The receiver is a radio frequency-chopped, crystal-video type with a bandpass filter followed by a radio frequency switch/detector module, a linear video amplifier and processing circuitry. The video amplifier output is routed to a comparator before being compared with a radio-frequency-chopping signal in the processing logic. This procedure detects any continuous wave signal and triggers a 3 kHz audio output, a logic output, an alert lamp drive and a logic blink signal (2 Hz). The

third unit is a control panel which includes the alert light indicator. Three versions of the system are available. The -44(V)1 incorporates the R-2097 receiver for coverage of the H/I bands: the -44(V)2 uses the R-2098 J band receiver, while the -44(V)3 covers both bands.

Receiver frequency and bandwidth: selectable
Sensitivity: -45 dBm (minimum)
Max rf input level: 1 W cw (integral limiter)
Receiver dimensions: 165 × 94 × 229 mm
Receiver weight: 1.8 lb (0.85 kg)

STATUS: in production.

Argo
Argo Systems Inc, 884 Hermosa Court, Sunnyvale, California, 94088-3452
TELEPHONE: (408) 737 2000
TELEX: 352077

AN/ALR-52 ECM receiver
Used by the US Navy, this is a multiband instantaneous frequency measuring receiver,

which is also used as a shipborne system under the designation AN/AWR-11. Typical coverage is 0.5 to 18 GHz using sets of receiver modules which cover octave bandwidths. It is a large system, providing for up to two operator positions, each with a control unit and display, although a single-operator version with direct link to digital hardware is also available. Capabilities are extensive and permit measurement of radar parameters, analysis of con-

tinuous wave and pulse signals, and direction finding measurements. The shipborne system is one of the major electronic countermeasures warning sets used by the US Navy, and airborne systems can only be installed in relatively large aircraft, such as the US Navy/Lockheed P-3 Orion.

STATUS: in service.

Bunker-Ramo
Bunker-Ramo Eltra, 4300 Commerce Court, Lisle, Illinois 60532-3601

AN/ALQ-86 ESM system
Bunker-Ramo developed this set and shared production work with Loral and Hewlett-Packard. It is used in the US Navy/Grumman EA-6A Intruder carrier-borne electronic warfare aircraft, in conjunction with the Loral AN/ALQ-53 electronic reconnaissance system,

and the AN/ALH-6 recorder, which is also made by Bunker-Ramo. A modified version of the system, with improved performance, increased reliability and maintainability, is being produced in conjunction with the EA-6B Advcap (advanced capability) programme.

STATUS: in production.

Cincinnati Electronics
Cincinnati Electronics, 2630 Glendale-Milford Road, Cincinnati, Ohio 45241
TELEPHONE: (513) 733 6100
TELEX: 214452
TWX: 810 464 8151

AN/ALR-23 and AN/AAR-34 infra-red warning receivers
Over 600 examples of these sensors are in service in US Air Force/General Dynamics F-111 fighter bombers. The system comprises two hemispherical detection heads with associated cryogenic-cooling units, a processor, control unit and cathode ray tube display. It provides passive detection of threats by sensing infra-red emissions characteristics of hydro-carbon exhausts. It is said to be insensitive to clouds, sun or ground-clutter, and provides automatic warning to the pilot and counter-measure command equipment. There is a multiple threat detection mode and a multiple discrimination mode. The operator can select variable field of view limits. Detector cooling is achieved by a closed-cycle cryogenic system. Detection performance is classified.

Cincinnati Electronics AN/AAR-34 infra-red warning receiver system

Dimensions: (detector head) 394 mm long × 164 mm dia (each)
(cryogenic cooler) 559 mm long × 164 mm dia (each)
(processor) 414 × 166 × 195 mm
(control unit) 63 × 145 × 66 mm
Weight: (detector heads) 25.3 lb (11.5 kg) (two units)
(cryogenic coolers) 15.5 lb (7 kg) (two units)
(processor) 19.1 lb (8.7 kg)
(control unit) 1.13 lb (0.5 kg)

Power: 1.24 kVA at 115 V 400 Hz, plus 50 W at 28 V dc
Environmental specification: MIL-STD-504
Reliability: demonstrated mtbf 150 hours (90% confidence)

STATUS: in service.

AN/AAR-44 infra-red warning receiver
The AN/AAR-44 system has been developed under US Air Force Avionics Systems Division

Cincinnati Electronics AN/AAR-44 infra-red warning receiver system

sponsorship to provide warning of surface-to-air missile attacks. Suitable for large transport aircraft and large helicopters, the system uses a conical detector head which is usually positioned on the rear underside of the aircraft fuselage. Warning data received by the crew includes missile position on a sector threat indicator and command and audio warnings. The system will also automatically initiate

countermeasures to neutralise threats. The system combines track and search functions and has multiple threat capability. It can discriminate between solar radiation, terrain and water backgrounds.

Dimensions: (sensor) 366 × 376 mm dia
(processor) 191 × 168 × 259 mm
(control/display unit) 104 × 145 × 79 mm

Weight: (sensor) 34.3 lb (15.6 kg)
(processor) 9.3 lb (4.2 kg)
(control/display unit) 1.4 lb (0.6 kg)
Power: 190 VA at 115 V 400 Hz, plus 25 W at 28 V dc

STATUS: in production for US Air Force.

Dalmo Victor

Dalmo Victor Operations, Bell Aerospace Division of Textron Inc, 1515 Industrial Way, Belmont, California 94002
TELEPHONE: (415) 595 1414
TWX: 910 376 4400

In early 1986 Dalmo Victor was bought by the Singer Company for $174 million.

AN/ALR-46(V) radar homing and warning system

The AN/ALR-46(V) can identify and analyse frequency-agile emitters and can process 16 emitters simultaneously in priority order and feed data to a jammer, anti-missile seeker, or a data collector system. It uses a signal processor, CM-442/ALR-46, which provides it with in-cockpit threat parameter programming and unambiguous identification. The system, the US Air Force's standard radar warning asset, is now undergoing an update to add the latest threats to its identifying and processing capabilities.

STATUS: in service.

AN/ALR-62 radar warning receiver

Developed for the US Air Force/General Dynamics F-111, FB-111 and EF-111A tactical and strategic fighters and jamming aircraft, this is the major sensor element in a US Air Force programme to update the electronic warfare equipment on these types. Other elements are believed to have been the AN/ALQ-137 internal jammer and AN/AAR-44 infra-red warning

receiver. The ALR-62(V) is a third-generation development that includes a complete antenna set, forward and aft receivers, a digital signal processor and cockpit-mounted threat indicator/countermeasures control units. The ALR-62(V)3 is used in the FB-111 while the ALR-62(V)4 is used in the EF-111A and the ALR-62(V)4 also constitutes a major portion of the EF-111A terminal threat warning system. The system monitors and analyses radar threats. At least 400 aircraft sets were delivered for F-111/FB-111 operations, and a further 40 sets for EF-111A use.

STATUS: in service.

AN/APR-39A threat warning system

The new lightweight (less than 7 kg) system is intended for use with light aircraft and helicopters, in particular those flying 'nap-of-the-earth' missions. It provides aural and visual warning of any radar energy incident on the host aircraft over the very wide band. Detected radiation is compared with the characters of up to 16 000 emitters contained within the system's library, and each emitter is assigned a priority in accordance with the magnitude of the threat. The system can then track up to eight of the highest priority threats, displaying threat type, bearing from the aircraft, and degree of danger on a cathode ray tube display.

The APR-39A is a development of the earlier APR-39, of which several thousand are in operation. The new equipment is based on digital technology, and has a claimed mtbf of 2200 hours. A version called Triton is optimised for use aboard small ships.

STATUS: US Marine Corps and US Army have 23 sets for assessment, the first being delivered in October 1984. The US Air Force also has an interest for its Sikorsky HH-60D Night Hawk helicopters.

AN/APR-39A(V) threat warning system

The new lightweight (less than 9 kg) digital system is developed for use with helicopters and light fixed-wing aircraft, in particular those flying nap-of-the-earth missions. It provides aural and visual warning of any radar energy incident on the host aircraft over a very wide frequency range. Detected radiation is compared with emitter characteristics contained within the system's threat library, and each emitter is assigned a priority in accordance with the magnitude of the threat. The system displays multiple threats. However, the highest priority threats are highlighted while keeping track of other emitters in the threat buffer. The system displays threat type, bearing from the aicraft, and degree of danger on azimuth indicator displays.

The APR-39A is an upgrade of the earlier analogue APR-39, and is designed for potential application on US Army Marine Corps and Air Force aircraft. The system has completed US Army testing and has a demonstrated mtbf of greater than 900 hours.

STATUS: production quantities scheduled for delivery to US Army, Marine Corps and Air Force starting in early 1988.

Eaton

Eaton Corporation, AIL Division, Commack Road, Deer Park, NY 11729
TELEPHONE: (516) 595 5000
TELEX: 510 227 6073

AN/ALQ-99 stand-off tactical jamming system

Eaton provides the tactical jamming section to the AN/ALQ-99 electronic warfare system which, launched in 1965, represents a major element of the defensive equipment on the US Navy/Grumman EA-6B Prowler and US Air Force/General Dynamics EF-111A Raven electronic warfare aircraft. Grumman is the prime contractor for both of these weapons systems, and they are further described under that company's entry. The ALQ-99 jammer is

claimed to be the first such system with real-time digital processing to determine the nature of threat signals and automatically assign jamming priorities.

In May 1975 an upgraded version of the system, designated AN/ALQ-99E and optimised for the US Air Force, was launched for the Raven. In October 1984 steps were put in hand to improve the system further under the terms of a $65.88 million contract awarded by the US Air Force to a team led by Eaton and comprising General Dynamics, Comptek Corporation, Delco Systems and Whittaker Tasker Systems. It was competing against a group comprising Teledyne, Raytheon and Intermetrics and led by Grumman. The contract covers development and production of equipment for the 42 Ravens. Modifications will affect virtually every part of the system. Thus the receivers will incorporate analogue/digital converters, new signal processors will operate faster and have bigger memories, and new exciters for the transmitters will provide a greater number of jamming modulations, each one compressed into a narrower bandwidth. Along with the exciters and transmitters will go improved antennas that will generate narrower beams for better energy concentration, though these will be the subject of separate contracts awarded in 1985. Each aircraft has five exciters driving ten transmitters, and the US Air Force plans to replace two of the exciters with new designs. Likewise each EF-111A has three digital computers, which will be replaced by systems with MIL-STD-1750A architecture.

STATUS: AN/ALQ-99 and -99E in service. Six sets of improved -99E are to be delivered by early 1986 for flight trials beginning in June of

that year. Production decision is scheduled for early 1987 leading to initial service equipment being supplied by mid-1989.

AN/ALQ-130 communications jamming system

Delivery of this system commenced in 1974 and it is now used on US Navy/Douglas A-4 Skyhawk, Grumman A-6 Intruder, and EA-6B Prowler, LTV A-7 Corsair, and McDonnell F-4 Phantom aircraft. The system's principal function is the disruption of enemy defence communication links and surface-to-air missile radar operations. Jamming may include broad-band and/or acoustic noise or spot-frequency jamming, but the former is more likely. Development was initiated as an AN/ALQ-92 update programme. It is internally-mounted on EA-6B, but may be externally carried on other types.

STATUS: in service.

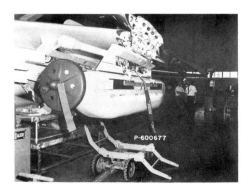

Eaton AN/ALQ-99 jammer pod installed on US Navy/Grumman EA-6B Prowler

US Navy/Grumman EA-6B Prowler carries Eaton AN/ALQ-130 communications jamming system internally

Eaton AN/ALQ-161 is being developed for US Air Force/Rockwell B-1 bomber

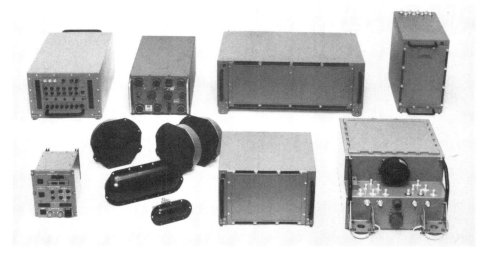

Eaton AN/ALR-77 ESM system for Lockheed P-3C Orion

AN/ALQ-161 ECM system for B-1B

Eaton is prime contractor for the most costly and capable ECM system to be built in the Western world – the AN/ALQ-161 suite for the US Air Force/Rockwell B-1B. A substantial part of the 40 per cent increase in EW funding requested by the US Defense Department in its Fiscal 1985 plan related to this project. The ALQ-161 was launched in 1972 under Eaton leadership for the Rockwell B-1A bomber, but the effort came to a halt when the aircraft was cancelled in June 1977. However the programme was reinstated in October 1981. Eaton accordingly resumed its task as ALQ-161 project leader, having participated in limited flight trials with one aircraft since 1979. While the system has undergone some redesign to incorporate new requirements and technology that has arisen during the dormant period, a basis of confidence was built up on the results of more than 400 hours of operation during 95

flights with a B-1A prototype. Principal goal of these early trials was to demonstrate that the system would operate in harmony with the B-1A's flight control system and OAS (offensive avionics system). The current equipment is being developed under restart and initial production contracts worth more than $1700 million to Eaton and its sub-contractors. Each ship-set is understood to cost around $20 million, approximately one-tenth the cost of a complete B-1B, and comprises no fewer than 108 line-replaceable units, with a total weight of 4700 lb (2132 kg) exclusive of cabling, displays and controls. Total cost of the ALQ-161 programme is estimated at more than $2000 million.

Operating frequency range is approximately 0.5 to 10 GHz, to cover Soviet early-warning

and ground-controlled interception radar, surface-to-air missile and interceptor radar frequencies. Jamming signals in the higher regions of the ECM spectrum are emitted from three electronically-steerable, phased-array antennas: one in each wing-glove leading-edge, and the other in the fuselage tail-cone. Each antenna provides 120° azimuth and 90° elevation coverage. Lower frequency signals are emitted from quadrantal-horn antennas mounted alongside the high-frequency equipment.

Major sub-contractors in the AN/ALQ-161 programme are Northrop, Litton Industries and Sedco Systems. All companies serve as sub-system managers and have responsibilities for receivers, data-processing and jamming techniques. Northrop provides the low-frequency jamming antennas, and Sedco is supplying the phased-array antennas.

Eaton is responsible for system integration and for the line-replaceable units, including 51 unique designs. Most of the line-replaceable units are about 1 to 2 cubic feet (0.03 to 0.06 m³) in volume and weigh between 40 and 80 lb (18 to 36 kg). The majority are readily accessible and can be easily removed or installed by one or two people. Total power required is about 120 kW.

Key improvements that have been introduced since the programme was restarted include a frequency extension into the K band to improve the warning and jamming performance; extension into the low-frequency domain (below 200 MHz) to improve the response of the warning system; introduction of a digital radio-frequency memory to permit the deception jamming of more advanced Soviet radars, notably those employing pulse-Doppler techniques; and introduction of a new tail-warning system. The latter was originally to have been the Westinghouse AN/-ALQ-153 radar for the B-1B, chosen in competition with Eaton's own AN/ALQ-154. Both systems had been flight-tested in 1978, and the ALQ-153 was selected to update the Boeing B-52G. However subsequent investigation by Eaton showed that the tail-warning function could be integrated with the ALQ-161 to a very large extent, with considerable savings in weight and cost. Eaton was accordingly awarded a $9.1 million contract in September 1983 to extend the ALQ-161 system to include this function. In another change, an IBM 101D computer now replaces the earlier Litton system as the central processor, making for commonality with a similar machine elsewhere in the aircraft. The function of the computer, software for which (based on Jovial) is to identify the function of every hostile radar, assess its potential threat, and assign a jamming priority. Three sets of phased-array antennas mounted in the wing leading edges and tail provide full 360° coverage in bands 6, 7 and 8, antennas for lower frequencies being located fore and aft in the airframe.

Eaton AN/ALQ-161 ECM system for Rockwell B-1B

Lockheed P-3 Orion flight-testing Eaton AN/ALR-77 ESM system

Close-up of Eaton AN/ALR-77 ESM system on Lockheed P-3 Orion port wing-tip

Some indication of the magnitude of the system is provided by the observation during early design that although the preferred subsystem configuration involved an increase in uninstalled weight of 125 lb (57 kg), it eliminated 750 lb (304 kg) of cabling.

STATUS: in production. Two complete systems were delivered for continuing trials with a modified Rockwell B-1A in the summer of 1984 and for the first production B-1B; the latter made its initial flight in October 1984. Production rate was built up to four systems a month by May 1986. Continued testing and evaluation will be supported by a comprehensive integration test-rig at Eaton's Deer Park, New York, plant. In August 1985, Eaton was awarded a $1800

million contract for the final 92 ALQ-161 shipsets, believed to be the largest ever EW contract. Problems in development were reported in mid-1986 leading to the first 22 B-1Bs not being equipped with the tail-warning system which is scheduled to enter service in mid-1987.

AN/ALR-77 ESM system

This programme was initiated by the US Navy in 1982 with a $11.2 million development contract to Eaton for the production of eight pre-production electronic support measure systems. Each system will use wingtip-mounted antennas in the Lockheed P-3C Orion and employ signal processing to provide the crew

with airborne and shipborne signal location and threat data. Four detectors on each wing tip give full 360° coverage, providing directional accuracies of between 1° and 2° rms. A further requirement is that the system should provide targeting data for radar-guided Harpoon missiles. Following tests a production run for 275 additional systems is expected. A technical innovation on the ALR-77 (and on the AN/ALQ-161 above) is the use of microwave integrated circuits, reducing size by a factor of ten. Flight trials of the ALR-77 from the Naval Air Development Center in Pennsylvania began in late 1985.

STATUS: under development.

E-Systems

E-Systems Inc, Memcor Division, PO Box 23500, Tampa, Florida 33630
TELEPHONE: (813) 885 7000
TELEX: 523455
TWX: 810 876 9174

Prototype installations under evaluation are as follows: Lockheed C-130 Hercules; de Havilland (Canada) Buffalo; Northrop F-5; Dassault Mirage III; Casa C101; Agusta A129; and Dassault Falcon.

STATUS: in production. First production order for 359 sets was placed in October 1975. In early 1984 deliveries totalled 5200 sets. In November 1983 E-Systems agreed to supply 230 sets of AN/APR-39(V)1 equipment to Standard Electrik Lorenz for the PAH-1 helicopter.

AN/APR-39 radar warning receiver

This system is used in US Army, Navy and Marine Corps helicopters and can determine the frequency, pulse repetition frequency, pulsewidth, persistence and threshold power level of surface-to-air missile and anti-aircraft gun-laying radars. Coverage is continuous across the E to I band and there is additional coverage in C, D and part of J band. The system employs an under-fuselage blade antenna and four spiral antennas, two dual-receiver units being associated with each pair of spiral antennas. The AN/APR-39(V)1 system was originally developed for the US Army, remains the only fully-developed radar warning system operational in that service, and is the standard radar warning receiver system on all Army helicopters including the Hughes AH-64A Apache, the first production aircraft of which made its initial flight in late 1983. It is also used on various US Navy and Marine Corps helicopters, as well as fixed-wing liaison and cargo aircraft and selected high-performance fighter aircraft operated by allied armed forces.

Current applications include: Bell UH-1 Iroquois; Bell OH-58 Kiowa; Hughes OH-6 (500MD); Bell AH-1 Cobra; Hughes AH-64 Apache; Boeing-Vertol CH-46 Sea Knight; Boeing-Vertol CH-47 Chinook; Grumman OV-1 Mohawk; North American OV-10 Bronco; Westland Lynx; MBB PAH-1 (BO105); Westland Gazelle; Beech RU-21 King Air; Fairchild Merlin; and Hawker Hunter.

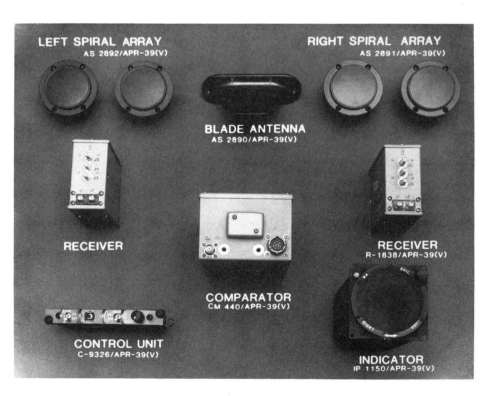

E-Systems AN/APR-39(V)2 radar warning system on US Army/Boeing Chinook helicopter

ESL

ESL Inc, a subsidiary of TRW Inc, 495 Java Drive, Sunnyvale, California 94088-3510
TELEPHONE: (400) 738 2888
TWX: 910 339 9526

General Electric

General Electric, Aerospace Electronic Systems Department, French Road, Utica, New York 13503
TELEPHONE: (315) 793 7000

General Instrument

General Instrument Corporation, Government Systems Division, 600 West John Street, Hicksville, New York 11802
TELEPHONE: (516) 933 3406
FAX: (516) 923 3432
TELEX: 311512
TWX: 510 221 1865

AN/ALR-66(V)1 ESM radar warning receiver

The AN/ALR-66(V)1 is the nomenclature for the upgraded AN/ALR-66. The system uses a powerful computer/processor which accepts up to 500 000 pulses per second from a mixed scenario of sea, air, and land-based radar emitters. It provides unambiguous emitter identification even in complex signal environments. Programmable alphanumeric symbology of up to three identifiers per emitter are presented on a polar high brightness crt display. Now available with eeprom memory, the user can reprogram the complete library on the flight-line or in-flight. This is done in less than 90 seconds with the computer memory loader, also built by General Instrument.

The AN/ALR-66(V)1 is used as the standard ESM system on the US Navy Lamps Mk I (SH-2) ASW helicopters, and is used for similar missions by other navies on HSS-2 and Sea King helicopters. The AN/ALR-66(V)1 is compact, lightweight, and easy to install. It is suitable for use on a wide variety of maritime patrol aircraft.

Receiver type: crystal-video high sensitivity receivers available
Frequency coverage: contiguous over E to J band
Warning/ID: all pulsed radars, including pulse-Doppler, cw, icw, LPI, 3-D, jitter/stagger pulse compression and frequency agile radars including known and unknown emitters
Azimuth coverage: full 360°
Emitter storage: in excess of 1800 emitter modes

AN/ALQ-151 airborne jammer

Although little is publicly known about this system produced by a TRW subsidiary, it is understood to be installed in US Army/Bell EH-1H ECM helicopters. Design commenced

Cross² ECM system

An advanced electronic countermeasures system based on new component technology has been produced by General Electric. It takes advantage of modern large-scale/very large-scale integrated circuit technology and uses monolithic microwave integrated-circuit com-

Reprogramming time: 90 s per platform, on the flight-line, by customer personnel, using the computer memory loader
Symbology: 1, 2 or 3 symbols per emitter as desired. Programmable by customer
Interface: RS-232 or MIL-STD-1553 as desired
Dimensions:
(receiver) 7.65 × 3.31 × 5.4 inches (194 × 84 × 137 mm)
(processor) 23.9 × 6.71 × 8.37 inches (607 × 170 × 213 mm)
(display) 3.25 × 3.25 × 9.25 inches (83 × 83 × 235 mm)
(indicator) 5.11 × 1.81 × 3.72 inches (130 × 46 × 95 mm)
Weight: 27 kg
Power: 400 W
Reliability: over 630 h mtbf demonstrated (MIL-STD-781)

STATUS: in production (ASU and AFP).

AN/ALR-66(V)3 surveillance and targeting system

The AN/ALR-66(V)3 surveillance and targeting system is the successor to the US Navy's operationally-deployed AN/ALR-66(V)2

in 1975 and is believed to have reached the full-scale production phase. In view of its recent development, this is probably an advanced deceptive jammer.

STATUS: in production.

ponents linked to silicon-diode phased-array antennas. The configuration is claimed to provide high effective radiated power, with minimum size and cost and sizeable cost-of-ownership benefits.

STATUS: under development.

system, and has been specifically designed for installation aboard the Lockheed P-3C Orion ASW aircraft.

The system detects, identifies and locates radars in the C to J band frequency range, and provides over-the-horizon (OTH) targeting through features such as ultra-high sensitivity receiving system, precision DF accuracy, positive emitter identification in high-density environments, interactive plasma display and control, self-protection to the aircraft, completely automatic operation, and complete built-in-test capability.

The basic modes of operation are:
Surveillance mode: the highest priority emitters are presented on the plasma display in positions corresponding to their range and bearing, utilising unique symbology which is either emitter- or platform-related.
Targeting mode: precise targeting data on any emitter is presented on the plasma display. This operating mode provides the OTH targeting data required for weapon activation.
Emitter waveform analysis: raw video is accepted from the computer-converter, and data is presented directly on the plasma display, permitting the operator to analyse emitter waveforms.

AN/ALR-66(V)3 surveillance and targeting system

AN/ALR-66(V)1 radar warning receiver

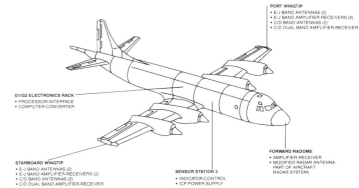

Diagram showing AN/ALR-66(V)3 layout in Lockheed P-3C Orion

Power requirement: less than 1 kVA
Total system volume: less than 4 ft³ (0.11 m³)
Number of units in total system: 19

AN/ALR-66(VE) radar warning receiver system

The AN/ALR-66(VE) is a lightweight rwr system designed for use in fighter aircraft and helicopters, providing enhanced performance and extended versatility due to the use of eeprom in the computer memory. The system is mechanically designed as a drop-in replacement/upgrade of older analogue and digital radar warning receivers used in combat aircraft such as the APR-25, 36/37, 39 and ALR-45 and 46. The ALR-66(VE) has emitter storage of more than 1000 radar signals, making it one of the most powerful and economical rwr systems available. The use of eeprom in the memory section of the unit allows complete memory reprogramming in 90 seconds on the flight-line by customer personnel. The system is capable of automatically recording all major parameters of detected radars for post-flight analysis.

The ALR-66(VE) is in operational use and has been proven in NATO and other countries. This system is available for a wide variety of fighter aircraft including the McDonnell Douglas F-16, F-4 and A-4, Northrop F-5 and F-20, Dassault Mirage and Sepecat Jaguar.

Receiver type: crystal-video high sensitivity receivers available
Frequency: contiguous over E to J bands. C/D band, omni or directional capability (option)
Warning/ID: all pulse radars; cw, icw and pulse-Doppler, LPI, 3-dimensional, jitter/stagger, pulse compression, and frequency agile radars including known and unknown emitter
Radar storage: more than 1000 emitters
Programming time: 90s per aircraft, on the flight line, by customer personnel using the computer memory loader
Symbology: 1, 2 or 3 symbols per emitter as desired. Programmable by customer
Interface: RS-232 or MIL-STD-1553 as desired
Mission recording: solid state, automatic in-flight radar parameter recording (optional)
Dimensions:
(receiver) 7.03 × 1.73 × 6.19 inches (179 × 44 × 157 mm)
(processor) 7.6 × 4.84 × 14.83 inches (193 × 123 × 363 mm)
(power supply) 7.97 × 10 × 4.12 inches (202 × 254 × 105 mm)
(display) 3.25 × 3.25 × 9.25 inches (83 × 83 × 235 mm)
(indicator) 5.11 × 1.81 × 3.72 inches (130 × 46 × 95 mm)
Weight: 29.3 kg
Power consumption: 400 W
Reliability: over 725 h mtbf demonstrated (MIL-STD-7811)

STATUS: in production.

AN/ALR-79 radar warning receiver

The AN/ALR-79 radar warning receiver is a newly nomenclatured US system, being a modified AN/ALR-66(V)1 system, featuring increased sensitivity, an IFM and an interface with several jammers. The AN/ALR-79 completely and automatically controls the operation of the jammers via frequency set-on and other interfaces. The AN/ALR-79 therefore offers enhanced capabilities for self-protection. It demonstrates the flexibility of the AN/ALR-66 design and the ease with which new capabilities and interfaces can be added to the AN/ALR-66 rwr family.

STATUS: in production and service.

AN/ALR-80(V) radar warning receiver system

General Instrument announced in June 1985 its latest fully digital radar warning receiver (rwr) system, the AN/ALR-80(V), designed for use in fighter aircraft and attack helicopters. This system is an outgrowth of the ALR-66(VE) rwr system. The AN/ALR-80(V) rwr system provides enhanced performance and flexible, modular versatility due to the use of eeprom in the computer memory and new powerful software. The system is mechanically designed as a drop-in replacement/upgrade of older analogue and digital radar warning receivers used in combat aircraft. The AN/ALR-80(V) has library storage capacity of more than 1800 radar modes, making it, the makers claim, the most powerful and economical rwr system available today. The use of eeprom in the memory section of the unit allows complete memory reprogramming in 90 seconds, on the flight-line, by customer personnel. The system is capable of automatically recording all major parameters of detected radars for post-flight analysis.

The AN/ALR-80(V) is General Instrument's answer to the need to protect fighter aircraft from the highly sophisticated radars used in today's weapon systems. The system is available for a wide variety of fighter aircraft including the McDonnell Douglas F-16, F-4 and A-4, Northrop F-5 and F-30, Dassault Mirage, and Sepecat Jaguar.

Receiver type: crystal-video high sensitivity available
Frequency: contiguous over C through J bands (C/D band: omni or directional). Legends vary for various systems, ie frequency vs frequency coverage, radar storage vs emitter storage, etc, order also varies
Warning/ID: all pulse radars; cw, icw and pulse-Doppler, LPI, 3-dimensional, jitter/stagger, pulse compression, and frequency agile radars, including known and unknown emitters
Radar storage: More than 1800 emitters

Programming time: 90 s per aircraft, on flight-line, by customer personnel, using the computer memory loader
Symbology: 1, 2 or 3 symbols per emitter as desired. Programmable by customer
Interface: RS-232 or MIL-STD-1553 as desired; for use with jammers, chaff/flare or data links
Mission recording: (optional) solid state, automatic in-flight radar parameter recording (>3 h)
Dimensions: basically similar to AN/ALR-66(VE) (see preceding entry)
Weight: 29.3 kg
Power consumption: 500 W
Reliability: over 725 h mtbf demonstrated (MIL-STD-781)

STATUS: in production.

ALR-606(V)1 ESM radar warning receiver system

The ALR-606(V)1 ESM system is a fully digital radar warning receiver. It is based on the AN/ALR-66(V)1, with which it shares the same hardware configuration. The ALR-606(V)1 is designed especially for the export market with the following in mind:
1. Item for item exact replacement of older analogue and digital rwr systems in previous generation aircraft, giving ideal upgrade capabilities for maritime patrol aircraft such as the Grumman S-2 Tracker.
2. Full digital signal processing, high intensity crt, E to J band coverage, alphanumeric symbology; it is ideally suited for such newer generation helicopters as the Aérospatiale Super Puma, Gazelle and Alouette III, Westland Sea King Lynx, and maritime patrol aircraft such as the Fokker F-27M, Embraer EMB-111, Dassault Atlantique, de Havilland Dash 7 and 8, etc.
3. Use of eeprom to provide customers with their own reprogramming capability for display symbology and program parameters.

The ALR-606(V)1 is in world-wide use in numerous types of aircraft and is available for export to most free-world countries.

Frequency: E to J band contiguous
Warning I/D: all pulse types; cw, icw and pulse-Doppler, LPI, 3-D, jitter/stagger, pulse compression, and frequency agile radars
Receiver type: crystal-video
Azimuth coverage: 360°
Processor memory: 18 000 words eeprom/ram/rom; 68 K total capacity
Symbology: Up to 3 symbols per emitter, programmable by customer
Interface: RS-232 or MIL-STD-1553 as desired
Weight: 27 kg
Reliability: over 630 h mtbf demonstrated (MIL-STD-781).

STATUS: in production.

AN/ALR-66(VE) is designed to replace older radar warning receivers in existing combat aircraft

AN/ALR-80(V) radar warning receiver

ALR-606(VE) radar warning receiver system

The ALR-606(VE) is a fully digital radar warning receiver. It is based on the proven ALR-66(VE) technology, with which it shares the same hardware configuration. The ALR-606(VE) is designed especially for the export market with the following in mind:
1. Item for item exact replacement of older analogue and digital rwr systems in previous generation fighter aircraft, giving ideal upgrade capability for Northrop F-5A to F, Lockheed F-104, McDonnell Douglas F-4 and A-4, LTV A-7 aircraft etc.
2. Full digital signal processing, high intensity crt, C to J band coverage, alphanumeric

symbology; ideally suited for new generation fighter aircraft such as the Northrop F-20, General Dynamics F-16, Dassault Mirage, etc.
3. Use of eeprom to provide customers with their own reprogramming capability for display symbology and program parameters.

The ALR-606(VE) is in world-wide use in numerous types of fighter aircraft and is available for export to most free-world countries.

Frequency: (C to J band) contiguous (C/D band) omni or directional
Warning/ID: all pulse types; cw, icw and pulse-Doppler, LPI, 3-D, jitter/stagger, pulse compression and frequency agile radars

Receiver type: Crystal-video higher sensitivity receivers available
Azimuth coverage: 360°
Processor memory: 18 000 words eeprom/ram/rom; 68 K total capacity
Symbology: up to 3 symbols, programmable by customer
Interface: RS-232 or MIL-STD-1553 as desired
Mission recording: solid state in-flight radar parameter recording (optional)
Weight: 29.3 kg
Reliability: over 725 h mtbf demonstrated (MIL-STD-781)

STATUS: in production.

Goodyear

Goodyear Aerospace Corporation, 1210 Massilon Road, Akron, Ohio 44315
TELEPHONE: (216) 796 2121
TELEX: 986439

AN/ALE-39 chaff dispenser

One of the leading in-service countermeasure dispenser types, the AN/ALE-39 replaces the AN/ALE-29 on most US Navy aircraft. It is in operation on Douglas A-4 Skyhawk, Grumman A-6 Intruder, EA-6B Prowler, LTV A-7 Corsair, McDonnell F-4 and RF-4 Phantom, Grumman F-14 Tomcat, Bell AH-1T, AH-1J, UH-1N, Boeing Vertol CH-46 and Sikorsky CH-53 helicopters, as well as the US Navy/McDonnell

Douglas F/A-18 Hornet and the US Marine Corps AV-8A Harrier. It has been selected for the AV-8B and is used on several non-US aircraft. Contracts for over 3000 sets had been placed by the end of 1985 and over 2000 sets are in service. Production is expected to continue through the 1990s.

The system accommodates three types of payload: chaff, infra-red flares and expendable jammers. Cartridges are loaded in groups of ten up to the normal system capacity of 60 (expansible to 300). Each cartridge is 1.4 inches (36 mm) in diameter and 5.8 inches (147 mm) long. Typical payload would be RR-129 chaff cartridges, Mk46 Mod 1C or MJU-8/B flares and AM-6988A POET expendable jammers.

STATUS: in production.

US Marine Corps AV-8B V/Stol fighter can carry Goodyear AN/ALE-39 chaff dispenser

Grumman

Grumman Aerospace Corporation, South Oyster Bay Road, Bethpage, New York 11714
TELEPHONE: (516) 575 0574

US Navy/Grumman EA-6B Prowler electronic warfare aircraft

This is a dedicated EW variant of the US Navy/Grumman A-6 Intruder and has been in low volume production (six a year) since the late 1960s. The Prowler's primary mission is to protect surface vessels and aircraft by jamming hostile radars and communications. Secondary missions include electronic surveillance, anti-ship missile defence, and training for the radar operators of ships and aircraft. The bulged fin fairing encloses sensitive surveillance receivers that can detect radars at long range, and the AN/ALQ-99 tactical jamming system is the principal EW equipment on both the EA-6B and EF-111A. The Grumman aircraft has a crew of four (one pilot and three ECMOs, or electronic countermeasure officers) and has less automatic capability than the larger EF-111A with its two-man crew. It is designed specifically to meet US Navy mission requirements, and carries much of its jamming equipment in up to five pods attached to the fuselage and wing pylons. The ALQ-99 has 10 times the jamming power of previous systems, according to Grumman. Each pod is equipped to jam one of seven frequency bands, and the aircraft can carry any combination of pods and fuel tanks, depending on the mission. Emitter information is fed to a central digital computer that processes data for display or recording. Detection, identification, direction-finding and activation of the jammer system can be accomplished manually or automatically. Principal EW systems are the Lundy ALE-29A chaff and flare dispenser, Sanders AN/ALQ-41 track-breaking systems (in conjunction with AN/APR-27 missile-alert warning receiver), Sanders AN/ALQ-9L communications jammer, and AIL AN/ALQ-99 tactical jamming system.

The EA-6B has been the subject of several improvement programmes. The EA-6B Excap (expanded capability) was the first, being

US Navy/Grumman EA-6B Prowler EW derivative of Grumman A-6 Intruder has Eaton AN/ALQ-99 ECM system in five external pods

delivered in 1983. This aircraft featured updated EW equipment to meet threat emissions in six frequency bands. Then followed the EA-6B Icap (increased capability) in mid-1976, with shorter processing times, new multi-format displays, better radar deception equipment (thought to be the AN/ALQ-49 system), and improved communication, navigation and IFF equipment.

The EA-6B Icap 2 is a further improvement, in which the exciters mounted in the external pods can now jam two different frequency bands simultaneously. The core capacity of the AN/AYK-14 computer has been increased, and there the system can now act in concert with equipment on other Prowlers such that three aircraft can together mount a co-ordinated countermeasures mission. The first Icap 2 aircraft was handed over to the US Navy in April 1984. Icap 2 is the third significant EW update for the Prowler. Grumman will build 40 EA-6B Icap 2s and probably modify a further 15 earlier versions.

The most recent initiative is the EA-6B Advcap (advanced capability), launched by the US Navy in late 1982 with a contract on Grumman. A substantial part of the work will be conducted by Litton-Amecom, which is developing a new receiver/processor system, based on the AN/ALR-73 passive detection system in the US Navy/Grumman E-2C Hawkeye aircraft. Flightworthy prototype equipment should become available in 1986 and lead to production awards by 1989. The total value of the Advcap

programme is expected to exceed $400 million (at 1983 prices) and should produce equipment with applications for the EF-111 programme. The commitment to an equivalent update of this programme announced in early 1984 may be assumed to incorporate some Advcap technology.

STATUS: in production.

USAF/Grumman EF-111A Raven electronic warfare aircraft

Grumman is prime contractor for the US Air Force/Grumman EF-111A Raven EW aircraft programme, which calls for extensive modification of existing F-111A tactical-fighter airframes, including the installation of new wiring, new environmental control system (which includes liquid cooling), 90 kVA power-generation system, and structural modification to the fin, weapon bay, rear fuselage, wing glove and environmental control system inlet. The EF-111A has a crew of two (pilot and electronic warfare officer) and is credited with a large automatic threat detection and counter-measure capability. Designed specifically to meet US Air Force mission requirements, its primary mission is the detection, tracking and interception of hostile air-defence forces.

Major components of the EF-111A tactical jamming system, which is largely adapted from

the EW suite on the Grumman EA-6B Prowler (see separate entry), are:
AN/ALQ-99E tactical jamming system
AN/ALR-62(V)4 radar warning system
AN/ALQ-137(V)4 self-protection system
AN/ALE-28 chaff flare dispensers
AN/ALR-23 infra-red tail-warning receiver
AN/ALE-28 ECM dispensers.

The total suite weighs about 6500 lb (2950 kg).

Where details are available these systems are described under manufacturer entries. Subcontractors to the EF-111A project include:
IBM, with its 4 Pi general-purpose central computer and Raytheon, with transmitters and exciters.

Two EF-111A prototypes were flight-tested in 1978, and the first operational EF-111A entered service at Mountain Home Air Force Base, Idaho, in October 1981. All 42 aircraft have been delivered. A detachment of EF-111As assigned to NATO was deployed to the UK during February 1984.

In October 1984 the US Air Force announced that it had chosen a team of companies, led by Eaton, to upgrade the AN/ALQ-99E tactical

US Air Force/Grumman EF-111A EW derivative of General Dynamics F-111A equipped with AN/ALQ-99 ECM

jamming system to increase effectiveness against early warning and ground-control intercept radars and surface-to-air missiles. Three functions will be improved: stand-off jamming, close-in jamming, and penetration/escort. The contract is worth $65.88 million. Contractors are Eaton, General Dynamics, Comptek Research, Delco Systems and Whittaker Tasker Systems. Delivery of the first

six prototype kits is scheduled for fiscal year 1986 and production of 38 kits (allowing for aircraft attrition) will continue between fiscal years 1987 and 1991. Technical changes will include a reduction in the number of receivers and exciters and the incorporation of preprocessors in the computing system.

STATUS: in service.

Honeywell

Honeywell Electro-Optics Division, 2 Forbes Road, Lexington, Maryland 02173
TELEPHONE: (617) 862 6222
TWX: 710 326 6389

AN/AAR-47 missile warning set

Honeywell has identified a large and growing business area in electronic warfare, and is investing heavily to enter the market. The company's first product is the AN/AAR-47 missile warning set which detects the infra-red signatures of approaching surface-to-air missiles and gives the pilot an indication of

range and bearing. Decoys, such as the AN/ALE-39 flare/chaff dispenser, can be automatically triggered and the system is also compatible with the AN/APR-39A radar warning receiver.

The feasibility of such a warning system was first demonstrated by Honeywell as long ago as December 1977, and, following development of the AAR-46 system in 1979, full-scale development of the AAR-47 started in March 1983. The AAR-47 is small, needs no cryogenic cooling and has a mtbf of 1500 hours.

Dimensions: (sensor) 4.75 inches dia × 7.8 inches (120 mm × 200 mm)
(processor) 8 × 10.1 × 8 inches (203 × 257 × 204 mm)

Weight: (4-sensor system) 31 lb (14 kg)
(sensor) each 3.4 lb (1.5 kg)
(processor) 17.4 lb (7.9 kg)
Power: (4-sensor system) 75 W, 28 V dc (4 W per sensor, 59 W for processor)
Azimuth coverage: 360° given by 6 sensors
System performance degrade: above 30 000 ft

STATUS: in development. A further system, the dual-mode missile warning system (DMMWS) is also being developed. This combines an infra-red detector with a pulse-Doppler radar for accurate determination of threat range.

IBM

IBM, Federal Systems Division, 6600 Rockledge Drive, Bethesda, Maryland 20817
TELEPHONE: (301) 493 0111

AN/ALR-47 radar homing and warning system

Developed for the US Navy/Lockheed S-3A Viking ASW carrier-borne aircraft, this is a comprehensive passive electronic warfare system which uses four cavity-backed planar spiral antennas in each wing-tip. The aerials are orthogonally directed to enhance monopulse direction-finding, ensuring that threat direction is measured very accurately. Associated with them are twin, highly-sensitive, narrow-band receivers, and a comprehensive processor. Manual or automatic system operation is possible, control being exercised over frequency-band limits, speed of tuning and signal selection. The processor indicates frequency-

F-4G Wild Weasel carries IBM AN/APR-38 warning equipment

scanning limits, scan speed, pulse-length, pulse repetition frequency and bearing limits of any detected radar transmissions.

STATUS: in production.

AN/APR-38 radar warning receiver

IBM supplies radar warning receiver equipment

used in the AN/APR-38 electronic surveillance measures system installed in US Air Force/McDonnell Douglas F-4G Wild Weasel aircraft. Details of the system are given in the McDonnell Douglas entry.

STATUS: in production.

ITT Avionics

ITT Avionics Division, 500 Washington Avenue, Nutley, New Jersey 07110
TELEPHONE: (201) 284 0123
TELEX: 133361
TWX: 710 989 1479

AN/ALQ-117 jamming system

Over 600 examples of this system have been produced for the US Air Force/Boeing B-52 strategic bomber fleet. An initial production contract was awarded in 1972, and installation took place during Rivet Ace programme lay-ups, in which all B-52G and -H variants were fitted with electro-optical viewing equipment in

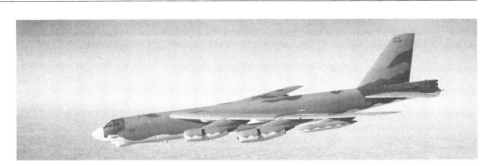

US Air Force/Boeing B-52G and -H strategic bombers carry ITT AN/ALQ-117 jamming system

chin blister installations. AN/ALQ-117 is an I to J band (8 to 20 GHz) system and may include

noise and deception jamming elements. The AN/ALQ-117 system is being upgraded and redesignated AN/ALQ-172 to extend the capabilities of US Air Force Boeing B-52s. Following a US Air Force production decision in July 1984, two sets of equipment were ordered for trials, and ITT Avionics has been gearing up to provide about 20 sets a month. The B-52G will have an updated system designated AN/ALQ-117 Pave Mint, while the B-52H will have the AN/ALQ-172.

STATUS: AN/ALQ-117 in service, AN/ALQ-117 Pave Mint and AN/ALQ-172 in early production.

AN/ALQ-136 advanced jammer system

This is an advanced helicopter countermeasures system developed under contract to the US Army. Engineering development models were produced in 1977 and production started in 1980. In late 1982 ITT Avionics was awarded a $64 million contract for further production equipment. The system is used in US Army combat helicopters, its primary role being to provide protection against radar-guided weapons. The system is being fitted to the US Army's Bell AH-15 Cobra attack helicopters, and will also go on the Hughes AH-64A Apache battlefield helicopter.

In 1982 the US Army funded a further development programme for an improved version for use aboard a special electronic mission aircraft (SEMA). This system is contained in an almost identical package to the helicopter version, weighs 62 lb (29 kg), and gives protection against radar-directed ground and airborne threats.

Weight: 42 lb (19 kg)
Volume: 0.7 ft³ (0.02 m³)

STATUS: in production.

AN/ALQ-165 ASPJ airborne self-protection jammer

The system is designed to operate with the AN/ALR-79 and -67 radar warning receivers. ITT Avionics and Westinghouse were jointly awarded contracts in late 1981 to produce the AN/ALQ-165 ASPJ system.

STATUS: under development (see ITT/Westinghouse entry).

AN/ALQ-172 jamming system

A revision of the AN/ALQ-117, see above.

ERWE enhanced radar warning equipment

This system was developed and manufactured for West Germany, the prime application being Air Force and Navy operated Panavia Tornados. The system uses an Itek computer for data collection, sorting, analysis, assignment of

Enhanced radar warning equipment for West German Air Force and Navy Panavia Tornados

priorities, and alphanumeric display function generation. An emitter library is installed and is accessible from the cockpit for threat updating. Co-production arrangements exist with AEG-Telefunken.

STATUS: in production.

ITT/Westinghouse

ITT Avionics Division, 500 Washington Avenue, Nutley, New Jersey 07110

Westinghouse Electric Corporation, Aerospace and Electronic Systems Division, PO Box 746 Friendship Site, Baltimore, Maryland 21203

AN/ALQ-165 ASPJ airborne self-protection jammer

In terms of quantity production, the AN/ALQ-165 jammer (formerly known as ASPJ) could be the most significant US electronic warfare system in the current decade. Production potential is estimated to exceed 3000 sets, and contract value, at 1984 prices, will exceed $2000 million. AN/ALQ-165 is a joint US Navy/US Air Force programme to provide a common active electronic countermeasures capability in Grumman F-14 Tomcat, A-6 Intruder, McDonnell Douglas F/A-18 Hornet, AV-8B, and General Dynamics F-16 aircraft. Other current aircraft remaining in front-line service after the above types have been equipped would also be eligible for this system. ALQ-165 is intended to replace the current Sanders AN/ALQ-126 and Westinghouse AN/ALQ-119 systems.

The joint-services specification for advanced self-protection jammers was developed around 1976. It was evident then that an internal electronic countermeasures system, small enough to fit into modern combat aircraft and with wideband jamming capability, was a logical future development. Even the US electronics industry, where electronic warfare capability far exceeds that of anywhere else in the Western World, recognised that the system would challenge the resources of individual companies. Accordingly, with direct encouragement from the services, industrial consortia were formed to bid for this contract.

In its original proposal the US Navy asked for the innovative application of available technology with miniaturised packaging techniques. The technical risk in the programme could therefore be very high, but the aim is to achieve previously unattainable capability in very small volume. This is necessary if a complete avionics suite is to be packed into increasingly small aircraft, notably the F/A-18, where accommodation for equipment is already at a premium. Some critics claim that this degree of miniaturisation could be too much of a technology and cost driver. There is little doubt, however, that ASPJ will set new standards in internal electronic countermeasures technology.

Three teams of contractors formed in 1977 to compete for ASPJ Phase 1 development contracts, resulting in two of them being selected in August 1979 to proceed with the first phase of full-scale engineering development. In further competition, ITT/Westinghouse was chosen in August 1981 over the Northrop/Sanders team to move into full-scale development. That team built 16 sets of equipment (nine development models and seven prototype sets) for a two-year evaluation beginning early in 1983. At that time the target cost was about $350 000, and initial reliability of the system

(already met by mid-1982) was to be not less than about 200 hours mean time between failures.

The transmitter will use dual-mode pulse/continuous wave, travelling-wave tubes, and a custom-built processor is in development using thick-film hybrid medium-scale integrated circuits. These offer higher packaging densities than conventional flat-pack or dual in-line circuits, and could be superseded by vhsic devices in the future. All waveforms and data for technique generation are contained within software, so that the system can be optimised with specific threats under given circumstances. Design of the receiver is based on projections as to what the signal environment will be in Central Europe in the 1990s. The travelling-wave tubes themselves represent about a quarter of the cost of an ALQ-165 set, while their reliability largely determines the reliability of the entire set. Accordingly the US Air Force has launched a programme to improve the reliability of these devices. Westinghouse says that ASPJ is not a high-risk development because it is based on components designed for the AN/ALQ-131. Built-in diagnostics should detect 96 per cent of all faults, and isolate 99 per cent of these to specific line-replaceable units.

A podded version has been developed for the AV-8B aircraft which does not have sufficient space to carry the system inside the airframe. Following collaborative production of the first 100 or so systems, ITT and Westinghouse will compete for full production awards expected in mid-1987. By this means the customer secures the benefits of dual-source procurement.

The basic system will consist of five line-replaceable units (two receivers, two transmitters and a processor), weighing 225 lb (102 kg). An additional augmentation set with high-band transmitter/receiver capability, and suitable for US Navy Grumman F-14A Tomcat or A-6 Intruder operation weighs 81 lb (37 kg). The system has been adopted for the F-16.

The ASPJ specification is summarised below. It also calls for all the equipment to fit in a 2 cubic foot (0.057 m³) volume space and to expand up to 35 GHz capability before 1990. Eventual expansion up to 140 GHz is a long-term goal. Some Soviet fire-control and interception radars are already reported to be operating up to 40 GHz.

McDonnell Douglas F/A-18 fighter carries AN/ALQ-165

Frequency coverage: 0.7-18 GHz
System response time: 0.1-0.25 s
Volume: <2.1 ft³ (0.059 m³)
Weight: 225 lb (102 kg). Augmentation system adds 85 lb (38.6 kg)

Receiver-processor
Dynamic range: 50 dB
Sensitivity: –71 dBm
Resolution: 5 MHz
Instantaneous bandwidth: 1.44 GHz
Pulsewidth: 0.1 μs (minimum)
Input pulse amplifier: 20 dBm (max)
False alarm rate: 5/h (max)
Signal detection capability: pulse, cw, pd, agile

Jammer
Peak power: 58-63 dBm
Pulse-up capability: 5-7 dB
Set-on accuracy: ±0.5 to ±20 MHz
Duty cycle: 5-10%
Jamming capacity: 16-32 signals
Modes: noise, deception

STATUS: under development. Prototype system was successfully demonstrated to US Navy and Air Force during late 1983. First full-scale development model was delivered in January 1984 to General Dynamics for evaluation in its F-16 avionics integration rig at Fort Worth. Forecasts are for more than 2500 sets

by 1992, with first deliveries scheduled for 1988. Some 14 engineering development sets are being built for flight-test, simulator, reliability and maintenance evaluation. First flight-test of an ALQ-165 took place in September 1985, aboard a US Air Force/General Dynamics F-16 fighter and flight tests fitted to a US Marine Corps AV-8B Harrier II are expected to start in October 1986.

Litton

Litton Industries Inc, Amecom Division, 5115 Calvert Road, College Park, Maryland 20740
TELEPHONE: (301) 864 5600
TWX: 710 826 9650

AN/ALR-73 ESM system

Developed to meet US Navy requirements for an effective passive detection system to supplement the early-warning radar used in Grumman E-2C Hawkeye aircraft, this highly capable electronic surveillance measures system is tailored to the role and configuration of the E-2C and its systems. It has an extensive array of antennas, considerable processing capability, and its main purpose is to alert the operators to the presence of electronic emitters at distances well beyond their maximum detection range.

The system uses 52 grouped antennas in four sets, one for each of several wavebands. Each complete set is positioned to look at a 90° sector, thus providing 360° azimuth coverage. The forward and aft antennas are in the fuselage extremities, and the sideways-looking aerials are in the tailplane tips. All receiver sets are under separate control, so wavebands are scanned independently and simultaneously in all sectors. Each antenna has dual processing channels and uses digital closed-loop rapid-tuned local oscillators. The latter provides instantaneous frequency measurement with fast time response and high-accuracy frequency determination.

Receiver outputs are collected at a signal preprocessor unit which performs pulse-train separation, direction-finding correlation, band-tuning and timing, and built-in test equipment tasks. Data is then in a form suitable for the general-purpose digital computer, which has overall control of electronic surveillance measures operations, and will vary frequency coverage, dwell time and processing time according to prescribed procedures. Control of these parameters is aimed at maximising the probability of intercepting signals in particular missions. Other on-board sensor data and crew inputs will determine the technique adopted. Data such as signal direction of arrival, frequency, pulsewidth, pulse repetition frequency, pulse amplitude and special tags is sorted by the computer and transmitted to the E-2C central processor.

Receiver and processor technology devel-

AN/ALR-45 radar warning receiver system

oped for the ALR-73 is the basis for the new receiver/processor group being produced by Litton, with Texas Instruments and ITT Avionics for the Grumman EA-6B Advcap tactical jamming aircraft (see separate entry).

STATUS: in production.

AN/ALQ-125 ECM system

Sometimes referred to as Terec (tactical electronic reconnaissance) this passive electronic system can locate and identify radar emitters, thus developing an electronic order of battle. It is installed in a limited number of US Air Force/McDonnell Douglas RF-4C Phantoms. All types of fixed and mobile ground-based radars are included in the system's threat library, and data can be recorded for replay or transmitted on data link. Up to 256 complex programmable PRI (pulse repetition interval) palleins are available by which agile emitters may be identified. Twenty-four systems have been procured by the US Air Force.

STATUS: in service.

LR-5200 ESM system

This highly capable Elint gathering system will provide real-time tactical intelligence for force-level and forward-area army commanders. It will indicate the location, composition, state of readiness and intent of an electronically-active opposing military force. It includes magnetic-tape recording of the database for post-mission

analysis or use in developing and maintaining the tactical electronic order of battle. The system can accommodate radar warning receivers and ECM jammers.

The LR-5200 system will incorporate all airborne, ground-processing and maintenance support equipment. Airborne equipment can be fitted internally or in a pod, and the system is sufficiently compact for installation in fighter airframes. The ground-processing/analysis facilities include an ESM data-analysis centre for permanent or transportable control centre installations. A remote data-terminal facility can be used on command-and-control aircraft, land vehicles, naval vessels or forward-area platforms. The system's modular construction allows configuration for various levels of performance.

Fighter aircraft installation
Frequency band: 0.5-18 GHz (up to 40 GHz provision)
Azimuth coverage: (acquisition) 360° (direction-finding) 240°
DF accuracy: 0.5-1° RMS
Location accuracy: 5% of range
Processor sensitivity: –65 dBm
Dynamic range: 60 dB
Weight: 225 lb (102 kg)
Volume: 5 ft³ (0.16 m³)
Reliability: 1000 h mtbf, 0.5 h mttr

STATUS: under development.

Litton Industries Inc, Applied Technology Division, 645 Almanor Avenue, Sunnyvale, California 94088-3478
TELEPHONE: (408) 732 2710
TWX: 910 339 9271

AN/ALR-45/-45F radar warning and control system

Standard US Navy tactical radar warning system in fleet use since the early 1970s. System was a replacement for the APR-25, which was the first standard tri-service rwr in inventory. ALR-45 is a crystal-video receiver utilising a hardwired programmable signal processor, four cavity-backed planar spiral antennas with preamplifiers and processor.

US Navy/Grumman E-2C Hawkeye carries Litton AN/ALR-73 ESM system

The AN/ALR-45F replaces the ALR-45 processor and azimuth indicator with a dual ATAC-16M computer-based threat processor and an alphanumeric microprocessor-based, MIL-STD-1553 compatible, display terminal. The processor features threat software reprogrammability, on-board avionics and EW suite communications, and interface control of such systems as HARM, APR-43, ALQ-126A/B, ALE-29/39, ALQ-162, ALQ-164 and ALQ-165. The processor and the display terminal are standard AN/ALR-67 weapons-replaceable assemblies and can be used interchangeably in the 45F configuration of McDonnell Douglas F-4J, and A-4M, LTV A-7E and AV-8C Harrier aircraft without kit wiring changes. Deliveries of the AN/ALR-45F version are scheduled to continue in 1986 following a $7.2 million contract awarded by the US Navy covering 50 systems for A-7E, F-45, A-4M and AV-8C carrier-based aircraft. Production of this batch, which follows on from the 55 set/$8.5 million buy in 1984, will continue until January 1987 and brings to 213 the number of AN/ALR-45Fs supplied to the US Navy.

STATUS: ALR-45 is extensively used by the US Navy and a number of NATO nations. Production has reached several thousands of sets.

The ALR-45 is being replaced by the US Navy with the AN/ALR-67(V) and the ALR-45F. The first three AN/ALR-45F systems were delivered in April 1983 as part of a $3.6 million contract and an additional 105 sets were delivered by September 1984.

AN/ALR-46 radar warning receiver

A significant US Air Force system, the ALR-46 is used in all front-line types except the McDonnell Douglas F-15 Eagle, General Dynamics F-111 and FB-111 types. It replaced APR-25/26 and APR-36/37 systems; in its AN/ALR-46(V)3 version it is virtually a digital version of the APR-36/37 the main factor in the change from analogue to digital being the addition of a Dalmo Victor DSA-20 processor. Replacement is under way, however, as the AN/ALR-69 radar warning receiver enters service. The AN/ALR-46(V)6 is available with an analogue processor.

The AN/ALR-46 is a 2 to 18 GHz system. It uses a wide-open, front-end, crystal video type receiver, with a digital processor which can be reprogrammed in the field to accept new threat data. In addition to providing cockpit visual and audio warning cues, it can directly control jamming systems. Export examples of AN/ALR-46(V)6 have been supplied to Iran, Korea, Malaysia, Thailand and Saudi Arabia, and AN/ALR-46(V)3 is used in Switzerland, Taiwan, Portugal, Egypt and the US Air Force.

STATUS: in service.

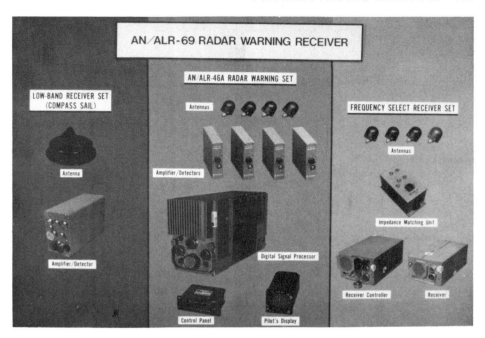

AN/ALR-69 radar warning receiver system

AN/ALR-67 radar warning receiver

Development began in March 1975 and the AN/ALR-67 is currently in the operational evaluation phase by the US Navy. It provides 1 to 16 GHz frequency coverage, and is being used initially on the McDonnell Douglas F/A-18 Hornet. The system features are similar to AN/APR-36/37 functions. There are four spiral antennas and an Itek Victor V processor is used for all threat assessments. Conventional cockpit displays and aural warnings are provided, and integration with active electronic warfare systems is possible.

STATUS: in production. Designed for installation in the US Navy F/A-18 Hornet aircraft and for the Grumman A-6E and F-14 and AV-8B

AN/ALR-68 radar warning receiver system

Harrier. Litton has received contracts exceeding $200 million in total for the AN/ALR-67(V) from the US Navy (including FMS sales to Spain and Australia) and Canada. A $7 million contract to provide a prototype installation in an F-14 aircraft was awarded in March 1985.

AN/ALR-68 radar warning receiver

This digital threat-warning receiver system is based on improvements incorporated in the AN/ALR-46 device. A major component of this development has been the introduction of an Applied Technology advance computer (ATAC) produced by Itek (now a division of Litton Industries).

AN/ALR-68 is a wide-open front-end, crystal video, field-programmable system which can detect threats from ground, shipborne and airborne radar signal sources. It provides audio and visual alarms to the aircrew and has in-cockpit threat-parameter programming facilities which permit hands-off tactical ECM operation. Design was undertaken for the West German Air Force and the equipment entered production in 1979. It is installed in Panavia Tornado aircraft.

STATUS: in production.

AN/ALR-68 advanced radar warning system

The West German Air Force advanced radar warning receiver system (ARWS) is now deployed operationally. It is a digital threat warning receiver system based on improvements to the ALR-46. Conversion to digital operation is achieved by the addition of the Applied Technology ATAC computer.

The ALR-68 is a wide-open crystal-video, field programmable system which provides in-cockpit threat-parameter programming and hand-off to tactical ECM systems. It uses a digital threat processor containing software provisions unique to the West German threat scenario, and is the first wide-open digital, general-purpose computer-based rwr to become operational in Europe. Installed in McDonnell Douglas F-4F and RF-4E aircraft of the West German Air Force. The ALR-68 is manufactured in association with AEG of Ulm, Federal Republic of Germany.

AN/ALR-69 radar warning receiver

An outgrowth of the earlier models of ALR-46 warning receivers, the AN/ALR-69 employs a frequency selective receiver system (FSRS)

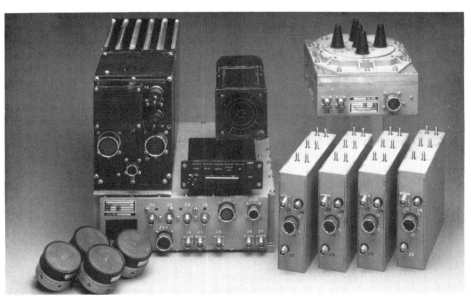

AN/ALR-67 radar warning receiver system

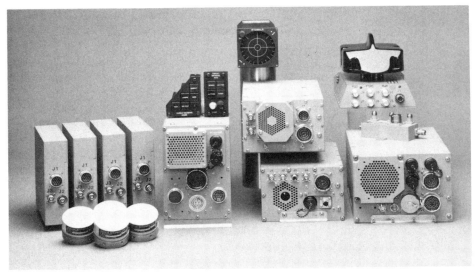

Litton AN/ALR-74 radar warning system

Units of Litton enhanced radar warning equipment

and a low-band launch alert receiver (Compass Sail) added to the basic ALR-46. Compass Sail detects and analyses SAM guidance beams to warn the pilot of the operational situation in which the threat missile system is tracking the target aircraft and also guiding a missile toward the aircraft. The FSRS detects and analyses higher frequency missile guidance radiations that are not associated with the tracking radar and gives relative bearing of the missile's approach. The system is designed to activate automatically ECM resources. The signal processor provides executive control for the FSRS. It accepts video inputs from five receivers and processed information from the FSRS, sorts and analyses the data, identifies and labels the radar signals found, tracks the status of these radar signals, and generates signals to provide threat warning to the operator. Functions performed by the FSRS include warning and direction finding on cw signals, accurate frequency measurements on pulse signals for ambiguity resolution, threat antenna scan type and rate analysis and jammer frequency set-on for jammer power management and blanking functions. Examples of the AN/ALR-69 system are installed in General Dynamics F-16s supplied to Denmark, Norway, the Netherlands, US Air Force and Pakistan.

In September 1983 a further update programme, valued at $24.2 million, was announced. The contract was let by the US Air Force and included options which could total a further $22 million. By September 1984 more than 2800 AN/ALR-69 sets had been delivered.

STATUS: in production and continuing. More than 3200 systems have been delivered to the

US Air Force and to NATO countries. The AN/ALR-69 is used on the F-16, Fairchild A-10 McDonnell Douglas F-4 and Lockheed C-130 aircraft.

AN/ALR-74 radar warning system

Affording self-protection for General Dynamics F-16 and other tactical aircraft, the ALR-74 was formerly known as the AN/ALR-67/69 Update and is designed with advanced technologies to achieve commonality between US Air Force and Navy radar warning systems. The system consists of amplifier-detector, processor, low-band receiver, frequency measurement receiver system, controller, control indicator and azimuth indicator.

STATUS: more than $111.0 million has been awarded to Litton for this system. Deliveries of prototype models were completed in late 1983 and flight testing on McDonnell Douglas F-4 and F-16 aircraft was completed in November 1984. First pre-production contract was awarded in December 1984 for initial F-4, RF-4 and F-16 requirements. Under the pre-production contract a number of hardware and software product improvements are being incorporated as required to meet the projected threats of the 1990s.

ERWE enhanced radar warning equipment

This system comprises a fully-programmable digital threat processor and three separate receiver elements which provide C to D band

coverage and both wide-band and narrow-band coverage for the E to J band. A crystal-video receiver is employed to provide wide-band coverage for rapid threat response, and a superheterodyne receiver is used to provide high sensitivity and selectivity. Both receiver sections employ the Applied Technology ATAC computer. The equipment provides a prioritised alphanumeric display and a high-intensity azimuth indicator and is capable of rapid change of threat parameters and priorities from mission to mission. ERWE was developed for the European threat scenario and is now operational on the West German Air Force's Tornado aircraft.

AN/ALQ-125 Terec tactical electronic reconnaissance system

Terec is essentially a passive system that detects, locates, identifies fixed and mobile ground-based emitters.

Terec is a US Air Force programme to equip 24 McDonnell Douglas RF-4C Phantoms with ESM receivers. These can 'listen in' to hostile emissions, measuring range and bearing. Information can be recorded, down-linked to a Terec portable exploitation processor, or transmitted to accompanying aircraft tasked to destroy the target.

STATUS: in limited production, 50 sets having been ordered by January 1985.

Lockheed

Lockheed Missile and Space Company Inc, 1111 Lockheed Way, Sunnyvale, California 94086
TELEPHONE: (408) 742 4321
TWX: 910 339 9227

PLSS precision location strike system

PLSS is designed to locate hostile ground-based radar emitters and to provide guidance for air strikes directed against them. It can be traced back to the Pave Onyx and Pave Nickel programmes of the mid-1960s, aimed at providing strike-guidance against missile radars in South-East Asia, Pave Onyx developed into ALSS, the airborne location and strike system successfully tested at the White Sands missile test range in 1972. ALSS did not become operational in South-East Asia, but three years later, in 1975, five modified US Air Force/

Lockheed U-2C high-altitude reconnaissance aircraft were deployed to the United Kingdom for a three-month evaluation under European conditions. In the same year the PLSS system received its first funding, the new equipment building upon ALSS by having wider coverage, better signal discrimination and processing, improved reliability, and able to operate against both continuous wave and pulsed signals.

In July 1977 an eight-company team led by Lockheed was chosen over another team headed by Boeing to begin full-scale development of the PLSS system, an activity planned to last 4.5 years. Flight trials began in 1983. Originally the equipment was to have been the principal sensor aboard the Compass Cope high-altitude reconnaissance remotely piloted vehicles, but when that programme was cancelled in the autumn of 1977 it was switched to the Lockheed TR-1, a substantially redesigned U-2. The near real-time system pinpoints the locations of emitters by triangulation. Three aircraft with PLSS equipment fly race-track

pattern more or less parallel to the edge of the battle zone or frontier, but in friendly skies. The position of each is continuously and accurately monitored by ground-based distance measurement equipment transponders. The bearing of emitters from each aircraft is measured, transmitted to the ground by data link and correlated there with the differences between the time of arrival of the signals at each aircraft.

STATUS: in development and flight trials.

TRAAMS time-reference angle of arrival measurement system

A time-reference angle of arrival measurement system similar to that in development for PLSS (see separate entry) is being developed by Lockheed under US Air Force Armaments Laboratory and US Navy sponsorship. This sensor system will allow a remote-piloted vehicle such as the Lockheed Aquila to home on to enemy communications emissions. Initial operation is to be across hf, vhf and uhf bands, and extension to microwave capability is under consideration.

STATUS: under development.

Lockheed-Georgia Company, 86 South Cobb Drive, Marietta, Georgia 30063
TELEPHONE: (404) 424 3689
TELEX: 542642

EC-130H Compass Call electronic warfare aircraft

This version of the Hercules transport is one of the two principal tactical jamming aircraft in the US Air Force inventory, the other being the Grumman EF-111A Raven. Their common purpose is to identify and disrupt enemy radars and communications. While the EF-111A can operate in any of three modes (stand-off, close-in, and the penetration escort of strike aircraft, the Lockheed EC-130H Hercules is too vulnerable to be exposed, and is limited to the stand-off mode.

The Compass Call aircraft was a response to an urgent request from Tactical Air Command for a communiction jammer. Development of equipment and operational tactics are the responsibility of the Tactical Air Warfare Center. The aircraft entered service in mid-1982 with the 41st Electronic Combat Squadron. It is distinguished from all other Hercules C-130 variants by the large vertical aerial forward of the vertical fin, and by two more under-wing antennas.

Loral

Loral Electronics Systems, Ridge Hill, Yonkers, New York 10710
TELEPHONE: (914) 968 2500
TELEX: 996597

AN/ALR-56 radar warning receiver

Together with the AN/ALQ-135 internal electronic surveillance measures set, this radar warning receiver suite forms the tactical electronic warfare system (TEWS) which is installed in all models of the US Air Force/McDonnell Douglas F-15 Eagle. Designed in 1974, the ALR-56 was claimed to be the first operational radar warning system with a digital processor to control power management functions, and capable of being reprogrammed in the field within minutes, using software techniques.

The radar warning receiver system comprises five antennas, a high-band tuner unit, low-band receiver/processor, power supply unit and cockpit control and display. There are four circularly-polarised planar spiral aerials (two facing forward, one in each wing-tip, and one rearward-facing at the tip of each fin) to provide 360° azimuth high-band coverage. Additionally, there is a blade antenna under the fuselage for lower hemisphere low-band coverage.

The system uses separate high and low-band receiver units. The high-band tuner unit provides digital control of the antennas, and is a dual-system capable of scanning a very wide portion of the electromagnetic spectrum. The single-conversion superheterodyne design provides high sensitivity, and the use of dual YIG radio frequency pre-selection affords excellent sensitivity and spurious signal rejection.

The low-band receiver unit is integrated with the system processor, which is a 32 K 16-bit word unit with a cycle time of 1.5 microseconds. With the comprehensive antenna/receiver-derived data the processor can conduct direction-finding operations, and low-band electronic tuning is carried out directly.

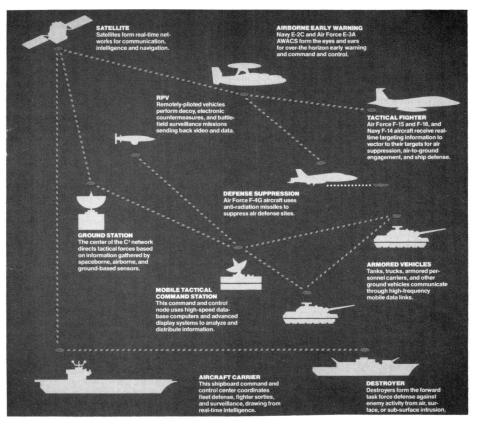

Loral interpretation of typical electronic warfare applications in modern operations

Pre-processor facilities, which convert intercepted signals into digital data for the central processor element, are included in the processor unit. A reprogrammable threat library is held in memory and can be easily changed.

A self-contained precision signal source within the system provides calibration and/or built-in testing over the entire tuning range, and there is provision for multiplexing important

receiver functions. The latter are controlled by serially-generated NRZ Manchester-coded data. The precise data rate also serves as the reference clock for the receiver frequency synthesiser.

Information provided in the cockpit includes data for a circular threat-evaluation display and audio signals. The display unit shows threat bearing and distance, and threat characteristics are presented by alphanumeric symbology. The display is designed to be legible in high ambient lighting and has pilot-selectable clutter-elimination programs. There are two cockpit control units. One unit combines basic radar warning receiver and tactical electronic warfare systems and TEWS control functions, and the other, smaller, unit is specifically for radar warning receiver use, with interrogated data readout facilities.

Loral was awarded a contract in late 1982 to begin the first stage of updating the system for the US Air Force/McDonnell Douglas F-15 fighter. The $22.3 million contract funds the development of an enlarged memory module for the signal processor, with up to five times the capacity of existing production equipment, the new equipment being designated AN/ALR-56C. A new processor occupying only two-thirds the volume of existing equipment will be put into production with the new memory module and long-term development is underway to improve receiver capability. An extension of the receiver system bandwidth, which will permit detection of millimetric radars, is being privately funded by Loral.

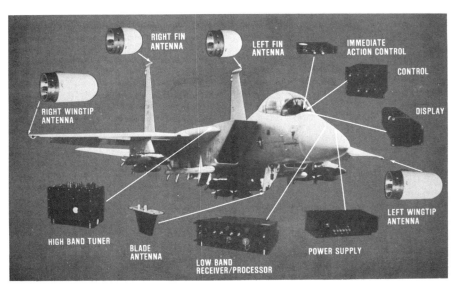

Loral AN/ALR-56 radar warning equipment on US Air Force/McDonnell Douglas F-15 Eagle

Dimensions and weight
(power supply unit) 445 × 267 × 156 mm; 26.4 lb (12 kg)
(wing-tip antennas) 89 dia × 144 mm; 1.8 lb (0.8 kg) each
(fin antennas) 89 dia × 100 mm; 0.9 lb (0.43 kg) each
(low-band antenna) 133 × 84 × 45 mm; 0.4 lb (0.2 kg)
(high-band receiver) 241 × 187 × 202 mm; 22 lb (10 kg)
(low-band receiver/processor) 451 × 404 × 165 mm; 53 lb (24 kg)
(display unit) 406 × 149 × 137 mm; 13.3 lb (6 kg)
(control unit) 165 × 111 × 114 mm; 2.4 lb (1.1 kg)
(immediate action controls) 29 × 76 × 127 mm; 0.7 lb (0.3 kg)
Total system weight: 138 lb (63 kg)
Total system volume: 2.41 ft³ (0.068 m³)
Installation: 13 lb (5.9 kg) for rf transmission lines
Power: 550 VA at 115 V 400Hz
3 to 7 A, 28 V dc

STATUS: in production: the total value of the programme is estimated by Loral at $300 million.

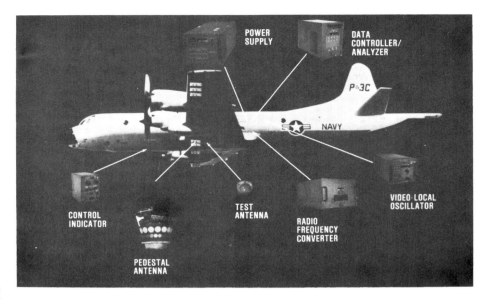

Loral AN/ALQ-78 ESM equipment on US Navy/Lockheed P-3C Orion

Radar warning system for helicopters
Loral is developing what it calls a 'new concept' in radar warning technology for large or expensive US Army helicopters. The new system, which can be integrated with the company's AN/APR-39(V)2 digital processor, has a unique receiver to handle next-generation threats. In conjunction with this effort is an internal research and development programme to integrate infra-red warning, chaff and flares management, and optical fire-control sensor cueing into this next-generation receiver system for future Army applications such as the LHX light armed helicopter.

AN/APR-38 ESM system for F-4G Wild Weasel
This significant US Air Force programme for the McDonnell Douglas F-4G Wild Weasel was sub-contracted to IBM (radar warning receiver), Texas Instruments (radar warning computer) and Loral (displays).

The Loral-furnished control/indicator set comprises seven items in the front and rear cockpits housing the pilot and system-operator. Data is assigned priorities by the central processor before transmission to the cockpit items. All displays are controlled from a special electronics unit, they comprise a plan-position indicator for each crewman, and a panoramic analysis and homing indicator/control unit for the system operator. The remaining units are a lighting controller, system and missile control panel and aircraft commander's warning panel. Alphanumeric displays are used to designate signals: 'A' corresponds to anti-aircraft gun radar, '2' to Soviet SA-2 tracking radar and so on. The 15 highest priority threats are assigned and labelled, and the most significant threat is placed within a triangular symbol. The operator can override this with his diamond-shaped 'Bear' symbol.

Automatic weapon-release is also handled by the system, blind attack being possible by sighting a green aircraft cursor on the gun-sight display over a red reticle positioned by the threat processor.

STATUS: in service.

AN/APR-43 radar warning receiver
An integrated antenna and radar warning receiver, the APR-43 adds continuous wave detection and C/D band direction-finding capabilities to the current family of radar warning receiver devices employed on front-line US Navy fighters.

AN/ALQ-78 ESM system
Used in US Navy/Lockheed P-3C Orion ASW aircraft the AN/ALQ-78 is a radar-warning system designed to detect anti-submarine and conventional electronic warfare threats. A high-speed rotating antenna in the aircraft belly operates in an omni-directional search mode and, using threat-processor control, it initiates automatic direction-finding as signals are acquired. Threat analysis is carried out within the equipment, and data is displayed to system operators in the aircraft.

STATUS: in production at a rate of some 12 sets a year since mid-1970s.

Rapport II ECM system
Rapport is unusual, perhaps unique, in being an American system procured by a European customer after competitive evaluation and development, not already being available 'off the shelf' for US applications. It is designed to protect aircraft against airborne and ground-based radar-directed weapons. The system continuously analyses, identifies and measures the bearings of threat radars, and can rank up to 14 such threats in order of priority according to reprogrammable data held in the system's memory. An alphanumeric display presents data to the pilot on the threat and its bearing from the aircraft, with audio 'alert' tones. Automatic noise or continuous wave pulse countermeasure action can then be initiated.

The Rapport II system (Rapport is the acronym for rapid alert and programmed power-management of radar targets) was the subject of a Belgian Air Force requirement for an ECM suite that could be carried internally in its Dassault Mirage V aircraft. Dissatisfied with equipment available at the time, during 1973 the Belgian Air Force began to search for an ECM suite for the new aircraft, and issued RFPs (requests for proposals) to US and European industry for a completely new system. Evaluation narrowed down to three US companies: ITT, Loral and Sanders. These three were funded by the Belgian Air Force to conduct installation and effectiveness studies, resulting in July 1975 in the award to Loral of a $4.5 million contract for the development, manufacture and installation in a representative aircraft of a Rapport II system. A critical design review was conducted in October of that year, and the system was shipped to Belgium early in 1977 for a comprehensive test and assessment programme by the Belgian Air Force later that year. As a result of these tests, in mid-1978 the Belgian Government placed a production contract on Loral worth $20 million. Loral provided offsets of 75 per cent of the contract value, representing 63 per cent to MBLE for development and production, and 12 per cent to SABCA. Foreseeing a likely application on the General Dynamics F-16 (chosen in June 1975 by Belgium, Denmark, Norway and Holland for their Lockheed F-104G replacement), options were also arranged for that

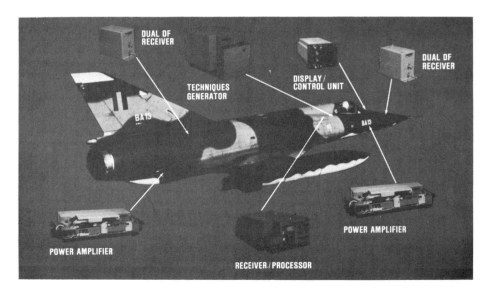

Loral Rapport II ECM equipment on Belgian Air Force/Dassault Mirage V

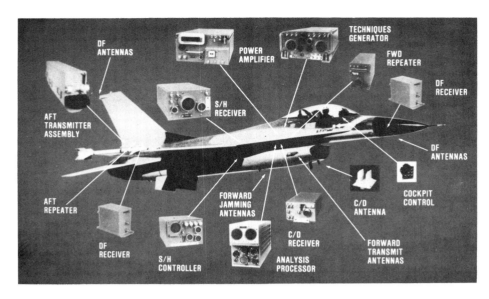

Loral Rapport III ECM equipment on Belgian Air Force/General Dynamics F-16

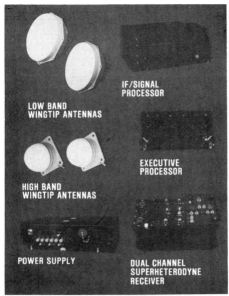

Loral supplies passive receiver equipment for use on Royal Air Force Nimrod MR2 and AEW3 maritime patrol and early warning aircraft

possibility. Installation of Rapport II on 50 Dassault Mirage Vs was completed during 1981.

Rapport II comprises seven line-replaceable units (Irus), together with associated receiver and jamming antennas. The Irus comprise a receiver/processor, techniques generator, control and display unit, two dual-channel direction-finding receivers and two dual-band power amplifiers. The system brought to ECM technology for the first time such features as solid-state power amplifiers and distributed processing, but most importantly it incorporated in the same suite of equipment both passive (radar warning) and active (radar jamming) functions. Such internal integration of tasks within the system obviated the need for space-consuming and expensive interface boxes.

Frequency range: E to J band (2-20 GHz)
System weight: 282 lb (128 kg)
System volume: 5.1 ft³ (0.14 m³)
Digital processor: TI 2520
Memory: 32 K bytes
Growth flexibility: MIL-STD-1553

STATUS: in service with Belgian Air Force.

AN/ALQ-178 Rapport III ECM system

US Air Force designation for the Rapport III system in production for Belgian and Israeli General Dynamics F-16 aircraft.

Closely following the selection and development of the Rapport II ECM suite for its Dassault Mirage V fighters, in August 1979 the Belgian Air Force decided in principle to adopt the system for its General Dynamics F-16s. The Belgian Government funded General Dynamics to conduct installation and effectiveness investigations of a system now known as Rapport III and sent two aircraft to Eglin Air Force Base for flight trials, concluded in March 1982. The Belgian Air Force continued in its commitment to the Rapport system despite initial delays in the flight programme that cut into the procurement budget and the decision of the Dutch Government in 1980 to acquire the Westinghouse AN/ALQ-131 system (on the grounds that the Loral equipment would not fit into the limited space available on its two-seat F-16Bs).

Meanwhile, in a move to investigate the potential of the Rapport III system as an interim measure until the availability of ASPJ (the advanced self-protection jammer, now the ITT/Westinghouse AN/ALQ-165), at the end of 1981 the US Air Force was directed by Congress to conduct a 'full and fair evaluation of the system' The US Air Force had itself adopted the AN/ALR-69 Compass Tie and the AN/ALQ-131 jammer for its own F-16s.

Rapport III is generally similar to its predecessor, though a number of refinements have been incorporated including a repeater, and the equipment for the Belgian Air Force has been designated Rapport III (I), the (I) standing for Improved. The Belgian Air Force planned to acquire 52 'full house' systems, incorporating both passive and active capabilities, and 56 passive-only systems, to equip 108 aircraft out of a planned fleet of 116. However budget difficulties have led to a postponement of the programme, estimated to be worth about $180 million. In December 1983 however Loral announced an $88 million order from a 'friendly foreign country', noting that it was the first production order. The country was later identified by other sources as Israel. However, press reports from Brussels say that Israel had taken over development of the system (or rather, financing continued development), and that Rapport III had been installed on Israeli Air Force F-16s at an early date. A contract worth $78 million may have been placed with Loral during 1978. The report says that the system 'performed admirably during the Lebanon campaign, with Israeli F-16s knocking out great numbers of Syrian surface-to-air missile sites and shooting down 82 Syrian aircraft against a loss of only two'. The Belgian Government, which had been kept posted of Israeli-finanaced development, was by September 1984 reported to be seeking ways by which its own Rapport III(I) system could be funded. Loral meanwhile has not confirmed any Israeli involvement.

Rapport III provides E to J band (2 to 20 GHz) detection with simultaneous fore and aft jamming. A 'look-through' facility permits rapid assessment of jammer effectiveness. Threat data generated by the system includes direction-finding, pulse repetition frequency, pulse-width and scan rate. Information is presented to the pilot on a circular cathode ray tube with alphanumeric detail.

System weight: 368 lb (116 kg)
Installed volume: 2.8 ft³ (0.079 m³) in avionics bay, 5.5 ft³ (0.156 m³) in tail compartment

STATUS: in production, and probably in service with Israeli F-16 fighters. A decision on a Belgian order is expected.

ARI-18240/1 ESM system

Loral is supplying an ESM system to the Royal Air Force for installation on the British Aerospace Nimrod AEW3 and (possibly) Nimrod MR2 aircraft. Sets of fore and aft facing cavity-backed planar spiral antennas are in wing-tip pods for a total of 16 aerial units. Also in each pod is a receiver unit, which extracts threat data from the radar warning receiver sensors for onward transmission to an intermediate-fre-

quency processor unit in the centre fuselage. Loral also supplies a power supply unit, and the absence of any further hardware suggests that all subsequent use of threat data is integrated into the basic aircraft systems.

Few technical details are available, although it seems probable that each antenna set covers a 90° azimuth sector and thus provides all-round electronic surveillance measure coverage. Both high- and low-band antennas are used and frequency coverage is probably E to J band (2 to 20 GHz).

STATUS: in production.

EW-1017 ESM system

In 1979 Loral launched a new electronic surveillance system under this designation, several units of which may be similar to those in production for the ARI-18240/1 system. With more accurate and sensitive radar receivers than previously, the system extends the range and performance of patrol and surveillance aircraft. Advanced digital microprocessing, an expanded computer memory and new interactive graphics display system enhance the system's ability to detect and analyse in real time electromagnetic activity over great distances from the host aircraft.

STATUS: in production: applications include Royal Air Force/British Aerospace Nimrod AEW3, and the German version of the Dornier/Breguet Atlantic maritime patrol aircraft.

Multibeam transmitter assembly

Loral is developing, as part of a company-funded activity, a distributed miniature travelling-wave tube transmitter system which, it is claimed, will considerably improve the efficiency and directional-jamming capability of an active electronic warfare system.

Demonstrations were conducted in early 1981 to show the feasibility of a multibeam transmitter system which would use a 10-element forward-looking array to provide a field of view measuring 120° in azimuth and 60° in elevation. This would use five transmitter elements on either side of the fuselage. A nine-element aft-looking array is proposed to complement this unit. Several miniature 40-watt travelling-wave tubes (Varian Associates VRT-6110-A3 devices in prototype units) are used to feed each array element. As a result of this redundancy, the system is claimed to be less likely to fail than a conventional system using only a single unit, and also to provide graceful degradation in the event of a failure, as the contribution of any one unit is relatively small. Operating range is 4.8 to 18 GHz.

In a total electronic surveillance measures system Loral is proposing that a superheterodyne receiver with crystal-video receivers should be used for direction-finding. The effectiveness of such a system against monopulse and simultaneous targets in different azimuth locations has been demonstrated. Total system power requirements are 3.5 kVA forward and 2 kVA aft in the type of installation described above, which would be suitable for the General Dynamics F-16.

A configuration that could be installed in the F-16 has been outlined by Loral and exemplifies the most compact internally-mounted electronic surveillance measures system the firm has yet contemplated. The two sets of forward-looking arrays and their associated crystal-video receivers would be installed in blisters on either side of the chin engine air-intake. The forward transmitter box would be located in space already reserved for possible electronic warfare use and connected to the aerials by coaxial cables about 7 feet (2.2 metres) in length. Loral has also stated that it will consider the applicability of this new technology for retrofit to aircraft such as the McDonnell Douglas F-15 Eagle, Grumman F-14 Tomcat or General Dynamics FB-111.

STATUS: under development.

Loral AN/ALQ-123 infra-red countermeasures pod on US Navy/McDonnell Douglas F-4 Phantom wing pylon

Loral Corporation, Electro-Optical Systems, 300 N Halstead Street, Pasadena, California 91107

AN/ALQ-123 infra-red countermeasure system

This infra-red jamming system is a pod-mounted unit with a ram-air turbine powered generator. It is designed to provide protection against SA-7 surface-to-air missiles and is suitable for Grumman A-6 Intruder and LTV A-7 Corsair types. Tail installations in other US Navy types have been proposed.

Jamming is achieved by modulating an infra-red source in a manner which will break the lock-on of an infra-red seeking missile. Development was completed in March 1973, and production commenced a few months later. The system is now used by the US Navy and several export customers.

STATUS: in service.

AN/ALQ-157 infra-red countermeasures system

Designed to meet US Naval Air System Command requirements associated with the protection of large helicopters, the AN/ALQ-157 consists of two infra-red transmitters, considered by Loral to be a key technology and proprietary to the company. A control power supply, electromagnetic interference filter assembly and pilot's control indicator. Pilot selection of any one of up to five pre-programmed jamming codes is possible. Main application is US Navy/Sikorsky CH-53A/D and Boeing Vertol CH-46E helicopters. Operational testing has been successfully completed and production began in 1983.

STATUS: Loral received an initial production contract, worth $1.5 million, in December 1983, the application being US Marine Corps CH-46 helicopters. The company is building 121 sets. Qualification and acceptance tests began in October 1984.

Loral AN/ALQ-157 infra-red jamming equipment attaches to tail rotor pylon of US Army/Boeing Vertol CH-47 Chinook helicopter

Lundy Electronics

Lundy Electronics and Systems Inc, Robert Lane, Glen Head, New York 11545
TELEPHONE: (516) 671 9000
TELEX: 126929
TWX: 510 223 0605

AN/ALE-43 chaff cutter/dispenser pod

This is a high-capacity chaff system which holds rolls of chaff material and cuts it to the appropriate dipole length during operation. Lundy claims to have overcome the difficulties associated with past chaff-cutters by developing a unique chaff-roving supply system.

The system has a chaff-roving hopper behind which is the chaff-cutter system. A remote processor controls operation of the cutter.

Chaff is drawn simultaneously from up to nine chaff-roving packages in the hopper. Each roving passes through a guide-tube which terminates at draw rollers and a cutting roller. As dipole lengths are cut, they are discharged into a turbulent airflow, and so distributed

efficiently behind the pod. Each cutter assembly consists of a drive motor, clutch/brake unit and three indexing-cutting rollers, the latter embodying blades which yield specific combinations of dipole lengths.

The system can be podded or mounted internally. The podded version can fit a variety of standard pylon attachments and is used on the Boeing B-52, Grumman EA-6B Prowler, McDonnell Douglas F-4 Phantom and EA-4A Skyhawk. Internally-mounted versions have been used on the McDonnell Douglas ERA-3B Skywarrior and Boeing NKC-135. The system is frequently used as a training aid, but its primary wartime functions would be for anti-ship missile defence, area-saturation operations (over battlefields), corridor seeding and aircraft self-protection.

Podded version
Pod length: 11 ft (3.37 m)
Max diameter: 1.58 ft (0.48 m)
Empty weight: 305 lb (139 kg)
Loaded weight: 625 lb (284 kg)
Chaff payload: 8x RR-179 roving packages
Max dispensing rate: 7.2 × 10⁶ dipole-inches/s

Internal version
Basic weight: 80 lb (37 kg)
Max loaded weight: 420 lb (191 kg)
Chaff payload: 9 × RR-179 roving packages
Max dispensing rate: 8.1 × 10⁶ dipole inches/s

Common characteristics
Max continuous dispense time: 660 s
On-time: select 1–9 s in 1 s steps or continuous
Off-time: select 1–9 s in 1 s steps
Operable altitude band: 0 to 50 000 ft
Power: 1.7 kVA at 115 V 400 Hz, plus 2.5 A at 28 V dc

STATUS: in production.

AN/ALE-44 chaff dispenser pod

This is a chaff or flare dispenser system suitable for supersonic aircraft, usually installed as a two-pod system with a control unit in the cockpit. Each pod houses two dispenser modules and a sequencer. The pods are lightweight units which can be installed on wing-tip, underwing or under-fuselage store locations. They have dual-channel dispensing capability and can be quickly reloaded with RR-129 chaff or Mk 46 infra-red flare cartridges. The cockpit unit permits selection of burst rates, burst interval, and units per burst. Flares and chaff can be dispensed simultaneously.

Pod length: 6.75 ft (2.06 m)
Pod section: 4.375 × 7.125 inches (110 × 180 mm)
Pod empty weight: 30 lb (13.6 kg)
Pod weight with 32 chaff cartridges: 44 lb (19.9 kg)
Pod weight with 32 flare cartridges: 50 lb (22.6 kg)
Control unit weight: 2.5 lb (1.1 kg)
Modes: 1 or 2 units per burst
Programs: 1, 2, 4, 8 or continuous bursts per program
Rate: 4, 2, 1 or ½ bursts/s

STATUS: in production.

Supersonic countermeasure pod
Developed by Lundy Technical Centre, this new system is suitable for a wide range of supersonic strike aircraft. It comprises a control unit and two dispensing pods, each pod holding two dispensing modules and a sequencer. Total system capacity is 64 RR-129 chaff dispensers or Mk 46 infra-red flare cartridges. Flight-qualified for supersonic flight, and with a low frontal area, the system is marketed as being suitable for installation in a pod or on a pylon. The control unit permits selection of burst rate, burst interval and units per burst. Specification is identical to AN/ALE-44 unit, also built by Lundy.

STATUS: in development.

Countermeasure dispenser
This is a lightweight system designed for the US Air Force and successfully flight-tested on Cessna O-2 and Bell UH-1 aircraft. It carries up to 20 Mk 50 infra-red decoy flares or chaff cartridges and its primary use is in forward air controller aircraft.

STATUS: under development.

Countermeasure system
A recent development has been a 14-cartridge chaff/flare dispenser which can be accommodated on virtually any surface by using a contoured mounting plate. Aircraft countermeasure loads can be tailored to missions by using different numbers of dispensers, and chaff cartridge capacity is claimed to be adequate to protect aircraft such as General Dynamics F-16 and McDonnell Douglas F-4 Phantom within 2 seconds of ejection, against radars in the E to J bands.

Dispenser
Dimensions: 710 × 130 × 150 mm
Weight: 31 lb (14.2 kg) with flares

Cartridge
Dimensions: 40 mm dia × 127 mm
Weight: (I/R) 0.7 lb (0.32 kg) (Chaff) 0.6 lb (0.3 kg)
Flare: 15 kW output, 2-5 μm range
Chaff: rcs of 20 m² within 1-2 s of ejection

STATUS: under development.

Magnavox
Magnavox, Government and Industrial Electronics Company, 1313 Production Road, Fort Wayne, Indiana 46808
TELEPHONE: (219) 429 6000
TELEX: 0232485

AN/ALR-50 radar warning receiver
This was part of a very substantial US Navy programme which has passed production phase but units are still widely used in the service. The system was a development of the early AN/APR-27 system, also a Magnavox-produced unit, and it in turn now seems likely to be superseded by the AN/APR-67. Operating frequencies were probably in the range 4 to 20 GHz (G to J band), and production was almost continuous throughout the early 1970s, with as many as 800 units, worth $22 million, being ordered in 1972. At least 1300 sets were delivered to the US Navy and used on Douglas A-4 Skyhawk, Grumman EA-6A Intruder, EA-6B Prowler, LTV A-7 Corsair, F-8J/RF-8G Crusader, McDonnell RF-4B/F-4N Phantom, Grumman F-14 Tomcat and North American RA-5G Vigilante.

STATUS: in service.

AN/ALQ-91 communications jammer
Developed for the US Navy as a countermeasure system in the vhf/uhf and hf communications waveband. It forms part of the tactical homing

US Navy/Grumman F-14 Tomcat fighters have Magnavox AN/ALR-50 radar warning receivers

and warning system for Grumman F-14 Tomcat fighters and Douglas A-4M Skyhawk light bombers. An order for 98 sets was made in 1971 and, although it remains in service, more advanced sets are now replacing it.

STATUS: being withdrawn from service.

AN/ALQ-108 IFF deception set
Representing a little-publicised sector of the electronic warfare market, this system is used in

US Navy's Grumman E-2C Hawkeye, Lockheed EP-3E Orion and S-3A Viking types to improve survivability in ASW and Elint operations. Production of about 300 sets is reported, most systems having been delivered in the early 1970s. It is likely that this remains an important system.

STATUS: in service.

McDonnell Douglas
McDonnell Douglas Corporation, Box 516, St Louis, Missouri 63166.
TELEPHONE: (314) 232 0232
TWX: 910 762 0635

AN/ALQ-76 jamming pod
This unit was derived from the earlier AN/ALQ-31 and is part of the AN/ALQ-99 system. It is a podded unit, which is superseding older types of noise and deception jamming systems. McDonnell Douglas undertook design for the US Navy, and was awarded prime contractor status, although development and production is with Raytheon. Operating frequencies are between 2 and 8 GHz, and 650 pod units will be supplied for use on Grumman EA-6A Intruder, EA-6B Prowler and Douglas A-4E Skyhawk. (see also under Raytheon).

STATUS: in service.

US Air Force/McDonnell Douglas F-4G Wild Weasel
The McDonnell Douglas F-4G Wild Weasel is a significant and current EW type which fulfils a large part of front-line dedicated and co-operative jamming duties with the US Air Force.

The Wild Weasel programme was initiated in 1965 following introduction of Soviet SA-2 radar-guided anti-aircraft missiles in North Viet-Nam to counter US bombing. In a top-priority effort, Republic F-105 Thunderchiefs and McDonnell Douglas F-4C Phantoms were modified to carry receivers that could detect transmissions from the SA-2 guidance radars. Having pinpointed the radars, the Wild Weasel aircraft could than attack and destroy them with strike anti-radar so that the SA-2s could not be fired. The F-4C had the Itek AN/ALR-46 radar homing and warning system.

Along with those hastily contrived adaptations, the US Air Force was taking a long-term

view of the surface-to-air missile suppression role, and these deliberations crystallised in the form of a much more optimised Phantom variant, the F-4G, adapted from the F-4D airframe. Continued choice of the F-4 as the test available Wild Weasel aircraft was due to its two-seat configuration, with an observer to operate the EW equipment; an ability to carry heavy loads, which meant that comprehensive ECM equipment and missile-suppression weapons could be carried on the same aircraft; high speed and long range; and the safety of two engines.

Two test aircraft were flown in operational tests during 1977, and the first F-4G was delivered in March 1978. The basic airframe now became the F-4E , which having wing leading-edge slats was more manoeuvrable. Production built up to three a month, and the last of 116 aircraft was built in 1982.

The principal ECM system is the AN/APR-38

radar homing and warning system, the separate elements of which are produced by Loral, IBM and Texas Instruments with McDonnell Aircraft as prime contractor and developer of software. The single 20 mm gun (a feature that characterised the F-4E) was removed and the space used for the new EW system. The APR-38 process incoming signals and, from a knowledge of range, speed and height, computes launch conditions for the missile selected. The aircraft carries AN/ALQ-119 jamming pods, also the AN/ALQ-131 self-protection jammer. The APR-38 employs no fewer than 56 antennas.

Three squadrons, including one for training, form the 37th Tactical Fighter Wing at George Air Force Base, and there are fully operational wings in the Philippines and West Germany. All aircraft have specially modified, smokeless J79

engines and are able to carry AGM-88 HARM high-speed anti-radiation missiles, APG-45 Shrike anti-radiation missiles, AGM-78 Standard ARM or AGM-65A/B Maverick, television guided missiles. An infra-red guided version of Maverick is to be introduced and, additionally, the aircraft can carry conventional Mk82 bombs, AIM-9 Sidewinder and AIM-7 Sparrow self-defence weapons.

In 1983 a $108 million contract was awarded to McDonnell Douglas by the US Air Force for the development of an improved Wild Weasel system. The company will conduct full-scale engineering development on modified directional-receiver systems to improve location and attack capability against radiating threats such as air-defence radar sites. Engineering development was due for completion in 1984 and the equipment entered production in 1986.

The main suppliers to the F-4G programme have been:
Lear Siegler (AN/ARN-101 nav/attack system)
IBM (radar warning system)
Texas Instruments (radar warning computer)
Loral (control indicator sub-system)
Tracor (AN/ALE-40 ECM dispensers)
Westinghouse (AN/ALQ-119 ECM pod)

More details of specific items are given, where available, in the appropriate company entries. A small number of F-4G Phantoms are believed to be operated by the Israeli Air Force.

STATUS: in service. The US Air Force operates 116 McDonnell Douglas F-4Gs, of which 96 are dedicated to combat units, the remainder being retained for training and attrition.

Motorola

Motorola, Government Electronics Division, PO Box 1417, Scottsdale, Arizona 85252
TELEPHONE: (602) 949 4176
TELEX: 667490

SPEW small platform electronic warfare tactical system

Under contract to the US Air Force Avionics Laboratory, Motorola is developing a compact, lightweight ECM system for unmanned vehicles to counter early warning and ground-controlled intercept radars in support of penetrating

aircraft. The company built a combined receiver/processor/transmitter and a four-element antenna for laboratory tests in 1984, and is currently developing flyable brassboard system hardware.

STATUS: in development.

AN/ALT-16A solid state amplifier

The high-powered solid state amplifier designed and produced for the AN/ALT-16A ECM system for the US Air Force is the first military-qualified solid state amplifier of its kind and size to operate over an octave bandwidth at its

required frequency and produce hundreds of watts. This advanced unit improves reliability by more than 10 to 1 and reduces specified weight approximately to one-third.

The solid state amplifier is functionally and physically interchangeable and is designed to replace conveniently a twt. It is designed for easy installation and maintenance and, in contrast to a twt, is much more reliable.

In addition to providing a significant reduction in input power requirements the solid state amplifier offers a lower life-cycle cost. The unit weighs 60 lb (27.5 kg)

STATUS: in production.

Northrop

Northrop Corporation, Defence Systems Division, 600 Hicks Road, Rolling Meadows, Illinois 60008
TELEPHONE: (213) 757 5181
TWX: 910 321 3044

AN/AAQ-4 infra-red countermeasures system

An infra-red countermeasures system first introduced as an internal configuration on US Air Force/McDonnell Douglas EB-66 aircraft for deployment in South-East Asia. Subsequently, the system was reconfigured for helicopter application and updated with multi-threat capabilities for redeployment on US Air Force/Sikorsky H-53s. In the helicopter configuration, the dual-transmitter provides protection on bands I and II without engine suppressors. The system electronically modulates a vistal caesium infra-red source to produce a highly-effective jamming signal.

Length: 33 inches (838 mm)
Diameter: 13 inches (330 mm)
Weight: 140 lb (63.5 kg)
Power: 4 kVA at 115 V 3-phase, plus 20 W at 28 V dc

STATUS: in service.

AN/AAQ-8 infra-red countermeasures pod

The AN/AAQ-8 is a multi-threat infra-red countermeasures system capable of operating in supersonic environment. This pod is a second-generation system updated to meet new and continuing threats, and has been extensively deployed on US Air Force McDonnell Douglas F-4, LTV A-7 and Lockheed C-130 aircraft. The pod can be configured with a ram-air turbine allowing protection independent of aircraft prime power and cooling resources.

Length: 7.5 ft (2.29 m)
Diameter: 10 inches (254 mm)
Weight: (MkVI) 235 lb (107 kg)
(MkV2) 264 lb (120 kg)

Power: 4 kVA at 115 V 3-phase, plus 20 W at 28 V dc

STATUS: in service. In 1985 the AN/AAQ-8 underwent trials fitted to a British Aerospace Buccaneer operated by the Royal Aircraft Establishment.

AN/ALQ-135 noise/deception jamming system

Otherwise known as the ICS (internal countermeasure set) the ALQ-135 is a component of the tactical electronic warfare system for the US Air Force/McDonnell Douglas F-15 Eagle fighter. It operates with the AN/ALR-56 radar-warning system and the AN/ALQ-45 countermeasures dispenser.

The AN/ALQ-135 is a noise/deception jamming system which uses dual-mode, pulsed/continuous wave, travelling-wave tube transmitters. All equipment is mounted internally and jamming system management is provided by the AN/ALR-56 radar warning receiver processor. No details of operating frequency range or installation data have been revealed, but it is thought to use a four-antenna array providing 360° azimuth jamming coverage.

The basic ALQ-135 consists of two line-replaceable units (lrus) plus appropriate waveguides and antennas. Each lru consists of a control oscillator and an rf amplifier that are mission interchangeable. Over the years the system has continued to evolve along with capabilities of the aircraft and changes in the

McDonnell Douglas TEWS tactical electronic warfare system includes Northrop AN/ALQ-135 noise/deception jamming system carried internally

threat. While maintaining commonality with the original system and support electronics, the AN/ALQ-135 has been updated to include high-band coverage and digital receiver/processor power-managed functions.

Initial development funding commenced in August 1974, and led to a $25 million production contract in September 1975. This has been followed by further production contracts totalling $800 million for 900 sets by the end of 1982. Additional countermeasures improvements are being developed for this system under a $60 million contract.

STATUS: in production.

AN/ALQ-136 ESM jammer

One of two compact internal radar-jammer systems (the other is AN/ALQ-162) currently available from Northrop. No technical data has been released and, as yet, it is uncertain whether this system will be committed to production. It is intended for use on aircraft too small to accommodate the new AN/ALQ-165 advanced self-protective jammer system.

STATUS: under development.

AN/ALQ-155(V) B-52 power-management system

The AN/ALR-155(V) provides integral set-on receivers for each jamming transmitter, plus increased effective radiated power density through accurate frequency set-on. The system is a power-management evolution of the ALT-28(V) active ECM set providing automated hand-off from the ALR-46 radar warning receiver with near instantaneous jammer response. It is computer-managed and field programmable and contains automatic frequency control in all modes and a wide variety of ECM techniques that are automated or manual (or semi-automated). The system has a 12-transmitter upload capability.

The ALQ-155(V) improvements are: frequency agility against multiple threats; pulse repetition interval trackers; cover pulse jamming techniques; false target generation through pseudo-random noise; hybrid IFM receiver and central receiver capability; programmable noise optimisation; compatibility with AN/ALQ-117 deceptive I/J band jammer; down-link jamming; increased pulse-up power for cw to pulse operations; electronically steerable antenna system; compatibility; coherent and incoherent jamming.

AN/ALQ-162 ESM jammer

Funded by the US Navy and Army, the AN/ALQ-162 is a small (36 lb, 16.4 kg) radar-jamming system which can be supplied with its own receiver/ESM management processor, or made compatible with many existing types of radar warning receiver processor systems. Internal or podded installations are proposed, and, is to be integrated with the AN/APR-43 Compass Sail/Clockwise rwr. The ALQ-162 has been mounted in wing pylons without taking up needed fuel or stores stations, inducing drag, or occupying valuable fuselage electronics space. The system makes use of advanced jamming techniques, is software programmable to meet new threats and includes built-in test devices to increase maintainability and can operate autonomously using its own receiver/processor, in a 'stand-alone' capacity or in conjunction with a variety of on-board radar warning receivers. Its reprogrammability provides the flexibility to accommodate a unique or rapidly changing threat environment. Current installation commitments of the AN/ALQ-162 include: the US Navy's LTV A-7E, McDonnell Douglas A4-M, RF-4B and F-4S, and AV-8C Harrier; the US Army's Bell EH-1X, Sikorsky EH-60, Grumman OV-1D and RV-1D, and Beechcraft RC 2D and RU-21; and NATO's McDonnell Douglas F/A-18, General Dynamics F-16, and Saab F-35 Draken.

It is claimed to use advanced jamming techniques and probably has dual-channel, pulsed/continuous wave, transmitter elements. A reprogrammable threat library/techniques generator system can be provided. The receiver/processor and power supply/transmitter units are built as two separate items but they may be installed as a single unit.

STATUS: under development. The first pair of export orders for ALQ-162 were announced in September 1984. Some 1500 sets are to be built for Canadian Defence Force/McDonnell Douglas CF-18 fighters and for Danish Air Force/Saab F-35 Draken and General Dynamics F-16 aircraft. In the Canadian aircraft the system will operate in conjunction with AN/ALQ-126B pulse-jammers to confer an 'all threats' capability similar to that planned for US Navy tactical aircraft.

AN/ALQ-171 ECM system

Possibly using AN/ALQ-162 technology, the AN/ALQ-171 ECM system is a lightweight self-contained receiver/processor/transmitter sys-

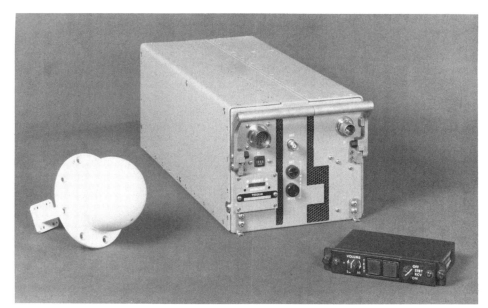

Northrop AN/ALQ-162 units

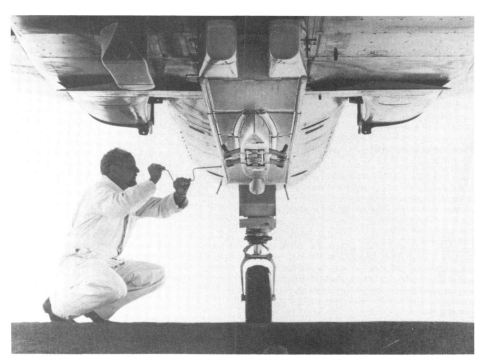

Northrop AN/ALQ-171 conformal ECM system under aircraft fuselage

tem proposed for conformal installation beneath any Northrop F-5 variant, including the F-20 Tigershark and proposed in pod form for McDonnell Douglas F-4, General Dynamics F-16, Douglas A-4 and Fairchild Republic A-10 aircraft. The unit extends along about one-third of the fuselage centre-line length, and is claimed to permit good aerodynamic performance and to be compatible with the use of any weapon or stores combination. A reprogrammable threat-processor is included in the system and modular design features have been embodied.

The ALQ-171 was launched in late 1976 when the Swiss Air Force approached Northrop to propose an ECM capability for its 66 F-5E and six F-5F fighters, also built by Northrop and delivered between 1978 and 1981. The system was tested by the US Air Force in its Electronic Warfare Environmental Simulator at Fort Worth, Texas, and completed flight demonstration and testing at Edwards Air Force Base, California, in January 1983. The programme was covered by a Swiss Air Force/Northrop agreement signed in September 1980 and run on an FMS (Foreign Military Sales) venture.

STATUS: under development.

AN/ALT-28 noise jammer

Produced for the US Air Force/Boeing B-52 strategic bomber fleet in the 1970s, this noise jammer has been combined with the AN/ALQ-155 system described above.

STATUS: in service.

Fit check of Northrop AN/ALQ-171 ECM pod on General Dynamics F-16

MIRTS modular infra-red transmitting system

This is an advanced subsonic, full-flight, internal infra-red countermeasures system under development for future deployment in a wide range of aircraft including helicopters. It utilises advanced jammer technologies including a variable optics/reflector design providing optimum aircraft infra-red signature coverage combined with state-of-the-art digital electronics and mode switching power supplies to enhance system reliability, maintainability, and versatility. MIRTS is presently being considered for installation on many different US Air Force and Navy aircraft including special VIP aircraft.

Transmitter-receiver
Dimensions: 228 × 240 × 635 mm
Weight: 52.5 lb (23.6 kg)
Power: 115 V 3-phase 400 Hz, 2.7 kVA
28 V dc, 5.6 W

Control unit
Dimensions: 190 × 259 × 318 mm
Weight: 22.8 lb (10.3 kg)
Power: 340 VA at 115 V 3-phase 400 Hz, plus
5.6 W at 28 V dc,

STATUS: under development.

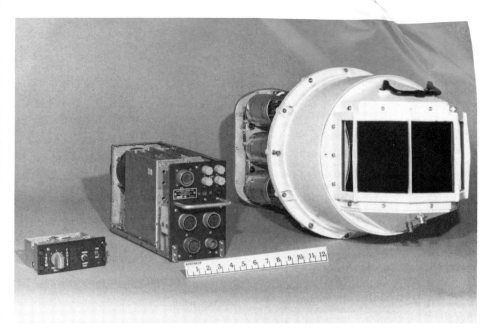
Northrop modular infra-red transmitting system

Perkin-Elmer

Perkin-Elmer, Electro-Optics Division, 100 Wooster Heights Road, Danbury, Connecticut 06810
TELEPHONE: (203) 797 6015
TELEX: 965954

Dole laser warning receiver

Dole (detection of laser emissions) is an experimental system which has demonstrated the feasibility of detecting laser emissions using a dispersed set of sensors on an airframe in the same manner as an all-aspect radar-warning receiver. Development was sponsored by the US Air Force Wright Aeronautical Laboratories, and flight evaluation conducted by the 4950th Test Wing. A combined radar/laser warning receiver set was evaluated using a Bell/Dalmo Victor AN/ALR-46A radar-warning receiver display to show threats determined by the radar and laser warning system processors.

STATUS: evaluation sets only.

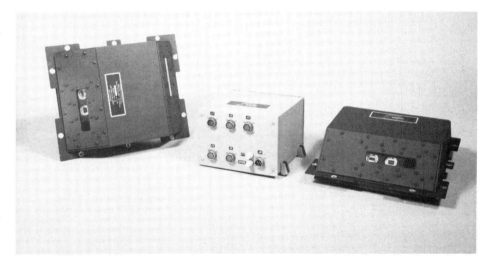
Perkin-Elmer AN/AVR-2 laser warning receiver set

AN/AVR-2 laser warning receiver

This is a laser-detection set which is integrated with the AN/APR-39 radar warning receiver system for integrated threat detection on US Army and Marine Corps helicopters. The

sensors detect incident laser radiation and identify type of threat through the video and audio channels of the radar warning receiver. The AN/AVR-2 system has broadband spectral response over a 360° field of view.

STATUS: operational testing completed. Scheduled to enter production in 1988.

Dolram multi-threat warning receiver

Dolram (detection of laser, radar and multimeter threats) is an integrated threat-detection and warning system which senses emissions over a broad band of frequencies. It has been designed to meet aircraft survivability requirements over the modern battlefield, where one or all of these types of sensor might be in use. Flight-tests completed in 1983 completed at Eglin Air Force Base under the direction of the 4950th Test Wing fully demonstrated the detection, location and identification of all threat emitters. The multi-sensor system uses integrated signal processing and a common display.

The system was developed from earlier demonstrations of a laser warning sub-system which had been integrated with an AN/ALR-46A radar warning receiver. With expanded laser warning capability and the addition of millimetre wave receivers, the system became Dolram.

STATUS: flight testing completed.

Perkin-Elmer Dolram warning system

Raytheon

Raytheon Company, 141 Spring Street, Lexington, Massachusetts, 02173
TELEPHONE: (617) 862 8600
TELEX: 923455

Raytheon handles many major EW sub-contracts. The company does not, however, hold prime-contractor status on any current large airborne EW programmes.

The most significant airborne EW system with which Raytheon is currently involved is the AN/ALQ-76 noise jamming pod (designed by McDonnell Douglas, but developed by Raytheon, which is now producing 680 production sets). This unit is used in the US Navy AN/ALQ-99 programme.

A high proportion of airborne jamming systems made in the USA uses transmitters and antennas produced by Raytheon, and the company has a substantial involvement in ground-based and seaborne electronic warfare systems. Since 1960 the company has built more than 2000 jammers.

AN/ALQ-42 ESM system

This electronic-support measures system is designed for the US Navy/Sikorsky SH-60B Seahawk (Lamps III) helicopter. It provides omni-directional azimuth coverage and has been configured to detect and track submarine transmissions. Four antennas are used, and the system will undoubtedly have a high-speed digital processor, probably incoporating a threat library to assist radar identification. No installation or performance details have been revealed.

STATUS: in production.

AN/ALQ-99 deception jamming system

This is a high-power jamming system designed for stand-off and escort jamming aircraft such as the US Navy/Grumman EA-6B Prowler and US Air Force/General Dynamics EF-111A Raven. Raytheon delivered its first airborne jamming set, an AN/ALQ-76, in 1965, and was awarded a contract for AN/ALQ-99 development in July of that year. This was developed from the earlier programme and entered production in 1968, supporting the EA-6B programme.

In May 1975 a follow-on contract from the US Air Force was awarded to support development of the AN/ALQ-99E for EF-111A operations. First production order was received in February 1979 and delivery commenced in December 1980. Total value of all AN/ALQ-99 systems built by early 1982 was $300 million.

The US Army has recently tested AN/ALQ-99E equipment in a Bell UH-1H helicopter, with an eye on potential future operation with the Sikorsky UH-60A Black Hawk. No full-scale development funds have yet been released for such a project.

The ALQ-99 on the EA-6B is carried in five external pods, each housing two high-power jamming transmitters, associated exciters, and steerable transmitting antennas (see also entries on EA-6B and EF-111A under Grumman). Various combinations of equipment can be chosen to cover the required frequency bands. The flexibility afforded by a digital computer driving control and display equipment permits the system to be used in manual, semi-automatic or fully automatic modes. In the first mode, operators search the frequency band assigned to them, identify threats and activate jammers. In the semi-automatic mode the computer identifies and indicates threats, permitting operators to initiate ECM activity. In the fully automatic mode, the computer sorts the detected signals in order of priority and activates the appropriate jammers.

The AN/ALQ-99E developed for the EF-111A has a number of improvements over the EA-6B system. They include a more efficient transmitter that can jam a greater number of radars, the ability of exciters to drive several jammers with different frequencies, and the option to select either directional or omni-directional signal propagation.

STATUS: in production.

Scarecrow jammer

Remotely piloted vehicles such as the Lockheed Aquila and Lear Siegler/Developmental Sciences Skyeye are potential applications for a new radar jammer, the existence of which was disclosed in 1984. The system weighs about 18 kg (excluding antenna) and has an output power of about 1 kW. The high-gain antenna for both transmission and reception.

STATUS: in development for the US Army, following award.

Raytheon Company, Electromagnetic Systems Division, Wayne, New Jersey 07470
TELEPHONE: (201) 628 0800

Escort electronic surveillance system

Raytheon has produced an advanced, low-cost radar detection system for maritime patrol aircraft, with emphasis on coastal protection and illicit surface-vessel acitivity such as drugs and arms smuggling. Escort (electromagnetic surveillance collector of radar transmissions) provides rapid and sensitive rf signal interception for real-time identification and classification. The system comprises a receiver, processor, magnetic tape unit and control/display with keyboard data entry, and is driven by wing-tip mounted antennas and antenna switching/pre-amplifier units.

The system provides 360° coverage in azimuth and ±30° field in elevation. Processed rf information is compared with known emitter types, characteristics of which are held in a library in the tape unit. Information made available to an operator in the aircraft can be shown on a television screen either as a situation display or in tabular form. In the first mode signals can be located as a bearing from the aircraft, as with a plan-position indicator display, and comments can be added to each track file. In the second mode the format can be tailored to the user's particular requirements.

The Escort system is generally applicable to adapted versions of business aircraft, but is particularly aimed at a variant of the Beechcraft B200 Super King Air known as the Maritime Patrol B200T. This version was chosen by Beech (a subsidiary of Raytheon) in April 1979 as the most suitable aircraft in its product line for conversion into what the company saw as a new requirement: low-cost airborne deterrence against all forms of illegal shipping movement. The substantially modified aircraft carries a comprehensive avionics suite for this purpose, including Escort and air-surface detection radar.

STATUS: a number of Beechcraft B200Ts have been sold, notably to Japan's Maritime Safety Agency, but it is not clear whether they carry Escort.

Sanders

Sanders Associates Inc, Daniel Webster Highway South, Nashua, New Hampshire 03061-0868
TELEPHONE: (603) 885 3660
TELEX: 094 3430
TWX: 710 228 189

AM-6988A POET expendable jammer

Although no details have been released, an expendable jammer which can be carried in a standard 1.4 inches (36 mm) diameter and 5.8 inches (147 mm) long cartridge is known to be in production and can be carried in many existing countermeasure dispensers. This jammer is specifically compatible with the AN/ALE-39 countermeasures dispensers.

STATUS: in production.

AN/ALQ-92 communications jamming system

This designation has been quoted as the communications jamming element of the US Navy AN/ALQ-99 tactical noise-jamming system. This particular unit may be installed only in

AM-6988A POET expendable jammer

the Grumman EA-6B Prowler carrier-borne electronic warfare aircraft, and is understood to use an antenna below the forward fuselage.

STATUS: in service.

AN/ALQ-94 noise deception jammer

This is an internally-mounted system used in US Air Force/General Dynamics F-111 FB-111 aircraft. Production of about 500 sets has been reported, and although still widely used it is being replaced by the AN/ALQ-137 system in the FB-111.

STATUS: in service.

AN/ALQ-126A defensive ECM system

Over 1200 ALQ126/126A systems were produced (see *Jane's Avionics 1985-86*), all of which have been converted to the -126A standard, which is in service with the US Navy and Marine Corps, McDonnell Douglas A-4, F-4 and RF-18, LTV A-7, and Grumman A-6, EA-6B and F-14 aircraft: the ALQ-126A has also been sold to the Royal Netherlands Air Force.

STATUS: in service but no longer in production.

AN/ALQ-126B ECM system

The ALQ-126B replaces the ALQ-100 and 126A systems in US military service, and is totally compatible with the latter. The initial production contract was awarded in August 1982 and deliveries started in May 1984: the -126B will be in production at least until 1990. The -126B equips the McDonnell Douglas F/A-18s of the Spanish, Canadian and Australian air forces.

The system is a multi-mode, power-managed, reprogrammable defensive electronic countermeasures system, the first such system in US Navy and Marines Corps service. Distributed micro-processors, microwave integrated circuits solid state amplifiers, a digital instantaneous frequency measurement receiver and advances in large-scale integrated circuits all provide state-of-the-art performance.

The ALQ-126B can act alone, or integrated with the ALR-45F/APR-43 or ALR-67 radar warning receivers, the ALQ-162 ECM system and the HARM missile, for example.

Dimensions: 411 × 270 × 609 mm
Volume: 2.3 ft³ (0.065m³)
Weight: (total) 190 lb (86.3 kg)
Power requirements: 3 kVA max

STATUS: in production and service.

AN/ALQ-137 ECM system

Forming, with the AN/ALR-62 radar warning receiver, a comprehensive internally-mounted electronic warfare system for the US Air Force/General Dynamics F-111 tactical fighter and FB-111A strategic bomber, this system is in full-scale production. It supersedes the AN/ALQ-94 noise/deception jammer, and is referred to as an advanced development, so probably retains a similar dual-channel,

Sanders AN/ALQ-126B deception jammer

US Navy/LTV A-7E Corsair IIs carried Sanders AN/ALQ-81 jamming system

continuous wave/deception, jamming capability, but with more advanced processing. Trials were conducted in 1974 and the set was first ordered in 1977 under a $24 million contract. The equipment may also be installed in US Air Force/General Dynamics EF-111A electronic warfare aircraft.

STATUS: in production.

AN/ALQ-144 infra-red countermeasures set

This is an electrically powered active IRCM system. While it was primarily designed to provide small and medium sized helicopters with protection against heat-seeking missiles, it does have applications on some fixed-wing aircraft. It is an omni-directional system consisting of a cylindrical source surrounded by a highly efficient modulation system to confuse incoming missiles. Since commencement of production in 1978, over 1300 systems have been delivered. The AN/ALQ-144 is installed on Bell AH-1, UH-1 and OH-58, Sikorsky SH-3 and UH-60, McDonnell Douglas AH-64 and Rockwell OV-10 aircraft. The set weighs 28 lb (13 kg). An approved export version (ALQ-144 (VE)) is also available.

STATUS: in production and service.

AN/ALQ-147 infra-red countermeasures set

An improved variant of the AN/ALQ-132 IRCM system developed by Sanders, the AN/ALQ-147 is designed for aircraft which cannot support the large power generation load

Sanders ALQ-144 infra-red countermeasures set mounted on upper fuselage of helicopter

associated with arc lamp or resistive source systems. Instead JP fuel is burned in a ram-air filled duct. The AN/ALQ-147 unit for the Rockwell OV-1 and Grumman RV-1 aircraft is mounted on the rear of the fuel tanks. Installation includes a filter which reduces visible emissions and makes the system suitable for night operations.

STATUS: in service.

AN/ALQ-149 communications countermeasures system for US Navy/Grumman EA-6B Prowler

The AN/ALQ-149 has been in full-scale development since 1983. Sanders Associates is serving as the lead of a three-company joint venture involving the Avionics Division and the Aerospace Optical Division of ITT.

The AN/ALQ-149 will jam hostile communications signals and long-range, early warning radars.

The system consists of two parts: an on-board portion (Sanders) containing receivers, operator controls and receiving antennas; and the external pod (or pods, depending on mission configuration) (ITT) containing the transmitters.

The on-board elements include an acquisition sub-system which has separate communications and radar intercept and processing elements that co-ordinate their activities with the AN/ALQ-99 radar jammer. The analysis sub-system acquires communication and radar signals and defines appropriate responses. Results are then transferred to the central processing sub-system which interfaces to other aircraft systems, including the on-board mission computer, thus enhancing the effectiveness of the EA-6B.

The jammer pod system has the capability and flexibility to jam simultaneously any threats identified, including prioritised combinations of communication signals and radar signals within the AN/ALQ-149's frequency range.

The original contract was awarded in 1983 for seven EDMs at a cost of $43 million.

STATUS: recently passed the critical design review. Release to manufacturing is expected.

AN/ALQ-156 missile detection system

The AN/ALQ-156 pulse-Doppler radar detects approaching missile threats with 360° of azimuth coverage, in time for an appropriate countermeasures response. This lightweight system features high reliability, minimal power consumption, solid state design and reprogrammability. Initial application was for the US Army/Boeing CH-47 helicopters. Continued development and testing has proved the AN/ALQ-156 to be an effective anti-missile defence for both fixed and rotary-wing aircraft. The system responds to surface-to-air and air-to-air missile threats by triggering the release of a decoy from an associated dispenser system.

Infra-red missiles are decoyed with use of a flare. Pulse-Doppler techniques are used to make the system immune to battlefield clutter down to nap-of-the-earth.

The AN/ALQ-156 consists of a receiver-transmitter unit, a control unit, and up to four antennas.

Weight: 50 lb (23 kg)

STATUS: in production and service.

AN/ALQ-164 ECM pod

This is the pod version of the AN/ALQ-126B DECM system with an integral Northrop AN/ALQ-162 cw jammer, designed specifically for the AV-8A/B/C Harrier. Two engineering development models were produced and successfully tested on AV-8C aircraft; further operational tests and evaluation were carried

out in mid-1985 leading to production. This system has potentially a wide application for fighter, attack, surveillance, transport and ASW aircraft.

Pod dimensions: 85 inches long × 16 inches dia (2160 × 410 mm)
Pod weight: 350 lb (160 kg)

Sperry

Sperry Corporation, Defense Products Group, PO Box 4648, Clearwater, Florida 33518
TELEPHONE: (813) 577 1900
TELEX: 523453
TWX: 810 863 5622

Jampac(V) ECM jamming system

This is a compact jamming system developed for light strike/attack aircraft and helicopter applications, and especially for export versions of aircraft such as the Northrop F-5. JamPac units can be attached to aircraft singly or in sets of up to three (typically) or four (maximum).

JamPac provides self-protection as well as combat system evaluation or ECCM training. Within each unit is a transmitter operating in the range of 0.8 to 15.5 GHz, using 30 per cent bandwidth voltage tuned magnetrons (VTMs). Two high-efficiency VTM transmitters can be accommodated in one JamPac. Power output per jammer is 150 to 400 watts cw and power-management receivers control jammer-frequency tuning. Where a radar warning receiver system is already installed in an aircraft, the power-management receivers can be deleted.

JamPac units are carried underwing or fuselage and should not interfere with the stores carriage. Aircraft performance deterioration with JamPacs attached is said to be negligible.

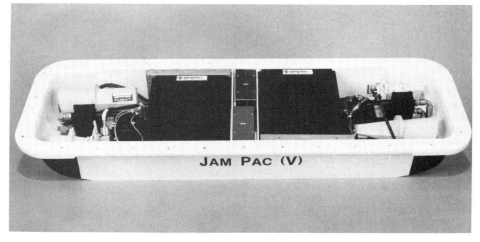

Internal view of Sperry JamPac (V) ECM jamming system showing two VTMs

Frequency range: 0.8-15.5 GHz with 30% bandwidth VTMs
Power output: 150-400 W cw
Jammers/JamPac: 1 or 2
Unit weight: 60 lb (27 kg)
Unit volume: 0.84 ft³ (0.027 m³)

STATUS: developed.

AN/ALQ-176(V) ECM pod

The AN/ALQ-176(V) is a pod-mounted jamming system designed, developed and tested to fulfil aircraft mission requirements for ECM support, stand-off, jamming (SOJ), and combat evaluation and training. The pod system comprises transmitter modules and complementary antenna, housed in cannisters to form a slim pod with a 10-inch (25 cm) diameter. The AN/ALQ-176(V) mounts on standard aircraft stores/munitions (either wing or fuselage) stations. The pod offers the option of using aircraft internal power or a ram-air turbine. The AN/ALQ-176(V) has been tested and evaluated favourably by the US Air Force, US Navy and the Canadian and Norwegian Air Forces.

The modular design of the AN/ALQ-176(V) pod allows flexibility in ECM configurations. A two or three-cannister arrangement is available. The two-cannister configuration [AN/ALQ-176(V) 1] houses up to three voltage tuned magnetron transmitters. The three-cannister version [AN/ALQ-176(V)2] provides housing for five transmitters. Dependent on frequency band and tube selection, each transmitter produces 150 to 400 watts cw (up to 30 per cent bandwidth) at efficiencies of greater than 50 per cent. The transmitters provide ECM across the

Sperry AN/ALQ-176(V)1 in two-canister configuration with ram-air turbine electrical supply and internal electronics tray showing modular design

specified threat frequencies. Transmitter protection circuits safeguard the system and provide fault status to the cockpit control panel.

The AN/ALQ-176(V) has been specifically engineered to allow rapid turnaround in meeting the demands of new threat frequencies. The AN/ALQ-176(V) can accommodate new demands through selection of different stored parameters or through easy replacement of the transmitter tube or antenna. Transmitter spares, maintenance costs, times and training skills required are greatly reduced by the use of standard transmitter design. In addition the design incorporates built-in tests and provides

for parameter/mode selection at the cockpit control panel.

The flexibility in ECM pod configuration and standard aircraft stores/munitions stations installation enables the AN/ALQ-176(V) ECM pod to be used on a wide variety of tactical and support aircraft.

Frequency coverage: VTM transmitters from 0.8 to 15.5 GHz
Transmitter power output: 150-400 W per tube cw (dependent on frequency and tube selection)
Modulation: noise (various)
Weight of transmitter module: 15 kg

Pod input/output power: (AN/ALQ-176(V)1) 2.6 kVA/1.2 kW
(AN/ALQ-176(V)2) 4.5 kVA/2 kW
Pod weight: (AN/ALQ-176(V)1) 100 kg
(AN/ALQ-176(V)2) 145 kg
Pod diameter: 25 cm
Pod length: (AN/ALQ-176(V)1) 199 cm
(AN/ALQ-176(V)2) 259 cm
Control: standard aircraft ECM cockpit control panel

STATUS: US Air Force has ordered 30 systems and additional orders have been placed by the Royal Norwegian Air Force.

Texas Instruments

Texas Instruments Equipment Group, PO Box 660246, Dallas, Texas 75266
TELEPHONE: (214) 480 1417
TELEX: 73324
TWX: 910 867 4702

AN/APR-38 radar warning computer

An unidentified model of the Texas Instrument airborne computer range is used as the radar-warning processor in the AN/APR-38 Wild Weasel system. The unit has a reprogrammable threat library.

STATUS: in service.

MPA-Jammer

In mid-1985 Texas Instruments was awarded a $5.9 million 30-month advanced development contract by the US Air Force's Aeronautical Systems Division to develop a monolithic phased array (MPA) jammer for airborne use. The use of advanced electronic technology will give the new system greatly enhanced reliability when operational.

STATUS: in development.

SAWS silent attack warning system

SAWS is an infra-red missile warning system, using advanced technologies such as infra-red search-and-track to detect and classify attacking missiles by their infra-red signature and

then alerting the pilot or automatically deploying countermeasures as appropriate.

At the end of 1985 Texas Instruments was awarded a $5.9 million contract by the US Air Force Wright Aeronautical Laboratories for the design and development of SAWS, leading to a flight qualified prototype system, including ground and flight-testing and system evaluation.

STATUS: in development.

EW suite for Tornado

In April 1986 Texas Instruments are reported as having won its first export order for electronic warfare equipment. Unspecified systems, with a contract value of $100 million are to be supplied to the Panavia Tornado programme.

Tracor

Tracor Inc, 6500 Tracor Lane, Austin, Texas 78721
TELEPHONE: (512) 926 2800
TWX: 910 874 1372

AN/ALE-29 ECM dispenser

Widely used on a variety of US Navy tactical aircraft. Each unit can accommodate up to 30 RR-129 or RR-144 chaff cartridges or Mk 46/Mk 47 infra-red decoy flares. The normal aircraft installation is two units. A cockpit-mounted programmer unit allows crew selection of countermeasure deployment, or automatic operation can be initiated by a direct link from threat-warning devices.

STATUS: in service.

AN/ALE-39 ECM dispenser

Tracor has been awarded a contract to produce this widely used dispenser at its Laredo facility. A full description of the system is given in the Goodyear entry.

STATUS: in production and service.

AN/ALE-38/41 bulk chaff dispenser

This bulk chaff system carries up to 300 lb (136 kg) of expendable countermeasure material in rolls (RR171-6). Metallised glass, aluminium chaff or aluminium-backed Mylar tape can be loaded.

The unit is contained in a podded installation and was initially developed for US Air Force/Republic F-105 Thunderchief and McDonnell Douglas F-4 aircraft. It was also adopted later for the Teledyne-Ryan AQM-34G/H Firebee RPV, which in operation as a chaff-seeding drone is known as Combat Angel. These units are currently qualified for use on F-4 Phantom, F-105 Thunderchief, LTV A-7 Corsair, Douglas A-4 Skyhawk, Northrop F5A/B and Saab J 35.

Examples have been supplied to Israel, West Germany, Republic of Korea, Denmark, Norway, Japan and other countries. Production is continuing for a variety of customers.

STATUS: in service.

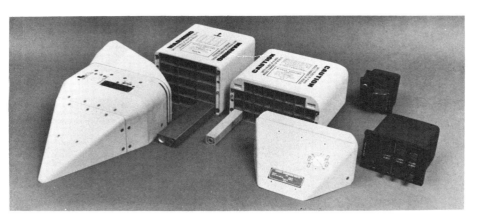

AN/ALE-40 chaff/flare dispenser for US Air Force/General Dynamics F-16 fighter

Tracor AN/ALE-40 chaff/flare dispenser for McDonnell Douglas F-4

AN/ALE-40 chaff dispenser

This first-generation system was developed to meet a US Air Force requirement calling for a high countermeasure payload capability, simple handling and loading procedures, improved decoy effectiveness over current systems, high

reliability, and a modular design which could be adapted easily to different aircraft types. Design commenced in 1974, and concentrated initially on a RR-170 chaff-cartridge-compatible unit for the McDonnell Douglas F-4 Phantom. In 1977 the system was tested with MJU-7/B infra-red flares.

The F-4 installation units are now designated AN/ALE-40(V)1, -2, -3. There are two small cockpit units, and four similar dispensers which can be attached on either side of the two inboard armament pylons. Each dispenser can accommodate 30 RR-170 chaff cartridges, or the outer modules can carry 15 MJU-7/B flares. With each rechargeable cartridge module is a streamlined nose-cap. Aerodynamic drag is said to be similar to that of a Sidewinder missile and launcher.

A smaller system, AN/ALE-40(N) was developed for the Royal Netherlands Air Force/ Northrop NF-5 fighter. In this variant there are two skin-mounted dispensers which attach to the rear fuselage, each of these carries 30 RR-170 chaff cartridges, or 15 MJU-7/B flares. Loaded weight is less than 32 kg. A similar unit has been flown on a Lockheed F-104 Starfighter.

The AN/ALE-40(V)4, -5, -6 is a variant for internal installation within the US Air Force/ General Dynamics F-16 aft fuselage. Capacity is the same as that of the AN/ALE-40(N). A similar-size system, designated AN/ALE-40(V)7, -8, -9, is available for internal mounting within the F-5E/F left wing root.

An extremely large system, AN/ALE-40(V)10 is an internal installation for the US Air Force/Fairchild A-10 aircraft. It comprises up to 16 dispensers holding as many as 30 cartridges each; 480 RR-170 chaff cartridges or 480 M-206 infra-red flares constitute the maximum capacity. Four dispensers are mounted in each wing-tip and each wheel-well.

Other variants of the AN/ALE-40 are operationally deployed on British Aerospace Hunter, Aeritalia G-91 and Dassault Mirage. All systems have cockpit control-units tailored to individual requirements, and which allow pilot selection of salvo/burst rate, ripple rate, and so on. All units show the number of cartridges remaining in each dispenser.

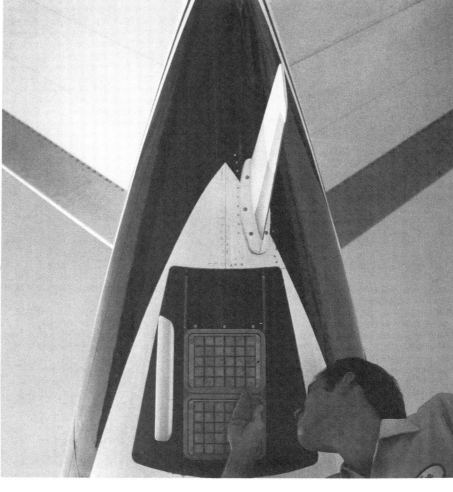

Tracor AN/ALE-40 chaff dispenser in rear fuselage installation

	F-4 pylon	F-16 internal	A-10 internal
Dispenser length:	(outboard) 528 mm	216 mm	193 mm
	(inboard) 348 mm		
Dispenser depth:	123 mm	269 mm	260 mm
Dispenser width:	241 mm	170 mm	147 mm
Load weight:	(120 chaff) 48.1 kg	(60 chaff) 22.5 kg	(480 chaff) 158 kg
	(60 chaff + 30 flares)	(30 chaff + 15 flares)	(240 chaff + 240 flares)
	58 kg	25.2 kg	168 kg
		(30 flares) 27.9 kg	(480 flares) 177 kg

Power (standby): 0.1A at 28 V dc
(dispensing at 100 ms intervals) 6A at 28 V dc

STATUS: in production.

AN/ALE-40 TACDS threat adaptive countermeasure dispenser system

The advanced-generation AN/ALE-40 unit is a 'form, fit and function' replacement for the AN/ALE-40(V) dispenser, and embodies solid-state microprocessor technology with threat-adaptive programs. The latter accepts information from the aircraft radar-warning receivers, air-data system and throttle transducers to determine the optimum deployment of expendables such as flares or chaff.

The following operating modes are available: off/standby, manual, semi-automatic and fully automatic. Mode selection is made by the crew on a cockpit control/display unit. An emergency jettison mode is available to dispense all expendables on command, the manual mode

requires pilot pre-selection, and automatic mode selects the appropriate expendables as well as the optimum dispensing rate. In semi-automatic mode the system requires the pilot to initiate an automatically optimised routine.

The dispenser includes growth capabilities that will allow threat characteristics to be rapidly updated, and is designed for rapid retrofit installation on existing AN/ALE-40-compatible aircraft. Engineering development is complete and production began in 1984.

STATUS: in production.

Quick Fix IIB countermeasures helicopter

Under a $51 million contract awarded in October 1984 by the US Army, three Tracor companies will combine forces to modify a standard US Army/Sikorsky UH-60A Black Hawk helicopter to incorporate electronic surveillance and active countermeasures

equipment. The aircraft will be designated Quick Fix IIB EH-60A and the three-year programme will include production and installation of equipment, integration and flight-test. The contract is being administered by the US Army's Electronics Research and Development Command, the responsible body being the Signal Warfare Laboratory in Warrenton, Virginia.

STATUS: prototype evaluation programme

AN/ALE-45 countermeasure dispenser

The ALE-45 second-generation threat-adaptive countermeasure dispenser, designed for the US Air Force/McDonnell Douglas F-15 Eagle fighter, provides upgraded, semi-automatic and automatic operation.

STATUS: in production.

AN/ALE-47 countermeasure dispenser

A follow-on from the AN/ALE-45, the ALE-47 it was the subject of a $3.9 million, 17-month development contract placed by the US Air Force Aeronautical Systems Division with Tracor in February 1984. It will be compatible, box for box, with the earlier AN/ALE-40 to permit retrofit.

STATUS: in development.

US Air Force/Navy

United States Air Force, Aeronautical Systems Division, Air Force Systems Command, Wright-Patterson Air Force Base, Ohio 45433

INEWS integrated electronic warfare system

Described as being 'a system of potentially revolutionary impact', the INEWS integrated electronic warfare system is a long-term joint-services programme being tailored by the US Air Force and US Navy to the next generation of

combat aircraft, notably the US Air Force's ATF advanced tactical fighter and the US Navy's advanced tactical aircraft (ATA). The Rockwell B-1B and Northrop Advanced Tactical Bomber would also be candidates. Its purpose is to protect these aircraft in the presence of the severe electronic environment expected in the next decade. In an address to the US Air Force's 1983 annual symposium, the head of the US Air Force Aeronautical Systems Division (which is responsible for INEWS), Lt Gen Thomas McMullen predicted that the new system would 'Go beyond the capabilities of current airborne electronic warfare defensive systems to provide integrated, multi-spectral warning and automatic countermeasures capability for the total electromagnetic threat. Advanced technology and integration concepts will be required to yield the necessary performance in the projected environment of high signal density'.

With an interest in the frequency range from dc to that of light, INEWS is specifically considering for the first time, taking the capabilities of an airborne EW system into the electro-optical region of the spectrum, and will be able to combat, for example, laser designator devices. It will exploit a new technology known as monolithic microwave integrated circuits (mmic) that permits operation at very low voltages. In conjunction with advanced antenna-array technology, mmic will enable the construction of solid-state jammers with an effective radiated power of more than 10 kW. They will be small and light and will be less expensive and more reliable than current travelling-wave tube amplifiers that need nearly a thousand times greater driving voltages. The system will also employ vhsic (very high speed integrated circuit) technology, under development by the US Defense Department and industry, to achieve faster processing speed and the miniaturisation necessary to accommodate it inside the aircraft rather than on external pods.

The capabilities of INEWS will extend beyond those of even the AN/ALQ-165 ASPJ airborne self-protection jammer (due to enter service in 1986), although the latter may eventually incorporate some of the technology under development for the new system. INEWS will be modular in nature, embodying Pave Pillar technology, so that equipment may be optimised to specific aircraft or missions. Unlike ASPJ and many of the earlier systems that had to be carried in external pods, INEWS will not be on 'add-on' or retrofit system. It will be a 'core' system, designed into the aircraft from the earliest stages, and will be permanently tied in with other systems and equipment by way of

a digital or optical data-bus. It is the degree of integration with other aircraft systems that sets INEWS apart from previous EW suites. Opinion among the US Air Force and the companies involved is that the system will demand a wider range of technological talent than any previous defence electronics system; the programme is very largely guided by competitors' proposals.

Six industrial teams submitted bids in February 1984 for the first phase of a competition to develop INEWS. They are: TRW, Westinghouse and Honeywell; Hughes Aircraft and Loral; Texas Instruments, ITT and Litton; Eaton, E-Systems and IBM; Sanders Associates, General Electric and Motorola; and Raytheon and Northrop. Five of the teams were chosen in July 1984 under $3 million contracts to proceed with Phase 1A definition studies lasting 18 months. Two teams will be chosen to go forward to a 31-month Phase 1B period, with an individual team allocation of $55 million, following a Defense Systems Acquisition Review Committee scheduled for March 1986. The programme is being administered by the US Air Force, Wright-Patterson Air Force Base, Ohio.

At this stage the total cost of INEWS cannot be gauged with any accuracy. Tentative figures varying all the way from $3500 million to $50,000 million (the latter accompanied by a research and development budget alone estimated at $1 000 million) have been surmised.

The objectives of Phase 1B, due to be completed by November 1988, include: advanced studies and analysis to determine how INEWS can be specifically adapted to the ATF and ATA projects; establishment of risk-reduction programmes on specific hardware; to develop hardware and software technologies; the start of lead-in planning for the full-scale development programme, due to start in November 1988, when one of the two Phase 1B teams will be awarded a Phase 2 contract, the overall aim being to reduce the number of outstanding problems at the critical design review.

INEWS is a sub-set of the US Air Force's Pave Pillar integrated avionics architecture programme (see separate entry), although there is little overlap in technology between the two programme offices.

Specifically, the five teams in Phase 1B were: Loral and Hughes Radar, with BDM, Sperry and

Boeing as sub-contractors; Raytheon and Northrop, with AT&T, GTE, Magnavox, General Research, Boeing and Tracor; TRW Westinghouse, with Honeywell, Perkin-Elmer and Tracor; ITT and Litton, with Texas Instruments and Rockwell; Sanders and General Electric, with Teledyne, Bell and Dalmo Victor.

These teams are listed, since, for the losers it is accepted by most observers that there will be little new airborne electronic warfare work before the year 2000, if at all. INEWS is so big and so comprehensive a project that it will be virtually the only airborne EW programme in the USA, apart from some minor updates, once it moves into full-scale development.

Work on the development of a INEWS-related computer, using vhsic technology began early in 1985 when Westinghouse and TRW were awarded contracts totalling $23 million. Using 1.25-micron vhsic technology, the computer will execute the MIL-STD-1750A computer architecture at 3 million instructions per second. The computer is also destined to form part of the Ultra-Reliable Radar (URR) project and a common signal processor concept. INEWS will also incorporate Ada and artifical intelligence.

The full-scale engineering development phase, which follows Phase 1B in November 1988, will last four years, and will involve flight testing breadboard systems, most probably on a McDonnell Douglas F-15 and/or General Dynamics F-16. A production contract, with each winning team member acting as an independent production source, is scheduled for 1993.

STATUS: in Phase 1A definition studies.

Silent attack warning system

In Summer 1985, General Electric, Honeywell and Texas Instruments were awarded US Air Force contracts for the initial 15-month design effort into the Silent Attack Warning System, a non-radiating system which will warn pilots of missile attacks. After this initial phase, one or two prototypes will be built.

In this phase, Honeywell was awarded $4.5 million, Texas Instruments $5.9 million and General Electric $1.8 million.

Westinghouse

Westinghouse Electric Corporation, Aerospace and Electronics Division, PO Box 746, Friendship Site, Baltimore, Maryland 21203
TELEPHONE: (301) 765 1000

AN/ALQ-101 noise/deception jamming pod

This is one of the US Air Force's most widely-used jamming pods, and it has also been supplied abroad. The project began in 1966 and was referred to as Sesame Seed. Five examples of a development pod, designated QRC-335A/101-1, were developed, each with fore and aft facing antennas and operating in the 2 to 8 GHz frequency range.

The first production units, AN/ALQ-101(V)3, were more powerful versions of the prototypes, but were soon followed into production by the AN/ALQ-101(V)4, which covered the 2 to 20 GHz frequency range. A further frequency band extension was added in Model AN/ALQ-101(V)6.

The final production version, and the most numerous unit in service, was the AN/ALQ-101(V)8. This was the first electronic countermeasures pod to use a gondola layout, where a trough compartment underneath the main body considerably increases the available volume in the pod, without significantly increasing the cross-sectional area. A further frequency-range extension was incorporated

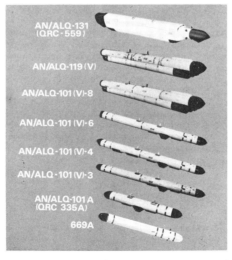

Westinghouse noise/deception jammers in US Air Force service

in the system, and many earlier production pods were later updated to this standard.

AN/ALQ-101 pods are also in service in the UK, Israel and West Germany.

Production was as follows:

QRC-335A/101-1	25 prototypes
ALQ-101(V)3	171 pods
ALQ-101(V)4	324 pods
ALQ-101(V)6	158 pods
ALQ-101(V)8	400 + pods

Westinghouse AN/ALQ-101(V)8 ECM pod

Pod length: (QRC-335A/101-1) 7.5 ft (2.3 m) (all later models) 12.75 ft (3.9 m)
Pod width: 9.8 inches (250 mm)

STATUS: in service.

AN/ALQ-105 noise jammer pod

This was a repackaged AN/ALQ-101 system which effectively split the original pod in half and positioned the two resulting units on either side of an EF-105 Thunderchief fuselage. About 90 examples of this system were produced.

STATUS: in service.

AN/ALQ-119 noise/deception jamming pod

Initiated as project QRC-522 in 1970 and now one of the most numerous jamming pods in service with the US Air Force, AN/ALQ-119 was one of the first dual-mode (noise and deception) jammers to appear. It was used initially on the McDonnell Douglas F-4 Phantom, but was subsequently adopted for the Fairchild A-10 and General Dynamics F-16 as well as by the F-111 and McDonnell Douglas F-15. The system has a three-band frequency-range transmitter which covers the terminal threat range. Both noise and deception jamming modes can be employed. Each pod has dual-mode travelling-wave tube emitter elements. The pod has the gondola cross-section introduced by Westinghouse on the AN/ALQ-101 pod.

Recent producton has been of the AN/ALQ-119(V)15 model, which is recognisable alongside earlier versions of the pod by the addition of a radome below the front end of the gondola portion. Earlier versions of the pod with the US Air Force and other air forces are being upgraded to this standard. Features of the (V)15 version include automatic control of power radiated, frequency selection and signal type. A shorter-body version, the AN/ALQ-119(V)17, is also operated by the US Air Force.

The total production run exceeds 1600 units at production rates of between 10 and 30 sets per month. Sets are operated by Israel and West Germany. It will be superseded, in US Air Force service, by the AN/ALQ-131 system.

With the appearance of new Soviet air-to-air and ground-to-air threats, the US Air Force in the early 1980s instituted a major improvement programme, designated AN/ALQ-119A (Seek Ice). Raytheon was appointed to strip out and replace much of the existing electronics and incorporate current Rotman lens technology. The changes will facilitate reprogramming and make for greater reliability and ease of servicing. To date several bench-test systems have been built. The several versions of ALQ-119A with the US Air Force are currently being superseded by the ALQ-131A modular jamming system.

STATUS: in service with F-4, F-16, F-111 and A-10 aircraft. The ALQ-119 is the mainstay of US Air Force's electronic warfare suites.

AN/ALQ-131 noise/deception jamming pod

The first development contract for this system, let by the US Air Force in 1972, called for an advanced noise/deception jamming pod covering five wavebands, and incorporating a digital processor for direct control of the emitter elements. The specification also required a modular design suitable for extension or contraction to meet specific application requirements. The standard pod appears to comprise a single processor and two independent fore and aft jamming sets. Motorola was sub-contracted to provide the two lower band jamming units, and Loral developed the processor. Total waveband capability may exceed the range 2 to 20 GHz.

Operationally the pod can be tailored to meet the waveband requirements of differing missions, and the processor is reprogrammable on the flight-line to take account of revised threat priorities, or new threats as they appear. The AN/ALQ-131 is claimed to be the first self-protection system to be programmable on the flight-line.

Westinghouse AN/ALQ-119 jammer on outboard pylon of US Air Force/Fairchild A-10 aircraft

Although essentially a podded system, the AN/ALQ-131 can also be arranged for internal installation. The pod is built in modular canisters which provide structural support, cooling and environmental protection to all associated equipment. The centre of each module is an I-beam which carries loads and cooling supplies. The latter uses a freon-to-air supply which has no moving parts and needs no electrical power for operation. By varying module combinations, 16 different equipment configurations can be accommodated. The main module in any configuration is the interface and control unit which contains the signal processor and threat library. New threat data can be loaded on the aircraft in less than 15 minutes. There is also a digital waveform generator which can produce up to 40 simultaneous waveforms for deception-jamming modulation. Reliability continues to grow steadily, field data indicating a mean time between failures of more than twice the specified figure.

Following intensive pre-production, the initial AN/ALQ-131 production contracts were let in 1976, and by 1980 production was being funded annually for well over 100 systems. Production rates of more than 13 a month have been achieved, and more than 650 of the 900-plus systems had been delivered by March 1986. Total sales exceeding 1500 systems are anticipated. In addition to the United States, the ALQ-131 has been bought by the Netherlands, Japan, Egypt and Pakistan.

Main current application, however, remains the US Air Force/General Dynamics F-16, on which it is likely to be superseded by the ITT/Westinghouse AN/ALQ-165 airborne self-protection jammer (ASPJ). The US Air Force initiated an advanced version of the ALQ-131, and production deliveries have begun. Performance will be further enhanced by the inclusion of an additional power-management module. The CPMS (comprehensive power management system) in this upgrade is derived directly from the US Air Force/US Navy ASPJ programme, and most of the CPMS equipment is identical with it. This CPMS completed acceptance testing and integration with the ALQ-131 in early 1985.

Aircraft using the AN/ALQ-131 include General Dynamics F-16, F-111, McDonnell F-4 Phantom, LTV A-7 Corsair, Fairchild A-10 and Lockheed AC-130 Hercules. As an indication of cost, an Egyptian contract for 40 pods placed in February 1983 was stated to be worth $29 million.

As an indication of the system's importance, it is one of a very small number of tri-service systems to be chosen as a flight-test article for vhsic technology. In fact a standard ALQ-131 jammer is currently being modified to carry a signal-processing computer incorporating vhsic chips, and will be the first airborne system to fly with these ultra-fast microcircuits. The US Air Force plans to have the system ready for test during 1987.

Pod length: 112.2 inches (2850 mm)
Pod depth: 25 inches (639 mm)
Pod width: 11.8 inches (300 mm)
Weight: 561 lb (255 kg)

STATUS: in production. The system is operational with Tactical Air Command and US Air Force Europe. The updated ALQ-131 Block II system completed functional performance verification in mid-1985. In early 1986 the US Air Force began a series of stringent reliability and maintainability tests on the Block II ALQ-131, following problems in similar areas with the Block I systems.

A detailed text programme on the Block I systems, involving 57 000 hours of testing up to October 1985, showed a mean time between repair of 45 hours, compared with a specified figure of 19.3 hours.

In 1983/84 Westinghouse received orders to produce 220 Block II systems for the US Air Force and 34 for the Royal Netherlands Air Force. The 1985 fiscal year procurement request was rejected by the US government because of the reported reliability problems and, as of March 1986, the 1986 fiscal year procurement was still being discussed – the US Air Force's request was for $124.6 million for 96 systems, only $48.6 million was authorised but $84.6 million was appropriated. A total of 50 Block II systems had been delivered to the US Air Force by March 1986.

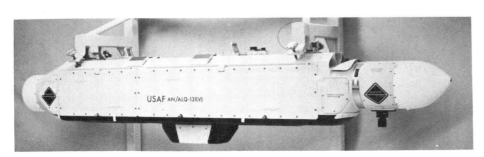
Westinghouse AN/ALQ-131(V) ECM pod

AN/ALQ-153 tail warning radar

Development of this system began in 1975, ground-tests were completed in 1976 and a competitive fly-off against a similar set took place in 1976 and 1977. The AN/ALQ-153 was selected in 1978 and was followed by full-scale production.

This is a pulse-Doppler radar that increases protection from rear-closing threats, either aircraft or missiles, and will be a prime sensor for countermeasure deployment and crew warning in the US Air Force/Boeing B-52G and -H Stratofortress aircraft.

It has been stated that the radar met or surpassed all technical requirements during evaluations, exhibiting no false alarm problems at low or high altitude. The unit demonstrated 99 per cent probability of detecting threats at or beyond specified range. Equally important, introduction of the ALQ-153 has been claimed to cut dramatically the false alarm rate.

STATUS: production complete for US Air Force/Boeing B-52G. Deliveries started in April 1980 and 321 Boeing B-52 aircraft were equipped by the end of 1985.

AN/ALQ-165 ASPJ airborne self-protection jammer

Westinghouse and ITT Avionics, as a joint-venture team, were awarded a contract to develop the AN/ALQ-165 system in late 1981.

STATUS: in development (see ITT/Westinghouse entry).

Data processing, management and displays

The McDonnell Douglas MD-80 twin-jet is the first aircraft to introduce a digital flight guidance system into commercial airline service. The standard flight deck features a performance management system and optional flight management system adds lateral guidance to the pitch capability for even greater fuel savings. An optional head-up display, shown stowed above the captain's windscreen, provides necessary operational data for landings, take-offs or go-arounds in either visual or instrument conditions. Airlines also can select an optional EFIS to replace electromechanical flight, navigational and engine instruments. In 1987 McDonnell Douglas will certify a new windshear sensing system which can be installed on any MD-80 or DC-9 twin-jet.

Data handling

INTERNATIONAL STANDARDS

Programming Languages

General-purpose processors have to be pro-grammed to do specific tasks and whereas, until about a decade ago, all airborne process-ing was so specialised that programming instructions were written in 'assembler' or 'machine' code (which are machine-dependent techniques, and very time-consuming), most processors today can be programmed using a high-order language.

The high-order programming language used to generate most software in the engineering industry, and familiar to many computer users, is **Fortran**. This is a recognised language in the USA (Fortran 78 is defined in MIL-STD-1753), but its aeronautical use has generally been restricted to non-airborne, software-develop-ment tasks.

Programming with high-order language has many advantages over lower-level languages.
(a) The program structure is clearly visible, both to the software designer and subsequent users, so modifications to it can be accommo-dated more easily.
(b) Most tasks can be accomplished with fewer statements, so increasing programming pro-ductivity.
(c) Programmers are easier to train and can become familiar with a high-order language much more rapidly than with any lower-level programming method.
(d) The program should no longer be machine-dependent. It can be loaded into any machine which has the appropriate compiler facilities, and therefore software development and testing can be conducted before airborne hardware is available, providing the host machine has similar operating capabilities.

Notwithstanding these advantages, and in apparent contradiction to (b) a high-order language program, after compilation, takes up considerably more machine storage space than a program written in a lower-level language; neither is it as efficient, in that it operates less rapidly. These disadvantages have been more than offset by the enormous increase in storage and operating speed made available in airborne processors in the last decade. High-order language software applications have therefore been very closely associated with technical improvements in processor hardware.

A great number of processor/software combi-nations are offered by the manufacturers of digital equipment used in commercial aircraft, largely because these are sold as discrete systems with relatively well-tried interfaces, or customised analogue discrete interfaces to other systems. Many companies, for reasons of commercial security and because programs are relatively small, find it advantageous still to use machine-dependent software techniques. This situation is unlikely to last indefinitely and commercial software rationalisation could follow the standardisation example being set by the military, where currently there are several languages (typically a nationally-preferred high-order language in each major nation), but an internationally-recognisable high-order language is being developed. A review of high-order languages in use in several nations, and a review of the new internationally-recognised high-level language are presented below.

France

French military suppliers are encouraged to prepare software in **LTR**. This is a high-order language recognised by the French Army, Air Force and Navy, and a root compiler is maintained by the French military authorities. It is available at no charge to suppliers who are required to furnish LTR software to French military projects. No notable users of this language exist outside France, but it will remain the preferred high-order language for future French military software.

Germany, Federal Republic

Suppliers of programmable military items to West Germany are encouraged to use **Pearl,** a high-order language developed and used ex-clusively in West Germany. It employs a slightly different approach to most other high-order languages, the specification having two sec-tions. One is the problem division, which defines all functions, algorithms, etc, and is designed to be compatible with all types of processors. The system division describes the communication with hardware and program interfaces, so it is machine-dependent.

Although support for Pearl is strong in West Germany, it is being superseded by the Ada high-order language.

United Kingdom

The preferred British high-order language is **Coral 66** (British Standard 5905). Although well documented and supported by the UK Ministry of Defence for Army, Air Force and Navy use, only a small proportion of Coral experience has been accumulated on airborne hardware. De-spite this, it is probably still the most well-established European high-order language.

The Royal Signals and Radar Establishment (formerly the Royal Radar Establishment) at Malvern, with Ministry of Defence support, has developed a support facility called Mascot (modular approach to software construction, operation and test). By pools/channels con-cepts it aids visualisation of programming requirements. It can also be used to make automatic checks, ensuring the consistency and completeness of programs. The portability of programs between processors is also con-siderably simplified.

Coral/Mascot is being superseded as the preferred high-order language by Ada.

United States

The bulk of all types of military digital processor experience has been with US projects, and the US Department of Defense operates a list of approved languages and control agents. These are:

Language	Control Agent
Fortran	Assistant Secretary of Defense (Comptroller)
Cobol	National Bureau of Standards
Tacpol	US Army
CMS-2	US Navy
SPL/1	US Navy
Jovial J3	US Air Force
Jovial J73	US Air Force
Ada	Department of Defense

The languages used most frequently for airborne applications have been CMS-2 and Jovial J3/J73. CMS-2 is used in current US Navy computers, eg AN/AYK-14 in the F/A-18 fighter and Lamps III helicopter and has, generally, a wide range of support facilities. Jovial probably represents the most exten-sively-used airborne high-order language, and Jovial J73 is defined in MIL-STD-1589B. Jovial was devised by Jules Schwartz of the US company Systems Development Corporation in the late 1950s/early 1960s. (Jovial is an acronym standing for Jules Own Version of the International Algebraic Language.) It is a language especially suited to real-time process-ing applications and was developed largely from Algol. Although now superseded by Ada, Jovial remains in service: the principal airborne applications at the present time are the General Dynamics F-16 fighter, and the Rockwell B-1B bomber. However, over the past few years Ada, a new high-order language, has been developed and is now the preferred computer language throughout NATO (including France and the USA).

Ada high-order language

Since January 1975, when the US Department of Defense formed a high-order language working group with representatives from each US military agency and various Department of Defense agencies, there has been strong pressure for the development of a high-order language, available for world-wide use.

Development continued between 1975 and 1978 via stages referred to as 'Strawman', 'Woodenman', 'Tinman', 'Ironman', and 'Steel-man'. After evaluation of languages available, world-wide, a predominantly French-used high order language, developed by CII-Honeywell-Bull, was selected as most able to meet the 'Steelman' requirements. This forms the basis of a new internationally-recognised high-order language, referred to as **Ada**. It is defined in MIL-STD-1815A. By registering the program name as a trademark, the US Department of Defense intends that only a limited number of language compilers will be approved, thereby ensuring inter-operability of programs between computer types; some 350 programming languages are currently in use in US military equipment.

The first reference document, *Reference*

Northrop F-20 is first aircraft to fly with operational mission computer programmed in Ada

Manual for Ada Programming Language, was published by the US Department of Defense in July 1980. This has been superseded by a definitive ASCII-standard manual and ISO-Standardisation documents were published in 1984. In late 1983 the US Under Secretary of State for Defense announced that Ada would be mandatory for 'embedded' computer applications in projects entering advanced development after January 1984.

An Ada language compiler developed by Rolm Corporation and Data General was the first to be approved by the US Department of Defense, and became available for the Rolm Hack/32 computer in late 1984. Certification involved 1776 separate tests. Several Ada compilers are now being marketed and, although the number of compilers is expected to rise rapidly, only a small number of them will have official approval. The remainder will be software research and development tools.

American development has concentrated on two independent projects, managed by the US Army and Air Force respectively. Both forces intended to produce approved software for major military system applications and dual team development was felt to be more likely to produce an optimum language than the cheaper, single-team alternative. The US Department of Defense intends that both teams should develop an Ada programming support environment (APSE) to assist program writing in the same way as Mascot supports Coral in the UK. The Army project, ALS (Ada language system), is being developed by SofTech, with Honeywell as the major sub-contractor. The Air Force project, AIE (Ada integrated environment), is being conducted by Intermetrics Incorporated, with Massachusetts Computer Associates as major sub-contractor.

By late 1983 it was clear that the ALS and AIE projects would not yield mutually-compatible software. As the Air Force project was the least advanced, its development team was ordered to study and incorporate interface compatibility requirements to preserve the objective of total software commonality. This led to formal definition of a Kernel APSE (KAPSE) in 1985.

The ALS project is using the Digital VAX family, Motorola M6800 microprocessor and Nebula family of computers as host machines. Equipment hardware addressed by the Air Force is the IBM 370/VM and all proposed MIL-STD-1750A computers. The following companies are supported by the US Department of Defense to develop Ada compilers and software support tools: Digital Electronic Cor-

poration, Intel, Rational Machines, Rolm Corporation, TeleSoft Incorporated and Western Digital Corporation.

The US Air Force plans to introduce Ada in four stages via the ALS project. It will apply Ada to VAX-hosted MIL-STD-1750A computers, such as the CDC Cyber and Zilog Z8000, and the first airborne equipment likely to have an Ada-programmed processor is AMRAAM, the Hughes AIM-20 advanced, medium-range air-to-air missile due to enter service in 1986 on the General Dynamics F-16 fighter. An operational system will then be developed alongside an existing Jovial J73 language, so that operational experience (without the associated risk in-service) is made available for existing sub-contractors. Thirdly, Ada will be used as the primary language on a small, low-risk, program. The target date for its introduction as the mandatory language for all airborne and associated processor applications is 1986/87.

Ada is likely to be made available for the US on 10 types of host computer, and 20 different target computers. Features of the language which will contribute to software interchangeability include: structured programming to divide programs into manageable sub-programs; language mechanisms that recognise the 'typing' or category of variables, eg integer, fixed-point, floating-point, Boolean, Programmer (a choice of categories is permitted); technique called 'packaging' which provides a non-volatile program portion for which entry and exit is through 'scaled' interfaces; a program module called a 'task' to define and control concurrent operations, and oversee program operation synchronisation; program portions called 'generic' units, which are easily repeatable when there are many similar operations.

Ada is not an easy language to learn; training in the US in 1983 comprised 30 to 45-day courses, using experienced programmers, over a six-month period. Training issues were highlighted in SofTech's training schedule for the US Army. Ada requires not only skilled programmers with, for example, Pascal or Coral expertise, but also an understanding of the trade-offs involved with equipment, and of performance constraints. It is also not clear that Ada will be really portable; whether for example the Ada processing of sensors on a US Air Force tactical fighter could be used by similar sensors on an Army tank; or even if nominally the same, it can be guaranteed identical in all respects. There is a body of opinion that holds the view that total portability cannot be justified

as a blanket requirement across the entire spectrum of potential applications.

Defence authorities in all Western European countries have indicated a willingness to adopt Ada, but the UK Ministry of Defence considered the target dates set by the US Department of Defense to be very optimistic.

Criticisms of the language stem partly from the enormity of the task that is being undertaken. Although Ada's structure was initially seen as able to cater for all the existing specialities, it is possible that it will not be sufficiently versatile to be made compatible with current intelligent knowledge-based system (IKBS) developments.

These new requirements may institute what could be a continuous development programme for Ada, resulting in several versions of the language. The approval authorities believe that although software development costs may escalate in the future, the in-service savings will considerably offset this development outlay, particularly because only one 'set' of recognised compilers will be available worldwide.

In the past years there has been an increase in activity in Ada validation, a number of companies within NATO countries having reached milestones in their respective development programmes involving this high-order language. A survey published in 1984 predicted that the market for Ada would grow to be worth $900 million by 1989, but that includes all military applications, not merely programming for avionics equipment.

Market leaders in the USA are TeleSoft, and Rolm, both of which have had Ada compilers validated, the former on a MC-68000 microprocessor, the latter in conjunction with MV-100 processors. Rolm has also announced the first 16-bit Ada compiler, targeted for the AN/UYK-64(V) computer.

In the UK, the Ministry of Defence has set a particularly tight timescale, 1987, for the implementation of Ada in all mission-critical software and, while this target might slip, Ferranti and CAP, for example, are collaborating to develop the TeleSoft compiler for use on the Ferranti Argus M700 processor. The development work was completed in early 1985 and the compiler has been submitted to the US Department of Defense for validation. In advance of this, however, is GEC Software, whose Verdix compiler is already Department of Defense validated.

West Germany, Spain, Israel and Denmark are also pressing ahead with the implementation of Ada. The West German company IABG obtained validation for its compiler towards the end of 1984; this programme has the support of the West German Defence Ministry.

Meanwhile the United States is well in the lead, and in September, 1984, the US Air Force/McDonnell Douglas F-15 became the first aircraft in the world to fly with a mission-critical system – in this case the flight-control computer – programmed in Ada. The F-15A Ada development is a part of the DEFCS (digital electronic flight control system) programme, a joint undertaking between the US Air Force's Flight Dynamics Laboratory and McDonnell Douglas (see also under Flight Control). The compiler was supplied by Zilog and built by Irvine Computers. The F-15's Lear Siegler flight control system uses four Zilog Z8002 microprocessors.

Finally, in December 1984, Northrop claimed that its private-venture F-20 fighter had become the first aircraft to fly with an operational mission computer programmed in the new language (the F-15 system being for trials only).

In September 1984 the McDonnell Douglas F-15 Eagle became the first aircraft to fly with mission-critical system programmed in Ada

Data Transmission Standards

As digital processors have replaced analogue equipment there has been increased digital data transmission to and from equipment. Whereas individual links (cables) are necessary to carry separate data in an analogue system, the same information can be sent as a sequence of data-words using digital data transmission. This has led to the definition of internationally-recognised digital data transmission standards to ensure that individually manufactured systems have compatible interfaces. A résumé of these standards, together with some of the more common analogue characteristics, is given below.

ARINC 407

This characteristic defines the standard four or five-wire connections between synchro transmitters and receivers.

ARINC 419

This document brings together the quite widely varying standards used in individual avionic systems.

ARINC 429

This is a single-source, multi-sink uni-directional data transmission standard. Digital avionics equipment suppliers are the almost exclusive users in the commercial airliner field, but where equipment commonality runs across into military applications, the standard can be applied to that category of aircraft. ARINC 429 was agreed among the airlines during 1977/1978, at the time that the specification for the first new-generation, digital airliners–the Boeing 767 and 757 – were being fixed, it describes the MK 33 digital information transfer system.

An ARINC 429 data-bus consists of a single, twisted, shielded-cable pair, with the shield grounded at both ends and at all breaks. One end of the data-bus is at the transmitting line-replaceable unit, and there may be up to 20 receiver ends, each configured as a 'stub' on the basic data-bus. Only one system may transmit on a particular data-bus, though many others may receive that information.

Information can be carried in one of two data-rate bands. The high-speed data-rate band is 100 K bits a second (±1 per cent) and the low-speed data-rate band is 12 to 14 K bits a second (±1 per cent on selected value).

Modulation is RZ-polar, which means that three states can exist: hi, null or lo. In any bit interval, a hi-state returning to a null-state represents a logic '1' and a lo-state returning to a null-state represents a logic '0'. Hi or lo states can be between ±13 volts to ±5 volts, and null can be between ±2.5 volts.

All information is sent as 32-bit words, with a parity-bit included (odd parity required). The identity of each word is transmitted first and a gap between words equivalent to at least 4-bit time is necessary.

In any 32-bit word bits 1 to 8 are a label, and bits 9 and 10 are a source/destination identifier. Bit 32 is the parity digit, and a sign/status matrix precedes this, being either 2 or 3 bits (bit numbers 31 and 30, or 31 to 29) depending whether data is ISO-alphabet, BCD or binary. Bits 11 to 28 or 29 (18 or 19 bits in total) are available to carry actual data. Special word formats can be provided, the most commonly used being latitude/longitude data. The ARINC standard defines such details together with the mandatory labels assigned to all data.

Each ARINC 429 transmitting unit on an aircraft has an associated data-bus. On the commercial aircraft such as the Boeing 767 and 757 and Airbus Industrie A310. Advantages of the ARINC 429 bus characteristic are that it is in large-scale use and is now a well known system, that many receivers can be loaded on to a single bus, that it uses relatively inexpensive components, and is simple to execute in terms of hardware. It is also inherently reliable. Disadvantages are that the system operates in

one direction only (each box acting only as a 'sender') so that twice the number of cables are needed to equal the performance of a 'two-way' system, such as MIL-STD-1553. Also, the system is considered by some to be dated by comparison with current digital technology. There may be as many as 100 ARINC 429 data-buses.

ARINC 453

This standard relates to a very high-speed data-bus used specifically to supply data from a weather-radar receiver unit to its associated display. Pulse characteristics are similar to ARINC 429, but users are not constrained to a 32-bit word format.

ARINC 547

This defines the characteristics of the signal between a VOR and the compass director, horizontal situation and radio magnetic indicators, and autopilot systems.

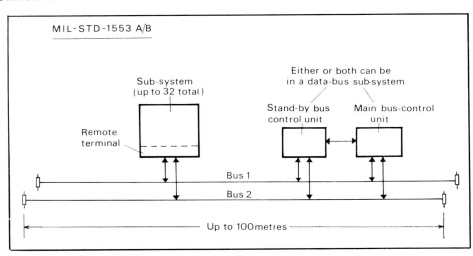

MIL-STD-1553 A/B

Sub-system (up to 32 total)

Remote terminal

Either or both can be in a data-bus sub-system

Stand-by bus control unit

Main bus-control unit

Bus 1

Bus 2

Up to 100 metres

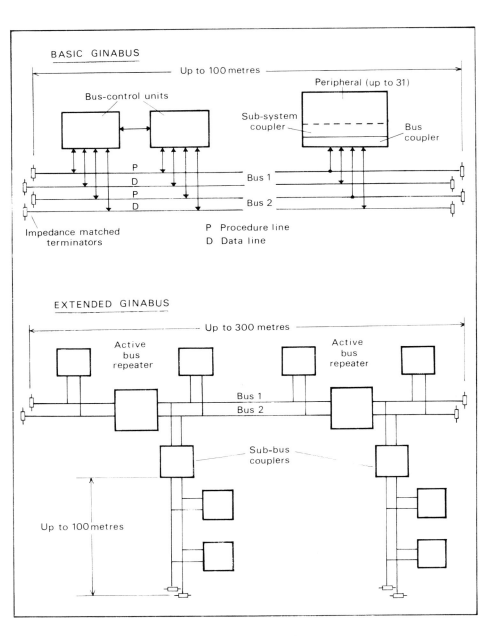

BASIC GINABUS

Up to 100 metres

Bus-control units

Peripheral (up to 31)

Sub-system coupler

Bus coupler

Bus 1

Bus 2

Impedance matched terminators

P Procedure line
D Data line

EXTENDED GINABUS

Up to 300 metres

Active bus repeater

Active bus repeater

Bus 1
Bus 2

Sub-bus couplers

Up to 100 metres

Boeing 767, one of several new commercial transport aircraft to use the ARINC 429 digital data-bus

ARINC 561
A description of the characteristics of an inertial navigation system or inertial navigation type of system

ARINC 568
This describes the signal format to be used in the transmission of distance measuring information.

ARINC 571
A description of the signal characteristics for an inertial sensor or platform, acting as the principal source of attitude and acceleration information.

ARINC 575
This specifies the format of output signals from a digital air data computer.

ASCB
The ASCB (avionics synchronous control bus) developed by Sperry Flight Systems is a 600 K bits a second data-bus system used in its own flight control equipment. The bus controller is in the autopilot control unit and information can be routed between sensors, flight-instrumentation processors and electronic flight instrumentation (EFIS) displays. Knowledge of the data-bus message structures is required to interface with Sperry installations on aircraft such as the Aérospatiale/Aeritalia ATR 42 and de Havilland Canada Dash 8 feeder-liners. It has been compared to a commercial version of MIL-STD-1553B but without many of the costly performance and environmental specifications of the latter. It is not yet clear whether this standard will gain wide-scale acceptance among either avionics manufacturers or airframe builders.

DEF STAN 0018-2
This is the UK Defence Standard Specification MIL-STD-1553B equipment. In terms of data-bus structure, word structure and protocol, the standards are virtually identical.

MIL-STD-1553A
This is a military, multi-source, multi-sink, bi-directional, very high-speed data-bus specification. Since formal definition in 1975 it has been widely used, and examples of its application can be found on the Space Shuttle, Rockwell B-1 bomber, General Dynamics F-16 and McDonnell Douglas F/A-18 fighters, and updated Boeing B-52 bombers.

A MIL-STD-1553A data-bus consists of a single twisted cable pair and, although not required by the specification, it has usually at least one layer of shielding, grounded at all ends and breaks. The data-bus is terminated by resistors and along its length equipment is connected by transformer-coupled stubs. Up to 32 stubs can be accommodated, and the total data-bus length must not exceed 100 metres. In practice fewer stubs and much shorter bus installations are used.

Information is carried at 1 MHz (1 M bits a second) data-rate, using Manchester bi-phase pulse-coded data. Transmitted pulses are between 18 and 27 volts peak/peak, and the detectable peak/peak voltage at any stub is in the range 6 to 9 volts.

All information is sent as 16-bit words, with a parity bit (odd parity required) added, and preceded by two synchronisation pulses which occupy a time equivalent to three bits; up to 50 K words a second can therefore be carried.

As any unit can transmit to any other unit, the potential for clashed signals is too high to allow uncontrolled use of the data-bus. A bus controller function is accordingly allocated to one stub, and this may be a dedicated unit, or a software function hosted in a powerful processor. The remaining stubs are referred to as remote terminal positions. The majority of traffic on a MIL-STD-1553A data-bus is carried by three types of data transfer. These are: remote terminal to remote terminal, remote terminal to bus controller, and bus controller to remote terminal.

Different data-transfer types are recognisable by the structure of messages, which can comprise three types of words: command status and data. Each type of word is characterised by its structure:

Command word: can only be generated by the bus controller. Synchronisation pulses are identical to status word. The 16-bit data section is divided into four groupings:
(a) 1 bit, set to zero or one, designates whether the terminal addressed will transmit or receive. (Depending on how this bit is set, the whole word is called a transmit command or receive command word.)
(b) 5 bits identify the remote terminal being addressed (address is in the range zero to 31 inclusive).
(c) 5 bits identify the sub-address associated with the remote terminal (again, the address range is zero to 31).
(d) 5 bits convey information on the number of words to be transmitted or received. This imposes a 31-word limit on any data-transfer.

Status word: can only be generated by remote terminals. Synchronisation pulses are identical to command word. The 16-bit data section includes 5 bits which identify the address of the originating remote terminal. The remainder of the bits is dedicated to various functions which illustrate (to the bus controller) various operating data; these are chiefly concerned with efficiency of operations.

Data word: can be sent in any message. Synchronisation pulses are opposite to command/status words and 16 bits can be used to convey any selection of data, eg BCD, binary, ISO-alphabet. The operating procedure for each data-transfer type is as follows:
Remote terminal to remote terminal:
Bus controller (BC) sends receive command word to appropriate remote terminal (RTr)
BC sends transmit command word to appropriate remote terminal (RTt)
RTt sends a status word, followed by the commanded number of data words.
RTr sends a status word.
Remote terminal to bus controller:
BC sends transmit command word to appropriate remote terminal (RTt)
RTt sends a status word, followed by the commanded number of data words.
Bus controller to remote terminal:
BC sends receive command word to appropriate remote terminal (RTr)
BC sends appropriate number of data words
RTr sends a status word.

MIL-STD-1553A requires a knowledge of the remote terminals in use, how much information will pass to/from each remote terminal and remote terminal or bus controller combination, and how frequently each transfer will take place. Based on this information the bus controller has to be pre-programmed to provide the desired mix of data-transfer types, and at the appropriate data-rates. Typical operations are based on a maximum data-rate interwoven with lower data-rates of exactly half, quarter-value etc; eg 50 Hz, 25 Hz, 12.5 Hz, 6.25 Hz.

Each unit on the data-bus has to know its remote terminal address. This can be generated automatically by socket pins as the unit is installed, or programmed in software.

Only a few MIL-STD-1553A data-buses are used in an aircraft. Usually one or two suffice to integrate a distributed-computing system, and there may be extra data-buses dedicated to special functions, for example, stores management and flight control

MIL-STD-1553B
All the procedures and capabilities of MIL-STD-1553A are retained in this subsequent standard, first issued in 1978, but two additional capabilities have been added:

Dynamic bus control: the bus controller function can be passed to different remote terminals. This can complicate some aspects of the software needed, but is designed to provide graceful degradation of data-bus operation in the event of the primary bus controller unit, or the host computer, failing.

Broadcast mode: it is often found that some information is being transmitted several times by one remote terminal to several other remote terminals. Using broadcast mode, the bus controller tells all the terminals to prepare to receive, and one transmission therefore reaches all destinations. This may be useful only in the case of non-critical data, as there is no remote terminal status word response to assure the bus controller that the data has been received.

MIL-STD-1773B
This is the fibre-optic version of MIL-STD-1553. The word structure, protocol and data-rate characteristics are unchanged. The specification is based on fibre-optic cabling requirements and is intended for military applications where electromagnetic pulse (EMP) protection is vital.

High-speed data-bus
In mid-1982 the baseline characteristics of a new, high-speed, data-bus system were defined by America's Society of Automotive Engineers and are under development by the SAE's High Speed Data-bus Committee. This forum will issue an initial specification which is expected to be introduced alongside MIL-STD-1553 later this decade.

A major objective is to provide a capability for integrating avionics sub-systems which use MIL-STD-1553 internally. Characteristics already under consideration include a distributed bus-control feature, 20 M bits a second data-rate (the eventual designated data-rate will depend on progress with current vhsic research), split-channel operation with dedicated procedure and data paths (as the French Digibus system), choice of fibre-optic or coaxial cables, 64 remote terminals, integrated data and audio capability and a maximum message length of around 4000 16-bit words (compared with only 32 words on the current MIL-STD-1553 specification).

Industry is being canvassed for opinions on these objectives and the desirability of any additional system features. Implementation is likely to depend on the availability of vhsic-type components. (Vhsic is the US tri-service very high speed integrated circuits programme now occupying a number of companies.)

Digibus data-bus

Originally called Ginabus (Gestion des Informations Numeriques Aéroportées – Airborne Digital Data Management), Digibus was designed jointly by ESD (Electronique Serge Dassault), Avions Marcel Dassault-Breguet and Sagem, between 1973 and 1976. It embodied experience gained in the S3 data-bus used in ballistic weapons designed in 1970-73, and in 1975-77 a French Navy standard data-bus (BSM – Bus Standard Marine) was derived from Ginabus for surface vessel and submarine applications.

Digibus is now the standard data-bus of all branches of the French military and is defined in Specification GAM-T-101, which the French are proposing as a NATO standard. The system operates at 1 MHz (1 M bits a second) and uses two twisted cable-pairs, each shielded with two mesh screens either side of a mu-metal wrap. One cable pair conveys data while the other carries protocol messages which have a similar structure to those of the command and status words in MIL-STD-1553. Bus ends are terminated with 75-ohm loads and nominal pulse characteristic is ±6 volts with a 1-pulse starting positive and a 0-pulse starting negative. A broadcast mode is available.

Up to 32 ports can be accommodated, including the bus-control units and 16-bit data words are used. Maximum bus length is 100 metres, but active bus-repeater units are in development to allow extension to 300 metres plus sub-bus couplers that can be used to graft sub-buses (each up to 100 metres long) onto the main installation. Additional ports can be added to the network, there being up to 31 ports available on each of the 31 possible sub-buses. Protocols include transmit, receive, explicit address and label-mode operation. There is an 'echo' on the data line which acknowledges data reception. Odd-parity protocol is used and typical operating speeds accommodate 30 to 40 K words a second.

The Digibus has been adopted on the Dassault Mirage III NG, IVP, F1 and 2000 variants as well as the Alpha Jet and Atlantic NG.

OBDH data-bus

This is a specialised data-bus specification developed by the European Space Agency (ESA) for use on scientific and communication satellites. Like Digibus, it uses segregated protocol and data wires, but the data-rate is only 550 K bits a second. Remote terminal and bus control equipment has been designed in France by Crouzet.

Wide-bandwidth tactical data-bus

A current US Air Force Avionics Laboratory research project is the definition of parameters to specify a wide-bandwidth data-bus for possible use on tactical aircraft to handle radar and electronic warfare sensor data. The requirements are also expected to define a system that will handle conventional narrow-bandwidth signals.

Dual-redundant data-bus installations

Almost all military aircraft which use, or are intending to use, a multi-source, multi-sink information-transmission system have specified dual-redundant data-bus installations.

In this configuration two identical data-bus systems are installed between dual-configuration remote terminals and bus controller units. If any single remote terminal, bus controller, or data-bus component fails, communication is assured via one route between any two points in the aircraft. More complex failure-survival capability can be incorporated by appropriate use of bus controller intelligence. The aim of a dual-redundant data-bus installation is to provide a high degree of data availability.

Computer Standards

Two major computer instruction-set architecture standards are in the process of implementation in the military industry. These will have far-reaching influences on future computer developments in both military and civil fields. A brief description of each is given prior to descriptions of hardware in production.

MIL-STD-1750 16-bit instruction-set architecture

The US Air Force has standardised an instruction-set architecture for use in all new computer procurements. The elements of this procedure are embodied in MIL-STD-1750. This does not require a standard computer system, but by standardising the techniques used to accomplish major computational functions, advantages will result from software life-cycle-cost savings and more frequent hardware competition, which should, in turn, lead to a more rapid infusion of electronic technology into airborne computing.

MIL-STD-1750 procedures are being used to develop a common set of support software applicable to all computers, irrespective of their manufacturer or date of manufacture. This has not been possible before, and cross-compilers have become commonplace to translate between machines using customised versions of the same programming language. As experience with MIL-STD-1750-compatible machines is built up, so the risks of operating errors will reduce as more mature techniques are continually being used. This should in turn reduce the risk involved and the time needed to develop software for new applications. Furthermore, software should become a standard item, allowing computer hardware to be updated for example, and merely transferring to new equipment the software developed for a previous generation machine. This should substantially reduce cost, software development being a labour intensive and hence very expensive item.

MIL-STD-1750 developments were begun in 1976 at Wright-Patterson Air Force Base, Air Force Avionics Laboratories, using a Strawman architecture ('Strawman' was the name given to a particular stage in the development of a high-order language, see Programming Languages section). The initial MIL-STD-1750 standard architecture-set was developed with industrial participation and officially released on 21 February 1979.

Current activity includes a MIL-STD-1750 instruction architecture-set User's Group, which is responsible for developing revisions to the initial release standard.

MIL-STD-1750 is a general-purpose register architecture, which can be used for both fixed- and floating-point operations. It is applicable to 16 and 32-bit fixed-point and 32 or 48-bit floating-point operations. A total repertoire of 100 instructions is envisaged, and this should be adequate for all foreseeable airborne computing tasks. The basic standard is satisfactory for a 65 536 word main-storage unit, but this is not a limitation. Applications with more storage requirements, up to 1 megaword limit, can be handled by expansible addressing facilities. The standard has deliberately refrained from defining programming protocol for input/output channels, thus assuring continuing use of channels and devices already in production.

US aircraft which use MIL-STD-1750 protocol include: Fairchild A-10, General Dynamics F-16 and F-111, and McDonnell Douglas F-4G (Wild Weasel) update/upgrade programmes, Lantirn (F-16 and A-10), Northrop F-20 development, Boeing KC-135 improvements, the new LHX combat helicopter project, and MATE (modular automatic test equipment) for the US Air Force/McDonnell Douglas C-17 advanced transport.

MIL-STD-1862 32-bit instruction-set architecture

The US Army initiated MIL-STD-1862 developments, and these are now being implemented by multi-service groups and made complementary to the 16-bit MIL-STD-1750 instruction-set architecture.

Development commenced in 1979, when the US Army commissioned Carnegie-Mellon University to develop a 32-bit instruction-set architecture called 'Nebula'. A preliminary issue of MIL-STD-1862 was made in March 1980, and since then the US Army has requested proposals for computers which are compatible with this standard: 32-bit microprocessor technology having now matured to full-scale production phase.

IBM has published the following comparisons of computer capability of these machines:

	1980	1985
	NATO AWACS	
	CC-2 computer	**AN/UYK-41**
Speed	2 Mips	3 Mips
Storage	2.5 megabits	2 megabits
Power	6500 W	100 W
Volume	37 ft^3	0.6 ft^3
	B-52 OAS*	**AN/UYK-49**
	computer	
Speed	400 Kips	500 Kips
Storage	250 kilobits	1 megabit
Power	550 W	20 W
Volume	0.9 ft^3	0.12 ft^3
	B-52 DBNS**	**single-module**
	computer	**computer**
Speed	400 Kips	500 Kips
Storage	128 kilobits	128 kilobits
Power	450 W	5 W
Volume	0.9 ft^3	0.02 ft^3
		(one card)

*Offensive Avionics System
**Doppler Bombing/Navigation System

. The objectives of the MIL-STD-1862 programme are the same as those for MIL-STD-1750. In summary, the standard should result in the availability of universally-applicable software support facilities, reduced software developments, risks and timescales and easy interchangeability of software between computers.

FRANCE

Crouzet

Crouzet SA, 25 rue Jules Vedrines, 26027 Valence Cedex
TELEPHONE: (75) 429 14 41
TELEX: 345807
FAX: 75552250

Alpha 732 general-purpose computer

This optimised processor unit has been developed for general application in Crouzet's products. Its operating speed is about 1 million operations a second and 1000 microprogrammed instructions can be accommodated, including real-time management instructions.

The Nadir 2 system includes an Alpha 732 processor comprising two 312 × 166 mm multilayer printed circuit boards, one devoted to the processor and the other to the memory unit. Another version, based on the same vlsi (very large-scale integration) components, uses leadless carrier chips implanted on 160 × 160 mm ceramic circuits. There are two such devices, one each for processor and memory functions. The Nadir 2 can be built-up, adding memory units, co-processors or further processor units to meet requirements.

STATUS: in production.

Crouzet Alpha 732 general-purpose computer

Electronique Serge Dassault

Electronique Serge Dassault, 55 quai Carnot, 92214 Saint-Cloud
TELEPHONE: (1) 46 02 50 00
TELEX: 250787

Electronique Serge Dassault is the leading manufacturer in France of computers for aircraft and missiles, and is presently mass producing within the framework of agreements with Sagem various ranges of computers intended for major military programmes such as the Dassault Mirage F1 and Mirage 2000 aircraft, strategic and tactical missiles, etc.

A new generation of computers was created in 1983, making wide use of advanced components and technologies, including custom lsi circuits in large hybrid modules.

ESD has also been selected to develop, in collaboration with CIMSA and Sagem, the 'French Military Computers' (CMF) programme of the next decade.

M182-84 on-board digital computer

Using the very latest hybrid micro-electronic technologies, the M182-84 computer has been designed for combat aircraft. It features exceptionally small weight and volume; modular construction for easy extension and maintenance; and a standard (Digibus) digital input/output unit and a specific input/output unit for adaptation to all weapon system requirements.

STATUS: M182-84 is in mass production for the export Mirage F1 and Mirage F1 CR aircraft.

UCM 2084 computer

The 2084 belongs to a series of general-purpose digital computers designed to operate with high reliability under extremely severe environmental conditions. It makes use of the latest technologies, including in particular microprocessors assembled in hybrid modules and offering an extremely high performance/volume ratio.

ESD 2084 processor

Within the volume of a ½ ATR case, the 2084 computer comprises essentially: an arithmetic unit; a working memory of up to 512 K 18-bit words; a fully comprehensive input-output system, including the simultaneous management of two multiplexed buses (of the Digibus type); and a power supply.

Extremely comprehensive basic software, such as the LTR compiler, real-time monitor, assemblers, micro-assemblers and programs to assist in the development and debugging of application programs and microprograms, has been developed for these computers.

Because of its modular construction and high computing capacity (350 Kops), the 2084 computer can perform the following functions: flight preparation (initial alignment, etc); inertial, Doppler and hybrid navigation; calculations for air-to-ground attack in all possible configurations; weapon system equipment monitoring and management of navigation and weapon system control units, and management of analogue and digital data transfers, including in particular Digibus transfers.

STATUS: The 2084 computer has been selected for the various configurations of the Dassault Mirage 2000 aircraft: 2000 DA, 2000 N, 2000 Export (two computers per aircraft).

At the present time, some 700 Type 2084 computers have been ordered, of which 300 had been delivered by mid-1986.

ESD 2084 XR general purpose computer

The 2084 XR computer belongs to the range of digital computers (84 series) especially designed to meet the requirements of combat aircraft and to operate with a high degree of reliability in harsh environment conditions.

Format: ½ ATR
Addressing capacity: 4 M words
Computing speed: better than 1 Mips
Execution time
(add): 0.833 μs
(multiply): 1.25 μs
(divide): 6.542–7.583 μs

UCM 3084 general-purpose computer

The UCM 3084 miniaturised computing unit is the latest general-purpose development of the 2084 XR and provides 10 times the computing power and 16 times the memory of the previous generation of computers. It is intended for applications where high computing capacity is required within a small volume. The main feature of the 3084 is that it can be configured to suit the physical constraints of the application. For airborne use the 3084 can be used for navigation, weapon system control, maintenance records and data transfer management, for example.

Multiprocessor configuration
Format: ¾ ATR
Addressing capacity: 4 M words per processor
Computing speed: better than 3 Mips

STATUS: flight trials began in the summer of 1985.

Computers for Franco-German helicopters

The computers proposed by Teldix and ESD for the bus management sub-system of the Euro-copter PAH-2, HAP and HAC-3G helicopters perform the functions of processing (system and vehicle monitoring, back-up navigation, etc), management of data transfers over Type 1553 buses and interfacing with equipments not connected to the buses.

These computers possess considerable processing capacity and fully comprehensive basic software comprising in particular a high-order language compiler.

The input-output system of these computers consists of a 1553B bus management unit and couplers for encoding and decoding analogue signals as well as for the transmission and reception of discrete signals. The computers make use of hybrid technology, rendering them highly reliable and compact.

MIL-STD-1553B aircraft multiplex data bus

Stemming from its Digibus experience, ESD also manufactures 1553B components based on custom lsi chips. These components are designed to make easy their use for developing high level couplers as for the Digibus.

MIL-STD-1553B remote terminal coupler: in a very small size (25 × 25 mm), it handles the complete remote terminal functions for a redundant or non-redundant 1553B bus. Coupled to any 8-bit or 16-bit processor through a shared memory, it operates without processor control after initialisation: data transfers, generation of interrupts and transfer reports associated with each data buffer.

MIL-STD-1553B management auxiliary (MGA 53): as the front end of a processor, the MGA 53 handles the bus controller or remote terminal functions coupled to the memory of the host processor, it generates the bus frame described by a channel program and processes data as the remote terminal coupler.

Digibus GAM-T-101 multiplex data-bus

With Digibus, ESD has been responsible for the majority of on-board multiplexed data bus in French military systems.

At the end of 1982, Digibus became a French inter-services standard denoted by specifications GAM-T-101.

During 1984, ESD introduced the notion of 'communicators' offering high-level services to the user (transport layer of OSI Model).

Today, many military programmes use the Digibus for data transfers between equipments. These include in particular Dassault Mirage 2000, Mirage F1 and Atlantique 2 aircraft, and in missile, naval (submarines and surface vessels) and on-ground applications.

A large set of Digibus couplers of two types (remote terminal and control unit) is being produced by ESD. They use hybrid technology with standard and custom lsi chips bonded on multilayer ceramic substrates.

ESD furnishes component kits mounted on customers' cards, high-level coupling cards integrated in customers' equipment or complete coupling equipments for sub-systems.

By the end of 1985 ESD had delivered approximately 10 000 remote terminal couplers and 1000 control unit couplers.

In addition to the above coupling units, ESD also manufactures ground support facilities such as a procedure generation, bus monitor and Digibus simulator.

STATUS: in production.

ESD 3084 processor

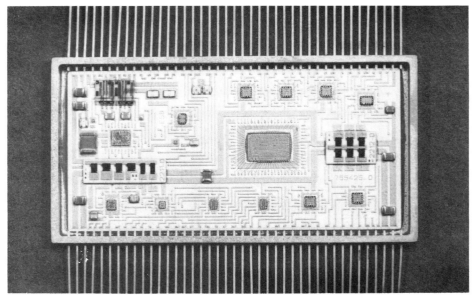

Digibus makes extensive use of lsi technology

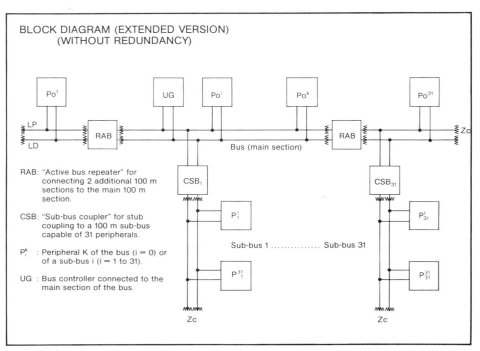

BLOCK DIAGRAM (EXTENDED VERSION)
(WITHOUT REDUNDANCY)

RAB: "Active bus repeater" for connecting 2 additional 100 m sections to the main 100 m section.

CSB: "Sub-bus coupler" for stub coupling to a 100 m sub-bus capable of 31 peripherals.

P_i^k : Peripheral K of the bus (i = 0) or of a sub-bus i (i = 1 to 31).

UG : Bus controller connected to the main section of the bus.

Schematic diagram of ESD Digibus data transmission system

Sagem

Société d'Applications Générales d'Electricité et de Mécanique, 6 avenue d'Iena, 75783 Paris Cedex 16
TELEPHONE: (1) 47 23 54 55
TELEX: 611890

MIL-STD-1553B equipment

Sagem has developed hardware and associated support systems in order to offer a full data-bus integration service using MIL-STD-1553B operating procedures. The design process begins with functional-partitioning to determine an acceptable system architecture. Selection is based on data-flow definitions according to mode of operation and data-frame information (for example, type of data exchange, data definition, data-rate) provided by the design support system.

Using advanced microelectronics based on thin-layer hybrid circuits, and gate-arrays, the company provides a wide range of MIL-STD-1553B bus-controller and remote terminal configurations. Depending on sub-system requirements, the systems can be in a stand-alone or embedded configuration.

STATUS: in production.

UCG 90 general-purpose computer

This unit is designed to provide all computations and data processing requirements for the weapon delivery systems of modern strike aircraft. It can conduct air-to-ground weapon delivery computation, head-up display and rangefinder (laser or radar) control, stores management and MIL-STD-1553B data-bus control. It uses a fast, parallel, micropro-grammed processing unit and an eprom memory.

The UCG 90 is built around the UTR 90 single-card computer which has a 830 000 operation per second capability and 1 M words of address. An advanced software environment

Sagem Copra general-purpose computer

has been developed, including the facilties for real-time troubleshooting, high-order languages and dynamic validation.

Computer type: binary, two's complement, fixed- and floating-point
Word length: 16-bit and 32-bit
Max address range: 1024 K words (program); 32 K words (constants)
Instruction-set: 130 (expansible to 183)
Typical execution time (16-bit)
(add) 0.6 μs
(multiply) 1.2 μs
(divide) 4.2 μs
Central memory: 128 K words (expansible to 256 K words)

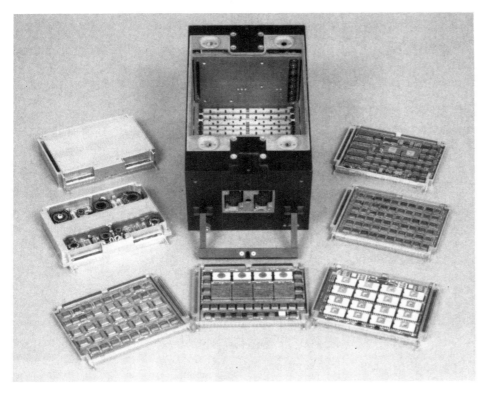

Sagem UCG 90 general-purpose computer

Sagem EBS 2501 magnetic bubble-memory recorder

Input/output options: discrete, dc analogue, synchro, ARINC, MIL-STD-1553B, Ginabus
Format: ¾ ATR Short
Weight: 22 lb (10 kg)
Software support: Pascal, Ada

STATUS: in production.

Copra general-purpose computer

Sagem, in co-operation with ESD, has developed this specially-configured computer for fault-tolerant operations. It has demonstrated a 97 per cent probability of correct operation during a seven-year mission, and a failure probability of 10⁻¹⁰ each hour of operation. It is intended for satellite, inertial navigation and new military aircraft control-system applications.

Computer type: binary, two's complement, fixed- and floating-point
Word length: 16-bit and 32-bit
Instruction-set: 150
Typical speed: 600 Kops
Central memory: 128 K bytes expansible to 1 M byte
Input/output options: redundant multiplexed serial bus
Volume: (DIL circuits) 14 litres
(gate arrays and hybrids) 5 litres
Power: 25 W at 28 V dc

STATUS: under development.

EBS 2501 magnetic bubble-memory recorder

This equipment consists of a controller, one to four holders and 16 M bit memory cartridges. It has been selected for the Dassault Mirage 2000N to store mission data including terrain contour and radar image correlation data bases. A version will also be used in the Dassault ATL2 to store mission data as well as main computer programs for restarting after inadvertent shutdown. Magnetic bubble-memory was chosen as it meets requirements for small installation and can withstand severe environmental conditions.

Memory capacity: 2-8 M bytes
Average access time: 20 ms
Transfer frequency: 80 K bytes/s
Interface options: HDLC, Digibus
Error rate: 10⁻¹⁰
Reliability: 10 000 h mtbf
Volume: (controller) 58 × 194 × 240 mm
(holder) 95 × 127 × 225 mm
(cartridge) 23 × 103 × 162 mm
Power: 70 W at 28 V dc

STATUS: under development.

Sfena

Sfena, Controls and Systems Division, Aérodrome de Villacoublay, BP 59, 78141 Velizy Villacoublay, Cedex
TELEPHONE: 46 30 23 85
TELEX: 260046
FAX: 46 32 85 96

Series 7000 general-purpose processors

Sfena Series 7000 computers are characterised by a modular and adaptable architecture, which is claimed to minimise hardware and development costs in any application, and avoid technical obsolescence by providing for continuous updating of standard modules. They use a flexible memory design, allowing a wide range of memory types, capacity and access-time characteristics to be matched to any application. The input/output system is completely modular, providing again for a high degree of customisation, and there is a highly-developed operating system to support all applications.

The basic central processing unit is the 16-bit CPU 7800, which is organised to operate with a memory bus and an input/output bus. A more recent development is the 7068 central processing unit, which is available for especially demanding tasks. Parallel processors are able to operate simultaneously with either central processing unit for applications requiring the very highest performance. The interrupt system also allows processing in up to eight hierarchic levels, with an extension to 16 levels.

Normal addressing capacity is 64 K words, extensible up to 256 K words. Mass memories can be added externally, and internal memories can be semi-conductor (with or without battery back-up), programmable read-only memory or reprogrammable read-only memory.

Software support is available in LTR high-order language.

	7800 CPU	7068 CPU
Operation	binary, two's-complement, fixed- and floating-point	
Word length		
fixed-point	16-bit	16-bit
floating-point	32-bit	32-bit
Max address range		
basic	64 K words	64 K words
option	256 K words	8 M words
Instruction set	Up to 150	Up to 150
Execution time		
load	0.9 μs	—
store	1.2 μs	—
fixed-point multiply	7.5 μs	—
floating-point multiply	13.2 μs	—
Input/output options	large range of possibilities, including ARINC and MIL-STD-1553	
Format	(4 cards) ½ ATR	(2 cards) ½ ATR
Power	6.3 A at 5 V dc	2.5 A at 5 V dc

STATUS: in production.

Thomson-CSF

Thomson-CSF, Division Equipements Avioniques, 31 rue Camille-Desmoulins, BP 12, 92132 Issy-les-Moulineaux
TELEPHONE: (1) 45 54 92 40
TELEX: 204780
FAX: (1) 45 54 09 45

Data analogue converter

This unit is designed to receive, convert, process and multiplex analogue and discrete input signals in order to present them in a digital format for an ARINC 429 bus application. It uses microprocessors and cmos components, and is suitable for civil and military applications. Interchangeability is provided by a personalised program (menu) which automatically identifies the aircraft type, switches inputs to the appropriate interfaces and connects the correct ARINC formatting process. Extensive internal monitoring and a built-in test function provide failure detection and easy on-line maintenance.

Format: 6 MCU
Weight: 8 kg
Power requirements: 115 V 400 Hz
Environmental specification: DO 160
Refresh rate: 16 times/s
Resolution: 4096 points
Inputs: 150 analogue and 200 discrete signals
Outputs: ARINC 429 serial data-words, BITE and validity signals

STATUS: in service.

Thomson-CSF data analogue converters

Thomson-CSF CIMSA 15M/125F processor

Thomson-CSF CIMSA 15M/05 processor

Thomson-CSF, Compagnie d'Informatique Militaire Spatiale et Aéronautique (CIMSA), 12 avenue de l'Europe, 78140 Velizy-Villacoublay
TELEPHONE: (1) 49 46 96 70
TELEX: 698529

15M/05 general-purpose processor

This unit has been designed for process-control or signal-processing applications. It is less capable than the 15M/125 which is also made by CIMSA. Typical avionic applications are radar, sonar or electronic warfare processing. M-rack and ATR-case versions of the processor are produced and can be used for ground-based or airborne applications. Software languages for which support is available are Mitra and LTR.

Computer type: binary, two's-complement, fixed- and floating-point

Word length: (fixed-point) 16-bit
(floating-point) 32-bit
Max address range: 32 K words (extensible to
64 K)
Instruction-set: 98
Approximate instruction run time
(ram or prom/ttl) 1.2 μs
(ram/hmos) 1.2 μs
(ram/mos (4 K/board)) 1.7 μs
(ram/mos (32 K blocks)) 1.8 μs
Micro-instruction run times: 110, 165 or 220 ns
Format: ½ ATR, 1 ATR or Rack 2 M
Power: 300 W at 24-48 V dc

STATUS: in production.

15M/125 F general-purpose processor

There are two versions of this processor. The
15M/125 is a central processing unit with main
memory extensible to 512 K words, and the
15M/125F is a version with a built-in fast
arithmetic operator. The unit is complementary
to the 15M/05 unit and is better-suited to high-
speed processing duties or tasks which require
a large amount of rapidly accessible memory.
Joint-service operations in weapon-command
and control, communications management,
navigational aids, detection systems (radar,
sonar, electronic warfare) and command aids
are possible.

Software support is available for Mitra and
LTR programming languages.

Computer type: binary, two's-complement,
fixed- or floating-point
Word length: (fixed-point) 18-bit (16-bit usable)
(floating-point) double-length
Max address range: 128 K words, extensible to
512 K words core or semi-conductor
Instruction-set: 128
Format: 1 ATR and ½ ATR
Power: 24-28 V dc or 115-200 V 400 Hz

STATUS: in production.

15M/125X general-purpose processor

This is the most recent addition to the CIMSA
range of general-purpose units and employs
many internal operating procedures common
to previous units, although it is a 32-bit machine
with up to 1 million operations a second
capability. A 1 M byte memory can be addressed
and versions of the processor are typically
installed in a 1 ATR case. LTR real-time
software support is provided, and numerous
applications are expected in high-performance
military equipment.

STATUS: under development.

Thomson-CSF CIMSA 15M/125X general-purpose processor

Thomson-CSF CIMSA 125MS general-purpose processor

125MS general-purpose processor

A development of the 15M/125 range, specifically packaged for a space application (Spacelab). It uses the 16-bit word central processing
unit of the 15M/125F and is installed with all
peripherals in a single box.

It is similar to 15M/125F except:

Instruction-set: 135
Addressable memory: 64 K words
Dimensions: 190 × 280 × 500 mm
Weight: 67 lb (30.5 kg)
Power: 390 W at 24-32 V dc

STATUS: in production.

GERMANY, FEDERAL REPUBLIC

Litef

Litef, Lorracherstrasse 18, Postfach 774,
Freiburg 7800
TELEPHONE: (761) 49011
TELEX: 0772669

LR-1432 digital computer

This fast, micro-programmed, general-purpose
computer uses state-of-the-art large-scale
integrated circuits and is designed for the
military. Fixed- and floating-point options are
available. The main memory holds both program and data in eprom/ram-type storage with
a core memory capacity of 64 K 16-bit words.

A bus interface is available to connect the

computer to a digital data-bus system. This is
fully compatible with dual-redundant MIL-STD-
1553B, allowing operation as bus controller or
remote terminal. Input/output circuits are provided for 16 various discrete signals. The whole
processor, including a power-supply unit and
64 K of storage, can be accommodated in a ½
ATR Short case.

The LR-1432 is in production as the main
computer for the tri-national Panavia Tornado
(1½ ATR) and the LR-1432K is used as a
navigation computer in the Franco-German
Alpha Jet (½ ATR).

Computer type: binary, fixed- or floating-point
Word length: (fixed) 16-bit
(floating-point) 32-bit

Max address range: 64 K words (extensible to
512 K)
Instruction-set: 69 (including floating-point)
Store access time (core)
(8/16 K module) 450 ns
(32 K module) 350 ns
Input/output options: various analogue con-
verters, plus digital to MIL-STD-1553, ARINC
(Dits), Panavia digital links, NTDS, serial
inputs/outputs (eg: CVR, TV-tab) and discretes
Format: ½ ATR Short, ¾ ATR Short or 1½ ATR
Short
Power: 28 V dc or 115 V 400 Hz

STATUS: in service.

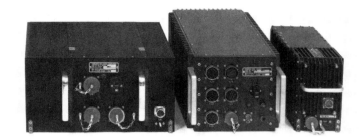

Three versions of Litef LR-1432 processor, left: Tornado main computer; centre: tank computer under evaluation; right: Alpha Jet navigation computer

This version of Litef LR-1432 is main computer used in Tornado

ISRAEL

Elbit

Elbit Computers Limited, Advanced Technology Centre, PO Box 5390, Haifa 31053
TELEPHONE: (4) 514 211
TELEX: 46774

ACE-3 computer

The Elbit ACE-3 computer has been selected to equip the Israeli Air Force's F-16C/Ds. The computer is designed to MIL-STD-1750A standards and is compatible with MIL-STD-1553B data-buses: it has 64 K words of memory and operates at 550 Kips. Extensive built-in test routines are claimed to be able to detect 97 per cent of all faults.

STATUS: in production.

ACE-4 computer

The Israeli Lavi combat aircraft will be equipped with the Elbit ACE-4 computer for centralised avionics management. The ACE-4 operates at 600 Kips and has 128 K bytes of store: it can control two MIL-STD-1553B data-buses and has a claimed mtbf of 2500 hours.

Dimensions: 124 × 194 × 224 mm
Weight: 5 kg

STATUS: in development.

Elbit ACE-3 computer to be installed in Israeli Air Force F-16C/Ds

Elbit ACE-4 computer to be installed in IAI Lavi

Elisra

Elisra Electronic Systems Limited, 48 Mivtza Kadesh Street, Bene Beraq 51203
TELEPHONE: (3) 787 141, 787 131
TELEX: 33553

MIL-STD-1750A computers

This company has developed a family of fast 650 000 operations a second (650 Kops) and flexible computers for both military and laboratory applications. The version designed for airborne use is the GPAP, packaged in a ½ ATR case. Its processor and 64 K memory occupy three out of eight available slots.

Elta Electronics

Elta Electronics Industries Limited, a subsidiary of Israel Aircraft Industries, Electronics Division, PO Box 330, Ashdod 77102
TELEPHONE: (55) 30333
TELEX: 31807

EL/S-8600 computer

There are several versions of the basic EL/S-8600 main computer, all of which use similar central processing units and architectures, and are customised to various military applications. The EL/S-8610 has 15 modules (each with two printed circuit boards) in a 1 ATR box, and the EL/S-8611 has only four modules, in a customised box. In addition to their airborne applications, other computers in the range are used for land and sea operations. Typical applications are inertial navigation and weapon-delivery systems, artillery tactical fire-control system, message-switching centres and mobile

Some military computers produced by Elta

communications control. Software support is available for 'C', Fortran-77, Iso-Pascal, APL and Cobol.

EL/S-8610
Computer type: binary, two's-complement, micro-instruction (15 register), fixed-point
Word length: 16-bit (32-bit instruction words)
Max address range: 64 K byte (expansible to 2 M byte with mapping unit)

Instruction-set: 90
Micro-instruction cycle time: 225 ns
Micro-instruction depth: 1024 words (expansible to 2048 words)
Typical execution time
(add/subtract) 0.66 μs
(load) 1.34 μs
(multiply) 6.56 μs
(divide) 9.26 μs

Input/output options: wide variety, customise to use
Dimensions: 394 × 257 × 194 mm
Weight: ≤47.3 lb (21.5 kg)

STATUS: in service.

ITALY

Selenia
Selenia, Defence Systems Division, PO Box 7083, Via Tiburtina Km 12,400, 00131 Rome
TELEPHONE: (6) 43601
TELEX: 613513

SL/AYK-203 and -204 general-purpose computers
These two families of high-power computers are based respectively on the Intel 8086 and 286 microprocessors, and are intended for real-time military applications. The modular architecture permits processing power, memory type and size, and input/output modules to be tailored to a wide variety of requirements. They are highly fault-tolerant, and comprehensive fault-detection circuits are incorporated.
Input/output modules options: MIL-STD-1553B/ARINC 429/Parallel and serial channels/tape recorder driver/display drivers, raster and stroke/analogue, digital, discrete, synchro, frequency multichannel/A/D and D/A converter multichannel

	SL/AYK-203/1	SL/AYK-204/1	SL/AYK-204/2	SL/AYK-204/3
Microprocessor type & instruction-set	8086	286	286	286
Number of microprocessors	1	1	2	3
Word length	16-bit data + 2-bit parity	16-bit data + 2-bit parity	16-bit data + 2-bit parity	16-bit data + 2-bit parity
Max addressable memory in modules of various ram/eprom/prom combinations	1 M byte	16 M bytes	24 M bytes	32 M bytes
Typical speed	0.3 Mips	0.8 Mips	1.6 Mips	2.4 Mips
Numeric co-processor	optional	optional	optional	optional

Typical physical characteristics: for cases completely fitted with modules and power supply 115 V 400 Hz or 28 V dc
½ ATR Short/6 modules/9.5 kg/80 VA
¾ ATR Short/10 modules/13 kg/130 VA
1 ATR Short/13 modules/16 kg/160 VA

STATUS: SL/AYK-203/1 in production; SL/AYK-204/1, -204/2, -204/3 in development.

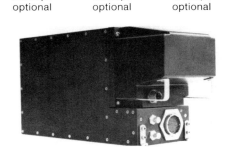

Selenia SL/AYK-203 airborne computer

SWEDEN

Ericsson
Ericsson Radio Systems, Airborne Electronics Division, PO Box 1001, S-431 26 Mölndal
TELEPHONE: (31) 671 000
TELEX: 20905

SDS80 standardised computing system
Designed for aircraft and army/naval applications, the SDS80 is intended as a low-cost general-purpose processor, cost savings extending over software development as well as expenditure on hardware. Current design consists of three modules: D80 computer, PUS80 program development system and Pascal/D80 high-order language. By adding D80 modules the system can be expanded from a relatively simple unit to an extremely high-performance embedded computer system. Using an inter-module bus, up to 15 processing units, input/output modules and bulk storage facilities can be connected. A low-cost microprocessor complement, the D80M, is under development, as well as Ada software compilers for the D80 computer and microprocessor.
The D80 computer has been selected as system computer for the radar, display and other systems in the Saab JAS39 Gripen combat aircraft and for the mid-life update

Ericsson D80 computer

programme of the British Royal Navy's Sea Harriers.
The company is producing an Ada environment, based on the D80 computer, under contract to the Swedish government. Designated SDS80/A, the system will include the Ada language and an Ada Integrated Development Environment. A prototype D80/A will be delivered early in 1987, with production deliveries starting later that year.

Word length: 32-bit
Typical speed: 1 Mips
Typical dimensions: 125 × 193 × 318 mm
Weight: 14.5 lb (6.5 kg)
Instruction time: from 125 μs
Interfaces: include MIL-STD-1553B
Power: 56 W (ac or dc)
Cooling: forced-air or fan

STATUS: in production.

UNITED KINGDOM

Computing Devices
Computing Devices Company Limited, Castleham Road, St Leonards-on-Sea, East Sussex TN38 9NJ
TELEPHONE: (0424) 53481
TELEX: 95568

ACCS 2000 general-purpose computer
The ACCS 2000 (airborne computing and communications system) is a family of micro-programmed computers designed to provide standard low-cost data processing for a wide range of vehicles and missions. The computers can be used as multi-processor units with an

extensive interface capability. There is a high degree of functional and mechanical modularity for flexible growth. The system architecture is not affected by modular configuration changes and is the feature most responsible for providing low-cost and versatility.
The general-processing module contains the microprogrammed control, arithmetic unit,

registers and bus interfaces. Supporting elements such as micromemory, real-time clocks, bootstrap memory and interrupt logic are contained within the processor support module. These units form a central processing unit for the family of computers.

An ACCS 2000 variant has been selected as the mission computer for the Royal Air Force's Harrier GR5. Contained in a standard ATR case, it includes the following modules: power-conversion, general processor, memory-control, two 64 K × 8-bit word core memories, input/output unit and three MIL-STD-1553 bus controllers.

Word length: 16-bit
Typical speed: up to 675 Kips
Input/output: 3 MIL-STD-1553B ports, up to 16 other channels
Memory: 32 K words 18-bit core modules (900 ns access time) 32 K words 18-bit semiconductor modules (400 ns access time)

STATUS: in production.

Computing Devices ACCS 2000 general-purpose computer

Ferranti

Ferranti Computer Systems Limited, Western Road, Bracknell, Berkshire RG12 1RA
TELEPHONE: (0344) 483 232
TELEX: 848117

FM1600 general-purpose processor

The FM1600 designation covers a family of general-purpose computers, of which the FM1600D is a unit developed for airborne use. It can be employed in avionics systems that require high operating speeds, sophisticated interface structures and rapid peripheral responses. An instruction-set particularly suited to highly complex real-time applications has been developed with over 330 instructions based on 60 basic forms including powerful arithmetic and logical shift and jump instructions. The power offered by the instruction-set facilities is increased by a three-address architecture and hardware implementation.

Input/output peripheral control facilities ensure rapid interrupt response and efficient peripheral handling. Peripherals are connected via high-speed serial, star-connected interface channels, and control is by means of a dedicated controller which can monitor up to 16 channels. Priority-scanning to assure short response time for high-priority peripheral equipment is available.

Memory is organised in blocks of different sizes, speeds and technology, covering ram, rom, and core storage. Each block can have up to four interface ports, of which one is allocated to the processor/memory highway.

Extensive software is available to support Coral 66 and Fortran IV operations. Airborne applications include the processor to the Thorn EMI Searchwater radar in Royal Air Force/ British Aerospace Nimrod MR2s and in the Royal Navy/Westland AEW Sea King helicopters.

Computer type: binary, two's-complement, fixed- and floating-point
Word length: (fixed-point) 24-bit (floating-point) 48-bit
Typical speed: 350 Kips

Max address range: ≤ 32 K words core (expansible)
Input/output options: up to 16 B-serial input/output channels
Dimensions: 393 × 191 × 318 mm
Weight: 44-55 lb (20-25 kg) depending on options
Power: 360 VA at 115 V 400 Hz

STATUS: in service.

Argus M700 general-purpose processor

Argus M700 is a military system that embodies all the advantages of a fully modular building-block approach to design. Argus system architectures can be tailored to specific requirements, using a wide range of standard computer modules. In this way equipment may be developed in which the different sub-systems have identical, fully-interchangeable circuit cards, the same programming language and the same software development system.

Processing power of existing systems is provided by the Argus M700/20, a central processor that combines a powerful architecture with an advanced implementation to provide high performance with reduced space and power demands. M700/20 features 16-bit word length, multiple operating modes, a sophisticated interrupt structure and hardware-implemented double-length arithmetic.

Argus M700/40 is a recent update and is the first processor in the Argus range to be implemented entirely in large-scale integrated form to increase significantly the power and versatility of Argus M700-based systems. The Argus M700/40 military microprocessor is fully compatible with all other Ferranti Argus M700

series products, and can therefore replace other M700 processors in existing systems as well as form the basis of new configurations.

Argus M700 computers may be connected to other elements of an overall system via a MIL-STD-1553B digital data-bus, while internal communication uses Eurobus 18/A.

All memory technologies can be incorporated in Argus M700-based computers, including random access memory, programmable read-out memory and core store. A maximum 256 K words can be connected to each Eurobus and a further 64 K words of private memory can be provided for each processor. Private memories, used to hold the main part of operational programs, are accessed via a special high-speed interface.

Software support is available for Coral 66 applications and is associated with Mascot modular software construction facilities. Argus M700 has been developed in close co-operation with UK Ministry of Defence research establishments and is a preferred computer for UK future defence programmes.

In mid-1983 the Argus M700 was allocated its own defence standard in recognition of its widespread use in defence computer systems. The hybrid M700/40 is the preferred processor for the British Aerospace Rapier low-level anti-aircraft weapon system.

Computer type: binary, fixed- and floating-point
Word length: (fixed-point) 16-bit (floating-point) 32-bit
Typical speed: (M700/20) 250 Kips (M700/40) 1.6 Mips
Instruction set: over 120
Input/output options: wide range of customised modules

Ferranti FM1600D processor

Two versions of Ferranti Argus airborne processor. Background cards comprise Argus M700/20, and hand-held in foreground is Argus M700/40

Typical dimensions: 380 × 190 × 193 mm
Weight: 28.6 lb (13 kg)
Power: 200 VA at 115 V 400 Hz

STATUS: in production.

F-100L microprocessor

This is a 16-bit microprocessor system, comprising a microcomputer set with memory interface, timer and interrupt controller, multiply/divide unit and clock-generation chips. It can accommodate a 32 K word address range and has comprehensive software support facilities including cross-product and resident development software, Coral 66, sub-routine library and hardware-test programs. Using these, single-unit or multi-processor systems can be developed.

The F-100L is part of an overall system concept. Its structure has been chosen to provide the facilities to enable fast, real-time systems, as well as less complex applications, to be developed in the simplest possible manner. To meet this aim a comprehensive instruction set is available, plus program interrupt and reset/start facilities. One of several addressing modes can be used and fast instruction times are possible.

F-100L development was sponsored by the UK Ministry of Defence, and the system meets many military requirements. Applications are under way for fire-control equipment in land/sea operations and for the digital control of aircraft turbine engines.

Operation: 16-bit, binary
Max address range: 32 K ram plus 1.2 M byte mass storage
Instruction-set: 153
Clock rate: 13.5 MHz
Memory access time: 1.2 μs/word
Execution time: (multiply) 9-13.5 μs (divide) 12-16.2 μs
Radiation hardness: immune to latch-up, high resistance to neutron and cumulative ionisation damage.

STATUS: in production.

F-100L processor hybrid FBH 5092

The Ferranti F-100L microprocessor, plus clock-generator, multiply/divide unit and two bus-buffer chips, are available as a single hybrid package, reducing space requirements by about 50 per cent compared with the volume needed to mount the same components on a printed circuit board. System reliability is increased by reducing the overall number of soldered joints and by testing the chip set prior to integration. The hybrid pack is also claimed to allow more heat dissipation with significantly lower average temperature and reduction of high-spot temperatures compared with printed circuit board-mounted assembly.

Dimensions: 81.3 × 22.9 × 4.6 mm
Operating temperature: (commercial) 0 to 70°C (military) –55 to +125°C

STATUS: in production.

F-200L microprocessor

The F-200L microprocessor is an enhancement of the F-100L microprocessor, but still remains compatible in executing a comprehensive real-time instruction-set that is orientated to programming in a high-level language. F-200L features a parallel data/address multiplexed bus and facilities for memory mapped input/output, direct memory access, and vectored program interrupts. The instruction-set incor-

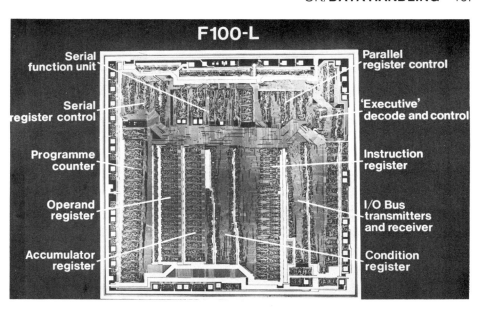

Close-up of F-100L processor chip

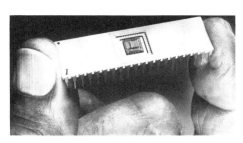

Ferranti F-100L military microprocessor in 40-pin pack

porates arithmetic and logical operations, along with jump, shift, bit manipulation and multi-length instructions. The F-200L provides an increased addressing capability, with increased performance and an on-chip multiply/divide facility. Supporting cards and components plus a range of support software enables both small and large systems to be constructed to suit the required needs.

Operation: 16-bit, binary
Operation speed: over 300 Kips
Address range: 64 K to 1 M word
Temperature range: –55°C to +125°C
Power: 1 W from single +5 V power supply

STATUS: in production.

MIL-STD-1750 processors

Ferranti's Navigation Systems Department based in Edinburgh offers systems containing the company's F-DISC processors which have been updated to include the MIL-STD-1750A instruction-set architecture.

STATUS: under development.

Ferranti Computer Systems Limited, Bracknell Division, Cwmbran Department, Ty Coch Way, Cwmbran, Gwent NP44 7XX
TELEPHONE: (0633) 84117
TELEX: 497636

MIL-STD-1553B interface

The Ferranti MIL-STD-1553B interface card set is designed to connect the Ferranti F-100 microprocessor to a MIL-STD-1553B/Def Stan 00-18 (Part 2) dual-redundant data-bus system.

This versatile design allows bus control and remote terminal functions to be accomplished with identical hardware. Interfacing uses microprogram control, enabling it to be adapted to meet the requirements of systems which will have different protocols, software interfaces or error-recovery procedures.

Software interfaces split the data into functionally separate areas, permitting comprehensive data buffering and minimal processor involvement in transfers to and from the databus. The interface is implemented on two standard-sized printed circuit cards and is fully compatible with all Ferranti F-100 systems. The logic card allows stub coupling or direct connection to the dual-redundant buses, and contains the 1553B bus driver/receiver along with encoder/decoder logic, 16-bit control/status shift registers and a message buffer. The interface and sequencer card is connected to the logic card and contains the 8-bit microprogrammed sequencer. The microprogram is held in a 40-bit word length prom, and addressing is carried out by the program controller.

Interface facilities include all standard 1553B input/output functions, the logic card providing the interface between the sequencer and the 1553B data-bus. Bus-control commands are held in memory before being transmitted onto the bus. All types of transfer, including bus control, are handled by the card set in bus-controller mode. In this mode the card set inspects returning status words and automatically initiates a re-entry sequence in the event of error-detection. Errors are reported back to the processor and initiate a program interrupt.

The card set also operates in all terminal modes, the interface between the card set and the sub-system being software defined and allowing the use of cyclic or single buffers. The microprogrammed sequencer enables the user to allocate areas of memory to specific functions, for example inputs from different subsystems will have unique areas of address space. Software interfacing minimises the rate and number of program interrupts.

Bus controller hardware has been supplied by Ferranti for the MIL-STD-1553 data-transmission rig operated at British Aerospace, Brough, and for the Nightbird project development at the Royal Aircraft Establishment, Farnborough.

STATUS: in production.

GEC Avionics

GEC Avionics Limited, Airadio Products Division, Christopher Martin Road, Basildon, Essex SS14 3EL
TELEPHONE: (0268) 22822
TELEX: 99225

Type AA27811 MIL-STD-1553B interface unit

This unit permits existing avionic equipment to interface directly with a MIL-STD-1553 bus without redesign. The unit can also act as a single terminal on the data highway while itself supporting up to three terminals, thereby reducing highway congestion. The design is sufficiently flexible to allow other dedicated data highway formats, such as ARINC 568 and Panavia 64 K bit, to be catered for. Data input is via a two-port MIL-STD-1553B multiplexed dual-redundant data-bus, or by three separate inputs to ARINC 429 standard. Six discrete signals can also be inputted, together with a variable dc signal at up to 40 V. The output formats are similar. The Type AA27811 is normally packaged in a box, but it is available as separate printed circuit boards for integration directly into avionic systems.

Dimensions: 299 × 43 × 127 mm
Weight: 1.32 kg
Power: 0.5 A at 27.5 V dc

STATUS: in production and service, and selected for the UK's EAP Experimental Aircraft Programme fighter demonstrator.

GEC Avionics MIL-STD-1553B interface unit

GEC Computers

GEC Computers Limited, Elstree Way, Borehamwood, Hertfordshire WD6 1RX
TELEPHONE: (01) 906 6933
TELEX: 22777

4080M military computer

A version of the GEC 4080M computer equips the British Aerospace Nimrod AEW3 as the main processor of radar track data.

The 16-bit central processing unit has 15 cards in the AEW variant, with 256 K of core storage and 512 K bytes of semi-conductor store.

Dimensions: 496 × 257 × between 190 and 225 mm high
Weight: 19 kg
Power requirements: 200 V 400 Hz 3-phase
Operating temperature: –26° C to +70° C

STATUS: in production. GEC Computers is believed to be studying the possibility of equipping the Nimrod AEW with a computer of larger capacity, based on an existing design used for commercial purposes.

Logica

Logica Space and Defence Systems Limited, Cobham Park, Downside Road, Cobham, Surrey KT11 3LX
TELEPHONE: (01) 637 9111
TELEX: 267413

Logos speech-recognition system

Logos is a family of continuous speech-recognition systems developed by Logica to control a variety of systems by the human voice. Logos represents a signficant technical advance over isolated and connected word-recognition systems, allowing continuous recognition of spoken phrases of any length from a predefined vocabulary. A set of word sequence-rules (syntax) may be entered into the machine to reduce the number of word choices at any point in the phrase. This not only reduces the computation power needed but in certain situations improves the recognition accuracy.

Two important factors determining the Logos configuration required for a particular application are the size of vocabulary needed and the maximum number of active words in any phrase. In addition, Logos includes acoustic analysis features to improve recognition performance under noisy conditions, such as aircraft cockpits. The system can be controlled by a visual display unit or host computer attached by a serial or parallel link.

Each user is required to 'train' the machine by speaking one example of each word in the required vocabulary. In the case of an airborne application the pilot, or crew member concerned, would perform this task so that a set of vocabulary words (templates) are recorded. These templates would then be transferred to the aircraft speech recognition equipment, probably in cassette form, and commands to operate would be compared to the recorded templates and actioned.

STATUS: in production. Logos was originally designed for the UK Government's Joint Speech Research Unit. Logica has signed licence agreements for the development of the Logos systems with several UK companies.

Smiths Industries

Smiths Industries Aerospace and Defence Systems Limited, 765 Finchley Road, Childs Hill, London NW11 8DS
TELEPHONE: (01) 458 3232
TELEX: 928761

MIL-STD-1553B equipment

Development work in MIL-STD-1553B by Smiths Industries has been undertaken at three levels: custom design has been used to produce compatible lsi components, these components have been used in a range of modules, and the modules have been incorporated into units and systems to perform various data-bus related functions. The lsi components allow remote terminals or bus controllers to be assembled with the minimum of hardware to interface any equipment with the data-bus and allow all options declared in the standard to be easily implemented. These components, which are fabricated in low-power cmos, are currently being applied in numerous aircraft and missile programmes.

A range of modules has been developed to allow system designers to optimise their equipment based on standardised interfaces. The range is based on a common 16-bit backplane bus and, as well as the MIL-STD-1553B remote terminal and bus controller, the range includes ac and dc analogue input/output, synchro and discrete (28 V) interfaces as well as more specialised designs for communicating with fuel probes, flux-valves and motors. New designs are being added to the range continually and most are in production.

A number of MIL-STD-1553B terminals, to cover varied customer requirements, have been developed based on a standard case, power supply and the range of modules described above. These terminals may include high-speed processing, input/output interfacing, data-storage, or a combination of these functions to provide the capability of customised mission and executive computing.

Smiths Industries is to develop the MIL-STD-1553B data-bus for the mid-life update programme of the British Royal Navy's Sea Harriers. The Smiths-developed bus control and interface unit, as well as controlling the bus, will also drive the aircraft's head-down display.

STATUS: in production.

MIL-STD-1553B interface

A custom-lsi interface unit has been developed by Smiths Industries in which each interface, comprising 25 lsi chips, replaces approximately 20 boards of conventional electronics necessary to obtain the equivalent performance. Four chips are custom lsi components and the full chip set provides a dual-redundant facility, meeting all current US and UK requirements. Custom-lsi components are high-speed silicon-gate cmos devices and a radiation-hard version using silicon-on-sapphire (sos) technology is also available.

Trials and installations at the Royal Aircraft Establishments at Bedford and Farnborough have provided the basis for continual expansion of MIL-STD-1553B systems horizons. The installation at RAE Bedford on board a British Aerospace 748 represents one of the world's largest flying 1553B test-beds. The system integrates equipment ranging from an inertial navigation platform through to laser ranging and head-up displays; it also provides a two-tier (avionics and utilities management) bus system.

A number of aircraft programmes have now selected Smiths Industries MIL-STD-1553B systems. These include the Indian Air Force's Jaguars, the Royal Air Force's Buccaneer update programme, the UK's Experimental Aircraft Programme and the Royal Navy's Sea Harrier mid-life update. The Sea Harrier system provides the latest in reconfigurable 1553B terminals. These terminals are integrated into the Sea Harrier head-up display, bus control and interface unit, air data computer and missile control system, all of which are being supplied by Smiths Industries.

STATUS: in production.

Display multi-purpose processor

This is a wholly modular, display processor unit which can be configured to meet customer requirements. It is based on an extensive library of standard electronic cards and a range of ATR/MCU case sizes. An internal data-bus allows the system to be updated or expanded easily and to incorporate new technology components. The unit is designed to meet military specifications and is suitable for head-up display, weapon-aiming, mission control and electronic flight instrument computing applications.

Central feature of the system is a single-card computing element which provides a 16-bit microprocessor, 4 K words of random access memory and 24 K words of ultra-violet eprom storage. Additional modules provide the following facilities: global memory, cursive or raster symbol generation, analogue inputs, discrete inputs, serial digital interface, and MIL-STD-1553B interface as either remote terminal or bus controller.

Format: 1/2 ATR (4 MCU) 124 × 194 × 318 mm
3/4 ATR (6 MCU) 190 × 194 × 318 mm
1 ATR (8 MCU) 256 × 194 × 318 mm
Weight: (4 MCU) 13 lb (6 kg)
(6 MCU) 20 lb (9 kg)
(8 MCU) 26.5 lb (12 kg)
STATUS: under development.

UNITED STATES OF AMERICA

Control Data

Control Data Corporation, Aerospace Division, 3101 East 80th Street, Minneapolis, Minnesota 55440
TELEPHONE: (612) 853 2622

AN/AYK-14 CDC 480 general-purpose processor

A member of the CDC 480 computer family, the AN/AYK-14 has been selected by the US Navy as its standard airborne computer.

The designation CDC 480 describes a family of modular, microprogrammable computers which represent low life-cycle cost, standard computing elements for a broad range of military systems and environments. Functional partitioning for the computer results in a set of standard modules which are configured and packaged to the user's specification and

requirements. Basic shop-replaceable assembly types provide a general complement of computer module building blocks.

The CDC 480 general-purpose capability can be used with 16 or 32-bit instruction-sets, as the modules can be configured as a 32-bit computer (horizontal expansion). The emulation task can also be divided between two processors (vertical expansion) to increase processing throughput. Configurations range from 16-bit single card IOP, to 32-bit high-speed processor with addressability to 512 K words and extensive input/output capability. Memory modules are available in configurations of 16 K or 32 K words. Memory types (core or semi-conductor) are also interchangeable.

The CDC 480 is adaptable to a variety of enclosures for airborne, shipboard and ground-based applications as a general-purpose processor, emulator, controller, dedicated processor or algorithm unit. Representative

applications include weapon delivery/fire control, guidance, communications, navigation, display sub-system control, radar or sonar processing system control, electronic counter-measures and electronic surveillance measures management, and digital flight control. It has been selected for the US Navy/McDonnell Douglas F/A-18 fighter (which has two CDC 480 mission computers), Sikorsky SH-60B Seahawk Lamps Mk III helicopter, Grumman E-2C Hawkeye, McDonnell Douglas/British Aerospace AV-8B, Grumman EA-6B Prowler and Lockheed P-3C Orion. Sperry has been selected as a second-source manufacturer for the AN/AYK-14.

On the F/A-18 fighter the AYK-14 is installed as a dual-redundant system, one computer serves navigation and engine control while the other acts in a mission-management role, handling and processing weapons and target information. Each computer can accommodate 64 K words of memory, and functions at 800 000 operations a second. Under Defense Department funding the AYK-14 will be upgraded to incorporate a new processor and input/output modules and so double the throughput. Application of vhsic technology is also foreseen and in mid-1985 the US Department of Defense awarded Control Data a $13 million contract to develop Ada software programmer support tools for the airborne AN/AYK-14 and the related shipborne AN/UYK-43/44 computers.

Computer type: binary, fixed- or floating-point
Word length: 16 or 32-bit
Typical speed: 300-800 Kips
Max address range: 512 K words
Execution time (high-speed floating-point)
(add/subtract) 2.2 μs
(multiply) 3.7 μs
Input/output options: MIL-STD-1553A (two buses)
32 I/O discretes
8 interrupts
NTDS fast, slow, ANEW and serial
Dimensions: 194 × 257 × 356 mm
Weight: 34 lb (15.5 kg)
Power: 350 W typical

F/A-18 Hornet avionics system incorporates two CDC 480 (AN/AYK-14) mission computers

STATUS: in production.

Cubic

Cubic Corporation, 9333 Balboa Avenue, San Diego, California 92123
TELEPHONE: (619) 277 6780
TELEX: 6831168
TWX: 910 335 2010

Data links for Joint-Stars

Cubic is developing the air-to-ground data-link for the US Army and Air Force's Joint Surveillance Target Attack Radar System (Joint-Stars – see separate entry), under a $25 million contract awarded at the end of 1985. The

equipment will be installed in a Boeing C-18 (modified 707), with development continuing into 1988.

STATUS: in development.

Delco Electronics

Delco Electronics Division, General Motors Corporation, Santa Barbara Operations, 6767 Hollister Avenue, Goleta, California 93117
TELEPHONE: (805) 961 5903
TWX: 910 334 1174

M362F general-purpose processor

This is a general-purpose unit which uses parallel, binary, floating-point, two's-complement, 16-bit processing. A typical instruction mix yields an operating speed of about 340 000 instructions a second (340 Kips).

The M362F can be tailored to particular

operations by specifying from a wide choice of memory types and standard input/output circuit modules. It is mechanised on two operating modules. The instruction repertoire can be varied by adding microprogram memory, or changing the existing microprogram memory, which consists of 10 integrated circuits. The M362F will operate with core and/or semi-conductor memory. The processor provides 72 basic machine instructions, including nine special-purpose types that are mechanised using the basic input/output instructions. Micro-instructions are held in a 512-word control memory, but this can be expanded to 2048 words, providing for such operations as byte (8-bit); control jumps, skips and transfers;

register-variable shifts, register/register floating-point arithmetic; logical immediate and register/register; and macro-instructions. The latter provides square root and trigonometric functions much faster than by using sub-routine executions.

Development of M362F software can be accomplished on support equipment such as mini-computer-directed systems, IBM 360/370 or similar facilities. Mini-computer equipment enables software development on a stand-alone basis in the laboratory.

The M362F is used as the fire-control computer in the General Dynamics F-16, and has a comprehensive set of Jovial or assembly-coded software facilities.

Computer type: parallel, binary, fixed- and floating-point, two's-complement
Word-length: 16-bit
Typical speed: 340 Kips
Max address range: 65 536 (64 K) memory locations
Instruction-set: 72 (expansible)
Execution time (fixed-point)

	Semi-conductor memory	Core memory
add/subtract/load	1.46 μs	2.4 μs
multiply (32-bit)	4.6 μs	4.6 μs
multiply (64-bit)	20.5 μs	22.7 μs
divide	8.2 μs	8.7 μs

Input/output options
analogue/digital/analogue converter
analogue input/output multiplexer
discrete input/output (28 V)
MIL-STD-1553 bus
multi-purpose serial-digital processor
Format: processor + 16 K core memory + 4 K rom in ½ ATR case
Weight: 14.2 lb (6.5 kg)
Power: 28 V dc

STATUS: in service.

M362S general-purpose processor
Related to the M362F, this is a 32-bit, high-speed, general-purpose processor. It uses microprogrammed, parallel, binary, fixed- and floating-point, two's-complement operations, and has a typical operating speed of 750 Kips.

The processor provides 96 basic machine instructions, microprogrammed in a 512-word control memory, with possible expansion to 1024 words. A random-access memory system is available and comprises semi-conductor cmos memory modules, memory controller and an error-detection and -correction unit. A total of 65 536 (64 K) memory locations (16-bit) can be accommodated.

Packaging is either for forced-air cooling, as in aircraft bays, or a radiant cooling frame, for space applications. The system has been selected for NASA's Inertial Upper Stage programme. A comprehensive set of Jovial and assembly-coded software facilities are available.

Computer type: binary, fixed- and floating-point, two's-complement
Word length: 32-bit
Typical speed: 750 Kips
Max address range: 65 536 (64 K) memory locations
Instruction-set: 96 (expansible)
Execution time

Fixed-point	Semi-conductor memory
add/subtract/load	0.88 μs
multiply (32-bit product)	2.72 μs
multiply (32-bit product)	4.29 μs
divide	5.66 μs
Floating-point	
add	2.6 μs
double precision add	4.2 μs
multiply	4.6 μs
divide	9.8 μs

Input/output options: analogue/digital/analogue converter
analogue input/output multiplexer
discrete input/output (5 or 28 V dc)
MIL-STD-1553 bus
multi-purpose serial-digital processors
Dimensions: 365 × 365 × 152 mm
Weight: 54 lb (24.5 kg)
Power: 220 W at 28 V dc

STATUS: in production.

M372 general-purpose processor
This is a development of the M362F used as the fire-control computer of the General Dynamics F-16. It has been designed for applications requiring extended performance in memory, input/output, throughput and computational

Delco M372 general-purpose processor

speed. Operating speed is typically between 570 and 720 Kips, and non-volatile, electrically-alterable memory of 32 K, 64 K, 128 K or 256 K words capacity can be accommodated and addressed. In addition to analogue and discrete input/output channels, up to three MIL-STD-1553 dual-redundant digital data-bus channels can be accommodated, and the computer executes the new US Air Force MIL-STD-1750 standard instruction-set architecture.

The system is supported by Jovial software, and packaging dimensions can be between ½ and 1 ATR, depending on the features incorporated. The M372 computer is used as the General Dynamics F-16's enhanced fire control computer, as the Martin Marietta Lantirn pod control computer, as the C-5B MADAR multiplexer/processor, and as the Sikorsky HH-60D helicopter's mission computer.

Computer type: binary, parallel, microprogrammed, fixed- and floating-point, two's complement
Word length: 16-, 32-, or 48-bits
Typical speed: 570–720 Kips
Max address range: 1 058 576 (1024 K) words
Instruction-set: MIL-STD-1750A, Notice 1, 262 instructions
Execution time

	Semi-conductor memory	Core memory
Fixed-point		
add/subtract	0.992 μs	1.288 μs
multiply	2.861 μs	3.143 μs
divide	8.111 μs	7.955 μs
Floating-point		
add/subtract	2.770 μs	3.868 μs
multiply	3.424 μs	3.518 μs
divide	10.643 μs	10.408 μs

Dimensions (typical installation): 112 × 170 × 190 mm
Weight: 16 lb (7.3 kg)
Power: 100 W, 28 V dc

STATUS: in production.

Magic IV general-purpose processor
The Magic IV computer is the fuel savings advisory and cockpit avionics system (FSA/CAS) computer for the US Air Force's C-135 and KC-135 aircraft. It is also used in the performance management system in the Boeing 747 and McDonnell Douglas DC-10 and MD-80 series.

The hardware used in this new processor is unrelated to earlier Delco processors. Magic IV is a high-performance, all large-scale integrated micro-computer system, which promises to

provide lower cost, reduced power, weight and size, greater modularity and reliability advantages over existing machines. Typical operating speed is 250 000 operations a second (250 Kops). There are three basic machines, the M4116, M4124 and M4132, available with 16-, 24- and 32-bit word architectures respectively. They are seen to be suitable for remote-terminal or sensor-orientated processor applications.

Almost exclusive use of large-scale integrated circuits has resulted in a great reduction in the number of components compared to a conventional processor. The large-scale integrated units are:

	M4116 (16-bit)	M4124 (24-bit)	M4132 (32-bit)
Cpu-control unit	1	1	1
Cpu-arithmetic unit	4	6	8
Input/output control unit	2	3	4
Memory controller	2	3	4
TOTAL	9	13	17

Additional large-scale integrated circuits may be needed to meet programmable communication-interface and digital-input requirements. Mean time between failure estimates ranges from 27 000 hours for a simplex configuration to 150 000 hours for a dual-redundant system. Nmos large-scale integrated circuits are used with typical component densities of 1000 gates and 5000 transistors per 250-mil chip, with pair-gate delays below 5 nanoseconds. Delco claims better nuclear-radiation tolerance than contemporary dynamic nmos circuits due to the static-logic design, substrate bias design and exclusive use of nor logic. Digital inputs compatible with ARINC 561, 575 and 583 can be provided.

Computer type: binary, parallel, fixed-point, two's-complement
Word length: 16-, 24- or 32-bit
Typical speed: 250 Kips
Max address range: 32 768 (32 K) words
Instruction-set: 89 (expansible)

Execution time	Semi-conductor memory
add/store/jump	2.5 μs
store (double-length)	3.5 μs
add (double-length)	5.5 μs
multiply	11 μs

Dimensions: 167 × 69 × 127 mm
Weight: 3 lb (1.4 kg)
Power: 25.3 W

STATUS: in production.

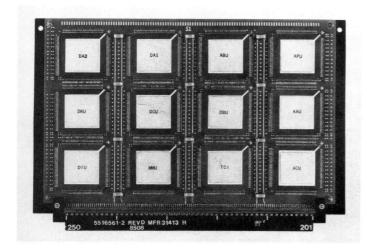

Delco Magic V computer cpu side (left) and memory side (right)

Magic V general-purpose processor

The Magic V processor is an all-vlsi implementation of a cpu and memory system that executes the MIL-STD-1750A, Notice 1 instruction set architecture. Designated M572, this single-card computer is being developed on contract for the EF-111 upgrade programme, for the Grumman F-14 and A-6F aircraft display system, and as both an engine monitor system and a turbine engine control processor for the LHX helicopter. It is also being employed as a mission computer on two classified programmes for the US Air Force.

The Magic V processor utilises 3-micron bulk cmos technology to configure a complete MIL-STD-1750A central processor in just 10 vlsi parts. An eleventh vlsi part provides extended memory management to address up to 1 million words, and a twelfth chip provides an IEEE-488 bus interface that enables external communication with the cpu and memory for monitoring performance and for developing software. These 12 vlsi chips are mounted on one side of a single ½ ATR size circuit card assembly.

The back side of the single circuit card assembly mounts the vlsi memory controller that interfaces the cpu with the memory devices mounted on this back side. Typical configurations of M572 include up to 192 K words of cmos ram plus a start-up rom, or various combinations of ram, eeprom, uvprom devices, all capable of being addressed by the programmable vlsi memory controller.

The M572 typically operates at 850 000 to 1 million operations per second running the DAIS instruction mix. A built-in feature of the M572 is the ability to couple multiple cpu/memory cards in a multiprocessor configuration. The EF-111A utilises six identical cpu/memory cards to yield five million instructions per second. The chip design and mechanisation provides extensive built-in test and fault management hardware and software, leading to very high fault detection and isolation, typically 98 per cent. A built-in self-checking capability provides virtually 100 per cent fault detection in dual cpu installations.

Computer type: general-purpose, microprogrammed, fixed- and floating-point, custom cmos vlsi, ttl compatible interfaces
Word length: 16-, 32-, and 48-bits; 48-bit logic unit
Instruction-set: MIL-STD-1750A, Notice 1; all specified options
Memory size: up to 192 K words cmos ram on ½ ATR size circuit card; up to 256 K words cmos ram on SEM-E size circuit card
Max address range: 1 million words

Addressing modes: direct, indirect, immediate, indexed, non-indexed, relative, base relative, BIT, byte
Input/output: all mandatory features and non-application-dependent MIL-STD-1750A options implemented, discrete and digital outputs/inputs, analogue outputs/inputs, and MIL-STD-1553B bus interfaces implemented in various system level applications
Test interface: built-in IEEE-488 interface for DMA operations, test communications, software development, and real-time performance monitoring
Execution time (semi-conductor memory)

	Single precision	Double precision
load	0.218 µs	1.356 µs
add/subtract	0.356 µs	3.013 µs
multiply	2.150 µs	4.082 µs
divide	4.772 µs	6.482 µs

Dimensions: 4.3 × 6.4 × 0.5 inches (109 × 163 × 13 mm) (cpu and memory)
Weight: 0.45 kg, with memory side fully populated
Power
(cpu) 2.2 W
(cpu and memory) 7W

STATUS: under development

Fairchild

Fairchild Communications and Electronics Company, 20301 Century Boulevard, Germantown, Maryland 20874
TELEPHONE: (301) 428 6000
TELEX: 892468
TWX: 710 828 9700

The company produces three MIL-STD-1553-based systems which perform different functions in data-bus equipment already in operational service. These are used in the US Air Force-General Dynamics F-16 and the Fairchild A-10 aircraft.

Master bus controller

This unit provides total control over data passing between sub-systems on a MIL-STD-1553A data-bus. It interfaces to dual-redundant buses and is programmed to adhere to MIL-STD-1553A protocols. The unit commands and controls all data transfers, provides for validity checks, detects unreliable remote terminals by analysing status words, recognises data-bus failures and automatically switches to the redundant data-bus in the event of a fault being detected. It also contains self-test features to ensure confidence in internal operations. The programming technique can accommodate additional remote terminals and messages with minimum changes to the software or hardware. There is also capacity for extra computational tasks within the microprocessor in addition to the bus-control function.

Dimensions: 170 × 343 × 193 mm
Weight: 14 lb (6.35 kg)
Power: 115 V ac 400 Hz 3-phase
Design and construction: MIL-E-5400
EMC test: MIL-STD-462
Environmental test: MIL-STD-810B

STATUS: in production.

Remote terminal

This is a compact system that performs analogue to digital conversion of aircraft control signals. It receives analogue information from controls and discrete switches, and digital data from systems such as radar altimeters and radar systems for transmission on a MIL-STD-1553A/B data-bus. The unit contains a multiplexed converter, which scales and formats data to comply with data-bus requirements and receives discrete switch analogue and digital inputs. Flexibility is incorporated by using microprocessor control and modular design, making the unit easily expansible and an effective system for the integration of analogue-control sub-systems onto a MIL-STD-1553A/B data-bus.

Dimensions: 170 × 343 × 193 mm
Weight: 12 lb (5.44 kg)
Power: 115 V ac 400 Hz 1/3-phase
Design and construction: MIL-E-5400
EMC test: MIL-STD-461
Environmental test: MIL-STD-810B

STATUS: in production.

Data transfer equipment

This is a general-purpose solid-state memory cartridge and memory-management system providing avionics computers and sub-systems real-time access to mission planning data. A multi-mode file-access system, under microprocessor control, performs all file-management functions, so reducing user overhead processing and simplifying interfaces. This system, in conjunction with a ground-based computer, permits the loading of pre-flight mission data, maintains in-flight memory access recordings and performs post-flight analysis. The system features high reliability, real-time data exchanges with the MIL-STD-1553 data-bus, and has a high capacity non-volatile memory cartridge. It uses the MIL-STD-1553 data-bus, MIL-STD-1750 hardware and meets MIL-STD-1589B (Jovial J73) software requirements. It comprises a cockpit unit and an interchangeable cartridge. In the US Air Force/General Dynamics F-16 the cockpit unit can be operated by the pilot and has random access storage of between 8 and 131 K 16-bit words. Addressing for up to 2 M words capacity is provided for future high-density cartridges.

Dimensions: 178 × 146 × 114 mm
Weight: 7 lb (3.18 kg)
Power: 110 V ac 400 Hz, 1-phase
Cartridge data capacity: 8 K 16-bit words
Cpu speed: 200 Kips

STATUS: in production. Early in 1985 General Dynamics ordered 780 shipsets of this equipment, value $15 million, for installation in F-16 aircraft.

General Electric

General Electric Company, Space Systems Division, Valley Forge Space Center, PO Box 8555, Philadelphia, Pennsylvania 19101
TELEPHONE: (215) 962 2000

Airborne computers

General Electric supplies airborne computer-based systems for both military and aerospace markets. The first application of a microprocessor to an airborne product built by the Aircraft Instruments Department was the signal-conditioning and distribution unit for the Rockwell B-1A in the early 1970s. With the revival of the aircraft as the B-1B a redesign of the system was to take advantage of newer technologies which include 16-bit I²L radiation hardened design microprocessor and large-scale integrated components. The system interfaces with 64 engine sensors plus 12 more from the aircraft and electrical multiplex sub-system.

The B-1B system includes a dual redundant signal processing and power supply to enhance single point failures, signal conditioning, sensor excitation, data conversion, linear computations, thrust computations, periodic and initiated self-test parameter compensations for adverse temperature and generate warning messages. Output information is transmitted via a dual redundant digital EMUX word-generated interface with aircraft avionics. The computer can perform 7 million instructions a second.

STATUS: under development.

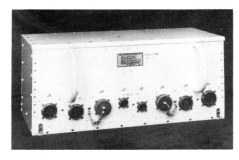

Signal conditioning and distribution unit for Rockwell B-1B

Kearfott

The Singer Company, Kearfott Division, 1150 McBride Avenue, Little Falls, New Jersey 07424
TELEPHONE: (201) 785 6000
TWX: 710 988 5700

SKC-3140 bus controller and main computer for AM-X

Two Kearfott SKC-3140 bus controllers and computers equip each Aeritalia/Aermacchi/Embraer AM-X aircraft, providing the main and back-up computations and data control.

A single ¾ATR lru, containing computer, bus controller and power supplies controls data from navigation and attitude sensors, radar, head-up display, weapon-aiming and communications controllers and the various control and display panels in the aircraft, and data is passed around the aircraft on a MIL-STD-1553B dual-channel data-bus.

Weight: 20 lb (9 kg)
Reliability: over 3600 h mtbf
Power requirements: under 200 W at 115 V 400 Hz

STATUS: in production.

AN/AYK-18 general-purpose computer

This unit is in production for the USAF/General Dynamics F-111 fighter avionics update programme. Two units are installed in each aircraft; one is the general navigation computer (GNC), and the second is designated the weapon delivery computer (WDC). In conjunction they are called the weapons/navigation computer (WNC) and when mounted in one rack with an advanced micro-electronics converter (AMC), the units are known as the digital computer complex (DCC).

The WNC has been designed as an upgraded central computer for the F-111 for use until the 1990s. These new processors, each based on a single SKC-3132 central processing unit, offer increased computational capability and reliability, and are compatible with existing and potential instruction-set architectures (CP-2/CP-2EX and MIL-STD-1750 respectively on the F-111).

The AMC is a programmable, multiplexed, solid-state data converter. It is divided into three sections, two of which provide instruction execution, serial digital data processing and analogue-to-digital conversion for the WNC inputs, and the third provides digital-to-analogue conversion for out-going signals. The unit is partitioned on 15 cards plus 1 built-in self-test card for the F-111D version, with an extra processing card for the F-111F.

The built-in test functions identify faults to shop-replaceable unit level. Error reports are displayed on latching indicators on the face of the DCC chassis. Flight personnel can confirm DCC operational status by means of the 7 WNC and 8 AMC indicators.

Computer type: binary, two's-complement, fixed- and floating-point
Word length: 16-bit
Max address range: 128 K words core (64K words available in F-111)
Operating speed: in excess of 450 Kops
Instruction-set: micro-coded, choice of CP-2EX or MIL-STD-1750
Micro-code memory: 4 K words
Input/output options: 24 inputs/16 output discretes, plus 8 external interrupts and two MIL-STD-1553B rt/bc ports
Dimensions: 193 × 258 × 471 mm
Weight: 48.5 lb (22 kg)
Power: 400 W MIL-STD-704
Built-in test: exceeds 95% capability to sru
Reliability: (calculated) 3026 h mtbf (specification) 2000 h mtbf

STATUS: in production.

Digital computer complex (DCC) for US Air Force/General Dynamics F-111

Kearfott AN/AYK-18 general-purpose computer

LTV

LTV Aerospace and Defense Company, Sierra Research Division, 247 Cayuga Road, PO Box 222, Buffalo, New York 14225
TELEPHONE: (716) 631 6200
TELEX: 6854308
TWX: 710 523 1864

Airborne data link

Late in 1985 LTV received a $34 million contract to develop an airborne platform and telemetry system. Radar information, detected by an electronically-steerable phased array antenna mounted in a de Havilland DHC-8, will be transmitted to a ground station, forming part of the Gulf Range Instrumentation System at Tyndall Air Force Base, Florida. The use of the link in this way extends the effective area of the range to 200 miles offshore.

M/A-COM

M/A-COM, Government Systems Division, 3033 Science Park Road, San Diego, California 92121
TELEPHONE: (619) 457 2340
TWX: 910 337 1277

LMC-1750A MIL-STD computer

The LMC-1750A was developed primarily as a satellite communications processor for an airborne application but is also available in two configurations for other customers. As a general-purpose processor it has a 1 M word internal memory and up to 440 input/output lines. The secure processor version offers multi-level security software, red/black data separation and Tempest hardening. The LMC-1750A has full MIL-STD-1750A architecture and runs on MIL-STD-1589B (Jovial J73) software.

Typical applications include communications processing, data processing in terminals, plasma display control and dynamic antenna steering and control.

Throughput: 750 Kops
Memory: 1 M byte, 16-bit words
Input/output: up to 440 lines using RS232, RS-422, MIL-STD-188C, A/D, D/A, etc. Compatible with MIL-STD-1553B data highways
Dimensions: (including mounting tray) 9.81 × 7.94 × 25.25 inches (249 × 202 × 641 mm)
Format: ¾ ATR chassis
Weight: 59 lb (26.8 kg)
Power input (max): 420 W
Power output (max): 250 W

Mitre

The Mitre Corporation, Burlington Road, Bedford, Massachusetts 01730
TELEPHONE: (617) 271 2425
TELEX: 923458

Airborne data link

A new generation airborne data link receiver has been developed by the Mitre Corporation as a low data-rate communications link from a satellite. The link is intended for a variety of requirements, including passing communications from air traffic control, meteorological information and the operating company, as well as providing public communications.

STATUS: in development. A series of successful flight trials were carried out in a Rockwell Sabreliner 65 during the summer of 1985.

Norden

Norden Systems Inc, subsidiary of United Technologies, PO Box 5300, Helen Street, Norwalk, Connecticut 06856
TELEPHONE: (203) 852 5000
TELEX: 965898
TWX: 710 468 0788

AN/AYK-42(V) processors

The most significant machine in this designated range is the Norden PDP-11/34M processor. All machines in the series are airborne processing units which are software and interface-compatible with Digital's PDP-11 system. Fully militarised, they are suitable for applications ranging from tactical avionics to complex command and control.

The PDP-11/34M unit includes a complete processor, memory, peripheral interfaces and power supply on a single chassis. Modular unit construction assures quick and easy replacements.

Features include a complete PDP-11 instruction set (over 400 instructions), 1 K word cache memory option, memory expansible up to 128 K words, using 16 K or 32 K word memory modules, 900 nanoseconds cycle time in core memory, memory management and protection, hardware multiply and divide, floating-point processor option, hardware stack-processing and an input/output rate up to 1.1 M words a second.

Another processor in this range, the Norden PDP-11/70M, can comprise up to four 1 ATR boxes, each dedicated to processing, power supply, memory (256 K words) and expander (extra input/output options) facilities respectively. This unit can perform up to 850 000 instructions a second. The Norden LSI-11M is a single-card version of the same processor design which operates at approximately 200 000 instructions a second. It has 4 K words of random access memory and can be associated with 16 K or 32 K word core storage modules.

The specification below refers to the Norden PDP-11/34M.

Computer type: binary, fixed- and floating-point
Word length: 32-bit or 64-bit
Instruction-set: > 400 instructions (as PDP/11)
Typical speeds: (without cache) 275 Kips (with cache) 400 Kips
Execution time
(add) 1.9 μs
(multiply) 9.0 μs
(divide) 12.9 μs
Input/output: 5-12 slots. Up to 1.1 M words/s
Memory: (core) up to 128 K words
Environmental specification: MIL-E-5400
Dimensions: 498 × 257 × 194 mm
Weight: 50 lb (22.7 kg)
Power: 410 W max

STATUS: in production.

Rockwell

Rockwell International, North American Aircraft Operation, PO Box 92098, 100 North Sepulvedra Boulevard, Los Angeles, California 90009
TELEPHONE: (213) 647 1000
TELEX: 664363
TWX: 910 348 7101

MIL-STD-1553 electrical multiplex system for B-1B

The data transmission (and power-distribution control) system used in the US Air Force/Rockwell B-1B is an improved version of the MIL-STD-1553 system which was operational on the previously developed B-1A prototypes. Built by Harris Corporation, Government Information Systems Division, the system will control the B-1B major sub-systems using techniques which eliminate approximately 32 000 wires and cables. The system is one-third lighter than an equivalent conventionally-wired system.

Terminal flexibility is such that extra units can be deleted or added for future interfacing with new inputs or outputs. As currently configured the system acquires, commands and controls more than 9000 inputs and outputs. It is designed to govern such functions as electrical power distribution to sub-systems and avionics equipment, landing gear, engine instruments, fuel, weapons and environmental-control systems, and aircraft lights. The system collects and conditions signals at remote terminals and transmits them between any two points in the aircraft over a common data-bus. The saving in wire length over a conventional system is about 80 miles.

The system also supervises all signal data, using a centralised controller. This unit routes data between any two points in the system, and can solve combinational, sequential or interlock equations to produce output commands. A significant cost saving has been achieved through the replacement of the hybrid integrated circuits used in the B-1A, with monolithic circuits in the B-1B system. The semiconductor devices are manufactured by Harris Corporation, Texas Instruments and General Instruments.

STATUS: under development.

Sanders

Sanders Federal Systems Group, 95 Canal Street, Nashua, New Hampshire 03061-2034
TELEPHONE: (603) 885 9760
TELEX: 943430
TWX: 710 28 1894

HMS-1750A processor

The HMS-1705A is a high speed processor designed to run the MIL-STD-1750A instruction-set architecture. It is designed to full military standards and is suitable for a wide range of applications, including airborne.

The MIL-STD-1750A set is executed at 1.75 million instructions per second (Mips) in the standard digital avionics information system (DIAS) mix, or 4 Mips in fixed point computations. The standard four-card configuration (the cards are 6 × 9 inches) includes 64 K of ram, with space for an additional 128 K words of memory and a variety of interfaces. The HMS-1750A can also be packaged in a ½ATR box, including power supplies.

Weight: 7 lb (3.2 kg)
Power requirement: 52 W at 5 V dc

STATUS: in production.

Sperry

Sperry Corporation, Sperry Aerospace and Marine Group, Defense Systems Division, PO Box 9200, Albuquerque, New Mexico 87119
TELEPHONE: (505) 822 5000
TELEX: 501386

MIL-STD-1553 data-bus

Initially developed for the Space Shuttle programme (where it was called a multiplexer/demultiplexer system) Sperry MIL-STD-1553 data-bus equipment is now available for the US Army/McDonnell Douglas Helicopters AH-64A Apache. The data-bus integrates the Martin Marietta tactical airborne designator/pilot night vision sensor systems (TADS/PNVS) with aircraft fire-control components. Primary bus control is vested in the fire-control computer, with the Sperry Type III MRTU (multiplex remote terminal unit) providing back-up bus control. Bus controller outputs are used to control weapons and to provide information to the cockpit. Features include built-in fault-detection and location, including a sub-system which alerts the flight-crew and maintenance personnel to the operational status of the aircraft avionics.

Sperry produces three types of MRTU each of which represents a different level of remote terminal capability:

Type I MRTU is the basic building block and operates between MIL-STD-1553 data-bus and aircraft systems.

Dimensions: 127 × 178 × 188 mm
Weight: 10 lb (4.54 kg)
Power: 25 W

Type II MRTU is specifically designed for installation in small or oddly-shaped spaces, such as the pylons of a helicopter.

Dimensions: 79 × 109 × 215 mm

Weight: 4 lb (1.8 kg)
Power: 10 W

Type III MRTU incorporates a high-speed digital computer (Sperry SD-175) and is designed to provide primary, back-up or shared bus control. It also provides functions such as fire-control, stores management, guidance and navigation, on a time-shared basis.

Dimensions: 127 × 178 × 260 mm
Weight: 12 lb (5.44 kg)
Power: 40 W

STATUS: in service.

Sperry Corporation, Commercial Flight Systems Division, PO Box 21111, Phoenix, Arizona 85306
TELEPHONE: (602) 869 2311
TELEX: 668419

AN/AYK-14 computer

Sperry has been chosen as a second source for the US Navy's standard general-purpose computer, the AN/AYK-14 (see entry under Control Data). The Sperry version incorporates 64 K word memory store units produced by Electronic Memories and Magnetics Corporation.

1416A general-purpose processor

This compact general-purpose digital processor provides highly reliable and powerful processing for general avionics, fire-control, communications, weapons and counter-measures control, and electronic warfare applications. The software base, Sperry U1616, is AN/UYK-20 compatible and host compilers are available for Coral 66, Fortran and CMS-2 high-order languages. There is a real-time executive facility for scheduling interrupt handling, error-processing and system initialisation.

Sperry 1416A general-purpose processor

Computer type: binary, fixed- and floating-point, two's-complement
Word length: 16-bit
Typical speed: 700 Kops
Max address range: 512 K words

Memory type	Access time	Cycle time
hmos ram	170 ns	300 ns
nmos ram	380 ns	500 ns
cmos ram	380 ns	500 ns
prom	230 ns	350 ns

Typical execution time
(add) 0.50 and 1.50 μs
(multiply) 2.75 and 3.50 μs
(divide) 6.25 and 7.25 μs
Format: 1 ATR
Weight: 33 lb (15 kg)
Power: 175 W at 28 V dc

STATUS: in production.

Teledyne

Teledyne Ryan Electronics, 8560 Balboa Avenue, San Diego, California 92123
TELEPHONE: (619) 569 2413
TELEX: 695028
TWX: 910 335 1174

TDY-750V micro-computer

Teledyne has developed a single-card micro-computer, designated TDY-750V, designed to MIL-STD-1750A with a projected speed of 1.5 million operations a second (1.5 Mops). First deliveries of this version were expected to be made towards the end of 1985, with a radiation

hardened variant, capable of 1.8 Mops, available during 1986. The system uses cmos silicon-on-sapphire circuits and will have 128 K words of memory on the same 6 × 9 inches (152 × 229 mm) card.

STATUS: in development.

US Air Force/Navy

United States Air Force, Aeronautical Systems Division, Air Force Systems Command, Wright-Patterson Air Force Base, Dayton, Ohio 45433
TELEPHONE: (513) 255 2725

Pave Pillar integrated avionics architecture

Under the direction of US Air Force's Avionics Laboratory, Pave Pillar is a development activity being conducted in conjunction with industry to integrate to a hitherto unprecedented extent the various electronic systems that will feature in the next-generation combat aircraft such as the US Air Force's ATF advanced tactical fighter and the US Navy's advanced tactical aircraft (ATA), and updates of current fighters such as the F-16. It will also incorporate new technologies in vhsic, sensors, voice command, colour graphics and algorithms, as well as fibre-optics and Ada-based software.

Behind Pave Pillar is the desire to improve the reliability of a total avionics suite, thereby improving the availability of aircraft on the flight-line; cut down the space occupied by electronics equipment, firstly by employing vhsic micro-miniature chip technology, and secondly by preventing wasteful duplication of sensors, processors, power supplies and other sub-systems common to many items of equipment; cut cost; and further reduce a pilot's

workload. Reliability and space-saving will be further enhanced by major reductions in the number of interconnecting cables and connectors; reductions of as much as 90 per cent are foreseen. The advent of vhsic circuits and high-speed digital data multiplexing would permit what are currently line-replaceable units (lrus) to be replaced with printed circuit boards or line replaceable modules (lrms). The architecture proposed would also permit built-in test circuits to check the entire system down even to the level of individual chips, cutting down or even eliminating the intermediate avionics shop facilities. The avionics would no longer be contained in lrus but in lrms.

The Avionics Laboratory has done a comparison, using the US Air Force/General Dynamics F-16, to demonstrate the advantages of Pave Pillar. The current F-16 contains 58 lrus in its avionics suite, has an avionics mean time between failures of 7.3 hours, and can fly an average of three missions a day. The same aircraft with Pave Pillar would have 58 lrms weighing only a fraction of that of the current avionics suite, mtbf would climb to around 35 hours, and availability would increase to about 4.5 missions a day. The 38-man avionics intermediate shop needed to maintain a 72-aircraft wing would be eliminated, and with it the need to deploy six Lockheed C-141 Starlifters with every location change. The proportion of maintenance time occupied by avionics as a proportion of that of the whole

aircraft would decrease from 17 to 5 per cent. The number of avionics personnel needed to maintain the avionics of a wing would be halved, from 142 to 71.

Seven companies (Boeing, Rockwell, Northrop, General Dynamics, Grumman, Lockheed and McDonnell Douglas) are currently working on $1 million Pave Pillar Phase 1 contracts which end in October 1986, at which time the most promising technologies will be transferred to the ATF programme.

Phase 1 includes hardware and software design and development of modular packaging concepts and other design work for testing in the later phases.

The ultimate Pave Pillar-derived system will be highly fault tolerant; that is the computer will be able to recognise and isolate faults and bring alternative or 'back-up' processing on-line instantly.

A common signal processor, for radar, electro-optics and electronic warfare functions is being developed and the integrated navigation module is based on ring-laser gyros: Westinghouse is developing the Ultra Reliable Radar (URR) with Pave Pillar compatible architecture. (See entries for URR and INEWS.) The ICNIA programme involving TRW and ITT/Texas Instruments is also related to Pave Pillar.

STATUS: US Air Force development effort – risk reduction for ATF programme.

Pilot's Associate

Pilot's Associate (PA) is one of the leading US Air Force research and development efforts involving artificial intelligence as a means of enabling the crew of an attack aircraft or helicopter to receive better and more effective information from the sensors. Like Pave Pillar, ICNIA and INEWS (see separate entries), PA will will greatly advance avionics technology and will have a number of applications, including the Advanced Tactical Fighter and LHX helicopter projects.

PA will provide the pilot with an optimum interface with his sensors and aircraft systems, including, in near real-time, data on tactics and mission planning and situation assessment. It will do this by combining under one software program the data from aircraft sensors and systems with an expert system, which will bring human knowledge and experience to bear on the immediate situation very many times faster than even the most adept pilot could make an assessment.

PA entered the research and development phase at the beginning of 1986 with the awarding of three-year cost-sharing contracts to two teams headed by Lockheed-Georgia and McDonnell Douglas. The former team, including General Electric (USA), Teknowledge, Goodyear Aerospace, Search Technology, Defense Systems and the Carnegie-Mellon University (a world leader in expert systems research) has a contract valued at almost $12.5 million, while the latter, which includes Texas Instruments, has a $8.4 million contract.

The initial phases of PA concentrate on five areas of research; systems status, mission planning, situation assessment, tactics and pilot interface.

The systems status function will monitor aircraft functions and diagnose faults and determine their effect on aircraft and mission performance. It might even be able to reconfigure the aircraft after battle damage to enable a recovery to base (see Self Adaptive Technology entry).

In mission planning, PA will take in navigation, threat avoidance fuel management and route replanning to give the pilot increased effectiveness to respond to changing situations. Indeed, situation assessment is the critical area in the PA programme, for if the software cannot define the situation clearly, then the rest of the algorithms will be even less well defined. Threats will have to be prioritised, taking into account the mission objective, and friendly forces will also have to be considered. Tactics will also be refined by the PA system computer to take account of the instantaneous situation, and provide advice to the pilot. In this respect the computer will have to be able to understand what the pilot intends to do, in basic terms, in particular situations.

The tactics function extends beyond the specific, single aircraft and co-ordination of effort between all aircraft participating in a mission is essential. The display of information to the pilot, and how much of the computer-derived knowledge is acted upon automatically, or presented to the pilot for his decision, is another crucial aspect of PA. Speech demands are one option here.

Following the current three-year phase, there will be a second phase lasting two years which will include advance development and real-time simulation: a formal demonstration of achieved progress is scheduled for 1988.

Data recording

CANADA

Canadian Marconi

Canadian Marconi, Avionics Division, 2442 Trenton Avenue, Montreal, Quebec H3P 1Y9
TELEPHONE: (514) 341 7630
TELEX: 05-827822
TWX: 610 421 3564

CMA-879 engine health monitor

The CMA-879 is a microprocessor-based data gathering system for use on aircraft or helicopters with up to three engines. It provides interfaces to existing sensors on the engine and aircraft systems, before performing preliminary data assessment and logging. The system can be used for the following real-time data acquisition applications: flight profile and engine usage recording, engine event detection, aircraft system event logging, time/temperature cycle recording, maintenance data recording or engine performance analysis. Each set of data is annotated with an event time from a clock, and an engine or aircraft identifier can be preset for identification during off-aircraft analysis. The sets of data can be collected at preset elapsed time intervals or continuously (using a high sampling rate). They can be triggered by a particular engine or system event, by a significant change in a parameter or group of parameters, or by combination of these factors.

Data is stored in a non-volatile memory and information can be extracted by a data collection unit without removing or opening the system. The data can be stored permanently in a mass storage cassette or analysed by the data collection unit. A remote data storage unit, which can be fitted in a survivable portion of the airframe, is also available. Maintenance needs no flight-line equipment, since extensive built-in test and automatic self-test can identify failures down to card-replacement level. Self-contained signal simulation and calibration modules check the unit after card replacements.

Dimensions: 127 × 198 × 350 mm
Weight: 4.1 kg
Input capacity: up to 200 channels per unit. Preferred input types include dc analogue (0-10 V) linear, ac analogue (±10 V) linear, discrete ground 28 V dc, or logic level, thermocouple (K or E), resistance bulb type 2, synchro, pulse (up to 100 000 pps), frequency (0.7-10 000 Hz) vibration, ARINC 429, pressure (variable reluctance)
Outputs: RS 232C or RS 423 serial port
Maintenance display: 4 × 7 segment led display, built-in test information
Memory: earom or bubble (192 K to 2 M byte)
Max sampling rate: every 100 ms
Nominal sampling rate: 1 K byte/engine/h
Environmental specification: MIL-E-5400 Class 1A

STATUS: under development.

CMA-776(M) engine health monitor

The CMA-776 status display system is a caution/warning system with optional check-list readout. The two most important current faults are displayed to the crew on a two-line gas-discharge display. Reduced pilot workload, increased safety, and reduced panel space are cited as advantages for the CMA-776. Engine performance monitoring can also be incorporated. By removing the control and display unit, the CMA-776 can become a pure engine health monitor, when it receives the designation CMA-776(M). All engine parameters are sampled at 30-second intervals, and all out-of-limit conditions are recorded whenever they occur. The result is extracted from the computer's solid-state memory by connection of a printer or digital cassette when the aircraft lands. Environmental and flight data is recorded alongside the engine parameter sampling.

STATUS: in production.

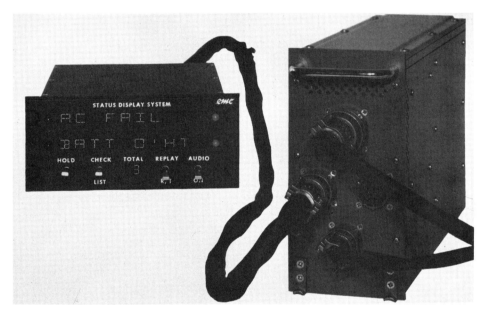

Canadian Marconi CMA-776 engine status display system

FRANCE

Electronique Serge Dassault

Electronique Serge Dassault, 55 quai Carnot, 92214 Saint-Cloud Cedex
TELEPHONE: (1) 46 02 50 00
TELEX: 250787

Airframe fatigue recorder

This device measures the steady-state load factors imposed on an aircraft during normal manoeuvres and the transient loading due to gusts imposed upon its structure. The system comprises an accelerometer and a separate counter and storage unit. The accelerometer is mounted as near to the aircraft centre of gravity as possible.

The accelerometer generates voltages proportional to the imposed loads, and these are compared in the counter and storage unit with reference voltages corresponding to pre-selected g-thresholds. A series of veeder-counters on the face of the storage unit display the number of counts at each g level.

Accelerometer
Effective measuring range: –10 to +10 g
Linear frequency range: 0-10 Hz

ESD structural fatigue recorder (SPEES)

Sensitivity to cross-acceleration: 0.02 g/g
Dimensions: 60 × 60 × 38.5 mm
Weight: 0.22 kg

Counter and storage unit
Number of thresholds: 8
Accuracy: 0.2%

Power: 10 W at 115 V 400 Hz
Dimensions: 180 × 85 × 76 mm
Weight: 1.4 kg

STATUS: in production. Over 1500 have been built and fitted to all French-built combat aircraft for national use and export. The system is manufactured under licence in West Germany for the Alpha Jet strike/trainer and Transall C-160 transport.

SPEES structural fatigue recorder

Launched during 1983, SPEES (Système pour l'Evaluation de l'Endommagement Structural) records the parameters of fatigue in the metal structure of an airframe. Aerodynamic sensors, accelerometers and strain gauges mounted on the airframe provide measurements that are processed and condensed into a format devised by CEAT (Centre d'Essais Aeronautiques de Toulouse). The data format reflects the principle that this type of structural damage, caused by loads imposed on the airframe over a period of time, is related to load variations rather than their absolute values.

The system does not duplicate the function of standard fatigue-meters, but its installation in

10 to 20 per cent of the aircraft in a particular fleet, all of which have fatigue-measuring devices, permits a more accurate and reliable correlation between SPEES data and fatigue-recorder readings.

The data format program is implemented in eprom, the data itself being recorded on a front-loading cassette. SPEES complements the fatigue recorder, providing a more efficient way of scheduling maintenance and of safeguarding an aircraft's service life.

Format: ¼ ATR
Weight: <3 11 lb (5 kg)
Power: 40 W at 200 V 3-phase 400 Hz

Capacity: 20 high-level (0 to +5 V) and/or low-level (0 to ± 16 mV) analogue signals, and 16 28 V discrete signals
Memory: 64 K reprom
Temperature range: –55 to +70° C

STATUS: available.

Enertec

Enertec, 1 rue Nieuport, BP 54, 78141 Vélizy-Villacoublay Cedex
TELEPHONE: (1) 39 46 96 50
TELEX: 698201

A member of the US-based Schlumberger group of companies, Enertec has specialised in the development and production of crash and maintenance flight recorders for civil and military aircraft.

Crash recorders impress digital, analogue or voice information continuously on magnetic tape in accordance with FAA Technical Service Order C51a. This standard protects the tape by specifying that the housing and transport mechanism be capable of withstanding the following conditions:
Shock: an acceleration of 10 000 metres a second for 5 milliseconds
Crushing: 22 700 N on 3 axes for 300 seconds
Dynamic penetration: 227 kg dropped from 3.05 metres on an impact area of 0.32 square centimetre
Fire: 1100° C over at least 50 per cent of outer surface for 30 minutes
Immersion of complete enclosure for 36 hours in a mixture of kerosene, Skydrol (hydraulic fluid) and fire-extinguisher fluid
Immersion in sea-water for 30 days
These crash recorders operate continuously, the direction of the tape drive being reversed

Enertec crash recorder: tape deck remains intact within mechanical and thermal protection after an accident

and track-changing being accomplished in 100 milliseconds.

PE 6573 digital flight data accident recorder
The PE 6573 was designed to meet ARINC 573 defining the flight-data and recording system characteristics for modern jet and turboprop transports.

Format: ½ ATR Long
Weight: (recorder) 12.6 kg
(anti-vibration mounting) 2.1 kg
Recording capacity and data-rate: 25 h, 768 bits/s
Replay data-rate: 4608 bits/s
Power supply: 115 V 400 Hz
Accident survival: to FAA TSO C51a

STATUS: in production.

PE 6010 and PE 6011 digital flight data accident recorders
The PE 6010 is a lightweight digital flight data accident recorder in which weight-saving is accomplished by a relaxation of the dynamic penetration requirement. In terms of accident protection, however, it meets all other requirements of FAA TSO C51a.

The PE 6011 is intended as a replacement for the PE 6010. It is more compact than its predecessor, meets TSO C51a completely with only a small weight increase, and has 50 per cent more capacity. The new design of tape deck is simpler, with attendant gains in reliability and maintainability.

Format: (PE 6010) ½ ATR Short (length 319 mm)
(PE 6011) ½ ATR (length 296 mm)
Weight: (PE 6010) 7.7 kg
(PE 6011) 10.2 kg
Recording capacity, data-rate:
(PE 6010) 16 h, 768 bits/s
(PE 6011) 8 h, 2308 bits/s
Replay data-rate: (PE 6010) 4608 bits/s
(PE 6011) 18 464 bits/s
Power supply: (PE 6010) 18-30.5 V dc
(PE 6011) 12-32 V dc
Accident survival: (PE 6010) TSO C51a (except penetration is static)
(PE 6011) TSO C51a

STATUS: the PE 6010 is in production and service with Dassault Mirage F1 fighters flown by the air forces of France, Greece, Egypt, Iraq, Kuwait, Morocco and Spain. It has been supplied for the Indian HAL Ajeet trainer and French Navy Super Frelon helicopters. The PE 6011 is in development for French Air Force/Dassault Mirage 2000 and 4000 fighters.

PE 6013 and PE 6015 digital flight data and voice accident recorders
These two recorders are developments of the PE 6010 and PE 6011 and employ their respective technologies. Both record voice as well as data.

Format: (PE 6013) ½ ATR Short
(PE 6015) ½ ATR (length 296 mm)
Weight: (PE 6013) 11.2 kg
(PE 6015) 11.2 kg
Recording capacity, data-rate:
(PE 6013) 90 minutes (voice) 6 h, 2304 bits/s (digital)
(PE 6015) 30 minutes (voice) 1 h, 2816 bits/s (digital)
Replay rate: (PE 6013) 13 824 bits/s
(PE 6015) 2816 bits/s
Power supply: (PE 6013) 12-32 V dc
(PE 6015) 20-30.5 V dc
Accident survival: (PE 6013) TSO C51a
(PE 6015) TSO C51a

STATUS: the PE 6013 has application to Dassault Mirage 2000 fighters and the PE 6015 has been fitted to prototypes of Panavia Tornado.

PC 6033 general-purpose digital cassette recorder
This device records serial data on a continuous basis over a long period of time. Typical applications are performance and maintenance recording, engine health monitoring, aircraft testing, and reconnaissance. The system complies with ARINC 591. Cassettes can be replayed at 80 times the recording speed.

Recording capacity, data-rate: 50 h, 138 M bits
50 h at 768 bits/s
25 h at 1536 bits/s
12.5 h at 3072 bits/s
Number of tracks: 12
Recording code: bi-phase L or M
Power supply: 19-32 V dc or 115 V 400 Hz
Error rate: <1 bit in 10⁵

STATUS: in production for and in service with European airlines including Lufthansa (A300, DC-10, B-747), Alitalia (A300, B-747, B-727), KLM (A310), Swissair (A310) and Austrian Airlines (A310) as part of AIDS (aircraft integrated data system) equipment.

Enertec PE 6011 crash recorder

Enertec PC 6033 performance/maintenance recorder

PS 6024 cassette memory system

This is a digital magnetic cassette tape recorder/reproducer designed for operation under severe environmental conditions. It comprises a cassette drive unit, microprocessor-controlled central processing unit, and electronic interface circuits. Its chief function is as an easily integrated mass memory unit for military computers with standardised interfaces.

Format: 3/8 ATR Short
Interfaces: V 24/RS 232, MIL-STD-1553B, Digibus, HDLC as required
Capacity: 6 M bytes formatted (1 K blocks)
Data density on tape: 1600 bits/inch
Max recording rate: 2.2 K bytes/s usable data
Format: Forward and backward serial recording in bi-phase code
Recording speed: 12 inches/s (300 mm/s)
Rewind speed: 50 inches/s (1270 mm/s)
Max length of data blocks: 2048 (2 K) bytes
Inter-block gap: 20 mm

STATUS: in production.

Enertec PS 6024 cassette memory system

PS 6028 cartridge memory system

A magnetic-tape mass-memory for military systems, the PS 6028 can operate in extremely severe environments. Main characteristics are: tape transport in a hermetically sealed, interchangeable cartridge; data capacity in excess of 6 M bytes with optional extension to 12 M; a formatter for handling data and detecting errors; and coupler options to permit interfacing with a variety of standard data-buses. The cartridge and memory are packaged into a 1/2 ATR Short case. The design and construction are appropriate to mounting in exposed locations in aircraft, vehicles and ships. Typical functions are program and data loading and archiving.

Capacity: > 6 M bytes, formatted in 1 K blocks. Optional extension to 12 M
Recording density: 2400 bits/inch
Peak recording rate: 7 K/s
Recording mode: serial, track by track to and from
Recording code: bi-phase level
Write and read speeds: 30 inches/s (760 mm/s)
Rewind speed: 90 inches/s (2290 mm/s)
Block length: ≤ 2048 (2 K)) bytes
Number of tracks: 8 (16 optional)
Inter-block gap length: 1.75 inches (45 mm)

STATUS: in production.

Sealed cartridge recorders

Enertec has developed a family of data storage devices for use in severe environments where conventional cassette recorders could not be used. Sealed cartridges are employed and typical applications include externally-mounted gun-pods, weapon pylons, remotely piloted vehicles, and in situations in which humidity, temperature extremes and contaminants such as mud, sand, lubricating oil and hydraulic fluid

Enertec ruggedised streaming tape drive

would damage the tape and transport mechanisms in conventional cassettes.

Central to this recorder range is the PA 3000 sealed cartridge, housing a tape deck, magnetic tape reels, record/replay heads, tape drive motor and electronics in a box 8.27 × 5.12 × 1.73 inches (210 × 130 × 44 mm) and filled with an inert gas.

STATUS: in service.

PV 6080 high-speed digital recorder

The PV 6080 is an airborne recorder-reproducer which allows recording of a digital data-stream up to 50 M bits per second during 30 minutes.

The PV 6080 utilises helical scan recording technology and the storage medium is an easy to handle cassette, similar to standard video cassettes. The PV 6080 provides also for two auxiliary analogue channels and one remote control interface. The microprocessor of the PV 6080 controls all functions and a self-test feature. Digital data processing using error correcting codes ensures high data integrity.

Other configurations allow for example recording of 5 M bits per second during 5 hours.

Enertec PS 6028 cartridge memory system

Enertec PV 6080 recorder

Jaeger

Jaeger, Division Aéronautique, 2 rue Baudin, 92303 Levallois-Perret Cedex
TELEPHONE: 757 31 35
TELEX: 620368

Flight-data acquisition unit

The Jaeger flight-data acquisition unit (FDAU), and its associated data entry panel are compatible with the digital flight data recorder requirements of ARINC 573 and 717.

The FDAU codes aircraft parameters into digital format and transmits them to the data recorder. The data entry panel allows the crew to enter flight number, time and events into the system. It also acts as a fault monitor.

STATUS: in production.

Jaeger flight-data entry panel

Omera

Société d'Optique, de Mécanique, d'Electricité et de Radio, Omera-Segid, 49 rue Ferdinand Berthoud, 95101 Argenteuil
TELEPHONE: (1) 39 47 09 42
TELEX: 696797

AA8-123 gunsight recorder

This camera is designed for use with TRT's VE 120 cathode ray tube sight. The system has been recently redesigned as a single unit combining the functions of both photocell assembly for automatic iris control and control box.

Capacity: 9 m × 16 mm film
Frame size: 10.4 × 7.5 mm
Rate: 16 frames/s
Film duration: 70 s
Firing marker: cannon, machine guns
Over-run timer: 0, 2.5 and 5 s
Weight: 1.13 kg
Dimensions: 147 × 110 × 65 mm

STATUS: in production and service.

AA8-130 gunsight recorder

This gun-camera consists of three parts; an optical camera, photocell unit for automatic iris control, and control unit. It has a large magazine, two film speeds, and can be fitted with a choice of two periscopes (with focal lengths of 32 and 50 mm) for installations where the camera cannot 'see' directly the outside world. This recorder is being replaced by the AA8-400 (see separate entry).

Capacity: 22.5 m × 16 mm film
Frame size: 10.4 × 7.5 mm
Rate: 5 and 16 frames/s
Film duration: 70 s
Firing marker: cannon, machine guns
Weight: 1.6 kg
Dimensions: 165 × 110 × 70 mm
Over-run timer: 0, 5, 30, 60 s

STATUS: in service.

AA8-400 gunsight recorder

The new AA8-400 cine camera gunsight recorder is a replacement for the AA8-130 system described above. It uses 16 mm film to photograph the outside world and the symbology superimposed on it by a head-up

display. Mounted above the head-up display, it can see through the combining glass by means of an integral periscope. Film is contained within a magazine that can be easily reached and replaced by the pilot during flight. The shutter operates at a fixed exposure time and is synchronised with the framing sequence of the cathode ray tube on the head-up display. The automatic iris is controlled by a photocell which sees the same view as the camera lens. A firing marker identifies the sequences filmed during actual and blank firings. A push-button injects test signals into the camera and its hybrid electronics.

Omera AA8-123 gunsight camera installed above and left of radar hood

Omera AA8-130 gunsight camera

Dimensions: 120 × 92 × 111 mm overall, including magazine
Weight: (camera) 0.8 kg
(magazine) 0.36 kg
Film: 16 mm standard, 0.15 or 0.11 mm thick; ASA 50, 100 or 200; black and white or colour
Magazine capacity: 15 or 20.5 m, darkroom loaded without tools
Lens: 30, 40 or 50 mm focal lengths, f/2.8-f/32
Image format: 14.5 × 7.5 mm
Exposure time: fixed, 1/50 s
Framing rate: 4 and 16 frames/s
Running time: 164 s at 16 frames/s with 20.5 metres of film

STATUS: in production.

OTA 204 head-up display video system

This system comprises the OTA 204 monochrome video recorder camera, using charge-coupled technology, and the Omera OEV 201 airborne magnetic tape recorder.

The principle is the same as that of the gunsight recorder described separately. The view of the outside world, and the symbology super-imposed on it by the head-up display are transformed into a video signal by the camera and passed to the tape recorder.

The video camera is of modular design and can be mounted as a single unit above the HUD or as two separate units, leaving only a small periscope above the HUD.

Camera
Type: ccd array, 600 × 576 pixels
Spectral response: 0.4 – 1.0 μm
Focal length: 16 mm
Dimensions: 120 × 94 × 113 mm
Weight: 0.5 kg
Tape recorder
Frequency response: 3 MHz
No of channels: 1 video, 2 audio
Recording time: 120 minutes
Power: 20 VA at 11 V ac 400 Hz single-phase
Dimensions: 218 × 270 × 110 mm
Weight: 6 kg

STATUS: in production and service.

Omera AA8-400 gunsight recorder

Omera OTA 204 video recorder system, with tape-recorder (left) and camera

Sfim

Sfim (Société de Fabrication d'Instruments de Mesure), 13 avenue Marcel-Ramolfo-Garnier, BP 74, 91301 Massy Cedex
TELEPHONE: (1) 69 20 88 90
TELEX: 692164

AP 1717 flight-data/crash recorder

Intended for smaller aircraft and helicopters, this system combines in one unit the data acquisition and tape-transport sections. Inputs are in accordance with the mandatory crash-recorder parameters specified by ARINC 573, with an option for recording some digital data. An AC 6445-A data entry panel is used for the insertion of time and other reference information. Data may be replayed without removing the tape from its housing.

Format: 3/4 ATR Short, alternatively 6 MCU or 1/2 ATR Long
Weight: 15 kg
Power: 60 W at 28 V dc

STATUS: in production.

AP 1750 voice and data/crash recorder

This system is designed for aircraft in which space or other limitations are such that separate voice and data recorders cannot be used. The recorder has one voice channel and a pulse-code modulated data-stream from a data acquisition unit.

Format: 1/2 ATR Short
Weight: 10 kg
Recording duration: (voice) last 90 minutes (data) last 6 h
Power: 40 W at 28 V dc

STATUS: in production.

ED 41xx data acquisition unit/data acquisition and flight management unit

This system is designed for the new generation transports with digital avionics and electronic flight-deck displays, and is contained in ARINC 600 housing. When fitted with the basic circuit cards appropriate to mandatory crash

Sfim AP 1717 flight-data/crash recorder showing acquisition and tape-transport sections

parameters, the unit can feed information to the crash-protected recorder. The addition of processing and input cards to the basic unit converts it into a data acquisition and management unit for aircraft and engine systems monitoring. Software for data acquisition and data processing are fully isolated. The device can be connected to a quick-access recorder as well as to a printer and to the flight-deck display.

AC 6510 flight-data entry panel

This flight-deck mounted unit has nine thumb-wheel selectors to enter flight and other reference information into the data acquisition unit. Communication between the flight data acquisition unit and the digital flight-data acquisition unit is via an ARINC 429 data-bus.

Format: 6 MCU
Weight: (basic) 6.5 kg (expanded) 7.5 kg
Power: 60 VA

STATUS: in production and service.

AC 6620 cockpit display unit

Documentary or calibration information may be transferred into the data acquisition unit by this device, or data played back from the crash recorder for checking purposes. A microprocessor permits the display of engineering parameters. The control/display unit also con-

trols an external printer, by which the entire recording system may be checked.

Dimensions: 102 × 56 × 30 mm
Weight: 2.5 kg
Power: 40 VA
Display format: 5 line × 16 character

STATUS: in production and service.

AC 6445 data entry panel and time display

This unit may be used in conjunction with the ED 3472 data acquisition unit. It generates and displays time in hours and minutes, allows identification of the flight by means of four thumbwheels, and also contains check buttons and warning lights.

Dimensions: 148 × 132 × 136 mm
Weight: 1.2 kg
Power: 4 W

ED 3333 data acquisition unit

For military aircraft applications, this system automatically increases the flight number recorded at the beginning of each flight. The system monitors the performance of sensors and transducers and signals a warning in the event of a failure, and transmits data to the crash recorder.

Format: 1/4 ATR Short
Weight: 3 kg
Power: 28 VA

STATUS: in production.

ED 3472 data acquisition and processing unit for ATR 42

This device has been designed for installations with weight or space constraints. It accepts the mandatory crash parameters, meets the requirements of ARINC 573, and can also take additional digital inputs from ARINC 429 data-buses. After processing, information is transferred to a crash-protected recorder. The system can be expanded by adding a microprocessor and associated electronics to the same box to monitor engine and flight parameters. A printer and quick-access recorder (such as Sfim AP17) may be connected to store maintenance data.

The system can store data for between 50 and 100 flights. Information can be replayed on the ground through a low-cost commercial micro-computer system such as an IBM PC or Apple 3.

Format: 1/2 ATR Short
Weight: 9.2lb (4.2 kg)
Power: 30 VA

STATUS: the system has been chosen as the standard mini-AIDS for the Aérospatiale/Aeritalia ATR 42, and is in production.

AD 3213-01 maintenance recorder

This cartridge-based system can record up to 45 hours of digital data in serial form at 768 bits a second. It contains a buffer memory that can transmit information in block form or continuously at rates of zero to 3072 words a second. The system may be connected to either a data acquisition or data management unit.

Format: 1/2 ATR Short
Weight: 6.5 kg
Power: 45 VA at 115 V 400 Hz
Temperature range: −15 to +55°C

STATUS: in production.

AD 3122-03 recorder

This cartridge-based digital recorder is designed for use with virtually any airborne

Sfim ED 41xx data acquisition/flight management system building blocks: left, flight data/management unit; above, AC 6620 cockpit display; below, AC 6510 data entry panel

Sfim AD 3213 recorder opened up to show tape cartridge

Sfim AD 3122 recorder showing electronics and tape cartridge

Sfim AD 36 loading unit

computer through an RS 232-C interface. It records serially via microprocessor control.

Format: ½ ATR Short
Weight: 6.5 kg
Power: 45 VA at 28 V dc or 115/200 V 400 Hz
Temperature range: −25 to +60° C

STATUS: in production.

AD 36 × 2-Ox cartridge recorder
The AD 36 family of recorders are intended for the loading and saving of computer data, and rugged construction permits the devices to be used in the most severe vibration and shock environment. These units are controlled completely by the computers into which they are plugged, operating in a simple start/stop mode.

Format: ⅜ ATR Short
Weight: 5.2 kg
Power: 40 VA at 28 V dc or 115/200 V 400 Hz
Temperature range: −25 to +60° C

STATUS: in production.

ARINC 717 aircraft integrated data AID system for A310
Sfim provides the aircraft integrated data system (AIDS), to ARINC 717 format, for the Airbus Industrie A310s operated by Air France and Sabena. The system is contained within a single box, with the mandatory crash parameters and general aircraft data being rigorously isolated from one another.

STATUS: in production.

SWEDEN

Saab Instruments
Saab Instruments AB, Box 1017, S-55111 Jonkoping
TELEPHONE: (36) 194 000
TELEX: 70045

Model 130S camera recorder
This device was specifically designed for Saab's RGS 2 weapon sight. Using 16 mm film, the recorder sits atop the sight, its low profile giving the minimum obstruction in the pilot's line of sight. It comprises a camera body, optics and film magazine. The latter faces the pilot, enabling him to replace it with a fresh one during flight. The magazine has a film-remaining indicator visible to the pilot and a test switch ensures operation before flight. The iris is manually selected to apertures between f/4 and f/22. The device has a mean time between failures of 3000 hours.

STATUS: in production.

Saab Model 130S sight recorder mounted on RGS weapon sight

UNION OF SOVIET SOCIALIST REPUBLICS

Aviaexport
Aviaexport, 32/34 place Smolenskaja-Sennaja, Moscow 121200
TELEPHONE: (095) 244 2686
TELEX: 7257

'Tester' flight/accident data recorder
This system is produced for light- and medium-size helicopters and fixed-wing aircraft, using as the recording medium metal tape 0.00006 inch thick. Digital data can be conserved at temperatures up to 500° C.

Weight: tape transport and conditioning box total 35 kg

Recording capacity: 128 inputs
Recording rate: 256 measurements/s
Storage: 25M bits (equivalent to a 3-hour flight)
Temperature range (working): −60 to +60° C
Shock: 12g

STATUS: believed to be in production.

UNITED KINGDOM

British Aerospace
British Aerospace Electronic Systems and Equipment Division, Downshire Way, Bracknell, Berkshire RG12 1QL
TELEPHONE: (0344) 483 222
TELEX: 848129

SCR 300 crash flight-data recorder
The SCR 300 is the latest in a series of airborne data recording systems that the division has produced for both military and commercial aircraft. The SCR 300 is also the smallest system produced by the company and believed to be the smallest such unit commercially available. Although the principal function is to preserve the flight record in the event of an accident, it is equally being used to collect data on engine health, navigation parameters and airframe fatigue as well as recording cockpit voice information.

The system comprises a crash-protected recorder and a data acquisition unit. The former is mechanically and thermally protected to FAA standards and has proved of value in a number of major accidents. It comprises a continuous, single-spool tape transport with six or eight tracks of information. Tracks may be recorded in series or parallel and in any combination required. The recording rate can be 128 data-words (12-bit) a second or 256 words a second giving a total recording time of up to 5 hours.

The data acquisition unit collects a variety of signals in digital, analogue or discrete format and converts them to a stream of serial digital data which it then transmits to the recorder. The number of parameters monitored depends on the aircraft type, timescale and sampling rate, but a typical system has more than 40 different digital, analogue and discrete signal inputs as well as an audio track.

A recent development has been to integrate the crash-protected recorder with a specialist health and usage monitoring computer, thus realising a significant reduction in weight, volume and cost. This enhances the routine value of installing a recorder whilst retaining its ultimate value as a post-accident database.

SCR300 recorder
Weight: 22 lb (10 kg)
Volume: 0.23 ft³ (0.007 m³)

Combined health and usage/protected recorder system
Weight: 22 lb (10 kg)
Volume: 0.32 ft³ (0.009 m³)

STATUS: in production and service on Indian Air Force Jaguars. In August 1983 it was selected, in SCR 300E form, for the EAP (Experimental Aircraft Programme) demonstrator. Later that year Indian Air Force announced contract for 40 SCR 3001, to be supplied as component kits and assembled by Hindustan Aircraft at Lucknow. In 1984 the system was selected to equip 13 types of aircraft and helicopters in British military service.

British Aerospace SCR 300 flight-data recorder: tape transport (left), electronics and data acquisition unit (right)

Ferranti

Ferranti Defence Systems Limited, Navigation Systems Department, Silverknowes, Edinburgh EH4 4AD
TELEPHONE: (031) 332 2424
TELEX: 727101

Video cameras

Ferranti has developed two airborne video cameras, the FD5000 and FD5500, to replace conventional wet-film cameras used for head-up display and gunsight recording. The FD5000 is a black and white camera and the FD5500 records in colour.

Each version has automatic iris control, but focusing is fixed and focal length adjustment is by manually changing lenses. A 525- or 625-line standard is selectable and the cameras feature a 380 × 488-pixel metal-oxide semiconductor sensor.

In late 1985 Ferranti was selected to supply a monochrome video camera, designated FD 5040 to the Swedish Saab JAS 39 programme. The camera, which will have an automatic iris control, will provide a 400-line black and white picture as part of the JAS 39's display and video recording system. Included in the contract is the provision of six cameras for prototype installation.

STATUS: under development.

FD 6000 Video recorder for Tornado F2

In March 1984 Ferranti was selected to develop a video recorder, based on the Philips 2000 commercial recorder, for installation in the Panavia Tornado F2. It is used in conjunction with the FD 5000 series of monochrome or colour video cameras.

The recorder will be housed in the cockpit and will automatically record the signals being relayed to the five monochrome principal displays in the aircraft. This information will have de-briefing, training or mission planning applications.

Dimensions: 295 × 288 × 161 mm including anti-vibration mounting
Weight: 8 kg
Video standard: monochrome, CCIR-compatible, 625 lines, 50 Hz
Video linearity: 10 grey scale minimum

STATUS: under development. The first order, for eight pre-production models was received in mid-1985, with deliveries starting in 1986: over 200 recorders will eventually be required by the Royal Air Force.

Ferranti FD 5000 pilot's display recorder

Engineering model of Ferranti FD 5040 camera selected for Saab JAS 39 Gripen

GEC Avionics

GEC Avionics Limited, Recording Systems Division, 2 High Street, Nailsea, Bristol BS19 1BS
TELEPHONE: (0272) 856 511
TELEX: 444791
FAX: (0272) 851 467

036/A01 cartridge recorder/reproducer

This bi-directional, four-track digital magnetic tape read-write unit is designed for use in helicopters, typically associated with sampling or program-loading applications in ASW sonics equipment. The panel-mounted, cartridge-loading unit houses the tape, record/replay head and tape-motion sensors, and is sealed against dust and moisture to provide a tape life of 5000 or more passes. The cartridge can be changed with gloved hands. The drive module provides tape motion in either direction at 30 inches a second record and reproduce, or 120 inches a second fast forward and reverse. Data can be recorded and replayed at data-rates equivalent to 1600 bits an inch or 6400 bits an inch over one or all tracks simultaneously. The drive module can incorporate an RS232 interface controller as needed.

Tape: 0.25 inch wide
Recording duration: 2 minutes

Tracks: 4
Head: single 4-track dual purpose record/replay
Data: (rate) 1600/6400 bits/inch
(capacity) 184 M bits (1 bit = 1 flux reversal)
Dimensions: 171.5 × 146 × 166 mm
Weight: 4 kg including cartridge

STATUS: in production.

Negretti Aviation
Negretti Aviation, 73/77 Lansdowne Road, Croydon CR9 2HP
TELEPHONE: (01) 688 3426
TELEX: 267708

Airframe fatigue meters
Negretti Aviation (a division of Negretti & Zambra) has been involved in the design and development of fatigue load meters for nearly 30 years. With the support of the Royal Aircraft Establishment (RAE), Farnborough, it now builds a family of such devices, based on RAE's fatigue-load measurement policy. The purpose of the devices is to record structurally damaging, low-frequency g events at the centre of gravity of the aircraft. These events are chosen as representing damaging amounts of mainspar deflection and a correlation has been established between them and damage to other critical parts of the airframe. The record enables a realistic update of remaining airframe life to be made, also of 'safe' failures associated with specific load event patterns.

Normal acceleration, or g, is measured using a special compound accelerometer designed originally by the RAE. This device is contained within the recorder or, where accessibility at the centre of gravity does not permit direct reading, remotely located.

Recent activity has focused on the development of solid-state units, based on microprocessors and non-volatile memories, and miniaturisation of the transducers themselves, with up to 40 per cent reduction in weight and volume.

Concurrently with the new designs, the existing range of devices has been updated by

the introduction of solid-state switching to replace the earlier mechanical relays.

STATUS: in production and service on wide range of commercial and military aircraft including, most recently, the tri-national Panavia Tornado. More than 31 000 devices have been produced.

Integrated monitoring systems
Negretti Aviation currently produces third-generation monitoring systems, consisting of an airborne computer unit and a data retrieval unit, which can service several aircraft, and an optional cockpit display unit. This suite of equipment allows for airborne recording and processing of engine, transmission or airframe data, and access during flight to recorded information, plus the ability to down-load the information at suitable intervals and allow comprehensive and accurate assessment of critical health and life-consumption parameters at the operator's maintenance facility.

Applications include single and multi-engined aircraft and helicopters. Fixed-wing installations may incorporate airframe fatigue and engine health monitoring. Helicopter versions are available with transmission and rotor head monitoring. Alternatively, the airborne computer unit may be configured to act as a data acquisition unit, supplying a remote crash-survivable recorder, as well as carrying out the airframe fatigue and engine health monitoring functions.

The systems are self contained, microprocessor-based and with a highly modular hardware and software structure, storing the solutions to complex algorithms (computed in a real time environment), in high density, non-volatile memory. As well as recording and processing specific parameters, the airborne computer unit can monitor limit exceedances, engine vibration and low-cycle fatigue, its modular design allowing rapid response to new applications.

The data retrieval unit is used to down-load, edit or clear the stored data from one or more aircraft at convenient intervals, independent of aircraft power. It allows on-the-spot analysis, using the keyboard and display, or a hard-copy read-out with the optional built-in printer. Alternatively, the data can be transferred to a ground-based processing facility, using a serial data link or a cassette tape.

The cockpit display unit comprises a 12-character, sunlight-visible display, and a multi-function keyboard, for the interrogation and control of the airborne computer unit during flight. It can also be used to enter data such as the aircraft all up weight or flight number, displaying engine parameters and manually initiating performance data sampling.

Dimensions: (105 series airborne computer unit) 5 × 4 × 7 inches (127 × 102 × 178 mm) (415 series airborne computer unit) 7.5 × 5 × 12.5 inches (191 × 127 × 318 mm) (data retrieval unit) 4 × 11 × 14 inches (102 × 279 × 357 mm)
Weight: dependent on application
Inputs: analogue or digital, optional MIL-STD-1553B interface

STATUS: in production and service.

Engine monitoring systems
This company builds third-generation engine monitoring systems suitable for all gas-turbine powered aircraft and helicopters, consisting of an airframe-mounted airborne computer unit (ACU), a data retrieval unit (DRU) which can service several computers, and an optional cockpit display unit (CDU).

The 105 Series ACU is primarily designed for single or twin-engined, fixed-wing aircraft, while the larger 415 Series ACU facilitates more comprehensive applications in multi-engined aircraft.

Both systems are self-contained and micro-processor-based, storing the solutions to complex algorithms, computed in a real-time environment, in high density, non-volatile memory. Hardware and software are both highly modularised, enabling rapid response to new applications, and permitting a wide variety of interface requirements, whether analogue, digital or data-bus. The combination of processing and discrimination of relevant data enables the operator to identify and analyse significant occurrences immediately while accurately logging routine performance levels to facilitate trend analysis and on-condition maintenance procedures.

The DRU is used to down-load, edit or clear the stored data from one or more aircraft at convenient intervals, independent of aircraft power, and allows on-the-spot analysis, using the keyboard and display, or a hard-copy read-out, with the optional built-in printer. Alternatively the data can be transferred to a ground-based processing facility, using a serial data link or a cassette tape.

Negretti Aviation engine monitoring recorder

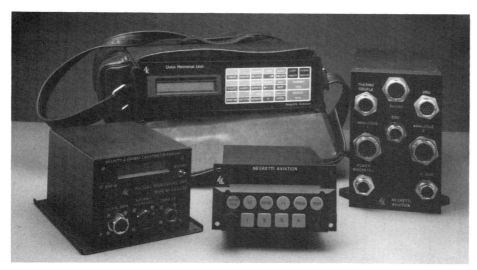
Negretti Aviation integrated monitoring system

Negretti engine monitoring computer

The CDU comprises a 12-character, sun-light-visible display and a multi-function keyboard, for the interrogation and control of the ACU during flight. It can also be used to enter data such as the aircraft all-up weight or flight number, displaying engine parameters and manually initiating performance data sampling.

Dimensions: (105 series ACU) 5 × 4 × 7 inches (127 × 102 × 178 mm)
(415 series ACU) 7.5 × 5 × 12.5 inches (191 × 127 × 318 mm)
(DRU) 4 × 11 × 14 inches (102 × 279 × 356 mm)
Weight: dependent on application

STATUS: in production and service.

Negretti engine monitoring system data retrieval unit

Normalair-Garrett

Normalair-Garrett Limited, Electronics Division, Clarence Street, Yeovil, Somerset BA20 2YD
TELEPHONE: (0935) 751 81
TELEX: 46132

1203V maintenance data recorder

Claimed by NGL to be the world's smallest airborne maintenance data recorder, the 1203V was chosen by Bendix Flight Systems to become part of that company's data recording system for the US Navy/McDonnell Douglas F/A-18 Hornet fighter.

The recorder is a sealed unit to ensure maximum reliability and a data-error integrity of 1 in 10^7, yet still allowing quick access for replay and data retrieval. Tape tension and motion are independently controlled via precision servo electronics, eliminating the need for pinch wheels and co-belts. Recording-head wear is kept to a minimum by the use of a glass-bonded ferrite/ceramic construction. Control electronics are mounted on 10-layer printed circuit boards, and cmos logic is used

for minimum power dissipation. A high standard of built-in test capability isolates faults to sub-assembly level and, by the use of automatic test equipment, faults can be diagnosed down to individual components.

STATUS: in production and service.

1403V large capacity data recorder

This is a 42-channel, two-head, write-only machine for information collection under severe environmental conditions. A significant feature of the device is that it generates a data-demand clock signal that may be used to select the correct speed automatically to match the data input rate required by the associated system. This results in constant tape density information, so reducing wastage. The recorder has a servo-controlled capstan and spools to ensure constant tape tension and spool brakes to prevent spool rotation during 'power off'.

Recorder electronics housed in a separate unit are mounted on printed circuit boards for maintenance and reliability, and are based on

cmos technology for reduced power dissipation. The system may be adapted to record FM analogue or digital data, and its capacity of 25 G bits ($25 × 10^9$ bits) doubled by a simple tape change.

STATUS: entering production.

1472V quick access recorder for aircraft condition monitoring

The 1472V is suitable for use in any monitoring/recording application, such as fatigue monitoring, engine health monitoring etc when data integrity is of paramount importance. It incorporates the very successful 1203V maintenance data recorder magazine using an identical sealed tape transport magazine (TTM) and simplified repackaged associated circuitry.

Data is received from a data acquisition unit or a multiplexed data-bus for recording. Control is achieved by a Zilog Z80A 8-bit microprocessor. Data can also be pre-recorded on the TTM by a ground station for uploading for initialisation of the aircraft system.

Input data is formatted maintaining header information and parity, and grouped into blocks containing checksum and status word data. The status word provides a means of communicating header validity, block length and data integrity to the replay station operator.

A non-volatile memory is provided to store the tape position when the 1472V is powered down. Flags are incorporated in order to communicate BIT failure or tape 85 per cent full (change TTM flag) to the aircraft system.

Data integrity is maintained by comprehensive 'power on' BIT and 'continuous' BIT check routines.

The unit operates from 28 V dc aircraft power.

NGL Type 1203V miniature recorder   *NGL Type 1403V large-capacity recorder*

STATUS: in development and on trial.

Type	Application	Recording mode	Capacity	Data-rate	Tracks	Unit dimensions (mm)	Power	Environment	Data integrity
1203V	Aircraft recording system	Digital	12 M bits	30 K bits/s	4-track switching	123 × 89 × 53	28 V dc 13 W	MIL-E-5400 MIL-STD-810B	BER $<1 × 10^7$
1472V	Aircraft recording system	Digital	11 M bits	8 K bits/s	4-track switching	192 × 158 × 104	28 V dc 25 W	MIL-E-5400 MIL-STD-810C	BER $<1 × 10^{-7}$
Solid-state data store	Aircraft recording systems	Digital (16 M bits)	4 M bits max	819 K bits/s	N/A	261 × 194 × 58	28 V dc 25 W	MIL-STD-810C	N/A (semiconductor store)

NGL 1472V data recorder

Solid-state data recorder

The solid-state recorder (SSR) is designed to provide non-volatile data storage in a removable medium, without moving parts, for use in airborne applications. The SSR achieves this aim by using the latest developments in electrically erasable programmable read-only memories (eeproms). The SSR is designed for the storage of raw or processed data for condition monitoring (eg maintenance recording) but it can also be used for up/downloading mission specific data. The SSR has a storage capacity of up to 2 M bytes and operates at 5 V.

The standard SSR interface is an asynchronous RS422 serial digital interface at baud rates up to 9.6 kHz and an 8-bit parallel interface. The SSR formats the received data into pages which are then written to the eeprom. The pages are formed in a non-volatile ram (novram) so that no data is lost at power down/interrupts.

The novram also stores the last eeprom address written to by the control electronics. At power up, recording starts from the address stored in the novram.

The SSR provides a 'replay' indication when 85 per cent of the available capacity has been used. The percentage capacity used for the 'replay' indication can be changed purely by a firmware change in the SSR's control electronics.

Replay of the data stored in the SSR is via either the asynchronous RS422 interface or the 8-bit parallel interface. The control electronics transmits the data in the same order as it was recorded. At completion of replay, the percentage capacity used is reset to zero.

The SSR performs two independent BIT routines: 'power up' BIT and 'continuous' BIT. A BIT failure is indicated on the SSR status output.

The SSR is designed to operate in conjunction with an NGL-designed power supply which operates with input power to MIL-STD-704C and BS 3G 100 Part 3 with voltages in the range of 16 to 32 V dc. When the input power falls below a preset level a signal is asserted by the power supply to the SSR to ensure retention of all stored data.

Additional interfaces to MIL-STD-1553B and ARINC 429 can be supplied to interface the SSR with an aircraft multiplex data-bus.

STATUS: in development.

Data storage unit

The data storage unit (DSU) provides 4 M bits of non-volatile memory in an abbreviated ¼ ATR case. The unit accepts commands and data from a MIL-STD-1553B multiplex data-bus and stores the data in a first line removable data storage magazine (DSM). The unit provides an RS422 serial data link to the tactical data store. The DSU firmware provides both sequential and random access to the non-volatile memory in the DSM. The main sub-assemblies of the prototype DSU are:

1553B RT: based on a single chip RT, the RT interfaces to the DSU processor via a dual-port ram.

Processor: cmos 16-bit microprocessor which has 75 per cent spare memory capacity and 70 per cent spare processing capacity in normal operation. This spare capacity can be used for pre-processing of data (eg data compression) to make more effective use of the non-volatile memory.

Data storage magazine: this is a 4 M bit non-volatile memory using 64 K eeproms arranged as 256 K × 16. Maximum write data-rate is at present 6400 16-bit words a second, which will increase to 25 600 words a second when second-generation 64 K eeproms are available and to 51 200 words a second when 256 K eeproms are available. The 256 K eeproms will also increase the capacity of the DSM to 16 M bits maximum. The DSM also stores the last sequential write address so that stored data will not be overwritten. The DSM utilises a novel self-latching handle enabling convenient insertion and removal even with a gloved hand.
RS422: RS422 interface to tactical data store.

Power supply: +28 V; MIL-STD-704D; 25 W max.

Tactical data store

The tactical data store (TDS) when used in conjunction with the data storage unit (DSU – see previous entry) is designed to provide an efficient means of up/downloading mission specific data into or from the aircraft systems. Since such data often originates from or is required in the operations briefing room, the TDS is designed to be conveniently small and light so that it can be hand carried by the aircrew and occupy a minimum of panel space in the cockpit.

To achieve this, the TDS is a slave unit of the DSU. All communication to the multiplex data-bus is via the DSU and power for the TDS is supplied from the DSU. This allows the TDS to be a small, low cost unit.

The TDS contains an RS422 interface, an 8-bit cmos microcontroller with associated eprom and ram and 1 to 2 M bits of non-volatile memory using eeprom. The unit mates via a nine-way connector into a panel-mounted receptacle.

STATUS: in development.

Penny & Giles

Penny & Giles Data Recorders Ltd, Recorder Tape Division, Airfield Road, Christchurch, Dorset BH23 3TJ
TELEPHONE: (0202) 481 771
TELEX: 41555

Airborne program loaders

Penny & Giles's computer-program loaders are based on the industry standard DC300A data cartridges and the expanded-capacity DC300XL units. They provide a range of features appropriate to military or commercial fixed-wing aircraft and helicopters, and land- and sea-based vehicles and vessels in severe environments where low error-rates are essential. Notable applications of these devices are the Royal Air Force/British Aerospace Nimrod MR1, MR2 and AEW3 aircraft.

PLU 2000 program loader

Apart from applications on Royal Air Force/British Aerospace Nimrod MR2 and AEW3

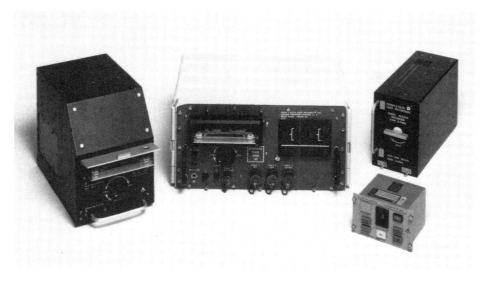

Penny & Giles airborne and ground program loaders

Penny & Giles ADL 7000 data logger

aircraft, this device was rushed into service along with the Thorn/EMI Searchwater radar, for Royal Navy/Westland Sea King helicopters during the Falklands crisis of 1982. The unit can accommodate up to three optional modular interfaces, has a capacity of 3.2 M bits expansible to 6.4 M bits, and can load up to 10 preselected programs. It is operated by a remote control unit.

PLU 3000 program loader
This device has all the features of the PLU 2000 with the addition of an integral control panel for the selection of interfaces and programs. An interim system in the development of the GPLU 2000 ground program loading unit, it has been used in trials with the Panavia Tornado.

STATUS: limited quantity production.

PLU 7000 program loader
This microprocessor-based system, represents a new generation of loading units. With significant advances in engineering and economics, though still employing the proven DC300A cartridge with 140 feet (43 metres) of tape, the system is under evaluation for some 20 to 30 projects, and the procurement cycle for Royal Air Force, Navy and overseas customers has begun. The PLU 7000 is a development of the Penny & Giles device designed and built for British Airways' Boeing 757s.

The system is offered in two forms: ARINC 600 in 4 MCU form factor and ARINC 404 in ½ ATR Short box. Capacity is 28 M bits, expansible as needed.

STATUS: under development.

ADL 7000 data logger/quick access recorder
This system, which can also be arranged for maintenance recording and performance monitoring, is software-based, has a capacity of 78 M bits, and comes in an ARINC 600 – 4 MCU or ARINC 404 – ½ ATR Short case. Full control and status indications are provided, and data can be replayed aboard the aircraft or on the ground as desired. The system is based on industry standard DC300A and DC300XL data cartridges.

STATUS: in production.

ADS 7100 airborne data store
Designed for general-purpose data storage with read/write applications, the ADS 7100 provides data-logging type facilities and a full, directly addressabie file structure for up to 800 files. The capacity is 28 M bits and the unit is similar to the ADL 7000 in packaging and environmental details.

STATUS: in production.

Penny & Giles ACD 701 cartridge drive unit

ACD 701 cartridge drive module
This module is designed to integrate into customers' own recording systems, has a 1600 bits an inch drive mechanism and is ruggedised for airborne use. The standard drive train includes servo, tape control, read/write and read decoder electronics. The capacity is 34 M bits unformatted.

STATUS: in production.

ADR 800 accident data recorder
This family of recorders, designed for civil and military fixed-wing aircraft and also used for helicopter flight trials, continues in service. It has a capacity of 138 M bits and can interface either directly or via ARINC 573. Packaged into a ½ ATR Short unit, with or without vibration isolators, the system is protected according to the requirements of TSO C51a. Operating temperature is between –65° to +70°C.

ADR 7000 accident data recorder
This is the generic designation of a new family of accident data recorders under study for current and future civil and military transports, general aviation aircraft, and helicopters.

Emphasis has been placed on survivability, low cost and weight, and versatility. The microprocessor-controlled system would be fully crash-protected, with integral or remote signal-conditioning, and an optional voice channel. The recording duration could be tailored to suit needs, and a sonar beacon would facilitate underwater detection.

STATUS: under study.

Accident data voice recorder for Harrier GR5
Penny & Giles, in conjunction with Plessey Radio as prime contractor, was chosen in December 1982 to provide the combined accident data and cockpit voice recorder for the Royal Air Force Harrier GR5. The system has also been proposed as a private-venture development for the US Marines AV-8B Harrier. Plessey will work on the data handling and normalising unit and Penny & Giles, the tape-transport system. Penny & Giles' activity grew out of work undertaken for the ADR 7000 accident data recorder (see separate entry) but with much more emphasis on the voice-recording aspect.

STATUS: under development.

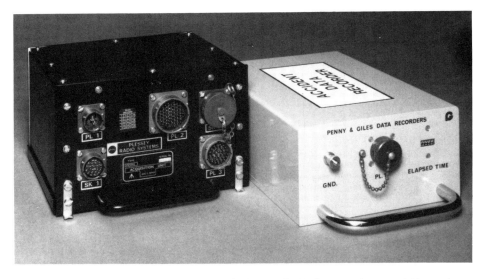

Penny & Giles accident recorder with (left) Plessey Radio Systems data acquisition unit

Plessey Avionics

Plessey Avionics, Martin Road, West Leigh,
Havant, Hampshire PO9 5DH
TELEPHONE: (0705) 486 391
TELEX: 86227

PVS 1573 flight-data acquisition system

This system is based on a range of Plessey-developed equipment that goes back a decade, and complies with ARINC 573. In its most basic form the PVS 1573 meets current and proposed CAA and FAA standards for accident data recording, but can be expanded by means of plug-in modules to become a full AIDS (aircraft integrated data system) suite.

The system comprises a Type PV 1585 flight-data acquisition unit, Type PV 1591 flight-data entry panel, Type PV 1587 data management and entry panel, and a Model 209 digital flight-data recorder with marker beacon for underwater recovery. The PV 1591 and PV 1587 are configured individually to customer needs. The Model 209 recorder is supplied by Lockheed Aircraft Service Company, though customers for the PVS 1573 may specify other approved equipment. However a particular advantage of the Model 209 is its ability to transfer its 25-hour record onto a dump recorder in only 16 minutes. The dump recorder is plugged into a test socket on the aircraft, enabling data to be retrieved without removing or disturbing the flight recorder itself.

STATUS: in production.

PRS 3500A data/voice accident recorder

This system is claimed to be the first to accept information from the new multiplexed data-bus based on MIL-STD-1553B. In May 1983 the company announced a £750 000 design and development contract aimed at bringing the system into service on the Royal Air Force Harrier GR5. The PRS 3500A consists of the Plessey PRS 3501A data acquisition unit and the Penny & Giles D50330 accident data recorder.

The system will provide 2 hours' continuous recording time, taking in up to 50 parameters at 240 words a second. The small size of the aircraft restricts equipment space, and digital transmission reduces the size and weight of wiring looms while increasing data integrity. Recorded data can be extracted in 6 minutes, using a portable transfer device. The system comprises two units: the flight-data recorder and a data acquisition unit.

Dimensions: (flight-data recorder) 115 × 172 × 440 mm
(data acquisition unit) ½ ATR Short
Weight: (flight-data recorder) 8.3 kg
(data acquisition unit) 4.5 kg

STATUS: in production. An order has been received for 60 systems for the Royal Air Force's Harrier GR5s.

PV 1820C structural usage monitoring system

Designed to monitor airframe fatigue by continuously measuring structural loads during flight, the PV 1820C is based on the company's earlier PVS 1820 engine usage life monitoring system. Data and the computed results are recorded on a cassette-loaded quick-access recorder for ground replay and analysis. The system comprises a PV 1820C structural monitoring unit, a PV 1819C control unit, quick-access recorder Type 1207-003, and a transient-suppression unit Type PV 1845. Microprocessor operation simplifies the measurement of stress by enabling the output from a number of strain-gauges (typically 16) to be monitored, along with other flight parameters, on a cassette recorder.

Plessey PV 1580 mandatory flight-data recorder with, left, PV 1587 data management and entry panel

Dimensions: (PV 1820C) 3.71 × 7.62 × 15.49 inches (94 × 194 × 394 mm)
(PV 1819C) 5.74 × 2.99 × 7.63 inches (146 × 76 × 194 mm)
(1207-003) 5.74 × 2.03 × 6.65 inches (146 × 51.6 × 169 mm)
(PV 1845) 3.74 × 2.08 × 7.48 inches (95 × 53 × 190 mm)
Weight: (PV 1820C) 11.96 lb (5.44 kg)
(PV 1819C) 3.49 lb (1.59 kg)
(Type 1207-003) 3.89 lb (1.77 kg)
(Type PV 1845) 2.5 lb (1.4 kg)
Sampling rate: Each parameter is defined by 10 bits of a 12-bit word, the 11th bit being a compression flag and the 12th being reserved for parity. The sampling rate for each quantity is programmable in binary steps from 1 to 128 times a second.
Data compression: programmable 16:1, 8:1 or 4:1
Self-test: comprehensive automatic self-test and fault diagnosis is built in.

STATUS: in production and service.

PV 1580 mandatory flight-data recorder

The PV 1580 is designed to full ARINC 573 requirements as a mandatory crash/flight-data recorder with world-wide aviation authority approval. It can be easily expanded into a full integrated data system.

Dimensions: (data management and entry panel) 7.48 × 5.74 × 4.48 inches (190 × 146 × 114 mm)
(acquisition unit) 7.59 × 4.88 × 19.48 inches (193 × 124 × 495 mm)
Weight: (data management and entry panel) 6.99 lb (3.18 kg)
(acquisition unit) 11.96 lb (5.44 kg)

STATUS: in production and service.

PV 1584 data recorder

A compact unit combining recorder and acquisition electronics within a single ½ ATR case, the PV 1584 is designed to replace ARINC 542 electromechanical recorders. Plessey claims that the system is unique in meeting the full requirements of the FAA and CAA (as they apply to this type of equipment) in a single unit of this size. The system is approved by all major certification authorities, including those of Australia, Canada, Germany, the UK and the USA. Its signal-handling capability meets all current requirements including those of the CAA's Specification 10 and the FAA's Part 121 section 343 with 10 per cent excess capacity.

The system can be used in conjunction with a copy recorder to extract data for analysis in situ without disturbing the equipment. A PV 1591 flight-data entry panel can be used for manual

Plessey PV 1584 data recorder

insertion of documentary information into the recording. Alternatively, a PV 1587 data management and entry panel (see entry below) can be installed to provide a two-way information interchange between the recorder and the flight crew.

The PV 1584 comprises a protected flight recorder at the front of the unit and a data acquisition electronics section at the rear. Design baseline is ARINC 573, input parameters being sampled at the rate of 64 a second, each sample being converted into a 12-bit binary word. The resultant digital data stream is routed to the recorder section and placed on a Mylar tape loop giving 25 hours' continuous recording. The sampling rate for given parameters can be varied as required. The entire capacity of the tape can be dumped in 16 minutes by means of a Lockheed 235 data-dump recorder. The unit contains an eprom giving eight programmes selected by changing wire links in the aircraft. The device can, therefore, be reprogrammed to meet requirements so that, for example, interchangeability could be maintained between eight aircraft types in an operator's fleet. Signal types comprise 26-volt synchro, dc ratio and absolute, ac ratio, potentiometer, variable resistance, and frequency (tacho or pulse probe). Discretes comprise dc series and shunt and ac series and shunt.

Format: ½ ATR Long
Weight: 25 lb (11.34 kg)
Power: 25 W at 115 V 400 Hz

STATUS: in production and service. During 1985 the PV 1584 was selected to equip the Royal Air Force's Nimrod aircraft and the British Aerospace ATP. It is standard equipment on the BAe 748, 146 and 125 and on the Shorts 360.

PV 1587 data management and entry panel

The PV 1587 data management and entry panel, normally mounted on the flight-deck, acts as a two-way communication link between the flight crew and an ARINC 573 flight-data acquisition system. The DMEP has a central processing unit with a microprocessor and 16 K of memory. The system processes the computations associated with out-of-limit checks on parameters, selects the desired parameters from the data stream for display purposes, and organises the routeing and storage of documentary data.

Dimensions: 7.5 × 5.75 × 4.5 inches (190 × 146 × 114 mm)
Weight: 7 lb (3.18 kg)
Power: 15 W at 115 V 400 Hz
Out-of-limit checks: no of checks 100; no of types of checks 12;
(types of checks) conditional on values of two other parameters; conditional on value of one other parameter; unconditional
Display: any parameter from the data system output of 512 words/s
Data insertion: four 4-bit bytes of documentary data

STATUS: in production and service.

Plessey PV 1587 data management and entry panel in Concorde supersonic transport

PVS 1820 engine usage life monitoring system

This flexible monitoring system incorporates a real-time data processing function for the detection and display of engine parameter exceedances, the computation of accumulated low cycle and temperature fatigue life, and continuous data recording for detailed analysis of engine design criteria.

The microprocessor-based system can include a display suitable for the cockpit or equipment bay giving parameter limit exceedance warning and immediate access to engine life data without reference to ground replay equipment. The system also can analyse retrospectively flight and engine data by an airborne cassette-loaded recorder and a ground-based data replay system.

Dimensions: (engine monitor unit) 194 × 90 × 321 mm
(visual display unit) 74 × 146 × 169 mm
(cassette recorder) 52 × 146 × 169 mm
Weight: (engine monitor unit) 4.44 kg
(visual display unit) 0.9 kg
(cassette recorder) 1.77 kg
Inputs: 16 spool-speed inputs, 18 discrete (on/off) signals and up to 40 analogue parameters. (Latter may include synchro, ac and dc voltage ratios, and potentiometer or variable resistance signals)
Sampling rate: programmable in binary steps from 1–156 parameters/s
Data-rate: programmable in binary steps from 32–512 words/s
Recorder duration: 32 samples/s for 12 h or variable in binary stages to 512 samples/s for 0.75 h
Power: 19 W at 28 V dc

STATUS: under development.

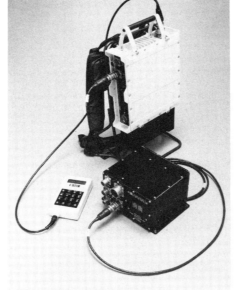

Plessey PRS 2020 engine monitoring system

PRS 2020 engine monitoring system

Developed from the PVS 1820 system, this design is lighter, smaller, uses less power and is compatible with MIL-STD-1553 (Def Stan 00-18 Part 3/1) data-bus interfaces. It also includes a 32-bit high-speed microprocessor, plus hybrid and uncommitted logic array-based electronics and bubble memory storage.

The PRS 2020 consists of several component units including:
PRS 2021 engine monitoring unit: a single-box data acquisition and processing unit for analogue signal conditioning, discrete digital conditioning and/or digital data-bus parameter recovery. The heart of this unit is a single very large-scale integrated hybrid circuit comprising a 32-bit 19-register central processor with 4 K bytes of random access memory and 8 K bytes of read-only memory reserved for applications programs.
PRS 2023 display: available for taking a 'quick look' at data, the display can be mounted on the engine monitoring unit or remotely. The format of the display, which uses 12 light-emitting diodes, can be defined by the customer
PRS 2026 data retrieval unit: a rugged battery-powered unit for extracting data from the engine monitoring unit for subsequent analysis, or for system test and calibration.

Plessey also offers the options of a bulk storage recorder, a portable ground-replay unit and comprehensive systems software packages.

Dimensions: (PRS 2021) 200 × 194 × 124 mm
(PRS 2026) 280 × 260 × 80 mm

STATUS: under development.

PA3520 aircraft integrated monitoring system

The PA3520 aircraft integrated monitoring system records engine lifing information and structural data and comprises the Plessey PA3521A data acquisition and processing unit and the Penny & Giles D50330 Mk 2 accident data recorder.

Plessey PVS 1820 engine usage life monitoring system

Plessey modular data acquisition system

Accident related data can be dumped directly into a portable recording unit for later analysis, thus avoiding the need to remove the data recorder itself. Engine and structural data can be dumped into a unit with a display unit, allowing instant viewing of critical data, or the information can also be stored for subsequent analysis. The PA3521A can accept discrete, shaft rotation and analogue inputs, or can interface with a MIL-STD-1553B data-bus.

Format: ½ ATR Short
Weight: 4.55 kg
Power: 28 W at 28 V dc

STATUS: in development.

Modas modular data acquisition system

Modas has been designed to be able to gather and record any type of signal to be found on an aircraft during flight trials. The system is expansible to meet specific requirements, up to 4096 input channels, and 131 072 channels per second being possible. Eight separate sampling programmes can be selected in flight, according to the demands of each trial.

A pulse coded modulation (pcm) digital technique is employed for recording data in either 14-bit parallel or simultaneous multi-track serial streams. A comprehensive ground replay facility is also available, and the Modas can interface with a telemetry system.

Dimensions/format and weight:
(acquisition/processing unit) ½ ATR; 10 kg
(control unit) 146 × 191 × 151 mm; 3 kg
(monitor unit) 146 × 191 × 166 mm; 3 kg
(small recorder) ¾ ATR Short; 13 kg
(large recorder) 533 × 360 × 200 mm; 39 kg
Power: 28 V dc or 400 Hz, 3-phase

STATUS: in service in a wide range of British and European aircraft.

Smiths Industries

Smiths Industries, Aerospace and Defence Systems Limited, 765 Finchley Road, Childs Hill, London NW11 8DS
TELEPHONE: (01) 458 3232
TELEX: 928761

Engine health monitoring system for helicopters

Smiths Industries is developing an engine health monitor for the Westland WG 30 helicopter that will measure several engine parameters, including speeds, temperatures and throttle movement from which engine usage can be determined from flight to flight.

STATUS: in development.

Engine life computer for BAe 146

Fitted to the British Aerospace 146 airliner, this system records engine parameters which are transferred after flight, via a data-transfer unit, to a ground-based computer for analysis. The resultant data is then used to establish engine performance and to predict maintenance requirements.

Low cycle fatigue counter

In recent years low cycle fatigue has been recognised as a major limiting factor in the lives of turbine engines. These powerplants benefit from being run at as near constant speed as possible, so that the mechanical stresses set up under changing conditions are kept to a minimum. With aircraft engines, particularly those for military types, benign environments and operating usage common to turbines for electrical generation or natural gas pumping is never possible; a typical flight may involve a number of power excursions, or low frequency cycles, between flight idle and maximum power, each such excursion from the cruise setting representing a cycle. Each time the throttle or power lever is moved, stress variations are set up in the rotating components that produce minute stress cracks and these increase according to usage. This cumulative damage, known as low cycle fatigue, is monitored by recording the number of times that an

engine has been cycled through specified rpm limits between overhauls. The results are compared with the damage produced by a 'standard cycle', and factored to allow for the scatter between individual engines of the same type.

The Smiths Industries low cycle fatigue counter, weighing 3.7 lb (1.7 kg), is housed in a ¼ ATR Dwarf case and accepts four inputs from tachogenerators or pulse probes in single or two-shaft engines. The functions can be extended to include temperature inputs so that resulting creep and fatigue damage can be taken into account. Additional inputs can be included to accommodate three-shaft engines.

The counter's four readouts show selected component lives, such as low-pressure and high-pressure compressors and their corresponding turbines. The unit computes the reduction of low cycle fatigue life using the same basic formula as that used by a stress analyst, but the data-processing uses a microprocessor in conjunction with a memory to compute low cycle fatigue from its stored programs. The damage cycles representing loss of engine life as a result of low cycle fatigue are displayed on electromechanical registers to give a running total. The device can handle up to four channels of information and display it on the front face of the unit. Built-in test indicators show input or equipment failures.

STATUS: in service to monitor Rolls-Royce/ Turbomeca Adour engines of Hawk trainers of Royal Air Force's Red Arrows aerobatic team. Formation aerobatics impose most severe usage on engines of any form of flying and, as a result of unavoidable constant throttle movement, low cycle fatigue is dominant failure mechanism, and so determines overhaul life.

Series 0600 and 0700 engine life computers

Extensions of the low cycle fatigue counter, these microprocessor-based systems grew out of Smiths Industries' engine life counter developed in conjunction with the UK Ministry of Defence (and described in *Jane's Avionics 1983-84*). They are considerably more versatile than the relatively simple fatigue counter, monitoring and processing parameters other than engine speed and temperature.

Series 0600
Format: ¼ ATR Dwarf Short
Weight: 4 lb (1.8 kg)
Inputs: up to six, selected from speed (as frequency), temperature, pressure and discretes (as voltages)
Computation: 16-bit processor, 8 K bytes eprom, 2 K bytes ram
Capability: Assembler programmable; provides analysis as required from low cycle fatigue life usage, thermal fatigue life usage, creep life usage, and limit exceedance, and BITE
Outputs and displays: 6 × 5-digit mechanical displays, two BITE indicators

Series 0700
Format: ¼ ATR Dwarf Short
Weight: 4 lb (1.8 kg)

Smiths Industries engine life counter

Inputs: up to 12, selected from speed (as frequency), temperature (as thermocouple or resistance bulb voltage), pressure (as voltage), digital (ARINC 429, RS 422), discretes (as switch positions), torque (as voltage), and vibration (as charge amplifier)
Computation: 16-bit processor, 32 K bytes eprom, 4 K bytes ram, 2 K bytes non-volatile memory, and memory expansion
Capability: Pascal programmable; provides analysis as required from low cycle fatigue life usage, thermal fatigue life usage, creep life usage, torque life usage, vibration, vibration tracking filters, limit exceedance, data formatting for bulk storage, and BITE
Outputs and displays: 5-digit led display, 2-digit memory location identifier, BITE indicator, RS 422 to tape recorder or remote display, RS 422 to data transfer unit, ARINC 429 digital data

STATUS: under development and evaluation.

Series 0800 engine life computers

Based on the technology and approach of the 0600 and 0700 series, the 0800 series of engine life computers has extended capabilities to include some airframe monitoring. The company envisages an eventual merging of engine and airframe monitoring systems and the 0800 is a move in this direction. The immediate aim is to develop a system to meet a potential Royal Air Force requirement for combined engine/airframe monitoring equipment.

Format: ¼ ATR Short
Weight: 6.6 lb (3 kg)
Inputs: up to 42 selected from engine/airframe idents, speed, temperature, pressure, digital (ARINC 429, RS 422, MIL-STD-1553B), DEF STAN 0018, discretes, torque, vibration, air data, and synchros
Computation: 16-bit processor, 32 K byte eprom, 4 K byte ram, 2 K byte non-volatile, 1 M bit bubble (optional), memory expansion
Capability: Pascal programmable; provides analysis as required from low cycle fatigue life usage, thermal fatigue life usage, creep life usage, torque life usage, vibration, vibration

Smiths Industries low cycle fatigue counter

tracking filters, limit exceedance. Provides record of engine ident, aircraft ident, incidents, time and date, raw data sets, and BITE
Outputs and displays: 5-digit led display, 2-digit memory location identifier, BITE indicator, RS

422 output to tape recorder or remote display, RS 422 output to data transfer unit, digital data to ARINC 429, MIL-STD-1553B, DEF STAN 0018

STATUS: under development.

Thorn EMI

Thorn EMI Electronics, Computer Systems Division, Wells, Somerset BA5 1AA
TELEPHONE: (0749) 72061
TELEX: 44254

Tape transport systems
Thorn EMI makes a range of tape transports, suitable for missile and aircraft applications

HER 200M
This system uses 60 metres of 0.5-inch tape, with five recording speeds from $^{15}/_{16}$ to 15 inches per second. The transport is supported on anti-vibration mounts within a cylindrical case, so no added shock insulation is needed.

Dimensions: 3.78-inch dia × 8.42 inches (96 × 214 mm)
Weight: 5.3 lb (2.4 kg)
Tape speeds: $^{15}/_{16}$, 1, $3^3/_4$, $7^1/_2$ and 15 inches/s
Tape running time: 630 s at $3^3/_4$ inches/s
Altitude limit: 82 000 ft
Max linear acceleration: 50 g
Power: 23 W 28 V dc

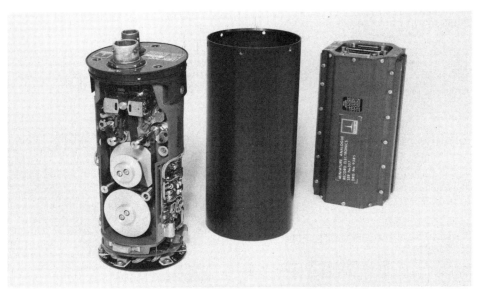

Thorn EMI Type HER 200M tape transport

HER 402 and 600
Two ruggedised tape transports operating to full military specifications and packaged in conventional rectangular boxes. Each offers six tape speeds (up to 30 inches per second), with up to 28 record tracks being possible using a pair of multi-track heads. The Type 402 takes 650 feet of tape, the Type 600 1400 feet.

Dimensions
(402) 11 × 6.25 × 6 inches (279 × 159 × 152 mm)
(600) 15 × 8.7 × 7 inches (381 × 221 × 178 mm)
Weight
(402) 17.6 lb (8 kg)
(600) 33 lb (15 kg)

Tape running time: (at $3^3/_4$ inches/s)
(402) 34.6 minutes
(600) 149.3 minutes
Max altitude: 70 000 ft
Max linear acceleration: 30 g
Power: (402) 34 W
(600) 54 W at nominal 28 V dc

Vinten Avionics

Vinten Avionic Systems Limited, 6 Wadsworth Road, Perivale, Greenford, Middlesex UB6 7JD
TELEPHONE: (01) 998 1011
TELEX: 935524

Formerly S Davall & Sons Limited, the company in January 1985 changed its name to emphasise both its membership of the Vinten group of companies and its main product-line.

Type 1002 pilot's display recorder
This device was introduced to fulfil the dual role of head-up and head-down recording. In the head-up display mode it 'sees' the outside world on which is superimposed flight and weapon-aiming commands and information, the record subsequently being used for training, de-briefing, and assessment of weapon-aiming accuracy and target damage. By a simple switching arrangement the camera can be used in a head-down mode to record information from the display unit of a surveillance or attack radar.
    The camera is mounted on the rear of the head-up display and consists of three units: the film transport itself, interchangeable lenses and an exposure control unit. The latter is mounted so that its optical axis coincides with the field of view axis of the display.

Film capacity and type: 9 m × 16 mm double-perforated cine with 20, 40, 80, 160, 320 ASA rating
Framing rate: 4/s or pulsing at 1/s
Running time: 5 minutes at 4 frame/s
Lens options: (head-up display) f/5.6-22, F=25.4 mm, fov 23° × 17°, f/5.6-22, F=65 mm, fov 9.2° × 6.6°
(head-down display) f/2-16, F=20 mm, fov 28° × 21°
Power: 28 V dc, 1.5 A start, 0.5 A running

Weight: (camera) 3.25 lb (1.48 kg)
(control unit) 0.95 lb (0.425 kg)

STATUS: in production and service.

Type 1192 recycling wire recorder
This device, based on a recording medium of stainless-steel wire, preserves the last 25 hours of flight-selected data for crash analysis. Wire has long been used as a recording medium and gives the maximum data storage for a given volume and weight. The Type 1192 mechanical transport can be used in conjunction with any data-processing and normalising system providing analogue or digital information. The system is designed to withstand 100 g, 5 millisecond shocks, the impact of a 500 lb (227 kg) steel bar dropped from 10 feet (3 metres), a static crush force of 5000 lb (2270 kg), exposure to flame at 1100° C for 5 minutes over half the outside area, and a combination of fuel, hydraulic fluid, lubricating oil, fire extinguisher fluid, and salt water for 36 hours.

Dimensions: 8-inch (203 mm) sphere reducing to 7-inch (178 mm) cylinder over central region
Weight: 16 lb (7.27 kg)
Recording medium: 0.0023-inch dia stainless-steel wire, breaking force > 12 oz (0.34 kg)
Data transfer rate: 1536 flux reversals/s
Power: 115 V 400 Hz 2-phase and 28 V dc

STATUS: in production and service.

Type 1200 cockpit voice recorder/reproducer
The Type 1200 voice recorder/reproducer is designed for mounting in the confined areas typical of fighter cockpits. When connected to the communications control system, it provides intelligible speech and record/replay facilities. The system has only two controls, a selector

switch for record/replay and a toggle switch for on/off, rewind, and fast forward. When a microphone in the recorder detects speech it switches on automatically and continues to operate until speech stops. The recorder then continues to run for a preset period, and switches itself off until the next speech sequence causes it to repeat the cycle. Cassettes may be replayed through the aircraft system or through a ground-based playback unit. The system employs standard Philips C60 and C90 audio cassettes. A version designated the Type 1200 Mk II can record and reproduce data as well as voice.

Dimensions: 5.8 × 1.8 × 6.6 inches (147 × 43 × 168 mm)
Weight: 3 lb (1.36 kg)
Tape cassette: Philips Compact C60 (30 minutes audio/track) or C90 (45 minutes track)
Tape erase: automatic prior to recording

STATUS: in production and service.

Type 1207 digital tape recorder
This device has been designed for use in military aircraft and helicopters. It records digital information on two or four input channels on standard audio-format (Philips type) cassettes. Tape speed is adjustable within certain limits by means of a preset control, but operation of the recorder is accomplished by means of a single on/off switch. Cassettes may be removed or inserted in flight without removing the recorder from its mounting.

Dimensions: 6.67 × 5.76 × 2.03 inches (169 × 146 × 51.6 mm)
Weight: 3.9 lb (1.77 kg)
Cassette type: Conrac CM 300
Recording duration: 1 h 20 minutes at 17.83 mm cm/s, 2 h at 11.88 mm cm/s
Recording density: 800 bits/inch

Series 2000 pilot's display recorder

Tape speed: 0.468-0.702 inch/s (11.88-17.83 mm/s)
Input format: ARINC 573-6 drive, 4 mA peak-to-peak
Power: 100 mA at 28 V dc

STATUS: in production and service.

Type 1267 digital tape recorder
The Type 1267 recorder is designed to provide very high data integrity in the most hazardous environments. Initially intended for airborne engine monitoring, it offers the tape transport stability essential for high-capacity digital recording and can support a variety of military and commercial applications. Size, weight and cost have been kept to a minimum, partly as a result of incorporating technology from the Type 1207 recorder.

Dimensions: 5.75 × 2.41 × 6.91 inches (145.1 × 61.1 × 175.4 mm)
Weight: 5.5 lb (2.5 kg)
Recording pattern: simultaneous parallel or serial serpentine with rapid track change; recycling available
Tape speed: 3 inches/s or 1.875 inches/s
Recording duration: 18.7 or 28.1 minutes at 3 inches/s; 30 or 45 minutes at 1.875 inches/s
Recording density: 4000 bits/inch max
Data capacity: 50 M bit at 2500 bits/inch; 80 M bit at 4000 bits/inch

STATUS: in service.

Type 1300 pilot's display recorder
The Type 1300 camera photographically records head-up display symbology superimposed on the pilot's view of the outside world. The record can be employed for the training of combat tactics, for de-briefing, or for assessment of weapon-aiming and target damage. Mounted on the side of the head up display it gives the least visual obstruction to the pilot and preserves a clear surface rear on the head up display combiner glass for control purposes. The system comprises a camera unit, film magazine, and interchangeable lenses. Automatic exposure control and variable shutter speed permit the use of film emulsion speeds within a wide brightness range. The exposure is

Type 2320 video cassette recorder

controlled by a built-in electronic sensing unit and emulsion speed is adjusted automatically by a coding feature in the film magazine that presets the appropriate controls.

Dimensions: 8.3 × 7.5 × 1.8 inches (210 × 190 × 45 mm)
Weight: 6.75 lb (3 kg)
Film type: 22 m × 16 mm × 0.1 mm double-perforated cine
Framing rate: 16 or 32 frames/s
Film movement indicator: visual indication
Exposure control: integral with camera, automatic over brightness range 78-5028 ft lambert
Running time: 3 minutes at 16 frames/s
Shutter speed: 1.25-10 ms
Event marker: remote operation
Lens (with periscope): f/2.8-22, F=25 mm, fov 16° × 22°; f/2.8-22, F=50 mm, fov 8° × 11°
Power: 28 V dc, 2 A starting, 1 A running

STATUS: in production and service.

Series 2000 pilot's display recorder
This new-generation family of cameras records photographically head-up display or gunsight symbology superimposed on the pilot's view of the outside world. The record can be used for training, de-briefing and assessment of target damage. Units can be built as 'flatpacks' or in more conventional form to overcome the space problems associated with the installation of this type of equipment in fighter aircraft. Conventional magazine loading permits rapid film replacement in flight. Manual and automatic exposure control options are available and a range of lenses and periscopes allows the system to be matched to individual requirements.

Dimensions: 4.2 × 2.7 × 1.6 inches (107 × 69 × 41 mm)
Weight: 2.2 lb (1 kg)
Film: 16 mm × 0.1 or 0.15 mm, 50, 100 or 200 ASA
Framing rate: 16/s and single shot; alternatives available
Running time: 3 minutes at 16 frames/s
Lens: f/2-f/16, F=25.4 mm, fov 18° × 13°
Shutter speed: 1/100s standard, other speeds available
Event marker: remote operated
Indicators: film contents and motion
Automatic exposure control: integral when requested

STATUS: in production and service.

Series 2300 video cassette recorders
These small, lightweight helical-scan devices are suitable for recording both video and data signals in a hostile environment and are produced in both top and front-loading formats.

The top-loading Type 2300 standard record only unit and Type 2301 record/replay unit offer an extremely cost effective solution for the recording of TV formatted signals, such as head-up display cameras and head-down displays, for post-sortie de-briefing and training purposes. Since standard VHS format is retained, ground replay can be on commercial video recorders and monitors.

The front-loading Type 2320 has been developed in order to provide a high-bandwidth recorder with fully released components. This recorder, which has both record and replay facilities, offers a selection of signal-processing electronics to match non-standard input information such as thermal imaging, radar and other electro-optical signals, in addition to conventional raster-scan television data. The unit is particularly suited to such tasks as tactical reconnaissance and surveillance where the need for data integrity is paramount. The Type 2320 range of units includes high data-rate analogue and digital versions.

All Series 2300 recorders have heaters to permit operation at low temperatures.

Dimensions: (Type 2300) 261 × 252 × 110 mm (Type 2320) 257 × 325 × 168 mm
Weight: (Type 2300) 6 kg (Type 2320) 12 kg
Recording system: rotary helical-scan
Video signal system: PAL colour/CCIR monochrome 625 lines, or NTSC colour/EIA monochrome 525 lines, plus other formats for non-standard signals
Recording time: (Type 2300) up to 180 minutes (Type 2320) 120 minutes nominal, dependent on format
Event marker: visual flag

STATUS: in production and service.

Series 2700 video cassette recorders
These miniature ruggedised units are intended for applications involving limited space and a harsh environment. They provide one video channel and one audio/digital channel using 0.5-inch VHS-C compact cassettes, the choice of which permits a great reduction in recorder volume by comparison with earlier systems while at the same time retaining commonality of recording format and allowing ground replay of cassettes on standard full-size VHS machines by means of a simple mechanical adaptor.

The Series 2700 is a top-loading unit of minimum height, the Type 2708 a front-loading unit particularly suited to cockpit panel-mounting. Both can be tailored to meet the recording needs of television-format signals such as those often employed in head-up and head-down displays, thermal imaging equipment, radar and other electro-optical sources, and high data-rate analogue and digital systems. All Series 2700 systems incorporate heating elements for operation at low ambient temperatures.

Dimensions: (Type 2700) 183 × 193 × 77 mm (Type 2708) 150 × 176 × 125 mm
Weight: 3.2 kg
Recording system: rotary helical-scan
Video system: PAL colour/CCIR monochrome 625 lines, or NTSC colour/EIA monochrome 525 lines, plus other formats for non-standard signals
Recording time: up to 40 minutes
Event marker: visual flag

STATUS: in production and service.

Type 3000 monochrome video sensor
A compact lightweight video camera employing solid-state imager technology, the Type 3000 is designed for combat aircraft and is specially suitable for the recording of head-up display/outside world information. The two-unit system comprises a camera head and electronics unit, both employing solid-state construction. The

Type 2708 miniature video cassette recorder

Type 3000 television sensor

camera head contains a burn-resistant ccd element lens with automatic exposure control for operation over a wide range of brightness and a facility for mounting optical filters. A range of periscopes permits the system to be optimised to widely different installations, minimising obscuration of the pilot's view. The electronics unit providing power, interfacing and video processing for the camera head, also doubles as the pilot's control unit. The equipment produces a standard video composite in 50- or 60-cycle formats which may be either recorded for later analysis or displayed on a television monitor. When used in conjunction with a suitable video cassette recorder the system can provide a long-duration record of head-up display symbology and outside world for pilot de-briefing, system failure and weapon-delivery analysis.

Dimensions: (camera head) 120 × 56 × 40 mm (electronics unit) 140 × 120 × 60 mm

Type 3100 colour video camera

Weight: 1.5 kg
Camera head sensor: solid-state imager
Field of view: 20° × 16° (nominal)
Refresh rate: 50 or 60 Hz
Power: 8 W at 12 V dc or 28 V dc
Signal/noise ratio: 48 dB

STATUS: under development.

Type 3100 colour video camera
A compact lightweight device, the type 3100 camera employs a single solid-state imager

with integral filter to produce a full colour picture, and is intended for military aircraft. A separate camera head, with remote electronics unit, permits installation in situations where space is restricted. The camera head itself comes in either standard or flat-pack form for convenience of installation, and contains the imaging chip, drive electronics and lens. The latter has an automatic exposure control, and a number of periscopes, including see-through types are available. The electronics unit provides power, interfacing and video processing for the camera head, and also doubles as the pilot's control unit; for this function the unit can incorporate video control circuitry. Dual output channels permit two independent video devices, such as a recorder and a monitor, to be driven simultaneously.

Dimensions: (standard camera head) 92 × 70 × 60 mm
(flat-pack head) 89 × 107 × 38 mm
Weight: (camera head) 0.8 kg
(electronics unit) 1.5 kg
Sensor: solid-state imager with integral colour filter mosaic
Refresh rate: 50 Hz (PAL) or 60 Hz (NTSC)
Field of view: 20° × 16° nominal
Power: 1 A at 28 V dc

STATUS: in production and service. First deliveries of the Type 3100 were made in mid-1985.

UNITED STATES OF AMERICA

Bendix
Bendix Corporation, Energy Controls Division, 717 North Bendix Drive, South Bend, Indiana 46620
TELEPHONE: (219) 237 2100
TELEX: 258426
TWX: 810 299 2514

PMUX propulsion data multiplexer
The Bendix propulsion multiplexer monitors and collects engine data including temperatures, pressures, speeds, fuel flow, variable geometry parameters within the powerplant, cockpit discrete (ie on/off) signals, and engine

identification. This compact engine-monitoring system provides data appropriate to examining engine condition trends, tracking limited-life equipment and improving engine life forecasts. The data generated by PMUX can improve engine performance evaluation, leading to reductions in fuel consumption and further savings from more extensive use of on-condition rather than scheduled maintenance. Additional savings can result from the early detection of potential failure of engine components and the elimination or reduction of secondary engine damage.

STATUS: under development.

Bendix propulsion data multiplexer

Collins
Rockwell International, Collins Division, 400 Collins Road NE, Cedar Rapids, Iowa 52498
TELEPHONE: (319) 395 4994
TELEX: 464421
TWX: 910 525 1321

Turbine monitoring system
At the end of 1984 Collins was awarded a $1 million contract by the US Air Force's Aeronautical Systems Division for the study of an integrated turbine engine monitoring system. With Systems Control Technology and Woodword Governor as sub-contractors,

Collins evaluated current US Air Force engine-monitoring equipment and technologies, and then develop hardware and software specifically for the new system, which will enter full-scale development in 1987.

STATUS: work completed in 1985. The development phase has yet to start.

Edo
Edo Corporation, Western Division, 2645 South 300 West, Salt Lake City, Utah 84115
TELEPHONE: (801) 486 7481/3846
TELEX: 388315

Model 655 head-up display television camera recorder
This device is designed for advanced military aircraft to record head-up display symbology superimposed on the outside world, and can also be used as flight-test instrumentation during equipment development. A substitute for the earlier-generation film-camera recorders, it has magnetic recording compatibility with many types of reproducer, and instant replay; no time is lost in film processing. For training and reconnaissance, a front-seat presentation can be provided at the rear-seat position.

As a video tape recorder, the system can record equipment anomalies and failures,

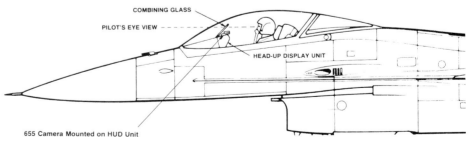

Installation arrangement of Edo Model 655 recorder in General Dynamics F-16 fighter

provide a basis for the assessment of weapon delivery and target damage, permit an evaluation of pilot proficiency, act as an intelligence gatherer, and help pilot and crew training. As an airborne video display generator the system provides real-time information to other processing and display equipment; the ability to provide a front-seat picture to the weapons

system operator or instructor in the rear seat can substantially augment the value of a two-seat aircraft for front-line service or training. The system can also act as a video communications link, providing, in conjunction with an S or L band transmitter, a data link to an airborne command post, sensor reporting post, flight-test centre, or ground-based intelligence

Edo Model 655 head-up display television camera recorder

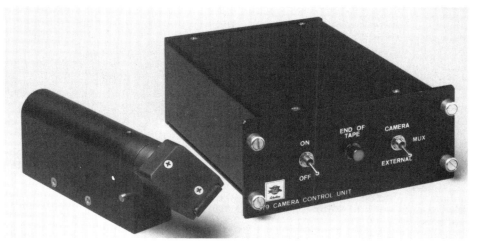

Camera and control unit for Edo Model 679 gunsight television recorder

organisation. The Model 655 vidicon unit is detachable from the body of the camera so that it can be used to record cathode ray tube data in other avionic systems such as a radar display or threat-warning panel.

Dimensions (overall): 7.5 × 3.75 × 6.57 inches (190 × 95 × 167 mm)
Weight: 3 lb (1.3 kg)
Lens: f/1.8, 18 mm, iris switchable to f/1.8 or f/8
Field of view: 22.5° × 17° at 4:3 image format; 20° × 20° at 1:1 image format
Interface: compatible with GEC Avionics head-up displays, and has a multiplexer that permits input from external video source to be recorded simultaneously with the HUD video.
Sensor: 0.625 inch FPS (focus, projection, scan) vidicon sulphide target or (optionally) silicon diode target
Bandwidth: flat to 10 MHz ± 1 dB, 12 MHz –3 dB
Resolution: 700 tv lines at 4:3 image format; 800 lines at 1:1 format
Synchronisation generator: EIA RS-170 or CCIR European
Horizontal line rate: 525 US or 625 European
Frame rate: 60 Hz US or 50 Hz European
Video output: to RS-170 or CCIR

STATUS: in production.

Model 679 gunsight television camera

This gunsight television camera is designed for installation in the cockpits of combat aircraft where size and mounting constraints may limit the use of other head-up display camera systems. There are always difficulties with the cockpit installation of 'outside world' cameras because they need to command the same field of view as the pilot but without obstructing his view. To reduce the size of the viewing system, the Model 679 is configured as two units: the television camera, and a camera control unit. The camera can be mounted just in front of the head-up display or above it, while the control unit can be sited in any convenient panel space.

The system can be used in conjunction with airborne video tape recorders, video displays and S or L band transmitters. The tape-recording system augments pilot training and permits his efficiency to be assessed, providing an estimation of target damage, and reconnaissance and intelligence-gathering. It can also pinpoint faults in weapon-aiming or other equipment. As a source of video information, the Model 679 can drive other displays and data-processing equipment, for example, to give the observer a 'front window' reproduction of the pilot's view in a two-seat combat aircraft.

Dimensions: (camera) 6.7 × 1.25 × 2.16 inches (170 × 32 × 55 mm)
(control unit) 6.5 × 5.75 × 2.5 (165 × 146 × 64 mm)
Weight: (camera) 0.68 lb (0.3 kg)
(control unit) 2.7 lb (1.2 kg)
Sensor: 0.625 inch FPS sulphide target vidicon; optional silicon diode target
Bandwidth: flat to 10 MHz ± 1dB, 12 MHz ± 3dB

Resolution: 700 lines tv 4:3 image format, 800 lines tv 1:1 format
Image format: 0.28 × 0.21 inches at 4:3; 0.25 × 0.25 at 1:1
Frame rate: 60 Hz (USA), 50 Hz (European)
Dynamic range: 10 000:1 sulphide target
Grey scale: 10 shades at 1 ft candle illumination
Lens: focal length 0.8 inch, f/1.8 and f/8 switchable; fov 22.5° × 17° at 4:3 image format, 20° × 20° at 1:1
Power: 16 W at 28 V dc

STATUS: in production and service. One application is in British Aerospace Sea Harriers in service with Royal Navy.

Model 1633 Night Sight low-light level television camera

This device is intended for use in harsh environments, such as combat aircraft, rockets or missiles, and spacecraft. Using state-of-the-art techniques for image intensification and special automatic gain-control circuitry, Night

Edo Model 1633 Night Sight low-level television camera

Sight can operate over brightness levels of more than 5 million to one. Bore-sighted to other imaging and fire-control equipment such as laser rangefinders and forward-looking infra-red scanners, the device can add a dimension to ground-attack aircraft operating at night or in bad weather. Its ability to distinguish objects in near or total darkness supplements the performance of the system with which it is used, providing real-time data to a display in the cockpit or for post-flight evaluation and training.

The Night Sight is based on the 4804 family of silicon intensified target vidicons, with full horizontal and vertical sweep failure protection supplied from the sweep circuits, and accepts standard 16 mm C-mount lenses. Qualification is to MIL-E-5400 and MIL-E-16 400.

STATUS: in production.

Model 1631 television camera

This harsh-environment tv camera is designed for combat aircraft, sounding rockets and armoured vehicles. It generates a high-quality 800-line resolution picture over –20° to +60° C, using a silicon diode target vidicon or other suitable vidicon.

STATUS: in production for sounding rockets, anti-aircraft gun mounts, remotely piloted vehicles, helicopters, tanks and other tracked vehicles.

Model 1953 low-light camera

The small low-light level airborne video camera is specifically designed for use on stabilised gimbal mounts on drones and RPVs, but is also suitable for manned aircraft applications where space is at a premium. The electronics unit is

Miniature camera head and control unit for Edo Model 2531 cockpit colour television camera

mounted separately. The camera uses a silicon intensified target (SIT) vidicon.

Dimensions: (camera head) 2.25-inch dia × 8 inches (57 × 203 mm)
(control unit) 7.75 × 3.35 × 5.22 inches (197 × 85 × 133 mm)
Weight: (camera) 1.875 lb (4.125 kg)
Power: 28 W at 28 V dc
Operating temperature: –20 to +71° C
Operating altitude: unlimited (sealed version)
Video characteristics: 4:3 aspect ratio, 16 mm format, 525-945 lines, 50/60 Hz, programmable, 15 MHz bandwidth

STATUS: in production.

Edo Model 1953 video camera and control unit

Model 2531 cockpit colour television camera

Designed for high-performance military aircraft, the Model 2531 comprises a miniature camera head and remote control unit. The sensing device can, therefore, be kept very small, causing minimum obstruction of the pilot's forward view when monitoring the outside world and head-up display. The Model 2531 is of all-solid-state construction based on an mos image sensor with convertibility to monochrome or colour tape recording systems. The camera has an automatic iris, adjusting to light levels from dusk to full sunlight. A multiplexer permits the input from an external source to be recorded simultaneously with the camera video on a single cassette recorder. An optional demultiplexer unit allows head-up display and external video to be replayed on separate recorders.

Dimensions: (camera head) 4 × 3 × 1.125 inches (100 × 76 × 29 mm)
(control unit) 4.75 × 6.37 × 3.75 inches (120 × 162 × 95 mm)

Weight: (camera head) 0.5 lb (0.22 kg)
(control unit) 4.75 lb (10 kg)
Lens: f/1.4-360, fl 25 mm, fixed focus
Video standard: NTSC or PAL colour
Resolution: 350 lines
Sensitivity: 1000 lux normal video (34 lux min scene illumination)
Power: 17 W at 28 V dc or 115 V 400 Hz 3-phase

STATUS: in production.

Endevco

Endevco Corporation, Rancho Viejo Road, San Juan Capistrano, California 92675
TELEPHONE: (714) 493 8181
TELEX: 685608

Microtrac engine vibration monitor

Microtrac is a microprocessor-based engine vibration monitor which is basic equipment on the Boeing 757 and 737-300 airliners and optional on the 747 and 767 transports. The use of microprocessors permits the device to be reprogrammed to suit most aircraft/engine combinations. Vibration is detected by piezoelectric accelerometers mounted at sensitive regions in the engines. The piezoelectric crystal generates a voltage proportional to the vibration level and the signal is passed to a processor in the avionics bay.

The Microtrac processor uses a narrow-band digital filter, controlled by the output from a tachometer on the engine, to isolate the vibration frequency. The narrow bandwidth means that the signal-to-noise ratio of the final vibration indication is high, and the filter's transient response is fast so that it can track the fundamental vibration frequencies of each of the engine's rotors during rapid accelerations.

The filter's accuracy is determined by the number of bits and computer program. Built-in test equipment is provided, and only one programmable read-only memory needs to be changed to adapt the system to a different engine type. Microtrac has an ARINC 429 digital data-bus output so the engine vibration can be received by any EICAS, ECAM, AIDS, or similar unit with a compatible interface. An analogue output to conventional instruments can be provided. The built-in test results can also be put on to the ARINC 429 bus, and up to 17 fault indications can be stored for display to maintenance crews.

In-flight engine balancing measurement is an optional feature with Microtrac. The amplitude and phase of the once-per-revolution vibration of the Microtrac accelerometer is presented to the microprocessor which selects the weight and mounting position required to balance the rotor more accurately. This data is stored in an expanded memory for later display to ground-crew, and it may be passed into the ARINC 429 bus for a direct cockpit readout. The in-flight balancing measurement can be accomplished more accurately than on the ground and

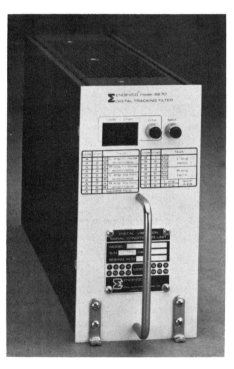

Endevco Microtrac narrow-band tracking filter

repeated ground run-ups are avoided. Time and fuel are therefore saved.

Format: 3 MCU
Weight: 7.75 lb (3.5 kg)

STATUS: in production.

Views vibration imbalance early-warning system

Endevco produces analogue engine vibration monitors for the Boeing 747, Canadair Challenger, Mitsubishi Diamond and British Aerospace 146 aircraft. All aircraft certificated to FAR Part 25 need engine vibration monitoring. Views is an adaptation of the Challenger system to suit the Garrett TFE731 engine.

Views uses a piezoelectric accelerometer mounted on the engine carcass. The solid-state computer amplifies and filters signals from the accelerometer and converts them into a cockpit readout of engine imbalance. There is an optional warning light which is activated when

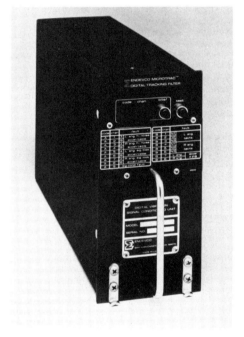

Endevco Microtrac engine vibration monitor

the engine vibration level is more than 1.1 inches a second for 3 seconds. Endevco stresses that the main function of Views is to establish vibration trends so that maintenance can be scheduled in time to prevent permanent damage. Views will also display the sudden increase in vibration that accompanies the loss of a fan, compressor or turbine blade.

Dimensions: (signal-conditioner) ¼ ATR case (display unit) 5.75 × 2.6 × 2.9 inches (146 × 67 × 73 mm)
Total weight: (2 engines) 12.7 lb (5.8 kg)
Display range: 0-2 inches/s per channel
Accelerometer: 1 Endevco Model 6222M8 per engine
Frequency response: 5-5500 Hz
Specification: FAR Part 25

STATUS: similar systems in production for Boeing 747, Challenger and Diamond.

Fairchild

Fairchild Weston Systems Inc, Data Recorders Division, PO Box 3041, Sarasota, Florida 33578
TELEPHONE: (813) 371 0811
TELEX: 052890
TWX: 810 864 0406

Model 100A and Model A100A cockpit voice recorders

Fairchild Weston claims that the 8000-plus cockpit voice recorders that it has supplied to 300 users worldwide represent more than 85 per cent of these devices in service. The Model 100A and Model A100A are identical except for the deletion on the latter of some little-used meter and test buttons that were a feature of the earlier system. Both meet FAA TSO-C84 and ARINC 557 requirements and provide a crash-protected, continuous 30-minute record of speech between crew members on the flight-deck and communications going through the public-address system. The system comprises two principal items: a recorder, with electronics and tape transport in a single box, and a control unit. The latter can have an integral microphone or it can accept signals from a remote unit. The recording medium is Mylar tape in a continuous loop and the claimed mean time between failures for the system is 6000 hours.

Dimensions: (control unit) ATA Panel, 5.75 × 2.25 × 2.5 inches (146 × 57 × 64 mm) (recorder) ½ ATR Short
Weight: (recorder) 23.9 lb (10.86 kg) (control unit) 1.2 lb (0.55 kg)
Power: 20 W maximum at 115 V, 400 Hz
Bulk erase: fail-safe, double electrical interlock, total erase level >35 dB
Frequency response: (tape transport) 150–5000 Hz to 3 dB levels

STATUS: in production.

Model F800 digital flight data recorder

This recorder is heavily based on Fairchild Weston's technology and experience built up during the production of some 8000 voice recorders, and is being built for the Beech C-12F Super King Air and Gates Learjet C-21A, both types having been ordered by the US Air Force to meet its OSA (operational support aircraft) mission. The particular voice recorder whose tape-transport system was chosen as the base for development into the F800 is the 100A/A100A (see separate entry) which has an average mean time between failures of 6000 hours and a mean time between overhauls of 8000 hours.

The requirement for the OSA aircraft called for mission and date identification, vertical acceleration (−3 to +6 g), heading from a compass system, airspeed and altitude, and elapsed time. Up to seven more parameters can be recorded, using synchro or voltage inputs that are transformed into digital information via a Fairchild F-8 microprocessor. The system is designed to interface with mini- or business-type computers for data readout and automatic testing without the need for the complexity and expense associated with larger computer centres.

The F800 system meets FAA TSO-C51a requirements and operates either as an ARINC

Fairchild Weston Model A100A cockpit voice recorder opened up to show tape transport and surrounding crash-protection packaging

Fairchild Weston Model F800 digital flight recorder opened up to show electronics cards and tape transport, but without crash-protection packaging. Tubular device on front end is the under-water acoustic locator beacon

573/717 recorder or as a replacement for an ARINC 542 metal-foil type. It can also function as a combination recorder, permitting inter-changeability in airlines which use both types of recorder. It is the flexibility of the system that has permitted its transition from commercial to military use, according to Fairchild Weston. The system is accommodated in a 29 lb (13 kg) box that fits into the same space as that occupied by the older foil-type recorders, and can withstand a temperature of 2100° F, and 1000 g impacts without damage to the tape. To aid retrieval of the unit at sea when only the approximate position of the aircraft is known, an underwater locator beacon is activated by immersion in salt water. The system will 'ping' for 30 days with new batteries.

The recording medium is a 0.25-inch wide, 450-foot (137-metre) long Mylar tape, with six recording tracks providing 25 hours' total recording time; by comparison the Model 100A voice recorder from which the F800 was developed has the typical CVR recording time of 30 minutes. Automatic track switching ensures continuous data recording. At a cost of between $8 000 and $16 000 depending on customer requirements, according to Fairchild

Weston, the system is cheaper to buy and has lower operating costs than the metal-foil recorders that it will be replacing.

Latest versions of the F800 are expanded to take in more parameters in accordance with proposed new FAA requirements. In addition, a new option incorporates the addition of a solid-state memory card for data monitoring and storage. The DMS option, as it is called, permits fast data retrieval on certain parameters which may call for urgent action, such as engine temperature exceedances and the shocks sustained during hard landings.

STATUS: in production and service. More than 600 F800 recorders are in service and the unit has been selected as standard on the Airbus A320, Fokker F50/F100, ATR 42, CN-235 and Embraer EMB-120, as well as the Boeing 737 family.

Model F880 cockpit voice/data recorder

This cockpit voice/data recorder meets the requirements of SAE AS 8030 and TSO C111 which specify parameters for such devices as they relate to the general aviation sector. The F880 combines cockpit voice and data record-ing functions in a single ½ Long ATR box. Two voice channels are provided, one to record pilot/co-pilot radio transmissions, the other to record cockpit noise from an area microphone, recording time is 30 minutes. A third channel records the last 30 minutes flight-data, and a fourth records the last eight hours of such information, sampled every 16 seconds.

STATUS: in development. Prototype units were tested in 1985.

Fairchild Weston cockpit voice recorder control units, panel-mounted and with integral truncated-cone microphones

Fairchild Weston Systems Inc, 300 Robbins Lane, Syosset, New York 11791
TELEPHONE: (516) 349 2200
TWX: 510 221 1836

AN/AXQ-15 Hitmore weapons system video recorder for helicopters

The Hitmore (helicopter-installed television monitor recorder) system, based on Fairchild Weston's solid-state camera, increases the proficiency of helicopter TOW missile operator by recording on video his handling of the weapon for immediate monitoring by an instructor aboard the aircraft, or post-flight playback.

The benefits, according to Fairchild Weston, are that no 'target conditioning' is needed, real-time impact-points can be observed and commented upon; the record can be played back directly on landing; 40 simulated firings can be accommodated on a single cassette; the reusable tape cuts costs; a short-term record of student's actions can be used to demonstrate performance; and the system is capable of modular growth. The overall objective is to improve the gunner's first-round kill probability.

Camera
Dimensions: 2.5 × 2 × 3.75 inches (63 × 51 × 95 mm)
Frame rate: 30 Hz
Format: EIA RS-170 compatible

Video monitor
Dimensions: 6 × 6 × 14 inches (152 × 152 × 357 mm)
Weight: 10 lb (4.55 kg)
Display area: 4.25 × 3.25 (108 × 83 mm)
Visual: 525-line scan, 8 shades of grey
Frame frequency: 30 Hz

Video recorder
Dimensions: 9.6 × 6 × 13 inches (243 × 151 × 330 mm)
Weight: 23 lb (10.3 kg)
Recording system: rotary two-head helical scan
Power: 30 W at 28 V dc

STATUS: in production and service.

Fairchild Weston AN/AXQ-15 Hitmore system

AN/AXQ-16(V) cockpit television sensor

The Fairchild Weston AN/AXQ-16V cockpit television sensor allows the recording of real-time gunsight or head-up display symbology and audio in fighters and tactical-strike aircraft. It comprises a small, solid-state television camera and an electronics unit.

The camera, or video sensor head, uses a charge-coupled photosensor array for high sensitivity and dynamic range. The principal optical element is a 31 mm, f/2.8 lens with automatic iris control and the camera can operate at low light levels.

The electronics unit can be used as a camera mounting platform or may be remotely sited. A military-qualified airborne video tape recorder stores the images. The camera can also drive a monitor to provide real-time displays for second crew members in two-seat aircraft.

The system can replace existing film cameras or interface with any head-up display, the separation of camera and associated electronics permitting the sensor to be kept as small as possible, thereby minimising obscuration of

Fairchild Weston AN/AXQ-16(V) cockpit television sensor showing sensor head and electronics unit

the pilot's forward view. Several variable-angle, customised sensor heads are available for different installations.

The tape recorder recommended for use with the AN/AXQ-16V is the TEAC V-1000 AB-R, a remotely controlled unit using 0.75-inch tape cassettes and with a recording time of 20 or 30 minutes. Dual audio tracks permit accompanying voice commentary. This recorder has been designed for extended standby operation with instant record start-up.

Dimensions: (sensor) 3.25 × 3 × 3 (83 × 76 × 76 mm) with height variable according to installation
(video tape recorder) 6.2 × 13 × 9.6 inches (157 × 330 × 244 mm)
(electronics unit) 7.4 × 4 × 1.46 (188 × 102 × 37)
Weight: (combined unit) 2.43 lb (1.1 kg)
(video tape recorder) 23 lb (10.45 kg)

Lens: Custom f/2.8, f.1. 31 mm, automatic iris
Refresh rate: 30 frames/s
Line rate: 15 750 lines/s
Power: 20 W at 115 V 400 Hz 3-phase

STATUS: in production. Television sensor has been tested in several US Air Force tactical aircraft and is specified for Northrop F-5, Grumman F-14, General Dynamics F-16, McDonnell Douglas F-15 and F/A-18, LTV A-7, Fairchild A-10 and Lantirn versions of F-16 and A-10. Contracts for F-15, F-16 and A-10 systems totalling 1500 systems and worth $15.2 million were placed by the US Air Force's Aeronautical Systems Division in 1981. It is employed in remotely piloted vehicles and for flight test instrumentation, and is marketed in the UK by GEC Avionics.

General Electric

General Electric Company, Avionic and Electronic Systems Division, Aircraft Instrument Department, 50 Fordham Road, Wilmington, Massachusetts 01887
TELEPHONE: (617) 937 4101
TELEX: 947406

EMSC engine monitoring computer system

General Electric manufactures the EMSC engine monitoring system computer for the General Electric F110 engine which has been chosen for the US Navy's F-14D Tomcat and US Air Force F-16 C/D fighters. The computer provides in-flight monitoring of engine

exceedance, fault or trends. Engine-related signals are acquired from the engine monitoring system processor via the MIL-STD-1553B engine signal data-bus. Diagnostic data is retained in non-volatile memory which annunciates this data visually to the cockpit.

A secondary function downloads the data via the RS223C series communication link to the data display and transfer unit for ground-support evaluation. This link also uploads information to the EMSC such as the aircraft engine diagnostic information, time data, aircraft serialisation and life usage data.

STATUS: in production.

General Electric engine monitoring system computer

Genisco

Genisco Memory Products Corporation, 10874 Hope Street, Cypress, California 90630
TELEPHONE: (714) 220 0720
TWX: 910 591 1898

AN/ASH-27 signal data recorder

This system was designed to record sensor information from MAD and sonobuoy equipment, voice, control and time information, and reference signals and computer commands, aboard US Navy/Lockheed S-3A ASW aircraft. The equipment operates automatically and can play back recorded data at any time during a mission.

The system comprises two units rack-mounted: an RD-349/ASH-27 magnetic tape transport and an MX-8959/ASH-27 interface unit, the latter containing the appropriate signal-conditioning circuits. Remote and locally controlled functions include record and play-back at 7.5 inches a second, fast forward and rewind at 240 inches a second, and stop. The transport uses two 14-track write heads and two 14-track read heads in conjunction with 1-inch tape. Two tracks are employed for internal reference, leaving 26 tracks available for data.

Separate servo-controlled torque motors drive the supply and take-up reels to maintain constant tension in the tape as it approaches and leaves the head. The transport is sized to provide 5.5 hours of continuous recording on 16-inch reels. An 80 dB/cycle signal-to-noise ratio permits a wide dynamic range.

Dimensions: (tape transport) 17.4 × 9 × 25.8 inches (442 × 229 × 656 mm)
(interface unit) 8 × 8.9 × 19 inches (203 × 219 × 483 mm)

STATUS: in production and service.

ECR-40 tape recorder

The ECR-40 is one of several tape cartridge recorders manufactured by Genisco suitable for airborne and other military uses, as well as commercial applications where a ruggedised design is called for. It emulates the industry standard nine-track parallel 1600 bpi format, with a 0.5-inch tape cartridge and housed in a standard ½ ATR box.

Format: ½ ATR
Weight: 35 lb (77 kg)
Power: 100 W in operation, 28 V dc

Genisco ECR-40 cartridge tape recorder

Altitude: up to 50 000 ft
Characteristics: 50 M bytes capacity on 650 ft 0.5-inch cartridge, 9-track parallel recording at 6400 bpi, 72 or 200 K bytes/s data transfer rate

STATUS: in production and service.

Hamilton Standard

Hamilton Standard, Division of United Technologies, Bradley Field Road, Windsor Locks, Connecticut 06096
TELEPHONE: (203) 623 1621
TELEX: 994439

Flight-data systems

Hamilton Standard produces examples of both types of information recording system currently operational in military and commercial aircraft. The first is the basic 25-hour capacity flight-data acquisition system (BFDAS) to monitor mandatory crash parameters, designed in accordance with ARINC 573 and 717 for mounting and electrical interfaces, and the second is a development with expanded capability to measure other parameters, and referred to as an aircraft integrated data system (AIDS).

The expanded systems have been developed principally for the wide-body transports, but may also find application in the new-generation narrow-body transports, such as the Boeing 767 and 757. The two types of system are built up from a family of building blocks, or modules, which comprise:

Flight-data entry panel (FDEP)
Data management unit (DMU)
Flight-data acquisition unit (FDAU)
Auxiliary data acquisition unit (ADAU)
Management control unit (MCU)
Digital flight data acquisition unit (DFDAU)
Recorders or printers to handle outputs

FDEP 100, 120, 121, 122, 123 flight-data entry panels

The FDEP permits flight-crews to feed commands or information such as flight number,

gross weight, and aircraft identification into the AIDS or BFDAS recording system. The information may also control various analysis routines in the AIDS processor. The unit has three configurations: thumbwheel-controlled unit for a BFDAS, push-button FDEP, and a standard AIDS unit with display and recall facilities. These units are part of Hamilton Standard's Mk II, III and IV BFDAS/AIDS systems installed on Boeing 737 and 747, Douglas DC-9-80 and DC-10, and Airbus A300 and A310 transports. Specific applications are FDEP 100: 737, 747 and DC-10; FDEP 120:, 747, DC-9-80, A300; FDEP 123: A310; FDEP 121: 767, 757.

FDEP 100 for Mk II AIDS
Dimensions: 5.8 × 4.5 × 9 inches (147 × 114 × 228 mm)
Weight: 7 lb (3.18 kg)
Power: 35 W

FDEP 120 for Mk IV AIDS
Dimensions: 5.75 × 2.25 × 6 inches (146 × 57 × 152 mm)
Weight: 4 lb (1.82 kg)
Power: 4 W

DMU 100, 101 data management unit

This unit is the 'brain' for the AIDS system, analysing the real-time information from the FDAU, the ADAU, or the DGDAU. It controls the digital AIDS recorder, having program logic that determines what information to record and when to do so. It also provides data to displays including the FDEP and airborne printer, and contains extensive built-in test equipment. This DMU is part of the company's Mk II and expanded Mk III AIDS system in the following aircraft: DMU 100: 747, DC-10, A300; DMU 101: A310.

Hamilton Standard flight-data acquisition system

	DMU 100	DMU 101
Format:	1 ATR Long	6 MCU
Weight:	36 lb (16.36 kg)	12.7 lb (5.7 kg)
Power:	180 W	55 W

FDAU 100 flight-data acquisition unit

Essentially a data gatherer, this unit contains the signal conditioning needed to rationalise the many types of signal from engine, airframe and systems sensors. Signals are multiplexed and digitised so that they can be recorded on the DFDR for accident investigation purposes. The unit is fully compliant with ARINC 573 and meets regulatory agency flight-data acquisition requirements. The unit is used on Boeing 727, 737, 747, McDonnell Douglas DC-10 and Airbus A300 airliners.

Format: ½ ATR Long
Weight: 18.5 lb (8.41 kg)
Power: 70 W

ADAU 100 auxiliary data acquisition unit

This microprocessor-controlled unit provides the multiplexing, signal conditioning and digitising of strain-gauge and thermocouple signals for use in engine condition monitoring equipment. The serial output data is sent to a DMU as part of an expanded AIDS system, and is controlled by software stored in a read-only memory. The unit also provides calibration and cold-junction compensation for the input signals. The ADAU is part of the company's Mk II AIDS system for the Boeing 747 and McDonnell Douglas DC-10 transports.

Format: ⅜ ATR Short
Weight: 9 lb (4.1 kg)
Power: 45 W

Hamilton Standard digital flight-data acquisition unit

MCU 110, 111 management control unit

This is the control analysis unit for the Mk IV AIDS system, providing a reduced or more specialised data-gathering capability. Using state-of-the-art microprocessors, this AIDS provides some of the benefits of larger systems although packaged into a smaller volume. The computer reads the FDAU input data and decides when it is to be passed to the quick-access recorder or optional printer. The MCU writes over data in the FDAU serial data stream in unused word locations sequence numbers, flight modes, and other specialised information to aid transcription and analysis on the ground. The unit is part of Hamilton Standard's Mk IV AIDS system on the 747, DC-10, DC-9-80 (MCU 110) and A300 (MCU 111) transports.

Format: ⅜ ATR Short
Weight: 8 lb (3.6 kg)
Power: 25 W

DFDAU 120 digital flight-data acquisition unit

Complying with ARINC 717 and performing the same functions as the FDAU for the mandatory flight recording systems on the new-generation transports built to ARINC 700, the unit contains a microprocessor permitting it to record some AIDS information in addition to its crash recorder functions. This unit is standard equipment on the Boeing 767 and 757 and is a basic option on the Airbus A310. It is also used as a building block for an expanded AIDS on the A310.

Format: 6 MCU
Weight: 15 lb (6.8 kg)
Power: 40 W

AIRS 100 accident information retrieval system

Hamilton Standard has developed under contract to the US Army a solid-state airborne flight-data recorder, a ground-readout unit, and a batch-process computer for data analysis. The recorder receives, conditions and digitises information from aircraft sensors and, under microprocessor control, stores it in a crash-proof, non-volatile solid-state memory device. The system can record 40 essential parameters, such as acceleration, air-speed, altitude, attitude, heading, engine power, and fault-warnings to the crew.

The heart of the system is a crash-survivable module housing the solid-state memory device and meeting the requirements of FAA TSO C51a. Depending on requirements, the device can store between 15 minutes and 4 hours of prior flight history. In the event of an accident the unit can be read out directly, or the memory chip can be removed for the subsequent extraction of flight information.

Dimensions: 6.5 × 6.8 × 6.5 inches (165 × 172 × 165 mm)

STATUS: in development for small aircraft types including helicopters, business jets and fighters.

JET

JET Electronics and Technology Inc, 5353 52nd Street, Grand Rapids, Michigan 49508
TELEPHONE: (616) 949 6600
TELEX: 226453

ETC-100A engine temperature cycle counter for F-20

Designed exclusively for the Northrop F-20 Tigershark/General Electric F404 powerplant installation, this engine-mounted unit has six event counters and two failure flags that are readily visible to maintenance personnel. The flags may be reset, but the event counters are cumulative and so provide engine life-cycle data when periodic maintenance is conducted. Provision for four additional failure flags is made in the existing equipment.

STATUS: under development.

Engine life monitors

JET has developed a range of engine life monitors which measure engine stress factors such as temperatures, speeds and throttle angle, from which the engine life used may be calculated. Initially, trials were conducted on J85-powered US Air Force T-38 jet trainers in 1983, but in 1984 JET introduced a version for the commercial market. The ELM-101 engine monitor has been developed specifically for use with the Lycoming LTS-101 turboshaft engine which powers the Bell 222, MBB/Kawasaki BK 117, Aérospatiale 350 and other helicopters.

STATUS: in production.

Lear Siegler

Lear Siegler Inc, Instrument Division, 4141 Eastern Avenue SE, Grand Rapids, Michigan 49508
TELEPHONE: (616) 241 7000
TELEX: 226430
ITT TWX: 4320036

Data transfer system

The DTS (data transfer system) is an automated mission planning and post-mission analysis system for aircraft with digital avionics. It was initiated by LSI in the 1970s to obviate information-transfer problems encountered during the pre-flight loading of mission-related information. Errors and delays in aircraft readiness were being caused by the manual insertion of pre-flight information such as target coordinates and waypoints via aircraft system keyboards. The company says that the introduction of the DTS has cut such pre-flight errors from more than 5 per cent to less than

0.1 per cent while at the same time reducing cockpit or flight-deck initialisation time from 30 minutes to less than 1 second.

In addition to the loading and retrieval of normal mission data, DTS equipment is now being used to initialise JTIDS (Joint Tactical Information Distribution System), Navstar GPS (satellite-based Global Positioning System), missile guidance control units and voice-control interactive devices. DTS equipment also provides immediate post-mission print-out and analysis of information, both operational and flight-test.

The LSI DTS is used on over 20 applications to date, including fighter, multi-engine transport, and helicopter aircraft.

A typical DTS comprises a solid-state data transfer module, a receptacle into which the module is inserted, and a ground-based mission-data terminal with a software database. The data transfer module (DTM) is a small, rugged, solid-state portable cassette or cartridge used to load or retrieve digital data. The

original production Model 3245A contained 8 K 16-bit words or read-write random access memory. An interchangeable low-cost DTM is now available that provides a variable memory, with options between 2 K and 128 K 16-bit words. Further memory expansion is planned for introduction in 1986-87.

The data transfer module receptacle, mounted in a convenient location in the aircraft, accepts the DTM via a safety spring-loaded door. Depending on user needs, the receptacle is modularly expansible to accommodate the electronics for a MIL-STD-1553 interface, a DTM data management microprocessor, more memory, or other electronics. They can be provided either within the receptacle or packaged remotely to conserve cockpit space.

DTS ground computer terminals incorporating user-friendly software are available for loading and retrieving digital data into and from the DTM. Initial mission data terminals (MDTs) were large dual-operator systems with alphanumeric displays which were introduced in the

Lear Siegler data transfer module receptacle

Lear Siegler interface receptacle unit

late 1970s. They incorporated a large bulk memory for automated pre-mission data loading and post-mission data retrieval. Inputs were directed from a number of sources providing disk storage, cathode ray tube display, and hard copy. All classified data was stored on a removable disk. The MDTs contain their own maintenance diagnostics for self-test and also test the DTM prior to data loading.

Second-generation mission data ground terminals (MDGTs) take advantage of technology gains since the original MDT introduction. The MDGTs are one-half the size and cost of the original terminals and offer colour graphics plus mouse and/or digitiser tablet enhancements. Improved memory capacities and throughput coupled with the above enhancements have permitted parallel software advances and expansion. Original data loading functions have been expanded to include automated flight planning and mission planning tasks. DTS ground terminal systems are packaged in shock proof, watertight cases for ease of transport.

STATUS: in production.

Standard flight-data recorder system

LSI is entering production on a solid-state standard flight-data recorder (SFDR) system developed under a US tri-service specification.

The solid-state SFDR is designed to replace older oscillograph or tape-based recorders. It is being initially introduced by the US Air Force on the General Dynamics F-16 tactical fighter. The SFDR will be used on a variety of tactical fighters, multi-engine aircraft, and helicopters.

The Model 6213A SFDR consists of two boxes: the signal acquisition unit (SAU) with auxiliary memory unit, and the crush-survivable memory unit (CSMU). The CSMU is also being used on Sweden's Saab JAS 39 fighter.

The SAU receives up to 200 discrete, analogue, and digital mux bus parameters which are converted into digital information, data compressed, and stored within the SAU's auxiliary memory unit. Daily monitoring and recording of general airframe and engine health such as structural loads and engine low-cycle fatigue is typical. Conversion and data management functions are accomplished by a MIL-STD-1750A processor. High-speed databuses transfer data between the acquisition unit and the recorder, and permit rapid readout onto a data logging system on the ground.

Selected aircraft parameters are sent to the small armour plated and insulated CSMU to ensure data recovery following a flight incident. The ruggedised CSMU is designed to withstand temperatures of up to 1350°C and mechanical shock up to 1700 g. The CSMU non-volatile memory has a life expectancy of 60 000 hours which together with the built-in test facility, permits it to be mounted in inaccessible regions of the airframe.

The SFDR is capable of both daily maintenance data recording and retrieval plus recording of essential crash data within the ruggedised CSMU.

Routine maintenance data is retrieved from the aircraft by means of flight line ground readout equipment (GRE). Retrieval data is then transferred to a support and test station (STS) for data decompression, analysis, and implementation. The STS can also load operational flight program (OFP) data in reverse order from the STS to the GRE in the

SFDR which in turn is connected to the mux bus.

STATUS: entering initial production for tactical military and commercial aircraft.

SAU
Dimensions: 6.2 × 7 × 7.25 inches (157 × 178 × 184 mm)
Weight: 14.35 lb (6.51 kg)
Memory: (program) 16 K words
(scratchpad) 2 K words
(non-volatile) 2 K words
(auxiliary) 256 K words, growth to 512 K
Environment: MIL-E-5400 Class II

CSMU
Dimensions: 3 × 3 × 4.6 inches (76 × 76 × 117 mm)
Weight: 3.48 lb (1.58 kg)
Memory: 28 K words, growth to 64 K
Recording time: (normal flight profile) 60 min
(active flight profile) 15 min
Environmment: (impact) 1700 g 6 ms
(penetration) 500 lb 15 ms
(static crush) 5000 lb 5 min
(fire) 1100°C flame, 30 min, and equivalent oven test
(fluid immersion) 40 h

Lear Siegler Inc, LSI Avionic Systems Corporation, 7-11 Vreeland Road, Florham Park, New Jersey 07932
TELEPHONE: (201) 822 1300
TELEX: 136521

AN/AYQ-8(V) Ulaids

The universal locator airborne integrated data system (Ulaids), designated AN/AYQ-8(V), is a comprehensive data recording and display system, with ground and airborne elements for improving mission effectiveness and maintenance diagnostics, and reducing aircraft downtime.

A complete system (Ulaids can be built up to suit specific needs) would comprise four sections: master monitor display, terminal and data entry panel; signal acquisition conditioning terminal (SACT); aircraft health monitor recorder and playback units; and flight incident recorder, playback unit and terminal.

Critical signals related to airframe and engine health are passed from the relevant monitor units to a SACT, for real-time conditioning, converting, the detection of exceedances and multiplexing. Data is then passed to the appropriate recorder, and all exceedances which would endanger the aircraft or affect mission effectiveness are displayed to the pilot. Corrective actions are displayed for each detected exceedance, on a priority basis. Any event, or a series of events can be recorded for ground-based analysis, by pilot action. Data can be recalled on the flight line, to speed turn-rounds, and further analysis can be done in depth for airframe and engine performance

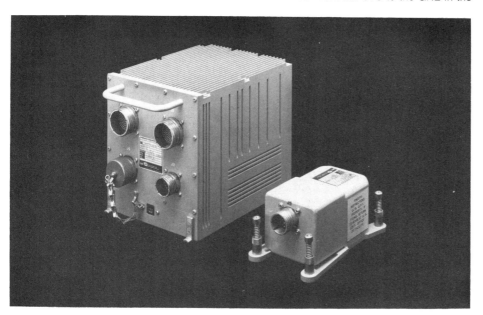

Lear Siegler standard flight-data recorder system with SAU (left) and CSMU (right)

trend, pilot performance and system monitoring, for example.

The AN/AYQ-8(V) has been test flown on an LTV A-7 aircraft, in which the flight recorder element monitored 76 flight critical parameters, while the whole Ulaids system monitored a total of 189 parameters throughout the aircraft.

STATUS: in development.

AN/ASH-28 signal data recorder

LSI provides a variety of airborne data acquisition systems for US Air Force and allied military aircraft, including General Dynamics F-16, McDonnell F-4G Wild Weasel, Boeing B-52, KC-135, and Lockheed C-141. The most recent development, the AN/ASH-28, records digital data on a 25-hour capacity quickly removeable cassette. The system is designed for the McDonnell Douglas F-15 Eagle, and is housed within a single box containing two linear accelerometers, one angular accelerometer and three rate-gyros, and interfacing with the F-15's MIL-STD-1553 digital data-bus. The company was awarded a contract worth more than $1 million in October 1983 for recorders to support an F-15 programme and expects to deliver nine units within a one-year period.

STATUS: in production.

Lockheed

Lockheed Aircraft Service Company, Division of Lockheed Corporation, PO Box 33, Ontario, California 91762-8033
TELEPHONE: (714) 988 2357
TELEX: 182974
TWX: 910 581 3409

Model 209 digital flight and maintenance data recorder

Since the introduction by Lockheed in 1958 of the first FAA-approved flight recorder on the Boeing 707, the company's recorders have become standard equipment on the world's airlines and more than 5000 have been built. The Model 209 digital flight and maintenance recorder, designed to accommodate the increased complexity of the widebody jetliners, was introduced in 1970 with the Boeing 747, and is now in service with most of the world's airlines. The manufacturer says that it has the highest demonstrated reliability of any flight-data recorder in current service. Apart from the widebodies, the Model 209 has been chosen by a number of airlines for their Boeing 767 and 757 transports and commuter and regional-transport aircraft such as the de Havilland Canada Dash-7 and Dash-8 and Saab 340. The Model 209 has also been selected by the US Air Force for its Lockheed C-130, C-5B and Boeing E-3A aircraft.

The Model 209 digital flight-data recorder was developed for use with the FAA-mandated EFDARS (expandable flight-data acquisition and recording system). Apart from its application as a mandatory crash recorder, a unique electronic motor speed control and high-speed playback facility permit the device to be used as a data recorder for maintenance purposes. Some 25 hours of digital information can be copied in a 20-minute turnround, using the companion LAS Model 235 copy recorder. As a crash recorder the device is contained within a titanium housing and contains mounting facilities for an optional, certificated underwater locator beacon.

Format: 1/2 ATR Long
Dimensions: 4.9 × 7.6 × 19.5 inches (124 × 193 × 493 mm)

Lockheed Model 209F digital flight recorder

Weight: 22.5 lb (10.23 kg)
Operating temperature: –55 to +70°C
Motor: ac induction type operating at 1/60 rated speed during recording to conserve life
Drive system: V-belt and pulley
Speed control: electronic system using optically coupled tachometer disc for lowest wow and flutter
Input signal: Harvard bi-phase, 768 bits/s
Recording medium and speed: Mylar tape, 0.37 inch/s record, 11.84 inches/s high-speed playback
Bit packing density: 2076 bits/inch at 64 words (12 bits each)/s
Recording duration: 25 h, continuously updated
High-speed playback time: < 20 minutes for total record
Power: 12 W at 115 V 400 Hz record, 30 W playback
Qualification: survivability to FAA TSO C51a, using titanium housing and dry thermal insulation
Reliability: 8000 h mtbf, 15 000 h or 5 y mtbo. BITE and status outputs according to ARINC 573

STATUS: in production and service with more than 2300 delivered to operators world-wide.

Model 209F digital flight-data recorder

Lockheed has introduced a second-generation version of the widely used Model 209 digital flight-data recorder (see separate entry). Designated Model 209F, the first units were delivered to airframe manufacturers in September 1982.

The Model 209F represents a two-year product-improvement effort by the company and incorporates state-of-the-art advances in electronics, mechanical design and production procedures. The result is a system with a claimed 8000 hours mean time between failures, a one-third increase over the 6000 hours of its Model 209E predecessor, which is also a claimed 'industry highest' figure.

As with the Model 209E, the -F can record more than 120 parameters, using Mylar tape on an endless loop that can accommodate the most recent 25 hours of flying.

Operating temperature: –55° to +70°C
Environment: tested by exposure to 1100°C for specified time, crush force of 5000 lb (2270 kg), impact shock 1000 g, immersion in sea water for 30 days, and damage resistance equivalent to 500 lb (227 kg) steel bar dropped on end from height of 10 ft (3 metres).

STATUS: in production and service.

Model 280 quick-access recorder

The Model 280 quick-access recorder was designed primarily for recording and storing aircraft maintenance data, and has a capacity of more than 51 hours. The system is built to ARINC 591 specifications, and uses a commercially proven cassette modified to accommodate reel brakes when it is removed from the recorder in order to avoid tape spillage. The system can be easily adapted for use with one, two, three, or four flight-data acquisition units, and can handle two such units without the use of a data management unit.

Format: 1/2 ATR Short or 1/2 ATR Long case
Weight: 17.5 lb (7.95 kg)
Input signal: Harvard bi-phase, 768, 1536, 2304 or 3072 bits/s
Playback time: < 32 minutes for total record

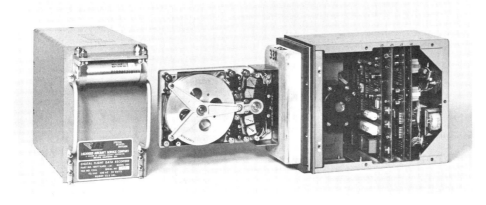

Lockheed Model 209 digital flight and maintenance data recorder

Bit-packing density: 2400 bits/inch
Power: 18 W

STATUS: in production.

Model 319 flight-data recorder system

This crash recorder system is specially designed for business and commuter aircraft where space and weight considerations are even more important than in commercial transport aircraft. Space is conserved by combining the functions of a data acquisition unit and recorder, normally accommodated in two ½ ATR Long boxes, into a single ½ ATR case with the same capability. The system can playback the entire 25 hours of data in less than 20 minutes at high speed without removing it from the aircraft.

Apart from the saving in volume, Lockheed says that the Model 319 is 30 per cent lighter and requires 80 per cent less power than equivalent air-transport equipment. The system records the parameters mandated by the FAA's FAR Part 121 airworthiness schedules for this class of aircraft in standard ARINC 573 format.

Dimensions: 4.9 × 7.6 × 19.5 inches (124 × 193 × 493 mm); ½ ATR Long
Weight: 25 lb (11.36 kg)
Operating temperature: –30° to +50° C
Input signals: 80 analogue, 30 discrete
Sample rate: programmable from 1 sample in 4s to 8 samples/s
Tape speed: 0.37 inch/s for record, 11.84 inches/s high-speed playback
Bit packing density: 2076 bits/inch at 64 words (12 bits each)/s
Recording duration: 25 h, continuously updated
Power: 25 W at 115 V 400 Hz
Qualification: survivability to FAA TSO C51a
Reliability: 6000 h mtbf, 15 000 h or 5 y mtbo

STATUS: in production.

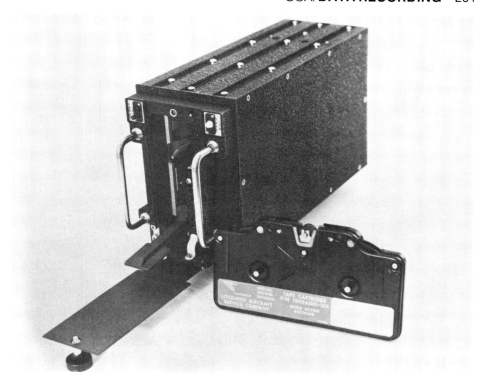

Lockheed Model 280 quick-access recorder and cassette

Lockheed Model 319 flight-data recorder

McDonnell Douglas

McDonnell Douglas Corporation, McDonnell Aircraft Division, PO Box 516, St Louis, Missouri 63166
TELEPHONE: (314) 232 0232
TWX: 910 762 9635

Aimes expert system

Flight-testing of the Aimes avionics integrated maintenance expert system, funded and developed by McDonnell Douglas began in January 1986 using an F/A-18 aircraft.

Aimes uses rule-based artificial intelligence to detect and isolate avionics failures down to electronic card level and is a development of the standard fault isolator in the F/A-18. It is one of the very first applications of artificial intelligence to enter flight-testing.

The system monitors the aircraft's mission computers for indications of a failure and records data from suspect boxes: queries are then generated by the artificial intelligence, dependent on the data and tests are performed to determine the validity of the queries. A conclusion as to the nature and location of the fault is then stored in a 256 K bubble memory for post-flight analysis.

Aimes uses a 80186 microprocessor, operating at 6 MHz, and interfaces with the MIL-STD-1553B digital data-bus. The expert system, which has forward and backward chaining and an in-built rule de-bugger, is written in PLM language.

At present, the prototype Aimes box can monitor only two avionics units in the aircraft, the communications controller and the digital display indicator, but further development of capability, including integrating Aimes into the data-bus, rather than having it as a monitor, is in hand. Another future plan is to rewrite the expert system in Ada. Eventually, McDonnell Douglas would like to make Aimes a part of the F/A-18 data storage set, being developed by Hamilton Standard for installation in aircraft built from 1987 onwards: the Aimes software could be ready for this application by 1988/89.

Dimensions: 6 × 6 × 9 inches (150 × 150 × 225 mm)

STATUS: Phase 1 flight trials will last through 1986.

Northrop

Northrop Corporation, Electronics Division, 2301 West 120th Street, Hawthorne, California 90250
TELEPHONE: (213) 600 3000
TWX: 910 321 3044

AN/USQ-85 TEMS turbine engine monitoring system

Northrop is producing an advanced diagnostic system, the turbine engine monitoring system (TEMS), which is designed to improve tactical aircraft operational readiness while reducing maintenance costs and turnaround time.

TEMS continually monitors jet engine performance while recording and storing the data used in determining aircraft engine conditions. It helps maintenance personnel verify the operational status of an engine, helps confirm problems noted by the pilot, and records engine performance trends which could lead to future problems.

TEMS is said to be the first turbine engine diagnostic system to qualify for use with US Air Force tactical aircraft, and has been designated AN/USQ-85.

Northrop is providing 234 TEMS shipsets, associated hardware and support to the US Air Force and Boeing Military Airplane Company for the US Air Force's Fairchild A-10 and Boeing KC-135R aircraft. The company is also producing an additional 166 spares and modules for these aircraft. Northrop is also providing two engineering models and nine pre-production units with aircraft fatigue monitoring capabilities to Grumman for testing on the F-14 fighter.

The prime objective of TEMS is to aid the military in implementing an 'on condition maintenance' concept, whereby engine inspection and repair is performed on an as needed basis, determined by electronic computerised diagnostics, rather than at conventional intervals based on flight hours or elapsed calendar time.

Northrop's system uses highly advanced

sensors placed at key points to monitor jet engine performance, thereby eliminating unnecessary jet engine removals and test cell run-ups. The system also reduces the amount of time and fuel required to trim or fine tune jet engines.

TEMS is adaptable to a variety of powerplants on other aircraft, both military and civilian, without hardware redesign, through incorporation of adaptive software. TEMS airborne data computation capability and memory capacity allows analysis of data to be made while the plane is airborne, presenting a 'go' or 'no go' status for maintenance decisions after landing, allowing rapid aircraft turnaround in combat conditions.

STATUS: in production. Deliveries began in 1984 and TEMS is operational in Boeing KC-135R aircraft. TEMS are also being delivered to Fairchild A-10 aircraft, beginning in February 1986.

Data recorded on Northrop AN/USQ-85 TEMS unit is dumped into data collection unit following sortie of Fairchild A-10 aircraft

Novatech

Novatech Corporation, 1200 Diamond Circle, Lafayette, Colorado 80026
TELEPHONE: (303) 666 4156
TELEX: 450184

Data acquisition system

This airborne recording system uses data compression techniques that analyse information on board the aircraft before storing it in the memory. Modular hardware and software permits maximum user flexibility. The standard system accommodates up to eight dual-channel input modules of high and low level signals and up to eight digital signals. The system incorporates internal clock and full self-diagnostic capabilities. It interfaces directly with ground-based computers either through the RS 232 port or by use of a memory module.

Format: standard ARINC 404A ½ ATR Short
Dimensions: 4.5 × 6.5 × 12 inches (114 × 165 × 305 mm)
Inputs: switch closure or digital logic levels; high-level signals (0-10 V or 4-20 mA); transducers with conditioning for thermocouples, strain gauges, pressure, displacement or acceleration sensors and most other aircraft instruments.
Environment: ruggedised for shock and vibration to MIL-STD-810C Class 2.

STATUS: in production.

Novatech data acquisition system with hand-held display/programmer

Sundstrand

Sundstrand Data Control Inc, Avionic Systems Division, Overlake Industrial Park, Redmond, Washington 98052
TELEPHONE: (206) 885 3711
TELEX: 286144
TWX: 910 449 2860

UFDR universal flight-data recorder

This 25-hour capacity, crash-protected digital flight data recorder has a variety of configurations. Together with optional accessory equipment, the system can be made completely interchangeable with existing flight-data recorders to ARINC 542, with digital recorders to ARINC 573/717, or expanded-parameter systems intended for CAA-regulated applications.

The recorder uses Kapton tape for durability, a single motor, and co-planar reel-to-reel tape transport. It can be hard-mounted in the aircraft, and considerable attention has been given to reducing the number of moving (and therefore wearable) parts. High-speed data retrieval permits the entire tape content to be transferred to a copy recorder in less than 30 minutes, so that the operation may be conducted during normal turn-round.

The system uses a patented feature called Checkstroke that monitors the data recorded each second and verifies its accuracy against the continuous stream of information received into the recorder's electronics. This built-in test feature ensures the validity of the recorded data.

Sundstrand universal flight data recorder with desk-top readout/test system

With the recent release of the FAA NPRM 85-1, new versions of the UFDR have been designed. The new H model is considered a six-parameter unit (meeting that portion of the NPRM) and the new G Mod 10 model meets the expanded 11-parameter portion of the NPRM.

The H Model is expansible to the G Mod 10 by simply adding one circuit card and a 5-volt dc power supply.

A new feature of these models is the addition of programmable inputs which allow the user to tailor the UFDR to his particular transducer types. Another plus is the increased accuracy of the system.

The following inputs are available for each model:

H Model	G Mod 10 Model
Vertical acceleration	Same as H Model
Heading	Pitch attitude
Pneumatic attitude	Roll
Pneumatic airspeed	Flap
Marker beacon	Engine 1 to 4
Time	Pitch trim position
Trip and date	1 spare programmable
2 programmable inputs (altitude and airspeed)	2 spare 0-5 V dc Total air temperature

Format: ½ ATR Long per ARINC 404
Weight: 30 lb (13.6 kg)
Input power: 115 V 400 Hz single-phase
Compatible transducers: synchros, 0-5 V dc 3-wire, 0-18 V dc 3-wire, 0-5 V dc 2-wire, 0-15 V 2-wire, ac ratio type 1, resolvers and selsyns
Recording capacity: 25 h at 768 bit/s
Input interface: (digital flight-data system) ARINC 573/717 data stream from remote flight data acquisition unit
(flight-data recorder) vertical acceleration, altitude, airspeed, heading, marker beacon and three discrete signals (on/off, to/from etc). Alternatively can be ARINC 573/717 input from a remote flight-data acquisition unit
(expanded-channel flight-data recorder) same as for flight-data recorder plus seven additional synchro inputs
Tape format: blocked incremental, bi-phase encoded, with 768 data bits, preamble, postamble and IRG
Tape type: Kapton, 0.25-inch magnetic
Tape speed: recording, incremental 5 inches/s = 0.414 inch/s net; playback 5 inches/s
Tape capacity: 450 ft for 25 h recording
Number of tracks: 8 bi-directional sequential
Playback time: 30 minutes with copy recorder, 2 h with sequential track transcription
Environmental conditions: RTCA DO-160, TSO-C51a
Crash survivability: TSO-C51a

STATUS: in production and service for over 20 years. In May 1984 the system was chosen for the US Air Force/McDonnell Douglas KC-10A Extender tanker. More than 2500 of these devices have been built or are on order.

SETS-II severe environment tape system

SETS-II is claimed to be the lightest and most compact digital data recording system currently available. It emphasises a very low error rate during recording and playback in the most harsh military air, sea and land environments, and comprises two modules: a tape drive system, with the recorder electronics and mechanical elements, and a tape unit containing the recording medium, magnetic recording

head tape, tension device and tape-end sensor. The system accommodates 300 feet of 0.25-inch tape, with a four-track single-gap read-write head, a recording density of 1600 bits an inch transfer rate of 48 000 bits a second (phase encoded) and total capacity of 23 million bits. The nominal error rate is 1 in 10^7 bits. Tape speed is 30 inches a second write and 120 inches a second search and rewind. Total system weight is 4.74 lb (2.02 kg).

STATUS: in development.

Micro-aircraft integrated data system

The micro-aircraft integrated data system (Micro-AIDS) has been developed to automate the various data-logging functions on an aircraft, including the gathering and processing of data concerning engine condition and operational performance monitoring, exceedances and fuel usage, etc. A single unit is used to gather the data which is stored in solid-state memory.

The Micro-AIDS operates on the ARINC 573 data-stream, and recorded data can be recovered by ground crew using a portable data retrieval unit.

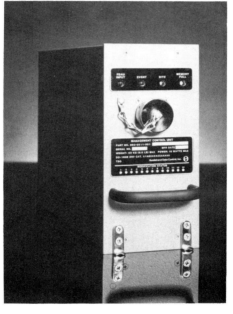

Sundstrand Micro-AIDS single-box recording system

Format: ARINC 600 2 MCU (3 MCU in expanded version)
Weight: 5 lb (2.3 kg) (8 lb (3.6 kg) expanded)
Power: 15 W at 115 V 400 Hz
Capacity: 25 flights

STATUS: in production.

Model AV-557C cockpit voice recorder

Latest in Sundstrand's family of cockpit/flight-deck voice recorders, the Model AV-557C provides a crash-survivable record of the last 30 minutes of flight-crew conversation by recording them simultaneously on four separate tracks. The new device succeeds the earlier Models AV-557A and AV-557B, offering improvements in reliability and maintenance cost by incorporating integrated circuits.

The specially coated polyimide-base magnetic tape recording medium is less sensitive to heat than conventional polyester film, so that data can be recovered after exposure to 240° C, compared with 110° C in conventional equipment. Recorder electronics are mounted on functional plug-in modules for ease of test and replacement. The tape has a clear window at each end so that the beginning and end of tape can be detected, whereby, capstan motor rotation is electronically reversed, and the record, erase and monitor heads are switched for bi-directional recording.

The recorder is used in conjunction with a microphone monitor that can be installed on the flight deck. It is housed within a unit that also contains an amplifier, filter, monitor indicator, push-test and bulk erase buttons, and a headphone jack for auditing recorded

Sundstrand AV-557C cockpit voice recorder

Tape-drive unit for Sundstrand SETS-II tape recorder

Sundstrand data transfer unit AN/ASK-7

information. An external microphone can also be plugged into the unit.

Format: ½ ATR
Weight: 23 lb (10.5 kg)
Tape type, drive and speed: 0.25 inch (6.35 mm) polyimide-base, ac hysteresis synchronous motors, 2.75 inches/s (70 mm/s)
Bandwidth: 300-5000 Hz ± 3dB
Maintenance period: 6000 h
Environmental:
(fire) meets requirements of TSO-C84 and TSO-C51a; temperature of 1100° C over 50% of equipment surface for 30 minutes
(salt water immersion) 30 days
(explosion) meets or exceeds DO 160 requirements

STATUS: in production and service; over 2000 have been built.

AN/ASK-7 data transfer unit

This device, originally designed for use with the Boeing AGM-86 air-launched cruise missiles, has now been adapted for the US Air Force/ Boeing B-52 and Rockwell B-1B aircraft.

The unit transfers large quantities of digital information from ground-based master systems to aircraft equipment. It will also monitor performance parameters as part of the central integrated test system on the B-1B. In the B-52s, they form a small part of the current OAS (offensive avionics system) update programme to improve the effectiveness of these 1950s-designed subsonic bombers.

Major features of the AN/ASK-7 include full record/replay capability, MIL-STD-1553 interface, error-correction code, remote electronics, extensive built-in test, and nuclear hardening.

The system comprises a control unit, containing most of the recorder electronics, and an instrument-mounted cassette drive unit, accommodating two cassettes at a time.

Weight: (cassettes) 7 lb (3.1 kg)
(cassette drive unit) 8 lb (3.6 kg)
(control unit) 15 lb (6.8 kg)
Power: 20 W at 115 V 400 Hz

STATUS: in production. A modified version has been TEMPEST-approved for use in the Integrated Operational Nuclear Detection System being developed by Texas Instruments.

Navigation and nav/attack

CANADA

Canadian Marconi

Canadian Marconi Avionics Division, 2442 Trenton Aenue, Montreal, Quebec H3P 1Y9
TELEPHONE: (514) 341 7630
TELEX: 05-827822
TWX: 619 421 3564

CMA-2000 Microlander microwave landing system

Development of the CMA-2000 continues, leading to qualification to ARINC-727 (MLS airborne receivers) scheduled for a late-1986 pre-production availability. Originally test flown in January 1985, Canadian Marconi development receivers continue test and evaluation in a joint Transport Canada/ Canadian Marconi flight programme based at Uplands Airport, Ottawa. Updates to these development receivers include the addition of highly integrated microwave modules produced at the company's plant in Montreal. Other flight evaluations of the MLS receiver have been made on-board a Canadian Forces' CC-130E Hercules, which was flown at 6° glideslope, and a de Havilland Dash 7 aircraft which was flown in conjunction with the September 1985 Montreal/Ottawa ICAO seminar on MLS. Further evaluations are planned for a Boeing 757 aircraft and a complex valley approach programme is underway in British Columbia.

Canadian Marconi production equipment will be to the ARINC-727 specification currently in the final stages of definition by the Airlines Electronic Engineering Committee (AEEC). The ARINC-727 MLS has a 3 MC case size and the Canadian Marconi design uses high levels of integration featuring surface mount technology and large-scale integration. As a consequence, power, weight, size and cost are minimised. The CMA-2000 has processing and memory capacity to be able to provide guidance for curved and segmented approaches currently being defined by Special Committee SC-151 of the Radio Technical Commission for Aeronautics.

STATUS: in development. Canadian Marconi will deliver pre-production ARINC-727 receivers in late 1986 and production equipment will be available in early 1987.

AN/APN-208 (V)/CMA-708B Doppler navigation system

A member of Canadian Marconi's fifth generation of this type of equipment, the AN/APN-

Canadian Marconi AN/APN-208(V)/CMA-708B Doppler system

208(V)/CMA-708B Doppler navigation system was designed to meet the requirements of ASW and search and rescue helicopters.

The system comprises four units: a transmitter-receiver-antenna, signal data converter, computer control indicator and an optional pilot steering indicator. The transmitter-receiver-antenna uses a Gunn diode rf source to generate four Janus-configuration beams, and FM/continuous wave modulation gives high accuracy and immunity from carrier noise, precipitation, surface spray, and reflections from nearby objects such as airframe structure and sling loads. The signal data converter contains processing circuitry and a Canadian Marconi mini-computer designed for airborne applications. The control indicator provides navigation information on two 14-character displays, and mode and status information on annunciators.

Dimensions: (transmitter-receiver-antenna) 439 × 113 × 439 mm
(signal data converter) 193 × 198 × 119 mm
(computer control indicator) 146 × 152 × 165 mm
(pilot steering indicator) 83 × 83 × 127 mm

Weight: (transmitter-receiver-antenna) 5.59 kg
(signal data converter) 9.77 kg
(computer control indicator) 3.54 kg
(pilot steering indicator) 1.04 kg
Max speed: (forward) –50 to 300 knots
(vertical) 5000 ft/minute
(lateral) 100 knots
Accuracy: (velocity) 0.25% in forward and lateral movement, 0.2% vertically
(navigation) 0.30% with no heading error
Rf power: 50 mW radiated at 13.325 GHz

STATUS: in production and service.

AN/APN-221 Doppler navigation set

Claimed to be the most powerful and cost-effective self-contained helicopter navigation system currently available, the AN/APN-221 radar set is supplied for night/adverse weather search and rescue helicopters. This fifth-generation system (it spans the technology range from thermionic valves and attitude-stabilised antennas to all-solid-state circuitry, fixed antennas and microcomputers) was derived directly from the AN/APN-208(V)/CMA-708B helicopter Doppler navigation

During 1985 Canadian Marconi's CMA-2000 Microlander underwent flight trials on board this Canadian Armed Forces' CC-130E

Canadian Marconi CMA-2000 Microlander airborne MLS receiver and display

Control/display unit for Canadian Marconi AN/APN-21 Doppler radar

system developed to meet the requirements of ASW and search and rescue helicopters, where versatility in operation and interface options were essential.

A specific application of the APN-221 Doppler navigation set, which was developed under the sponsorship of the US Air Force Systems Command, Aeronautical Systems Division, is the Pave Low III/Sikorsky HH-53 medium lift helicopter. The operational requirements of this aircraft called for an extension of the AN/APN-208(V)'s capability to include guidance in poor weather or darkness through the use of advanced computer-access techniques and cockpit displays.

The four-unit APN-221 comprises a four-beam lightweight antenna, ¾ ATR Short signal-data converter with 16-bit microcomputer, 6-line by 12-character control and display unit, and a steering/hover indicator. It provides three pilot-selectable navigation coordinate systems, with automatic conversion from one to another. They are latitude/longitude, worldwide alphanumeric UTM, and arbitrary grid. Data can be stored for up to 15 mission waypoints, 10 targets of opportunity, or 25 library waypoints (including Tacan beacon locations). There are also three pilot-selectable search patterns with automatic turn-point computation and navigation/guidance outputs: creeping line, expanding square, and sector. The system provides aircraft velocity outputs to the flight control system, enabling coupled hover manoeuvres and automatic approaches to the hover to be conducted.

Dimensions: (antenna) 439 × 439 × 113 mm
(signal data converter) 194 × 198 × 319 mm
(control/display unit) 146 × 114 × 165 mm
(steering/hover indicator) 83 × 83 × 127 mm
Weight: (antenna) 5.59 kg
(signal data converter) 9.45 kg
(control/display unit) 3.54 kg
(steering/hover indicator) 1.14 kg
Transmission: 4-beam Janus, time-shared (200 ms/cycle), 3 ×6.7 degree beam-width from Gunn diode, 13.325 GHz with FM/cw modulation optimised for flight envelope
Velocity range: (forward) –50 to 300 knots
(lateral) 100 knots
(vertical) ±5000 ft/minute
Accuracy: (forward) 0.3% speed along velocity vector ±0.2 knots
(lateral) 0.32% speed along velocity vector ±0.2 knots
(vertical) 0.2% speed along velocity vector ±20ft/minute
Inputs: pitch/roll attitude and heading and mode control and clock pulses from INS or central computer
Microcomputer: (architecture) word-addressed, 16-bit parallel microprogrammed, single/multi-precision
(instruction-set) 70, Hewlitt-Packard compatible
(addressing) immediate, direct, indirect up to 32 K words
(memory) ram/rom up to 32 K words
(software) Assembler/Fortran mixed, complete Hewlitt-Packard support software library

STATUS: in production.

CMA-734 omega/vlf systems

The CMA-734 series is designed for use in applications where compact size and low weight are required. It has a choice of absolute or relative mode and continuous self-monitoring (built-in self-test). Four versions of the system are currently available.

CMA-734 standard: with an INS-type control/display unit, nine-waypoint storage, and separate optional vlf receiver (¼ ATR).

Number of waypoints: 9
Outputs: cross-track deviation, to/from indication, waypoint alert, DR warning signal, failure warning signal to autopilot, steering commands
Inputs: heading, speed
Power: 28 V dc at 50 W (typical)
Units: antenna coupler unit, receiver/processor unit, control-display unit, plus optional vlf if required
Weight: 24 lb typical (may vary, depending upon configuration)
Dimensions: (receiver/processor) ½ ATR Short (control/display unit) 4.5 × 5.75 × 6.1 inches (114 × 146 × 155 mm)

STATUS: in production. In 1985 the Canadian Department of National Defence ordered eleven CMA-734s, valued at C$250 000, to equip Canadair Challenger aircraft.

CMA-734 Arrow: features a liquid-crystal display, storage for up to 99 flight plans of up to nine waypoints each, and a built-in vlf receiver. The system can be hard-mounted in most helicopters.

Number of waypoints: 891
Outputs: } as for CMA-734 standard
Inputs: }
Power: 28 V dc at 40 W
Units: antenna coupler, receiver/processor (built-in vlf), control/display
Weight: 18 lb (with E-field antenna)
Dimensions: (receiver/processor) ½ ATR Short (control/display unit) 3.75 × 5.75 × 2.38 inches (95 × 146 × 60 mm)

STATUS: in production.

CMA-734 Alpha: featuring storage for up to 4000 waypoints (in flight-plans) and up to 1800 data points with identifiers. The control/display unit has a two-line display, one line of which is alphanumeric, and all-push-button entry/control operation.

Number of waypoints: 4000
Outputs: } as for CMA-734 standard
Inputs: }
Power: 28 V dc at 50 W (typical)
Units: antenna coupler, receiver/processor (built-in vlf), control/display unit
Weight: 21 lb typical (may vary, depending upon configuration)
Dimensions: as for CMA-734 standard

STATUS: in production.

CMA-734 sensor system: an omega/vlf sensor for use with flight and navigation management systems. The vlf receiver is built-in, and ARINC 429 digital data-bus interfaces are provided. An 'Arrow' control/display unit may be added as an option.

Number of waypoints: 891
Outputs: ARINC 429
Inputs: dual ARINC 429
Power: 28 V dc at 30 W (without optional CDU)
Units: antenna coupler, receiver/processor
Weight: 13.6 lb (6.7 kg) without antenna coupler or control/display unit
Dimensions: (receiver/processor) ½ ATR Short

CMA-771 omega/vlf systems

Systems in the CMA-771 series conform to ARINC 599 standards as well as to those of DO-160, DO-164, and TSO-C94. They also include ARINC 429, 561, and 575 digital data-bus circuitry to permit integration with a wide range of other systems such as Doppler, radio, INS, digital air data, and FMS. High-level discrete outputs and synchro signals are also available to drive HSIs. Two versions of the CMA-771 are currently available.

CMA-771 standard

Number of waypoints: 9
Inputs: heading, speed
Power: 115 V ac, 400 Hz, 75 VA (typical)
Units: antenna coupler, receiver/processor (built-in vlf), control/display unit
Weight: 33 lb (15 kg) typical (may vary, depending upon configuration)
Dimensions: (receiver/processor) ¾ ATR Long (control/display unit) 4.5 × 5.75 × 6.1 inches (114 × 146 × 155 mm)

STATUS: in production.

CMA-771 Alpha: features enhanced storage (up to 4000 waypoints) and a two-line display, one line of which is alphanumeric. The system also has a much greater computing power than the standard version, with high growth capacity.

Number of waypoints: 4000
Inputs: heading, speed
Power: 115 V ac, 400 Hz, 75 VA (typical)
Units: antenna coupler, receiver/processor (built-in vlf), control/display unit
Weight: 32 lb (14.5 kg) typical (may vary, depending upon configuration)
Dimensions: (receiver/processor) ¾ ATR Long (control/display unit) 4.5 × 5.75 × 6.1 inches (114 × 146 × 155 mm)

STATUS: in production.

CMA-764 omega/vlf sensor system

The CMA-764 omega/vlf sensor system (OSS) is a compact, lightweight system housed in a 2 MCU (approximately ¼ ATR) box. It features digital synthesised tuning of four omega plus one vlf channel, and has ARINC 429 data-bus

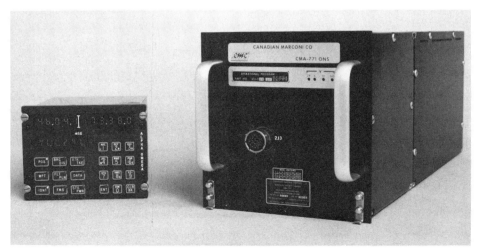

Canadian Marconi CMA-771 Alpha omega/vlf navigation system

interfaces. It is intended for use with advanced flight and navigation management systems, either as the only long-range sensor, in conjunction with inertial-type sensors, or with a second CMA-764.

Number of waypoints: 891
Outputs: ARINC 429

Inputs: triple ARINC 429
Power: 28 V dc at 20 W (without optional control/display unit
Units: antenna coupler, receiver/processor
Weight: 7 lb (3.2 kg)
Dimensions: 2 MCU

STATUS: in production, selected for Beech Starship 1.

FRANCE

CEIS Espace

Compagnie pour l'Electronique, l'Information et les Systèmes Espace, ZI de Thibaud, rue des Frères-Boudes, 31084 Toulouse Cedex
TELEPHONE: (61) 44 39 31
TELEX: 521039

Alpha-Nav area navigation system

This area navigation system, introduced during 1983 for light aircraft, is based on a plotting table, keyboard and microcomputer. Co-ordinates of waypoints are plotted on the table and then keyed into the computer. The system's memory can accommodate the positions of two waypoints (which can be outside the limits of the plotting table) and a DME. Aircraft speed, wind speed and direction, and map scale are also taken into account.

For radio navigation it is sufficient to enter bearing, or range and bearing information to obtain actual position coordinates with reference to the plotting table, as well as compass course, drift correction, time and distance to waypoint, and groundspeed. For dead-reckoning navigation, this information is obtained once the aircraft position has been entered.

STATUS: in production.

Crouzet

Crouzet SA, 25 rue Jules-Védrines, 26027 Valence Cedex
TELEPHONE: (75) 42 91 44
TELEX: 345807

Nadir Mk 1 Doppler navigation computer

The Nadir system comprises a processor and a control/display unit, and generates navigation guidance information from a Doppler velocity sensor. It can compute in geographical or UTM coordinates, store the positions of up to nine waypoints, compute windspeed and direction, and preserve the navigation data between the two points after switch-off. Nadir can also provide special functions, such as time-to-go to a waypoint or destination, provide a hover-hold signal to the autopilot of a helicopter, or correct the gimbal error of a gyro-compass.

STATUS: in production. The system is fitted to French Navy ASW helicopters.

Nadir Mk 2 navigation/mission management system

This is a multi-purpose processing and display system that can store details of up to 100 waypoints, and the characteristics of up to 100 VOR/DME stations. It is not limited to Doppler but can operate with many other navigation sensors. Nadir Mk 2 comprises a 4 MCU central processing unit and a general-purpose control/display unit. It can provide the following functions:
(1) Navigation management based on Doppler,

Crouzet Nadir Mk 1 system in Gazelle helicopter. Control/display unit is at lower centre

omega, VOR/DME, heading sensor, inertial sensor and air data inputs.
(2) Flight management, with guidance for optimum cruise conditions, fuel and weight management, and engine monitoring.
(3) Air data computations involving speed, altitude and outside air temperature.
(4) Interface with autopilot, radar, and navigation indicators.
Nadir Mk 2 may be used in a two-system configuration, in which one computer is responsible for navigation and flight management, while the second deals with weapons and aircraft management.

STATUS: in production.

Sextan H40 Doppler/inertial navigation system

The Sextan H40 is a hybrid Doppler-monitored inertial navigation system employs Sfena ring-laser gyros to give the short term qualities of a strapdown inertial system with the long-term ones of a Doppler radar. The system comprises a navigation/air data unit, vibrating-quartz pressure-sensor unit, and solid-state three-axis magnetometer. The navigation/air data unit contains the Sfena inertial measurement sensors (three 12 cm path-length ring-laser gyros and three accelerometers), and the Crouzet processing unit, Alpha 732, with a one million operations per second processing speed capability.

Equinox ONS 200A omega/vlf

Designed for dual installation in commercial aircraft, the ONS 200A omega comprises three units: an antenna coupler, receiver/processor and control/display unit. The system operates in conjunction with the worldwide vlf coverage provided by the eight omega stations.

Number of waypoints: 29
Inputs and outputs: As for ARINC 599

Receiver/processor
Type: 16-bit microprocessor, 16-bit address bus and data-bus, cycle time 0.5 μs
Memory: (program) eprom 18 K words 16 bits (computation) ram 4 K words 16 bits, including 256 words in protected memory
Dimensions: ½ ATR Short, 384 × 194 × 126 mm
Weight: 7.8 kg

Control/display unit Type 20
Display: 15 numerals and 16 status lights
Dimensions: ATI 146, 146 × 114 × 164 mm
Weight: 2 kg

One configuration of Crouzet Nadir Mk 2 control/display unit

Navigation/air-data unit for Crouzet Sextan H40 system

Crouzet ONS 200A omega/vlf

Antenna: 4 types available
(Type 10) H-field, to ARINC 580, aerodynamic profile, 311 × 235 × 48 mm, weight 3.3 kg
(Type 30) E-field, 370 × 170 × 86 mm weight 0.7 kg
(Type 40) H-field 'brick' type for installation inside wing tips, 176 × 160 × 44 mm, weight 1.6 kg
(Type 50) E-field coupler inserted in aircraft's ADF antenna

STATUS: in production.

Equinox ONS 300 omega/vlf

This is a more advanced version of the ONS 200 with fully digital inputs and outputs.

STATUS: in production for Airbus A300, and chosen by Garuda and Tunis Air.

Equinox OSS T400 omega system

OSS T400 is the designation of a family of omega navigation systems for commercial and military fixed-wing aircraft and helicopters. Each system comprises a control/display unit mounted on the flight-deck, a remote receiver/processor unit, and antenna. A choice of four types of antenna coupler permit optimised performance to be achieved. The sensor unit can be coupled directly to a flight management system via ARINC 429 connections.

STATUS: in development. The system has been chosen for the Dassault Falcon 900.

Equinox ONS 500 and 600 omega/vlf

These are versions of the ONS 200A modified for military use. The antenna-coupler units are the same, but the Type 500 receiver/processor has 5 K words of 16 bits for the memory program, while the Type 600 operates with 4 K 16-bit words. They operate in conjunction with the Type 10 and Type 20 control/display units. The Equinox systems are compatible with Crouzet's Navstar GPS (see separate entry).

STATUS: in production. One customer is the Irish Air Corps, which uses ONS 500 systems in its five Aérospatiale SA 365 Dauphin 2 helicopters.

Navstar GPS receiver system

Crouzet is planning to produce a complete range of equipment to receive transmissions from the US Defense Department's Navstar GPS navigation satellites. A prototype receiver

Crouzet Navstar GPS system comprises, left to right, control/display unit, receiver/processor, antenna coupler, and antenna

Receiver/processor unit for Crouzet OSS T400 omega system

Crouzet ONS 500 and 600 omega/vlf

was shown at the 1984 Expo Naval in Paris. These systems can measure position to within 10 to 20 metres, and the groundspeed of the vehicle to within 0.1 metre a second. The receiver system comprises an antenna and pre-amplifier, receiver/processor, and control/display unit. The receiver uses an Alpha 732 processor operating at 1 million operations a second and communicates with other equipment via an ARINC 429 bus. It can be coupled to an inertial navigation system. The Navstar GPS receiver developed by Crouzet will be compatible with Crouzet's Equinox omega navigation systems.

STATUS: in development.

Jaeger

Jaeger, Division Aéronautique, 2, rue Baudin, 92303 Levallois-Perret
TELEPHONE: (7) 37 71 20
TELEX: 62368

Digital VOR/DME

The Jaeger VOR/DME instrument takes navigation data, derived from suitable ground beacons and processed by a flight management system such as the Collins Proline using the ARINC 429 digital data-bus. It requires a 28-volt dc input and weighs less than 2.1 kg.

STATUS: in production.

Jaeger digital VOR/DME instrument

LMT

LMT Radio Professionnelle, 46 quai Alphonse-Le-Gallo, 92103 Boulogne-Billancourt Cedex
TELEPHONE: (1) 608 60 00
TELEX: 202900

Airtac airborne Tacan beacon

This system is designed for installation aboard tankers or mission-lead aircraft so that they can be located by other aircraft. It operates in the same way as a standard Tacan beacon except that it provides only 15 Hz information, in line with antenna size requirements. The system is therefore compatible with any standard Tacan that can operate with 15 Hz only, such as AN/ARN-52, Mitac and Deltac. Traffic capacity of the Airtac is identical with that of a normal Tacan beacon in that it can reply to a maximum of 100 aircraft.

The system comprises a transmitter-receiver, a remote control box, an electronically scanned antenna that produces a 15 Hz modulation, and an adapter driven by a compass signal controlling the electronic rotation of the antenna, so that the beaming information provided to the interrogating aircraft is always related to magnetic North, irrespective of the heading of the beacon aircraft.

Hemispherical antenna housing LMT Airtac system mounted under fuselage

LMT NRAP-1A Microtacan

Dimensions: (transmitter-receiver) 220 × 420 × 340 mm
(aerial) dia of hemisphere 280 mm
Weight: (transmitter-receiver) 15 kg
(antenna) 5 kg
Number of channels: 20 at 63Y, 64 at 126X
Peak power output: 100 W
Receiver sensitivity: –93 dBm

NRAP-1A Microtacan

This system provides data in digital form and, with modification, in analogue form, for the type of indicators used with the US AN/ARN-52 system. It makes extensive use of semiconductor components and integral linear and logic circuits, and contains only four thermionic valves.

Digital system
Format: 1/2 ATR
Weight: 10 kg
Modes: ground-to-air and air-to-air, channels X and Y

Digital/analogue system
Number of channels: 252 (126 each for channels X and Y)
Sensitivity: better than –87 dBm
Output power: 1.5 kW minimum, using broadband power amplifier

Accuracy: (distance) ±280 m
(bearing) ±20.5°

STATUS: in production under Franco-German collaborative agreement.

Deltac Tacan

This system was designed for the Dassault Mirage 2000. Based on a 1/2 ATR box, the fully-solid-state transmitter-receiver is of modular design (including the rf generator), has built-in self-test facilities, does not need forced-air cooling, and contains a microprocessor for data-filtering and area navigation computation.

Weight: 10 kg
Acquisition time: (bearing) 5 s
(distance) 1 s
Tracking rate: (bearing) 20°/s
(distance) 4500 knots
Memory: (bearing) 3-5 s
(distance) 8-12 s
Accuracy: (bearing) ±21°
(distance) ±20.1 n mile
Modes: air-to-air, transmitting and receiving
Power output: 1 kW peak
Power: 75 VA at 115 V 400 Hz, 50 mA at 28 V dc
Sensitivity: –90 dBm

STATUS: in production.

LMT Deltac Tacan system

Sagem

Société d' Applications Générales d'Electricité et de Mécanique, 6 avenue d'Iéna, 75783 Paris Cedex 16
TELEPHONE: (1) 47 23 54 55
TELEX: 611890

Uliss inertial navigation and nav/attack systems

These modular systems all employ high-accuracy inertial components: two dynamically-tuned gyroscopes and three dry accelerometers, a microprocessor-controlled computer working at 360 000 operations a second with eprom memory and highly integrated electronics using large-scale integrated and hybrid circuits.

Uliss systems fall into two categories, in both of which the main inertial navigation unit is contained within a 3/4 ATR Short case. In the first category are navigation versions, with a position accuracy of better than 1 nautical mile an hour. In the second category the navigation function is combined with the computation necessary for weapon delivery. Alignment time is 90 seconds for stored heading and 5 to

10 minutes for self-contained gyrocompassing. Standard interfaces permit the systems to be linked to other equipment via Gina or MIL-STD-1553B data-buses or ARINC serial data lines. A sophisticated failure-detection system can detect faults at module level with 93 per cent confidence, isolate them and signal their presence on a magnetic annunciator without external test equipment.

Specific systems and applications are:
Uliss 45: Boeing KC-135 tankers of French Air Force
Uliss 47: Dassault Mirage F1 CR (reconnaissance) and Mirage F1 EJ for Jordan
Uliss 52: Mirage 2000 (DA and N fighters), French Air Force
Uliss 52X: Dassault Rafale experimental prototype (two systems)
Uliss 53: Dassault ATL2 for Aéronavale
Uliss 54: Mirage IV of Armée de l'Air
Uliss 60: flight-tests carried out in Mirage, with 0.25 nautical mile typical error at end of mission
Uliss 80: Dassault Super Etendard for Argentina
Uliss 81: Dassault/Dornier Alpha Jet and Mirage V for Egypt, and Alpha Jet for Cameroon
Uliss 82: Sepecat Jaguar for India
Uliss 90: Jaguar for India

More than 80 per cent of the components and sub-assemblies are common to all members of the Uliss family, the principal differences residing in specific interfaces and computation functions. Details of five members of the Uliss family follow.

Uliss 45 inertial navigation system

This system has been optimised for high accuracy in long-range navigation and certain other special applications. Its interfaces comply with ARINC 561 and embody significant flexibility, for example, in order to communicate with two DME or Tacan receivers with Kalman filtering for better accuracy. The equipment is used on long-range transport aircraft, and as an accurate position and velocity reference for flight-development purposes.

Dimensions: (navigator) 420 × 194 × 191 mm
(control/display unit) 209 × 114 × 127 mm
Weight: (navigator) 16 kg
(control/display unit) 3 kg
Accuracy: 1 n mile/h
Power: 250 VA at 115 V 400 Hz

STATUS: in production for Boeing KC-135 tankers of French Air Force.

Uliss 52 inertial navigation system

This navigation system is designed for high-performance combat aircraft. Avionics information and commands are distributed by a digital multiplexed data-bus. It comprises three units: a UNI 52 inertial navigator, a PCN 52 control/display box and a PSM 52 mode selector, fitted with an automatic insertion module allowing information to be fed in such as flight plan, system data, and maintenance information.

Dimensions: (navigator) 386 × 194 × 191 mm
(control/display unit) 208 × 114 × 146 mm
(mode selector) 148 × 38 × 127 mm
Weight: (navigator) 15 kg
(control/display unit) 3 kg
(mode selector) 0.9 kg
Accuracy: 1 n mile/h cep
Power: 200 VA at 115 V 400 Hz

STATUS: in production for Dassault Mirage 2000.

Uliss 60 inertial navigation system

Uliss 60 incorporates the very accurate S060 dry gyroscope and is pin-to-pin interchangeable with other Uliss systems.

STATUS: in production.

Uliss 81/82 inertial nav/attack system

The Uliss 81/82 combines in a single box all the functions of an inertial navigation and a fire-control system. It comprises three units: a UNA 81 inertial/attack box, a PCN 81 control/display unit, and a PSM 81 mode selector.

The system provides aircraft position, velocity, and attitude information; computation of navigation and steering information to way-points; position updating by navigation fixes; weapon delivery computation (ballistic, determination of release point, ripple spacing of weapons, safety pull-up information, and head-up display information for target acquisition and commands for 'blind' release of stores); attack modes; air data computations and multiplex bus control for Ginabus or MIL-STD-1553B.

Dimensions: (navigator) ¾ ATR Short, 386 × 194 × 191 mm
(control/display unit) 216 × 116 × 153 mm
(mode selector) 151 × 41 × 135 mm
Weight: (navigator) 16 kg
(control/display unit) 3.5 kg
(mode selector) 1 kg
Accuracy: position 1 n mile/h cep, weapon-delivery 5 mrad
Power: 220 VA at 115 V 400 Hz

Sagem Sirsa inertial reference navigation system

STATUS: in production for Dassault/Dornier Alpha Jet and Dassault Mirage V.

Uliss 90 inertial navigation system

Sagem pioneered the development of integrated inertial navigation and attack systems with the Uliss 80 series. These systems now equip many front-line fighters like the Sepecat Jaguar for India, and the Dassault Super Etendard, Alpha Jet NGEA and Mirage 3 and 5

Sagem MSD 20 inertial navigation system for armed helicopters

operated by several international air forces. As there is a demand for more and more operational capability from such systems, Sagem has developed an improved single-card embedded computer (UTR 90) which demonstrates a 830 000 operations per second throughput and has an addressing capacity of 1 Mwords while maintaining an upward software compatibility. An advanced software development environment has been created along with UTR 90, including facilities for real-time trouble-shooting, higher-order languages, dynamic validation.

This computer fitted to Uliss 90 is interchangeable with the original UT 382 50 single-card Uliss computer.

STATUS: in production.

MSD 20 strapdown inertial navigation unit

In conjunction with a Doppler radar, the MSD 20 system makes up a Doppler-monitored INS suitable for the new generation of armed helicopters. It is particularly useful for low-level operation, where it is difficult to take external position 'fixes'. The inertial elements comprise two tuned-rotor GSD gyroscopes and three gas-damped, dry accelerometers. The computing system contains 8- and 16-bit microprocessors for logical and mathematical processing respectively, and the reprogrammable memory can accommodate 56 K words. The system is contained in a 191 × 193 × 305 mm box. Attitudes are measured to an accuracy of 0.1°, heading to 0.5° and navigation to 0.5% of distance travelled.

STATUS: under development.

Sirsa ring-laser gyro inertial reference system

Sirsa systems, which are the result of a co-operation between Sagem (prime contractor) and Singer Kearfott, are a new family of ring-laser gyro inertial systems intended for commercial and military transport aircraft. Sirsa systems basically comply with existing ARINC recommendations, but can incorporate 'growth options' to satisfy particular customer needs.

Sirsa 738: is an air data inertial reference system (ADIRS) complying with ARINC 738 recommendation.
It provides:
Inertial data: such as present position (1 nautical mile an hour class), ground speed, track, heading, drift angle and attitude
Air data parameters: as per ARINC 738 eg baro

Uliss 81 inertial nav/attack system for Alpha Jet NGEA

altitude, true airspeed, derived from air data module (ADM) measurements sent to, and processed in the ADIR unit

Main growth option allowed for in the Sirsa 738 by incorporation of GPS inertial Kalman filtering, from measurements of a GPS receiver either remotely located, or embedded in the ADIR unit.

Dimensions: 318 × 322 × 194 mm
Weight: 20 kg

STATUS: under development.

Sirsa 561-3: is the most powerful model in the Sirsa 561/571 family, whose other members are Sirsa 561, 571, 561-1 and 561-2, all intended as substitutes to conventional (gimballed) ARINC 561/571 inertial navigation systems. Each of these systems complies with ARINC 561

mechanical and electrical standards with additional capabilities such as multi-DME Kalman filtering (Sirsa 561-1) or flight-plan storage in a magnetic bubble memory card (Sirsa 561-2) or both (Sirsa 561-3), which results in a mini-FMS function.

Modular interface cards and software packages are available for each option. Input/outputs in the most comprehensive version (Sirsa 561-3) are:

Attitude angles

DME inertial horizontal situation parameters such as present position, ground speed, track, heading, drift angle

Horizontal air routes and radio-beacon insertion (FMS loader type), storage (magnetic bubble memory module) and retrieval (alphanumeric control and display unit)

Horizontal guidance parameters such as

desired track, track error, distance to go, time to go, cross track error, steering signal

Efficient display through colour cathode ray tube

Inertial-GPS Kalman filtering is also proposed as a customer option.

Dimensions: 510 × 210 × 194 mm
Weight: 20 kg

STATUS: in development. A 3-month flight trial of a Sirsa mounted in an Air France Boeing 747 was completed in August 1985. It achieved 700 hours of operation, including 500 of navigation, without any failure and without re-calibration.

The maximum error on any one flight was 17.2 nautical miles after eight hours, compared with the ARINC 704 specification of 20 nautical miles accuracy.

Sfena

Société Française d'Equipements pour la Navigation Aérienne, Aérodrome de Villacoublay, BP 59, 78141 Vélizy-Villacoublay Cedex
TELEPHONE: (1) 46 30 23 85
TELEX: 260046
FAX: (1) 46 32 85 96

Ring-laser gyros

In anticipation of a growing requirement for military and commercial aircraft, Sfena in 1984 set up a dedicated research, development and production plant at Chatellerault in central France to begin manufacture of ring-laser gyros for inertial reference or inertial navigation systems.

Sfena is currently producing two ring-laser gyro units with optical path-lengths of 12 and 33 cm.

With its 12 cm ring-laser gyro, Sfena is developing mid-course guidance/control systems for tactical missiles (pre-production systems for the new ANS anti-ship missile) and hybrid Doppler/inertial systems for helicopters in co-operation with Crouzet (Sextan system for the Franco-German PAH-2 combat helicopter).

With its 33 cm ring-laser gyro, Sfena is manufacturing stand-alone inertial systems: Pulsar, designed for transport aircraft has been undergoing flight evaluation on a Caravelle airliner since late 1985; and the Totem demonstrator system for military aircraft will fly in 1986 on a Dassault Mirage 3. Sfena is also producing an inertial system for the Ariane IV satellite

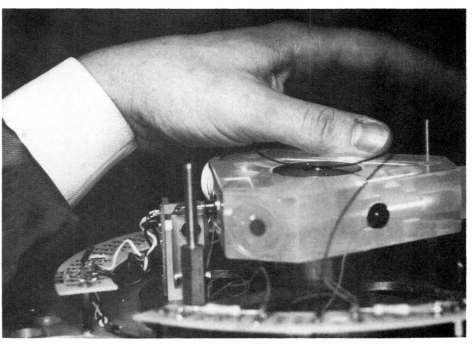

Sfena 33 cm path-length ring-laser gyro

launch vehicle, qualification of which was achieved in 1986.

Although Airbus has chosen Honeywell to build ring-laser gyros for the A320, Sfena is planning to fly an ARINC system aboard an

A300 airliner. Production of inertial reference and navigation system equipment is being done in conjunction with Crouzet.

STATUS: in development.

Sfim

Société de Fabrication d'Instruments de Mesure, 13 avenue Marcel-Ramolfo-Garnier, 91344 Massy Cedex
TELEPHONE: (1) 69 20 88 90
TELEX: 692164

28SH strapdown inertial navigation system

The 28SH system is designed to meet the requirements of ASW, anti-tank, combat, scout and observation helicopters, in conjunction with a Doppler radar to minimise long-term errors. Other inputs comprise magnetic heading and barometric and radio height information. The strapdown system employs high-precision gyros rigidly mounted to the structure of the containing box.

The system is contained in three boxes: an inertial reference unit (iru), a navigation and control/display unit (cdu), and a flux valve heading sensor. The inertial reference unit carries three two-axis tuned rate-gyros. The computing section employs cmos prom for protection against power-supply transients, and earom for storage. Sfim claims that the 'GAM' gyros employing Hookes' joint principle

Inertial reference unit for Sfim 28SH system

are the cheapest and simplest such devices to produce, as they eliminate all the mechanical components associated with conventional gimbals, and there are no bearings, motors or slip-rings.

Control/display unit and flux valve heading sensor for Sfim 28SH system

Readiness: 90 s after switch-on
Interfaces: most information available in ARINC 429, MIL 1553, or Ginabus data-bus format, though certain data can be supplied in analogue form to feed flight or other instruments
Accuracy (all accuracies to 95% cep):
(heading) 0.3°
(pitch and roll) 0.2°
(velocities) 1.5 m/s in horizontal axes, 1 m/s vertically
(present position) 0.3%

Alignment: automatic, needing no external reference
Computer: (instructions) 84
(memories) prom (24 K words of 16 bits), ram (8 K words of 16 bits), earom (2 K words of 16 bits)
Dimensions: (Iru) 155 × 190 × 193 mm
(cdu) 130 × 145 × 205 mm
(magnetic sensor) 125 × 89 × 81 mm

Weight: (Iru) 6 kg
(cdu) 2.7 kg
(magnetic sensor) 0.7 kg
Power: (total) 105 W at 28 V dc
Certification: meets French standards 7304, 2021D, 2025, and 510 for combat helicopter category

STATUS: the system has been successfully flight-tested by official centres in France, UK and West Germany, and is in production for helicopter applications and aircraft refit programmes.

Socrat

Socrat (Nouvelle Société), 2-4 rue Kuss, 75640 Paris Cedex 13
TELEPHONE: (1) 45 89 89 54
TELEX: 206420

Series 6200 VOR/ILS and marker receivers

Designed for use in military aircraft, the Series 6200 uses solid-state electronics to minimise space requirements. The VOR/ILS band is covered with 200 channels at 50 kHz spacing and frequency selection is by rotating knob with mechanical display. The glideslope receiver has 40 channels.

Dimensions: (receiver) 61 × 198 × 352 mm
(control box) 63 × 57 × 143 mm
(beacon receiver) 124 × 25 × 83 mm
Weight: (receiver) 4.6 kg
(control box) 0.55 kg
(beacon receiver) 0.5 kg

STATUS: in production.

Series 8900 VOR/ILS and marker receiver

The Series 8900 is intended for use in military aircraft, and is available in four combinations of VOR/ILS, marker and 30 Hz receiver, housed in a ¼ ATR box, or in special configuration to suit the application. Each covers the ILS/VOR band with 200 channels at 50 kHz spacing, with 40 channels in the glideslope receiver.

TRT

Télécommunications Radioélectriques et Téléphoniques, Defence and Avionics Commercial Division, 88 rue Brillat-Savarin, 75640 Paris Cedex 13
TELEPHONE: (1) 581 11 12
TELEX: 250838

TAC-200 Tacan

This low-cost digital Tacan builds on the company's experience in digital avionics including radio altimeters, distance-measuring equipment and air traffic control transponders. Software analysis of the timing and amplitude of all received pulses permits good co-channel and adjacent channel performance with no echo susceptibility.

The transmitter is fully-solid-state, and signal-processing by microprocessor software instead of discrete circuits reduces the number of components. Extensive use of integrated circuits keeps the power consumption low. The two-unit system comprises the IRT-200 transmitter-receiver and the INT-200 distance-measuring equipment indicator.

Acquisition time: 1 s
Interrogation rate: 20 (track), 40 (search), automatically varied by microprocessor

Sensitivity: better than –90 dBm
Accuracy: (distance) 0.02 n mile
(bearing) 1°
Memory time: 10 s

STATUS: in production and in service with French Air Force and Navy Embraer Xingu and Aérospatiale Epsilon communications and training aircraft and Aérospatiale AStar helicopters.

TDM-709 distance measuring equipment

Primarily designed for transport aircraft, the

TRT TDM-709 digital DME transmitter-receiver

TDM-709 DME system measures the slant range to a selected DME beacon. It also provides distance information on a number of DMEs to a navigation computer. This ARINC 709 system is microprocessor-controlled, permitting software analysis of timing and amplitude of received pulses, giving good co-channel and adjacent channel performance, and suppressing echoes. The system extensively uses software for output filtering and extrapolation. Five stations can be tracked simultaneously and the positions of the nearest 20 stations indicated. The TDM-709 features a uhf pre-amplifier for improved sensitivity.

Format: 4 MCU
Weight: 5 kg
Accuracy: ±100 m
Reliability: mtbf >7000 h (estimated)
Power output: 700 W
Power: 35 W at 115 V 400 Hz
Sensitivity: better than –92 dBm

STATUS: in production and service.

TDM-709P distance-measuring equipment

TRT is developing a precision DME system, derived from the TDM-709, to be used in conjunction with the company's microwave landing system. The principal operating difference is the guidance accuracy, which is 7 metres compared with a 100 metres navigation accuracy of the TDM-709.

Format: 4 MCU
Weight: 5 kg

STATUS: under development.

Microwave landing system

Based on its background of Category III radio altimeter work, TRT is developing a microwave landing system for commercial aviation. Guidance information is detected by a rapid Costas-loop technique which is relatively immune to noise. After processing, signals are passed to a course-deviation indicator. In order to provide comprehensive self-test, 50 per cent of the software and 10 per cent of the equipment and circuits are devoted to built-in test equipment.

STATUS: system had been bench tested by early 1984. Complete MLS/DME installation was test-flown at end of 1984.

Principal units of TRT TAC-200 Tacan navigation system

GERMANY, FEDERAL REPUBLIC

Becker

Becker Flugfunkwerk GmbH, Niederwaldstasse 20, Postfach 1980, D-7550 Rastatt
TELEPHONE: (7222) 121
TELEX: 781271

NAV 3000 VOR/ILS system

The NAV 3000 is a series of VOR/ILS receivers which can be used in many airborne applications: the sets cover the VOR/ILS frequency band and there is also a glideslope receiver available. The family features push-button operation and liquid-crystal display readout of selected frequency. The receiver can interface with a number of commercially available indicators, or with those produced by Becker.

Dimensions: (control box) 146 × 47.5 × 225 mm (associated RMI display) 82.5 × 82.5 × 129.5 mm

STATUS: in production.

NAV 3301 VOR/LOC system

The NAV 3301 is a small navigation set, with liquid-crystal display readout of frequency covering the VOR band with 200 channels at 50 kHz spacing. The set is designed mainly for light aircraft and gliders, and is the same size as the associated Becker AR 3201 radio transceiver.

Dimensions: (control box) 60.6 × 60.6 × 212.5 mm

Becker NAV 3301 VOR/LOC system

STATUS: in production.

Litef

Litef (Litton Technische Werke), Lörracher Strasse 18, Postfach 774, D-7800 Freiburg
TELEPHONE: (761) 49010
TELEX: 772669

Doppler system for Alpha Jet

This West German-based Litton subsidiary has built 190 sets of its own design of Doppler radar for the German version of the Franco-German Dassault/Dornier Alpha Jet strike/trainer. The system is used in conjunction with a Teledyne Ryan Electronics AN/APN-220G Doppler velocity sensor, built under licence in Germany.

With the addition of aircraft attitude and heading as correction factors, the system computes a number of quantities for navigation. A unique filter permits a very accurate instantaneous velocity to be derived for weapons delivery. The computer provides signals to the control/display unit for the following outputs: present position, up to 10 waypoints or target coordinates, four Tacan coordinates (including channel number and station height), five targets of opportunity, automatic correction for magnetic variation, fix and update facilities (using Tacan or known points), ground speed and drift, wind speed and direction, range and bearing to waypoints or targets, time-to-go to

Litef Doppler system for West German Alpha Jet

waypoints or targets, automatic back-up calculation for failures, and diagnostic test results.

Accuracy: 0.4% cep
Range: 40-600 knots, 0-50 000 ft over land with pitch and roll up to 45°. Accurate navigation between ±80° latitude
Weight: (sensor) 9 kg
(processor) 6.7 kg
(control/display unit) 2.8 kg

Power: (total) 186 W at 28 V dc
Reliability: (predicted) (sensor) 8000 h
(computer) 6000 h
(control/display unit) 12 000 h

STATUS: in production for West German version of Alpha Jet.

ISRAEL

Elbit

Elbit Computers Limited, Advanced Technology Center, PO Box 5390, Haifa 31051
TELEPHONE: (4) 514 211
TELEX: 46774

Update WDNS weapon delivery and navigation system for F-4

After many years in service, the navigation and weapon delivery equipment on Israeli Air Force/McDonnell Douglas F-4 Phantoms

(AN/ASN-46, AN/ASN-63 and AN/ASQ-91) was concluded to have marginal performance, low mean time between failures, high maintenance costs, and high mean time to repair. Elbit was contracted to upgrade the entire system, which it has now done in conjunction with Kearfott. The new system is fully digital, gives improved navigation and weapon delivery accuracy, flexibility, reliability and maintainability without modifications to the aircraft.

The Update WDNS, as the system is called, provides guidance for visual and blind navigation and weapon delivery, using an inertial measurement unit as the principal sensor. The

system also accepts inputs from the central air data computer, fire control radar, and various infra-red and optical target trackers. It operates in air-to-air, air-to-ground and navigation modes. In the first application it continuously computes missile firing envelope, in the second it provides information for ccip (continuously computed impact point), dive-toss, toss, and strafe attack, or for the release of special ordnance (for example 'smart' bombs). For navigation it computes and displays steering commands, and updates position, height and speed information.

The Update WDNS comprises eight line-replaceable units. They are the mission computer (the central digital computer for the system); aircraft interface unit; nav/attack data panel, essentially a control unit; navigation display unit, providing information on a 2.4 × 3 inch cathode ray tube; a pilot control panel, enabling the pilot to select radar or system information on his radar display, and to designate waypoints or targets; a filter regulator to supply regulated dc power during periods of normal power interruptions up to 5 seconds; an inertial measurement unit; an all-attitude reference platform; and a relay switching unit that passes signals to other aircraft systems.

Unit	Dimensions (mm)	Weight (kg)
(mission computer)	301 × 279 × 193	19.5
(aircraft interface unit)	361 × 239 × 203	14.5
(nav/attack data panel)	223 × 146 × 142	2.9
(navigation display unit)	240 × 127 × 160	4.5
(pilot control panel)	52 dia × 107	0.22
(filter regulator)	254 × 76 × 76	1.8
(inertial measurement unit)	334 × 216 × 207	10
(relay assembly)	209 × 109 × 119	2.72

STATUS: in service.

Navigation accuracy: 0.8 n mile/h
Weapon delivery accuracy: 7 mrad cep
Reliability: 300 h mtbf predicted
Total system weight: 56.14 kg

LCWDS low-cost weapon delivery system

Elbit's LCWDS is a compromise between the full-performance WDNS weapon delivery and navigation system produced by the company and the simple reflector gunsights still extensively used on light strike/trainer aircraft. The system is based on digital computation, improving aircraft capability and pilot's performance, and is particularly suitable for trainers and light attack aircraft and interceptors.

The system interfaces with other sensors and instruments and comprises a central mission computer, single combiner-glass head-up display and control unit. Specific interfaces include flux-valve, angle-of-attack, air data computer and gyro reference sensors, armament programmer and stick and throttle switches, and director instruments such as HSI and ADI. The mission computer is based on a 16-bit nmos microprocessor working in conjunction with a 32 K program memory that can be expanded up to 64 K. The three modes of operation comprise air-to-ground, air-to-air, and navigation. Each may be divided into submodes. In air-to-ground they are ccip toss manoeuvre, and manual. In air combat they are hot-line (snap-shoot) and lead-computing sight. The navigation mode displays steering commands to pre-selected way-points and calls for a Doppler system or sensor.

Elbit also has what it calls a minimum system, which is a very low cost version without the head-up display. This configuration uses the original aircraft gunsight to present air-to-air and air-to-ground aiming points in, respectively, ccip and snap-shoot and lead-computing modes.

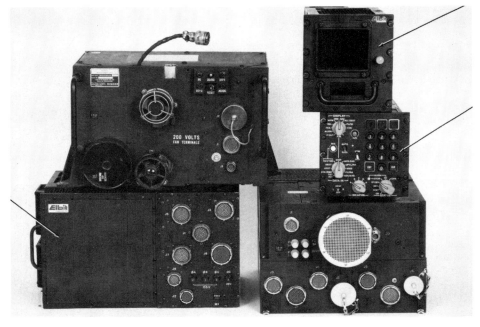

Five Elbit Update WDNS units. Arrows point (lower left) to aircraft interface unit, (top right) navigation display unit, and (lower right) nav/attack data panel

Dimensions: (mission computer) 250 × 195 × 140 mm
(head-up display) 320 × 120 × 150 mm
(control unit) 167 × 175 × 127 mm
Weights (mission computer) 10 kg
(head-up display) 5 kg
(control unit) 3 kg
Power: (total) < 390 W
Accuracy: (air-to-air) 3 mrad
(air-to-ground) 10 mrad (both figures depend on accuracy of aircraft gyro reference)
Reliability: 800 h specification
Velocity: 0-650 knots
Angular rates: (pitch and yaw) 90°/s
(roll) 360°/s
Altitude: 0-50 000 ft

STATUS: in production.

System 81 weapon delivery and navigation system

System 81 was designed to provide pilots of single-seat combat aircraft with such a standard of accuracy that they can navigate to the target at low level and deliver weapons with a high assurance of destruction on the first pass. It comprises three units: central computer, head-up display and control terminal unit (ctu). These communicate with other systems on the aircraft to derive navigation commands and weapon delivery information for air-to-air and air-to-ground attack. The equipment is virtually automatic in operation, being designed on the American Hotas (hands on throttle and stick) principle whereby the pilot is relieved of nearly all 'housekeeping' duties that distract his attention, such as switch selection and data-entry during flight. System 81 is part of the wider family of Israel Aircraft Industries' WDNS (weapon delivery and navigation system) 341 equipment.

In the air-to-air mode using missiles, the system computes the launch envelope and displays opportunity information on the head-up display. With guns, it provides hotline (snap-shoot) commands and lead-computation. For the air-to-ground mode it generates ccip(continuously computed impact point), designated toss and toss commands, together with data for the release of 'smart' bombs. For en route navigation, System 81 permits the co-ordinates of up to 40 waypoints to be stored before flight.

The central computer comprises a processing unit and memory, symbol generator, analogue/digital and digital/analogue converter unit, core store and power supply. The control and display unit is used to enter data into the central computer and extract it by a keyboard and three six-figure counters. The head-up display has a single combiner glass.

Dimensions: (central computer) 273 × 382 × 195 mm
(cdu) 175 × 167 × 127 mm
Weight: (central computer) 20 kg
(cdu) 2.5 kg

Elbit LCWDS system: from left, mission computer, head-up display and control unit

Components of Elbit System 82 without associated head-up display

Computer memory: 16-bit word length, 48 K × 16-bit eprom, 6 K × 17-bit ram, 2 K × 17-bit non-volatile ram
Registers: 17 × 32-bit registers (for arithmetic unit), and 7 × 16-bit registers
Accuracy: 0.6 n mile/h en route navigation, 6 MR cep for weapon delivery
Reliability: 500 h mtbf. Built-in self-test
Power: 40A at 115 V 400 Hz 3-phase
Cdu interface: 1 M bit/s 16-bit words

STATUS: in service on Israeli Air Force/ McDonnell Douglas A-4, F-4 and IAI Kfir aircraft.

System 82 weapon delivery and navigation system

This is an upgraded version of the System 81. The growth potential in the central computer, amounting to 40 per cent spare computational time, has allowed the addition of a sophisticated stores management system, an armament control and display unit and back-up computer. The stores management system replaces the weapon selector, fuze selector, weapon data insertion panel, gun switch, missile switch, all stores switches, release programme selector and jettison switch (which were previously separate items) with a single unit. The back-up computer is based on an 8086 processor and assists with the computation and display of lead angles for air-to-air combat and trajectories for air-to-ground delivery.

Dimensions: (stores management computer) 321 × 194 × 172 mm
(back-up computer) 241 × 194 × 140 mm

Elbit System 82 in improved IAI Kfir C-7 fighter

Weight: (stores management computer) 7 kg
(back-up computer) 9 kg
Memory: (stores management computer) 58 K eprom, 3 K ram, 1 K non-volatile ram
(back-up computer) 16 K eprom, 1 K ram

STATUS: in production for McDonnell Douglas A-4, F-4 and IAI Kfir aircraft as part of update programme. System 82 is part of family of Israel Aircraft Industries' WDNS 391 equipment.

UNITED KINGDOM

British Aerospace

British Aerospace plc, Electronic Systems and Equipment Division, Downshire Way, Bracknell, Berkshire RG12 1QL
TELEPHONE: (0344) 483 222
TELEX: 848129

Ring-laser gyro inertial navigation system

In January 1985 British Aerospace was one of two UK companies to receive £1 million UK Ministry of Defence contracts as a step towards the development of military inertial navigation (IN) equipment using ring-laser gyros (RLGs). Potential applications for such equipment include a Royal Air Force/Panavia Tornado mid-life update, the Puma helicopter replacement, and UK versions of the EFA European fighter aircraft. British Aerospace has delivered a pair of production-standard systems for flight trials at the Royal Aircraft Establishment, Farnborough. They will use triangular-type RLGs, with optical path-length of 30 cm, to meet the medium-accuracy requirement for

this class of aircraft, which calls for drifts of not more than 1 nautical mile per hour. The system itself is packaged into a single 20 kg box and is compatible with MIL-STD-1533B digital data-bus, but can have alternative interfaces, eg for the serial data link in the Tornado or for an ARINC 429 system.

The accelerometers are mounted rigidly to the inside of the case, and their signals have to be corrected by computer to compensate for changing aircraft attitude to derive inertial velocities.

Advantages claimed for laser-gyro IN equipment compared with current electromechanical systems are lower first cost resulting from fewer components and therefore lower labour costs, lower maintenance costs as a result of greater reliability and smaller costs when attention is needed, and substantially reduced warm-up times.

Earlier, in 1978, the Division secured the British Ministry of Defence contract to design and build the first UK RLG aircraft IN equipment.

In October 1981 a prototype RLG inertial navigation system began airborne trials in a

Comet IV at Britain's Royal Aircraft Establishment, becoming the first such equipment of British origin to fly and the first such European system to be built. It marked a milestone in the Royal Aircraft Establishment's laser-inertial development programme, and represents more than three years' work by British Aerospace. A second system was delivered in December of that year.

The original development contract called only for trials in the Comet. But with results from both prototype systems well inside the target accuracy an extension of the flying programme was considered justified. On a Britannia flight-trials aircraft the full polar capability of the British Aerospace laser gyro system was demonstrated successfully by flying directly over the North Pole, during which mission it continued to operate continuously and correctly. Gyrocompassing at high latitudes (78° N), without modification to the equipment, was also completed successfully. A terminal error of 10 miles after 9 hours in the air was typical of the performance achieved. This corresponds to an accuracy of about twice that called for by the US Air Force specifications for medium-accuracy inertial navigation equipment.

In July 1985, the Division was awarded the contract to develop the main navigation system of the EH 101 naval helicopter based on its RLG aircraft INS. This success in the joint Anglo-Italian collaborative programme represents a UK world-first in that it is the first laser INS to be fitted in a helicopter anywhere in the world and it is the first military laser INS system to be designed and built outside the USA.

In November 1985, British Aerospace announced it had teamed with Teldix of West Germany to make joint bids for future European RLG contracts.

STATUS: in pre-production.

Terprom

Terprom (terrain profile matching) provides highly accurate navigation, precise terrain following and exact ground proximity warning with stealth. Using a digital map and taking inputs from the aircraft's radar altimeter and navigation systems Terprom matches measured profiles with the stored database to provide a high degree of navigational accuracy. This positional knowledge allows Terprom to examine the database ahead of the aircraft in order to carry out terrain following or to give ground proximity warnings.

The use of Terprom eases crew workload and permits automatic terrain-following, ground proximity warning and precise navigation with little or no forward emissions, and with no requirement for external navigation aids.

More than 100 sorties have been flown on a British Aerospace Jetstream based at the College of Aeronautics, Cranfield, demonstrating navigation, terrain following and ground proximity warning, flights were also carried out with Terprom driving the aircraft via the autopilot.

Terprom has flown more than 40 sorties on two fast jet aircraft, the General Dynamics F-16 and the Panavia Tornado. Navigation has been demonstrated at low level and high speed which is fully automated and considerably more accurate than existing systems.

Tornado prototype P12 was used for Terprom trials. Equipment is mounted internally

Format: ¾ ATR

STATUS: further flight trials planned using F-16 and Tornado aircraft during 1986.

British Aerospace plc, Brooklands Road, Weybridge, Surrey KT13 0SJ
TELEPHONE: (0932) 53444
TELEX: 27111

Sea Harrier mid-life update programme

The British Aerospace Sea Harriers operated by the Royal Navy are to have their avionics updated in a mid-life update programme (MLU) during the second half of the 1980s.

A BAe125 aircraft is to be modified and used as a flying test-bed, with the revised Sea Harrier cockpit displays and controls installed in place of the conventional co-pilot's seat.

The aircraft will be equipped with Ferranti Blue Vixen radar, and other elements of the MLU, including a Smiths Industries head-up display and weapon aiming computer (Hudwac), MIL-STD-1553B data-bus interface and control, digital air data computer and missile control system. The Marconi Guardian ESM system is also a part of the MLU.

Buccaneer update

British Aerospace is prime contractor for the updating of the avionics in 60 of the Royal Air Force's Buccaneers. Update radar (Ferranti Blue Parrot), navigation (Ferranti FIN 1063) and electronic warfare equipment (Marconi Guardian ESM) will be fitted over the next few years.

Mock-up of British Aerospace laser platform

British Aerospace ring-laser gyro sensors being tested

Ferranti

Ferranti Defence Systems Limited, Navigation Systems Department, Silverknowes, Ferry Road, Edinburgh EH4 4AD
TELEPHONE: (031) 332 2411
TELEX: 727101

FIN 1000 series inertial navigation systems

FIN 1000 is the designation of a family of inertial navigation systems based on the Ferranti miniature inertial platform and using mostly floated rate-integrating gyros and pendulous accelerometers. The group includes particular systems optimised for long-term high accuracy and for rapid-reaction alignment. Versions announced include:-
FIN 1010 Developed for the Panavia Tornado, this system has all-digital interfaces and a nominal accuracy of 1 nautical mile an hour.
FIN 1070 A version of the Tornado system, this has analogue interfaces and comprehensive route navigation, and was chosen for the Mitsubishi F-1 advanced trainer.
FIN 1012 Fitted to Royal Air Force/British Aerospace Nimrod MR2 and AEW3 aircraft, this system is optimised for long-term accuracy.
FIN 1064 This system was chosen in the early 1980s to replace the Navwass nav/attack equipment in Royal Air Force/Sepecat Jaguar as part of the mid-term refit. The FIN 1064 acts not only as the main navigation system but is also the weapon delivery and mission computer, information being loaded by way of a portable data store. It employs 64-bit parallel processing and has comprehensive interfaces with many other aircraft systems. The first set was delivered in 1980 to British Aerospace, Warton,

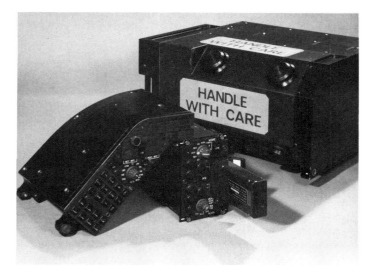

Ferranti FIN 1064 IN system configured for Sepecat Jaguar

Ferranti FIN 1064 control/display unit installed on port cockpit coaming of Sepecat Jaguar

for rig-testing, and several sets were installed in Jaguars for flight-trials beginning in 1981. In March of that year Ferranti was contracted to build 100 sets, equal to about two-thirds of the Royal Air Force's fleet of Jaguars.

Principal advantages over the earlier system are a reduction in size and weight, and an increase in storage capacity. Five boxes are replaced by one, with a two-thirds reduction in volume and a 50 kg weight-saving. The four-fold increase in capacity, from 16 to 64 K, means that the aircraft can accommodate all its roles without the need for reprogramming. This requirement has become increasingly important as the different roles have expanded and new and more complex equipment became available.

FIN 1063 Closely based on the system now being retrofitted on UK Jaguar, the FIN 1063 is part of the avionics update programme for the Royal Air Force/British Aerospace Buccaneer S2B aircraft. A new control/display unit will be tailored to the Buccaneer's cockpit, and special interface electronics are being developed to match its analogue instrumentation. An instruction to proceed was given by British Aerospace to Ferranti in February 1985.

FIN 1031 Navhars navigation heading and attitude reference system This is the system for the Royal Navy/British Aerospace Sea Harrier. It is similar to the FIN 1064 equipment used in the Royal Air Force Jaguars, but has dynami-

cally tuned Oscillogyros in the platform. Navhars is normally used in the Doppler-monitored mode since Sea Harriers are required to operate from ships that do not have an at-sea alignment reference.

The Navhars system combines a high-quality inertial platform with a powerful, dedicated minicomputer to produce information for navigation, weapon-aiming and aircraft management. A key feature is the choice of a pair of two-axis dry-bearing Oscillogyros, developed by Ferranti, in place of the three single-axis devices traditionally employed for platform stabilisation. They use rotating beams as the gyroscopic mass. Changes in angle between the planes of rotation of the beams and the platform reference surface are used to measure platform displacement in two axes. Other aircraft sensors provide Doppler radar velocities, true airspeed, and flux valve magnetic heading.

The 8/16P mini-computer was developed by Ferranti to process the data generated by the platform and other aircraft equipment. A 2-minute self-alignment can be accomplished on land or at sea. During sea alignment certain compensations are made for ship's motion. Once airborne, alignment is further refined over a period of 5 minutes by damped, second-order erection techniques. Doppler velocities are corrected for sea-state.

The system can display on command up to

10 waypoints, present position in latitude and longitude or grid coordinates, range and bearing to waypoint or destination, destination, speed, and track, magnetic or true heading, on-top and Tacan position-fixing, wind speed and direction, drift-angle and ground speed, and fuel remaining.

Inertial platform unit
Dimensions: 8.35 × 8.46 × 13.1 inches (212 × 215 × 332 mm)
Weight: 26.25 lb (11.9 kg)

Processor unit
Dimensions: 10.28 × 7.84 × 15 inches (261 × 199 × 381 mm)
Weight: 30 lb (13.64 kg)

Control/display unit
Dimensions: 5.77 × 6 × 5.47 inches (147 × 152 × 139 mm)
Weight: 5.5 lb (2.5 kg)
Power: 200 V 3-phase 400 Hz + 28 V dc for switching and lighting

STATUS: in production for Royal Navy Sea Harriers.

FIN 1070 Ferranti has supplied a system for the United Kingdom's EAP demonstrator that made its first flight in June 1986.
FIN 1075 In March 1985 Ferranti was instructed to proceed with the production of an inertial navigation system for the Royal Air Force's Harrier GR5s. Designated FIN 1075, the system is form, fit and functionally interchangeable with the Litton AN/ASN-130 navigation system being built for the McDonnell Douglas AV-8B. The UK system is based on the floated rate-integrating gyro platforms fitted to Nimrod, Buccaneer, Tornado and Jaguar aircraft, and will interface with the GR5's digital data-bus, avoiding the need for a dedicated control/display unit. The first of eight pre-production sets for systems integration at British Aerospace was delivered in mid-1986, and production will be phased over a three-year period.

FIN 1100 strapdown inertial reference system

A fully aerobatic, Schuler-tuned attitude reference, motion-sensing and navigation system for aircraft, ships and land vehicles, the FIN 1100 IRS is contained in a ½ ATR box weighing less than 7 kg, consuming 70 watts, and has a projected reliability of 2700 hours. The system is based on three high-grade 1° an hour rate-integrating gyros and three precision force-feedback accelerometers, with a microcomputer to execute the algorithms necessary to compute outputs for attitude, rate, linear velocity, acceleration, and displacement. It is governed either by a simple rotary switch or by

Ferranti Navhars three-unit inertial navigation system for Royal Navy Sea Harrier

Ferranti FIN 1100 strapdown inertial reference system

Left to right: Ferranti FIN 1110, FIN 2000 and FIN 1100 systems

a separate control/display unit, and contains extensive built-in test circuitry.

STATUS: system has been selected for the Casa C-101 DD trainer.

FIN 1110 2 GINS inertial navigation system

This system has been developed by Ferranti as a private venture, the aim being to produce a low-cost, lightweight and compact inertial navigation system (INS) for the foreseen market in helicopters and non-aerobatic transport aircraft. The system is contained in a ½ ATR Short box weighing 7 kg and costing about one-third the price of a full inertial navigation system. Complexity is avoided by the use of only two gimbals (hence the designation 2 Gimbals INS) in place of the four needed in aerobatic aircraft. At the same time the system has a strapdown azimuth sensor pack. The adoption of a gimballed platform rather than a strapdown system for this purpose has several advantages. The most significant is the rapid and independent alignment capability, the system finding true north to an accuracy of 0.1° within a few minutes of powering up. The system is thus freed from the inherent errors caused by variations in the earth's magnetic field. Continuing alignment in the air permits operation from moving platforms such as ships.

An original concept was to incorporate sufficient computing power to permit integration with an external sensor. Since Doppler is widely used in military helicopters, this was chosen for initial trials. Working closely with UK Doppler manufacturers, Ferranti modelled the potential errors in both inertial navigation and Doppler, the outcome being a 15-state Kalman filter, providing a much greater degree of accuracy than could be obtained by either Doppler or INS in isolation. However, once the filter has refined the error estimates, the system can operate using only occasional bursts of Doppler, an advantage for covert or 'stealth' operations when any emission of radiation is undesirable. Another refinement is the ability to recognise the step changes that occur with Doppler, typically in the scale factor between land and sea. Parallel studies have also been conducted in conjunction with the GPS satellite navigation system, Tacan, omega, Tercom, and air data equipment, and even a land vehicle's odometer.

FIN 1110 is the basis of the integrated navigation system used on the Hughes 530MG Defender light armed helicopter, in which it is coupled with a Racal Avionics Management System. Another FIN 1110 set is currently flying at Royal Aircraft Establishment, Farnborough, on a Sea King helicopter to assess the potential navigation improvement and better radar-scanner stabilisation.

STATUS: selected for Hughes 530MG helicopter.

Ferranti FIN 2000 gimbal system

FIN 2000 inertial navigation system

A prototype of this fully aerobatic inertial navigation system, which incorporates a new miniature gimballed platform, was shown at the 1980 SBAC display at Farnborough. Smaller and lighter than previous Ferranti equipment, the system is designed for retrofitting on existing aircraft or for new designs in which space is at a premium.

One of the design goals was a very rapid response time: the FIN 2000 can be aligned and ready for navigation within 2 minutes of switching on. It is suitable for operation world-wide, either as a self-contained navaid or in conjunction with other navigation equipment. All navigation and attitude information is available in digital form from the inertial navigation unit, and can be supplied to the pilot on an optional control/display unit.

The inertial navigation unit contains the inertial platform, a powerful dedicated computer, and the associated interface electronics. The miniature inertial platform is stabilised by a pair of two-axis dry-bearing oscillo-gyros, and inertial velocities are derived from three accelerometers mounted on it. Platform stabilis-ation and the derivation of velocities are controlled by the computer, which also pro-cesses the inertial information to provide present position, velocities, and attitude outputs. The extensive use of digital techniques permits considerable flexibility in meeting individual customer needs.

Number of waypoints: 10
Alignment: 1 mode for all conditions. 2-minute readiness time from switch-on, with indication of alignment status
Outputs: present position in lat/long, attitude, ground speed and track; range, bearing and time-to-go to destination; drift angle and heading
Test feature: system checking via bit modes
Accuracy (design goals): position 1 n mile/h (1850 m/h) rms, velocity 2.5 ft/s (0.76 m/s) RMS, heading 0.25° RMS, attitude 0.15° RMS, all in first 2 hours of operation

Inertial navigation unit
Dimensions: 7.67 × 7.87 × 15.15 inches (195 × 200 × 385 mm)
Weight: 8.16 lb (18 kg)
Power: 160 W at 28 V dc

Control/display unit (optional)
Dimensions: 5.78 × 6.02 × 5.5 inches (147 × 153 × 140 mm)
Weight: 1.13 lb (2.5 kg)
Power: 40 W at 28 V dc

FIN 3020 laser-gyro strapdown inertial navigation unit

The FIN 3020 inertial navigation unit is a ¾ ATR single-box ring-laser gyro, strapdown inertial navigation and attack device. An autonomous source of accurate navigation, velocity, attitude and body rate data and designed for high performance military aircraft, it was developed from the prototype FIN 3000 that began flight trials in late 1982. The major difference between the FIN 3000 and FIN 3020 is the size, with latest-technology electronics replacing off-the-shelf circuit cards and sensor repackaging that has permitted the original-size ring-laser gyros to be accomodated in a single ¾ ATR box.

The system consists of a strapdown sensor cluster, containing Ferranti Type 160 ring-laser gyros and accelerometers, plus interface elec-tronics, navigation/sensor computers and power supply. The Type 160 ring-laser gyro has a triangular path length of 43 cm, larger than that used by any other manufacturer in a similar sized inertial navigation unit; the longer path length provides greater gyro performance and potential for future use than alternative smaller gyros. The special accelerometers are mounted in an orthogonal assembly of advanced mech-anical and electrical design. The main interface conforms to MIL-STD-1553B, but additional or alternative interfaces, including the Panavia standard serial link, ARINC 429, and a variety of analogue and synchro interfaces, may also be incorporated. FIN 3020 uses two Ferranti F-DISC computers, the latest addition to the DISC family. The F-DISC computer has suf-ficient spare capacity to perform many other functions, such as weapon-aiming, Kalman filtering, flight management, integration with other navigation systems, and as a controller for the data-bus. The sensor computer uses a special microprogram permitting high-fre-quency motion compensation. The navigation computer can be microprogrammed to produce a MIL-STD-1750A instruction-set.

Format: ¾ ATR
Weight: 20 kg
Power: 160VA at 115 V 400 Hz
Reliability: 2500 h mtbf
Position error: 1 n mile/hour (cep)
Reaction time: 4 minutes to full accuracy, with gyro-compass align; 1 minute with memorised heading align

Accuracy (attitude) 0.1°, (heading) 0.10°, (H velocity) 3 knots, (V velocity) 2 knots
Linear acceleration: ± 12*g* along any one axis
Roll rate: ± 400°/s
Calibration interval: 1 year

STATUS: two prototype FIN 3000 engineering models are flying, achieving around 1 nautical mile an hour accuracy in manoeuvring flight. Pre-production models for delivery to Royal Aircraft Establishment, Farnborough, are in preparation. By January 1985 the prototype FIN 3000 had flown 1660 hours without failures. Flight trials included simulated air combat in British Aerospace Buccaneer aircraft.

FINAS inertial nav/attack system

Under the designation FINAS (Ferranti inertial nav/attack system), the company is marketing a set of equipment comprising a FIN 2000 dry-gyro inertial platform, a Type 4500 head-up display with front-mounted control panel, and a Type 105D laser rangefinder. The system dispenses with the panel-mounted control unit associated with the FIN 2000 platform; all functions are selected from the front panel of the head-up display.

STATUS: private venture development; no applications have been announced.

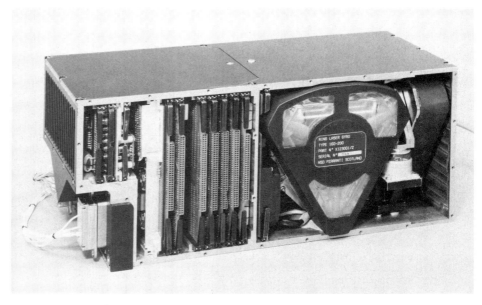

Ferranti FIN 3020 unit showing two of three ring-laser gyros

Ferranti FINAS system

GEC Avionics

GEC Avionics Limited, Airadio Products Division, Christopher Martin Road, Basildon, Essex SS14 3EL
TELEPHONE: (0268) 22822
TELEX: 99225

AD 380 and AD 380S automatic direction finders

Latest in a number of automatic direction-finding systems developed and produced by GEC Avionics, the AD 380 is designed to ARINC 570 and covers the frequency range 190 to 1799 kHz in 0.5 kHz steps while its variant, the AD 380S, covers 190 to 1599.5 kHz. The latter, however, additionally covers the international maritime distress frequency of 2182 kHz with the ability to tune to ± 0.5 kHz on either side of the nominal frequency, rendering it particularly suitable for search and rescue. This difference apart, both systems are designed to the same standard.

AD 380 systems have automatic, crystal-controlled frequency selection and are of all-solid-state construction with instantaneous electronic tuning. Built-in test facilities are incorporated.

A range of controller options are available, permitting single, dual or programmable operation. Standard controllers come in three versions, all of which provide selection of frequency and mode ADF, antenna, test or beat frequency oscillator, together with volume control. The decade frequency selectors display the operational frequency in 1-inch (25.4 mm) high numerals. Frequency controls comprise three concentric knobs which operate the logic frequency circuitry.

The first type of controller is the G4032E, a single-frequency version, and the second type, the G4033E, is used for tuning two receivers. Mode facilities in each case are selected by toggle switches, and each controller has a test-button. The third controller type is the G4034E, designed for rapid retuning of a single receiver to one of two frequencies which can be then selected by operation of a transfer switch when a white bar is displayed across the figures of that frequency not in use. Mode facilities are selected by a rotary switch.

A programmable controller, the AA-3809, conforms to the dimensions, form-factor and electrical requirements of ARINC 570. It provides a four-channel pre-select facility and can control both versions of the AD380. Six push-buttons permit instantaneous selection of frequencies previously entered into the memory store, which retains information when the equipment is unpowered.

The displayed frequency is always that to which the receiver is tuned and the operator can change stored information while in flight or select the 'N' button which allows normal operation of the controller. A brightness control, which is independent of the main aircraft panel lighting system, is incorporated. Normal mode facilities are also available for pilot operation.

Format: (receiver) 1/4 ATR Short
Weight: (receiver) 10 lb (4.5 kg)

STATUS: in production and service.

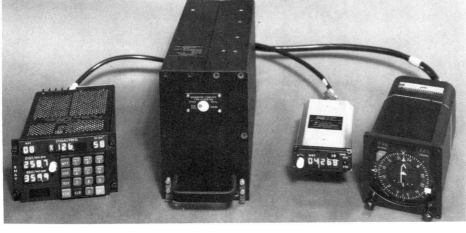

GEC Avionics AD2620 integrated navigation and control system

A620C navigation system

The AD620C is a microprocessor-governed integrated navigation and control system based on the earlier AD260 navigation computer. It provides not only navigation facilities but also control of sensors such as Tacan and VOR/DME and, in addition, can compute true airspeed (TAS) from basic air data information. TAS provides smoothing for Tacan and VOR/DME signals and dead-reckoning navigation in the absence of Tacan and VOR/DME. It can operate in polar or cartesian coordinates and will automatically tune a Tacan or VOR/DME when waypoints are selected. Other facilities include full slant-range correction and outputs to an automatic flight control system or weapon-aiming system. 'Fix' and 'mark' facilities are also available. Data, entered by a control and display unit keyboard, is retained when power is cut off.

The system comprises three units, a navigation computer unit (ncu), a control and display unit (cdu), and a remote readout unit (rru). Each controls a microprocessor to allow simple interfacing with its companion units to form an ARINC 429 digital data highway. The inclusion of a microprocessor in the control and display unit also permits a rapid display response to keyboard input while at the same time allowing display formatting to take place within the control and display unit itself. The three units have considerable spare computer capacity for future development and growth.

Features of the navigation computer unit include an expanded memory capacity for both programme and in-flight data acceptance, improved interface circuit design for better performance and higher integrity, and a new form of mechanical construction which makes each module a self-contained entity and allows testing down to component level with simple equipment. A built-in test programme monitors all input and output signals as well as the computer software.

Dimensions: (ncu) 7.6 × 5 × 12.5 inches (194 × 127 × 318 mm)
(cdu) 3.75 × 5.75 × 6.5 inches (95 × 146 × 165 mm)
(rru) 1.2 × 3.2 × 5.7 inches (31 × 80 × 144 mm)
Weight: (ncu) 7.7 lb (3.5 kg)
(cdu) 4 lb (1.8 kg)
(rru) 0.75 lb (0.34 kg)

STATUS: in production and service.

AD2770 Tacan

The AD2770 Tacan navigation and homing aid is suitable for all types of aircraft. This system is now used on the majority of front-line aircraft in service with UK forces. It provides range and bearing information from any selected ground Tacan station or from any suitably equipped aircraft, and is available in a number of forms offering outputs in digital (ARINC 429) or analogue form or a combination of the two. Output signals may be provided to drive range/bearing or deviation indicators on a pilot's panel or to interface directly with a computer.

The system has full 252-channel X and Y mode capability and operates up to 300 nautical miles range with a range accuracy of better than 0.1 nautical mile. Bearing accuracy is said to be better than 0.7° on normally strong input signals. It comprises two units, a transmitter-receiver, hard-mounted in the avionics bay, and a panel-mounted remote control unit. A switching unit, also installed in the avionics bay, is required when two antennas are fitted. A mounting tray with a cooling air blower is necessary if a cooling air supply is not available from the aircraft's own air-conditioning system.

The transmitter-receiver section with a digital interface only is contained in a ¾ ATR Short case with a front doghouse. Versions with analogue outputs are accommodated in a case of longer dimensions to house the additional circuitry. Signal-processing circuitry is largely

digital in the interests of system reliability, and continuous integrity monitoring techniques eliminate the risk of erroneous outputs.

Range and bearing analysis and output formats are prepared in a general-purpose computer module called the analyser. This allows both range and bearing signal processing to use the same circuitry, with a consequent reduction in the number of components. The range system uses a parallel search method, said to be unique, which by making use of all signal returns achieves a very rapid lock-on.

The AD2770 system operates in three modes, receive (giving bearing information only), transmit-and-receive, and air-to-air (providing range information only). Transmitter frequency range is from 1025 to 1150 MHz with an output of 2.5 kW peak pulse power. Receiver frequency coverage is from 962 to 1213 MHz. The tracking speed range is from 0 to 2500 knots.

The design of the AD2770 system is flexible both electrically and mechanically, and alternative configurations with appropriate form factors, output characteristics and mechanical and electrical interfaces can be provided for new aircraft or for retrofit.

Dimensions: (transmitter-receiver standard unit) 7.6 × 7.5 × 15 inches (194 × 191 × 380 mm)
(control unit) 2.25 × 5.75 × 3.25 inches (57 × 146 × 83 mm)
(antenna switch) 2.7 × 5.1 × 2.2 inches (69 × 130 × 56 mm)
Weight: (transmitter-receiver standard unit) 31 lb (14 kg)
(control unit) 1 lb (0.45 kg)
(antenna switch) 0.55 lb (0.25 kg)

STATUS: in production and service. By early 1985 more than 800 systems had been ordered.

AD2780 Tacan

A follow-on from the AD2770 series supplied for the Panavia Tornado and other front-line types, the AD2780 is also proposed for military applications and is, says GEC Avionics, lighter, smaller, and less expensive than its predecessors, with extensive use of large-scale integration and microprocessor technology. The system provides slant range and relative bearing to a standard Tacan station, range rate (which approximates to ground speed when not used in conjunction with the company's area navigation system), time-to-go to waypoint or station, ARINC 429 serial data output and 252 channels in X and Y modes. Outputs of range are available in digital format to ARINC 429, and in analogue to dial-and-pointer displays. The system has been chosen for the UK's EAP demonstrator.

Frequency range: (transmitter) 1025-1150 MHz
(receiver) 962-1213 MHz
Range rate output: 0-999 knots with accuracies ±15 knots for 0-300 knots, and ±5% for 300-999 knots
Time-to-station output: 0-99 minutes

Transmitter-receiver for GEC Avionics Type AD2780 Tacan

GEC Avionics AD 620K navigation computer unit

Dimensions: 5 × 6 × 12.5 inches (127 × 153 × 318 mm)
Weight: 7.7 lb (3.5 kg)
Tracking speed: 0-1900 knots, 0-20°/s
Memory: (range) 10 s
(bearing) 4 s

STATUS: in production and service. At the end of 1985 the Royal Netherlands Air Force ordered 32 AD2780 sets to equip its MBB BO 105 helicopters. This was followed by an order to equip the Royal Air Force's Tucano trainers being built by Shorts.

AD 620K integrated navigation system for MB 339K

Designed for the single-seat Aermacchi MB 339K Veltro light strike aircraft, the AD 620K is a development of the earlier AD 620C navigation system provided for the two-seat Aermacchi MB 339A trainer. The AD 620K has an improved performance compared with earlier versions owing to the addition of an AD 660 Doppler velocity sensor and associated processing and displays. The system operates in conjunction with an inertial platform, nav/attack information and guidance being presented on a wide-angle head-up display. The navigation computer can accept data from a wide variety of navigation sensors together with an attitude and heading reference, to compute present position.

STATUS: the system was selected for the MB 339K in October 1984 and is in production. The AD 620C was chosen for the MB 339.

GEC Avionics Limited, Airport Works, Rochester, Kent ME1 2XX
TELEPHONE: (0634) 44400
TELEX: 96333

AD660 Doppler velocity sensor

This Doppler sensor, claimed to be the world's smallest such system, is packed into a single unit containing the antenna, transmitter-receiver, tracker, and digital and analogue inputs and outputs, all under microprocessor control. The transmitter is a Gunn diode generating about 200 mW of power at 13.325 GHz. The signal is frequency-modulated, the actual frequency being selected by the microprocessors to give the best return signal in any set of conditions, eliminating for example 'height-hole' effects. The system employs four beams, any three of which provides satisfactory performance, the fourth being used to enhance operation at extremes of altitude and to provide a self-check function. The standard output for ground speed and drift angle is an ARINC 429 serial digital data highway, but ARINC 582 and 561 digital and ARINC 407 synchro outputs can also be provided.

Dimensions: 379 × 237 × 132 mm

STATUS: in production. The AD660 is a standard option on Boeing 727 and 737 transports, and has been adopted by British Airways and Lufthansa among other operators.

In October 1984 it was selected by Casa for its C-101 DD Aviojet advanced trainer/ground-attack aircraft.

GEC Avionics AD660 Doppler velocity sensor

Plessey Avionics

Plessey Avionics Limited, Martin Road, West Leigh, Havant, Hampshire PO9 5DH
TELEPHONE: (0705) 486 391
TELEX: 86227

GPS receivers

Plessey has been developing global positioning system receivers in conjunction with Magnavox and the UK Ministry of Defence since 1977.

Members of the Series 9000 family are suitable for all applications, including airborne. The PA 9020 is a two-channel version for use in helicopters; the PA 9050 is a full five-channel system for aircraft use. Both can interface with the aircraft's other avionics.

Dimensions: (PA 9050 receiver) 193 × 90 × 316 mm

STATUS: in development.

Plessey PA 9050 GPS receiver

Racal Avionics

Racal Avionics Limited, Burlington House, Burlington Road, New Malden, Surrey KT3 4NR
TELEPHONE: (01) 942 2464/2488
TELEX: 22891

Area navigation system

Certificated by the UK Civil Aviation Authority in 1981, this system is in widespread use by helicopter operators supporting the North Sea oil industry. It comprises a control/display unit (cdu) and a navigation unit, and accepts information from any or all of the following sensors: VOR/DME, Decca Navigator, Loran, omega/vlf, Doppler velocity and air data. The cdu shows the present position in latitude and longitude, Decca coordinates, and bearing and distance to any waypoint from the aircraft or from another waypoint.

The Racal Avionics RNav system can store the details of a 100 waypoints with alphanumeric designators together with information on up to

30 routes which can contain 250 route segments. Waypoint positions and routes can be entered before flight by a data transfer device or a tape loader. Existing routes can be edited or new routes created in flight using cdu keyboard procedures; for example, a search and rescue steering profile can be added to an existing route. The system is powered continuously by an internal battery, so that information is not lost from the memory when the aircraft is powered down. Full autopilot and flight director facilities enable en route and search and rescue flights to be conducted automatically. Serial data outputs allow navigation-related information to be shown on an EFIS or weather-radar display.

The complementary Mk 32 Decca Navigator receiver and associated antenna amplifier is used with this system.

This RNav equipment forms the core of a search and rescue hover guidance system recently certificated by the UK's Civil Aviation Authority; it is claimed to be the only civil system approved for use in fog or at night, and is used by most helicopter operators in the North Sea.

Type 80791A control/display unit
Dimensions: 146 × 114 × 220 mm
Weight: 2.5 kg

Type 80792A navigation computer unit
Format: 4 MCU
Weight: 4.5 kg

STATUS: in production.

RNS 5000 area navigation system

This advanced, second-generation area navigation system meets the standards set down in the UK Civil Aviation Authority's proposed

Control/display for Racal Avionics RNS 5000 area navigation system

Minimum Navigation Performance Specification. It is designed for fixed-wing regional carriers and commuter aircraft and helicopter operators, where its accuracy would be expected to reduce operating costs by virtue of the fuel savings possible with precision navigation. A key feature of the RNS 5000 is its ability to use DME/DME position-fixing as a basis for high accuracy.

The system is suitable for retrofit applications, and comprises a control/display unit that can show eight lines of data on a cathode ray tube and a navigation unit computer that accepts information from the following navigation sensors: dual DMEs, VOR, Loran or omega/vlf, Doppler velocity, inertial navigation and air data. Provision is also made for GPS (the satellite-based global positioning system) and MLS (microwave landing system). Automatic tuning of the VOR/DME receivers is based on aircraft position, track, and the geometry of the station being used.

Front-panel display for Racal Avionics area navigation system

Racal Avionics TANS system Type G indicator

Racal Avionics Type 9308 ground speed/drift presentation

Racal Avionics Type 9306 hover meter

Antenna unit for Racal Avionics Doppler Type 71

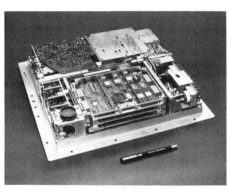

Antenna unit for Racal Avionics Doppler Type 80E

Antenna unit for Racal Avionics Doppler Type 92

The RNS 5000 can store the coordinates of 5000 waypoints with alphanumeric designators and 300 routes, representing a database sufficient to store the route structure of a major airline. Waypoint positions and stored routes can be updated by a tape-loader.

Full autopilot/flight director facilities are provided in the lateral and vertical planes and the system can command a parallel track, with left or right offset distances selected by means of the cdu. Serial outputs enable navigation-related information to be shown on an EFIS or weather-radar display. A fuel management function shows the predicted fuel quantity at destination and intermediate points, as well as time and distance remaining against fuel reserves. Racal claims that the system in airline service could pay for itself through fuel savings in only 18 months.

Control/display unit
Dimensions: 146 × 114 × 165 mm
Weight: 2.5 kg

Navigation computer unit
Format: 4 MCU
Weight: 4.5 kg

STATUS: in production. System has been ordered by Embraer for its Brasilia commuter aircraft, and is in operation with British Airways and Brymon Airlines.

Doppler velocity sensors

Five current models comprise the Doppler 71 and 72 antenna units, for helicopters and fixed-wing aircraft respectively, the new Doppler 80 (helicopter and light aircraft), and Doppler 91 and 92 antenna units for helicopters and fixed-wing aircraft respectively. By January 1984 over 3000 of the earlier Doppler 71 and 72 devices had been produced.

The Doppler 91 and 92 units, first shown at Farnborough in 1984, use the best features of

Control unit for Racal Avionics RNS 252 navigation system

the previous models and incorporate microprocessor technology and new manufacturing techniques. The Doppler 80, with printed antennas, Gunn diode rf source and switched beams, is for low-level helicopter operations, where low weight is particularly important. The Doppler 91 and 92 units use waveguide antennas and varactor multiplier transmitters for high accuracy at greater altitudes. These two units can directly replace the Doppler 71 and 72, fitting into the same mounting trays.

Antenna units for helicopters have a speed range of –50 knots to 300 knots forwards, and 100 knots laterally. The fixed-wing sensors have a corresponding speed range of –50 knots to 1000 knots and 200 knots laterally. Velocity data can be provided in either analogue or ARINC 429 digital format, and a MIL-STD-1553 data-bus may be specified. The microwave signals produced by the Doppler 90 series are specially tailored to reduce errors created by heavy rain, snow and hail.

The transmission characteristics of the Doppler 91 and 92 units can be remotely controlled, that is, a low-power 'stealth' mode can be selected for minimum detectability or the transmitter switched off if the beam goes above the horizon, for example when the aircraft is banking. They transmit information to other aircraft systems in ARINC 429 serial digital data form, but the MIL-STD-1553B remote terminal format can also be supplied.

A recent addition to the Racal Doppler family is the 80E. Retaining the weight and cost advantages of the earlier members, the 80E has a new microprocessor-based tracker/output interface module, giving better flexibility and self-test facilities.

Dimensions: (Doppler 71/72) 406 × 406 × 127 mm
(Doppler 80) 356 × 381 × 80 mm
(Doppler 91/92) 358 × 391 × 118 mm
Weight: (Doppler 71/72) 16.5 kg
(Doppler 80) 8.6 kg
(Doppler 91/92) 11 kg
Power: (Doppler 71/72) 115V 400 Hz
(Doppler 80 and 91/92) 28 V dc

STATUS: in production. The RDN 80B system is produced under licence by Electronique Serge Dassault. The Doppler 91 has been selected for the EH 101 helicopter.

Navigation computers for Doppler velocity sensors

Racal Avionics produces two types of computer and control unit to process and present information from its family of Doppler velocity sensors. They are the TANS tactical area navigation system and the PBDI position and bearing distance indicator. Both are single units that process, interface, control and display Doppler velocities, heading and attitude.

Position bearing and distance indicator for Racal Avionics Doppler 80

Antenna unit for Racal Avionics Doppler Type 80

TANS displays present position in latitude and longitude, grid coordinates, bearing and distance to a waypoint and between waypoints. In conjunction with an air data system, it can provide reversionary navigation information.

Two versions of the PBDI are available, one providing navigation information in latitude and longitude, the other as grid coordinates.

TANS and PBDI can store the coordinates of up to ten waypoints or destinations; the PBDI incorporates a digital steering indicator, while TANS needs an external analogue steering indicator.

A newer device, the Racal Navigation System RNS 252, is a navigation computer incorporating all the facilities of the TANS and PBDI. The system has a display comprising four lines of 12 characters using green light-emitting diodes. In addition to the navigation capability in conjunction with conventional gyro compass and vertical gyro, the RNS 252 can interface, via an ARINC 429 serial data-bus, with a strap-down attitude and heading reference system to give Doppler-monitored inertial navigation data.

All three devices can drive the Racal Avionics automatic chart display.

Dimensions: (Type 9447 TANS computer) 146 × 162 × 247 mm
(Type 80475 PBDI) 106 × 106 × 195 mm
(Type 80895 RNS 252) 146 × 123 × 163 mm
Weight: (Type 9447 TANS computer) 6 kg
(Type 80475 PBDI) 2 kg
(Type 80895 RNS 252) 3.5 kg
STATUS: in production.

Ground speed/drift meter Type 9308

This standard aircraft indicator-size unit shows ground speed within 3.5 knots at 100 knots or 5 knots at 300 knots. Drift is registered within 0.5° of true value. If the signal is lost, the last-measured ground speed and drift are displayed.

Hover meter Type 9306

This standard aircraft indicator-size unit displays along-heading velocity over the range -10 to 20 knots with an error not greater than 1 knot, and vertical velocity range ±500 feet a minute with a maximum error of 40 feet a minute.

STATUS: in production. Type 71 Doppler is installed on Royal Air Force/Aérospatiale-Westland Puma, Sikorsky-Agusta SH-3D and Royal Navy and British Army Lynx helicopters.

Smiths Industries

Smiths Industries, Aerospace and Defence Systems Limited, 765 Finchley Road, Childs Hill, London NW11 8DS
TELEPHONE: (01) 458 3232
TELEX: 928761

Navstar GPS receiver systems

Smiths Industries has formed a collaborative agreement with Polytechnic Electronics plc and Interstate Electronics Corporation to design, develop and market GPS equipment. The PEL XR-1 GPS receiver, which is believed to be the only UK receiver in production, is a single-channel C/A code receiver designed to demonstrate, evaluate and test the Navstar GPS system. The XR-1 has sufficient capacity to be reconfigured to meet many users' requirements. The system is modularised around logical vlsi 'building blocks' and thus emulates the structure of production sets which will be available for use with the full satellite network. The XR-1 has been successfully demonstrated in a number of countries. The XR-3 receiver is under development as a single-channel (expansible to five channels), C/A code system which will be available with one or more modules to meet multi-user needs. As a core unit printed circuit boards can be assembled either as a stand-alone receiver with navigation programme and multiple interface options, or as a separate board sub-assembly for integration with a host system. The system is being developed for land, sea and air applications, and has been proposed to meet the GPS requirement for the EH 101 helicopter programme.

STATUS: XR-1 in production. XR-3 under development.

STL

Standard Telecommunication Laboratories Limited, London Road, Harlow, Essex CM17 9NA
TELEPHONE: (0279) 29531
TELEX: 81151

Navstar GPS receiver system

STL is the technology research centre for STC and is under contract to the UK Ministry of Defence. The company is developing a Navstar GPS satellite navigation receiver system independently of the two principal companies in this field, Collins and Magnavox. The organisation is building a number of systems for tri-service testing and evaluation implemented initially in medium-scale integration but later to incorporate large-scale integration technology. Technical authority for the UK effort is vested in the Royal Aircraft Establishment, Farnborough.

The first breadboard model was built in the late 1970s. Using just a single channel, it became the basis of a successful bid for the UK Ministry of Defence development contract against other UK manufacturers. At the time of the 1984 Farnborough Air Display the system was said to be three to four years away from full production.

STATUS: in initial development. Low-cost, multi-channel system began trials at Royal Aircraft Establishment during 1983.

UNITED STATES OF AMERICA

Aeronetics

Aeronetics, 2100 Touhy Avenue, Elk Grove Village, Illinois 60007
TELEPHONE: (312) 437 9300
TELEX: 687 1468

Radio magnetic indicators

Aeronetics produces a family of radio magnetic indicators under the designation 3100 and 3300 for medium and heavy twins and helicopters, using remote-mounted signal-processing, and a 7100 series panel-mounted instrument for single, light and medium twins, and light helicopters.

3100 Presents aircraft heading on servo-driven compass azimuth card read against a fixed lubber line. Each of the two pointers provides information from VOR or ADF. The instrument is packaged in a 3 ATI case. The solid-state electronics are made up into plug-in circuit modules.
3300 This system permits an interface with the new-format navigation receivers.
7100 This instrument interfaces directly with the latest panel-mounted ADF and navigation receivers built by Bendix, ARC, Collins, King and Narco, without the need for adaptors or an external power source. The system comprises a full dual-switch radio magnetic display indicator and an electronics converter unit.

STATUS: all in production.

BDI-303A digital radio magnetic indicator

This system, introduced in September 1984, is intended to complement the new digital avionics suites now being widely adopted in general aviation. It has a traditional VOR/ADF display of bearing, DME distance and compass information, while accepting ARINC 429 format data from external sensors. It also incorporates an emergency compass mode with direct input of flux-detector.

Arnav

Arnav Systems Inc, 16100 SW 72nd Avenue, Portland, Oregon 97224
TELEPHONE: (503) 684 1600
TWX: 510 600 2185

AVA-1000 Loran C

The AVA-1000 is claimed to be the only panel-mounted Loran C with TSO approval, with a capacity for holding data on up to 200 waypoints, and to use plain-language displays. The system has a supplemental type certificate for IFR flight in Alaska and areas of the USA approved for Loran C navigation. The system has completed TSO testing in accordance with FAR Part 21. Remaining flight-testing was completed in the company's Piper Aerostar corporate aircraft.

In 1983 the company introduced the AVA-1000R, a remote-mounted Loran system tailored to turboprop, jet and commuter aircraft. With alphanumeric presentation and capacity for 200 waypoints, it is TSO certificated and cleared for IFR operation in areas of Loran coverage.

Dimensions: 158.8 × 82.6 × 271.8 mm
Weight: 2.2 kg
Scaling: (en route mode) ±25 n miles each CDI dot equivalent to 1 n mile
(approach mode) ± 1.25 n miles, each CDI dot equivalent to 0.25 n mile
Update rate: position update each 0.5 s
Resolution: (distance) 0.1 n mile
(crosstrack) 0.1 or 0.01 n mile (selectable)
Accuracy: meets or exceeds en route accuracy requirements of AC 90-45A in Loran C ground-wave coverage areas

R-20 Loran C

The R-20 is a new high-performance panel-mounted VFR Loran C system for the general aviation market which is claimed to contain more computing power than any other comparable system and to require less pilot monitoring.

It has a capacity for 200 waypoints and a liquid-crystal-type display with a built-in CDI presentation. A high degree of automation obviates the need for the pilot to insert data manually for secondary stations or magnetic variation.

STATUS: in production.

R-21/NMS Loran C

The R-21/NMS incorporates, in one low-cost unit, a number of functions not found on any other Loran C or vlf/omega at any cost.

Nav management functions can be used, without interrupting normal navigation, to calculate direction and speed of actual winds aloft as well as true airspeed. When primed with fuel remaining and fuel-burn rate, the R-21/NMS will automatically compute and display absolute range. Vertical Navigation (VNav) information can also be computed to determine accurately the optimum descent/climb locations. There is automatic magnetic variation correction and the unit has three automatic notch filters to insure signal integrity anywhere in the world. There is also an optional extended range feature which allows navigation outside the normal areas of coverage.

Dimensions: 3.25 × 6.25 × 10.7 inches (83 × 159 × 272 mm)
Number of waypoints: 200
Waypoint entry: alphanumeric
Power: 10-45 V dc, 19 W
Weight: 2.1 kg
Warranty: 18 months

R-40 Loran C

The R-40 is TSO and IFR certified for areas of the USA including Alaska. It features three-line light-emitting diode readouts in a clear, alpha-numeric, plain language display. The system has 200 waypoints and 20 course legs may be programmed. Parallel track, magnetic variation and propagation error are automatically corrected. An extended range feature provides for increased Loran coverage by as much as 400 nautical miles outside of the normal Loran chain coverage by use of a skywave signal reflected off the ionosphere.

Nav management functions can be used, without interrupting normal navigation, to calculate direction and speed of actual winds aloft as well as true airspeed. When primed with fuel remaining and fuel-burn rate, the R-40/NMS will automatically compute and display absolute range. Vertical navigation (VNav) information can also be computed to determine accurately the optimum descent/climb locations.

Dimensions: 3.25 × 6.25 × 10.7 inches (83 × 159 × 272 mm)
Weight: 2.2 kg
Display: 5 × 7 dot matrix
CDI sealing: approach ±1.25 n miles-0.25 n mile per dot
En route ±5 n miles per dot
Notch filters: 9 internal
Distance resolution: accurate to within 0.1 n mile
Warranty: 18 months
Power: 10-45 V dc, 19 W

R-50 Loran C

This new IFR Loran C is a development of the successful AVA-1000, and the performance has been improved to permit operation in areas of poor Loran coverage, such as the North Slope of Alaska, Bermuda, the Caribbean and parts of the mid-West. The panel-mounted system can accommodate up to 200 waypoints, and present position is shown on an alphanumeric 5 × 7 dot-matrix display. Outputs are provided to an autopilot, a CDI, and to equipment using RS 232 signal format. Also designed for general aviation, the system is FAA certificated and has TSO approval.

STATUS: in production.

R-60 Loran C

The R-60 is a remote-mounted version of the R-50, designed for Dzus rail installation on transport aircraft and helicopters. The receiver (contained in a ½ ATR case) and control/display unit together weigh 3.4 kg. Other details are as for the R-50.

STATUS: in production.

Computer and control/display unit for Arnav AVA-1000 Loran C

Presentation of Arnav R-20 Loran C control/display unit

Presentation of Arnav R-60 Loran C control/display unit

Bendix

Allied Signal Inc, Bendix Avionics Division, Teterboro, New Jersey 07608
TELEPHONE: (201) 393 2142
TELEX: 134414
TWX: 710 990 6136

Avionics management system

This avionics management system provides the military pilot with an integrated approach to the control and display of avionics equipment and information. By combining the various func-

tions required to manage the cockpit, the mission and the performance of the aircraft, the system reduces the proliferation of control and display equipment and enhances the pilot's ability to function effectively in a complex and hostile combat environment. The system brings together in one box the control and information presentation for a variety of CNI (communication, navigation and identification) equipment. It communicates with that equipment by way of a MIL-STD-1553 digital data-bus or through remote terminal units.

Functions that the AMS can accomplish include flight and mission management, control

of avionics equipment, monitoring of flight and navigation information, performance management, and tactical navigation. In the performance management mode, for example, AMS presents real-time guidance to a helicopter pilot on engine torque available, or the torque needed to hover, climb, or provide the best range or endurance. The information is based on data entered into the two mission computers and made available to the pilot on four interactive pages of a control/display unit.

STATUS: system began flight trials in 1985.

Mock-up of Model 360 troop-carrying helicopter flight-deck. Two Bendix mission computers and display units occupy console and central area of flight-panel

ADF-2070 automatic direction finder

The Bendix ADF-2070 is a panel-mounted unit designed for the general aviation sector. It is claimed to be able to receive signals from exceptionally long ranges and has two sensitivity settings, extended range reception and conventional ADF which is used primarily on approach. This extended range facility is provided by a 'coherent detection' feature which results in good reception characteristics with high immunity from thunderstorm and other static interference. Continuous digital tuning ensures lock-on to the desired frequency.

The system also features a blade antenna which serves both the communications radio and the ADF systems. The ADF sensor is installed in the base of the blade and feeds signals to the receiver via a small amplifier which is mounted adjacent to the antenna blade but within the aircraft skin. It is claimed that this configuration provides nearly twice the gain of other combination-antenna units, results in a reduction in cable length, and no impedance-matching requirement.

Provision is made for correction of quadrantal error either on the ground or while airborne. The receiver output can drive either a standard ADF indicator with rotatable azimuth card or an HSD-800 horizontal situation indicator.

Allied Signal Inc, Bendix Avionics Division, 2100 NW 62nd Street, Fort Lauderdale, Florida 33310
TELEPHONE: (305) 776 4100
TELEX: 514417
TWX: 510 955 9884

MLS-20A microwave landing system receiver

Bendix has been involved in the development of microwave landing system technology and equipment since 1967, and pioneered the time-reference scanning-beam (TRSB) principle that was later adopted by the International Civil Aviation Organisation as the international standard for this high-frequency landing aid. The MLS-20A is a 200-channel angle receiver with associated control/display unit and omni-directional antenna, and is based on custom large-scale integration and microprocessor technology. It is compatible with conventional analogue flight instruments, is approved to FAA-TSO-104 and meets the requirements of the new international standards. The system has a coverage of 20 nautical miles and registers from 0° to 20° in elevation and up to 60° either side of the extended runway centre-line. It is designed to work with all MLS ground stations transmitting the FAA-ER-700-08C signal format.

Dimensions: (receiver) 5 × 3.9 × 11.75 inches (128 × 100 × 301.2 mm)
(control/display unit) 5.75 × 2.625 × 6.5 inches (147.4 × 67.3 × 166.7 mm)
Weight: (receiver) 9 lb (4.1 kg)
(control/display unit) 4 lb (1.82 kg)
Number of channels: 200
Frequency range: 5031.00–5090.70 MHz
Power: 1.5 A at 28 V dc

STATUS: in production. The system is installed on Panavia Tornado aircraft of the Italian Air Force. Over 100 MLS 20A systems had been delivered by late 1985 and it was, at that time, the only MLS receiver certified by the FAA for IFR use.

MLS-21 microwave landing system receiver

Bendix announced the MLS-21 microwave landing system receiver in September 1985 at which time over 100 orders had already been received. The receiver is compatible with ARINC 727, can automatically tune DMEs and can accept analogue or digital inputs and outputs: it is also compatible with the Bendix EFS-10 and other electronic flight instrument systems.

The MLS-21 can be used in either automatic or manual modes: in the latter the pilot can select ±60° in azimuth and up to +20° in elevation for the approach. The maximum selectable angles can be limited to suit the capabilities of the aircraft.

The MLS 21 can receive 200 MLS channels in the frequency band 5031 to 5090.7 MHz (channels 500 to 699).

Dimensions: (receiver) 102 × 124 × 330 mm (control/display unit) 63 × 80 × 60 mm
Weight: (receiver) 2.9 kg
(control/display unit) 0.86 kg
Power: 0.75 A at 28 V dc

STATUS: in production.

Bendix panel-mounted ADF-2070 ADF indicator

Bendix MLS 21 microwave landing system receiver, control/display unit and antenna

Dimensions: 1.75 × 6.25 × 9.2 inches (45 × 159 × 234 mm)
Weight: 5.79 lb (2.63 kg)

STATUS: in production and service.

RN-260B (AN/ARN-127) VOR/ILS system

This widely used VOR/ILS system comprises a compact, solid-state 200-channel VOR/LOC remote-mounted receiver, a 40-channel glide-slope receiver and marker-beacon receiver, and a panel-mounted control/display unit. All three receivers operate independently. The system has been adopted by the US Air Force, Army and Coast Guard.

Dimensions: (receiver) 183 × 130 × 319 mm (control unit) 146.1 × 66.7 × 114.3 mm
Weight: (receiver) 4.5 kg
(control unit) 1 kg
Power: 25 VA at 26 V 400 Hz plus 1.5 A at 28 V dc

STATUS: in production and service.

RIA-35A ILS receiver

The Bendix RIA-35A is an ARINC 710-compatible ILS receiver, part of the CNI700 digital

Bendix RN-260B (AN/ARN-127) VOR/ILS system

nav/com family. It uses dual micro-processors and custom large-scale integration design, with continuous monitoring of all unit sub-assemblies.

In September 1985, Bendix announced that the RIA-35A had been selected as standard on the Gulfstream G-IV aircraft: it had previously been selected by 55 airlines worldwide.

STATUS: in service and production.

Collins

Collins Avionics Division, Rockwell International Corporation, 400 Collins Road NE, Cedar Rapids, Iowa 52498
TELEPHONE: (319) 395 1000
TELEX: 464421
TWX: 910 525 1321

Digital RMI/DME indicators

Designed to interface with both the current ARINC 700 sensors and the older-generation analogue avionics, this new series of combined radio magnetic and distance-measuring (RMI/DME) indicators forms a family of instruments that share many common features, including electrical, servo, thermal, packaging and lighting methods. Use of a patent digital encoder/driver module, common to each instrument and each channel within it, permits packaging techniques that greatly improve the effectiveness of the heat-transfer arrangements. The encoder/driver module consists of an 11-bit digital encoder, dc motor and associated drive circuits, all under microprocessor control. Each of the six channels is isolated from its neighbour for integrity, and each has its own single-chip micro-computer. Each channel monitors its own faults and activates its own flag or shutter and output signal.

The members of the digital RMI/DME instrument family comprise:-
RMI/733A radio magnetic indicator: a compact lightweight VOR/ADF selectable three-servo instrument. A version designated RMI-733A is a three-servo instrument designed to display information from an ADF receiver.
RDMI-743 radio distance magnetic indicator: this instrument features liquid-crystal DME readouts for better reliability and readability. The three-servo unit is selectable to either VOR/ADF or VOR only.
RDMI-743A radio distance magnetic indicator: with a magnetic-wheel DME display, the RDMI-743A is a four-servo VOR instrument.

STATUS: all in production.

51Z-4 marker-beacon receiver

The Collins 51Z-4 marker-beacon receiver automatically provides aural and visual indication of passage over airways and instrument landing system marker-beacons. The system is approved for Category II approaches in a number of Collins' all-weather avionic system

certifications. Operating at a frequency of 75 MHz, the receiver sensitivity can be varied between two pre-adjusted levels through a cockpit HI-LO switch. The HI position is used to gain early indication of a marker beacon, the LO position is then subsequently used closer to the beacon for a sharper position fix. Alternatively, the receiver sensitivity is continuously variable from the cabin or flight-deck by means of a potentiometer.

The system is of all-solid-state construction and contains triple-tuned circuitry for the rejection of spurious signals generated by television and FM broadcast transmitters. It is designed for three-lamp indication but may be easily modified for single-lamp operation by removal of a resistor and wiring the three-lamp outputs together. In either type of operation, outputs can operate two sets of indicator lamps in parallel.

An optional self-test facility causes the internal generation of 3000, 1300 and 400 Hz marker signals which are detected in sequence by the receiver. The indicators light in order and the corresponding aural tones are also generated.

The unit is suited to retrofit installation since it is mechanically and electrically interchangeable with a number of other marker-beacon receivers, including the Collins 51Z-2 and 51Z-3 units. An associated marker-beacon antenna, the Collins 37X-2 system, is also available. Designed for operation with the 51Z-4 and other compatible receivers, the antenna is plastic-filled and sealed to reduce the effects of precipitation static. This unit, which weighs less than 1 lb (0.45 kg), can be mounted without cutting into the airframe structure, and has negligible drag.

Format: 1/4 ATR Short Low
Weight: (without self-test option) 3 lb (1.36 kg) (with self-test option) 3.23 lb (1.47 kg)

STATUS: in production and service.

DME-42 Proline DME system

The DME-42 is an all-digital DME system, which can provide complete information on up to three DME stations using a single receiver. It is designed for business or commuter aircraft where only a single DME facility is needed, but where additional information is useful and it can directly replace the earlier DME-40 and -41 systems. Station information is presented on the associated IND-42 display.

The system can provide the display with data to show distance-to-station (up to 300 nautical miles), time-to-station (up to 120 minutes) and ground speed (up to 999 knots).

Dimensions: (DME-42) 1/2-ATR Short Dwarf (IND-42) 42 × 86 mm
Weight: (DME-42) 2.4 kg
(IND-42) 0.41 kg
Power: 0.8 A at 28 V dc plus 0.3 A for display

STATUS: in service and production.

VIR-32 Proline navigation receiver

The VIR-32 is the first digital VOR/ILS navigation receiver designed for business aircraft and can directly replace the VIR-30A or be installed in an all-digital aircraft. Several versions of the VIR-32 are available, but all are identical in format, the variations being selected by making the appropriate wiring connections. The system can receive 20 VOR/localiser and the associated 40 glideslope channels.

Dimensions: 3/4-ATR Short Dwarf
Weight: 2 kg
Power: 1.5 A at 28 V dc
Operating altitude limit: 70 000 ft

STATUS: in service and production.

DF-206A automatic direction finder

The DF-206A is a lightweight and rugged ADF, to full military avionic standards, operating from 100 kHz to either 2200 or 3000 kHz, with 500 kHz steps. An optional plug module makes it compatible with the MIL-STD-1553B digital data-bus, and the combining of loop and sense antennas helps reduce system size. The use of modern electronics means that the DF-206A is half the size and weight of the earlier ARN-89 system, for example, while having a designed mtbf of 4000 hours.

Dimensions: (receiver) 79 × 127 × 279 mm (control) 146 × 57 × 96 mm
(antenna) 216 × 43 × 419 mm
Weight: (receiver) 2.5 kg with -1553B capability;(control) 0.7 kg
(antenna) 1.4 kg
Power: 18 W at 28 V dc and 7.8 W at 26 V ac

STATUS: in production.

AN/ARN-144(V) VOR/ILS receiver

The ARN-144(V) VOR/ILS receiver is said to be the first such system to be compatible with the MIL-STD-1553B digital data-bus. All the standard VOR, localiser, glideslope and ILS beacon facilities are available with 160 VOR channels and 40 localiser/glideslope channels being selectable at 50 kHz spacing.

A number of configurations are produced to meet specific military applications. For example the R-2274(V)4 is used on the General Dynamics F-16 and Northrop F-20 aircraft, while the R-5094/ARN-514 is standard on the McDonnell Douglas F/A-18. A number of different control panels are also available.

In mid-1985 Collins received a $4.6 million contract to supply the US Air Force with over 5000 AN/ARN-144(V) sets for use on several types of military transport aircraft.

Dimensions: 104 × 127 × 304 mm
Weight: 3.6 kg
Power: 25 W at 28 V dc

STATUS: in production and service.

TCN-40 lightweight Tacan

The TCN-40 brings digital Tacan technology to training, communications, utility, and other secondary-role aircraft and helicopters. While fulfilling the need for lower weight, size, and cost, the basic navigation functions are comparable with those of the company's full-capability systems. According to Collins, digital technology has the following advantages: instantaneous channelling with no mechanical misalignment problems; nominal search and lock-on times of 1 and 3 seconds respectively for distance functions and bearing; accuracies of 0.1 nautical mile for distance measurement and 1° and 2.5° for OBI and RMI bearings; elimination of the 40° lock-on error by the use of special monitoring circuits; and the avoidance of interference with IFF and air traffic control transponders by the use of mutual suppression.

In addition to a multi-function digital distance display and ARINC-standard two-out-of-five cockpit control, the TCN-40 system comprises two remote units: the DME-40 distance measuring and BRG-40 bearing computers. Considerable output versatility is incorporated.

Dimensions: (DME-40) ARINC ½ ATR Short Low; 4.9 × 3.5 × 14 inches (124 × 89 × 356 mm) (BRG-40) identical with DME-40
Weight: (DME-40) 7.17 lb (3.26 kg) (BRG-40) 6 lb (2.72 kg)
Transmitter frequency band: 1025-1150 MHz
Receiver frequency band: 962-1213 MHz
Number of channels: 252

Transmitter power: 300 W peak
Receiver sensitivity: -82 dBm
Memory after signal loss: (distance) 12 s (bearing) 3 s
Range: 250 n miles
Track rate: 999 knots, 20°/s
Qualifications: FAA TSO C66a, RTCA DO-138

STATUS: in production and service.

AN/ARN-139(V) air-to-air Tacan

The AN/ARN-139(V) transforms an aircraft into a flying Tacan station by providing air-to-air bearing and distance information, inverse Tacan operation, and selectable range ratios. These capabilities are added to all the standard features of the AN/ARN-118(V) Tacan. Rendezvous with an ARN-139(V) equipped aircraft or ship simplifies critical missions by providing a reliable readout of bearing and distance during the approach.

The ARN-139(V) is derived from the ARN-118(V), of which more than 20 000 units have been built. The ARN-139(V) provides distance and bearing transmission, and inverse Tacan operation. The latter function allows a tanker aircraft, for example, to read the bearing and distance to a Tacan-equipped aircraft in need of air refuelling. Inverse Tacan also enables a pilot to determine the bearing to a DME-only ground station. The selectable range-ratio capability permits the pilot, by means of a switch, to limit all replies to within four times the range of the nearest aircraft, or to concentrate on aircraft more than 30 times the distance of the nearest aircraft.

Dimensions: (AN/ARN-139(V)) 10 × 7.7 × 19.4 inches (254 × 196 × 494 mm) (C-10059/A and C-10994 control units) each 5.75 × 3 × 3.2 inches (127 × 76 × 81 mm)
Weight: (AN/ARN-139(V)) 69.125 lb (31.3 kg) (control units) each 2 lb (0.9 kg)
Transmitter frequency band: 1025-1150 MHz
Receiver frequency band: 962-1213 MHz
Transmitter power: 500 W minimum, 750 W typical
Receiver sensitivity: -92 dBm
Modes: x/y channels, inverse beacon, inverse air-air, inverse transmit-receive, inverse receive
Range: 390 n miles
Track rate: 3600 knots, 20°/s
Accuracies: (distance) 0.1 n mile (digital) 0.2 n mile (analogue) (bearing) 1° (digital) 1.5° (analogue)
Reliability: mtbf designed for 1000 h

STATUS: in production. Recent application is US Air Force/McDonnell Douglas KC-10 Extender. AN/ARN-118(V) was awarded original US Air Force production contract in 1975, and

is standard equipment with that service and with US Coast Guard. It is also becoming standard for US Army and Navy and has been chosen by over 35 countries.

ADF-700 automatic direction finder

Designed in accordance with ARINC 712, the ADF-700 automatic direction finder was introduced in 1980. It is based on experience gained with the earlier DF-203 and DF-206 receivers. These technical advances were pioneered by Collins Air Transport Division and incorporated into draft ARINC characteristic 712, which also provided for ARINC 429 interfaces and an integral loop/sense antenna.

In addition to mechanical and reliability improvements introduced by the new characteristic, a significant performance improvement came with the change from analogue to digital technology. In the ADF-700 all bearing signal baseband processing is performed digitally. The reference and bearing signals are converted into digital form using a 12-bit cmos analogue-to-digital converter, and these are subsequently handled by an Intel 8086 16-bit microprocessor. The ARINC 429 input/output functions are performed by an Intel 8049 processor in conjunction with a Collins universal asynchronous transmitter-receiver.

The advent of third-generation microprocessors permits the introduction of digital ADF signal processing to improve accuracy and reliability, a significant advantage being the elimination of mechanical adjustments. The system incorporates improved self-test capabilities as a result of using the 8086 microprocessor to control a test sequence incorporating a digital test signal synthesis. The same

Collins TCN-40 Tacan system with channel selector, range and bearing readout, and receiver

End-face of Collins ADF-700 receiver

power supply sub-assembly is also used in the Collins VOR-700 and ILS-700, leading to reductions in spares' inventories and maintenance costs.

Format: 2 MCU per ARINC 600
Weight: 6.6 lb (2.9 kg)
Power: 115 V 400 Hz 26 VA
Modes: ANT-aural receiver, ADF navigation, cw/mcw
Tuning: ARINC 429 dual serial bus
Frequency range: 190-750 kHz
Channel spacing: 0.5 kHz
Bearing accuracy: better than 0.9° with ARINC 712 antenna in 35 μV/m field, exclusive of antenna error

STATUS: in production.

VOR-700 VOR/marker-beacon receiver

The VOR-700 vhf omni-directional range/marker-beacon receiver incorporates the newest digital technology combined with the background derived from previous industry standards such as the Collins 51RV-2, 51RV-4 and 51Z-4 VORs. The system was designed in accordance with ARINC 711.

All bearing signal baseband processing in the VOR-700 is accomplished digitally. The 30 Hz reference and variable signals are converted into digital form using a 12-bit cmos analogue-to-digital converter, and are thereafter handled by an Intel 8086 16-bit microprocessor. The ARINC 429 input/output functions are performed by an Intel 8048 microprocessor in conjuction with a Collins universal asynchronous transmitter-receiver large-scale integration-based circuit. Additional functions, such as self-test, auto-calibration and monitoring, are also conducted digitally.

Digital processing improves the accuracy of measuring bearings by reducing the effects of temperature variation and ageing. Implementation of 30 Hz bandpass filters in firmware, compared with previous analogue methods, permits improved tracking of ground-station modulation frequency variations, increased navigation sensitivity, and better rejection of undesired components in the modulation of received signals.

VOR-700 parts count has been reduced by 40 per cent compared with the most recent analogue-technology VOR systems such as the 51Z-4 marker beacon receiver and the VOR portion of the 51RV-4. The system is fully compliant with ARINC 711 and extends many parameters of previous-generation equipment. In particular, bearing-measurement accuracy is five times better and there are 40 per cent fewer adjustments.

Format: 3 MCU per ARINC 600
Weight: 8.9 lb (3.9 kg)

Receiver unit for Collins VOR-700 system

Control/display unit for Collins LRN-85 omega/vlf navigator

Power: 115 V 400 Hz 30 VA
Frequency range: (VOR) 108-117.95 MHz (marker beacon) 75 MHz
Channel spacing: (VOR) 50 KHz

STATUS: in production.

LRN-85 omega/vlf navigator

Designed for the worldwide navigation of commercial, corporate and military aircraft, the LRN-85 is claimed to be one of the most comprehensive omega/vlf systems available. By comparing all available signals from omega and vlf ground stations against a precise rubidium frequency standard, the system can navigate a direct route between departure and arrival points, eliminating dependence on Vortacs, and providing continuous position information to the crew.

As well as maintaining 100 per cent duty cycle on the omega transmitters, the system receives both data sidebands from the seven worldwide vlf stations.

A notable feature of the system is the mass memory, which can store the coordinates of all the world's Vortacs and major airports, and the waypoint memory, with capacity to hold the coordinates of up to 100 pilot-entered waypoints. For the majority of airports, therefore, the crew need only enter the identifying codes of the start and destination airports, and the LRN-85 computes a direct course between them. Should this still not be adequate to define a route or series of regularly used routes, the crew can insert the appropriate waypoints and

destinations and designate them by alpha-numeric codes.

Two dedicated micro-computers provide the LRN-85 with power and flexibility to drive or command many functions, for example, provide desired track, cross track, and distance-to-go to a horizontal situation indicator; automatic and manual waypoint advance; roll command with turn anticipation; data-bus feed for automatic transmission of waypoint information to a second LRN-85 or other navigation system; internally computed magnetic variation; true and magnetic display of bearing, desired track, and wind speed and direction; computed offset waypoint; automatic computation of diurnal shift; and test mode for station deselect and for computer diagnostics.

The system comprises five units: E-field vlf blade antenna or H-field loop, antenna coupler, optional equipment unit (containing the atomic frequency standard, a battery, and additional interfaces), receiver/processor, and control/display unit. In its ARINC 599 version, the optional equipment unit is incorporated into the receiver/processor.

Dimensions: (control/display unit) 5.75 × 4.5 × 6.25 inches (146 × 114 × 159 mm) (receiver/processor) 7.5 × 7.67 × 19.67 inches (190 × 193 × 498 mm)
Accuracy: 2 n miles cep with minimum of 2 usable stations
Qualification: ARINC 599

STATUS: in service, no longer in production.

Navstar GPS receiver system

A competition to build Navstar GPS receiving equipment, involving Collins and Magnavox, began in 1974. Collins' Government Avionics Division successfully participated in the concept validation phase by building what was called the generalised development model under contract to the US Air Force Avionics Laboratory. The seven-year competitive development phase to build Navstar GPS receiving equipment for the US military concluded in April 1985 with the choice of Collins as the contractor. In that month the US Air Force Space Division authorised Collins to proceed with a $61.6 million initial programme covering integration and further development. The company believes that initial production could be worth $434 million over the next five years, with perhaps as many as 6100 sets being built for users. Total Department of Defense requirement may reach 21 000 sets by the end of the century.

On 22 May 1983 a Rockwell Sabreliner 65 corporate jet fitted with Collins Navstar GPS receiving equipment became the first aircraft to fly the Atlantic (Cedar Rapids to Paris) using

Navstar GPS aerials can just be seen on top of the fuselage of this F-16, just behind cockpit; second aerial is on nose wheel door

satellite navigation. The flight took five days owing to the limited availabilty of the prototype six-satellite chain, designed at that time to give maximum coverage at the US Air Force test site at Yuma, Arizona.

The system aboard the Sabreliner comprised a single antenna and electronics unit, two single-channel GPS receivers, a navigation computer and a control/display unit. It is reported that the system gave a position accuracy of 24.6 feet with respect to a predetermined point on the destination airfield after its five-sector, 4228 nautical miles flight.

Air, land and maritime users will be able to assemble the systems most suited to their needs from a series of common equipment and software modules. For example, the Grumman A-6E, General Dynamics F-16, and Boeing B-52 would share three units in common: a controller to direct the antenna towards the satellite, a controlled-reception antenna, and a five-channel receiver. The A-6E and F-16 would have a fixed-reception antenna, and the A-6E would further have an interface and a control/display unit. For all vehicles, the Collins system has in common 82 per cent of all software, 75 per cent of line-replaceable units, and 94 per cent of all shop-replaceable units.

For tactical aircraft GPS will facilitate rendezvous, target positioning, weapons delivery, and recovery of aircraft. In addition it will have an advanced anti-jamming capability for reliable and high-quality performance under the most severe conditions. The GPS tactical air equipment will permit continuous signal tracking during all manoeuvres, and will be integrated into other systems so that GPS information can improve mission effectiveness.

The air-breathing segment of the US strategic deterrent will likewise benefit from an improvement of its penetration and survival capabilities. The worldwide common grid position system available from GPS will improve strategic targeting and accuracy of weapons delivery. Stand-off weapons such as the US Air Force/Boeing AGM-86B air-launched cruise missile will also be more effective as a result of accurate and continuous position-updating.

In December 1982 Collins signed agreements with several major non-US electronics companies to promote its user equipment for the Navstar GPS system. Among them were Bell Telephone Manufacturing Co of Belgium, Fabbrica Italiana Apparecchiature Radio-elettriche of Italy, and GEC Avionics in the UK.

Designed for the military, many questions must be resolved before the system will be made available for commercial use.

STATUS: in full-scale development of user equipment for GPS Joint Program Office at Space Division of US Air Force Systems Command. First set of units, single-channel manpack and five-channel airborne receiver, were completed in autumn 1982. System began field tests in December 1982 in full-scale development trials under $75 million contract calling for production of 55 sets for use with various agencies to gain user experience. First installation on a high-performance aircraft (a General Dynamics F-16 fighter) was completed in July 1984, and flight-testing began at Yuma Air Force Base in early 1985. Another set has been installed in a Boeing B-52 strategic bomber. Competition between Collins and Magnavox to build operational receiving equipment continued until spring 1985, when Collins was selected for first procurement of equipment. Airadio Products Division of UK company, GEC Avionics, is to promote Collins GPS airborne system under agreement signed in November 1982. A small GPS receiver, measuring 7.7 × 7.4 × 4.8 inches (195 × 188 × 122mm), known as Navcore 1 was reported to be in production in late 1985. During late 1985 a Collins Navstar GPS was used in trials over the Atlantic Ocean to determine the feasibility of using GPS, via a satellite communications link to provide automatic aircraft position reporting over oceans.

Near-flush antenna for Collins Navstar GPS five-channel receiver on upper fuselage of Boeing B-52 trials aircraft

Collins' Navstar GPS receiving equipment for the Sikorsky UH-60 Blackhawk

Collins CP-1516/ASQ automatic target hand-off system and control/display unit

CP-1516/ASQ automated target hand-off system

The CP-1516 is a battlefield mission management system used in conjunction with a control and display unit and up to four standard hf, vhf or uhf radios to provide a tactical command control, communications and information (C³I) network. The digital communication network can provide for stores management, target hand-offs and other similar functions to be passed to airborne forces in short radio bursts which are difficult for the enemy to detect or jam.

The CP-1516 features a recall capability for 12 previously received messages and allows the transmitting of pre-formatted messages or free-text messages using an alphanumeric keyboard. Non-volatile memory in the unit retains all critical information in the event of a power loss. In addition, the CP-1516 maintains the current status of up to 10 active airborne missions and two preplanned missions.

Various control-display unit options are available for data entry and display. The CP-1516 is fully compatible with the McDonnell Douglas AH-64 Apache data entry panel and TADS/PNVS display, the Bell OH-58 AHIP control-display and mast-mounted sight display and the JOH-58 light combat helicopter control-display unit. The Collins CMS-80 night-vision goggle compatible control-display is also compatible with the CP-1516.

Modern electronic battlefield systems including Sincgars, PLRS/JTIDS hybrid (PJH), Tacfire communications, Comsec and all MIL-STD-1553 avionics including digitally-generated map displays are completely compatible with the CP-1516.

The CP-1516, which was first demonstrated in May 1985, is suitable for use in conjunction with the Collins CMS-80 avionics management system, and has been selected for the US Army's OH-58D and JOH-58 combat helicopters.

The computer within the CP-1516 incorporates 8 K bytes of ram, 2 K bytes of earom and 196 K bytes of program memory, with an expansion capability of another 80 K bytes. The system is compatible with MIL-STD-1553B and B data-buses.

Dimensions: 136 × 165 × 203 mm
Weight: 4.5 kg
Power: 40 W max (dc)

STATUS: in development.

Delco

Delco Systems Operations, General Motors Corporation, 6767 Hollister Avenue, Goleta, California 93117
TELEPHONE: (805) 961 5011
TWX: 910 334 1174

Carousel IV inertial navigation system (AN/ASN-119)

During the late 1960s the Carousel IV inertial navigation system was the subject of the largest ever single military procurement of such equipment, when it was chosen by the US Air Force for its fleet of Lockheed C-5A Galaxy and C-141 StarLifter transports and Boeing KC-135 tankers. It had earlier been chosen as standard fit for the Boeing 747, when it was fitted as a three-system installation to become the first certificated commerical inertial navigation system. A guaranteed mean time between failures greater than 1250 hours was an important factor in Boeing's choice of the Carousel IV. Some 7000 sets have now been delivered to support 30 military programmes and more than 60 airlines, and the Carousel IV has a demonstrated mean time between failures of more than 3000 hours on commercial aircraft.

Improved versions of the Carousel IV are being built for the Boeing E-3A Sentry AWACS under the designation AN/ASN-119, the Titan III ICBM missile fleet, numerous helicopter applications, and the airborne element of the US Army's Guardrail intelligence programme. The AWACS installation comprises two ASN-119 platforms, operating in conjunction with a single Northrop AN/ARN-129 omega system and a Teledyne AN/ADN-213 Doppler velocity sensor. The combination provides an accuracy of less than 1 nautical mile in a 10-hour mission.

Each system comprises three elements: an inertial navigation unit with gyros, accelerometers and computing functions, a control/display unit, and a mode selector unit. A battery unit to maintain operation during power transients is optional.

Dimensions: (inertial navigation unit) 215.9 × 259.1 × 510.5 mm
(control/display unit) 114.3 × 146.1 × 152.4 mm
(mode selector unit) 146.1 × 38.1 × 50.8 mm
Weight: (inertial navigation unit) 25 kg
(control/display unit) 2.1 kg
(mode selector unit) 0.45 kg

STATUS: in production.

Carousel-Six inertial navigation system

The Carousel-Six is an ARINC 561 INS combining the platform technology used in the Carousel IV with an expanded system computer and cathode ray tube control/display unit. The cdu includes a microprocessor and has 8-line by 16-character symbology. The expanded

Control/display presentation for Delco Carousel VI inertial navigation system

memory uses semi-conductor devices. Designed primarily for corporate aircraft it has been certificated on the Gulfstream, Dassault Falcon, Canadair Challenger and Boeing 727 types and has found applications in both civil and military aircraft, both as new equipment and as a replacement for existing Carousel IV equipment.

To assist flight safety, the system can provide a wind-shear detection function and has growth provision for fuel-consumption and range management. The system is designed to use the Vandling NDB-2 navigation database, can programme flight-plans by alphanumeric identifiers, can store up to 20 flight-plans and up to 75 additional pilot-defined waypoints, and can select routes by city pair or numbered designator. Expanded interfaces allow the system to communicate with other equipment, notably the Global Series IIIA and -B, Collins LRN-85 and FMS-90, Canadian Marconi CMA-771, Sperry and Bendix radars, and with VOR beacons.

As with the Carousel IV, the -Six retains Delco's method of achieving high accuracy by rotating the stable platform on which the attitude and accelerometer sensing devices are mounted. In addition to cancelling errors, it provides automatic calibration of the inertial devices.

Input signals comply with ARINC 407 (synchro), ARINC 419/429 (digital 2-wire), ARINC 561 (digital 6-wire), ARINC 568/709 (DME), ARINC 547/711 (VOR) and ARINC 545/565/575 (air data). Outputs are to synchro, 2- and 6-wire digital standards.

Dimensions: (inertial navigation unit) 215.9 × 259.1 × 495.3 mm
(control/display unit) 114.3 × 146.1 × 152.4 mm
(mode selector unit) 146.1 × 38.1 × 50.8 mm
(battery unit) 160 × 124 × 320.5 mm
Weight: (inertial navigation unit) 23.6 kg
(control/display unit) 3.27 kg
(mode selector unit) 0.45 kg
(battery unit) 7.7/12.3 kg (15/30 minute life)
Power: 115 V ac 400 Hz, 1023 W (warm-up), 253 W (operational)

STATUS: in production.

Low-cost inertial navigation system

The low-cost INS development program continues Delco's Carousel-class systems activity into applications that require somewhat lower performance at substantially lower cost. Typical navigation accuracy rating for the LCINS is 2 to 4 nautical miles per hour. The LCINS is a strapdown configuration, utilising Incosym Inc two-degree-of-freedom gyros and companion 2-axis accelerometers. In this configuration, the entire inertial reference assembly is substantially reduced in size, yet still incorporates the novel Carouselling technique where one set of instruments is rotated continuously to offset bias errors and to enable pre-flight instrument

Delco Carousel IV inertial navigation system

calibration/compensation. A digital microprocessor performs all the measurement data processing, instrument torquing computation, scaling, attitude, and navigation functions.

Steering commands and other autopilot interfaces are provided.

Dimensions: 7.5 × 5.2 × 11 inches (190 × 132 × 279 mm)

Weight: 17 lb (7.7 kg)
Power: 100 W (28 V dc)

STATUS: in development.

Emerson

Emerson Electric Company, Electronics and Space Division, 8100 West Florissant Avenue, St Louis, Missouri 63136
TELEPHONE: (314) 553 3232
TELEX: 209903
TWX: 910 761 1126

AN/ARN-89B automatic direction finder

This ADF comprises four line-replaceable units: loop antenna, impedance-matching amplifier, receiver and control unit, and is used in conjunction with a sense antenna and bearing indicator. The system can operate-manually or automatically on any AM or continuous wave signal between 100 and 3000 kHz. In the Compass mode it provides automatic bearing indication by tuning to the appropriate station frequency. The 'loop' mode permits manual direction-finding through a null-tone determination of bilateral bearing. An 'antenna' mode enables the system to be used as an AM or continuous wave communications receiver. A second control unit and associated transfer unit enables the system to be operated by a crew-member elsewhere in the aircraft.

Dimensions: (antenna) 11.75 × 11.75 × 1 inches (298.5 × 298.5 × 25.4 mm)
(impedance-matching unit) 2 × 2.5 × 1.5 inches (50.8 × 63.5 × 38.1 mm)
(receiver) 5 × 11 × 5.5 inches (127 × 279.4 × 139.7 mm)
(control unit) 5.75 × 3.75 × 6 inches (146.1 × 95.3 × 152.4 mm)
Weight: total 12.75 lb (5.79 kg)
Power: 1.3 A at 28 V dc
Frequency range: 100-300 kHz

Emerson AN/ARN-89B automatic direction finder system four line-replaceable units

Bearing accuracy: 3°
Reliability: >2660 h

STATUS: in production, principally for US Army helicopters and fixed-wing aircraft, also

for Canadian Armed Forces/Lockheed CP-140 Aurora ASW aircraft, and ships and tanks. System has been in production, in ARN-89 and -89A forms, since 1971, and to date more than 2000 systems have been built.

Foster

Foster AirData Systems Inc, 7020 Huntley Road, Columbus, Ohio 43229
TELEPHONE: (614) 888 9502

LNS616 area navigation system

With the launch of the LNS616 in mid-1983, Foster provided general aviation with a long-range, self-contained navigator within a single panel-mounted unit, requiring only a remotely-mounted Loran C aerial and receiver (there is no need for an additional panel-mounted Loran control unit). The system integrates Loran C, Vortac and area navigation methods (RNav) with microprocessor technology to give continuous, accurate navigation. It conforms to FAA

Foster single-unit LNS616 area navigation system

Circular AC90-45A in the USA, including the mid-continental area which has poor Loran coverage.

The LNS616 has applications in three areas: long-range (in general aviation terms) navigation involving sectors of 600 to 2000 nautical miles; shorter sectors, where the destination is at some distance from a navigation beacon or other position-fixing source; and as an aid in the terminal area by providing greater flexibility to accommodate local air traffic control directions.

Foster has discarded push buttons for programming in the LNS616. They are replaced by two concentric rotary knobs to programme the information display and control a cursor. A seven-function rotary switch selects the required operational mode: VOR/LOC reference, programming, navigation, time/wind, GRI display and auxiliary. A 'page' pushbutton provides the display of additional information levels in each selected mode. Four pushbuttons permit the pilot to review or programme flight-plans, waypoint addresses, VNav (vertical navigation) parameters or aircraft present position. The pilot is alerted to important computations by a 'message' pushbutton and annunciator, which is also used to erase flight plans or reference information or to insert waypoints. The display itself comprises two rows of 12 dot-matrix characters and symbols.

The system can store 26 non-volatile flight-plans, and their organisation is assisted by a 100-waypoint memory and 250 pilot-programmable navigation references. In the VNav mode, the LNS616 provides pilot-programmable flight-path angles (0° to 9.9°), altitude accuracy and resolution to within 100 feet and altitude computation to 65 000 feet.

Format and dimensions: (Model RNC601 control/display unit) 3 ATI panel mounted (Model LR651 remote-mounted Loran C receiver and mounting tray) 133.6 × 105.4 × 280.4 mm
Weight: (control/display unit) 1.48 kg (Loran C receiver) 2.73 kg

STATUS: in production. The LNS 616 was selected in spring 1985 by the US Forest Service to equip 19 Beech 58P Baron aircraft used as tanker support in fire-fighting tasks.

RNAV 511 area navigation system

A low-cost full-time navaid for general aviation, the RNAV 511 is claimed to have the lowest work-load factor of any system available, and interfaces with most commercially available VOR/DMEs, compass director indicators, horizontal situation indicators, and autopilots. Designed in accordance with the recommendations of an FAA study on RNav utilisation, the system meets all criteria for light aircraft avionics of its type for single-pilot IFR and VFR operation.

The system is based on two-waypoint storage as being optimum for light aircraft operation, especially by one pilot. Microprocessor computation provides continuous updates of range and bearing to the active waypoint, but an alternative display of ground speed and time-to-waypoint may be selected, providing a useful mode for flight-planning and fuel management.

For the private pilot, RNav applications vary from simplifying cross-country flights and finding airports and airfields without co-located navigation beacons, to skirting with minimum fuel and time penalties terminals and restricted zones and penetrating complex terminal areas, RNav instrument approaches, or simply as range advisories during ILS, ADF, VOR, and MLS procedures.

Dimensions: (panel-mounted control/display unit) 6.25 × 2.375 × 8.75 inches (159 × 60 × 222 mm)
(steering adaptor) 6 × 4 × 1.375 inches (153 × 102 × 35 mm)
Weight: (control/display unit) 2.5 lb (1.1 kg)
(steering adaptor) 1 lb (0.4 kg)
Power: 0.7 A at 28 V dc
Max waypoint offset: 199 n miles
Max distance to next waypoint: 199 n miles

STATUS: in production.

RNAV 612A area navigation system
Designed for medium-sized piston, turboprop and turbine-powered fixed-wing aircraft and helicopters, the RNAV 612A is a high-performance, microprocessor-controlled area navigation system, developed from the RNAV 612, that can accommodate up to four waypoints. It was introduced in September 1983, meeting the IFR requirements of FAA Advisory Circular AC90-45A and containing a TSO-approved VOR/LOC converter. In the normal RNav mode, the system provides linear (constant course width) steering signals to the horizontal situation and compass director indicator flight instruments. Steering sensitivity is selectable between 'en route' (± 5 nautical miles full-scale) and 'approach' (± 1.25 nautical miles full-scale).

Digital bearing and distance to waypoint are displayed continuously, while groundspeed and time to the next waypoint, or present position with respect to the selected Vortac, can be shown by pressing a button. In the VOR/LOC mode, standard track-angle VOR and localiser steering drive signals are provided to the HSI and CDI, while magnetic bearing and DME distance is continuously displayed. As in the RNav mode, groundspeed and time to next Vortac, and aircraft present position, are displayed by pressing a button.

The incorporation of a Foster 612 RIU remote interface unit, improves system effectiveness in the following ways: the addition of Range Monitor (a Foster trademark) which, by switching between Nav 1 and Nav 2 beacons, reduces the possibility of pilot disorientation at critical navigation situations; a sine/cosine drive signal to the radio magnetic indicator showing the next waypoint or Vortac beacon; provision for those panel interfaces requiring sine/cosine-type drives to interface with radio magnetic indicators; DME slant range correction, which eliminates the altitude error from all RNav computations; and distance drive signals for flight directors and VNav equipment made by other manufacturers.

STATUS: in production.

Foster RNAV 612A area navigation system

Foster 612 remote interface unit

RNAV 612 area navigation system
This system, introduced in 1979, has all the capabilities of the RNAV 612A together with four additional features:
'Auto waypoint', whereby the equipment automatically establishes an unlimited number of successive waypoints, within reception range of any Vortac, along the route
A fifth waypoint which can be selected for programming or navigation by the 'Auto waypoint' push-button
'Load present position' capability which automatically establishes a present position waypoint at the location where the VOR/DME button was pressed. This equipment can also store for later use the address of a particular runway threshold or an important location in search and rescue operations
Crosstrack offset steering (up to 20 nautical miles to the left or right of the course) can be selected in 0.1 nautical mile increments in the RNav mode, thereby avoiding bad weather or restricted areas along the planned route. It may also have applications in precision aerial mapping and search and rescue.

A revolutionary feature of the RNAV 612, at the time of its introduction, was plug-in interchangeability with the King KNC 610 single-waypoint RNav system.

STATUS: in production. System was CAA approved in May 1982.

AD611 area navigation system
This designation describes a family of modular RNav systems that can be specified or built up according to need, and aimed at the professional market. They are assembled from various combinations of four types of identically sized unit: range selector, waypoint setter, and the horizontal display and data entry modules that together comprise the memory waypoint system.

The simplest system is the one-waypoint RNav, comprising a range/mode selector and a manual waypoint setter. As with all Foster RNav equipment, range and bearing to the waypoint are entered by thumb-wheels, easing operation

Foster AD611 area navigation system

in rough air. Despite its simplicity, the one-waypoint system has found wide acceptance, particularly among operators with limited panel space.

Since these manual waypoint setters have no memory, the two next most advanced variations (with two and three waypoints) are obtained by adding, respectively, one and two more such units. The two-waypoint system is acknowledged to be the minimum needed for most RNav operations, while the combination handling three meets the challenge of IFR flying.

A 10-waypoint system, claimed to be the most straightforward RNav available, uses a range display mode selector, and the two units of the non-volatile memory waypoint option.

Finally, an 11-waypoint system can be assembled by adding a manual waypoint setter to the 10-waypoint system just described.

STATUS: in production. In 1980 US Navy installed 177 AD611/A-T sets on its Beech T-34C trainers. RNav has enjoyed growing popularity for civil applications since mid-1970s, but this turboprop T-34C is believed to be first military primary trainer with this capability. AD611/A-T is specialised version of AD611.

VNAV 541 and 541/A vertical guidance systems
In order to initiate climb or descent to a new level at the correct time and range, the company provides the VNAV 541 and 541/A vertical navigation guidance computers. These devices compute the triangulation problem involving height and distance to the destination airfield, a new flight level, or waypoint-crossing altitude, to provide a commanded flight profile. The pilot enters the required altitude at the airfield or flight level, selects the flight-path angle (1.5°, 3°, or 4.2°), and then flies the aircraft so that the altimeter readings match the demanded readouts on the VNav; the system provides the basic vertical guidance.

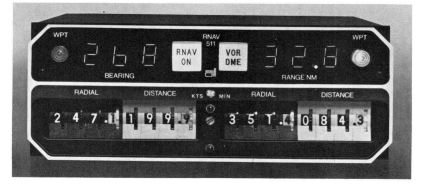

Foster RNAV 511 area navigation system

Descent angle: (VNAV 541) 1.5° cruise descent, 3° approach descent, designed for unpressurised aircraft
(VNAV 541/A) customer-specified cruise descent, 3° approach descent for pressurised aircraft and jets

Dimensions: 3.25 × 1.3 × 5.5 inches (82.7 × 33 × 140 mm)
Power: 2 A at 28 V dc

STATUS: in production.

Global Systems

Global Systems Inc, 2144 Michelson Drive, Irvine, California 92715
TELEPHONE: (714) 851 0119
TELEX: 681340
TWX: 910 595 2451

GNS-500 vlf

In 1974 the company introduced the GNS-500, a system that combined vlf reception with microprocessor data-handling to provide 10 position updates per second. This facility was particularly suited to the new generation of long-range business aircraft since pilots could now programme their flights by entering the coordinates of departure points, waypoints, and destinations.

GNS-500A omega/vlf

In 1976 the omega network of eight very-low-frequency transmitters was introduced, and in May of that year the GNS-500A was approved by the FAA for IFR en route navigation in US national airspace. In anticipation of this development, Global had been perfecting the GNS-500A, which could receive the new stations as well as the vlf transmitters.

GNS-500A Series 2 omega/vlf

In response to customer requests, improved equipment was introduced to offer non-volatile retention of waypoints and present position, automatic computation of magnetic variation, and a continuous clock to maintain Greenwich Mean Time and date.

Number of waypoints: 10
Outputs: position, track, distance and time to destination, ground speed, horizontal situation indicator, wind speed and direction
Inputs: heading, true airspeed, ARINC 571 bus interface with inertial navigation, radar, area navigation, and navigation-management systems
Power: 5 A at 28 V dc
Units: control/display, receiver/computer, optional equipment
Weight: 38 lb (17.27 kg)
Approval: North Atlantic, US domestic en route IFR, as sole over-water navaid, and in remote areas world-wide

Control/display presentation for Global Systems GNS-500A Series 2

Global GNS-500A Series 3 control/display unit

GNS-500A Series 3 omega/vlf

The Series 3 was introduced in 1980 to answer the need for more waypoints and a faster and clearer data presentation from any combination of up to five letters and numbers, on nine flight plans of up to 20 waypoints each. Information is presented on a sunlight-readable cathode ray tube providing a 14-character by 8-line matrix. In the Series 3 system all data is organised into one of three sections: navigation, data, and flight plan. Each section in turn contains information laid out on a number of 'pages'.

Number of waypoints: 127
Outputs: position, track, distance and time to destination, ground speed, horizontal situation indicator, wind speed and direction
Inputs: heading, true airspeed, ARINC 571 bus interface with inertial navigation, radar, area navigation, and navigation management systems
Power: 5 A at 28 V dc
Units: control/display, receiver/computer, optional equipment
Weight: 39 lb (17.72 kg)

GNS-500A Series 3B omega/vlf

This refinement was introduced in September 1981, and features trip planning, fuel planning, and offset waypoint facilities. In the trip-planning mode, the system computes distance and time for each leg as well as total flight distance and time. When the fuel-planning mode is selected, the system computes time and distance to the reserve point based on ground speed and manually inserted values for fuel remaining, fuel flow, and desired reserve. It also displays fuel efficiency in terms of nautical miles flown per 100 lb (45.36 kg) of fuel consumed. The offset-waypoint facility allows the crew to set up an artificial waypoint based on distance and bearing from a real waypoint, so avoiding the need to overfly the actual waypoint.

GNS-500A series 3C omega/vlf

In what it promotes as 'the EFIS connection', Global and the major manufacturers of electronic flight instrument systems (Bendix, Collins, Sperry, Smiths Industries and Thomson-CSF) have been working together to bring compatibility between the two types of equipment so that EFIS displays can now show navigation information. As a result, software changes in the GNS-500A (resulting in a version called the Series 3C) now permit EFIS-compatible navigation data to be passed to electronic or electromechanical flight instruments, generally via an ARINC 561 bus. Apart from EFIS compatibility, the system offers

offset waypoint indication, fuel planning, and variable-turn anticipation.

The offset waypoint facility introduced in the Series 3B is expanded so that the waypoint is indicated by an alphanumeric identifier followed by a star (for example, LAX* for Los Angeles International airport). The system defines such a waypoint in terms of bearing and distance from the real waypoint. The offset waypoint page now appears any time such a waypoint is called up, even if the same point had previously been defined. This ensures verification of the bearing and distance selected each time an offset is flown.

For variable-turn anticipation, the system now accomplishes a leg change based on ground speed and turn angle instead of beginning the manoeuvre 36 seconds before the estimated time of entry on to the new leg. The time to change a leg can vary from 8 to 96 seconds, and the new method minimises the overshoot associated with large turn angles.

GNS-500A Series 4 omega/vlf

The Series 4 is the latest improvement to the GNS-500A family offering further operational features: it can directly replace earlier models. The Series 4 operates with the Global NDB-2 navigation database and can implement 127 waypoints and nine flight-plans with up to 20 waypoints each. The system, which comprises three units, is fully compatible with ARINC 561 and 571.

Weight: (total system) 17.5 kg
Power: 7.5 A at 28 V dc

STATUS: the Series 4 is the latest to enter production. Earlier series continue in service.

GNS-1000 flight management system

In July 1983 Global announced the launch of the GNS-1000 flight management system. The GNS-1000 will be the 'lead system' for the new generation of equipment, with varying levels of capability.

Global had recognised that the earlier GNS-500 was running out of development potential; it accommodated virtually all the growth functions envisaged in the early 1970s, but had no spare processing capability or connector pins. The GNS-1000, therefore, represents a fresh approach and it will not be possible to upgrade the GNS-500 equipment to the new standard.

A typical two-channel GNS-1000 installation comprises two ½ ATR Short flight management computers, two control/display units, two ½ ATR Short receiver/processor units, two H-field antennas, and a single shared NDB-2 navigation database. The new system achieves position-sensing and computation and it also performs automatic fuel-planning functions and radio tuning. Other facilities, such as vertical navigation, are being introduced as part of growth development.

One problem that confronts manufacturers of equipment that communicate with many different systems is that of ensuring all-round compatibility between older aircraft with analogue navigation, communication and flight control installations, the more recent mixed analogue digital types, and the all-digital/aircraft. Global has overcome this difficulty with what it calls a 'configuration module'. This is a separate rack-mounted box that provides, in code form, a description of all the equipment in that particular aircraft that have to communicate with the GNS-1000.

In early 1986 the FAA approved the installation of a Global Flight Information System in a IAI Westwind 2 aircraft. This system incorporates a data link receiver into the GNS-1000 so that the crew can receive weather reports, routine flight information and air-to-air messages, routed from a central computer in California and originating from any compatible source.

STATUS: in production, replacing earlier GNS-500 range.

Multiport NDB-2 navigation data bank

The Multiport NDB-2 is a version of the solid-state, mass-memory device that holds Jeppesen navigation information in digital form for updating cockpit and flight-deck navigation systems. The Multiport function enables the device to feed in information to six navigation systems in sequence, eliminating the wiring currently needed to switch information from one system to another. NDB-2 is a unique system, maintaining in a non-volatile, reprogrammable memory the coordinates, frequency, magnetic variation and elevation for vhf navaids, together with the coordinates of high and low points along routes, standard instrument departures and stars, outer markers and runway thresholds.

All data in the system is supplied by Jeppesen Sanderson, and can be updated every 28 days on the same revision cycle as the charts themselves. The memory module is removed and inserted in an updater, a computer unit installed at turbine service centres worldwide, and the information is revised within a few minutes.

The Multiport NDB-2 interfaces with the Global GNS-500A, Litton LTN 72RL, Delco Carousel VI, Canadian Marconi Alpha omega/ vlf, Bendix NMS-46A, and Honeywell Lasernav systems.

STATUS: in production.

Gould

Gould Inc, NavCom Systems Division, 4323 Arden Drive, El Monte, California 91731-1997
TELEPHONE: (818) 442 0123
TELEX: 215380
TWX: 910 587 3428

AN/ARN-84 MicroTacan

Operational features of the ARN-84 Tacan include x-y position fixing, air-to-ground link with any standard Tacan station, air-to-air bilateral ranging, receive-only (bearing and beacon identification), radio frequency data link, air-to-air bearing and inverse mode. The system provides automatic antenna switching together with digital and analogue range and bearing outputs to an aircraft computer. The solid-state design gives a level of reliability previously unmatched according to Gould, and there are no 40° false lock-ons even in the absence of North burst data. The range is 300 nautical miles.

STATUS: in production. System is fitted to US Navy/McDonnell Douglas F/A-18 Hornets.

AN/ARN-130 MicroTacan

This system incorporates features that significantly improve the utility of the MicroTacan series, including two-way air-to-air bearing and range measurement, full y-mode operation, and solid-state coupler. Precision digital and analogue outputs are simultaneously available from the transmitter-receiver, and a separate converter is not needed for retrofits. A single-tube power has increased the reliability and reduced power source dissipation so that the mean time between failures is now 1500 hours, and there is continuous in-flight performance monitoring. The range is 399 nautical miles.

STATUS: in production and service.

SDM 77 distance measuring equipment

Introduced in late 1984, this panel-mounted DME features a liquid-crystal display, and all computing is microprocessor-controlled. The system measures range up to 199 nautical miles, and acquires ground speed 8 seconds after lock-on to a station, with a simultaneous readout of time-to-go, up to 99 minutes. An audio volume control associated with station identification is mounted on the front face. A version designed for remote control of the navigation function, and eliminating the need to tune both NAV and DME, is designated SDM 77A. A further version, the SDM 77B, provides for both remote and local control.

Transmitter-receiver and control unit for Gould AN/ARN-84 MicroTacan

Dimensions: 6.687 × 1.57 × 10.0 inches (169.8 × 39.87 × 254 mm)
Weight: (SDM 77A) 3.0 lb (1.36 kg)
(SDM 77B) 3.2 lb (1.45 kg)
Number of channels: 200
Frequency: (receiver) 978-1213 MHz
(transmitter) 1041-1150 MHz
Power: 500-250 mA at 11–32 V dc
Range: 0-199 n miles
Range accuracy: (0-99.9) ± 0.1 n miles, (100-199) 1 n mile
Ground speed: 399 knots
Ground speed accuracy: 15% after 8 s, 5% after 45 s

STATUS: in production.

Harris

Harris Corporation, PO Box 94000, Melbourne, Florida 32902
TELEPHONE: (305) 727 9100
TELEX: 803719
TWX: 510 959 6291

DTM/D digital terrain management/display

The DTM/D is being test-flown on the US Air Force AFTI (advanced fighter technology integration) F-16 aircraft. It combines a digital map-generation system, developed by Harris, with a colour multi-function display from Bendix, and is expected to provide a navigation system for combat aircraft 10 times more accurate than conventional systems, but with only minimal electromagnetic emissions and, at the same time, leaving the aircraft's radar free for other purposes.

Much as the British Aerospace Terprom system (see separate entry), the DTM/D correlates radio altimeter information with a digitally-stored map, providing for covert navigation which will, according to the USAF, enhance terrain masking tactics and will give the pilot the ability to look beyond hills in front of and to the side of his aircraft. This will allow him to work out his best route well in advance and be able to relate constantly his position in relation to the threats and targets.

STATUS: flight-testing in AFTI F-16 continues.

Honeywell

Honeywell Inc Commercial Aviation Division, Avionics Systems Group, 5775 Wayzata Boulevard, St Louis Park, Minnesota 55416
TELEPHONE: (612) 542 5133

Ring-laser gyro systems

In January 1981 Honeywell delivered to the Boeing 757 and 767 programmes the world's first production laser gyro inertial reference system. These two airliners represent the first production applications of the ring-laser gyro (RLG), heart of the laser inertial reference system. The Honeywell Avionics Division at Minneapolis was then the only plant in the world in quantity production of RLGs, a position that stems from a Boeing contract for the delivery of 400 inertial reference systems up to the end of 1984, with an option on a further 200. Each system consists of three ARINC 700 series digital inertial reference units.

Honeywell has been developing RLGs for more than 20 years, and the first-generation strapdown systems employing them began flying in 1974. Standards of performance equivalent to that of INS equipment was first demonstrated in flight trials at Holloman Air Force Base in 1975, and have since been confirmed in flight tests by the US Navy, NASA, McDonnell Douglas, Boeing, and Honeywell.

Environmental immunity during simulated combat manoeuvres and carrier operations has been demonstrated. More than 600 hours in nine different aircraft types had been flown up to the end of 1980, by which time Honeywell had completed its RLG production plants, a $15 million investment.

A Honeywell system installed aboard an Air France Boeing 747 for flight trials demonstrated in early 1981 an accuracy of 0.26 knot on its initial flight (7 hours 51 minutes) from Chicago to Paris. By comparison, the system specification called for 2 knots. Residual velocity on landing at Paris was only 3 knots.

In August 1981, the company's Military Avionics Division delivered to McDonnell

Douglas for test and evaluation on the AV-8B the first military laser gyro INS. In 1980 the US Air Force had completed testing a Honeywell RLG navigation system on a Lockheed C-141 transport at the Central Inertial Guidance Test Facility at Holloman Air Force Base, New Mexico. Test results indicated a circular error probability of 0.88 nautical mile an hour and a velocity error of 2.9 feet a second.

By late 1983 production of ring-laser gyro units had settled at its present rate of 200 a month (with a yield of 90 per cent), and in August 1984 Honeywell celebrated both the production of its 5000th ring-laser gyro and the millionth flight hour. These RLGs are used both in conventional inertial navigation systems and in the much newer IRS inertial reference systems.

Apart from commercial and military aircraft applications, Honeywell also produces laser gyros for missiles and spacecraft.

Honeywell ring-laser gyro inertial navigation systems were selected in 1985 for retrofit into the West German Air Force's F-4F Phantoms. In September 1985, 459 Honeywell commercial RLG systems had been installed in 153 airliners: the RLGs had amassed over 10 million operating hours with a demonstrated mtbf of 57 000 hours.

Honeywell family of triangular ring-laser gyros now equip many types of military and civil aircraft, missiles and land vehicles

Laser inertial reference system

The world's first production ring-laser gyro IRS inertial reference system was chosen by Boeing as part of the avionics package common to both the 767 and 757 airliners. It was also selected for the new Boeing 737-300, and as an option on the Airbus A310.

The strapdown configuration is so called because the gyro-stabilised platform of current conventional inertial navigation and attitude-reference systems is replaced by three ring-laser gyro units mounted rigidly and at right-angles to one another. The laser gyro detects and measures angular rates of motion by measuring the frequency difference between two contra-rotating laser beams made to circulate (hence the term 'ring') in a triangular cavity by mirrors. When the units are at rest the distances travelled by each beam are the same, and the frequencies are the same. When the unit rotates, one path lengthens while the other shortens and so a frequency difference is established proportional to the rate of rotation of the unit. The difference is measured and

processed digitally in ARINC 704 format as aircraft attitude in pitch, roll, and yaw.

Since the accelerometers are mounted rigidly in the box, their signals are related to aircraft axes, and have to be processed to convert them to the external inertial reference frame necessary to provide navigation and flight control information and guidance.

Honeywell claims that the laser gyro inertial reference system will cost significantly less than an ARINC 561 inertial navigation system or any other combination of devices offering the same functions and performance. A strapdown system has no moving parts to wear, fail, or become misaligned; no gimbals, torque motors, spin-motors, slip-rings, or resolvers, and no scheduled maintenance, realignment, or recalibration requirements are anticipated. A typical installation comprises three inertial reference units (containing the sensing and computing elements) and a display unit. The Honeywell laser device is contained within a low-expansion, triangular glass block, with a 34 cm path length. It has a claimed mean time between failures of 8000 hours.

Outputs: primary attitude information to displays and automatic flight control systems (afcs), linear accelerations, velocity vector and angular rates to afcs, wind-shear detection and energy management, magnetic heading for displays and afcs, and long-range navigation data
Specification: ARINC 704
Format: 10 MCU
Weight: 20.1 kg
Power: 91 W
Accuracy: (10 h flight) position 2 n mile/h (95% probability), velocity 12 knots (95% probability)
Reliability: mature mtbf predicted to average > 4500 h
Self-test: BIT (initiated and continuous) detects 95% of failures with 95% confidence level

STATUS: in production for Boeing 767, 757 and 737-300, and is optional system on Airbus A310.

Lasernav laser inertial navigation system

Introduced during early 1983, Lasernav (a Honeywell trademark) exploits the strapdown inertial reference system developed for the Boeing 767 and 757, but also has facilities that make it suitable for long-range business and corporate jet aircraft.

In addition to normal navigation functions, Lasernav can, if required, replace all customary attitude and heading sensors, including compass system components such as the flux detector, resulting in a reduction of up to 14 separate boxes. This is said to result in a weight saving of up to 60.3 kg, and a volume reduction of up to 50 per cent by comparison with equivalent dual installations in other systems.

Lasernav comprises two units: an inertial navigation unit and a control/display unit. The memory can store the coordinates of up to 255 waypoints, in 20 routes and with up to 20 waypoints per flight-plan, for immediate recall. In dual installations each system can store a different flight-plan, but can share it with the other systems if required. For international routes, or long over-water sectors, the system can use a Global NDB-2 database, and only the departure point and destination coordinates need to be inserted. Lasernav then computes a great-circle route, selects and identifies air traffic reporting points, and lists bearing and distance to the nearest VOR/DME on the 8-line by 14-character control/display unit.

A notable advantage is the sharp reduction in alignment time; the interval from switch-on to 'nav ready' is as little as 2.5 minutes at the equator and 10 minutes at 60° latitude.

Honeywell laser gyro-block test area

Dimensions: (inertial navigation unit) 322 × 324 × 193 mm
(control/display unit) ARINC 561
Weight: (inertial navigation unit) 21.1 kg
(control/display unit) 3.2 kg
Power: (total) 146 W at 115 V 400 Hz plus 28 V at 28 V dc

STATUS: in service.

Lasernav II navigation management system

Introduced in late 1984, Lasernav II is an inertial system for airlines and general aviation that combines self-contained, strapdown laser inertial position and aircraft motion sensors with externally-sensed radio signals to provide a single, efficient, integrated guidance package. The notable advantages of previous Honeywell laser inertial systems are maintained in Lasernav II: 2.5 to 10 minute alignment time, dependent upon latitude; three to four times the reliability of conventional systems; reduced size, weight and power consumption. The system can also be mounted in an unpressurised environment.

Position data obtained from VOR/DME and omega/vlf stations is blended with inertial position information using high-speed digital computing techniques. DME/DME updating is obtained by way of an auto-tuning function that requires no pilot inputs. VOR/DME or omega/vlf updating is obtained by manually tuning the radios. Triple inertial navigation system mixing combines inertial data from two other Lasernav II systems to calculate a composite inertial position.

In addition to normal navigation functions, Lasernav II can replace all conventional attitude and heading sensors, including vertical and directional gyros, flux valves, and compass controllers. In a typical dual installation, the total box count can be reduced by up to 14 separate boxes, resulting in weight-savings of up to 133 lb (60 kg) and a greatly simplified installation.

Lasernav II comprises: a navigation management unit and a control/display unit. An internal non-volatile memory can store 20 flight plans of up to 20 waypoints each. An NDB-2 worldwide database (required for DME auto-tuning) facilitates the automatic flight-planning feature that requires the pilot to input only the departure and destination points. A great-circle route is computed, intermediate waypoints are selected, and range and bearing to nearest VOR/DME is displayed, all automatically.

Dimensions: (navigation management unit) 322 × 324 × 193 mm
(control/display unit) ARINC 562
Weight: (navigation management unit) 22.1 kg
(control/display unit) 3.2 kg
Power: (total) 160 W at 115 V 400 Hz plus 19 W at 28 V dc

Accuracy: (inertial position) 2 n miles/h with 95% probability
(velocity) 8 knots with 95% probability
(heading) 0.4° (pitch and roll) 0.1°

STATUS: in production. Installations approved include Gulfstream II and III, Dassault Falcon 50, and Canadair Challenger CL-600 and CL-601 business jets, and McDonnell Douglas DC-8-71 and Boeing 737-200 airliners. In July 1985 Honeywell announced that the US Air Force was to fit Lasernav II sets in seven C-20A Gulfstream III aircraft operated by the 89th Special Air Mission.

Laseref inertial reference system

A derivative of the Lasernav inertial navigation system, Laseref (a Honeywell trademark) is intended as an attitude and heading reference (or primary sensor) for flight and navigation management equipment on top-of-the-line business aircraft such as the Canadair Challenger Gulfstream III and Dassault Falcon 50. Sharing many of the Lasernav modules and assemblies, Laseref is claimed to be the only laser-based inertial navigation sensor designed specifically for general aviation.

The new solid-state, strapdown sensor generates present position, ground speed, heading, and windspeed and direction to flight management systems and navigation equipment, and attitude and heading to flight instruments, weather radar stabilisation, and autopilots. It replaces vertical and directional gyros, compass systems, fluxgate sensors and other independent navigation equipment with self-contained sensors and computing circuits to provide digital outputs, together with the ARINC 407 synchro outputs still needed by current-generation avionics, in ARINC 429 format. As with Lasernav II, alignment time is greatly reduced in comparison with gimballed inertial systems; typically 2.5 to 10 minutes depending on latitude.

The Laseref system comprises two components: an inertial reference unit, containing

Honeywell Laseref inertial reference system

the sensing and computing elements, and a mode-select unit that provides power to the former and governs its operation. Data insertion is through the control/display unit of the flight management system used in conjunction with the Honeywell system. The company offers an optional device, the inertial sensor display unit, as an alternative means of initialisation. This unit has a small display for reading out inertially computed present position, groundspeed, wind data and heading. Comprehensive built-in test equipment is provided; the system performs a rapid pre-flight self test and monitors its operation throughout flight. Like Lasernav II, Laseref can be installed in an unpressurised bay.

Dimensions: (inertial reference unit) 322 × 324 × 193 mm
(mode select unit) 146 × 38 × 63.5 mm
(inertial sensor display unit) 146 × 114 × 167 mm
Weight: (inertial reference unit) 21.1 kg
(mode select unit) 0.45 kg
(inertial sensor display unit) 2.3 kg
Power: (total) 137 W at 115 V 400 Hz or 28 V dc

STATUS: in production.

Laseref II inertial reference system

Laseref II performs the same functions as Laseref and uses the same inertial sensor assembly, but in addition is designed to interface with new-generation digital avionics such as flight management systems, using the aircraft standard communications bus (ASCB). The Laseref II may also interface with aircraft having the ARINC 429 digital data-bus. Laseref II has been selected as the standard factory IRS installation on the Dassault Falcon 900.

Dimensions: (inertial reference unit) 322 × 324 × 193 mm
(mode select unit) 146 × 38 × 62.5 mm
Weight: (inertial reference unit) 21.1 kg
(mode select unit) 0.45 kg
(inertial sensor display unit) 2.3 kg
Power: (total) 137 W at 115 V 400 Hz or 28 Vdc

STATUS: in production. Laseref II is standard in a dual configuration on the Gulfstream IV, Dassault Falcon 900 and Canadair Challenger 601. Other applications include the BAe 125-800 and de Havilland Dash 8.

ADIRS air data inertial reference system

Honeywell's new ARINC 738 air data inertial reference system is an air data computer combined with an inertial reference system (IRS).

The ADIRS provides complete ARINC inertial reference system outputs including primary attitude and heading, body rates, acceleration, ground speed, velocity and aircraft position and ARINC 706 air data outputs, which include altitude, true airspeed, Mach number, air temperature and angle of attack.

Each ADIRS is equipped with three air data inertial reference units, one control display unit, and eight air data modules mounted remotely, adjacent to the pitot and static pressure sensors (see below). The ADIRS is manufactured at Honeywell's Commercial Aviation Division in Minneapolis.

Air data inertial reference unit (ADIRU)

The air data reference electronics and the laser gyro inertial reference system are packaged in a 10 MCU box which weighs 43.5 lb (19.7 kg) and requires a nominal power of 109 watts. The unit meets the functional requirements of ARINC 738 and the environmental requirements of D0-160B. Aircraft maintenance is simplified by extensive reporting of ADIRS Iru's status to the A320 centralised fault data system.

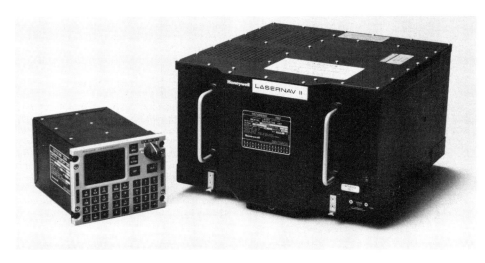

Honeywell Lasernav II inertial navigation system for general aviation

Air data module (ADM)

The air data module weighs 1.4 lb (0.64 kg) and requires a nominal power of 1.8 watts. It is packaged in a module which is 6 inches (152 mm) long, 3 inches (76 mm) wide and 2 inches (51 mm) high. The unit meets the environmental requirements of DO-160B. It features a state-of-the-art solid-state pressure transducer.

Control display unit

The control display unit is packaged in accordance with ARINC 738 (6.75 inches (171 mm) high – form factor B), weighs 5 lb (2.27 kg) and requires a nominal power of 5 watts (exclusive of warning lights). The unit meets the functional requirements of ARINC 738 and the environmental requirements of DO-160B. It features a liquid-crystal display.

Military Avionics Division, Aerospace and Defense, Honeywell Inc, Honeywell Plaza, Minneapolis, Minnesota 55408
TELEPHONE: (612) 378 4180
TELEX: 290631

Honeywell's H-423 RLG-based inertial guidance system was recently selected as the form, fit and functions standard by US Air Force

AN/ASN-131 SPN/GEANS precision inertial systems

The SPN/GEANS, standard precision navigator/gimballed electrically-suspended gyro aircraft navigation system, now designated AN/ASN-131, was developed primarily under sponsorship of the US Air Force Avionics Laboratory at Wright-Patterson Air Force Base, Ohio. It has been selected by the US Air Force for upgrading the navigation capability of the entire B-52 bomber fleet and is now in production at Honeywell, St Petersburg, Florida, for that and other programmes.

The basic SPN/GEANS system consists of an inertial measurement unit (IMU), an interface electronics unit (IEU) and a complete software library. For a stand-alone system capability, these two units are supplemented with a digital computer unit (DCU). The IMU contains, in addition to the velocity measuring unit (VMU) and two ESGs, temperature control electronics, accelerometer pulse rebalance and ΔV readout electronics, precision timing reference, gimbal control electronics and built-in test functions (BITE), as well as serial digital data-bus communication electronics. The IEU provides power conversion, control and sequencing electronics, additional BITE circuits, and a common serial data-bus interface with other units of the inertial system and other subsystems.

The ESG has only one moving part, a suspended hollow beryllium ball, which is combined with two optical pick-offs to give error and timing signals. These in turn are used to drive the IMU platform gimbals to maintain a stable reference base for the accelerometers. Three highly accurate, single-axis accelerometers (contrasted with the two-degree-of-freedom ESGs) are used within the VMU which is mounted on the stable platform inner element. These accelerometers, oriented in an orthogonal triad configuration, measure

accelerations directly and provide the incremental velocity pulses to the computer, which it uses in its software algorithms to calculate velocity and position parameters.

The ESG system requires little or no reliance on other navigation aids for most aircraft applications and thus can be described as self-contained.

SPN/GEANS is being deployed throughout the entire US strategic aircraft fleet and widespread use is also expected for long-range reconnaissance and patrol missions, as well as specialised cargo and transport usage in both military and civil applications and in tactical military aircraft.

In addition to the traditional function of position determination, or basic navigation, the higher accuracy outputs of velocity and attitude data have opened up a new realm of possibilities for stabilisation and/or motion compensation of other (non-inertial) sensors. These include high precision radars, sonars, lasers, optical and electro-optical devices.

Laser inertial and global positioning system

Honeywell is to market a combined laser gyro based inertial and Navstar global positioning navigation system being developed by Stanford Telecommunications Inc for use in commercial aircraft.

H-423 strapdown inertial guidance system for military aircraft

In August 1985, the US Air Force selected the H-423 as the standard inertial navigation system for its fighter, transport and helicopter aircraft (Lockheed C-130, General Dynamics F-111, McDonnell Douglas RF/F-4 retrofit, F-15E and F-15A-D retrofit, Sikorsky HH60A, LTV A-7 retrofit and possibly Fairchild A-10). Production deliveries under the contract began in early 1986.

The Honeywell system was selected based on the RLG production maturity, system

performance and Honeywell's extensive past RLG experience. Prior to the US Air Force award, Honeywell was already under contract with Northrop and McDonnell Douglas to provide RLG navigation systems for the Northrop F-20 and F-15E aircraft, respectively.

The H-423 was developed according to the US Air Force SNU84-1 specification which was the RLG version, an update to ENAC77-1. The system is a self-contained unit comprising three Honeywell GG1342 RLGs, three solid-state accelerometers and associated electronics along with a complete MIL-STD-1750A navigation processing package. It also contains built-in test circuitry to achieve a greater than 95 per cent fault detection.

Under the US Air Force contract, Honeywell is guaranteeing the H-423 will achieve 2000 hours mtbf in a fighter/helicopter environment and 4000 hours mtbf in a transport environment. Honeywell claims the H-423 has the highest reliability and maintainability, and lowest life cycle costs ever achieved by a military aircraft inertial navigation system allowing the US Air Force to go from a three-level to two-level maintenance system. These performance and reliability advantages have been the primary drivers for all US Defense Services now going to the RLG technology for all future aircraft inertial navigation systems.

Honeywell was also selected by Sweden for incorporation in its new fighter aircraft, the JAS 39 Gripen. In addition, it was selected by the Federal Republic of Germany for its F-4F modification programme. The many nations presently conducting aircraft modification programmes are expected to put RLG systems on their aircraft.

Dimensions: 3/4 ATR: 18.1 × 7.6 × 7.89 inches
Weight: 48 lb (22 kg)
Power: 140 VA
Interface: dual 1553B digital data-bus
Accuracy: ≤0.8 n miles/h with full performance from 22 second stored heading alignment
Specification: SNU84-1, FNU85-1

STATUS: in production.

Hughes

Hughes Aircraft Company, Radar Systems Group, PO Box 92426, Los Angeles, California 90009
TELEPHONE: (213) 648 2345
TWX: 910 348 6681

AETMS airborne electronic terrain map system

The subject of a research contract from the US Air Force, AETMS is essentially a moving-map display in which cartographic details are generated by feeding into its memory a

digitised terrain database appropriate to the area over which an aircraft is flying. The system relies on the database that the US Defense Mapping Agency began preparing during the late 1970s. The terrain information can be processed to provide the most appropriate data for terrain-avoidance and navigation guidance, and then presented to the pilot on a standard colour or monochrome cathode ray tube-type display.

Synthetic maps of this kind avoid the maintenance and unreliability problems associated with conventional electromechanical moving maps. The information can be presented either as a conventional plan view, shaded to

represent relief, or as the view seen by the pilot flying a low-altitude mission. Tactical and navigation symbols can be added as required.

With currently available technology, AETMS can store enough data on commercial disks to produce detailed maps covering up to 250 000 square miles (648 000 km²). The system operates in conjunction with an AN/AYK-15A computer. Terrain elevation and slope gradients can be updated and displayed at a rate of 1.2 million samples a second, with 256 × 256 pixel resolution. Hughes is studying other storage devices, including bubble- and solid-state memories.

Hughes, in competition with Westinghouse

and Harris, has been working on a development of the system, called ITARS (integrated terrain access and retrieval system). Phase 3 of this programme calls for manufacture of three flight-trials units.

STATUS: project succeeded by ITARS (see following entry).

ITARS integrated terrain access and retrieval system

ITARS is the follow-on programme from AETMS, (see previous entry), Hughes being selected to continue the development, under a $3.2 million, four-year US Air Force Aeronautical Systems Division contract awarded in October 1985. Texas Instruments had been the other developer of AETMS.

ITARS will display colour-coded surface features and man-made structures. Pilots can elect to see the information in look-down or look-ahead views or on the head-up display.

ITARS automatically will share its stored data with other systems aboard the aircraft to aid in navigation, terrain-following and terrain-avoidance, weapon delivery, sensor blending, mission planning and threat avoidance. It eliminates the need for a pilot to input that data manually.

The Hughes modular design makes ITARS flexible enough for a variety of applications, ranging from helicopters and small tactical fighters to strategic and reconnaissance aircraft, depending upon the mission, the aircraft, and the branch of service.

ITARS will enable military pilots to 'pre-fly' a mission in a simulator to acquaint them with the terrain and location of probable threats, and to permit them to plan their routes to limit exposure to detection and enemy fire.

The system's modular design permits it to be

Synthetic image of Mt St Helens, Washington. The different tones represent vegetation, snow, earth and rock

tailored to the needs of the user, with the initial version supporting up to 10 000 square miles (25 900 km²) of terrain data, and the system can be expanded to include terrain data covering 250 000 square miles (647 500 km²). Digital map information being incorporated in the

system by Hughes will come from a database produced by the Defense Mapping Agency.

STATUS: in development, first of two productionised ITARS will be delivered for flight development at the end of 1986.

King

King Radio Corporation, 400 North Rogers Road, Olathe, Kansas 66062
TELEPHONE: (913) 782 0400
TELEX: 42299

Silver Crown KNS 80 integrated navigation system

The KNS 80 is a single 6.75-inch (172 mm) wide by 3-inch (76 mm) high panel-mounted unit. It comprises a 200-channel VOR/LOC receiver, 200-channel digital DME, 40-channel glideslope receiver, and digital RNav computer that can store up to four VOR/LOC frequencies and waypoints. Information is displayed on a full-width light-emitting diode display, and the system can operate in conjunction with an ARINC horizontal situation indicator or course deviation indicator. Extensive use is made of large-scale integrated (lsi) technology, the single lsi chip accomplishing the work that would have required 40 conventional integrated circuits. As with other members of the Silver Crown family, the KNS 80 is aimed at the lower end of the general aviation sector.

Dimensions: 160 × 76 × 305 mm
Weight: 2.7 kg
Power: 25 W at 11-33 V dc

RNav section
Distance to next waypoint: 199.9 n miles in 0.1 n mile increments, 0.1° angle increments selectable on display

STATUS: in production.

Silver Crown KNS 81 integrated navigation system

A development of the KNS 80, the KNS 81 integrated navigation system can accommodate up to nine waypoints and embodies a number of new features: a remote-mounted DME enabling a KDI 572 indicator to be positioned directly in front of the pilot, providing simultaneous digital readouts of distance, ground-speed and time to either a Vortac station or an RNav waypoint; simultaneous display of all waypoint details (bearing, distance and frequency) on the KNS 81 panel for easy programming and updating of navigation information; a 'radial' push-button permits a rapid bearing check to a chosen Vortac or RNav waypoint, displayed on the DME panel indicator in place of groundspeed and time to next waypoint; a 'check' push-button permits a rapid cross-check of bearing and distance from the Vortac without disturbing other navigation instrument settings; a radio magnetic indicator output to a KI 229 or KNI 582 indicator gives an accurate bearing to a selected RNav waypoint or Vortac station.

Dimensions: 160.3 × 5.1 × 291.2 mm
Weight: 2 kg
Power: 15 W at 11-33 V dc

RNav section
Number of waypoints: 9
Distance to next waypoint: 199.9 n miles in 0.1 n miles increments; 0.1° angle increments

STATUS: in production.

Gold Crown KNR 665 integrated area navigation system

A member of the Gold Crown family of avionics for the upper end of the general aviation spectrum (the business and corporate turbo-props and jets), the KNR 665 can operate in conjunction with the KFC 300 flight director and autopilot system.

The KNR 665 can memorise and display up to 10 waypoints and/or Vortac stations. Waypoint information in Jeppesen format includes Vortac frequency, two courses (outbound and reciprocal) and waypoint bearing and distance. All waypoint information is displayed simultaneously for ease of reference. Individual parameters for any waypoint can be changed without resetting other data; the information is inserted by push-button and checked on a 'scratch-pad' before entering it into the system. An autocourse facility provides Air Traffic Control clearance to a particular waypoint by pressing the 'autocourse' button, whereby the route to that particular waypoint is instantly computed and displayed. The new course is automatically displayed on the horizontal situation indicator. The system can be checked by a push-button.

STATUS: in production.

KTU 709 Tacan

Particularly suited to corporate aircraft, where weight, cost and power requirements are

Presentation of King KNS 80 integrated area navigation system

King KNS 81 area navigation system in RNav mode

notably important, the KTU 709 Tacan transmitter-receiver is based on King's extensive DME experience, and the extensive use of large-scale integration (lsi) technology; the system specifically is based on the new-generation KDM 706 distance-measuring equipment. This Tacan provides bearing, slant range, range rate and time-to-station or waypoint information to a KDI 572 control/indicator unit. Transistors provide a 250-watt peak-to-peak output for a typical range of 250 nautical miles and the system covers 252 channels; all tuning is done electronically, using a digital frequency synthesiser designed around a King-developed lsi chip.

Dimensions: 3 × 5 × 10.25 inches (76.2 × 127.0 × 260.4 mm)
Weight: 5.8 lb (2.6 kg)
Reliability: 2000 h design
Number of channels: 250
Frequency: (transmit) 1025-1150 MHz (receive) 962-1213 MHz

STATUS: in production.

KNS 660/650 integrated navigation system

The KNS 660, and its slightly less comprehensive KNS 650 version, are total navigation management computer systems representing a major development effort by King. Aimed at the turbine-powered corporate and commuter sector of general aviation, KNS 660/650 is the designation for a family of great-circle navigation management systems to work in conjunction with King's Gold Crown III avionics and other compatible units. The system was first shown at the Paris Air Show in June 1983.

The KNS 660 has its own database, ARINC 429 input/output formats, analogue processing, air data function, vertical navigation computation, frequency management and optional omega/vlf sensor. The KNS 650 produces basic steering commands from Vortac range and bearing signals and contains its own navigation and glideslope receivers. The KNS 660 has a choice of two control/display units, the 6-inch (152 mm) KCU 568, and the 4.5-inch (114 mm) KCU 567, for installations with limited space.

The KNS 650 is designed to use the smaller unit. The KNS 660 will handle signals from VOR/DME, omega/vlf, Loran C, INS, GPS (the satellite-based global positioning system), AHRS (attitude heading and reference system) and compass. King's KTU 709 Tacan system can replace the DME if required.

There is now available an optional receiver/processor providing a choice between omega/vlf and Loran C. A potential growth area is MLS.

STATUS: production began in July 1984. First customer was Swedair, whose 10 Saab-Fairchild 340s will have KNS 660. The reason for this choice was given as the system's ability to show distance-to-airport with internal omega/vlf sensor, and to provide ILS guidance from Gold Crown III vhf/nav receiver used on this aircraft.

King is also developing a global positioning system (Navstar) receiver as part of the KNS-660 family. Flight trials began in early 1986.

Lear Siegler

Lear Siegler Instrument Division, 4141 Eastern Avenue SE, Grand Rapids, Michigan 4960P
TELEPHONE: (616) 241 7000
TELEX: 226430
TWX: 810 273 6929

Nav/attack update for RNZAF A-4s

In late 1985 Lear Siegler signed a $62 million contract to update the McDonnell Douglas A-4 and TA-4 aircraft of the Royal New Zealand Air Force with advanced avionics including a

MIL-STD-1553B digital data-bus, cockpit management system, inertial navigation system, head-up display, weapons control and display and radar.

Litton

Litton Aero Products, 6101 Condor Drive, Moorpark, California 93021-2699
TELEPHONE: (805) 378 2000
TELEX: 662619
TWX: 910 494 2780

LTN-72 inertial navigation system

The LTN-72, introduced in 1972, is a self-contained, all-weather, worldwide navigation system for commercial aircraft that is independent of any ground-based aids. Designed to incorporate area navigation facilities, the INS provides continuous position, navigation, and guidance data.

The system comprises three units: mode selector (msu), control/display (cdu), and inertial navigation box (INU). The first is used to energise and align the system prior to flight and to select navigation or attitude reference modes of operation. The cdu permits the crew to enter present position and waypoint coordinates, select track steering, and display information generated by the system. The INU houses the

gimbal structure with its gyros and accelerometers, associated electronics, power supply, and data converter. Ease of maintenance has been emphasised: for example, the principal mechanical elements (gyros and accelerometers) can be removed and replaced in 20 minutes using only screwdrivers. The gimbals are cantilevered, permitting the servo electronics to be mounted directly on the platform. This permits the use of flexible leads instead of slip-rings in some cases, improving reliability. The platform has only two slip-rings compared with four on a conventional platform. The LTN-72 has very extensive self-test and failure-detection facilities; the system complies with ARINC 561 in that the probability of an undetected failure in attitude during the last 30 seconds before touchdown is less than 1 in 10^6. Again, an analogue output test feature permits tests not only of the INS but also of the flight instruments by driving them to various test readings.

Inertial navigation unit
Dimensions: 10.2 × 8.625 × 19 inches (267 × 219 × 507 mm)
Weight: 59 lb (26.8 kg)

Control/display unit
Dimensions: 5.75 × 4.3 × 6.2 inches (146 × 114 × 157 mm)
Weight: 5 lb (2.3 kg)

Mode selector unit
Dimensions: 5.75 × 1.5 × 2 inches (146 × 38 × 51 mm)
Weight: 1 lb (0.5 kg)

Number of waypoints: 9 plus remote entry capability
Waypoint offset capability: up to 399 n miles worldwide
Display: 7-segment incandescent numerals for all ARINC terms, together with display of INS parameters on flight director, horizontal situation, or remote indicators
Inputs: self-contained inertial guidance. Avionics interface is designed to ARINC 561 and 575, and compatible with all flight directors and autopilots
Outputs: actual track, track angle error, cross-track and desired track, plus access to computer during flight for great-circle distance computations
Power: 115 V 400 Hz

STATUS: in production. The Anglo-French Concorde is one application of LTN-72.

LTN-72R inertial/area navigation system

The LTN-72R inertial navigation system is a development of the LTN-72 with automatic radio position update, automatic omega position update, and triple-system mixing capability. It may be operated in the area navigation mode, using range and bearing information from selected Vortac (VOR/DME) stations or from a combination of omega transmitters, providing very high accuracy independent of time.

The system uses newly developed gyros, platform, and accelerometers, and a new, expansible C-4000 digital computer.

Number of waypoints: 9, plus remote entry capability
Waypoint offset capability: up to 400 n miles world wide
Display: as for LTN-72

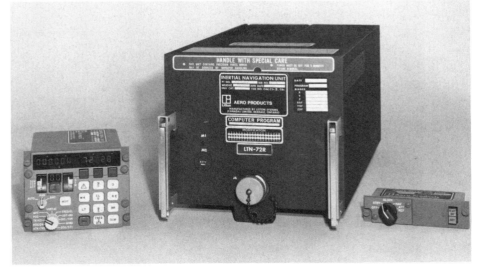

Litton LTN-72 control and display unit, inertial navigation unit and master control box

Litton LTN-72RL control/display unit with multi-line light-emitting diode symbols

Inputs: as for LTN-72 plus data on up to 9 VOR and DME stations
Outputs: the system is certificated for en route, terminal, and approach categories. It provides crosstrack, ground speed, position (both updated and 'raw'), waypoints, Vortac position, frequency, and elevation, update code, magnetic variation, distance and time-to-go, wind speed and direction, desired track, status, self-test, and malfunction codes
Memory: uv-erom (ultra-violet erasable read-only memory) for optimum protection with flexibility for change
Power: 115 V, 400 Hz
Weight: 61 lb (27.7 kg)

STATUS: in production.

LTN-72RL inertial/area navigation system

The LTN-72RL is an advanced, worldwide inertial navigation system that can automatically update itself by radio navigation fixes. It has a control/display unit that functions as an 'intelligent' data terminal and incorporates a 5-line by 16-character light-emitting diode display for presentation of operator-entered or computer-processed data. Waypoint data and VOR/DME locations can be pre-stored in the computer, and the system contains an algorithm (mathematical model) of magnetic variation that can be used to compute magnetic heading, track, and desired track, independently of the aircraft compass system; this algorithm is limited to latitudes between 60°N and 60°S. The pre-stored database contains information specified by the operator on selected vhf navaids, airports, and some high-altitude waypoints. This bulk data is programmed in read-only memory.

A section of electrically alterable memory is allocated for particular waypoints or fixes not contained in the standard databases; up to 160 routes, with an average of 20 waypoints per route, can be stored in this way and recalled for use at any time. The total number of waypoints in all routes is limited to 3200.

STATUS: in production. Certificated in July 1981 for Saudia Airlines Boeing 747s.

LTN-90-100 ring-laser gyro inertial reference system

The systems in this series, designated from LTN-90 to LTN-100, are the latest in Litton's navigators, being based on the RLG technology. The LTN-90 RLG inertial reference system was specified by Airbus Industrie for the A300-600 and A310 airliners. The first Litton system began flight trials in May 1979 aboard an A300 and, in November 1982, became the first laser-based gyro navigation system to receive air-worthiness approval (FAA TSO). The ARINC 704 strapdown system is now cleared for unrestricted, worldwide use as a sole means of navigation and attitude measurement under the FAA's FAR 121 Part G and FAR 25 regulations. The A310 installation comprises three parallel

Litton LTN-90-100 inertial reference system

Airbus A310 has triple-redundant Litton LTN-90 installation

LTN-90 systems and is certificated as the primary attitude source and as the sensor for position and velocity information required for navigation (the actual navigation computation is conducted in the aircraft's flight management system). The system is additionally certificated to provide aircraft rotation rates and accelerations to the flight control system.

With some 4000 hours accumulated by the nine systems in three A310 development aircraft up to the time of certification, the LTN-90 was showing average position error rates of less than 1 nautical mile an hour and average ground speed errors at the end of a flight of less than 5 knots; reliability was also being acclaimed by Airbus Industrie.

Litton is now supplying the LTN-90-100 with advanced electronics and an improved RLG system for digital-computer-based applications and the LTN-91 inertial reference system with

additional outputs. Although most production has been in support of the airline industry, Litton is considering the top end of the general aviation market, where long-range corporate aircraft such as the Dassault Falcon 50 and Gulfstream III now demand a navigation capability up to airline standard.

The LTN-90 comprises an inertial reference unit, a mode selector unit, and an inertial sensor display unit. Heart of the system, and contained within the inertial reference unit, are the RLGs that measure rotation accelerations and rates about the three aircraft axes, and the three single-axis accelerometers that measure accelerations and rates along the aircraft axes. The RLG system and accelerometers are mounted at right-angles to each other and are rigidly secured to the case. Unlike the Honeywell RLGs, which are based on a triangular light path, the Litton LG-8028 units designed specially for the LTN-90 use a square-path configuration, with a 28 cm path-length. The reason for this, says Litton, is that for the same scale factor a square is smaller than a triangular gyro, resulting in a more compact overall sensor assembly. The square gyro is said to produce less backscatter because of the improved angle of incidence (45° for square versus 30° for triangle), which makes for lower random noise.

The substantially greater data processing power required by strapdown systems over conventional gyro-based inertial navigation equipment is provided, in the case of the

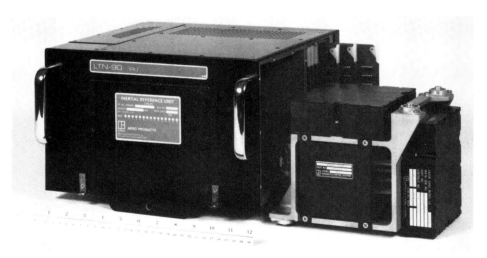

Inertial reference unit of Litton LTN-90

LTN-90, by three Zilog 8000 microprocessors, each of which is assigned a particular task. This computing power is also employed to correct temperature variations. Traditionally, gyroscopes are susceptible to changes in temperature, causing inaccuracies, and the normal approach to dealing with this effect is to maintain a constant temperature by trickle heating. In the LTN-90 the effect is compensated by mathematically modelling the way in which the characteristics of the system and accelerometers vary with changing temperature, and then by applying a correction to their output signals.

Easy maintenance is a key advantage claimed by Litton. It is achieved by functional partitioning, so that each circuit board has all the components needed to support a major activity, more effectively isolating faults and simplifying trouble-shooting. This partitioning is achieved by reducing the number of electronic plug-in modules from between 15 and 20 to 7. Maintenance is further improved by the provision of special connectors along the edge of each circuit board, permitting the system to be tested in situ.

Litton describes the LTN-90 as a sensor rather than (in the case of navigation) as a complete system, and this is the way in which it is used in the A310. However, the company offers this inertial reference system as a complete system for installations in which further integration (by means of a flight management system) is not desired. With inputs only from an air data computer, the LTN-90 should provide accurate outlets of attitude, heading, present position, drift angle, ground track, flight-path angle, ground speed, vertical velocity and wind speed, as aircraft angular rates and linear accelerations. Alternative control/date and mode selector unit configurations can be supplied to optimise specific aircraft requirements.

Inertial reference unit
Weight: 44 lb (19.9 kg)
Dimensions: 7.64 × 12.69 × 12.52 inches (194 × 322 × 318 mm)
Power: 110 W
Cooling: ARINC 600
Mtbr: 2500 h

Mode selector unit
Weight: 1 lb (0.45 kg)
Dimensions: 1.5 × 5.75 × 2 inches (38 × 146 × 51 mm)
Power: negligible
Cooling: none
Mtbr: 50 000 h

Inertial sensor display unit
Weight: 5 lb (2.27 kg)
Dimensions: 4.5 × 5.75 × 6 inches (114 × 146 × 152 mm)
Power: 15 W
Cooling: none
Mtbr: 15 000 h

Performance: 95%
Heading: 0.4°
Pitch: 0.1°
Roll: 0.1°
Position: 2 n miles/h
Ground speed: 8 knots
Flight-path angle: 0.4°
Body rates: 0.1%

Third-generation Litton INS, AN/ASN-109, for US Air Force/McDonnell Douglas F-15 fighter

Body accelerations: 0.01 g
Reaction time: 10 minutes

STATUS: in production for Airbus A310 and A300-600 and US Navy/Boeing E-6A Tacamo. By mid-June 1983 Litton had orders and options for 700 sets. Only one A310 customer had chosen non-Litton system.

LTN-96 ring-laser gyro inertial reference system for C-23A

A version of the LTN-90 designated LTN-96 was chosen by Shorts for the C-23A Sherpa (a variant of the company's 330 light transport), selected in March 1984 by the US Air Force as its EDSA European distribution system aircraft. The LTN-96 will use Tacan for position updating and can store up to 99 waypoints.

LTN-92 inertial navigation system

The LTN-92 provides proven state-of-the-art technology in a standard ARINC 561 1 ATR box. The inertial sensors and much of the electronics are the same as found in the LTN-90-100 inertial reference unit which is currently flying on the Airbus A310, Gulfstream III, Canadair Challenger and other high technology aircraft. The LTN-92 has already been chosen for installation on the Cathay Pacific Boeing 747-300s and -200s as well as retrofits on their Lockheed L-1011 aircraft.

The LTN-92 INS is comprised of three separate units: the inertial navigation unit (INU), the control/display unit (cdu) and the mode select unit (msu). Three 28 cm ring-laser gyros and a triad of force-rebalanced accelerometers comprise the instrument cluster of the INU. The instrument electronics digitises the instrument control signals and the instrument outputs for easy microprocessor interface. There are three microprocessors used to process the instrument data; perform input and output interface functions and perform the navigation calculations. Because no heat is required and the latest technology parts are used, power consumption is reduced to a maximum of 175 watts total operating power.

The cdu has a 5-line light-emitting diode dot matrix display and keyboard providing the INU interface with the crew. The msu controls the operational mode of the INS.

The LTN-92 is pin for pin compatible with existing ARINC 561 INS systems and ARINC 571 ISS systems. In addition there are three 429 high-speed digital output buses, nine 575/429 low speed input buses, 12 programmable synchro outputs, four synchro inputs, four analogue dc or ac two-wire outputs, DME pulse-pair input and 2 × 5 radio tuning line interfaces. This allows the LTN-92 to interface with existing avionics equipment installations as well as being a part of new installations.

Other capabilities are programmed into the LTN-92 to make it a versatile navigation unit. The system will accept RNav, Tacan, Omega, triple INS mixing or manual position updates for improved performance. With an external database or using the internal 16 K × 16 eerom data storage the crew may programme a flight plan and the INS will steer the aircraft along the entered flight plan. An interface to a weather radar system is provided for modification of the flight plan during flight. Intersystem communication is utilised to check system performance and sending flight plan and initialisation data between systems. An extensive software built-in test monitor program of more than 150 performance checks is continually run to assure the validity of the computed data.

The LTN-92 has been designed for future growth capability. Space has been provided for incorporation of an air data computer, GPS or omega dual-system capability.

LN-39 inertial navigation unit

Litton's LN-39 INU is the sensing and data processing device that has been chosen by the US Air Force as the basis of the standard inertial navigation system on the Fairchild A-10. The contract, including options, calls for production of 1100 sets. The unit contains a P-1000 platform, stabilised by two G-1200 gyros and mounting three A-1000 accelerometers, and an LC-4516C general-purpose computer. The P1000 platform is an advanced version of the system employed in the Northrop F-5, McDonnell Douglas F-15 and F/A-18 fighters, and in the cruise missile programme. An optional control/display unit controls the system and selects and displays data and status/failure information. Particular emphasis has been placed on reliability, and this is accomplished partly by the use of large/medium-scale integrated circuits and hybrid components, allowing a significant parts-count reduction.

In February 1983 the LN-39 was chosen as the INU for the US Air Force/General Dynamics F-16 under a $16 million plus contract awarded to the Litton Industries Guidance and Control Systems Division. The contract calls for production of 89 sets, together with options for an

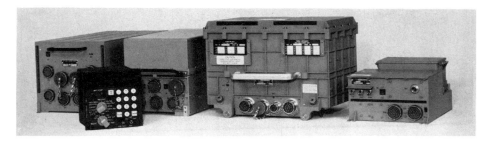

Litton AN/ASN-92 second-generation INS for US Navy Grumman F-14

Litton LTN-92 inertial navigation system

additional 700 systems and 2000 systems for other aircraft applications. This contract, part of the MSIP (multi-stage improvement programme) for the General Dynamics F-16, displaces Singer Kearfott (the inertial navigation system supplier for the F-16A and -B) from the F-16C/D.

Inertial navigation unit
Dimensions: 7.5 × 7.6 × 18.1 inches (191 × 193 × 460 mm)
Weight: 38.2 lb (17.36 kg)

Control/display unit
Dimensions: 5.75 × 6 × 7.3 inches (146 × 152 × 185 mm)
Weight: 8.2 lb (3.73 kg)

Power: (running) 180 W
(starting) 550 W
Accuracy: (gyro compass) 0.8 n mile/h cep, 2.5 ft/s
(stored heading) 3 n miles/h cep, 3 ft/s
Align time: (gyro compass) 8 minutes at 21°C
(stored heading) 1.5 minutes at 21°C
LC-4516C computer: 16-bit single or 32-bit double precision, 65 536 words, semi-conductor ram/rom or eprom 24 K words total
Mtbf: Specified goal, 740 h over full military environment

STATUS: in production for US Air Force's Fairchild A-10 and General Dynamics F-16 and FB-111 aircraft. A total of 2267 units are on order with 1340 delivered so far. Anticipated additional orders should continue LN-39 production for F-16 until 1992.

Litton Industries, Guidance and Control Systems, 5500 Canoga Avenue, Woodland Hills, California 91365
TELEPHONE: (213) 715 2020
TELEX: 698483
TWX: 910 494 2091

LN-92 inertial navigation system
The LN-92 is Litton's version of the US Navy AN/ASN-139 (CAINS II) ring-laser gyro INS. It is designed for operation in high-performance carrier-based-aircraft environment. The accuracy specifications are better than 1 nautical mile an hour cep and 3 feet per second velocity

with a 4-minute reaction time. The unit is a form, fit and function alternative to the Litton AN/ASN-130, a tuned-rotor gyro INS.

The gyros used in the LN-92 are the Litton LG-9028 28 cm ring-laser gyro used in conjunction with the A-4 accelerometer triad. This system will incorporate MIL-STD-1750 processors for navigation and signal data processing. The reliability of this unit will permit the Navy to employ a two level maintenance concept.

LN-93 US Air Force standard ring-laser gyro INU
The LN-93 is the Litton US Air Force Standard ring laser gyro INU. This unit completed all formal qualification testing at CIGTF. A production contract was awarded in August 1985. Units will be delivered for the Lockheed C-130 and McDonnell Douglas RF-4C aircraft in 1986,

and a system variant will be delivered for the McDonnell Douglas F-15 later. The LN-93 is a function alternative to the Litton LN-93 Standard INU currently onboard Fairchild A-10, General Dynamics F-16C/D and other US Air Force and Army aircraft.

LN-95 inertial measurement unit for AINS
The Litton Alternate Inertial Navigation System (AINS) is a pure inertial hybrid strapdown system using three 40 cm ring-laser gyros. The LN-95 is a simple, compact, easy to assemble inertial measurement unit capable of operating under any small ICBM scenario.

AN/ASN-130A inertial navigation system for AV-8B
At the 1983 Paris Air Show Litton announced that its AN/ASN-130A inertial navigation system had been chosen for the US Marine Corps' AV-8B Harrier, representing a continuation of the procurement process for this system begun earlier with the US Navy/McDonnell Douglas F/A-18 fighter and with the US Marine Corps/ Grumman EA-6B ECM aircraft. The choice of a conventional 'wheel and gimbal' system is significant since it had been in competition over the past year aboard development AV-8Bs against the AN/ASN-140 laser-ring gyro system by Honeywell.

Litton system AN/ASN-130 for US Navy F/A-18 and McDonnell Douglas AV-8B fighters

Litton LN-95 inertial measuring unit

During the airborne trials the Litton system demonstrated both the 30-second AHRS mode as well as the four-minute gyrocompass mode in more than 500 flights. Reliability was considered excellent (the predicted mean time between failures is 1950 hours) while navigation accuracy remained at less than 1 nautical mile an hour. Designed as a high-performance, rapid-reaction system for US carrier-based aircraft, the AN/ASN-130A is said to embody the highest reliability of any Litton inertial navigation equipment. This is accomplished by a large reduction in the parts count (the number of components used), by the adoption of hybrid circuits, advanced gyro and accelerometer design, and by a simpler and less critical platform configuration.

The AN/ASA-130A is a complete system contained within a single weapons replaceable assembly, the principal elements being a P-1000C platform, G-1200 gyros, A-1000 accelerometers, and LC-4516C digital computer. The latter is a 16-bit general-purpose device with 30 K memory operating at 238 000 operations a second. The current AV-8B avionics suite uses about 70 per cent of this memory, leaving a large margin for new functions such as forward-looking infra-red, side-ways looking radar, low-light television and special flight paths may be computer-controlled for best area coverage in these reconnaissance missions.

Weight: 35 lb (15.9 kg)
Volume: 0.7 ft^3 (0.02 m^3)
Power: (start) 800 W
(run) 185 W, 115 V 3-phase 400 Hz
Accuracy: 1 n mile/h cep
Reliability: 1950 h predicted mtbf

STATUS: in production. Litton has more than 700 of these systems under contract.

LTN-211 omega/vlf

In 1979 the FAA certificated the Litton LTN-211 omega system, in dual installation form, as meeting all requirements for sole navigation aids over the North Atlantic on Olympic Airways' Boeing 707s, Icelandair's Boeing 727s, and the McDonnell Douglas DC-8s of Airlift International and Intercontinental Airways. Previously omega had been authorised only in conjunction with other navigation systems, for example, Doppler.

The LTN-211 airborne equipment in conjunction with the omega worldwide network of very-low-frequency stations, or the US Navy's chain of vlf transmitters, provides a limited-error display of guidance information for long-range, great-circle navigation. It is programmed to adopt under certain circumstances a back-up, dead-reckoning mode of operation based on available aircraft velocity and heading. Switchover to the dead-reckoning mode is accomplished automatically when the number and quality of signals received falls below the requirement needed for acceptable position tracking and navigation.

When the initial position, time and date have been entered, the system automatically chooses the stations to be used on the basis of measured signal/noise ratios of the transmissions being received at the time. This method yields the best position accuracy since it considers both the quality of the stations selected as well as their propagation stability, and uses those stations least likely to be affected by diurnal signal changes. The basic LTN-211 uses signals of the three standard omega frequencies, 10.2, 11.33, and 13.6 kHz. Capability to process the fourth omega frequency, 11.05 kHz, is an available option. In the omega/vlf system, this fourth frequency is used in conjunction with a vlf converter to process signals from the US Navy transmitters.

The system comprises three units: receiver/processor, control/display, and antenna coupler.

Receiver/processor
Dimensions: 7.5 × 7.6 × 19.6 inches (191 × 194 × 498 mm)
Weight: 26 lb (11.8 kg)
Power: 62 W

Control/display unit
Dimensions: 5.75 × 4.5 × 6.2 inches (146 × 114 × 158 mm)
Weight: 3.8 lb (1.72 kg)
Power: 24 W

Antenna
Dimensions: 18 × 10.5 × 1.75 inches (457 × 105 × 45 mm)
Weight: 8 lb (3.63 kg)
Power: 1 W
Number of waypoints: 9
Inputs: signals from all 8 omega stations and the US Navy vlf chain in ARINC 599 format including true airspeed input for computation of wind speed and direction
Computer: TMS 9900 second-generation microprocessor, with 16 bits a word, and directly addressing 24 576 words of ultra-violet read-only memory reprogrammable memory
Outputs: ground speed, track, heading, drift angle, cross-track, track-angle error, present position in latitude/longitude, waypoints, distance and time-to-go, wind, desired track, station status and signal quality, GMT, date, self-test, and malfunction codes. All outputs are to ARINC standard. Digital outputs, together with autopilot command steering signals and synchro outputs to the horizontal situation indicator can be provided
Failure detection: C9000 computer/processor has extensive BIT facilities within itself, as well as controlling the BIT section of the LTN-211 system. Probability of BIT system detecting a fault anywhere in the navigation system is more than 99%

STATUS: in production. LTN-211 equipment has been installed aboard Boeing 707, 727, 737, 747, McDonnell Douglas DC-8, DC-10, Gulfstream II, III, Lockheed C-130, P-3, L-1011, Jetstar, Dassault Falcon, British Aerospace HS-125, Cessna Citation and Fokker F 27.

LTN-700 GPS navigation set

The LTN-700 GPS navigation set is claimed to be the first GPS navigation set specially designed for commercial aviation applications. Design specifications for the LTN-700 are based on ARINC and RTCA documents.

The LTN-700 implements a single-channel fast-sequencing L1 C/A-code architecture. For medium dynamic applications (<4 g), the single-channel fast-sequencing configuration provides the performance of a five-channel continuous tracking configuration at the cost of a single-channel sequential tracking configuration. The LTN-700 design philosophy is to minimise rf signal processing in favour of digital signal processing and to minimise hardware digital signal processing in favour of software digital signal processing resulting in a significant reduction of costly rf components while improving performance, and results in improved reliability.

There are two basic phases of air navigation: approach/landing and en route/terminal.

The 100-metre 2 dRMS accuracy of the SPS (standard positioning service) will allow the LTN-700 to meet the current and projected future requirements for all en route/terminal and non-precision approach operations.

For medium dynamic applications (<4 g), the LTN-700 provides similar performance to a five-channel set:
- 40 metres cep position accuracy
- 0.1 metre per second cep velocity accuracy
- 100 nanosecond 1-sigma time accuracy
- 270 second 95 per cent reaction time
- continuous data demodulation from four satellites simultaneously
- processing of four measurement pairs (pseudorange and deltarange) each second.

The LTN-700 is designed to utilise the SPS C/A code signal to provide three-dimensional position and velocity, and time outputs with the following accuracies:
Position: 40 metres cep (100 metres 2 dRMS)
Velocity: 0.1 metre per second (0.2 knot) cep
Time: 100 nanoseconds 1-sigma.

The dynamic capability of the set is:
Velocity: 1200 knots
Acceleration: 4 g
The set will provide the above accuracies under these dynamic conditions. The interference immunity of the set is 24 dB I/S (interference-to-signal power ratio). Thus, the set will continue to provide the above accuracies when subjected to a narrow-band interfering signal 24 dB above the received GPS signal level.

The designed reaction time is 270 seconds 95 per cent (cold power on to output of first fix).

Litton LTN-700 GPS navigation computer

The set utilises battery backed-up low power cmos memory for storage of GPS almanac data, waypoints, present position and other critical data. A battery powered real-time clock (RTC) is used to maintain time. Thus, the operator is only required to enter initialisation data when the set is first installed.

The set is packaged in an ARINC 600 6 MCU case (approximately ¾ ATR Short). This is the same form factor as the Department of Defense Phase II GPS user equipment medium dynamic set intended for installation in transport/tanker/reconnaissance aircraft and helicopters.

The environmental specifications are based on Radio Technical Commission for Aeronautics (RTCA) Environmental Conditions and Test Procedures for Airborne Equipment (DO-160A). The decompression limit is equivalent to an altitude of 40 000 feet (12 000 metres) MSL. The overpressure limit is equivalent to an altitude of -15 000 feet (-4600 metres) MSL. The set is designed to operate in the vibration environment typical of rack-mounted, non-vibration isolated equipment installed in fixed-wing-turbojet engine aircraft. This is DO-160A classes B and 0.

Dimensions: 7.5 × 12.76 × 7.64 inches (191 × 324 × 194 mm)
Weight: 22 lb (10 kg)
Temperature limits: (low operating) -15°C (high operating) +70°C (low ground survival) -55°C (high ground survival) +85°C
Power: 155 V ac 400 Hz

Complete Litton LTN-311 omega/vlf system: from left, control/display unit, receiver and aerial

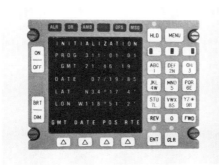

Litton LTN-311 control/display unit

Litton LTN-3100 control/display unit

LTN-311 omega/vlf/GPS

The LTN-311 is the first omega navigation system designed to incorporate GPS, and it may be a pin-for-pin replacement unit for aging LTN-211 omega systems. It provides over 300 waypoints and adapts to special user software.

The compact control/display unit provides an active display screen formatted into 8 lines of 16 led characters, 15 inches (381 mm) high. The display incorporates an alphanumeric keyboard, four soft keys, and a full menu-driven scenario. The top line of the screen displays the page function and the bottom line the soft key legends. Red leds may be substituted for marine application. The enhanced colour and user-friendly, menu-driven scenario combine with the other features of the cdu to lessen operator workload and the possibility of operator error. The cdu is self-leading needing no pilot's guide to operate the system.

The LTN-311 RPU is an advanced development based on Litton's extensive design experience. The power supply is an ac/dc non-interruptible power supply module; the computer and memory are combined on a single card which fulfils both the memory and processing roles. The RPU has a spare card slot, which can incorporate the Litton designed, single-card Micro Nav GPS. The Micro Nav provides increased navigation capability as well as dual-system operation at minimal cost. The RPU can communicate with ACARS via an ARINC 429 port. Alternatively, this port could be used for access to a data bank or to a recording device.

A significant feature of the LTN-311 is the incorporation of extensive BITE. The alphanumeric led cdu allows plain language reporting of all detected malfunctions along with recommended maintenance codes. All multi-box radio based products of the past have had major non-verified removal problems. The LTN-311 speaks to the pilot and maintenance engineer in English language to break through this barrier. This improved BITE system is designed to maximise the mtbr/mtbf ratio, minimise spares requirements, increase line maintenance efficiency, and reduce training requirements.

LTN-3100 omega/vlf

Combining an existing omega/vlf receiver with new software and panel-mounted flight-deck display, the LTN-3100 is aimed at the general aviation sector, where low cost is particularly important. Based on sensor technology used on Litton's LTN-211 omega/vlf system, the LTN-3100 received an FAA supplemental type certificate in early 1983 following a development programme which began in July 1979. The system comprises an antenna, receiver/processor, and control/display unit. It uses VOR/DME or DME/DME as the primary en route navigation facility, and the system gives them priority when signals are at acceptable strength. Elsewhere it automatically selects the best omega/vlf signals. An optional vertical navigation system can be used to drive compass director and horizontal situation indicators, and requires a servoed altimeter input.

The system provides navigation and status information on a two-line 16-character liquid-crystal display. It can store up to 99 waypoints in a non-volatile memory, making use of an earom chip. These waypoints can also be stored as routes, a maximum of 16 such routes containing up to nine waypoints being available from the memory. In all respects the LTN-3100 is a full-capability, long-range area navigation management computer meeting all current requirements for area navigation en route and terminal accuracy.

STATUS: in production.

Micro Nav GPS

Litton Aero Products has design drawings for a single-card GPS receiver that may optionally be integrated with its own INS, ONS and AHRS navigation systems that will provide enhanced, accurate updating and dual system capabilities. This miniaturised receiver card will be integrated with technology components to produce a stand-alone Micro Nav navigation system small enough to fit in the console of most aircraft.

STATUS: in development. Production deliveries start late 1986.

LTV

LTV Aerospace and Defense Company, Sierra Research Division, 247 Cayuga Road, PO Box 222, Buffalo, New York 14225
TELEPHONE: (716) 631 6200
TELEX: 6854308
TWX: 710 523 1864

Station keeping system for C-130H

LTV is supplying Lockheed with a further 16 station keeping navigation systems which facilitate close formation flying in stores and troop dropping exercises.

STATUS: in production.

Magnavox

Magnavox, Government and Industrial Electronics Company, 2829 Maricopa Street, Torrance, California 90503
TELEPHONE: (213) 618 1200
TELEX: 696101

Navstar GPS receiver

For seven years up until April 1985, when the US Department of Defense announced its choice of Collins, that company and Magnavox been in competition to build the air, sea and land-receiving equipment for the Navstar GPS satellite navigation system.

Subsequently, Magnavox has begun a co-operative development, called Euronav (see following entry), with TRT and LMT of France, SEL of West Germany, Elmer of Italy and others. Euronav is a fully military, multi-purpose, multi-channel GPS receiver/processor unit designed primarily for use by European NATO forces. It will be considerably smaller than the planned US equipment and is being designed to the same performance specifications.

Magnavox GPS receiver

Euronav GPS system

Magnavox announced the new Euronav GPS system in early 1986 and claims that it represents a new generation, as well as a new concept, in high-precision military GPS receivers. The Euronav receiver incorporates numerous technical advances over other military GPS receivers currently available. Due to its small size and weight, reduced power consumption, and advanced performance characteristics, Euronav systems will be able to fill many operational navigational needs which cannot be satisfied by earlier systems. The Euronav equipment is designed with only four plug-in printed circuit boards, a substantial simplification compared to other military quality GPS receivers. Euronav meets SS-US-200 and NATO specifications.

Each partner in the Euronav project has the option to manufacture the jointly designed hardware locally or to purchase equipment from Magnavox. The common design enables each member of the consortium to offer the most cost effective GPS product tailored to their respective customers' needs, whether for air, ship or ground application.

STATUS: initial deliveries are expected in early 1987.

Narco

Narco Avionics Inc, 270 Commerce Drive, Fort Washington, Pennsylvania 19034
TELEPHONE: (215) 643 2900
TELEX: 846395

Centerline II RNAV 860 area navigation system

The RNAV 860 area navigation system has extended the company's range of Centerline II avionics designed for the singles and light twins sector of the general aviation market. Introduced in 1982, the system operates in conjunction with Narco's DME-890 distance-measuring

Narco RNAV 860 area navigation presentation

equipment and NAV 824 and NAV 825 200-channel VOR/LOC and VOR/LOC glideslope receivers.

'En route' and 'approach' modes are selected by push-button, as well as ground speed and time to next waypoint, and immediate indication of position with respect to the Vortac (VOR/DME) being used can also be given.

The RNAV 860 can function as an area navigation system without a left/right course-deviation display; digital steering readouts enable the pilot to follow VOR radials or any selected track.

Dimensions: 158.8 × 60.3 × 222.3 mm
Weight: 1.2 kg
Waypoint offset: (max distance) 199.9 n miles; (increments) 0.1°/0.1 n mile
Distance to next waypoint: 199.9 n miles

STATUS: in production.

Northrop

Northrop Corporation, Electronics Division, 2301 West 120th Street, Hawthorne, California 90250
TELEPHONE: (213) 757 5181
TWX: 910 321 3044

NAS-21 astro/inertial navigation system

Northrop's involvement with stellar/inertial navigation equipment goes back to the late 1940s, when the first-generation jet bombers being planned or in production required highly accurate, long-range navaids independent of ground beacons or give-away airborne transmissions. The long-term accuracy of the early inertial systems was far from adequate to meet navigation and weapon delivery needs, and so manufacturers developed ways to update them in flight. Northrop's answer was a star-tracker that could pin-point the position of the aircraft by reference to selected bright stars, and use this information to correct the gradual drift of the inertial navigation system.

A particular advantage of the star-tracking method was that the much higher cruising altitudes of the jet bombers put them above almost all cloud and haze, and so the stars were always visible.

Whereas the emphasis has now changed to low-level delivery, almost always below cloud levels (at least in Europe), stellar-monitored inertial systems continue to have wide-scale applications, for example, in long-range military aircraft and strategic missiles and, in any case, the NAS-21 can function entirely satisfactorily beneath as much as 95 per cent cloud cover.

The NAS-21 provides continuous position, velocity, and attitude information, and can interface with Doppler, Tacan, GPS ground-mapping radar, altimeters, and missile systems. It comprises four units: an astro-inertial instrument, digital computer, control/display box, and power supply. The astro-inertial unit is the movement and position sensor, containing a three-gimbal, stable reference platform, with two-degrees-of-freedom star tracker, two gyros, and three accelerometers mounted on it. The tracker conducts a square search over large angular regions of the sky, picking up and identifying an average of three stars a minute by day or night. The computer is a fully qualified, general-purpose military instrument designed for airborne applications needing high data rates. The ephemerides (positions and visibilities) of 61 stars are stored in the computer's memory and permit celestial navigation anywhere in the world.

The system is being flown aboard the first three US Air Force/Rockwell B-1B bombers as part of the navigation accuracy information system which monitors the performance of the standard inertial navigation system. It will view

Northrop NAS-21 stellar/inertial system on test

the sky, and lock on to selected stars, through a transparency behind the flight deck. The system has also been test flown to qualify it as a master reference to check the performance of the B-1B's own navigation system. The company has built three sets of equipment for this purpose.

A new, radiation-hardened astro-inertial system, the NAS-27, is under development.

Accuracy: air or ground-aligned: astro-inertial (occasional star tracking), better than 1000 ft cep over 10 h; pure inertial (no star tracking), less than 0.5 n miles/h

Attitude readout: better than 25 arc seconds
Velocity: 0.5 ft/s per axis typical error
Weight: 147 lb (66.8 kg)
Volume: 3.25 ft³ (0.35 m³)
Reliability: 800 h mtbf

STATUS: in production.

ONI

ONI Offshore Navigation Inc, PO Box 23504, New Orleans, Louisiana 70183
TELEPHONE: (504) 733 6790
TELEX: 200435

ONI-7000 Loran C

ONI's Type 7000 Loran navigator is claimed to represent a breakthrough in signal-processing technology, enabling it to operate in regions where other systems are unusable. The company claims it is the only system capable of providing coast-to-coast coverage over the USA.

The ONI-7000 automatically tracks up to eight stations in as many as four chains simultaneously, thereby providing a degree of redundancy not previously available. It also incorporates a number of features to reduce pilot work-load, including a flight-planning computer mode, automatic waypoint sequencing, pre-stored routes and automatic correction of magnetic variation.

Waypoints can be defined by latitude and longitude, Loran time differences or as an offset radial/distance from a previously defined waypoint. Any number of routes can be defined and stored in non-volatile memory, using up to 200 stored waypoints. The control/display unit employs fibre optics for better oblique and daylight visibility. Outputs are provided to CDI and remote annunciators, and optional outputs comprise signals to HSI, RMI, radar, and TAS. Optional inputs include compass, IAS, and encoding altimeter.

The system consists of three units: antenna, receiver/computer and control/display unit.

In April 1983 ONI announced FAA approval of the system under TSO C-60a requirements. As a result it was expected to be certificated in Beech Baron and King Air, Cessna P210 and Conquest, IAI Westwind and Aérospatiale Twin Star and Twin Dauphine, Sikorsky S-76 and Bell 222 helicopters.

Dimensions: (receiver/processor) 7.5 × 7.625 × 12.57 inches (191 × 194 × 320 mm)
(control/display unit) 5.75 × 4.5 × 6.5 inches (146 × 114 × 165 mm)
(antenna) 6.5 × 3.5 inches (base) × 14.5 inches long (146 × 89 × 368 mm)
Weight: (receiver/processor) 13.1 lb (5.95 kg)
(control/display unit) 5 lb (2.27 kg)
(antenna) 1.5 lb (0.68 kg)

STATUS: production deliveries began in September 1981 to helicopter and fixed-wing aircraft operators in USA, Canada, and Mexico. System has received FAA's TSO approval and is cleared for en route and terminal IFR operation in several aircraft types in US airspace.

In October 1985 the ONI-7000 was the first Loran C system to be approved for non-precision approaches by the FAA. The first approach under the new approval was flown using a Beech King Air 200 at Hanscoms Field, Massachusetts, in November 1985. Several other approaches have since been approved for the ONI-7000.

Flite-Trak 100 telemetry system

In 1979 ONI proposed a joint programme to the FAA whereby they could evaluate the use of Loran C to pinpoint the position of helicopters operating beyond the reach of air traffic control surveillance radars in the crowded airspace of the Gulf of Mexico. The objective was to

ONI-7000 Loran C navigator

telemeter to the air traffic control centre at Houston position information derived from Loran C equipment aboard helicopters servicing the many offshore oil installations in the area. This information was then to be displayed on computer-graphics terminals adapted to resemble as far as possible conventional air traffic control radar screens.

The US Government financed development of the FAA displays, while ONI launched and

ONI Flite-Trak 600 terminal for air traffic control centres and private operators

funded its own airborne telemetry equipment, at the same time developing a commercial version of the FAA system to enable helicopter operators to track their own fleets. The FAA system became operational during 1981, and ONI began delivering equipment to the commercial companies participating in the FAA's evaluation programme. In November 1981 ONI delivered the first Flite-Trak 600 despatcher terminal with colour graphics display to the Transcontinental Gas Pipeline Company. That operator is using the system in conjunction with its automated Gulf of Mexico communications network to control its fleet of 28 helicopters. Mobil Oil, Chevron Oil and Tenneco Oil have also installed Flite-Trak equipment. Chevron's system has a master terminal in New Orleans and two remote terminals at other Gulf Coast bases operating off the central computer. Shell Oil is using the system in its North Sea oil operation to improve search and rescue effectiveness. Aramco has installed a Flite-Trak 700 system incorporating a two-way data link, which has improved its crew scheduling, maintenance and parts-inventory control.

In 1984 the FAA replaced its equipment with a Flite-Trak 600 in the Houston ATC centre. China's Civil Aviation Administration has adopted Flite-Trak, operating on hf frequencies,

Flite-Trak 100 telemetry system

to control helicopter traffic in the Hainon/Pearl River offshore areas, and base stations are being installed in Zuhai, Zhangiang and Guangzhou during 1985.

In 1984 ONI introduced the Flite-Trak 110 airborne data link system, with colour cathode ray tube and full alphanumeric keyboard. Like the Flite-Trake 100 system, it operates in the vhf and hf bands over ACARS, SITA or private radio network.

Flite-Trak 100 airborne telemetry units interface with a number of navigation systems including Loran C, omega/vlf, INS and Decca.

Singer Kearfott

The Singer Company, Kearfott Division, 1150 McBride Avenue, Little Falls, New Jersey 07424
TELEPHONE: (201) 785 6000
TELEX: 133440
TWX: 710 998 5700

AN/ASN-128 Doppler navigation system

Singer Kearfott's AN/ASN-128 is the US Army's standard lightweight airborne Doppler navigator and comprises three units: receiver-transmitter-antenna, signal data converter and control display unit. A steering hover indicator can also be included as an optional extra. With inputs from heading and vertical references, the system provides aircraft velocity, present position, and steering information from ground level to above 10 000 feet.

Volume: (velocity sensor including signal data converter) 1037 inches^3 (16 993 cm^3)
(control display unit) 224 inches^3 (3670 cm^3)
Weight: (velocity sensor including signal data converter) 23 lb (10.43 kg)
(control display unit) 7 lb (3.8 kg)
(hover indicator) 2 lb (0.9 kg)
Propagation: 4-beam configuration operating FM/continuous wave transmissions in K band. Beam-shaping eliminates need for land/sea switch. Single transmit-receive-antenna uses full aperture in both modes to minimise beamwidth and reduce fluctuation noise
Number of waypoints: 10
Self-test: localisation of faults at line-replaceable level by BITE
Reliability: (complete system) >2000 h mtbf
(velocity sensor) >4500 h mtbf

STATUS: in production. In service with US Army Sikorsky UH-60A, Bell AH-1S, McDonnell Douglas AH-64A and Boeing CH-47D helicopters: also in Royal Australian Air Force Bell UH-1H and CH-47D, in Hellenic Air Force UH-1H, in Taiwan Army CH-47B and in Spanish Army MBB BO 105 and CH-47B helicopters. System, with hover indicator, is being built for West German Army/MBB BO 105, PAH-1 and Sikorsky VBH helicopters. Also built under licence in Japan for JASDF AH-1S and CH-47D.

AN/ASN-137 Doppler navigation system

The AN/ASN-137 is the multiplexed version of the AN/ASN-128 (see entry above). It is compatible with the MIL-STD-1553B data-bus, interfaces with the ASN-99 projected map display and has ARINC 575 or 429 outputs.

The ASN-137 uses the same antenna as the ASN-128 and has a control and display unit of the same size, and with the same front panel as the -128.

Additional features in the ASN-137 include hover bias correction for precision hovering, 12-point magnetic deviation entry and the addition of latitude/longitude and UTM grid zone outputs.

The AN/ASN-137 is installed on the US Army's Bell OH-58D and the US Navy's Sikorsky HH-3 and HH-60 helicopters. It has been selected for retrofit on the McDonnell Douglas AH-64A.

Weight: 26 lb (12 kg)
Power: 90 W at 28 V dc
Reliability: 2600 h mtbf (calculated)

STATUS: in production and service.

Singer Kearfott AN/ASN-128 Doppler navigation system: from left, velocity sensor (transmitter-receiver), control display unit and signal data converter

SKD-2110 Doppler navigation system

The SKD-2110 is a complete Doppler navigation system in one line-replaceable unit, emulating the two-unit AN/ASN-137 (see separate entry), providing all the customary navigation functions.

Dimensions: 369 × 342 × 50 mm
Weight: 5.5 kg
Reliability: 8500 h mtbf

SKD-2125 Doppler navigation system

The SKD-2125 is an enhanced lightweight Doppler navigation system suitable for fixed-wing aircraft, capable of operating up to 50 000 ft and 500 knots. It is a development of the AN/ASN-128 (see separate entry), having a larger antenna, increased transmit power and better receiver characteristics.

STATUS: in service in Lockheed C-130 aircraft.

AN/ARN-138 microwave landing system

The AN/ARN-138 airborne microwave landing system has been developed to the full military specifications of the US Navy and Marines and is suitable for carrier operations, use at remote sites and in conventional transport aircraft.

The glidescope can be selected by the pilot between 2° and 15° and approaches can be made down to Category II levels (100 feet cloudbase and 0.25 mile visibility). Range can be determined to within 15 feet using the associated L band Tacan system. Total system volume is 0.28 ft^3 and weight is 10.5 kg.

STATUS: in production.

AN/ARN-145 Micro-pcas

The AN/ARN-145 is a tactical microwave landing system primarily intended for military helicopters. It uses the scanning beam principle in the Ku band and can facilitate approaches down to Category II levels (100 feet cloudbase and 0.25 mile visibility), with range accuracy of 25 feet from an associated Tacan beacon. Glideslope can be selected by the pilot between 3° and 12°. Total system volume is 0.37 ft^3 and weight is 7.8 kg.

STATUS: in production.

SKN-2416 inertial navigation system for F-16

Singer Kearfott's SKN-2416 inertial navigation system equips the US Air Force F-16s. This was the third major contract for the SKN-2400/2600 class of inertial navigation and nav/attack

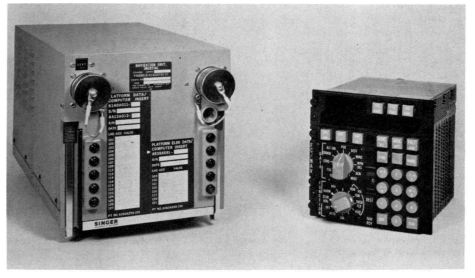

Singer Kearfott SKN-2416 INS for GD F-16 fighter

systems: in 1973 France had chosen it for the Dassault Super Etendard, and in the following year Sweden followed suit for the Saab JA 37.

In July 1975, Belgium, Denmark, the Netherlands and Norway chose the General Dynamics F-16 which aircraft are also equipped with the SKN-2416.

The SKN-2416 is a prime sensor for aircraft velocity, attitude and heading, and also the principal source of navigation information. It comprises three units: an inertial navigation unit (INU), fire-control/navigation panel, and an emergency battery and mounting.

The INU is a self-contained navigator consisting of an inertial platform and 10 electronics cards packaged into a single line-replaceable unit. The heart of the unit is an inertially stabilised platform with four gimbals for all-attitude operation, vertical and azimuth gyros, and three sub-miniature, pendulous linear accelerometers. The INU operates in conjunction with a cockpit-mounted control unit for entering, changing, or recalling digital computer data; displaying navigation, weapons delivery, and operational status data; controlling electrical power to the INU, fire-control computer, and target identification laser; and providing control facilities to the INU.

The system can be updated or driven with present position information (latitude and longitude), velocities in INS coordinates, gimbal angles about INS axes, and gyro torquing rates (all in MIL-STD-1553 format). Outputs, also on the data-bus format, are: present position (latitude, longitude, altitude), aircraft attitude (pitch, roll, heading true or magnetic), aircraft velocity (horizontal and vertical), and steering information (track angle error).

There are five operating modes: stored heading alignment, normal alignment, navigation, attitude reference, and calibration. The first of these is a rapid-align mode which permits entering navigation to an accuracy of better than 3 nautical miles an hour circular-error probability within 90 seconds provided the aircraft has not been moved during shutdown (the computer stores the last heading recorded before shutdown). In the normal mode the system aligns itself within 8 minutes, with a corresponding accuracy of 1 nautical mile an hour.

INU
Dimensions: 7.5 × 7.5 × 15.2 inches (191 × 191 × 386 mm)
Weight: 33 lb (15 kg)
Interface: MIL-STD-1553 data-bus
Performance: accuracy better than 1 n mile/h
Rapid-align time: 9 minutes at 0° F (−17.8° C)

Fire control/navigation panel
Dimensions: 5.75 × 6 × 7.3 inches (146 × 152 × 185 mm)
Weight: 8.25 lb (3.75 kg)

INU battery and mounting
Type: NiCd
Endurance: up to 10 s during interruption of primary power supply. Battery charged through INU during normal operation

STATUS: in production for General Dynamics F-16A and F-16B. (Litton won competition to supply LN-39 system as follow-on equipment for F-16C/D.)

SKN-2440 HAINS high accuracy inertial navigation system for B-1B

For its Rockwell B-1B the US Air Force has chosen Singer Kearfott's SKN-2440 high accuracy inertial navigation system (HAINS). This system uses as its starting point the SKN-2416 inertial navigator set for the US Air Force/General Dynamics F-16 fighter, a fully mature design with more than 2000 delivered to date, and Singer Kearfott's SKN-2430 Standard INU, which has been qualified to US Air Force technical exhibit ENAC 77-1.

Singer Kearfott SKN-2440 INS equips US Air Force Rockwell B-1B bomber

Singer Kearfott's approach to the development of the HAINS system has been to 'design in' the basic performance improvements needed to achieve the navigation accuracy specified for the new bomber. The system incorporates inertial components of higher accuracy, software enhancements to compensate for certain error sources, and a high-speed mini-computer that can accommodate a much greater computational load. Taken together, these improvements provide a quantum performance increase over that demonstrated by the production General Dynamics F-16 inertial system. The performance has been demonstrated during laboratory and flight tests conducted by the US Air Force Central Inertial Guidance Test Facility at Holloman Air Force Base, New Mexico.

The system comprises two line-replaceable units: the inertial navigation unit, and an INU rack. A completely self-contained navigator, the INU is the prime sensor on the B-1B for aircraft velocity, attitude and heading, and is also the prime source of navigation information. Navigation data is developed from self-contained inertial sensors comprising one vertical and two horizontal accelerometers and two two-axis displacement Gyroflex gyroscopes. These elements are mounted on a four-gimbal, gyro-stabilised stable platform, with the accelerometers (which are maintained in a known reference frame by the gyroscopes) as the primary source of information. The system provides pitch, roll and heading in both analogue and digital form for display, radar stabilisation and other purposes, and can be aligned on the ground and in the air.

Dimensions: 19.1 × 7.63 × 7.53 inches (485 × 194 × 191 mm)
Weight: 38 lb (17.3 kg)
Digital interface: MIL-STD-1553B

STATUS: in production.

SKC-3007 navigation/weapon-delivery computer for A-7

This computer has been designed to upgrade the performance of the US Air Force/LTV A-7D and A-7K aircraft. Intended to serve this aircraft through the 1990s, the system by comparison with the earlier nav/weapons computer will have improved computational capability, better reliability, easier maintenance through extensive built-in test to eliminate sophisticated intermediate level test equipment, be capable of operating with existing (TC-2A) and future (TC-2A (ex) and MIL-STD-1750A) instruction-set architectures with memory expansible to 256 K. In addition it was required to be form, fit and functionally interchangeable with its AN/ASN-91 predecessor. Software interchangeability was also an important operational

goal, and was demonstrated in 38 flights with an A-7.

The most important task is to solve a real-time ballistic problem for weapons, so permitting wide latitude in approach to a target. The system provides for low altitude night attack (LANA) using a forward looking infrared (Flir) and automatic terrain following (ATF) making it possible to perform low altitude navigation, target detection and attack during day or night under the weather visual conditions. During en route cruise the system computes and outputs present position and guidance information to the destination or target; it can store the coordinates of up to 19 waypoints. Nine additional waypoints or targets can be entered. The system drives the head-up display via a MIL-STD-1553B digital data-bus, and also the moving-map display.

Dimensions: 12.90 × 9.48 × 18.40 inches (327.7 × 240.8 × 467.4 mm)
Weight: 72.9 lb (33.1 kg)
Power: 431 W
Reliability: 2560 h mtbf predicted

STATUS: in production on A-7 and other special operating forces contracts.

SKN-4020 CAINS II ring-laser gyro inertial navigation system

The Singer Kearfott SKN-4020 was selected by the US Navy for their carrier aircraft inertial navigation system (CAINS II-AN/ASN-139) and is currently building systems for the full scale engineering development phase of this programme.

SKN-4030 ring-laser gyro inertial navigation unit

Conforming to the US Air Force Technical Specification SNU-84-1 for standard form, fit and function inertial navigator, the system comprises an inertial navigation unit, containing three ring-laser gyro sensors, three accelerometers and associated electronics and a control display unit. Navigation accuracies of better than 0.5 nautical miles an hour and velocity accuracies significantly better than the SNU-84-1 requirements of 2.5 feet per second in each axis were demonstrated with systems during tests at the Central Inertial Guidance Test Facility (CIGTF) at Holloman Air Force Base, New Mexico.

Dimensions: (inertial navigation unit) 7.5 × 7.6 × 18.1 inches (190.5 × 190 × 459.7 mm)
(control display unit) 5.7 × 6 × 7.3 inches (144.8 × 152.4 × 459.7 mm)

STATUS: private venture initiative, in development.

Sperry

Sperry Corporation, Aerospace and Marine Group, Commercial Flight Systems Division, Flight Systems, PO Box 21111, Phoenix, Arizona 85036
TELEPHONE: (602) 869 2311
TELEX: 667 405

Sperry MLZ-900 microwave landing system: from left, receiver, control unit, antenna

IONS-1020 and IONS-1030 inertial omega navigation system

Development and production of the IONS series of systems has been discontinued. (See *Jane's Avionics 1985-86*).

MLZ-900 microwave landing system receiver

This system was originally designed by Hazeltine as the Model 2800. Sperry took over production of the system under a licensing agreement and by mid-1983 had begun production. The MLZ-900 comprises a receiver/computer, antenna and control unit, permitting users of ground-based time-referenced scanning-beam MLS transmitters to select approach courses of up to 60° off the runway heading, and glide-slope angles up to 20°. The system can drive existing flight-director and horizontal-situation indicators as well as EFIS.

The FAA is installing the MLZ-900 in 10 aircraft for the FAA-sponsored MLS programme at Richmond Airport, Virginia.
STATUS: in production. The system was initially certified in a Cessna Citation II in November 1985 and has been installed in a Bell 222, Agusta A109, Westland 30, and Sikorsky S-61 and S-76 helicopters as well as Citation II and III, Falcon Jet, Fokker 50 and 100, and Saab 340.

GZ-810 GPS sensor

The GZ-810 GPS sensor is used to provide Navstar global positioning system data to a Sperry flight management system (see separate entry). It provides all the standard GPS navigation functions within a single ¼ ATR Short box, weighing 8 lb (3.6 kg) and requiring 40 W of power at 28 V dc.

STATUS: in production.

300 series avionics

The 300 series is designed to give single and light twin-engined aircraft a comprehensive and integrated avionics suite. The system comprises the following units:

Close up of MLZ-900 display

RT-385A nav/com: a 720-channel vhf radio and 200-channel VOR/ILS receiver
IN-380 navigation indicator: for VOR navigation and ILS approaches
R-546E ASF receiver: with digital tuning and associated IN-346A indicator
RT-359A transponder: certified to 15 000 ft and compatible with most altitude encoders
SDM-77 DME: with lcd.

Also compatible with 300 series equipment are the Sperry SMA-90 audio panel, and the 200A autopilot.

400 series avionics

The 400 series is designed for high performance single and twin-engined aircraft. It comprises the following units:

RT-485B nav/com set, with 720-channel radio and 200-channel navigation receiver facilities, led displays, and three nav/com station memory
R-446A ADF receiver and associated **IN-346A** indicator
IN-404A RMI, **R-443B** glideslope and **R-402** marker beacon receivers and **RT-459A** transponder
EA-401A encoding altimeter, operational up to 35 000 ft
RTA 477A DME with led display
RN-479A 10-waypoint RNav system, with led display.

The 400 series is also compatible with the SMA-90 audio panel/amplifier, the 400B autopilot and YD-840B yaw damper.

1000 series avionics

The 1000 series is a complete avionics suite for the business aircraft. It is standard equipment on the Cessna Conquest I and II and optional on the 402, Chancellor and Golden Eagle: it features led displays throughout the system's units which include:

RT-1038A communications transceiver, **R-1048** navigation receiver and **C-1046A** ADF receiver, with digital readouts of frequency and pre-selection facility to enable a series of frequencies to be stored and rapidly recalled to use
IN-480AC navigation indicator, **R-1043A** glideslope receiver, associated **IN-1004B** RMI, **EA-801A** encoding altimeter, **SXP 1060** transponder and **RTA 1077A** DME receiver. As with the 400 series, there is a 10-waypoint RNav set in the 1000 series (in this case the **RN-1079A**) and the system is complete with either a 800B or 1000 series integrated flight control system.

Teledyne

Teledyne Ryan Electronics, 8650 Balboa Avenue, San Diego, California 92123
TELEPHONE: (619) 560 6400
TELEX: 695028

AN/APN-217 Doppler velocity sensor

This transmitter-receiver is the primary navigation sensor for the US Navy's Sikorsky SH-60B Seahawk Lamps Mk III helicopter, and deliveries of a slightly modified version, the AN/APN-217(J), have been made to Japan for its Sikorsky HSS-2 Sea Kings. The US Navy is also planning to install this sensor on its Sikorsky MH-53E and RH-53D minesweeper helicopters. The US Marine Corps plans to install the APN-217 on its Sikorsky CH-53 helicopters.

The AN/APN-217 is a microprocessor-controlled, solid-state, single-unit radar with a unique continuous wave space-duplex design

that avoids modulation losses, eliminates altitude 'holes', and bypasses the limitations associated with modulated systems. At low altitudes the APN-217 is claimed to be 100 times more sensitive than modulated Doppler navigation radars. A combination of six special features prevents acquisition or tracking of antenna vertical sidelobe returns, making the system suitable for coupled transitions from forward flight to hover and vice versa.

Dimensions: 16.9 × 16.3 × 6.9 inches (429 × 414 × 175 mm)
Weight: 34 lb (15.5 kg)
Outputs: dc signals to afcs and flight instruments such as hover indicator, digital outputs to MIL-STD-1553 provide velocities to navigation computer, and one version of APN-217 has digital outputs to ARINC 575
Accuracy: (digital) 0.3% of ground speed, 0.3 knot in heading and drift, 35 ft/minute vertically. (analogue) 2% of ground speed, 0.5 knot in heading and drift (afcs)
5% ground speed, 1 knot in heading and drift
70 ft/minute vertically (flight instruments)

Teledyne Ryan AN/APN-217 Doppler velocity sensor

Control/display unit presentation of Teledyne Ryan AN/APN-218 Doppler

Display: outputs are compatible with a number of displays, including CP-1251 cdu, and with ground speed/drift and hover instruments
Controls: land/sea select. System in SH-60B is controlled via MIL-STD-1553 data-bus
Power: 55 W at 28 V dc
Speed range: –40 to 350 knots along track, ±100 knots drift, ±5000 ft/minute vertically
Reliability: demonstrated mtbf 3750 h
Self-test: BITE detects 98% of predictable failure modes

STATUS: final development in late 1981. Deliveries of Lot 1 production began in late 1982.

AN/APN-218 Doppler velocity sensor

This Doppler velocity transmitter-receiver is a high-performance, nuclear-hardened radar chosen by the US Air Force as its common strategic Doppler (CSD) navigation system for fixed-wing aircraft. Competitive development of the APN-218 began in 1976. Production awards followed in 1978 and deliveries began the following year.

The performance and reliability of the system stems from two characteristics that it shares with the parent APN-213 Doppler: use of a continuous wave space-duplex transmit-receive system that is typically 7 to 10 dB more efficient in its use of transmitter power than a modulated system, and a high-gain, narrow-beamwidth planar array antenna for high accuracy and sensitivity.

The features that commended it to the US Air Force as its CSD are nuclear-hardness, high performance, choice of ARINC 575 or dual-redundant MIL-STD-1553 data-bus interface, and integral radome.

The system can be used as a source of velocity information to other equipment, or in conjunction with a ground speed and drift indicator or computer display unit as a self-contained navigation system. The optional cdu combines the functions of navigation computer and control/display unit, and contains an incandescent alphanumeric display panel and

Teledyne Ryan AN/APN-218 Doppler velocity sensor

a keyboard for entering data and selecting operational modes. Up to 10 waypoints can be accommodated, and a non-volatile scratch-pad memory holds critical information during power transients or interruptions.

Dimensions: (sensor) 28.2 × 25.4 × 6.7 inches (716 × 645 × 170 mm)
(gdsi) 5.75 × 3 × 6.1 inches (146 × 76 × 155 mm)
(cdu) 5.75 × 6 × 6.5 inches (146 × 152 × 165 mm)
Weight: (sensor) 70.2 lb (31.9 kg)
(gdsi) 3.4 lb (1.5 kg)
(cdu) 8.5 lb (3.9 kg)
Speed: 96-1800 knots
Altitude: 0-70 000 ft
Beam geometry: 4 beams time-shared
Accuracy: 0.14% rms
Terrain: land/smooth sea
Power: 170 VA at 115 V 400 Hz
Transmitter power: 1.5 W
Frequency: 13.325 GHz
Operational reliability: >3000 h
Self-test: BIT both continuous and commanded

STATUS: APN-200/213 and APN-218 are in production; latter as retrofit for US Air Force/ Boeing B-52 KC-135 aircraft. System is also fitted to US Air Force/Lockheed C-130 and C-141 and to General Dynamics FB-111 aircraft. Modified version of AN/APS-218, designated AN/APN-230, has been chosen for US Air Force/Rockwell B-1B bomber. AN/APN-200/ 213 is fitted to Boeing E-3A Sentry (AWACS) and Lockheed S-3A Viking and has been chosen for Grumman EA-6A.

AN/APN-233 (220) Doppler velocity sensor

This system can be used either as a single-unit velocity sensor providing outputs to other aircraft systems, or with a computer/display

Doppler velocity sensor for AN/APN-233

unit (cdu) and HSI as a self-contained navigation facility. As a velocity sensor to provide data to the navigation and weapons-delivery systems, it was chosen by the West German Air Force for its Dassault/Dornier Alpha Jet strike/ trainers. Versions of the equipment have also flown on RPVs such as Teledyne's BGM-34C, and on helicopters, and was designed for applications in which size, weight, performance, and reliability are critical factors. The APN-220 family evolved from a small, lightweight Doppler sensor originally designed for the US Army and was subsequently qualified by that service and by the US Air Force and the West German Air Force.

An optimum velocity range and near-zone rejection are offered for each application.

System characteristics of typical fixed-wing application
Dimensions: (sensor) 16.78 × 11.46 × 4.45 inches (426 × 291 × 113 mm)
(cdu) 6 × 5.75 × 6.5 inches (152 × 146 × 165 mm)
Weight: (sensor) 21.3 lb (9.66 kg)
(cdu) 8.5 lb (3.86 kg)
Output: heading, vertical velocity, and ground-speed/drift to aircraft systems, eg afcs, or to CP-1251 and HSI
Number of waypoints: 5:10, entered via front-panel keyboard
Range: (typical system) –40 to 600 knots forward, 150 knots in drift and 5000 ft/minute vertically. Height, up to 50 000 ft
Accuracy: (over land) 0.25% + 0.2 knot (over sea) 0.3% + 0.2 knot
Power: sensor 28 W at 28 V dc
(cdu) 30 W at 28 V dc
Reliability: 2600 h mtbf demonstrated in Alpha Jet
Self-test: BITE diagnostic program locates faults at first-line level to 95% confidence

STATUS: in production. System has been selected for US Navy/Grumman C-2A Grey-hound and US Marine Corps/Rockwell OV-10D

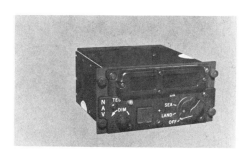

Ground speed/drift indicator for Teledyne Ryan AN/APN-218

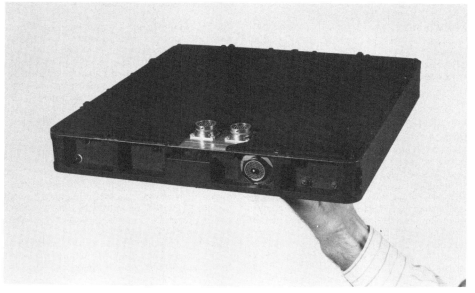

Teledyne Ryan Model 2000 Doppler velocity sensor

Bronco observation-post aircraft. In former application Teledyne has to deliver 43 sets between 1984 and 1989 to equip second batch of C-2As to supplement 12 already in service.

AN/APN-231 radar navigation system
In final development during early 1984, the AN/APN-231 will be the primary navigation system for the US Navy/Grumman EA-6A aircraft. It also integrates with other equipment including the attitude/heading reference system, air data computer, search radar, ECM sets and flight instruments. The system comprises an AN/APN-200 Doppler velocity sensor, a CP-1573/APN-231 computer display unit and a CV-3780/APN-231 signal data converter.

Dimensions: (velocity sensor) 149.9 × 632.5 × 652.8 mm
(computer display unit) 152 × 146 × 165 mm
(data converter) 226 × 259.1 × 426.7 mm
Weight: (velocity sensor) 20 kg
(computer display unit) 3.6 kg
(data converter) 18.2 kg
Range: 50-999 knots forwards, 0-200 knots in drift, 0-5000 ft/minute vertically
Height: up to 40 000 ft
Accuracy: (velocity sensor) 0.13% + 0.1 knot

STATUS: production deliveries began in mid-1984.

Model 2000 Doppler velocity sensor
The Model 2000 is designed to full military standards and is compatible with the MIL-STD-1553B digital data-bus. The use of advanced micro-electronics leads to a predicted mtbf exceeding 8500 hours as well as particularly good accuracy. The system can accommodate velocities between –55 to +350 knots, drift velocities up to 100 knots and vertical velocities of up to ±5000 feet per minute. Accuracies are specified as being within 0.2 per cent of actual velocity horizontally and 0.05 per cent vertically.

Dimensions: 13.5 × 14.6 × 1.9 inches (342 × 370 × 48 mm)
Weight: 11 lb (5 kg)
Power: 20 W at 28 V dc

STATUS: in production.

Texas Instruments
Texas Instruments Equipment Group, PO Box 660246, Dallas, Texas 75266
TELEPHONE: (214) 995 2011
TELEX: 73324
TWX: 910 867 4702

TI 9100 Loran C Navigator
With the TI 9100 Loran C Navigator, Texas Instruments has brought the benefits of this precision, long-range hyperbolic radio navigation system to the general aviation community. Loran C stations serve coastal and inland waters in the United States (including Alaska) and Canada, as well as major areas of Europe, the North Atlantic and North Pacific. The coverage is now so widespread, claims Texas Instruments, that it is feasible to use Loran C for general aviation. The reliability and accuracy of the TI 9100 has been recognised, and it has been certificated by STC and TSO for en route and terminal operations in US national airspace Loran C coverage areas. The company claims that the TI 9100 was the first of its type to be certificated.

The TI 9100 is a microprocessor-based self-contained panel-mounted unit that can perform 14 distinct navigation functions together with other tasks. It provides point-to-point navigation direct and can store up to nine waypoints, which can be entered in any of four ways: lat/long, range and bearing from another waypoint, two bearings from two other waypoints, or two time differences. Apart from its own display, the system can drive three high- or low-level CDIs or HSIs for simultaneous cross-track error display. Cross-track error sensitivity for the CDI is 0.01 nautical mile. Full-scale deflection is 5 nautical miles en route or 1.25 nautical miles on approach.

Data provided by the TI 9100 includes:
Cross-track error and ground-track angle Distance and direction (left or right) off course between the selected FROM and TO waypoints, and the ground-track angle in degrees.
Range and bearing Distance in nautical miles and bearing in degrees from the present position to the next selected waypoint.

Range and radial Distance and bearing (radial) from the next waypoint to the present position.
Course and distance Course and distance between two selected waypoints or the total distance between them if a number of waypoints have been selected.
Estimated time en route Average ground speed in knots and ETE to the next waypoint.
Latitude and longitude Aircraft present position, and also used to enter estimated present position at power-up on the ground.

The system operates automatically with all existing Loran chains, choosing the best stations, and can be updated easily to meet changes or additions to the networks.

Autopilot interface: An optional steering output is available for use in conjunction with most standard autopilots
Dimensions: (receiver) 6.25 × 3.25 × 11.6 inches (159 × 83 × 295 mm)
(pre-amplifier) 4.25 × 2.63 × 1.69 inches (108 × 67 × 43 mm)
Weight: (receiver plus pre-amplifier) 7.2 lb (3.2 kg)
Accuracy: meets en route and terminal requirements of FAA AC 90-45A in primary Loran C coverage areas without propagation corrections

	Cross-track error	Along-track error
En route	1.5 n miles (2.8 km)	1.5 n miles (2.8 km)
Terminal	1.1 n miles (2 km)	1.1 n miles (2 km)
Approach	0.3 n mile (0.5 km)	0.3 n mile (0.5 km)

Repeatability: 0.1 n mile (0.2 km) typical in primary Loran C coverage
Power: 11-33 V dc 24 W

STATUS: in production.

TI 9100A Loran C Navigator
Similar to the TI 9100 in accuracy, reliability and installation arrangements, the TI 9100A has a greater capacity, so that the coordinates of up to 40 waypoints can be stored. These waypoints are internally arranged to form four flight-plans, each having ten waypoints. Reported price for the system in 1984 was $6495.

STATUS: in production.

TI 91 Loran C Navigator
Introduced in 1982, the TI 91 is a development of the TI 9100. It combines the features of the earlier model with a remote-mounting facility, standard autopilot interface and an optional ARINC 419/429 interface whereby Loran information can be overlaid on the picture of any Sperry or Bendix data/navigation/weather radar

Control panel of Texas Instruments TI 9100 Loran C

Texas Instruments TI 91 Loran C system showing receiver, control unit, display and antenna

display. The system can also drive a second display unit, so that pilot and co-pilot can have their own weather and navigation pictures.

Dimensions: (receiver) 124 × 180 × 330 mm
(control unit) 82 × 82 × 58 mm
(display) 101 × 51 × 157 mm
(pre-amplifier) 135 × 61 × 33 mm

Weight: (receiver) 3.9 kg with tray
(display) 0.39 kg
(control unit) 0.25 kg
(pre-amplifier) 0.2 kg

STATUS: in production.

Tracor

Tracor Inc, Aerospace Group, 6500 Tracor Lane, Austin, Texas 78725
TELEPHONE: (512) 929 2233
TELEX: 776410
TWX: 910 874 1372

Tracor Model TA 7800 omega/vlf system

Model TA 7800 omega/vlf

Designed for the airlines, the Model TA 7800 can receive vlf signals simultaneously with omega transmissions, identifying the best combination of signals, and displaying the most reliable navigation mode. The system can operate in its primary hyperbolic or circular mode, in a secondary rho-rho mode, or (in the unlikely event that signals degrade to the extent that only two omega and/or vlf stations are usable) a tertiary rho-rho mode.

The antenna and receiver electronics are sensitive enough to respond to signal strengths below normal thresholds. A unique adaptive noise-blanker and frequency cross-coupler combine to improve the signal/noise ratio up to 20 dB. Another feature of the Model TA 7800 permits it to 'work through' modal interference events, a propagation difficulty with omega that can cause serious navigation errors. Appropriate circuits detect the presence of such interference and if necessary suppress signals from the offending station.

Receiver/processor unit
Format: 3/4 ATR Long
Weight: 29 lb (13.15 kg)
Power: 55 W typical, choice of 28 V dc or 115 V 400 Hz
Cooling: convection
Reliability: >4000 h mtbf
Altitude: tested to 55 000 ft
Certification: TSO C94

Control/display unit
Dimensions: 4.5 × 5.75 × 6.25 inches (114 × 146 × 159 mm)
Weight: 4 lb (1.82 kg)
Power: 9 W typical
Environment: sealed unit
Display: sunlight readable
Reliability: > 10 000 h mtbf
Certification: TSO C94

Antenna coupler unit
Configurations: 3 'H'-field: ARINC 599, internal brick, or external teardrop, 3 'E'-field omega/ADF, omega/ADF Boeing configuration, or omega blade
Certification: TSO C94

STATUS: in service.

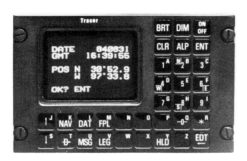

Control/display unit for Tracor 7880 omega/vlf

TA 7880 omega/vlf navigation system

The TA 7880 is the first omega/vlf navigation system to offer a control/display unit with full colour capability providing efficient transfer and display of vital navigation data. The TA 7880 comes with 352 non-volatile, pilot-definable waypoints which are contained within up to 99 flight plans of up to 99 waypoints per plan. The TA 7880's optional database stores as many as 30 000 waypoints and can be easily updated on a microfloppy diskette using Tracor's disk loader.

Dimensions: (control/display unit) 5.75 × 3.75 × 8 inches (146 × 95 × 203 mm)
(navigation processor) 4.9 × 7.5 × 12.7 inches (124 × 191 × 323 mm)
(antenna coupler) variable
Weight: (control/display unit) 4.5 lb (2.04 kg)
(navigation processor) 8.25-9.5 lb (3.75-4.31 kg)
(antenna coupler) 1.8-6.2 lb (0.82-2.81 kg)

STATUS: in production.

TA 7885 multi-sensor navigation management system

The Tracor TA 7885 offers all of the features of the TA 7880 omega/vlf navigation system, plus the ability to interface with: VOR/DME (autotune), IRS/INS, omega/vlf, Loran-C, and GPS. The system includes a standard 30 000 waypoint database, and interfaces directly with flight control systems, autopilots, radar navigation displays, and cabin information displays.

Dimensions: (control/display unit) 5.75 × 4.5 × 7.75 inches (146 × 114 × 197 mm)
(navigation processor) 4.9 × 7.5 × 12.7 inches (124 × 191 × 323 mm)
(antenna coupler) variable
Weight: (control/display unit) 5.6 lb (2.54 kg)
(navigation processor) 10.5 lb (4.76 kg)
(antenna coupler) 1.8-6.2 lb (0.82-2.81 kg)

STATUS: in production.

TA 7900 ARINC 599 omega/vlf system

The Tracor TA 7900 is a new generation ARINC 599 system which significantly reduces hardware weight and power consumption. An optional 30 000 waypoint database and an optional VOR/DME option offer a multi-sensor management system. Design provisions include Tracor's GPS module.

Format: (receiver/processor) 3/4 ATR Long
Dimensions: (control/display unit) 5.75 × 4.5 × 7.75 inches (146 × 114 × 197 mm)
(receiver/processor) 7.75 × 7.6 × 19.6 inches (197 × 193 × 498 mm)
(antenna coupler) variable
Weight: (control/display unit) 5.6 lb (2.54 kg)
(receiver/processor) 16 lb (7.26 kg)
(antenna coupler) 1.8-6.2 lb (0.82-2.81 kg)

STATUS: in production.

Trimble

Trimble Navigation Limited, 585 North Mary Avenue, PO Box 3642, Sunnyvale, California 94086
TELEPHONE: (408) 730 2900
TELEX: 6713973

Global positioning system

Trimble is developing a range of low-cost GPS receivers suitable for a wide variety of uses including airborne applications. Flight trials have been conducted using a Beech Bonanza. The system used, designated 10X, uses both GPS and Loran signals to give precise navigational data.

Universal Navigation

Universal Navigation Corporation, 3545 West Lomita Boulevard, Torrance, California 90505
TELEPHONE: (213) 325 4081
TELEX: 705806
TWX: 910 349 6205

Universal UNS-1 control and display unit

UNS-1jr Navigation computer

First shown at the 1984 NBAA Convention, the UNS-1jr navigation computer is derived from the company's UNS-1 flight management system (see separate entry on page 300), employing the same control/display unit and operating programme. It is designed for general aviation, specifically for light twins, turboprops and helicopters, but could also serve as a second or third navigation unit on a larger installation. The UNS-1jr provides the same guidance as the UNS-1 in lateral, or area, navigation, but does not have the capability of computing aircraft performance, coupled vertical navigation or fuel management. Neither does it have the extended Jeppesen database of the UNS-1. Accordingly, it is much smaller, the navigation computer being contained within an 8 lb (3.6 kg) 2 MCU box, which does, however, have the standard Jeppesen database. Recent software developments include the availability of heading and approach modes, regional databases, frequency management and storage for 200 routes.

STATUS: in production. The UNS-1A will be available in late 1986 with a full alphanumeric keyboard and larger display.

UNS-OSS omega/vlf

This sensor was designed to operate in conjunction with the company's UNS-1 flight management and UNS-1jr navigation management systems, working either alone or in combination with inertial or Loran C sensors. The system comprises an antenna and receiver, the control/display unit of the UNS-1 or UNS-1jr systems providing the necessary management functions. New receiver techniques together with digital processing combine to reduce the size and weight of the UNS-OSS system over previous omega/vlf sensors. Operation is automatic, control being exercised via the UNS-1 or UNS-1jr. All eight omega and ten vlf stations are monitored.

Format: ARINC 600/2 MCU
Weight: 6.5 lb (3 kg)
Power: 15 W at 28 V dc

STATUS: in production.

UNS-LCS Loran C

UNS claims that its Loran C sensor is the smallest and lightest such device currently available. It was designed to operate in conjunction with the company's UNS-1 flight management and UNS-1jr navigation management systems. Working alone, or in conjunction with inertial or omega/vlf sensors, this device further improves the accuracy of the UNS-1 and UNS-1jr systems. The system comprises only a single receiver box and antenna, the UNS-1 or UNS-1jr providing the control/display unit. It uses all available Loran C transmissions to compute a best position, computer-controlled notch filters reducing interference from any nearby frequency but non-Loran signals.

Format: ARINC 600/2 MCU
Weight: (receiver) 7 lb (3.2 kg)
Power: 20 W at 28 V dc

STATUS: in production. In January 1986 the Universal Loran receiver was certified by the FAA for IFR en route and terminal operation throughout virtually all of the continental USA. Certification followed a 30-hour flight trials programme.

Flight management and control

CANADA

Canadian Marconi

Canadian Marconi Company, Avionics Division, 2442 Trenton Avenue, Montreal, Quebec H3P 1Y9
TELEPHONE: (514) 341 7630
TELEX: 05-827822
TWX: 610 421 3564

CMA-923 flight advisory computer

This flight advisory computer is designed to optimise fuel use by calculating and displaying the best combination of airspeed/Mach number and altitude for the aircraft's gross weight and environmental conditions. Using ground speed and fuel-flow inputs the system displays specific range (miles per pound or kilogram of fuel) for manual fuel-flow fine-tuning. The flight-planning mode allows calculation of best cruise flight-level and second-segment climb for current gross weight and environmental conditions. The display can indicate V_1, V_R and V_2 for take-off, or optimum climb, long-range cruise or high-speed cruise parameters. Towards the end of a flight the top-of-descent point, V_{REF}, and runway-length required for landing are also displayed. All parameters are calculated using data taken from the aircraft manuals and are developed in accordance with FAA operating limit conditions.

The single unit system accepts outside air temperature, altitude and true airspeed inputs, in either ARINC 565 (analogue) or ARINC 429

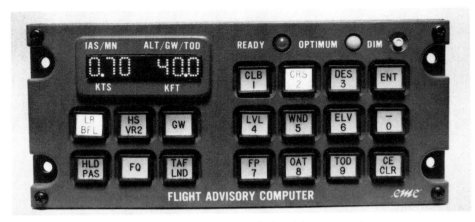

Canadian Marconi CMA-923 flight advisory computer

(digital) forms, plus individual engine fuel flow inputs as 0-5 V signals. There is an optional ARINC 429 ground speed input.

Entries which are made by the crew include: runway length and elevation, desired flap setting, aircraft payload, flight distance, predicted wind vector, air temperature, destination conditions and elevation, and any air traffic control altitude restrictions. Before flight the display shows the maximum take-off weight limits, the optimum flight-level, fuel quantity, time of flight, and take-off velocity data. During

climb, cruise and descent the crew is presented with optimum air speed/Mach number information or the system can be interrogated at any time to indicate fuel quantity and time-remaining data, or current gross weight. Fuel alerts are provided when the flight time remaining is less than 45 minutes.

Dimensions: 145 × 67 × 165 mm
Weight: 1.6 kg

STATUS: in production.

FRANCE

Elecma

Division Electronique de la Snecma, 22 quai Galliéni, BP 305, 92156 Suresnes Cedex
TELEPHONE: (1) 4772 81 84
TELEX: 620385

RN-1435 full-authority digital engine control

Elecma has delivered three prototype RN-1435 full-authority digital engine controllers to Turboméca. The system has been tested on the Turmo engines of a Super Puma, and tests on other types of helicopter engines are envisaged. The RN-1435 is being developed under a French Government contract in close co-operation with engine manufacturer Turboméca, which designed the fuel-flow regulator. The computer provides the various speed regulation phases of the free turbine and gives overspeed protection. On a twin-turbine helicopter, the two computers can exchange data

to balance the loads on the engines and to provide redundancy and failure protection.

STATUS: under development.

RN 1493 digital engine control unit for helicopters

Elecma has delivered 35 prototype units of its RN 1493 full-authority digital engine control units for in-flight testing of the Turboméca TM 333 and TM 319 turboshaft engines which power the Dauphin and Ecureuil helicopters. The unit is also being adapted, in conjunction with Lucas Aerospace, for use on the RTM 322 engine. The unit has also completed flight tests on the Epsilon, used with the TP 319.

Dimensions: 276 × 185 × 127 mm
Weight: 5.5 kg
Processor: Motorola 680 g series

STATUS: in production.

Elecma RN 1493 full-authority digital engine control unit for helicopter turbine engines

RM 1627 full-authority digital engine control

The RM 1627 full-authority digital engine control is now in production and is fitted on the M53-P2 engine of the Dassault Mirage 2000.

Sagem

Société d'Applications Générales d'Electricité et de Mécanique, 6 avenue d'Iéna, 75783 Paris Cedex 16
TELEPHONE: (1) 4723 54 55
TELEX: 611890

SNL 10 engine control computer

The SNL 10 determines the N_1 limit rotation speed of the General Electric CF6-50 and the EPR on the Pratt & Whitney JT9D-59 engines

used on the Airbus A300. It provides automatic engine control at the authorised maximum rating via auto-throttle control. The N_1 limit is computed from aerodynamic parameters (Mach number and altitude) and in accordance with the flight phase (take-off, take-off at adapted thrust, go-around, maximum continuous, climb or cruise). The SNL 10 includes a general-purpose micro-programmed computer and input and output interfaces.

STATUS: in service.

Sagem SNL 10 engine control computer

Sfena

Société Française d'Equipements pour la Navigation Aérienne, Controls and Systems Division, Aérodrome de Villacoublay, BP 59, 78141 Vélizy-Villacoublay Cedex
TELEPHONE: (1) 46 30 23 85
TELEX: 260046
FAX: (1) 46 32 85 96

Airbus A300 airliner has Sfena automatic flight control system

Automatic flight control system for A300

Although collaboratively developed with other nations involved in the European Airbus consortium, Sfena leads the automated flight control system (afcs) development programme and has sub-contracted system components to Smiths Industries (UK) and Bodenseewerk (West Germany). Sfena was appointed prime contractor in November 1969.

The predominantly analogue, dual-dual fail-operational afcs developed by these three companies has been installed in all Airbus A300 airliners except the forward-facing crew compartment models, and the A300-600, which have a digital afcs, described below.

The A300 afcs is designed to provide automatic control of all stages of flight from initial climb to touchdown. It is certificated for operations to Category IIIA weather minima. Sfena points out that the afcs is closely linked to the aircraft flight controls and uses sensors which are available in other systems in addition to dedicated sensors. External references are provided by inertial navigation or heading-reference units, air data computers, ILS/VOR receivers and radio altimeters.

The afcs is divided into six independent sub-systems:

Autopilot/flight director All processing functions in this sub-system are duplicated, and processors are used which have two computational paths, so that there is an output for both a real and an 'imaginary' servo. The system can operate as a simplex self-monitoring system in non-critical operating modes, and as a dual-dual fail-operational system in safety-critical modes. In cruise only one set of computers is used, but the second set is continuously synchronised and can be selected as a replacement if there is a malfunction. Both sets operate in unison during automatic-landings.

Each computer set comprises a longitudinal (pitch) and lateral (roll and yaw) computer, plus a logic computer. The latter generates operating

logic and engagements, including indications on the cockpit control unit and mode indicators. This is a digital electronic unit which uses a 4 K 8-bit word read-only memory. Basic autopilot modes provided by the other computers, and sequenced by the logic computer, include pitch and roll/heading hold.

Other autopilot modes from the pitch computer are altitude, speed and vertical-speed hold, plus altitude acquisition with arm and capture-phase facilities. The lateral computer provides heading pre-select, VOR/LOC arm, capture and track, and navigation mode facilities. The latter include back-beam approach using flight-director guidance. Facilities jointly provided by the two computers are turbulence, auto-approach, automatic-landing (flare and roll-out), overshoot, and control-wheel steering modes.

When cruise modes are used the computers automatically limit the amplitude and speed of commands, compare the amplifier outputs in both lanes of each computer, and compare the actual servo output with the output of a pseudo-servo, or electronic 'image'. Mechanical limiting is also incorporated in the control-surface servos.

In the automatic-landing mode, both sets of computers are engaged and the configuration produces two direct servo-motor drives, and two 'imaginary' servo drives. A majority-vote assures continued operation of the system after an initial failure, and passive disconnection of the system after a second failure.

In addition to the processors a control unit in the flight-deck glareshield provides all the controls for engagement of the autopilot/flight-director modes and autothrottle. Each pilot also

has a failure and performance indicator which shows autopilot and autothrottle warnings, the system modes in use, and the redundancy level available during approach and landing. Two sets of dynamometric rods, comprising two rods per control column, are used to sense crew inputs during control-wheel steering phases.

Pitch trim This also uses a dual-dual fail-operational processor configuration. The system minimises residual longitudinal control forces during periods of control activity by varying the setting of the variable-incidence tailplane, when the autopilot is in use. During manual flight the crew can place the trim datum at any desired position. The system also incorporates Mach-trim compensation. Each active lane drives an independent trim actuator, and if the active lane is a defective lane the second lane will take over. All the above operations are handled by two identical pitch-trim computers. Two additional computers, which are for pitch compensation, produce additional pitch trim demands, using triple angle-of-attack data. When necessary, these units disengage the trim computers. A small control unit installed on the flight deck permits selection of either or both lanes of the overall trim system.

Yaw damper This sub-system is also based on a dual-dual layout, so that it will survive an initial failure and fail passively after a second failure. A control-panel allows selection of either system lane and each lane has an identical set of lateral accelerometer and yaw-rate gyro units, in each of which are two similar sensors. There is a connection from the lateral computer to each yaw-damper computer so that roll angle can be used to co-ordinate yawing manoeuvres.

Autothrottle This is a single-lane, self-monitoring system, analogous to one dual-lane set of autopilot/flight director/processors. It should disengage automatically on detection of a failure. The system has inputs appropriate to drive speed select, N_1 limit, go-around and automatic-landing flare modes. A single actuator drives separately clutched coupling units on each engine throttle, and dynamometric rods assure easy pilot-override when necessary. Autothrottle mode controls are incorporated in the glareshield control panel.

Test This is a sub-system dedicated to fulfilling the double function of fault location during flight for maintenance purposes, and monitoring of safety circuits. The detection of a defective unit is ensured by a continuous reading, in-flight and on the ground (the latter on request), of the 'go/no-go' signals exchanged between the afcs units and peripherals. This detection function is completed by a display of resulting failure codes, initiated on request. The safety-monitoring function consists of automatically testing the afcs sub-system comparators during a ground test.

The test system is duplicated to maintain integrity and complete segregation between the sub-systems. In flight, the system is continuously checked by a self-test programme

Airbus A300 autothrottle and autotrim equipment by sub-contractor Bodenseewerk

which shares processor time with the monitoring programme.

All test facilities are included in two identical digital computers. A test control and display panel is on the flight deck.

Speed reference system This simplex, self-monitoring facility generates the pitch commands required during take-off and go-around (after autoland disconnect) manoeuvres, and commands the autothrottles in N_1-limiting mode during abnormal angle-of-attack manoeuvres, or if windshear is encountered during approach. Take-off references are to hold V_2+10 knot or 18 degrees pitch attitude and on go-around it will hold $V_{ref}+10$ knots or 18 degrees pitch attitude. In the event of an engine failure it will hold V_{ref}.

Computer units
2 longitudinal computers each ¾ ATR Long
2 lateral computers each ¾ ATR Long
2 logic computers each ¾ ATR Long
2 pitch trim computers each ½ ATR Long
2 pitch compensation computers each ⅜ ATR Short
2 yaw damper computers each ½ ATR Long
1 autothrottle computer ½ ATR Short
2 test computers each ½ ATR Short
1 speed reference system computer ⅜ ATR Short

Control units
1 glareshield control panel
2 mode indicators
1 trim engagement unit
1 yaw damper engagement unit
1 test control and display panel

Other units
2 pitch dynamometric rods (on control columns)
2 roll dynamometric rods (on control columns)
2 trim actuators
2 lateral accelerometer units
2 yaw-rate gyro units
1 autothrottle actuator assembly
2 autothrottle coupling assemblies
2 autothrottle dynamometric rods
1 two-axis accelerometer unit

STATUS: in service.

Digital automatic flight control system for A300/A310/A300-600
New digital processors which replace the largely analogue elements of the existing automatic flight control system in the Airbus Industrie A300 wide-body airliner have been developed and put into service, so bringing the standard of this afcs up to that of an almost wholly digital system. This is available in all A300 production with the forward-facing crew compartment. The new system was first flown on the A300 in December 1980, and entered service with Garuda Indonesian Airways in January 1982. It is certificated for Category IIIB operation. The newer A310 version, the first of which entered service in April 1983, has the new digital automatic control system as standard.

The digital automatic control system provides flight augmentation functions (pitch trim in all modes of flight; yaw damping, including automatic engine-failure compensation when the autopilot is engaged, and flight-envelope protection); has a comprehensive complement of autopilot and flight-director modes which permit automatic operations from take-off to landing and rollout; a thrust-control system which operates throughout the flight envelope, has a derate capability, contains protection features against excessive angle-of-attack; and a fault-isolation and detection system for line maintenance.

Design has been in accordance with ARINC 600 and 700 characteristics, and has led to the adoption of ARINC 429 data-buses between the automatic control system processors and sensors. Four to six processors are used, comprising two flight-augmentation computers, one or two flight-control computers (the second unit being necessary only if Category III automatic-landing capability is required), together with a thrust-control computer. A second thrust-control option is available, and the system also includes a flight control unit providing pilot interface with the autopilot/flight-director and autothrottle functions, and a thrust-rating panel which allows crew access to thrust-limit computations.

A further new item of equipment is an engagement unit, with pitch trim and yaw damper engage levers, autothrottle-arm and engine-trim controls, which is mounted in the flight-deck roof panel. Two pitch and roll dynamometric rods are also used as control-wheel steering sensors. There are two trim actuators, an autothrottle actuator, two coupling units (one on each engine), and two further dynamometric rods connected to the throttle-control linkage.

Flight-deck controller equipment has been revised and a new autopilot/flight-director and autothrottle mode selector is installed in the centre glareshield. Variable data can be entered by rotating selector knobs, and shown by liquid-crystal display readouts. The various modes are engaged by push-buttons. Modes available are altitude capture and hold, heading select, profile (capture and maintain vertical profiles and thrust commands from flight-management system), localiser, landing and speed reference. Autothrottle modes include delayed-flap approach, speed/Mach-number select, and engine N_1 or engine pressure ratio selection.

The thrust-rating panel is mounted above the centre pedestal. This has comprehensive controls permitting thrust-levels to be selected, depending on operating mode, and providing for selection of such facilities as derated-thrust take-off.

The fault isolation and detection system has a dedicated maintenance-test panel which, on a 2-line by 16-character display, provides written alert messages based on fault information from automatic-testing activities. This is conducted in all line-replaceable units, and includes fault-isolation, tests to check for correct operation after maintenance action and on-ground auto-

matic-landing availability checks. Up to 30 faults, from six flights, can be stored and retrieved.

Production of the new system is jointly conducted by Sfena (as prime contractor) and Smiths Industries (UK) with Bodenseewerk (West Germany).

Computer units
2 flight control computers (10 MCU size)
1/2 thrust-control computer (8 MCU size)
2 flight augmentation computers (8 MCU size)

Control units
flight-control unit (glareshield)
thrust-rating panel (centre panel)
FAC/ats engagement unit (roof panel)
Maintenance/test panel

Other units
2 pitch dynamometric rods
2 roll dynamometric rods
2 trim actuators
autothrottle actuator
2 engine coupling units
2 engine dynamometric rods

STATUS: in service.

CGCC centre of gravity control computer
The CGCC controls the in-flight centre of gravity of the Airbus A310 aircraft in order to optimise performance and reduce fuel consumption.

The calculation is achieved by inputs from the fuel quantity indicating system; the fuel feed system (i.e. configuration) and associated systems such as the air data computer and flight management.

The computer is designed in accordance with ARINC 600 standards and as a self-monitored duplex computer system in a single box.

Format: 2 MCU
Weight: 3.7 kg
Power: 30 W 28 V dc

STATUS: in revenue flight service, certificated on A310-300 aircraft.

AP 205 autopilot
This system was designed at the request of the French Air Force for the French version of the Sepecat Jaguar. It is a simple and rugged system allowing pilots to fly hands-off while conducting other operational tasks.

Autopilot facilities include, in the pitch axis, pitch attitude- and altitude-hold modes, and in the roll axis, heading-hold mode. Trim functions provide automatic pitch trim and manual trim-assist in roll. Further modes for navigation or guidance purposes can be added. Modular construction is claimed to minimise maintenance actions, and the system is designed to be retrofitted into aircraft not equipped with autopilot. The first autopilot was delivered to the French Air Force in November 1981.

Dimensions: 124 × 210 × 253 mm
Weight: 5.4 kg
Power: <40 VA 200 V ac 400 Hz 3-phase
<1 VA 26 V ac 400 Hz 3-phase
<15 W 28 V dc

STATUS: in service.

AP 305 autopilot/flight-director system
This system is derived from an earlier Sfena system, the Tapir, which equipped civil-operated Fokker F27 and French Air Force Nord 262 light transports. For French Navy Nord 262s a radio-altimeter low-altitude mode has been added for low-flying operations over the sea. It is an analogue system with sideslip suppression, turn co-ordination and Category I

Airbus A310 has Sfena digital automatic flight control system

auto-approach capability. Flight-director facilities include altitude capture and hold, selected-heading capture and hold, VOR/ILS capture and tracking, attitude-hold, and appropriate operation indications and alerts.

Dimensions
(system computer) 421 × 125 × 194 mm
(control panel) 180 × 146 × 57 mm
(mode selector) 135 × 146 × 27 mm
There is also a sideslip detector and four servo actuators.
Weight: (excluding actuators) 8.6 kg
(including actuators) 21.6 kg
Power: <80 VA 115 V ac 400 Hz single-phase
<150 W 28 V dc

STATUS: in service.

AP 405 autopilot and autothrottle system

The AP 405 has been specially designed to meet the requirements of the French Navy for the low-altitude and high-speed operations of its carrier-borne Dassault Super Etendard, and can monitor radio height above sea level. Precise angle of attack-holding is provided by the autothrottle, including during carrier approaches when precise control is a deciding factor in achieving consistency and safety of operations.

Autopilot functions are pitch-hold, altitude capture and hold, heading hold (with control-wheel steering override), semi-automatic pitch trim and the provision for instinctive disconnect to make rapid flight-path changes. At very low altitudes there is also a radio-altitude hold mode with continuous flight-path monitoring. The autothrottle system can capture and hold a selected angle of attack.

Dimensions
(system computer) 409 × 198 × 94 mm
(control/indicator panel) 115 × 26 × 150 mm
(autothrottle actuator) 120 × 120 × 200 mm
(angle-of-attack selector panel) 37 × 45 × 50 mm
Total weight: 7.9 kg
Power: <60 VA 115 V ac 400 Hz single-phase
<1 VA 26 V ac 400 Hz single-phase
<60 W 28 V dc

STATUS: in service.

AP 505 autopilot

The AP 505 has been designed for Mach 2+ combat aircraft. Pilots of the Dassault Mirage F1, which is equipped with the system, are claimed to be satisfied with its reliability and ease of use. The system can be switched on before take-off, and engaged or disengaged by a hand-grip trigger in the control column. In basic mode it maintains the longitudinal attitude

held when the pilot releases the stick trigger, and either the heading or bank angle, depending on whether re-engagement is effected at a bank angle of less or more than 10°. Autopilot modes provide automatic flight at a pre-selected altitude, heading or VOR/Tacan/ILS radio-aid bearing. Limit on attitude hold facilities are ±40° in pitch, and ±60° in roll. More than 600 sets have been built and supplied in eight countries.

Dimensions
(system computer) 190 × 202 × 522 mm
(function selector unit) 132 × 142 × 27 mm
(heading selector unit) 99 × 35 × 80 mm
Total weight: 14.9 kg
Power: <100 VA 200 V ac 400 Hz 3-phase
<6 VA 26 V ac 400 Hz single-phase
<15 W 28 V dc

STATUS: in service.

AP 605 autopilot

The AP 605 is a digital flight control system developed for the Dassault Mirage 2000 supersonic combat aircraft. The high-capacity Sfena Series 7000 computer allows future extensions to the basic autopilot, to provide new modes suited to the particular missions flown by the aircraft. Flight control signals which pass between the autopilot and the Mirage 2000 fly-by-wire flight control system use a digibus serial data transmission system.

The AP 605 provides semi-transparent control, meaning that it is engaged or disengaged simply by releasing or taking hold of the control stick, which has a trigger-switch in the hand-grip. There is a high degree of internal monitoring, and computer design and organisation have been configured to reduce on-board maintenance. In basic mode the autopilot will hold pitch angle to any value within the range ±40°, or bank angle in the range ±60°. Additionally there are altitude capture and hold modes, preset altitude acquisition and fully-automatic-approach capability down to 200 feet. The nuclear M2000 version autopilot includes a terrain-following mode.

Sfena AP 605 digital flight control system developed for Dassault Mirage 2000

Dimensions
(system computer) 194 × 124 × 496 mm
(control unit) 24 × 146 × 115 mm
Total weight: 12.7 kg
Power: <100 VA 200 V ac 400 Hz 3-phase
<1 VA 26 V ac 400 Hz single-phase
<30 W 28 V dc

STATUS: entering service.

AP 705 autopilot

The AP 705, equipping the Dassault ATL2 ASW aircraft, is designed to provide precise flight-path control and a high level of safety at very low altitudes above the sea. The system can hold a course while maintaining altitude. These levels of performance are made possible by the quality of inertial data available on the aircraft, and by the versatility of the microprocessor-based computer. Built-in automatic testing enables the operation of the system and its safety devices to be checked before take-off, and also facilitates on-board maintenance.

The three-axis autopilot includes pitch trim and can drive a flight-director system. Autopilot modes permit pressure-altitude hold, glide-slope-beam tracking (in Category I weather minima) and radio altitude hold over the sea at very low-levels in reduced visibility. Lateral modes provide for holding the heading or course at the time of engagement, heading hold and homing or tracking on radio navaids, or navigation waypoints.

Dimensions
(system computer) 384 × 256 × 194 mm
(control unit) 200 × 164 × 67 mm
(servo-actuator) 185 × 183 × 101 mm
Total weight: 23 kg
Power: <50 VA 200 V ac 400 Hz 3-phase
<150 VA 28 V ac

STATUS: in production.

B 39 auto-command autopilot

Introduced into French Air Force/Dassault Mirage III interceptors since 1975 to replace older auto-command systems, the B 39 is now available for retrofit to other versions of the same aircraft type. It is a relatively simple system, but uses modern integrated circuit techniques to confer benefits in terms of performance, safety and maintenance standards. The new auto-command computer is physically interchangeable with the original equipment.

Autopilot functions are reduced to attitude and altitude hold modes. Attitude hold includes short-term capability, stability augmentation, and uniform artificial-feel load against load factor irrespective of flight conditions.

Dimensions: 264 × 200 × 140 mm
Weight: 6.5 kg
Power: <25 VA 200 V ac 400 Hz 3-phase
<1 A 28 V dc

STATUS: in service.

FDS-90 flight-director system

The FDS-90 system comprises an attitude director indicator and a navigation coupler/computer unit.

The attitude director indicator is a 4-inch (102 mm) unit which can operate autonomously using self-contained gyros and power inverters. It uses a ball-type real-world display and has a three-cue command capability. There are annunciators for go-around, decision height and flight-director mode monitor. The gyro can be caged.

The B152 nav coupler/computer is a small, panel-mounted unit which interfaces the attitude director indicator to navigation equipment. It has eight buttons which permit selection of different flight-director operating modes. These are:

Dassault Mirage F1 fighter has Sfena AP 505 autopilot

HDG: captures and tracks the heading selected on the horizontal-situation indicator
V/L: captures and tracks VOR and ILS localiser beam
BC: tracks the back-course localiser
ALT: maintains the altitude existing at the time of selection
GS: captures and tracks an ILS glideslope beam
VS: maintains the vertical speed that exists at the time of engagement
IAS: maintains the airspeed that exists at the time of engagement
FD: removes the flight-director bars from view.

There is also a button on the collective-pitch control to initiate the go-around mode and establish the correct pitch angle for climb-out. An additional safety feature is that, irrespective of the mode selected, any relevant deviation outside preset limits for more than 10 seconds causes the flight-director annunciation on the automatic director indicator to flash.

H140 HSI
Format: 4 ATI standard case
Weight: 2.5 kg

B152 Unit
Format: 3 ATI standard case
Weight: 1.3 kg
Power: <1 A 28 V dc
<0.25 A 26 V ac 400 Hz
Lighting power: <0.5 A 28 V dc
Equipment is fully TSO'd, and meets environmental category D0160

STATUS: in service.

Ministab stability-augmentation system

This is the most basic of three related helicopter stability/flight control systems produced by Sfena. It consists of three independent channels for roll, pitch and yaw. Each incorporates a computer with integral rate-gyro in series with the control linkage and having approximately 10 per cent control authority.

Upstream of the actuator a magnetic-brake/force-gradient assembly is used for stability-augmentation system control activity isolation, artificial feel and stick trim, and detection of pilot control inputs. In manoeuvring flight the pilot's control inputs are detected and integration terms removed to eliminate any resistance from the stability augmentation system. Dynamic stability is still ensured by pure rate terms, and integration is restored when the manoeuvre is complete. This technique provides good damping and positive static stability.

An inexpensive static transducer can be added to give long-term altitude hold.

Ministab installation weight and performance is specific to the helicopter type.

STATUS: in service.

Duplex Ministab stability-augmentation system

In duplex Ministab there are two cyclic channels which are monitored to detect any potential hardover commands. This feature extends the attitude hold qualities of the basic Ministab system, so permitting true 'hands-off' operation, a considerable reduction in pilot workload, and fulfilling all requirements for single-pilot IFR operation.

Monitoring in the cyclic channels (pitch and roll) is achieved by a separate computer which generates an 'image' of the real actuator position, and inhibits demands if a discrepancy is detected. Hardover commands are eliminated in this way, and any system failure causes reversion to natural helicopter stability.

The installation consists of five Ministab computers driving three actuators (the anti-torque channel is unmonitored because hardover faults are not critical). Altitude hold

Sfena Ministab stability-augmentation system was designed for such helicopters as Aérospatiale SA 340-series, manufactured in Britain as Westland Gazelle

through the pitch channel is an important part of the system, and heading hold through the roll channel is optional. 'Beep' trims are provided in pitch and roll.

The duplex Ministab has been demonstrated to meet all the requirements for CAA certification of single-pilot IFR operations.

STATUS: in service.

Helistab stability-augmentation system

Busy missions which include safety-critical operations require two pilots. Although sophisticated flight control systems can be used, fail-operational requirements are not essential, so Sfena has developed a system which offers substantial automatic capability with relatively low cost.

Helistab is a modular system consisting of a basic three-axis Ministab system, plus an autopilot computer and associated equipment. In normal operation sufficient Ministab functions are in use that this constitutes an adequate reversionary mode in the event of an autopilot failure.

The autopilot replaces the rate-gyro integrated terms by vertical and directional-gyro inputs, providing references for automatic control relative to attitude or heading. Basic autopilot functions are long-term pitch and roll attitude hold, long-term heading hold in cruise and hover, and auto-trim in pitch. Functions which can be added include turn co-ordination and heading capture in cruise, automatic

hands-off recovery from unusual attitudes, barometric altitude or airspeed hold in cruise, Doppler-based longitudinal and lateral speed control in hover, and collective-to-yaw coupling.

Helistab is transparent, therefore the pilot's control inputs are detected and the attitude hold terms removed to keep the autopilot from resisting pilot control in manoeuvring flight. The peripheral equipment used by Helistab makes the installation helicopter dependent.

STATUS: in service.

AFDS 95 flight control system

A two-, three- or four-axis modular autopilot, the AFDS 95 is designed for IFR operation with light and medium helicopters. In its most basic version, the system has an analogue computer to drive the electromechanical actuators in the control circuits and the trim and artificial feel actuators. With the addition of inputs from a vertical gyro it provides stabilisation and long-term attitude and heading hold, and can provide altitude hold and heading select functions if this information is available. More comprehensive versions incorporate a panel-mounted digital computer, with eight push-buttons, for selecting the more advanced control modes. The fourth axis facility is intended for operators needing automatic transition and hover capability. The system is approved to FAA TSO C9C.

STATUS: in service.

Sfena AFDS 95 autopilot system

IFS 86 integrated flight system

The Sfena IFS 86 integrated flight system has been designed specifically for commuter and corporate aircraft. It is a low cost, highly flexible system with many optional features so that it can be tailored to suit many applications. It can interface directly with communication and navigation receivers, such as those produced by Collins, King and Bendix. Data from the IFS 86 feeds instruments and displays using a standard ARINC 429 data-bus.

The primary flight display of the IFS 86 shows altitude, airspeed and vertical speed on a colour raster display: there are no 'conventional' displays in the system. All such basic data is processed in a single flight-data computer, while navigation and auto-pilot related data is handled in the flight-guidance computer. Each of these computers has built in a fault detection and isolation system. With all the options incorporated the IFS 86 includes 23 line-replaceable units, weighs 72 kg and requires 640 watts' power. In the basic version these figures are reduced to 18, 60 kg and 585 watts. The system was initially evolved for the Aérospatiale/Aeritalia ATR 42 42-seat feeder-liner.

STATUS: in development.

Artist's impression of Sfena IFS 86 helicopter instrument panel

Sfim

Société de Fabrication d'Instruments de Mesure, 13 avenue Marcel-Ramolfo-Garnier, 91344 Massy Cedex
TELEPHONE: (1) 69 20 88 90
TELEX: 692164

AFCS 85 autopilot/flight-director system

The basic AFCS 85 system is a simplex series three-axis autopilot with long-term hands-off VFR (visual flight rules) capability, including heading hold over the entire flight envelope and baro-altitude or airspeed hold in cruise.

Adding the FDC 85 flight-director coupler, and pitch and roll autopilot monitor, gives single-pilot IFR capability by providing facilities such as automatic lateral and vertical guidance and by driving command bars on the attitude director indicator.

Automatic trim is standard and VFR versions with FDC 85 or altitude, airspeed, and heading-select modes are available. The system has been chosen for the Aérospatiale Alouette III, Gazelle, AStar and Twin Star helicopters. More than 800 systems had been ordered by early 1986.

STATUS: in production.

AFCS 155 autopilot/flight-director system

The AFCS 155 is a series duplex, fail-passive, autopilot. Optimised for both single and dual-pilot IFR operations, covering the whole flight envelope from hover to VNE, it is IFR-certificated on the Aérospatiale Super Puma and the Dauphin helicopters and can be integrated with Sfim-developed couplers. These include the FDC 85 (three- or four-axis) and FDC 155 four-axis system. In the duplex configuration each channel has its own power supplies, sensors and interconnections.

Basic functions include long-term attitude- and heading-hold, turbulence compensation, collective-link mode and auto-trim. All autopilot configurations satisfy single-pilot IFR requirements and there are three upper modes: heading select, altitude- and airspeed-hold.

'Fly-through' or 'transparent' handling characteristics allow the pilot to make quick attitude- and heading-changes while benefiting from dynamic damping. On releasing the controls the autopilot returns to long-term stabilised flight at the previously set attitudes. If new attitude settings are required the pilot can either use the 'stick-release' button on the cyclic pitch grip, or he can change the settings at a slow rate by using the stick-top four-way 'beep' trim button. The beep trim button is also used to alter slowly the reference airspeed

Sfim FDC 155 helicopter flight control system all-mission coupler

when the automatic airspeed hold mode has been selected. All versions of the system include automatic trim which keeps the stick centred so that the autopilot has full authority.

Additional flight-director coupler facilities provide IFR automatic navigation, radio-navigation and approach capabilities, including steep-approach MLS beam captures and track. There are also additional modes for anti-submarine warfare and offshore or search-and-rescue operations. The latter includes automatic pattern flying, automatic up and down transitions to and from selected radio altimeter height, Doppler/radio height or sonar cable hover, and low Doppler-speed automatic hold. Compatible couplers include FCS 85 three- and four-axis navigation and approach couplers (latter unit has Category II capability), and FCS 155 all-mission coupler. More than 700 AFCS 155 systems had been ordered by early 1986.

Weight: (autopilot computer) 8 kg
(servo amplifier) 4.2 kg
(autopilot control box) 1.6 kg
(4 actuators) 1 kg each
(3 trim servos) 1.3 kg each
(baro-sensor) 1.5 kg
(flight-director computer) 2 kg
(flight-director coupler box) 2 kg
(collective pitch motor) 2 kg

STATUS: in production.

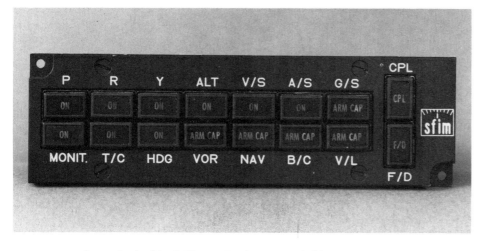

Controller for Sfim 85T31 single-pilot automatic flight control system

AP 146 autopilot

This autopilot is designed for anti-submarine aircraft equipped with MAD systems. It provides accurate radio-height hold, with filtering to compensate for swell up to sea state 5. Safety circuits inhibit active or slow failures in the pitch channel for extended low-altitude operation.

STATUS: in service.

AP 155 autopilot

Optimised for single-pilot IFR operation, the AP 155 is a two-channel system. It continues to maintain IFR performance after one failure.

STATUS: in service. The system is installed on Aérospatiale AS 332 Super Puma and AS 365-N Dauphin helicopters.

AP 85 autopilot

The AP 85 autopilot is a single-channel system, and being of modular layout can be easily tailored to suit a variety of light and medium helicopters. Configurations can vary from two-axis VFR systems to suites for three-axis, single-pilot operation, with coupled flight director. The CDV 85 is a flight-director and coupler for cruise and let-down which can be used for manual or automatic approaches, and permits single-pilot IFR operation with the Aérospatiale TwinStar and Dauphin 2 helicopters.

STATUS: in service on Aérospatiale AS 316/319 Alouette III, AS 341/342 Gazelle, AS 350 Ecureil/AStar and AS 355 TwinStar helicopters.

Thomson-CSF

Thomson-CSF, Division des Equipements Avioniques, 178 boulevard Gabriel-Peri, 92240 Malakoff
TELEPHONE: (1) 46 55 44 22
TELEX: 204780

Engine interface unit

The engine interface unit EIU operates during ignition, start-up and indication of engine parameters. It gathers analogue, discrete and digital data originating from the cockpit and other systems (central fault detection and isolation unit, Fadec, landing gear, slat position) and after suitable processing transmits them to the Fadec. This data is passed in the shape of digital and discrete (ARINC 429) signals.

Each A320 will be equipped with two EIUs, which are identical and interchangeable.

Format: 3 MCU per ARINC 600
Power: 28 V dc

STATUS: in development.

GERMANY, FEDERAL REPUBLIC

Bodenseewerk

Bodenseewerk Gerätetechnik GmbH, Postfach 1120, D-7770 Überlingen
TELEPHONE: (7551) 811
TELEX: 733924

Digital thrust-control computer

Bodenseewerk is responsible for the thrust-control computer section of the digital flight control system on the Airbus A310 twin-engined wide-body airliner. The French company, Sfena, is in overall charge of the programme, and the UK firm, Smiths Industries, is also involved. The thrust control computer is an extension of Bodenseewerk's activities in that area, which started with the autothrottle on Lufthansa's Boeing 707 fleet, and continued with the Airbus A300 autothrottle. The company delivered the first A310 unit in June 1981.

The thrust-control computer controls the fuel flow, and hence engine thrust, according to the aircraft speed selected by the pilot. It calculates the maximum admissible engine rpm or pressure ratio, and controls it during take-off according to circumstances (runway length, gradient, atmospheric conditions). Furthermore, it provides protective functions ensuring the maintenance of four safety limits: maximum flight speed, maximum admissible engine speed, minimum flight speed, and maximum angle of attack.

The digital electronics are housed in an ARINC case. The computer is a redundant system using up-to-date technology microprocessors, with dissimilar software to guard against the danger of a software failure not being detected. The mechanical clutch units and the actuator are identical to those of the A300.

STATUS: in production for A310.

Digital engine control unit

A digital engine control unit for the Turbo-Union RB.199 turbofan engine used in the Panavia Tornado is being developed by Bodenseewerk in close collaboration with MTU. It will provide full-authority engine control and improved engine response without exceeding parameters. It will also feature built-in advanced self-test for pre-flight and in-flight checks.

The system comprises two self-monitoring lanes for dry engine operations, a third lane being dedicated to afterburner control. The decu interfaces ensure that it is fully interchangeable with the existing analogue main engine control unit of the RB.199.

STATUS: in pre-production phase.

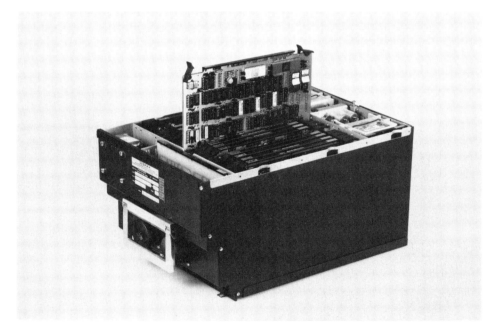

Bodenseewerk digital thrust-control computer used in Airbus Industrie A310

Bodenseewerk digital engine control unit for Turbo-Union RB.199

ITALY

OMI

Ottico Meccanica Italiana SpA, Via della Vasca
Navale 79, 00146 Rome
TELEPHONE: (6) 547 881
TELEX: 610137

FQG-28 fuel quantity gauging system

This is a capacitance sensor system developed
for SIAI-Marchetti S205, S211 and S260 light
aircraft. The system uses an electronic unit to
analyse sensor data in such a way as to correct
for tank shape. This method of characterising
probe data by software allows for considerable
commonality of equipment between aircraft
types, and consequently is less expensive than
a system with mechanical sensor compensation. The basic system can be adapted to

many aircraft types. It measures quantity to
within 2 per cent and provides a low-level
indication through warning lamps. Mean time
between failures exceeds 5000 hours.

Dimensions: (electronic unit) 160 × 102 ×
33 mm
(indicators) each 60 × 51 mm dia
(probes) to suit application
Weight: 2.1 kg
Power: MIL-STD-704C

STATUS: in production.

UNITED KINGDOM

Dowty

Dowty Electronics Limited, Dowty Group, Arle
Court, Cheltenham, Gloucestershire GL51 0TP.
TELEPHONE: (0242) 521 441
TELEX: 43176

Higher harmonic control system

Dowty announced in 1984 that it was collaborating with Westland to perfect a system to reduce
vibration in helicopters. Known as the higher
harmonic control system, it will compensate for

out-of-balance forces acting on rotors by
modulating their pitch angle over the course of
each rotation.

STATUS: in development.

DSIC

Dowty and Smiths Industries Controls Limited
(DSIC), Arle Court, Cheltenham, Gloucester-
shire GL51 0TP
TELEPHONE: (0242) 527 888
TELEX: 43176
FAX: (0242) 582 579

DSIC was formed in 1977 and promotes the
aero engine control activities of the Dowty
Group and Smiths Industries. DSIC work-
centres (namely Dowty Fuel Systems, Dowty
Electronics and Smiths Industries at Basingstoke) have supplied equipment for more than
100 different types of military and civil aircraft.
The world's first full authority electronic engine
control to be certified for civil flight (for
Concorde's Olympus 593 engines) was designed and developed by Dowty Electronics in
the 1960s. More recently, in 1982, DSIC had
further success with the world's first flight of a
single engined V/Stol aircraft equipped with a
full authority digital engine control (Fadec).
The hydromechanical element of the control
was designed and developed by Dowty Fuel
Systems with Smiths Industries responsible for
the electronics. DSIC has now received production contracts for this equipment.

Control system for vectored-thrust powerplants

Initial tests began in 1979 and continued with
flight-tests conducted in a British Aerospace
Harrier GR3 V/Stol aircraft. Most conventional
flight conditions had been demonstrated by
late 1982 with hovering flight-tests in 1983.
Flying commenced in 1984 with a type-
approved system. Further testing continued
both in the UK and US until 1985.

The Digital Engine Control System (DECS)
for the Rolls Royce Pegasus engine completed
type-testing in early 1986; certification of the
engine with the new control system was
achieved in May 1986.

The production of DECS, which comprises of
a hydromechanical fuel metering unit and two
digital engine control units, has begun and
initial deliveries were made early in 1986 to
meet orders for engines which will equip the
Harrier aircraft of the Royal Air Force, US
Marine Corps and Spanish Navy.

Current orders, which take production
through to 1987, are valued at £30 million. DSIC
says that potential orders, for delivery up until
1991, should amount to £85 million.

STATUS: in production.

Engine control for RPVs

DSIC's involvement in the production of modular digital engine control systems also includes
programmes for gas turbine powered unmanned aircraft. DSIC received a contract to
manufacture engine controls for the Teledyne
CAE 373-8 engine during 1985 and is currently
working on other similar projects. This follows
earlier success with the engine control system
for the Hindustan Aeronautics PTAE-7 engine.

The control systems' modular construction
enables modifications to be incorporated
quickly to meet changes in engine specification,
installation, fuel type or aircraft operating
characteristics. The control modules are fully
interchangeable, compact, and can be installed
in the nose cone of an engine if necessary.

STATUS: in production.

Advanced military engine control systems

Extensive research and development work is
being carried out on a new generation of digital
afterburner control systems which are being
developed for advanced military engines of the
1990s which will supersede the RB.199 variants.
Also in conjunction with Hindustan Aeronautics
DSIC is jointly developing a full authority digital
control for an engine being conceived by the
Gas Turbine Research Establishment (GTRE)
of India.

The equipment for the GTRE project comprises digital electronic control unit, the main
engine control, the reheat control (including
pumps) and nozzle/variable actuation systems.

STATUS: under development.

N_1/N_2 limiter

The DSIC N_1/N_2 electronic limiter is currently in
service with several major airlines and can be
supplied on new engine equipment or on a
straight replacement basis for older types of
limiter.

The main purpose of the unit is to provide
engine speed limiting to preset datums, thus
protecting the engine against overspeeds which
arise from slam accelerations, control system
failures and certain engine failures.

STATUS: in production.

BVCU bleed valve control unit

The DSIC electronic BVCU has a main lane
with back-up and a two-digit BITE. The BVCU
provides automatic control of the intermediate

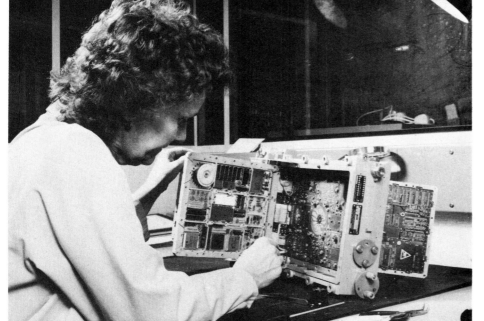

DSIC digital electronic control unit for Pegasus engine digital engine control system

pressure (IP) and high pressure (HP) compressor bleed valves. It replaces a complex system of variable guide vanes, thereby improving engine fuel consumption and preventing engine surge.

STATUS: in production.

V.2500 engine accessories
DSIC has been awarded contracts for the following accessories on the new International Aero Engines V.2500 engine:

Variable stator vane actuator (VSVA)
The VSVA positions four variable stages of engine vanes – the inlet guide vanes and three sets of stator vanes.

Electronic engine control (EEC)
The EEC supplies an electrical current to the torque motor which in turn operates a spool valve. The spool valve switches metered fuel flow to either side of the actuator piston. Actuator piston feedback to the EEC is provided by a dual-wound LVDT assembly.

STATUS: in production.

DSIC nose cone-mounted electronic fuel control system and alternator for an RPV

GEC Avionics
GEC Avionics, Airport Works, Rochester, Kent ME1 2XX
TELEPHONE: (0634) 44400
TELEX: 96333

Flap/slat control computer for Airbus A310 and A300-600
The flap/slat control computer is part of a fully-monitored, fail-operative, duplex system. It is a digital system using dissimilar redundancy to satisfy high-integrity requirements. Built-in safety features include protection following a mechanical jam, breakage or runaway.

STATUS: in production.

Flap/slat control computer for Airbus A320
This unit, for the latest member of the Airbus family, is based on the same techniques of dissimilar redundant hardware and software as the earlier devices but takes advantage of the most recent generation of microprocessors. This has resulted in reduced weight and volume.

STATUS: in development.

Automatic flight control system for British Aerospace One-Eleven
Several automatic flight control system variants are in service, ranging from three-axis auto-stabilisation units with autopilot approach facilities meeting ICAO Category I requirements to a fail-passive automatic-landing system which includes automatic-throttle control and a minimum decision height of 60 feet (18.3 metres).

STATUS: in service.

Automatic flight control system for Concorde
GEC Avionics jointly developed with Sfena (France) the Concorde automatic flight control system (afcs), which comprises eight separate systems with no fewer than 38 individual units, and provides fully-automatic control from climb-out to automatic landing. The system is duplex-monitored and fail-operational to meet the operational requirements of ICAO Category

Airbus A300/A310 flap/slat control computer

IIIA, and is cleared for automatic landing with 4.5 metres decision height and 200 metres runway visual range. It was the first afcs to use a push-button controller fitted into the flight-deck glareshield.

STATUS: in service.

General-purpose three-axis auto-stabiliser
Design and development by GEC Avionics of a general-purpose three-axis auto-stabilisation system, with integral gyro pack and self-test, was started in 1978 and the system is now in quantity production for an undisclosed customer. The entire system is enclosed by a single ½ ATR box.

STATUS: in production.

Auto-stabilisation system for Harrier
Designed and developed specifically for V/Stol operations, this system provides three-axis stability augmentation during vertical take-off, transition, wing-borne flight, hover and landing. Over 400 systems have been produced for all UK, US and Spanish armed forces customers of the British Aerospace Harrier/AV-8A (which does not have an autopilot). Rate gyros and self-test facilities are included in each processor.

Individual line-replaceable units are: pitch/roll auto-stabilisation computer, yaw auto-stabilisation computer and lateral accelero-meter unit.

STATUS: in service.

GEC Avionics auto-stabilisation system for Royal Air Force Harrier, together with additional computer and sensor unit, constitutes autopilot for Royal Navy Sea Harrier version

GEC Avionics advanced subsonic aerial target autopilot

Autopilot for Sea Harrier

The Royal Navy/British Aerospace Sea Harrier is the first Harrier variant to have an autopilot to reduce the pilot's workload. Designed to work in conjunction wiith the Harrier auto-stabilis-ation system, the autopilot currently provides roll and pitch attitude hold, heading hold, barometric height hold, and self-test. An autotrim function is being added as part of the current Sea Harrier mid-life update programme.

STATUS: in service. Mid-life update in progress.

Control systems for RPVs and drones

GEC Avionics currently produces flight control equipment for target and surveillance RPVs, both for the air vehicles themselves and their associated ground stations. One of the longest-running programmes is the manufacture of the automatic flight control system, with auto-stabilisation, for the Australian Jindivik target drone. Command signals from a ground-control station are received on the aircraft via radio link, and summed with auto-stabilisation signals generated on-board. The original Mk L5 system, comprising six types of unit, has given way to the Mk 4 system, which has a flight control computer, accelerometer, and three-axis rate-gyro unit. More recently the company has begun volume production of a flight control system for its Phoenix RPV, chosen in February 1985 by the British Army for battlefield surveil-lance and real-time targeting.

The company also supplies flight control equipment for the Falconet subsonic target vehicle produced by the UK's Flight Refuelling. The ground command and control unit is closely coupled with the range tracking radar to reduce the complexity of the airborne equip-ment. GEC Avionics builds a universal drone pack that has been used on drone conversions of the Hawker Siddeley Sea Vixen. Fitted to the ejector-seat rails, the pack comprises computer, accelerometer, controller and three-axis rate-gyro unit.

The company also produces the height-keeping computer for Flight Refuelling's sea-skimming towed target. It commands the target to maintain a pre-determined radio height irrespective of altitude changes by the towing aircraft.

Flight control system for Lynx

GEC Avionics has produced several variants of the basic Lynx automatic flight control system. Facilities common to all variants include pitch and roll attitude stabilisation with yaw-rate stabilisation in heading hold. Army versions of the Lynx also have a barometric height hold, using collective and lateral acceleration control. Two naval versions provide radio-altitude height hold using collective control, or radio-altitude acquire and hold with automatic transition to and maintenance of the hover for sonar use in anti-submarine warfare operations.

STATUS: in service.

Australian-designed Jindivik drones have GEC Avionics flight control equipment

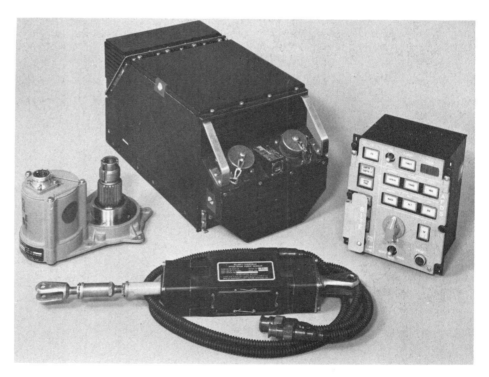

Panavia Tornado equipment by GEC Avionics includes control panel, computer, pitch stick-force sensor and drive motors

GEC Avionics Flight control system for Westland/Aérospatiale Lynx helicopter includes computer, controller and drive units

Full flight regime autothrottle for 747

The GEC Avionics autothrottle is installed on Boeing 747 airliners, and is in service with more than 40 airlines. As well as the production line in Rochester, the Atlanta facility of GEC Avionics Inc in the USA also supplies autothrottles. The full flight-regime autothrottle system computer is programmed with the aircraft's engine pressure ratio, speed and Mach number control laws which permit it to control the throttles automatically in one of those three modes, selectable on a control/display unit on the flight-deck.

For take-off, climb and go-around, the system controls engine pressure ratio. Speed or Mach control may be selected in cruise, and speed control is used in the descent, hold and approach phases. During autopilot-coupled landing the throttles are retarded automatically. All four throttles are controlled to the same position.

The system accepts angle of attack, flap position, total air temperature, air speed, Mach number and aircraft attitude as inputs. As well as the three primary modes, sub-modes are provided for aircraft and engine protection. These are: control to a safe minimum speed, flap placard-speed protection, engine over-boost protection, and constant throttle hold during the critical take-off regime. In the engine pressure ratio mode the crew can select the engine pressure ratio for take-off, climb, cruise, or go-around, and the appropriate engine pressure ratio is acquired and maintained, though the crew can modify its limit if required. In the speed mode a commanded speed set on the aircraft's autopilot control panel is acquired

GEC Avionics full flight regime autothrottle used in Boeing 747

and maintained, while in the Mach mode the existing Mach number is kept constant. In the latter two modes, engine pressure ratio protection is provided as a sub-mode.

As well as governing the 747's throttles the full flight regime autothrottle system sends the selected air speed for display on the airspeed indicators, and generates a fast/slow indication on the attitude director indicators. Total air temperature and engine pressure ratio limits are displayed to the crew and the target engine pressure ratio is shown on each engine

pressure ratio indicator. Built-in test equipment is provided and the system status and warning indications are incorporated on the flight mode annunciator panels.

The autothrottle system increases engine life and saves fuel, responding rapidly to changes in the aircraft configuration or flight conditions and avoiding unnecessary throttle excursions. Pilot work-load is also reduced.

STATUS: in production.

Louis Newmark

Louis Newmark, Aircraft and Instrument Division, 80 Gloucester Road, Croydon, Surrey CR9 2LD
TELEPHONE: (01) 684 3696
TELEX: 264004

Louis Newmark activities in helicopter auto-pilots stems from the mid-1950s, when the company undertook to manufacture under licence the Lear Siegler autopilot for the Sikorsky S-58, production of which was being licensed by Westland from Sikorsky under the UK designation Wessex HAS1. The UK company then followed up with its own design of autopilot for the Wessex HAS3.

FN31 flight control system

The Type FN31 is a versatile autopilot for the Royal Navy/Westland Sea King helicopter incorporating stability-augmentation, heading hold, barometric height hold, radio height hold, automatic transitions, hover control, and plan-position control. It is a simplex system derived from the company's FN30 duplex system, currently in service with the Royal Navy.

FN31 flight control system facilities are engaged on the centrally-mounted pilot's controller unit. This indicates the modes selected and disengagement or engagement of any of the system's four channels can be effected. Servo-amplifier output selectors actuate variable-orifice valves in the auxiliary servo unit, and these act in series with the pilot's own demands, but with limited authority. Extended control authority is by spring mechanisms in the auxiliary servo-unit and by automatic inching of the cyclic trim during automatic transitions.

Selection of the stabiliser facility gives a combination of attitude and heading hold, with three-axis damping of aircraft motion. The flight attitude of the helicopter becomes stabilised relative to the cyclic column position, and heading hold takes effect whenever the pilot completes any heading manoeuvre. A heading

Royal Navy Sea King HAS5 has FN31 flight control system

trim control is provided for controlling the helicopter during flat turns.

In cross-country flying, barometric altitude and stabilisation facilities may be engaged together or separately. By pressing a manoeuvre button, situated on the hand-grip of the collective lever, the pilot can disengage altitude hold temporarily, and fly the helicopter to a different altitude.

Radio altitude hold can be engaged separately. When this mode is engaged the helicopter is stabilised relative to an inertial height, which is derived from radio altitude and vertical acceleration data. This stabilises the helicopter when flying over undulating terrain. With this mode engaged the pilot can make controlled changes in altitude by turning the 'set radio height' control knob to other settings.

When the pilot engages 'trans down', the helicopter begins a controlled descent to hovering altitude, at the same time decelerating

to zero ground-speed. Forward speed information is obtained from a Doppler-radar sensor. The helicopter hovers at an altitude which is preset by the pilot on a 'set hover height' control. When 'trans up' is engaged the helicopter climbs from the hover altitude to the altitude set on 'set radio height', and accelerates to the ground-speed set on the 'set exit speed' control. Both transition manoeuvres are programmed for completion in minimum time.

In anti-submarine warfare operations, at the completion of a transition to hover, a sonar can be lowered, and as it enters the water the pilot can select 'cable hover', after which plan-position and height are controlled relative to the submerged sonar. On retrieving the sonar the pilot re-selects 'Doppler hover' in preparation for an up-transition. In rescue operations the pilot can set the hover switch at 'Doppler' and engage an auxiliary hover-trim control, which allows the winch-operator to command small increments of fore-aft and lateral ground-speed.

Four attitude indicators are included in the flight control system. Attitude signals for the indicators, and for other systems in the aircraft, are obtained electrically from two vertical-gyro sets and a repeater platform. Each vertical-gyro unit transmits synchro signals to the main indicator of one pilot and the standby indicator of the other pilot, ensuring that neither pilot suffers complete loss of attitude information through a single system failure. As the accuracy of transmitted information depends on having accurate alignment between the repeater platform-unit and the associated vertical-gyro unit, these are mounted on a common base-plate.

Dimensions and weight: (pilot's controller) 240 × 147 × 117 mm, 6 lb (2.7 kg)
(channel selector) 147 × 150 × 144 mm, 3.5 lb (1.5 kg)
(stick canceller) 90 × 58 × 51 mm, 0.5 lb (0.2 kg)
(altitude controller) 230 × 141 × 135 mm, 2.5 lb (1.2 kg)
(amplifier) 531 × 286 × 104 mm, 22 lb (10 kg)
(2 vertical-gyro units) 230 × 160 × 134 mm, 8.5 lb (3.8 kg) each
(repeater platform) 410 × 205 mm dia, 19.5 lb (8.9 kg)

(2 5-inch attitude indicators) 141 × 141 × 230 mm, 4.8 lb (2.2 kg) each
(2 3-inch attitude indicators) 83 × 83 × 250 mm, 2.8 lb (1.2 kg) each
(rate gyro unit) 103 × 103 × 51 mm, 1 lb (0.5 kg)
Power: 220 VA 200 V ac 400 Hz 3-phase
100 W 28 V dc

STATUS: in service.

LN400 flight control system

The LN400, in production for the Westland 30, is designed for single-pilot IFR operation and provides duplex stability-augmentation for pitch, yaw, roll and collective axes. Automatic trim in pitch and roll ensures mid-point operation of the series actuators, so ensuring full authority at all times. The system incorporates collective-acceleration control, heading hold and yaw-trim facilities. A radio or baro-metric-height hold can be provided as an option. The system comprises a computer, controller and sensor.

Computer unit (NDN 8919-01) This is contained in a standard ½ ATR Short case with front plug connectors. The computing circuits are of modular form using plug-in circuit boards for each lane. Simple manipulation of control-law parameters is effected by appropriate linkages on the printed circuit boards. Duplex integrity is assured throughout, each lane of each channel containing an independent stabilised power supply. The lane isolation is maintained by the embodiment of separate front plug connectors.

The duplex stabilisation boards of each channel are identical, each board containing the necessary electronics for interfacing with the appropriate sensor, processing the control-law signals and providing the series-actuator drive circuitry and control logic. In addition, simplex boards contain the heading-hold circuitry and provide the parallel-actuator drive for automatic trim in the pitch and roll axes.

Growth capability in the computer unit allows for an optional barometric or radio-height hold facility.

Pilot's controller (NDN 8921-01) This unit provides push-buttons to engage and disengage both lanes of pitch, yaw, roll or collective channels. There are two dual-purpose meters that may be used to monitor each of the four channels during flight or alternatively (as part of the built-in test equipment) to diagnose channel or lane faults.

The auto-stabilisation functions engage as soon as power is available, and illuminate the collective channel push-buttons. On achieving remote-gyro verticality, pitch, roll and yaw channels of both lanes are engaged by depressing the 'ase engage' button. When both lanes of all channels are engaged, feedback signals

Pilot's controller for Louis Newmark LN400 flight control system has engagement buttons for each axis of two-lane autopilot, and other facilities

Westland 30 Civil Lynx has Louis Newmark LN400 automatic flight control system

from the series actuators are compared in each channel. If a pre-determined level of disparity is detected between a pair of channel lanes, the relevant fault light illuminates, and the central-warning system is informed. The pilot then takes appropriate action, based on built-in test indications, to disengage one or both channels.

The selected heading may be fine trimmed by using a 'beep' switch on the controller, and there is space for an optional height hold control.

Sensor unit (NDN 8923-01) This contains duplex yaw-rate gyros, duplex vertical accelero-meters and has growth capability for additional sensors. Each sensor is excited from separate power supplies and electrical connections are via two connectors, thus maintaining lane isolation. An external three-wire heading signal is required to complete the total system.

Dimensions and weight: (computer unit) 125 × 194 × 319 mm; 12 lb (5.5 kg)
(pilot's controller) 137 × 95 × 165 mm; 5.3 lb (2.4 kg)
(sensor unit) 130 × 65 × 212 mm; 4.1 lb (1.9 kg)
Power: 48 VA 115 V ac 400 Hz single-phase 210 W 28 V dc

STATUS: in production and service.

LN450 digital flight-path control system

The size and complexity of the North Sea oil and gas activities have given rise to the need for a dedicated search and rescue service. Recognising this need, Bristow Helicopters joined with Louis Newmark and HM Coastguard to develop an aircraft and system optimised for the task.

The outcome of this activity is a Bell 212 transport helicopter fitted with a Darnell forward-looking infra-red system (to detect survivors in the water), a Racal-Decca Doppler navigation system to measure speed and drift, and the Louis Newmark LN 450 flight-path computer. With this combination, the aircraft after detecting and locating a survivor can carry out an automatic, pre-programmed recovery manoeuvre bringing the aircraft into wind over the survivor so that the winch-operator can begin work.

The LN 450 is a digital four-axis micropro-cessor-based system providing special-manoeuvre information or commands to the flight-director or autopilot. It can be used either in coupled or uncoupled mode. It can be installed either as a single-channel system for en route flying, or as a two-channel system for search and rescue work. Principal crew interfaces are a pilot's controller and a computer unit.

Pilot's controller (NDN 9642) This incorporates mode-dedicated push-button switches, each of which illuminates when the appropriate logic

circuitry is selected. Digital light-emitting diode displays are provided for 'set hover height', 'radio height' and 'indicated airspeed hold' inputs. The system incorporates high-integrity self-testing and continual amplitude- and rate-monitoring, with fault indicators on the controller.

Duplex or simplex inputs from standard aircraft sensors are acceptable. The following coupled/uncoupled modes are available: survivor overfly, winch-man hover-trim, transition up/down, radio or barometric height hold, vertical speed hold, indicated airspeed hold and heading hold. There are navigation mode selections for ILS/LOC, VOR, area-navigation, microwave landing system, back-course approach and go-around. Except for go-around, all navigation modes include capture and track logic. There is growth capability to incorporate an ASW mode.

Computer unit (NDN 9634) Each computer unit contains a signal conditioner, micro-processor and power supplies. In each unit the sensor analogue input signals are conditioned, multi-plexed, and fed into the microprocessor, where autopilot and flight-director information for all four axes is computed in accordance with the programmed control laws. The autopilot outputs control the helicopter through the existing stability-augmentation system, while the flight-director outputs drive command bars on the cockpit attitude-director indicator units. To maintain maximum integrity the system is arranged so that only two-channel outputs for either autopilot or flight-director are active from each computer unit. All outputs are fed back for signal self-monitoring, and should a discrepancy in excess of a pre-determined threshold occur, the computer unit automatically disconnects the system then reverting to simplex. If the microprocessor cannot establish in which the fault has occurred, the pilot must ascertain the corrective action. Louis Newmark and Bristow Helicopters commenced specialised joint flight-development of a LN450 system on a Bell 212 helicopter in mid-1982.

Dimensions and weight (duplex system)
(2 computer units) 91 × 193 × 320 mm; 7.7 lb (3.5 kg) each
(pilot's controller) 146 × 162 × 165 mm; 6.1 lb (2.8 kg)
(a simplex installation has only one computer unit, and the pilot's controller weighs 3.3 lb (1.5 kg))
Power: 0.4 VA 115 V ac 400 Hz single-phase 63 W 28 V dc

STATUS: in production. CAA certification for system in Bell 212 helicopter in IFR role was awarded in August 1983. This event is claimed as a 'world first'. Certification for Westland S-61N, with auto-hover capability, followed in late 1984. The system is a standard fit on the Westland 30.

Flight control system for EH 101
Louis Newmark is involved in a design study for the Anglo-Italian EH 101 helicopter. The flight control system is expected to be based on a four-axis digital triplex design.

Nine EH 101 prototypes are to be built, the first of which will fly in 1987 with an LN 450 system.

Lucas Aerospace
Lucas Aerospace Limited, Electronic Systems and Equipment Division, Electronic Systems, York Road, Hall Green, Birmingham B28 8LN
TELEPHONE: (021) 777 3222
TELEX: 336755

SDS-200 engine control computer
In 1981 Lucas completed a four-year development programme for its SDS-200 Fadec system, culminating in bench-testing on Avco Lycoming's 800 shaft horsepower advanced technology demonstrator engine. The SDS-200 is a single-channel device which controls fuel flow and intake geometry for helicopters. Lucas is continuing to develop it with the objective of full production.

Dimensions: 7 × 3.5 × 6.5 inches (178 × 89 × 165 mm)
Weight: 5.5 lb (2.5 kg)
Mounting: on engine

STATUS: under development.

SDS-300 engine control computer
Avco Lycoming has chosen the SDS-300 for the latest versions of its LTS 101 family of turboshaft engines, starting with the LTS 101-

Lucas Aerospace SDS-200 engine control computer

750 series. The system modulates fuel flow via a hydromechanical metering system governing compressor and turbine speeds, limiting exhaust-gas temperature, and featuring automatic starting and surge recovery. In multi-engined installations the SDS-300 performs torque limiting and load sharing.

During 1984 the Lucas SDS-300 Fadec became the first such system to be certified for use on a helicopter, in conjunction with the Avco Lycoming LTS 101 engine powering the

ESC-102 engine supervisory control unit for Rolls-Royce RB.211-535

US Coast Guard/Aérospatiale HH-56A helicopters. The SDS-300 unit has been developed in the Lucas facility in New Jersey, USA.

Dimensions: 7 × 3.5 × 6 inches (178 × 89 × 152 mm)
Weight: 5 lb (2.3 kg)
Mounting: airframe or avionics bay
Qualification: FAR Part 33

STATUS: in production for Avco-Lycoming LTS 101-750 shaft-turbine engine for light helicopters.

SDS-400 engine control computer
This is similar to the SDS-300 in concept and has been developed for the auxiliary power unit of the McDonnell Douglas AV-8B, also supplied by Lucas. As well as replacing the hydromechanical fuel control, the SDS-400 sequences and monitors the auxiliary power unit, permitting in-flight power generation for the first time.

Dimensions: 6.5 × 5 × 4.5 inches (165 × 127 × 114 mm)
Weight: 4.5 lb (2 kg)
Mounting: airframe

STATUS: in production for US Navy.

SDS-500 engine control computer
Lucas Aerospace is to provide the SDS-500 full authority digital engine control system for the Rolls-Royce Turboméca RTM 322, and the unit is also in development for the auxiliary and emergency power unit in the Swedish JAS 39 Gripen fighter and Aeritalia/Aermacchi/Embraer AM-X aircraft. The system incorporates a 16-bit microprocessor and Lucol software.

STATUS: in development.

Lucas main engine control unit for Turbo-Union RB.199 powerplants of Panavia Tornado

CUE-400 main engine control unit for Turbo-Union RB.199

DECU-500 digital main engine control unit

The RB.199 engines for fighter versions of the Panavia Tornado are the first application for the Lucas military Fadec. This system weighs the same and occupies the same space as the analogue main engine control unit currently fitted in Tornado and described below, but Lucas claims it has better control accuracy and system reliability. The ability to make changes at software level is also of great importance.

The digital engine control unit has the built-in test equipment and self-monitoring common to digital systems, and transfer to the safety channel is done at module level. Thus working modules from both channels can be combined rather than using one channel or the other. The engine is monitored via an interface to the air data system.

A testing programme has been successfully completed on powerplants installed in test facilities at Rolls-Royce in Bristol, and the equipment has been specified for the Tornado F2 air-defence version of this aircraft.

Flight trials of the Tornado unit began in February 1984 and since then Lucas DECUs

have flown in full control of both engines on the Tornado, including dry and re-heat thrust, without manual back-up. Production deliveries started in late 1984, for use on engines fitted to the Royal Air Force's Tornado F2.

Tornado unit dimensions: 12 × 7.5 × 7.5 inches (305 × 191 × 191 mm)
Weight: 33 lb (15 kg)
Mounting: avionics bay

STATUS: in production.

CUE-400 main engine control unit

The Turbo-Union RB.199-augmented turbofan engines which power the Panavia Tornado are controlled by this analogue electronic unit. Dry-thrust operation uses one of two channels with automatic channel switching if the built-in test equipment detects a failure. The second channel has a separate simplex reheat control fitted with built-in safety networks to prevent a dangerous situation arising from a failure.

The main engine control unit uses twin electromechanical metering devices to modulate the fuel flow according to its calculations,

and maintains the engine within its operating envelope. The system operates three actuators for reheat control, two to control the fuel and one to operate the variable nozzle.

Dimensions: 12 × 7.5 × 7.5 inches (305 × 191 × 191 mm)
Weight: 33 lb (15 kg)

STATUS: in production.

Self-adaptive full-authority digital engine control

Lucas is conducting a feasibility study for the US Army to find out the potential improvement for helicopter engine controls arising from the self-adaptive capability of digital systems. The company is looking at improving the turbo-shaft's power response and stability and improving the helicopter's performance and safety when a defect arises in the control systems, engine or airframe.

STATUS: feasibility study.

Lucas Aerospace/GEC Avionics

Lucas Aerospace Limited, Electronic Systems and Equipment Division, Electronic Systems, York Road, Hall Green, Birmingham B28 8LN
TELEPHONE: (021) 777 3222
TELEX: 336755

ESC-102 engine supervisory control

Lucas Aerospace is the prime contractor to Rolls-Royce on this joint design with GEC Avionics. The controller, which is in production, is used on the Rolls-Royce RB.211-535C turbofan installed in the Boeing 757 airliner. The companies share equal responsibility for production.

The controller consists of two electrically independent controls in a single case, each section with its own processor, input and output modules. The unit can be engine-

mounted or located in an avionics bay. The prime control channel optimises powerplant performance by maintaining accurate engine pressure-ratio control under all flight conditions. This channel interfaces with the air data computer, flight-deck instrumentation and on-board maintenance systems via ARINC 429 data-buses, and directly with engine-parameter transducers. Control is by a continuous trim function on the output stage of the hydro-mechanical fuel control unit, thus setting and maintaining the commanded engine pressure-ratio.

The second control channel provides an automatic engine over-speed limit and temperature protection. This channel has separate engine parameter transducer inputs and interfaces with the outputs of the hydromechanical fuel control unit. It is also linked to a fault-indication display which takes data from a comprehensive built-in test facility.

ESC-102 engine supervisory control unit for Rolls-Royce RB.211-535

Dimensions: 12 × 8 × 5 inches (305 × 203 × 127 mm)
Weight: 17 lb (7.7 kg)

STATUS: in service.

Racal

Racal Avionics Limited, Burlington House, Burlington Road, New Malden, Surrey KT3 4NR
TELEPHONE: (01) 942 2464
TELEX: 22891

RAMS Racal avionics management systems

These are members of a family of systems with facilities for control, display, interfacing and processing that can be configured to meet a variety of avionics management requirements. Multi-sensor navigation, communication frequency and mode control, flight performance monitoring including fuel management, monitoring of engine parameters, integration of external sensors relating to aircraft mission, are all examples of functions provided by RAMS. The purpose of the equipment, which can be configured to meet the requirements of military and commercial fixed-wing aircraft and helicopters, is to reduce pilot or crew work-load and increase safety.

The main interface between crew and the aircraft systems being controlled is a high-brightness cathode ray tube control and display unit, with alphanumeric keyboard and dedicated function keys. Associated with this are processor interface units designed for the specific applications in hand. Interfacing can be analogue or digital, with a variety of industry standards such as ARINC 429 and MIL-STD-1553B.

Of the four general configurations that comprise the RAMS family, two have ARINC 429 interfacing and two have MIL-STD-1553B. The two ARINC 429 members are the simplex RAMS 1000 and the duplex RAMS 2000, the corresponding MIL-STD-1553 equivalents being the simplex RAMS 3000 and the duplex RAMS 4000. Duplex configurations offer greater redundancy, interfacing and processing capabilities. A portable data-transfer device, with solid-state memory, is used to enter information and instructions from a ground loader into the RAMS database. A variety of monochrome and colour displays with associated symbol generators, liquid-crystal display remote indicators, stiff stick controllers and auxiliary control panels are available to make up any required configuration.

Type 5402 control/display unit
Dimensions: 146 × 162 × 247 mm
Weight: 6.5 kg

Type 5401 processor interface unit
Dimensions: 124 × 184 × 362 mm
Weight: 9.2 kg (typical)

Type 5405 data-transfer unit
Dimensions: 146 × 38 × 160 mm
Weight: 0.6 kg

STATUS: in production. A notable application of RAMS is the Hughes 530MG Defender light armed helicopter, which first flew in May 1984. This has a RAMS integrated display and

Integrated crew-station on Hughes 530MG helicopter with RAMS multi-function display, top, and control/display unit. Device at upper left is a swing-out sight for TOW missile

control/display unit, and a Racal Avionics Doppler 80E velocity sensor linked to it by means of a MIL-STD-1553B digital data-bus. The Royal Navy has also adopted RAMS for its Westland Lynx helicopters, together with 12-inch (305 mm) tactical displays. A system has also been installed on a Royal Aircraft Establishment Puma helicopter for general trials and assessment work. RAMS has now been selected for the new MBB BK 117 A-3M military helicopter, first seen at the 1985 Paris Air Show.

Smiths Industries

Smiths Industries, Aerospace and Defence Systems Limited, 765 Finchley Road, Childs Hill, London NW11 8DS
TELEPHONE: (01) 458 3232
TELEX: 928761

SEP10 automatic flight control system

The SEP10 has been selected for the British Aerospace 146. It provides three-axis control or stabilisation and incorporates a two-axis (pitch and roll) autopilot, elevator trim, flight director and yaw damping facilities. It uses simple, well-proven control laws and the minimum of sensors. There is also a 'transparent' control facility which allows the pilot to temporarily disengage the autopilot clutches and sensor chasers, and to manoeuvre the aircraft manually, so adjusting the datum of the basic and manometric autopilot modes.

The autopilot is based on rate-type control laws. Pitch and roll-rate signals are derived from ARINC three-wire attitude references, thus eliminating the need for rate gyros. Other ARINC standard interfaces accept a wide range of sensor inputs, including those from barometric and radio-navaid sensors, and allow systems to be tailored to suit operator's needs. The autopilot computer uses digital computing techniques to provide outer-loop control and to organise the mode logic, and has capacity to accommodate optional facilities. Analogue computing is used for the inner-loop stabilisation computing, servo-drive amplifiers and safety monitors.

The system can be supplied with either a parallel-acting yaw damper, which uses a rotary servomotor to drive the rudder and rudder pedals, or a series yaw damper which drives a linear actuator in series with the rudder control run. In each case, the yaw damping system is self-contained and consists of an analogue yaw computer, sensor and the relevant actuator or servo motor.

Flight-director computations are performed within the digital section of the autopilot computer, which can supply commands to V-bar or split-axis flight-directors. The flight-director and autopilot share common mode-selection and outer-loop guidance but, if desired, they can be operated independently.

Emphasis has been placed on maintainability and ease of testing, both for the installed system and for individual units in the workshop. Routine testing is designed to confirm correct functioning of safety devices, the tests being performed by operating a test button in conjunction with buttons on the mode selector. Modular construction has been used extensively to ensure that faulty equipment can be corrected and recertified easily and quickly.

The following descriptions of individual line-replaceable units outline the operation of a full SEP10 system.

Components of Smiths Industries SEP10 automatic flight control system installed in British Aerospace 146 aircraft

Autopilot controller The autopilot controller, in addition to providing autopilot and yaw damper engage or disengage controls, also includes pitch-rate and roll-angle selectors, and the elevator and rudder trim indicators. Engagement of the autopilot and yaw damper is confirmed by the illumination of a legend within each selector, and by the illumination of engage monitor lights on the mode selector. Pitch control uses a spring-centred lever which has a non-linear feel so that minor adjustments can be made instinctively. Roll control is accomplished by rotation of the control knob, which remains offset by a displacement proportional to the roll angle demanded in the basic mode, but returns automatically to the central position on selection of an alternative mode.

Autopilot controller for SEP10 automatic flight control system

Mode-selector There are 11 push-button switches, each illuminating as mode-indicators, for the selection of both autopilot and flight-director functions. Control mode engagement is confirmed by the illumination of a green triangle on the appropriate button, and for modes that include both arm and engage facilities, an amber triangle is illuminated during the arm phase, changing to a green triangle when the mode is engaged. A turbulence facility is included to soften flight disturbances in turbulent air. This reverts the autopilot to the basic stabilisation mode and at the same time reduces the overall gain of the system. Autopilot and yaw damper engagement lights are provided so that the full engagement state of the system can be seen on the mode selector. There is also provision for remote mode indication.

Autopilot computer The autopilot computer receives both analogue and logic information from sensors, controllers and selectors, and processes them to formulate the pitch and roll axis demands and the flight-director commands. The majority of autopilot computing is performed digitally, although analogue techniques are used to provide pitch and roll-stabilisation and authority limitations. Correct functioning of the computer safety circuits is verified by a test facility at a convenient remote station.

Yaw computer This unit takes short-term damping information from a yaw-rate gyro, while the lateral accelerometer senses acceleration along the lateral axis for slip-skid pre-

SEP10 mode selector

British Aerospace 146s have Smiths Industries SEP10 autopilot

vention. A suitable series yaw damper is available to suit customer demand.

Altitude selector The altitude selector provides facilities for altitude pre-select and vertical speed hold modes as well as the normal altitude alerting functions. Altitude information is obtained from either a servo-altimeter or an air data source. Selected altitude is displayed on a counter readout. A warning flag obscures this display in the event of a power failure or absence of altitude valid signal, and a test facility allows checking of the associated audio-visual signals and altitude pre-select function.

Altitude selector for Smiths Industries automatic flight control system

Air data unit Where there is a requirement for a Mach-hold facility to secure better fuel economy, the basic airspeed sensor can be replaced with an air data unit providing the necessary extra outputs.

Monitor computer For operation to Category II weather minima, this unit completes the performance monitor functions necessary to provide a fail-safe pitch channel. It independently monitors autopilot pitch, localiser and glide-slope deviation and provides outputs that can be used to disconnect the autopilot and provide warnings to the pilot. The computer is completely independent of the autopilot, and a self-test facility allows a check to be made on the correct operation of all the monitoring functions.

STATUS: in production.

Automatic flight control system for helicopters

In association with Agusta-OMI a defect survival automatic flight control system for helicopters has been developed using a dual-redundant digital processor configuration. It provides auto-stabilisation in pitch, roll and yaw and a full range of autopilot modes suitable for military or civil applications. It has been designed to meet onerous naval operating requirements including anti-submarine warfare and is also suitable for civil passenger transport, cargo and off-shore operations. The key certification requirement is the ability to control a helicopter, with a single- or two-pilot crew, in day or night operations over a wide range of weather conditions. For less severe applications, simplified versions of the system are available.

Major equipment elements are the two computers, the pilot's control unit and the dynamic sensor units containing accelerometers and yaw rate sensors as well as a hover trim control unit (if required). Data communication between the units and associated sensor systems is fully digital to ARINC 429. The flight control computers are arranged to drive dual-series actuators and parallel actuators connected to the pitch, roll, yaw and collective controls.

Two identical lane flight control computers perform all the processing for inner loop auto-stabilisation and outer loop autopilot functions.

Triplex yaw rate, normal and lateral acceleration signals are provided for the auto-stabilising function by the dynamic sensor unit. Three independent sources of pitch and roll attitude are also fed into each computer.

In addition to providing automatic control in pitch, roll, yaw and collective axis, the flight control computers output flight director commands for display to the pilots.

The Flight Control System has been selected by EH Industries for the EH 101 helicopter.

STATUS: pre-production.

FMCS flight management computer system

Conforming to the full ARINC 702 specification and a standard option on the Airbus A310 and A300-600 aircraft, this FMCS is the prime interface between crew and aircraft and enables optimum performance to be achieved from take-off to final approach. Main functions include flight planning, navigation, performance optimisation, flight-guidance (with coupling to autopilot and autothrottle) and display processing. The operational procedures create a working routine which is easy to implement and is similar for all phases of flight, optimising the factors affecting flight profile to give greater economy of fuel consumption, flight-time and aircrew work load.

The system design is based on a parallel multiprocessing arrangement of microprocessors within the flight management computer unit. This technique permits high processing capability and gives the flexibility to accommodate future expansion of functions and procedures. Two sets of dual 16-bit microprocessors, one dedicated to navigation the other to performance functions, provide overall throughput of over one million operations a second. Additional microprocessors are dedicated to input/output and database control functions. A bubble memory provides 256 K words of memory for navigation database storage. There is provision for up to 56 discrete inputs and 16 discrete outputs, plus 32 input and 12 output ARINC 429 channels. The system contains its own built-in test routines which constantly monitor system operation and fault detection.

The crew interface is with the control/display unit which has a 14 by 24 character cathode ray tube (crt) format. The bottom line can be used for scratch-pad entries. A full alphanumeric keyboard is provided, together with function keys and 12 line-select keys adjacent to the crt. Self-contained built-in test provides a cued step-by-step test of all push-buttons, annunciators and the crt display. For routine

Control/display unit (cdu) for the Smiths Industries flight management system

operations, most of the information is defaulted from the navigation database requiring a minimum of manually entered data.

Dimensions: (computer unit) 8 MCU (control/display unit) 10.5 × 9.0 × 5.75 inches (267 × 229 × 146 mm)
Weight: (computer unit) 28 lb (12.7 kg) (control/display unit) 13.9 lb (6.3 kg)
Power: (computer unit) 200 W (control/display unit) 87 W

STATUS: in production. Airline customers include Kuwait Airways, Saudia, British Caledonian, Air France, Sabena, Nigeria Airways, Air India, Cyprus Airways and Singapore Airlines. The Smiths FMCS has also been chosen for the Boeing E-6A of the US Navy.

Performance command system

This inexpensive performance command system has been designed to provide optimisation of trip-cost. It is largely based on Smiths Industries (Harowe Division) components and is housed in a 3 ATI case. The computer/totaliser continuously monitors economy-sensitive flight parameters (such as height and speed) and can be integrated with other Harowe systems. These include fuel quantity (Model 2120 fuel quantity indicators), airspeed and altitude (Model 2084 Mach/airspeed indicator), total air temperature (Model 2262 indicator), DME analogue pulse-pairs, and air-conditioning and anti-ice bleed-switch positions.

Before flight the pilot enters the aircraft's zero-fuel weight, economy index, cruise altitude, destination altitude and predicted wind data. During cruise the pilot selects the appropriate flight mode to obtain continuous airspeed and engine pressure-ratio command information. The pilot can also select DME-based wind data or a display of total fuel quantity, top-of-descent and optimum cruise information. Early flight trials showed a 2.8 per cent fuel saving compared with a ground-based computer generated flight plan.

Dimensions: 83 × 83 × 203 mm
Weight: 2.2 kg
Power: 10 W at 115 V 400 Hz ac

STATUS: in production. The system was certificated in October, 1983 on board a Boeing 737 of Air California and in August 1984 the system was given an FAA supplemental type certificate, the trials for which were done on board a Boeing 727-200 of Air Alaska.

STS 10 full flight regime autothrottle

Designed for the Boeing 737-300 and now installed as standard equipment, Smiths Industries autothrottle has been developed from the highly successful system supplied to Boeing for the 727-200 and 737-200 aircraft. It interfaces with flight management systems, digital air data systems, inertial reference systems and digital autopilots and uses advanced digital techniques for higher reliability, easier maintenance and lower cost of ownership.

The system comprises a digital computer with independent electromechanical drive to each throttle lever. The computer, which is housed in a single ½ ATR Long box, accepts analogue and digital information from sensors and systems on board the aircraft. After processing this data the computer generates outputs to drive servo-actuators which adjust the position of each throttle lever independently to achieve optimum engine performance. A further output from the same computer drives the fast/slow indicators on the ADIs.

The autothrottle includes a number of unique features designed to enhance performance and promote flight safety. A particular feature of the system is the ability to override the actuator drive and adjust the throttle levers manually, without the pilot applying more force than he would normally use in manual operation.

Smiths Industries autothrottle computer

To achieve precise control throughout the full flight regime, the autothrottle computer continuously monitors all the necessary engine and aircraft parameters and adjusts the thrust in accordance with the prevailing flight conditions. Protection is included to prevent exceeding predetermined N_1 engine limits and maximum aircraft incidence.

The system includes damping controls which are designed to minimise throttle activity during normal flight conditions. If excessive gust rates are detected, these controls are modified to allow a faster throttle response in compensating for windshear effects.

If a large change in vertical wind speed occurs during the approach a command is inserted which enables the system to achieve the required level of thrust more quickly.

Provision has been made within the autothrottle computer to allow for future growth and to interface with different engine variants.

STATUS: in production; standard fit on 737-300 airliners.

Fuel gauging and management

Smiths Industries has supplied analogue fuel gauging systems of varying complexity for more than 75 types of fixed-wing aircraft and helicopters. Of these aircraft the following are still in production: Airbus A300, British Aerospace HS 125, HS 748, Harrier/Sea Harrier/AV-8A, Hawk, Casa C-212, Embraer Bandeirante, GAF Nomad, Shorts Skyvan, Sepecat Jaguar, Dassault/Dornier Alpha Jet, Saab Viggen, Agusta A109, and Aérospatiale/Westland Gazelle and Puma. The A310's digital system is described separately below.

The Smiths fuel gauge is an ac or dc driven moving-coil or servo indicator which may have twin pointers to show the contents of pairs of tanks. A fuel-level warning device set by the pilot is optional. A more advanced installation may have digital readings on light-emitting diode displays and may incorporate a summation facility to give total fuel level.

The company has found that metallic tubular tank probes are the best contents' sensors, combining the most reliable results with low weight. They are protected against corrosion and are anodised to reduce the risk of microbiological contamination.

Any fuel system is tailored to the individual aircraft and in the design stages Smiths Industries undertakes a computer study of tank geometry and the effect of changes in wing loading. The company offers a self-compensating probe which incorporates capacitors to allow temperature variation in the fuel. Infra-red compensation may be provided. The fuel level is detected by a float, capacitance measurement or thermistors (which provide the most compact solution).

As well as the measurement of fuel contents, the company offers several optional fuel management functions. Automatic refuel/defuel is provided by pre-selectors at the refuelling point. Instead of relying on the flight-deck indicators, which are too far away to be of use

Autothrottle actuator assembly

on large aircraft, the refueller sets the amount of fuel required and the pumps are stopped automatically at the correct level.

Another fuel management function is centre of gravity indication. Before take-off the pilot sets into the system the actual centre of gravity and it monitors fuel consumption from the tanks throughout flight, giving a continuous centre of gravity readout. Allied to this are systems which control fuel distribution between tanks to keep the centre of gravity within limits as fuel is burned off. Smiths also has a system that pumps fuel between tanks in flight to counteract trim changes, either for vertical take-off or supersonic aircraft. Such a system has been applied to the Harrier family of fighters.

The company also provides fuel flow-rate indication equipment to show the mass of fuel consumed by one or more engines and fuel remaining. Current aircraft weight may also be computed and shown.

Typical configurations
High-performance strike aircraft: Servo indicator, centre of gravity position indicator, separate bridge amplifier (weighing 0.7 lb, 0.3 kg), and self-compensating probes. The system is designed for severe conditions including large temperature variations in the tanks.
Executive jet: Moving-coil indicator, separate bridge amplifier, and uncompensated probes. No fuel management is needed.
Transport aircraft: Moving-coil indicator with combined amplifier, refuelling pre-selector, infra-red compensated probes and separate immersed reference unit. Full temperature and density compensation is needed, and fuel management speeds up the refuelling operation.
Helicopter: Servo indicator with combined amplifier, infra-red compensated probes, and separate immersed reference unit.

STATUS: in production.

Digital fuel gauging
Smiths Industries has developed the digital fuel quantity indication system (FQIS) for the Airbus A320, the contract for production units was signed in December 1984.

The A320 FQIS incorporates advanced digital computing technology, bringing improved accuracy and reliability to the system compared with the analogue equipment employed on current Airbus aircraft (see separate entry).

Built-in test, failure recording and a recall facility are also included.

STATUS: in production for the Airbus A320.

Digital fuel management

Smiths Industries has developed and flight-tested a digital fuel management system which can be applied to any civil or military aircraft. The only specific application to date is for the Airbus A310 airliner, for which a digital fuel quantity indication system is produced by Smiths in collaboration with Intertechnique of France.

The digital system reduces attitude errors and performs temperature compensation within the processor so that linear tank probes can be used. Smiths says that fewer probes are needed, thereby increasing accuracy. Fuel is sampled by a densitometer as it is on-loaded. The computer is connected to the aircraft's attitude sensor so that changes from straight

Smiths Industries fuel systems components for British Aerospace 146 aircraft

and level flight may be taken into account when measuring fuel levels.

Built-in test equipment is included and the system is duplex. Input and output data is sampled and abnormal measurements are rejected, so indicator fluctuations and errors are reduced. The processor can control the refuelling operation, including fuel distribution between tanks and cut-off.

The system specific to the A310 and the forward-facing crew cockpit A300 uses digital fuel readout and shows aircraft weight and total fuel weight. Intertechnique designed the probes, which are characterised to suit the individual aircraft and to compensate for tank shape variations. Smiths Industries makes the contents and totaliser indicators, whereas production of the processors is shared between the two companies.

Flexibility is always one of the advantages of digital systems and this one may be adapted to changes in operating procedures and tank configurations by modifying the software. The system is also suitable as a retrofit in existing A300s without changing the tank probes since the software can correct the differences.

STATUS: in production.

Series 200 engine limiter

This system monitors jet-pipe temperature and controls it by limiting fuel flow. Its applications include the Rolls-Royce Viper turbojets powering the Aeritalia MB.326, British Aerospace 125, and Soko Jastreb aircraft. Other engines to have had Series 200 limiters include Avon, Orpheus and Pegasus.

Dimensions: 6.75 × 6 × 2 inches (170 × 150 × 50 mm)
Weight: 4.12 lb (1.87 kg)
Temperature range: 3 datums between 350° and 1000° C can be externally selected
Accuracy: ±2.5° C

STATUS: in production.

Series 500 engine limiter

The Series 500 is used for the Rolls-Royce Pegasus engine powering British Aerospace Harrier and AV-8B aircraft. It limits jet-pipe temperature and compressor shaft speed.

Dimensions: 6.75 × 6 × 2 inches (170 × 150 × 50 mm)
Weight: 4.5 lb (2 kg)
Temperature: 4 datums between 350° and 800° C can be externally selected, each capable of fine adjustment by ±7.5° C

Smiths Industries Series 800 temperature limiter for Rolls-Royce/Allison TF41 engine

Speed range: Preset between 10 400 and 12 700 rpm
Accuracy: ±2° C and ±1% of rpm

STATUS: in production.

Series 600 engine limiter

This jet-pipe temperature limiter is used for civil and military Rolls-Royce Spey engines on aircraft including the Nimrod MR2, British Aerospace One-Eleven, Buccaneer and Trident, Gulfstream II and Fokker F28. It has also been selected for the Italian/Brazilian AM-X aircraft and Rolls-Royce Tay engines for the Gulfstream IV and Fokker 100 aircraft. The Series 600 has the facility to compensate for changes in intake temperature.

Dimensions: 8.5 × 6 × 2.5 inches (215 × 150 × 65 mm)
Weight: 4.1 lb (1.88 kg)
Temperature range: Preset between 450° and 650° C. Setting may be fine-tuned by ±25° C
Accuracy: ±1.5° C

STATUS: in production.

Series 800 engine limiter

The Series 800 limits jet-pipe temperature, compressor delivery temperature and mass flow, and is engine-mounted on the Rolls-Royce/Allison TF41 Spey engine which powers some versions of the LTV A-7 Corsair. A time-profiled datum shift is provided in which the upper temperature limits are increased temporarily during sudden power demands such as take-off or go-around.

Dimensions: 10 × 7.5 × 3.25 inches (250 × 190 × 80 mm)
Weight: 8 lb (3.63 kg)

Temperature range: Preset between 0 and 1000° C
Accuracy: ±2.5° C and ±0.25% of rpm

STATUS: in production.

Series 900 engine limiter

Rolls-Royce and other engine manufacturers are evaluating the Smiths Industries Series 900 system which may be used as a jet-pipe temperature and/or mass-flow limiter. The principal benefit of this limiter is its low-cost, low-density design which uses single-sided circuit boards to accommodate growth. Double-sided printed circuit boards may be used if space is at a premium.

Dimensions: 6.5 × 4.75 × 2.75 inches (150 × 115 × 65 mm)
Weight: 2.2 lb (1 kg)
Temperature range: Preset between 0 and 1000° C and fine-tuned to ±25° C
Accuracy: ±2° C over ambient range –26° to 70° C
±4° C over range –55° to 125° C

STATUS: development completed.

Temperature monitor units

Smiths Industries engine temperature monitor units are designed to provide reliable warning to the crew if an engine over-heats or exceeds a critical temperature limitation whether during light-up or normal running. An indication of turbine air cooling flow failure is also given.

Dimensions: 101 × 76 × 38 mm
Weight: 0.6 kg
Power: 115 V 400 Hz or 28 V dc

STATUS: in production.

UNITED STATES OF AMERICA

Aero Systems

Aero Systems Aviation Corp, 5415 NW 36th Street, PO Box 52-2221, Miami, Florida 33152-2221
TELEPHONE: (305) 871 1300
TELEX: 808125
FAX: (305) 884 1400

CD-3000 fuel management computer

The CD-3000 fuel management system computer processes information on aircraft fuel-flow and ground speed to calculate and display specific range (nautical miles per 1000 lb of fuel) data. This is maximised for best fuel economy by adjusting speed and altitude while observing immediate optimum changes in the display. In addition to displaying specific range, the unit can also provide a readout of aircraft gross weight, fuel weight remaining, endurance range (in both distance and time), fuel flow, fuel used and ground speed.

Reliability and ease of operation are assured by all-solid-state technology, including microprocessor control and incandescent digital displays. There is also a built-in emergency memory power supply which can safeguard data for up to 48 hours. Units can be programmed to produce data calibrated in pounds or kilograms.

Two versions of the system are available, one for aircraft up to 100 000 lb (45 360 kg) all-up weight, the other providing for up to 1 000 000 lb (453 600 kg) all-up-weight. Aero Systems has developed a series of fuel-flow interfaces which permit installation with a variety of linear, pulse, synchro and second harmonic (magnesyn) fuel-flow meters. In addition the system requires a ground speed reference, which can be from omega-vlf, inertial navigation system or DME, and either digital or ARINC 568 format. Installation is claimed to take less than 40 man-hours.

Supplemental type certificates have been obtained for installation in McDonnell Douglas DC-10 and DC-8, Boeing 707 and 727, British

Aero Systems CD-3000 fuel management computer

Aerospace 125 Series 700, Lockheed Jetstar II, North American Sabreliner and Gulfstream III, Challenger 600 and 601 and Falcon 50. Equipment has also been installed on Lockheed C-130, Boeing 737, McDonnell Douglas DC-9, Gulfstream II, Gates Learjet Series 35 and

Fairchild-Swearingen Metroliner. The Australian Department of Transportation has issued notice of acceptance for any Australian-registered aircraft.

Installation: (panel space) 76 × 146 mm
Weight: 1.2 kg

STATUS: in production.

AiResearch

AiResearch Manufacturing Company, Electronic Systems Division of The Garratt Corporation, 2701 East Elvira Road, Tucson, Arizona 85706
TELEPHONE: (602) 573 6000
TELEX: 165936

GEMS energy management system

The GEMS energy management system is claimed to be the first airline-standard equipment of its type for business aircraft. It is intended for use on all Garrett TFE731 equipped aircraft. The device is being offered either as a factory option or as a retrofit on aircraft currently in service.

GEMS is intended to simplify flight planning and ensure optimum conditions for fuel economy, time or cost as well as providing engine power-setting trim through the fuel control computers that govern the Garrett TFE731 powerplants. This device allows accurate and automatic trimming of thrust levels once the power lever angle is manually established within the GEMS trim authority range.

The system comprises an energy management computer and is controlled through the UNS-1 display and control/display unit or its own control/display unit. Peformance data can also be displayed on weather radar displays. The computer carries the store of information including engine data, derived aircraft data and the cruise control and flight manual data. Five policies or modes are available for flight profile planning and reflect the system's flexibility. They are: minimum time with range assurance, maximum range, minimum overall cost based on fuel and time costs, maximum endurance, and manual speed selection. Once the policy is selected GEMS determines the best altitude, speed and power setting for climb, cruise and descent modes and displays this information on the control and display unit. GEMS is the only electronically coupled autothrottle available for business jets.

Dimensions: (computer) 7.5 × 9 × 14.5 inches (191 × 229 × 368 mm)
(control/display unit) 4.5 × 5.75 × 8 inches (114 × 146 × 203 mm)
Weight: (computer) 22.1 lb (10.02 kg)
(control/display unit) 4.8 lb (2.18 kg)

STATUS: in production. Flight trials completed and system certificated on LearJet models 35, 35A, 36, 36A and 55.

Engine performance reserve controller

The Garrett APR automatic engine performance reserve controller is used on aircraft powered by the Garrett TFE731 turbofan engine. The controller detects a loss of thrust on take-off from either engine through the comparison of

selected rotor speed. The high-speed rotor (N_2) is automatically selected during automatic power reserve operations and a 5 per cent difference is considered the nominal threshold. Following detection of a power decay, the system automatically boosts the remaining engine's maximum thrust by raising the high-pressure spool maximum speed by 1 per cent and the maximum operating temperature by 25° C. Manual APR control is available to give thrust boost should the system fail.

Weight: 4.75 lb (2.2 kg)
Dimensions: 160 × 74 × 269 mm
Power: 14 W at 28 V dc

STATUS: in production for TFE731-powered aircraft.

Full-authority engine control for Garrett ATF-3 engine

The controller for the Garrett ATF-3 turbofan engine, which powers the Dassault Falcon 200 executive jet and HU-25A Guardian, executes closed-loop control of the powerplant. Its solid-state electronic design includes a continuous integral monitoring system. Failure detection is included, and manual back-up control is selected automatically if there is a problem. Dual redundancy provides closed-loop protection against hard-over failures. External test points are provided for rapid fault isolation. Inputs include fan speed and low-pressure and high-pressure compressor speeds, inlet pressure and temperature, turbine discharge temperature, and power-lever position. The unit controls the engine fuel valve and bleed valve positions.

Dimensions: 8 × 17 × 7 inches (203 × 431 × 178 mm)
Weight: 19 lb (8.6 kg)
Cooling: natural convection
Inputs: 10
Outputs: 9

STATUS: in production.

Full-authority digital engine control for Garrett TFE731 engine

Garrett has developed a full-authority engine fuel control system for the TFE731 turbofan, operating according to an engine acceleration/temperature schedule. It provides closed-loop exhaust-gas temperature control during acceleration, as well as on-speed governing and integral bleed-air control based on exhaust-gas temperature. Over-temperature protection is given at all times. The equipment has the same features as those for the ATF-3: solid-state design, failure detection with automatic switch to manual backup, dual redundancy, continuous integral monitoring, and external test points. It also includes fault annunciation, fault detection and fault isolation capabilities, to line-replaceable unit level.

Dimensions: 338 × 197 × 96 mm
Weight: 6.3 kg
Cooling: natural convection
Inputs: 10
Outputs: 7

STATUS: in production for TFE731-5 power-plant.

Full-authority engine control for F109 engine

A full-authority digital engine control unit is under development for the Garrett F109 engine fitted to the new US Air Force/Fairchild T-46 trainer aircraft. The microprocessor-based engine computer unit provides automatic starting, trimming and safety limiting. Built-in test facilities check the controller and engine sensors, accommodate certain failures and can isolate faults to line-replaceable unit level. A data-logging feature monitors and records 12 engine parameters for maintenance analysis.

Dimensions: 12.8 × 7.6 × 5.0 inches (321 × 191 × 137 mm)
Weight: 15.2 lb (6.9 kg)
Power: 28 V dc

STATUS: in qualification test.

Integrated engine computer for TPE331-14 powerplant

The digital integrated engine computer is made by Garrett for its TPE331-14 turboprop engine. It provides automatic start sequencing, variable red-line temperature indications, torque indication, closed-loop torque and temperature limiting, data-logging, built-in test and fault-isolation. The unit's redundant torque indication ability permits manual mode dispatch even after a computer failure. There are serial outputs from a non-volatile memory of the number of engine hours operated, time spent above specified maximum temperatures and number of starts. A personality module is used to adjust the software to a specific engine so that performance degradation will be referenced to an actual, rather than theoretical, baseline.

Dimensions: 5.5 × 12.38 × 7.13 inches (140 × 314 × 181 mm)
Weight: 11.6 lb (5.3 kg)
Temperature range: –55 to 71° C
Power: 28 V dc

STATUS: in production.

Full-authority engine control for Garrett TPE331 engine

The Garrett full-authority engine controller for the TPE331 turboprop has the following features: temperature-limited automatic starting; isochronous propeller governor speed control; closed-loop torque and temperature limiting; single red-line exhaust-gas temperature correction; continuous speed and fuel governing; power-level matching; and simplified engine rigging.

Dimensions: 7.6 × 5.4 × 11.85 inches (193 × 137 × 301 mm)
Weight: 10.2 lb (4.6 kg)
Cooling: natural convection and radiation
Power: 28 V dc

STATUS: in production for TPE331-8 and -12B powered aircraft.

T800/ATE 109 engine control unit

This control is under development for the LHX helicopter engine of the Allison/Garrett team. It features a MIL-STD-1750A microprocessor, ADA language and a MIL-STD-1553 data-bus.

IAI Westwind 2 has Garrett TFE731 fan engines fitted with Garrett engine performance reserve controller system

Engine power trim system

This is a combined autothrottle and flight management system suitable for commercial aircraft and designed to provide fuel-savings and crew workload benefits for a low initial cost. It can also protect engines against temperature and engine pressure-ratio overshoots. Garrett has produced a system which simplifies crew procedures by dividing work between pre-flight planning (which on completion is loaded into the system memory) and in-flight control, which is conducted in real time taking account of influences as they arise. The ground computer supplies optimised cruise altitude and Mach number data. An advantage of leaving this function on the ground is that operators can subsequently modify planning policy with no change to airborne hardware, software or operating procedures.

For operation the crew presses a button to identify the appropriate flight phase: take-off, climb, cruise, long-range cruise or go-around (there are also two spare button locations), and to dial in a reference temperature for normal or reduced power take-off. The system indicates engine pressure-ratio setting, and has an engine number selector which provides individual pressure-ratio indications if engines are operating with different bleed-air conditions. A single button engages automatic engine power trimming to the climb schedule or holds constant Mach number. A second button can be used to engage a further operating mode which trims fuel consumption by small Mach-number adjustments as the aircraft weight decreases. The system has built-in test facilities and alerts the crew immediately when a fault is detected.

Installation comprises a cockpit control and display unit, a power trim computer and a throttle control gearbox for each engine. Internal data transfer is conducted via an ARINC 429 data-bus and rack-mounted equipment meets ARINC 600 requirements. The equipment is FAA-approved for use on Boeing 727-200 aircraft.

STATUS: in production.

Propeller synchrophaser

The Garrett synchrophaser controls the phase relationship between the blades of two aircraft propellers. Noise and vibration can be minimised and the objectional beat frequencies associated with propellers operating at different rotational speeds be eliminated. The synchrophaser is designed to operate with twin turboprops and electronic engine fuel controls. It works with existing speed detection systems and requires no rigging adjustments, calibration, or indexing. It is compatible with engines operating closed-loop on torque and temperature. The system uses a master-and-slave engine concept with a full 120° phase angle authority for three-bladed propellers.

Dimensions: 2.3 × 5.05 × 8.05 inches (57 × 128 × 204 mm)
Weight: 1.75 lb (0.79 kg)
Power: 28 V dc with MIL-STD-704 Category B transient immunity
Authority limit: 2.4% max speed separation
Steady-state phase-angle accuracy: $< 5°$ (typical)
Operating range: cruise and max continuous speeds

Vibration: RTCA DO-160-PAR. 8 curves
Electromagnetic interference: MIL-STD-6181D

STATUS: in production.

Engine synchroniser

When two common-structure mounted engines operate at slightly different rotational speeds, a third beat frequency is set up in the structure which can cause discomfort to passengers. Garrett is producing an electronic engine synchroniser to reduce this problem. It interfaces with the engine through the existing electronic fuel control, and requires no modifications to the aircraft wiring. The synchroniser may be applied to two, three, or four engines without modification. It works by nominating one engine as master and adjusting either fan (N_1) or compressor (N_2) speeds of the other powerplants so that all speeds are closer to each other. The trim authority varies with power lever setting. The flight deck switches allow on/off and N_1/N_2 selections, and a synchronisation indicator is optional.

Dimensions: 9.4 × 7.3 × 3.6 inches (239 × 186 × 91 mm)
Weight: 4.6 lb (2.1 kg)
Max acquisition time: 30 s
Synchronisation accuracy: (N_1 control) 11 rpm (N_2 control) 18 rpm

STATUS: in production.

Astronautics

Astronautics Corporation of America, 4115 N Teutonia Avenue, PO Box 523, Milwaukee, Wisconsin 53201-0523
TELEPHONE: (414) 447 8200
TWX: 910 262 3153

Pathfinder P3B general aviation autopilot

Astronautics produces a range of general aviation autopilots under the trade-name Pathfinder, of which the P3B is one embodying almost all of the components available in this family of modular autopilots. The all-electric P3B is a three-axis system which provides roll, yaw and pitch stabilisation, heading selection and control, VOR/LOC tracking with automatic cross-wind correction, automatic unlimited-angle VOR/LOC intercepts, manually-selected standard-rate turns, pitch stabilisation in all attitudes, altitude hold with automatic pitch-trim, automatic or manual glideslope intercept, glideslope tracking with automatic pitch trim and manually-operated electric pitch trim.

STATUS: in service.

Flight-director system

The company has developed a three-cue system which gives full IFR capability to helicopters. It combines a versatile flight-director computer with a three-command-bar attitude director indicator, dual-bearing pointer horizontal-situation indicator and a multi-mode computer controller. In IFR conditions, the system provides ILS, VOR, and ADF approach capability, including the option to include steep-angle ILS approaches.

The three-cue system adds a collective-command steering bar to the pitch and lateral command steering bars used in two-cue systems. Continuous altitude, airspeed, vertical speed, ADF, VOR/ILS, and (optional) Doppler and altitude-alert inputs permit the flight-director computer to respond to pilot-selected flight modes. A pilot can execute automatic

Astronautics three-axis helicopter autopilot system

VOR/ILS intercepts, glideslope intercepts, vertical and airspeed holds, deceleration rates and altitude holds. The computer provides automatic intercept and tracking of VOR and ADF courses, glideslope and localiser, and automatic initiation of deceleration ILS approaches. Pilot and co-pilot displays and controls are effected by optional slaved horizontal situation indicators and transfer controls.

The indicators are 5-inch (127 mm) units, hermetically sealed with a dry nitrogen-helium atmosphere. Direct-current servos used in these units are claimed to result in considerably less heat dissipation, less power drain, higher torque and greater reliability than typical ac-servoed units.

The mode controller has four rotary switches for mode, nav select, vertical speed and airspeed selections. An optional remote dual-course selector can be added to supplement the basic controller.

The flight-director computer accepts inputs from the mode controller and an array of sensors. It computes the pitch, lateral and collective commands required to adhere to selected and/or scheduled flight parameters,

and displays the commands on the attitude director and horizontal situation indicators.

Dimensions and weight: (ADI) 127 × 133 × 194 mm; 7 lb (3.2 kg)
(HSI) 127 × 108 × 174 mm; 6 lb (2.7 kg)
(flight-director computer) 59 × 194 × 319 mm; 6.5 lb (2.9 kg)
(mode controller) 146 × 105 × 127 mm; 3.5 lb (1.6 kg)
(remote course selector) 146 × 32 × 165 mm; 1.5 lb (0.7 kg)
Power: 45 VA 115 V ac 400 Hz
0.5 A 28 V dc
8 VA 5 V ac (for lighting)

STATUS: in production.

Three-axis autopilot for Hughes 500

Astronautics has developed an autopilot for light helicopters which has been FAA certificated and is available to Hughes 500D, 500E and 530 users. The autopilot provides full three-axis control to reduce pilot fatigue. There are seven basic operating modes, plus hands-off stabilisation. In basic attitude-retention mode, the helicopter can be flown hands-off not only in straight-level flight but also during climbs, descents and turns. The desired attitude will be held even in autorotation, and all turns are automatically co-ordinated. In altitude-hold mode the desired altitude is maintained to within ±20 feet. The altitude can be captured with vertical velocities as high as 1000 feet a minute. Following high vertical velocity climb/descent the helicopter will smoothly change pitch attitude and capture the selected altitude. In heading hold mode the pilot may select the desired heading either before or after the mode is engaged. All turns to the new heading are automatically co-ordinated at a bank angle of 20°. Hands-off hover capability is provided either in or out of ground effect. Heading hold may be engaged any time during hover. Any new desired heading is entered by moving the 'bug' on the heading gyro, and yaw damping is provided throughout the autopilot flight envelope.

STATUS: in service.

Bendix

Bendix Avionics Division, Allied Signal Inc, 2100 NW 62nd Street, Fort Lauderdale, Florida 33310
TELEPHONE: (305) 776 4100
TELEX: 514417
TWX: 510 955 9884

FCS-870 automatic flight control system

Bendix has combined integrated circuit technology with several new automatic flight control features to produce a system which offers optimum performance over a wide range of general aviation aircraft. The FCS-870 is an autopilot, with flight-director and independent yaw damper options which form the basis of many options. It is designed for installation in a broad range of aircraft types, from heavy singles through most turboprop powered types. The system meets or exceeds the TSOs for these classes of aircraft.

The complete FCS-870 consists of the cockpit instruments (including flight controller, attitude director indicator, horizontal-situation indicator and mode annunciator) and a remote-mounted computer amplifier and servos. The yaw damper option adds a side-slip sensor, a panel-mounted turn-and-slip indicator, and a remote yaw servo.

Flight controller This is a small panel-mounted unit used to select the desired operational modes of the system. All nomenclature on the panel is back-lighted for easy night-viewing.

Mode annunciator This can be mounted in any convenient 'head-up' panel location so that the pilot can monitor the autopilot or flight-director functions in use. The unit also alerts the pilot to some fault and armed-system conditions.

Computer amplifier This is the main flight control system unit and is an all-solid-state device which houses all the lateral and longitudinal computational circuitry, power supplies and altitude transducer. It also contains the calibration circuits which ensure compatibility with specific aircraft sensor and output requirements, plus an additional circuit board providing input signals for flight-director operations. Lateral and longitudinal data circuits are segregated on opposite sides of the unit, so reducing the amount of inter-wiring, and augmenting reliability and serviceability. Relays have been eliminated and all heat-generating components are near the outside of the unit to improve cooling.

Servos Advanced design servos provide the greatest torque required for the highest-performance aircraft likely to use the system. Three similar units are used for pitch, roll and trim control.

Yaw damper The independent yaw damper provides positive turn co-ordination and rudder control in all flight conditions. In twin-engined types the yaw damper is claimed to provide substantial assistance in maintaining directional control during an engine-out sequence.

Instrumentation The company recommends integration with the following Bendix flight instruments:

DH-886A director horizon indicator (4-inch model) or
DH-841V director horizon indicator (3-inch model), plus
HSD-880 horizontal-situation indicator (4-inch) or
HSD-830 horizontal-situation indicator (3-inch).

Specific features of the flight control system are: command-turn (half- or full-rate co-ordinated turns can be initiated by rotating a knob); full-pitch integration (provides smooth capture of desired altitude, and eliminates stand-off errors); automatic altitude pre-select; pitch synchronisation (keeps elevator surfaces aligned with trim to eliminate disengagement disturbances); control-wheel steering (manoeuvre to desired attitude with button depressed, and then release to leave flight control system maintaining pilot's demand); coupled go-around, automatic cross-wind correction; all-angle intercepts; pitch-rate command and manual or automatic glideslope capture.

Weight: (basic autopilot) 19.7 lb (8.94 kg) (with 3-inch FD/HSI) 31.1 lb (14.11 kg) (with 4-inch FD/HSI) 37.65 lb (17.08 kg) (optional yaw damper) 7.35 lb (3.33 kg)
TSO compliance: C9C, C52A, DO160.

STATUS: in service.

Series 3 FCS-60 digital flight control system

The designation FCS-60 embraces a family of four flight control systems with performance to suit different categories of regional commuter and business aircraft. The new member of the Series 3 family was announced in April 1984 and, along with two new digital weather radars and a navigation system, will complete Bendix Series 3 avionics suite for general aviation, based on ARINC 429 data-bus.

The FCS-60 is basically a three-axis system under microprocessor control. The simplest system available has one channel per axis, weighs 31 lb (14 kg), and is driven by a single air-data sensor. Navigation and sensor units are controlled by a single fail-passive flight controller, the outputs from which operate pilot and co-pilot flight-director instruments. The system can be expanded to dual-simplex or duplex configuration, the second of which may be certificated to Category II operation. Both have dual air-data sensors, and the second version has dual drive motors and three microprocessors. The system provides pitch stabilisation with automatic trim, roll stabilisation, heading hold, yaw damping and turn co-ordination. The yaw damper can be used independently of the other channels. Additional modes provide lift compensation during turns, compensation for turbulence, and pilot-commanded inputs.

STATUS: in production.

Bendix Energy Controls Division, Allied Signal Inc, 717 North Bendix Drive, South Bend, Indiana 46620
TELEPHONE: (219) 237 2100
TWX: 810 299 2514

EH-L2 digital electronic fuel control

Bendix has developed this full-authority digital engine control for future airliners, and the system has been test flown on a Boeing 747 during an evaluation known as the electronic propulsion control system programme. The EH-L2 is designed to give improved performance over current systems, while retaining good fuel consumption. A version of the control is being adapted for testing in the Ategg advanced technology engine gas generator programme at Pratt & Whitney's Government Products Division.

The EH-L2 comprises a full-authority digital electronic controller and complementary fuel-flow handling unit. It incorporates micro-electronic-based computation including selective or complete redundancy as appropriate for high reliability and fault tolerance. The EH-L2 has built-in high accuracy quartz pressure sensors, and Bendix has proved its tolerance to shock, vibration, electromagnetic interference, electromagnetic pulse interference, and temperature extremes.

The company has also developed an advanced system for digital full-authority control of complex variable-cycle engines. The design is intended for future advanced military aircraft, which are likely to have variable-cycle engines to suit supersonic flight, while improving subsonic performance. This engine control has also been used by Detroit Diesel Allison for its JTDE joint technology demonstrator engine and Ategg programmes.

STATUS: under development.

Bendix EH-L2 digital electronic fuel control system

Boeing

Boeing Military Airplane Company, Box 3707, Seattle, Washington 98124
TELEPHONE: (206) 237 1716
TELEX: 329430

Future flight management concepts

Early in 1986 Boeing was awarded a cost sharing contract by the US Air Force to develop flight management concepts which will enhance mission effectiveness and survivability of strategic bombers and airlift aircraft. Boeing will be assisted by Lear Siegler.

Current and future strategic aircraft will face increasingly demanding mission requirements. Operating with today's flight management systems, particularly in high threat areas, requires excessive crew workload and severely limits overall operational effectiveness. To take advantage of new offensive and defensive technology to improve the effectiveness of strategic aircraft will require the development of significantly improved flight management concepts.

The contract will be carried out in two phases. Phase One will be concerned with the development of system requirements. Boeing will be responsible for aircraft and mission selections, mission segment analysis and navigation and trajectory tracking requirements.

Lear Siegler will be responsible for trajectory generation, threat management and mission planning requirements.

Having identified these requirements Phase Two will develop concepts of mission planning; trajectory optimisation and control including time based trajectory control, fuel optimisation, threat response and avoidance, target acquisition, weapon delivery and air drop. In addition the incipient problems of integration and prioritisation of various requirements for fuel optimisation, survivability, time constraints, ride quality and multi-aircraft co-ordination will be addressed:

Century

Century Flight Systems Inc, Municipal Airport, PO Box 610, Mineral Wells, Texas 76067
TELEPHONE: (817) 325 2517
TELEX: 706207

Century I autopilot

This is an all-electric, rate-based, lightweight single-axis wings-level/heading system. An electric actuator in the aileron circuit provides the control power for attitude stabilisation and pilot-commanded, knob-controlled turn-rates of up to 200° a minute. A tilted rate-gyro inside a standard 3-inch (76 mm) case senses roll-rate and rate-of-turn, for both instruments and servo, and the system can be slaved to VOR/LOC. Century I can also be used as a back-up to the Century IIB, -III or -IV vacuum/electric systems, sharing the same roll-servo.

Dimensions: 3.5 × 3.5 × 7.64 inches (89 × 89 × 194 mm)
Weight: 7 lb
Power: 1.25 A at 14 or 28 Vdc

STATUS: in production.

Century IIB autopilot

A single-axis development of the Century I, the -IIB has two panel-mounted instruments, a directional gyro with a heading bug and an attitude gyro, as sensors, and additional navigation options. The system includes a solid-state computer, control panel and roll servo providing lateral stabilisation, roll command and heading select. An optional radio coupler permits heading select, VOR capture and tracking, increased sensitivity for VOR approaches, and localiser capture. An HSI may be substituted for the directional gyro.

Dimensions: (directional gyro) 3.37 × 3.25 × 6.37 inches (85.6 × 82.6 × 161.8 mm)
(control unit) 3.75 × 2.00 × 4.25 inches (92.3 × 50.8 × 108 mm)
Weight: 9 lb (4 kg)
Power: 2 A at 14 V dc or 1.5 A at 28 V dc

STATUS: in production.

Century III autopilot

This two-axis autopilot comprises directional- and attitude-gyro panel-mounted instruments, control unit, computer/amplifier, and electric servos. It is claimed to be approved for more makes and models of aircraft than all of its competitors combined, and to be the preferred altitude hold system for dealer-installed, retrofit installations in a wide range of singles and twins. The standard features include separate roll and pitch engagement, altitude hold, pitch and roll command, and automatic or manual electric pitch trim. The optional radio navigation coupler is identical with that for the Century IIB.

Dimensions: (control unit) 5.0 × 2.25 × 2.5 inches (127 × 57.2 × 63.5 mm)
(panel instruments) as for Century IIB

Weight: 19.5-24 lb (8.9-10.9 kg) depending on options
Power: 4.5 A at 14 or 28 V dc

STATUS: in production.

Century IV autopilot/flight-director

The two-axis Century IV autopilot has built-in VOR/LOC/GS couplers and outputs to drive a single-cue, V-bar attitude-director. Altitude hold, automatic and manual electric pitch trim, and go-around modes are standard features, as is the Century NSD-360A HSI. The programmer is back-lighted to serve as an annunciator on mode engagement, thus saving space. Options include a separate panel-mounted annunciator, and a modified computer compatible with ARINC HSIs.

Dimensions: (programmer) 5.95 × 2.08 × 2.80 inches (151 × 52.8 × 71 mm)
(annunciator) 3.1 × 1.1 × 1.6 inches (77.7 × 26.9 × 39.4 mm)
(NSD-360A) 3.2 × 3.06 × 8.75 (81.3 × 77.2 × 222.5 mm)
Weight: 6.6-34 lb (12.1-15.5 kg)
Power: 6 A at 14 or 28 V dc

STATUS: in production.

Century 21 autopilot

This system represents an advance on previous Century autopilots in having the lateral, 'wings level' servo slaved to both roll position and roll rate, the attitude and rate sensors being contained within the panel-mounted ADI and HSI instruments. Response rates are tailored to match input commands and operating modes. Features include roll stabilisation, heading select, built-in VOR/LOC radio coupling for intercept and track, and dual-mode (heading and approach) operation. The combined programmer and annunciator has front-panel access for adjusting the rate and position

Century 21 autopilot comprises HSI and ADI indicators and programmer/annunciator

Beech Bonanza is typical application of various Century autopilots

sensors for the autopilot. Options include the Century NSD-360A slaved or non-slaved HSI or similar compatible instrument as an alternative to the standard directional gyro. 'Stand alone' manual electric trim is certificated in some aircraft.

Dimensions: (directional gyro) 3.2 × 3.02 × 7 inches (81.3 × 76.7 × 177.8 mm)
(attitude gyro) 3.2 × 3.02 × 7 inches (81.3 × 76.7 × 177.8 mm)
(programmer) 6.24 × 1.83 × 11.18 inches (158.5 × 46.4 × 284 mm)
Weight: 12.5-13.4 lb (5.7-6.1 kg)
Power: 1.2 A at 14 or 28 V dc

STATUS: in production.

Century 31 autopilot

Century 31 is the company's most recent two-axis autopilot, and blends position and rate signals from the attitude and directional gyros for smooth pitch and roll control, autopilot commands being tailored to the particular mode in use. Features include roll stabilisation, heading select, pitch command, altitude hold, proportional automatic and manual electric pitch trim, VOR/LOC/GS radio coupling, and dual-mode (heading and approach) operation. Adjustments to the autopilot functions can be made through the front panel of the programmer. The system includes a patented velocity servo amplifier in the variable-speed servo system, providing trim response appropriate to the situation, for example, small changes during cruise giving place to more rapid response during go-around after a baulked landing. The Century NSD-360A slaved or non-slaved HSI or other compatible flight director may be substituted for the attitude indicator.

Dimensions: (directional gyro and attitude indicator) as for Century 21
(programmer) 6.25 × 2.35 × 12.5 inches (158.75 × 59.7 × 317.5 mm)
Weight: 22-24.8 lb (10-11.27 kg)
Power: 2.3 A at 14 or 28 V dc

Century 41 autopilot/flight-director

The first Century autopilot to feature digital processing, the Century 41 is a two-axis system with built-in VOR/LOC/GS couplers and outputs to drive a single-cue, V-bar flight director.

It employs both position and rate signals to command the flight control servos. Other features include synchronised pitch attitude and altitude hold modes, a pitch modifier, automatic and manual electric trim and automatic pre-flight check schedule. VOR/LOC/GS capture intercepts are tailored to ground-speed, intercept angle, wind direction and distance from the ground-station. The VOR tracking circuitry incorporated gain-reduction to ensure 'soft' passage around the station.

When using the go-around mode, the pilot merely has to press a single button, 'clean up' the aircraft and add power; the autopilot flies to a calibrated pitch-up attitude appropriate to the single-engine safety speed for that particular type of aircraft, and turns on to a new heading preset by the pilot. The NSD-360A slaved or non-slaved HSI may be substituted for the standard directional gyro.

Dimensions: (directional gyro and attitude indicator) as for Century 21
(programmer) 4.62 × 1.97 × 2.43 inches (117.35 × 50.04 × 61.72 mm)
Weight: 27.5-34 lb (12.5-15.45 kg)
Power: 2.3 A at 14 or 28 V dc

STATUS: in production.

Century 41 autopilot with HSI and ADI indicators, mode programmer and annunciator

Collins

Collins Avionics Division, Rockwell International, 400 Collins Road NE, Cedar Rapids, Iowa 52498
TELEPHONE: (319) 395 1000
TELEX: 464421
TWX: 910 525 1321

AP-105 autopilot

Available for aircraft in the executive-jet category, the AP-105 provides three-axis automatic flight control. The system offers a full complement of navigation and vertical mode options, including altitude pre-select, and can present all computed steering data on flight-director displays for manual control guidance, or to monitor the autopilot performance.

The AP-105 is designed to exceed automatic and manual Category II approach requirements. Glideslope scheduling using radio-altimeter information is incorporated, and an all-angle VOR/LOC capture capability is included, plus automatic back-course operation. The glidescope function can automatically capture from above or below, either before or after localiser capture. Vertical speed hold, airspeed hold, Mach hold and altitude hold modes can be integrated to fly departure, en route and arrival segments. Automatic mode changing is standard and automatic elevator trim permits ripple-free transitions between modes, and autopilot disconnects.

The system comprises an autopilot amplifier, controller, mode coupler, mode selector, airspeed sensor flight computer, altitude selector, yaw-damper and three servos.

A control-wheel synchronisation facility is engaged by depressing a button on the yoke, and the autopilot then follows pilot control inputs, taking up the desired attitude when the wheel is released. The rudder channel can be operated in manual flying modes to enhance turn co-ordination, or engaged for all flying modes. In established attitudes the aircraft is gyro-mode stabilised, but displacement of pitch or roll rate manoeuvre-controls permits manual override. Attitude rates are used as a basis for all control inputs and automatic lift-compensation is provided in the basic control laws.

Weight: (total) 23.4 kg

STATUS: in production.

AP-106A ProLine autopilot

This three-axis automatic flight control system, a member of the company's ProLine avionics range, can be used in a wide range of small twin-turboprop aircraft types, and integrated with the Collins FD-112 flight-director. Facilities are similar to those provided by the AP-105 autopilot and include: Category II glideslope steering, all-angle course capture, coupled go-around, airspeed hold, linear VOR coupling, RNav-type linear VOR-deviation steering, and control wheel synchronisation.

Weight: (total) 6.9 kg

STATUS: no longer in production, but remains in service.

APS-65 autopilot

Introduced in 1982, the APS-65 was claimed to be the first digital autopilot offered for turboprop types with dual microprocessor computation. The basic operating modes are: roll hold, pitch hold, heading hold, navigation mode, approach mode, indicated airspeed hold, vertical speed hold, climb and descent, altitude select and go-around. Additionally, when climbing, the autopilot can be programmed to fly the aircraft at optimum efficiency, thus providing fuel-savings.

A new control technique for the pilot to use when introducing pitch, altitude, airspeed and vertical speed hold changes has been incorporated. This employs a rocker switch that can be operated to select precise alterations. Altitude can be 'nudged' in increments of 100 feet, pitch in increments of 1°, airspeed in increments of 2 knots, and vertical speed in increments of 100 feet a minute. The new technique has been incorporated to reduce pilot workload and permit smoother flying. Operational safety is enhanced through the dual processor configuration, which ensures that no single equipment or software fault can result in exceeding preset limits or a malfunction in more than one axis of control. Built-in monitoring assures automatic disengagement when a fault is detected, and appropriate diagnostics assist in fault detection in any element. The system is lighter and has a lower parts count than comparable analogue systems, and higher reliability is attributed to these features.

The autopilot is suitable for the heavier business turboprop types such as the King Air, and can be integrated with electromechanical or electronic flight director systems and navigation systems produced by Collins.

STATUS: in production. First to be certificated was the Beech King Air in April 1983. By September 1985 the APS-65 had been chosen for 16 types of turboprop business aircraft, one of the most recent being the Piper Cheyenne IV.

Collins APS-65 digital autopilot controller

The King Air was noteworthy as the first general aviation all-digital turboprop to be certificated. Its Collins avionics suite included APS-65 autopilot, EFIS-85 electronic flight instrument system, ADS-80 air data system, and nav/com equipment.

APS-80 autopilot

Designed for high-performance executive and corporate jets, the APS-80 autopilot provides three-axis flight control from initial climb to final approach. It has been developed from experience accumulated with the FCS-110 all-weather landing/automatic flight control system. A fully-integrated installation would use a Collins FDS-85 (or FDS-84) flight-director system. The APS-80 incorporates many integrated-circuit components and has automatic test functions. Smooth autopilot control during aircraft configuration changes is claimed through the use of torque-rate limiting servos. Each servo is designed to fail-passive on detection of a 0.6 g pitch disturbance caused by servo failure, or if preset values of roll angle and roll rate are exceeded.

Category II approach performance is available, including radio altimeter-based flight-path scheduling to extend performance further if necessary. An internal comparator replaces external monitoring systems to maintain acceptable roll and pitch attitude, localiser and glideslope deviation, and heading and radio altimeter data during the approach.

Climb, en route and descent profiles can be flown using autopilot modes. These incorporate gain-programming control laws, automatic turn co-ordination, lift compensation, command smoothing, all-angle adaptive VOR/LOC capture, turbulence mode, vertical synchronisation and half-bank angle limiting for high-altitude manoeuvring. The autopilot is compatible with a wide variety of navigation systems, including VOR/DME, inertial navigation system and omega/vlf.

The system comprises an autopilot amplifier, computer, control panel, two primary (pitch and roll) servos and mountings, and a yaw damper with computer, servo and mount.

Weight: (autopilot) 12.4 kg
(yaw damper) 4.8 kg
(dual flight-director system) 27.7 kg

STATUS: in production.

APS-85 digital autopilot

Introduced in March 1984, the APS-85 autopilot completes the ProLine 2 family of digital avionics for general and business aviation. APS-85 is an entirely digital, fail-passive autopilot with dual-redundant flight-guidance computers. It can be certificated to Category II operation, and Category IIIA landings will be possible with some growth. The system was flown during 1983 and early 1984 on a company Sabreliner test bed and on prototype Saab-Fairchild SF 340 commuter aircraft; it has been chosen as the standard autopilot on the latter and also on the British Aerospace 125-800 and Learjet 55/35 business aircraft.

Apart from its digital nature, the APS-85 has a number of features not found on earlier general aviation autopilots by this company, including the APS-80. For example it includes climb, cruise and descent modes. When selected in the climb mode, the system flies the aircraft according to an airspeed or Mach number schedule appropriate to that type of aircraft, and its weight at take-off. Likewise in the cruise mode the autopilot commands speed in accordance with the aircraft performance/altitude charts. In the descent mode the system sets up a rate of descent tailored to the aircraft manual and, with appropriate sensors, will be able to detect and compensate for wind-shear. Three diagnostic modes – report, input and output – are incorporated. In the first, faults in the flight control computers are displayed to the crew. In the second, the system reads out information from outside entering the system, for example from the air data system or control-surface position sensors. In the final mode, the system provides particular flight control computer outputs for examination. The diagnostics on the APS-85 are self-contained, and do not require additional test equipment.

Collins says that the new system will operate more smoothly, cost less and will have greater reliability than the same manufacturer's system it replaces. It weighs 42 per cent less, occupies 55 per cent less space, and has 80 per cent fewer wires and cables.

STATUS: first aircraft to be FAA-certificated with the APS-85 in December 1984, was a British Aerospace 125–800 business jet. It was also the world's first all-digital business aircraft. System is now in production. The 125–800 rework was done by Arkansas Modification Center at a reported cost of around $2.2 million, which also included installation of Collins EFIS-85 display, weather radar, and attitude/heading reference. The APS-85 has also been selected for the 30 and 50 series Learjets.

FCS-80II ProLine flight control system

This comprises the Collins APS-80 autopilot and FDS-85 flight-director system. Each system is described separately.

FCS-110 flight control system for L-1011

This analogue flight control system, a combined project involving both Collins and Lear Siegler, has been installed in the majority of Lockheed L-1011 TriStar airliners. A comprehensive set of autopilot modes, including Category IIIB automatic-landing capability, is incorporated. It is installed in conjunction with the Collins FD-110 flight-director system. All work on the autoland system was done by Lear Siegler, which was also responsible for other elements of the system. In conjunction with a Hamilton-Standard flight management system, the FCS-110 permits flight-path optimisation. Two-dimensional optimisation was certificated in 1972, and '3-D' in 1977. Prior to the abandonment of the L-1011 programme around 1980 Lockheed had been experimenting with a '4-D' system on its demonstration TriStar.

STATUS: in service.

FCS-240 flight control system for L-1011-500

This advanced digital flight control system was designed to replace the FCS-110. It is a dual-

Collins APS-85 autopilot includes FCC-86 computer, mode select (right) and annunciator (right) panels, and representative servo actuator

Main flight panels of British Aerospace 125-800 with Collins APS-85 autopilot

dual integrated autopilot/flight-director system and was initially tailored to the Lockheed TriStar-500 wide-body transport. Compared to its analogue predecessor the number of computing units has diminished, from seven to two, and benefits are weight savings, improved reliability, and lower cost of ownership. Category IIIB automatic-landing capability is included with a comprehensive set of conventional autopilot modes.

STATUS: in production.

ACS-240 manoeuvre load control system for L-1011

The Collins ACS-240 is the world's first production digital active flight control system,

and was designed to operate in conjunction with the FCS-110 or FCS-240 flight control systems fitted to the Lockheed L-1011 TriStar. The purpose of the ACS-240 is to alleviate unacceptable loads due to gusts on long-span versions of Lockheed's wide-body transport. In November 1979 Lockheed test-flew an L-1011-500 TriStar with wing-tip extensions of 4.5 feet (1.3 m). Owing to the lower induced drag with these tips, fuel savings of up to 3 per cent and a ride-quality improvement of 8 per cent, were achieved. However the greater wing-root bending moment during turbulence would have reduced the fatigue life unacceptably, and so Collins was contracted to provide a control system operated by acceleration sensors that would drive both ailerons upwards in unison at the onset of a gust, shifting the centre of lift inboard and so reducing its moment. The system, weighing 89 lb (40 kg), was made a retrospective modification, at customer option, on all L-1011-500s. It uses AED, a structured high-order language developed by SofTek.

STATUS: in service on some TriStar 500 airliners.

FCS-700 flight control system for 767/757

This fully-digital triplex autopilot and flight-director system, with fail-operational automatic-landing capability, was designed for the new Boeing 767 and 757 transports.

Development sprung from the FCS-111X experimental flight control system which Collins and Boeing evaluated during the late-1970s. The fully-digital architecture permits integration of comprehensive self-test and failure-protection monitoring, improved system performance monitoring, built-in growth and flexibility, and improved equipment test and serviceability features. Large-scale integrated circuit technology is used to confer a 3:1 ratio mass reduction compared with the experimental FCS-111X.

Autopilot modes include control-wheel steering, automatic cruise hold/select modes for heading, altitude, vertical speed and airspeed/Mach number, approach modes with back-course capability, and automatic approach and landing. The latter facility includes automatic flare control, roll-out guidance and coupled go-around. Computed flight-director steering is available in non-coupled flight, including take-off. Turn co-ordination and dutch-roll damping is provided. Dual or triplex control computer configurations can be used.

Steering commands from a navigation computer, using a wide variety of navigation sensors, will permit completely automatic control of pre-planned vertical and lateral flight profiles. Large-scale integrated circuit technology is used in ARINC 429 bus interface drives and digital multiplexers, and in several areas of each flight control computer, which is based on the Collins CAP-6 processor configuration. Computer operating speed is 300 000 operations a second and high-order language programming is used throughout the system.

Maintainability improvements are claimed from the MCDP-701 maintenance central display panel which uses a microprocessor to perform control, display and data management tasks. In-flight system failures are indicated by the maintenance control display panel and fault data is stored for up to ten flights.

STATUS: in production for Boeing 767 and 757 airliners.

APS-841H MicroLine autopilot

This pitch/roll autopilot system, a member of the company's MicroLine avionics family, is available for light helicopters such as the Aérospatiale AStar, Bell JetRanger variants, and Hughes 500D. It integrates with radio and navigation system inputs to provide hands-off

TriStar 500 has Collins FCS-240 flight control system

Computers and glareshield controller for Collins FCS-700 automatic flight control system used in Boeing 757 and 767

operation during climb, en route and approach phases of flight. Full-time attitude stabilisation can be selected alone or integrated with coupled-navigation modes. Automatic trim capability is available and permits manual flight-path adjustments without system disengagements.

Uninstalled two-axis autopilot and guidance control system mass, excluding flight instrumentation, is in the order of 10.5 kg. The panel-mounted computer-controller measures 3.25 × 3.25 × 9 inches (83 × 83 × 229 mm). Both 115 V ac and 28 V dc power supplies are required. The system is approved under FAA TSOs.

STATUS: no longer in production, but remains in service.

HFCS-800 flight control system

The HFCS-800 helicopter flight control system can be configured for single- or dual-pilot operation and offers what the company calls total-mission IFR capability. Operating modes

covering operations from start-up to shutdown have been incorporated, with special emphasis given to autopilot assistance and control at low-speeds and in hover.

Two separate sub-systems are used, one for automatic flight control (afcs) and the other for the flight-directors (FDS). The afcs provides automatic stabilisation and control, while the FDS computes automatic path-steering for any desired manoeuvre. The sub-systems may be purchased jointly or separately, and they meet all the requirements of FAA TSO C9A and C52A.

The following list of operating modes provides an indication of the system's performance:
Airspeed hold: maintains indicated airspeed, and pilot may 'beep' airspeed after initial selection. May be used if airspeed is above 35 knots (1 knot = 1.85 km an hour).
Vertical speed hold: holds barometric vertical speed if airspeed is above 60 knots.
Altitude hold: holds barometric altitude if airspeed is above 60 knots.
Heading select: captures selected HSI heading if airspeed is above 35 knots.
Navigation: captures and tracks selected VOR, LOC, Doppler, omega or RNav course if airspeed is above 35 knots.
Approach: captures and tracks course, providing three-dimensional control, from VOR, ILS, LOC (back-course), MLS or RNav sensors. May be used above 35 knots airspeed and up to 12° glideslope.
Airspeed/vertical speed: holds both parameters if airspeed is above 35 knots.
Navigation transfer: in two-pilot operation transfers control to co-pilot's side.
Hover augmentation: holds hover position and radio altitude.
Transition-to-hover: decelerates aircraft to hover at 50 feet (15 metres) radio altitude.
Go-around: accelerates to climb at 70 knots. May be used after take-off for departure.

Boeing 757 has Collins FCS-700 digital flight control system

The system has been selected by the US Coast Guard for installation in the Aérospatiale HH-65A (SA 336G) Dolphin helicopter.

Weights
Automatic flight control system
(afcs computer) 14.2 lb (6.43 kg)
(afcs panel) 3.3 lb (1.5 kg)
(servos) 2.1 lb (0.96 kg) each (three or four installed)
(feel/trim units) 3 lb (1.36 kg) each (two installed)
(yaw servo) 4.6 lb (2.1 kg)
(collective servo) 4.6 lb (2.1 kg)
(vertical gyros) 6.8 lb (3.1 kg) each (two installed)
(airspeed sensor) 1.1 lb (0.5 kg)
(yaw rate gyro) 2 lb (0.9 kg)

Flight-director system
(flight-director computer) 10.8 lb (4.9 kg)
(flight-director panel) 1.3 lb (0.6 kg)
(attitude control) 3.3 lb (1.5.kg)
(attitude-director indicator) 6.9 lb (3.1 kg) each (two installed)

STATUS: in production.

FCC-105-1 automatic flight control system

Based on the company's FCC-105 system, this flight control system is applicable to the Sikorsky S-76 and H-76 helicopters and has been in production since 1977. It offers a wide range of features up to full three-axis stability-augmentation with turn co ordination, pitch, roll and yaw attitude and altitude/airspeed hold modes. Each system amplifier has three independent channels (pitch, roll and yaw) contained in individual modules which can be selected on or off separately. This building-block principle permits customers to select as many features as necessary to meet their particular requirements.

The system comprises a stability-augmentation system amplifier, yaw switch, heading-hold amplifier, airspeed switches, control panels, linear electromechanical actuators, indicator panel, rate-gyros, cyclic switches and airspeed-hold amplifier.

Dimensions (main processor): 190 × 318 × 113 mm
Weight (main processor): 7.5 lb (3.4 kg)
Power: 35 W

STATUS: in production.

FCC 110 auto-stabilisation system

This is a dual-redundant auto-stabilisation system for the US Army/Hughes AH-64 Apache helicopter. The system comprises two rate gyros and two analogue computer units. Airspeed sensors and the gyros provide inputs to the computers which calculate the appropriate stabiliser angle for a mission configuration, taking account of longitudinal stability demands. Other features include pitch-axis control augmentation, and trim features which reduce pilot workload and improve control characteristics during low-altitude, high-speed, combat operations.

Dimensions: (one unit) 178 × 203 × 279 mm
Weight: (one unit) 8.2 lb (3.75 kg)
Power: 40 W per unit

STATUS: in production. First production Apache was accepted by US Army early in 1984.

Collins radar display is featured in this IAI Westwind 2 corporate jet

CMS-90 avionics management system

Introduced at the turn of the decade, the CMS-90 multi-function control/display unit has been combined with a variety of other systems to provide the crew interface. It can access, control and show the status of up to 21 separate avionics systems through a MIL-STD-1553 digital data-bus.

By early 1981 various versions of the systems had been chosen for the Sikorsky HH-65A helicopters of the US Coast Guard, HU-25A of the US Coast Guard, and the HC-130 Hercules, also for the same customer. It has also been selected for the US Air Force Fairchild A-10A Thunderbolt aircraft. At the Paris Air Show that year it was demonstrating applications in inertial navigation display, hover-guidance for helicopters, and target assignment for tactical aircraft.

In December 1981 Delco Electronics ordered CMS-90 under a $15 million contract as part of its commitment to provide 300 sets of fuel saving advisory system equipment to the US Air Force, part of a programme to upgrade its fleet of Boeing KC-135s. Deliveries for this application began in January 1983 and will last until the end of 1986. The FSAS system on the KC-135 is expected to save between 2 and 4 per cent of the fuel used by advising the crew of the most efficient speed, engine pressure ratio, altitude and descent profile.

The US Army has selected the CMS-90 for Special Operations and Special Electronic Mission Aircraft (SEMA). In addition the CMS-90 has been chosen for the AH-64A Apache.

Most recently CMS-90 was chosen as part of the ICNI (integrated communications/navigation/identification) system for the US Air Force/General Dynamics FB-111. The system

Display presentation of the Collins CMS-90 as ICNI management system on General Dynamics FB-111

provides control/display facilities for uhf, hf, secure voice, ILS, Tacan, IFF, and the INS. This application marks the first installation of CMS-90 into a high-performance supersonic aircraft, and its particular benefit is a reduction in display area from the 135 square inches (871 cm²) needed for previously separate systems, to 41 square inches (265 cm²).

STATUS: in production.

Delco

Delco Systems Operations, Delco Electronics Corporation – A Subsidiary of GM Hughes Electronics Corporation, 6767 Hollister Avenue, Goleta, California 93117
TELEPHONE: (805) 961 5903
TWX: 910 334 1174

This company was formerly known as Delco Systems Operations, General Motors Corporation; as Delco Electronics, Santa Barbara Operations; and as AC Electronics, General Motors Corporation

Performance management system

The Delco performance management system was initially developed for use in the Boeing 747 as a retrofit system and is now an option in new 747 deliveries.

It is a two-box system with the addition of an engine interface unit and a switching unit. The first device processes engine discrete signals, such as fuel flow, into a format usable by the computer. The switching unit handles all switching of autopilot, autothrottle and instrument signals from standard aircraft configuration to performance management control.

The performance management system achieves closed loop control of the aircraft by providing commands to the autopilot for vertical steering, and either directly or via the inertial navigation system to the autopilot for lateral steering, and the autothrottle for engine thrust control.

Delco has also adapted the performance management system for all models of the McDonnell Douglas DC-9, though the engine interface unit is not required, and in the case of the MD-80 the switching unit is not needed. The system has also been adopted for the McDonnell Douglas DC-10 and Boeing 727 airliners.

Dimensions: (computer unit) ½ ATR Long (control/display unit) 146 × 114 mm high (engine interface unit) ¼ ATR Long (switching unit) ½ ATR Long
Weight: (computer unit) 29 lb (13.2 kg) (control/display unit) 6 lb (2.7 kg) (engine interface unit) 4 lb (10 kg) (switching unit) 22 lb (10 kg)
Power: (computer unit) 200 W (control/display unit) 38 W (engine interface unit) 4.2 W (switching unit) 110 W

STATUS: in production.

Installation of Delco performance management system in Boeing 747

Fuel savings advisory and cockpit avionics system

Delco provides the fuel savings advisory and cockpit avionics system (FSA/CAS) for the upgrading of the current Boeing C-135 and KC-135 fleet of about 700 aircraft.

In this system, a crt control and display unit replaces one of the standard inertial navigation system units to simplify crew management tasks; in particular, all basic data for flight use can be entered at this one point. In addition to the usual engine and navigational information, a new fuel-management panel and centre of gravity display have been installed for inputs of fuel state, disposition and usage. This latter function is invaluable in tanker operations and ensures efficient fuel allocation and usage. The panel is linked to other avionic systems by a MIL-STD-1553B data-bus.

The FSA provides commands to the pilots during climb, cruise and descent using flight path optimisation algorithms and flight manual

data for lift, drag and thrust. An additional mode computes all required take-off and landing parameters based on crew inputs for present conditions.

Many KC-135s are being refitted with GE/Snecma CFM56 engines, which provide extra thrust over earlier types. FSA/CAS is integrated with this engine, offering more economic operation.

Dimensions: (FSA computer) ½ ATR long (control/display unit) 5.75 × 7.11 inches (146 × 181 mm) (bus controller) 8.15 × 5.19 inches (207 × 131 mm) (fuel panel) 18 × 4.85 inches (457 × 123 mm) (fuel management computer) 9.25 × 9.8 inches (235 × 249 mm) (remote display unit) 3.26 × 3.26 inches (83 × 83 mm)
Weight: (FSA computer) 26.2 lb (11.9 kg) (control/display unit) 10 lb (4.5 kg) (bus controller) 9 lb (4.1 kg) (fuel panel) 27.5 lb (12.5 kg) (fuel management computer) 48 lb (21.8 kg) (remote display unit) 2.1 lb (0.95 kg)

Delco performance management system for McDonnell Douglas MD-80

Power: (FSA computer) 95 W (control/display unit) 44 W (bus controller) 34 W (fuel panel) 92.8 W (fuel management computer) 139.5 W (remote display unit) 10 W

STATUS: in production.

General Electric

General Electric Company, Aerospace Electronic Systems Division, PO Box 5000, Binghamton, New York 13902
TELEPHONE: (203) 373 2121

General Electric flight control set for McDonnell Douglas F-15

AN/ASW-38 automatic flight control set for F-15

The ASW-38 provides the control functions for the US Air Force's McDonnell Douglas F-15 Eagle. It provides three-axis command augmentation to improve the fighter's handling qualities over the very wide speed/height envelope. The analogue system incorporates two channels in each of the three axes, also supplied by dual sensors. Three-axis fail-safe trim control and stall-inhibit functions are also provided. The set comprises two computers (one for pitch, the other for roll and yaw, a three-axis dual-channel rate sensor box, two-axis dual-channel accelerometer box, dual-channel pressure sensor unit, stick-force sensor, and pilot's controller.

Dimensions: (computers) 6 × 6 × 15 inches (152 × 152 × 381 mm) (rate sensor) 4.5 × 4.5 × 7 inches (114 × 114 × 178 mm) (acceleration sensor) 4 × 2.3 × 4 inches (102 × 57 × 102 mm) (pressure sensor) 4.5 × 3 × 5 inches (114 × 76 × 127 mm)

Weight: (total set) 44.5 lb (17.8 kg)
Power: 265 W
Reliability: 905 h predicted, 1100 h actual over 10 000 flight hours

STATUS: in production for current versions of F-15 fighter. The F-15E dual-role version in development will have a digital flight control system by Lear Siegler.

Stability-augmentation system for A-10

Stability-augmentation for the US Air Force/Fairchild A-10 in pitch and roll is provided by a two-axis, two-channel system that not only improves handling qualities but makes for more accurate weapon delivery. Each channel in a given axis is independent of the other, except for monitoring cross-links, and both are disengaged if significant desynchronisation occurs. Separate 'engage' switches permit single-channel operation, however. The system also provides turn co-ordination, compensation for the pitching moment caused by operation of the speed brake or recoil from the General Electric GAU-8 30 mm cannon. The pitch and yaw monitors can also be self-tested. The system comprises a single-box computer and pilot's control panel. The computer contains six rate-sensors, four of which are solid-state devices manufactured by General Electric.

Dimensions: (computer) 6.1 × 7.7 × 12.5 inches (155 × 196 × 318 mm)
(control panel) 5.8 × 3.4 × 6.5 inches (146 × 86 × 165 mm)
Weight: (total) 15.9 lb (7 kg)
Power: 73 W

STATUS: in production.

Fuel management system

General Electric's mass-flow measurement system, introduced in 1981, senses the mass-flow rate of fuel to the propulsion system or in refuelling operations to another aircraft. The output can be displayed in terms of flow rate and/or fuel consumed, or used as an input into other system elements. Recent technology improvements include the design of a flow-transmitter in which the impeller is driven by the flow of fuel, eliminating the impeller motor. An

General Electric pitch- and yaw-stability augmentation system for Fairchild A-10 with flight control unit (left) and computer

extension of this approach led to the production of a flowmeter in which fuel provided fluid torque to drive the rotor. The output is then a pulse which is proportional to the mass flow rate. No input power is required.

STATUS: in production.

Liquid-level measurement system

General Electric has developed a liquid-level transducer which provides an output proportional to the level of fluid in a tank. The primary application is the measurement of oil, but hydraulic fluid and fuel quantities may also be measured. Civil applications which utilise General Electric fuel management systems or liquid level measurement systems and, in some

cases both, include Airbus A310; Boeing 707, 727, 737, 747, 757 and 767; McDonnell Douglas DC-8, DC-9 and DC-10.

In the military field, aircraft using these systems comprise the Northrop F-5E/F, Lockheed P-3C Orion, Grumman A-6/EA-6B, McDonnell Douglas F-15 and F/A-18, and General Dynamics F-16. The Boeing KC-135 Stratotanker and McDonnell Douglas KC-10 Extender also use them for air-to-air refuelling, and the Lockheed SR-71 reconnaissance aircraft uses the system to measure fuel quantity. Development activities include the completion of a fluidic transmitter and work on an advanced non-intrusive flow transmitter.

STATUS: in production.

US Air Force/Lockheed SR-71, world's fastest operational aircraft, has General Electric fuel quantity measurement system

Global

Global Systems Inc, A Unit of Sundstrand, 2144 Michelson Drive, Irvine, California 92715
TELEPHONE: (714) 851 0119
TELEX: 681340
TWX: 910 595 2451

AFIS airborne flight information system

Global has expanded the capabilities of its GNS-1000 flight management system (which see) into a full airborne flight information system, with comprehensive facilities for flight-plan creation before flight and amendment in flight. Weather updates can be received and flight-related messages be sent or received. Before flight the pilot accesses the Global Data Center through the Global Portable Computer, which is a part of the AFIS system, for en route weather and other flight information. This information is recorded on a mini-computer disk, which is later loaded into the AFIS system via a data transfer unit. During flight the comparison of planned and actual flight path can be viewed at any time on a control and display unit.

Global airborne flight information system

An air-to-ground ground-to-air data-link can be used to update the flight plan at any time in flight in the event for example of changing weather or air traffic control requirements.

Messages can also be passed on this link, which is part of the Global Data Center.

STATUS: in development.

Hamilton Standard

Hamilton Standard, Division of United Technologies, Bradley Field Road, Windsor Locks, Connecticut 06096
TELEPHONE: (203) 623 1621
TELEX: 994439

FCES 101 flight control system for F-20

Hamilton Standard builds the digital flight control system for the Northrop F-20 Tigershark. The dual-redundant comprises a computer, cockpit control unit, air data unit, lateral and vertical accelerometers, and pitch, roll and yaw rate gyros.

Dimensions: volume 1200 in³ (19 700 cm³)
Weight: 37 lb (16.8 kg)
Power: 158 W

STATUS: a limited programme has been underway to provide equipment for four development F-20 aircraft and supporting spares and test-rigs.

Northrop F-20 has Hamilton Standard digital flight control

ASE 6, 19, 21 and 23 autostabilisation systems

Developed from early electromechanical two-axis stability-augmentation systems (first operational in 1958 and claimed to have been the world's first light-helicopter stability-augmentation systems), these units still serve in many Sikorsky S-61, S-62 and S-64 helicopters, and remain in production.

Typical computer dimensions: 266 × 180 × 318 mm
Weight: 22-35 lb (10-16.5 kg)
Power: 65 W

STATUS: in service.

FCC-100 automatic flight control system

This system was designed specifically for the US Army/Sikorsky UH-60 Blackhawk helicopter. It meets requirements calling for quick and easy field maintenance and repair and is claimed to have been the first application of a dual digital processor-based stability-augmentation system for a production helicopter. Features include three-axis stability augmentation with turn co-ordination, pitch, roll and yaw attitude and altitude/airspeed hold modes.

Dimensions: 356 × 305 × 217 mm
Weight: 18 lb (8 kg)
Power: 75 W

STATUS: in production.

FCC-105 automatic flight control system

This automatic flight control system introduced modern digital technology features in the Sikorsky HH-60, CH-53E, SH-60 and MH-53E helicopters. It is a dual-redundant system which provides three-axis stability augmentation, turn co-ordination and hands-off flight. It can also include a stick-force feel system, and has four degrees-of-freedom facilities and hover augmentation. The most recent version, on the Sikorsky MH-53E helicopter, includes MIL-STD-1553B digital data interfaces.

The baseline 53E computer used in the Sikorsky SH-60B includes automatic approach to hover, three-axis Doppler and radio-altitude coupling, with provision for tactical navigation computer coupling.

Dimensions: 190 × 432 × 266 mm
Weight: 25-29 lb (11.5-13.5 kg)
Power: up to 83 W

STATUS: in production.

Flight management systems

As part of the Hamilton Standard organisation, Arma produced the first flight management system to be certificated for airline service, and deliveries began in 1977. Designed around the Lockheed L-1011 TriStar, it is now in service with British Airways, Delta, Gulf Air, Pan Am and Saudia Airways and, under the designation FMS 500, is standard equipment on all L-1011-500 aircraft.

Arma produces a family of equipment based on modular techniques that allow upgrading of a basic performance system to the full flight management system standard. In the case of the FMS 500 the navigation element was developed from the earlier Mona (modular navigation) system.

The flight management system provides navigation control by interfacing with the aircraft's navigation computer and position-sensing equipment, and can handle INS and DME information for optimum updating. En route waypoints are stored in the flight management system computer via a control/display unit and the system automatically steers the aircraft from one waypoint to the next by providing the appropriate commands to the navigation computer and thence to the autopilot. This method achieves completely automated 'three-dimensional' flight from shortly after take-off through to the destination terminal area.

The systems interface with all major aircraft components including engines, navigation sensors, autopilot, autothrottle, air data computer, and designated flight instruments.

Dimensions: (computer unit) 1 ATR Long (control/display unit) 4.5 × 5.75 × 9 inches (114 × 146 × 229 mm)
Weight: (computer unit) 45 lb (20.4 kg) (control/display unit) 8.5 lb (3.86 kg)
Power: (computer unit) 400 W (control/display unit) 45 W

STATUS: in production. Derivatives of system are being developed for other commercial and military aircraft.

PMS 500 performance management system

The PMS 500 has been designed for the Lockheed L-1011 TriStar providing outputs to drive the autothrottle system and pitch computers to maintain the best vertical profile. In this system there are no navigation inputs. The PMS 500 automatically controls engine exhaust pressure ratio, altitude and airspeed via the autothrottle and pitch interfaces to match

commanded values. The equipment guides the aircraft in the vertical plane from take-off, in the climb, during cruise and descent.

Dimensions: (computer unit) 1 ATR Long (control/display unit) 4.5 × 5.75 × 9 inches (114 × 146 × 229 mm)
Weight: (computer unit) 40 lb (18.1 kg) (control/display unit) 8.5 lb (3.9 kg)
Power: (computer unit) 300 W (control/display unit) 45 W

STATUS: in service.

Pilot's performance system

The basic element of the 500 Series family is the pilot's performance system, which comprises a control and display unit and avionics bay mounted computer unit. As in other systems initial pre-flight data is entered via the control and display unit. After selection of the appropriate flight mode, for example, climb, cruise or descent, the computer calculates the altitude, airspeed and engine thrust to provide optimum fuel savings based on variables such as ambient temperature, aircraft gross weight, wind speed and direction and remaining distance to go. These performance parameters are then displayed continuously on the control and display unit and the aircraft is flown to match them.

Dimensions: (computer unit) 1 ATR Long (control/display unit) 4.5 × 5.75 × 9 inches (114 × 146 × 229 mm)
Weight: (computer unit) 37 lb (16.8 kg) (control/display unit) 8.5 lb (3.9 kg)
Power: (computer unit) 275 W (control/display unit) 45 W

STATUS: under development for Boeing 747, though principal market segment is considered to be all large transport aircraft and helicopters.

Hamilton Standard PMS 500 performance management system for Lockheed L-1011

Hamilton Standard AIC-12 air-intake control for McDonnell Douglas F-15 fighter

Hamilton Standard EEC-103 supervisory control has been chosen for Pratt & Whitney JT9D engines in Boeing 767

Weight: 21 lb (9.5 kg)
Power: 50 W

STATUS: in production for Airbus A310 and Boeing 767 airliners.

AIC-12 air intake control for F-15

Hamilton Standard has delivered more than 3000 air-intake controls for the US Air Force/ McDonnell Douglas F-15 Eagle. The system controls the F-15's intake ramps to achieve subsonic flow to the engine via a normal shock wave, minimising the energy loss due to the shock. The system also controls the bypass doors which allow the correct flow of air into the compressor at the right pressure.

A single control unit measures free-stream and intake static and total pressures, total air temperature and the aircraft's angle of attack. This data is used by the microprocessor element to calculate the optimum inlet throat area, and control signals are generated to drive the ramps and bypass doors via servo-actuators. A by-product of the calculation is the aircraft Mach number, which is fed to the EEC-90 electronic supervisory control detailed below.

Dimensions: 11 × 9 × 7.5 inches (279 × 229 × 190 mm)
Weight: 16.2 lb (7.3 kg)
Power: 65 W

STATUS: in production for F-15.

EEC-81 temperature limiter for F-14

The Hamilton Standard EEC-81 temperature limiter is an analogue electronic unit which has been in production for the Pratt & Whitney TF30 engines powering the US Navy/Grumman F-14 Tomcat since 1969. More than 1800 units have been delivered. The unit receives turbine inlet temperature signals from engine thermocouples and compares them with preset datum temperatures. The resulting temperature-error signal drives a fuel control trim motor.

Dimensions: 10 × 4 × 6 inches (254 × 102 × 152 mm)

Hamilton Standard provides EEC-81 temperature limiters for twin Pratt & Whitney TF30 engines in Grumman F-14

Weight: 5.9 lb (2.7 kg)
Power: 30 W

STATUS: in production.

EEC-90 engine supervisory control

Hamilton Standard claims that the US Air Force/McDonnell Douglas F-15 was the first high-performance military aircraft to have microprocessor-based engine controls. The EEC-90 was designed for the Pratt & Whitney F100 augmented turbofan engine that powers the F-15 Eagle and the General Dynamics F-16, and more than 3000 were delivered in the first ten years of production which began in 1972. Flexibility, accuracy and growth potential are cited by Hamilton Standard as advantages over earlier analogue systems.

The EEC-90 uses a general-purpose digital processor to compare the aircraft and engine performances with stored data which gives the desired performance under the relevant operating conditions. The system then drives the engine's hydromechanical fuel control and changes engine geometry schedules, such as exhaust nozzle position, to achieve the desired performance. The EEC-90 is mounted on the engine and is protected against adverse temperature, pressure and vibrations. Its function is essentially one of fine-tuning engine performance.

Dimensions: 17.2 × 11.6 × 6.2 inches (437 × 295 × 157 mm)
Weight: 24.5 lb (11.1 kg)
Power: 160 W

STATUS: in production.

EEC-103 engine supervisory control

Pratt & Whitney's JT9D-7R4 commercial turbofan engine comes with a Hamilton Standard EEC-103 supervisory control, so the digital unit · is to be found on the Boeing 767 and Airbus Industrie A310 and A300-600 airliners. It is similar in function to the EEC-90 but has a smaller volume and its function is mainly to control fuel; there are few, if any, variable-geometry control features. Built-in test equipment is incorporated and fault information is passed into the ARINC 429 data-bus to assist fault diagnosis. Reliability of the system on 53 Boeing 767s with the JT9D-7R4 at the end of 1984 was running at an average of one failure in 33 000 hours; highest reported figure was 50 000 hours on the United Airlines fleet of 19 Boeing 767s. The EEC-103 was one of the first such systems to begin reliability testing (in 1980) on the new company environmental test facility.

Dimensions: 14.75 × 13.9 × 3.75 inches (375 × 353 × 95 mm)

EEC-104 full-authority engine control for 757

The dual-redundant EEC-104 control system has been chosen for Pratt & Whitney's PW2037 turbofan, which is optional on some versions of the Boeing 757. Pratt & Whitney powered 757s began commercial services at the end of 1984 with Delta Air Lines. It is based on EEC-103 technology.

This system applies full electronics technology (ie without mechanical reversion) with signals derived from an engine-mounted digital control system that employs advanced gate arrays, semi-conductors and environmental packaging as protection against extreme temperature variations, vibration and degradation by fuel, hydraulic fluid and engine oil. The system represents a co-ordinated design effort by Pratt & Whitney, Hamilton Standard, Mostek and Microelectronics Center (all subsidiaries of United Technologies). It is intended to provide specific engine pressure ratios at pre-determined power-lever positions, ranging from take-off through intermediate thrust, derated power, to reverse limit. The digital control eliminates the power overshoots common to conventional equipment and which results in a shortening of engine life.

The full-authority engine control allows 'set and forget' operation, since the throttle setting is a reflection of a thrust demand. The EEC-104 automatically trims fuel flow to the engine so no throttle 'tweaking' is needed. As well as controlling fuel flow the dual-channel unit takes care of engine starting, variable stator-positioning and active clearance control. A constant-speed idle function is provided and the system controls engine acceleration and deceleration. Protection against over-heating and over-speeding is provided. The result is simpler operation, more consistent and accurate thrust settings and improved engine life.

Dimensions: 5 × 13.5 × 19.2 inches (127 × 343 × 488 mm)
Weight: 37.1 lb (16.8 kg)
Power: 50 W

STATUS: in service.

EEC-118 MACS multiple-application control system

Hamilton Standard has drawn on its long experience in engine control to produce a standard electronic engine control which is suitable for turboprop and turboshaft engines producing up to 5000 shaft horsepower (3700 kW), and turbofans rated at up to 5000 lb (22.2 kN). MACS is an integrated electronic and hydromechanical unit designed to reduce operating cost and increase engine performance and life.

Hamilton Standard EEC-104 full-authority digital engine control system under construction

Hamilton Standard PSC-101 synchrophaser synchronises propeller phasing on Lockheed C-130 Hercules

these limits are stored in a semiconductor memory for later diagnosis. This is intended to help solve any problems as they arise. Data accepted by the propulsion multiplexer includes engine pressures, temperatures, fuel-flows, vibration levels, oil condition and positions of variable-geometry features such as intake guide-vanes and active-clearance controls.

Dimensions: 16 × 14 × 4 inches (406 × 356 × 102 mm)
Weight: 18 lb (8.1 kg)
Power: 24 W

STATUS: in production.

Hamilton Standard EEC-118 multi-application engine control system

The system has been chosen for the Pratt & Whitney Canada JT15D (Mitsubishi Diamond, Peregrine and SIAI-Marchetti S.211), the PW100 (de Havilland Canada Dash 8, Embraer EMB-120 Brasilia and Aérospatiale/Aeritalia ATR-42) and the Rolls-Royce Gem (Westland 30, Lynx and Agusta A 129 helicopters). For the turbofan, MACS provides automatic thrust ratings, climb without throttle movement, engine synchronisation and over-speeding protection. Turboprops also have torque limit display, propeller speed control and automatic trim. The EEC-118 offers fast-response power-turbine governing, torque sharing and limitation, and temperature limitation.

Dimensions: 7.5 × 7.5 × 2.1 inches (191 × 191 × 53 mm)
Weight: 4.5 lb (2 kg)
Power: 11 W

STATUS: in production.

Digital engine control systems

Hamilton Standard has in hand the following family of digital engine control units:-
EEC-131 is intended for the Pratt & Whitney PW4000 engine destined for the Boeing 747 and 767 and the Airbus A300 and A310 airliners:

this unit is in development, it is essentially an upgraded EEC-104.
EEC-132 will be used on the Pratt & Whitney PT6B-35 engine and commercial flight-tests on a Sikorsky S-76 helicopter took place in 1984.
EEC-150 is an improved version of the EEC-104, and is being developed for use on the IAE V 2500 turbofan engine. This is a five-nation collaborative powerplant involving two of the 'big three' engine companies, Pratt & Whitney and Rolls-Royce. The engine has also been chosen for the Airbus Industrie A320 airliner.
EEC-133 For the military market, Hamilton Standard is developing this unit for the PW3005 engine.
EEC-106 The only full authority single-channel unit in the product range, this system is being evaluated, along with a mechanical reversionary system, for the Pratt & Whitney F100 engine of the McDonnell Douglas F-15 fighter under the US Air Force's DEEC (digital electronic engine control) programme.
Engine control for LHX Hamilton Standard and the UK's Lucas Aerospace are making a joint proposal to both Pratt & Whitney and Avco Lycoming for digital engine control equipment in support of the LHX light utility and attack helicopter. Both engine companies are in competition to build the powerplant for this aircraft.

EDM-110, EDM-111, and EDM-112 PMUX propulsion multiplexers

Hamilton Standard supplies propulsion multiplexers for the Pratt & Whitney JT9D-7R4 (EDM-110) and General Electric CF6-80 (EDM-111) big turbofan engines, and the PW 2037 (EDM-112) 37 000 lb (16 800 kg) 165 kN thrust medium-size turbofan. The applications for these are the Boeing 747, 767 and 757, and Airbus A310 aircraft. The propulsion multiplexer collects analogue data from the engine and air data system, converts it to digital format, samples it, and changes it into a multiplexed data stream. The data is then transmitted to a ground-based diagnostic unit.

The propulsion multiplexer is programmed with the engine's mechanical and thermo-dynamic limits, and parameters going beyond

PSC-100, PSC-101, PSC-102 solid-state propeller synchrophasers

The first type of propeller synchrophaser built by Hamilton Standard employed thermionic tube technology and is still used aboard the Lockheed C-130 Hercules and P-3 Orion and CP-140 Aurora aircraft. The unit synchronises all four engines at 100 per cent rpm and controls the propeller phase to minimise sound level.

The solid-state synchrophasers which have superseded them have the benefits of later technology: less weight, volume, power consumption and crew workload, and greater reliability. Aircraft noise levels are also lower because the phasing is controlled more accurately. As well as performing the functions of the earlier synchrophasers, the solid-state device also provides dynamic speed control for power changes and aircraft manoeuvres, again reducing noise. The solid-state PSC-100 is in production for the Grumman E-2 Hawkeye aircraft and C-2 Greyhound, while the PSC-101 is built for the civil and military Hercules. The PSC-102 was proposed for the shelved Lockheed L-400 Twin Hercules project.

Weight: 10 lb (4.5 kg)
Power: 42 W

STATUS: in production.

PSC-103 digital propeller synchrophaser

Hamilton Standard is developing this digital synchrophaser for the next generation of commuter airliners and it has been selected for the de Havilland Canada Dash 8 and Embraer EMB-120 Brasilia. Its functions are similar to those of the solid-state analogue synchrophasers described above, but will have improved control accuracy and reliability. The digital unit is also considerably smaller and lighter and consumes less than half the power of the earlier units.

STATUS: under development.

Honeywell

Honeywell Inc, Military Avionics Division, PO Box 31, 2600 Ridgway Parkway, Minneapolis, Minnesota 55413
TELEPHONE: (612) 378 4141
TELEX: 290631

Flight control system for F-14

Honeywell builds the three-axis analogue auto-pilot system for the US Navy/Grumman F-14 Tomcat. The system has three channels for pitch, two successive failures giving fail-opera-tional/fail-safe, and two channels each for roll and yaw, giving fail-safe reversion after a failure in those axes. The system is not connected in any way to the wing-sweep mechanism.

STATUS: still in production for F-14.

Automatic flight control system for C-5B

Honeywell provided the flight control system for the US Air Force/Lockheed C-5A Galaxy. It was claimed to be the first operational flight control system to provide complete Category III automatic landing and rollout guidance with fail-safe performance. Honeywell is providing an improved (though still analogue) five-box 'form, fit and function' version of the original flight control system in the C-5B; the computer boxes fit into the same racking as the C-5A, have the same connectors, and indeed are functionally interchangeable with the flight control boxes in the earlier aircraft. Common-

US Air Force/Lockheed C-5 freighter has Honeywell flight control system

ality of the two analogue systems in order to exploit the heavy investment in ground-test and avionics maintenance equipment was a major factor in the decision to go for a 'minimum-change' system. The principal purpose of its upgrading is to improve reliability and ease of maintenance. The system incorporates three-axis stability-augmentation and autothrottle.

STATUS: first sets were delivered to Lockheed in April 1985.

Dafics digital automatic flight and inlet control system for SR-71

As part of a major programme to improve performance and reliability and reduce pilot workload, Honeywell was contracted in mid-1978 to supply a completely new digital flight and propulsion control system for the US Air Force/Lockheed SR-71 aircraft. The first ship-sets of what is called Dafics (digital automatic flight and inlet control system) for this pro-gramme were delivered in mid-1983.

Right at the beginning of its career it was appreciated that the wide extremes of speed and altitude available to the SR-71 – greater than with any other aircraft – together with the aerodynamic characteristics of its thin delta wing would entail in some areas of the flight envelope handling characteristics unacceptable for manual operation. In short, the aircraft under some circumstances was bound to lack adequate stability in all three axes and would need both an autopilot for long-duration missions and a sophisticated autostability system to shelter the pilot from these handling inadequacies.

Honeywell built the original flight control system for the SR-71, which entered service in January 1966. It was basically a three-axis eight-channel analogue system that managed the flying controls. A separate control system was implemented in the original aircraft to manage the inlet spikes of the engines to maintain the entry shock waves in the correct position. (This function was integrated with the flight control in the new digital Dafics system.) Positioning of these spikes is critical to half an inch. Also critical, in this case to the fuel economy of the SR-71, is the centre of gravity control whereby fuel is transferred between tanks in order to balance the aircraft nearly on its centre of lift. By doing this, the trim drag caused by having to carry a degree of up-elevon during cruise is greatly reduced. The task of managing these and other 'housekeep-ing' systems, together with the need to monitor navigation accuracy at 30 nautical miles a minute and operate the mission sensors, imposes a heavy work-load on the two-man crew. Accordingly, with the arrival of reliable digital technology, the US Air Force decided to capitalise on its benefits in order to make the SR-71 easier to fly and more efficient, though the mechanical signalling of commands to the actuators is retained.

The SR-71 Dafics system is based on work done originally by the US Air Force Flight

US Air Force/Lockheed SR-71 has Honeywell Dafics flight control system

Dynamics Laboratory around the mid-1970s and the flight-control system designed by Honeywell for Sweden's Saab Viggen fighter. It has three channels, each with self-monitoring capability, so that it remains operational after any two failures. The principal computing element in each channel is a Honeywell 5301 processor.

There are three levels of redundancy (two fail-operate, and finally fail-safe), and a comprehensive built-in test schedule employs in-line monitoring techniques. The system is said to generate a 10 per cent fuel saving just as the result of being able to control more accurately the positioning of the engine inlet spikes.

STATUS: it is believed that some 20 of the original 32-strong fleet of SR-71s have now been modified with the Dafics system.

JET

JET Electronics and Technology Inc, 5353 52nd Street, Grand Rapids, Michigan 49508
TELEPHONE: (616) 949 6600
TELEX: 226453

FC-530 autopilot for Learjet 30

Available for new installation or retrofit to Gates Learjet Series 30 corporate and business aircraft, the FC-530 autopilot was certificated in September 1982 and employs many features which have previously been available only for the Learjet Series 55. JET developed a new flight control system, the FC-550, for the Learjet 55, and the airframe manufacturers then requested a version of the system that could be offered as a retrofit to augment the performance of the earlier Learjet 30 series aircraft.

All autopilot controls and annunciations are grouped on a new glareshield controller. New

facilities for the Series 30 include altitude preselect, full and half-bank mode selections, electrical pitch disconnect facility and illuminated daylight-readable controller indicators. The system integrates directly with aircraft gyro-sensors, primary navigation receivers and flight-director instruments.

The total system comprises the following units:
AC-530 autopilot controller
CA-530 computer/amplifier
Air data unit
AV-500 normal accelerometer
SA-200F pitch servo actuator
SA-200A roll servo actuator
RG-227 rate gyros
FU-500 position sensors
YD-530 yaw damper

STATUS: in production.

JET FC-530 autopilot controller designed for Learjet Series 30

FC-535 autopilot for C-21A Learjet 35 OSA

A development or variant of the FC-530, this autopilot has been specified for the US Air Force/Gates C-21A Learjet 35 OSA (operational support aircraft) programme. This newly configured hybrid digital system has dual controllers and computers.

STATUS: in production for 80 C-21As being supplied to the US Air Force under a leasing arrangement with Gates Learjet.

King Radio

King Radio Corporation, 400 N Rogers Road, Olathe, Kansas 66062
TELEPHONE: (913) 782 0400
TELEX: 42299

Silver Crown KAP 100 autopilot

This is a panel-mounted single-axis, 'wings level', digital flight control system which includes a KG258 horizontal-reference indicator and KG107 directional compass. An optional slaved compass system can be substituted for the latter item. Options include manual electric trim, control-wheel steering and a yaw damper. Lateral modes include: heading select, navigation tracking, approach and localiser backcourse modes. The system was announced in June 1982 and is certificated on several single-engined aircraft types.

KAP 100 with KG 107 directional-gyro, no yaw damper
Weight: 10.9 lb (4.9 kg)
Power: 14 V dc 3.1 A
1.6 A at 28 V dc

STATUS: in production.

Silver Crown KAP 150 autopilot

This is a panel-mounted two-axis digital autopilot providing pitch and lateral control facilities.

It is integrated with standard flight-instrument packages which are available from the company's range of products. Slaved compass and remote mode annunciator options are provided. Autopilot modes include: pitch hold, heading select, altitude hold, navigation tracking, approach, glideslope and localiser back-course, vertical trim and control-wheel steering. A yaw damper is available as an optional extra. The system is certificated on several single-engined aircraft types. It was announced in June 1982.

KAP 150 with KG 107 directional-gyro, no yaw damper
Weight: 18 lb (8.2 kg)
Power: 14 V dc 5.1 A
2.5 A at 28 V dc

STATUS: in production.

Silver Crown KFC 150 autopilot/flight-director system

This is a KAP 150 autopilot with a 3-inch (76 mm), air-driven gyro, flight instrument

package. Single-cue V-bar presentation is used on the flight director. Also included in the package is a KCS 55A slaved compass system with KI 525A pictorial navigation indicator. The KFC 150 provides the following modes: pitch attitude hold, altitude hold, flight director, heading select, navigation, approach, glideslope, back course, vertical trim and control wheel steering. The system was introduced in June 1982.

KFC 150 without yaw damper
Weight: 25.3 lb (11.5 kg)
Power: 14 V dc 8.7 A
4.4 A at 28 V dc

STATUS: in production.

Silver Crown KFC 200 autopilot/flight-director system

This automatic flight control system is suitable for a wide range of single- and twin-engined light aircraft. It comprises a two-axis autopilot with flight-director instruments, and can be configured as the full KFC 200 or the lower-cost KAC 200 system. Both variants have a two-axis autopilot, the low-cost option providing wings-level and pitch attitude hold with altitude, navigation, approach, localiser, back-course and heading select modes. The larger variant additionally includes go-around and control-wheel steering modes.

Flight-instrumentation options include the KG 258 flight command indicator and KI 525A horizontal situation indicator for the low-cost KAP 200, or the same horizontal situation indicator with a KI 256 flight command indicator in the more comprehensive KFC 200 system. Manual electric pitch-trim facilities are included, and options include slaved gyro and yaw damper installations.

KFC 200 without yaw damper
Weight: 28 lb (12.8 kg)
Power: 15.5 A at 14 V dc, or 9.5 A at 28 V dc

STATUS: in production.

Gold Crown KFC 250 autopilot/flight director

This system is effectively the computation and control function of the Silver Crown KFC 200 in conjunction with the 4.25-inch (108 mm) KFC 300 flight director. A solid-state computer

King Radio KFC 150 autopilot can be integrated with flight-director instruments for wide range of light aircraft

King KFC 300 system, top to bottom, mode annunciator, horizontal situation and compass director indicators and mode selector

King KFC-400 digital flight control system

generates flight-director commands in parallel with three-axis autopilot control signals. The system comprises a KAP 315 mode annunciator, KCL 310 flight director, KPI 552 pictorial navigation indicator (essentially a horizontal situation indicator), KAS 297 altitude selector, KC 290 model selector and KC 291 yaw mode controller.

Weight: 44 lb (20 kg)
Power: 80 VA at 115 V 400 Hz
11A at 28 V dc
42 VA at 26 V 400 Hz

STATUS: in production for high-performance piston twins and medium-size business turbo-props.

Gold Crown KFC 300 autopilot/flight director

King's top-of-the-range flight control system is designed for operation in conjunction with its KNR 665A digital area navigation system, and is intended for corporate and regional transport aircraft. The flight-director and autopilot are fully integrated, so that the crew can fly the aircraft manually by reference to the former, or can couple its commands to the autopilot and simply monitor the director instruments. Vertical navigation in one option operates in conjunction with RNav (area navigation) to provide the desired altitude or clearance throughout flight.

Weight: (less KNR 665A) 74 lb (33.64 kg)

STATUS: in production.

KFC-400 digital flight control system

Intended for turboprop and turbine-powered general aviation aircraft, the KFC-400 was announced in early 1985. The company's first all-digital autopilot, it is a dual-channel system designed to fail passive. The system is micro-processor-driven, and when used in conjunction with King's KNS 660 navigation and frequency management can provide full three-dimensional flight guidance, with automatically scheduled climb and descent profiles.

The system includes a new digital air data computer providing altitude, airspeed, vertical speed and Mach number via an ARINC 429 digital data bus, and it can also drive a new range of electromechanical flight instruments. Continuous fault-monitoring is incorporated.

STATUS: in production. The system was test-flown on the company's Beechcraft King Air 200, and was being installed in a Piper Cheyenne 400 LS during late 1984 for initial certification.

KFC-325 flight control system

The KFC-325 is a three-axis digital flight control system designed to suit high performance turboprop aircraft. An extensive preflight test of full time monitors ensures system integrity while airspeed compensated control maintains proper response to changing aircraft configurations. Manual electric trim speed is also adjusted to aircraft speed.

The remote mounted flight computer contains four microprocessors, with one dedicated to each of the following functions: roll, pitch, yaw damp and logic. In addition to providing these computations the microprocessors provide extensive preflight test of pitch, roll and accelerometer monitors which ensure system integrity during flight operations.

The KFC-325 is configured with a 4-inch (102 mm) electromechanical ADI and HSI with growth provisions for interface with the Bendix EFS-10 electronic flight instrument system EADI and EHSI displays. The electromechanical instruments include the KCI 310A rotating sphere flight command indicator and the KPI 553A pictorial navigation indicator with readout of DME distance, groundspeed and TTS plus radar altitude and distance from 1000 feet AGL to touchdown.

In addition to such standard modes as altitude hold (ALT), heading select (HDG), Nav (VOR/RNav), approach (APR), glideslope (GS), reverse localiser (BC), control wheel steering (CWS), indicated airspeed (IAS) hold, and yaw damp (YD); the KFC-325 also has the standard comfort modes of soft ride and half bank. Altitude/vertical speed preselect is optional. The optional KAS 297C altitude and vertical speed selector is a panel mounted unit which can interface with a KEA 346 counter drum pointer servoed altimeter.

The KFC-325 also includes as standard equipment the KDC 222 air data sensor which provides altitude and airspeed as well as normal and side slip inputs to the flight control system. The servos use capstan assemblies which may be left in the aircraft should servo repair be required, thus allowing the aircraft to remain rigged.

STATUS: in production.

KFC-275 flight control system

The KFC-275 digital flight control system is designed primarily for piston twins. This system uses the same KCP 220 autopilot computer with four microprocessors and the same KDC 222 air data sensor as the KFC-325. However the KFC-275 uses a different flight instrument system, 3-inch (76 mm) electro-mechanical instruments including the KI 256 V-bar flight command indicator and the KCS 55A slaved pictorial navigation indicator (PNI) system.

The KMC 221 mode controller provides mode selection and annunciation for most modes along with the KAP 185A mode annunciator. Annunciations are provided for both armed and coupled modes when appropriate.

Like the KFC-325, the KFC-275 can be configured with optional altitude/vertical speed preselect. However, there is a choice of either the KEA 130A three-pointer encoding altimeter or the KEA 346 counter drum pointer servoed altimeter to go with two different versions of the KAS 297 altitude/vertical speed preselector.

STATUS: in production.

King KFC-275 three-axis digital autopilot

Lear Siegler

Lear Siegler Inc, Astronics Division, 3400 Airport Avenue, PO Box 442, Santa Monica, California 90406-0442
TELEPHONE: (213) 452 6000
TWX: 910 343 6478

Experimental digital flight control system for IFFC F-15

In an important demonstrator programme the US Air Force is showing the benefits of coupling the automatic flight control system, fire-control radar and head-up display to reduce the target tracking time for gunnery and weapon delivery. The benefits are substantially greater weapon-delivery accuracy, and better survivability since the attacking aircraft is exposed to counter-attack for a much shorter time. The IFFC (integrated flight/fire-control) system, as it is called, is based on a specially modified McDonnell Douglas F-15B two-seat Eagle. In simulated attacks on targets, the pilot makes his approach in the usual way, but by switching in the IFFC system the complex gunnery or stores-delivery computations are fed from the attack radar to the head-up display and flight control system, which sets up a fast and accurate intercept path, providing the correct lead angle in much shorter time than possible simply by pilot judgment.

Initial and successful trials were conducted at Edwards Air Force Base, California using the standard General Electric flight control system. Follow-on trials, beginning in June 1983, employed a four-channel Lear Siegler digital system for faster processing and greater ease of optimising aircraft response; in fact the control laws may be changed during flight, for example to 'tune' the aircraft for air combat and ground attack in the same mission. The system initially employed Pascal as the software language, and is claimed by McDonnell Douglas to be the first use of a high-order language in an aircraft flight control suite, resulting in greatly reduced programming costs. It was then reprogrammed for Ada, the US Defense Department's new standard language for all future military systems, to assess the performance of the software throughout the flight envelope. In September 1984 the trials F-15B became the first aircraft to fly with a mission-critical system based on the new language, and represented a milestone in the then two-and-a-half-year old DEFCS digital electronic flight control system programme conducted jointly by the Flight Dynamics Laboratory of the US Air Force Wright Aeronautical Laboratories and McDonnell Douglas. A total of 14 flights, concluding in November 1984, was planned to verify the Ada-based flight control system. Results from the IFFC part of the programme are being assessed to determine the implications in cost and scheduling of incorporating it in the US Air Force fleets of F-15 and General Dynamics F-16 fighters, currently operational versions of which have analogue flight control equipment.

STATUS: experimental programme to test advantages of integrating fire- and flight control systems on combat aircraft.

Digital flight control system for F-15E

Lear Siegler is to build the digital triplex automatic flight control system for the F-15E derivative of the McDonnell Douglas Eagle. Whereas the standard F-15 has a General Electric analogue flight control system, the additional requirements of deep penetration, with emphasis on terrain-following and terrain-avoidance, allied with new sensors, calls for a system of greater performance and reliability. Digital technology here permits control laws to be refined through software changes with an ease and rapidity not possible with analogue systems. An advantage of digital flight control

equipment is that the control laws can be optimised both for air combat and for high-speed, low-level flight. Lear Siegler is also responsible for the digital pressure sensors used to provide speed and aerodynamic pressure sensing for the system. A particular challenge is verification of the validity of the signals from these devices. The system is based on MIL-STD-1750 instruction-set architecture, and uses Fairchild 9450 microprocessors.

STATUS: in development. Preliminary design review of the system was held in January 1985, and production was planned to begin in mid-1985. A total of 516 F-15E aircraft will be bought between 1986 and 1991.

Rudder augmentation system for KC-135R

Lear Siegler provides the rudder augmentation system for some 300 of the US Air Force fleet of 650 Boeing KC-135 tankers scheduled to be retrofitted with the General Electric/Snecma CFM56 turbofan engines. The new engines generate much more thrust than their predecessors, and failure of an outboard engine at take-off could lead to catastrophic loss of control before the crew could react. The Astronics Division rudder augmentation system is essentially a high-authority single-channel yaw damper, based on a rate-gyro, that monitors engine rpm differences on the outboard engines and provides a correction signal to the rudder servo if the difference exceeds a given threshold value. The system is nuclear-hardened.

STATUS: in production.

VDA versatile drone autopilot for BQM-34 RPV

The Lear Siegler VDA versatile drone autopilot has been extensively used in the Teledyne-Ryan BQM-34 Firebee RPV, one of the very few remote-piloted vehicles to have seen operational service. Examples have also been installed in QT-33 drone versions of the Lockheed T-33 trainer, and in a NASA-developed three-eighths scale F-15 model which was used to assess some aspects of full-size aircraft behaviour, notably spinning.

The VDA is a modular, single-box, autopilot which contains all sensors, processing and output control circuits. Sensors included are a vertical gyro, altitude transducer, airspeed transducer or Mach computer, and a yaw-rate gyro. To the single, basic electronic module can be added customised modules to suit different vehicles and requirements. The extra modules permit subsonic, supersonic, low-altitude, advanced manoeuvrability, proportional command, evasive flying and pre-programmed mission manoeuvres. It is claimed to be suitable or adaptable for a wide-range of vehicles and missions.

The self-contained air data equipment uses solid-state electronics and is supplemented by digital synchronisers and manoeuvre programming devices. The VDA can hold g-levels to within ±0.25 g up to airframe limits and within 3 seconds of command, hold high-g while climbing, diving or holding altitude, maintain low-level flight (50 feet (15 metres) above sea level) and accommodate pre-programmed manoeuvres, plus proportional or augmented-proportional control.

Dimensions: 165 × 203 × 457 mm
Weight: 28 lb (13 kg)
Power: 150 VA at 115 V ac plus 30 W at 28 V dc

STATUS: in production and service.

Autopilot for Skyeye RPV

Lear Siegler produces the autopilot for the Skyeye R4E-40 RPV built by its subsidiary Developmental Sciences. Skyeye can fly a

65 kg payload for 8 hours in tactical reconnaissance missions. Some two or three squadrons, each with six aircraft, are operational in Thailand, and the US Army Aviation Command has bought a number for evaluation.

STATUS: in limited production.

Lear Siegler Inc, Instrument Division, 4141 Eastern Avenue, Grand Rapids, Michigan 49508
TELEPHONE: (616) 241 7000
TELEX: 226430
TWX: 810 273 6929

Performance data computer system

Lear Siegler's performance data computer system has been developed jointly by Boeing and the Instrument Division of Lear Siegler for the Boeing 727 and 737 airliners. This collaborative programme involved energy management studies, ground-based flight simulation and crew inputs from a variety of airlines.

The crew operates the system through a control and display unit, command 'bugs' on the combined airspeed indicator/Machmeter, engine pressure ratio gauges and the mode-annunciator panel. Modes are provided for take-off, climb, cruise, descent, holding, go-around, engine loss, and the indication of fuel reserves at both the original destination and the alternative airport. Each mode has a variety of options which are available as data 'pages' in the control and display unit. The cruise mode, for example, has four pages detailing conditions applicable to economy, long-range, manual (that is, crew-selected) speed and thrust limit.

Lear Siegler claims an exclusive function in the 'look-ahead' capability, which is the ability of the system to provide a display of information relevant to other flight phases for reference purposes without erasing or disrupting the current commanded 'bug' settings on the airspeed indicator or engine pressure ratio gauges.

The performance data computer system receives its data from the navigation and central air data systems, as well as from the fuel totaliser unit, autopilot, autothrottle, anti-ice valve control, total air temperature probes, and the cabin air-conditioning and pressurisation mode control. These inputs are processed in the performance computer and provided in the appropriate form as flight information in the control and display unit.

The system contains its own built-in test circuits which constantly monitor system operation and fault detection.

By late 1985 over 1200 systems had been produced and orders received from over 80 airlines. Lear Siegler monitoring of fuel savings with 14 airlines indicated fuel cost reductions of 2 to 6 per cent.

A similar system, the Lear Siegler fuel savings advisory system, is used in US Air Force Lockheed C-5 and C-141 transports.

STATUS: in production.

Flight management computer system

This ARINC 702 system is committed to production for the new Boeing 737-300 short/medium-haul airliner, which is now in service.

The system has four microprocessors with 1024 K words of addressable eprom and memory. They are dedicated to navigation, performance, input/output and ARINC 429 interfacing functions. The TI 990 instruction-set is used and a section of random access memory is accessible by all four processors. A 192 K words memory section is dedicated to the operator's navigation database and can be structured according to the customer's specification. A wide range of options are available from a simple listing of navaids and airports to a detailed coverage of the airline's operating

Lear Siegler performance data computer system

Lear Siegler flight management computer system for Boeing 737-300

Lear Siegler flight management system control/display unit

routes, including such data as Sids, Stars and gate assignments. Lear Siegler has an arrangement with Jeppesen Sanderson whereby the latter provides a regular navigation database.

The crew interface is via a cathode ray tube-based control and display panel on which 14 by 24 character lines of information can be displayed. Aircraft lateral and vertical profile data is presented and critical information is highlighted by reverse-video presentation. The bottom of the cathode ray tube can be used as a scratch-pad for crew entries.

Interfaces are available to the autopilot and autothrottle for automatic profile tracking, and fuel savings in the range 4 to 14 per cent are predicted in normal operations.

Dimensions: (display) 10.5 × 5.75 × 9 inches (267 × 146 × 229 mm)
(computer) 12.6 × 10.1 × 7.6 inches (320 × 257 × 193 mm) (8 MCU)

Weight: (display) 18 lb (8.2 kg)
(computer) 35 lb (15.9 kg)
Power required: (total) 250 W

STATUS: in service and production.

Military flight management system
The MFMS military flight management system is intended for military transport aircraft. The system includes a fuel savings computer (FSC), control and display unit (cdu) and a display interface control unit (DICU). The computer has both C-141 and C-5 software supported in a single computer program. Identical hardware is used in both aircraft.

An aircraft wiring jumper selects aircraft type and version when the computer is installed. An lsi installation kit is used for rapidity and ease of

aircraft modification. Software is written in high level language with full documentation and support tools. Full military data and publications are provided. The system has an mtbf guarantee of 2000 operating hours.

Operational aspects include climb, cruise, descent and transition coupling to the auto-throttle and autopilot. The system also provides advisory information on an alphanumeric cdu that interfaces with the aircraft's INSs for enhanced navigation. The MFMS features flight management, navigation, fuel savings, Tacan and waypoint database, take off and landing computations, colour graphic tactical aids, moving map displays, enhanced airdrop capability, time-of-arrival control, altitude alerting, wind shear prediction and detection as well as audio alert tones.

Tactical functions are provided to enhance the mission performance of the aircraft and crew. Performance advisory aids including maximum climb, obstacle clearance, V_{MO}/M_{MO}, buffet margins, rapid descent and optimum approach data all of which help to improve tactical efficiency with reduced crew workload.

A built-in test system verifies overall system operational status to the flight crew and offers a method of fault-isolating lrus without support equipment.

STATUS: in service.

McDonnell Douglas
McDonnell Douglas Corporation, McDonnell Aircraft Division, PO Box 516, St Louis, Missouri 63166
TELEPHONE: (314) 232 0232
TWX: 910 762 0635

Tactical flight management system
McDonnell Douglas is developing what it calls the tactical flight management system for the planning of air-to-air and air-to-surface missions in fighter aircraft. Research started in 1981 under a $1 million contract from the US Air Force and the concept-development phase concluded in the autumn of 1984 with the testing of the system in a simulator by US Air Force pilots. The system links the aircraft's mission computer with the flight control system

and the engines. After feeding in such data as target location, danger areas and weapon-load, the system can then automatically direct or fly the aircraft to the target, having computed the optimum mission flight-plan.

In the summer of 1984 McDonnell Douglas said that the system could be operational in three to five years.

STATUS: in development.

NASA
National Aeronautics and Space Administration, Dryden Flight Research Facility, PO Box 273, Edwards Air Force Base, California 93523
TELEPHONE: (805) 258 3311

Hidec highly integrated digital engine control system
As part of an effort to make more efficient use of the information from engine and airframe sensors, NASA in January 1985 launched its Hidec programme, based on a modified McDonnell Douglas F-15 fighter as the test vehicle. The goal is to increase the performance of both engine and aircraft by more sophisticated control laws that permit operation closer to operational limits; in the case of the engine,

allowable temperatures and pressures and the surge line, and for the airframe, the flight envelope. Normally the engine operates some way from the surge line (the combination of speed and pressure) in order to prevent compressor stall and possible damage, and this margin is commanded, via the fuel-control system, to increase at high angles of attack and high altitude, ie under combat conditions. While permitting the engine to be run safely, available performance is lost owing to inadequacies in the engine control system.

Hidec takes in air data information (angle of attack, speed, temperature, altitude) and other parameters, and anticipates changing conditions through commands transmitted via the pilot's controls, and computes a set of signals to command operation of the engines' digital control system. NASA anticipates gains of 6 to 8 per cent in thrust and fuel efficiency at high

subsonic speeds. The system is software-based, and so not available for analogue (ie hydromechanical) engine control equipment. The Hidec programme has three phases. In the first (January to April 1985) the equipment and flight-test recorder were tested. In the second, lasting from July to September 1985, the adaptive stall-margin control system was tested. The third phase, beginning in January 1986, will refine the control system and take it further into the dynamic situations associated with combat manoeuvres.

Hidec is based on the DEFCS digital electronics flight control system built by McDonnell Douglas as a replacement for the standard analogue system on F-15.

STATUS: experimental programme.

Safe Flight

Safe Flight Instrument Corporation, 20 New King Street, PO Box 550, White Plains, New York 10602
TELEPHONE: (914) 946 9500
TELEX: 137464

SCAT speed command of attitude and thrust system

Safe Flight has pioneered the development of wing-lift instrumentation and technology and has now produced the speed command of attitude and thrust (SCAT) system that provides pitch guidance during the take-off, initial climb and go-around phases of flight and thrust guidance for landing approach. When combined with Safe Flight's Autopower (automatic throttle system), it can offer the standard of speed stability appropriate to Category III approach and landing requirements.

SCAT drives the pitch-command bar of the flight director during these low-speed phases of flight. If an engine fails just after take-off, the crew is immediately presented with the pitch attitude giving the best rate of climb. It does this by matching the aircraft's polar diagram (its lift/drag characteristic) against the thrust level available.

Emphasis is placed on the system's ability to combine target angle of attack and acceleration along the flight-path to optimise the crew's performance during these critical periods which demand precise attitude and speed control if the situation is to be stabilised. It is particularly useful in restrictive runway operations, noise-abatement procedures, and at times of forecast wind-shear.

Format: 3⁄8 ATR Short
Weight: 8.5 lb (3.9 kg)

STATUS: in production, in service and recently certified on the Cessna Citation III.

Performance computer system

Safe Flight's performance computer system is designed to provide increased accuracy in obtaining the best balance between fuel economy and airspeed, while reducing crew workload in the operation of business jets. The system computes and displays the speed appropriate to maximum specific range under the prevailing conditions. Alternatively long-range speed, percentage of maximum specific range being achieved, optimum cruise altitude or several other parameters may be selected as the baseline parameter. Inputs from various aircraft sensors, together with stored performance data are used in these computations, which take into account all external factors such as the effects of fuel burn or change of wind velocity, appropriate to the type of aircraft. The computer is based on an Intel 8085 microprocessor and has 28 K bytes of read-only memory and 2.25 K bytes of random access memory. The computer can supply data to compatible autothrottles, or thrust-management systems produced by Safe Flight, which are controlled by means of a control/display unit.

Format: (computer) 3⁄8 ATR Short
(control/display unit) 2.25 × 5.75 × 4.75 inches (57 × 146 × 121 mm)
Weight: (computer) 7.5 lb (3.4 kg)
(control/display unit) 1.2 lb (0.55 kg)
Power: 115 V ac 400 Hz

STATUS: in production and service.

Thrust management computer

A thrust management system, introduced by Safe Flight in 1983, operates throughout the flight regime. By monitoring all important engine parameters, the system gives protection against exceeding engine limits. It can compute

Safe Flight performance computer system computer and control/display unit

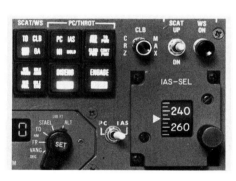

Typical presentation of Safe Flight's speed select feature

the correct power setting (N₁ or engine pressure ratio) for take-off, climb, cruise and go-around. It also monitors N₁, N₂ and turbine exhaust temperature settings for cruise to maintain efficient operation within approved limits. For autothrottle-equipped aircraft the computer controls the throttles from brake release to the landing flare.

Format: (computer) 3⁄8 ATR Short
(thrust target indicator) 1.4 × 2.5 × 5.6 inches (36 × 64 × 142 mm)
Weight: (computer) 7 lb (3.2 kg)
(thrust target indicator) 0.75 lb (0.34 kg)

STATUS: in production.

Airspeed select feature for SCAT autopower systems

An airspeed select feature is now available for the Safe Flight SCAT Autopower system which enables the pilot to program the automatic throttles to hold appropriate speeds for all phases of flight, allowing economical cruise performance, variable holding and approach speed scheduling. Any speed selected by the pilot is maintained by the SCAT system, but a safety feature precludes the aircraft flying

Safe Flight SCAT computer

Control unit for Safe Flight thrust management computer

below the speed appropriate to the maximum permitted angle of attack.

STATUS: in production. Certificated for use on McDonnell Douglas DC-9 and Boeing 727 aircraft.

Silver Instruments

Silver Instruments, 16100 SW 72nd Avenue, Portland, Oregon 97224, USA
TELEPHONE: (503) 684 1600
TELEX: 510 600 2185

Fueltron fuel management system

Fueltron is a complete fuel management computer system for improving economy, convenience and safety on twin and single-engined general aviation aircraft and helicopters powered by piston and turbine engines. Large digital displays enable rapid, precise setting of engine power for any flight condition, together with more accurate mixture-leaning than with an exhaust-gas temperature readout. The system is available in a large number of configurations. It can pay for itself in 300 to 500 hours of operation in a typical twin, through savings in fuel, time, and engine wear. FAA-approved installation kits are available for more than 210 aircraft models including Aérospatiale, Bell, Hughes and Sikorsky helicopters, and aircraft produced by Beech, Cessna, Piper and Gulfstream, as well as for the Douglas DC-3/C-47.

All electronics are in a single, panel-mounted unit. Incandescent displays give digital indication of fuel remaining (which may be translated into time remaining based on flow rate), and flow rate for the engine or engines. The computer calculates how much fuel has been burned by integrating fuel flow throughout the flight, and fuel used can be displayed instead of fuel remaining. A warning light indicates when time remaining falls below 48 minutes.

Fueltron receives its data from the digital output of a flow transducer located in the fuel line to each engine. This output, which is directly proportional to fuel flow, is combined in the Fueltron computer with precise time data from a quartz crystal clock. The system's accuracy is claimed to be 1 per cent or better, and Silver says that fuel economy can be improved by up to 10 per cent when setting engine power and mixture by Fueltron rather than exhaust-gas temperature. Fuel flow for a given climb rate can be reduced, and up to several times more spark-plug life can be achieved.

Indicator and computer
Dimensions: 3.25 × 3.25 × 6.14 inches (83 × 83 × 156 mm)
Weight: 1.7 lb (0.77 kg)
Power: 10-30 V dc; 1.4 A at 14 V, 0.7 A at 28 V
Accuracy: ±1%
Environmental: TSO C44a
Mtbf: ≥3000 h

Standard flow transducer
Weight: 0.31 lb (0.14 kg) per engine
Flow range: 1.5-60 US gal/h (5.7-227 litres/h) per engine
Temperature range: –65 to 125° C
Life expectancy: >10 000 h

Turbine aircraft PT Series flow transducer
Weight: 9 oz (0.25 kg) per engine
Flow range: 8 ranges from 1000–20 000 lb/h max
Temperature range: –90 to 300° C
Life expectancy: 10 000 h

STATUS: in production for wide variety of general aviation aircraft.

Fuelgard fuel management system

Fuelgard, a simplified version of Fueltron, is suitable for light single-engined aircraft. An incandescent display gives a digital readout of fuel flow with 2 per cent accuracy. A switch

Silver Instruments Fueltron fuel management system display

allows an alternative display of fuel used. Like Fueltron, Fuelgard allows accurate mixture-leaning for piston engines, and fuel savings are expected to be about 10 per cent. The total fuel used remains in the equipment's memory, a useful facility where more than one flight is made before refuelling. Fuelgard can be installed in fixed-wing aircraft or helicopters with a single reciprocating or turbine engine.

Dimensions: 2.9 × 1 × 4.4 inches (74 × 25 × 112 mm)
Weight: 0.9 lb (0.4 kg)
Voltage: 14 V or 28 V dc
Flow rate range: 1.5-60 US gal/h (5.7-227 litres/h)
Accuracy: ±2%
Approval: TSO C44a

STATUS: in production.

Simmonds

Simmonds Precision, 150 White Plains Road, Tarrytown, New York 10591
TELEPHONE: (914) 631 7500

PAS performance advisory system

PAS is a performance advisory system that uses a microprocessor to monitor, store and process real-time aircraft performance data. It consists of three items: a display unit, a control unit, and a computer.

The crew enters all necessary flight-plan data via the control unit keyboard as part of the pre-flight checks. This flight-plan is followed in the air and a crew member turns the control unit selector to each of the seven flight modes to read off the recommended engine pressure

Simmonds performance advisory management system

ratios, pitch attitude and indicated airspeed setting.

PAS receives signals such as pressure, altitude, Mach number, true air speed, true air temperature, fuel quantity, engine anti-ice and bleed valve position, flap angle and landing gear position. This information is used to update continuously the displayed information.

The display unit has a two-position toggle switch. On the Boeing 727 when the switch is in one position the display window reads centre and pod engine pressure ratios, commanded pitch attitude and commanded indicated airspeed. With the switch in the alternative lower position the display reads pod and centre engine pressure ratio limits, optimum and maximum altitude. Simple software changes adapt the system for any type of aircraft.

The parameters displayed on the upper scale are commands that result in the optimum operation for the selected mode. The lower scale parameters provide information for planning and comparison of various alternatives such as change of flight-plan, diversion or extended holding.

The display of optimum altitude and maximum altitude are of particular benefit, allowing the crew to accommodate quickly air-traffic control requests for flight-level changes. Optimum altitude is based on considerations of cost while maximum altitude considers the operational limits of thrust, buffet boundaries, and speed. The control and display unit receives directions from the mode selector on the control panel. There are eight positions: pre-flight, take-off, climb, cruise, turbulence, descent, holding and approach.

The performance advisory system computer is self-teaching and self-testing. It rejects entry data that does not fall within established parameters and it shuts down automatically on detection of failure. A DME lock-on facility is offered as a customer option.

STATUS: operational on some commercial Boeing 727s and military transports.

Monopole torque system

The monopole torque system is based upon four major components: a torque shaft, monopole sensor, engine mounted signal conditioner, and suitable indicator to display the instantaneous torque and speed measurements.

The shaft consists of two intermeshed gears used to measure the amount of "twist" on the drive coupling. During shaft rotation the gears are monitored by a single variable reluctance sensor whose output is presented as a series of pulses. The phase relationship between adjacent pulses varies linearly with torque level, while the frequency of the pulses provides a measure of shaft speed.

The signal conditioner converts the inputs to a linear dc voltage representative of the actual torque and speed of the rotating assembly.

System accuracy: better than ±1%
Reliability: 250 000 h mtbf

STATUS: in production.

Simmonds Precision Products Inc, Instrument Systems Division, Panton Road, Vergennes, Vermont 05941
TELEPHONE: (802) 877 2911
TWX: 510 299 0039

Fuel gauging systems and fuel level sensing

Simmonds Precision products include fuel tank sensors, cockpit and refuelling indicators, signal conditioning units and in-tank wire harnesses. These products provide the cockpit crew, ground maintenance and refuelling crew with information on the quantity of fuel contained in each tank, as well as total fuel onboard. Accurate gauging is provided over a range of aircraft attitudes, fuel types and fuel temperatures. Associated features, usually particular to the aircraft model, provide low fuel warning, eg 'bingo fuel', lateral and/or longitudinal assymetrical fuel imbalance warning and control, refuelling sequencing and automatic fuelling shut-off.

Fuel-gauging systems continue to evolve taking advantage of digital technology and solid state displays. These computerised and microprocessor-based fuel-gauging systems provide improved accuracy, higher reliability and improved maintainability and self-test features than earlier analogue and electro-mechanical systems.

Simmonds' gauging systems are fitted on a wide range of commercial and military aircraft of which the following are some examples – General Dynamics F-16, Northrop F-20 Tigershark, McDonnell Douglas F-15 Eagle and F/A-18 Hornet; the Sikorsky CH-53 and the UH-60A Blackhawk; Boeing 727 and 737, Lockheed L-1011; Panavia Tornado and the NASA/Rockwell Space Shuttle.

Simmonds also produces a range of low fuel level sensing systems. These may be used as separate stand-alone systems, as a back-up or to provide system redundancy for the main fuel gauging system. Level sensing systems are provided for the F-15, F-16, F/A-18, Airbus A320 and the Space Shuttle.

STATUS: in production.

Simmonds monopole torque system

Fuel management systems

Simmonds Precision has generated many patents since 1950 for its aircraft fuel gauging and management systems and has produced systems and components for more than 50 000 military and commercial aircraft. Based on this knowledge and experience, Simmonds introduced the world's first military computerised fuel management system on the Rockwell B-1B in 1974. The centre of gravity monitoring and automatic transfer of fuel from tank to tank optimises the aircraft's centre of gravity and weight distribution for improved fuel and flight efficiency as well as maintaining the aircraft weight distribution in a safe flight configuration.

Modern fuel management systems are responsible for controlling the fuel flow and tank sequencing. Discrete outputs or multiplexed data bus outputs are used to control the respective fuel pumps and fuel sequencing valves. Software algorithms and self-test are used to ensure correct actions are being carried out and that circuits are responding. Reversionary modes are automatic upon component failure or power loss.

Centre of gravity is controlled and implemented utilising information of fuel quantity and its location as well as stores and their

location to determine the appropriate moment arm. The centre of gravity is then calculated and adjusted by fuel transfer until the calculated centre of gravity matches the optimum or demanded position. The system's ability to measure and manage centre of gravity and control fuel movement can also be used to minimise the potential effects of battle damage. By monitoring any sudden changes in centre of gravity, fuel-flow variation or fuel pressure fluctuations, a damaged area can be isolated. The system will then automatically transfer the remaining fuel to an undamaged tank and close appropriate valves to retain system integrity.

Modern fuel management systems can also provide the function of thermal management enabling the fuel to be used as a heat sink. This can be effected by using the fuel management system's ability to move or circulate cooler fuel towards certain heat exchangers and hotter fuel towards the engine inlet. This is achieved when the management system is complemented by a fuel system architecture that takes maximum advantage of various heat exchangers aboard the aircraft. By continuously monitoring the fuel quantity and temperature in each tank, the system computes the best recirculation pattern in all flight regimes.

Simmonds has produced fuel management systems that comply with today's military standards utilising MIL-STD-1750A processors

Simmonds fuel management computer

and MIL-STD-1553B data-bus interfaces to link the fuel system to integrated displays, the aircraft avionics and other aircraft subsystems. Computer software is written in Jovial J73 higher order language.

STATUS: in production.

Computerised fuel measurement system

A fuel measurement system introduced by Simmonds for the new Boeing 737-300 airliner fulfils several fuel management sensor and

Fuel quantity in General Dynamics F-16 fighter is measured by Simmonds fuel gauging equipment

Simmonds fuel measurement indicator for Boeing 737-300

processing requirements in addition to providing fuel-quantity indications. The system is continuously self-calibrating, using microprocessor-based electronics to conduct system-level tests and to effect corrections. It can also test for, and indicate, up to nine fuel system faults, and is said to be easier and cheaper to maintain than conventional equipment. It can

override compensator faults, including those caused by water contamination.

The software can be modified to meet specific airline or airliner requirements, and is also available for existing models of the Boeing 727 and 737 and McDonnell Douglas DC-9 airliners. Mean time between failures is claimed to be in excess of 13 000 hours.The system

provides low-level warning. Load pre-select and all-digital volumetric top-off units are available, although minor modifications eliminate the need for these latter units.

STATUS: in production.

Sperry

Sperry Corporation, Aerospace and Marine Group, Defense Systems Division, PO Box 9200, Albuquerque, New Mexico 87119
TELEPHONE: (505) 822 5000
TELEX: 660319

SAAHS stability augmentation/ attitude hold system for AV-8B

This digital system was the first of its type to be applied to any of the Harrier series of aircraft. It comprises a limited-authority stability-augmentation system with some conventional autopilot functions to reduce pilot workload. The single-channel system operates on all three axes, and is accompanied by a mechanical back-up for reversionary operation. Extensive self-monitoring circuitry switches out the system in the event of a fault, so that it fails passive. Monitoring is accomplished in three ways: equipment, software monitoring of equipment, and software monitoring of performance. In addition a comprehensive pre-flight schedule is provided.

Stability-augmentation modes comprise three-axis rate damping in both vertical and cruise flight, and in transition between these phases. It also includes rudder/aileron and stabilator/aileron interconnects to improve turn co-ordination. Autopilot modes include attitude, altitude and heading hold, automatic trim, control-column steering, and airspeed limit schedules. A rudder-pedal shaker alerts the pilot to the onset of potentially dangerous side slip during the hover.

Principal element of SAAHS in a 13.5 lb (6.1 kg) digital flight control computer with 2901-bit slice, 16-bit processor working at 470 000 operations a second, with ultra-violet erasable programmable read-only memory. The flight control programme occupies 20 K words of memory, of which 9 K are taken by the control law code and 11 K by the built-in test schedule. The other units comprise a three-axis rate-sensor, lateral accelerometer, stick sensor, and forward pitch amplifier. In December 1982

the system demonstrated a complete 'hands-off', automatic vertical landing.

Weight: (total) 27 lb (12.3 kg)

STATUS: in production for US Marine Corps/ McDonnell Douglas AV-8B close-support V/Stol aircraft.

Avionics system for QF-100 drone

Sperry Flight Systems has developed a suite of drone avionics for installation in US Air Force/North American F-100 Super Sabres to transform them into full-scale aerial targets under the designation QF-100. Initial operations by the US Air Force began in June 1983.

The targets are used for air-to-air and ground-to-air missile evaluation and combat-crew training and to meet the demands of this range of operations each has a digital flight control system which interfaces with much existing ground-control and test equipment.

The digital processor in the aircraft is based on Sperry SDP 175 equipment with air data sensors, analogue/digital and digital/analogue converters, power supplies and interface electronics. All this equipment is installed in a single line-replaceable unit, compared to four units in a broadly equivalent analogue system. The digital signalling techniques have facilitated the incorporation of a large amount of system testing capability, aimed at meeting pre-mission test-stand and airborne system test-set requirements. These reduce the time needed to complete tests and improve fault detection.

STATUS: in service.

DAFCS digital automatic flight control system upgrade for B-52/KC-135/C-130

In September 1984 Sperry was awarded a $99.7 million contract by Tinker Air Force Base, Oklahoma, for the design, development and delivery of a new digital autopilot for the Boeing B-52 and KC-135 fleets. The contract also called for delivery of one prototype system for the Lockheed C-130. This system will provide

an order of magnitude improvement in the reliability and maintainability of the autopilot of all three aircraft.

The DAFCS digital automatic flight control system comprises three line replaceable units (Lrus) which replace numerous Lrus of the existing systems. These new Lrus are the digital amplifier unit (DAU), the air data sensor unit (ADSU) and the status test panel (STP). The DAU is the heart of the autopilot system and is a dual, Z-8002 processor design. A separate add-on task is provided by a MIL-STD-1750 processor card which can be added later. A 1553B input/output card will be demonstrated as part of the DAU but will not be delivered under this contract. Significant growth in the processor capability and memory is a US Air Force requirement. The ADSU provides the DAU with the necessary sensor information for the airspeed/altitude/Mach hold modes of the autopilot. The STP provides maintenance with an interface during diagnostic testing as well as reading the results of the numerous built-in tests.

Boeing will provide the aircraft modification design, flight testing and airframe modifications for the B-52 under a separate US Air Force contract. Sperry is responsible for the entire KC-135 programme, including the provision of airframe modification kits.

Under contract at present are over 1000 systems – approximately 270 for the B-52 and 740 for the KC-135. Flight tests of both aircraft began in March 1986. Flight tests of the C-130 DAFCS are expected to be authorised for late 1986.

STATUS: in development.

Sperry Corporation, Aerospace and Marine Group, Commercial Flight Systems Division, PO Box 21111, Phoenix, Arizona 85036
TELEPHONE: (602) 869 2311
TELEX: 667405

Flight management system

The advent of the Boeing 757 and 767 airliners boosted the development of many types of commercial aircraft avionics, including flight management systems. Sperry has produced a digital flight management system (FMS) as standard features for these aircraft, and also optional for the Airbus A310. More recently, in November 1984, Sfena announced its choice of Sperry to provide the FMS for the Airbus A320. In this system, the FMS is integrated into the same boxes as the flight control computers.

The Sperry flight management system features vertical navigation which, in conjunction with inputs from the central air data computer and pre-programmed information, provides climb, cruise and descent guidance options. Lateral navigation produces steering data in conjunction with INS, radio navigation aids and pre-programmed information. It may be coupled to the autopilot for ease of operation, though an interlock with the radio altimeter prevents coupling below 1500 feet on some aircraft. Autothrottle may also be coupled.

The crew can communicate with the system via a control and display unit and keyboard, and there is a scratch pad capability. The flight management system will then determine the best method of achieving a selected profile and will carry it out via the autopilot and autothrottle. In cruise it controls airspeed and altitude and can present a continuous indication of optimum

US Marine Corps/McDonnell Douglas AV-8Bs have Sperry autopilot

Sperry flight management system for corporate operators and regional airlines

Sperry flight management control/display unit for Airbus A310

altitude. When interrogated, it indicates the savings or penalty incurred with any proposed altitude change.

A magnetic disc memory stores all performance data, navigation aids, route profiles and airport information (an advanced system will feature bubble memory). Consequently the initial pre-flight checks require only that the crew verifies the correct date of data and engine identification and selects the route by inserting an identifier number, from on-board documentation, which 'loads' the sector. If the chosen route is not a regular one, or if it is not available in the data bank, then the waypoint co-ordinates may be loaded manually via the keyboard.

The INS inputs may be updated from radio navigation aids and, with the flight-director in operation, command information may be monitored on the attitude deviation indicator. The flight management system also features automatic tuning for the radio navigation equipment.

When this flight management system is used to retrofit older analogue-based aircraft a data adapter interface is fitted.

Certification programmes are being conducted in the USA with United Airlines McDonnell Douglas DC-10 and Boeing 747 airliners, and a dual system installation has been retrofitted into the British Airways 747 fleet. Sperry has now signed an agreement with Boeing to provide the advanced flight management computer (AFMC) with thrust management functions for the 747-400 and with McDonnell Douglas for the MD-80.

STATUS: in production. Now selected for MD-88 with certification expected in late 1987.

Flight management system for general aviation

Capitalising on the heavy investment put into and experience gained with equipment designed for the airlines, Sperry at the October 1984 NBAA Convention introduced a flight management system for general aviation. It is aimed specifically at the new generation, top-of-the-range business jets such as the Gulf-stream IV, Dassault Falcon 900 and Canadair Challenger 600/601, and commuter aircraft exemplified by the Aérospatiale/Aeritalia ATR-42, Saab/Fairchild SF340 and British Aerospace ATP Advanced Turboprop. Sperry claims it to be the first system to offer full lateral navigation, vertical navigation, and performance management for such aircraft, and emphasises its speed and ease of operation.

The system comprises an NZ 600 or NZ 800 navigation computer, PZ 800 performance computer, and a CD 800 control/display unit. The NZ 600 provides 112 K bytes of memory, the NZ 800 320 K bytes, both devices being the latest type electronically eraseable programmable read-only memory (eeprom). It can store permanently flight-plans containing up to 1000 waypoints, for example a combination of 100 plans each with 10 waypoints or 50 plans with 20 waypoints. These flight-plans would be based on a ground data-base such as Lockheed's Jet Plan, the information being carried on 3.5-inch floppy disks.

In September 1985 the Company announced the development of an interface between the Lockheed Jet Plan service and the FMS. By use of a single portable computer and the FMS data loader, this interface eliminates the need for manual input of flight plan and meteorological data into the FMS. The performance computer provides take-off computations, recommended flight-levels and performance data that can be used in an advisory capacity, eg to provide optimum speeds, altitudes and routes for best economy.

Two versions of the control/display are under development, one monochrome and the other colour; the first costs less, but the second is more compatible with the colour EFIS displays now proliferating even in general aviation aircraft, and colour effectively adds another dimension to the quality of information displayed. A notable advance in control/display unit technology is the replacement of the traditional line cursor by a line selector, so that the control/display unit is nearly akin to a touchscreen device.

Sperry has recently announced plans to offer a GPS global positioning system airborne receiver for use with the FMS. This sensor will be designed for the business aviation and regional airlines market. Currently Sperry is working with Motorola's Radar Operations on sensor development. It is also intended to market an omega/vlf navigation sensor with the FMS.

STATUS: the Sperry FMS is expecting FAA certification in the second quarter of 1986 on the ATR 42, Falcon 900 and Cessna Citation III. The British Aerospace 125-800 received certification early in 1986. The system should also be certificated for the Gulfstream IV and Canadair CL-601-3A by the end of 1986.

Performance management system

Sperry has developed a performance management system for the Boeing 727, 707, 737, and McDonnell Douglas DC-8 and DC-9 transports. It is said to require few changes in retrofitting and does not need any modification to existing sensors and aircraft instrumentation.

The performance management system is a three-box system consisting of computer, control and display unit and indicator. The computer integrates all performance, auto-throttle, thrust rating and air data functions. The control and display unit provides crew interface and displays performance data on a cathode ray tube with sub-mode selection via a conventional keyboard. The indicator provides the crew with a convenient display of the active thrust and speed commands and may be mounted adjacent to the engine pressure ratio gauges for ease of monitoring.

The system optimises the aircraft's performance throughout the vertical flight envelope. Four function keys are fitted at the lower edge of the control and display unit and seven adjacent mode-select keys are used to select climb, cruise, descent, hold and approach. These modes provide full flight control including sub-modes for take-off and go-around.

The performance management system features an 'engine out' mode which provides logic to the computer that will sequence on the indicator engine data.

The indicator is based on a Sperry 3-inch (76 mm) cathode ray tube. It is divided into five data fields with up to 16 characters per field, each field being separated by lines and legends to denote command information. These areas indicate mode, engine pressure ratios, limits and speeds. A manual speed control knob allows the pilot to select an air speed to comply with air-traffic control instructions or any other required deviation from the programmed schedule.

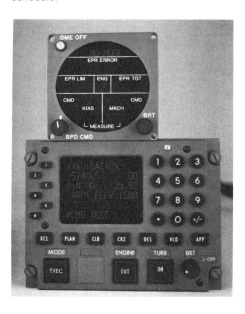

Sperry performance management system

The indicator also carries a 'DME off' lamp which warns the crew when the DME-dependent ground speed calculations are invalid.

A windshear detection and alert feature was certificated by the FAA in the latter part of 1985 for Boeing 737-20 aircraft. Certification is also in progress for a similar windshear detection capability on Boeing 727 aircraft. Certification assistance was provided by launch customer Piedmont Airlines.

A firm order for PMS/windshear has also been received from Varig-Cruzeiro Airlines of Brazil. Evaluation is currently under way by the Brazilian flag carrier—VASP. Development work has also begun on a separate, stand-alone windshear detection system to interface with a wide variety of commercial aircraft.

A highly instrumented and computer-monitored fuel savings evaluation programme conducted in 1984 aboard Frontier Airlines Boeing 737-200 aircraft in revenue service produced average fuel burn reductions of 3.8 per cent.

Dimensions: (computer) ¾ ATR Long
(control/display unit) 5.75 × 4.5 × 10.875 inches
(146 × 114 × 276 mm)
(indicator) 3 × 7 inches (76 × 178 mm)
Weight: (computer) 28 lb (12.7 kg)
(control/display unit) 4 lb (1.8 kg)
(indicator) 3.3 lb (1.5 kg)
Power: (computer) 58–79 W
(control/display unit) 16 W
(indicator) 12.5 W

STATUS: in production.

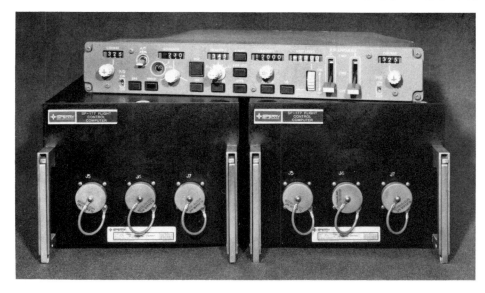

Sperry SP-177 automatic flight control system provides Category IIIA autoland capability in Boeing 737-200

SP-150 automatic flight control system for 727

This automatic flight control system embodies an automatic-landing function, and is used extensively in Boeing 727 airliners. It superseded the earlier SP-50 automatic-landing system, and employs more solid-state components for rate-sensing and integration. The system has a dual-channel configuration which meets FAA Category IIIA requirements (50 feet decision height, 700 feet runway visual range).

A wide-range of stabilisation, attitude hold and external-sensor steering modes are available in the basic autopilot. Additionally there is provision for area navigation steering and for a radio-altimeter input so that control law gains can be varied during the approach. Fail-operational performance is assured by the dual-computer configuration and a dual-channel yaw damper is also part of the overall flight control system.

Computational equipment used in the system includes two pitch computers (each ½ ATR), two yaw damper couplers (each ⅜ ATR) and a single roll-control computer (¾ ATR).

STATUS: in service.

SP-177 automatic flight control system for 737

This is an integrated digital/analogue system for the Boeing 737 twinjet airliner and providing Category IIIA automatic-landing capability. Main elements of each installation are two flight control computers and a glareshield-mounted controller. These are associated with an AD-300C attitude director indicator and an RD-800J horizontal situation indicator for each pilot. The system can be integrated with a performance management system and autothrottle.

Each computer performs pitch and roll computation for autopilot and flight-director functions. The system has been configured so that pilot involvement is minimised and to ensure that the flexibility of the system is available with simple and logical crew control operations. The pitch axis uses pitch attitude and rate, altitude rate, vertical acceleration and longitudinal acceleration for stabilisation. Vertical acceleration is blended with the altitude-rate signal from the air data system to provide filtered altitude rate for altitude acquisition, altitude hold, vertical speed and glideslope control. Radio altitude and radio-altitude rate are used to command automatic-landing flare. Additionally, there is a flight-director take-off mode which, based on flap setting and an angle-of-attack sub-mode, keeps speed above stall.

The design aims to provide high maintainability and reliability, has built-in self-test features, and incorporates the minimum number of line-replaceable units. Operation of the built-in test facilities can be selected from the flight-deck, and provides a readout on a performance data computer system display.

This system is being installed in current production Boeing 737-200s.

Format: (flight control computer) 1 ATR Long (2 installed)
System weight: 81 lb (38 kg)
Mode control panel: 440 × 74 × 348 mm
Digital interfaces: ARINC 429
Power: 160 W

STATUS: in service.

SFS-980 digital flight-guidance system for MD-80

Used exclusively on the McDonnell Douglas MD-80 (formerly DC-9 Super 80) transport, the SFS-980 is a new automatic flight-guidance system based on relatively few line-replaceable units, especially in respect of the range of functions performed. In addition to a comprehensive selection of conventional autopilot/flight-director operating modes the system is FAA-cleared for Category IIIA automatic-landing operations (50 feet decision height, 700 feet runway visual range) and has a full-time autothrottle. The large scale use of digital computing has led to the installation of a comprehensive built-in self-monitoring and maintenance diagnosis capability. Major line-replaceable units and functions are described in more detail below.

Digital flight-guidance computer (dfgc) There are two identical 1 ATR Long (29 lb, 13.15 kg) dfgc units, each autonomous with respect to the other, and individually capable of handling all system functions, including fail-passive

Sperry SP-150 autopilot system and control unit

Sperry SFS-90 Digital flight-guidance computer for McDonnell Douglas MD-80 airliner

McDonnell Douglas MD-80 family has Sperry SFS-980 flight-guidance system; this is an example of new MD-83, which entered service in March 1985

automatic landing. Within each unit analogue/digital conversions take place, and the complete system occupies 31 board places, against a maximum capacity of 51 boards. Each digital processor has 30 K words of read-only memory and 4 K words of random-access memory. In addition to all autopilot/flight-director processing, each unit also generates thrust-rating indication signals, plus maintenance data storage and status test-panel data. The aircraft can have (optionally) a head-up display for take-off and go-around data presentation, and guidance signals for this unit are also generated in each dfgc.

Flight-guidance control panel (fgcp) This unit fits in the centre glareshield and contains mode selection and control functions for full-time autothrottle, both flight-directors, the autopilot and altitude alerting. Pilots may select which dfgc controls all functions. Autothrottle speed/Mach, selected heading, vertical speed and selected altitude readouts are shown on seven-segment incandescent lamps which may be dimmed by a control knob on the bottom of the panel. The speed/Mach knob is a three-position control, while the heading selection is one of two concentric knobs. The outer knob provides selection of maximum bank angle for all autopilot/flight-director lateral modes except loc. The inner knob is a four-position device providing heading and autopilot/flight director heading mode selection. The three-position alt knob provides altitude selection and autopilot/flight director altitude pre-select mode arming.

Vhf/nav control panels Two panels on either side of the fgcp allow selection of Vortac station frequencies and courses. Displays are incandescent-lamp readouts.

Flight mode annunciator (fma) One fma on each pilot's panel provides instrument failure warning for ILS, attitude, heading, automatic-landing, auto-trim or instrument-monitor functions. They indicate which autopilot or flight-director system is engaged, and warn of autothrottle or autopilot disconnects. The units also annunciate which autothrottle, flight-director and/or autopilot modes are armed, and in which mode the system is currently controlling.

Autopilot duplex servo drive Three duplex servos drive the ailerons, elevators and rudder. The duplex servo functions only during ILS, land or go-around mode. A linear actuator provides normal yaw damping functions in other flight regimes.
 Each servo has two separate dc electric motors whose outputs are summed in a differential gear-train with a single output. Each servo sends position and rate-feedback signals to the dfgc, where servo 'models' monitor actual servo operation. The duplex servo design provides fail-passive protection against hardover manoeuvres. A fault in one servo channel is cancelled by the mechanical velocity summing conducted in the dual servo differential.

Autothrottle/speed control system Conducted in the dfgc, this function provides fast/slow attitude-director indicator commands throughout the entire flight regime. During take-off and go-around it provides pitch guidance for the flight-director, autopilot and optional head-up display. The speed control system provides a speed margin above stall (alpha speed) for all autothrottle modes, in addition to the autopilot stall-protection feature.
 The autothrottle system automatically prevents excursions beyond maximum operating airspeed and Mach values, slat and flap placard speeds, and engine epr limits. It also keeps the aircraft at or above alpha speed for prevailing flap/slat position and angle-of-attack, using limit data stored in the dfgc solid-state memory.
 Automatic reserve thrust enhances safety, operational economy and noise reduction. In the event of an engine failure, as indicated by a difference of more than 30 per cent between engine N_1 (fan) speeds, simultaneously with slats-extended indication, the good engine thrust is automatically boosted by about 4 per cent. Automatic reserve thrust also allows use of less-than-maximum certificated thrust for take-offs and go-arounds without corresponding reduction in take-off gross weight, with benefit to fuel and maintenance costs.

STATUS: in service.

SP-300 digital automatic flight control system for 737-300

This is essentially an all-digital version of the hybrid analogue/digital SP-177 automatic flight control system designed for earlier versions of the Boeing 737-200. The previous system employed analogue circuitry to compute the safety-critical Category IIIA automatic landing functions, and digital circuits for en route flight control. The configuration of the SP-300 meets the requirements for fail-passive Category IIIA automatic landing and independent computation for Category II flight-director approach. To meet the safety criteria for approach and landing, the system has a 'dual-dual' configuration: each of the two control channels has two processors, different from one another and with different software. In this way the possibility common-mode failures and software errors is reduced to a very low level.
 The mode-control panel on the glare shield provides centralised control for all autopilot, flight-director and autothrottle functions.

STATUS: in production. Certification of all modes was completed in December 1984.

SPZ-1 autopilot/flight director for 747

The SPZ-1 system, with minor changes, is fitted to all versions of the Boeing 747 airliner. It comprises the pitch and roll functions and associated electromechanical ADI and HSI flight directors. Boeing is the design authority, Sperry making the equipment to the Boeing's specifications and drawings. The current SPZ-1 is a three-channel system, pitch and roll computations being accommodated in three separate boxes. The two sets of flight director instruments are normally driven by different computers, but can be switched as necessary in the event of a failure. Each of the three channels has its own set of attitude, air data and other sensors, and the entire system is designated as fail-operational (defined as the situation whereby no single failure will cause the performance of the system to fall below the limits required by the autoland manoeuvre).

Sperry SPZ-1 autopilot is standard on all versions of the Boeing 747

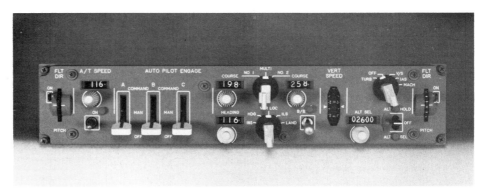

Glareshield controller for Sperry SPZ-1 autopilot/flight-director system

Three SPZ-1 configurations have been developed. As originally planned, the system incorporated two channels in pitch and roll, and was designated as a fail passive (no single failure will cause a 'hardover' control demand) Category II system. The two channels are compared one with another by means of a single box monitor and logic unit, or MLU. This system also incorporates a one-box automatic stabiliser trim. From this was developed the basic three-channel fail-operational system, and with it the 747 became the first US transport aircraft to be certificated for Category IIIA autoland. The upgrading was accomplished partly by adding the third autopilot and partly by changes in the mode-select panel, and by the addition of three landing control logic units, or LCLUs, that replace the MLU. Finally, in 1976, the autoland capability was extended to include automatic rollout control along the runway, representing Category IIIB. This capability was gained by replacing the LCLU with a landing rollout control unit, or LRCU, and by adding a microprocessor-controlled built-in test schedule.

The performance of the SPZ-1 has since been enhanced by the introduction of the AAIP analogue autoland improvement programme. This was launched in 1977 with the initial aim of minimising the number of disconnects being experienced at ILS capture. It was expanded in 1980 to include optimisation of the ILS tracking capability, and touchdown footprint, go-around, and landing flare performance.

System configuration

Box	No of Units	Weight	Format
Roll computer	3	16.9 lb (7.7 kg)	⅜ ATR
Pitch computer	3	17.8 lb (8.1 kg)	⅜ ATR
Autostabiliser trim unit	1	10.0 lb (4.5 kg)	⅜ ATR
LRCU	2	31.0 lb (14.1 kg)	¾ ATR
LCLU	2	23.0 lb (10.4 kg)	¾ ATR

STATUS: all three configurations in production and service.

Sperry Corporation, Aerospace and Marine Group, Avionics Division, PO Box 29000, Phoenix, Arizona 85038
TELEPHONE: (602) 863 8000
TELEX: 668419

SPZ-200A autopilot/flight-director system

This autopilot/flight-director system is available for corporate business aircraft such as the Beech King Air and British Aerospace Jetstream 31. It is available with either 4- or 5-inch (102 or 127 mm) instruments and includes an air data

system. Options include vertical navigation control.

STATUS: in production.

SPZ-500 automatic flight control system

This is an integrated autopilot/flight-director system suitable for corporate aircraft that provides all the necessary pilot interface controls, air data computation and control servos to fly automatically a selected flight profile. It integrates with companion flight-director and air data systems.

The autopilot is a full-time system which provides continuous control through all phases of climb, cruise and descent, and with a full complement of lateral and vertical modes. These may be flown automatically by engaging the SPZ-500, or manually by following computed steering commands presented on the flight-director instruments. The latter also enables the pilot to monitor autopilot performance.

The SPZ-500 contains an air data system to provide information over a wide range of flight profiles.

The SPZ-500 is a full three-axis autopilot with a yaw damper. It has acceleration and rate-limiting circuits to provide smooth autopilot performance without compromising positive control action. Turn entry and exit is smooth, and by programming the roll-rate limit as a function of selected mode, rates are matched to the required manoeuvres.

Control of the autopilot for basic stabilisation and attitude-command is provided through the autopilot controller. Engaging the system with no flight-director mode selected causes the aircraft to maintain the existing pitch attitude, roll to wings level and then hold the existing heading. With a navigation or vertical-path mode selected, engaging the autopilot automatically couples the selected mode. When the autopilot is engaged the yaw damper is automatically engaged to provide yaw stabilisation through control of the rudder. When the autopilot is not engaged, the yaw damper may be used separately to assist the pilot during manual flight.

The autopilot controller includes the turn knob and pitch wheel, allowing the pilot to insert pitch and roll commands manually. The amount of bank or pitch change is proportional to the command selected. The soft-ride engage button reduces autopilot gains for smoother operation in turbulence, and the automatic elevator trim annunciators show any out-of-trim condition. The autopilot may be pre-flight checked with the test button.

Touch control steering allows the pilot to take control of the aircraft momentarily without disengaging the system. The pilot can push the button on the control wheel and manually change the aircraft flight-path. While the touch control steering button is pressed, the autopilot

synchronises on the existing aircraft attitude. On releasing the button the system holds the new attitude and resumes the coupled-flight mode.

Several flight-director instrumentation sets, either 4- or 5-inch units (102 or 127 mm), can be integrated with the flight control system. If the Sperry ADZ-241/242 air data system is installed this includes Sperry air data instrumentation. SPZ-500 variants are available for the Dassault Falcon 10, 20 and 50 business jets.

Dimensions and weights
Autopilot
(controller) 67 × 146 × 114 mm, 1.5 lb (0.68 kg)
(computer) 194 × 71 × 321 mm, 6 lb (2.72 kg)
(3 servos & mounts) 100 × 129 × 224 mm, 16.3 lb (7.35 kg)
(normal accelerometer) 51 × 25 × 61 mm, 0.3 lb (0.14 kg)
(trim servo) 56 × 84 × 175 mm, 5.5 lb (2.5 kg)

Air data system
(computer) 193 × 124 × 361 mm, 11.5 lb (5.22 kg)
(altimeter) 83 × 83 × 159 mm, 2.5 lb (1.13 kg)
(VNav computer/controller) 38 × 83 × 272 mm, 2 lb (0.91 kg)
(vertical speed indicator) 83 × 83 × 140 mm, 2 lb (0.91 kg)
(Mach/airspeed indicator) 83 × 83 × 186 mm, 2.8 lb (1.27 kg)

Flight-director system
(ADI (AD-650B unit)) 129 × 129 × 223 mm, 7 lb (3.18 kg)
(HSI (RD-650B unit)) 103 × 129 × 208 mm, 7.3 lb (3.31 kg)
(remote controller) 38 × 146 × 66 mm, 0.8 lb (0.36 kg)
(computer) 194 × 71 × 321 mm, 5 lb (2.27 kg)
(mode selector) 48 × 146 × 114 mm, 1.3 lb (0.59 kg)
(rate gyro) 46 × 52 × 95 mm, 1 lb (0.45 kg)

STATUS: in service.

SPZ-600 automatic flight control system

The SPZ-600 automatic flight control system is designed specifically for the new generation of long-range, high-performance, business-jet aircraft and is claimed to be business aviation's only completely automatic fail-passive/fail-operational flight control system.

It is a complete dual-channel system from the sensors to the servos. All channels (roll, pitch and optional yaw) are fully-operational at all times and the performance of each is continuously monitored and compared. If a failure that would result in a hardover manoeuvre occurs in any channel, it is immediately shut-down, resulting in single-channel operation in that axis only. Performance status is continuously displayed on the autopilot status panel, as well as the autopilot master warning annunciator. Sperry says that the unique monitoring system and duplex servo design, have been well tried in air-transport aircraft. Each of the duplex servos incorporates two independent servo motors that operate from signals applied by their own autopilot channels. The common tie to a single control surface is accomplished through a mechanical differential gear mechanism.

In the event of a system fault, the master warning annunciator flashes amber, and may be cancelled by pressing the annunciator. The status panel then indicates that the system has automatically disconnected the malfunctioning channel and is single-channel in that axis only. The status panel also enables the manual selection of single-channel operation in any axis, and has provisions for testing the dual autopilot channels and monitoring circuits prior to flight.

Control of the autopilot for basic stabilisation and attitude command is provided through the autopilot controller. Engaging the system with no flight-director mode selected causes the aircraft to maintain the existing pitch attitude,

Examples of British Aerospace Jetstream 31 corporate aircraft have Sperry SPZ-200A automatic flight control system

Flight-director displays (HSIs and ADIs), computers and controllers are part of Sperry SPZ-600 automatic flight control system

roll to wings-level, and then hold the existing heading. With a navigation or vertical path mode selected, engaging the autopilot automatically couples the selected mode. When the autopilot is engaged, the yaw damper is automatically in use. When the autopilot is not engaged the yaw damper may be selected separately to assist the pilot during manual flight. The autopilot controller also has a turn knob and pitch wheel, which allow the pilot to insert pitch and roll commands manually. The amount of bank or pitch change is proportional to the command selected. A soft-ride engage button reduces autopilot gains for smoother operation in turbulence, and a couple button selects which flight-director is driving the autopilot.

Touch control steering allows the pilot to take control of the aircraft momentarily without disengaging the system. He pushes the touch control steering button on the control wheel and manually changes the flight path as desired. While the button is pressed, the autopilot synchronises with the existing aircraft attitude, and on releasing the button the system holds the new attitude or resumes the coupled-flight mode.

The flight-director system uses standard 5-inch (127 mm) instruments and there is a choice of displays, with either split-cue or V-bar directors, and vertical-scale differences. The ADZ-242 air data system includes a full set of appropriate instruments, including altitude-alert controller and true airspeed/temperature indicator. SPZ-600 variants are available for the Fokker F27, British Aerospace 125-700, Canadair Challenger, Aérospatiale/Aeritalia ATR 42, and Cessna Citation aircraft.

Dimensions and weight:
Autopilot
(controller) 69 × 146 × 114 mm, 1.5 lb (0.68 kg)
(status/switching panel) 48 × 146 × 114 mm, 1.6 lb (0.73 kg)
(2 duplex servos & brackets) 106 × 205 × 298 mm, 20.8 lb (9.44 kg)
(computer) 194 × 71 × 321 mm, 5 lb (2.27 kg)
(yaw actuator) 54 dia × 232 mm, 2 lb (0.91 kg)
(normal accelerometer) 31 × 25 × 61 mm, 0.3 lb (0.14 kg)

Flight-director system
(ADI (AD-650B unit)) 129 × 129 × 224 mm, 7 lb (3.18 kg)
(HSI (RD-650B unit)) 103 × 129 × 203 mm, 7.3 lb (3.31 kg)
(remote controller) 38 × 146 × 66 mm, 0.8 lb (0.36 kg)
(computer) 194 × 71 × 321 mm, 5 lb (2.27 kg)
(mode select) 47 × 146 × 114 mm, 1.3 lb (0.59 kg)
(rate gyro) 46 × 52 × 65 mm, 1 lb (0.45 kg)

Air data system
(computer) 193 × 125 × 361 mm, 11.5 lb (5.22 kg)
(altimeter) 83 × 83 × 159 mm, 2.5 lb (1.13 kg)
(vertical speed indicator) 83 × 83 × 146 mm, 2 lb (0.91 kg)

(Mach/airspeed indicator) 83 × 83 × 186 mm, 2.8 lb (1.27 kg)
(VNav computer/controller) 38 × 87 × 500 mm, 2 lb (0.91 kg)

Gyro references
(vertical gyro) 157 × 165 × 238 mm, 7.3 lb (3.31 kg)
(directional gyro) 191 × 154 × 229 mm, 4.7 lb (2.13 kg)
(fluxvalve and compensator) 121 dia × 73 mm, 1.4 lb (0.64 kg)

STATUS: in service.

SPZ-700 autopilot/flight-director for Dash 7

Chosen as the standard autopilot/flight-director combination for the de Havilland Dash 7 transport, the SPZ-700 is a full-time, three-axis system for Category II approaches. It provides all usual vertical and lateral flight-director modes, together with RNav and MLS guidance. These vertical modes are altitude hold, altitude select, vertical speed hold, IAS hold, while the MLS mode provides for the steep approach path appropriate to Stol aircraft. The lateral modes are standard heading, navigation, ILS, back course, VOR approach and RNav.

The system is controlled via a single control panel, that annunciates functions engaged. Manual demands can be fed in through the pitch trim wheel and a proportional roll knob. For manual flight a yaw damper can be chosen, independently of the autopilot.

STATUS: in production.

SPZ-4000 automatic flight control system

Specifically designed for turboprop aircraft, this system provides the accuracy, smoothness, and reduced weight and volume advantages attributable to digital control systems. It also

offers more control authority than an analogue system and has comprehensive self-test capability. Autopilot, flight-director and air data functions are integrated into a single system.

The functions of the three-axis autopilot and flight-director are combined in a single flight control computer, and lightweight flight control servos are used. Engaging the autopilot without selection of flight director mode causes the aircraft to maintain the existing pitch attitude, and to roll wings level on the existing heading. With a navigation or vertical-path mode selected, engaging the autopilot automatically couples the selected mode. The yaw damper is automatically engaged in autopilot operations, and can be manually engaged to assist in 'hands-on' flying.

Control functions include pitch wheel and turn knob inputs for pitch and roll commands. There is a pre-flight autopilot check facility and an annunciator to draw attention to out-of-trim conditions.

STATUS: in production.

SPZ-8000 digital autopilot

The SPZ-8000 is Sperry's second all-digital general aviation flight-control system, the first being the SPZ-7000 certificated in Aérospatiale SA-365N and Sikorsky S-76 helicopters. Designed with emphasis on corporate aircraft such as the British Aerospace 125-800, the new system is unusual among its kind in having a bi-directional data-bus – the Sperry ASCB Avionics Standard Communications Bus – and apparently being designed for use with a flight management system accommodating both lateral and vertical guidance. Sperry believes that its ASCB system provides both the higher 'refresh' rates needed for flight-control applications and a greater flexibility in use than the 'one-way' ARINC 429 standard. The system is built around two FZ-800 flight computers, and is fail-operational.

STATUS: in production. First aircraft to be certificated was the de Havilland Canada Dash 8. This was followed by the British Aerospace 125-800, Aérospatiale/Aeritalia ATR-42, Cessna Citation III and the Dassault Falcon 900. The Gulfstream G-IV is expected to follow suit during 1986.

Automatic flight control system for A109

The Agusta A109 system has been designed to reduce pilot workload and improve reliability and safety at a realistic cost. A typical installation includes two autopilot (helipilot) systems (combined in series for redundancy) and may be used with or without a flight-director system.

This duplex system consists of one directional and two vertical gyros, controller panel, two computer units and five series actuators, two each for pitch, roll and one for yaw control. A trim computer and two trim actuators, for

Sperry SPZ-8000 is company's first all-digital autopilot for fixed-wing aircraft

pitch and roll adjustments, are also part of the system.

A flight-director facility can be incorporated but requires the addition of a mode controller, navigation-receiver inputs, plus attitude-director and horizontal-situation indicators. Flight-director functions include heading, vertical speed, barometric altitude and airspeed select, go-around procedure selection, and VOR, ILS, MLS, omega/vlf, RNav system, and Loran or Tacan coupling.

STATUS: in service.

Automatic flight control system for Bell 212 and 412

The Bell 212 system has been designed to minimise pilot workload, improve reliability and contribute to flight safety, without incurring high costs. An installation includes two autopilots, known as helipilot systems (combined in series for redundancy) and may be used with or without a flight-director system.

A single helipilot system consists of dual Tarsyn gyros which combine both DG and VG inputs, controller panel, computer unit and series actuators, two each for pitch and roll and one for yaw control. A trim computer and two trim actuators, for pitch and roll adjustments, are included.

Addition of a flight-director facility requires a mode controller, navigation-receiver inputs plus attitude-director and horizontal-situation indicators. Flight-director presentations include heading, vertical speed, barometric altitude and airspeed select, go-around procedure selection, and VOR, ILS, MLS, omega/vlf, RNav and Loran or Tacan coupling.

STATUS: in service.

Automatic flight control system for AS 350/355

Tailored to the Aérospatiale AS 350/355 Ecureuil helicopter family, this is a combined autopilot/flight-director system. Single-pilot IFR operation is possible with this equipment.

The autopilot (helipilot) is a three-axis stabilisation system with sufficient authority to control the helicopter throughout the entire flight regime. Autotrim is also available to enhance performance and to reduce pilot workload. The system comprises the helipilot computer, a trim computer, three series actuators, two trim actuators, three actuator-position indicators, a controller, C-14 gyrocompass and GH-14 gyro-horizon/attitude-director indicator.

Autopilot operating modes are divided into three categories. The stability-augmentation system modes provide transient motion-rate damping, and pilot cyclic inputs determine aircraft attitude-rate responses. Attitude hold modes permit gyro-stabilisation to any pre-selected attitude, and pilot cyclic inputs are used to determine aircraft attitude response. Flight-director coupling modes permit hands-off flying and flight-director display on the attitude-director indicator.

The flight-director system performs the computation necessary to bracket and hold the flightpath selected by the pilot. Facilities effectively anticipate manoeuvre requirements and cause automatic flight-path transfer between modes. All command information is presented on the attitude-director indicator. The system comprises a flight-director computer, airspeed sensor unit, altitude sensor, navigation receiver, fluxvalve and compensation for a C-14 gyro-compass, mode selector, and RD-550 horizontal-situation indicator. A radio-altimeter can also be included as a flight-director sensor. The flight-director system shares with the autopilot the CH-14 gyro-horizon/attitude-director indicator to display information to the pilot.

Flight-director operating modes permit selection of heading, vertical speed, barometric

Sperry SPZ-7000 digital automatic flight control system installed in Sikorsky S-76 helicopter

Sperry SPZ-7000 digital flight control system and associated flight-director equipment installed in Aérospatiale SA 365C Twin Dauphin helicopter

altitude, airspeed and go-around procedures. External references which can be coupled to the system include VOR, ILS, MLS, omega/vlf, RNav systems, Loran or Tacan.

STATUS: in service.

SPZ-7000 digital flight control system

In development since the turn of the 1980s decade, the SPZ-7000 is Sperry's entry in the market for digital helicopter autopilots. The system was certificated aboard the Sikorsky S-76 Mk II in November 1983 (now also certificated on S-76B model), and in December 1984 was specified by the airframe company as the factory standard option for this upgraded

version of the corporate helicopter; in the S-76 Mk II the dual-channel system with Helcis II flight/director, permits IMC (instrument meteorological conditions) operation with one pilot. The system has also been certificated on the Aérospatiale SA-365N helicopters. The system has also been selected for the Bell 400, a new twin-turbine, seven-seat helicopter to be built by Bell in Canada, and for the Bell 222UT utility transport helicopter; the aircraft was certificated with this system in January 1985. The system is currently also flying in an Aérospatiale 365N Dauphin.

The SPZ-7000 is Sperry's fourth-generation helicopter autopilot, and is claimed to be the first digital four-axis (pitch, roll, yaw and collective) system to receive civil certification. Originally designated DFZ-700, the fail-passive system is microprocessor-based, has two flight computers providing full autopilot and stability augmentation, dual flight-director mode-selectors, automatic pre-flight and en route testing features, and comprehensive diagnostic circuits.

A version of the system designated SPZ-7300 was chosen by Agusta for search and rescue versions of the A109 and AB 212 (licence-built version of the Bell 212), and in an AB 412, also for search and rescue duties. The SPZ-7300 is essentially an analogue system with a digital flight-path computer added to provide important new functions for the search and rescue mission. In addition to the standard flight-director modes, it generates a two-stage decelerating approach to the hover, hover augmentation, and the ability to hover while coupled to a Doppler navigation reference for better hover performance.

STATUS: all versions in production.

SPZ-7000 digital flight control system for SA 365N

This advanced digital system for the Aérospatiale SA 365N Dauphin helicopter combines autopilot and flight-director functions and caters for dual gyro-compass sensors and displays, with both attitude-director indicators and horizontal-situation indicators for each crew member.

Instrument-panel units consist of flight-computer controllers, mode selectors and altitude pre-select/air data displays. The flight-computer modes cover all conventional stability-augmentation, attitude-hold and flight-director coupling capabilities, plus hover augmentation, collective coupling, deceleration and radar-altitude hold. Flight-director modes cater for on-board sensor coupling to heading, vertical speed, barometric altitude, airspeed, pre-selected altitude, go-around and deceleration scheduling. External sensor data can be accommodated from VOR, ILS, MLS, omega/vlf, RNav systems, Loran and Tacan. The altitude pre-select panel permits the pilot to enter the desired altitude, and can be used as a temporary display of air data information.

The system is a complete four-axis stability-augmentation/autopilot, with dual flight computers, dual flight-director mode selectors, three-cue attitude-director indicators, and incorporating automatic test and system diagnostics and monitoring features. It is microprocessor-based and fail-passive. After first failure there is no aircraft reaction, and in the case of a second failure a 3-second delay can be accommodated. First flight of the SA 365N system took place in 1982, initial delivery of a single-pilot IFR system following in February 1983.

STATUS: in service.

S-TEC

S-TEC Corporation, Route 4, Building 946, Mineral Wells, Texas 76067-9990
TELEPHONE: (817) 325 9406

S-TEC, with a workforce of some 50 people, produces a range of off-the-shelf autopilots, with varying refinements, for light aircraft.

System 40/50 autopilots

The basic system 40 'wing leveller' autopilot consists of a combined programmer/computer/annunciator in a single box a turn co-ordinator, and an electrically operated roll servo. It permits both straight and turning flight, while inputs from a radio beacon provide the means for VOR tracking for en route navigation, localiser tracking for approach and reverse or back-course tracking for go around. The system can be upgraded to include a heading mode by adding the optional directional gyro. The System 40 costs $2695, to which may be added $1150 for the directional gyro. Installation charges are extra, depending on aircraft type.

The System 50 has all the foregoing functions, but adds an altitude-hold mode and elevator trim indicators. The optional directional gyro, can be added to provide heading hold. An accelerometer confers pitch stability and automatically disconnects the system in the event of a pitch axis fault. In addition the system incorporates a pre-flight test feature that tests internal limiter circuitry. The basic System 50 costs $4595, with installation costs in addition. Both systems are FAA TSO-approved.

STATUS: both systems in production and service.

System 60 autopilots

The System 60 is a rate autopilot, that is it uses rate of change of aircraft motion instead of attitude to generate demands to the flying controls. Launched in 1978, the System 60 is a single or two-axis system comprising an electric turn co-ordinator, 3-inch (76 mm) air-driven directional gyro, mode programmer/annunciator, pitch and/or roll guidance computers and servos, masterswitch and control wheel disengage switch, and altitude transducer.

STATUS: in production.

Universal

Universal Navigation Corporation, 3545 West Lomita Boulevard (Unit B), Torrance, California 90505
TELEPHONE: (213) 325 4081
TWX: 910 349 6205

UNS-1 integrated flight management system

This integrated flight management system computes lateral area and vertical navigation track guidance on a world-wide basis while providing real-time fuel management information. The UNS-1 centralised computer accepts data inputs from up to five external long-range sensors, range information from a DME set, as well as air data computer and engine fuel-flow inputs. All position data is separately tracked in the computer and is used to develop a best-computed position. Outputs are displayed on the control and display unit and made available to flight-directors.

The system comprises a control and display unit and a navigation computer unit (ncu) in those configurations that interface with an external DME unit. In configurations featuring an internal DME a signal switching unit (ssu) is added. The ncu uses a standard data base that can store up to 6900 locations in the data base with an additional 200 pilot-definable positions, and there is storage capacity for up to 100 routes. Sufficient capacity is available for further expansion to 30 000 locations, of which 18 800 would form the main database. Stored information is updated in the cockpit, by using a bar-code reader with printed bar-code entries or by using a standard 3.5-inch floppy disk and a disk drive unit.

Typical data input configuration could use up to three INS sets, two omega/vlf or Loran-C sensors, fuel flow from four engines and an air

Universal UNS-1 flight management system

data computer to supply true airspeed, total air temperature and baro-corrected altitude information. Automatic scanning of DME-DME range is a segregated operation which produces latitude and longitude positions corrected for slant-range error. A dual installation is possible, each interfaced with all sensors, and a built-in battery prevents memory loss when the unit is removed from the aircraft.

Dimensions: (cdu) 146 × 114 × 171 mm
(ncu) 4 MCU
(ssu) 2 MCU

Weight: (cdu) 2 kg
(ncu) 4.9 kg (standard database)
5.2 kg (expanded database)
(ssu) 3.4 kg
Interfaces: (inputs) six ports with ARINC 429, 571 or 575 format
(outputs) ARINC 429, 571 or 575 format
Environmental specification: DO 160A Cat B2E1/A/MNO/EXXXXXZAAAA
Power: 45 W at 28 V dc

STATUS: in production.

US Air Force

United States Air Force, Aeronautical Systems Division, Flight Dynamics Laboratory, Wright-Patterson Air Force Base, Ohio 45433
TELEPHONE: (513) 255 2725

Self-adaptive technology for flight control damage compensation

The US Air Force is attempting to program computer-based flight control systems to compensate for severely degraded aircraft handling characteristics resulting from in-flight damage. Aeronautical Systems Division (ASD), Wright-Patterson Air Force Base has awarded several contracts for work on a self-repairing flight control system which would compensate for damage resulting from hostile action or collision and has begun flight testing initial computer software.

Century Computing received a contract from ASD's Flight Dynamics Laboratory to integrate current laboratory facilities and the computer software necessary for conducting appropriate simulations. The laboratory is one of ASD's four Air Force Wright Aeronautical Laboratories.

Lear Siegler, Charles River Analytics and Scientific Systems Inc have received contracts for development of reconfiguration concepts which would enable maximum performance of aircraft flight control systems after actuator failure or control surface damage. Initial devel-

opment contracts also have been awarded to Grumman and General Dynamics for implementation, demonstration and evaluation of the control reconfiguration concepts.

Under the nine-year, $45 million programme, Charles Stark Draper Laboratory will develop computer software to assess maintenance concepts, mission planning, control structure and reconfiguration strategies. Honeywell will develop an expert system for maintenance diagnostics to be included in the self-repairing system.

Other contracts were awarded to Analytical Methods, for development of an aerodynamic model to predict stability and control parameters of an aircraft with damaged flight control surfaces; Calspan for partial flight test and demonstration, completed in 1985, of control reconfiguration concepts on the TIFS (total in-flight simulator) aircraft; General Electric for flight control mixer software and knowledge-based flight control diagnostics; McDonnell Douglas for a combined inertial guidance and flight control sensor set with enhanced reliability; and Universal Energy Systems for technical support necessary to assess key reconfiguration technologies.

A typical 'real-life' scenario in which the system would be useful involves combat loss of a fighter aircraft's wing. Currently, to compensate for loss of control on one side of the aircraft, a pilot would have to move the control column so that remaining roll-control surfaces

(ailerons and stabilators) could maintain stable flight. New control coupling characteristics might then appear, for example, any roll inputs might then generate pitch and yaw movements. Handling the aircraft might become so difficult that, in the stress of combat, the pilot would not have the opportunity to adjust. Problems with navigation, aircraft management and possibly injury, would be too great for the pilot to handle and the only way out would be to eject. However, the aircraft itself might have some control margin left and could theoretically be recovered to a base.

The programme now under way seeks to determine how the response of the flight control system to pilot demands could be immediately modified so as to remove, as far as possible, the now difficult handling qualities. It will include investigation of solutions to extreme scenarios, such as cases where damage is so great that the aircraft would be uncontrollable even with the pilot implementing previously determined instructions on emergency manipulation of flight control surfaces. Engineers will analyse the feasibility of reconfiguring aerodynamic surfaces that normally work in phase with one another (such as the left and right landing flaps, or the left and right leading edge flaps) to operate anti-phase, generating additional aerodynamic moments to counter the unbalanced moments caused by damage.

Work has begun on the simulation facility which will permit laboratory engineers and scientists to look at reconfiguration techniques by bringing together test systems and the great computational power which will be available through the facility. Some flight testing of initial software has also been carried out.

An initial flight control 'mixer' developed by General Electric was tested during 1985 aboard the laboratory's NC-131H TIFS. The mixer is computer software which may be used in a self-repairing mechanism to monitor sensor inputs and determine if all flight control surfaces are healthy enough to take actions directed by the pilot. If that is not the case, the mixer decides which flight control surfaces to move instead to achieve the desired aircraft response. The mixer flown aboard the TIFS is more simple than the versions which will eventually be developed under the programme.

Further flight tests of the complete self-repairing system are scheduled to begin in 1989. A single-axis (simple roll, pitch or yaw) demonstration of the system will be performed in a high performance aircraft to validate the reconfiguration concept, and should run until 1990. A multi-axis (combinations of roll, pitch and/or yaw) demonstration, also scheduled to begin in 1989 aboard a high performance aircraft, will validate total system concepts and will run until 1992.

STATUS: research programme.

This F-15 fighter lost most of one wing in an in-flight accident, but managed to recover the aircraft safely

Advanced flight control

(including fly-by-wire, fly-by-light and "fly-by-speech")

This section describes flight control equipment, based on electrical or electronic signalling of crew demands, that is required to operate continuously during all flight phases and which is in development, prototype or production form. (Automatic flight control and auto-stabilisation systems which have only a part-time function are not included here but will be found in the preceding section.)

Fly-by-wire

With few exceptions fly-by-wire (fbw) systems, are multiplexed to maintain high reliability. Even so, they are conventionally associated with a parallel, mechanical back-up system to provide limited 'get-you-home' control (there may be speed or altitude restrictions) in the event of complete electrical or electronic system failure.

In recent years flight control equipment with a much lower probability of failure (typically less than 10^{-6} per hour or one total failure in a million flying hours) has entered service or is undergoing tests. It is designed for a new generation of combat aircraft in which traditional levels of natural stability have been exchanged for a high level of instability in order to exploit substantial performance, weight and cost benefits. The degree of instability now required to realise benefits fully is such that a fast-acting, full-time computer-based system is essential to control and protect the aircraft at all times. The performance requirements are so demanding that no mechanical reversionary system can provide even an emergency degree of control. Total failure of the flight control system is followed within perhaps less than a second by loss of control and structural break-up, the pilot being unable to react sufficiently quickly to the situation. The terms 'fully fbw', 'full fbw' and 'full-time fbw' are now being used to describe such systems and to differentiate between them and earlier equipment with dissimilar redundancy, for example, mechanical reversion.

Fully fbw flight control equipment for commercial aircraft is now being developed, following statistically acceptable military experience, and most major airframe manufacturers have already assessed its feasibility and are conducting rig or airborne trials. Elements under test include side-stick controllers (notably in an Airbus A300) and fault-tolerant processors and the Airbus A320, due to fly in 1987, will have side-stick control. Guidelines on reliability are specifying failure probabilities of between 10^{-8} and 10^{-9} per hour (that is, a total failure every hundred million or thousand million flying hours). Multiplex channels with dissimilar electronic or software redundancy to avoid common-mode failures may be required for commercial aircraft.

Industry has, therefore, come to recognise two categories of electrical/electronic flight control: fbw, where failures can be accepted because the aircraft can be controlled (albeit with perhaps some degradation in performance) by mechanical means; and fully fbw, in which total failures cannot be accepted. The distinction is made throughout this section, and means that credence can be given to some earlier aircraft with electrical/electronic signalling and processing without implying the kind of performance and safety margins that have become associated with the blanket description 'fbw'.

The distinction between the terms 'digital' and 'fbw' (or 'full fbw') also needs to be kept clear. Early fbw or full fbw systems were analogue in nature, the change to digital technology coming later. However a parallel development is now occurring in non-fbw aircraft, in which analogue equipment is being replaced by digital systems. Although feasible, it would not be economic to replace a 'conventional' flight-control system with an fbw (or fully fbw) system because the benefits of the latter cannot be fully exploited with an existing airframe. Thus the analogue flight control system on the US Air Force/McDonnell Douglas

F-15 Eagle fighter, designed in the mid-1960s, will be replaced by a digital one in the new F-15E multi-mission fighter for improved weight, cost, space and reliability considerations, but it will not feature either fbw or full-fbw technology.

A major development now clearly in prospect for the next generation of military aircraft is the integration of flight control and propulsion management functions such as the scheduling of engine intake-ramp and after-burner nozzle movement. Such a system is being developed in the USA, where a McDonnell Douglas F-15 fighter is being modified to incorporate two-dimensional thrust-vectoring nozzles to earlier Stol performance from short runways.

Fly-by-light

A new technology which entered flight-test in 1985 is fly-by-light (fbl), in which digital signals are transmitted as light pulses along optic fibres. There is a close 'read-across' between fbl and fbw; equipment now being tested corresponds in performance and safety standards to fully fbw in electronic flight control. However, fbl is so recent that it is able to rely heavily on electrical/electronic signalling philosophy and technology to the extent that (at least on production aircraft) it will bypass the need for a mechanical reversion system. There is, therefore, no need to make the distinction between fbl and fully fbl.

Fly-by-speech

Also included in this section are the experimental fly-by-speech (fbs) systems, entering flight-test from 1985, in which the pilot commands certain aircraft functions by voice input. A computer recognises certain sounds and can, for example, control radio channel or waypoint changes, the firing of missiles or even the flight-path itself.

It is certain that this marks the start of the development of a technology which will lead to full voice-control of a combat aircraft within a decade.

FRANCE

Crouzet

Crouzet SA, 26 rue Jules-Védrines, 26027 Valence Cedex
TELEPHONE: (75) 42 91 44
TELEX: 345807

Fly-by-speech system

The second stage of development of Crouzet's speech recognition system, based on isolated word recognition was completed in June 1984, trials having been conducted on a Dassault Mirage III at the Bretigny flight test centre.

Some 40 flights were carried out, involving 14 pilots, and demonstrated the feasibility of

controlling such functions as radio frequency changes, weapon management selections and radar mode control.

A further set of flight tests, involving an Aérospatiale/Westland SA 330 Super Puma helicopter, were completed in November 1985.

The next phase started in October 1984 on the Mirage 2000 simulator at Istres and involves the use of multi-word commands, and this phase will be flight tested on a Mirage IIIB in 1986. Operational units are scheduled to be available in 1987.

STATUS: in development.

Crouzet is developing a voice activated flight control system

Dassault

Avions Marcel Dassault-Breguet Aviation, 33 rue du Professeur-Victor-Pauchet, 92420 Vaucresson
TELEPHONE: (1) 741 79 21
TELEX: 203 944

Fbw systems for Mirage III, 5 and 50

In common with most high-performance military aircraft of its era the Dassault Mirage III was developed with electrically-signalled flight control information and some electronic processing (analogue) for improved capability. There is also a mechanical back-up. Powered-control

operation without fly-by-wire signalling will allow the aircraft to be recovered from most flight conditions although its handling qualities are degraded. The Mirage III system was retained in the Mirage 5 and the Mirage 50.

STATUS: in service.

Fbw system for Mirage IV

The Dassault Mirage IV is aerodynamically similar to the Mirage III but is considerably larger. Having been designed during the same decade it incorporated a similar fbw system. Philosophy and operation modes are understood to be approximately the same as those described for the Mirage III.

STATUS: in service.

Fbw system for Mirage F1

The Dassault Mirage F1 first flew in 1966 and uses a flight control system which is essentially like that of the Mirage III in that it uses electrical-signalling, some electronic processing and powered controls with a mechanical back-up. The fbw system configuration was altered to match the flight control surface layout of a conventional aircraft planform.

STATUS: in service.

Fbw system for Mercure

The Dassault Mercure airliner first flew in 1971 and uses electrical-signalling in conjunction with a mechanical back-up system to convey flight control information from the flight-deck and systems to the flying surfaces. The basic aircraft is conventionally stable and the flight control system includes auto-stabilisation. Control can be maintained in the event of a total fbw system failure, although the handling qualities would be somewhat degraded.

STATUS: in service.

Fully fbw systems for Mirage 2000 and 4000

A fully fbw analogue system was specified for the Dassault Mirage 2000 fighter, which has a similar configuration to that of the Mirage III but it is conventionally unstable. The fully fbw flight control system provides an appropriate degree of stability and mixes control inputs from the pilot and systems so that airframe stresses and aerodynamic limits (including those during supersonic flight) are not exceeded. The aircraft's available performance can, therefore, be fully exploited with better manoeuvrability than with a conventional fbw system.

The wing trailing-edge control surfaces (elevons) are used for control in the pitch and roll axes and are signalled electronically without any mechanical back-up. There are two surfaces on each wing, each served by a twin-body servo-jack and each body taking demands from two of the four available electronic channels. The single rudder is signalled via a triple fully fbw arrangement. A 'fifth channel' emergency system, which uses battery-supported signalling to provide the pilot with a degree of actuator control, is also employed, suggesting that in some portions of the flight envelope the aircraft has a reasonable degree of natural stability.

The system has quadruplex sensors associated with the pilot's controls, and multiple gyros, accelerometers, and air data sensors. These feed in to a quadruplex processor arrangement which uses hybrid computing techniques. Each processor is largely analogue, with digital computation of control-gain parameters. Dassault is responsible for most flight control system development and provides interfaces which enable automatic flight control system demands to reach the flying controls.

The Mirage 2000 system was used on the Mirage 4000 demonstrator, which first flew in 1979. This remains a prototype application whereas the Mirage 2000 is in full-scale production, entering service in July 1984.

STATUS: in production.

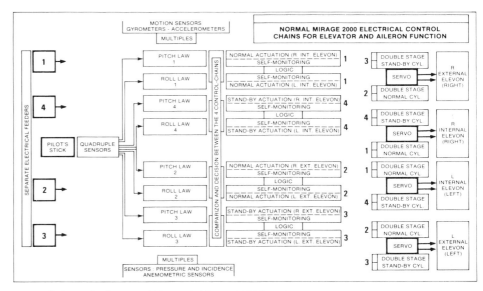

Schematic diagram of Dassault Mirage 2000 fully fly-by-wire flight control system

---

Sfena

Société Française d'Equipements pour la Navigation Aérienne, Controls and Systems Division, Aérodrome de Villacoublay, BP 59, 78141 Vélizy-Villacoublay
TELEPHONE: (1) 46 30 23 85
TELEX: 260046
FAX: (1) 46 32 83 96

FMGS flight management and guidance system for A320

The Airbus A320 will be the first commercial transport aircraft to employ digital fly-by-wire technology in the primary flying controls (though the A310 uses fbw to command the flap, slat and airbrake secondary-control movement). In the new transport miniature sidesticks will command pitch and roll attitude. The sticks are centred by simple springs providing 'return to neutral' forces independent of speed or altitude. The rudder continues to be mechanically operated via conventional rudder pedals.

Elevators and ailerons are each controlled by four independent channels, contained in two 8 MCU boxes, the quadruplex system commencing with eight electrical movement LVDT (linear variable differential transformer) pick-offs (four an axis) in the sidestick controllers. Each pilot will have his own sidestick, mounted on the side console, the captain operating his with the left hand, the first officer with the right.

While two of the three primary flying controls are activated by fully fbw equipment, and thus have no manual reversion, total failure of the aileron and elevator channels can be compensated by way of the mechanically operated tailplane trim and rudder; tailplane trim replaces

Layout of Airbus A320 flight deck, with captain's sidestick controller on left-hand console. Vacant space left by removal of conventional control column permits folding table to be fitted

elevators, and rudder stands in for ailerons. As with the A310, fbw will also be used for flaps, slats and speed-brakes, and in the form of Fadec (full-authority digital engine control) will also be used for engine management. Two flight-augmentation computers in 8 MCU boxes provide dual-channel command signals for yaw-damping rudder trim, rudder travel limit system, and flight-envelope protection. The entire system is monitored by fault detection and isolation software, with warning indications and appropriate vital actions and responses being automatically displayed on the EFIS cathode ray tube displays. Airbus is sufficiently confident about the integrity of the system that it believes the reversionary mode will never be employed operationally.

Some 75 hours of flying with sidestick controllers has been accomplished on the Airbus A300 demonstrator aircraft, largely to assess pilot reaction and to refine their design.

In November 1984 Sfena announced that it had chosen US company Sperry as its principal partner in the programme, replacing UK firm Smiths Industries, its associate on the A300 and A310 flight control systems. Sperry will provide the flight management system, which will be integrated with the Sfena flight control system into a single computer unit. For commercial aircraft, it will be the first time that flight control and flight management functions have been brought together in a single box. The aim is to reduce complexity, expand the use of the flight

SFENA flight control unit as fitted to Airbus A320

management system and reduce fuel consumption.

West Germany's Bodenseewerk will continue its association with Sfena on the A320.

STATUS: in full-scale development. First production set is scheduled for delivery to Airbus in early 1987.

Thomson-CSF

Thomson-CSF, Avionics Systems Division, 178 boulevard Gabriel-Péri, 92242 Malakoff Cedex
TELEPHONE: (1) 46 55 44 22
TELEX: 204780

ELAC elevator and aileron computer

Thomson-CSF is responsible for the design and production of the elevator and aileron computer (ELAC) sub-system which forms part

of the fly-by-wire system of the Airbus A320. Its main function is to control electronically the elevators and ailerons of the A320, the power required to displace the movable surfaces being provided by the aircraft's hydraulic system.

In order to carry out its functions the computer receives data from the various aircraft sensors such as the sidestick, movement sensors, accelerometers, etc, as well as from systems such as the inertial reference system, automatic pilot, radio-altimeters etc.

The architecture of the ELAC sub-system is based on the use of two redundant computers which continuously self-monitor themselves. These concepts have already been tried and tested on the Concorde and Airbus A310 through which programmes Thomson-CSF has acquired its experience of fbw controls.

Format: 6 MCU per ARINC 600
Weight: 8 kg
Power: 28 V dc

STATUS: in development.

GERMANY, FEDERAL REPUBLIC

MBB

Messerschmitt-Bölkow-Blohm GmbH, Helicopter and Aircraft Group, Military Aircraft Division, Postfach 80 11 60, D-8000 Munich 80
TELEPHONE: (89) 600 00
TELEX: 5287975

Experimental digital flight-control system for future combat aircraft

This was an experimental fully fly-by-wire flight control-system, flown between December 1977 and April 1984. In all, about 160 flights were completed in five different aircraft configurations: initially with the stability of the basic air-vehicle, and eventually progressing to a negative stability margin approaching 20 per cent.

The project was authorised in 1974 by the Federal Ministry of Defence, with the intention of developing and testing an advanced fail-safe digital flight control system. Instability was introduced by adding a Lockheed F-104 horizontal tailplane aft of the cockpit, and a jettisonable tail-trimming mass of up to 750 kg attached to the lower rear fuselage.

A quadruplex digital processing system was used, each channel using a 31 lb (14 kg), 16-bit, 18 K word computer supplied by Teledyne. A strapdown navigation and angle of attack and sideslip measurement system using skewed sensors, also by Teledyne was incorporated, and it was claimed to use only half the components of a conventional equivalent sensor arrangement.

Lockheed F-104G CCV experimental aircraft in final configuration

CCV Phase 4, between 1983 and 1984, was to develop and flight-test back-up software, a digital autopilot integrated into the flight control system and a laser gyro inertial navigation system, carried in a centre-line pod. Trials were completed in 1984.

STATUS: experimental system, flight-testing completed.

Panavia

Panavia Aircraft GmbH, Arabellestrasse 16, D-8000 Munich 86
TELEPHONE: (89) 92171
TELEX: 529825

Fbw system for Tornado

The Panavia Tornado flight control system uses triplex electronic signalling and processing, with quadruplex actuation, to provide a high

degree of manoeuvrability throughout the entire flight envelope and has a mechanical back-up system for emergency control, though with degraded handling qualities. The fly-by-wire processing is incorporated in the command and stability augmentation system (csas), and the automatic flight functions, which include safety-critical capability (for example, automatic terrain-following), are integrated with the automatic flight control system. These are separately configured but very closely allied systems. The triplex csas components are

associated with a duplex spin prevention and incidence limiting system (spils) which ensures that the crew can fly the aircraft to its structural and aerodynamic limits without the risk of loss of control.

Design, development and production of the systems have been a tri-national venture between GEC Avionics in the UK, Bodenseewerk in West Germany and Aeritalia in Italy.

Command stability and augmentation system
The csas is an analogue, fbw manoeuvre

Panavia Tornado spils computer opened up to show four-board, front-connector layout

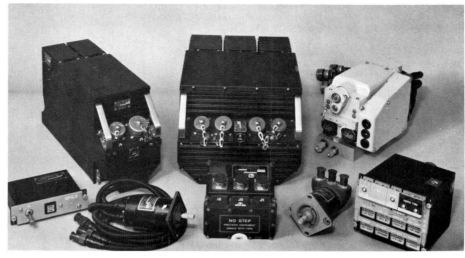

Panavia Tornado csas equipment includes computer units, control panels, triplex-position transmitters, triple gyro-packs and quadruplex first-stage actuators

demand system. It provides electrically-signalled pitch, roll and yaw control and automatic stabilisation of aircraft response to pilot command or turbulence. Gain-scheduling improves handling qualities and control stability over the entire flight envelope and operates in conjunction with the spin prevention and incidence limiting system. 'Carefree' manoeuvring allows the exploitation of the aircraft's full lift capability under all flight conditions without the risk of structural damage resulting from a pilot's control demand for control-surface movement that would exceed the design strength of the airframe.

Individual line-replaceable units are the csas pitch computer, csas lateral computer, csas control unit, pitch, roll and yaw rate gyros, and pitch, roll and yaw position transmitters.

Spin incidence and limiting system

Spils is a duplex analogue system which operates in conjunction with the csas to achieve maximum aircraft lift in low-level flight. It limits aircraft incidence, irrespective of the pilot's demands, when maximum safe angles of attack are reached. Individual line-replaceable units are the spils computer and spils control unit.

Automatic flight director system

The digital autopilot and flight director system (afds) automatically controls the flight-path in all modes, including terrain-following, and sends signals to the director instruments enabling the crew to monitor autopilot performance or to fly the aircraft manually. Pitch autotrim is also incorporated. The duplex self-monitoring processor configuration provides high-integrity automatic control, permitting low altitude cruise with appropriate safety margins. The flight director remains available after most single failures. Autopilot manoeuvre-demand signals are routed to the control actuators through the command and stability-augmentation system. A 12-bit processor is used with 6 K words of stored program and 1 K words of data store. Typical computing speed is around 160 Kops and program cycle time is 32 milliseconds. Individual line replaceable units are the two afds computers, afds control unit, autothrottle actuator and pitch and roll stick-force sensors.

STATUS: in production.

ISRAEL

IAI

Israel Aircraft Industries Limited, Ben Gurion International Airport, Tel Aviv 70100
TELEPHONE: (3) 97 13 11
TELEX: 371133

Autocommand Mabat flight control system

This is an analogue flight control system used in the IAI Kfir fighter and developed from a Sfena unit used in the Dassault Mirage III fighter. Autocommand is a pitch-only stabilisation system which alleviates undesirable transient handling characteristics (including during the transonic region of the flight envelope); improves dynamic stability with consequent benefits in target pursuit, aiming, firing and weapon-release accuracy; provides stick-force/g scheduling; and provides attitude hold for a limited period after stick release, and automatic barometric height hold (except during transonic flight).

Autocommand Mabat is a single line-replaceable unit comprising three interconnected modules: an analogue processor, air data system and power supplies. The main unit is the processor, which uses dc operational amplifier technology. Integrated circuit ac/dc and dc/ac conversion units, and self-test circuits are incorporated, and the unit uses multi-layer printed circuit boards and a motherboard to reduce internal wiring to a minimum.

Pitch auto-stabilisation is provided to requirements set out in MIL F-8785, with a stick-force/g characteristic set by the manufacturer, and pitch-attitude hold in the ±20° pitch range.

Israel's two-seat Kfir TC2 has IAI Autocommand Mabat analogue flight control system

After releasing the stick maximum pitch angle change is less than 5° per minute. The maximum change of altitude due to a malfunction cannot exceed 10 metres during the first second following the failure, and any normal acceleration produced is within a specified envelope. Self-test can be initiated by the pilot and if the test-button lamp does not light within 10 seconds this is taken as an indication of failure. System engagement/disconnection can take place anywhere within the flight envelope. Automatic disconnection will occur only if a malfunction occurs in the compensator system, the autocommand safety system, or if the number one hydraulic system disconnects and the preliminary servo changes over to the number two. Manual disconnect is achieved via several separate override or disconnect buttons, or by exerting a high stick-force.

Altitude hold is available only if pitch stabilisation has been engaged. It is usable throughout the flight envelope, except in the transonic region (Mach 0.95 to 1.15). In level flight altitude errors are within ±30 feet or 0.1 per cent of altitude, or double these values at 30° roll angle, and not outside a linear extension to 150 feet (46 metres) error at 60° roll angle. Aircraft oscillations are limited to periods in excess of 20 seconds, and maximum normal acceleration is less than 0.1 g. The hold mode can only be engaged if normal acceleration is less than 0.5 g. Automatic disconnect will occur if rate of height change exceeds 2000 feet a minute (610 metres), when the speed is between Mach 0.95

and 1.15, or if the undercarriage is extended.

Sensors and output devices associated with the unit include a rate gyro unit, accelerometer unit, left- and right-hand compensation servos, pitch pre-servo, stick dynameter and trim motor. A range of dedicated cockpit controls and annunciators are also installed.

System performance conforms to MIL 18224C and MIL F-8785, and electrical characteristics are to MIL-STD-704A

Dimensions: 306 × 198 × 195 mm
Weight: 10 kg
Power: 100 mA 115 V ac 400 Hz single-phase; 1 A 28 V dc
Reliability: >1400 h mtbf

STATUS: in service.

JAPAN

Mitsubishi

Mitsubishi Jukogyo Kabushiki Kaisha, 5-1 Marunouchi 2-chome, Chiyoda-ku, Tokyo 100
TELEPHONE: (3) 212 3111
TELEX: 22282/22243

Fully fbw system for T-2CCV

A specially modified example of its T-2 advanced jet trainer to incorporate CCV technology has been built by Mitsubishi for the Technical Research and Development Institute, a part of Japan's Defence Agency. It was handed over to the TRDI in March 1984 to begin a two-year flight-tests programme in support of research into this aspect of aircraft design.

The T-2CCV has a triplex redundant digital fly-by-wire flight control system operating the conventional control surfaces (trailing edge flaps, rudder and leading stabiliser) and a vertical canard surface to provide side force control. It also has horizontal canards making it aerodynamically unstable.

The T-2CCV programme was launched by

Mitsubishi T-2 CCV has two sets of forward-mounted canard surfaces for direct side-force control

the TRDI in 1978 and the first flight was conducted by Mitsubishi in August 1983. Five control modes are to be investigated including control augmentation, relaxed static stability,

manoeuvre load control, direct lift control, and direct side-force control.

STATUS: under development.

SWEDEN

Saab-Scania

Saab Aircraft Division, S:581 88 Linkoping
TELEPHONE: (13) 181 365
TELEX: 50040

Fbw system for AJ 37 Viggen

The AJ 37 all-weather attack, JA 37 interceptor versions and related trainer and reconnaissance models of the Saab Viggen all use a standard electrically-signalled fly-by-wire flight control system which provides auto-stabilisation during some flight modes. The aircraft itself is aerodynamically stable and a mechanical back-up system provides control after a total fbw system failure.

STATUS: in service.

Experimental fully fbw system for AJ 37 Viggen

An experimental fully fbw system has been developed and flown as part of the Elektriskt Strysystem (ESS) development programme in preparation for the use of this technology on the Saab JAS 39 Gripen which is due to enter service with the Royal Swedish Air Force around 1990. It is a triplex digital flight control system developed by Honeywell, which will eventually have all-electronic sensors and signalling, although it was controlled manually during its first flight in the ESS Viggen test vehicle in 1983.

Initial testing to demonstrate the technology was due to last six months, following which the aircraft was flown with the centre of gravity moved progressively aft, to positions where the

conventional aircraft would be unstable. The experimental system has used elevon and rudder surfaces whereas the Gripen will have a moving foreplane as a part of the pitch and roll control system. It is also likely that the system will incorporate weapon-aiming modes for use with the external ordnance and internal gun proposed for the Gripen.

STATUS: flight-tested.

UNITED KINGDOM

Cranfield

Cranfield Aeronautical Services Limited, College of Aeronautics, Cranfield, Bedfordshire MK43 0AL
TELEPHONE: (0234) 750 111
TELEX: 825072
FAX: (0234) 751 871

Astra Hawk and VAAC Harrier

The Astra Hawk and VAAC Harrier projects are two British aircraft, both with digital fly-by-wire flight control systems, but designed to serve different purposes: they both made their first flights, having been modified at the College of Aeronautics at Cranfield, in the summer of 1986.

The Hawk, known as the Advanced Systems Training Aircraft (Astra), is being prepared as an airborne simulator for the Empire Test Pilots' School (ETPS) at Boscombe Down. It is certainly unique in Britain, if not in the world, in

having variable stability, controlled by a computer operated by the rear cockpit pilot. The front cockpit, where the trainee sits, has a control column connected via a computer and electric wiring to the flying controls' hydraulic actuators. In the rear cockpit, the instructor has a control and display unit on which he can set the aerodynamic stability derivatives of almost any kind of aircraft: he retains the conventional flying control system unaltered from that of a standard Hawk.

The trainee can then experience the characteristics of, say, an unstable fighter, a light transport, or, more specifically, a Tornado or Jaguar: he can, of course, also fly an 'electronic' Hawk. He makes conventional movements of stick and rudder and the computer, operated by the instructor's inputs, in turn commands the flying controls and, through the electronic artificial feel system also installed in the aircraft, the stick forces felt by the trainee, to replicate the desired aircraft type.

A computer-controlled safety monitor checks that the student's inputs will not cause the Hawk to fly outside its approved flight envelope, and, if such a trend is developing, prevents the manoeuvre by controlling the flow of hydraulic fluid at the actuators. The use of this safety monitor, and the presence of the rear-seat pilot with conventional controls, means that the Astra Hawk has only a simplex (single-channel) fbw system, greatly reducing system cost and complexity.

The Astra Hawk's front cockpit has also been redesigned and modified by Cranfield's workshops to include a Smiths Industries head-up display and multi-function colour head-down display, as will be installed in the Hawk 200. It also has a side-stick controller, only the second aircraft in Britain to have such a device, which can be used as an alternative to the standard control column.

After a four-phase flight-test programme at Cranfield the aircraft will be delivered to the

ETPS, where it will be used for pilot training, although there maybe some research could be done. For example, in the design of digital flight control systems, the 'transport delay' – the time delay between pilot command and control surface deflection – is critical, and the Astra Hawk could be used to determine the effects of such delays on pilotability.

The VAAC (vectored thrust aircraft advanced flight control) Harrier is a converted two-seat version, and is configured with the rear cockpit equipped with fly-by-wire flight control, while the front cockpit, with full instrumentation, remains standard. The rear cockpit has a Smiths head-up display, and, apart from engine and very basic flight instruments has also been equipped with displays which show the positions of the flight controls. The fbw flight computer has also been supplied by Smiths: it has full authority in the pitch plane, and of throttle and nozzles, it also controls flaps and has limited authority in roll and yaw.

The aircraft is being prepared for use at the Royal Aircraft Establishment at Bedford: it was flown following conversion at Cranfield late in 1985 to check the basic characteristics were unchanged, and, after a six-month period of calibration and other engineering work, it also flew under 'electronic control' in the early summer of 1986; after a short period of basic evaluation, it was to be delivered to Bedford, where it will be used to evaluate control systems as part of the national programme to develop advanced V/Stol aircraft, in conjunction with an advanced simulator at Bedford which will be operated 'in parallel' with the aircraft trials.

GEC Avionics

GEC Avionics Limited, Airport Works, Rochester, Kent ME1 2XX
TELEPHONE: (0634) 44400
TELEX: 96333

Flight control system for AM-X

GEC Avionics is co-operating with Aeritalia in design and development of a fly-by-wire flight control system for the Italian/Brazilian AM-X strike aircraft. The system comprises two dual-redundant flight-control computers each based on 16-bit microprocessor hardware and incorporating fail-safe software. The system commands the movement of seven control surfaces, and incorporates automatic pitch, roll and yaw stabilisation. Analogue computing is used for the actuator control loops, the pilot-command paths and rate-damping computation. Digital computing handles gain scheduling, electronic trim and integration of the airbrake. Redundant microprocessors in the flight control computer units monitor system performance and are associated with built-in test facilities designed to provide a high confidence and rapid comprehensive system check. Testing is initiated by the pilot prior to flight and is conducted automatically thereafter. In addition to the flight control computers, the fbw system also includes pilot control position sensors, three-axis rate gyros and air data components.

STATUS: in flight-test.

Experimental fully fbw system for Jaguar

Development of an integrated flight control system for the British Jaguar fbw technology-demonstrator programme commenced in 1977 and first flight was accomplished in October 1981. This project is the world's first complete all-digital fbw system (albeit experimental), operating without mechanical or electrical reversionary controls. The system has been manufactured to production standards and provides a ready basis for future fbw applications. The principal system contractors are GEC Avionics and Dowty Boulton-Paul, which provided the servo-actuators.

Individual line-replaceable units are four flight control computers, two actuator drive and monitor computers, pilot's control panel, five control surface position transmitters, a lateral

Italian/Brazilian AM-X has GEC fbw flight controls

Jaguar fbw installation contains two actuator drive and monitoring computers (left) and four flight control computers (right)

British Aerospace/GEC Avionics fully fly-by-wire Jaguar with experimental forward strokes to augment instability

Airship Industries Skyship 600 has GEC Avionics fly-by-light system

Fibre-optic-based fly-by-light flight control system for Airship Industries Skyship 600

accelerometer and three quadruplex gyro packs.

A target loss-probability of 10^{-7} per hour for the complete system led to adoption of a quadruplex system configuration. Where essential data is measured, quadruplex sensors are used also, and their outputs are routed via optically-isolated lanes for consolidation in each flight control computer. Processing is accomplished using 16-bit words and operating with a 28 K word program and 6 K words of stored data. Typical processor operating speed is 800 000 operations a second and computing takes place in cyclic frames of between 5 and 80 milliseconds duration. Failure-survival relies on majority-voting of the quadruplex components.

Flight control computer outputs to the tailplane (pitch/roll), rudder (yaw) and spoilers (roll) drive two-stage hydraulic actuators. Two hydraulic supplies are connected to either two or three sub-actuator units at each control-surface location, providing the degree of redundancy necessary for the actuator to survive a single hydraulic power-supply failure without generating catastrophically large actuator movements. Each flight control computer is supplied from up to four independent electrical sources, including a battery-backed supply.

GEC Avionics supplies the quadruplex rate-gyro packs and duplex lateral accelerometer units which are used as sensors. A diagnostic display unit stores failure information for ground maintenance personnel.

Initial trials were completed during winter 1981/82, and there followed an intensive ground evaluation of the effects on system performance of lightning strike, using high-current pulses. Later flight-test objectives included the evaluation of handling with and without external stores, and the determination of stall departure and spin-prevention control characteristics. Unstable aircraft configurations were evaluated in 1983, and the 1984 flight-test programme began in March with the aircraft modified to incorporate wing-root leading-edge strakes to provide aerodynamic instability (previously, 500 lb (227 kg) of ballast in the rear fuselage had been added to give a degree of mechanical instability). The maximum degree of instability investigated during their trials was about 10 per cent (that is, the centre of gravity was some 10 per cent behind the mean centre of pressure). By comparison, the standard Jaguar has a positive stability margin of about 8 per cent. The fast-acting control system, in conjunction with incidence-limiting circuitry, permitted very high pitch and roll rates to be attained. During the programme stall departures, spin-prevention and 'carefree' handling were all demonstrated. The programme was concluded in July 1984 after 75 flights.

STATUS: experimental system to gain flight experience for future combat aircraft, in particular the UK's EAP experimental aircraft programme and EFA European fighter aircraft.

Digital fully fbw system for EAP
In preparation for the European Fighter Aircraft project, Britain is investigating the appropriate technology under an activity known as EAP (experimental aircraft programme). A single prototype to embody technology and equipment flew for the first time in 1986, and has a quadruplex digital fully fbw flight control system based very largely on that of the fbw Jaguar (see separate entry). Four flight control computers are arranged in a 'dual-dual' configuration, experience with the 'dual-triplex' Jaguar system having shown that the degree of redundancy required to remain operational with any two failures could be achieved with the simpler arrangement. The highly unstable EAP demonstrator aircraft uses fbw technology for the primary flying controls (flaperons, rudder and canard surfaces), and for manoeuvre flaps, nosewheel steering and intake ramp movement. Fbw may also be employed to operate the afterburner nozzles. Apart from the flight control computers, the system also includes a control column sensor assembly, rudder-pedal sensor (supplied by Penny & Giles), four aircraft-motion sensors (supplied by German company Litef), four actuator drive units (from Bodenseewerk), and a pilot's control unit. The processors are designed to give a high throughput, and their instruction set is optimised for flight control applications. They compute appropriate aircraft response to meet pre-determined control laws, and represent the interface between pilot inputs, aircraft motion sensors, air data information, and the electro-hydraulic actuators that drive the flying control surfaces.

STATUS: in flight-test.

Fbl system for Skyship 600
Fly-by-light, or fbl, is a development of the fbw principle, enabling the flight-control system to operate even in the presence of high levels of electromagnetic interference, and virtually eliminating the risk of lightning damage to the flight control system in aircraft constructed of composite materials. During the early 1980s GEC Avionics built a demonstration rig representative of one lane of a triplex flight control system and containing a force-stick control column, rate gyro, flight-control computer actuator and actuator drive unit. It uses fibre-optic data transmission from the flight control computer to the actuator, a hydraulic-to-electric generator to provide electrical power at the actuator, and monitoring equipment in the actuator which sends data, via a fibre-optic link,

to a monitor processor within the flight control computer.

In 1982, GEC Avionics discussed a contract with UK company Airship Industries for an optically signalled flight control system to be fitted for evaluation to one of its Skyship airships, paving the way for an operational system. The system was subsequently adopted for the Airship Industries Skyship 600, which made its first flight in March 1984 and was certified with mechanical flight controls in mid-1985 and with fbl shortly thereafter, thus becoming the first aircraft with optically-signalled primary flying controls. In addition to the benefits described, the installation of an fbl system paves the way for an autopilot.

Airships have non-rigid structures which continuously flex and distort in turbulent air, so that the control surfaces are also in continuous movement even though the yoke and pedals in the cabin may be held stationary. The pilot's task in counteracting the effect of these deflections can therefore be tiring, especially over long periods. With new and demanding applications in prospect for small airships, such as ASW, early warning, oil-rig and fishery protection, the emphasis is now on upgrading their performance and equipment to free the crew from the traditional piloting duties in order to concentrate on other tasks.

It is not considered possible to replace the traditional mechanical cable and pulley controls by a fbw system owing to the risks associated with lightning strikes; unlike aircraft, airships are largely non-metallic, and the cables carrying the signals could also double effectively as lightning conductors. However, control equipment based on optical signalling would be immune to such a risk, and GEC Avionics has produced a duplex system with two lanes of sensors and servos. While quadruplex (four-lane) signalling is essential on conventional aircraft to outvote a faulty channel before it can command catastrophically sudden or large control surface movements that would over-stress the airframe, similar control-surface runways on an airship would result in a leisurely response, giving adequate time for the pilot to identify the malfunctioning channel and switch it out. Duplex signalling in this case has been adequate.

STATUS: in service in Airship Industries Skyship 600.

UNITED STATES OF AMERICA

Bendix

Bendix Flight Systems Division, Route 46
Teferboro, New Jersey 07608
TELEPHONE: (201) 288 2000
TELEX: 34414
TWX: 710 990 6136

Experimental fully fbw system for F-16/AFTI

Bendix has developed a hybrid digital/analogue fully fly-by-wire flight control system for the advanced fighter technology integration (AFTI) programme, directed by the Air Force Systems Command Flight Dynamics Laboratory at Wright-Patterson Air Force Base. The programme uses an extensively modified General Dynamics F-16 to investigate how new propulsion, aerodynamic-control, and systems technology can be brought together to enhance fighter performance. The aircraft has two additional ventral control surfaces beneath the engine intake, and an extended dorsal fairing in which much of the experimental equipment is installed.

Objectives of the development programme include demonstration of the integration of a digital flight control system with advanced fighter operation concepts such as direct-force and weapon-line control, and automatic manoeuvring-attack systems. Pilot/vehicle interface evaluation will form a large part of the total development task. The flight control system is designed to provide six degrees-of-freedom commands, using multimode control laws in triple-redundant computers. Basic processor is a Bendix BDX-930, and double fail-operational probability is less than one chance in 10^7 flying-hours. To achieve this objective, in addition to the three digital lanes in the flight control system, there is also an independent analogue lane, normally disconnected, which is associated with each digital channel. Within each computer, self-test facilities isolate failures on up to 95 per cent of occasions. A failed digital computer is disconnected immediately, and automatic switchover to the independent analogue back-up is completed. Pilot engagement of the latter, using a side-stick mounted switch, is also possible.

The first flight of the AFTI/F-16 took place at Carswell Air Force Base, Fort Worth, on 10 July,

1982, and the aircraft began flight-tests from NASA's Dryden Center six days later. Initial flight-development software did not include some of the advanced modes which was flown later in the development programme. A two-phase, 275-sortie, flight-test schedule is now in progress. Phase 1 included 125 sorties and is designed to explore thoroughly flight control capability. The 150-mission second phase will evaluate the control system modes in conjunction with advanced sensors in weapon-release trials. It is anticipated that a 10 to 1 improvement in ground-target tracking capability will be demonstrated.

Dimensions: (flight control computer) 139 × 203 × 406 mm
(actuator interface unit) 266 × 139 × 369 mm

STATUS: experimental system in flight-test.

Digital fully fbw system for F-16C/D

In January 1984, as part of a major improvement programme General Dynamics chose Bendix to build a flight control computer to drive the quadruplex digital fully fbw flight control system for later versions of the F-16C and F-16D now-standard US Air Force lightweight fighter. Prototype equipment was delivered in 1985 and it will be incorporated into production aircraft in 1988, converting the F-16 from an analogue to a digital fully fbw fighter. Bendix, which began work on digital flight control systems aimed at the F-16 as early as 1978, won the award in competition with General Electric, Lear Siegler (which builds the analogue flight control system in the F-16A/Bs and some C/Ds), GEC Avionics and Hamilton Standard. It is planned to equip 600 aircraft but the total procurement may be for as many as 2500 aircraft (the companies which tendered for the contract quoted for that number). An important reason for the shift from analogue to digital is the need to improve the performance of the aircraft to exploit effectively with the Martin Marietta Lantirn night-targeting system.

The system has evolved from the equipment developed by Bendix for the US Air Force/General Dynamics AFTI advanced fighter technology integration programme, based on an F-16 airframe. A fundamental change from the flight control system in the AFTI aircraft is

Bendix single-box digital flight control system for later General Dynamics F-16 fighters

however the progression from three digital plus one analogue channels per axis to four digital channels, the configuration favoured by General Dynamics for operational reliability (the system has to remain operational after any two failures). The four digital channels use MIL-STD-1750A architecture utilising Jovial high-order language, and each has a processor, memory, input/output functions, discrete failure logic, MIL-STD-1553B digital data-bus and serial link to other channels. The computer is designed to remain fully operational following any two consecutive failures within the quadruplex part of the system. The single unit, weighing 50 lb (22.7 kg) and housed in a 1 ATR Long box, has the equivalent function of the four separate units on the AFTI aircraft and occupies about half the space. In all other respects the system meets the fit, form and function requirements that permit it to replace directly the earlier analogue system.

Computing and processing technology is based on large-scale and very large-scale integration, and gate arrays. It is designed to accommodate further growth, incorporating vhsic components, without the need for new software. Flight-critical functions are hard-wired into the system for the highest integrity, but less critical signals communicate with other equipment via the digital data-bus. The system has 48 K words of prom memory, 2 K words of ram scratch-pad memory, 8 K words of input/output scratch-pad memory, and a 2 K word memory to record faults.

Benefits of the all-digital system over the previous analogue one are given as better reliability, lower power density, smaller number of components, greater ease of 'tuning' to meet changes, and lower life-cycle costs. Cost per ship-set is said to be less than $100 000.

Bendix consider this new flight control system to be a strategically important step forward, since it can be seen as the basis for future integrated flight-critical, full-authority flight control systems, incorporating thrust/side-force vectoring, terrain-following and fire-control.

Dimensions: 20.25 × 10.75 × 8.75 inches (514 × 273 × 222 m)
Weight: 50 lb (22.7 kg)
Volume: 1 ft³ (0.03 m³)
Reliability: demonstrated mtbf > 2150 hours
Power: 200 W

STATUS: in development for later US Air Force/General Dynamics F-16C and F-16D single-seat and two-seat fighters, beginning with aircraft scheduled for delivery in October 1988.

AFTI/F-16 has distinctive canards, or foreplanes, operated by Bendix experimental flight control system

Boeing Vertol

Boeing Vertol Company, PO Box 16858, Ridley Park, Philadelphia, Pennsylvania 19142
TELEPHONE: (215) 522 2700
TELEX: 845205

Adocs advanced digital optical control system

The first production application of fly-by-light technology in the USA is likely to be a combat helicopter. The features of this technology which make it suitable for this role are the invulnerability of optical cables to electro-

magnetic interference, their ease of repair; a dramatic reduction in complexity of flight control systems, and superior flying qualities. US industry, with Boeing Vertol as the prime contractor, commenced flight-testing in November 1985, of a Sikorsky UH-60 Black Hawk helicopter fitted with the advanced digital optical control system (Adocs) based on fibre-optic technology. It will be the foundation of a production system to support the LHX light combat helicopter required by the US Army for the 1990s.

The system is based on optical position transducers, fibre-optic cabling, computers and pilot controllers. The transducers are coded plates attached to the flying control actuators, and pilot's controls, whose movements are measured by fibre-optic filaments. The latter comprise strands of silica. There are no electrical signals in the sensing and transmission elements and so they are invulnerable to electronic countermeasures or the strong electromagnetic pulses generated in a nuclear explosion. Information in serial word form passes along the fibre cables to a computer, and the same cable can pass signals in the reverse direction to act as signal commands, giving a weight and volume saving.

The flight control processor is a three-channel system, each channel containing two microprocessors representing the electronic analogue to the mechanical link between the rotor and the pilot's controls, and a third which handles automatic flight control functions (autopilot stability-augmentation). The primary flight control system remains operational throughout two failures, although the automatic flight control system can support only one failure because data from sensors outside the system is shared between the processors and cross-compared to eliminate errors. The preferred control solution for the single-pilot LHX is a four-axis side-arm controller providing pitch, roll, yaw and collective pitch to the right hand, leaving the left hand free to operate other equipment. Such a system was tested by Sikorsky during 1982 on a modified Canadian

US Army/Sikorsky UH-60A Black Hawk has been adapted as Adocs test-bed

National Research Council/Bell UH-1 helicopter. Bendix is teamed with Boeing Vertol to provide the triple-redundant servo-actuators that control the helicopter's rotor. These actuators are commanded by optical signals, with optical error and feedback signals.

The UH-60 test-bed has an Adocs system for one pilot and a standard control system for the other. An engineer's station behind the pilot is fitted to record and display information, and monitor the transition between conventional and optical systems.

Boeing Vertol will fly the experimental UH-60 for 100 hours, after which the Army will evaluate the aircraft for 15 hours at various centres. The

aim is to produce a battle-worthy system for the nocturnal/poor-weather, nap-of-the-earth missions envisaged for the LHX helicopter. The company estimates that Adocs will reduce the weight of a flight control system by 25 per cent for an aircraft of LHX size (9000 lb, 4000 kg), rising to about 50 per cent for a helicopter of about 40 000 lb (18 100 kg), corresponding to the projected V-22 Osprey multi-mission aircraft.

STATUS: in flight-test development: first flight took place in November 1985 and testing is scheduled to last until mid-1986.

General Electric

General Electric Company, Aerospace Control Systems Department, Aerospace and Electronic Systems Division, PO Box 5000, Binghamton, New York 13902
TELEPHONE: (607) 770 2000

Digital flight control system for F/A-18

While the US Air Force/General Dynamics team chose a fully fly-by-wire and non-reversionary, though analogue, flight control system for their F-16, the US Navy/McDonnell Douglas design group went for a digital fbw system with mechanical redundancy for the then new F/A-18 project.

The General Electric solution for the F/A-18 flight control requirements is a digital, four-channel fbw system operating the primary

flying controls – ailerons, stabilators and rudders – and leading-edge and trailing-edge flaps and nosewheel steering. The system incorporates 32 servo loops to drive and control these functions. At the same time the conventionally mounted control column (in contrast to the F-16's sidestick controller) has a mechanical link to the tailerons for reversionary pitch and roll control. All fbw computations are accomplished by four digital computers operating in parallel, accepting inputs from LVDTs (linear variable differential transformers) sensing control column and rudder-pedal movement, and also from analogue motion sensors, and formulating commands to the redundant electro-hydraulic servo-actuators driving the control surfaces. The system therefore remains operational after two failures. It has a high degree of integration with other aircraft equipment, communication being accomplished by means of a MIL-STD-1553 data-bus.

Two special display modes are used in conjunction with the system. The first is a flight control failure matrix, the second provides recovery guidance during spins, and both use the same cathode ray tube unit. In the first mode, the display is selected to show details of the fault after the pilot's attention has been drawn to its presence by an indication on the central warning system. The display signals an 'x' at the position of the fault on a schematic diagram of the system painted on the screen. In the second mode, the display shows the position of the control column needed to recover from a spin, assumed to exist when yaw rates greater than 15° a second occur simultaneously with speeds of 125 knots (232 km per hour) or below. This mode is selected automatically, having absolute priority over other displays when this combination of speed and yaw rate occurs. The two modes originally were provided for flight-test purposes, but have been retained in production aircraft following recommendations from Naval Air Test Center and McDonnell Douglas test pilots.

The system comprises two flight control computers (each incorporating two General Electric micro-programmable digital processors specially designed for flight control applications), two rate-sensor boxes, two acceleration-sensor boxes, an air data sensor box, radar pedal force sensor, and pilot's control panel.

The F/A-18 is the first aircraft in which major performance shortfalls have been rectified by software changes rather than expensive modifications to equipment or structure. To prevent the aircraft from falling back on its tail on a pitching deck the main wheels are mounted well behind the centre of gravity. This gave rise to a bad mismatch between the aerodynamic pitching moment generated by the stabilators and the weight moment about the main landing gear, resulting in excessive nosewheel liftoff speeds. There was consequently only a brief time between nosewheel and main gear liftoff, resulting in dangerously abrupt rotation as the aircraft left the ground. To overcome this

General Electric flight control set for McDonnell Douglas F/A-18 Hornet fighter

difficulty, an additional pitching moment was obtained by arranging the two rudders to toe-in, during take-off, via a relatively simple software change. Additional moment was generated by re-scheduling the leading-edge flap deflections at take-off. By these two ploys, nosewheel liftoff speeds were reduced by 25 to 35 knots (46 to 65 km per hour) depending on gross weight. Again, software changes calling for differential movement of the trailing-edge flaps, and later the leading-edge flaps (and

associated with local mechanical stiffening of the wing) were employed to improve the roll rate at high speed and low altitude; aeroelastic twisting of the thin wing under these circumstances was found during flight test to work against the roll demand of ailerons and stabilators, reducing the roll rate below the specified figure. Yet another software modification led to a reduction in the undesirable coupling between roll and yaw, which was putting unacceptable loads on the vertical fins.

All of these refinements were accomplished at very little cost in computer capacity. The nosewheel lift-off speed reduction accounted for only 50 words of code (40 for rudder toe-in, 10 for leading-edge flap re-scheduling); improved roll-rate took 120 words (50 for differential leading-edge flaps and 70 for trailing-edge flaps); and implementation of the cross-axis commands to reduce fin loads took 50 words.

STATUS: in production for F/A-18.

Honeywell

Honeywell Inc, Military Avionics Division, PO Box 312, 2600 Ridgway Parkway, Minneapolis, Minnesota 55413
TELEPHONE: (612) 378 4178
TELEX: 290631

Fully fbw system for X-29A

The Grumman X-29A demonstrator is a research aircraft built under contract to the Defense Advanced Research Projects Agency (DARPA) to provide data on a wide range of aerodynamics and control technologies.

The first X-29A made its initial flight on 14 December 1984 at Edwards Air Force Base, where the trials programme is being conducted by NASA's Ames-Dryden Flight Research Facility. The aircraft is unusual in having a forward-swept wing, and unique in being designed for supersonic flight in this configuration. Fore and aft stability has been traded for performance and structural benefits to the extent that the X-29A is the most unstable of any of the currently flying fully fly-by-wire operational or research aircraft; specifically, the instability at take-off (ie with full fuel), is about 35 per cent. In other words, the centre of gravity of the aircraft is some 35 per cent of the mean aerodynamic chord behind the centre of lift at speeds less than Mach 1. As the aircraft accelerates through Mach 1 the centre of pressure moves back over the wing to coincide more or less with the centre of gravity and so trim drag becomes essentially zero, making for the most efficient combat conditions. The instability of the X-29A at low speeds has been compared with that of an arrow fired backwards.

The flight control installation is a three-box system, each box containing full multi-axis digital control computations plus a back-up analogue system. The boxes measure some 7.5 × 7.5 × 25 inches (191 × 191 × 635 mm), and cross-channel monitoring is used to detect faults. Presence of a fault is given immediately to the pilot on a flight control panel, and he is also advised by ground-control from real-time telemetry analysis. While the system automatically reconfigures itself into a hybrid digital-analogue mode in the event of a fault, the pilot can switch the analogue system in at any time as a confidence check. The system remains fail-operational after any failure, and fail-safe after two. The volume of information from the sensors and the speed with which it has to be handled calls for the use of dual processors in each flight control computer. One of the processors is associated with input signals from sensors and the pilot's controls, the other handles commands to the control surfaces and contains the appropriate control laws. Each processor has a dedicated 8 K memory that, if necessary, can be expanded to 16 K, and commands to the control surface actuators are updated 40 to 80 times a second. The control surfaces themselves are the most advanced of any aircraft; in the pitch plane they comprise the all-moving foreplanes, full-span flaperons on the wings, and flaps on the wing trailing-edge extension, all moving in accordance with control laws optimised over the flight envelope to generate the least drag in any manoeuvre. Much of the background for the system is attributable to Honeywell's digital flight control system for the JA 37 Viggen fighter, now built by Saab-Scania.

The biggest single task in the flight control system has been software development to

Unique Grumman X-29A research aircraft has Honeywell digital fully fbw flight control system

optimise aircraft response over the flight envelope. Honeywell is responsible for writing the software, which is then checked out by Grumman on its simulator at Bethpage. Basis of the software were the results from drop-tests of a free-flight model conducted in 1981 at NASA's Langley Center. The software is all written in machine code, resulting (for a research aircraft, at least) in far better efficiency than a high-order language.

STATUS: research vehicle for developing and testing technologies for future combat aircraft.

Lear Siegler

Lear Siegler Inc, Astronics Division, 3400 Airport Avenue, PO Box 442, Santa Monica, California 90406-0442
TELEPHONE: (213) 452 6000
TWX: 910 343 6478

Analogue fully fbw system for F-16

The flight control suite for the US Air Force/General Dynamics F-16 fighter is the world's first fully fly-by-wire system to enter service. It was originally designed for the General Dynamics YF-16 entry in the US Air Force's LWF light weight fighter competition, launched in April 1972 and was carried over into the production version of the aircraft. It is the first operational system in which all pilot commands to the primary flying controls are transmitted as electrical signals, there being no rods or cables or any means of mechanical reversion.

Design was put in hand during 1969; at that time digital technology was still in its infancy, digital systems were larger than analogue, and failure mechanisms largely unexplored. The system is therefore entirely analogue in nature, with four channels in each of the three axes providing the necessary redundancy to meet the failure criterion of being able to continue flight after two failures. This criterion is equivalent to one total failure in 10 million flying hours. Heart of the system is a 20 kg box containing the four flight control computers,

each rigorously isolated from electrical contact with one another. Each has its own power supply, and individual input signals from the pilot's side-stick controller, sensors and air-data system. Each computer processes its signals and provides independent output commands to the servo actuators that move the flying control surfaces.

The F-16 is the first aircraft to benefit in cost, weight and size from having negative or 'relaxed' stability (about 3 to 4 per cent) in the pitch plane, and the flight-control system has to provide appropriate artificial stability to mask the dangerous 'raw' handling characteristics of the F-16 and make it acceptable to the pilot. The system also provides what has come to be called 'carefree manoeuvring', by imposing limits on aircraft response so that the pilot cannot inadvertently impose commands that would cause the aircraft to suffer structural damage or to stall and spin under excessive loads. The flight control system directly commands the three primary flying controls (flaperons, tailerons and rudder), sets the flaps for take-off and landing, and moves the wing leading-edge flaps to maintain optimum aerodynamic efficiency during high-g combat. In the interest of simplicity, since the F-16 is a relatively short-range short-endurance day fighter, no autopilot function is provided; the pilot has to fly the aircraft from take-off to touch-down through the digital flight control system.

For the first time in an operational aircraft, a

console-mounted side-stick takes the place of a conventional centre-mounted control column. This stick permits small movement (about 0.125 to 0.25 inch, 3.2 to 6.4 mm) about the pitch and roll axes, detected by two sets of four LVDT (linear variable differential transformer) transducers that convert this movement into electrical signals. Conventional rudder pedals are fitted, but again these simply operate on four LVDTs providing quadruplex electrical command signals through the computers to the rudder servo-actuator.

Fault isolation within the computer box, and subsequent reconfiguration of the computers to compensate for the new situation, are achieved by the use of analogue output voters that interface sensors and computations with movement of the servo-actuators. A 63-function self-test schedule is run automatically as part of pre-flight checks. By early 1985 about a million operating hours among eight air forces had been accumulated. During this time no confirmed instances of total system failure had been recorded.

Dimensions: 18 × 10 × 8 inches (457 × 254 × 203 mm)
Weight: 44 lb (20 kg)

STATUS: in production for F-16A and -B (single- and two-seat versions) and initial production for the later, improved F-16C and -D. Deliveries are scheduled to run past 1990 under current contracts. Some 2000 sets had been supplied by January 1985.

Experimental digital fully fbw system for F-16

As prime contractor for the analogue fully fbw system on all General Dynamics F-16A and -B fighters, Lear Siegler proposed a digital, quadruple system for advanced versions of the F-16C and -D to be bought by the US Air Force. The system was designed to meet the form, fit and function specification as a replacement for the earlier automatic flight control system suite, and design work began in January 1980. Flight-testing of a single unit, in an otherwise analogue-based system, commenced at Edwards Air Force Base in mid-1982. Initial reports claim that in a wide range of manoeuvres and flight conditions, including high angle of attack, 'zero' airspeed and high-*g* manoeuvres to the airframe limits, there were no functional differences between analogue and digital processing. A continuing evaluation, substituting further digital processors for analogue units, then demonstrated how the advantages of digital technology over analogue could be realised without the risk levels currently associated with the former.

The competition for an all-digital fully fbw system was won by Bendix in early 1984, but the Astronics Division equipment will be integrated with MIL-STD-1750 microprocessors and flown on continuing trials under General Dynamics contract.

STATUS: experimental system in flight-test.

Digital fully fbw system for Lavi

The fully digital flight control system for the Israeli Aircraft Industries Lavi has quadruplex redundancy, and comprises two boxes, with two digital channels built into each box. The twin-box configuration hinges on the survivability issue, which is given great emphasis: if one is damaged the other will provide sufficient control authority to regain base. Each digital channel has associated with it an analogue channel that can take over its function in the event of failure. Design total failure rate is not greater than 1 in 10^7 hours.

STATUS: in development. The programme was initiated in October 1982, and production deliveries are to begin in 1988. Six prototype Lavi aircraft are being built, the first scheduled to fly in 1986.

Digital fully fbw system for JAS 39

The digital fbw flight control system for Sweden's Saab JAS 39 Gripen has a triplex layout and is built into a single box. Nominal failure rate (ie a fault situation resulting in loss of the aircraft) with the system is expected to be about 1 in 10^6 hours, but Lear Siegler work is

First production fully fly-by-wire flight control system is Lear Siegler equipment on F-16

showing a degree of reliability half an order of magnitude greater than this, ie 5 in 10^7 hours.

Astronics Division is responsible not only for design and construction of the flight control computer boxes, but also for testing and integration of the complete system, including the Moog flying control servo-actuators. These are single-stage, direct-drive devices, using samarium-cobalt electric motors to operate the main hydraulic power valves, and the JAS 39 is the first aircraft to use this new technology.

STATUS: in development. The programme was initiated in April 1983 and production is to begin in 1990.

Digital fully fbw system for AT-3

In January 1985 Lear Siegler announced that it has been selected to design and build the triplex digital fully fbw flight control system for the AT-3 basic/advanced trainer being developed by Taiwan's Aero Industry Development Centre (AIDC). The Lear Siegler system was in competition with Bendix, which had offered a version of the digital flight control system chosen for versions of the General Dynamics F-16. AIDC has a contract with General Dynamics for technical assistance, and the Astronics system will incorporate a blend of the technologies and configurations employed in the Saab JAS 39 and McDonnell Douglas F-15E flight control suites.

Artist's impression of IAI Lavi fighter, which has Lear Siegler flight control system

STATUS: in development as a four-year programme for the Chinese Nationalist Air Force AT-3 fighter. Lear Siegler plans to build some 200 systems.

Summary of recent flight control systems proposed or in development by Lear Siegler

	F-16 digital (proposal only)	Lavi	JAS 39
System type	Fully fbw	Fully fbw	Fully fbw
Lear Siegler responsibility	Flight control system computer	Flight control system computer	Total system
Status	Research	Production	Production
Start date	January 1980	October 1982	April 1983
Digital channels	4	4	3
Independent backup	1	2/2	3
Control surfaces	5	7+2	7+3
Cross-channel data links	None	Serial	Serial
MIL-STD-1553 bus	None	2	2
Channel synchronisation	No	Yes	No
Box size	18 × 10 × 8 in (457 × 254 × 203 mm)	2 (15.3 × 13 × 7.7 in, 389 × 330 × 196 mm)	17.6 × 12.6 × 7.6 in (447 × 320 × 193 mm)
Total available card area	960 in² (0.62 m²)	1920 in² (1.24 m²)	1260 in² (0.81 m²)
Available area/channel	240 in² (1548 cm²)	480 in² (3097 cm²)	420 in² (2710 cm²)
Card size	6 (7 × 7 in, 178 × 178 mm)	10 (7 × 9 in, 178 × 229 mm)	10 (7 × 8 in, 178 × 203 mm)
Power consumption	150 W	2 × 180 W	309 W
Weight	44 lb (20 kg)	2 × 47 lb (21 kg)	44 lb (20 kg)
Reliability (mtbf)	2181 hours	1327 hours	1769 hours
Fault tolerance	2 fail operate	2 fail operate	—
Analogue inputs/channel	88	128	128
Memory (total/used)			
prom	16/12 K	24/18 K	32/20 K
ram	2/1.2 K	3/2 K	4 K
Language	LSI HOL	C	LSI HOL

Lear Siegler Inc, Instrument Division, 4141 Eastern Avenue SE, Grand Rapids, Michigan 49508
TELEPHONE: (616) 241 7000
TELEX: 226430
TWX: 810 273 6929

Voice-controlled interactive devices

LSI's Instrument Division is developing a voice-controlled interactive device, or VCID, which was qualified during 1982 as the first flight-worthy voice-command system to fly in US tactical aircraft. The aircraft carrying the experimental VCID is the US Air Force AFTI Advanced Fighter Technology Integrator programme, a heavily modified General Dynamics F-16 fighter equipped with digital flight controls and other equipment to investigate and demonstrate the technology foreseen as being essential for the next generation of US fighters. The voice-command system was employed to initiate and control symbology on multi-function displays without the need for the pilot to remove his hands from the control column or throttle box under combat manoeuvring conditions calling for high g, or having to divert his attention from the outside into the cockpit.

The Lear Siegler system is 'personalised', that is to say, the voice recognition element depends on matching word commands in the pilot's voice to the same words recorded by that pilot under 'ready-room conditions' on a solid-state cassette contained within the system. In this way the likelihood of error is considerably reduced. These commands are words and phrases such as 'COMM 1', '322.20', 'radar range 20', 'weapon status'.

The vocabulary and syntax rules are established by the operational requirements and are entered into the system by peripheral keyboard display in the ground support equipment. For this operation, the system is in a pre-flight 'edit' mode. In the 'train' mode following entry of the vocabulary, the pilot/operator is prompted to speak each word in the vocabulary a number of times which establishes a reference template for each word and word transitions. When all templates have been prepared they are transferred into a portable solid-state memory, the data transfer module, in a cassette format. The pilot carries the cassette to the cockpit, inserts it

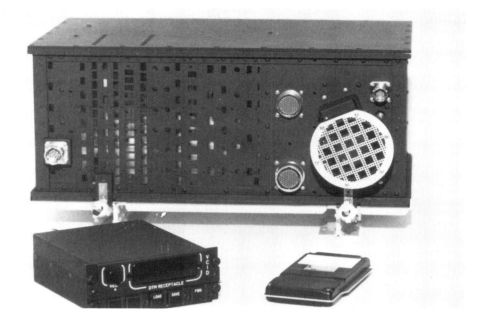

Lear Siegler VCID comprising data-processor (top), voice-command cassette (right) and cassette receptacle (left)

into the panel-mounted controller and loads the system. The pilot's voice templates are now stored in the system and available for use as the voice commands when required.

A six-month trial with the system began in December 1982 with variations of flight conditions, such as g-loading, and also with different background noise levels and quality, to see how the system retained its performance under degraded conditions. During the tests, the equipment was required to display on a cathode ray tube screen whatever command the pilot spoke. Results varied from 100 per cent correct score with low background noise to about 80 per cent as the noise level increased. Adequate performance is essential under high-g conditions, when accurate manual selection of controls or switches may be difficult or impossible due to heavy workload or g forces acting on the hands to make them nearly immobile and when the pilot's voice quality may be substantially changed.

Development and flight-testing of the Lear Siegler continuous speech recogniser are continuing during 1986 in the US Air Force's AFTI/F-16 and a UH-60 helicopter of the US Army.

In September 1984, Lear Siegler signed a licence agreement with Verbex, a division of Exxon Enterprises, transferring Verbex's speech recognition technology to Lear Siegler. This gave Lear Siegler access to the Verbex Model 3000 proprietary continuous voice-recognition technology to supplement its own development programme.

The Lear Siegler VCID comprises a data-processor, a cassette or module holding pilot commands as a 'comparator', and a panel-mounted controller/input into which the cassette is plugged.

STATUS: research tool for initial application in tactical aircraft.

McDonnell Douglas Helicopters

McDonnell Douglas Helicopters Inc, Centinela and Teale Streets, Culver City, California 90230
TELEPHONE: (213) 305 3400
TELEX: 182436

Hughes Helicopters changed its name in August 1985 to McDonnell Douglas Helicopters.

Experimental higher harmonic control system for helicopters

In late 1982, as part of a research programme jointly funded by the National Aeronautics and Space Administration, the US Army and Hughes Helicopters, a Hughes OH-6A helicopter was first flown with a fly-by-wire flight control system which damps high-frequency vibration by modulating rotor activity with small amounts of blade-incidence movement. Initial trials were conducted with approx 0.65° incidence authority, but up to 2° were available for later tests. The system uses vibration sensors positioned on the airframe and microprocessor-based computing elements which determine the amount of blade feathering necessary to compensate for the vibration. Wind-tunnel testing by NASA's Langley Center at Hampton, Virginia, preceded flight tests, and early flights used an open-loop system prior to full closed-loop operations. Preliminary testing was completed during March 1983. Using a Sperry Flight

McDonnell Douglas OH-6A helicopter fitted with experimental higher harmonic flight control system

Systems micro-computer in conjunction with a three-axis accelerometer mounted under the pilot's seat, demonstrated that closed-loop operation of the system could reduce vibration at around 40 knots (74 km an hour) (the speed for the most obtrusive effects) by up to 80 per cent.

The company's engineers refer to the HHC higher harmonic control system as a limited-authority fbw flight control system. It is

regarded as potentially valuable in improving passenger ride in commercial helicopters, or for improving optical sensor-stabilisation in military applications. Flight-testing resumed in 1984 at Mesa, Arizona with an improved version of the control system, which was claimed to show an 80 per cent or more reduction in airframe vibration levels at all speeds. In manoeuvring flight reductions of 70 per cent or greater were demonstrated.

STATUS: under flight-test development by US Army Applied Technology Laboratory.

LHX flight development helicopter

In October 1985, the first flight took place of a McDonnell Douglas Helicopters AH-64A Apache modified as part of the US Army's LHX development programme.

Apart from testing various avionic and other systems destined for the LHX, the AH-64A is equipped with four sidearm controllers and electronic rudder pedals. The side-arm controllers operate a full authority fly-by-wire system, with the McDonnell Douglas main simulator facility in St Louis being used extensively in developing the flight control laws. The fbw system is being developed primarily by Honeywell, partner to McDonnell Douglas in the LHX competition.

STATUS: in development.

Voice-control for F-16

The US Air Force has started flight trials of an interactive voice system (IVS) in the advanced fighter technology integration (AFTI) F-16 (see separate entry).

The IVS is being used to control various avionics functions and displays; the changing of waypoints, weapons and radio frequencies are typical examples.

In the current system the pilot has to pre-record a number of commands on a tape cassette which is then carried in the aircraft. In flight, the pilot's actual voice commands are

McDonnell Douglas Helicopters AH-64 test helicopter

compared by the IVS computer with the 'template' held on the cassette. The IVS also incorporates a synthetic speech module, so that the system can correctly reply to the pilot's commands.

The development work will continue with investigations into the effects of high-*g* and high noise levels on the system's ability to recognise speech. Two IVSs, developed by Lear Siegler and Texas Instruments, are in the programme.

STATUS: in development.

US Air Force

US Air Force, Aeronautical Systems Division, Wright-Patterson Air Force Base, Ohio 45438
TELEPHONE: (513) 255 2725

Voice recognition system

A voice-activated electronic system to be integrated with aircraft communications and navigation equipment is being developed for operational use by the US Air Force Aeronautical Systems Divison (ASD). A contract for the voice system has been awarded to SCI Systems Inc, Huntsville, Alabama. Based on technology developed by Votan of Fremont, California, the SCI system will feature connected voice recognition, enabling pilots to speak normally without pausing between each

word. The system is expected to have greater accuracy and improved error/correction over a previously tested system.

As in all voice-recognition systems currently under development, the user records words as templates and the operational system compares incoming speech with that recorded and identifies the similar word. As part of its in-house programme, the ASD Flight Dynamics Laboratory will attempt to apply to the voice system some dynamic training work currently being carried out by Honeywell under contract. With a dynamic retraining capability, the system will make new templates for words the pilot pronounces differently, possibly due to stress or fatigue. The dynamic retraining technique has been described as an adaptive, template-updating algorithm. The system monitors each recognised word on a 'goodness of fit'

score, and words with poor fit records are tagged for retraining. The next time a tagged word is recognised and confirmed by the pilot it is saved as an additional template for that word. If additional recognition of the word indicates better fit scores for the new template, it is saved and the old template is discarded. The US Air Force in-house programme is also looking at voice synthesis and voice feedback.

The SCI voice system was scheduled to commence flight tests in January 1986 and will use a MIL-STD-1553 data-bus to ensure integration, operational and expansion capabilities.

STATUS: flight tests will take place in a US Air Force/Boeing C-135C and are scheduled to run for three years.

Cockpit displays, instruments and indicators

CANADA

Canadian Marconi

Canadian Marconi Company, Avionics Division, 415 Legget Drive, PO Box 13330, Kanata, Ontario K2K 2B2
TELEPHONE: (613) 592 6500
TELEX: 05-34805

CMA-2014

This system features a colour crt display and full alphanumeric keyboard. It uses software keys on each side of the display for maximum operational flexibility. The system can be used as a central controller for a number of systems via ARINC 429 data buses. It can also be interfaced with Canadian Marconi's latest Omega/vlf systems – CMA-771, DMA-734 and the CMA-764.

Canadian Marconi's new CMA-2014

CMA-823 multi-purpose display

The alternative to a cathode ray tube display for many applications, the CMA-823 needs only 6.5 inches (165 mm) behind the instrument panel or console, and does not require a symbol generator or any other remote-mounted unit. The display itself is shown up in yellow-orange on a thin-film electro-luminescent flat panel, using 320 by 240 lines (76 800 pixels), at 68 lines an inch. The system is based on the Intel 8086 microprocessor, and interfaces directly with MIL-STD-1553B or ARINC 429 digital data-bus. The display is visible in direct sunlight, and is night-vision-goggles compatible. There are five mode-selector keys, a page forward/backward key, ten software-controlled keys, and a brightness control.

Dimensions: 7.75 × 6.11 × 6.5 inches (197 × 155 × 165 mm)
Weight: 10 lb (4.5 kg)

Canadian Marconi status display system

Canadian Marconi CMA-730 engine instruments for AH-64A Apache helicopter

Status display system

Canadian Marconi's status display system (SDS) is designed to replace current master warning and caution panels by incorporating warnings and faults on a single display. It summarises the warnings, thus reducing crew distraction and workload. Major functions include: automated display of the regular aircraft checklist, with items being deferred for later recall as required; fault storage, in which a separate memory records equipment faults during flight; engine data maintenance recording under which a maintenance memory takes samples of engine parameters and associated flight parameters, if available, during flight for later replay by ground maintenance personnel on the SDS display or as hard copy via a printer.

The SDS display comprises two lines each of 10 alphanumeric 16-segment characters with two 7-segment numeric characters. Red or amber status lamps indicate the degree of importance, and the brightness levels are adjustable to meet day or night viewing conditions. Among growth features to be offered is an emergency checklist facility that displays the appropriate actions to be taken as soon as a major fault is sensed by the system.

Discrete inputs can comprise any combination of standard applied 28-volt dc levels. Outputs comprise a logic-level drive for a master caution light with five additional logic-level outputs. Programs permit the coverage of up to 120 warnings or cautions; these are arranged in order of priority according to the aircraft manufacturer's specification. Up to 150 checklist items may be contained at current storage capacity. Features include a fault cancellation, storage and recall facility and the automatic display of checklist titles.

The system has dual redundant processing and power supplies and a channel failure indicator. Additional inputs for maintenance storage and replay include engine over-speed, over-temperature or over-limit discrete signals; flight inputs from an ARINC 575 data system; periodic sampling of N_1, N_2, exhaust gas temperature and fuel flow. Total capacity is in excess of 1050 items or 90 hours of flight recording of engine and flight parameters with no-flight fault discretes, at current recording rates.

Format: (signal data converter) ⅜ ATR Short (control/display unit) 5.75 × 2.62 × 8 inches (146 × 67 × 203 mm)
Weight: (signal data converter) 6 lb (2.72 kg) (control/display unit) 2.5 lb (1.13 kg)

STATUS: in production.

CMA-730 engine instruments

Canadian Marconi's vertical-scale solid-state engine instruments have been installed in a number of corporate and military aircraft and helicopters. These include the Dassault Falcon 10 and 20, Canadair Challenger, and Gulfstream II in the executive category, and the Canadair CF-5 and Grumman OV-1D Mohawk military aircraft. Helicopters to be equipped with the instruments include the Bell 214ST, Hughes AH-64A Apache, Sikorsky UH-60A Black Hawk, Aérospatiale Puma, and Westland Sea King (a particular example at Britain's Royal Aircraft Establishment).

The Falcon 10 instrumentation comprises five vertical-scale instruments which give read-outs of fan speed, exhaust-gas temperature, compressor speed, fuel flow, and fuel quantity. Each value is indicated by a vertical column composed of small lights which may be coloured appropriately; blue for low, amber for medium, and red for danger. A digital reading below each column can be incorporated if desired. Other functions such as engine oil temperature and pressure can be given, and helicopter applications use vertical scales for rpm and torque readings.

An additional instrument is the selectable parameter digital display, which gives digital readings of a chosen function. This may be fan speed, compressor speed, exhaust gas temperature, fuel flow, fuel used, or fuel quantity. This digital readout complements the vertical analogue display and is claimed to increase monitoring precision. It provides redundancy for the vertical scales. The function to be displayed is selected by a rotary switch, and a light indicates which function is active. Used with the sensor interface circuits which provide dc input signals, the display contains a digital voltmeter, binary-coded decimal encoder and sever-segment decoder drives. The level of brightness is chosen with a manual dim control, and this selected level sets the mean operational threshold for the automatic dimming circuit. Optional over-range warning lights record critical values that have been exceeded.

The company has built redundancy into the instrument by using parallel solid-state circuits and dual power supplies. Reliability is claimed to be better than that of mechanical instruments since there are no moving parts. Increased resolution can result from the use of complementary digital readouts, and scale expansion and contraction are available. Parallax compensation and viewing angle reading errors are avoided, and the displays are visible in bright sunlight, with a high contrast ratio between display and panel. The electronic elements consist of plug-in modules with printed circuit-boards, and no hermetic sealing is needed. Individual light-sources can be replaced within 10 minutes. Even if a segment fails, the readability remains unimpaired.

Measurement accuracy: ±0.5% full-scale deflection
Response: 1 s average
Design standard: (civil) DO-138 (military) MIL-E-5400 Class 1A
Approval: TSO

STATUS: in production.

CMA-776/CMA-887 engine performance monitor

When applied specifically as an engine para-meter recorder, the CMA-776 status display automatically records engine inputs and air data throughout all phases of flight. Automatic sampling during engine start, take-off, climb, cruise, and landing are supplemented by automatic over-limit detection and recording, and measurement and storage of spool-down times. Cruise recordings are qualified for stability within operational limits. Typical para-meters recorded include fan speeds, inlet turbine temperature, fuel flow, engine pressure ratio, vibration, rpm and torque. Additional ARINC 429 digital air data inputs include outside air temperature, altitude, airspeed and Mach number.

Data is removed on the ground using the CMA-887 Data Transfer Unit (DTU). Baseline engine performance is established and all parameters are resolved to constant power reference levels and altitudes within the DTU. Subsequent flight data is compared against baselines to produce graphic performance trend plots. Deviations from baseline perform-ance are readily detected on performance trend plots' output by connecting the DTU to a commercial keyboard printer (TI-703).

STATUS: in production.

Canadian Marconi CMA-776 engine status display system

Computing Devices

Computing Devices Company, PO Box 8508, Ottawa, Ontario K1G 3M9
TELEPHONE: (613) 596 7000
TELEX: 05-34139

AN/ASN-99 projected-map display

This projected-map system, with company designation PMS 4-1, has an interface for use with a navigation computer and comprises two units, a projected-map display PMD 2-2 and an electronics assembly unit EAU 1-2. The display unit contains 35 feet (10.7 metres) of maps in 35 mm colour-film format in a replaceable cassette, film drive system, optical elements, display screen, and controls. The assembly unit contains the electronics for processing the digital bit-serial and word-serial input com-mands from the aircraft navigation computer.

There are eight operating modes: Normal, North-up, Manual, Data, Test, Decentre, Hold, and Landing. In Normal mode, the projected-map image is orientated with aircraft track vertically up. Aircraft position is shown at the centre of the circular display or at the origin of the radar graticle depending on the Decentre button illumination. In North-up, the map is orientated with true north vertically up. On selection of Manual, the maps can be slewed in

the event of computer navigation failure. The Data mode causes key chart information to appear on the screen. When the Test position is selected a test pattern appears on the screen. The Decentre facility causes the aircraft position to move from the centre of the circle to the origin of the radar graticle, or vice versa. The Hold position stops map movement, and the film can then be slewed to a desired position. Selection of the Landing mode results in the automatic display of the nearest airfield ap-proach chart.

Map scales and coverage: 1/500 000 and 2 million, 1200 × 1200 n miles and 2500 × 1500 n miles
Capacity: 200 frames of 35 mm film
Accuracy: displayed position error 0.0276 inch cep (0.142 mile at 1/500 000 scale)
Brightness: visible in 10 000 ft-candles ambient illumination at 31° to the normal at the screen
Film slew rates: frame advance (row change) 3s max, data chart access 3s max
Reliability: mtbf 1000 h
Service life: 10 000 h or 10 years
Dimensions: (display) 7 × 6 × 15.6 inches (177 × 152 × 397 mm)
(electronics unit) 9.8 × 5.75 × 15.25 inches (249 × 146 × 387 mm)
Weight: (display) 21 lb (9.55 kg)
(electronics unit) 21.8 lb (9.91 kg)

Power: 294 VA at 115 V 400 Hz 3-phase, 140 VA at 21 V 400 Hz, 2 VA at 5 V 400 Hz, 0.2 A at 28 V dc
BIT: go/no-go check provided by test pattern. Continuous automatic checking of both units by monitoring circuits in the electronics unit, and appearance of a fail flag in the event of a fault. Fault isolation of failed unit by latching indicators on the electronics unit on selection of test mode

STATUS: in production. More than 1500 sets supplied to number of countries for tactical aircraft, helicopters, and military transports.

PMS 5-6 projected-map display

This system is similar to the PMS 4-1, but takes its signals from air-navigation sensors and ground-based aids. Scale combinations are 1/50 000 and 1/250 000, which are particularly useful in helicopter operations. The full map display is augmented by 16 data-charts. Simul-taneous map and numerical display of present position permit the crew to pass back to base accurate co-ordinates or to update computed present position.

STATUS: in production for Italian Air Force/ Aeritalia G222 transport.

Presentation of Computing Devices projected-map display and overlaid annotations

Electronics assembly for Computing Devices projected-map display

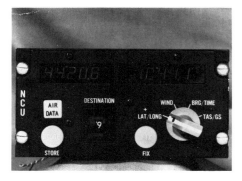

Control unit for Computing Devices projected-map display

Litton Canada

Litton Systems Canada Limited, 25 Cityview Drive, Rexdale, Ontario M9W 5A7
TELEPHONE: (416) 249 1231
TELEX: 06-989406
TWX: 610 492 2110

Led flat-panel displays

Litton Systems Canada has developed a number of flat-panel light-emitting diode (led) display systems. This developed technology has been applied to the military market with alphanumeric and vectorgraphic displays in aircraft data entry and control/display units, and is used in multi-function programmable keypads. Both monochromatic display surfaces at 4096 pixels per square inch and multi-colour presentation of 2600 pixels per square inch are available. All displays are sunlight readable and in the case of monochromatic green, compatible with second- and third-generation night-vision goggles.

STATUS: displays and multi-function keypads in production.

Litton Canada typical led flat-panel display

FRANCE

Badin-Crouzet

Badin-Crouzet, Aérodrome de Toussus-le-Noble, BP 13, 78117 Chateaufort
TELEPHONE: (1) 39 56 80 30
TELEX: 697771

Cabin pressure indicator

This is an integrated display of attitude, vertical speed and cabin differential pressure, using dial-and-pointer indications. The unit's electronics operate on inputs from solid-state sensors. Alarm signals operate if the maximum allowable values are exceeded on either the cabin altitude or the inside/outside differential pressure. The integrated unit simplifies installation on the flight-deck and improves the reliability compared with that of using three separate instruments.

STATUS: in production for the Aérospatiale/Aeritalia ATR-42.

Crouzet

Crouzet SA, 25 rue Jules-Védrines, 26027 Valence Cedex
TELEPHONE: (75) 79 85 11
TELEX: 345807
FAX: 75 55 22 50

Type 120 air data computer

The modular design Type 120 air data computer is intended for use on the new generation of transport aircraft and has already been chosen for the KC-135s of the French Air Force.

The computer comprises two identical and interchangeable absolute pressure sensors for measuring static and total pressures; a cpu board including arithmetic unit, memories and sensor measurement circuits; an input/output board for all analogue and digital interfacing and encoding and a power supply board.

A large number of built-in tests are available and permanent monitoring of all functional circuits takes place during flight, the results of which are expressed as maintenance words stored in a protected memory or despatched on the digital lines.

A type 130 series has been proposed for use on fighter aircraft with a MIL 1553B or Digibus interface.

Pressure sensors are the Crouzet Type 51 featuring vibrating quartz blades.

Dimensions: 2 MCU as per ARINC 600
Weight: 12.3 lb (5.6 kg)
Power: 115 V, 400 Hz single-phase
Consumption: 25 VA
Mtbf: greater than 10 000 h of operation

STATUS: in production.

Crouzet Type 120 air data computer

Laser airspeed measurement system

Crouzet in collaboration with several French Government research agencies has been testing a combined airspeed/ground speed measurement system based on carbon-dioxide laser radiating at 10.6 microns. Carried by an Aérospatiale SA330 Puma helicopter, the system measures the reflection of radiation from air particles to measure airspeed. The system could also be used as a ground-based or airborne wind-velocity sensor for warning of local wind-shear and for artillery shell trajectory correction.

Voice warning unit

A spin-off from the work undertaken by Crouzet in automatic speech-processing, the voice warning unit alerts the flight-crew to the onset of a failure or alarm using microprocessor based speech-synthesis techniques. Crouzet believes that, with the increasing complexity of aircraft equipment and alarm systems, it is becoming increasingly difficult for the crew to associate traditional warning horns immediately and correctly with the particular failure. The voice warning system is thus designed to reduce this reaction time by identifying immediately the nature of the failure.

This work is now resulting in a synthesised voice alarm warning system (VAWS) based on compressed speech and which will be fitted in the ACX. This has resulted in a small unit (1 litre) providing alarm warnings in plain language. It is based upon a similar system that has been used on the Dassault Mirage III NG.

A multi-function keyboard will be fitted in the cockpit and consists of a programmable dialogue facility between the pilot and the radio communication, radio navigation and IFF systems. Its use may be extended to other functions such as maintenance.

The module measures just 127 × 76 × 55 mm and communicates via Digibus. It carries a 17-character scratch pad display; a 12-key digital keyboard for data insertion and a keyboard with 9 variable label, alphanumerical or graphical keys.

STATUS: in development.

Sfena

Sfena (Société Française d'Equipements pour la Navigation Aérienne), BP 59, Aérodrome de Villacoublay, 78141 Velizy-Villacoublay Cedex
TELEPHONE: (1) 46 30 23 85
TELEX: 260046
FAX: (1) 46 32 85 96

Sfena gyro horizons

Sfena Series 705 gyro horizons have been installed on virtually all commercial transports built during the last decade, and more than 12 000 examples are in service with approxi-mately 150 airlines and 30 air forces. The main application of the instrument is as a stand-by attitude reference device.

The new Series H3XX gyro horizons are 3 ATI-sized and weigh about 1.5 kg. They make extensive use of modern alloys, have built-in static inverters and gyro-speed monitors, and Sfena-patented simplified erection and anti-spin devices. The basic versions include H301 (drum display), H321 (sphere display) and H341 (sphere display with ILS capability).

A related gyro horizon is the 4ATI-size H150 stand-by attitude indicator selected by Boeing for the 757 and 767 airliners. Other current production applications include H301 for Airbus Industrie A300-600 and A310, Socata Epsilion and Casa CN-235; H321 for Aérospatiale 365N helicopter and Embraer EMB-312; and H341 for Saab-Fairchild SF340 airliner. Recently Airbus Industrie has selected the H321 for A320 and Boeing the H341 for 737-300 EFIS configur-ation.

Attitude director indicator

Sfena produces a wide range of attitude director indicators for both civil and military use. Current military programmes include Transall C-160, Dassault-Breguet ATL 2 Atlantique, Aérospatiale Nord 262; and Aéro-spatiale SA 341 and SA 342 helicopters. Civil standard attitude direction indicators are fitted in Concorde, Airbus Industrie's A300/A310, and Fokker F27.

The current range provides a wide choice of instruments, and features flight-director com-mand bars, failure-flag indicators and warning lights for decision-height alerting. Some instru-ments include a basic yaw indicator for additional assistance in asymmetric flight.

Weight and balance and tyre-monitoring system

This capacitative-balance system measures aircraft take-off weight and computes centre of gravity to within 1 per cent. It also monitors tyre pressure to provide advance warning of poss-ible blow-out. Weight on the main gear and nosewheel is computed from measuring the degree of shear stress in the axle-beam, while tyre distress is calculated from measurement of pressure or diameter as torsion in the axle beam.

STATUS: evaluated with success on Airbus A300-B4 and A310, and proposed for A320 and Boeing 767.

Sfena main landing gear sensor installation

Sfim

Sfim (Société de Fabrication d'Instruments de Mesure), 13 avenue Marcel-Ramolfo-Garnier, 91344 Massy Cedex
TELEPHONE: (1) 69 20 88 90
TELEX: 260046

All-attitude flight control indicators

Sfim has developed a range of spherical indicators on which all-attitude information is displayed to give the pilot heading, roll and pitch information on a single dial without freedom limits around the three axes.

Some versions are fitted with command bars to display signal information from navigation aids or landing systems. Four types of instru-ments are available and equip Dassault Mirage F1 and 2000; Sepecat Jaguar and Dassault Super Etendard; French Navy/Westland Lynx helicopters; and Soko Galeb, VTI/CIAR Orao and FMA Pucará aircraft.

A helicopter version is designated 11-2 and features a manual pitch setting mode of ±10°. The various features of the series are:

Type	810	811	816
Roll	Yes	Yes	Yes
Pitch	Yes	Yes	Yes
Heading	Yes	Yes	Yes
ILS/VOR/Tacan	Yes	No	Option
To-from	Yes	No	*
Beacon	Yes	No	Option
Side-slip	Yes	Yes	Yes
Failure detection	Yes	Yes	Yes
Warning flag	Yes	Yes	Yes

* separate unit, on option

Type 810 and 811
Dimensions: 97 × 97 × 203 mm
Power: 26 or 115 V 400 Hz plus 28 V dc

Type 816
Dimensions: 81 × 81 × 208 mm
Power: 26 V 400 Hz plus 28 V dc

STATUS: in production.

ISG 80 attitude indicator

The ISG 80 is a complete flight control indicator in an 80 mm by 80 mm instrument panel indicator. The panel-mounted unit includes an all-attitude vertical gyro assembly and rep-resents roll and pitch attitude, and aircraft heading (from an external source) on a 75 mm diameter spherical display. There is also a ball side-slip indicator and provision for VOR/ILS

Sfim ISG 80 all-attitude and flight control indicator

Sfim IS 816 attitude indicator

deviation presentation. The system is fully monitored and has built-in pre-flight test facilities. A separate electronics unit contains the power supply, servo and computation circuits, together with failure detection circuits.

Dimensions: (instrument): 210 × 83 × 83 mm; (electronics unit): 144 × 85 × 77 mm
Weight: (instrument): 2.8 kg; (electronics unit): 1.2 kg
Run-up time: up to 90 s
Accuracy: 0.5°

Maximum angular rates: (heading and roll) 400°/s; (pitch) 90°/s
Power: 30 VA at 115 V 400 Hz plus 35 W at 28 V dc

STATUS: in production.

Type 550-2 high accuracy AHRS
The 550-2 attitude heading reference system, of a Sfim design of a corrective computation card,

is claimed to be the only such instrument to guarantee a vertical accuracy of 0.3° (1 sigma) for all flight manoeuvre conditions. The 550-2 two-gyro platform is the ideal low-cost solution for back up of inertial nav-attack systems or as the main gyro system for trainers.

Dimensions: (gyro unit): 6 litres, 6.5 kg; (electronic unit): 5 litres, 4.6 kg
Power: 115 V 400 Hz less than 1 A

STATUS: in production.

Thomson-CSF
Thomson-CSF, Division des Equipements Avioniques, 178 boulevard Gabriel-Peri, 92240 Malakoff Cedex
TELEPHONE: (1) 46 55 44 22
TELEX: 204780

Icare map display and Mercator remote map reader for Mirage 2000N
Thomson-CSF's involvement with head-down displays stems from the cathode ray tube imaging device designed in the 1950s as a radar display for the Vautour night-fighter. The system was the first French equipment of its kind and it led to the development of the moving-map display, produced for the French Air Force, in which navigation and weapon-delivery information were superimposed on optically-projected map images, the maps being stored on film and moved across the display area in accordance with aircraft movement.

Thomson-CSF, under a French Ministry of Defence contract has developed for the Dassault Mirage 2000N aircraft an integrated electronic map system with two full colour display units collecting data from the fully electronic Mercator remote map reader (RMR), as well as from radar and aircraft symbol generation.

In the US two Mercator RMRs are presently involved in flight trials by the US Air Force, while in France the first Icare systems will enter operational use in 1986 with the French Air Force.

A derivative of these electronic map displays is now under development for use in helicopters and is designated Helicare. Thomson-CSF is also working on complementary system configurations specially adapted to perspective displays of digital terrain modelling.

STATUS: in production.

Forward view repeater display
Thomson-CSF has developed what it terms a forward view repeater which reproduces the pilot's field of view forward for the benefit of rear-seat occupants. It was developed in response to conclusions that some shortfalls in the instruction of pilots in advanced training aircraft resulted from the instructor's inability to

scan the view directly ahead because it was obscured by the pupil's ejector seat head-rest. The system is based on a video camera that films the head-up display and outside world from the front compartment and reproduces it on a television monitor in the rear compartment.

STATUS: in production.

EFIS electronic flight instrument system for A310
In 1979 Thomson-CSF, in conjunction with West Germany's VDO Luftfahrtwerke, won the competition against Collins and Sperry to supply the electronic flight instrument colour cathode ray tube display system for the Airbus Industrie A310 airliner.

Airbus A310 flight deck

Thomson-CSF Mercator remote map reader

Airbus A310 flight deck

Mock-up of Airbus A320 flight deck

Thomson-CSF representative family of instruments

The electronic flight instrument display designed for the A310 uses six identical 6.25 by 6.25 inch (159 by 159 mm) shadow-mask colour cathode ray tubes. Each pilot has on the flight panel in front of him a pair of EFIS instruments: a primary flight display cathode ray tube which replaces the conventional electromechanical attitude director indicator, and a navigation display cathode ray tube which supplants the earlier horizontal situation indicator and weather-radar displays. Each pair of cathode ray tubes is driven by a single symbol generator, and a third such system acts as a 'hot' stand-by. Also associated with each EFIS pair is a control/display unit. Two more displays on the centre panel represent what is called the ECAM (electronic centralised aircraft monitor) display, presenting information on aircraft systems in any phase of flight and schematic diagrams of the hydraulic or electrical systems, for example, to supervise their operation. One of them is normally reserved for warnings, the other for systems. The ECAM system operates independently of the pilots' EFIS instruments, being driven by two symbol generators (one operational, the other a 'hot' spare) and a single control unit mounted on the throttle-box. All six cathode ray tubes are interchangeable reducing the number of spares required. Frame-repetition frequency is 70 Hz.

Thomson-CSF has continued to refine the system. Colour stability and visibility of the displays in high ambient lighting conditions have been demonstrated, the latter at 100 000

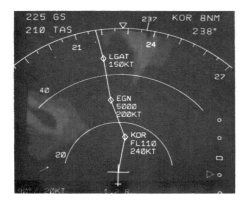

Presentation of Thomson-CSF electronic horizontal situation indicator showing planned track between bad weather and waypoints

lux instead of the 85 000 lux specified. Nine colours are used on the primary flight display, seven are employed for symbology, to avoid confusion, and two (blue and brown) represent sky and earth. A uniformly smooth sky, unmarred by the raster scan, was achieved by slightly out of focus imaging and has proved particularly acceptable to the eye. The three-dimensional effect familiar to pilots who have used electromechanical primary flight displays has been achieved by masking certain symbols when they would normally be occulted by others mounted further forward in the instruments. Thomson-CSF and Airbus Industrie have devised, therefore, a presentation that is instantly recognisable as that of a typical primary flight display, but one that can also utilise the vast quantity of newly available digital data circulating within an aircraft on the data-bus. Airbus Industrie and Thomson-CSF have added the following information to the periphery of the instrument: a moving speed scale along the left-hand edge with its standard symbols generated by computer; a selected pressure altitude readout along the right-hand edge; an autopilot and autothrottle mode annunciator along the top edge; and radio altimeter heights along the lower edge of the instrument.

For the navigation display, classic compass-card symbology has been retained, again in conjunction with the flexibility that digital sensing and processing affords. Its use is innovative in that it creates new symbology to suit each phase of flight, in the form of an electronic map display with five distance scales marking the course to be followed with radio waypoints and their identifying codes. On selection by the pilot, a three-colour weather map can be superimposed on the navigation map, resulting in the saving of the panel space that would be required by a separate weather cathode ray tube, and the more comprehensible integration of weather and navigation information. The warning and system status cathode ray tubes on the centre panel show information previously unavailable to flight crews. In addition to special alarms, failure warning is given by presenting the crew with synoptic displays of failed systems, and indications of vital actions to be followed to meet the difficulty and the effects on other systems.

The possibility of common-mode faults affecting the software in the two sets of flight-director equipment is eliminated by using

different software for each. The two sets of software were developed by two teams working independently of one another, in different localities. Software was implemented in machine-code.

STATUS: in production. In 1981 Thomson-CSF delivered sets to forward-facing crew cockpit simulator at Toulouse, to flight-control test-rig, and to first A310. Approximately 12 sets were completed by the end of 1981, 16 in 1982, and about 30 were completed in 1983. Production runs at some ten sets a month. At the end of September 1985, 17 A300-600s were being operated by four airlines and 72 A310s were being operated by 15 airlines.

EIS electronic instrument system for A320

In August 1984 Thomson-CSF was chosen to develop and build the electronic flight instrument system for the new Airbus Industrie A320 airliner. The system for the new transport is an integration of the EFIS and ECAM cathode ray tube (crt) suites installed on the earlier A310, of the crt instrument suite that is standard on the Airbus A310, though with a number of important differences.

As with the A310 there are six crt displays, but their disposition on the instrument panels are different. While on the earlier aircraft the flight director and navigation director are stacked vertically, with the two ECAM instruments side-by-side on the centre panel, on the A320 the pilots' crts are ranged side-by-side, with the ECAM displays situated one above the other. The crts themselves are larger, 7.25 by 7.25 inches (184 by 184 mm), compared with 6.25 by 6.25 inches (159 by 159 mm), so that more symbology can be displayed without congestion. Each crt can show six colours. The upper ECAM, for the first time, shows primary engine parameters, replacing the ten conven-

Artist's impression of Airbus A320

tional electromechanical dial-type instruments on the A310. The lower ECAM will normally display systems information, such as electrical or hydraulic circuits to show the location of faults. Confidence in the electronic display generated by A310 experience has resulted in the decision to delete most of the traditional electromechanical, dial-and-pointer instruments, retaining just a few as a back-up in the most important functions.

As with the A310 system, all six crt units are identical and interchangeable with one another. However, whereas the A310 employs five symbol-generators to drive them, the A320 uses only three, and with more complex functions they are called display management computers. All three are identical to one another. The crts and computers are grouped within very advanced architecture permitting extensive redundancy; in the event of a crt, sensor or computer failure the system automatically reconfigures itself, with top priority given to flight commands

and engine warning. The greater level of redundancy and self-test is at least in part responsible for the improved autoland performance – Category IIIB on the A320 compared with Category II on A310. At the same time there has been a significant weight decrease, from 153 kg to 115 kg. As with the A310 system, elimination of software errors in the pilots' flight director instruments is accomplished by the use of two independent development teams. However, unlike the earlier equipment, software is implemented in high-order language.

STATUS: in final development, with development equipment scheduled for delivery from 1985.

Digital engine control indicators

These families of instruments cover the various parameters required for engine control (N1, N2, epr, egt, ff, rpm, torque, itt etc). These instruments feature advanced design including the use of solid-state high contrast lcds, microprocessors and cmos technology to create, it is claimed, reliable, standardised and flexible instruments that may be used in a wide variety of aerospace applications.

In addition to the fundamental engine control indicators, the company product line also covers a number of other associated indicators including APU controls, hydraulics, cabin pressure and electrical power.

STATUS: in production and service.

TRT

Télécommunications Radioélectriques et Téléphoniques (TRT), Defence and Avionics Commercial Division, 88 rue Brillat-Savarin, 75640 Paris Cedex 13
TELEPHONE: (1) 45 81 11 12
TELEX: 250838

TRT AHV-8 radio altimeter

AHV-6 radio altimeter

This is claimed to be the most widely used radio altimeter of its generation and installed on more than 40 aircraft types in service with 20 military forces throughout the world. There are three sensor variants, AHV6-011, -311 and -611, all with identical basic installation characteristics. A choice of 3 ATI or smaller indicators is provided. Shockmounts for fixed-wing or helicopter applications are available. The equipment is all-solid-state and has no warm-up time. Analogue outputs conform to ARINC 552.

Typical installation
Dimensions and weight: (transmitter-receiver unit) 96 × 124 × 343 mm; 4 kg
(indicator) 3 ATI; 1.2 kg
Max altitude: 2500 or 5000 ft (10 000 and 15 000 ft limits in option)
Accuracy: 1.5 ft or 2% of altitude, above 100 ft ± 24%
Rf power: 250 mW typical (FM/continuous wave)
Power: 115 V 400 Hz

STATUS: in production.

TRT AHV-6 radio altimeter

AHV-8 radio altimeter

This lightweight unit satisfies the increasing demand for a height sensor for an automatic-landing system, helicopter stabilisation and ground-proximity warning system coupling. It conforms to ARINC 522 and a version is available for missile applications. It is an all-solid-state model which does not require warm-up time. Over nine different commercial aircraft types are fitted with the AHV-8 radio altimeter. It has been specified for the Aérospatiale/Aeritalia/ATR 42.

Typical installation
Dimensions and weight: (receiver-transmitter unit) 230 × 90 × 90 mm; 2.2 kg
(antenna) either 70 mm or 178 mm dia; 0.4 kg
(indicator) 3 ATI; 1.2 kg
Max altitude: 5000 ft
Accuracy: 0.5 ft or 2%
Rf power: 200 mW typical
Outputs: 3
Power: 25 VA at 28 V dc or 115 V 400 Hz

STATUS: in production.

AHV-9 radio altimeter

This is a high-range reconnaissance sensor with a digital output suitable for recording and

TRT AHV-9T radio altimeter is used in Panavia Tornado

digital indications. The system is solid-state and has no warm-up time.

It can be used as a primary sensor system for automatic landing systems certificated for Category II conditions.

Typical installation
Dimensions and weight: (transmitter-receiver unit) 97 × 124 × 383 mm; 5 kg
(antenna) 70 × 154 × 113 mm; 0.5 kg
(indicator) 2 ATI; 0.8 kg
Max altitude: 32 000 ft (64 000 ft option)
Accuracy: 0.5 ft or 2%
Rf power: 600 mW (FM/continuous wave)
Outputs: 4
Power: 70 VA at 115 V 400 Hz

STATUS: in production.

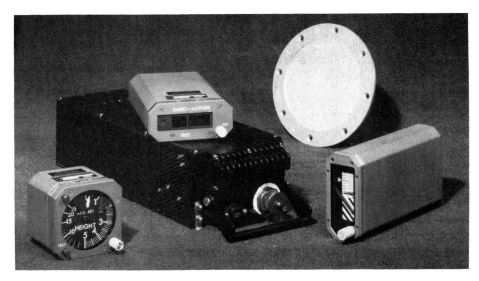

TRT AHV-9 radio altimeter

AHV-9T radio altimeter

This is a version of the AHV-9 unit designed for the Panavia Tornado. Low and high altitude bands can be used, the variable output being available from a microprocessor-based receiver unit. The system features cmos integrated electronics, stripline technology antenna, and has built-in fault detection capability.

Dimensions and weight: (transmitter-receiver unit) 109 × 154 × 324 mm; 4.5 kg
(antenna) 78 × 88 × 33 mm; 0.2 kg
(indicator) 61 × 61 × 158 mm; 0.9 kg
Max altitude: 50 000 ft
Accuracy: 1 ft ±2%
Rf power: 60 mW (FM/continuous wave)
Outputs: 5 digital, 3 analogue
Power: 70 VA at 28 V dc 400 Hz

STATUS: in production.

AHV-12 radio altimeter

Designed for use on the Dassault Mirage 2000, this system incorporates the most recent digital

TRT AHV-12 digital radio altimeter

developments. Its main features are a wide altitude range, high accuracy and high integrity level.

Typical installation
Dimensions and weight: (transmitter-receiver unit) 193 × 90 × 315 mm; 5 kg
(antenna) circular, rectangular or small size
(indicators) ARINC 429 (digital), ARINC 552 (analogue)
Altitude range: 0 to 70 000 ft
Accuracy: ± (1 ft plus ±1%)
Power: 45 VA at 115 V ac 400 Hz

STATUS: in production.

AHV-10 integrated radio altimeter

So called because it combines in one box transmitter-receiver circuits and display, this has been developed for general aviation aircraft and helicopters. TRT claims high accuracy through effective tracking of the nearest point on the ground so as to eliminate averaging errors. A narrow receiver bandwidth and continuous tracking are used to protect against electromagnetic interference. By integrating transceiver and indicator in one 3 ATI case weight is minimised and inter-unit wiring eliminated. The instrument interfaces with the ANT-133 or AHV-6-123 antenna types and conforms to ARINC 552.

Dimensions: (indicator) 3 ATI
(antenna) ANT-133 108 × 46 mm; AHV-6-123 108 × 35 mm
Weight: (indicator) 1.3 kg
(antenna) ANT-133 0.12 kg; AHV-6-123 0.12 kg
Power: 10 VA at 28 V dc

Altitude range: 0 to 2500 ft
Accuracy: 0.5 ft ± 2%
Rf power: ⩾ 50 mW

AHV-16 radio altimeter

This microprocessor-based instrument has a reprogrammable memory and is designed for business and commuter and light military aircraft. It has both digital and analogue outputs to ARINC 552 and 429 standards. It can interface with existing digital and analogue avionic equipment, and with electromechanical instruments as well as electronic flight instrument systems.

Weight: 4.84 lb (2.2 kg)

STATUS: the AHV-16 has been selected for the British Aerospace ATP.

AHV-20 radio altimeter

This is a miniature radio altimeter for helicopters, fixed-wing aircraft and missiles.

Typical installation
Dimensions and weight: (transmitter-receiver unit) 157 × 81 × 81 mm; 1.2 kg
(antenna) 178 mm dia; 0.5 kg
(indicator) 3 ATI; 0.6 kg
Max altitude: 2500 ft (5000 ft option)
Accuracy: 0.5 ft ±2%
Rf power: 50 mW
Power: 10 VA at 28 V dc

STATUS: in production.

AHV-530 radio altimeter

Developed to ARINC 707 and 600 specifications, this unit is suitable for Category II and III autoland applications in commercial aircraft. It is an all-solid-state, digital system with a microprocessor-based computer and a total of only 450 components. There is direct access to all components, and automatic fault detection and identification is provided by software. The unit is currently available for the Airbus A300-600 and A310, Boeing 757 and 767 airliners, and is used by 17 airlines. It is also specified for the Fokker 100 airliners of Swissair.

Typical installation
Dimensions and weight
(transmitter-receiver unit) 3 MCU; 4.5 kg
(antenna) horn or microstrip
(indicator) ½ 4 ATI; 1.2 kg
Accuracy: 1 ft ±2%
Rf power: 70 mW
Power: 17 VA

STATUS: in production.

TRT AHV-530 digital radio altimeter

TRT AHV-530A has all AHV-530 functions and is interchangeable with existing AHV-5 radio altimeter

TRT AHV-10 radio altimeter

AHV-530A radio altimeter

This is designed to incorporate a large proportion of basic AHV-530 radio altimeter technology and to be interchangeable with the previous generation AHV-5 unit. It had been ordered by 25 airlines by late 1983. Potential applications include MD-80s. The analogue outputs are configured to ARINC 552 requirements.

Typical installation
Dimensions and weight
(transmitter-receiver unit) 5 kg
(antenna) 178 mm dia; 0.4 kg
(indicator) 3 ATI; 0.8 kg
Accuracy: 0.5 ft ±2%
Rf power: 70 mW
Power: 28 VA

STATUS: in production.

APS-500 ground-proximity warning system

The TRT APS-500 ground-proximity warning system (GPWS) complies with ARINC 594 recommendations and is compatible with all ARINC 552/552A radio altimeters. TRT has for many years specialised in the manufacture of radio altimeters, and the APS-500 complements the company's own extensive range of these instruments, so that customers can have an integrated system provided by one manufacturer. The APS-500 system is of fully digital, solid-state design based on a microprocessor and, says TRT, may be considered as a second-generation GPWS.

Operational modes are Mode 1: excessive rates of descent; Modes 2 and 2B: excessive closure rate to terrain; Modes 3A and 3B: negative climb rate following take-off and altitude loss under 700 feet (210 metres); Mode 4: insufficient terrain clearance on approach with incorrect configuration; Mode 5: excessive glideslope deviation.

In all modes the warning takes the form of an imperative aural alarm (the characteristic 'whoop, whoop') and the words 'pull up' are repeated, with the exception of Mode 5 in which the word 'glideslope' is annunciated.

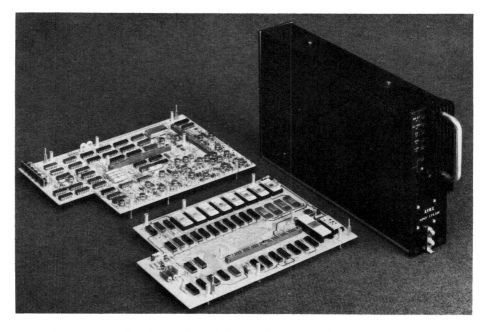

TRT APS-500 digital ground-proximity warning system, showing circuit boards

At the heart of the APS-500 is a single-chip, 16-bit microprocessor with a solid-state memory using random access memory devices for data storage and reproms for program store. Complementary metal-oxide silicon (cmos) integrated circuitry is widely employed throughout. A time-shared operating sequence allows self-monitoring at a rate of 30 times a second, irrespective of mode. The system employs 32-bit words for critical algorithms which require high accuracy; the filtering and integration routines. This factor, together with the high sampling rate, reduces the probability of spurious alerts. Decision algorithms are adapted to each warning mode requirement.

The APS-500 comprises a wired chassis with plug-in interconnecting printed circuit board, an input-output peripheral plug-in printed circuit board, a digital computer plug-in printed circuit board, a front panel with solid-state displays, and the power-supply section.

Built-in self-test facilities are provided for in-aircraft test and servicing, and automatic test equipment is available for bench repair and maintenance.

Only 50 per cent of the 16-bit memory in the APS-500 GPWS is at present employed, so there is considerable growth potential to meet any future mode-change requirements.

Format: ¼ ATR Short
Weight: 5.7 lb (2.6 kg)

STATUS: in service.

GERMANY, FEDERAL REPUBLIC

Conrac

Conrac GmbH, Postfach 60, Industriestrasse 18, D-6992 Weikersheim
TELEPHONE: 793 4675

Speed reference and stall warning system

This provides a pilot with guidance on optimum pitch attitude during take-off and approach.

The indicator displays the ratio actual airspeed/stalling speed and the percentage of total available lift being used at any instant.

The system has an externally mounted angle-of-attack sensor, the output signal of which is fed to the system's computer where it is related to flap position. The pilot is warned of impending stall by a stick shaker. The solid-state electronics consist of a power-supply module, dc amplifiers, and a number of shaping circuits.

Dimensions: (computer) 3.5 × 3.5 × 7.5 inches (89 × 89 × 190 mm)
(flap position transmitter) 2 inches dia × 1.5 inches (50 mm dia × 38 mm)
Weight: (computer) 2 lb (0.9 kg)
(flap position transmitter) 0.5 lb (0.35 kg)
(sensor) 1 lb (0.45 kg)
(indicator) 1 lb (0.45 kg)
(stick shaker) 5 lb (2.27 kg)
Power: 28 V dc

Diehl

Diehl GmbH & Co, Heinrich-Diehl-Strasse 2, D-8505 Röthenbach
TELEPHONE: (911) 59771
TELEX: 623942

Warning panels for Alpha Jet

A series of warning panels with the type numbers 3189-12, 3189-21, 3189-31, 3190-21, 3191-12 and 3191-31 have been developed by Diehl and are designated for French, Belgian and West German versions of the Dassault-Breguet/Dornier Alpha Jet. The panels provide visual and aural warning of system failures to front and rear cockpits and also contain switches for control of battery, generators and inverter.

In the event of the failure of a system which is not critical to aircraft safety, the panel illuminates an amber master warning light which is external to the main unit while simultaneously illuminating a captioned amber light (indicating which system is failed) on the warning unit's front panel. If the failed system is flight-critical, a red master-warning and a red panel caption lamp illuminate and an aural alarm is also sounded. An external switch enables the master warning lamps and the aural warning to be cancelled and the system to be reverted to standby. The captioned lamp, however, remains lit until the system fault has been rectified.

The brightness level of all warning lamps is adjustable to meet day and night ambient lighting. A test facility permits functional checks of all electronic circuitry and warning lights. The 3190-21 model provides critical failure (red) warning on six channels but has no non-critical (amber) warnings. All other models provide red warning facilities on 5 channels and amber warnings on 15 channels.

Dimensions: (3189-12, 3189-21, 3189-31) 115 × 118 × 147 mm
(3190-21) 91 × 88 × 124 mm
(3191-12, 3191-31) 115 × 118 × 125 mm
Weight: (3189-12, 3189-21, 3189-31) 1.15 kg
(3190-21) 0.55 kg
(3191-12, 3191-31) 0.85 kg

STATUS: in production and service.

Diehl warning system panels for Dassault-Breguet/Dornier Alpha Jet

Kollsman

Kollsman System-Technik GmbH, Kirschstrasse 20, D-8000 Munich 50
TELEPHONE: (89) 812 2073
TELEX: 5215170

Digital air data computer

Developed to meet the demand for better control of aircraft performance in order to offset increasing fuel costs, this general-purpose system was designed to ARINC 706-2 and is suitable for commercial aircraft and helicopters.

A range of air data inputs is processed including total and static pressures, total temperature, and two sources of angle of attack data. The computer is based on five modules and employs vibrating quartz barometric sensors.

The outputs conform to the ARINC 429-5 digital databus specification, and comprise altitude and altitude rate, true or calibrated airspeed, Mach number, static and dynamic pressures and angle of attack. In addition to supplying the primary flight instruments, the output data can also be transmitted to other avionics equipment such as autopilot, auto-

throttle speed control, and performance command and flight management systems. The principal processing element is a 16-bit microprocessor with 1 K words of random access memory and 10 K words of read-only memory.

Format: 1/2 ATR Short (4 MCW)
Weight: ≤ 11 lb (5 kg) depending on configuration
Reliability: 6000 h mtbf design
Power: 50 W at 115 V 400 Hz

STATUS: in production.

Litef

Litef, Lörracherstrasse 18, Postfach 774, Freiburg D-7800
TELEPHONE: (761) 49010
TELEX: 772669

LTR-81 attitude and heading reference system

This is a strapdown-type reference system with two degrees of freedom tuned rotor gyroscopes. It is suitable for commercial airliner applications, and is claimed to be about one-half of the cost of a laser-ring gyro-based system. Installation and interface functions meet ARINC

requirements, and equipment entered service in December 1983 aboard Alitalia-operated McDonnell Douglas MD-80 airliners. The LTR-81 has been selected as standard equipment for the Fokker 50 and 100 aircraft as well as for the British Aerospace ATP.

STATUS: in production. (See also under Litton (USA) entry, page 368).

Teldix

Teldix GmbH, Postfach 105608, Grenzhöferweg 36, D-6900 Heidelberg 1
TELEPHONE: (6221) 5120
TELEX: 461735

KG10-1 map display for helicopters

This stowable system has been developed to ease crew work-load in all-weather helicopter operations, particularly at low level. It is also suitable for low-speed fixed-wing aircraft fitted with a source of navigation data. A notable feature of the KG10-1 is that it can use unprepared standard maps, folded as convenient, and clamped between the edge frames of the display area. Navigation information to the system can be taken from a variety of sources in the aircraft, such as Doppler, and is converted into signals to energise x and y light-spot drive-motors by means of a microprocessor chip in the display unit. All electronics, including the M6800 microprocessor, are carried on a single circuit board in the display unit, adjacent to a control/display panel. Power

supplies, navigation information, and built-in test equipment signals are fed to the unit by a free multicore cable connected from the aircraft's navigation computer.

There are four operational modes: nav, fix, map, and grid. In the first, the light-spot moves under the map in accordance with information furnished by the navigation computer. In the second, the crew can correct the position defined by the light-spot by external position fixes. The third mode is chosen when a map is inserted and the system has to be aligned or 'zeroed'. In the fourth mode, the grid co-ordinates of any point on the map can be displayed, eg for reporting.

Dimensions: 365 × 250 × 45 mm
Weight: 2.4 kg
Map display area: 230 × 210 mm
Position display: light-spot
Power: 10 W at 28 V dc
Appropriate UTM charts: 1:50 000, 1:100 000, 1:250 000, 1:500 000, 1:1 000 000
Navigation computer interface: ARINC 575

STATUS: in production.

Teldix KG10 portable map display

ISRAEL

Elbit

Elbit Computers Limited, Advanced Technology Center, PO Box 5390, Haifa 31051
TELEPHONE: (4) 514 211
TELEX: 46774

Multi-function display

This is essentially a monochrome armament control and display panel for combat aircraft with limited cockpit space, showing video information from a variety of sensors on a 4 by 4 inch (102 by 102 mm) cathode ray tube. The display cathode ray tube is surrounded on the face of the panel by 28 push-buttons for the selection of different functions.

Display size and type: 4 × 4 inches (102 × 102 mm), P43 phosphor, raster scan selectable at 875 and 525 lines, 60 Hz
Brightness: 200 ft-lamberts (685.2 cd/m²) minimum
Contrast: 7 shades of grey
Contrast ratio: 7:1 minimum at 1000 ft-candles
Video bandwidth: 30 Hz – 20 MHz to 3 dB points
Dimensions: 175 × 147 × 249 mm
Weight: 6.5 kg
Power: 20 W at 28 V dc, 70 W at 115 V 400 Hz
Reliability: 1500 h mtbf per MIL-HDBK-271A

STATUS: in production for US-supplied McDonnell Douglas F-4 Phantoms and indigenous IAI Kfir fighters.

Display system for Lavi

Elbit is providing three head-down displays for Israel's IAI Lavi aircraft. Two of them are colour presentations and the third black and white. Data-sharing between them will ensure display redundancy. Control-column, throttle and display keyboard are all encoded in the display computers, which may themselves have a back-up function to the main aircraft computer. The system also includes a sophisticated head-up display.

Easily discernible in high brightness or low ambient light levels, the advanced, integrative displays will greatly simplify the pilot's workload, projecting exactly the data required at any given moment.

Elbit display computer/symbol-generator for Lavi

Elbit multi-function display

Mock-up of Lavi cockpit

ITALY

Litton

Litton Italia SpA, Via Pontina Km 27 800, 00040 Pomezia, Rome
TELEPHONE: (6) 911 921
TELEX: 610391

Lisa-X strapdown attitude and heading reference system

This is a small, lightweight system using microprocessors and advanced, dry-tuned gyroscopes. The gyros are not freely mounted in gimbals but are rigidly fixed to the case of the instrument. An integral unit incorporating both gyros and electronics, Lisa provides full dynamic analogue and/or digital outputs for pitch, roll and heading attitudes and rates, and fore and aft, lateral and vertical acceleration. The system can interface with omega, Navstar GPS, Tacan or VOR/DME. Associated with Lisa are an optional cockpit control unit and compass-adjust unit. It has been successfully tested in a SIAI-Marchetti S-211 and Aermacchi MB-326 and MB-339, is in production for the S-211, and is under evaluation for the Italian/Brazilian AM-X aircraft. The unit provides a wide range of interface possibilities, including multiple synchro outputs, high-speed multiplex data-bus and custom analogue or digital outputs. Lisa uses dry-tuned rotor gyroscopes and is

Litton Italia Lisa strapdown attitude and heading reference system

suitable for retrofit as an attitude and heading reference system or inertial guidance reference in advanced ground, marine and airborne applications.

Size: 190 × 165 × 300 mm
Weight: 17.8 lb (8 kg)
Power: 28 V dc 90 W

STATUS: in production.

Stand-by attitude and heading reference system for Tornado

Litton Italia is prime contractor for the attitude and heading reference system used in the Panavia Tornado. The equipment was designed and developed by the company and is an all-attitude, four-gimbal platform system with floated two-degrees-of-freedom gyroscopes. Microprocessor-controlled electronics provide a 14-channel interface capability. The system may operate in true-north directional gyro mode with full aircraft motion and latitude compensation, or in slaved or compass mode using a precision three-axis magnetic detector for north reference. Built-in test facilities contribute to the overall aircraft safety requirements demanded during automatic terrain-following operations.

Size: 250 × 240 × 370 mm
Weight: 40 lb (18 kg)

STATUS: in production.

Mini-Lisa strapdown attitude and heading reference system

One of the latest units in the Lisa range. It has been designed for minimum weight to ARINC 429 and MIL-STD-1553B. Omega navigation inputs are combined with Schuler-tuned inertial data to give accurate position and velocity air inputs to navigation and weapon systems.

Weight: 11 lb (5 kg)
Power: 60 W at 28 V dc

Lisa-2000 strapdown attitude and heading reference system

Using already-developed Lisa technology, this system has been designed for smaller aircraft as a single-box attitude and heading reference system. It is also applicable as a stand-by unit for larger aircraft. The system has been tested to typical military flight profiles, including aerobatics, to verify accuracy and reliability. There are three outputs in heading and two each in pitch and roll.

Weight: 18 lb (8 kg)
Power: 90 W at 28 V dc

STATUS: in production.

Litton Italia stand-by attitude and heading reference system for Panavia Tornado

Cali avionics computer

This is intended to complement the Mini-Lisa attitude and reference system (see separate entry) for application as a nav/attack management system for small aircraft. The system has a high data throughput and large memory, enabling it to perform real-time weapon delivery functions with head-up display, omega and barometric/laser ranging. The system incorporates a microprocessor and plug-in modules. An omega option extends the unit's capabilities so as to form the basis of a navigation-control computer sub-system. A projected development will cover global positioning system satellite navigation functions.

Input/output: digital and/or analogue, ARINC 429, MIL-STD-1553B, and synchro
Format: 1/2 ATR Short
Weight: 15.5 lb (7 kg)
Power: 60 W at 28 V dc

Litton Cali avionics computer

JAPAN

JAE

Japan Aviation Electronics Industry Limited, 21-6 Dogenzaka, 1-chome, Shibuya-ku, Tokyo 150
TELEPHONE: (3) 463 3111
TELEX: 2423345

JRA-100 radio altimeter

This is a lightweight design for helicopters, business and light aircraft. It displays radio height from zero to 2500 feet. A decision-height marker is selectable by means of a control knob which is also used to select the self-test function. An output is available for autopilot, ground-proximity warning and flight-director systems. The transmitter-receiver and antenna are in an integrated unit thereby obviating the need for external screened cables.

Dimensions: (transmitter-receiver-antenna unit) 320 × 120 × 70 mm
(indicator) 83 × 83 × 110 mm
Weight: (transmitter-receiver-antenna unit) 4.5 lb (2.3 kg)
(indicator) 2 lb (0.9 kg)

Frequency: 4300 MHz
Pitch/roll limits: 30° all directions
Max altitude: 2500 ft
Accuracy: (0–500 ft) 3 ft or 3%
(> 500 ft) 5%
Power: 32 W at 28 V dc

STATUS: in production.

JSN-8 strapdown laser attitude and heading reference system

This high-performance system is suitable for military and civil applications. The sensors are three Honeywell laser gyroscopes with three JAEL-manufactured JA-5 accelerometers. All the sensor equipment and associated processing is enclosed in a single unit, and there is a separate control panel. In its current form the equipment operates solely as an altitude and heading reference system, but it can be developed to include inertial navigation functions in due course, and so already has inertial navigation type interfaces.

It can operate in high angular-rate conditions, up to 400° a second roll-rate, and does not

suffer any performance degradation due to acceleration. It has a short reaction time and does not require a fluxvalve device. Reliability is quoted to be in excess of 2300 hours. In-flight alignment can be accomplished aided by such systems as Doppler, Tacan or Navstar, and magnetic heading can be provided as an output.

Dimensions: (attitude and heading reference unit) 250 × 330 × 180 mm
(control panel) 100 × 146 × 76 mm
Weight: AHRS 43.6 lb (19.8 kg)
Control panel: 1.7 lb (0.8 kg)
Input/output: ARINC 429 or MIL-STD-1553 or special (digital)
Synchro attitude signal outputs, and analogue turning-rate signal (analogue)
Valid signal and bit command output (discrete)
Alignment time: (normal) 2.5 minutes
(stored heading) 1 minute
Environmental spec: MIL-E-5400T Class 2X
Electromagnetic interference spec: MIL-STD-461A, 462
Power: 30 W at 115 V 400 Hz, 115 W at 28 V dc

STATUS: under development.

Tokyo Aircraft Instrument

Tokyo Aircraft Instrument Company Limited, 35-1, Izumi-Honcho 1-chome, Komae-shi, Tokyo 201
TELEPHONE: (3) 489 1121
TELEX: 2422246

IDS-8 horizontal situation indicator

This is a conventional course indicator with heading and course setting knobs mounted at the bottom of the bezel. Course and 'miles' indication are provided by a veeder counter and signals are provided for autopilot guidance.

Instrument failure is indicated by red flag monitors. The instrument is currently fitted on the P-2J and PS-1 anti-submarine and maritime surveillance aircraft.

STATUS: in production.

SWEDEN

Ericsson

Ericsson Radio Systems AB, Airborne Electronics Division, Torshamnsgatan, 21-23 Kista, S-163 80, Stockholm
TELEPHONE: (8) 752 1000
TELEX: 13545

EP-12 display system for Viggen

The integrated display system for the intercepter version of the Saab Viggen fighter collects, processes and displays all flight, navigation, radar and tactical data. It comprises five main units: head-up, multi-sensor and tactical displays, waveform generator and power supply. The three displays utilise cathode ray tubes with contrast enhancement tech-

niques to provide clear symbology even in bright sunlight. The multi-sensor display supports all-weather surveillance and interception, while the tactical display shows a track-oriented moving map superimposed with navigation and tactical data, updated at between 10 and 50 Hz depending on the amount of data.

An additional mini-size display, consisting of a 1 inch (25 mm) crt with a magnifying lens, can be mounted on top of the instrument panel. The magnified and collimated picture is ideal for TV and IR imagery. Also associated with EP-12 system is a cassette recorder to sample data at 4 Hz for the purpose of de-briefing and training.

STATUS: in production and service.

EP-17 display system for JAS 39

The cockpit of the Saab JAS 39 Gripen will feature advanced electronic information presentation on four display units: three head-down and one head-up. Conventional dial-and-pointer instruments will be retained only as back-ups for some of the more critical parameters. All four displays are computer-controlled, permitting the presentation to be tailored for every type of mission. It also provides considerable flexibility under emergency conditions by re-directing information between displays.

The three head-down units are the flight data display, the tactical display and the multi-sensor display. The flight data display provides the pilot with attitude, air data, engine data etc.

The integrated electronic map on the tactical display shows geographical features and obstacles, such as radio towers and masts, with the tactical situation superimposed. Map scale and display are automatically selected to suit different phases of a mission. The multi-sensor display presents a computer-processed radar picture, or optionally infra-red or thermal imagery, superimposed with target symbology.

All head-down displays are dimensionally and mechanically identical and measure about 6 × 4.7 inches (152 × 120 mm). The large

Ericsson multi-sensor display (centre) and tactical display (right) in Saab Viggen interceptor

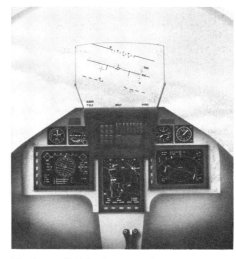

Mock-up of JAS 39 cockpit flight panels with (left to right) flight display, tactical display and multi-sensor display. The wide-field diffractive optics head-up display shows flight data and tactical information

diffractive-optics head-up display which has been developed in co-operation with Hughes Aircraft, presents a large, bright picture to the pilot.

The EP-17 has two display processors, each driving two displays and communicating with other aircraft equipment by means of MIL-STD-1553B data-buses. Each processor contains an Ericsson SDS-80 computer, the standard computing element in the aircraft. Specially developed, high performance graphics processors provide excellent dynamics in the imagery.

The head-down displays employ raster-generated symbology, while stroke-writing is used for the head-up display. Updating is accomplished at 30 Hz. Recording and replay facilities are similar to EP-12.

STATUS: in development for the Swedish Air Force/Saab JAS 39 Gripen fighter.

UNITED KINGDOM

British Aerospace

British Aerospace, Electronic Systems and Equipment Division, Downshire Way, Bracknell, Berkshire RG12 1QL
TELEPHONE: (0344) 483 222
TELEX: 848129

MAHRS 702 microflex attitude and heading reference system

This is a low-cost strapdown system designed for helicopters and light combat aircraft. The system uses British Aerospace microflex gyros as the primary sensors to measure body rates and the outputs from these are processed using a third-order integration routine to solve navigation equations. Each gyroscope is a miniature angular-rate sensor with a flexibly-mounted wheel. Small displacements of the wheel during angular movement result in countering precessional torques generated in a servo loop which keep the gyro rotor stable relative to its driving shaft. The currents measured at the torquer are a direct measurement of vehicle body rates.

The microprocessor-based computer includes strapdown routines, microflex and

British Aerospace microflex attitude and heading reference system

magnetometer calibration and navigation programs. The system accepts either omega position samples or Doppler velocities from other vehicle sensors to derive track and ground speed.

STATUS: under development.

SGP 500 twin-gyro platform

This is a three-box system comprising attitude and heading reference unit, electronic unit and a platform control unit, providing both heading and local vertical outputs. When integrated with a Doppler system, such as the Racal Type 72 and a TANS navigation computer, second-order levelling is used to ensure vertical accuracy during turns or high-*g* manoeuvres. The reference unit has twin interchangeable gyro assemblies utilising the Rotorace principle to minimise friction. The control unit is designed to ARINC 306, while the electronics unit contains the functions necessary to control the platform and the heading computer. Corrections are applied for earth's rotation and meridian convergence.

Dimensions: 250 × 193 × 124 mm
Weight: 6 kg
Outputs: (pitch) to 0.2° accuracy (2σ)
(roll) to 0.2° accuracy (2σ)
(heading) to 0.5° accuracy (2σ)
(accelerations) to 0.01 *g*
(angular rates) to 0.2°/s (up to 400°/s)
Output format: ARINC 429, synchro or 1553B
Environment: BS3G100

RAI 4 remote attitude indicator

This indicator works on a synchro output from a vertical gyro. The indication is essentially a sphere which moves relative to a fixed aircraft symbol (pitch) and to a peripheral bank scale (roll). The sphere has 360° of freedom in roll and ±100° in pitch.

Format: 4 ATI × 8.3 inches (210 mm)
Weight: 4 lb (1.82 kg)

British Aerospace CGI-3 indicator

Power: 11 VA at 115 V 400 Hz and 28 V dc plus 5 V ac/dc (lighting)

STATUS: in service.

CGI-3 compass gyro indicator

The CGI-3 is a panel-mounted instrument containing all the elements (apart from the flux valve and a small annunciator) of a gyro magnetic compass and an RMI. Although it is ideal for installation in helicopters and executive aircraft, the CGI-3 has been fully type tested and cleared for operation in a severe military environment.

The gyro uses a wheel and gimbal assembly and the readout of heading is by means of a rotating dial, with 5° graduations, which moves in the conventional sense with changes in aircraft heading.

The transistorised slaving amplifier contained within the instrument case provides the necessary energising voltage for a British Aerospace magnetic detector unit (flux valve) and then amplifies and discriminates the signals received

Indicator panel of British Aerospace's Views system

back from this unit in order to maintain the indicated heading in accordance with the direction of the earth's magnetic field.

Dimensions: 3 × 3 inches (76 × 76 mm)
Weight: 5.5 lb (2.5 kg)

VIEWS vibration indicator engine warning system

Vibration signals emanating from rotor imbalance, bearing wear, foreign object damage, etc, are detected by piezoelectric accelerometers. The level of engine vibration is continuously monitored and the measured signals are displayed on a panel-mounted indicator. They are also simultaneously compared with preset levels which have been determined as being related to an engine fault condition and if this preset level is exceeded a warning lamp on the indicator unit signals the presence of an unsafe or potentially dangerous engine condition and a discrete output feeds the aircraft central warning system. Each indicator is capable of displaying two input channels simultaneously.

Cargo Aids

Cargo Aids Limited, Brett Drive, Bexhill-on-Sea, East Sussex TN40 2JP
TELEPHONE: (0424) 216 611
TELEX: 95285

Underslung load measurement system

The transport of loads slung from hooks under the aircraft rather than internally stowed is convenient and rapid and has long been an established practice. From a military point of view its great advantage is that loads can be attached and released quickly, exposing the helicopter to hostile action for the minimum time. However it is not always practicable to weigh such loads beforehand, and some means of protecting the aircraft is therefore necessary to prevent structural damage, particularly during flight manoeuvres.

The underslung load measurement system designed by Cargo Aids for all marks of Aérospatiale Puma, Westland Wessex and Westland Sea King helicopters currently operated by the Royal Air Force and Royal Navy can be used by any aircraft with load-weighing requirements of up to 10 000 lb (4536 kg). The system comprises a load-measuring device and a cockpit-mounted indicator. The former fits on to the standard cargo hook suspended from cables attached to the aircraft. The load itself is measured by strain-bolts, transducers that transmit a digital signal to the indicator. The latter gives a readout on large green light-emitting diode displays for clarity and compatibility with night goggles.

STATUS: in production.

EEL

EEL Limited, Kings Building, Castle Street, East Cowes, Isle of Wight PO32 6RH
TELEPHONE: (0983) 291 515
TELEX: 869180

EEL airborne FM telemetry torquemeter

This system provides a direct, accurate and reliable measurement of the torque in a helicopter's transmission system. The indicated torque is independent of any correction factors for air temperature, torquemeter shaft temperature, engine and gearbox life and efficiency, fuel characteristics or aircraft operating conditions.

The system consists of four component parts: rotating assembly; fixed antenna and head amplifier; signal processor and indicators. It operates without physical contact between the rotating shaft and the stationary structure and is said to have an overall accuracy of within ±1% of full-scale torque. BITE is provided within the torquemeter equipment to enable system integrity and calibration accuracy to be checked and faults to be located.

Power: 28 V dc 1.5 A max

Ferranti

Ferranti Instrumentation Limited, Aircraft Equipment, Lily Hill House, Lily Hill Road, Bracknell, Berkshire RG12 2SJ
TELEPHONE: (0344) 3232
TELEX: 848117

Ferranti Instrumentation manufactures a range of artificial horizons for use as main attitude indicators or as stand-by instruments. In the latter mode most of them are driven directly from the aircraft dc supply via an integral static inverter. Some Ferranti instruments also feature pitch and roll signal pick-offs.

The FH30/32 series is becoming a standard fit on Royal Air Force aircraft such as the British

Aerospace Jet Provost, Hawk, Panavia Tornado and Lockheed C-130 Hercules, and Boeing-Vertol Chinook HC.1 helicopters. A retrofit programme is underway to replace the existing Ferranti instrument used in the Jaguar with an FH31F. Ferranti's horizons have also been supplied to Finland for the Valmet L70, and to Australia and New Zealand for the Aerospace CT4.

Ferranti FH31F is fitted in Royal Air Force/ Sepecat Jaguar strike aircraft

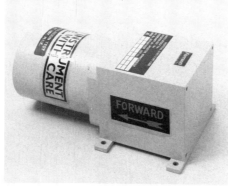

Ferranti FS 60 vertical gyro

A larger instrument (case size 4 inches, 102 mm) from the FH42 series has been adopted as the main instrument in the British Aerospace Sea Harrier and for a retrofit programme for the Royal New Zealand Air Force Strikemasters. A similar instrument has been selected for the new Australian Consortium basic trainer.

The company also manufactures a range of 2-inch (51 mm) instruments, known as the FH20 series. These ac driven instruments are fitted to all McDonnell Douglas F-4 Phantoms and Tornado GR.1 strike aircraft. A new dc driven range is now available. The type FH22 is the basic version and the type FH23 is available with pitch and roll potentiometers.

Type FS 60 vertical gyro

This is an electrically driven, air-erected, two-gimbal, gyroscope unit which measures and transmits aircraft attitude by means of potentiometers on the pitch and roll axes. An integral inverter generates an ac supply from an external low voltage dc source. Components from the earlier FH 22 and FH 31 series of instruments are used in the FS 60 to give a potential high reliability. Ferranti's patent 'pulse torque' erection system enables the erection rate to be reduced for minimum 'in flight' errors.

Dimensions: 140 × 64 × 62 mm
Weight: 2 lb (0.9 kg)
Environment specification: BS 3G100
Gyro drift: 0.5°/minute max
Erection rate: 2°/minute nominal
Power: 28 V dc

Ferranti Defence Systems Limited, Display Systems Department, Silverknowes, Edinburgh EH4 4AD
TELEPHONE: (031) 332 2411
TELEX: 727101

Comed 2036 combined map and electronic display

The increasing effectiveness of surface-to-air guns and missiles obliges strike and interdiction aircraft to operate at the lowest possible height above ground, making use of natural features such as valleys to shadow them from detection radars. But by so doing the pilot's 'look ahead' range is severely diminished and turning points, fix-points, and targets may be obscured by the same topography that affords protection. Head-up guidance provides a good method of giving aircraft-management information, but does not help with navigation. A topographical map is required for this purpose, and in recent years this has taken the form of a projected moving map display, driven by the navigation

system. The aircraft symbol is at the centre of the display and the map is moved across the display area at a speed proportional to the map-scale and aircraft speed.

The maps used with current displays have a major disadvantage in that it is not possible to annotate them with planned route, target information and other tactical data; this has to be marked up on a hand-held map as an adjunct to the moving map. Ferranti, however, has developed the Comed (combined map and electronic display) system, in which the map is electronically annotated with intelligence or navigation data appropriate to the particular mission. Using no more space than a conventional moving map, Comed provides a colour topographical map display annotated with dynamic navigation information as required; this can include aircraft track and commanded track, present position, en route fix-points, target position, locations of known hostile detection or anti-aircraft devices, and tactical information such as the delineation of forward edge of the battle area. In addition, the system can print out alphanumeric information such as time-to-go to fix-point or target.

Comed's cathode ray tube and projection facilities also permit it to perform other operational tasks. These can include the display of high-resolution, high-contrast symbology in raster form from radar, low-light television or forward-looking infra-red sensors. Electronics countermeasures threats can also be shown up. Tabular displays of weapons status, destination co-ordinates, or other tactical information can be shown on tabular 'forms' projected from images stored on the film. The system can be programmed with a library of aircraft and engine check-lists. Comed can also act as a back-up primary flight information display, particularly for the horizontal situation and attitude director command indicators.

Comed interfaces with the main aircraft navigation computer via a MIL-STD-1553 serial digital data link. From a knowledge of film-strip layout, aircraft present position and demanded scale, the main computer calculates the appro-

priate map-drive words, and transmits them to Comed by the data link. Within the display the information is converted into a form suitable for driving the map servoresolvers.

Comed uses standard 35 mm colour film up to 57 feet (17 metres) long. Typically this provides coverage of an area 10 000 square nautical miles at a scale of 1/250 000, plus selected target areas at 1/50 000 and sufficient film frames for tabulated displays. Film replacement can be done through a side access-panel in under a minute.

The system has been ordered by the Indian Air Force for its Sepecat Jaguars, made under licence by Hindustan Aeronautics. The first production Comed system was delivered during early 1983. Over 500 systems have been ordered for the McDonnell Douglas F/A-18 Hornets being supplied to the US Navy and the armed forces of Australia, Canada and Spain. They are delivered to Bendix Flight Systems Division as part of the horizontal situation indicator for that aircraft.

Dimensions: (cathode ray tube face) 5.5 × 5.5 inches (139.7 × 139.7 mm)
(unit) 7.1 × 8.58 × 23.78 inches (180.3 × 217.9 × 604 mm)

STATUS: in production. In May 1986 Ferranti announced that Comed had been selected for the Tornado ECR version ordered by the West German Air Force. Deliveries of some 40 systems, plus spares, will begin in 1988.

Digital map display

Ferranti is currently developing a full-colour three-dimensional digital map, with several aspects which will provide a hitherto unprecedented level of flexibility. Features on the presentation can be selectively displayed or erased. Hazards such as terrain and obstacles above aircraft altitude can be made to stand out in contrasting colour. Safe areas occasioned by terrain masking can be depicted, and areas may be viewed from different angles and altitudes. Data updating will be rapid and simple. The map will be displayed on a Ferranti colour electronic display (see above) which incorporates a multifunction keyboard and electronically generated symbology.

STATUS: in development.

Romag remote map generator

In August 1982 Ferranti announced a new and advanced remote map generator, called Romag, for aircraft display systems. The equipment generates map symbology for cockpit presentation, but can be remotely located in the aircraft, so alleviating space problems in combat aircraft. It is based on the well-proved technique of storing maps in film-strip form, using the film and film-traction unit currently in production for the Panavia Tornado, McDonnell Douglas F-18 Hornet and Indian Air Force Jaguar.

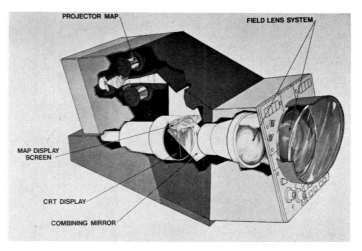

Optical arrangement of Ferranti Comed display

Film is selected and positioned by high-speed servos driven by present-position signals from the aircraft's navigation system, using a MIL-STD-1553B digital data-bus. Maps at several scales may be stored on the film and selected when required. The frame is scanned by a colour or black and white television camera and the picture transmitted to a cockpit display. The system can drive at least two displays, so that the rear occupant of a two-seat fighter could have his own display with no other weight and space penalty apart from an extra display unit. Another possibility is that Romag could present a negative image for night use. This form of presentation would allow the brightness of the display to be adjusted as desired.

Early in 1986 it was announced that for the first time in the UK, an avionics shadow mask crt colour display unit had been flown in a military fighter – a Hunter. Using a Ferranti CED 2054 display in combination with the Romag remote map generator, the trials were conducted by the Royal Aircraft Establishment as part of the Nightbird programme to evaluate electro-optic sensors.

Format: ¾ ATR Short
Weight: 26 lb (12 kg)
Outputs: 525-line 60 frames/s RS 170 standard, or 625-line 50 frames/s CCIR, 4:3 aspect ratio. Red, green, blue synchronised on green or monochrome

STATUS: in production.

Ferranti Romag produced display and installation of CED 2054 in RAE Hunter

Programmable display generator

Designed for use with the new military and commercial cockpit and flight-deck displays, this 7 kg ½ ATR unit can be remotely located from the system being driven, making for ease of installation. Based on microprocessor and matrix-storage techniques, the unit provides outputs for three independent colour or shaded monochrome raster-scan formats, which can be combined, superimposed or mixed to form composite displays. A fourth output channel provides calligraphic (cursive) symbology.

The symbology is generated by appropriate software, permitting simple modifications and allowing ready adaptation to particular requirements. A comprehensive library of symbols has been developed, offering easy and clear interpretation of data; complex shapes can be colour or monochrome shaded. The generator is designed to operate in conjunction with the MIL-STD-1553B data-bus, but can be adapted for any other data transmission system.

Format: ½ ATR Short
Weight: 15.4 lb (7 kg)
Power: 115 V 400 Hz or 28 V dc, 120 W
Raster standard: 625-line, CCIR
Raster resolution: 768 × 576
Cursive resolution: 1024 × 1024
Interface: MIL-STD-1553B dual-redundant

STATUS: under development.

Type 2060 monochrome electronic display

The Type 2060 monochrome electronic displays are small high-brightness raster-presentation devices designed for use where space is at a premium. Units are currently available with screen diagonals ranging from 105 to 280 mm. Different aspect ratios are selectable. 115-volt 400 Hz or 28-volt dc options are available as are 525- or 625-line variants. The displays can be supplied with or without a passive contrast enhancement filter matched to the cathode ray tube phosphor. The filter can be either bonded to the cathode ray tube or mounted away from the cathode ray tube face to provide optimum visibility in specific conditions. All units are compatible with Ferranti special automatic test-set facilities, but can be maintained and serviced using only standard laboratory equipment.

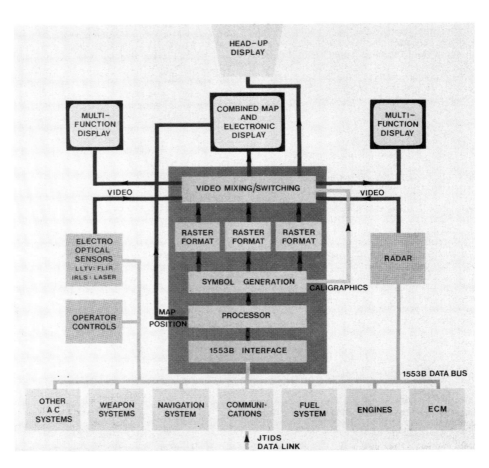

Interface between Ferranti programmable display generator and other aircraft systems

Currently the company is marketing the MED 2065 which measures just 250 mm deep with a usable flat screen area of 127 × 127 mm. As well as head-down applications for radar or sensor imagery, the MED 2065 will be offered as a rear seat monitor for head-up display and gunsight video cameras.

Video input is to CCIR Standard 625 lines at 50Hz with 525 lines at 60 Hz as an option.

Type 2051 colour electronic displays

The Type 2051 range of military standard colour electronic displays has evolved from proven electronic hardware modules used in Ferranti's Comed display. A feature of these modules is their small size. A major part of the development programme involved the design of a special anti-vibration mounting system for the shadow-mask cathode ray tubes used in military aircraft.

All Type 2051 devices incorporate high-efficiency passive contrast enhancement filters, in addition to active adaptive electronic controls to ensure adequate display visibility under all conditions. They operate in modes which include multi-stand raster, variable-speed stroke, and hybrid (combinations of raster and stroke). They are designed to operate with or without state-of-the-art push-or-press-button multifunction front panels designed by Ferranti to user specification.

Colour displays are currently being produced for several evaluation programmes. CED 2054, a high brightness unit with 0.3 mm triad dot pitch has successfully undergone evaluation flight trials in a Hunter aircraft. A second phase of trials with a 0.2 mm shadow-mask is scheduled to begin during early 1986. Useful screen area aspect ratios of 1:1 and 4:3 with various diagonals are under development. This parameter determines width and height of the line-replaceable unit. Line-replaceable units are typically 25 mm wider and 25 mm higher than the useful screen dimensions.

Currently, CCIR white raster brightness levels measured directly at the cathode ray tube faceplate are greater than 856 cd/m² (250 footlamberts). Cathode ray tubes now being developed for Ferranti displays will be capable of at least twice this brightness.

GEC Avionics

GEC Avionics Limited, Airport Works, Rochester, Kent, ME1 2XX
TELEPHONE: (0634) 44400
TELEX: 96333/4

AHRS attitude and heading reference system

This microprocessor-based strapdown guidance and navigation system incorporates three high performance GI-G6 rate integrating gyros and three linear accelerometers for inertial measurement in three axes. The system has been specifically developed for Spearfish, the Royal Navy's new heavyweight torpedo, but it is equally suitable for many other land, sea and air applications.

By selecting the appropriate gyroscope from the GI-G6 range the AHRS can be configured for optimum performance in any given application.

The self-contained system replaces a number of discrete units such as directional and attitude gyros, stabilisation gyros and accelerometers, weapon-aiming gyros, and accelerometers and inertial navigation platforms. It provides digital outputs of attitude and heading, and three-axis rate and acceleration.

Dimensions: 11.6 × 6.5 × 7.4 inches (295 × 165 × 188 mm)
Weight: 15 lb (6.8 kg)
Power: 115 V 400 Hz or 28 V dc

STATUS: in production.

Digital air data computers

GEC Avionics produces a series of digital air data computers of modular design based on microprocessor technology. Plug-in modules permit any output interface combination for compatibility with other equipment, while the chassis design simplifies servicing and repairs. Maintainability is aided by built-in test and the absence of lifed components. Rectification usually consists of diagnosis and module replacement without need of adjustments.

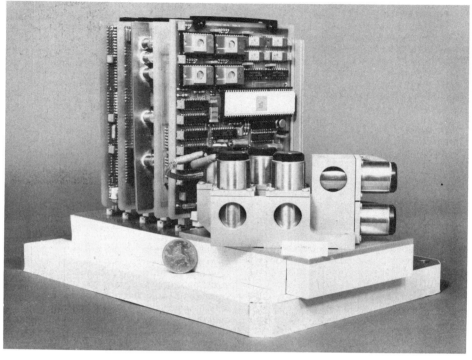

GEC Avionics attitude and heading reference system

IS 03-004 digital air data computer

This is a specially developed helicopter air data system for the updated US Army/Bell AH-1S Cobra light attack helicopter. It uses a swivelling pitot-static probe enabling speed to be measured in all axes from zero to maximum in any direction. The sensor information is processed by a 16-bit micro-computer and then transmitted to an analogue-presentation airspeed indicator and other systems in serial digital format. Manually initiated and automatic built-in test equipment is incorporated.

The outputs of the system are longitudinal indicated airspeed and true airspeed, lateral indicated airspeed and true airspeed, vertical true airspeed, downwash true airspeed, pressure altitude, static air temperature, and system-status discretes. The system is also used on the Agusta A-129.

Airspeed and direction sensor
Dimensions: 3.8 × 12.5 × 9.7 inches (97 × 317 × 246 mm)
Weight: 2.4 lb (1.1 kg)

Electronics processor unit
Dimensions: 4.2 × 8.2 × 7.5 inches (107 × 208 × 190 mm)
Weight: 7.4 lb (3.4 kg)

Low-airspeed indicator
Dimensions: 3.25 × 6 × 3.25 inches (83 × 152 × 83 mm)
Weight: 11.4 lb (5.2 kg)
Power: 34 W at 28 V dc and 140 VA at 115 V for de-icing

STATUS: in production.

IS 31-019-01 performance indicating system for helicopters

This equipment carries out performance and flight management computations for helicopters. A control and display unit provides the pilot interface, and a unique swivelling pitot-static sensor enables the system to supply data throughout the entire helicopter flight envelope. An omni-directional low airspeed indicator and vertical-strip torquemeter are also available.

The system interfaces extensively with other sensors and can provide measurements of peak-performance index, engine torque, power and payload margin, all-up weight, optimum cruise and climb speeds, VNE, take-off performance, icing detection and de-icing system control, engine-out performance, range and endurance. The system can also output three-axis airspeed signals down to hover and

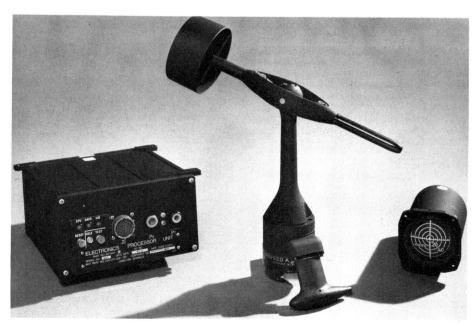

GEC Avionics digital air data system for Bell AH-1S helicopter

provide accurate height rate in addition to normal air data parameters.

A non-volatile memory can store information for post-flight retrieval and record engine-fatigue criteria. The system utilises a 16-bit microprocessor and interface options include analogue, ARINC 429 and MIL-STD-1553B. Automatic, manual or external electronically-initiated bit are included.

Airspeed and direction sensor
Dimensions: 97 × 317 × 246 mm
Weight: 1.1 kg
Power: 140 VA at 115V 400 Hz (de-icing)

Control and display unit
Dimensions: 124 × 153 × 140 mm
Weight: 1 kg

Torque meter
Dimensions: 76 × 152 × 165 mm
Weight: 0.7 kg

Processor unit
Format: 1/2 ATR Short
Weight: 6 kg
Power: 48 W

STATUS: under development.

ADC 81-01-08/32 analogue air data computer

This twin-capsule analogue air data computer meets full UK military standards and is currently supplied for all versions of the Sepecat Jaguar.

Format: 1/2 ATR
Weight: 18 lb (8.2 kg)
Power: 50 W

STATUS: in production.

DADC 50-048-01 digital air data computer

This digital air data computer interfaces with a head-up display/weapon-aiming system and uses vibrating cylinder transducers, a 16-bit microprocessor and has both digital and analogue outputs of air data parameters. Manually initiated and automatic built-in test equipment are incorporated.

Apart from the motherboard there are only four modules. The chassis is of a non-standard direct-mounting design.

Dimensions: 8.5 × 7 × 7.5 inches (216 × 178 × 190 mm)
Weight: 8 lb (3.6 kg)
Power: 35 W

STATUS: in production.

DADC 50-074-01 air data computer

This fully independent twin-channelled digital unit incorporates four solid-state transducers, twin power supplies, separate processing and a variety of output interfaces including ARINC 429 and MIL-STD-1553B. In addition to auto-matic built-in test, fault diagnosis and initiated bit can be achieved manually or through the MIL-STD-1553B data-bus with 95 per cent confidence. The chassis uses a system of thermal ladders and cold-wall cooling tech-niques, and includes a time totaliser.

Format: 1/2 ATR Short

STATUS: under development.

SADC 50-027-08 air data computer

The 50-027-08 triplex transducer unit is a secondary air data computer which has been specially designed for all versions of the Panavia Tornado. It is a three-channel device which uses six transducers (three pitot, three static) to produce analogue and discrete outputs for the aircraft systems. All computing is analogue, outputs are both triplex and duplex and the system incorporates automatic built-in test equipment. It meets full US military specifications.

Format: 3/8 ATR
Weight: 14 lb (6.4 kg)
Power: 50 W

STATUS: in production.

Standard central air data computers

At the end of 1981 GEC Avionics was contracted to develop a range of Standard Central Air Data Computers (SCADCs) to upgrade US Air Force and Navy aircraft. The intention of the pro-gramme is to update to a common standard the great variety of air data systems currently installed in some 30 variants of ten different types in US military service. These are McDonnell Douglas F-4 Phantom and A-4 Skyhawk; Grumman E-2 Hawkeye, C-2 Grey-hound and A-6 Intruder; General Dynamics F-111; Vought A-7 Corsair; Boeing KC-135 Stratotanker; and Lockheed C-141 StarLifter, C-5 Galaxy, and S-3 Viking.

To satisfy these requirements SCADC has been developed in five configurations, each having more than 80 per cent core commonality with the others. One or two special-to-type modules are included in each configuration to accommodate unique aircraft interfaces. The high commonality of designs is expected to produce significant life-cycle cost and mtbr (mean time between removals) benefits. The equipment is designed to the latest software, computing and data-transmission specifi-cations used in US military aircraft. Over 95 per cent failure detection and isolation is predicted for the built-in test systems, and ground-crew initiated tests will enable 98 per cent of all line-replaceable unit failures to be detected and isolated. Following completion of all devel-opment flight and ground tests, SCADC is now in quantity production with over 1100 units ordered by June 1985.

STATUS: in production.

The five configurations are as follows:

SCADC unit CPU 140/A This flies on 15 US Air Force and US Navy aircraft types, is designed to MIL-STD-1553B and includes an altitude en-coder and altimeter power switching.

Dimensions: 7.5 × 7.5 × 9.5 inches (191 × 191 × 242 mm)

Weight: 21 lb (9.4 kg)
Power: 61 W

SCADC unit CPU 141/A Fitted on the US Air Force/Lockheed C-5A, C-141A, C-141B military transports, this system is designed to MIL-STD-1553B and includes an altitude encoder and altimeter power switching.

Dimensions: 7.5 × 5.0 × 19.5 inches (191 × 127 × 496 mm)
Weight: 27 lb (12 kg)
Power: 79 W

SCADC unit CPU 142/A This equips six versions of the US Air Force/General Dynamics F-111, and is designed to MIL-STD-1553B with altitude encoder.

Dimensions: 7.5 × 14.0 × 19.0 inches (191 × 356 × 483 mm)
Weight: 31 lb (14.3 kg)
Power: 64 W

SCADC unit CPU 143/A This computer has been designed to MIL-STD-1553B for eight ver-sions of the McDonnell Douglas F-4 Phantom. An ARINC 429 serial digital interface is an alternative to a MIL-STD-1553B interface.

Dimensions: 7.5 × 17.0 × 12.0 inches (191 × 432 × 305 mm)
Weight: 32 lb (14.3 kg)
Power: 47 W

SCADC unit CPU 152/A This two-channel unit is fitted on all variants of the US Navy S-3 aircraft and includes dual MIL-STD-1553B navigation system and cockpit instrument interfaces.

Dimensions: 8.75 × 5.6 × 17.4 inches (222 × 142 × 442 mm)
Weight: 33 lb (15 kg)
Power: 76 W

Head-down displays

With the background of such multi-national projects as Panavia Tornado and British Aero-space Nimrod, GEC Avionics has developed a range of multi-sensor, multi-function television tabulator, E-scope and other displays. The systems comprise a cathode ray tube driven by a waveform generator embodying all raster scan techniques. The waveform generator produces synthetic symbols by what the company calls a time-shared, digital technique to give high accuracy and resolution of

GEC Avionics standard central air data computer

GEC Avionics head-down display

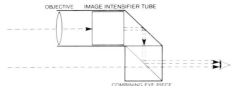

Optical path employed in Cats Eyes

modulated video signals with a minimum of components; a notable feature is the elimination of the 'staircase' effect in raster-generated graphics. The synthetic video is directly compatible with standard television signals from a variety of sensors, so that video signals can be mixed to produce an overlay of data and symbology on a pictorial display.

The waveform generator is completely digital in operation and receives data in standard serial digital form. By storing this information it can synthesise continuously video signals to drive one or more displays. The signal range can be modified or extended to meet individual requirements, particularly those suited to all-raster applications. The display unit provides a high-contrast television picture of sensor and computed data and has a multi-function keyboard permitting the operator to communicate with the aircraft computer. A filter on the face of the cathode ray tube improves the contrast under high ambient illumination and photosensors adjust the brightness of the display in accordance with the ambient lighting.

STATUS: in production and service. Other applications, as well as Tornado and Nimrod MR, include gunship version of Lockheed C-130 Hercules, British Aerospace Hunter, and aircraft such as British Aerospace Buccaneer, which use Anglo-French Martel television-guided air-to-surface missiles.

Reversionary flight/engine instrument display

GEC Avionics has produced a set of complementary flight and engine instruments display based on liquid-crystal display and fibre-optic technology that can function after a major failure of power or standard instruments. The engine instruments display shows analogue/digital engine readings of fuel quantity, engine speed and turbine gas temperature, while the flight instrument display provides speed and altitude, and pitch and roll attitude.

Information to the liquid-crystal displays comes by way of a high-integrity fibre-optic data link, so that they are independent of all other instruments or displays. The system is currently beginning tests at British Aerospace Brough Division's Avionic Systems Demonstrator Rig. It comprises two units, each 4 inches high by 3.25 inches wide and 9.5 inches deep (102 by 83 by 241 mm). Each contains a reprogrammable memory so that display sym-

bology can be tailored to particular requirements. The units are made by the company's Powerplant Systems Division at Rochester.

STATUS: in assessment.

Digital moving-map display

Recent advances in bulk-data storage techniques, combined with developments in electronic colour head-down displays, are being exploited by GEC Avionics to produce synthetic map presentation, that can be shown on a shadow-mask, raster-scan or flat-screen cathode ray tube.

The map data uses innovative digitising techniques to transform existing paper maps into a digital display format that can be fed into a standard colour head-down display. The display can then be overlaid with navigation instruments, route details, mission briefing notes or other written information. The system is contained in a ¾ ATR Short line-replacement unit mounted away from the densely equipped cockpit area.

The map can be scaled exactly to coincide with a radar ground-map, and can be rotated, scrolled, and have its scale changed. it can be re-programmed in the field to accommodate mission planning and updating.

STATUS: in development.

Tyre temperature measurement

Using a radiometer mounted on an airliner's undercarriage bogie to sense the temperature within the carcass of the tyre, the GEC Avionics system helps reduce the risks associated with tyre over-heating. The cockpit display gives a direct readout of this temperature in a format to suite the particular application.

STATUS: in development.

Graviner

Graviner Limited, Poyle Road, Colnbrook, Berkshire SL3 0HB
TELEPHONE: (02812) 3246

Digital fire detection system

Graviner introduced what was claimed to be the world's first digital fire detection system for aircraft engine bays at the 1984 Farnborough Air Show. While still relying on the Graviner Firewire for the sensing of temperature, the use of a single integrated circuit for the analysis of data reduces system weight and increases reliability. The integrated circuit is programmed with a number of differing waveforms against

which the received data from the various sensors is continually referenced to determine normal operation, overheat or fire. Spurious warnings are claimed to be almost totally eliminated by the application of digital technology.

STATUS: in development.

Smoke detector system

The Graviner smoke detector uses a light scatter principle to detect smoke in airliner toilets and galleys, providing a suitable warning at either the cabin crew station or on the flight-deck. It is designed to comply with the RTCA DO160A environmental standards. The detector

is a self-contained unit within which a light-emitting diode projects a hollow beam of light onto a light absorbing surface. A detecting photo cell is positioned on this material, at the centre of the hollow beam, so that in normal circumstances no light falls on it. The presence of smoke in the region causes scattering of the light beam, so that radiation now falls on the photocell. At an obscuration level of 10 per cent the warning is automatically triggered. To optimise reliability and minimise false warnings, the light and the photocells are pulsed in unison and smoke must be detected for ten consecutive pulses for the alarm to be triggered.

STATUS: in production.

MEL

MEL, Manor Royal, Crawley, West Sussex RH10 2PZ
TELEPHONE: 0293 28787
TELEX: 87267

Mission display units

MEL has been selected to equip the EH-101 helicopter with pilot's mission display units.

The company has delivered some 5000 similar cathode ray tube displays for military applications in recent years and is currently developing a new range of flat, thin crt displays based on channel electron multiplier tech-

nology. This is said to offer the advantages of low power demands, low volume, high reliability and ruggedness over rival flat-panel technologies, while retaining the necessary brightness and performance characteristics of conventional crt-based systems.

STATUS: in development.

Negretti Aviation

Negretti Aviation Ltd, 73-77 Lansdowne Road, Croydon CR9 2HP
TELEPHONE: (01) 688 3426
TELEX: 267708

Engine monitoring systems

This company builds third-generation engine monitoring systems suitable for all gas-turbine powered aircraft and helicopters, consisting of an airframe-mounted airborne computer unit (ACU), a data retrieval unit (DRU) which can

service several computers, and an optional cockpit display unit (CDU).

The **105 Series** ACU is primarily designed for single · or twin-engined, fixed-wing aircraft, while the larger **415 Series** ACU facilitates more comprehensive applications in multi-engined aircraft.

Both systems are self-contained and micro-processor-based, storing the solutions to complex algorithms, computed in a real-time environment, in high density, non-volatile memory. Hardware and software are both highly modularised, enabling rapid response to new applications, and permitting a wide variety

of interface requirements, whether analogue, digital or data-bus. The combination of processing and discrimination of relevant data enables the operator to identify and analyse significant occurrences immediately while accurately logging routine performance levels to facilitate trend analysis and on-condition maintenance procedures.

The DRU is used to down-load, edit or clear the stored data from one or more aircraft at convenient intervals, independent of aircraft power, and allows on-the-spot analysis, using the keyboard and display, or a hard-copy readout, with the optional built-in printer. Alterna-

Negretti engine monitoring computer

Negretti engine monitoring system data retrieval unit

tively the data can be transferred to a ground-based processing facility, using a serial data link or a cassette tape.

The CDU comprises a twelve-character, sunlight-visible display and a multi-function keyboard, for the interrogation and control of the ACU during flight. It can also be used to enter data such as the aircraft all-up weight or flight number, displaying engine parameters and manually initiating performance data sampling.

Dimensions: (105 series ACU) 5 × 4 × 7 inches (127 × 102 × 178 mm)
(415 series ACU) 7.5 × 5 × 12.5 inches (191 × 127 × 318 mm)
(DRU) 4 × 11 × 14 inches (102 × 279 × 356 mm)
Weight: dependent on application

STATUS: in production and service.

Louis Newmark

Louis Newmark Limited, Aircraft and Instrument Division, 80 Gloucester Road, Croydon, Surrey CR9 2LD
TELEPHONE: (01) 684 3696
TELEX: 264004

Type 5755 attitude indicator

Newmark has developed 3-, 4- and 5-inch (76, 102 and 127 mm) versions of an attitude indicator. The 5-inch (127 mm) version, the Type 5755, has been built in the greatest quantity and has been fitted in Westland Sea King HAS.2s of the Royal Navy. A derivative, the

Newmark Type 5755 attitude indicator

9222, which features cross command bars, has been adopted for the Sea King HAS.5.

The attitude sphere is free to rotate through 360° in pitch and roll and is coloured so that the upper hemisphere is grey to represent the sky and the lower hemisphere is black. A power-failure flag is fitted and a slip indicator is mounted at the lower edge of the instrument. A pitch and roll trim knob is fitted to facilitate trimming of the sphere. An elapsed-time indicator at the rear of the instrument enables an accurate check to be kept on the unit's total running time.

Dimensions: 5.25 × 5 × 6.75 inches (133 × 127 × 171 mm)
Weight: 4.75 lb (2.2 kg)
Power: 115 V 400 Hz

6000 series attitude and heading reference system

This is a lightweight, compensated, first-order, all-attitude, attitude and heading reference system suitable for use in a close-support aircraft. It comprises three major units: a displacement gyroscope assembly (DGA), electronic control amplifier (ECA) and compass system controller (CSC). The DGA contains a directional and vertical gyro and attitude gimbals which assure all-attitude alignment. Accelerometers are also incorporated to detect fore-aft and lateral accelerations to aid accurate gyro erection during violent manoeuvres.

The system is technically similar to the Lear Siegler 6000 AHRS.

Dimensions:
(DGA) 5.9 × 5.5 × 10.2 inches (150 × 140 × 260 mm)
(ECA) 6.1 × 7.4 × 9.1 inches (155 × 188 × 231 mm)
(CSC) 5.75 × 1.5 × 4 inches (146 × 38 × 102 mm)
Weight: 31 lb (14.1 kg) inc mounting bases
Environment spec: MIL-E-5400 Class II
Gyro drift: (vertical) 3°/h
(directional) 0.5°/h
Power: 90 W at 115 V 400 Hz single-phase

STATUS: in production.

Strapdown heading reference system

This system comprises a sensor unit and an indicator. It is self-contained and provides a true-north indication in a form suitable for airborne guidance or land navigation. The sensor unit contains a two-axis tuned-rotor gyro, associated electronics, attitude-compensation accelerometers and microprocessor-based computing. The system can be aligned in 5 minutes. In normal operation it functions automatically, but can be re-aligned at any time by pressing a button. The operator can also manually slew the indicator heading card.

Dimensions: (sensor unit) 4.5 × 5.8 × 6.9 inches (115 × 148 × 178 mm)
(indicator) 3.9 × 3.9 × 3.9 inches (100 × 100 × 100 mm)
Total weight: 10 lb (4.5 kg)
Accuracy: 0.5° to true north
Drift rate: <1°/h
Power: 75 W at 24 V dc

STATUS: under development.

Page

Page Aerospace Limited, Forge Lane, Sunbury-on-Thames, Middlesex TW16 6EQ
TELEPHONE: (0932) 787 661
TELEX: 27520
FAX: (0932) 780349

Voice, tone and display warning systems

Page Aerospace designs and manufactures a wide range of alerting systems for civil and military aircraft. These include centralised and de-centralised warning systems, synthesised programmable tone and voice systems, flight mode annunciators, attention-getters and indicator modules.

Currently, the majority of British Aerospace aircraft are fitted with Page warning systems, as are the Panavia Tornado and the Jaguar. Page has also been selected to supply the central warning system for the Shorts produced version of the Tucano.

In recent years the company has developed a screen finishing process which results in high levels of readability even in extreme brightness.

These screens are now used in the central warning panel installations of the SF-340 and British Aerospace 125 and 146 systems.

For night flying operations the company has developed a night vision goggle compatible range of screens and bezels for displays and instruments. These screens effectively filter out the infra-red content from illuminated cockpit devices and minimise 'flooding' of the goggles.

Other areas of research cover high quality synthesis of audio tones and voice. Units are currently being evaluated by the Royal Aeronautical Establishment, Farnborough, and aircraft manufacturers in order to assess the optimum man-machine interface.

These systems allow any desired human or robotic, male or female, voice sound in any accent or language to be provided. Additionally, the word vocabulary, tone characteristics, message and tone priority order, intersound gap, repetition rate and volume are pre-programmable to individual customer requirements.

Voice synthesis is implemented by modelling the human voice tract using linear predictive coding techniques to recreate the original sound from pre-determined speech parameters. This design approach offers two important advantages over any other type of voice system — its extremely efficient use of memory capacity and the ability to vary intonation and concatenation of the message.

Page central warning system for Mk 60 series export Hawk

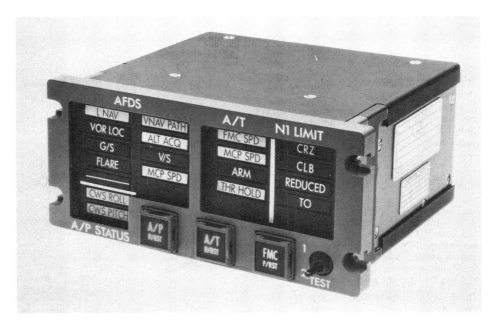

Page flight mode annunciator for Boeing 737-300

Plessey

Plessey Avionics Limited, Martin Road, West Leigh, Havant, Hampshire PO9 5DH
TELEPHONE: (0705) 486 391
TELEX: 86227

PVS 1712 radar altimeter

This J band pulse radar altimeter is designed for applications ranging from helicopters to high-performance fixed-wing aircraft. The use of a high operating frequency has enabled the design of a single unit system, containing transmit, receive and electronic functions. The unit uses a pulse leading-edge tracking technique to measure precisely the time interval for radar pulse travel. It is an all-solid-state design and requires no warm-up. Very short (4 nanoseconds) pulses are used. The system includes automatic error-correcting circuitry which fully compensates for errors due to internal delays and their time/temperature drift.

The transmitter is a 5-watt Gunn diode oscillator and a pseudo-homodyne receiver is used with a low-noise video pre-amplifier and main amplifier, the gain of the latter being geared to expected signal level at the range tracking point.

Plessey PVS 1712 radar altimeter

Typical installation
Dimensions: 218 × 116 × 76 mm
Weight: 2 kg
Altitude range: 0–1000 ft

Accuracy: ±(2% height +2 ft) at 500ft; ±(5% height +2 ft) above 500 ft

STATUS: in service.

PA 5000 compact radar altimeter

This simple integral unit operates in J band. By using advanced microwave field effect transmitter and stripline technology, the company claims enhanced electronic countermeasures resistance and better fade immunity for helicopters in low altitude hover, and the ability to detect hazards such as power cables. The system can be used up to 5000 feet. As height increases the transmitted pulse-width is widened to maximise the echo. The design allows for covert helicopter operations, when the pulse-width can be fixed and the transmitted power reduced below the nominal 1 watt so as to reduce detectability by electronic warfare sensors. Dual leading-edge tracking is used to ensure a response to the nearest object and to give accurate height ranging down to zero feet.

A number of functions are available to meet specific operating requirements. They include height range, track rate, search rate, digital message format, self-test, transmitter power control, and altimeter inhibit. All functions are under software control. Modular design enables the basic PA 5000 system to be adaptable to a 0 to 1000 feet, 2500 feet or 5000 feet range.

Dimensions: 9.25 × 3.56 × 2.37 inches (235 × 90 × 60 mm)
Weight: 3.3 lb (1.5 kg)
Power: 20 W at 19–32 V dc
Accuracy: ± (3% height + 2 ft)

STATUS: under development.

PVS 1820 engine usage life monitoring system

This flexible monitoring system incorporates a real-time data processing function for the detection and display of engine parameter exceedances, the computation of accumulated low cycle and temperature fatigue life, and continuous data recording for detailed analysis of engine design criteria.

The microprocessor-based system can include a display suitable for the cockpit or equipment bay giving parameter limit exceedance warning and immediate access to engine life data without reference to ground replay equipment. The system also can analyse retrospectively flight and engine data by an airborne cassette-loaded recorder and a ground-based data replay system.

Dimensions: (engine monitor unit) 194 × 90 × 321 mm
(visual display unit) 74 × 146 × 169 mm
(cassette recorder) 52 × 146 × 169 mm
Weight: (engine monitor unit) 4.44 kg
(visual display unit) 0.9 kg
(cassette recorder) 1.77 kg
Inputs: 16 spool-speed inputs, 18 discrete (on/off) signals and up to 40 analogue parameters. (Latter may include synchro, ac and dc voltage ratios, and potentiometer or variable resistance signals)
Sampling rate: programmable in binary steps from 1–156 parameters/s
Data rate: programmable in binary steps from 32–512 words/s
Recorder duration: 32 samples/s for 12 h or variable in binary stages to 512 samples/s for 0.75 h
Power: 19 W at 28 V dc

STATUS: under development.

PRS 2020 engine monitoring system

Developed from the PVS 1820 system, this design is lighter, smaller, uses less power and is compatible with MIL-STD-1553 (Def Stan 00-18 Part 3/1) data-bus interfaces. It also includes a 32-bit high-speed microprocessor, plus hybrid and uncommitted logic array-based electronics and bubble memory storage.

The PRS 2020 consists of several component units including: **PRS 2021 EMU**, a single-box data acquisition and processing unit for analogue signal conditioning, discrete digital conditioning and/or digital data-bus parameter recovery. The heart of this unit is a single very large-scale integrated hybrid circuit comprising a 32-bit 19-register central processor with 4 K bytes of random access memory and 8 K bytes of read-only memory reserved for applications programs.

PRS 2023 display available for taking a 'quick look' at data, and can be mounted on the engine monitoring unit or remotely. The format of the display, which uses 12 light-emitting diodes, can be defined by the customer

Plessey PVS 1712 indicator installed in an AAC Gazelle

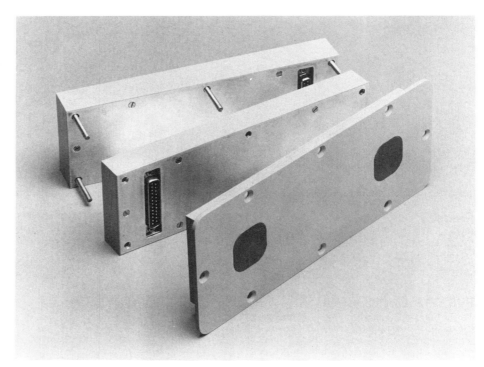

Plessey PA 5000 radar altimeter

Plessey PVS 1820 engine usage life monitoring system

PRS 2026 data retrieval unit (DRU), a rugged battery-powered unit for extracting data from the engine monitoring unit for subsequent analysis, or for system test and calibration.

Plessey also offers the options of a bulk storage recorder, a portable ground-replay unit and comprehensive systems software packages.

Dimensions: (engine monitoring unit) 200 × 194 × 124 mm
(data retrieval unit) 280 × 260 × 80 mm

STATUS: under development.

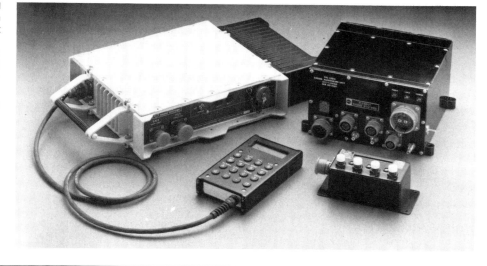

Plessey PRS 2020 engine monitoring system

Racal Acoustics

Racal Acoustics Limited, Beresford Avenue, Wembley, Middlesex HA0 1RU
TELEPHONE: (01) 903 9322
TELEX: 926288

AVAD automatic voice-alert device

This versatile voice-alerting system stores pre-recorded human speech in digitised form and operates under microprocessor control to assemble messages from words or phrases held in a vocabulary store. Since the voice is not synthesised, it can be in any language, either male or female, to provide the appropriate degree of stress and urgency for any situation. Racal says that high quality reproduction ensures that the voice remains recognisable and intelligible under all conditions.

Message priority order, repetition rate and volume are programmable to individual requirements to provide the maximum information to aircrews without disruption of their primary tasks. Racal claims that existing discrete audio warning systems can be replaced by AVAD with a minimum of installation cost and complexity.

There are three units in the current AVAD range, the V694, V695 and V691.

V694

The AVAD V694 is designed to integrate and control audio alerts keyed by signals from system sensors and push buttons. The message format is completely flexible and the system conforms to ARINC 577 audible warning systems and ARINC 726 flight warning computer systems for civil aviation applications.

Approximately 64 seconds of pre-recorded speech can be encoded and stored as a vocabulary of messages, phrases, words and tones. Messages can be constructed using words and phrases drawn from the unit memory under software control. This allows the total duration of all messages to exceed the vocabulary by a considerable amount. Up to 16 message channels are based on positive or zero volt keying. Four channels can respond to variable dc voltage inputs offering a sensitivity of up to 256 logic switching steps across the input range. A stabilised 5-volt dc output is available for use as a reference for an external potentiometer. Typical variable readouts might be aircraft altitude, electrical system voltages or engine pressure ratios.

Unit operation and vocabulary store may be defined by the user, and program software allows control of each message cycle. Each message has normal, re-grade and test priority values assigned to it. The output sequence is then controlled by the re-grade and test inputs.

There are two audio outputs, one for a telephone or headset and the other for driving a loudspeaker direct. The nominal level of each output can be adjusted by individual potentiometers offering a 20 dB range, with the relative volume of each message or message-format under software control. The complete message format, including repetition rate, pauses, message inhibit/de-inhibit and tones, may be similarly controlled. A key input is provided which inhibits any message in progress without affecting other keyed messages. A message will remain inhibited until its sensor input is removed and re-applied. Alternatively, the inhibit may be removed after a specified time, if the warning message keyline remains activated during this period.

Full test/reset facilities provide fast checking of unit operation. Correct operation of the microprocessor and amplifier circuitry and control program status are continuously self-monitored. In the event of a fault condition, both audio outputs are automatically inhibited and a 'fail' output is provided to activate an indicator to warn of unit failure.

V695

The V695 is a smaller four-channel unit with microprocessor control. It offers all the programming flexibility of the V694 but has a storage capacity of 9 seconds.

V691

The V691 is a single output device for applications in which only one warning such as 'fire' or 'pull up' is required. A second keyline may be used to provide a 'system enable/disable' input.

All the above units provide a bandwidth of 100 Hz to 4 kHz. Audio distortion is less than 4 per cent. Volume is preset and adjustable by 20 dB. The V694 is also programmable to different levels for each message and variation during messages. The operating temperature range of all units is from –40 to +70° C.

Dimensions: (V694) 6.49 × 4.52 × 2 inches (165 × 115 × 51 mm)
(V695) 5.28 × 2.4 × 1.26 inches (134 × 61 × 32 mm)
(V691) 4.33 × 1.96 × 1.1 inches (110 × 50 × 28 mm)
Weight: (V694) 2.1 lb (0.95 kg)
(V695) 0.48 lb (0.2 kg)
(V691) 0.3 lb (0.14 kg)

STATUS: in production.

Racal AVAD automatic voice-alert devices: V694 (left) and V691 (right)

Racal Avionics

Racal Avionics Limited, Burlington House, 118 Burlington Road, New Malden, Surrey KT3 4NR
TELEPHONE: (01) 942 2464
TELEX: 22891

Type 1655 automatic chart display

The automatic chart display is a simple and economical means of fixing aircraft position on a standard aeronautical chart. Position is indicated by the intersection of cross-wires that move in accordance with commands from a navigation computer such as the Racal TANS tactical area navigation system, the PBDI (position, bearing and distance indicator) or the new RNS 252 area navigation system. The chart area measures 254 by 254 mm, and five chart scales can be selected from 1:50 000 to 1:1 million. The system can be held on the thigh with a leg harness, rested on the knee, or table-mounted.

Dimensions: 306 × 306 × 67 mm
Weight: 3.2 kg

STATUS: in production.

Racal Avionics airborne chart display

Smiths Industries

Smiths Industries, Aerospace and Defence Systems Limited, 765 Finchley Road, Childs Hill, London NW11 8DS
TELEPHONE: (01) 458 3232
TELEX: 928761

HRA miniaturised radar altimeter

This unit meets requirements for a highly accurate measurement and flight-deck indicator device on civil and military aircraft. The outputs are suitable for use with flight control systems, including during automatic landing. Other applications include terrain-following and avoidance, reconnaissance and anti-submarine warfare. A pulse radar technique is employed with short pulse duration and pulse leading-edge tracking, and automatic receiver gain-programming at low altitudes. The system can operate over all types of terrain.

Typical installation
Dimensions: 203 × 150 × 96 mm
Weight: 2.9 kg
Operating frequency: 4300 MHz
Pulse repetition frequency: 10 kHz
Radiated power: 100 W peak
Altitude range: 0-5000 ft (1000 ft option)
Accuracy: ± 3 ft plus ±3%

Smiths Industries radar altimeter electronics unit

Smiths Industries radar altimeter indicator

Manoeuvre limits: (pitch) ±235° within stated accuracy; (roll) ±235° within stated accuracy

STATUS: under development.

General-purpose radar altimeter

Used in wide range of operational conditions, this system is in service with a large number of aircraft, particularly transport types. Operation is very similar to that of the HRA miniaturised radar altimeter.

Typical installation
Format: ½ ATR Short
Weight: 6.1 kg
Operating frequency: 4300 MHz
Pulse repetition frequency: 10 kHz
Radiated power: 100 W (peak)
Altitude range: 0-5000 ft (1000 ft option)
Accuracy: ± 5 ft plus ±3%
Manoeuvre limits: (pitch) ±30° within stated accuracy; (roll) ±30° within stated accuracy

STATUS: in service.

Compact radar altimeter

This is a lightweight pulse radar altimeter developed for applications requiring low volume installation and is particularly suitable for executive aircraft, low-cost military aircraft and missiles. It incorporates a high proportion of integrated and solid-state circuitry, as well as the operational features of other Smiths Industries' radar altimeters. A combined receiver/transceiver indicator version (for manned aircraft) and a stand-alone receiver-transceiver version (for missiles) are available.

Typical installation
Dimensions and weight: (manned a/c) 3 ATI × 195 mm; 1.4 kg
(missile) 3 ATI × 140 mm; 1.1 kg
Operating frequency: 4300 MHz
Pulse repetition frequency: 7 kHz
Radiated power: 100 W (peak)
Altitude range: 0-5000 ft (1000 ft option)
Accuracy: ± 5 ft plus ±2%
Manoeuvre limits: (pitch) ±30° within stated accuracy; (roll) ±30° within stated accuracy

STATUS: under development.

EFIS electronic flight instrument systems

Smiths Industries SDS 200 colour EFIS is fitted to British Aerospace ATP (Advanced Turboprop) aircraft.

The EFIS is a dual-channel system with four 6 × 5 inch (152 × 127 mm) cathode ray tubes, two primary flight displays and two navigation displays. Two symbol generators and one control panel complete the system.

The Smiths Industries SDS 200 EFIS features a new and improved architecture providing high system reliability by quadruplex operation and comprehensive data monitoring. This is a unique feature which provides improved system integrity and automatic cross cockpit monitoring.

Full reversionary capability is provided so that system performance is retained even after a total failure of one symbol generator. Furthermore, no single failure within the system will cause the loss of primary flight or navigation data from either side of the flight deck.

Smiths Industries SDS 200 EFIS equips British Aerospace ATP airliner

Three displays of Smiths Industries SDS 200 EFIS: left to right primary flight, navigation (map mode) and navigation (compass rose)

By using the most advanced electronic technology each display is provided with its own dedicated digital graphics generator to provide high definition, flicker free, display formats using both raster and stroke writing techniques.

The display unit power dissipation is typically only 35 watts, about half the power of earlier generations of equipment. This provides a display unit which does not require either cooling air or fans, leading to a simpler and less expensive installation and improved dispatchability.

The symbol generators are housed in a 6 MCU case with a power consumption of 140 watts.

All primary flight and navigation data is received using the industry standard ARINC 429 data-bus.

Weather radar data is received using ARINC 453 protocol.

The system is engineered to be fully compliant with RTCA DO 160 and DO 178A.

All units are interchangeable and suited to either fixed wing or helicopter installations.

Display units are available in a range of sizes up to 8 inches (203 mm) square.

Further identical units can be added to upgrade a four-tube EFIS to a six-tube EIS to include engine instrumentation, status displays, caution/warning displays and checklists.

STATUS: in production.

Smiths Industries multi-purpose display as specified for AV-8B, F/A-18 and Hawk 100/200

Multi-purpose displays

Smiths Industries' longstanding experience of advanced crt-based display systems has lead to the development of the latest generation of full colour multi-purpose displays suitable for a range of military aircraft applications. These systems offer a considerable advantage, in terms of both clarity and quantity of information presented, over the more traditional monochrome displays presently in use with the current generation of combat and tactical/reconnaissance aircraft. Smiths Industries currently has two multi-purpose displays undergoing trials, each has a distinctive set of operational characteristics suitable for varying aircraft system requirements. A third has been selected for the night attack versions of the US Marine Corps AV-8B, and the US Marine Corps/US Navy F/A-18.

2000 Series multi-purpose display

The 2000 Series multi-purpose display is capable of presenting data in either cursive, raster or hybrid form, in up to 14 colours, on a 7 × 6 inch (178 × 152 mm) display with a usable screen area of 6 × 4.5 inches (152 × 114 mm). The MPD works in conjunction with an external symbol/vector generator, transforming the outputs from the latter into visual formats for the display of synthetic radar and TV style data, as

Smiths Industries SDS display unit

Smiths Industries SDS 200 symbol generator

well as functioning as an interactive control and display unit.

Pilot and operator interaction is achieved via 24 momentary action keys located on the front panel around the crt face, all of which can be illuminated at low intensity during night-time operation. Ambient light is monitored by two sensors on the front panel and display brightness is controlled automatically around the level predetermined by the pilot.

The 2000 Series display is fully ruggedised, complying with all UK NATO and US specifications and is currently undergoing trials as part of the British Aerospace Experimental Aircraft Programme.

Dimensions: 7.05 × 8.1 × 17.17 inches (179 × 206 × 436 mm)
Weight: not exceeding 23.9 lb (10.86 kg)
Power: 125 VA max at 115 V/200 V, 400 Hz 3-phase plus 10 W max at 28 V dc
Usable screen area: 6 × 4.5 inches (152 × 114 mm)
Number of colours: any 14
Reliability: predicted mtbf 2500 operating hours

STATUS: undergoing trials.

2100 Series multi-purpose display

The 2100 Series is the latest full colour multi-purpose display to be developed by Smiths Industries. Incorporating a number of advanced electronic techniques, including a new high brightness, high resolution, flat screen, shadow mask, cathode ray tube, it will be capable of being viewed both in high ambient daylight conditions and at night in conjunction with night vision goggles.

The MPD is capable of presenting data in cursive, raster or hybrid formats on a usable screen area of 5 × 5 inches (127 × 127 mm). In cursive modes, the 2100 displays the output from an external graphics generator in a range of pre-determined colours. In raster modes TV-style data from a range of sources such as remote map readers, EO sensors and scan converted radar outputs, can be displayed in any one of four raster standards. When operating in hybrid modes, cursive symbology is written during the video frame blanking period thus presenting a raster display overlaid with highly accurate and independently controllable cursive symbology. An NTSC encoded output of the RGB Video is also generated and available for use by an external video recorder.

Pilot and operator interaction is achieved via 20 momentary action keys and four rocker switches located around the display face.

Ambient light is monitored by two sensors on the front bezel and display brightness is controlled automatically with respect to a predetermined level.

The 2100 Series complies with all relevant UK, NATO and US military specifications and has been selected by McDonnell Douglas for the Night Attack versions of the US Marine Corps AV-8B Harrier II and the US Marine Corps/US Navy F/A-18 Hornet.

Dimensions: 7.07 × 8.11 × 17.17 inches (180 × 206 × 439 mm)
Weight: 24 lb (10.9 kg)
Power: 200 V 3-phase 400 Hz 180 VA
Usable screen area: 5 × 5 inches (127 × 127 mm)
Reliability: >3500 h

STATUS: in development.

3000 Series multi-purpose display

The 3000 Series multi-purpose display uses the latest large-scale integration and hybrid packaging techniques to combine a graphics generator, amplifiers, high-voltage supplies and a full colour cathode ray tube into a single unit 6.25 inches (159 mm) square and 12.5 inches (318 mm) long. The unit has a built-in MIL-STD-1553B interface, making it a completely 'stand-alone' unit connecting directly onto the aircraft data-bus.

Smiths Industries SDS 200 control panel

Cockpit panels of British Aerospace Hawk 200 single-seat light-strike demonstrator fighter, with Smiths Industries head-up display, up-front controller (immediately below head-up display) and 3000 series multi-purpose display (white surround)

Smiths Industries Series 2000 multi-purpose display

The unit integrates fully into the nav/attack system and can display all primary flight and navigation data and control and display any parameter within the nav/attack system. Pilot or operator interaction is achieved via thirteen 'soft' keys whose function is determined by the selected display format. Two uncommitted keys can be specified as 'soft' or dedicated as required. Ambient light is monitored by two sensors on the front panel and display brightness is controlled automatically around the level predetermined by the pilot.

The 3000 Series display is qualified to UK military standards and complies with NATO and US specifications. It has been selected for the British Aerospace Advanced 100/200 Series Hawk aircraft.

Dimensions: 7.72 × 7.09 × 12.48 inches (196 × 180 × 317 mm)
Weight: not exceeding 22 lb (10 kg)

Power: 180 VA at 200 V 400 Hz 3-phase
Usable screen area: 5 × 5 inches (127 × 127 mm)
Number of colours: any 15
Reliability: predicted mtbf 2500 operating hours

STATUS: undergoing trials.

Display multi-purpose processor

This is a wholly modular, display processor unit which can be configured to meet customer requirements. It is based on an extensive library of standard electronic cards and a range of ATR/MCU case sizes. An internal data-bus allows the system to be updated or expanded easily and incorporate new technology components. The unit is designed to military specifications and is suitable for computing applications in head-up displays, weapon-aiming, mission-control and electronic flight instruments. The central feature of the system is a single-card computing element which incorporates a 16-bit microprocessor, 32 K words of random access memory and up to 128 K words of ultra-violet eprom storage. Additional modules provide global memory, cursive or raster symbol-generation, analogue and discrete inputs, serial digital interface, and MIL-STD-1553B interface as either remote terminal or bus-controller.

Latest applications which demonstrate the versatility of the DMPP series of processor units include a bus control and interface unit for the Sea Harrier FRS2, a bus interface unit and weapon aiming computer for the advanced 100/200 series Hawk, and waveform generators for Royal Air Force Harrier and Jaguar aircraft.

Format: ½ ATR (4 MCU) 124 × 194 × 318 mm; ¾ ATR (6 MCU) 190 × 194 × 318 mm
Weight: (4 MCU) 13 lb (6 kg); (6 MCU) 20 lb (9 kg)

STATUS: in production.

'Get-you-home' panel

Smiths Industries is developing what it calls a 'get-you-home' panel that provides all the essential information needed by the pilot for use in standby mode or to fly the aircraft back to base in an emergency. The device is a flat panel measuring 8 inches (203 mm) square by 1 to 2 inches (25.4 to 50.8 mm) deep that pulls out from a recess on the flight-deck. It provides dial-and-pointer or vertical-scale information in the form of a light-emitting diode matrix. The display is driven by two processors that share the task so that failure of one of them does not affect operation.

A typical aircraft installation would comprise two identical displays, one for flight and engine management (with, for example, speed, altitude, rate-of-climb, angle of attack and engine speed), and one for navigation and flight-director information.

STATUS: in development.

Led dot-matrix displays

In parallel with the development of cathode ray tube based electronic flight instrument system displays, Smiths Industries has produced a family of engine performance indicators using a hybrid dc-servo pointer/light-emitting diode readout. The servo-driven pointer moves over a conventional circular scale, while an emissive electro-optic drum counter constructed from a dot-matrix of high brightness light-emitting diodes offer a display which is fully compatible with crt and light emissive displays now being introduced in civil and military aircraft.

The readout is of the 'rolling-drum' type, with values appearing to roll smoothly upwards or downwards. The electronic readout is controlled by a microprocessor and an optical cell adjusts the brightness to match ambient levels.

The light-emitting diode instruments have been chosen for the Boeing 737-300, Airbus A310 and Saab JAS 39. They display N_1 and N_2 engine speeds, exhaust gas temperature, engine fuel and fuel used and, in some cases, engine pressure ratio.

This type of display has also been chosen for the McDonnell Douglas MD-88.

STATUS: in production.

Led standby engine display panel

Smiths Industries has been selected to supply a standby engine indicator display panel for the Boeing 767.

The display of engine parameters is by a light-emitting diode in a seven-bar format and incorporates four numeric readouts for each engine — N_1, N_2, egt and epr. Eight engine displays are housed in a 3 ATI case.

STATUS: in production.

Smiths Industries standby engine display panel for Boeing 767

Smiths Industries barometric standby display panel for EAP

Led barometric standby displays

Smiths Industries has supplied to British Aerospace a solid-state set of barometric displays for the Experimental Aircraft Programme (EAP). The unit is driven by the DADC and features he ght (with baro set), airspeed, Mach and vertical speed. The parameters are displayed in analogue and digital form using radially positioned light-emitting diodes for the pointer and an led dot matrix for the digital counter.

STATUS: in production.

Strapdown attitude and heading reference system for RPVs

This system has been designed for use with the flight control and navigation equipment carried by RPVs (remote-piloted vehicles). Extensive use is made of plastic-moulding techniques for mechanical components, and the electronic components incorporate thick-film, hybrid circuits. This attitude heading and reference system comprises a sensor unit, computer, magnetometer, and air data unit. The sensor unit generates analogue signals proportional to angular rates and linear accelerations by means of three 900 Series servoed accelerometers. The three-axis magnetometer provides the long-term drift correction required by the sensor unit. The 8086-based computer derives its heading and attitude information from the sensor unit. The air data unit provides basic pressure measurements from which barometric height and indicated airspeed are derived.

Flight control, including height and airspeed acquisition, hold, and stabilisation is provided by up to seven control surfaces driven by mark/space position command signals. A fully autonomous navigation function provides dead-reckoning navigation between waypoints.

Accuracies: (heading) ± 2°
(attitude) ± 1°
(rates) ± 2°/s
(navigation) typical 1 km in 20 km

Electronic systems for rate integrating gyros

These systems have been developed for a range of integrating gyros. They feature high bandwidth, high resolution and low drift over a wide range of inputs rates, and incorporate built-in test functions.

Units are available for sensing along one, two or three axes. The servoed electronics can be mounted either separately or as an integral part of a gyro assembly.

Input range: up to 350°/s continuous, or up to 500°/s intermittent
Drift rate: *g* insensitive to 3°/h, or *g* sensitive to 8°/h
Bandwidth: up to 100 Hz unfiltered
Power: 26 V 400 Hz and 40 V dc

300 RNA series horizontal situation indicators

This range of horizontal situation indicators (HSIs) is intended for civil and military fixed-wing aircraft and helicopters. Each instrument consists of a main frame, synchro frame and electronics. Large-scale integrated circuits are used for signal-processing, and synchros are used to drive the various displays.

The instruments can be used in NAV, TAC, APP or ADF modes. The range display is a four-digit electronic module at the top of the instrument face reading up to 999 nautical miles. Another numeric display is used to indicate the setting of the command track pointer. A knob is provided for selecting the relevant runway QDM on the command track pointer. These instruments provide complete ILS information as well as Tacan and ADF displays.

The 304/305 RNA HSIs are used in the British Aerospace Nimrod and 748 aircraft as integral parts of the SFS6 flight systems. The 309 RNA HSI equips Royal Navy/Westland Sea King helicopters and Royal Air Force/British Aerospace/Hawk trainers. The 306/307 RNA is used in the Sepecat Jaguar and British Aerospace Strikemaster. The 330 series HSI is used in the Royal Air Force's Panavia Tornado aircraft and Chinook helicopters.

FADS flexible air data system

FADS is a high performance distributed air data system, ideally suited for all military and civil applications, making use of spare computing capacity elsewhere on the aircraft. Using cmos technology it requires minimal power throughout. Modular in construction it offers various outputs, including ARINC 429 or MIL-STD-1553B and has full built-in test facilities.

Dimensions: 3.6 × 5.5 × 9.5 inches (91 × 140 × 241 mm)
Weight: 8.25 lb (3.7 kg)
Power: 20 VA

STATUS: in production.

Digital air data computer for civil aircraft

These air data systems are suitable for all civil aircraft and utilise the latest state-of-the-art vibrating cylinder sensors and ULA cmos technology to minimise weight and lower power demands. Data outputs conforming to ARINC 429 specification are provided and full in-flight monitoring and maintenance initiated built-in test facilities are included. The units meet the requirements of ARINC 706 specifications.

Format: 3/8 ATR
Weight: 8.71 lb (4 kg)
Power: 20 VA

STATUS: in production.

Digital air data computer for military aircraft

The digital air data computer comprises, where required for Vtol applications, two independent processor channels — a primary, with inputs of pitot and static pressure and a secondary channel with static only.

Depending on specification requirements, high or low range total pressure vibrating cylinder sensors are used. Outputs are provided in MIL-STD-1553 form. They feature comprehensive built-in test facilities which include

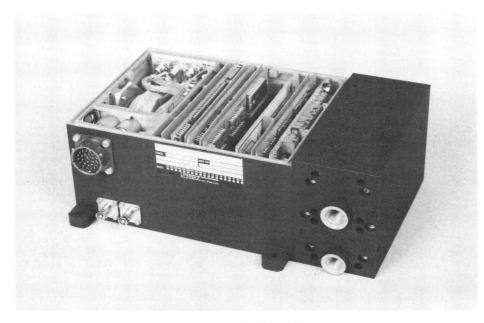

Smiths Industries Flexible Air Data System

light-emitting diodes (leds) in the instrument and simulating the rolling motion of a mechanical counter, both sense and rate indications are presented instinctively. Being light-emissive, the leds are compatible with the new flight-deck cathode ray tube digital displays while the presentation is familiar and flicker-free.

Compared with led-bar instrumentation, this device is much less affected by single led failures, and very accurate interpolation of data is possible due to the apparent motion in the display. A 7 by 5 matrix, 4.5 mm high is used, with three blank rows of leds between each figure, but a 9 by 5 matrix, giving 5.2 mm high characters and with reduced character spacing, is an alternative. In place of a mechanical failure flag the led matrix displays the word 'fail' if an indication is outside limits or invalid.

To minimise the number of moving parts a microprocessor is used to convert the input signal to digital form to drive the counter and then convert the digital signal back to analogue

continuous in-flight monitoring of the system. Performance and maintenance initiated BITE is capable of locating faults down to module level.

Format: 3/8 ATR
Weight: 11 lb (5 kg)
Power: 35 VA

STATUS: in production.

Speed computing display

This combined Mach/airspeed indicator measures and displays both indicated and flight-manual limit speeds. Analogue pick-offs on the capsule mechanisms provide outputs for computing Mach number which is presented on a two-digit numeral counter. With indicated airspeed, limit speed data and Mach number derived from integral capsules, this is a self-contained instrument displaying IRS, V_{MO}/M_{MO} and Mach number without the need for an air data computer.

Range: (airspeed) 60–450 knots
(Mach number) 0.4–0.99 Mach
(command speed) 100–450 knots
(altitude) –2000 to +50 000 ft
Format: 3 ATI case to ARINC 408
Weight: 3.5 lb (1.5 kg)
Power: 28 V dc

Mach/airspeed indicators

This is a range of indicators for both military and civil aircraft with accuracies to TSO C46a. The counter-pointer displays have a numeric readout of Mach number and a pointer on scale readout of airspeed. Mach inputs are to ARINC 545. Additional features include a limit-speed pointer and switch, airspeed warning, and a speed error output for use with an autothrottle system.

Format: 3 ATI ARINC 408 or 3 in STANAG
Weight: 4.5 lb (2 kg)
Range: (airspeed) 60 to 750 knots ASI
(Mach number) 0.4 to 2.5 Mach
(altitude) 1000–60 000 ft
Power: (Mach servo) 26 V 400 Hz
(speed switch) 28 V dc
(autothrottle) 26 V 400 Hz

STATUS: in production for Boeing 737, McDonnell Douglas MD-80, BAe 146 and Saab Draken and Viggen aircraft.

Led dot-matrix engine indicators

Smiths Industries has developed a hybrid counter-pointer engine indicator using a microprocessor to provide a presentation similar to that of a conventional mechanical drum counter. The instrument is more accurate, has fewer moving parts than a conventional device and has higher reliability. By using an array of

Smiths Industries engine indicators for Airbus A310 and A300-600 aircraft

Smiths Industries air data computer for civil aircraft

Smiths Industries digital air data computer for military aircraft

Smiths Industries engine instrument for British Aerospace 146 and 125-800

Smiths Industries Model 2300 fuel quantity indicator

Smiths Industries engine instruments for Boeing 747

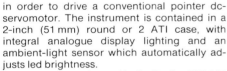

Smiths Industries engine indicators for Boeing 737-300

in order to drive a conventional pointer dc-servomotor. The instrument is contained in a 2-inch (51 mm) round or 2 ATI case, with integral analogue display lighting and an ambient-light sensor which automatically adjusts led brightness.

The instrument is used in the Boeing 737-300 to display N_1/N_2 engine speeds, EGT and fuel flow/fuel used for the CFM56 power plants. It is supplied as an option to Airbus A310 and A300-600 aircraft and has been selected by British Caledonian, Cyprus, Singapore Airlines and Air India. Pan-Am have selected Smiths Industries for equipping their A310 with a new set of engine indicators for the Pratt & Witney 4000 engined aircraft.

Counter-pointer dc-servo engine instruments

Smiths Industries produces a wide range of these instruments in sizes and displaying parameters appropriate to a wide range of applications. Examples are being supplied for use in Boeing 747 and British Aerospace 146 airliners.

Digital fuel gauging for Boeing 747

The Civil Systems Division of Smiths Industries in Malvern, Pennsylvania, USA, is to supply its Model 2300 digital engine fuel quantity indicators as standard equipment on the Boeing 747, starting in mid 1987. The first customer will be KLM airlines.

UNITED STATES OF AMERICA

Aero Mechanism

Aero Mechanism, 20327 Nordhoff Street, Chatsworth, California 91311
TELEPHONE: (818) 709 2851
TWX: 910 495 1730

8100 Series encoding altimeters

Intended for general aviation applications, this family includes three-pointer, counter-pointer and drum-pointer types. Altitude is optically encoded in 100 feet increments with a digital output. The series is designed for a range of

–1000 to 35 000 feet. The 8100 Series can be provided with extended scale (800 to 1050 mb) which complies with UK requirements.

Format: 3 ATI
Weight: 2 lb (0.9 kg)
Power: 14 or 28 V dc

STATUS: in production.

AM 275 altitude alert system

This is designed to monitor continuously and compare actual altitude with a desired altitude. A knob below the sunlight-readable six-digit

display operates the desired altitude-setting function. The display includes three visual commands: descent, climb and level over the range –1200 to +62 000 feet.

Format: $1/2$ ATI × 6.5 inches (165 mm)
Weight: 0.8 lb (0.36 kg)
Power: 0.7 A at 11 to 32 V dc

Aeronetics

Aeronetics, 2100 Touhy Avenue, Elk Grove Village, Illinois 60007
TELEPHONE: (312) 437 9300
TELEX: 6871468

Model 9130 horizontal situation indicator

This 4-inch (102 mm) horizontal situation indicator is suitable for medium-twin, light jet and helicopter applications. It provides simultaneous indications of selected VOR, RNav and localiser course, course deviation, and selected heading. During an ILS approach the glideslope deviation can be shown also. Range and course readouts use 7-segment pinlite displays and there is an ADF bearing pointer option.

Aeronetics also manufactures remote gyro and magnetic flux detector equipment which is compatible with this instrument.

Format: 4 ATI case
DME input: ARINC 568 serial code or King serial
Altitude range: –1000 to +50 000 ft
Power: 28 V dc

STATUS: in production.

Model 8000 HSI slaved compass system

This combines continuously slaved heading information with VOR, ILS and ground speed indications in one display so as to simplify navigation and IFR flight. The system provides heading, course and heading bootstrap outputs to other systems. The system uses the Model 8140 magnetic flux detector to sense the direction of the Earth's magnetic field; its output drives the stabilised Model 8100 directional gyro.

The Model 8130 horizontal situation indicator used with this slaved compass system is suitable for helicopter or high performance fixed-wing applications. An automatic emergency mode indicator advises the pilot when the system is in that mode.

Model 8130
Format: 3 ATI
Weight: 3 lb (1.4 kg)
Power: 28 V dc

Model 8100
Weight: 3.4 lb (1.5 kg)
Power: 28 V dc

Model 8140
Weight: 3 lb (1.4 kg)

AiResearch

AiResearch Electronic Systems Division, The Garrett Corporation, 2701 East Elvira Road, Tucson, Arizona 85706-7124
TELEPHONE: (602) 573 6000
TELEX: 165936

AiResearch has developed a series of air data computers for a variety of supersonic and subsonic aircraft, drones and spacecraft. They receive pitot and static pressures and air temperature and provide output information for use in navigation, fire-control, powerplant, cockpit display and other aircraft sub-systems, and flight instruments.

Digital air data computer for AV-8B

In common with all AiResearch air data computers, this model uses the quartz pressure transducer. The unit is similar in construction to that used on the US Air Force/Fairchild A-10 aircraft. It features a non-volatile memory for the retention of fault messages, even if power is turned off.

Dimensions: 5.03 × 7.63 × 14.06 inches (129 × 193 × 355 mm)
Weight: 16 lb (7.2 kg)
Inputs: 10
Outputs: 78

STATUS: in production for the US Marine Corps/McDonnell Douglas AV-8B V/Stol close-support-aircraft.

Digital air data computer for B-1

This computer was developed for the US Air Force/Rockwell B-1 supersonic bomber and features a digital mos large-scale integrated circuit processor and digital pressure transducers. It has an altitude reporting facility and the built-in self-test system provides continuous failure monitoring.

Dimensions: 6.25 × 7.67 × 19.67 inches (158 × 195 × 500 mm)
Weight: 27.7 lb (12.6 kg)
Inputs: 19
Outputs: 92

STATUS: developed for US Air Force/Rockwell B-1A supersonic bomber programme which was cancelled in 1977 and reinstated in late 1981 as B-1B.

Digital air data computer for JA 37

This is an all-digital computer with similar features to those of the B-1 air data computer and has bi-directional digital communication with the supplied sub-systems. It contains both static and dynamic self-test facilities.

Dimensions: 6.2 × 8.4 × 17.3 inches (157 × 213 × 439 mm)
Weight: 25.1 lb (11.4 kg)
Inputs: 16
Outputs: 37

STATUS: in service with Sweden's Saab JA 37 interceptor.

Air data computer for F-14

Although similar in basic design to other members of the AiResearch ADC family, the system for the US Navy/Grumman F-14 carrier-based air-superiority fighter has additional functions. These include control of the variable-geometry wing sweep angle, provision of a flight-parameter and status display, and the generation of certain commands for the engine thrust autopilot.

Dimensions: 7.75 × 5.7 × 20 inches (197 × 145 × 508 mm)
Weight: 27.5 lb (12.5 kg)
Inputs: 70
Outputs: 60

STATUS: in production and service.

Air data computer for A-10

Designed to full military environment and MIL-STD-1553 digital data-bus standards, this air data computer for the US Air Force/Fairchild A-10A close-support-aircraft features a digital pressure transducer and a ferro-resonant power supply. Modular programming is provided and the memory has a 6 K word read-only memory and a 256-word random access memory.

Format: ½ ATR
Weight: 13.9 lb (6.3 kg)
Power: 36 W

STATUS: in production for US Air Force/Fairchild A-10 close-support aircraft.

True airspeed computer

This device provides outputs of true airspeed for RNav, INS and display applications. It features a vector pressure-ratio transducer which facilitates computation of true airspeed using only one device. This servoed force-balance unit maintains a high level of performance under all aircraft operating conditions.

An accuracy of ±4 knots is obtained over a true airspeed range of 95 to 500 knots. A static air temperature output is available as a customer option.

Format: ⅜ ATR
Weight: 9.2 lb (4.2 kg)
Power: 30 VA at 115 V 400 Hz

STATUS: in production.

Standard central air data computers

In an attempt to rationalise their many weapons systems, and so cut acquisition and maintenance costs, the US Air Force and US Navy have co-ordinated the development of four standard central air data computers (SCADCs) having a high degree of commonality, and each applicable to many types of aircraft. AiResearch and Britain's GEC Avionics were awarded development contracts and evaluation of a substantial number of prototype systems supplied to the US Air Force and US Navy was completed in 1984: GEC Avionics was subsequently awarded the production contracts.

AiResearch had earlier secured contracts from two aircraft manufacturers to supply SCADC computers to Lockheed, for the C-5B Galaxy, and to Grumman for the C-2A Greyhound carrier on-board delivery aircraft.

The AiResearch SCADC configuration uses a non-volatile memory for storing air data and sub-system fault information. There is a high standard of fault-detection and high reliability ensures compatibility with a two-level maintenance strategy. AiResearch has developed four standard computers.

CPU-140/A
Applicable to C-2A, E-2C, A-4M, OA-4M, EA-6A, EA-6B, KA-6D, A-6E, A-7E, TA-7C and KC-135R
Dimensions: 143 × 140 × 242 mm
Weight: 9.4 kg
Power: 60 W at 115 V 400 Hz

CPU-141/A
Applicable to C-5A, C-5B and C-141B
Format: ½ ATR
Weight: 14.4 kg
Power: 72 W at 115 V 400 Hz

CPU-142/A
Applicable to F-111A, FB-111A, F-111D, F-111E, F-111F and EF-111A
Dimensions: 143 × 350 × 414 mm
Weight: 15.5 kg
Power: 77 W at 115 V 400 Hz

CPU-143/A
Applicable to F-4C, F-4D, F-4E, F-4J, F-4S and RF-4C
Dimensions: 143 × 416 × 305 mm
Weight: 16.6 kg
Power: 56 W at 115 V 400 Hz

STATUS: in production/development.

Engine performance indicators

Garrett makes a series of engine performance indicators for business, commuter and military aircraft. Typical examples are the British Aerospace HS 125-700, Gates Learjet 35/36 and 55, Lockheed JetStar 2, Casa CN235 and Taiwanese XAT-3. The indicators provide a

Typical Garrett engine performance indicators as fitted in Gates Learjet 35 executive jet

continuous analogue display of critical engine variables, such as fan and core rotation speeds, turbine temperature and fuel flow. Configurations with circular or square cross-section are available.

Solid-state circuitry is employed throughout. The indicators are unaffected by the large fluctuations in electrical supply during engine starts.

Power: 120 mA nominal, 450 mA max at 12–34 V dc, plus 250 mA at 5 V dc for lighting
Input signals: (rotation) monopole pulse (turbine temperature) chromel-alumel thermocouple per NBS Monograph 125
(fuel flow) second harmonic selsyn mass flow transmitter, 115 V ac 400 Hz reference
Rpm accuracy: (0 to 55° C ambient) ±0.25% at 100%
(–30 to 70° C ambient) ±0.5% at 100% rpm

Temperature accuracy: (0 to 55° C ambient) ±5° C at 900° C
(–30 to 70° C ambient) ±10° C at 900° C
Fuel flow accuracy: (0–55° C ambient) ±1% of full-scale
(–30 to 70° C ambient) ±2% of full-scale
Rpm range: 0–110%
Turbine temperature range: 100 to 1000° C
Fuel flow range: 0–2300 lb/h
Response: 3 s full-scale slew
Temperature range: (Operating) –30 to 55° C
(short-term) –30 to 70° C
(ambient extreme) –65 to 70° C
Shock: 6 g for 0.011 s
Vibration: 5–3000 Hz, 0.02 in amplitude, limited to ±1.5 g

STATUS: in production.

Propeller synchrophaser

The Garrett synchrophaser controls the phase relationship between the blades of two aircraft propellers. Noise and vibration can be minimised and the objectional beat frequencies associated with propellers operating at different rotational speeds be eliminated. The synchrophaser is designed to operate with twin turboprops and electronic engine fuel controls. It works with existing speed detection systems and requires no rigging adjustments, calibration, or indexing. It is compatible with engines operating closed-loop on torque and temperature. The system uses a master-and-slave engine concept with a full 120-degree phase angle authority for three-bladed propellers.

Dimensions: 2.3 × 5.05 × 8.05 inches (57 × 128 × 204 mm)
Weight: 1.75 lb (0.79 kg)

Power: 28 V dc with MIL-STD-704 Category B transient immunity
Authority limit: 2.4% max speed separation
Steady-state phase-angle accuracy: <5° (typical)
Operating range: cruise and max continuous speeds
Vibration: RTCA DO-160-PAR. 8 curves
Electromagnetic interference: MIL-STD-6181D

STATUS: in production.

Engine synchroniser

When two common-structure mounted engines operate at slightly different rotational speeds, a third beat frequency is set up in the structure which can cause discomfort to passengers. Garrett is producing an electronic engine synchroniser to reduce this problem. It interfaces with the engine through the existing electronic fuel control, and requires no modifications to the aircraft wiring. The synchroniser may be applied to two, three, or four engines without modification. It works by nominating one engine as master and adjusting either fan (N_1) or compressor (N_2) speeds of the other powerplants so that all speeds are closer to each other. The trim authority varies with power lever setting. The flight deck switches allow on/off and N_1/N_2 selections, and a synchronisation indicator is optional.

Dimensions: 9.4 × 7.3 × 3.6 inches (239 × 186 × 91 mm)
Weight: 4.6 lb (2.1 kg)
Max acquisition time: 30 s
Synchronisation accuracy: (N_1 control) 11 rpm (N_2 control) 18 rpm

STATUS: in production.

Astronautics

Astronautics Corporation of America, PO Box 523, 4115 North Teutonia Avenue, Milwaukee, Wisconsin 53201-0523
TELEPHONE: (414) 447 8200
TWX: 910 262 3153

Turn co-ordinator

The Astronautics turn co-ordinator is a standard presentation flight instrument displaying rate-of-turn and bank information. The unit is designed to meet the requirements of FAA/TSO C3b. It is powered by an Astronautics brushless dc motor. The environmental specification includes –40° C to 70° C and a pressure altitude of 20 000 feet.

Dimensions: 3.25 × 3.25 × 6.25 inches (82 × 82 × 158 mm)
Weight: 1.6 lb (0.72 kg)
Power: 14–28 V dc

STATUS: in production and in service.

E-scope radar-repeater display for Tornado

This display portrays video signals from the Texas Instruments terrain-following and attack radar in the Panavia Tornado IDS variant. In the terrain-following mode it shows a hyperbolic-shaped symbol representing the ground ahead, above which the aircraft symbol remains if the autopilot is functioning correctly. Topographical information can be shown in the ground-mapping mode. For check purposes, an operator-initiated test pattern can also be brought up on the screen.

STATUS: in production and service.

Radar/infra-red display for EF-111A

Part of an update for the US Air Force/General Dynamics EF-111A Raven, this display replaces an earlier one used in conjunction with the General Electric AN/APQ-113 forward radar on this electronic warfare aircraft. The direct-view storage tube permits good readability in direct sunlight. As an infra-red display, the system is compatible with the AN/ALR-23 receiver/countermeasure set. Symbology is electronically generated and stroke-written on the cathode ray tube.

STATUS: in production and service for the 42 General Dynamics EF-111As operated by the US Air Force.

Digital ECM display

This advanced display, originally designed for the US Navy/Grumman EA-6B aircraft, has such versatility that it can be used for a wide variety of airborne functions. It can, for example, accomplish computations and display for navigation and weapon-delivery. The unit contains

E-scope display for Tornado by Astronautics

Astronautics radar/infra-red display for EF-111A

Astronautics digital ECM display

a signal-data converter with a 16 K word digital computer.

STATUS: in production and service. In addition to EA-6B, system was selected for US Air Force/General Dynamics EF-111.

Moving-map display for A-10 and F-16

Astronautics provides for the General Dynamics F-16 and Fairchild A-10 a moving-map display based on specially prepared film covering the mission area and stored within it. The display is driven from the aircraft navigation system, the 35 mm film of a standard aeronautical chart passing through the viewing area at a rate proportional to the scale chosen (typically 1:50 000 or larger) and aircraft speed. Target, drop-zone, navigation or particular information relating to a mission can be overlaid on the map as desired, and JTIDS or other alphanumeric data can also be displayed. For night operation map negatives can be screened.

A MIL-STD-1553 digital data-bus interface is provided for communication with other equipment, for example, the navigation computer. A remote map reader with a colour output has also been developed.

STATUS: system underwent flight-trials on an A-10, and is currently under development for A-10 and F-16.

137150-1 display processor for 530MG helicopter

This display processor designed for the McDonnell Douglas Helicopters 530MG Defender generates independently programmable symbology for the multi-function displays, control/display unit, and the TOW missile display. Heart of the display system is a 16-bit high-speed general-purpose processor programmed in high-order language that manages all video, digital and analogue inputs and outputs, performs real-time computations and drives two independent symbol generators, each controlling raster and stroke displays.

Dimensions: 9.4 × 5.7 × 10.6 inches (240 × 145 × 270 mm)
Weight: 17 lb (8 kg)

Astronautics video display for McDonnell Douglas Helicopters 530MG Defender

Astronautics display processor for McDonnell Douglas Helicopters 530MG Defender

Inputs: high-speed parallel (digital), symbol brightness (dc analogue), 525 and/or 625-line video per EIA standard
Outputs: 4 raster video: 1 at 525/875 line, 2 at 525 per RS330, 1 at 875 per RS343
Symbology: azimuth and target; direction scales altitude and rate-of-climb; pitch, roll, TOW-firing symbology; navigation; and multi-function display switch annotation.

STATUS: in production.

Area navigation control/display for L-1011 TriStar

This system, installed on forward sections of Lockheed L-1011 throttle boxes so that it can be viewed and operated by both pilots, comprises a centrally-mounted electronic automatic chart, with identical control/display units (cdus) situated alongside, and a control unit and an electronic unit for the chart display. The cdus present pages of alphanumeric information as required, while the electronic chart shows pictorially the route being flown, with positions marked of waypoints, destination airfields and navigation beacons.

The control unit has a scale selector that can shrink or expand the navigation situation shown on the chart, a slew control that can move the display bodily about the screen or rotate it (the alphanumeric symbols, however, remaining upright), mode selector permitting north-up, track-up, or look-ahead display, and an RNav switch that can select data from one or other of the navigation computers on the aircraft. The chart can store 1000 symbols and display 500 of them at any one time, and can also act as a back-up to the cdus, presenting commanded alphanumerics as required.

The cdus can display 12 lines of information with 17 characters per line, and can access and display data stored in an aircraft computer, such as route details, flight plans, en route nav-aid frequencies, and terminal area procedures, for example, standard departures and approaches.

The system incorporates continuous failure monitoring and self-test, and the computing section furnishes a self-test pattern for complete 'end-to-end' checks.

Dimensions: (control unit) 5.8 × 2.7 × 6.3 inches (147 × 69 × 160 mm)
(chart) 8.5 × 9.1 × 15.1 inches (216 × 231 × 384 mm)
(electronic unit) 4.9 × 7.7 × 14.5 inches (124 × 196 × 368 mm)
(cdu) 5.8 × 9 × 12 inches (147 × 229 × 305 mm)
Weight: (control unit) 1.5 lb (0.68 kg)
(chart) 27 lb (12.25 kg)
(electronic unit) 14 lb (6.35 kg)
(cdu) 18 lb (8.17 kg)
Power: (chart) 200 VA
(electronic unit) 90 VA
(cdu) 90 VA

STATUS: in service.

Video display for AH-64 helicopter

This device is a high-resolution, high-brightness cathode ray tube for displaying flight, navigation and weapon data. Image information is passed to the visual display unit in the form of analogue composite video. Pitch and roll trim controls are provided on the front panel, together with a turn and slip indicator. Built-in test circuitry provides go/no-go status for the vdu and assists in check-out and fault location. The display has been chosen for the US Army AH-64 Apache helicopter.

Dimensions: 7.3 × 6 × 12.5 inches (186 × 152 × 318 mm)
Weight: 14.5 lb (6.59 kg)

STATUS: in production.

Multi-function display

ACA's multi-function display is a high resolution, high brightness single-unit system that accepts composite video data from external sensors and puts it on view on a raster-scanned cathode ray tube. Data can be put up as full grey scale pictures or two-tone (black and white) alphanumerics. Push-button switches on the front panel select the display required from external sources. Each switch is illuminated for identification in darkness or low light levels, and the display has an automatic brightness control to compensate for changing ambient conditions. There are also manual brightness and contrast controls. Built-in test circuits provide go/no-go indications of the unit's health and also other facilities to assist check-out using an external test-set. The unit is self-contained, with its own power supplies and cooling provisions.

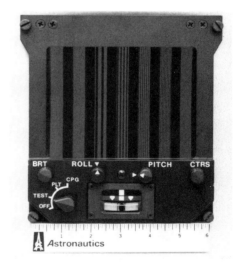

Astronautics multi-function display on AH-64A Apache helicopter

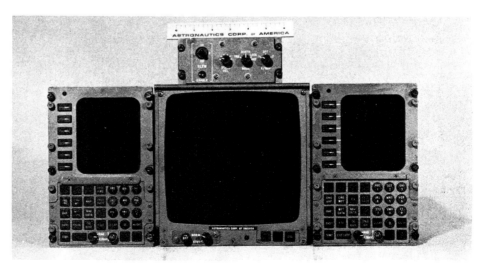

Astronautics area navigation electronic chart for L-1011, flanked by two control/display units

The system can be used to display television images from the Hughes AGM-65 Maverick air-to-ground missile so that the pilot can guide it to the target, duplicate the head-up display (HUD) scene for a HUD camera, and present radar information or data from a forward-looking infra-red system. It is also suitable to perform as a primary ADI or HSI flight instrument.

Dimensions: 7.9 × 6.6 × 12 inches (200 × 168 × 305 mm). Usable screen area 5 inches square (127 mm square)
Weight: 14.5 lb (6.59 kg)

STATUS: in production for Agusta A.129 Mangusta (Mongoose) helicopter; flown on this aircraft during 1983.

ACA 126370 horizontal situation indicator for SH-60B

This is a 5-inch (127 mm) instrument featuring standard ARINC 407 inputs and outputs with digital interface and extensive built-in test equipment. It was designed for the US Navy/Sikorsky Seahawk SH-60B Lamps (light airborne multi-purpose system) helicopter programme and is currently in production and operational. A mean time before failure figure of 3000 hours is claimed.

STATUS: in production and service.

ACA 113515 horizontal situation indicator for helicopters

This has proved to be a popular instrument for helicopters and has course-bar indicator, to/from flag, glideslope pointer, two bearing pointers and course-set knob with associated course-selection window.

STATUS: in production and operational on Bell 212, 214, 412, Sikorsky S-76, S-61, Aérospatiale Super Puma and Agusta AB212, AB412 helicopters.

4-inch horizontal situation indicator for Hawk

British Aerospace is the customer for this instrument, which is fitted to export versions of its Hawk light strike/trainer. Major features include a course-bar indicator, to/from indicator, glideslope pointer, two bearing pointers, course-set knob, course-selection window, and digital readout of range which is compatible with ARINC 568 digital input. The instrument meets full MIL-Spec standards.

STATUS: in production and service.

Astronautics 4-inch horizontal situation indicator for British Aerospace Hawk

ACA 123790 horizontal situation indicator

This is a 4-inch (102 mm) horizontal situation indicator to military specifications and fitted in the Bell AH-1S Cobra helicopter. It has similar features to other members of the company's family of instruments, including a range readout compatible with ARINC 582.

STATUS: in production and service.

ACA 130500 horizontal situation indicator

Similar in presentation to other company horizontal situation indicators, this 3-inch

Astronautics ACA 126460 horizontal situation indicator

(76 mm) diameter instrument is being supplied for the US Marine Corps/McDonnell Douglas AV-8B V/Stol aircraft.

STATUS: in production and service.

ACA 126460 horizontal situation indicator

Another 3-inch (76 mm) horizontal situation indicator differing only in minor detail from the ACA 130500, this instrument has been supplied for the Northrop F-5 and General Dynamics F-16 fighters.

STATUS: in production and operational.

Digital air data computer

This air data computer meets full US military standards. Present aircraft applications are the F-16, A-4, Mirage and a number of additional military aircraft. It is microprocessor-based using precision solid-state vibrating-quartz pressure-transducers, with analogue potentiometers and synchro outputs as well as dual-redundant serial digital data-buses.

Built-in test equipment allows a high degree of self-diagnosis and the computer has considerable growth potential. It has been designed also as a standard air data computer for retrofit and new aircraft programmes.

STATUS: in production and operational.

Analogue engine instruments

Astronautics produces circular-scale engine instruments for various US military aircraft.

Astronautics ACA 126370 horizontal situation indicator

Astronautics ACA 113515 horizontal situation indicator

Temperature, engine pressure ratio, jet nozzle position and fuel-flow are provided and the company makes a mechanical event timer which counts the number of times three specified engine temperature limits are exceeded.

Thermocouple temperature indicator
(temperature range) 0 to 1200° C
(diameter) 2 inches (51 mm)
(length) 6 inches (152 mm)
(weight) 1.5 lb (0.68 kg)
(power) 10 W 115 V, plus 2.5 W at 5 V for integral lighting
(specification) MIL-STD-I-27552B (USAF)

Synchro temperature indicator
(temperature range) 0 to 1200° C
(diameter) 2 inches (51 mm)
(length) 4.25 inches (108 mm)
(weight) 1 lb (0.45 kg)
(specification) MIL-STD-I-25685

Synchro thrust indicator
(engine pressure ratio range) 1.2 – 3.4
(other indications) cruise and take-off epr in windows
(diameter) 2.25 inches (57 mm)
(case) 2.375 inches (60 mm) square
(length) 4.75 inches (121 mm)
(weight) 1.5 lb (0.68 kg)
(specification) MIL-STD-I-25859A (USAF)

Nozzle position indicator
(nozzle range) 5 positions between open and closed
(diameter) 2 inches (51 mm)
(length) 1.75 inches (44 mm)
(weight) 0.6 lb (0.27 kg)

Fuel-flow gauge
(fuel-flow range) 0 – 5000 lb/h (0 – 2268 kg/h)
(diameter) 2 inches (51 mm)
(length) 6.125 inches (155.6 mm)
(weight) 1.5 lb (0.68 kg)
(power) 8 W 115 VA plus 2 W at 5 V for integral lighting
(specification) MIL-STD-I-27182A (USAF) or MIL-STD-26299D

Engine over-temperature timer
(display) indicates duration of limit exceeded and number of occasions for each temperature specified
(dimensions) 3.5 × 3 × 5.5 inches (89 × 76 × 131 mm)
(weight) 2.25 lb (1.1 kg)

STATUS: in production and service.

EPI engine performance indicator

The engine performance indicator (EPI) is a dichroic liquid crystal display monitoring and displaying five critical engine parameters. It is designed to achieve full dual-redundancy, to be of minimum weight, volume and power demands, to be easily maintainable and have high reliability. In bright ambient light the EPI is illuminated by a reflective mode, at lower light levels the display is illuminated by a high intensity source.

The EPI is designed to operate over a temperature range of –55°C to 85°C and within an altitude band from sea level up to 70 000 feet. It meets the requirements of MIL-E-5400T and MIL-STD-810C.

STATUS: in production.

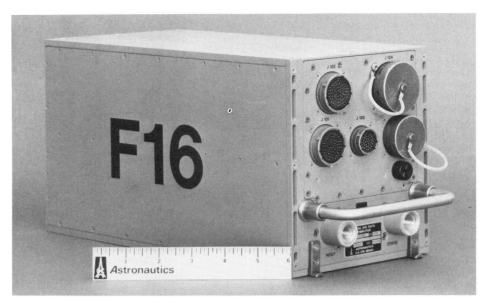

Astronautics digital air data computer

Astronautics engine performance indicator

Astronautics analogue engine instruments (left to right): over-temperature timer, exhaust-gas temperature indicator, and converter

---

Bendix

Bendix, Allied Signal Inc, Flight Systems Division, Teterboro, New Jersey 07608
TELEPHONE: (201) 288 2000
TELEX: 134414
TWX: 710 990 6136

Digital display indicator for F-20

Two of these indicators comprise the primary head-down flight and radar/systems information displays in the Northrop F-20 Tigershark. The two indicators form a dual-redundant system; if one display fails, the other can be commanded to screen whatever the pilot

needs. The system comprises the two monochrome cathode ray tubes displays, a single large-scale integration-based programmable display processor and a data-entry panel.

The pilot can select any one of ten modes from the menu. The latter can show EHSI (electronic horizontal situation indicator) information, INS (inertial navigation system) alignment, waypoints, control/navigation data, communications information, IFF (identification friend or foe), radar, head-up display, stores' disposition, and test functions. Information and symbology is painted on a high-brightness, high-contrast P43 phosphor display with a viewing area of 4.75 by 4.75 inches (121 by 121 mm), using time-shared stroke and 525-line raster writing. The system can, therefore, combine television and other electro-optic information with symbol overlay such as waypoint and target-location data.

Dimensions: (display indicator) 6 × 6 × 13.8 inches (152 × 152 × 351 mm)
(display processor) 8.6 × 8.75 × 12.5 inches (218 × 222 × 318 mm)
(data entry panel) 3.7 × 4.6 × 5.6 inches (94 × 117 × 142 mm)
Weight: (display indicator) 17.3 lb (7.86 kg)
(display processor) 28 lb (12.7 kg)
(data entry panel) 0.7 lb (0.31 kg)
Power: 340 W at 115 V 400 Hz 3-phase
Display characteristics: hybrid, stroke and raster, with resolution of 1000 tv lines
Reliability: target mtbf (display indicator) 4000 h (display processor) 3000 h

STATUS: in final development and limited production as company risk venture for Northrop F-20 Tigershark.

Moving map display for F-15
This remote map reader provides a continuous colour or monochromatic high-resolution video signal to a multi-function display. It uses existing 35 mm film strips housed in self-indexing, interchangeable cassettes which automatically engage and align when inserted into the unit. The system can be mounted in the equipment bay, so reducing pressure on cockpit space. The map images appearing on the screen are therefore synthetic, as opposed to the real images projected onto the screen of conventional moving map displays.

The centre of the map images is automatically aligned with the aircraft present position, using data from the navigation system. The unit incorporates a built-in test facility.

Dimensions: 7.6 × 7.5 × 12.56 inches (193 × 191 × 319 mm)
Weight: 27 lb (12.3 kg)

STATUS: in production for US Air Force/ Sikorsky HH-60D helicopter and McDonnell Douglas F-15 Eagle.

Series 3 electronic flight instrument systems for general aviation
Bendix announced a family of integrated avionics EFIS (electronic flight instrument system) displays for general aviation in 1982, and the concept began to crystallise at the 1983 Paris Air Show. Series 3, as the system is now called, comprises an EFIS display, a multi-function cathode ray tube, communication and navigation equipment and an ARINC 429 digital data-bus. The aim was to systematise the comparatively disorganised avionics situation in this field to enable equipment made by different manufacturers to communicate by adopting airline-style ARINC 429 architecture.

At the centre of the Series 3 system is the EFIS display, which is to be produced in three sizes: 6.25 inches square (158 mm square), 5 by 6 inches (127 by 152 mm), and 4.75 by 5 inches (120.7 by 127 mm) and with what Bendix claims to be a unique stroke-writing method. Typical configuration (at least for the higher-performance turbine types) would be two cathode ray tubes each for pilot and co-pilot and two shared EICAS (engine indication and crew alerting system) displays for systems and navigation.

STATUS: in production.

EFIS electronic flight instrument system for air transport
Bendix has developed an EFIS system for the airlines in which cathode ray tube and stroke/raster generator is combined into a single unit, rather than having two cathode ray tubes share one generator. The system is said to weigh about half as much as conventional EFIS devices, to consume about a third less power and occupy a third less space. It will also, reportedly, cost less.

Horizontal situation indicator
Designed for common installation with Bendix VSI and mode selector equipment, this instrument is suitable for a wide range of civil/military helicopter or fixed-wing applications. It can display aircraft heading, two simultaneous bearings, distance and course deviation. There are several electrical outputs for use in other aircraft navigation systems.

Dimensions: 4.25 × 5 × 12.6 inches (108 × 127 × 321 mm)
Weight: 6.8 lb (3.1 kg)
Power: 11.5 VA at 115 V 400 Hz plus lighting

STATUS: in production.

Vertical situation indicator
The Bendix VSI is designed for installation with the company's horizontal situation indicator and a common-mode selection panel. It has conventional ADI functions and uses a spheroid pitch/bank background. Roll and pitch trim is provided, and indications of glideslope, localiser and navigation warnings are standard. There is also an inclinometer and a rate of turn pointer.

Dimensions: 133 × 127 × 314 mm
Weight: 5.9 lb (2.7 kg)
Power: 8.5 VA 115 V 400 Hz plus lighting

STATUS: in production.

Bendix Avionics Division, Allied Signal Inc, 2100 NW 62nd Street, Fort Lauderdale, Florida 33310
TELEPHONE: (305) 776 4100
TELEX: 514417
TWX: 510 955 9884

Enhanced T/CAS 2 threat-alert/collision-avoidance system
Currently under test by the FAA on a Boeing 727 is a Bendix system that monitors the position and closing characteristics of other aircraft to provide collision avoidance information and guidance to the flight crew. Designated Enhanced T/CAS 2 to differentiate it from the Minimum T/CAS 2 being tested by Sperry Dalmo Victor, the system will begin operational tests in 1986. The FAA has ordered seven such systems as part of its LIP limited installation programme; it will be the first time aircraft with T/CAS equipment will have been able to interact with one another in an operational environment.

The system interrogates transponders on other aircraft to measure bearing, distance and altitude separation, and closing rates in these axes, and provides warning information and commands to the flight crew. In contrast to the Minimum T/CAS 2, the Enhanced T/CAS 2 will compute collision-avoidance commands in both vertical and horizontal planes; this is partly the result of using a phased-array antenna that provides more accurate bearing and rate-of-change of bearing. This antenna also generates a much narrower beam so that fewer aircraft are interrogated at a time, helping to obviate confusion in high-density terminal areas. Logic circuits determine potential conflicts in order of increasing risk, and direct the antenna to scan more frequently those that pose the greatest danger. Manoeuvre commands are generated by algorithms developed by Mitre Corporation, and the system was to be tested during 1985 in the Los Angeles area.

STATUS: evaluation programme.

Boeing
Boeing Military Airplane Company, Box 3707, Seattle, Washington 98124
TELEPHONE: (206) 237 1710
TELEX: 329430

Pictorial format displays
In an attempt to replace traditional alpha-numeric or symbolic information presentations, Boeing Military Airplane Company is studying ways, under contract to Wright Aeronautical Laboratories, to simplify aircraft cockpit and flight-deck displays. These advanced systems, known by Boeing as pictorial format displays, exploit emerging technologies in equipment and computer-generated imagery to lessen the pilot's workload.

In autumn 1982 five such displays, using colour cathode ray tubes, were integrated into a simulated cockpit representative of a typical single-seat fighter to solicit opinions from US Air Force and Navy pilots during ground-based but otherwise realistic 'missions'. This exercise spurred new developments in threat-warning

Pictorial format cockpit of the future, using 'expert systems' techniques

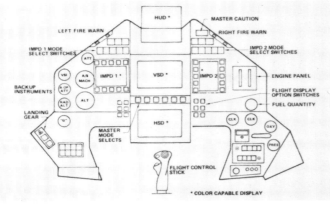

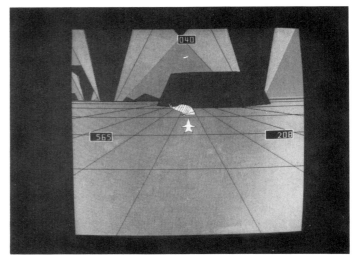

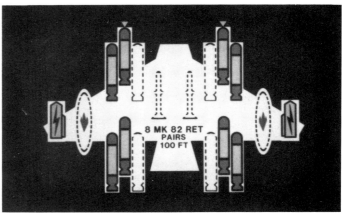

Top left and lower left, layout of five Boeing pictorial format displays in simulated cockpit. Top right, perspective viewpoint showing projected flight-path following route around nearest hill. Lower right, possible format describing stores disposition

systems and displays. The five cathode ray tubes comprised the head-up display (the primary flight instrument), the vertical situation display, the horizontal situation display, and two interactive multi-purpose displays. In June 1984 Boeing was awarded a follow-on contract by the US Air Force to expand this technology for use in two-seat combat aircraft. In July 1985 Boeing was awarded an add-on contract from the US Air Force to incorporate 'expert-systems' artificial intelligence into the multi-

crew pictorial format cockpit, already being developed under a previous contract.

Boeing will design, develop and implement a simulation to demonstrate pilot decision aided by an expert system. These systems use high performance computer programs to assist the pilot by actually making decisions in limited task areas.

The displays will be tailored to specific mission phases and controlled by an expert system called the Crewstation Information

Manager. The programme will also demonstrate the use of voice inputs to perform various cockpit management and control functions. The technology will be designed to include all-weather requirements for a wide range of aircraft and missions, including tactical fighters, bombers, support aircraft and helicopters.

STATUS: research tool.

Bonzer

Bonzer Inc, Division of Terra Corporation, 3520 Pan American Freeway, Albuquerque, New Mexico 81707
TELEPHONE: (505) 884 2321

Impatt radar altimeter

This is a low-cost, lightweight, radar altimeter suitable for general aviation. It features a combined antenna and transmitter-receiver unit, and an independent indicator. The antenna uses microstrip, solid-state technology, and a miniaturised pulse-measurement processing system. The servo-driven indicator incorporates a decision-height selector which gives visual and audio indications on reaching the preset altitude. An additional module provides interfaces for other systems, such as autopilot/flight-director, ground-proximity warning system, and recorders.

Dimensions and weight: (transmitter-receiver-antenna unit) 102 × 133 × 33 mm; 0.5 kg (indicator) 3 ATI; 1 lb (0.5 kg)

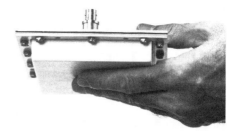

Bonzer Mini-mark radar altimeter

Bonzer Impatt radar altimeter

Frequency: 4300 MHz
Max altitude: 2500 ft
Accuracy: (40–100 ft) 5 ft
(100–400 ft) 5%
(> 400 ft) 7%
Power: 14 or 28 V dc
STATUS: in production.

Mini-mark radar altimeter
This was developed as a very low-cost radar altimeter suitable for general aviation. The system uses a combined antenna and trans-mitter-receiver unit and can have a decision-height unit added to give visual and audio indication of having reached a preset altitude.

Dimensions and weight: (transmitter-receiver-antenna unit) 89 × 127 × 153 mm; 0.9 kg (indicator) 51 mm dia × 89 mm deep
Frequency: 4300 MHz
Max altitude: 1000 ft
Accuracy: 7%
Power: 14 or 28 V dc

STATUS: in production.

Clifton
Clifton Precision, Box 4508, 2734 Hickory Grove Road, Davenport, Iowa 52808
TELEPHONE: (319) 383 6000
TELEX: 468429

Type AAU-34/A standby altimeter
This servo controlled automatic pressure stand-by altimeter is designed to MIL-A-83419B and to interface with air data computers conforming to MIL-C-27889 (CPU-46) and MIL-C-38240 (CPU-66). The operating range is –1000 to 80 000 feet.

Dimensions: 3.25 × 3.25 × 6.5 inches (82 × 82 × 164 mm)
Weight: 3.7 lb (1.6 kg)
Power: 25 VA at 115 V 400 Hz

Collins
Collins Avionics Division, Rockwell International, 400 Collins Road NE, Cedar Rapids, Iowa 52498
TELEPHONE: (319) 395 1000
TELEX: 464421
TWX: 910 525 1321

Integrated display system for Boeing 747-400
Collins is developing a new integrated electronic colour display system for the Boeing 747-400. (A Collins digital flight control and central maintenance computer will also be standard on the aircraft.)

The display system will feature six 8 inch (21 cm) square colour cathode ray tube displays. Any one of these units can be selected to show either electronic flight information system (EFIS) or electronic indication and crew alerting system (EICAS) data.

The new technology will give the 747-400 only 38 percent of the cockpit lights, gauges and switches as compared with current 747 aircraft and will lead to better aircraft availability, according to Boeing and Collins.

STATUS: in development.

FD-108/109 flight-director system
Essentially similar systems, the FD-108 uses 4-inch (102 mm) instruments and the FD-109 attitude director and horizontal situation indicator uses 5-inch (127 mm) instruments. Two categories of each are available, suffixed Y and Z, the latter incorporating air data and additional features. Either flight-director system can be integrated with many types of modern autopilots, and both versions are used on current production airliners.

The attitude director indicator has a V-bar display and a flat-tape background which provides linear pitch information through ±90°. Outputs are provided for annunciations of all submodes and approach progress indications. A conventional compass-rose horizontal situation indicator is provided with an electronic digital display of distance-to-go and a mechanical course readout.

STATUS: in production.

FD-110 flight-director system
The FD-110 system uses 5-inch (127 mm) attitude director and horizontal situation indicator instruments. Several configurations are available, making them suitable for individual airliner, company and autopilot combinations. Configurations can use cross-pointer or V-bar attitude director indications, various ARINC system interface compatibilities, or area navigation and inertial navigation system integration. The equipment is used in many current airliner types.

STATUS: in production.

FDS-84 ProLine flight-director system
Featuring separate 4-inch (102 mm) attitude director and horizontal situation indicator displays, the FDS-84 system is compatible with Collins FCS-80 and FCS-105 autopilot/flight control systems. It is a comprehensive system suitable for high-performance business and executive aircraft, and commuter airliners.

The attitude director indicator uses a flat-tape attitude display background and has V-bar steering command symbology. A radio-altimeter readout is optional in this unit. The horizontal situation indicator uses a compass-rose presentation with an electronic readout which can show distance, time-to-go or ground speed information. The equipment is fully-compatible with ARINC standard VOR, distance measuring equipment, inertial navigation system and omega/vlf sensors.

ADI-84 attitude director indicator (4-inch)
Dimensions: 4.2 × 4.2 × 9.4 inches (106 × 106 × 238 mm)
Weight: 5 lb (2.27 kg)
Power: 1 A at 26 V ac 400 Hz
Temperature range: –20 to +70° C
Altitude range: –1000 to +50 000 ft
TSOs: C3b, C4c, C52a

HSI-84 horizontal situation indicator (4-inch)
Dimensions: 4.2 × 4.2 × 8.8 inches (106 × 106 × 224 mm)
Weight: 5.3 lb (2.4 kg)
Power: 165 mA at 26 V 400 Hz
200 mA 28 V dc
Temperature range: –20 to +70° C
Altitude range: 0 – 50 000 ft
TSOs: C6c, C52a, C66a

REU-84 remote electronic unit
Dimensions: 3.25 × 4.67 × 5.2 inches (83 × 118 × 131 mm)

Weight: 1.1 lb (0.5 kg)
Power: 110 mA at 28 V dc
Temperature range: –40 to +70° C
Altitude range: 0–50 000 ft
TSOs: C6c, C52a, C66a

STATUS: in production.

FDS-85 ProLine flight-director system
Based on 5-inch (127 mm) attitude director and horizontal situation indicator instruments, and compatible with the Collins APS-80 autopilot, this system is suitable for high-performance business aircraft and commercial airliners. It is the company's most comprehensive mechanical flight-director system. Flat-tape attitude indication, V-bar steering commands and electronic readout of distance, time-to-go, speed or elapsed time are standard. A separate heading/course control panel is used.

ADI-85 attitude director indicator
Dimensions: 5.2 × 5.2 × 8.1 inches (131 × 131 × 206 mm)
Weight: 7 lb (3.2 kg)

HSI-85 horizontal situation indicator
Dimensions: 5.2 × 4.4 × 9.1 inches (131 × 112 × 229 mm)
Weight: 7.2 lb (3.3 kg)

HCP-86 heading/course control panel
Dimensions: 5.75 × 1.5 × 6 inches (146 × 38 × 152 mm)
Weight: 1.4 lb (0.63 lb)

Overall system
Temperature range: –20 to +70° C
Altitude range: 0–35 000 ft
Cooling: convection
Power: 44 VA at 26 V 400 Hz
260 mA at 28 V dc
10 W at 5 V ac/dc for lighting

STATUS: in production.

FIS-70 ProLine flight instrumentation system
Compatible with the Collins APS-80 or AP-106A autopilots, the FIS-70 comprises 4-inch (102 mm) attitude director and horizontal situation indicator instruments and is suitable for high-performance turboprop and business jet aircraft. It is essentially a low-cost version of the FDS-84 system, and excludes the digital distance/course readouts, and radio-altimeter

displays options. All other FDS-84 features are incorporated.

ADI-70 attitude director indicator
Dimensions: 4.2 × 4.2 × 8.25 inches (106 × 106 × 210 mm)
Weight: 4.7 lb (2.1 kg)
Temperature range: −15 to +71° C
Altitude range: −1000 to +50 000 ft
TSOs: C3b, C4c, C52a

HSI-70 horizontal situation indicator
Dimensions: 4.2 × 4.2 × 9 inches (106 × 106 × 229 mm)
Weight: 4.9 lb (2.2 kg)
Temperature range: −20 to +70° C
Altitude range: 0–35 000 ft
TSOs: C6c, C52a
Total power:
1.04 A at 26 V ac 400 Hz
0.2 A at 28 V dc
0.34 A at 28 V for lighting.

STATUS: in production.

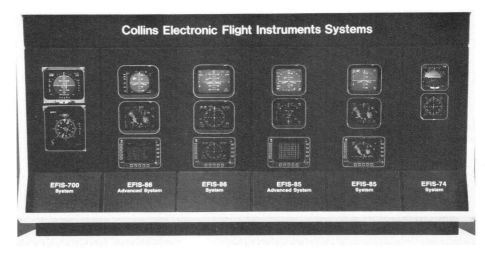

Complete line-up of Collins EFIS displays

EFIS electronic flight instrument systems

Collins has developed four families of cathode ray tube based electronic flight-deck instrument systems: EFIS-700, a four-tube 6 by 7 inch (152 by 177 mm) installation for the Boeing 767 and 757 airliners; EFIS-86, with five 6 by 5 inch (152 by 127 mm) tubes; EFIS-85, having five or three 5 by 5 inch (127 by 127 mm) cathode ray tubes; and EFIS-74, with 4 by 4 inch (102 by 102 mm) displays. EFIS 'firsts' claimed by Collins are certification of the first airline system (aboard the Boeing 767), of the first general aviation installation (EFIS-74), and of the first five-cathode ray tube system. By October 1984 EFIS-type displays had been chosen for 45 aircraft types, of which 40 installations had been certificated.

Collins EFIS-700 electronic display as attitude director indicator

Collins EFIS-700 electronic display as horizontal situation indicator with expanded forward sector

EFIS-700 electronic flight instrument system for 767 and 757

In December 1978, Boeing placed with Collins the world's first production orders for primary flight instruments based on cathode ray tubes rather than electromechanical indicators. They were designed initially for the Boeing 767 and 757.

The EFIS-700 electronic flight instrument system comprises an EADI (electronic attitude director indicator) and an EHSI (electronic horizontal situation indicator) for each of the two pilots in these transports, and each pair of instruments has an associated mode-control panel. They provide all the functions associated with earlier electromechanical ADI (attitude director indicator) and HSI flight-director instruments, and in addition show map and flight-plan data, weather patterns, radio height, automatic flight control modes and flight-path information, all on 7 by 6 inch EHSI (178 by 152 mm) and 5 by 6 inch EAD1 (127 by 152 mm) cathode ray tubes.

These instruments utilise bright, three-gun, robust shadow-mask cathode ray tube technology permitting no fewer than eight colours: the traditional red, blue and white associated with cathode ray tubes is augmented by magenta, yellow, green, cyan and white (not normally used). When used in conjunction with a contrast enhancement filter, the high resolution cathode ray tubes provide bright displays that are readable under all flight-deck lighting.

Each of the two pairs of EFIS instruments in an aircraft is driven by its own symbol generator, but a third, standby, generator is retained as a 'spare' that can be switched in as necessary on failure of a dedicated unit. These symbol generators utilise the Collins CAPS-8 sixth-generation data-processor, implemented with the latest medium-scale integration and bipolar bit-slice large-scale integration 2900 devices.

The electronic attitude director indicator presents primary attitude information, together with pitch and roll steering commands. Secondary data is also shown, such as ground speed, autopilot and autothrottle mode and many others. In order to keep the display uncluttered, information is switched out as soon as it is not needed; for example instrument landing system and radio height symbols are absent during cruise, appearing only during the final approach.

The electronic horizontal situation indicator depicts the horizontal position of the aircraft in relation to selected flight data and a map of the navigation features in the vicinity of the aircraft at any given time. Aircraft track, trend vector information and desired flight plan are also displayed. This allows rapid and accurate manual or automatic flight-path correction. Other information can also be displayed, such as wind-speed and direction, vertical deviation

Layout of flight panels in Boeing 767 and 757. Each pilot has a set of two EFIS displays, mounted one above the other. The two EICAS displays on the centre panel can be monitored by both pilots

from a selected profile, and time to the next navigation waypoint. Weather patterns can also be superimposed on the navigation picture.

During 1984 a third-generation 6 by 7 inch (152 by 178 mm) EFIS-700 system with six cathode ray tube displays was selected for the Fokker 100 airliner. In this system the symbol-generator electronics are incorporated in the display units themselves.

STATUS: following competitive selection of Collins' EFIS in 1978 for Boeing 767 and 757, the company was contracted to build 600 sets of equipment for deliveries which began in mid-1982.

EICAS engine indication and crew alerting system for 767 and 757

The Collins EICAS engine indication and crew alerting system was chosen by Boeing in March 1980 as standard equipment on the 767 and 757 airliners. The EICAS system, developed by the company's Air Transport Division, comprises two multi-colour cathode ray tube displays, two computers and a single selector panel. The display unit is identical to the electronic horizontal situation indicators on the pilot's display panels, though rotated through 90° in its function as an engine indicating system.

Each aircraft has two EICAS cathode ray tubes mounted one above the other on the centre instrument panel where they can be monitored by the two pilots. For those aircraft operated by three-man crews, a third cathode ray tube is installed on the flight-engineer's station. On the centre panel the top EICAS cathode ray tube is programmed to display primary engine information (engine pressure ratio, fan speed and exhaust gas temperature) as electronic symbols representing traditional circular scale and pointer instruments, together with cautionary information (for example, wheel-well overheat showing dangerously high tyre temperatures, or failure of a yaw damper in the flight-control system). The lower EICAS display shows lower priority information such as compressor speed, fuel flow and oil tempera-tures, pressures and tank contents.

In the case of failure of one EICAS display, priority information automatically switches to the other cathode ray tube, and the dual-redundant computer installation permits both cathode ray tubes to be driven from one unit.

The multi-colour cathode ray tube displays in the EICAS configuration measure 7 by 6 inches (178 by 152 mm) and are driven by one of the two computers, the other acting as a 'hot' spare.

STATUS: first engineering model for 757 was delivered to Boeing in mid-1980 and company has contracts for 300 sets of 757 equipment and same number for 767.

EFIS-85 electronic flight instrument system

As a result of its work on television-type flight-instrument displays for commercial aircraft such as the Boeing 767 and 757, Collins has developed equivalent systems for business, corporate and commuter aircraft. A prototype system was shown at the National Business Aircraft Association's (NBAA) Convention in September 1980, and two months later at the Commuter Airline Association of America simulated instruments were exhibited on the Embraer 120, Saab-Fairchild 340 and Short 360 stands.

A full EFIS-85 system comprises dual 5 by 5 inch (127 by 127 mm) electronic attitude director indicators and electronic horizontal situation indicators (one set of equipment for each pilot), a multi-function display on the centre panel to be shared by the pilots, and mode controls. The electronic attitude director indicator and electronic horizontal situation indicator cathode ray tubes (crts) can display all the information traditionally associated with electromechanical flight-director instruments and, in addition, weather-radar patterns, navi-gation maps, performance data and navigation waypoints. Presentation is very flexible: the EHSI-85 can show conventional horizontal situation indicator information on a circular scale or it can expand just the forward sector of the display and show weather maps.

The EFIS-85 crt incorporates a three-gun assembly, a shadow-mask, a faceplate with phosphor coating and a glass envelope to enclose the elements. The in-line electron gun

Collins EFIS-85 has been chosen for Westwind Astra

assembly provides improved convergence and mechanical rigidity and the high-resolution shadow-mask gives four to six times better resolution than that of a domestic television set because the phosphor dots are so much closer together. The displays use both stroke and raster writing; the high-intensity stroke writing of symbols, in conjunction with contrast-enhancement filters, enables displays to be read even in full sunlight. Primary colours are red, blue and green with easy synthesis of several derivative colours, including white.

A Collins multi-function crt display was being considered for the display of tabular navigation data from the FMS-90 or the LRN-85 in addition to the use of the EFIS system to show some of this information.

STATUS: in production. Collins claims EFIS-85 is first such system with five crts to be certificated. Aircraft used for approval trials was Dassault Falcon 100, and certification was announced in December 1982. Beech King Air 200, first turboprop with all-digital avionics, was certificated in April 1983. In June 1983 system was approved by UK's Civil Aviation Authority for British Aerospace HS.125-700; EFIS-85 has also been chosen for Westwind Astra business jet. In June 1984 EFIS-85 was certificated for the Cammacorp DC-8 Super 70. This was the first occasion on which 5-inch (127 mm) crts had been approved for an air transport appli-cation. The installation comprises four EFIS crts (two for each pilot), but no central electronic displays. The system is now being offered on the Learjet Model 55.

EFIS-74 electronic flight instrument system

EFIS-74 is for the general aviation single and light-twin turboprop market and comprises 4 by 4 inch (102 by 102 mm) EHSIs (electronic horizontal situation indicators) designed to work in conjunction with the company's ADI-84 attitude director indicator, and an information display/radar navigation centre, the Collins IND-270 cathode ray tube, which is used in conjunction with the comany's WXR-270 weather radar. The system is completed by a DCP-270 display control panel.

EFIS-74 was first shown in 1982 and is claimed to be the first business aircraft EFIS to be certificated.

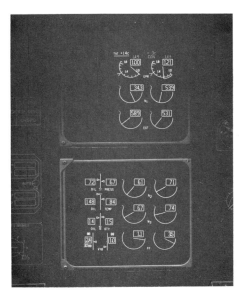

Collins EICAS engine instruments display

Cockpit of Beech King Air 200, the first all-digital turboprop aircraft. It has, among other systems, Collins EFIS-85 and APS-65 digital autopilot

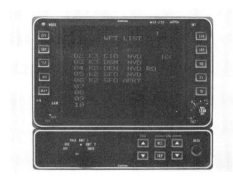

Collins information display/radar navigation centre for EHSI-74

The IND-270 cathode ray tube can present on pilot demand up to 128 pages of easily-accessed, pilot-programmable text, and navigation information from a Collins LRN-85 long-range navigation system. The equipment is programmed by a portable data-reader that can load the system with performance (for example, cruise/consumption) tables, emergency checklists and other alphanumeric information. Chapter-by-chapter indexing facilitates input and retrieval. In addition to the information stored within the DCP-270, many pages of data are available from the LRN-85.

STATUS: in production and in service. Certificated on Commander 690, King Air 200, Cessna 441, King Air F90, Bonanza, Beech 100 and Mitsubishi MU-2. A dual installation has been selected for the Learjet Model 35A.

EFIS-86 advanced electronic flight instrument system

In 1983, Collins introduced the EFIS-86 Advanced, an electronic flight instrument system with five 6 by 6 inch (152 by 152 mm) or 5 by 6 inch (127 by 152 mm) cathode ray tube (crt) displays for top-of-the-range GA and commuter aircraft. They comprise dual EADIs (electronic attitude director indicators), dual EHSIs (electronic horizontal situation indicators) and a single MFD (multi-function display). The two pilots each have an EADI and an EHSI, and

Collins 5 by 6 inch (127 by 152 mm) EADI (above) with primary airspeed display and EADI with weather imagery represent EFIS-86 system for Dassault Falcon 50 and 200

share the centrally-mounted MFD display, which Collins claims to be unique to its own EFIS system.

The high-resolution crts combine raster and stroke writing. The EADI provides conventional ADI information, together with airspeed, airspeed trend and multi-source vertical and lateral deviation information. The addition of air data information is facilitated by the increase in the display area of the EFIS-86 which ensures that the display remains legible. The large EHSI makes it possible to consolidate all conventional HSI, navigation and weather information directly in front of the pilot, thereby reducing the need to scan other instruments (that is, to integrate the weather radar display mentally). Three EHSI modes can be selected: full compass rose (as in conventional HSI), expanded sector display, or sector display with weather radar paints.

The system is linked with long-range navigation and flight management systems to show additional information, such as an expanded display of HSI, radar and navigation data from the flight instruments, either in combination or separately for more detailed examination. The display can store and show pre-selected lists of waypoints or up to 100 pages of pre-programmed data, such as checklists and emergency procedures, written in chapter form for ease of input and retrieval. The display also can act as a standby flight instrument in the event of an EADI or EHSI crt failure.

Earlier in 1983 Collins had exhibited at the Paris Air Show a less comprehensive version of the system, the EFIS-86, with just four crts (two EHSIs and two EADIs). One version was shown on two examples of the Saab-Fairchild 340. This particular system has display formats unique to the SF340, with symbology designed for the European commuter market.

An EFIS-86B has been certificated on the Canadair Challenger CL-601 and features an unique airspeed readout and airspeed trend predictor.

STATUS: in production. EFIS-86 systems have been installed in Falcon 50 and 200, Gulfstream III and Saab-Fairchild 340. In particular, EFIS-86 Advanced has been certificated on the Dassault Falcon 50 and 200, with airspeed indication on the ADI as the primary speed indication. The aircraft and their EFIS systems have been approved to Category II operation, the first such approval for an EFIS-equipped business jet. New display symbology has been developed for two Dassault aircraft with speed and height scales, and autopilot mode annunci-

ation. Other systems are flying on Embraer EMB-120.

Collins ProLine Concept 4

Introduced at the 1985 NBAA convention, the Concept 4 equipment represents a totally new line of advanced avionics for business jets. The heart of the new system is an integrated centre of avionics intelligence — Collins control central. This unit performs all flight control functions, processing inputs from sensors distributed throughout the aircraft and relaying the appropriate message to the autopilot servoes, flight management system and displays.

The electronic flight instruments have also achieved a new level of avionics integration. Display tube and driver circuitry are integrated into a single, compact 6 by 7 inch (152 × 178 mm) unit. No remote display driver unit is required. Displays include dual primary flight displays and navigation displays, engine indication and crew advisory system (EICAS) and multi-function display. If an abnormal engine condition or emergency develops, the display automatically re-formats to alert the pilot to the situation.

Any abnormalities detected by EICAS, or within the avionics system, are retained in memory and can be displayed on the multi-function display both when the aircraft is in the air or on the ground. This centralised fault isolation and identification system is said to be a unique feature of Collins Concept 4 avionics.

STATUS: in production.

Starship 1 integrated avionics system

Concept 4 technology (see above) is the basis of the advanced integrated avionics system specially developed for the Beech Starship 1. This innovative system uses twelve colour cathode ray tube (crt) displays, two monochromatic crt displays and two plasma displays, which virtually eliminate all of the conventional instrumentation and warning systems. The system covers primary flight displays, navigation displays, multi-function display, EICAS, air data system, weather radar, centralised radio-tuning units, comm/nav/pulse sensors, aircraft data acquisition system and engine data concentrators.

The avionics cooling, power distribution and physical layout have all been specially devel-

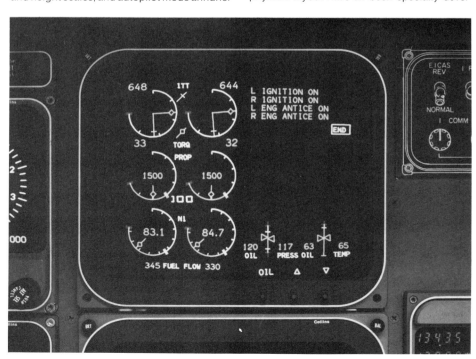

Collins Concept 4 EICAS

oped over a 2 year design period for the most efficient integration into the airframe. The manufacturer claims that dispatch and system reliability, ease of operation and on-board diagnostics have all been designed to new, higher levels for Starship 1. Flight deck ergonomics are enhanced by the location of Collins avionics controls adjacent to the displays. All radios can be tuned by just one knob on a radio-tuning unit, or the pilot may opt to tune via the flight management system keypad.

Flight trials began in May 1985 in a Beech King Air 300.

STATUS: a production shipset was delivered to Beech in August 1985.

Military electronic flight instrument systems

Collins began developing and flight-testing electronic displays in the early 1970s, with the aim of saving space in the crowded cockpits and flight-decks of military aircraft, and reducing the crew's workload. These electronic flight displays allow the integration of flight control, navigation, air data and forward-looking infrared/television sensor information in a space saving as well as a comprehensible form.

HSVD-800 horizontal situation video display

This high-resolution, monochromatic display is intended for tactical situations and can operate in six modes: conventional horizontal situation indicator, stroke-written electronic map, forward-looking infra-red or other raster video, radar map, hover and data presentation. The system accepts inputs in synchro or analogue form, or in MIL-STD-1553, MIL-N-49098 or ARINC 429 format. It is sunlight-readable and compatible with night-vision goggles.

STATUS: FAA-certificated on US Coast Guard/ Aérospatiale HH-65D Short Range Recovery Helicopter.

DU-776 electronic colour map/navigation display

This system combines colour stroke and colour raster writing to overlay flight information such as target assignment and navigation waypoints on a topographical display. The two writing methods may be used separately to optimise particular types of display. The Collins multibandpass filter ensures good sunlight readability but the colour rendering is maintained at low brightness levels for night operation. The stroke-writing presentation is controllable through a MIL-STD-1553 digital data-bus. The display is used in conjunction with a user-programmable symbol generator. A notable advantage is that radar, attitude director or horizontal situation indicator, map, forward-looking infra-red, search and rescue guidance, engine instruments and monitoring and crew-alerting functions can be integrated into one display.

STATUS: under development.

CAI-701 electronic caution annunciator indicator

The CAI-701 cathode ray tube instrument was chosen by Boeing as standard equipment for customers preferring electronic displays on their 767 and 757 airliners. The units use bright-display technology to provide three primary colours (red, green and blue), permitting the traditional combination of red and amber for caution annunciation, and a full colour spectrum, including white, for message display.

Central to the operation of the unit is a high-speed controller that executes a set of instructions to update the cathode ray tube display.

The new Collins displays have symbol generator built-into the instrument

Collins avionics in the Beech Starship 1

The unit accepts data in alphanumeric form from a high-speed ARINC 429 digital data-bus and combines symbol generator and display in one box. The cathode ray tube is fed from a triple-gun assembly so that in the event of one gun failing the system reverts from colour to monochromatic, though with full retention of information.

1D-1805/AJN-18 digital horizontal situation indicator

Claimed to be the first instrument of its type to use digital techniques, this instrument was developed originally for the McDonnell Douglas F-15 fighter. It receives data in both analog and digital formats. Five digital servo systems are used to position the azimuth card, the course pointer, heading marker, and the two bearing pointers. Positional feedback is by serialising optically encoded data from each servo. The instrument can interface with the 9119/AJN-18 flight-director adaptor unit so as to provide a digital link with other command outputs from a number of different navigation sensors, air data computers, and ADF, Tacan and ILS receivers.

Dimensions: 5.0 × 4.25 × 6.5 inches (127 × 108 × 165 mm)
Weight: 5.7 lb (2.6 kg)
Power: 115 V 400 Hz

STATUS: in production.

HSI-85 horizontal situation indicator

This indicator displays numeral counter DME, glideslope or vertical navigation deviation, and elapsed time, in addition to the standard HSI displays. Multiple-mode communicators, bright coloured markings and a mechanical course readout are used to enhance readability.

STATUS: in production.

FD-108 and FD-109 horizontal situation indicators

These indicators are part of the Collins FD-108 and 109 flight-director systems, with respectively 4 and 5-inch (102 and 127 mm) case sizes. Both types provide conventional displays of horizontal situation data with numeric readouts of DME and course. The instruments interface

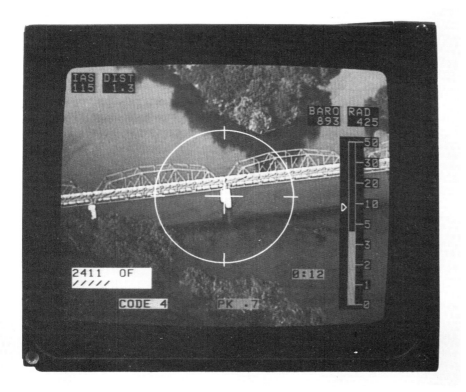

Colour video image from a television camera is overlaid with both raster-scan and stroke-written symbology on this Collins DU 776 military electronic display

with the attitude director, altitude controller, flight computer, mode coupler and the mode selector. They can be coupled to an inertial navigation system to show true course and heading.

AHS-85 strapdown attitude and heading reference system

Designed for business aircraft and regional airlines, this unit is aimed at a market largely ignored before 1983. The emphasis is on simplicity and all sensing and processing functions are grouped in one unit. The system uses a sensor package which eliminates all common high failure-rate components and uses rotors that spin at one-seventh the speed of conventional gyros. The primary sensors are two piezoelectric-equipped multi-sensors (each with four independent pick-offs) which determine two axes of angular rate and linear acceleration data, thus providing one axis of redundant information for system monitoring. These devices are complemented by a flux detector mounted independently in a portion of the airframe without magnetic disturbances. The replacement of any sensing element does

Collins AHS-85 piezoelectric multi-sensor

not require adjustment to the installation and compass compensation remains unchanged if the AHS-85 is replaced.

Processed data is compatible with flight instruments, autopilot systems, weather-radar

stabilisation and flight-data recording. True airspeed can be accepted from a remote air data system to provide additional information outputs. The system provides digital outputs, three-axis body rate and acceleration data, and does not suffer directional gyro gimbal errors.

Format: 1/2 ATR Short (excluding circular flux detector)
Weight: 15 lb (6.6 kg)
Accuracy: (pitch and roll) 5° nominal, 1° during manoeuvres
(heading) 1° nominal, 2° during manoeuvres
Power: 45 W

STATUS: in production.

ALT-50/55 radio altimeters

These two radio altimeters, one with a range of 2000 feet to zero, the other 2500 feet to zero, have been designed for business aircraft. Both types provide decision-height annunciation for Category II landings. Their decision-height annunciators can be set at any desired altitude, and both instruments can interface with high-performance flight directors and autopilots. The indicator incorporates a self-test button alongside the circular scale which is expanded over the range 500 feet to ground-level. An alternative indicator is offered with a numeric readout of radar altitude and decision height.

Dimensions: radio (transmitter-receiver) 3.5 × 3.5 × 12.5 inches (90 × 90 × 322 mm)
(indicators) 3 ATI × 4 inches (102 mm) deep
(antenna) 7 inches (178 mm) dia, 2.75 inches (70 mm) deep
Weight: (transmitter-receiver) 5.6 lb (2.54 kg)
(indicator) 1.5 lb (0.68 kg)
(antenna) 1.25 lb (0.56 kg)
Power: 28 V dc

AM 275 altitude alert system

AM 275 altitude alert system monitors continuously and compares actual altitude with a desired altitude. A knob below the sunlight-readable six numeral display operates the desired altitude setting function. The display includes three visual commands: descent, climb and level over the range –1200 to + 62 000 ft.

Format: 1/2 ATI × 6.5 inches (165 mm)
Weight: 0.8 lb (0.36 kg)
Power: 0.7 A at 11-32 V dc

860F-4 digital radio altimeter

This employs large-scale integrated technology and microprocessors but has an analogue output for driving existing non-digital instruments. It has been designed to ARINC 552/552A and can be used as a reference for Category IIIA automatic landings. Collins says that the digital technology gives a 30 per cent reduction in parts and 40 per cent saving in weight compared with the earlier non-digital 860F-1 system. The equipment can be used from –20 to 2500 feet. The outputs comply with ARINC 429 (digital) and ARINC 552/552A (dc analogue).

Format: 1/2 ATR short
Frequency: 4300 MHz
Weight: 12 lb (5.4 kg)
Power: 50 VA at 400 Hz 115 V

ADS-80/82 digital air data system

Collins has developed the ADS-80/82 air data system as part of the ProLine range of avionics equipment. The system processes data derived from pitot and static pressures and interfaces with the aircraft flight control and flight-director systems.

To adapt the system to a variety of aircraft a single plug-in module is mounted at the rear of the computer. Operating parameters are coded in the module according to aircraft type.

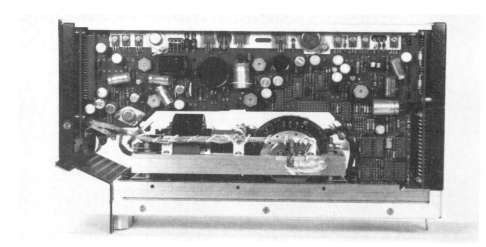

Collins AHS-85 attitude and heading reference system

Optional output instruments include a true airspeed/temperature indicator, Mach/airspeed indicator, altitude pre-selector/alerter, radio altimeter, encoding altimeter and vertical-speed indicator. These indicators are in 3 ATI format. An ARINC 575 true airspeed output is available for use in inertial or omega/vlf navigation systems.

Dimensions: ½ ATR
Weight: 6.6 lb (2.99 kg)
Power: 23 W

Conrac

Conrac Corporation, SCD Division, 1600 South Mountain Avenue, Duarte, California 91010
TELEPHONE: (213) 359 9141
TELEX: 675468

Television display for RF-5E

This 5 by 5 inch (127 by 127 mm) high-resolution, high-contrast monochrome raster display is part of the reconnaissance system for the US Air Force/Northrop RF-5E. The unit acts as a remote viewfinder for the television camera carried in the aircraft's nose. It is a dual-in-line system, that is, the display unit carries both the cathode ray tube tube and the electronics modules.

Display: P.43 phosphor writing on 4.8 × 4.8 inch (122 × 122 mm) aperture viewing area, operating at 525 or 875 lines, remotely selectable
Frame rate: 30/s
Field rate: 60/s
Horizontal scan frequency: 15.75 kHz or 26.25 kHz
Brightness: 250 ft-lamberts
Contrast ratio: 50:1 in low ambient light, 7:1 at 10 000 ft-candles
Grey scales: 10 shades of grey at 10 000 ft-candles
Resolution: minimum horizontal; 1000 tv lines at centre, 800 at corner
Reliability: 1500 h mtbf
Weight: 17 lb (7.7 kg)

Altitude range: –1000 to +50 000 ft
TAS range: 50 – 600 knots

STATUS: in production.

Turbine monitoring system

At the end of 1984 Collins was awarded a $1 million contract by the US Air Force's Aeronautical Systems Division for the study of an integrated turbine engine monitoring sys-

Dimensions: 5.95 × 5.95 × 15 inches (151 × 151 × 381 mm)

STATUS: in limited production for Northrop RF-5E.

Monitor display for AH-64A helicopter

Conrac builds a small-screen, high-resolution monitor display for the US Army/Hughes AH-64A attack helicopter. It is used in conjunction with the Martin Marietta TADS target acquisition designation sight.

STATUS: in production.

Stall-warning system

This is a low-cost version of the flight-optimiser and stall-warning system used on many commercial and military jets, and is particularly suitable for business-jet aircraft. It has a 2-inch (51 mm) stall-margin indicator, stick-shaker capability, an external angle-of-attack sensor and a computer. The system permits aircraft to be flown at optimum conditions throughout critical flight phases, such as approach and take-off, regardless of gross weight, centre-of-gravity position or ambient temperature. The sensor is rugged and simple, consisting of a vane connected to a potentiometer with an associated flap-position input providing compensation for various flap angle selections. The computer is an all-solid-state unit using dc

tem. With Systems Control Technology and Woodword Governor as sub-contractors, Collins will evaluate current US Air Force engine-monitoring equipment and technologies, and then develop hardware and software specifically for the new system, which will enter full-scale development in 1987.

STATUS: study contract.

power supplies, and the standard indicator continuously displays the ratio V/Vs. The system is designed to provide a steady indication, even in rough air, but to respond rapidly to changes in flight conditions. The computer will also permit activation of a stick-shaker through a 28-volt dc warning output.

Angle-of-attack sensor
Vane rotation: 50°
Vane sensitive region: ±25°
Power: 28 V dc (heater)
Dimensions: attaches via 83 mm dia plate
Weight: 0.45 kg

Computer (Model 54301)
Dimensions: 88 × 88 × 190 mm
Accuracy: ±1% of true V/Vs at V/Vs = 1, reducing linearly; to ±5% of true V/Vs at V/Vs = 2
Weight: 0.9 kg
Power: 5 W at 28 V dc

Indicator
Dimensions: 126 × 50 mm dia
Weight: 0.45 kg

Stick-shaker
Weight: approx 2.3 kg
Power: 28 V dc

Flap position transmitter
Dimensions: 38 × 50 mm dia
Weight: 0.23 kg

STATUS: in service.

Dataproducts

Dataproducts New England Inc, Barnes Park North, Wallingford, Connecticut 06492
TELEPHONE: (203) 522 3101
TWX: 710 476 3427

Ice-detection system

Dataproducts New England Aerospace Division has developed an ice-detection probe which employs what is claimed to be the most sensitive method of ice-accretion detection currently available. The probe operates in a cyclic fashion using the thermal characteristics of the ice which forms on it combined with the heat of fusion effect as the ice is formed. The

signal produced may be used to initiate a warning signal or to activate automatically the aircraft's de-icing system. Between detection cycles, the probe is cleared of ice by an integral heater circuit. It is claimed that an ice-accretion thickness of 0.005 inch (0.12 mm) can be detected within 5 seconds. A test facility for airborne confidence checking and for ground servicing purposes is incorporated. The standard ice detector is designed to sense ice formation on fuselage and nacelle air intakes for turbine engines, but other versions are available for carburettor ice detection on piston-engined aircraft and for other ice-sensitive areas.

The system has been selected as standard equipment for the Lockheed C-130 Hercules

and is being installed as a retrofit to all US military C-130s. Dataproducts is also producing ice detectors for the B-1B, Cessna Citation II, the T-46A and the F-16C/D.

STATUS: in production and service.

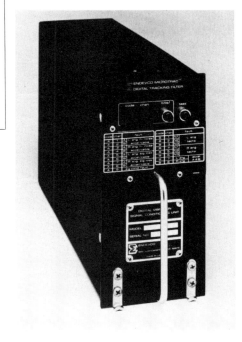

Endevco

Endevco Corporation, 30700 Rancho Viejo Road, San Juan Capistrano, California 92675
TELEPHONE: (714) 493 8181
TELEX: 685608

Microtrac engine vibration monitor

Microtrac is a microprocessor-based engine vibration monitor which is basic equipment on the Boeing 757 and 737-300 airliners and optional on the 747 and 767 transports. The use of microprocessors permits the device to be reprogrammed to suit most aircraft/engine combinations. Vibration is detected by piezo-electric accelerometers mounted at sensitive regions in the engines. The electrical signals are processed by the monitor in the avionics bay.

The Microtrac processor uses a narrow-band digital filter, controlled by the output from a tachometer on the engine, to isolate the vibration frequency. The narrow band-width means that the signal-to-noise ratio of the final vibration indication is high, and the filter's transient response is fast so that it can track the fundamental vibration frequencies of each of the engine's rotors during rapid accelerations.

The filter's accuracy is determined by the number of bits and computer program. Built-in test equipment is provided, and only one programmable read-only memory needs to be changed to adapt the system to a different engine type. Microtrac has an ARINC 429 digital data-bus output so the engine vibration can be received by any EICAS, ECAM, AIDS, or similar unit with a compatible interface. An analogue output to conventional instruments can be provided. The built-in test results can

Endevco Microtrac engine vibration monitor

also be put on to the ARINC 429 bus, and up to 17 fault indications can be stored for display to maintenance crews.

In-flight engine balancing measurement is an optional feature with Microtrac. The amplitude and phase of the once-per-revolution vibration of the Microtrac accelerometer is presented to the microprocessor which selects the weight and mounting position required to balance the rotor more accurately. This data is stored in an expanded memory for later display to ground-crew, and it may be passed into the ARINC 429 bus for a direct cockpit readout. The in-flight balancing measurement can be accomplished more accurately than on the ground and repeated ground run-ups are avoided. Time and fuel are therefore saved.

Format: 3 MCU
Weight: 7.75 lb (3.5 kg)

STATUS: in production.

General Electric

General Electric Company, Avionic and Electronic Systems Division, Aircraft Instruments Department, 50 Fordham Road, Wilmington, Massachusetts 01887
TELEPHONE: (617) 937 4101
TELEX: 947 406

Engine performance indicators

General Electric produces a wide variety of integrated numeric/analogue displays for various applications. Displays utilising General Electric developed dichroic liquid-crystal displays are provided for Boeing 757, 767 and KC-135R aircraft. Other displays utilising magnetic wheels for numeric presentation are provided for McDonnell Douglas F/A-18, AV-8B, and other military aircraft. McDonnell Douglas DC-10, Lockheed L-1011, Boeing 747, and Airbus A300 transport aircraft also utilise General Electric displays.

Vertical tape displays were supplied to the US Navy for the Grumman A-6E Intruder and EA-6B Prowler attack and electronic warfare aircraft. Derivative models in the four-tape configuration are being provided for the Boeing 747 and Lockheed C-5B transports.

General Electric has supplied instrumentation to measure almost all primary and secondary engine characteristics: engine pressure ratio, fan and compressor rotation speeds, exhaust-gas temperature, turbine inlet temperature, torque, propeller rpm, fuel flow, pressure (fuel, oil, air, and hydraulic), temperature (oil and fuel), and oil and fuel quantity. The vertical-scale instruments incorporate two, three, or four tapes per parameter.

STATUS: in production.

DJ 288 engine performance indicator

General Electric produces the DJ 288 standby engine performance indicator for the US Navy/McDonnell Douglas F/A-18 Hornet, and has designed a similar system for the McDonnell Douglas AV-8B Harrier II. Five engine parameters are displayed electromechanically for the F/A-18's two engines. Digital values of rpm, exhaust-gas temperature and fuel flow are given, below which are pointer displays of nozzle position and pressure. The DJ 288 can use ac, dc, or pulse-pair types of sensors, and General Electric is seeking to expand its applications. The system replaces ten conventional 2-inch (51 mm) dial instruments with a 3.7 by 5.75 inch (94 by 146 mm) display.

The digital displays use magnetic wheels driven by latch-decoder drivers. These provide a preset, variable update rate. The analogue presentations are controlled by dc torque motors using the servo-driven nulling position to ensure smoothness of display. The self-contained power-supply operates on a high-efficiency fly-back converter principle incorporating overload protection and operating over a wide input voltage range. The internal built-in test circuitry is powered by an external command signal. The 'canned' and 'display' built-in test sequence provides a visual demonstration check of the correct operation of more than 90 per cent of the component failures. The 'canned' built-in test provides an accuracy check on all channels, which is held for a nominal 8 seconds, following which the magnetic wheels sequence through all indexable digits at the rate of one numeral per second. The unit continues to cycle through these canned and display built-in test sequences until the command signal is removed, at which time normal operation is resumed.

Dimensions: 3.7 × 5.75 × 7.8 inches (94 × 146 × 198 mm)
Weight: 5.4 lb (2.4 kg)
Input power: 15 W at 28 V dc, plus 1.8 W for lighting
Operating temperature range: –54 to 71° C
Vibration: 10 g sinusoidal

Electromagnetic interference: MIL-STD-461 and 200 V/m (RS03)
Mtbf: 2700 h
Bit failure detection rate: > 99%
Rpm range: 2–110% ±1%
Exhaust-gas temperature range: 0 to 999° C ±5° C
Fuel-flow range: 300–15 000 lb/h ±100 lb/h
Nozzle-position range: 0–100% ±5%
Oil-pressure range: 0–200 psi ±10 psi
Full-scale response: (rpm, exhaust-gas temperature, fuel flow) 2 s

STATUS: in production.

Integrated engine instrumentation

General Electric claims to have pioneered the concept of integrated engine display systems, now standard options on the Boeing 747, McDonnell Douglas DC-10, Airbus A300, and Lockheed L-1011 TriStar airliners. The displays are offered in round or vertical configurations. The two 2-inch (51 mm) round and 2 ATI configurations are common for the airliner applications, while early examples of the vertical tape displays were supplied to the US Navy for the Grumman A-6E Intruder and EA-6B Prowler aircraft.

The round instruments typically have pointer and electromechanical counter displays. They are servo driven, by standard ac servos or by General Electric's Accutorque drive. This is a gearless, direct-drive torque motor used for pointer displays, either alone or in combination with digitally encoded individual magnetic counter wheels when high accuracy is needed. Accutorque may also be used with a minimum of gearing to provide a numerical readout on a mechanical counter, with no complex analogue to digital conversion or digital circuits. General Electric has supplied this type of instrumentation to measure almost all primary and secondary engine characteristics: engine pressure ratio, fan and compressor rotation speeds, exhaust-gas temperature, turbine inlet temperature, torque, propeller rpm, fuel flow, pressure (fuel, oil, air, and hydraulic), temperature (oil

General Electric engine performance indicators for US Air Force/Rockwell B-1B bomber

General Electric engine data panel for McDonnell Douglas F/A-18 Hornet

and fuel), and oil and fuel quantity. The vertical-scale instruments also use Accutorque, and may incorporate two, three, or four tapes per parameter.

Pointer/counter electromechanical displays
Dimensions: 2.27 × 2.27 × 6 to 13 inches (58 x 58 × 152 to 330 mm) (depth depends on complexity)
Weight: 1.2 to 1.8 lb (0.5 to 0.8 kg)
Accuracy: typically ±0.5% full-scale; ±0.1% full-scale for some digital instruments
Response time: typically 3 or 4 s full-scale

Vertical scale instruments
Dimensions: (dual-tape) 1.75 × 5.75 × 6 inches (44 × 146 × 152 mm) or 1.25 × 5.5 × 8 inches (32 × 140 × 203 mm) typical
3- and 4-tape instruments higher and deeper
Weight: (dual-tape) 2 lb (0.9 kg)
(3-tape) 3 lb (1.4 kg)
(4-tape) 4 lb (1.8 kg)
Accuracy: typically 0.5% full-scale
0.3% for engine pressure ratio
Response: (cold start) < 3 s
(after warm-up) < 1 s
Mtbf: (dual-tape) 3500 h demonstrated

STATUS: in production.

Solid-state engine instruments
The latest General Electric development in this area is the incorporation of liquid-crystal displays for the digital information, replacing electromechanical counters. Both the Boeing 767 and 757 use General Electric liquid-crystal displays for the standby engine instrumentation. The displays need very low power (measured in microwatts) and are lighter and more reliable. The data is displayed in white on a black background. General Electric has achieved completely modular construction, with separate mechanical and electrical assemblies for motor/potentiometer, gear-box, set counter, numeric display/printed circuit board, and plug- n circuit modules.

Solid-state liquid-crystal display instrumentation
Format: 3 ATI
Weight: 2.3 kg
Accuracy: 0.1 to 0.5% depending on parameter
Response time: less than 1s
Power: 28 V dc

STATUS: in procuction.

Flat-panel and solid-state displays
Active-matrix liquid-crystal displays incorporating thin-film transistor switches to control individual liquid-crystal display picture elements (pixels) are being developed by General Electric's Aircraft Instruments Department for the large-area cockpit and flight-deck displays that are foreseen as needed in the next-generation aircraft. General Electric has built 200 by 200 pixel, 2-inch (51 mm) square demonstration displays employing an amorphous thin-film silicon transistor switch at each pixel electrode to switch vertical columns of data lines and horizontal rows of scan lines, so providing a fully addressable matrix. Colour is achieved by the integration of red/green/blue filters, accurately registered with the pixels.

From early development work, General Electric researchers concluded that a liquid-crystal display with a thin-film transistor at each pixel can perform as well as, or better than, a cathode ray tube.

Large-area fixed-format dichroic liquid-crystal display instruments are also available, using the same formulations as currently in production for the Boeing 767 and 757 transports and KC-135 tanker. They can be arranged to suit a variety of applications including alphanumerics, bar-charts and simulated dial-and-pointer presentations.

STATUS: flat-panel displays in pre-production.

Gould
Gould Inc, Navcom Systems Division, 4323 Arden Drive, El Monte, California 91731-1997
TELEPHONE: (818) 442 0123
TELEX: 215 380
TWX: 910 587 3428

AN/APN-232 Cara combined altitude radar altimeter
The Cara system is an advanced technology, all-solid-state combination of a 0 to 50 000 feet, frequency-modulated, continuous wave conventional pressure-based altimeter and a radar altimeter operating at a nominal 4.3 GHz. The Cara programme to replace height-measuring equipment on most US Air Force operational aircraft with a single type of equipment was initiated by the Warner Robins Air Logistics Centre in early 1982. At that time Gould was awarded a $27.1 million initial contract. Reduction of life-cycle costs was a priority, and in the early stages of the programme the US Air Force adopted Gould's proposed two-level support policy. This stated that circuit module would be replaced in the field, according to built-in test indication, and failed units could be discarded at the air logistics centre level rather than being repaired. The proposal bypassed the need to repair modules, at squadron level, thereby eliminating field-level test equipment. Life-cycle costs are expected to be a third less than those of current systems, saving an estimated $123 million.

STATUS: in production as its standard altimeter for US Air Force. US Navy bought four Cara

Gould AN/APN-232 Cara system

pre-production system for independent evaluation. US Army has contracted for two sets of equipment. Flight tests by US Air Force began early in 1983. First production sets were delivered during 1984 to General Dynamics for F-16 fighter.

Hamilton Standard
Hamilton Standard, Division of United Technologies Corporation, Bradley Field Road, Windsor Locks, Connecticut 06096
TELEPHONE: (203) 623 1621
TELEX: 994439

Engine pressure-ratio transmitter
Hamilton Standard introduced this high-temperature four-bellow engine pressure-ratio transmitter (EPRT) in 1965, following production of earlier systems for the US Air Force's Convair B-58 Hustler supersonic bomber and France's Aérospatiale Super Caravelle airliner and Dassault Mirage family. Over 2000 EPRTs have been delivered for the Pratt & Whitney TF30-powered General Dynamics F-111.

Four bellows are positioned at 90° intervals in a cruciform layout in the engine intake and exhaust. A change in pressure tends to move the bellows, which are then maintained in their original position by a feedback servo system. The signal required to return the bellows is also used to drive a cockpit dial which reads the ratio of exhaust pressure to inlet pressure. The configuration minimises the effect of temperature changes.

Diameter: 5.4 inches (137 mm)
Length: 7.7 inches (196 mm)
Power: 35 W at 115 V, 400 Hz

STATUS: in service.

EPR-100 engine pressure-ratio transmitter
Essentially the civil version of the EPRT

described above, this system entered production for the Pratt & Whitney JT9D-powered Boeing 747 in 1968 and over 3000 have been delivered. The EPR-100 works in the same way as the EPRT, measuring the ratio of turbine exhaust pressure to compressor inlet pressure.

Dimensions: 9.75 × 7.4 × 5.75 inches (248 x 188 × 146 mm)
Weight: 5.5 lb (2.5 kg)
Power: 115 V 400 Hz

STATUS: in production.

EPR-101 engine pressure-ratio transmitter

This transmitter works in a similar manner to the EPRT and EPR-100, but calculates the pressure ratio electronically rather than mechanically. It can be configured to suit the application and has been supplied to the McDonnell Douglas DC-10, Airbus A300 and Lockheed L-1011 TriStar airliners. The CF6-powered DC-10 has three engine pressure-transducers and an ambient ram pressure-inducer. The DC-10 with JT9Ds has two extra ambient transducers. The A300 systems have no ambient pressure transducers.

The unit for the TriStar is called the integrated exhaust pressure-ratio transmitter (IEPRT) because the exhaust pressure-ratio is determined by integrating two separately measured pressures.

Dimensions: (DC-10 and A300) 4.9 × 3.25 × 8.08 inches (125 × 83 × 205 mm)
(TriStar) 6.5 × 8.75 × 10.2 inches (164 × 222 × 260 mm)
Weight: (DC-10 and A300) 3.95 lb (1.78 kg)
(TriStar) 7.25 lb (3.3 kg)
Power: 115 V, 400 Hz

STATUS: in production.

EPR-102 engine pressure-ratio transmitter

This solid-state integrated exhaust pressure-ratio transmitter has been chosen for the RB.211-535 turbofan engine that powers some versions of the Boeing 757 airliner. The EPR-102 calculates engine intake pressure for the engine control system and engine pressure ratio for the cockpit indicator. Vibrating cylinder pressure transducers convert the frequency of vibration to a digital pressure reading according to calibration equations stored in programmable read-only memories. Effects on the

calibration are taken into account using built-in temperature measurements.

An Intel 8085 microprocessor calculates the engine pressure ratio from these values, and converts it, together with intake pressure, into ARINC 429 data-bus format for transmission around the aircraft. The software, contained in programmable read-only memories, is trimmed to individual engine characteristics by a five-bit code.

Dimensions: 9.1 × 5.7 × 3.6 inches (231 × 145 × 91 mm)
Weight: 5.6 lb (2.5 kg)
Power: 13 W

STATUS: being flight-tested and in early production.

ETT-100 tachometer transmitter

Hamilton Standard supplies this tachometer transmitter which measures fan rotational speed on the Pratt & Whitney JT9D big-fan engines powering the Boeing 747, McDonnell Douglas DC-10 and Airbus A300. It is mounted on the fan casing and comprises an eddy-current detector which generates a small voltage as each fan blade passes. The signals are amplified and the time interval between pulses gives the revolutions a minute.

Dimensions: 3.75 × 4.9 × 2.75 inches (95 × 125 × 70 mm)
Weight: 1.12 lb (0.51 kg)
Power: 2.8 W

STATUS: in production.

EDM-110, EDM-111, and EDM-112 PMUX propulsion multiplexers

Hamilton Standard supplies propulsion multiplexers for the Pratt & Whitney JT9D-7R4 (EDM-110) and General Electric CF6-80 (EDM-111) big turbofan engines, and the PW 2037 (EDM-112) 37 000 lb (16 800 kg) 165 kN thrust medium-size turbofan. The applications for these are Boeing's 747, 767 and 757, and Airbus Industrie's A310. The propulsion multiplexer collects analogue data from the engine and air data system, converts it to digital format, samples it, and changes it into a multiplexed data stream. The data is then transmitted to a ground-based diagnostic unit.

The propulsion multiplexer is programmed with the engine's mechanical and thermodynamic limits, and parameters going beyond these limits are stored in a semiconductor

memory for later diagnosis. This is intended to help solve any problems as they arise. Data accepted by the propulsion multiplexer includes engine pressures, temperatures, fuel-flows, vibration levels, oil condition and positions of variable-geometry features such as intake guide-vanes and active-clearance controls.

Dimensions: 16 × 14 × 4 inches (406 × 356 × 102 mm)
Weight: 18 lb (8.1 kg)
Power: 24 W

STATUS: in production.

Helmet-mounted display

Hamilton Standard has developed a helmet-mounted display which uses two 1 inch (2.5 cm) cathode ray tubes to provide high resolution information to an observer, derived from onboard computers and sensors. The night viewing sensors allow the observer to view outside the aircraft, or the aircraft's instruments even in total darkness.

The display has been developed as part of the US Army's Advanced Rotorcraft Technology Integration (ARTI) programme and a system was delivered to Sikorsky in mid-1986 for trials.

Hamilton Standard helmet-mounted display

Honeywell

Honeywell Military Avionics Division, 2600 Ridgeway Parkway, Minneapolis, Minnesota 55413
TELEPHONE: (612) 378 4141
TELEX: 290631

Air data computers

Honeywell supplied the first digital air data computer for commercial transports, with a unit for the McDonnell Douglas DC-10 widebody. The analogue HG180 equipment still remains in production and is fitted in the Boeing 707, 727 and 737. Two improved air data computers have been developed based on the latest electronic systems and techniques.

HG280D80 air data computer

Intended for the McDonnell Douglas DC-10 and MD-80 transports, this microprocessor-based ARINC 576 type digital system can replace the earlier HG280D5 computer. The transducers use proved D5 sensors, with high accuracy and a fast warm-up time. Reliability is claimed to be excellent, with a mean time between failures prediction of 11 000 hours.

The central processor memory and input/output functions have been reduced from the 360 integrated circuits on nine boards of the earlier D5 to 22 circuits on a single card. Built-in test includes both continuous monitoring and manually activated self-test. The D80's non-volatile memory automatically stores information for five flights, and data may be retrieved by selecting a single switch on the front panel. If a failure has occurred either of two warning lights will illuminate.

Honeywell HG280D80 air data computer

Honeywell HG480B air data computer

The D80 can also accommodate limited power interruptions. A capacitatively maintained supply voltage and a cmos memory can retain critical data during power losses of several milliseconds. When power resumes the device automatically restores all outputs based on this stored data.

HG280D5
Format: 1/2 ATR
Weight: 18.2 lb (8.3 kg)
Power: 101 W

HG280D80
Format: 1/2 ATR
Weight: 13 lb (5.9 kg)
Power: 30 W

STATUS: in service.

HG480B air data computer
This digital computer meets ARINC 545 and replaces the analogue HG180 on Boeing 727 and 737 transports. This unit incorporates ARINC 429 buses.

A single-board micro-computer controls the functions, and built-in test (which includes both continuous monitoring and manually activated self-tests) reduces maintenance. Ground built-in test has been expanded and includes q-pot stimulation and transmission of fixed values on all signal output lines.

The HG480 is available in four versions. The basic configuration allows for differences in aircraft wiring but is functionally identical for what are designated the B1, B2 and B3 versions. Designed to operate with conventional analogue equipment, the system also provides outputs for digital instruments and digital autopilot. The B4 version is designed primarily to interface with an all-digital flight control system but can still operate with synchro-driven instruments.

Format: 1/2 ATR
Weight: 16.8 lb (7.6 kg)
Power: 44 W

STATUS: in production and in service.

HG580 air data computer
This advanced device was originally designed as a competitive bid for the Boeing 757. Although unsuccessful in this application, development is continuing for future aircraft use.

STATUS: under development.

ADIRS air data/inertial reference system
This system is an example of a new but growing move to bring together two systems dissimilar in nature but providing complementary data. Recent improvements in electronic component design have reduced the volume of circuitry needed to process a given function. For example, Honeywell inertial reference units (IRUs) currently flying on a variety of aircraft now contain two empty card bays or seven card slots. Similar reductions in the size of transducers and electronics have greatly reduced the size of air data computers (ADCs), and Honeywell has an advanced ADC comprising only two cards in a 4 MCU box, also containing two dual transducers. The result is that current technology ADC components can be accommodated within the existing Honeywell IRU. A low-risk modification to the existing IRU consists of deleting the ADC power supply, installing the remaining ADC circuit cards and components to fit one of the empty card bays in the IRU, sharing the existing IRU power supply and ARINC connector with the ADC, and installing connectors for pitot and static pressures on the front face of what is now an ADIRS system. Apart from the power supply and connectors, the two functions remain isolated from one another. The proposed aircraft installation would comprise three ADIRS boxes per aircraft. Honeywell claims the following savings over separate IRUs and ADCs: 74 per cent less by weight, between zero and 25 per cent less power required, 50 per cent less installed volume, twice the reliability, and 50 per cent greater redundancy.

STATUS: in development. System has been proposed for new Airbus A320 150-seat transport.

Altitude alerting device
In conjunction with an air data computer and servo altimeter, this altitude deviation indicator gives a warning when the aircraft is approaching or deviating from a preset altitude, latches being activated at a nominal 375 feet before reaching the set altitude and when at a nominal 1000 feet after deviating from it.

Dimensions: 5.9 × 3.2 × 1.5 inches (150 × 81 × 38 mm)
Weight: 2.0 lb (0.9 kg)

STATUS: in service.

Honeywell, Commercial Aviation Operations, 5775 Wayzata Boulevard, St Louis Park, Minnesota 55416
TELEPHONE: (612) 542 1533
TWX: 910 6576 2692

Honeywell radar altimeters are highly accurate, time-based and employ leading-edge track processing. They are currently in use in a variety of applications ranging from space installations to missiles and aircraft, many with extreme performance requirements and specifications such as nuclear hardening, and operation at high altitudes and hypersonic speeds. Accommodations have been made for the special requirements of rotary-wing aircraft. Systems are available which provide the minimum rf power necessary to obtain the signal-to-noise ratio needed; they tolerate low frequency vibration; and contain integrated signal processing for advanced hover performance. Tested and proven in the full range of environmental conditions, including heavy rain or snow and variations in temperature, Honeywell radar altimeters are claimed to be reliable at all altitudes up to 70 000 feet and to have the unique capability of meeting customer requirements during severe pitch and roll.

The basic Honeywell radar altimeter system is a lightweight, high-resolution, short pulse system that automatically locates terrain returns and provides a continuous, selective, precision, leading-edge track of the signal. It is composed of a receiver-transmitter, two antennas, and one or more height indicators. It is available in standard off-the-shelf configurations, or in special configurations designed to meet specific customer needs, and in most instances, without an increase in package size.

Honeywell has produced over 40 000 radar altimeters. They are standard on US Army and Navy aircraft. Honeywell is also the primary supplier to the US Air Force with altimeters in volume production for both strategic and tactical missiles and a variety of target drone programmes. Quality of Honeywell produced systems is ensured with automatic test equipment designed to check altimeters during the manufacturing process and at depot level military facilities.

7600/7500 series
Honeywell's 7600 and 7500 radar altimeter systems are the smallest available in the DoD inventory. They use the same proven concepts as the other Honeywell radar altimeter systems; however, all circuits have been re-designed to make maximum use of available integrated circuits. Other design improvements have been combined with time-proven techniques and several thousand HG7500 systems have been produced for target drones and commercial airline application.

Honeywell claims that their 7600 series system represents the industry's first integrally-packaged radar altimeter. The basic system consists of a receiver-transmitter-indicator (HG7600), two antennas (LG81L) and an optional height indicator. It is designed for instrument panel mounting in military and commercial aircraft. The 7500 series system is functionally the same as the 7600 series, but the receiver-transmitter unit and the height indicator are packaged individually.

Typical installation (7500 series)
Dimensions and weights: (transmitter-receiver) 1/2 ATR Short; 3.4 kg
(indicator) 83 × 83 × 101 mm; 0.6 kg
(antenna) 107 × 139 × 4 mm; 0.45 kg
Attitude range: 0-2500 ft

Honeywell 7500 radar altimeter

Accuracy: ±2 ft plus ±2%
Power: 20 VA at 115 V ac 400 Hz

STATUS: in service on MQM-107B and several DoD target drones and commercial airliners

HG7700 series

The HG7700 is specifically designed for the low cost, high-performance requirements of tactical missiles. Honeywell claims that the system is the result of independent development efforts and incorporates the latest accuracy, reliability and performance improvements at the lowest cost possible. It requires reduced power and has a highly-producible design. Predicted mtbf for the HG7700 is 7450 hours. The US Air Force and Rockwell Missile Systems Division have selected Honeywell's HG7700 for the AGM-130A Missile Programme.

STATUS: in development for AGM-130 missile.

AN/APN-194 radar altimeter

The APN-194 is claimed to be Honeywell's most versatile altimeter system. It is standard on all Navy fixed-wing and high performance aircraft, including the F-14. Functions include low altitude warning, radar altitude warning set input, ac rate, ac altitude errors and landing gear warnings. The system furnishes both analogue and digital outputs and measures altitudes from 0 to 5000 feet. Predicted reliability of the system is 4500 hours and the company guarantees 2500 hours mtbf. The use of high Rel MIL-SPEC parts increases the reliability of the APN-194. In addition, it has an adjustable low-altitude warning, indicated by a lamp built into the height indicator. A low-altitude warning discrete is also provided. The system contains circuitry that permits the altimeter to be checked before and during flight.

More than 8000 have been produced and installed in various applications since Honeywell introduced the system as a form, fit, and function replacement for the AN/APN-141(V) in the early 1970s. Although the system's technology and construction are very similar to that of other Honeywell altimeters, recently developed circuits have been included to ensure optimum system performance. The system includes the receiver-transmitter, as many as four analogue height indicators, and two high-gain antennas.

Typical installation
Dimensions and weights: (transmitter-receiver unit) 82 × 97 × 185 mm; 4.4 lb (2 kg)
(indicator) 1.6 lb (0.7 kg)
Frequency: 4400 MHz
Pulse repetition frequency: 20 kHz
Max altitude: 5000 ft
Accuracy: 3 ft or 4%
Pulse width: 0.02 or 0.20 μs
Transmit power: 5 W
Power: 25 W at 115 V 400 Hz

STATUS: in service as standard on all US Navy fixed-wing and high-performance aircraft.

AN/APN-194(V) radar altimeter indicators

Honeywell has a range of six different indicators which can be used with the AN/APN-194(V) system, having pointer/scale indications from 0 to 5000 feet (expanded from 0 to 200) or from 0 to 1000 feet. The choice extends to mounting, bezel or clamp; or lighting, red or white' and power supply, 115 watts 400 Hz or 28 volts dc.

All indicators include a low-height warning light, a pust-to-test button which initiates the built-in test function, and an OFF flag when

Honeywell AN/APN-209 radar altimeter

radar height exceeds 5000 feet, the power fails, or signal tracking is unreliable.

Dimensions: 104 mm diameter × 130.5 mm
Weight: 1.6 lb (0.7 kg)

STATUS: in production.

AN/APN-171 radar altimeter

For 20 years, US Navy rotary-wing aircraft have been flying with the APN-171 aboard as their standard altimeter. Many functions are available with this system, including low altitude warning, radar altitude warning set input, ac rate, ac altitude errors and landing gear warnings. Working with the Navy, Honeywell has maintained the APN-171 F[3], while incorporating the technology to reduce costs and improve performance.

Three basic Honeywell AN/APN-171(V) radar altimeter systems are currently available: the first uses the HG9010 receiver-transmitter for a 0 to 1000 foot system; one uses the HG9025 receiver-transmitter for a 0 to 2500 foot system; and the other uses the HG9050 receiver-transmitter to provide a 0 to 5000 foot system. All systems are available with standard or special output signals to represent a particular altitude range.

STATUS: in service as standard on all US Navy rotary-wing aircraft. Modification kit available for existing systems upgrade.

AN/APN-209 radar altimeter

The APN-209 was Honeywell's first radar altimeter with both electronic digital and analogue readouts, and is compatible with night vision goggles. The system is standard on all US Army helicopters, and features both high and low altitude set as well as an integrated indicator, receiver and transmitter. For more flexibility in installation, a version of the APN-209 with the receiver-transmitter separate from the indicator is under production.

STATUS: in service as standard on all Army helicopters.

AN/APN-222 radar altimeter

The APN-222 is one of Honeywell's high-altitude altimeters, measuring distances up to 70 000 feet. The system is highly accurate and reliable at both low and high altitudes and is operational on the RF-5E. Modifications were made by incorporating MIL-STD-1553B data-bus, resulting in the HG7197, which is qualified for use on the F-16 and has demonstrated its ability to meet US Air Force F-16 radar altimeter requirements.

Although the technology and construction utilised is very similar to that incorporated in

Honeywell's APN-194 and missile altimeters, recently developed circuits have been included to ensure optimum system performance at high altitudes. The APN-222 system includes the receiver-transmitter, a new analogue/digital height indicator, and two high-gain planar-array antennas.

Typical installation
Dimensions and weights: (transmitter-receiver unit) volume 2.75 litres; 6.5 lb (2.9 kg)
(antenna) 176 × 254 × 7 mm; 1 lb (0.5 kg)
(indicator) volume 0.92 litre; 2.5 lb (1.1 kg)
Power: 15 W at 28 V dc
Frequency: 4300 MHz
Pulse repetition frequency: 5 kHz
Max altitude: 70 000 ft
Accuracy: 5 ft or 2%
Transmit power: 500 W
Power: 35 W at 28 V dc

STATUS: in production. In service on RF-5E and a modified version is qualified for the F-16.

AN/APN-224 radar altimeter

The APN-224 was developed specifically for the Strategic Air Command's B-52, and meets the nuclear hardening and high reliability specifications of that aircraft. The system's performance and ability to withstand severe environments also led to its selection by the US Air Force for the B-1B. The APN-224 is also a strong candidate for other US Air Force applications such as the A-10, F-15E and the Air National Guard F-16.

As with other Honeywell radar altimeter systems, the APN-224 has a continuous failure monitoring capability. Its technology and construction is based on the APN-194, enhanced with recently developed circuitry to ensure optimum performance.

STATUS: in service on B-52 and B-1B.

Cruise Missile Radar Altimeter

The Cruise Missile Radar Altimeter (CMRA) was developed specifically for the Cruise Missile Programme, including the alcm and Tomahawk missiles. Honeywell's CMRA is a derivative product. A variety of features from other Honeywell altimeters are coupled with state-of-the-art advancements. The system has the capability to perform terrain correlation and navigation functions.

STATUS: qualified by US Navy and Air Force and in production for the Cruise Missile Programme. In service on Tomahawk and alcm missiles.

Ideal Research

Ideal Research Inc, 11810 Parklawn Drive, Rockville, Maryland 20852
TELEPHONE: (301) 984 5694
TELEX: 904059

MIAMI ice-detection system

The microwave ice accretion measurement instrument (MIAMI) permits the accurate measurement of ice accumulating in critical areas of the aircraft.

As an ice warning system, the MIAMI alerts the pilot to the earliest initiation of ice growth. As little as 0.003 inches (0.076 mm) of ice can be detected, illuminating a warning light on the annunciator panel. Then the pre-programmed microprocessor takes over and computes both icing rate in inches per minute and indicates the ice thickness in the cockpit digital display.

If required, the system can be used to activate/de-activate the de-icer equipment on aircraft and missile systems and works equally well with all types of de-icing or anti-icing equipment.

The system transducer element consists of a resonant surface waveguide. The resonant frequency of this varies according to the amount of ice accreted and so a relationship between ice thickness and frequency shift can be established. It is said that this type of transducer has the advantage of not requiring an external, and frangible, probe and the transducer may be profiled to conform to the contour of the mounting surface.

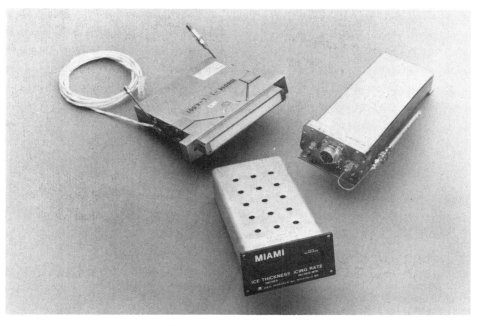

Ideal Research MIAMI System

The transducer can be mounted anywhere on the aircraft, including rotors or wings, and the system can be protected against sand or rain erosion. The system is microprocessor controlled and a single microprocessor unit can control any number of transducers.

STATUS: in production.

Intercontinental Dynamics

Intercontinental Dynamics Corporation, 170 Coolidge Avenue, PO Box 81, Englewood, New Jersey 07631-9990
TELEPHONE: (201) 567 3600
TELEX: 135323

Following the acquisition of Intercontinental Dynamics Corporation (IDC) by Sun Chemical Corporation, the IDC product line of air data instruments, cabin displays and related equipment will be marketed under a new name – Kollsman/IDC

Model 41707 reversionary altimeter

This instrument displays computed altitude derived from a central air data computer (CADC), and is housed in a 3 ATI case that also includes an integral precision pressure altitude mechanism as a standby. Both the computed and the standby altitudes read from –1000 to 53 000 feet.

The standard barometric setting by means of a knob in the lower right-hand corner operates two readouts, one displaying inches of mercury, the other millibars. An additional knob, located in the lower left corner, is used to select either air data computer altitude or the standby barometric altitude; the former is indicated by the legend CADC and the latter by PNEU.

The IDC pressure-sensing mechanism is augmented by a follow-up servo which operates on the altitude input from the CADC. The servo positions the mechanism so that the altitude displayed by the instrument agrees with the data from the CADC.

An electronic monitor continuously compares the integral pressure sensor output with that from the CADC. Should these not agree, or if there is an electrical failure or the CADC output is lost, the system automatically switches to the pneumatic mode. This is indicated by a yellow flag with the legend PNEU over the zero position on the circular scale. The accuracy in the standby mode is claimed to be within specification TSO C10b. The system uses the now standard counter-pointer presentation with co-acting numeral drum counters and a single pointer moving over a circular scale.

STATUS: in production.

IDC alerter for Cheyenne IV

Altimeter/altitude alerter for Piper Cheyenne IV

IDC has developed a combined altimeter and altitude alerter specifically for the Piper Cheyenne IV. It is an IDC 28704 altimeter with altitude reporting output and the Model 45100 altitude alerter. The latter provides barometric corrected pressure altitude deviation for the autopilot when in altitude-hold mode.

Type 35405 Mach/TAS indicator

This instrument, incorporating speed-limit indications has multiple presentations. A pointer shows indicated airspeed over the range of 60 to 120 knots, while a striped pointer shows maximum allowable airspeed. Both pointers operate from pitot and static pressure inputs and do not require electrical power.

The instrument also includes a triple-drum display of true airspeed over a range of 120 to 590 knots and a second triple-drum display shows Mach number. Both drum displays are positioned by the air data computer. If power is lost the true airspeed and Mach counter are covered by a flag. The type 35405 can interface with a flight-director system to drive the fast-slow bug on the altitude director. It can also provide the speed reference for an autothrottle

system. The reference speed is selectable through a remote switch and is indicated by an orange coloured command 'bug'. There is also V_{REF} 'bug' with control knob on the instrument face.

Format: 4 ATI (ARINC 408)
Weight: 5.2 lb (2.3 kg)
Power: 5 VA at 400 Hz 26 V

Type 35415 Mach/TAS indicator

This instrument has been designed for aircraft not having an air data computer. An input from a Rosemount outside air temperature probe is operated on by the integral Mach computing circuit to drive the true airspeed readout. The readout is obscured in the event of Mach circuit failure. This also occurs below Mach 0.52 and in the event of a power failure. Similarly the true airspeed display is obscured if the power fails or falls below 120 knots. The maximum allowable and airspeed pointer are activated by the pitot/static system and continue to operate even in the event of a power failure. The instrument interfaces with a flight-director and with an autothrottle system. There are two control knobs on the instrument face: an up-

IDC Type 39939 airspeed/Mach indicator

Intercontinental Dynamics radio/barometric 'talking altimeter' and Machmeter

down command which sets an orange 'bug' at a desired approach speed, thereby setting the speed datum for the flight-director and auto-throttle; the other knob sets the VREF 'bug'.

An optional fast-airspeed low-altitude amber warning light illuminates above 250 knots indicated airspeed if the aircraft is below 10 500 feet.

Range: 60–420 knots indicated airspeed, 0.52–0.99 Mach, 120–590 knots true airspeed and –1000 to +53 000 ft altitude
Format: 4 ATI (ARINC 408)
Weight: 5.2 lb (2.3 kg)
Power: 5 VA at 26 V 400 Hz

Type 39939 max allowable airspeed/digital Mach indicator

This indicator was designed to meet the demands of pilots who prefer reading Mach number using a two digit counter instead of the traditional Mach sub-dial as read against airspeed pointer.

In operation, the max allowable airspeed/digital Mach indicator accepts pitot and static pressures and positions the pointers. These pressures also drive electrical induction potentiometers which provide precise a/s and alt voltages. These voltages are used to compute Mach. The digital counter displays Mach numbers from 0.55 to 0.95. Whenever the instrument ceases to compute Mach values a black and white flag obscures the Mach counters. Indicated and maximum allowable airspeeds are read against a fixed dial which is extended at the lower speeds for enhanced readability.

A manually operated command marker allows the setting of approach speed and this provides electrical signals to the flight-director fast-slow bug and/or auto throttle. The unit also includes a VREF marker as an integral part of the indicator which is set by a bezel-mounted knob. Reminder bugs are also available.

Range: 60–400 knots airspeed, –1000 to +53 000 ft altitude, 0.55–0.95 Mach
Power: 26 V, 400 Hz 5 VA max
Input: aircraft pitot and static pressures
Output: ac airspeed error signal 100 mV/RMS/knot at 140 knots, dc airspeed error signal 200 mV dc/knot at 140 knots, fast-slow valid +28 V dc
Dimensions: ARINC 3 ATI
Weight: 2.9 lb (1.3 kg)

STATUS: in operation on Learjet, Gulfstream IV and Challenger CL600 aircraft.

Radio/barometric altimeter with voice terrain advisory

Termed a 'talking altimeter', a radio/barometric altimeter with voice terrain advisory keeps the flight-crew informed of vertical flight progress with particular reference to altitudes within 2500 feet of the ground. The unit combines the functions of standard barometric instruments with those of radio altimeters, but its notable quality is that it provides voice output warning of potential hazards in certain discrete flight regimes. In some cases, the unit may replace conventional ground-proximity warning systems whose functions they tend to cover. Radio/barometric instruments typically provide warnings in low-altitude flight whenever the aircraft-ground separation altitude decreases, during en route flight, and advise of increasing ground-proximity while flying within 2500 feet of rapidly rising terrain.

Intercontinental manufactures a range of radio/barometric models with differing options, though the principles of operation are broadly similar. The visual instrument presentation is closely allied to that of a conventional barometric altimeter with drum and pointer scales. A window showing radio altitude is set into the face of the instrument. Models giving either three- or four-digit radio altitude are available. The radio-altitude window digits read altitudes to above 2500 feet, after which the window goes blank.

Following take-off the voice-warning facility remains silent unless the aircraft begins to descend or it traverses rapidly rising ground, in which case a voice warning 'terrain' would be annunciated. Similarly, during en route flight, the same warning would be announced if the aircraft descended below 2500 feet or if rapidly rising ground were below.

Before entering the final approach pattern at the destination airfield, the crew can programme the radio/barometric altimeter for the correct decision height. Using a decision-height control on the cockpit instrument, the pilot illuminates the radio altimeter readout digits and by rotating the control knob can set the desired decision height into a memory, as indicated by the digits. During approach, the voice output will announce 'terrain' as the aircraft descends through 1000 feet and continues to announce radio altitude at 100-feet decrements. At decision height, the word 'minimum' is announced.

The system also provides a glideslope deviation warning mode in which any significant departure below the correct path causes the call 'glideslope' to be announced at 2-second intervals. If the divergence continues, the call is repeated, at twice the previous volume level and 1-second intervals. These 'glideslope' annunciations are interrupted by altitude calls whenever the two coincide. Glideslope calls 'cease' once the aircraft descends through decision height, whereupon the 'minimum' call is heard. The minimum (ie decision height) call takes priority over all other announcements, but the altitude calls continue as the aircraft continues to descend below decision height.

During all procedures the barometric altitude drum continues to function normally and the radio-altitude digital readout shows increasing or diminishing altitude below 2500 feet, in 10-feet (3-metre) increments.

Various models of the Intercontinental radio/barometric system, with differing options, are available. Some units provide digitised data output for automatic in-flight altitude reporting, altitude-change rate signals for vertical-speed indicators, and barometrically corrected altitude synchros for altitude alerters. Decision height warning lights, which illuminate coincidentally with the call 'minimum', are incorporated in certain units. Barometric settings may be presented in windows showing inches of mercury, millibars or both. Self-test facilities, for in-flight or ground use, are also incorporated. Intercontinental equipment models can interface with all existing ARINC and many non-ARINC radio altimeter transmitter-receivers. Instruments cover altitude ranges from –1000 to +50 000 feet and can be lit in either white or blue/white according to the

Intercontinental engine performance indicators as fitted to the Gulfstream II

configuration selected. The systems comprise panel-mounted instruments in standard ARINC 408 cases and remotely located radio altimeter-converter units which supply radio altitude information to the indicator units. Aural output may be channelled through standard inter-communication systems using cabin speakers or pilot headsets.

Format: (panel unit) 3 ATR and 4 ATR cases (converter) ⅜ ATR Short
Weight: (total system) 8.3 lb (3.8 kg)

STATUS: in production and service.

Engine instrumentation for Gulfstream II

Intercontinental Dynamics produces a range of vertical strip advanced engine instruments for the Gulfstream Aerospace Gulfstream II. They incorporate solid state circuitry, the output from which lights up a string of incandescent filaments. illuminating upwards from zero until the correct readout is shown. The counter then returns to zero and re-energises the filaments again, repeating the procedure 50 times a second. There are 100 filaments on the large displays, 30 on the smaller ones. Failure of one filament does not affect the remainder of the display. These instruments are available for engine temperatures, speeds and pressure ratio, fuel flow and quantity, and oil pressure and temperature. A numeric display can show any selected parameters.

Dimensions: (large units) 6.3 × 1.25 inches (160 × 32 mm)
(small units) 2.5 × 1.25 inches (63 × 32 mm)
Weight: (large units) 2 lb (1 kg)
(small units) 0.5 lb (0.2 kg)

STATUS: in production.

JET

JET Electronics and Technology Inc, 5353 52nd Street, Grand Rapids, Michigan 49508
TELEPHONE: (616) 949 6600
TELEX: 226453

RAI-303 remote attitude indicator

This instrument operates on synchro inputs from a vertical gyro or attitude reference system. The indicator displays attitude by means of a sphere moving with respect to a fixed peripheral dial and aircraft symbol. Loss of power is indicated by an OFF flag. A solid-state monitor samples and compares the input signal from the vertical reference unit with the output of the servo control transformers for the roll and pitch axes.

Dimensions: 3.25-inch square case to MS 33556 × 8.5 inches long (83 × 210 mm)
Weight: 3 lb (1.36 kg)
Power: 12 VA at 115 V ac 400 Hz

STATUS: in production.

ADI-350 attitude director indicator

This is a self-contained gyro indicator with flight-director needles, rate-of-turn and slip indication. Synchro pick-offs are provided for remote indicators, radar stabilisation and flight control functions. Inputs are needed to operate the flight-director needles and the rate-of-turn indicator. Applications include: Grumman A-6E, KA-6D and EA-6B, McDonnell Douglas F/A-18, Canadair CP-140, and Rockwell B-1.

Dimensions: 3.23-inch square × 9 inches (82.04 × 228.60 mm)
Weight: 6 lb (2.72 kg)
Power: 115/208 V 400 Hz 3-phase or 28 V dc

STATUS: in production.

AI-804 2-inch standby gyro horizon

This is a standby gyro horizon for high-performance general aviation aircraft. The self-contained package eliminates the need for additional electronic components. After complete loss of external power attitude information remains available for 9 minutes.

Dimensions: 2 ATI
Weight: 2.5 lb (1.1 kg)
Power: 115 V 400 Hz or 28 V dc

STATUS: in production for Lockheed JetStar, Gulfstream I, II and III, Learjet 20, 30 and 50, IAI Westwind 1 and 2, Dassault Falcon, Cessna 400 Conquest and Citation, Rockwell Sabreliner, Canadair Challenger, British Aerospace HS-125, Mitsubishi MU-2, DH Canada DHC-7 Dash, Fairchild Merlin, Beechcraft King Air, Piper Cheyenne, Embraer Xingu fixed-wing aircraft and Sikorsky S-76 helicopter.

AI-904 3-inch standby gyro horizon

This 3-inch (76 mm) standby unit was designed specifically as an attitude reference indicator for high-performance aircraft. It is currently being used in a number of military and commercial aircraft around the world. Over 9 minutes of attitude information is available after complete loss of all electrical power.

Dimensions: 3 ATI
Weight: 4.2 lb (1.9 kg)
Power: 115 V 400 Hz or 28 V dc

STATUS: in production.

ESIS emergency standby instrument system

JET Electronics and Technology has now introduced a new emergency standby instrument system (ESIS) for all aircraft fitted with EFIS cockpits, and claims to be the only supplier who offers all seven components in the ESIS.

The system comprises four cockpit-mounted units – a 2-inch (51 mm) attitude indicator; digital airspeed indicator; digital altimeter and heading/navigation system which is fed by a remote directional gyro. The sixth component is an air data computer while the seventh is a completely independent emergency power supply.

It is noteworthy that the FAA has recommended that EFIS-equipped aircraft carry a suitable 'standby' to provide 30 minutes of independently powered basic flight information for emergency use.

STATUS: in production.

AHR-102 attitude and heading reference system

This is a low-cost system which outputs pitch, roll, heading and turn-rate data. It comprises a reference platform, electronics control unit, compensator and control unit.

Model 102A does not have an earth rate corrector whereas Model 102B does have this device.

Dimensions: (platform) 6.7 × 6.7 × 12.5 inches (170 × 170 × 317 mm)
(electronics unit) 5.2 × 7.4 × 9.2 inches (132 × 188 × 234 mm)
(compensator) 1.8 × 1.8 × 1.8 inches (46 × 46 × 46 mm)
(control unit) 5.8 × 1.5 × 3.7 inches (147 × 38 × 94 mm)
Weight: (platform) 12.3 lb (5.58 kg)
(electronics unit) 6.0 lb (2.7 kg)
(compensator) 0.75 lb (0.3 kg)
(control unit) 0.6 lb (0.27 kg)
Power: 115 V 400 Hz

VG-501 vertical and DG-501 directional gyros

These are lightweight, digital, remote gyros incorporating spool-type rotors and simplified electronics. JET claims that the design eliminates power-consuming electromechanical units.

Dimensions: (DG-501) 6.75 × 5 × 5.5 inches (171 × 127 × 139 mm)
(VG-501) 6.6 × 4 × 4.8 inches (167 × 101 × 122 mm)
Power: 26 V 400 Hz and 40 V dc

VG-204 vertical gyro

JET claims that this modular, vertical gyro has a life expectancy four times that of conventional gyros. It has been designed to MIL-E-5400 for use in aircraft, helicopters, drones and remote-piloted vehicles. Outputs can be provided for radar stabilisation. The system is used in Sikorsky S-76, Hughes 500 MD, Bell 214 ST and 412 helicopters and Beech T-34C fixed-wing aircraft.

Dimensions: 6 × 4 × 4 inches (152 × 100 × 100 mm)
Weight: 5 lb (2.25 kg)
Power: 115 V 400 Hz

VG-208 vertical gyro

This is a similar design to that of the Model VG-204. It is used in the Dornier 228, Beech King Air, Piper Cheyenne, Embraer Bandeirante twins and MBB BK-117 helicopter.

JET AI-804 standby horizon indicator

Engine trend monitors

The ETM-600 is a microprocessor-based equipment designed for use with the Pratt & Whitney PT-6 engine. It determines the engine's current status and predicts future trends, indicating normal or accelerated degradation. Data is automatically stored in a removable memory module which is downloaded into a personal computer through a ground-based reader. An STC has been issued to the ETM-600 for use on the Beech King Air 200, B200 and B200C. Additional trend monitors are currently under development.

STATUS: in production.

JET ETM-600 engine trend monitor

Kaiser Electronics

Kaiser Electronics, Kaiser Aerospace and Electronics Corporation, 300 Lakeside Drive, Oakland, California 94612
TELEPHONE: (415) 835 5554
TELEX: 335326

With a workforce of over 1300, Kaiser Electronics is solely dedicated to design and production of cockpit display systems for combat aircraft. Major products include head-up and multi-function displays, in monochrome or colour, integrated helmet displays, advanced display processors and display system integration. Kaiser's first major programme, won in 1960, was the AN/AVA-1 vertical situation display and radar data converter for the US Navy's A-6A Intruder, of which 800 were built. This programme established the value of crt displays for flight information in aircraft cockpits, the evolution of which continues today. Kaiser display systems are featured in most US combat aircraft, a number of international aircraft and the Space Shuttle.

AN/OD-150(V) integrated cockpit display for F/A-18

In 1976 Kaiser was awarded contracts to develop and produce the head-up display and the multi-purpose display group for the US Navy/McDonnell Douglas F/A-18 Hornet. The three multi-function head-down displays operate in stroke, raster or hybrid and provide modes for weapon management, air-to-air and ground-map radar (with synthetic aperture and Doppler beam-sharpening capabilities), horizontal situation and tabular alphanumeric modes. Contact push-buttons around the edges of the displays are for mode selection and data call-up. The dual-combiner glass head-up display provides an instantaneous field of view of 18° by 18° and a total field of view of 20°.

STATUS: in production for US Navy/US Marines F/A-18 Hornet which entered service during 1982 and for the Hornets ordered by Australia, Canada and Spain.

AN/AVA-12 attitude display for F-14

The AN/AVA-12 is the primary attitude and director display for the US Navy/Grumman F-14 Tomcat. An unusual feature is that the vertical situation display cathode ray tube, though electrically independent of the head-up display, is mechanically integrated with the latter in a common box. The vertical display, or attitude director, uses a high-brightness cathode ray tube to produce 525-line raster television and stroke writing for the retrace. Another unusual feature is that there is no separate combiner glass for the head-up display, the inside face of the windscreen acting as the combiner element.

STATUS: in production and service.

AN/AVA-1 electronic attitude display for A-6

The AN/AVA-1 attitude indicator is of historical significance in being the very first aircrew display to use a cathode ray tube for flight information. As the pilot's primary flight instrument on the US Navy/Grumman A-6, it provides attitude, navigation and weapon delivery information and terrain-clearance commands.

STATUS: in service, 850 produced.

Display system for F-15E

Kaiser is building the head-up/head-down displays for the two-seat McDonnell Douglas F-15E, chosen by the US Air Force in early 1984 as its new dual-role fighter. Each aircraft will have four 6 by 6 inch (152 by 152 mm) cathode ray tube-type head-down displays (two in each cockpit) and one holographic head-up display (HUD). The latter will employ both raster-scan and stroke-written symbology to accomodate the forward-looking infra-red (Flir) imagery from the Martin-Marietta Lantirn system and the Hughes Aircraft AN/APG-70 radar. The Kaiser wide field of view head-up display employs a holographic single combiner-glass, permitting a smaller and lighter support structure with less obscuration of forward view.

The holograms for this HUD are made by an associate company, Kaiser Optical Systems Inc, of Ann Arbor, Michigan. With proprietary software developments this company has established a special hologram development and production facility. Current US Air Force plans call for 396 F-15E aircraft.

STATUS: in development.

F-14D/A-6F display suite

In 1985 Kaiser was awarded the entire display suite for the US Navy's F-14D/A-6F update programme. This is a hallmark programme, in that the display products for the two aircraft are

AN/AVA-12 vertical situation display by Kaiser Electronics for F-14 Tomcat fighter

Kaiser Electronics attitude display for Grumman A-6

identical. Furthermore, the multi-function panel displays were mandated by the Navy to be Kaiser's multi-purpose display repeater indicator (MDRI) displays already present in the Navy's inventory for the AV-8B and F-16 aircraft. Commonality of this kind is being sought throughout the Navy and this programme is setting the pace for cockpit displays.

In addition to the MDRIs, the display suite consists of an advanced display processor (DP) and a wide angle, low profile HUD. The F-14D complement of equipment includes one HUD, two DPs and three MDRIs. The A-6F will have one HUD, two DPs and five MDRIs.

The F-14D/A-6F head-up display features an advanced, extremely compact optics design, that provides a 30° total field of view and dual, flat holographic combiners for increased brightness and better see-through vision.

The display processor will use custom gate arrays and a single card MIL-STD-1750A processor and can simultaneously drive up to six cockpit displays. Reserve capacity and significant growth are inherent in the design.

STATUS: in development, with production deliveries beginning in 1989.

AH-1W control/display subsystem
Both the control/display subsystem (CDS) and full function signal processor (FFSP) are produced by Kaiser. The HUD is identical to that of the Army's AH-1S, which Kaiser also supplies. In contrast, the FFSP is an entirely new design, which features dual 68000 processors and software designed to DoD-STD-1679A requirements. The processor is programmable and includes the capabilities for vectors, circles, arcs and rotation.

STATUS: The US Marine Corps has ordered 44 AH-1Ts and is expected to retrofit its fleet to the AH-1W configuration.

Kearfott
Kearfott Division (Aerospace and Marine Systems), The Singer Company, 1150 McBride Avenue, Little Falls, New Jersey 07424
TELEPHONE: (201) 785 6000
TWX: 710 988 5700

AN/USN-2(V) strapdown attitude reference system
In February 1985, the Kearfott attitude heading reference system (AHRS) was selected as the US Tri-Service standard attitude heading reference system (SAHRS). This is a three-axes strapdown system, using ring laser gyroscopes.

The unit provides vehicle attitude, acceleration and navigational data.

Dimensions: 11 × 7 × 7 inches (279 × 178 × 178 mm)
Weight: 28 lb (12.7 kg)

STATUS: in production.

King
King Radio Corporation, 400 North Rogers Road, Olathe, Kansas 66062
TELEPHONE: (913) 782 0400
TELEX: 42299

Accuracy: (0–100 ft) 5 ft
(100–500 ft) 5%
(>500 ft) 7%
Power: 6 VA at 28 V dc

STATUS: in production.

Accuracy: (0–500 ft) 5%
(>500 ft) 7%
Power: 24 VA at 28 V dc

STATUS: in production.

KRA10A radar altimeter
A low-cost system suitable for independent use or in combination with King Silver Crown avionics equipment and tailored to general-aviation requirements. There is a standard facility for presetting decision height, which produces a visual and aural warning on reaching the set altitude, and antennas suitable for flat and sloping skin installations are available. The KRA10A is an all-solid-state system with short warm-up time and can be fitted with an auxiliary output to interface with flight-director and autopilot installations.

Typical installation
Dimensions and weights: (transmitter-receiver unit) 79 × 89 × 203 mm; 2 lb (0.9 kg)
(antenna) approx 100 × 100 mm aperture; 0.9 lb (0.4 kg)
(indicator) 100 × 83 × 83 mm; 0.9 lb (0.4 kg)
Max altitude: 2500 ft

KRA 405 radar altimeter
Part of the King Gold Crown avionics range, this is an all-solid-state radar altimeter suitable for twin-engine general aviation types. The KRA 405 interfaces with King KPI 553A HSI and KFC 300 autopilot to give smooth tracking of the glide-slope beam. It can provide indications from 2000 feet above ground level, and a usable output is available from 2500 feet above ground level for ground-proximity warning system operation. Separate transmit and receiver horn antennas are used.

Typical installation
Dimensions and weights: (transmitter-receiver unit) 83 × 133 × 296 mm; 6.3 lb (2.9 kg)
(antennas) each 178 mm dia; 2.6 lb (1.2 kg) total
(indicator) 83 × 83 × 170 mm; 1.7 lb (0.8 kg)
Frequency: 4300 MHz
Max altitude: 2500 ft

KEA 346 encoding altimeter
This servoed altimeter reads up to 50 000 feet and provides an output to King's KAS 297/297A altitude pre-select units (see below). The display combines pointer and five-digit numeric readouts along with barometric setting displays in millibars and inches of mercury.

Format: 3 ATI (ARINC 408)
Length: 8.9 inches (227 mm)
Weight: 3 lb (1.36 kg)
Power: 28 V dc

KAS 297 and 297A altitude pre-select units
These units can be used with the King KFC 200 flight-control system for general aviation and the KEA 346 altimeter to give altitude pre-selection and alerting. Selected altitude is displayed by five-digit numerals. The pre-select units also interface with the King KVN 395 RNav unit.

Kollsman
Kollsman Instrument Company, Daniel Webster Highway South, Merrimack, New Hampshire 03054
TELEPHONE: (603) 889 2500
TELEX: 943537

Electro-luminescent displays
For nearly 40 years scientists have attempted to imitate fireflies and glow-worms by producing electro-luminescent devices. In the 1960s evaporated thin-film phosphors were developed that were characterised by low brightness and short life. During the past ten years significant progress has been made by a number of companies, including Kollsman, through materials and process improvements. In particular, ac thin-film electro-luminescent displays are seen to be viable alternatives to servo-mechanical displays, light-emitting diodes, liquid-crystal displays and cathode ray tubes. Matrix

displays are now possible with resolutions greater than those of conventional television tubes and brightness levels greater than 1000 foot-lamberts (3426 cd/m²) are now practicable, compared with 100 foot-lamberts (342.6 cd/m²) of a typical television tube. Lifetimes greater than 40 000 hours have been achieved and efficiencies are said to be greater than those of more traditional displays, with outputs of two to three lumens per watt. Kollsman says that the manufacturing method that it has developed gives a cost-effective display, competitive with other equivalent systems.

In the manufacturing process various layers of thin films are vacuum-deposited on a glass substrate. The first layer, deposited directly on the rear surface of the glass (the front surface forms the face of the instrument) is a transparent conductor which provides one of the electrical connections. The second and fourth layers are insulators needed to prevent electrical breakdown. The third layer is the phosphor, the source of light. The fifth layer is a light-absorbent or black' material to improve contrast. The final layer is the back conductor and

Kollsman vertical-strip engine instrument display, configured for twin-engined aircraft, shows percentage rpm, exhaust-gas temperature and fuel flow rate

the entire assembly is sealed to prevent contamination from the external environment. Light is produced in the phosphor layer by applying an alternating voltage to the conducting layers, emission occurring as a result of the excitation of activator ions by injected electrons.

One of the major considerations in using any type of display in an aircraft instrument panel is its legibility in direct sunlight. In the Kollsman system this is achieved by preventing the sun's radiation from being reflected from the display.

Reflection can be reduced to 1 per cent by using a 'black' layer which not only absorbs most of the sunlight but permits lower brightness levels to be used, reducing the power needed to drive the system and increasing its reliability.

Based on this technology the company is developing vertical-scale and round-dial instruments and flat screen, alphanumeric displays. A vertical scale instrument can offer 35 elements per inch, integral lighting, sunlight readability, manual dimming, silk-screen scales for simple customising, modular design, and dc, analogue, or parallel binary interfaces with other equipment and sensors.

STATUS: under development.

Lear Siegler

Lear Siegler Inc, Astronics Division, 3400 Airport Avenue, Santa Monica, California 90406
TELEPHONE: (213) 452 6000
TWX: 910 343 6478

Model 6000 attitude and heading reference system

This is called a fifth-generation system by Lear Siegler, and about 2500 examples of the Model 6000 AHRS were produced between 1974 and 1982. There is a licence production agreement with Louis Newmark (UK) and technical features of the system are described in that company's entry.

Applications for the equipment are: the US Air Force/Fairchild A-10, Dassault-Breguet/Dornier Alpha-Jet, Fairchild T-46A, British Aerospace Hawk, SIAI-Marchetti S-211, and Sikorsky HH-60D helicopter. Retrofits are being conducted on McDonnell Douglas A-4 Skyhawk, Northrop F-5B, Convair F-106, Boeing B-52 Stratofortress, Lockheed C-141 StarLifter, Saab J-35 Draken, Kaman HSS-2B helicopter, Shin Meiwa PS-1 flying boat, and US Air Force QF-86 drone version of the F-86 Sabre.

Flight-tests and special purpose applications have been conducted with a further eight aircraft types.

STATUS: in production.

Model 2580D intelligent control/display unit

This unit is designed to simplify the avionics-management tasks common to high-performance military fixed-wing aircraft and helicopters. Data handling is accomplished through a dual MIL-STD-1553B digital bus, 16-bit processor and 24 K by 16 bit eprom, and 8 K by 16-bit random access memory. The processor and memory operate on 8086 instruction-set, and a Jovial compiler is available. The cathode ray tube screen measures 4.2 by 3.4 inches (107 by 86 mm) and has a format of 10 rows times 24 characters and a 128-character set of alphanumerics and special symbols. It is associated with single-function alphanumeric keys, of which 23 can be employed to define user functions.

Functional versatility is attained by having spare card slots for different applications and substantial spare processing power. The system is particularly valuable for applications calling for integrated control of communication and navigation systems within a single unit.

Dimensions: 5.7 × 7.875 × 9.5 inches (145 × 200 × 240 mm)
Weight: 14 lb (6.4 kg)
Self-test: bite facility

STATUS: in production.

Lear Siegler Model 6000 attitude and heading reference system

---

Litton

Litton Industries Inc, Aero Products Division, 6101 Condor Drive, Moorpark, California 93021
TELEPHONE: (805) 378 2000
TELEX: 662619
TWX: 910 494 2780

LTR-81-01 attitude and heading reference system

This has been certificated for the McDonnell Douglas MD-80s series. Alitalia MD-80s are being retrofitted with triple LTR-81-01s in place of earlier conventional vertical gyro and directional gyro systems. The system has also been selected for the MD-80s operated by Alisarda, and is standard on British Aerospace ATP. Litton claims that the LTR-81-01 offers reduced maintenance costs compared with conventional attitude and heading reference systems. It is a single-package replacement in ARINC 705 format. A mtbf of more than 11 000 hours has been established on MD-82 aircraft. The LTN-81 incorporates two K-273 two-degrees-of-freedom tuned rotor gyroscopes specifically developed by Litton's West German subsidiary, Litef, for strapdown applications.

LTR-81-02 all-digital system has been selected for the Fokker 100. It is identical to the -01 except the -01 contains two additional analogue ouput cards.

Litton LTR-81-01 selected for British Aerospace ATP

STATUS: in production. Certificated for Category III A autoland on MD-82 aircraft.

In March 1986 Litton announced that LTN-81-01 had been selected to equip the new

McDonnell Douglas MD-88 airliners ordered by Delta Air Lines: deliveries to Delta start in October 1987.

AD2610 ground-proximity warning system

The AD2610 ground-proximity warning system (GPWS), warns of potentially hazardous ground-proximity for Modes 1, 2A, 2B, 3, 4 and 5. It provides swept audio tone warning plus the voice command 'pull up' when activated, with the exception of Mode 5 when the command 'glideslope' is substituted. Outputs for visual warning lights are also incorporated. The system is digitally based and microprocessor controlled. Any changes required by regulatory authorities may be accommodated with minimal disruption to aircraft operations.

Two levels of self-test are incorporated: a crew-operated ground check is used for preflight serviceability testing, while during flight, automatic self-monitoring confirms performance. Analysis of in-flight warnings and isolation of faulty systems, whether in the ground-proximity warning processor or in the associated sensor systems, are facilitated by latching magnetic front-panel indicators.

In flight, the GPWS indicator logs a system failure while the remaining indicators log warnings for the five modes. On the ground, the GPWS indicator logs a computer failure and the remaining indicators record which sensor (flaps, gear, radio height, baro rate or glide-slope) has failed. The system has a guaranteed mean time between failures of 10 000 hours.

An optional item, the barometric altitude rate computer (BARC) (AA26101-1) manufactured by Harowe Systems of the USA, is available for use with the AD2610.

Format: (GPWS) ¼ ATR Short
(BARC) ¼ ATR Short Dwarf
Weight: (GPWS) 6.5 lb (2.95 kg)
(BARC) 3 lb (1.35 kg)

STATUS: in production and service.

Loral

Loral Electronic Systems, Ridge Hill, Yonkers, New York 10710-0800
TELEPHONE: (914) 968 2500
TELEX: 996597

AN/ASA-82 anti-submarine warfare display

This tactical information system is the primary data display for the US Navy/Lockheed S-3A Viking, being the link between the four-man crew and the array of electronic sensors. It presents high-speed, high-density data in the form of alphanumerics, symbols, vectors, conics and other appropriate formats from acoustic and non-acoustic submarine detection devices such as sonobuoys.

The system comprises a general-purpose digital computer and display generator driving five display units. The tactical co-ordinator (who directs the anti-submarine warfare mission in the S-3A) and the sensor operator are provided with identical displays showing forward-looking infra-red raw radar, scan-converted radar and tactical data, magnetic anomaly detector and acoustic data. Taken together, this information enables the tactical co-ordinator to anticipate the impending situation. The sensor operator has in addition an auxiliary readout unit that presents a continuous display of the acoustic environment, which it is his job to monitor. The pilot's display cathode ray tube shows overall command and control information that permits him to monitor the entire tactical situation by way of, for example, sonobuoy position, time-to-waypoints, aircraft position and track, navigation and predicted target positions. The lower part of the display area presents various cues and alerts and information describing the appropriate sequences of action. The co-pilot's display repeats the tactical situation presented on the pilot's cathode ray tube, but in addition can show outputs from non-acoustic sensors. The pilot's and co-pilot's cathode ray tubes are readable in an ambient brightness of 8600 foot-candles, and the symbology is shown in eight shades of grey on all cathode ray tubes except that of the pilot which has two shades. The total weight of the AN/ASA-82 system is 268 lb (122 kg).

STATUS: in service with Lockheed S-3A Vikings of US Navy and continuing in production for Lockheed CP-140 Aurora version of P-3 Orion for Canadian Armed Forces.

Integrated tactical display

A multi-purpose television and random-write alphanumeric, graphic and video display for military applications, this system is essentially a colour version of the AN/ASA-82 described above. It is suitable for 'stand alone' operation by virtue of a built-in TI-9900 microprocessor, display generator and 8 K by 16 random access/read-only refresh memory. It presents on a 16-inch (406 mm) screen, information from electro-optical, radar and acoustic sensors in different formats, each of which can be overlaid with tabular, graphic situation, television raster and raw radar symbology.

Loral says that the system shows significant improvements in performance, reliability, maintainability, and potential for growth over current displays. Local data storage, an internal refresh function and the ability to program through the microprocessor all help to off-load the host computer, increase the quantity of data that can be displayed and allow a more flexible operation of the equipment. The use of colour reduces clutter in high-density displays and increases operator comprehension.

Active display size: 9 × 12.7 inches (229 × 323 mm)
Spot size: 15 mils
Addressibility: 1024 × 1024
Random write speed: 100 000 inches/s
Linearity: 1%
Dimensions: (display) 18 × 14.5 × 22 inches (457 × 368 × 559 mm)
Weight: (display) 73 lb (33 kg)

STATUS: in limited production for some aircraft types with classified electronic warfare functions.

AN/APR-38 control indicator set for Wild Weasel F-4G

Loral produces the control indicator set for the AN/APR-38 reconnaissance tactical system for the F-4G, consisting of four displays, a digital display processor, and associated controls. The system analyses and presents data from radar emitters in four types of display: plan position, panoramic frequency analyser, real-time pulse video and homing (elevation and azimuth).

The plan position control indicator selects emitters by type, or combination of types, and displays their ranges and bearings from the

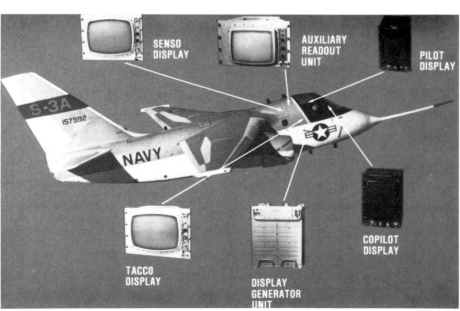

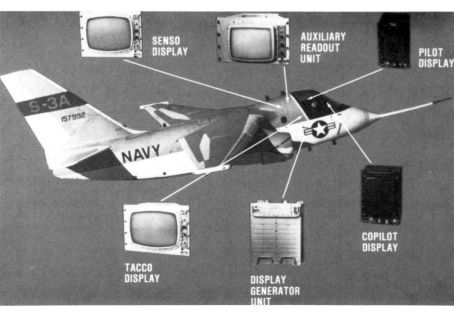

Seven principal units of AN/ASA-82 tactical display in US Navy /Lockheed S-3A Viking sub-hunter

Loral tactical integrated display

aircraft. It then indicates the range and bearing of the radar posing the greatest threat and superimposes on this the destruction footprint of the Standard anti-radiation missile so that the crew can judge whether or not they are within launch range. The panoramic analysis and homing unit contains two displays, one of amplitude versus frequency, the other showing emitter and missile situation relative to the aircraft or, alternatively, real-time signal amplitude versus time. It also controls access to the missile for reprogramming.

Computer: serial, 400 kHz bit rate, 400 word/frame, 16 bit word-length, 50 Hz refresh rate

Ppi display: range versus bearing marks at 10, 25, 50, 100, and 200 n miles
System weight: 100 lb (45.45 kg)
Reliability: 177 h mtbf

STATUS: in production and service.

McDonnell Douglas

McDonnell Douglas Aircraft Co, Box 516, St Louis, Missouri 63166
TELEPHONE: 314 234 7021

Wind-shear detection systems

McDonnell Douglas has been conducting research into wind-shear problems since 1973. Now the MD-88 will be the first of the MD-80 series to be equipped with an on-board wind-shear detection system, currently being developed by McDonnell Douglas.

The system will detect differences in windspeed on the ground and that at the aircraft's height on the approach. When these differences become excessive the pilot will be alerted.

On take-off the system will give both visual and aural warnings of wind-shear conditions. A separate wind-shear computer will be directly linked to the auto-throttles and autopilot to provide an automatic response and ensure that the aircraft maintains the correct pitch and speed to safely pass through the shear.

Certification is expected in 1988.

Norden

Norden Systems Inc, Subsidiary of United Technologies Corporation, Norden Place, Norwalk, Connecticut 06856
TELEPHONE: (203) 852 5000
TELEX: 965898
TWX: 710 468 0788

Multi-sensor/vertical situation display for General Dynamics F-111D

Under a $38 million contract from the Sacramento Air Logistics Command, Norden is building high-brightness, daylight-visible displays as part of an avionics upgrading programme for the US Air Force/General Dynamics F-111D. These integrated displays provide the crew with flight-direction, navigation and weapons-delivery information. The company produced displays for the F-111 programme as part of the original avionics installation in the late 1960s, and also upgraded the scan converter for the integrated displays system on the aircraft.

STATUS: in December 1983 a prototype system was delivered to the US Air Force for acceptance tests. Delivery of production displays began in December 1984 and have continued ahead of schedule.

Pacer Systems

Pacer Systems Inc, 900 Technology Park Drive, Building 8, Billerica, Massachusetts 01821

OADS omni-directional air data system

Developed for helicopter applications, this system consists of a mast-mounted sensor, connected cable and air data computer, and provides an airspeed output over the range 0 to 200 knots irrespective of direction. Additional outputs include forward, rearward and sideways speed components, air density, altitude, air temperature and pressure. The original Pacer OADS was used in the Bell X-22 ducted-propeller, tilt-engine V/Stol research aircraft built in the early 1970s. The current system is in production for the US Army/Hughes AH-64A Apache anti-tank and US Coast Guard Aérospatiale HH-65 Dolphin search and rescue helicopters.

Dimensions: (sensor) 10 × 3 × 3 inches (254 × 76 × 76 mm)
(computer) 7.25 × 5 × 9.5 inches (184 × 127 × 241 mm)
Weight: (sensor) 2.2 lb (1 kg)
(computer) 6 lb (2.7 kg)

STATUS: in production.

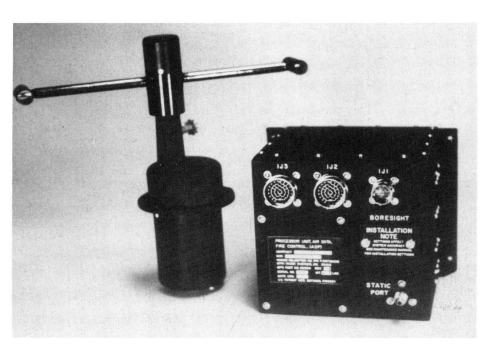

Pacer omni-directional air data system

Omni-directional airspeed indicator

This is a standard 3.25-inch (82.6 mm) instrument for use with the OADS system (see separate entry) in helicopters. There are three meters movements per axis: two bar pointers and an arrow pointer. A circle of dots surrounds the zero airspeed index for indicating 10 knots omni-directionally. A moving horizontal bar indicates a forward or rearward speed component against a scale in 5- and 10-knot increments. A vertical bar moves against a horizontal scale to indicate the left or right airspeed component when the helicopter is moving sideways. The other indications include absolute velocity, an excessive airspeed warning light, certificated and normal operating envelopes. Density altitude readout is an optional feature. The company has developed an alternative presentation in which the dynamic components of the display move behind a fixed horizontal reference line.

Dimensions: 3.1 × 6.4 inches (80 × 160 mm)
Weight: 1.6 lb (0.7 kg)

STATUS: in production for Sikorsky S-61 and Bell 212 helicopters.

Radio Systems Technology

Radio Systems Technology Inc, 13281 Grass Valley Avenue, Grass Valley, California 95945
TELEPHONE: (916) 272 2204

Warning panel

The warning panel brings a centralised warning facility to light and business aircraft. Such conditions as low fuel contents, abnormal battery bus-bar voltage and low oil or vacuum pressure triggers both audio and visual signals to attract the pilot's attention. The warning panel is supplied in kit form for do-it-yourself assembly.

Rosemount

Rosemount Inc, 12001 West 78th Street, Eden Prairie, Minnesota 55 334
TELEPHONE: (612) 941 5560
TELEX: 290183

Model 32 angle of attack indicators

There are two basic versions of the Model 32. For business and commercial aircraft the indicators relate angle-of-attack (AOA) to airspeed by expressing a ratio of actual airspeed to stall speed. For military applications AOA is measured in arbitrary units from 0 to 30. Both types of indicator can be provided with additional markings for displaying the most efficient angle-of-attack for different operating modes.

Type 853 orthogonal airspeed system

This air data transducer and computer combination is designed specifically for helicopter and Vtol aircraft, in which the flight vector (the direction of movement) can make very large angles with the fore and aft axis, and is also available for fire control and flight-test applications requiring extremely accurate data. Components are available with options that permit the system to be tailored to specific applications.

The basic sensor measures two orthogonal airspeed vectors so that if it is mounted vertically it can detect fore/aft and left/right velocities and if mounted laterally it will detect vertical and fore/aft velocities. The sensor is a probe approximately 18 mm in diameter and 0.3 metre long. The processor can be used with a cross-pointer indicator or a display which provides a digital velocity readout and a needle indication of vector alignment.

STATUS: under development.

871/872 Series ice detection system

The 871/872 series ice detector was developed to meet the need for a reliable device to alert the crew to ice formation. Used extensively on military jets and helicopters, the detector is now in use on commercial aircraft and has also been adapted for stationary applications, such as ground turbine installations and antenna towers. Other versions have been developed for icing-rate measurement for helicopters. Using a collector probe vibrating at its resonant frequency, the detector can indicate either ice accretion of a predetermined thickness or the rate of accretion of ice on the probe. Accurate positioning of the probe and the high collection efficiency of the detector ensure that an ice warning is given before accretion occurs on the aircraft or structure itself. The device meets the requirements of MIL-D-8181 for a Type 1 detector.

STATUS: in production.

Icing-rate indication system

Rosemount's model 871FN ice-detector employs an aspirated head to induce an airflow over the sensing probe at a rate which remains constant over a helicopter's operating airspeed range. The indicator instrument used in conjunction with the detector is the 512AG icing-rate meter panel. This incorporates a liquid water content indicator with press-to-test warning facility and a failure warning flag.

STATUS: in production and service.

Pitot heat monitor

Rosemount's pitot heat monitor is a solid-state device designed to meet FAA criteria. The unit monitors currents used by an aircraft's pitot heater and if the heater is selected to 'off' or becomes inoperative generates a signal which operates visual warning indicators on the flight-deck. Dual independent current sensors are incorporated so that the unit can monitor two heaters simultaneously, thereby reducing costs. A comprehensive failure analysis suggests that mean time between failures is in excess of 100 000 hours.

STATUS: in production.

Safe Flight

Safe Flight Instrument Corporation, PO Box 550, 20 New King Street, White Plains, New York 10602
TELEPHONE: (914) 946 9500
TELEX: 137464

Windshear warning recovery guidance

Safe Flight's airborne windshear warning system provides a voice-alert to the crew of high-performance aircraft at the start of an encounter with hazardous low-level windshear. The system is operative during take-off and approach and is a computer-based device which, using conventional sensing elements, resolves the two orthogonal components of a wind gradient with altitude and provides a threshold alert that an aircraft is encountering a potentially hazardous situation. The vectors concerned are horizontal windshear and downdraught drift angle.

Horizontal windshear is derived by subtracting ground speed acceleration from airspeed rate. The latter term is obtained by passing airspeed analogue data from the airspeed indicator or the air data computer through a high pass filter. Longitudinal acceleration is sensed by a computer integral accelerometer, the output of which has been summed with a pitch attitude reference gyro to correct for the acceleration component due to pitch. A correction circuit is employed to cancel any errors due to prolonged acceleration. This circuit has a 'dead band', equivalent to 0.2° of pitch, which prevents correction for airspeed rates of less than 0.1 knot a second. Summed acceleration and pitch signals are fed through a low-pass filter, the output from which is summed with the airspeed rate signal to give horizontal wind-shear.

The vertical computation (downdraught drift angle) is developed through the comparison of measured normal acceleration with calculated glidepath manoeuvring load. Flightpath angle is determined by subtracting the pitch-attitude signal from an angle of attack signal sensed by the stall-warning flow sensor. This is fed to a high-pass filter, from there to a multiplier to which the airspeed signal has been applied. Thus the flightpath angle rate, corrected for airspeed, provides the computed manoeuvring load term. This is compared in a summing junction with the output of a normal computer-integral accelerometer and the failure of the two values to match is the indication of acceleration due to downdraught. The acceleration, when integrated, is the vertical wind velocity and is further divided by the airspeed signal to compute the downdraught angle.

The outputs of both horizontal and vertical channels are determined solely by the atmospheric conditions, and ignore manoeuvres that do not increase the total energy of the aircraft. Windshear correctly compensated by increased engine thrust shows no change in airspeed in the presence of an inertial acceleration as thrust is applied. If, however, the shear goes uncorrected, an acceleration (or deceleration) becomes apparent. Similarly, in the case of the vertical component, a vertical displacement compensated by the crew shows a positive flightpath angle rate (in a downdraught) with less than the computed incremental normal acceleration. Correspondingly, in a downdraught for which the downdraught drift angle is allowed to develop, a negative angle rate at a near-constant 1 g results in the same computation and output. This is important to the crew as it eliminates the possibility that their actions

Safe Flight windshear computer

in anticipating or countering windshear might well mask the condition as far as the warning system is concerned.

Both downdraught drift angle and horizontal windshear signals are combined and the resulting output fed through a low-pass filter to the system computer. This provides two output signals, a discrete alert and, through a voice-generator, an audio alert. Warning output is set at a threshold of –3 knots a second for horizontal shear and –0.15 rad downdraught drift angle or for any combination of the two components which acting together would provide an equivalent signal level. According to Safe Flight, any wind condition requiring additional thrust equivalent to 0.15 g to maintain glidepath and airspeed will result in a non-stabilised approach.

The company has paid particular attention to the elimination of spurious alerts and claims that none had occurred up to early 1986. A cross-over network is employed to sense zero cross-overs of the combined windshear warning signal and this is sampled every 25 seconds. If the warning signal does not pass through a band close to zero, the network automatically provides failure indication, alerting the crew to the fact that the unit is inoperative. A self-test function activated by the pilot is also built into the system.

Format: ¼ ATR
Weight: 6 lb (2.72 kg)

STATUS: in production and service. Most recently certificated systems have been for British Aerospace 125-800, Cessna Citation III and Falcon Jet Falcon 50.

RGS recovery guidance system

Safe Flight also offers recovery guidance, an optional enhancement of the basic windshear warning system. With the combined windshear warning/recovery guidance system (WSW/RGS), as soon as the warning occurs, continuously computed pitch guidance for recovery is displayed on the flight-director command bars. The system also provides pitch guidance for

take-off and go-around on the same instrument, to maintain pilot familiarity with use of the system and promote confidence in it. The pilot does not have to change his flight scan or depart from accustomed procedures during the crucial emergency escape manoeuvre. The company maintains that by following the command bars, the best possible climb profile, maximising the performance capabilities of the aircraft will be achieved.

The system is armed automatically, even if the flight director is turned off, by the windshear warning system's alert output, but only becomes operative, displaying pitch guidance for recovery, when the pilot activates the GA switch. System logic may be programmed to perform these switching and display functions automatically, if desired.

Safe Flight's RGS displays pitch attitudes up to, but not in excess of, the stick shaker target. However, the RGS can be programmed to display stick shaker target information on the ADI slow/fast indicator, alongside the pitch

Safe Flight I-500 speed indexer

guidance display on the command bars. With this optional function, when the pilot activates the GA switch for recovery guidance, the slow/fast indicator changes from a speed mode to shaker mode, with the slow bar representing shaker target. The pilot then has a continuous visual indication of his margin to stick shaker. Internal system monitoring and a self-test function ensure system reliability.

STATUS: recovery guidance is presently available in combination with the windshear warning system in a single ¾ ATR box, or, where Safe Flight's SCAT (speed command of attitude and thrust) system is desired (or already installed), through tie-in of the wind shear warning and SCAT system computers.

I-500 speed indexer

The Safe Flight I-500 speed indexer provides for light transport aircraft a simple head-up display showing the current speed situation with respect to a reference supplied by an associated flight computer which is linked to three lights by a controller. 'F', 'O', or 'S' lights illuminate according to whether the aircraft is fast, on-speed, or slow, compared to the reference. The amber (over speed) illuminates at more than 6 knots over speed, green and amber are seen between 6 and 2 knots fast, green only at within 2 knots of reference speed, green and red between 2 and 6 knots slow and the red (slow) lamp flashes if the speed is more than 10 knots slow. The indicator is mounted on the glareshield.

Dimensions: (indexer) 2.6 × 1.6 × 2.3 inches (66 × 41 × 58 mm)
(controller) 7.1 × 2.25 × 3.5 inches (180.3 × 57 × 89 mm)
Weight: (indexer) 0.2 lb (0.09 kg)
(controller) 1.1 lb (0.5 kg)
Power: 250 mA at 28 V dc

Sanders

Sanders Associates Inc, Daniel Webster Highway, South Nashua, New Hampshire 03061
TELEPHONE: (603) 885 2810

Electronic display units for B-1B

Sanders Associates has supplied a number of electronic display units (EDUs) for the Rockwell B-1B. It is a random position display capable of presenting graphic information in response to X and Y deflection of signals and an unblank signal, Z.

The indicator uses a 12-inch (305 mm) diagonal, square crt with a usable display area of 8 × 8 inches (203 × 203 mm). The deflection system is capable of linear response at writing-rates of up to 125 000 inches per second. Small signal bandwidth is greater than 3 MHz. Positioning moves can be made and settled to within 0.05% of full screen in less than 35 microseconds. These displays depict threat and panoramic information which enables the B-1B's defensive systems operator to analyse threat situations quickly and to assign countermeasure steps.

Sanders display for Rockwell B-1B

Sperry

Sperry Corporation, Aerospace and Marine Group, Commercial Flight Systems Division, PO Box 21111, Phoenix, Arizona 85036
TELEPHONE: (602) 869 2311
TELEX: 667405

Attitude director indicators

Sperry has been involved in flight-director displays since the first Sperry Zero Reader cross-pointer command instrument. Now the company offers a family of instruments from which customers may choose installations

spanning the range from light aircraft to commercial transports. Several types of attitude director indicators allow 'customising' or even variation of presentation from pilot to co-pilot instrument layout.

HZ-6F attitude director indicator

This instrument provides attitude reference data, flight-director, and radio displacement data for aircraft control in pitch and roll during all phases of flight. Computed attitude commands are displayed on the horizontal and vertical flight-director bars and are slaved to the

flight-director controller. When the controller is switched off the command bars are biased from view.

All the other displays provide constant cues unless built-in fault circuits detect a malfunction. In most failure cases pointers are biased from view and an appropriate warning flag appears.

The HZ-6F is electronically self-contained with integral power supply, amplifiers, servos and monitors, and a self-test feature ensures pre-flight display integrity. Loss of either power or an attitude signal will cause the sphere to move to an obviously erroneous position.

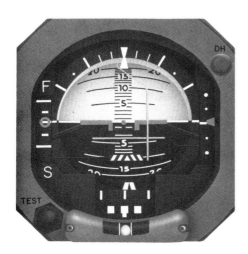

Sperry HZ-6F attitude director indicator

Dimensions: 5 × 5.25 × 8.5 inches (127 × 133 × 216 mm)
Weight: 9.5 lb (4.3 kg)
Power: 11.1 W at 26 V 400 Hz

STATUS: in production.

AD-300B attitude director indicator

Although broadly similar to the HZ-6F, the AD-300B does not display radio altitude.

Format: 5 ATI
Weight: 11 lb (4.9 kg)
Power: 15 W at 26 V 400 Hz

STATUS: in production.

AD-300C attitude director indicator

While still based on the earlier models, the AD-300C displays attitude, flight-director commands, instrument landing system deviation and speed commands on a 5-inch (127 mm) indicator.

Dimensions: 5 × 5 × 9.4 inches (127 × 127 × 239 mm)
Weight: 11 lb (4.9 kg)
Power: 15 W at 26 V 400 Hz

AD-350 attitude director indicator

The AD-350 offers a display of attitude, flight-director commands, instrument landing system deviation, speed command, rate of turn and radio altitude in two versions. Differing slightly in symbology, the AD-350-903 has two decision height warning lights while the AD-350-904 has just one. The instruments are otherwise identical.

AD-350-903/904
Dimensions: 5 × 5 × 9 inches (127 × 127 × 229 mm)

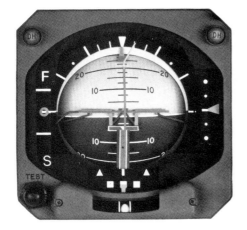

Sperry AD-350 attitude director indicator

Weight: 10.5 lb (4.8 kg)
Power: 11 W at 26 V 400 Hz

STATUS: in production.

AD-650/600 attitude director indicator

This family of attitude director indicators represents a completely new design incorporating micro-computer and micro-servo technology to replace earlier conventional mechanisms, torque motors and gears. They are lighter, run cooler and are more reliable. Both single-cue or cross-pointer displays are available in these 5-inch (127 mm) instruments. The AD-650 features digital readouts of decision height and radio altitude, these facilities being absent on the AD-600. These are specified for a number of corporate business aircraft.

The AD-600 series also provides a symbolic rising runway combined with an expanded-scale localiser which depicts the aircraft's position relative to the runway during the last 200 feet of descent. The glideslope pointer may be on the right, as in the AD-650A, or on the left as in the AD-650B. The AD-650H is a version optimised for helicopter operations.

Dimensions: 5 × 5 × 8.75 inches (127 × 127 × 222 mm)
Weight: 7 lb (3.2 kg)

STATUS: in production.

AD-550/500 attitude director indicator

The 500 series instruments are 4-inch (102 mm) versions of the 600 series. They have all the features and technology of the larger indicators and are compatible with Sperry flight-director systems and radio altimeters.

Dimensions: 4 × 4 × 9 inches (102 × 102 × 229 mm)
Weight: 4.7 lb (2.1 kg)

STATUS: in production.

AD-800 attitude director indicator

This attitude director indicator is an advanced technology, digitally interfaced unit offering the full range of flight data and cues. Flight-director, speed command, glideslope, expanded localiser and radio altitude are presented on high-torque meter movements which are software monitored for improved reliability.

A state-of-the-art micro-computer handles all data and can update the display at least 30 times a second to ensure smooth operation of all servo and meter-driven elements.

Format: 5 ATI
Weight: 10.2 lb (4.6 kg)
Power: 17.4 W at 115 V 400 Hz or 26 V 400 Hz

STATUS: in production.

GH-14 attitude director indicator

The GH-14 can offer several presentations from basic attitude indicator to full flight-director indicator. It features the well established Sperry spherical attitude presentation with full 360° of freedom in roll and ±80° in pitch. The pitch display is expanded to provide 1.4° of sphere movement for every degree of aircraft pitch movement.

The basic GH-14-100 at one end of the range is a simple 4-inch (102 mm) gyro horizon which can be compared with the GH-14-50 at the other end. The latter has rising runway presentation, expanded localiser, pitch-attitude trim, cross-pointer flight-director bars, flight-director flag and pitch and roll outputs. Each configuration is available for either 0° or 8° panel tilt.

Internal failure-monitoring is standard and when in operation any failure is indicated by a gyro-warning flag.

It is suitable for a wide range of aircraft, either as a primary instrument or standby, and is also suitable for helicopters.

Dimensions: 4 × 4 × 9 inches (102 × 102 × 229 mm)
Weight: 4.5 lb (2 kg)
Power: 26 V 400 Hz

STATUS: in production.

HAI 5 heading and attitude indicator

This instrument displays heading, pitch and roll information in a conventional and non-ambiguous form and it is particularly suited to helicopters where panel space is usually limited. The information is taken in three-wire synchro form from a gyro-magnetic compass and vertical gyroscope or from other sources, such as an inertial platform. The associated servo amplifiers are contained within the instrument case and integral lighting is provided.

Pitch and roll information is presented in conventional spherical form except that the sphere is slightly barrel-shaped to provide the maximum display area and reduce any 'tunnelling' effect. There is full 360° freedom in roll, while the pitch scale is expanded in the approximate ratio 4:3 and is limited in travel to about ±100°.

Heading is shown on a separate dial which rotates on the outside of the pitch/roll sphere. This dial has 2° graduations and moves in a conventional sense to indicate changes in aircraft heading. A dial-mounted 'set heading marker' can be positioned relative to the dial using the knob at the lower left-hand corner of the bezel. Once adjusted it rotates with the dial to give a constant indication of the required heading.

A ball side-slip indicator is provided at the bottom of the presentation. Two warning flags, for 'attitude' and 'heading' signal indication are provided. A flag indicating directional-gyro mode selection is installed for use with a compass system. There is a press-to-test button to check the attitude display and steering pointer indications of ILS localiser or VOR deviation can also be included.

Dimensions: 130 × 130 × 215 mm
Weight: 3 kg
Follow-up rates: (heading) 50°/s; (pitch) 200°/s; (roll) 400°/s
Power: 28 V dc

STATUS: in service.

Sperry HAI 5 heading and attitude indicator

Sperry MD-80 EFIS

Sperry RD-350J horizontal situation indicator

Sperry RD-650 horizontal situation indicator

Electronic flight instrument system for MD-80 series aircraft

These instruments give a pictorial presentation of all primary flight instrument data, displaying attitude director indicator and horizontal situation indicator information on separate screens. These easily installed direct replacements for electromechanical indicators will also display weather information, advisory maps and radio altitude data. This compact, easy-to-read system is specifically programmed with standard commercial display formats for civil aviation applications.

Dimensions: 5.06 × 6.06 × 10.5 inches (129 × 154 × 267 mm)
Weight: 10.5 lb (4.8 kg)
Power: 115 V, 400 Hz, 65 W

STATUS: in production. Sperry is also developing a new EFIS system for the MD-88 aircraft featuring 5 × 6 inch (127 × 152 mm) colour cathode ray tube displays.

ARINC 700 symbol generator

Sperry's symbol generator, designed to interface with other ARINC 700 equipment in commercial transport aircraft, takes in information from a variety of sensors and computers and provides outputs to colour displays. Converting basic aircraft parameters to a pictorial display requires a significant data processing capability, and Sperry's answer is a 16-bit processor designed for display applications with a special memory hierarchy scheme. The system incorporates the company's raster/stroke display technology stemming from a Sperry patent more than a decade ago on hybrid symbol generation. Bold symbology, such as sky/ground shading and weather radar returns, is painted using raster-scan techniques. Smooth moving high-resolution graphics, however, such as a navigation display calls for stroke writing. A special timing scheme permits the raster and stroke symbol generators to drive two displays simultaneously, each with its own writing characteristic. When one display is being stroke-written, the other is in the raster-scan mode. The methods are then switched to generate each composite display at a frequency of 80 times per second so as to eliminate flicker.

Another refinement is symbol priority. It is accepted that certain information needs to remain always in view without interference by other symbology. To ensure this, the company has developed an innovative technique providing several levels of priority. Symbols move smoothly into and out of these priority areas as if from behind mechanical windows, resulting in a smooth, uncluttered presentation.

STATUS: in production.

Horizontal situation indicators

Sperry produces a wide range of horizontal situation indicators to suit all types of operation and aircraft. These instruments may be part of integrated flight systems where they are compatible with and complementary to other instruments.

RD-350J horizontal situation indicator

This horizontal situation indicator provides heading, two DME distances, radio navigation information via a displacement bar, indications of selected course and heading, and to/from indications for VOR operations. Flags give indication of failures.

A Nav Mode annunciator takes the form of a rotary display in the centre of the compass card, controlled by VOR/LOC valid, LOC tuned, to/from and back course selected signals. The annunciator displays a symbol to indicate selected modes and can also indicate when radio data is invalid.

Dimensions: 5 × 5 × 8.9 inches (127 × 127 × 226 mm)
Weight: 9.2 lb (4.2 kg)
Power: 6.4 W at 26 V 400 Hz

STATUS: in production.

RD-450 horizontal situation indicator

This instrument gives a pictorial presentation of aircraft lateral position and integrates navigation deviations and compass into one instrument. It provides a repeated heading display, manually-selected heading and course facilities, and an indication of flight-path deviations.

Dimensions: 4 × 4 × 9.75 inches (103 × 103 × 248 mm)
Weight: 3.6 lb (1.6 kg)
Power: 2 W at 26 V 400 Hz

STATUS: in production.

RD-550 horizontal situation indicator

This instrument gives a pictorial presentation of the aircraft's lateral position, integrating navigation deviations and compass information into one instrument. The indicator provides course readout and DME range indications on

Sperry RD-550 horizontal situation indicator

digital displays and has mechanical presentation of vertical and course deviations. Remotely signalled or manually selected heading and course information can be displayed, together with glideslope deviation. Course and distance displays can be orientated for left or right-hand installations.

Dimensions: 4 × 4 × 9 inches (103 × 103 × 230 mm)
Weight: 5.9 lb (2.65 kg)
Power: 13 W at 26 V 400 Hz

STATUS: in production.

RD-650 horizontal situation indicator

The 4 inch × 5 inch (102 mm × 127 mm) RD-650 has luminous digital presentations of course and distance, which may be supplied with or without heading reference and course control knobs. A choice of remote controls for these horizontal situation indicators is available. They can be separate for each instrument in a dual flight-deck installation, or can have a single control that synchronises pilot and co-pilot heading reference while still providing independent course control.

Dimensions: (RD-650A) 5 × 4 × 8 inches (127 × 102 × 203 mm)
Power: 115 V 400 Hz

STATUS: in production.

RD-700 series horizontal situation indicators

The RD-700 series has been developed for applications in new aircraft or retrofit installations. High torque, low power-flag and shutter movements eliminate sticking displays and low-power devices coupled with open card construction result in low heat dissipation and power demands. All indications are conventional in presentation and numeric indicators feature standard veeder readouts. The following list summarises the presentation and displays of each instrument:

RD-700 horizontal situation indicator

Displays all standard navigation radio inputs, compass system and ARINC 561 INS data including digital readout of drift angle and ground speed.

Format: 5 ATI
Weight: 8.5 lb (3.9 kg)
Power: 115 V 400 Hz or 26 V 400 Hz

RD-700A horizontal situation indicator

All standard navigation radio and compass data are presented together with dual digital DME readout. The RD-700A HSI features new flag, shutter and annunciator mechanisms and improved packaging to enhance reliability.

Sperry claims a 44 per cent reduction in power consumption compared with earlier designs.

Format: 5 ATI
Weight: 8.5 lb (3.9 kg)
Power: 115 V 400 Hz or 26 V 400 Hz

RD-700C horizontal situation indicator

In addition to radio and compass navigation data this instrument includes ARINC 561 INS data and digital readout of time and distance to waypoint, and ground speed.

Format: 5 ATI
Weight: 8.5 lb (3.9 kg)
Power: 115 V 400 Hz or 26 V 400 Hz

RD-700D horizontal situation indicator

Although very similar to the RD-700C this presentation does not include the time-to-waypoint counter.

Format: 5 ATI
Weight: 8.5 lb (3.9 kg)
Power: 115 V 400 Hz or 26 V 400 Hz

RD-700F horizontal situation indicator

Displays all standard data including to/from, drift angle and digital readout of ground speed and distance to waypoint.

Dimensions: 5 × 5 × 8.5 inches (127 × 127 × 216 mm)
Weight: 8.5 lb (3.9 kg)
Power: 115 V 400 Hz or 26 V 400 Hz

RD-700G horizontal situation indicator

In addition to all standard navigation, radio, compass and ARINC 561 INS inputs, this instrument presents digital readout of drift angle, distance to waypoint and ground speed, though the drift and ground speed presentation is at the lower area of the indicator rather than the more usual upper region of the instrument.

Format: 5 ATI
Weight: 8.5 lb (3.9 kg)
Power: 115 V 400 Hz or 26 V 400 Hz

RD-700M horizontal situation indicator

In addition to standard information displays the RD-700M uses new, 11-position low-power mag-wheels for dual DME displays. A new thermal design includes a more efficient heat-sink mounting for high-power components and

Sperry RD-700M horizontal situation indicator

Sperry RD-850 horizontal situation indicator

heat-sensitive capacitors. Electronic and mechanical sections are segregated for maintenance access, while open-board packaging in the electronic section facilitates troubleshooting.

Dimensions: 5 × 5 × 8.6 inches (127 × 127 × 218 mm)
Power: 115 V 400 Hz or 26 V 400 Hz

RD-800 horizontal situation indicator

This instrument features three digitally-driven servoed displays in conjunction with two four-digit gas-tube displays showing time and distance to waypoints. Microprocessor control gives improved versatility in navigational data processing.

Format: 5 ATI
Weight: 8.8 lb (4 kg)
Power: 115 V 400 Hz or 26 V 400 Hz

RD-800J horizontal situation indicator

In this application the readout of true airspeed and ground speed is provided by conventional counter displays and for ease of interpretation the command bars are colour-identified.

Format: 5 ATI
Weight: 8.8 lb (4 kg)
Power: 115 V 400 Hz or 26 V 400 Hz

RD-850 horizontal situation indicator

The RD-850 horizontal situation indicator features the most up to date applications of instrument technology including microprocessor control. Coloured display elements are included together with distance-to-go and ground speed counters. Automatic direction-finder annunciators are fitted in the lower instrument area.

Format: 5 ATI
Weight: 10.2 lb (4.7 kg)
Power: 115 V 400 Hz or 26 V 400 Hz

Sperry Corporation, Aerospace and Marine Group, Defense Systems Division, PO Box 9200, Albuquerque, New Mexico 87119
TELEPHONE: (505) 822 5000
TELEX: 660319

ADZ air data system

Sperry produces a wide range of aircraft flight instruments and supplies complete packages of air data equipment for aircraft including the air data computer and a full set of associated instruments.

The ADZ air data system uses an AZ 241 or 242 computer depending on whether the

aircraft is a turboprop or jet. These are now complemented in production with the AZ-600 (turboprop) and AZ-800 (jet) digital air data systems. All systems use the Sperry patented vibrating diaphragm pressure sensor to provide outputs for altitude, airspeed, vertical speed, true airspeed, true air temperature and total air temperature information.

If required, a single computer can handle a dual flight director installation. Consequently, no matter which director is driving the autopilot the full complement of modes is available.

AZ-241 (turboprop)
Dimensions: 7.6 × 3.1 × 14.2 inches (193 × 79 × 361 mm)
Weight: 8.6 lb (3.9 kg)

AZ-242 (jet)
Dimensions: 7.6 × 4.9 × 14.2 inches (193 × 124 × 361 mm)
Weight: 11.5 lb (5.2 kg)

AZ-600 (turboprop)
Dimensions: 7.6 × 3.625 × 14.25 inches (193 × 92 × 362 mm)
Weight: 8.2 lb (3.7 kg)

AZ-800 (jet)
Dimensions: 7.6 × 3.625 × 14.25 inches (193 × 92 × 362 mm)
Weight: 9 lb (4.08 kg)

STATUS: in production for wide range of executive turboprop and jet aircraft, one of the latest is the Canadair Challenger.

Military air data computers

Sperry is the world's largest supplier of military digital air data computers. Equipment is in production for the McDonnell Douglas F-15 Eagle and F/A-18 Hornet, the General Dynamics F-16 and the Boeing KC-135R. All use a patented vibrating diaphragm pressure transducer. Sperry says that these are the only air data computers requiring no sensor re-calibration. The US Air Force is claimed to do no maintenance on the F-15 system, which is designated 'discard on failure'. Only seven assemblies were replaced, according to Sperry, during the first six years of F-15 equipment production.

STATUS: in service.

Digital air data computer

Sperry produces an all-digital ARINC 706 standard air data computer for Airbus Industrie A310 and Boeing 757 and 767 airliners. These all-solid-state systems use large or medium-scale integrated technology, cmos master transmitter-receiver chips and a Z8000 microprocessor. Built-in test sequences run continuously in flight to provide a 95 per cent fault detection capability, with a 99 per cent probability of correct fault identification. Test failure information is stored in the non-volatile memory for post-flight analysis. External sensor failures can also be detected and signalled visually. The system can accommodate 20 per cent increase in output parameters, more than 50 per cent computational growth and more than 100 per cent memory growth. The unit incorporates Sperry-patented transducers and has ARINC 429 transmitter and receiver interfaces.

Format: 4 MCU
Accuracy: (airspeed) ±2 knots at 100 knots, ±1 knot at 450 knots; (altitude) ±15 ft at sea level, ±80 ft at 50 000 ft; (Mach) ±0.003 Mach at 0.8–0.9 Mach in the range 25 000–45 000 ft
Weight: 12.5 lb (5.7 kg)
Power: 25 W
Reliability: 15 000 h mtbf
Failure memory: 6 failures/10 flights

STATUS: in service.

Sperry digital air data computer for (left to right) McDonnell Douglas F-15 Eagle, and F/A-18 Hornet and General Dynamics F-16 Falcon fighters

Sperry multi-function display for General Dynamics F-16 fighter

SI-800 Mach/airspeed indicator

This instrument, based on a Z-80 microprocessor, operates with computed airspeed, Mach number and maximum allowable airspeed (V_{MO}/M_{MO}) data from an air data computer and selected airspeed signals from a mode control panel through a multiple input-integrated input device using ARINC 429 master chips. Airspeed is displayed by pointer on dial over the range 60 to 450 knots. In addition there is a red and white striped V_{MO} pointer and four memory 'bugs' which can be moved around the periphery of the speed scale. Airspeed is also displayed on a three-digit counter, and commanded speed by a red 'bug'. Mach number is displayed in the upper counter display over the range 0.4 to 0.999 in 0.001 increments. Each display has an associated failure flag.

Format: 4 ATI S7.0
Weight: 5.3 lb (2.4 kg)
Power: 14 W at 115 V 400 Hz

VS-800 vertical speed indicator

A Z-80 microprocessor uses signals from an inertial reference system to drive a dc servo-driven display pointer presenting instantaneous vertical speed. The scale is marked at 100 feet a minute intervals over the expanded range 0 to 1000 feet a minute and at 500 feet a minute intervals from 1000 to 6000 feet a minute. Software adjusts pointer movement to compensate for dial non-linearity.

Format: 3 ATI S7.0
Weight: 3 lb (1.3 kg)
Power: 9.9 W at 115 V 400 Hz

BA-800 barometric altimeter

This combines a Z-80 microprocessor with electromechanical technology for processing altitude data received via an ARINC data-bus from an air data computer.

Altitude is presented on a counter-pointer display over the range –1000 to 50 000 feet. The numeric display is given by a continuous motion five-digit, four-drum, mechanical counter array. Below 10 000 feet the 10 000 counter shows green to alert the pilot. Reference barometric pressure is displayed both in inches of mercury and millibars. A set altitude marker 'bug' is provided. Sperry claims that the incorporation of microprocessor technology improves the dynamic response, accuracy and comprehensiveness of fault isolation compared with non-digital instruments.

Format: 3 ATI S8.0
Weight: 4 lb (1.8 kg)
Power: 11 W at 115 V 400 Hz

Air data instruments

Associated with the Sperry ADZ-241/242 air data system are the BA-141 altimeter, SI-225 Mach-airspeed indicator, and VS-200 vertical

speed indicator. The altimeter is of the counter-pointer type with numeric height readout; the Mach airspeed indicator has a numeric readout of Mach number.

Format: 3 ATI
Weight: (BA-141) 2.5 lb (1.1 kg)
(SI-252) 2.8 lb (1.2 kg)
(VS-200) 2 lb (0.9 kg)
Power: 26 V 400 Hz

SRS 1000 attitude and heading reference system

This strapdown attitude and heading reference system (AHRS) uses state-of-the-art technology to meet modern AHRS requirements with very low life-cycle cost. A 15-second reaction time is claimed, independent of ambient temperature or vibration.

There are two units: the attitude heading reference unit and compass controller unit. The reference unit accepts magnetic compass, air data and Doppler inputs to provide attitude, heading, body rates and accelerations, ground speed and drift angle information to other aircraft systems. The compass controller provides systems information and control functions to the crew.

An inertial measurement unit within the reference unit contains two flexure suspended gyros, and two toroidal accelerometers for the X and Y axes, plus a standard force-feedback accelerometer for the Z axis. The system has been selected as standard equipment on the Airbus Industrie A300 and A310 wide-body airliners.

STATUS: in production.

Multi-function display for F-16

In June 1981 Sperry won a $4 million contract to develop an advanced multi-function display system for the US Air Force/General Dynamics F-16. The system forms part of the F-16 MSIP (multi-staged improvement programme) to update the aircraft with latest new-technology equipment.

Type 1020 attitude and heading reference system

The system comprises two 4-inch (102 mm) square cathode ray tube displays with a digital, programmable display generator for each aircraft. The two-seat F-16Ds that make up about 15 per cent of the Fighting Falcon fleet have four displays, but still feed from the one signal generator. The displays present the pilot with navigation, radar and weapon-aiming information. They use raster-scan techniques to provide clear and sharp presentations of sensor information, symbol overlays and alpha-numerics in all ambient light conditions. The displays are capable of providing both 525-line and 875-line rates. Each display has 24 push-button keys arranged around the bezel for interactive control and data entry.

The programmable display generator uses advanced task-sharing techniques for high-speed operation and symbol-generation. It is a software-based device with two processors and 64 K words of core memory, permitting rapid changes of mode and the use of several sets of symbology. Data from the F-16's fire-control computer is fed to the system via the MIL-STD-1553 digital data-bus while video information comes from its electro-optical sensors.

The system's digital design permits rapid reprogramming and this flexibility includes the ability to integrate future avionics, including the Lantirn (low-altitude navigation and targeting infra-red for night) and joint tactical information and distribution systems (JTIDS). The display system incorporates comprehensive test and diagnostic routines and there are growth provisions for the generation of colour symbols and for the MIL-STD-1750A instruction set architecture. A variation of this display is also being used in the F-111 avionics modernisation programme.

Display colour: phosphor green P43
Symbol contrast ratio: 7:1 at 10 000 ft-lamberts (34 260 cd/m²)
Brightness: (display screen) 3000 ft-lamberts (10 278 cd/m²)
(filtered) 300 ft-lamberts (1028 cd/m²)
Line width: 0.0075 inch (0.19 mm)
Linearity: 2%
Usable screen area: 4 × 4 inches (102 × 102 mm)
Jitter: (symbology) < 2.5 mils
(raster video) 3 mils
Video inputs: 9 total
(electro-optical weapons) 525 line and 875 line
Video outputs: 6 total, 2 independent channels, 3 outputs/channel
Dimensions: (display) 5.6 × 5.6 × 12.1 inches (143 × 143 × 308 mm)
(display generator) 6.3 × 7.6 × 17.66 inches (159 × 194 × 448 mm)
Weight: (display) 13 lb (5.9 kg)
(generator) 28 lb (12.7 kg)

STATUS: in service on all F-16C/D aircraft. The total contract awarded by General Dynamics calls for the manufacture of up to 2250 sets.

Multi-purpose colour display for F-15

The multi-purpose colour display (MPCD) developed for the US Air Force/McDonnell Douglas F-15 fighter as part of a continuing MSIP (multi-stage improvement programme) enhancement for the C and D versions is a very high resolution 5 by 5 inch (127 by 127 mm) shadow-mask system with stroke, raster, or combined stroke/raster writing in any of 16 colours as determined by software in the symbol generator. Raster symbology is a full-colour representation of a colour sensor's output. Functions will include built-in test for a number of aircraft systems, graphic representation of aircraft stores configuration (replacing the electromechanical armament control panel on current McDonnell Douglas F-15s), display of video from electro-optic sensors and weapons, and interface for secure tactical information from a JTIDS communications system.

McDonnell Douglas has contracted with Sperry for a second version of the colour display for their F-15E programme. The F-15E fleet, which will incorporate air-to-ground and air-to-air mission capabilities, will be equipped with three Sperry MPCDs on each aircraft. Sperry is also under contract for the display processor for F-15E aircraft. The equipment will not only power the aircraft's three colour displays, but all the displays on the aircraft. Production of the display processor is scheduled to begin in early 1986.

The display's delta gun shadow-mask tube uses dynamic and static convergence for maximum symbol fidelity. A bandpass filter and 17 000 inches a second writing speed facilitates viewing in high ambient light levels, or even direct sunlight. Two cathode ray tube versions are available, one having 0.008 inch phosphor dot spacing, said by Sperry to be only two-thirds the dot size and spacing of other shadow-mask cathode ray tubes designed for airborne applications.

The Sperry programmable signal data processor interfaces with other equipment and sensors, generating monochrome and colour symbology for other displays. It comprises two elements: a three-channel symbol generator and a general-purpose processor. Self-test circuits monitor continuously all critical signals and failures are flagged on a built-in test indicator.

Dimensions: 6.75 × 7.07 × 15.25 inches (172 × 180 × 387 mm)
Weight: 26 lb (11.8 kg)
Screen size: 5 × 5 inches (127 × 127 mm)
Linewidth: typical 0.016 inch

Sperry multi-function cathode ray tube for F-15 Eagle fighter

Sperry ANMI for McDonnell Douglas F-15 fighter

Video bandwidth: dc 10 MHz
Television scan resolution: 70-line pairs/inch

STATUS: in production for US Air Force/McDonnell Douglas F-15 fighter. Contract includes options for up to 513 production display sets, the first such equipment with colour.

Air navigation multiple indicator for F-15

The ANMI air navigation multiple indicator is one of two elements of the vertical situation display in the US Air Force/McDonnell Douglas F-15 Eagle fighter, the other being a programmable signal data processor. Together they provide the pilot with navigation and radar information. The 4-inch (102 mm) square ANMI uses advanced filter and display techniques to produce high-context monochromatic symbology readable in direct sunlight.

STATUS: the first ANMI was delivered in 1972, and the system continues in production. Sperry has delivered more than 1500 such systems.

Multi-function display for AHIP

The Sperry multi-function display is one of the elements in the full integrated cockpit avionics installation designed for AHIP, the US Army's helicopter improvement programme to upgrade a significant number of its scout and reconnaissance helicopters and improve their battlefield effectiveness. AHIP is under assessment, with five Bell OH-58 Kiowas modified to take the new avionics and other systems.

Each AHIP aircraft contains two such displays. They receive data from two master controller processor units, either of which can drive both displays for redundancy purposes. The symbology is produced by an 875-line

Sperry multi-function display for AHIP demonstrator programme

Sperry remote frequency display for AHIP

raster scan video writing on a 4:3 aspect-ratio screen. The monochrome P43 phosphor and contrast enhancement filter, together with 0.008-inch spot size, generate a crisp picture with high resolution and easily visible in direct sunlight. Display modes and sensor inputs are selected via 14 push-button switches on the bezel surrounding the screen.

Dimensions: 7.08 × 8.10 × 15.25 inches (180 × 206 × 387 mm)
Weight: 20.3 lb (9.2 kg)
Screen size: 6.4 × 4.8 inches (162 × 122 mm)
Contrast: > 11:1 at 2000 ft candles, > 4:1 at 10 000 ft candles
Power: 56 W

STATUS: in production.

Remote frequency display for AHIP

The operational status of each of the five radios carried aboard the OH-58D prototype helicopters (see description under above entry) is displayed by means of this remote presentation system. It shows, on five lines of liquid-crystal readout, the radio number, operator in control, frequency, whether coded or in plain language, and two alphanumeric characters for channel identification.

The display receives its information through a serial digital interface with either of the two master controller processor units, the choice of unit being left to the operator. Data for each window remains displayed until updated. If the system does not receive an update within 2 seconds, the screen goes blank as a warning to the operator. The screen is designed for reflective daylight viewing and is back-lighted for night operation. Characters are painted in white on a black background.

Dimensions: 5.75 × 3.9 × 4 inches (146 × 99 × 102 mm)
Weight: 1.8 lb (0.8 kg)
Contrast: 20:1 in direct sunlight, 16:1 when back-lighted
Back-lighting: electro-luminescent
Power: 3 W normal, 53 W with heater

STATUS: in production.

Multi-purpose stroke display

The Sperry Multi-purpose stroke display is currently operational on a number of aircraft including the SH-3, SH-2F and EA-6B. The 6.5 inch × 8.5 inch (165 mm × 216 mm) crt is driven by a self-contained display generator thereby deleting the requirement for peripheral display hardware. The display requires no cooling air. The stroke formatting provides clear and crisp symbology in high ambient light conditions. A variation of this display, characterised by a stroke and raster capability, is now being produced for the US Navy's CV-Helo (SH-60F) anti-submarine warfare helicopter. The CV-Helo display can simultaneously provide raster formats (TV, flir, etc) along with stroke symbology overlay.

Sperry digital map display

Display colour: phosphor green P43
Symbol contrast ratio: 2.5:1 at 10 000 foot-candles
Usable screen area: 6.5 × 8.5 inches (165 × 216 mm)
Weight (SH-2F): < 28 lb (12.7 kg)
Size (SH-2F): 9 × 11 × 15 inches (229 × 279 × 381 mm)
Power consumption: 70 W nominal
Symbology rate: 28 320 inches/s
Cooling: none

STATUS: in service.

Digital map display set for the AV-8B

In July 1985, Sperry won a contract from McDonnell Douglas Corporation to design and develop a digital map system for use in the AV-8B Harrier II Night Attack Programme. Full-scale development has begun on the digital map display set which consists of a digital map computer and a digital memory unit. It will provide pilots with a moving map display that includes digitised paper maps and Defense Mapping Agency terrain information presented on a colour display.

STATUS: under development.

'Smart' multi-function display

Sperry is developing a 6 inch × by 6 inch (152 × 152 mm) 'smart' multi-function display for use as a primary flight instrument on military aircraft. Functions normally provided by a display controller are incorporated into the display unit eliminating the need for a peripheral controller. Format generation uses high-level macro commands which may be stored as tables and then compiled along with the associated aircraft dynamic data to generate the actual displays. Multi-function display modes and sub-modes are completely programmable and are stored in a non-volatile memory within the display. This system allows rapid switching between formats with minimal bus loading because only format designations and dynamic data are sent over the 1553 bus.

The cathode ray tube assembly uses a delta gun, taut shadow-mask tube with an aluminised tri-colour phosphor dot triad on a black matrix. Convergence is controlled statically and dynamically over the entire viewing area. The stroke-writing rate of this hybrid display is software controlled with a symbol generation speed of over 35 000 inches a second. Synthetically generated raster for background and stroke symbology fill is written at a rate of 180 000 inches a second. Display brightness and contrast is automatically controlled by two bezel-mounted light sensors and a remote light sensor.

Size: 8 × 8 × 15 inches max (203 × 203 × 381 mm max)
Weight: 36 lb max (16.3 kg)
Screen size: 6 × 6 inches (152 × 152 mm)
Line width: 20 mils max; 14 mils nominal: 11 mils minimum
Operating modes: stroke, hybrid (stroke/raster)
Digital interface: MIL-STD-1553B-RT
Raster type: RS-170- 525 lines

STATUS: in development.

Sperry Corporation, Aerospace and Marine Group, Avionics Division, PO Box 29000, Phoenix, Arizona 85038
TELEPHONE: (602) 863 8000
TELEX: 668419

EDZ-800 electronic flight instrument system for general aviation

Television-screen flight instrument and systems displays, initially developed for airliners such as the Boeing 767 and 757 and the Airbus Industrie A310, have started to appear at the upper end of the general aviation industry rather sooner than many observers had expected. Sperry's effort in the general aviation electronic flight instrument system field is represented by the EDZ-800 family of attitude director indicators and horizontal situation indicators replacing the conventional electromechanical attitude and horizontal situation indicators, and available with 5- and 6-inch (127 and 152 mm) displays. Initially, Sperry duplicated the symbols and formats of the earlier electromechanical instruments but, in addition, routed weather radar information to the horizontal situation indicator and some symbols appropriate to other sensors have also been added. Thus a dedicated weather radar display is no longer needed, producing an immediate and substantial space-saving in the overcrowded flight-decks of general aviation aircraft. The system also has great capacity for growth and flexibility for changes in presentation to meet the needs of integrated performance or flight management systems.

Although the two electronic flight instrument system (EFIS) displays on each pilot's flight panel are normally dedicated to specific functions (either horizontal situation or attitude director indication), it is possible to combine these displays electronically so that a single cathode ray tube can show basic information from both in the event of the failure of one display. The EFIS attitude director indicator

display also permits unwanted information to be automatically dropped from view. For example, the glideslope and localiser symbols disappear when the navigation receivers are tuned to stations en route and the radio altimeter readouts vanish above the system's maximum operating altitude. The horizontal situation indicator is also very flexible and can operate in three navigation modes. In the first of these, the display can show weather radar returns on an expanded, partial compass card presenting just the forward 90° sector. In the second mode, the weather is removed, leaving just a partial compass card with navigation mapping or standard courses and heading information. The third presentation gives a standard, full compass card horizontal situation indicator with all appropriate alphanumeric navigation data. Separate electronic flight instrument system control units for the two pilots permit the different modes to be selected.

Production began early in 1982 for the first two applications, the Gulfstream III and Canadair Challenger.

In November 1982 the first such installation, in a Canadair CL-600 Challenger, was certificated. Installations certificated since then are CL-601 Challenger, Gulfstream II, IIB and III, Dassault Falcon 50, Cessna Citation I, II and III, King Air 200, 300 and F90, Mitsubishi Diamond 1 and II, Gates Learjet 55, British Aerospace 125-700/800, Jetstream 31, Dash 8, ATR42 and Boeing 727-100 aircraft, and Sikorsky S-76 and Westland 30-160 helicopters. Falcon 900 certification was expected in March 1986.

A second-generation of Sperry EFIS, incorporating 8 × 8 inch (203 × 203 mm) colour cathode ray tube displays, has been chosen as standard equipment on the Gulfstream IV. The Gulfstream IV will use six 8 × 8 inch cathode ray tubes, perhaps a unique combination of number and size of displays on a general aviation aircraft.

STATUS: in production.

ED-603/803, ED-611/811 electronic flight instrument systems

Sperry has introduced a new generation design in the electronic flight instrument product line intended for the corporate aircraft and regional airline industries. Available in 5- and 6-inch sizes (127 and 152 mm) the new product is the result of a new symbol generator and software changes. The generator has increased stroke writing and memory capabilities, while software

Flight panels in Gulfstream G-IV with six second generation 8 x 8 inch (203 x 203 mm) Sperry cathode ray tubes

Sperry EHSI for corporate aircraft and regional airliners

updating has provided a variety of cosmetic improvements to the display.

The electronic attitude director indicator (EADI) features an enlarged sphere presentation with linear pitch tape, improved single cue aircraft symbol and stroke-filled single cue command bar. A digital T-bar airspeed and/or angle-of-attack presentation have also been added. Other improvements include a stroke-filled roll pointer, larger roll indices, shorter horizon indices and colour reversal on glideslope, angle-of-attack, rate of turn and expanded localiser scales. In addition, the 5-inch (127 mm) EADI is now available in a truncated sphere presentation.

The electronic horizontal situation indicator (EHSI) features a stroke-filled lubber line, colour reversal on the glideslope scale and weather radar mode annunciation.

The ED-603/ED-803 are analogue systems while the ED-611/ED-811 are digital.

STATUS: in production.

ASZ-800 electronic air data system for general aviation display

Sperry has integrated into two electronic displays all the flight-deck air data readouts appropriate to heavy turboprops and corporate business aircraft. They simplify the crew monitoring tasks, reduce panel space needs

and add a number of capabilities unavailable in conventional electromechanical instruments. These include a programmable reference speed window on the airspeed display and an integrated altitude alert/pre-select on the altimeter. The standard vertical-case cathode ray tubes can be programmed for either conventional circular-scale or vertical-tape format presentations.

The circular-scale airspeed indicator incorporates a pointer and counter-drum airspeed display and dedicated Mach number, true airspeed and total or static air temperature. Programming the V-speed window permits the pilot to enter the appropriate reference speeds for each flight based on aircraft performance curves and weather. V_i, V_r, and V_{ref} speeds can be entered via a set/select button and value knob on the instrument bezel. A red V_{mo} arc appears on the periphery of the airspeed scale.

Like the cathode ray tube airspeed indicator, the altimeter incorporates several related functions to simplify readability. The electronic symbology duplicates the conventional scale and drum height readout with dedicated windows for barometric pressure in millibars or inches of mercury. It also includes a vertical speed readout with a maximum digital reading of 500 feet a minute (152 metres a minute), though the pointer window can read up to 9999 feet a minute (3047 metres a minute).

Apart from these scale and drum formats, two types of vertical scale presentations are also available for each of these two air data instruments.

STATUS: in production.

AA-200 radio altimeter

Part of the Sperry Stars avionics range for general aviation and helicopter applications, this is an all-solid-state system with 3-inch (76 mm) indicator and dual horn antennas. It is compatible with ground-proximity warning systems and is available with a choice of indicators. One version has an expanded low-altitude scale (below 200 feet) and is suitable for helicopters. Decision height can be preset and both visual and audio warnings are provided on reaching this altitude.

Typical installation: transmitter-receiver unit, 2 horn antennas and an indicator

Frequency: 4300 MHz
Max altitude: 2500 ft
Accuracy: (0–500 ft) 3 ft or 3%
(> 500 ft) 4%
Power: 1.5 A at 22–31 V dc

STATUS: in production.

AA-300 radio altimeter

This consists of RA-315 and RA-335 indicators, RT-300 transmitter-receiver and AT-220, 221, or 222 antenna. The system interfaces with the AD-650 and AD-550 series attitude director indicators and with the VA-100 voice advisory unit. The RT-300 transmitter-receiver is a solid-state unit offered in three optional configurations for different outputs: the –901 provides an auxiliary output conforming to the first 500 feet of the ARINC characteristics for use in driving the rising runway bar of the attitude director indicator; the –902 provides an auxiliary output shaped to the full ARINC characteristics for use by ground-proximity warning devices and autopilots; the –903 is intended for hovering applications with helicopters.

The RA-315 indicator has a servo mechanism-controlled pointer display of radio altitude up to 2500 feet. Below 500 feet the scale is expanded to enhance readability. There is an adjustable decision height 'bug' and an amber decision height warning lamp. The RA-335 is similar to the RA-315 but is configured for helicopters, having a range of 0 to 1500 feet. Below 200 feet the scale is expanded to improve readability.

Format: (RA-315 and -335) 3 ATI × 45 inches (1143 mm)
(RT-300) 4.09 × 4.56 × 11.07 inches (104 × 116 × 281 mm)
(AT-220) 2.5 × 6.25 × 5.6 inches (63 × 159 × 142 mm)
Weight: (RA-315 and -335) 1.5 lb (0.7 kg)
(RT-300) 4.5 lb (2 kg)
(AT-220) 0.75 lb (0.3 kg)
Accuracy: (RT-300) 0–100 ft ± 3 ft, 100–500 ft ± 3%, 500–2500 ft ± 4%
(RA-315) 0–100 ft ± 5 ft, 100–500 ft ± 5%, 500–2500 ft ± 7%
(RA-335) 0–100 ft ± 5 ft, 100–500 ft ± 5%, 500–1500 ft ± 7%
Power: 0.5 A at 21–32 V dc

Sperry Dalmo Victor

Sperry Dalmo Victor Inc, PO Box 52004, Phoenix, Arizona 85072
TELEPHONE: (602) 869 2311
TELEX: 667405

Minimum T/CAS 2 threat-alert/collision-avoidance system

Receptive to the safety issues raised by the growing density of air traffic, Sperry and Dalmo Victor in 1983 formed a joint company to produce and market airborne collision-warning and avoidance equipment. The initial result of this pooling of resources is a system called Minimum T/CAS 2, so designated to differentiate it from the more capable but more expensive Enhanced T/CAS 2 under development by Bendix. Dalmo Victor brings to the

joint company its expertise in electronic warfare (notably in radar warning systems) and is responsible for design, while Sperry contributes its knowledge of airline operations and procedures to manufacture and marketing. Minimum T/CAS 2 was scheduled to begin an eight-month flight trials programme aboard a Boeing 727 transport of Piedmont Airlines early in 1985.

Minimum T/CAS 2 is basically an airborne interrogator transmitting signals into four quadrants around the 727 in a horizontal plane. It computes bearing (to 10° accuracy), range and range closure rate, and altitude and vertical separation closure rate for any other aircraft, and displays on a colour weather radar display information on potential conflicts. If the risk continues to increase the system computes and commands evasive manoeuvres in pitch on a modified vertical-speed indicator.

In preparation for a follow-on series of tests

planned for 1986/87, the FAA in December 1984 ordered from Sperry Dalmo Victor nine Minimum T/CAS 2 systems. In contrast to the 1985 trials, which involved non-co-operative or passive 'targets', the new series will investigate the interaction between aircraft that all have T/CAS equipment. These tests will involve three airlines: United, Republic and Piedmont, and will be known as the LIP limited installation programme. The principal aim will be to see how effectively the system can separate co-operative 'targets', and in particular whether the commanded manoeuvres can under some circumstances actually increase the risk of collision.

STATUS: evaluation programme.

Sundstrand

Sundstrand Data Control Inc, Overlake Industrial Park, Redmond, Washington 98052
TELEPHONE: (206) 885 3711
TELEX: 152308
TWX: 910 449 2860

Digital ground-proximity warning system

The Sundstrand digital GPWS is designed for service with aircraft equipped with ARINC 700 avionics. It provides the flight-crew with back-up warning for six potentially dangerous situations. Alerts and warnings are provided by steady or flashing visual indications and by

audible warnings. Each audio warning is also annunciated to identify the particular situation. The system is unusual in that the annunciated warnings give a relatively precise description of the hazard, rather than the 'pull up' and 'glideslope' warnings provided by most systems, and for this reason the full range is quoted:

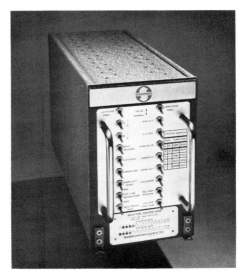

Sundstrand digital ground proximity warning system computer

Mode 1 - excessive descent rate
  alert: 'sinkrate, sinkrate'
  warning: 'pull up, pull up'
Mode 2 - excessive closure rate
  alert: 'terrain, terrain'
  warning: 'pull up, pull up'
Mode 3 - alert to descent after take off
  alert: 'don't sink, don't sink'
Mode 4 - alert to insufficient terrain clearance
  alert: 'too low terrain, too low terrain'
  alert: 'too low gear, too low gear'
  alert: 'too low flap, too low flap'
Mode 5 - alert to inadvertent descent below glideslope
  alert: 'glideslope, glideslope'
Mode 6 - alert to descent below minimums
  alert: 'minimums, minimums'

Alerts and warnings repeat continuously until the aircraft is manoeuvred out of the situation. A particular advantage of the variety of voice alerts is that its operationally orientated warnings permit confirmation by cross-checking of the panel instruments. Diagnosis of flight 'squawks' can thus be quickly carried out and

corrected. The speed of the ground-proximity warning system envelopes has been increased, providing longer warning times.

The system contains a number of features to assist in test maintenance and repair procedures. These include a non-volatile memory which stores both steady-state or intermittent faults occurring over the last ten flight sectors and which can be erased only when the unit is removed from the aircraft for bench work. The accepted test procedure is programmed within the computer and a simple test fixture is all that is required to re-address computer output data back into the computer itself. An alphanumeric display on the front of the unit can be used to isolate faults and indicate specific line replacement units which require replacement. On the bench, faults can be isolated to board level.

The GPWS complies with ARINC 600 standards, and its sub-components are grouped by circuit function on plug-in/fold out removable printed circuit boards with easily removed captive hardware. Latitude and longitude is used to modify warning boundaries at certain locations to reduce nuisance probability, or increase available warning time.

Automatic altitude call-outs are available on some models.

Format: 2 MCU

STATUS: in production and service.

Mark II ground-proximity warning system

Sundstrand's Mark II GPWS computer is designed for aircraft wired to ARINC 594 standard and is suitable for service in a wide cross-section of commercial, military or business aircraft. The Mark II model is claimed to be the first GPWS to use a Mach/airspeed input and therefore to have a much faster response time than previous GPWS computers. It was the first such system to offer voice-alerts which specifically identified each warning mode, and the first to offer a warning mode for minimum approach conditions.

Warning modes for the Mark II unit are generally the same as for the Sundstrand digital GPWS (see previous entry) although there are

Sundstrand stall-warning computer

differences between the warning times and the warning envelopes themselves.

Format: ¼ ATR Short
Weight: 8 lb (3.63 kg) max

STATUS: in production and service.

Stall-warning computer

The Sundstrand stall-warning computer provides output signals to such devices as stick-shaker and stall-warning horn, based on the aircraft's angle of attack, flap/slat settings and rate of increase of angle of attack. Self-monitoring circuits minimise false alarms and failures. Front panel leds allow isolation of misaligned and faulty sensor inputs.

Format: ⅜ ATR Short
Weight: 8.75 lb (4 kg)

STATUS: in production and service.

Terra

Terra Corporation, 3520 Pan American Freeway, Alberquerque, New Mexico 81707
TELEPHONE: (505) 884 2321

TRA 1000 radar altimeter

This system, suitable for executive and light aircraft, comprises a transmitter-receiver, antenna unit and decision height system. It operates in a restricted altitude range, and requires visual monitoring during operation. It is designed to alert the pilot when he descends through a pre-selected altitude (decision height). At this point the DH light illuminates and there is an audio warning.

Typical installation
Dimensions and weight: (transmitter-receiver) 89 × 127 × 152 mm; 0.9 kg
(indicator) 89 × 51 mm dia; 0.22 kg
(decision height system) 32 × 19 mm dia; 0.27 kg
Operating frequency: 4300 MHz
Power output: 3 W (peak)
Altitude range: 80-1 000 ft
Accuracy: better than ± 27%

STATUS: in service.

TRA 2500 radar altimeter

This is a lightweight radar altimeter for general use on executive and light aircraft. The com-

bined transmitter-receiver-antenna unit can be fuselage or wing-mounted and used during all flight phases. There is a decision height system which illuminates a light and produces an audio warning when the aircraft passes a preset radar height during the approach.

Typical installation
Dimensions and weights: (transmitter-receiver-antenna) 101 × 132 × 32 mm; 0.45 kg
(indicator) 3 ATI × 127 mm; 0.45 kg
Operating frequency: 4300 MHz
Altitude range: 40-2 500 ft
Accuracy: (40-100 ft) ±5 ft; (100-500 ft) ±5%; (500-2500 ft) ±7%

STATUS: in service.

Head-up displays and weapon aiming sights

FRANCE

Crouzet

Crouzet SA, 25 rue Jules-Védrines, 26027
Valence Cedex
TELEPHONE: (75) 79 85 11
TELEX: 345807
FAX: (75) 55 22 50

Weapons sight for helicopters

This electromechanical head-up weapons sight shows a collimated reticle which can be moved between –10° and +7° in elevation and ±6° in azimuth, angles compatible with the requirements for air-to-air and air-to-ground weapons launch. The device is mounted from the canopy frame of the helicopter, weighs 2 kg and has a field of view of 7.5°.

The system underwent flight trials in 1984 at the Cazaux Flight Test Centre as part of the competition to select a sighting system for the French Army Light Aviation Corps' Gazelle helicopters. The sighting system has also been selected by Aérospatiale for incorporation in its Ecureuil, Puma. Gazelle and Dauphin military helicopters, for export.

The sighting system is based on a modular concept; the basic component is the Sight Head (single or dual axis monocular system) which can be used for both day and night weapon firing in conjunction with 3rd generation micro-channel night vision goggles. Sight recording by ccd camera has been validated for both training and operational firing and is available as an option. The physical features of the Sight Head, such as its low weight, small size, simple and strong high-performance optical system (Angenieux lens, diode array on microelectronic support), give it the capacity to produce a remarkably high-quality image, both by day and night, that cannot be matched by a tube system.

The second main component of the sighting system is a Control Unit combining command and calculation functions. This unit includes the necessary controls for moving the sight head, symbology animation controls, firing tables stored in its internal memory, weapon firing computation system (guns and air-to-air missiles) and sighting telescope interface. The links with the missile system and telescope are designed in particular for export requirements (including coupling with Crouzet's Nadir computer for target designation). The system is equipped with a built-in automatic testing facility, with status information being displayed on the sight head.

To meet the weaponry requirements of French Army-operated Gazelles which are equipped with guns and Mistral missiles, Crouzet proposed a multi-purpose sighting and fire control system comprising the above-mentioned components that have already been qualified for the Gazelle. With this system, the pilot can fire the various weapons, including rockets, without any manipulation other than manual selection of functions on the front panel of the control unit.

STATUS: in service.

Crouzet sight-head mounted on canopy frame of light helicopter

Crouzet helicopter weapon sight-head

Sfim

Société Fabrication d'Instruments de Mesure, BP 74, 13 avenue Marcel-Ramolfo-Garnier, 91344 Massy Cedex
TELEPHONE: (1) 69 20 88 90
TELEX: 692164

Sfim's involvement with sight systems dates from 1972 when, in conjunction with the government agency APX and Sfim's subsidiary Sere Bezu, it began to produce APX-Bezu sights for firing helicopter-carried anti-tank missiles. Since that time it has produced a family of sighting systems for aircraft and land vehicles. Production of all types stands at about 1650 units sold to 30 different countries.

APX M 334 sights for helicopters

The designation APX M 334 covers a variety of roof-mounted sighting systems for helicopters. All are single eyepiece devices, and have two magnifications (3x and 10x) with corresponding fields of view (300 or 90 mrad). The line of sight is stabilised within 0.1 mrad and controlled in elevation and azimuth by precessing the stabilising gyroscopes of the mirror, with continuously variable sweeping speeds.

The sight has been fitted to such helicopters as Aérospatiale Gazelle 341 and 342, Super-Frélon SA 321; Westland Wasp, Lynx, WG 13; Bell 204, 205, 206/OH 58, 212; Sikorsky SH-3D and MBB BO 105.

The principal requirement of the M 334 series is to provide an operational system to detect a target at a range of about 30 km, and recognise and identify a tank at 5 km. An optional special TV camera arm allows the recording of observed images by a video recorder.

Variants of the basic system have specific capabilities:

M 334-04 Athos: observation and firing of manually guided missiles or seekerhead missiles

M 334-25: observation sight with range-finding and targets localisation. When used in conjunction with a navigation system it constitutes OSLOH III system giving target absolute position. When used in connection with an aiming collimator for targets designation it constitutes OSHAT system for air-to-ground gun and rocket firing

M 397 HOT: observation sight able to fire HOT automatic infra-red guided missiles, produced

Sfim Viviane day/night helicopter sight has been selected by the French Army to equip its Aérospatiale Gazelle helicopters armed with HOT missiles

with French-German collaboration by Euro-missile

M 334-05 Visigoth: observation of terrain and targets. Range-finding and illumination of targets up to 10 km. Firing of laser guided missiles.

M334-05 Visigoth
Magnification: × 3.2 and × 10.8
Field of view: 17.2° and 5.2°
Rangefinder wavelength: 1.06 micron
Rangefinder energy pulse: 100 mJ
Rangefinder pulse-width: 15 ns
Weight: 36 kg

STATUS: all members of the M334 are currently in production and service.

V3G sights for day/night helicopters

Based on previous developments of optronic stabilised platforms flown as early as 1976, Sfim has designed a sight, which incorporates an appreciable weight saving and a performance at least equal to that of previous systems, to meet the requirements of modern combat helicopters.

Sfim's 3rd generation sight (V3G) can be chin, roof, or mast-mounted. Depending upon which sensors are fitted, it can perform several functions: day and night observation; range-finding; anti-tank automatic guided missiles firing by day or night; and air-to-air heatseeker missiles. This series includes Viviane, selected for the addition of night capability to Euro-missile's HOT antitank missile system.

Sfim APX M334 sight

STATUS: prototype.

Thomson-CSF

Thomson-CSF, Division Equipements Avion-iques, 178 boulevard Gabriel-Péri, 92240 Malakoff
TELEPHONE: (1) 46 55 44 22
TELEX: 204780

Combat aircraft head-up displays VE 120, VE 110, VE 130, VEM 130

Thomson-CSF has developed a full family of head-up displays for tactical aircraft of various sizes and sophistication. They include advanced software for each phase of a flight allowing accurate attacks as well as navigation and piloting.

The VE 120, chronologically the first of the family, was configured to the Dassault Super-Etendard. The VE 110, especially designed for small cockpit tactical or trainer aircraft such as the most recent version of the Dassault Alphajet, has the most compact pilot's display unit (PDU).

The VE 130, with a larger PDU, is fitted to Dassault Mirage F1 and both versions of the Mirage 2000. A derivative of the VE 130 is the VEM 130 which includes an additional capability for television-type raster Flir image presentation.

The instantaneous fields of view of these dual-combiner displays are typically about 20° in azimuth and 15° to 18.5° in elevation, depending on aircraft type. The total field of view of these compact head-up displays is about 20°, with in some cases the possibility of growth to 24°.

Combination head-up displays plus collimated level displays

Thomson-CSF is developing monochrome wide field of view collimated displays, which will be soon available for association with either VEM 130 or holographic head-up displays.

VH 100 head-up display

Thomson-CSF is developing a family of head-up displays as a part of advanced weapon control systems in helicopters, in particular the air-to-ground firing of Stinger or Matra Mistral missiles.

Thomson-CSF head-up displays. Left to right: VE 130, VE 110 and VE 120

Thomson-CSF holographic head-up display for Dassault Rafale

Thomson-CSF VEM 130 head-up display

Thomson-CSF VH 100 head-up display for helicopters

The VH 100 has been flight tested fitted to a Gazelle in France, and on OH-58 and Bell 406 helicopters in the USA. In September 1985 Thomson-CSF announced that the VH 100 had been selected to equip the US Army's OH-58 helicopter missile sight sub-system programme. In all, 828 helicopters are to be retro-fitted.

STATUS: in development.

Type 196 gunsight

For many years Thomson-CSF has produced electromechanical gunsights with half-silvered mirrors to reflect target-position reticles and computed lead-angles into the pilot's field of view. The company has concentrated on reducing their size and increasing their field of view, the latter requirement being prompted by the increasing manoeuvrability of fighters. The most recent of these devices, the Type 196, was first shown in 1970 and appears to be the last of its kind to be produced by Thomson-CSF.

Gradual improvements in the development of gunsights have been maintained: in 1952 the Type 52 gunsight had two reticles with a field of 6°, whereas the Type 196 has eight reticles and a field of 10° by 8°.

STATUS: production decreasing as head-up displays attain more prominence, even in less sophisticated strike/trainers. Since 1950 over 5000 sights have been built for aircraft including Dassault Mystere IV, and Mirage, Anglo-French Jaguar, Franco-German Alpha Jet, Italy's Aermacchi MB326, and Japan's Mitsubishi F-1 and T-2 bomber and strike/trainers.

TMV 980 A head-up/head-down display

The TMV 980 A is an integrated head-up/head-down display system for the Dassault Mirage 2000 multi-role fighter. It comprises a digital computer and processor to generate the display symbology and help with flight and weapon-aiming computation, and three display units: a VE 130 cathode ray tube head-up display (see separate entry above), a VMC 180 interactive multi-function head-down colour display, and a VCM 65 complementary monochrome cathode ray tube for electronic support measures information.

The head-up display has a high resolution, high brightness cathode ray tube with a collimating optical system based on a 130 mm lens providing a wide total field of view. The binocular instantaneous field of view is increased in elevation by the use of a twin-glass combiner that transmits 80 per cent of the light incident upon it. Automatic brightness control, with manual adjustment, permits symbols to be read in an ambient illumination of 100 000 lux. The system provides continuous computation of tracer line in the air-to-air mode and impact and release points in the air-to-ground mode.

The main head-down display presents, in red, green, and amber on a 127 by 127 mm cathode ray tube radar, information such as a radar map, 'synthetic' tactical situation, and range scales; raster images from television or forward-looking infra-red sensors; tactical data from the system itself or from an external source.

A helmet-mounted sight may be integrated into the TMV 980A unit to improve target discrimination and off-boresight target designation, and another option is the substitution of the VE 130 head-up display by a VEM 130 system.

Weight: (electronic unit) 9 kg
(head-up display) 13 kg
(VMC 180 head-down display) 14 kg
(VCM 65) 4 kg

STATUS: in production for the Dassault Mirage 2000 fighter, the first squadron of which was formed in 1984.

Helmet mounted sights and displays

Thomson-CSF is developing helmet mounted displays for a wide variety of operational requirements for helicopters and for combat aircraft.

For advanced helicopters, Thomson-CSF is developing a very wide field of view helmet visor display.

Holographic head-up display for Rafale

For several years Thomson-CSF has been conducting research into holography, and the first application will be a holographic head-up display for the French Air Force/Dassault Rafale advanced fighter demonstrator programme. The French avionics company is making the system entirely in-house, its new special-purpose test and assembly building at Le Haillan having been completed in late 1984. Some assistance is also being given by the Thomson Group's central research facility in Paris. Prototype holographic head-up displays are currently flying in a Dassault Mystère 20 and a Dassault Mirage F1. Rafale flew in the summer of 1986. The field of view of these head-up displays is 30° in azimuth and 20° in elevation.

Thomson-CSF helmet-mounted sight

GERMANY, FEDERAL REPUBLIC

Teldix

Teldix GmbH, PO Box 105608, Grenzhöferweg 36, D-6900 Heidelberg 1
TELEPHONE: (6221) 5120
TELEX: 461735

Reflective-optics head-up display

Teldix, in a technology-demonstrator programme with Carl Zeiss, has sought to overcome the principal drawback of traditional head-up displays – restricted fields of view – by replacing the conventional lens and flat combiner glass with a suitable reflection-optics system. The improved field of view, says Teldix,

would result in a wider weapons-delivery envelope so that, for example, weapons could be released in a much sharper turn than now possible and their release ranges extended.

In the wide-angle head-up display, or W-HUD as Teldix calls it, light from the head-up display's cathode ray tube is reflected by a concave mirror onto a large aspherical combiner that can still show only modest weight increases over current systems. In a typical installation the field of view could be increased from present-day values of 20° to 25° in azimuth and elevation to 40° to 60° in azimuth and 25° to 30° in elevation. The optical geometry of the system requires a special cathode ray tube with a spherical screen, but the electronic picture could be produced by a standard symbol generator.

Display format: stroke symbols and alphanumerics for flight-control, navigation, weapons delivery. Blending with tv raster optional
Combiner dimensions: 400-500 mm wide × 200-250 mm high, depending on cockpit configuration
Line width: (stroke) 1.5 mrad
Aiming accuracy: 2 mrad near optical axis
Brightness: (stroke) 5000 cd/m²
(tv raster) 1500 cd/m²
Self-test: BITE
Weight: 115 kg

STATUS: technology development and demonstrator programme financed by West German Ministry of Defence. System was first shown and demonstrated publicly at 1981 Paris Air Show. Prototype flight models are expected to be delivered in 1986.

Teldix is partner of Smiths Industries and OMI in design and production of head-up display for the Panavia Tornado

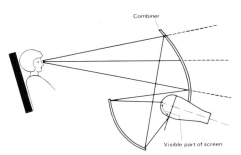

Ray geometry for Teldix reflective-optics head-up display

Curved combiner glass of Teldix reflective-optics head-up display

ISRAEL

Elbit

Elbit Computers Limited, Advanced Technology Center, PO Box 5390, Haifa 31053
TELEPHONE: (4) 517 111
TELEX: 46774

Helmet mounted sight

Elbit has developed a helmet mounted sight which will allow the pilot of an attack aircraft to visually cue his weapons.

The system comprises four units: helmet, sight and motion sensor; cockpit mounted raiator, used in conjunction with the sensor to derive head position; control panel; and computer.

STATUS: in development.

Night vision goggles/head-up display

Elbit is developing a head-up display system which projects flight data into the view of a pilot's night vision goggles, enabling the pilot to keep watch outside the cockpit while remaining aware of his speed, altitude, etc.

The night vision capability is unaffected and the addition of the HUD symbology capability adds only 15 g to the goggles basic on-helmet weight of 135 g.

STATUS: in development.

IAI/Tamam

Tamam Precision Instrument Industries, Subsidiary of Israel Aircraft Industries Limited, Industrial Zone, PO Box 75, 56000 Yahud
TELEPHONE: (3) 357 211
TELEX: 371114

Gyro-stabilised sighting system

This system, designed for light aircraft, helicopters and armoured vehicles, is a dual-magnification periscope, the optical head being outside the vehicle, free to scan in elevation and azimuth, and gyro-stabilised to eliminate image

movement due to aircraft movement or vibration. It can be controlled by a joystick slaved to an external designator or the designator can be slaved to the sight. The coating of the stabilised mirror is effective over a wide spectral range, from visual to near infra-red, and so can be used with laser or infra-red devices in conjunction with visual equipment.

A television camera permits viewing by both eyes, and enhances penetration of haze, and a video recorder allows a flight to be run through for post-mission critique or training purposes. An infra-red goniometer can be attached to the sight for accurate launch and automatic guidance of missiles against armoured targets.

Likewise, a laser rangefinder permits target ranging and the system can be integrated with a low-light television camera for passive, undetectable night vision.

Scan angles: 360° azimuth, –30° to +40° elevation
Stabilised tracking rate: up to 10° in both axes
Magnification: × 2.5, × 210
Field of view: 20°, 5° (depending on magnification)
Electrical pick-off accuracy: 0.1-0.2 mrad
Weight: 25 kg

STATUS: in production.

SWEDEN

Saab Instruments

Saab Instruments AB, Box 1017, S-551 11 Jönköping
TELEPHONE: (36) 19 40 00
TELEX: 70045

RGS 2 lead computing optical sighting system

The Saab RGS 2 forms the new generation of optical sighting systems. Following a modular design concept and using servo techniques, all but the most essential elements of the optical system have been eliminated from the sight head and housed in a separate box called the computer and gyro unit. Likewise, most of the

sighting system controls are found on a separate module, the control unit, which is normally installed adjacent to the weapon fire control system in the aircraft. These design innovations result in very small sight head base dimensions, a feature which allows the RGS 2 sight head's easy installation into even the tightest cockpit and the most crowded instrument panel. Despite these small base dimen-

*Twin RGS2 sight head installation in SAAB 105
side-by-side two-seat trainer*

Saab Instruments advanced RGS 2 configuration with automatic flight parameter compensation

Saab Instruments Helios lightweight modular helicopter sight

sions, the sight head is designed to provide a full two-eyed viewing capability through the combining glass.

Apart from small dimensions the modular design concept has other practical benefits. For optical sights in dual command trainers it is important that the aiming marks of the two sight heads for the instructor and student pilots have identical and synchronised deflections. In the RGS 2 system the separate computer and gyro unit can be used to control two sight heads in parallel through the output servo loop circuit.

Instead of controlling a second sight head in a twin sight arrangement the same principle can be used for slaving a laser rangefinder or an IR missile or radar homing device to the optical sight-line of the sight head.

The RGS 2 is a lead computing rate gyro controlled sighting system. The aiming mark is controlled by electrical signals in a servo system. These signals are outputs from the computer and gyro unit, which sends the relevant weapon aiming control signals to the servo loops based on the weapon aiming relations. The deflection of the aiming mark is accomplished by tilting a spherical mirror, which is part of the stiff closed loop servo system. The mirror is therefore very accurate and the system does not suffer from adverse temperature effects and long term drift.

In the basic configuration, the computer works with input signals from the three rate gyros for pitch, roll and yaw rates together with additional manually preset quantities for sight sensitivity and the weapons characteristics. In this configuration the RGS 2 is a self-contained system. In a more advanced configuration the system includes a roll angle input to the computer to stabilise the aiming mark to true vertical reference, and an altitude input to compensate for altitude variations.

The RGS 2A retains all the functions of the basic RGS 2 configuration in both air-to-air and air-to-ground modes, but also features automatic flight parameter compensation for enhanced ground attack capability. As well as extended weapons delivery functions the RGS 2A can be configured to display steering commands and alphanumeric characters, as for instance in an aircraft navigation mode.

The Saab RGS family comprises:

RGS 1	fixed type optical sight
RGS 2	basic universal gunsight
RGS 2	training configuration with two sight heads
RGS 2A	with CCIP modes for enhanced ground attack capability
RGS 2R	with CCRP modes using laser rangefinder
RGS 2H	optimised for helicopter use
RGS 2A-HUD	with head up display capabilities.

STATUS: the RGS 2 is certified for 15 different aircraft types; in operational service for eight years.

Helios helicopter observation system

The Helios is a roof-mounted observation system for scout/rescue helicopters. The system has 3x and 12x magnification, high resolution optics (by UK Company Pilkington) and high stabilisation accuracy. The eye-piece arm protruding into the cabin can be stowed close to the roof when not in use and the eye-piece height is adjustable to suit the observer. The system itself comprises five units: roof-mounted sight, electronic unit, control panel, control unit and line of sight indicator.

The Helios has provisions for laser range-finder/designator and thermal imaging system for night vision. It can accept commands from laser-warning receiver, hostile fire indicator, radar warning receiver, helmet sight or avionics systems for accurate target positioning, and generate outputs for weapons aiming. A camera port for a film or ccd camera is included to allow recording of the screen.

STATUS: in production.

Helitow helicopter weapon sight

The Helios system (see above) was the basis for development of the Helitow sight sub-system. Together with the Emerson Helitow missile launching and guidance sub-system it forms a state-of-the-art helicopter system for firing TOW missiles. The sight is fully prepared for the introduction of IR vision and laser rangefinder/designator. A Saab BT49H TOW missile simulation system can be attached for cost effective training.

STATUS: in production for the Swedish Army.

UNITED KINGDOM

British Aerospace

British Aerospace, Army Weapons Division, Six Hills Way, Stevenage, Hertfordshire SG1 2DA
TELEPHONE: (0438) 312 422

Roof sight for Lynx

In March 1986 British Aerospace announced it had been awarded a £60 million contract to update the roof sight used for firing TOW missiles from British Army Air Corps' Lynx helicopters. The modifications include provision for a thermal imager to give the sight day/night capability. The thermal imaging sub-system has been developed by Rank Pullin.

British Aerospace won the original contract to supply the sights, which are to a Hughes design, in 1978 and over 130 units have been delivered.

STATUS: in development.

Ferranti

Ferranti Defence Systems Limited, Electro-optics Department, Robertson Avenue, Edinburgh EH11 1PX
TELEPHONE: (031) 337 2442
TELEX: 72529

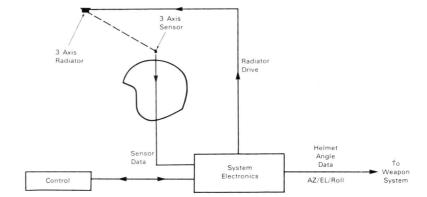

Computation of sight angles in Ferranti helmet-sight

Helmet mounted sights

In October 1980 Ferranti delivered to the Royal Aircraft Establishment (RAE) at Farnborough an advanced, comprehensive helmet-mounted display system for development and evaluation. It was tested in a RAE flight simulator and later flown aboard a Westland Sea King helicopter there for airborne trials.

The system was developed as a research tool by the company's Pointing and Displays Group at Edinburgh under a UK Ministry of Defence contract. Ferranti considers the potential application of these devices to fixed-wing aircraft to be very important, particularly in association with the new generation of advanced, short-range air-to-air missiles. The substantially greater manoeuvrability of the new, 'agile' fighters now in the design stage, together with new missiles such as the UK's Asraam (advanced short-range air-to-air missile), results in large off-boresight angles which may require a much larger field of view than accessible by conventional head-up displays.

This helmet-mounted equipment is intended for both fixed-wing aircraft and helicopters in conjunction with sighting and surveillance equipment for weapons or sensor control, but also has sea and land applications. It is adapted for operations in fast-patrol boats, large cruisers (as an alternative target acquisition system for sophisticated weapons), or as part of a low-cost protection system for armed merchantmen. In land applications it is suited to visual target acquisition and for the pointing of guns and missiles. It can be vehicle-mounted or infantry operated and used for battlefield surveillance and air defence.

The system comprises an electronics unit driving a three-axis magnetic field generator under the direction of a pilot's control unit, a small magnetic-field sensor mounted on the helmet to measure the direction of the pilot's line of sight, and an optical sight. The electronics unit can operate two heads. The latter has a simple display format with 'see through' capability. It presents an aiming mark in the form of an open cross, together with four arrow signs positioned and individually illuminated so as to provide a directional cueing facility. The symbols are optically presented, and are collimated. The cueing arrows, when slaved to the aircraft's nav/attack system, can direct the pilot's head towards the target; they could also be commanded by the navigator or observer in a two-seat aircraft to direct the pilot's arrows towards a target.

The accuracy of acquisition and designation is adequate to differentiate between individual but closely separated targets. The designation accuracy is said to be some ten times better than that achieved by verbal instructions between crew members.

As an alternative to the optical sight, a miniature cathode ray tube can be mounted on the helmet. This has optics giving a wider field of view for presentation of more complex imagery.

The electronics unit contains the power supply and signal conditioning, digital processing, and interface electronics needed to measure continuously the pilot's line of sight in azimuth, elevation, and roll with respect to the fore-and-aft axis of the aircraft or missile.

The control unit provides on/off switching, a function switch for mode-changing, display-lighting intensity control for independent adjustment to the brightness of the sighting reticle and discrete cueing signs, and a light to indicate that helmet sensor alignment is required.

The cathode ray tube and associated wide-band video amplifier are designed to give the best imagery for general surveillance and target discrimination. The television raster can be matched to existing or upcoming sensor line-standards. The picture is roll-stabilised to avoid pilot disorientation. The field of view is 40° wide, and imagery within it is collimated. A high data-refresh rate is provided for compatibility with rapid head-slewing, so that the picture is space-stabilised. Imagery is presented in four grey scales.

Electronics unit
Format: ½ ATR Short
Weight: 24 lb (10 kg)

Control unit
Weight: 1.7 lb (0.77 kg)

Magnetic field generator
Weight: 0.125 lb (57 g)

Helmet sensor
Weight: 0.04 lb (18 g)

Optical sight
Weight: 0.375 lb (170 g)

STATUS: system is in production for the British Aerospace Tracked Rapier ground-to-air missile system. It has been extensively tested in a helicopter simulator, and subsequently flown as part of the first visually-coupled system in Europe in an MBB BO105 helicopter. It has also been flown in additional trials aboard a Westland Sea King helicopter at RAE Farnborough.

Isis weapon-aiming sights

Ferranti's Edinburgh factory first produced gyro gunsights in 1943. It has since developed a family of optical weapons sights under the name Isis, for the aiming of guns, rockets and bombs in the close-support role, and for guns and missiles in air-combat.

The Isis sight can also use range information from a laser rangefinder, and can serve as a low-cost alternative to a cathode ray tube-type head-up display by displaying pointing information from a laser seeker such as the Martin Marietta Pave Penny or Ferranti LRMTS. The company claims that adequate accuracy is coupled with low maintenance cost and high reliability.

Operation is essentially the same for all members of the Isis family. A two-axis rate-gyro generates a lead-angle proportional to the rate-of-turn and a mirror attached to the rotor reflects the image of an illuminated reticle pattern through a collimating lens onto a combining glass and into the pilot's field of view.

In the ground-attack role the reticle is depressed below the armament datum through an angle that compensates for the combined effects of the weapon's ejection velocity, angle

Ferranti helmet-mounted sight system showing sensors and display projector

D-195R, fitted to British Aerospace Hawk trainer, is typical member of Ferranti Isis range of weapon sights

of attack of the airframe in the intended release conditions, the gravity 'droop' of the weapon during its flight, and the surface wind if known. These angles may be set up by the pilot to suit the intended attack speed, dive angle, and slant range at weapon release, or he may prefer to use a pre-determined set of parameters, one set for each type of weapon carried.

In air combat the unit measures the rate of turn of the sight-line and scales it to target range, thus producing a first-order lead-angle solution.

To assist steering and enable the pilot to make allowance for sudden manoeuvres by the target most sights include a second, fixed reticle representing the gun axis.

Typical sight, Isis D-195R Mk 3
Number of units: sight head, control unit, gyro interface unit, and throttle unit
Weight: (sight head) 9 lb (4.1 kg)
(control unit) 3 lb (1.36 kg)
(gyro interface unit) 3.6 lb (1.63 kg)
(throttle unit) 0.5 lb (0.2 kg)
Sight head dimensions: 9.54 × 9.89 × 5.20 inches (243 × 251 × 132 mm)

STATUS: in production for British Aerospace Hawk strike/trainer. Wide range of other sights in production including GSA Mk 3 lead-computing gunsight and Isis M 126WL light-weight sight for helicopters and light aircraft.

Ferranti Defence Systems Limited, Display Systems Department, Silverknowes, Edinburgh EH4 4AD
TELEPHONE: (031) 332 2411
TELEX: 727101

Type 4510 head-up display
Ferranti's first head-up display, the two-unit Type 4500, had its public début at the 1982 Farnborough Air Show. Initially a cursive-only head-up display, the FD 4500 became the basis for the FD 4510 designed for retrofit application on light strike and combat aircraft. The FD 4510 has both cursive and raster displays, the latter being selected by a switch on the up front control-panel. The FD 4500 and 4510 are physically identical and were based on the optical and symbol generation technology previously used on the Comed system (see separate entry in previous section). The extensive production runs of Comed had removed any design problems, while the weapon-aiming, interfacing, and air data techniques were derived from a series of complete weapon systems designed and built by Ferranti. The FD 4510 head-up display provides a full site of navigation symbology with steering and

Ferranti AF532 sight unit installed on roof of British Army/Westland Gazelle helicopter

location cues available at all times, and also generates weapon-aiming symbology, either automatically or by selection, for all known air-to-air and air-to-ground weapons. Weapon release can be triggered manually or automatically at the pilot's discretion. The control-panel allows the pilot to set-up initially the navigation system, select modes and display them during flights, and then change to a raster display when the mission sensors (such as forward-looking infra-red) come into play.

The pilot's display unit features a 24° circular field of view and is scaled 1:1 with the outside world. In order to achieve satisfactory brightness for the raster requirements, and to ensure night-vision goggles compatibility, the FD 4510 uses a P53 phosphor; this also permits good sunlight readability in the cursive mode. It also allows permanent fitting of an anti-reflection filter (if this becomes desirable in a specific installation) without compromising the brightness requirement in any light conditions.

The interface unit, contained in a ¾ ATR Short box, provides navigation and weapon-aiming solutions, waveform-generation in both

cursive or raster formats, and can act as a 1553B digital bus controller if required.

The head-up display forms part of the Fastac navigation system being fitted in the Spanish CASA 101 DD, an advanced trainer being used to prepare student pilots for the McDonnell Douglas F/A-18 Hornet fighter ordered by Spain, and is an integral part of various total system options at present being considered in the retrofit market worldwide. The Ferranti head-up display fits straight on to existing gunsight mounts and can, because of its compact size, meet the installation requirements of many types of light attack aircraft. A notable potential application is India's fleet of some 400 Soviet-designed MiG-21 fighters, the originals of which have relatively simple electro-mechanical gyro gunsights.

Ferranti Instrumentation Limited, Aircraft Equipment Department, Lily Hill House, Lily Hill Road, Bracknell, Berkshire RG12 2SJ
TELEPHONE: (0344) 424 001
TELEX: 848117

AF500 series roof observation sights
Since 1967 Ferranti and the optical company, Avimo of Taunton, have collaborated in the design of a family of sight-line stabilised weapon-aiming devices, under the general designation AF500. The AF532 roof-mounted helicopter sight is the latest joint project in this field, superseding the AF120 sight, introduced in 1970 and still operational in the British Army Air Corps' Westland Scout helicopters. The new sight is half the weight of its predecessor, but confers a greatly improved optical performance. A version of the AF532, the AF580, has been built for evaluation by the US Army aboard a Bell OH-58C helicopter during 1983 and 1984.

This gyro-stabilised, monocular, periscopic telescope is designed for the gunner/observer in reconnaissance helicopters, particularly when scouting targets for anti-tank helicopters.

The device has a built-in interface for a laser designator and rangefinder which (if of Ferranti design) can be installed at first-line level without the need for setting up or initial adjustment. It can also be adapted for night-vision equipment, helmet sights, and weapons, and there are facilities for attaching a television

Ferranti Type 4510 head-up display with integral mode controls and dual combiner glass

recording camera for training or intelligence-gathering. The design of the optical system is such that the varying eye positions in different helicopter installations can be easily accommodated.

The gyro-stabilised head protrudes above the roof forward of the rotor mast, while the down-tube and eyepiece extend downwards from the roof so that the eyepiece falls into a comfortable viewing position. The down-tube is adjustable in height and retracts sideways, locking close to the roof when not in use. A control handle is extended by the operator and adjusted, in tilt, so that it can be used with his right forearm resting on his knee. A horizontal thumbstick is used to steer the sight-line and a direction indicator, to show its direction relative to aircraft heading, is mounted on the glareshield in front of the pilot.

The sight provides a stabilised image of the chosen field of ×2.5 magnification for search and × 10 for identification and laser operation. The sight line may be steered through ±30° and ±120° in pitch and yaw planes respectively.

STATUS: in August 1982 a representative production standard AF532 sight was delivered for flight trials. System is now in production for British Army Air Corps' Westland SA.341 Gazelle light helicopters following a contract awarded in February 1984. In February 1983 AF580 system was delivered to US Army at Fort Rucker for trials evaluation. These trials were extended in 1984, when the company supplied a further two units, designated AF580, together with a Ferranti laser ranger and target designator. The systems were installed in two Bell OH-58 Kiowa helicopters, integrated with a Collins ATMS automatic target hand-off system, providing information for a reconnaissance helicopter to a gunship helicopter.

GEC Avionics

GEC Avionics Limited, Airport Works, Rochester, Kent ME1 2XX
TELEPHONE: (0634) 44400
TELEX: 96333/4

Hudsight head-up display for F-16

The head-up display for the General Dynamics F-16 fighter, ordered by the US Air Force, four European nations and several other countries is the latest in a line of GEC Avionics head-up displays that go back to the strike-sight of 1960 designed for the Royal Navy's Buccaneers.

The Hudsight system comprises an electronic unit (effectively the head-up display's 'brain'), a rate sensor unit and a pilot's display unit. The electronic unit is a small digital computer with a 16 K eprom memory. The system has spare capacity which has been used by General Dynamics to broaden the scope of the F-16's fire-control computer. The head-up display communicates with the Westinghouse AN/APG-66 attack radar and the fire-control computer, air-data computer and inertial navigation unit through the MIL-STD-1553 data-bus.

The pilot's display unit comprises the cathode ray tube assembly, the optical train and the control panel all mounted on a rigid chassis; the optical combiner glass, provided by UK company Pilkington PE, is sufficiently strong to withstand the aerodynamic loads resulting from damage or loss of the canopy. Also incorporated in the optical module is a standby sight which is independent of the principal electronic circuitry. The rate-gyro unit generates the pitch, roll and yaw rates and normal acceleration needed for accurate air-to-air weapon delivery.

The system operates in three modes: cruise flight management, ground attack, and air combat. The two combat modes are subdivided into further modes, each designed to provide the most effective information and guidance based on data from the F-16's fire-control computer. In the cruise mode the head-up display shows speed, height, heading, vertical velocity, velocity vector, climb/dive, Mach number, maximum available g, and range and time to destination. In the ground-attack mode there are five types of display: continuously computed impact point, strafe, dive-toss, electro-optical, and low-altitude drogue delivery for retarded weapons. Air-to-air symbology is divided into four configurations for the built-in gun and various types of air-combat missiles: snap-shoot (in which the system provides the guidance for transient gun-firing opportunities), smooth tracking (again for the gun), air-to-air missiles, head dogfight. In order to keep himself out of trouble from enemy fighters the pilot needs to monitor constantly his speed and height, and so in all these air-to-air modes energy-management scales may be selected in place of airspeed and altitude scales.

The system has built-in test facilities to isolate faults to a particular line-replaceable unit and predicted mean time between failure for the head-up display is now 870 hours.

The F-16 head-up display is the basis of a complex international production agreement.

GEC Avionics head-up display dominates General Dynamics F-16A instrument panel with its prominent combiner-glass and selector panel

Twin lines for head-up display production for the US Air Force and export aircraft have been established in the UK and the USA (at the Atlanta plant of GEC Avionics Inc), while production for the NATO F-16s of Norway, the Netherlands, Denmark and Belgium is largely in those countries. The overall production and maintenance activity is under the control of the GEC Avionics Central Management Team at Rochester in the UK.

STATUS: in production.

Wide-angle head-up display for F-16C/D

In March 1983 GEC Avionics announced a $50 million order to begin production of a new, wide-angle non-holographic head-up display for the US Air Force/General Dynamics F-16C/D fighter programme. This head-up display is based on development work undertaken for the US Air Force's AFTI (Advanced Fighter Technology Integration) programme.

The new head-up display provides the latest electronically-generated symbols, thrown up on a total field of view of 25°, which is much wider than that attained with previous head-up displays. The instantaneous field of view (that is the field seen by the pilot without moving his head from side to side) is 21° in azimuth and 15° vertical. GEC Avionics believes that this field of view is the most that can be achieved with conventional, refractive-optics head-up displays. The system uses the same electronics unit as that developed for the Lantirn (low altitude navigation and targeting infra-red for night) system, and the symbology and raster-scan (television-like) pictures are particularly suited to guidance and target acquisition at night or in poor weather. The system is claimed to represent the first applications of MIL-STD-1750A processor architecture, MIL-STD-1553B digital data transmission, and Jovial 73(MIL-STD-1589B) high-order language.

Pilot's display unit
Dimensions: 25 × 6.4 × 6.7 inches (635 × 163 × 170 mm)
Weight: 47.6 lb (21.6 kg)
Power: 98 W (including 25 W for standby sight)
Predicted mtbf: 2000 h

Electronics unit
Dimensions: 13.28 × 7.1 × 7.5 inches (337 × 180 × 191 mm)
Weight: 31 lb (14.1 kg)
Addressable memory: 64 K words (48 K eprom, 16 K ram)
Predicted mtbf: 1000 h

STATUS: in production for US Air Force/General Dynamics F-16C/D fighters. A second $75 million production contract was announced in August 1985.

Lantirn holographic head-up display for F-16 and A-10

Notwithstanding the capability and growth potential of the standard General Dynamics F-16 head-up display, the US Air Force in the mid-1970s issued requirements to industry for an even more advanced system. These, in

F-16Cs and -Ds have wide-angle GEC Avionics head-up displays

Installation of GEC Avionics Monohud

Two versions of Monohud that have been built for trials; left, for NASA, and right, for UK's Royal Aircraft Establishment

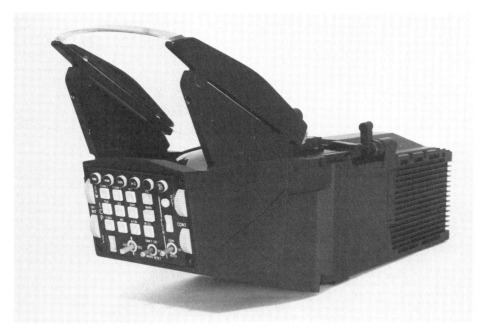

GEC Avionics wide-angle head-up display for US Air Force/General Dynamics F-16C/D

July 1980 was awarded a General Dynamics' development contract and initial production options totalling $103 million. By May 1982 the US Air Force had options on 603 holographic head-up displays for F-16s and 250 Fairchild Republic A-10s as part of the Lantirn programme. The system combines the wide-angle display geometry made possible by using hologram technology with a company developed method of combining raster and cursive-scan symbol writing that greatly reduces the amount of equipment needed for day/night head-up display. The optical train uses three combiner glasses, each of which is a sandwich, the hologram being imprinted onto a gelatin 'filling'. It provides a field of view of 30° laterally and 18° vertically. On the electronics side, it incorporates MIL-STD-1589B high-order language, MIL-STD-1750A airborne instruction-set architecture, and MIL-STD-1553B digital data-bus standards.

STATUS: programme is sponsored by Aeronautical Systems Division of US Air Force Systems Command for versions of the General Dynamics F-16 and Fairchild Republic A-10 equipped with the Lantirn system. First pre-production Lantirn head-up display was delivered to US Air Force in March 1982 and flight evaluation with F-16 began a few months later. Similar head-up display was delivered for trials with A-10. Flight trials with both aircraft have been satisfactorily completed.

conjunction with improvements in other areas, would provide the basis for a substantial upgrading of the F-16's effectiveness at a relatively early stage in its career under the designation, MSIP, for multinational staged improvement program. The specific improvement sought was the adoption of a new imaging capability known as Lantirn (low-altitude navigation targeting, infra-red for night) permitting the F-16 to operate in all weathers and at night.

It was soon realised that the field of view needed for this new capability would be far greater than that available with existing head-up display technology. Conventional head-up displays with lateral and vertical fields of about 13.5° and 9° respectively could be expanded to 20° and 15° using standard optics, but this would still be less than the field size that the US Air Force has set as its ultimate objective.

GEC Avionics started developing a head-up display based on holographic techniques, using the principles of diffractive optics, and in

Monocular head-up display

Designed for installation in existing cockpits, the monocular head-up display (Monohud) is a fully capable system in miniature. It offers flight and navigational modes to provide the pilot full flight parameters head-up during low level flight and landing, as well as computed weapon aiming solutions when used as a sighting system on military aircraft. It has been designed for applications in which size and weight constraints play a major role in the choice of systems. The Monohud system consists of a pilot display unit, an electronics unit (EU), a high voltage power supply unit (HVPSU) and a control panel.

The PDU can be fitted with minimal modification of the airframe. It allows an unobstructed view of instrumentation and the outside world in both the stowed and operational positions. In the operational position the PDU is situated approximately 8 cm in front of the pilot's eye and thus a 30° × 24° field of view is achieved. The PDU control panel includes manual controls for symbol brilliance and declutter.

The EU is compatible with a comprehensive range of discrete, analogue, digital, synchro, ARINC 429 and MIL 1553B inputs. In addition to the input interface, processor and symbol generator, the ½ ATR electronics unit contains the low voltage power supply and BIT. Built-in test eliminates the need for external support equipment at organisational and intermediate levels. The computed mtbf is greater than 2500 hrs.

Dimensions: (PDU) 76.2 × 76.2 × 190 mm
(EU) 320 × 124 × 193 mm
(HVPSU) 178 × 153 × 102 mm
(control panel) 57 × 146 × 79 mm
Weight: (total system) ±12.7 kg
Field of view: 30° vertical × 24° azimuth
Contrast ratio: 1.2 : 1 against 10,000 ft L
Reliability: 2500 h mtbf
Environmental qualification: MIL-E-5400T, Class 1B

Head-up display and weapon aiming computer for F-5

In both the air-to-air and surface-to-air areas rapid technological advances, especially in electronics and guidance systems, have quickened the pace of avionics obsolescence in a number of highly capable combat aircraft with years of airframe and engine life remaining. GEC Avionics has developed a head-up display and weapon aiming computer (HUDWAC) for the Northrop F-5E and F that, as an avionics up-grade, allows the performance and manoeuvrability of the F-5 to be combined with the advantages of head-up, eyes out flight.

The F-5 HUD utilises technology and hardware from other GEC Avionics HUD programmes, and performs air-to-air and air-to-surface weapon aiming calculations as well as delivering symbology for flight and navigation modes. The HUD is comprised of a pilot display unit (PDU), an electronics unit (EU) and a weapon data input panel (WDIP).

The PDU fits in an existing mounting tray. The PDU's 25° total and 15.75° vertical by 16.73° azimuth instantaneous fields of view are achieved without impinging on the F-5's ejection plane. A 16 mm cinematic camera is included in the PDU, but a TV camera is optionally available. The PDU control panel includes switches for controlling symbol brightness, standby sight selection, and symbol declutter. Data entry is via a key pad on the same control panel.

The EU will fit in all existing configurations of the F-5E/F, in the same location as the LCOSS gyro lead computer which is removed for HUDWAC installation. It contains a power

Installation of GEC Avionics holographic head-up display display unit in General Dynamics F-16 cockpit as part of Lantirn system

GEC Avionics HUDWAC for F-5

supply, integral cooling fans, and the circuit cards for processing, symbol generation, and interfacing with the F-5's avionics systems. The HUDWAC in its baseline configuration interfaces with the F-5's power supply, CADC, AHRS, fire control system, AN/APQ-153, -157 or -159 radar, and LN-33 INS. Optionally, the HUDWAC will interface with the AN/APG-69(V) radar and the LN-39. The system software provides weapon aiming calculations for weapons certified for F-5 carriage and is written to display symbology which complies with MIL-STD-1787. The system has the capability to expand and handle the AIM-9P-4 missile line of sight and off-boresight aiming.

The system operates in three modes: navigation, surface attack and air combat. In navigation and a landing sub-mode the display symbology shows speed, altitude, heading, velocity vector, pitch, bank, mach number and g. Inertial navigation information, when avail-

able, is also displayed. Surface attack modes include continuously computed impact point (CCIP) and continuously computed release point (CCRP) displays. There are three air-to-air modes: missiles; lead computing optical sight (LCOS) (for tracking 20 mm cannon attacks); and dogfight, which combines missiles, LCOS and the snap-shoot, or continuously computed impact line (CCIL), on the same display.

The system has built-in test facilities to isolate faults to a particular line replaceable unit. Predicted mtbf for the HUDWAC is greater than 750 hours. The HUDWAC's support equipment includes a semi-automatic system test set which uses a 16 bit microprocessor.

STATUS: full scale development. Initial customer delivery mid-1987.

Smiths Industries

Smiths Industries, Aerospace and Defence Systems Limited, 765 Finchley Road, Childs Hill, London NW11 8DS
TELEPHONE: (01) 458 3232
TELEX: 928761

Type 1402 head-up display

The Smiths Industries Type 1402 wide field of view head-up display uses diffractive optics in an optical configuration known as the 'Z-HUD'.

This arrangement, coupled with the use of holographic optical elements give, according to the company, excellent parallax performance, good image positioning accuracy and a large eye-movement box.

The Type 1402 combines cursive symbology for flight parameters and weapon aiming, together with a raster video picture derived from an infra-red sensor. The brightness controls for the two images are independent and a video recording camera can be fitted as an integral part of the HUD.

Total field of view: 30° horizontal × 22° vertical
Binocular instantaneous field of view: 30° horizontal × 18° vertical
Raster format: 525 or 625 lines, 25 frames/s
Power: 120 Va ac, 2 W dc
Weight: 21.8 kg

STATUS: in development.

Smiths Industries Type 1402 head-up display

5-50 Series head-up display/weapon-aiming computer

The 5-50 head-up display/weapon-aiming computer is based on a single advanced digital computer, similar to that supplied by Smiths Industries for the Panavia Tornado, that can readily interface with any of the head-up displays manufactured by the company. It provides all necessary flight, navigation and weapon-aiming symbology, and performs accurate weapon-aiming computation. As with the earlier 4-40 system, the symbology presented to the pilot at any particular time is limited to information appropriate to the mode of flight at that time, so minimising display clutter.

Though designed to interface with other digital systems, the 5-50 can accommodate analogue inputs, for example from sensors fitted to non-digital aircraft, by means of an interface matching unit. This arrangement, says Smiths Industries, is particularly useful during preliminary flight trials, or as an economic way of providing a suitable interface when only a small production run is envisaged.

Using UK or MIL-standard symbology, the 5-50 system can operate in a number of modes associated with waypoint and target acquisition, and weapon-aiming. These modes are for air-to-ground, continuously computed release point and continuously computed impact point, and for air-to-air, continuously computed impact line and weapon release. Additional weapon-aiming modes are depressed sight-line, director-guided aerial gunnery and toss-bombing. A breakaway cue is provided in most bombing modes to indicate that immediate recovery from the delivery manoeuvre is necessary to avoid impact with the ground or damage caused by flying through the blast envelope.

Symbology: MIL-D-81641(AS), STANAG 3648, or as required
Inputs: 8 serial digital data channels and 32 discrete commands
Memory: up to 9 K words
Speed: (add/subtract) 3.75 μs
(multiply) 8.63 μs
(divide) 9.13 μs
Dimensions: 5.05 × 7.74 × 14.97 inches (128 × 197 × 380 mm)
Weight: 21 lb (9 kg)
Accuracy: (boresight) 0.05 mrad
(symbol positioning) 0.22 mrad up to 5° off axis, 0.32 mrad 5-10° off-axis

STATUS: no longer in production but remains in service.

6-50 Series head-up display/weapon-aiming computer

The 6-50 builds on previous developments by the addition of facilities to include air data and navigation computation. The system incorporates many detailed improvements over earlier head-up display systems and is an economic, low-weight device suitable for a wide range of aircraft. It is designed to interface with the data-bus systems now coming into general use in the newer aircraft, but can also be adapted as a retrofit for earlier, 'analogue' aircraft which are being upgraded for mission enhancement by the addition of modern avionics. There are many variations in packaging but in its simplest form the 6-50 comprises a display/multi-purpose processor driving a pilot's display unit through a control panel.

The processor generates the symbology for the display unit, performs the weapon-aiming computations, and can be used for a variety of air data and navigation tasks. The designation 6-50 embraces not only a range of packaging options, but also of performance and capability. The most comprehensive system could be contained in a 1 ATR (8 MCU) box weighing 12 kg and capable of generating additional symbology for a raster display and of interfacing with a wide range of sensors via a MIL-STD-1553 data-bus. Extensive weapon-aiming computation facilities and cursive display symbology are also provided in a 3/4 ATR box (6 MCU) weighing 9 kg. At the other end of the scale, weapon-aiming computing and symbology for aircraft requiring only basic head-up display/weapon-aiming computer capabilities can be housed in a 1/2 ATR (4 MCU) unit of 6 kg.

Variations are also provided for the pilot's display unit. The simplest has a 102 mm collimating lens providing a 25° field of view for 11.9 kg. Another variant, in line-replaceable form and designed to MIL specifications for the McDonnell Douglas AV-8B Harrier, has a 114 mm collimating lens, a 22° fov, and weighs 17 kg.

The system has a precision dual combining-glass assembly and an electronically depressible standby sight. Special care in the

Smiths Industries 6-50 head-up display/weapon-aiming computer

mechanical design has resulted in a unit with the same rigidity as earlier units but weighing 20 per cent less. Compensation for windshield distortion is applied electronically and optically to suit the particular characteristics of the aircraft. The cathode ray tube is protected from the damaging effects of solar radiation by ultraviolet and infra-red filters in the optical train, and cathode ray tube brightness, display accuracy and deflection amplifier performance are monitored by a high-reliability built-in test system.

The standby sight is a precision light-emitting diode matrix on a glass substrate, the brightness of which can be adjusted by varying the intensity of the light-emitting diodes.

STATUS: the system is no longer in production, but remains in service with the US Marine Corps.

Head-up display/weapon-aiming computer for Tornado

The head-up display/weapon-aiming computer designed for the Panavia Tornado can readily interface with any of the head-up displays manufactured by the company. It provides all necessary flight, navigation and weapon-aiming symbology, and performs accurate weapon-aiming computation. As with earlier systems, the symbology presented to the pilot at any given phase of flight is limited to information appropriate to that phase, so minimising display clutter.

The Tornado system operates in a number of modes associated with waypoint and target acquisition, and weapon-aiming. These modes are (for air-to-ground) continuously computed release point and continuously computed impact point, and (for air-to-air) continuously computed impact line and weapon release. Additional weapon-aiming modes are depressed sight-line, director-guided aerial gunnery, and toss-bombing. A breakaway cue is provided in most bombing modes to indicate that immediate recovery from the delivery manoeuvre is necessary to avoid impact with the ground or debris damage caused by flying through the blast envelope.

Symbology: MIL-D-81641(AS), STANAG 3648, or as required
Inputs: 8 serial digital data channels and 32 discrete commands
Memory: up to 9 K words
Speed: (add/subtract) 3.75 μs
(multiply) 8.63 μs
(divide) 9.13 μs
Dimensions: 5.05 × 7.74 × 14.97 inches (128 × 197 × 380 mm)
Weight: 21 lb (9 kg)
Accuracy: (boresight) 0.05 mrad
(symbol positioning) 0.2 mrad up to 5° off-axis, 0.2 mrad 5-10° off-axis

STATUS: electronic unit for this system is in production for Panavia Tornado.

DMPP Series head-up display/weapon-aiming computers

The display/multi-purpose processor HUDWAC builds on previous developments (notably the 5-50 and 6-50 series computers) by the addition of facilities to include air data and navigation computation and MIL-STD-1553B data-bus compatibility. The system incorporates many detailed improvements over earlier head-up display systems and is an economic low-weight device suitable for a wide range of aircraft. While compatible with the new data-bus systems, it can also be adapted as a retrofit for earlier analogue aircraft which are being enhanced by the addition of modern avionics. There are many variations in packaging, but in its simplest form the system comprises a display/multi-purpose processor driving a pilot's display unit. A control panel enables pilot selection of operating modes and control of brightness and other functions.

The processor generates the symbology for the head-up display itself, performs weapon-aiming computations, and can accomplish a variety of air data and navigation tasks. The designation DMPP embraces a range not only of packaging options but also of performance and capability. The most comprehensive system built to date (for the Indian Air Force Jaguar) has been supplied in a 1⅛ ATR (9 MCU) box weighing 13 kg, generating additional symbology for a raster display, and interfacing with a wide range of sensors via a MIL-STD-1553 data-bus. Redesign of circuit boards, using the latest component technology, is currently underway such that this type of unit will soon be available in a ¾ ATR box (6 MCU) weighing 9 kg. Weapon-aiming computing and symbology for aircraft requiring only basic computation can be housed in a ½ ATR (4 MCU) unit weighing 6 kg.

Variations are also provided for the pilot's display unit. The simplest has a 102 mm collimating lens providing a 25° field of view and weighing 11.9 kg. Another variant, in line-replaceable form and designed to MIL specifications for the US Marine Corps/McDonnell Douglas AV-8B Harrier, has a 114 mm collimating lens, a 22° fov, and weighs 17 kg. This particular system has a precision dual combining-glass assembly and an electronically depressible standby sight. Special attention to the mechanical design has resulted in a unit with the same rigidity as earlier units of comparable performance, but weighing 20 per cent less. Compensation for windshield distortion is applied electronically and optically to suit the particular characteristics of the aircraft. The cathode ray tube is protected from the damaging effects of solar radiation by ultraviolet and infra-red filters in the optical train, and cathode ray tube brightness, display

accuracy and deflection amplifier performance are monitored by a high-reliability built-in test system. The standby sight is a precision light-emitting diode matrix on a glass substrate, the brightness of which can be adjusted by varying the intensity of the light emitting diodes.

A variant of this unit is now being offered with a redesigned optical system to provide a wider instantaneous field of view and a raster video capability. The wider field is achieved by employing a redesigned twin-combiner assembly in conjunction with a 165 mm diameter output lens truncated at the fore and aft edges to 114 mm. This has the effect of increasing the instantaneous field of view by approximately 50 per cent. A similar truncated-optics technique is utilised in another recent development – the so-called 'low profile' head-up display. The shallow depth of this unit makes it most suitable for the retrofit market.

The principal reason for maximising the instantaneous field of view is the growing requirement to provide head-up displays with a raster video picture of the outside world derived from a forward-looking infra-red sensor. Both of the above truncated-optics designs will have this capability. In the raster mode, cursive symbology is written in the frame flyback raster during night operations. The normal cursive symbology format is retained for daylight use.

Another raster head-up display, using conventional refractive optics, is supplied to the Indian Air Force Jaguar as part of the so-called DARIN system. Because the raster display of radar returns has to be available in daylight, a special opaque blind is incorporated in the combiner assembly. The blind is retractable when the raster mode is not in use during daylight hours. This unit is a variant of the head-up display which Smiths Industries supplies for the British Aerospace Sea Harrier, in service with the Royal Navy and the Indian Navy.

Among the options being offered on new designs of refractive-optics head-up displays are holographic plane combiners. A holographically-produced gelatin element is sandwiched between the combining surfaces such that its reflective wavelength is identical to the wavelength of the cathode ray tube phosphor. This has the effect of creating a near-perfect mirror to the cathode ray tube symbology and an almost perfect window to all other light wavelengths. In practice, a typical transmissivity of 90 per cent can be achieved and a reflectivity of 80 per cent. This compares with values of 70 per cent and 30 per cent respectively using traditional combiner coatings.

The major benefit of using holographic combiners is the increase of contrast between the display and the outside world as seen by the pilot through the combiners. At the same time, the increase in reflectivity permits the bright-

Smiths Industries is prime contractor for Panavia Tornado head-up display, produced in collaboration with Teldix of West Germany and OMI of Italy

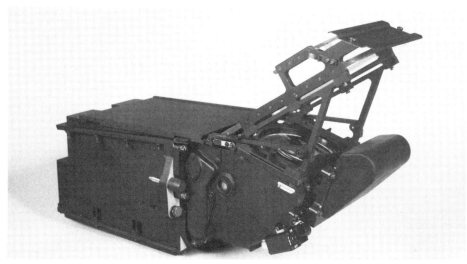

Smiths Industries raster head-up display for Indian Air Force Jaguar

ness of the cathode ray tube to be reduced, thus prolonging the life of the tube.

A variant of the unit used in the Royal Navy's Sea Harriers was selected in 1985 to form part of that aircraft's mid-life up-date programme. The new system's electronics unit includes the reversionary MIL-STD-1553B bus controller and a high power executive processor.

Diffractive-optics head-up displays

In 1981 Smiths Industries supplied a prototype diffractive-optics head-up display, or DHUD, for flight trials on a Jaguar aircraft of British Aerospace at Warton. The purpose of DHUD technology is to maximise the field of view of the display for compatibility with electro-optical sensors. The trials, with the DHUD working in association with a low-light television sensor, represented the world's first system demonstration at night to yield comprehensive analytical data on DHUD performance in a combat aircraft, according to Smiths. The field of view of this early model compassed

some 30° in azimuth and 20° in elevation, representing a three-fold increase over most conventional refractive-optics HUDs.

Modular electronics, incorporating a raster/cursive symbology facility with the cursive symbology being displayed during the frame flyback of the raster video display under night conditions), plus the wide field of view made possible by diffractive-optics geometry, permit compatibility with a broad range of electro-optic sensors, particularly forward looking infra-red devices.

Three other prototype models, each using different holographic optical element configurations, have been produced since. The most recent has a 40° field of view in azimuth, clearly capable of producing a 'wide-screen' video picture of the outside world.

At various times Smiths Industries has cooperated with Hughes Aircraft and Flight Dynamics in diffractive-optics head-up display projects.

STATUS: in development.

A prototype of Smiths Industries diffractive-optics head-up display

UNITED STATES OF AMERICA

Astronautics

Astronautics Corporation of America, 4115 N Teutonia Avenue, PO Box 523, Milwaukee, Wisconsin 53201-0523
TELEPHONE: (414) 447 8200
TWX: 910 262 3153

Model 131A head-up display

This device was produced for the US Army/Bell AH-1G light attack helicopter. It is located in the forward (gunner's) compartment.

Exit aperture: 3 inches (76 mm)
Total field of view: 20°
Instantaneous field of view: 8.5° monocular (15.5° binocular) at 18.33 inches from combiner glass
Standby reticle: fixed, red colour
Computer: microprocessor, 500 ns/instruction, full arithmetic capability
Memory: 4 K – 16-bit words rom, 256 – 12-bit words ram

STATUS: in service.

Astronautics Model 131A head-up display

Flight Dynamics

Flight Dynamics Inc, PO Box 1079, 5289 NE Elam Young Parkway, Hillsboro, Oregon 97123
TELEPHONE: (503) 640 8955
TWX: 910 460 2200

Holographic head-up displays for commercial transports

Flight Dynamics has identified a potential market for head-up displays in those airlines operating older aircraft whose autopilots have limited automatic-landing performance. Head-up displays are especially beneficial for package-parcel express carriers that have to maintain demanding overnight delivery schedules in all weather and for airlines utilising hubs in poor weather. Flight Dynamics is accordingly developing and certificating head-up displays for manual Category IIIA approaches with a 50-feet (15-metre) decision height to assist such operators. Manual Category IIIA landings will be flown with either single or dual head-up display installations, according to Flight Dynamics. In the first case one pilot will 'fly' the head-up display, the other monitoring ILS deviations on the conventional head-down ADI and HSI instruments. In a dual installation the second pilot will monitor flight-path deviations on his head-up display. To prevent common-mode failures (which could result in identical but wrong commands to both pilots) the second pilot's display symbology would be driven by data by-passing the head-up display computer. Category IIIA landings will be based on Category II ILS sensors, with flare indication partly based on radio altimeter data down to visual contact with the runway.

The head-up display could also be employed in conjunction with automatic landing systems to augment Category II and Category III operations. For example, a single system (costing around $200 000) could be used as an

Flight Dynamics head-up display with combiner glass lowered for use in Boeing 727 trials aircraft

independent monitor in aircraft with a fail-passive automatic landing system to improve safety and pilot acceptance.

Flight Dynamics has designed two basic versions of its head-up display. Irrespective of model, each head-up display comprises four units: an optical system with overhead projector and combiner glass, a computer, electronics drive unit, and pilot's control unit. The

company's first system, called the Transport HUD is aimed at Boeing 727, 737, 757 and 767, McDonnell Douglas DC-10, Lockheed C-130/L-100 Hercules, and Airbus A300, A310 and A320.

In August 1983 this system was certificated for Category I operation on the 727, and approval was given for Category II during January 1984. The company had hoped to

Components of Flight Dynamics head-up display

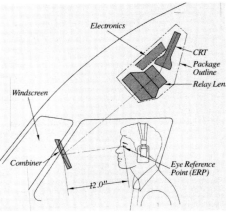

Ray geometry of Flight Dynamics holographic head-up display system

reach Category IIIA by the end of 1983, but development of the flare-cue software took longer than anticipated, and the 727-100 being leased by Flight Dynamics had to be returned to its operator, Piedmont Airlines. However FAA Category IIIA approval for the 727 installation was granted in July 1984. In a parallel programme, Flight Dynamics was planning to complete and fly, for development and certification, a prototype head-up display for the McDonnell Douglas DC-9. Certification was to be accomplished in conjunction with Airborne Express, which operates a fleet of ten McDonnell Douglas DC-9-32s on its overnight parcel-delivery service, but an order was never placed and the project lapsed.

The second type, known as the Compact HUD, would have its optical system packaged into a smaller unit for the more restricted flight-decks of DC-8, DC-9, 747 and general aviation aircraft such as the de Havilland Dash 7 and Dash 8 and Canadair Challenger, though retaining the basic performance of the first type.

Flight assessment of the two systems began in 1982. Evaluation equipment delivered to Boeing and NASA-Ames Research Center in early 1983 was being used for simulator studies to assess the dispersion along the runway of manual Category IIIA touchdowns, and certification would depend in part on the statistical results of several hundred 'landings'.

For best results, says the company, a transport aircraft head-up display should be conformal, the angles in the display symbology

being the same as those of objects in the outside world. Thus glideslope and horizon on the display would line up with their real counterparts. To accommodate the large angles needed in low-speed flight, such as during approach in strong cross-winds, the display itself has to be rather larger than those common in military aircraft and this calls for the use of holographic technology. By comparison with a typical 20° × 15° military head up display, the Flight Dynamics holographic system gives respectively 30° × 24° in an instantaneous, binocular field. The hologram is formed in a 1-mil (0.025 mm) thick layer of gelatin sandwiched between two layers of glass that form the combiner, and curved to provide the appropriate optical characteristics. The outer surfaces of the glass lenses are flat so that there is no distortion or curvature of the field when seen through the composite sandwich.

Flight Dynamics has also been pursuing prospective military applications. In combat aircraft, the wider fields of view made possible by holographic optics is required to exploit the wide fields of view covered by the new electro-optical devices such as forward-looking infra-red. Flight Dynamics at one time had a teaming arrangement with the UK company, Smiths Industries, the aim being to develop a holographic head-up display for the US Air Force/McDonnell Douglas F-15E Strike Eagle multi-mission version of this air-superiority fighter. (In the event, the head-up display competition for this aircraft was won by Kaiser.)

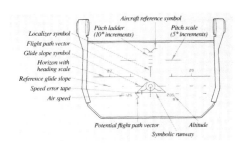

Typical symbology of Flight Dynamics head-up display in IMC approach mode

Model 1000 head-up display for airliners

Flight Dynamics began certifying its Model 1000 wide angle HUD to Category 3A approach standards in the Boeing 737-200 in mid-1985. It had earlier been certified for manual approaches to Category 3A minima in the Boeing 727-100.

The company has sold two systems to Boeing for cockpit research, two to Lockheed for test bed work on board a C-130, and other systems for use in simulators.

STATUS: in development.

General Electric

General Electric Company, Aircraft Equipment Division, PO Box 5000, Binghamton, New York 13902
TELEPHONE: (607) 770 2000

AN/ASG-26A lead-computing optical sight

General Electrics has developed from its earlier equipment an improved lead-computing optical sight that enables air-to-air combat to be conducted without the need for continuous target tracking. This capability is incorporated in the system for the US Air Force/McDonnell Douglas F-4E. The equipment comprises a head-up display, two-axis lead-computing gyroscope, gyro mount and lead-computing amplifier. For airborne targets the system displays gun and missile fire-control information by means of a servoed aiming mark. Against ground targets the pilot adjusts the aiming

mark manually to control gunnery, rocket and bombing displays.

The aircraft's own manoeuvres generate rate and acceleration signals in the gyro lead-computer, range-to-target is measured by radar, and angle of attack, air density and airspeed (needed for trajectory correction) are supplied by the air data computer. With these parameters fed into the system, the aiming reference is displaced so as to produce the appropriate lead angle and gravity corrections. Analogues of roll angle and range are also projected on to the combining glass. In ground attack modes other sensors generate corrections for drift, and offset bombing is also possible.

STATUS: in service, but no longer in production. General Electric has been engaged on lead-computing optical sight programmes for F-101, Lockheed F-104, F-105, General Dynamics F-111, and F-5 aircraft. For F-5E company provides AN/ASG-29 lead-computing sight (see below).

AN/ASG-29 lead-computing optical sight

This sight has been specially produced for the US Air Force/Northrop F-5E. The AN/ASG-29 has a family relationship with the AN/ASG-26A but comprises only two units, a pilot's display and lead computer. The system provides guidance for air-to-air and air-to-ground weapons delivery.

Total field of view: (azimuth) 12° (elevation) 14°
Instantaneous field of view: 7.5°
Collimating lens aperture: 4 inches (102 mm)
Weight: (sight head) 6.9 kg
(lead computer) 7.9 kg
(mounting base) 0.8 kg
Reliability: mtbf claimed to be over 300 h

STATUS: in service but out of production.

Honeywell

Honeywell Inc, Military Avionics Division, PO Box 312, 2600 Ridgeway Parkway, Minneapolis, Minnesota 55413
TELEPHONE: (612) 378 4178
TELEX: 290 631

Visually coupled systems

The Honeywell electro-optical and magnetic helmet sights have evolved with the related objectives of simplifying target acquisition for airborne weapons, and also improving this function by utilising the wide search-angle and flexibility of the human eye. This is accomplished by arranging for the pilot's line-of-sight to be determined in relation to aircraft axes. This data can be used as pointing information for a wide range of aircraft systems (AI radar, cameras, infra-red seekers, etc) and for weapon aiming.

The Honeywell helmet sight system consists of the following major elements: helmet-mounted unit (HMU), sensor surveying units (SSU), sensor electronics unit (SEU), and controls mounted on the instrument panel and control stick.

The system operates as follows: two sensor surveying units, rigidly mounted to the airframe and aimed in the pilot's direction emit fan-like beams of infra-red light rotating at constant velocity. The light beams sweep over a reference photo sensor and two pairs of helmet-mounted photo sensors (one pair on each side of the pilot's helmet). The time intervals between the pulses from the helmet photo sensors and the reference pulses are a measure of the angular position of the pilot's head relative to the XYZ reference axes of the aircraft. The photo sensor outputs are transmitted to the sensor electronics unit where the angular computations are performed and converted into azimuth and elevation information. A collimated reticle image, aligned with the sensor pairs, is reflected off the parabolic visor and into the pilot's right eye. Additional digital computations such as fire control and missile

launch envelopes may easily be incorporated in the sensor electronics unit with little or no increase in system complexity. In versions being supplied to the US Navy, a Honeywell HDC-202 digital computer is included.

AMHMS

AMHMS advanced magnetic helmet mounted sight system has been fully qualified for US Air Force applications through a contract with the US Air Force Aeronautical Systems Division and is undergoing flight evaluation in the AFTI-16 aircraft. The system uses a visor projected reticle and a line-of-sight computation, based on magnetic fields surrounding the pilot's helmet, to allow off-boresight target engagement by merely turning the head to place the reticle on the target.

VTAS

A total of 500 electro-optical helmet sight systems were supplied to the US Navy for fitting in F-4J aircraft. The VTAS visual target acquisition system is used for off-boresight targeting of the Sidewinder air-to-air missile.

IHADSS

The latest version of the Honeywell electro-optical system is the IHADSS integrated helmet and display sighting system which is now in production for the AH-64A advanced attack helicopter. Both the pilot and co-pilot/gunner are provided with helmet units and controls to allow independent and co-operative use of the system. The IHADSS sight component provides off-boresight line-of-sight information to the fire control computer for slaving weapon and sensor to the pilot's head movements. Real world sized video imagery from the slaved and gimballed infra-red sensor is overlaid with targeting as well as flight information symbology, and projected on a combiner glass immediately in front of the pilot's eye. The

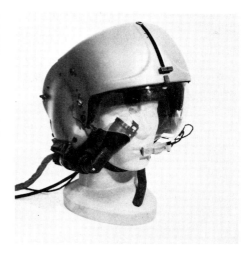

Honeywell IHADSS helmet sight/display

IHADSS allows night nap-of-the-earth flight at below treetop altitudes without reference to cockpit instruments, and rapid target engagement.

Advanced concepts

Honeywell is developing advanced helmet-mounted sight and display concepts which include wide fov monocular displays, NVG compatable displays, NBC mask compatable displays, eye tracking, and improved head tracking methods for both aircraft and ground vehicle applications.

STATUS: the company has delivered 500 VTAS systems for the F-4J aircraft. The IHADSS is currently in production for the Hughes AH-64A helicopter and is also to be fitted into the CH-53E, HH-60D, A-129 and PAH II aircraft. The AMHMS has been military qualified and is undergoing flight evaluation in the AFTI-16 aircraft. It is also in use in the helmet mounted oculometer system for the tactical combat trainer at the US Wright Patterson AFB.

JET

JET Electronics and Technology Inc, 5353 52nd Street, Grand Rapids, Michigan, 49508-0239
TELEPHONE: (616) 949 6600
TELEX: 226453

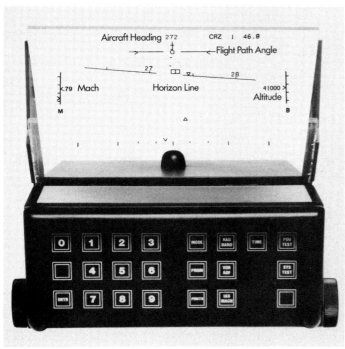

JET Series 1000 head-up display

Series 1000 head-up display for business aircraft

JET's head-up display for business aircraft, said to be the first such system designed for this application, was announced in September 1985.

Signals from standard aircraft sensors are processed by the HUD computer (a separate ½ ATR box) and displayed on the combiner, which has a 15° vertical × 30° horizontal field of view. A key parameter is aircraft acceleration, which gives the pilot an indication of windshear. The brightness of the display is automatically adjusted to suit the ambient conditions.

The display has four modes: take-off/go-around, cruise, approach and ILS. Critical reference speeds and flight path angles can be entered on the display control panel.

The Series 1000 is offered on the Pilatus PC-9 trainer as an option.

STATUS: in production.

Hughes Aircraft

Hughes Aircraft Company, Radar Systems Group, PO Box 92426, Los Angeles, California 90009
TELEPHONE: (213) 648 2345
TELEX: 910 348 6681

Diffractive-optics head-up display for JAS 39

In January 1983 Hughes Aircraft announced that it had been contracted by SRA Communications of Sweden to develop and produce a diffraction-optics head-up display for the Saab JAS 39 Gripen. SRA Communications is a member of the Ericsson Group and also one of the five-member JAS Industry Group set up in 1980 by the Swedish aerospace industry to develop and build the new aircraft. The initial head-up display (HUD) contract is worth more than $10 million.

Sweden is the first country to award a production contract for a HUD incorporating diffraction-optics technology. The advantages of this technology are claimed by Hughes to be a key factor in providing the JAS 39 with its capability in the air-to-air, air-to-ground, and reconnaissance roles.

Diffractive-optics HUDs have two principal advantages. First, by comparison with conventional systems, they have a much wider field of view, typically 30° by 20° compared with 20° by 15°. They are thus more suited to the new generation of combat aircraft, in which weapon-aiming symbology can make large angles with the flight vector. The wide field will also be useful in night operations to display data from electro-optical sensors such as forward-looking infra-red.

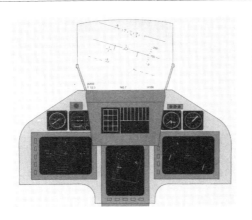

Artist's impression of flight panels in JAS 39 Gripen cockpit showing Hughes diffractive-optics head-up display

Secondly, the combiner glass on which the symbology is superimposed on the outside world acts as a mirror reflecting only a narrow band of light. The transmission index is about 85 per cent compared with 50 to 70 per cent. The symbology is also bright enough to stand out in direct sunlight without having to operate the cathode ray tube at such high power levels that their life is shortened.

The system also provides resistance to glare, reflections and spurious sun images – the latter particularly important, since bright sunlight can create 'hot-spots' on the display that prevent the pilot from seeing the symbology. The Hughes design based on proprietary technology using holography and lasers, employs a single combiner glass and so also eliminates

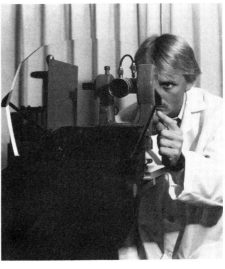

Hughes engineer tests first flight-test model of diffraction-optics head-up display for JAS 39

the bulky support structure necessary in HUDs that support two or more combiner glasses.

STATUS: Hughes delivered first holographic HUD for flight-test in November 1984 and production is expected to continue through 1990s. Similar equipment is being evaluated by US Air Force and Navy and other countries.

Hughes is also developing a diffractive HUD for the Israeli Lavi. The first pre-production HUD for the JAS 39 and the first flight-test Lavi HUD were delivered to the respective customers in mid-1986.

Kaiser Electronics

Kaiser Electronics, Kaiser Aerospace and Electronics Corporation, 300 Lakeside Drive, Oakland, California 94612
TELEPHONE: (415) 835 5554
TWX: 910 338 0196

Head-up display/weapon-aiming system for Alpha Jet

Kaiser supplies the head-up display/weapon-aiming system for the West German version of the Franco-German Alpha Jet. The three-unit system comprises a computer/symbol generator, sweep driver unit and pilot's display unit.

All or part of the same system are used in the MT-4, F-4EJ, C-1 and CCV programmes in Japan and the Taiwanese AT-3. In the Italian MB-339 application, the Kaiser HUD has dual flat holographic combiners.

STATUS: in production and service.

Head-up display for AH-1S helicopter

The head-up display for the US Army/Bell Helicopter AH-1S Cobra helicopter is a lightweight, low-profile conventional cathode ray tube that superimposes aiming information for the multi-barrelled gun and Hughes TOW anti-tank missile on to the pilot's forward field of view. The programming flexibility and spare capacity of the microprocessor-controlled symbol-generator permits other functions to be incorporated, such as the derivation of flight commands for nap-of-Earth flying and laser tracking and pointing information.

STATUS: in service.

Inertial navigation head-up display for A-10

This updated version of the Kaiser head-up display for the US Air Force/Fairchild Republic

Kaiser Electronics symbol generator, combined head-up display/control unit and sweep driver for Alpha Jet

Pilot's display unit for Kaiser Electronics head-up display in Cobra AH-1S helicopter

A-10 incorporates inertial navigation data from the A-10's inertial navigation system. The head-up display contains a MIL-STD-1553 multiplex data-bus system which can compute total velocity vectors and other algorithms.

STATUS: in service; 650 delivered.

F-4E wide field of view holographic head-up display

A wide field of view (WFOV) holographic head-up display (HHUD) is now in production for the F-4E aircraft, under a contract from an international customer. This is the first WFOV HHUD to enter prcduction for any aircraft. The F-4E HHUD features a single curved holographic combiner that offers a 20° × 30° total and instantaneous field of view. The HHUD operates in raster and/or stroke, and the display processor, which utilises a 1750A Processor, is packaged with the HUD.

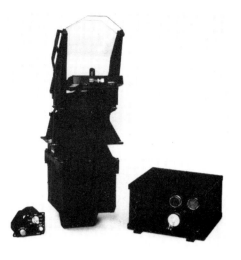

Kaiser Electronics head-up display for Fairchild A-10

McDonnell Douglas

McDonnell Douglas Astronautics Company, PO Box 516, St Louis, Missouri 63166
TELEPHONE: (314) 233 9377
TWX: 910 762 0635

Mast-mounted sight for AHIP helicopter

Following a flight-assessment programme with a single prototype, the US Army in October 1984 ordered into production the McDonnell Douglas mast-mounted sight for its AHIP (Army helicopter improvement programme) Bell OH-58 helicopters. The initial contract is for 16 units plus two spares, but the company estimates a potential outlet for more than 1000 systems for this and other aircraft. Mounted on the rotor mast, but above the rotor itself, the sight unit acts like the periscope of a submarine, in this case permitting the aircraft to conceal itself by using natural cover, but allowing the crew to peer over the cover to observe enemy dispositions and forces.

McDonnell Douglas, the MMS contractor, integrates their unique stabilisation platform and electrical systems with a sensor suite purchased from Northrop. Northrop announced a $33.4 million contract from McDonnell Douglas in September 1985 for the production and integration of 44 MMS sensor suites. The sensors include a low-light television, a thermal imaging sensor, and laser rangefinder/designator for Copperhead and Hellfire anti-tank missiles. Both companies are responsible to Bell, which is building 578 specially modified OH-58 (military JetRanger) helicopters to act as scouts and escorts for the Army's fleet of Hughes AH-64A Apache battlefield helicopters. A notable feature of AHIP is the automatic hand-off system by which targeting information from one aircraft can be transmitted to another or to ground-based weapons. AHIP itself is the precursor to the much more ambitious LHX light battlefield helicopter, to enter service around 1993/4.

Mounted over the main rotor drive shaft, the performance of such a system would be severely degraded by the high vibration levels in this area (a characteristic of all helicopters) were it not for the 'soft mount' devised by McDonnell Douglas, which provides a high degree of isolation; performance of the anti-vibration system is such that target bearing can be measured to within 20 milliradians, (about 1°). The over-riding need to keep down weight above the rotor has produced a sensor package weighing only 160 lb (73 kg); equipment-bay systems add another 90 lb (41 kg).

The television camera has a silicon-vidicon 'dawn-to-dusk' capability with an 8° field of view for target acquisition and a 2° field for recognition. The forward-looking infra-red sensor has a 120-element common-module detector array and two field of view, 10° and 3°.

McDonnell Douglas mast-mounted sight display is on left of this Bell OH-58D helicopter cockpit. View is adjusted by controls on stick held by pilot's right hand

Bell OH-58Ds with McDonnell Douglas mast-mounted sights

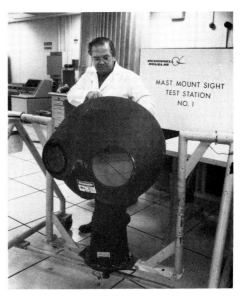

Carbon-epoxy spherical container housing the mast-sight electronics, sensors and anti-virbration mounting

McDonnell Douglas has equipped this McDonnell Douglas Helicopters 530F with its mast-mounted sight

Television and laser systems share a common optical path to minimise the number of components. A video tracker and digital scan-converter in the data-processor together permit the incorporation of other features such as auto-track, frame-freeze and point-track. In the cockpit are two multi-function displays for the presentation of video from the MMS and flight guidance and communication navigation information. The sensors themselves are contained within a 25.5 inch (650 mm) carbon-epoxy sphere with a sight-line 32 inches (810 mm) above the plane of rotor rotation. Data is channelled into the cockpit area via cables through a 0.9 inch (23 mm) tube running inside the drive shaft.

The MMS is applicable to a variety of helicopter platforms, including the McDonnell Douglas Helicopter's 500 series, the Sikorsky H-76, the Bell 406, the Agusta A129, and the MBB BK117. The MMS can be used to provide these platforms with both 'AHIP type' scout features and light attack capability, using Hellfire and/or TOW missiles. The Apache helicopter itself (which does not have a mast-mounted system, the Martin-Marietta TADS/PNVS system being nose-mounted) is regarded as a potential future application. Surface

vehicles, too, could benefit from a system on a telescopic mast.

STATUS: in MMS-equipped Bell Model 406LM Long Ranger helicopter first flew in August 1983. Five prototype units were delivered to the US Army by June 1984. Development testing began the following month and was concluded by September of that year. Operational testing commenced in October 1984 and was concluded in January 1985. Production and delivery of the 16 initial units was to begin in 1985. The US Army is currently funded for 578 AHIP/OH-58As, but expectation is for an eventual total of 720. The second production lot of 44 units was funded in 1985. In August 1985 McDonnell Douglas began flight testing the mast-mounted sights on a McDonnell Douglas Helicopters 530F.

Polhemus

Polhemus Navigation Sciences Division, McDonnell Douglas Electronics Company, PO Box 560, Colchester, Vermont 05446
TELEPHONE: (802) 655 3159
TWX: 510 299 0046

Advanced helmet-mounted sight

In 1979 a contract for the development of an advanced helmet-mounted sight was awarded jointly to Polhemus Navigation Sciences and Honeywell by the US Air Force and it became the basis of the current work in this field by the Vermont company.

Polhemus designs, develops and manufactures third-generation helmet-mounted sights specifically for tactical military aircraft. They are based on a company-patented method of measuring sight angles by the use of magnetic sensors that is produced under the trademark Spasyn.

A transmitter mounted in the cockpit generates a magnetic field some distance around it, which is sensed by a pick-up installed under the pilot's helmet visor. Signals are processed by a computer in an associated electronics unit and converted into angles representing the direction of the pilot's line of sight in relation to a particular datum, usually the aircraft's reference frame.

Coincidence between the helmet-aiming axis and the pilot's line-of-sight is attained by a small helmet-mounted optical generator that projects a virtual image (that is, located at infinity so that the pilot does not have to refocus his eyes) on to the helmet visor; the pilot then aligns the aiming reticle with the target. The reticle can also be used to provide cueing information that permits the pilot to be signalled or directed from an external source of information, for example, from a radar, forward-looking infra-red, or radar-warning device or by a weapons-systems operator. The same system, in conjunction with a more elaborate display, can provide the pilot with flight-director symbology or imagery and information from other aircraft sensors. The complete system, irrespective of complexity, comprises a magnetic-field radiator and tuning unit, helmet-mounted sensor and memory unit, electronics unit, control panel and reticle display unit.

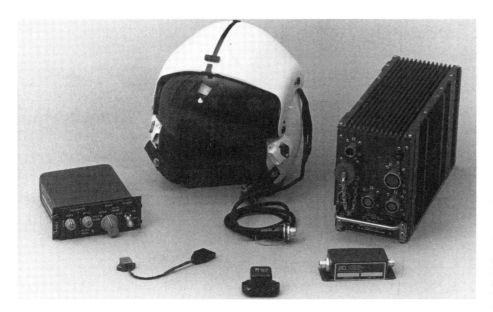

Polhemus helmet sight: left to right, control panel, magnetic-field sensor, helmet, radiator and tuning unit, and electronics unit

Dimensions: (radiator) 2.25 × 1.27 × 1.3 inches (57 × 32 × 33 mm)
(sensor) 1.13 × 0.83 × 0.55 inches (29 × 21 × 14 mm)
(electronics unit) 4.83 × 7.59 × 14.25 inches (123 × 193 × 362 mm)
(radiator tuning unit) 4.50 × 1.75 × 1.25 inches (114 × 44 × 32 mm)
(control panel) 5.75 × 2.25 × 7.67 inches (146 × 57 × 195 mm)

Weight: (radiator) 0.2 lb (0.09 kg)
(sensor) 0.03 lb (13 g)
(electronics unit) 19.9 lb (9.04 kg)
(radiator tuning unit) 0.4 lb (0.2 kg)
(control panel) 2.2 lb (1 kg)

STATUS: in production.

Sundstrand

Sundstrand Data Control Inc, Overlake Industrial Park, Redmond, Washington 98052
TELEPHONE: (206) 885 3711
TELEX: 286144
TWX: 910 449 2860

Electronic head-up displays

The electronic head-up display in the form chosen by McDonnell Douglas as a standard option for the MD-80 transport, provides a much larger number of modes than its electro-mechanical counterparts. Lateral guidance through localiser deviation and course error, together with airspeed, are provided during the take-off roll. At rotation it continues to give lateral guidance, now via magnetic heading, and vertical guidance by means of pitch attitude, plus airspeed. During the subsequent climb-out it shows attitude information in pitch, roll and heading together with height, speed and take-off/go-around command. In a missed approach situation it gives pitch and roll attitude, and heading, as well as height, speed and 'go-around' command. The HUD is driven from the third channel of the automatic flight control system, and acts as an autoland monitor.

Guidance during approach and landing falls into four modes. Above the decision height the system gives lateral guidance via localiser deviation and course error, flight-path guidance through glideslope deviation, pitch, roll and yaw attitudes, and speed and height. Between decision and flare heights the same information is presented, together with decision height and flare height messages. Between flare height and touch-down the information remains the same, except for the addition of safe roll-out limits. During the roll-out phase lateral guidance is given through localiser deviation and course error.

The display unit fits into an overhead cavity in the roof panel where it is normally stowed and swung out for use. It is fed with signals from a digital processor and controlled by a panel-mounted unit.

Pacific Southwest Airlines (PSA) is so far the only US carrier to be operating with a HUD. The airline began equipping with the MD-80 in 1981, but lack of manpower in training and economic restraints, permitted only a slow familiarisation programme, and not until November 1984 did the airline formally introduce the HUD into its

Sundstrand head-up display for McDonnell Douglas MD-80

standard flight procedure. The FAA was also moving cautiously, and imposing stringent training requirements on PSA; crew training was delayed for 18 months until a HUD could be introduced into the MD-80 simulator at San Diego, California. Though each aircraft has two displays (one for each pilot), PSA's policy is to permit only one crew-member at a time to use the device. In PSA aircraft the captain remains 'head-up' during the approach, monitoring the HUD and preparing to make the transition from instrument to visual guidance as the aircraft breaks cloud. The HUD system is reported to be invaluable in Category III operations at PSA's major hubs, Los Angeles and San Fransisco.

Pilot's display unit
Calibration: calibrated during manufacture and contains automatic boresight feature
Dimensions: 14 × 15 × 9 inches (356 × 381 × 229 mm)
Weight: 20 lb (9.1 kg)
Reliability: estimated 4000 h
Image generation: cathode ray tube 2 × 2 inches (51 × 51 mm) PI phosphor

Image brightness: 15 000 ft-lambert (51 390 cd/m²) max
Power: 30 W
Field of view: (horizontal) 30° (vertical) 26°
Accuracy: 3.5 mrad over entire field
Display integrity: fail-safe through auto alignment and self-test. Display illumination interrupted upon fault detection

Processor: stored program digital processor with 20 Hz sampling and computation iteration rate, 50 Hz cathode ray tube image refresh rate
Inputs: attitude, angle of attack, speed
Self-test: equipment and software completes self-checking sequence with each computation iteration. Display flashes to warn of faults
Weight: 15 lb (6.8 kg)

Control unit
Functions: flight-path angle select, press-to-test, barometric altitude select
Reliability: estimated mtbf 36 700 h
Face size: 5.75 × 1.9 inches (146 × 48 mm)

STATUS: in production.

Stores management

FRANCE

Alkan

R Alkan et Cie, rue du 8 Mai, 94460 Valenton
TELEPHONE: (1) 43 89 39 90
TELEX: 203876

T 900 C and T 905 C armament control and monitoring systems

The Alkan T 900 C and T 905 C are armament control and monitoring systems (ACMS) designed for fixed-wing aircraft and helicopters equipped with twin rocket-launching installations. They assist the gunner/weapon-system operator in monitoring a mixed weapons load, and deliver electrical firing impulses to the weapons according to preset schedules.

An ACMS comprises a panel-mounted firing-selection unit connected to the firing trigger and two decoder units on the launchers. Using a microprocessor in the controller the crew can select the desired type of rocket to be launched, the firing mode, time interval and the number of rounds per salvo. The quantity and type of ammunition available is recorded and displayed throughout the sortie and automatic load balancing between launchers is provided, irrespective of rocket types.

Decoders of 12, 22 or 36 channels can be provided to control launching systems with 3 to 36 tubes. Systems may be optionally extended to control up to four launchers.

All conventional types of rocket may be controlled, and include air-to-air, hollow-charge, smoke or flame dispensing, incendiary or electronic countermeasures.

The principal difference between the T 900 C and the T 905 C systems is that the former exercises control over four and the latter over three different types of rocket. The number of types of munition may be varied according to mission requirements, the total number being limited only by launcher capacity.

Firing interval may be selected at 0.2, 0.5, 1, 2 or 5 seconds and there are six choices of the number of rounds in each salvo: 1, 2, 4, 8, 12, or the full complement of rockets may be selected.

The system, which may also be used on armed ground vehicles and surface vessels, has been selected for the French Army's Aéro-spatiale/Westland Gazelle helicopters and is also in use on MBB BO105 helicopters.

Dimensions: (T 900 C) 5.5 × 4.1 × 5.2 inches (140 × 104 × 132 mm)
(T 905 C) 5.7 × 3 × 6.7 inches (146 × 76 × 171 mm)

Weight: (T 900 C and T 905 C) 2.2 lb (1 kg)

STATUS: in production and service.

T930 weapon management system

The T930 stores management system is intended for maritime patrol aircraft. Up to eight wing-mounted stores can be controlled by either the pilot from the flight-deck or the tactical officer stationed in the cabin. The two control panels are linked, with software controlling the dialogue and ensuring that the correct authorisation and information procedures are followed. The two crew-members each have a weapon inventory panel displaying the current weapons load. Either member can select the munitions to be released, but the tactical officer can only activate his selection after receiving authority on his panel from the pilot. One crew member also has a weapon-arming panel containing the gun- and rocket-arming and bomb-fuzing selections.

STATUS: in production; selected for the Fokker F27 Maritime Enforcer.

Alkan T 900 C armament control and monitoring system

Alkan T930 weapon inventory panel

Alkan T930 weapon arming panel

TRT

Télécommunications Radioélectriques et Téléphoniques (TRT), Defence and Avionics Division, 88 rue Brillat-Savarin, 75460 Paris Cedex 13
TELEPHONE: (1) 45 81 11 12
TELEX: 250838

Dot and Telecontrol systems

Dot and Telecontrol are described by TRT as target-designation and revectoring guidance systems for long-range, naval surface-to-surface missiles. Location of targets by the radars of the missile-launching vessel, however, is limited by the line-of-sight horizon to less than 20 nautical miles. For long-range attack it is necessary to use an aircraft, normally a helicopter, to acquire the target, transmitting its co-ordinates to the launching vessel. This is accomplished by the Dot system, which transmits data relating to one or more targets from the aircraft to a computer aboard the vessel.

The Telecontrol system is used to update the missile's course after launch, according to the target position information transmitted by the Dot system.

Details of frequencies employed, channel availability and modulation modes used have not been released, but both systems are believed to operate in the uhf band.

STATUS: in production.

Control and display units for TRT target designation and revectoring system for long-range missile control

---

GERMANY, FEDERAL REPUBLIC

Teldix

Teldix GmbH, Postfach 10 56 08, Grenzhofer-weg 36, D-6900 Heidelberg 1
TELEPHONE: (6221) 5120
TELEX: 461735

Missile control system

The Teldix missile control system has been specifically designed to control intelligent missiles fitted to aircraft such as the West German Navy's Panavia Tornados. The missile control system serves as a computing and interface system between the aircraft's main digital computer with its associated peripheral systems including stores management, inertial navigation, maintenance panel, warning system and others, and the relevant input/output circuits of the missile.

The missile control system comprises three units: the missile control panel, missile control unit and launcher decoder unit. The missile control panel contains a number of controls and displays which enable the operator to monitor the status of the missiles and select

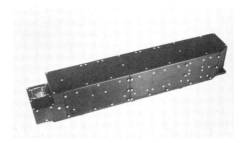

Teldix missile control launcher-decoder

those to be fired and the firing mode. The missile control unit contains the interface circuits that convert and process externally-fed data, feeding it to the missiles via the MIL-STD-1553B data-bus.

Interface between the missile control unit and individual missiles is provided by the launcher decoder unit and one of these is included in each launcher. The launcher decoder units contain digital/analogue and analogue/digital data conversion between the missiles and their power supply, or circuitry for direct communications via the -1553B data-bus.

STATUS: in production.

Teldix missile control unit

Teldix missile control panel

ISRAEL

Elbit

Elbit Computers Limited, Advanced Technology Center, PO Box 5390, Haifa 31051
TELEPHONE: (4) 514 211
TELEX: 46774

Stores management system

Elbit's stores management system is computerised, flexible, easy-to-manage and has dual-mode operation for maximum integrity against the accidental release of stores. Computed information is shown on an armament control and display panel, essentially a multifunction cathode ray tube unit. The pilot prepares several weapon-release schedules, according to his pre-flight briefing, so that during the attack phase he merely has to select the appropriate one for the task. The system comprises four units, an armament control and display panel, stores management computer, video electronic unit and (optionally) a video recorder.

The stores management system replaces the switches and selectors which would normally control guns, weapons and fuzes, weapon data, stores release programming, weapon-data insertion panel and stores emergency jettison. Pre-flight selections comprise: master mode for the aircraft's weapon delivery and navigation system, which takes in air-to-air combat, air-to-ground missiles, gun-firing, selective jettison, and navigation; weapon delivery mode (ccip, dive/toss, toss, and direct); type of weapon to be used; release mode (single weapon, pair, or salvo), release interval, and fuze mode (nose, tail, or nose and tail).

The cathode ray tube continuously displays the inventory (number, location and type) of stores being carried on the mission and is updated as stores are released or jettisoned. It also shows the information appropriate to the particular release programme, and warns of incorrect programming, equipment failure or 'beyond limits' condition. The system computer can also control a videotape recorder for post-

mission analysis or debriefing after a training flight.

Computer
Dimensions: 321 × 194 × 172 mm
Weight: 13.5 kg
Redundancy: two independent computing systems, dual power supply
Memory: 58 K eprom, 3 K ram, 1 K non-volatile ram
Input/output: 160 discrete inputs, 120 discrete outputs

Armament control and display panel
Dimensions: 172 × 147 × 267 mm
Weight: 7 kg
Scanning: 525 or 875 lines
Bandwidth: 20 MHz
Contrast ratio: 7:1

STATUS: in production and service, for and with McDonnell Douglas A-4 Skyhawk and F-4, IAI Kfir and in other aircraft exported from Israel.

SMS-86 stores management system for Lavi

Elbit was selected during early 1985 to develop a stores management system for Israel's Lavi fighter.

The system, which is fully computer-controlled, comprises two units. The stores management processor (SMP) includes a MIL-STD-1750 computer with an operational speed of 600 Kips and 128 K memory: it also has two MIL-STD-1553B data-bus interfaces. The armament interface unit (AIU) includes a stores interface compatible with MIL-STD-1760.

Dimensions: (SMP) 124 × 194 × 250 mm
(AIU) 124 × 194 × 400 mm

Weight: (SMP) 6 kg
(AIU) 10 kg
Power: 230 W max, 28 V dc and 115 V ac 3-phase
Reliability: (designed) better than 900 mtbf

STATUS: in development.

PWRS-1 programmable weapon release system

The PWRS-1 is a microprocessor-based weapon release system which enables the pilot to programme the automatic release of weapons, selecting the required type and quantity and, time interval and order of release. Up to nine weapons stations can be accommodated. The system comprises an electronics unit and control unit. The electronics unit, which can be sited at any convenient location in the aircraft, performs the necessary computations and generates the appropriate electronic commands for weapons release. The type, quantity and spacing of weapons are selected on the cockpit-mounted control unit. The central processor, within the electronics unit, is based on an Intel 8080 microprocessor and can be reprogrammed within an hour to account for differing aircraft performance or weapon ballistics.

Dimensions: (electronics unit) 184 × 185 × 128 mm
(control unit) 146 × 148 × 38 mm
Weight: (electronics unit) 2.6 kg
(control unit) 0.6 kg
Power supply: 15 W at 115 V 400 Hz

STATUS: in production.

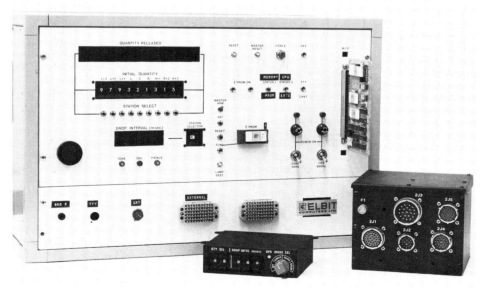

Elbit's stores management system display showing INV (inventory) mode, with store types and positions on aircraft. Pattern indicates that four AIM-9 Sidewinder missiles and four Mk 82 and SN 82 bombs are carried on wing pylons, and ECM pod on fuselage centre-line station

Elbit PWRS-1 control unit (left) and electronics unit, in front of system's test set

SPAIN

CASA

Construcciones Aeronáuticas SA (CASA), Space and Systems Division, Avenida John Lennon, Getafe, Madrid 28008
TELEPHONE: (1) 695 06 00
TELEX: 47937

SCAR armament control system
CASA's SCAR (sistema de control de armamento) armament control system provides full weapon selection and control facilities for up to six underwing carriage stations and a centrally located gun pod. The system's functions include: selection of weapon station for use in a mission; selection of weapon type fitted at a station; mode selection for bomb arming; time interval selection for bomb release in ripple mode; air-to-air and air-to-ground gunnery selection; emergency mode selective jettison; weapon availability and status; stores selected and stores remaining indication and status; gun ammunition counter indication.

Although designed principally for the CASA C-101 strike-trainer, the SCAR system is also suitable for other types of attack aircraft. As well as providing weapon selection and control facilities for bombs, rockets, guns and combined stores dispenser/launcher units, SCAR also supplies signals for control of weapon sighting and gun-camera systems.

The equipment comprises a central processor, usually located in an aircraft's nose bay,

and a main unit, a monitoring unit and an emergency unit, all installed in the cockpit. In the case of trainer aircraft, additional monitoring and emergency units, identical to those placed in the attack aircraft, are mounted in the student's cockpit.

The main unit is used for all control and selection functions, with the exception of emergency jettison for which the emergency unit is used. Monitoring units indicate the status and type of stores available and the quantities of stores/ammunition remaining or expended. In the training role, the functions of the student's portion of the system can be completely inhibited by the instructor pressing a button. An 'error' light indicates a system malfunction or a mis-selection of a store or weapon. In the case of mis-selection, the system, with the exception of emergency release, is automatically rendered inoperative. The emergency unit is hardwired to ensure complete reliability and contains a test-lamp pushbutton for status checking of all indicator lamps in the system.

SCAR is a microprocessor-controlled system employing digital techniques and TTL integrated circuitry. Operational parameters can be modified via software changes and an earom memory ensures that stored data is retained even when the system is unpowered. The system is of modular design and conforms to the recommendations of MIL-STD-1472 and MIL-HDBK-244. It can be readily adapted to different types of aircraft by changing module

input/output interfaces. An interface to a centralised control/display system (WAC, MFD, HUD) has been incorporated via discrete disturbance-free output lines. Compatibility to previous self-supporting system is maximised as the new interface is provided through the central processor connector previously dedicated to the front cockpit monitoring unit, redundant in the new configuration.

The new interface lines provide not only all status information included in the substituted monitoring unit, gun rounds inclusive, but also ripple interval status and certain control capabilities for remote gunnery selection (HOTAS interface). Power requirements are 2.5 A maximum for an illuminated panel system or 1.6 A maximum for a non-illuminated system, both at 28 V dc.

Dimensions: (central processor) 7.08 × 8.26 × 4.72 inches (180 × 210 × 120 mm)
(main unit) 7.87 × 2.75 × 3.97 inches (200 × 70 × 101 mm)
(monitor unit) 3.93 × 1.65 × 2.95 (100 × 45 × 72 mm)
(emergency unit) 2.75 × 1.57 × 3.18 inches (70 × 40 × 81 mm)
Weight: (central processor) 7.26 lb (3.3 kg)
(main unit) 3.56 lb (1.62 kg)
(monitor unit) 1.01 lb (0.46 kg)
(emergency unit) 0.61 lb (0.28 kg)

STATUS: in production and service.

UNITED KINGDOM

Base Ten

Base Ten Systems Limited, 11 Eelmoor Road, Farnborough, Hants GU14 7QN
TELEPHONE: (0252) 517 665
TELEX: 858888
FAX: (0252) 517 841

Stores management systems
Base Ten is a major supplier of stores management and weapon control equipment on Europe's Panavia Tornado multi-role aircraft, responsible for 18 separate electronic sub-assemblies. Its parent company, Base Ten of Trenton, New Jersey, is also heavily involved with this aircraft.

Gun control unit for Tornado
The UK company builds the electronic control units for the Mauser 27 mm cannon in the

Tornados of the three partner air forces, governing rate of fire, auto-recocking and purging. It provides a 'rounds remaining' indication to the pilot, and has comprehensive built-in test equipment.

Sidewinder control unit for Tornado
Base Ten has further developed the standard US system used to control the Sidewinder air-to-air missile, to meet UK requirements for Royal Air Force Tornados. The microprocessor-based missile unit interfaces with the stores management system to provide semi-automatic control of the missile. Features include both normal and off-boresight search and lock, simultaneous missile running, rapid acquisition, automatic selection of optimum missile in the presence of obscuration by the host airframe, and extensive built-in test equipment.

The current UK missile unit is being devel-

oped to enable it to control four missiles. Base Ten's parent company in the USA is providing a new control unit for the operation of four Sidewinders on the US Air Force's A-10s.

Stores management system for UK avionics demonstrator rig
Base Ten built the stores management system for the British Aerospace avionic systems demonstrator rig at Brough, North Humberside. The purpose of this MIL-STD-1553B rig is to test and evaluate advanced systems for performance, compatibility, integrity and reliability in support of future UK involvement in military aircraft. The system in this application provides the interface to a simulated weapon load, but with 'real' inputs and outputs, incorporating the new NATO standard weapons interface specified by MIL-STD-1760/STANAG 3837AA.

Computing Devices

Computing Devices Company Limited, Castleham Road, St Leonards-on-Sea, East Sussex TN38 9NJ
TELEPHONE: (0424) 53481
TELEX: 95568

SMS 2000 series stores management system
The Computing Devices SMS 2000 stores management system controls a complete aircraft inventory of releasable stores and armament including bombs, guided weapons, drop tanks and guns and rockets. It comprises cockpit control panels, which also indicate the status of the aircraft's armament and a digital stores management computer.

The system permits in-flight pre-selection of a weapon delivery package to suit a particular target or conditions. Control facilities are also

provided for selection or release modes (single or multiple stores), ground spacing optimisation in relation to aircraft speed and fuzing. Duplex selective and emergency jettison for the entire stores load is also permitted.

Hang-ups due to battle damage or for any other reason are displayed on the control panel and in such cases the system controls release of the remainder of the stores within centre of gravity and lateral balance limits. An optional serial data interface, conforming to MIL-STD-1553, permits the system to operate in conjunction with a navigation/attack computer. This interface is standard in the latest variant, the SMS 2001, which also incorporates up-to-date, micro-electronics technology.

The SMS 2000 is of modular construction and contains built-in self-test circuitry. For off-aircraft servicing and repair, an automatic test equipment is available. This, initially, enables a fault to be isolated to sub-module or circuit-card level and subsequent determination of the

precise component fault when the suspect module has been removed. The level of integrity of the system ensures that no single failure results in an inadvertent release of stores.

The system is designed for a wide range of light tactical attack aircraft and the SMS 2001/A variant has been selected as standard equipment for export models of the British Aerospace Hawk.

Dimensions: (control panel) 5.5 × 7.1 × 6.3 right, 3.5 inches left (140 × 180 × 160 right/90 mm left)
(stores management computer unit) 8.6 × 5.1 × 12.2 inches (220 × 130 × 310 mm)
Weight: (control panel) 6.16 lb (2.8 kg)
(stores management computer unit) 15.18 lb (6.9 kg)

STATUS: in production and service. An improved version, the SMS 2112, which offers additional features, is now available for specific aircraft applications.

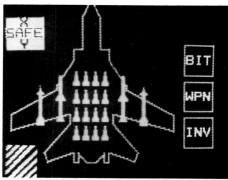

Typical weapons configuration display for Computing Devices in-flight display and control system

Computing Devices SMS 2000 stores management system with British Aerospace Hawk

SMS 3000 series stores management systems

SMS 3000 stores management systems are based largely on the experience gained with the company's SMS 2000 equipment but offer greater flexibility in heavier aircraft. They are intended for use with current and projected weapon systems, are microprocessor-controlled and have an expanded memory to accommodate future requirements.

The SMS 3000 series has a dual mode of operation which has two-fold intent: that of maximising safety by reducing the risk of accidental stores release, and of improving mission reliability. This involves the employment of a dual redundant channel system which operates in serial form to release stores in the safety mode. When in combat, one of two parallel systems can be used to effect release. The systems provide all the facilities of the 2000 series equipment, but have greater inbuilt flexibility. In display terms, for example, the control panels inform the aircrew of hazards such as missile misfire, unsafe height, unsafe spacing etc. There is provision for expansion by means of a range of modular stores interface units to match the increased processor capacity.

STATUS: in production and service.

SMS 4000 series stores management systems

The SMS 4000 series is the most recent addition to the company's range of stores management systems. Like its predecessors, the basic 4000 system is designed to cover a full range of weapon types, but its capability has been extended to take into account new-generation weapons such as the ASRAAM (advanced short-range air-to-air missile) weapon. New technology has also been employed to provide an integrated cockpit display and control suite, the size and position of which may be varied to suit particular aircraft installations. The display sequences have been optimised for specific mission requirements, differing store types, and are of enhanced ergonomic design for the reduction of crew workload.

The display and control system is especially flexible in operation. A colour or monochrome display unit provides the crew with status information including details of stores selected, fuzing, spacing and release modes. A video

insert can be included to show opto-electronic sensor inputs for weapon designation, while simultaneous display of the status of the selected weapon system is presented. The same unit also has a medium-range air-to-air combat display which depicts the location and hit-probability of hostile aircraft. The display unit uses internally-generated graphics, apart from inputs from weapon sensors which are digitally processed prior to presentation, providing a wide range of display and control options. Further automation results from the provision of synthetically-generated voice outputs to the crew and a voice-command recognition system, both of which may be of particular value in air combat.

The SMS 4000 system includes facilities to ensure inter-operability between all NATO aircraft and stores, and systems designed to the latest military standards including the following: (i) MIL-STD-1760 and STANAG 3837AA aircraft-store electrical interface; (ii) MIL-STD-1750A standard airborne computer instruction-set architecture; (iii) MIL-STD-1815A Ada standard high-order language; and (iv) MIL-STD-1553B and DEF/STAN 00-18 wire and fibre-optic data-buses. Maximum use of modular hardware and software design packages give optimum flexibility to accommodate all future aircraft requirements, and simplex or reconfigurable duplex system processing ensures that aircraft mission safety and success requirements are met and helps in minimising the need for support equipment and maintenance crew training.

The stores management processor manages the release for all weapon interfaces at a high level of integrity. Missile initiation, target acquisition and designation, weapon separation and system operational status are all processor-monitored and controlled. The processor also acts as a bus controller for other system equipment and can provide weapon aiming calculations and in-flight missile guidance commands. Conditioning, control and distribution of weapon system equipment and armament power supply is also performed by the processor unit.

Mission data interchange equipment enables appropriate pre-take-off briefing information to be entered into the processor before the mission. This may include an inventory of stores carried, attack profiles, weapon delivery packages, mission-specific software, information on friendly forces deployment, and expected threats. A post-mission de-briefing

facility is also incorporated. This records actual mission sequences together with information on equipment performance and maintenance data to aid rapid post-mission evaluation.

A system would typically comprise up to four electronic boxes: display and control units the processing system and interface equipment. The processor would normally be located in the aircraft's avionics bay and specific interface boxes such as store interface units would be carried close to the stores which they control for maximum integrity. A range of modular interface units permits the host aircraft to carry any single type or a mixture of stores. These are designed to withstand the harsh physical and electrical conditions found in the limited space available in wings, pylons and carriage equipment.

Dimensions: variable according to aircraft type
Weight: variable according to aircraft type

STATUS: under development.

In-flight display and control equipment

Computing Devices is currently developing a range of colour and monochrome display systems with finger-operated panel overlays to serve as an easy-to-operate, high-integrity, aircrew-to-weapon interface. Digitally-processed imagery from sensors associated with either weapons or reconnaissance equipment may be displayed together with internally-generated graphics. The finger-position overlay provides a unique control method. The display sequences may be optimised for specific applications, missions, crew ergonomics and workloads.

In the stores management role, these systems can provide 'attack packages' showing weapon types, quantities and type selected. Fuzing and release details may also be displayed and stored for immediate recall. Selective and combat jettison release packages are readily available for emergency use.

Processed missile sensor information can be shown on an inset display, allowing target location, recognition and hit-probability assessment for air-to-air and air-to-ground situations.

The equipment may also be applied to recognition and reconnaissance roles since the same technology can be used to process and display time images and information from a wide range of electro-optic and other sensors. Such systems could include polar, side-looking and synthetic-aperture radar, downward and forward-looking infra-red sensors, and low light television, permitting positive, rapid friendly/hostile identification. The high resolution video information output, with associated synchronisation pulses and digital data, can also be recorded.

STATUS: under development.

GEC Avionics

GEC Avionics Limited, Airport Works, Rochester, Kent ME1 2XX
TELEPHONE: (0634) 44400
TELEX: 96333

Stores management system for Tornado GR1

The stores management system developed by GEC Avionics for the Panavia Tornado GR1 aircraft can handle 26 types of weapon, controlling arming, fuzing, firing or release functions from its 35 outputs. The system also initiates the timed and sequenced emergency and selective jettison of all weapons and other stores such as fuel and equipment pods. This is a controlled output of 29 pulses over 2.2 seconds. A comprehensive 'bogus' weapon inventory is also included for training purposes.

Among the principal design aims of the Tornado stores management system were: meeting the specified rates of safety, jettison availability, mission success, an unprecedented degree of electromagnetic compatibility hardness and a significant reduction in crew workload. The system was also engineered to provide a minimum of interconnect wiring between the 12 units.

To achieve these targets, the company developed a dual-channel, digital-architecture system employing mini-computers and a digital data transmission system to provide the required integrity. Wherever possible, cmos circuitry is used, on account of its low power consumption, to eliminate the need for cooling and to meet the electromagnetic compatibility specification. Both channels are fully synchronised and perform in a consolidated mode until

channel unserviceability is detected, when authority is then given to the serviceable channel to proceed independently.

The level of redundancy ensures that the channels only close down for safety-critical failures and that the emergency jettison system is not only separated from processor control but made duplex to avoid any consolidation requirements.

Weapon control is split into defensive and offensive weapons. The pilot's control panel takes the defensive weapons and the navigator's control panel has the others. Crew workload is minimised by eliminating from the weapon package selection process both safety critical and non-applicable options.

(Stores management systems for the Tornados operated by the West German and Italian Air Forces are supplied by US electronics company Base Ten.)

STATUS: in production.

Modular stores management system

GEC Avionics continues to develop a family of stores management designs, called MSMS (modular stores management systems). Individual systems are based on a standard set of modules, packaged to satisfy specific customer needs, capable of controlling all existing weapon types, and expansible to meet the evolving requirements of MIL-STD-1760/STANAG 3837AA. From detailed studies carried out on numerous aircraft and weapon types, standard module designs have been rationalised to provide high commonality across a wide range of stores management system configurations. This concept allows MSMS to be easily reconfigured to meet the needs of both new-build and aircraft modernisation programmes.

Typically, the MSMS standard modules provide more than 80 per cent of the hardware in every stores management application. The range of 'pool' modules available cover: general applications such as processor/store functions, 28 V drive and input conditioning functions; Weapon specific such as air-to-air control functions, air-to-surface, and pre-setting functions; peripheral units such as data-bus interface modules, display drive modules; and future development, including MIL-STD-1760 control functions.

The principal features of this new family include proven functional units reducing technical risk and timescales; the practicability of producing tailor-made systems, providing low programme costs; and product support and maintenance commonality between aircraft types.

In conjunction with standard 'pool' hardware modules, the concept has been extended to embrace associated standard software modules. This, together with commonality of test facilities at module level, has a beneficial effect on costs when compared with new special-to-type designs for each application. The MSMS permits complete flexibility in the choice of control interfaces. Either existing aircraft bus interfaced control and display avionics may be used, or dedicated SMS control panels may be provided. The low crew workload features devised for the Panavia Tornado SMS have been continued and developed for use with MSMS.

Unit volumes are of course entirely dependent upon the specific installation, and the standard module sizes have been carefully chosen to allow for installation in a wide variety of unit shapes and sizes.

STATUS: available.

Plessey Avionics

Plessey Avionics, Martin Road, West Leigh, Havant, Hampshire PO9 5DH
TELEPHONE: (0705) 486 391
TELEX: 86227

Weapon control system for Harrier

Until the mid-1960s weapon arming and release systems mainly comprised manually-actuated switches controlling relay networks which completed the stores arming and release/firing sequence. The advent of new combat aircraft capable of accommodating a huge range of stores mounted on a variety of underwing and/or fuselage stations, raised a requirement for stores management methods to meet the control problems caused by asymmetric release of weapons, drop-tanks, or other disposable items. These difficulties were compounded by the diminishing size of airframes while the stores themselves remained largely unchanged, thus increasing their influence.

In the United Kingdom initial research into the problem was undertaken by the Royal Aircraft Establishment, Farnborough which later co-opted Plessey Avionics and Communications to develop an electronic weapon control system. This programme resulted in the system for the British Aerospace Harriers operated by the Royal Air Force, Royal Navy and Indian Navy, for which contracts well in excess of 200 units have been placed.

With the Harrier's weapon load distributed between wing and fuselage pylons, it is important that lateral balance is maintained as weapons are released, particularly during low-level strike attacks. The Plessey weapon control system computes and executes a release sequence which maintains aircraft balance under all conditions.

The system is digital-based and makes extensive use of silicon integrated circuitry. Logic circuits are of diode-transistor type and the number of relays has been reduced to the

Plessey weapon control system installed in British Aerospace Harriers

minimum required to handle the heavy switching-currents which occur at the time of weapon release.

When stores are loaded onto the aircraft, thumbwheel switches are manually set on the weapon control system's control panel to provide the pilot with an indication of the external stores status. During the loading operation the groundcrew also move the role-selection switches on the panel to an appropri-

ate setting for the type of stores carried. Two modes of operation, manual or automatic, are possible. In manual mode, the pilot during run-up to the target chooses, via pylon selector switches, which weapons are to be released. If two stores are carried on one pylon, then use of a single/double selector determines whether one or both stores will be dropped. Selection of fuzing is the next operation and then, if the choice of pylon, store and fuzing is valid for the weapon state and maintenance of lateral balance, a store indicator light will confirm that the selection is correct. Actual release is made by depression of a late-arming switch situated remotely from the weapon control system panel as a safeguard against premature release. The weapon control system prevents store release if the store indicator lamp remains unlit or if the stability of the aircraft will be jeopardised. Release intervals of 20, 40, 80, 160, and 320 milliseconds are selectable from a rotary switch on the weapon control system panel.

For automatic release, the auto/manual switch interfaces the weapon control system with the aircraft's nav/attack computer providing it with data on stores status, types selected and spacing required. Release interval then comes under computer control. In either mode, the pilot may use individual or 'clear aircraft' overriding jettison control switches to jettison stores separately or as a single salvo in an emergency.

During rocket or guided weapon release or during gun firing, the weapon control system ensures that power is supplied to the engine igniter circuits to safeguard against flame-out caused by aerodynamic flow disturbance or gas ingestion.

Built-in test circuits for functional and power supply checking are also controlled from the weapon control system panel. All switches can be operated with gloved hands and vital

controls are protected against inadvertent operation.

In two-seat Harriers a monitor unit in the rear cockpit duplicates the front cockpit jettison facilities and enables the control panel to be monitored during operation.

The system's capacity permits mixed loads to be carried and controlled on each pylon. The range of stores includes single or twin bombs, flares, one or two rocket launchers, two gun pods, external fuel tanks or guided weapons. Bomb fuzing signals can be transmitted to the five pylons.

Dimensions: (weapon control system unit) 4.5 × 8.25 × 12.625 inches (114 × 210 × 321 mm) (monitor) 4.5 × 8.25 × 6.3 inches (114 × 210 × 160 mm)
Weights: (weapon control system unit) 10.25 lb (4.65 kg)
(monitor) 5.25 lb (2.38 kg)

STATUS: in production and service.

Weapon control system for Jaguar

The Plessey weapon control system for the Sepecat Jaguar is a development of that provided for the Harrier (see previous entry) and employs the same digital, solid-state electronics technology. In the single-seat Jaguar the system is divided into two major elements, the main unit housing the control

logic and the cockpit-mounted control unit which contains the indicator lamps, the pilot-operated armament switches and the individual jettison switches. In the two-seat variant, a monitor unit is fitted in the rear cockpit for training purposes.

As well as providing all the functions of the Harrier system, the Jaguar unit has greater reliability as the result of wider use of solid-state logic. It provides fully duplicated release and jettison of individual stores, dispenses with autoselector switches on tandem store carriers, and provides a more advanced method of rocket firing control. It may be operated as an autonomous system or used in conjunction with the Jaguar nav/attack computer. Orders for Jaguar weapon control systems exceed 250 units.

In May 1985 Plessey announced a £1 million contract to supply weapons control panels to equip 31 Jaguars being built by Hindustan Aeronautics for the Indian Air Force.

Dimensions: (control unit) 6 × 5.75 × 6.16 inches (152 × 146 × 157 mm)
(logic unit) 6.125 × 13.75 × 16.75 inches (155 × 349 × 425 mm)
(monitor) 3.375 × 5.75 × 5.125 inches (85 × 146 × 130 mm)
Weights: (control unit) 3.875 lb (1.76 kg)
(logic unit) 32 lb (14.54 kg)
(monitor) 1.25 lb (0.56 kg)

STATUS: in production and service.

UNITED STATES OF AMERICA

Base Ten

Base Ten Systems Inc, PO Box 3151, 1 Electronics Drive, Trenton, New Jersey 08619
TELEPHONE: (609) 586 7010
TELEX: 510 685 5692

Stores management system for Tornado

Base Ten builds the SMS for the IDS interdictor/strike versions of the Panavia Tornado operated by the West German and Italian Air Forces; it also supplies the Sidewinder missile control unit for the former service. Largely as a result of Tornado involvement, the US company has greatly expanded its defence activities, notably with the MIL-STD-1553B weapons system on the private-venture Northrop F-20 fighter.

Base Ten stores management system for Tornado. Left to right: pylon decoder units, pilot's control unit, weapon-processor, observer's control unit, pylon decoder units. Foreground: pylon decoder unit

Bendix

Allied Signal Inc, Bendix Flight Systems Division, Teterboro, New Jersey 07608
TELEPHONE: (201) 288 2000
TELEX: 134414
TWX: 710 990 6136

ARCADS armament control and delivery system

The Bendix ARCADS system has been developed to provide symmetrical-release external stores management for the US Army/Bell AH-1J light attack helicopter. It permits selective programming and controlled release of stores

from between one and four external stations, and can be indexed and programmed for up to 12 separate stores. Control functions comprise stores selection, quantity for release, rate of release, mode of release (single, pair or all) and emergency jettison. The system displays to the flight crew the type and quantity of stores that have been selected; the desired rate of release (slow or fast); the selected mode of release (single, pair or all); stores low quantity; stores empty; and co-pilot override. A complete system comprises seven main units: a cockpit stores control panel, an emergency jettison control unit, an emergency jettison switch, two wing rocket delivery units and two wing adaptor cable assemblies. The units are of modular

construction and registration modules are readily replaceable to allow for changes of weapon complement. The system contains safety interlocks and built-in safeguards to ensure symmetrical release of stores.

Dimensions: (stores control panel) 5.75 × 4.5 × 6.5 inches (146 × 114 × 165 mm)
(emergency jettison select) 1.5 × 5.75 × 5.1 inches (38 × 146 × 129 mm)
(wing rocket delivery) 5.4 × 7.6 × 2 inches (137 × 193 × 51 mm)
Weight: (stores control panel) 5.3 lb (2.4 kg)
(emergency jettison select) 1.2 lb (0.54 kg)
(wing rocket delivery) 5.2 lb (2.36 kg)

STATUS: in production and service.

Boeing

Boeing Military Airplane Company, PO Box 7730, Wichita, Kansas 67277
TELEPHONE: (316) 526 2121
TELEX: 417484
TWX: 910 741 6900

Stores management system for B-52

In early 1984 Boeing Military Airplane Company was awarded a $22.6 million contract by the US Air Force Aeronautical Systems Division for work on the B-52 bomber ICSMS integrated conventional stores management system. The programme will provide uniform physical, electrical and software characteristics for all

future weapons specified for the B-52. Since 1981 Boeing has been working on related programmes; the medium-range air-to-surface missile and the joint tactical missile system.

Flight tests began during 1985 and it is anticipated that 69 B-52G bombers will be equipped with the new weapons and the ICSMS by the late 1980s.

Control Data

Control Data Corporation, 8100 34th Avenue South, Minneapolis, Minnesota 55440
TELEPHONE: (612) 888 5555

Advanced stores management system

In 1982 Control Data Corporation announced the award of a joint US Air Force/Navy contract worth $3.3 million to design and develop an advanced stores management system for fixed-wing aircraft. The contract called for a 12-month definition phase, followed by an 18-month development phase during which an advanced development model was built for a 7-month evaluation at Eglin Air Force Base, Florida beginning in mid-1985.

The programme was to be executed jointly by Control Data's Minneapolis-based Government Systems Division and by its British-based component, Computing Devices Company of St Leonards-on-Sea. The latter has developed a number of advanced stores management systems (see entry under the UK). It is likely that technology derived from these will be incorporated in the new US Air Force/Navy system.

STATUS: under development.

Fairchild

Fairchild Communications and Electronics Company, 20301 Century Boulevard, Germantown, Maryland 20874
TELEPHONE: (301) 428 6000
TELEX: 892468

AN/ASQ-165 ACIS armament control indicator set

The Fairchild AN/ASQ-165 armament control indicator set (ACIS) is an airborne ordnance stores management system developed for the US Navy/Sikorsky SH-60B Lamps III helicopter. The system controls one or two Mk 46 Mod 0/1/2 torpedoes, one or two Practice Multiple Bomb Racks with Sound Underwater Sources and up to 25 air-launched sonobuoys. Interfaces for control functions and for the ACIS microprocessor control system are provided for two BRU-14A bomb racks, the sonobuoy launcher, and a MIL-STD-1553 data-bus which interfaces in turn with an AN/AYK-14 airborne computer.

The system consists of a C-10488/ASQ-165 armament control indicator (ACI) for the Airborne Tactical operator or the helicopter pilot, and a CV-3531/ASQ-165 armament signal data converter (ACDC) which contains logic, interlock, driver and relay circuits. The microprocessor controls all tasks such as inventory usage, functional status and built-in self-test, with the exception of the jettison function, which is hard-wired. Jettison functions are redundant and the circuitry ensures that no single failure will cause inadvertent release or inability to jettison. Communication with all parts of the system apart from the jettison function is provided by the data-bus.

The ACI control panel functions include: master arm voltage switching for positive armament safety; torpedo selection, arming and launch; torpedo search depth, mode, ceiling and course programming; torpedo program status indicators (three); sonobuoy selection, arming and launch; and jettison left store, right store, all sonobuoys or all ordnance.

The AN/ASQ-165 precludes unintentional launching from single-point failure situations. The system features low-power Schottky TTL 54-LS series logic and tri-state, sunlight-readable switch/indicators. With the exception of some launch and jettison circuits, the self-test facility is quoted as being able to detect more than 98 per cent of faults. The system is compatible with automatic test equipment at both unit and module level. Failure rates for the ACI and the ASDC are stated as better than one in 25 700 hours and 3850 hours respectively.

Dimensions: (ACI) 8.63 × 5.75 × 6.69 inches (219 × 146 × 170 mm)
(ASDC) 7.62 × 5.88 × 18 inches (193 × 149 × 457 mm)

Weight: (ACI) 4.5 lb (2.04 kg)
(ASDC) 23 lb (10.45 kg)

STATUS: in production and service.

Armament control system for A-10A

Fairchild's armament control system (ACS) provides the pilot with stores status information and programmed control over the operation and release modes of all of them. The ACS comprises three separate types of line replaceable unit. The armament control panel (ACP) permits the pilot to communicate with the system, the interstation control unit (ICU) carries out signal processing, and the station control units (SCUs) act as the interfaces between the system and the weapon stations. A complete system consists of one control panel, one interstation and 11 station control units (seven type A, four type B).

The ACS allows the pilot to select the weapon stores at any of the 11 weapon stations at any time. The GAU-8A gun and camera may be selected for operation, together with several types of weapon fuzing option, by means of a switch. Single pairs, ripple-single and ripple-pairs are the possible weapon release mode choices. Selective rack and missile jettison are the choices available with the emergency jettison control. A built-in timing and sequencing system generates the appropriate time intervals between releases, ensuring safe release intervals for weapons and the appropriate weapon station priorities. Video signal channelling assists in the release of electro-optical weapon stores. A front panel displays the current status of each of the eleven stations and the quantity of ammunition remaining in the GAU-8A gun. The system can operate over a temperature range from –55 to +71° C. Microprocessor-controlled test equipment is used for flight line and intermediate level support.

Dimensions: (ICU) 9.5 × 15.3 × 8.5 inches (241 × 388 × 216 mm)
(ACP) 6.5 × 9.25 × 8 inches (165 × 235 x 192 mm)
(SCU types A and B) 3.75 × 5.6 × 2.8 inches (95 × 142 × 71 mm).
Weight: (ICU) 32.35 lb (14.7 kg)
(ACP) 10.75 lb (4.88 kg)
(SCU type A) 2.2 lb (1 kg)
(SCU type B) 1.4 lb (0.63 kg)

STATUS: in production and service.

AN/AWG-15 armament control system for F-14A

The Fairchild AN/AWG-15 armament control system (ACS) was designed and developed for weapon stores control to be implemented from a control indicator by an aircrew member or in conjunction with an airborne missile control

system (AMCS) computer. The system comprises a control indicator (CI), a power switching unit (PSU), and a set of command signal decoders (SCDs), one of which is located at each weapon station.

The ACS provides the preparation and control functions before releasing Phoenix, Sparrow and Sidewinder missiles or bombs, rockets and flares. It also controls and fires the M61-A1 gun.

For air-to-air functions, the system provides missile options, control/speed gate selection, inventory/ready control, weapon/station control and release by the pilot or other aircrew member. For air-to-ground functions, using bombs, rockets or flares, it provides weapon and station selection, inventory/ready control, mode selection (for example attack, delivery, fuzing and weapon mix) and release by the pilot. It also provides jettison command functions. Protective logic circuitry is incorporated to prevent inadvertent weapon firing or release.

The system has a high level of inbuilt redundancy with built-in test for 90 per cent fault detection and automatic isolation of faulty circuits, sub-units or components.

Dimensions: (CI) 12.5 × 11.5 × 10 inches (318 × 292 × 254 mm)
(PSU) 8 × 7 × 16.5 inches (203 × 178 × 419 mm)
(SCDA) 2.5 × 4 × 8 inches (64 × 102 × 203 mm)
(SCDB) 2 × 4 × 6 inches (51 × 102 × 152 mm)
Weight: (CI) 13.5 lb (6.13 kg)
(PSU) 20 lb (9.09 kg)
(SCDA) 1.88 lb (0.85 kg)
(SCDB) 1.1 lb (0.5 kg)

STATUS: in production and service with US Navy/Grumman F-14A Tomcat deck-landing fighter.

AN/AWG-15 technology improvement programme

In May 1985 Fairchild was awarded a full-scale development contract by Grumman for the AN/AWG-15 technology improvement programme (TIP).

It is expected that all the US Navy's F-14As will be modified, the total cost being estimated at $15 million. The TIP increases reliability and maintainability of the system, and will permit the carriage of new weapons on the aircraft.

STATUS: in development.

Lear Siegler

Lear Siegler Inc, Avionic Systems Division, 7-11 Vreeland Road, Florham Park, New Jersey 07932-0760
TELEPHONE: (201) 822 1300
TELEX: 136521
TWX: 710 986 8504

AN/AYQ-9(V) stores management system for F/A-18A

The AYQ-9(V) stores management system provides fully computerised control of stores release to optimise accuracy and minimise crew workload. Although available for a variety of attack aircraft, it is standard equipment on the US Navy/McDonnell Douglas F/A-18 aircraft, and is claimed to be the most advanced

system of its kind. Functions in the F/A-18 include maintaining an inventory of stores types, locations, quantity, status and special conditions (for example, hung store or locked pylon) for correct sequencing and display on cockpit cathode ray tube; preparation and activation of store and store-release mechanism, and maintenance of correct aircraft balance during release sequence; deactivation of

AN/AYQ-9(V) stores management system which equips McDonnell Douglas F/A-18

Lear Siegler helicopter integrated armament management system

stores after completion of attacks; and self-testing to diagnose failures in the system.

Thirty-three types of weapon on the F/A-18 can be controlled by the AYQ-9(V) including Sidewinder, Sparrow, Maverick, Harpoon and HARM missiles, several types of bomb (nuclear, freefall, retarded and 'smart'), forward-looking infra-red cameras and the aircraft's guns. Eleven types of mine can also be carried.

The stores management system consists of the processor and control units, together with decoders at each of nine weapons stations. It interfaces with the AN/AYK-14 mission computer via a MIL-STD-1553B digital data-bus, and a dedicated low-power bus connects the processor with the various decoders.

The AN/AYQ-9(V) provides for the operational readiness assessment of each store and weapons station, prepares the suspension and release equipment (including power-up, alignment and fuzing) and activates the stores sequencing, arming and release, including the off-boresight guidance of Sparrow and Sidewinder missiles.

The system also provides a stores inventory facility, by type, number, location, etc, to permit correct release and also for the display of requisite data to the pilot. There is extensive built-in test.

STATUS: in production for F/A-18.

IAMS integrated armament management system for helicopters

The Lear Siegler IAMS integrated armament management system is designed to bring advanced stores management capability to modern attack helicopters such as the Hughes AH-64 Apache and Sikorsky UH-60A Blackhawk. A simplified version of the stores management system developed for the F/A-18 (see above), it is also suitable for fixed-wing light-attack aircraft and for new projects or retrofits. The cockpit control and display panel incorporates the weapon release computer, while the high-current switching and weapons interfacing is performed in a stations interface unit mounted in the aft fuselage. IAMS has demonstrated compatibility with 35 types of weapon, including air-to-air and anti-radar missiles, gun and rocket pods, flare and sonobuoy dispensers and rack ejectable stores such as bombs, mines and torpedoes. The IAMS software allows this list to be modified to suit any particular requirement. For pilot training purposes realistic effects and displays can be produced without the need to carry stores.

During flight the pilot can check the stores inventory or change the weapon/fuze type or rocket fuze range for the selected attack programme. Any one of four preset programmes can be selected by a button on the pilot's control stick for each type of air-to-ground weapon carried. A rocket automatic programmes sequence (RAPS) facility allows the pilot to select three rocket attack programmes and have them automaticaly executed with one press of the firing button.

IAMS has extensive built-in test and a pilot training facility for simulation of weapons release.

Testing began in early 1985 aboard a Sikorsky H-76 helicopter. Sikorsky and Lear Siegler are co-operating on the marketing of the system.

STATUS: development completed. Flight-tested on Sikorsky H-76.

AN/AYQ-13 stores management system for AV-8B

The AN/AYQ-13 stores management system is a development of the AYQ-9 system (see separate entry) optimised for the US Marine Corps/McDonnell Douglas AV-8B and the Royal Air Force/British Aerospace Harrier GR5 aircraft.

As well as the primary functions for the AYQ-9 system described above, the AYQ-13 interfaces with and controls a 25 mm gun and provides for the control of stores at seven stations. Additionally, the system allows for the emergency jettison of all, or a selection of the stores being carried.

For the AV-8B the AYQ-13 is compatible with 38 types of store, including Maverick and Sidewinder missiles, laser-guided, cluster, practice, anti-tank, fire and conventional bombs, rocket launchers, decoy dispensers, and the aircraft's gun and external fuel tanks.

Additionally the AYQ-13 maintains an inventory of stores carried by type, location and quantity: it also takes account of special conditions such as minimum release interval, hung store, weapon gone and rack locked. This is to provide correct release sequencing and the required data for the cockpit display. The system also controls the preparation of the suspension and release equipment, power-up, alignment and fuzing and the dynamic power management of the SMS: it also provides for instant deactivation of stores on landing or when a change of circumstances in-flight so demand. The AN/AYQ-13 has a comprehensive built-in test facility and a pilot training mode for simulation of weapons release. The system comprises nine weapon replaceable assemblies; seven station control units; an armament control panel; and a stores management processor.

STATUS: in production.

Radio communications

Over the past few years most manufacturers of airborne radio equipment have adopted light-emitting diode displays for frequency, channel number, etc. This Cessna 210 is equipped with twin nav-com systems and a simple flight management system with leds, although the transponder retains conventional mechanical numbers. The aircraft also features the latest Bendix colour-display weather radar

CANADA

Aero Enterprises

Associated Aero Enterprises Limited, Division of Aircom Electronics Limited, 9501 Ryan Avenue, Dorval, Montreal, Quebec H9P 1A2
TELEPHONE: (514) 636 3874
TELEX: 05-822663

Ssb 10/100 and ssb 10/100 H hf radios

The ssb 10/100 is a single unit, lightweight all-solid-state hf radio for civil aircraft. The ssb 10/100 H is a similar radio for helicopters. Each has a 100 watts pep and operates in the simplex mode.

The radio unit plugs directly into instrument or console panels permitting quick installation. It weighs 2.72 kg and covers the band from 2 to 9 MHz with ten fixed channels. The power requirement is 8 A at 24 V dc.

The ssb 10/100 belongs to a family of radio equipment designed particularly for use in remote areas of Canada. The range includes 60-watt and duplex versions, matching antennas, and marine, ferry and land mobile radios.

Dimensions: 290 × 160 × 60 mm

STATUS: in production and service.

Canadian Marconi

Canadian Marconi, Avionics Division, 415 Legget Drive, PO Box 13330, Kanata, Ontario K2K 2B2
TELEPHONE: (613) 592 6500
TELEX: 05-34805

CMA-872 and CMA-873 voice message systems

The Canadian Marconi voice message systems are single-box devices using advanced microprocessor and memory technology to store and reproduce high-quality, digitally-produced speech messages.

The CMA-872 VMS can store 40 three-word messages or about 30 seconds of continuous speech. The smaller CMA-873 Micro VMS can store 15 three-word messages or 13 seconds of continuous speech. Each system incorporates extensive self-test facilities. High quality speech reproduction is ensured by using a 30 MHz

Canadian Marconi CMA-873 Micro VMS

digitising sampling rate. Any type of voice or language can be stored and reproduced. Each message can be preceded by an alerting tone and repeated as often as necessary. The messages can be sorted in order of priority so that, in the event of coincidental inputs, the most important message is reproduced first.

Canadian Marconi CMA-872 VMS

Dimensions: (CMA-872) 146 × 114 × 121 mm (CMA-873) 64 × 75 × 67 mm
Weight: (CMA-872) 1.8 kg (CMA-873) 0.45 kg
Power: (CMA-872) 14 W at 28 V dc (CMA-873) 2.5 W at 28 V dc

STATUS: in production.

Garrett

Garrett Manufacturing Limited, 255 Attwell Drive, Rexdale, Ontario M9W 5BB
TELEPHONE: (416) 675 1411
TELEX: 06-989142
Garrett manufactures aircrew personal locator beacons and survival radio sets, as well as transceivers for use at air radio facilities ground stations.

AN/PRQ-501 personal locator beacon

The AN/PRQ-501 is a lightweight emergency beacon/transceiver which serves the purposes of a radio beacon on which search and rescue forces can home onto downed aircrew, and for voice communication between the two parties.

It operates on 243 MHz in the beacon mode, and 243 and 282.2 MHz for both transmit and receive voice, with a maximum output power of 325 mW in the beacon mode and 200 mW on voice.

Dimensions: 149 × 79 × 38 mm
Weight: approx 0.625 kg

STATUS: in operational service with the Canadian Armed Forces.

Survival radio set

The Survival Radio Set (SRS) is a lightweight emergency beacon/transceiver which functions as both a radio beacon on which search and rescue forces can home onto and locate downed aircrew, and for voice communications

between the two parties. The SRS is activated automatically during ejection, by means of an acceleration switch, or can be manually activated. It is normally carried in a pocket of the aircrew's lifevest. Operational transmit frequencies in both the beacon and voice modes are 121.5 and 243 MHz, reception is a single channel at 243 MHz. Transmitter output power is 500 mW on 121.5 MHz and 200 mW on 243 MHz.

Dimensions: 150 × 75 × 50 mm
Weight: 0.5 kg including lithium battery pack

STATUS: in operational service with the Royal Swedish Air Force.

FRANCE

Electronique Aérospatiale

Electronique Aérospatiale, BP 51-93350, Le Bourget Principal
TELEPHONE: 48 62 51 51
TELEX: 220809

BCC 2125, BCH 2725 and BCT 2525 control units

These are panel-mounted digital control units for remote control of electronic equipment on commercial aircraft. The BCC 2125, BCH 2725 and BCT 2525 are for vhf and hf communication and ATC respectively. They are based around a number of printed circuit boards that are common throughout the range. Other units are also available for VOR/DME, VOR/ILS and ADF control.

The display is by non-emissive dichromic liquid-crystal displays, white on black background. The units conform to ARINC 720, 306, 716 and 429.

Dimensions: 146 × 66.6 × 140 mm
Weight: (BCC 2125) 1.08 kg (BCH 2725) 1.03 kg (BCT 2525) 1.36 kg

STATUS: in production and service.

RMP 850 radio management panel

The RMP 850 is a digital radio management panel which allows direct frequency control of all radio communications and radio navigation equipment on board. It is designed for commercial aircraft applications and the installation normally includes two or three RMP 850s. Interconnection between the panels and the radio equipment allows any set to be tuned from any RMP. The display is by non-emissive dichromic liquid crystal displays, white on black background. The panel consists of four printed circuit boards, one power supply, one input/output, one processing, and one for interface to the ARINC 429 bus. Frequency ranges covered are:

RMP 850 radio management panel

hf – 2 to 24 MHz with 1 kHz spacing
vhf – 118 to 136.975 MHz with 25 kHz spacing
VOR – 108 to 117.95A MHz with 50 kHz spacing
ILS – 108 to 111.95 MHz with 50kHz spacing
adf – 190 to 1750 kHz with 0.5 kHz spacing

Dimensions: 146 × 85.72 × 140 mm
Weight: 1.5 kg max

ERC 740 series and ERC 741 vhf radios

Electronique Aérospatiale's ERC 740 series is a family of vhf transmitter-receivers which provide air-to-air and air-to-ground communication for military aircraft. This family, which uses the same basic systems technology and common components across the range, comprises three major transmitter-receiver units, ERC 740, ERC 740E and ERC 740EF, and a range of remote controllers. The latter can be matched with the transmitter-receivers to build up any particular installation to customer requirements.

The ERC 740, which is an AM-only variant, covers the frequency band 116 to 150 MHz in which it provides 1360 channels at intervals of 25 kHz. Another AM-only transmitter-receiver, the ERC 740E covers a wider frequency range, from 100 to 150 MHz, in which it provides 2000 channels, also at 25 kHz increments.

The most comprehensive member of the family is the ERC 740EF, an AM/FM unit covering the 100 to 160 MHz band in which 2400 channels, again at 25 kHz spacing, are provided. Nominal transmitter output power is 20 watts but a selectable low power output of 5 watts is also available.

The ERC 741 vhf radio is an AM system which conforms to the same general specification as the ERC 740 equipment but also provides an azimuth homing capability.

Developed as a private venture in the early 1970s, the ERC 740 radios entered production later in the decade and have been ordered in quantity by the French Armed Forces. The ERC 740EF system has also been ordered by the Finnish Air Force to meet its requirement for vhf communication facilities on its British Aerospace Hawk aircraft. Initially these radios were supplied directly by Electronique Aérospatiale, but later-delivered Hawks will be equipped with Finnish-assembled radios manufactured under a licence agreement.

Dimensions: 319 × 128 × 100 mm

STATUS: in production and in service.

Electronique Aérospatiale TVU 740 transmitter-receiver

RTM 10 vhf/uhf transceiver

The RTM 10 is an airborne vhf/uhf transceiver for air-to-ground and air-to-air communications. It covers the frequency range 100 or 118 to 155.975 MHz for vhf and 225 to 399.975 MHz for uhf, with a guard frequency of 243 MHz. Channel spacing is 25 kHz giving 1520 or 2240 channels for vhf, and 7000 channels for uhf. RF output power is 20 watts, modulated 85 per cent, with a 5-watt reduced power output capability if required.

RTM 10 vhf/uhf transceivers

The transceiver incorporates six plug-in modules; transmitter, main receiver, guard receiver and varicap, audio interface, synthesiser and chassis/power supply. The equipment has been designed for ease of retrofit and various adaptors allow its installation without removal of existing aircraft wiring. The RTM 10 conforms to ARINC 404 and 429, MIL-STD 461B/462B, Air 7304, and Air 2021E requirements.

The RTM 10 is normally remotely-controlled by a BCU 750 control unit or by a centralised command unit. The BCU 750 provides manual frequency selection and 100 preset channels operation.

STATUS: in production.

RTV 10 vhf transceiver

The RTV 10 is a version of the RTM 10 for use in the vhf range only. It covers the range 100 or 118 to 155.975 MHz at 25 kHz spacing, with a guard frequency of 121.5 MHz. All other details are as for the RTM 10.

STATUS: in production.

RTU 10 uhf transceiver

The RTU 10 is a version of the RTM 10 for use in the uhf range only. It covers the frequency range 225 to 400 MHz at 25 kHz spacing, with a guard frequency 243 MHz. All other details are as for the RTM 10.

STATUS: in production.

TR 800R vhf transceiver

The TR 800R is a remote control vhf transceiver developed by Electronique Aérospatiale as one of its new Migrator series of systems. It is intended primarily for use in business aviation, more particularly in twins, trijets and helicopters. The transceiver is rack-mounted and covers the frequency range 118 to 135.975 MHz in 25 kHz steps giving 720 channels. This range can be extended to 143.975 MHz, giving 1040 channels, or from 116 to 149.975 MHz (1360 channels) for use on military trainers. For military use the equipment is known as the TR 800RM. It can be remotely controlled by any control unit operating in ARINC 410 two-out-of-five code, but is offered by the manufacturer with either control unit BC 600 or BCC 600, both forming part of the Migrator series.

The TR 800R employs a frequency synthesiser with single digital phase-lock loop, driven by a single crystal. The use of a modulation compressor on the micro input and a compressor on the audio frequency output enables a constant audio level to be maintained whatever the signal received by the antenna, or the percentage of modulation. An automatic squelch system takes the interference level of

the receiver input into account. A fully transistorised power amplifier is incorporated and will withstand all standing wave ratio conditions and all phase relations occurring in the antenna circuit. Output power of the equipment is 20 watts nominal.

The TR 800R has been approved in Eurocae Category 1 by STNA allowing its use in commuter airline aircraft. It also meets FAA TSO C37b and C38b requirements.

Dimensions: 276 × 135 × 62 mm
Weight: 2.5 kg

ECMRITS TRM 900 and TRU 900 transceivers

ECMRITS is a military communication system designed for high performance aircraft. The system consists of two transmitter/receivers and their associated remote control units, and one secure communication circuit. Voice and data link communications are made in the uhf and vhf frequency bands. Communication security is provided by one or more of the following ECCM modes; voice or data encryption, frequency hopping, spread spectrum.

The system architecture use of several data interfaces between units allows very flexible operation of the two transmitter/receivers and provides back-up capabilities in the event of failure. The system interfaces the aircraft high speed data-bus (DIGIBUS or MIL-STD 1553B).

The TRM 900 is the vhf/uhf transceiver and is compatible with communication protection systems operating in voice or data link modes using encryption, frequency hopping and spread spectrum techniques. Frequency ranges covered are 118 to 149.975 MHz on vhf, and 225 to 399.975 MHz on uhf, with channel spacing of 25 kHz giving 1280 vhf and 7000 uhf channels. The transceiver incorporates seven modules; transmitter, main receiver, guard receiver, AM modulator, FM modulator, synthesiser, power supply/chassis. It is controlled by a programmable digital control unit Type BCU 750, or a centralised command unit. Operating modes cover analogue voice communication (A2-A3), ciphered digitised voice communication (A9), data link (F9), ciphered or unciphered digitised voice in frequency hop mode. Optional modes include homing reception and F3 transmission/reception.

The TRU 900 is the uhf transceiver, details of which are similar to the TRM 900. It covers the frequency range from 225 to 399.975 MHz, giving 7000 channels at 25 kHz spacing. Optionally the range at the top end can be extended to 409.975 MHz. A guard frequency at 243 MHz can be provided. The TRU 900 is controlled by a programmable digital control unit Type BCU 750/753, or by a centralised control panel.

STATUS: in production and in service.

TVU 740 and 741 vhf/uhf radios

The Electronique Aérospatiale TVU 740 and 741 radios are combined vhf/uhf systems for air-to-air and air-to-ground communication in high performance fixed-wing aircraft and military helicopters. Both systems are essentially similar in design and construction and each provides 7000 channels in the uhf band spanning the range 225 to 400 MHz. In vhf, however, the TVU 740 covers the band 116 to 150 MHz, in which 1360 channels are provided, but the TVU 741 offers a selection of 1960 channels in the wider frequency range of 100 to 150 MHz. In all cases, the channels are spaced at a 25 kHz interval and AM modulation is standard.

Frequencies are generated by synthesis techniques and transmitted power output is selectable at either 5 or 20 watts. Azimuth homing facilities are provided in both systems.

Both the TVU 740 and 741 systems are among Electronique Aérospatiale's earlier range of synthesised airborne radios and the

Electronique Aérospatiale ERC 741 transmitter-receiver

BCU 750

BCU 753

BFA 900

TVU 740 has been extensively employed in French military service, particularly aboard Aérospatiale Gazelle, Puma and Super Frelon helicopters and earlier versions of Dassault Mirage series fighter aircraft. It has also been selected for service by a number of non-French military arms, one notable export application being for the Yugoslav Air Force, which uses the TVU 740 as the standard communication system in its Soko Galeb and Jastreb strike-trainer aircraft. Developed in the late 1960s and entering service during the following decade, the TVU 740 and 741 systems have become widely accepted in a number of military roles. They are no longer in quantity production, having been largely superseded by other equipment later developed by the same company.

STATUS: in service.

TRU 750 uhf radio

The Electronique Aérospatiale TRU 750 is a uhf airborne communication system covering the frequency band 225 to 400 MHz, in which it provides 7000 AM channels, up to 26 of which may be pre-selected, at incremental spacings of 25 kHz. The system, which is remotely controlled, can also incorporate an independent guard receiver which covers the band 238 to 248 MHz but which is tunable at intervals of 50 kHz. Transmitter power output is selectable at either 5 or 20 watts nominal. Frequencies are synthetically generated.

The system commenced development in the mid-1970s, entering production later in that decade. It is intended as a replacement for earlier quantity-produced equipment for which a specific retrofit market is envisaged.

STATUS: in production.

TRU 900

KFA 900

TRM 900

Units of ECMRITS military communications system

Electronique Aérospatiale ERC 740E transmitter-receiver

LMT

LMT Radio Professionelle, 46 quai Alphonse Le Gallo, BP 402, 92103 Boulogne Billancourt Cedex
TELEPHONE: (1) 46 08 60 00
TELEX: 202900

LMT is a subsidiary of Thomson-CSF

3520 hf/ssb radio

The 3520 is an hf/ssb system which has been designed for use in a variety of military and commercial applications, both fixed-wing and helicopters. It is a lightweight equipment consisting of the 3520A transceiver, 3520B remote control, 3596A antenna coupler and a 3520C adaptor for retrofit purposes. The hf band is covered from 2 to 29.999 MHz, providing 280 000 channels at 100 MHz.

Operating modes are usb, ame, cw and data. The transceiver is contained in a 6 MCU box and meets the requirements defined by ARINC 719. The system complies with ARINC 600, and may be associated with any control unit with a serial data output for remote control, in accordance with ARINC 429. It can be operated using single or dual controls and a digital pre-selector, Type 3597A, can be supplied for dual installation applications.

The transceiver is a latest technology, fully-solid-state unit, and includes a 200- or 400-watt amplifier. It is compatible with all antenna couplers. A comprehensive fault identification (BITE) increases maintenance efficiency and the modular construction minimises repair time by fast access to all circuit modules.

Frequency stability: $\pm 1 \times 10^6$
Mtbf: 1500 h in military airborne environment
Power supply: 115 V 400 Hz, 3-phase, 600 VA in transmit.
Optional 28 V dc with PA, 200 W pep

STATUS: in production and in service since 1984 in the French Air Force for retrofit and new fixed-wing and helicopter installations. The system is designed for use in place of the 618-T.

3527 hf/ssb radio

The 3527 is an hf/ssb radio system designed for airborne use. It has been developed primarily for AEW applications (French Navy ATL2 programme).

The communication system comprises the 3527F transceiver, 3527H remote control, 3596A antenna coupler, 3597A digital pre-selector and the 3598A fsk modem. The pre-selector filter is used to provide simultaneous operation of two 3527H units on the same aircraft, one for transmission and the other for reception.

The fully solid-state transmitter supplies a modulated peak power of 400 watts. The transceiver is driven by a digital synthesiser tuned for setting to 280 000 channels spaced at 100 Hz intervals. The antenna coupler accommodates wire antennas from 10 to 30 metres long. The fsk modem handles messages of 50 to 100 bauds according to a five-moment code.

A comprehensive fault identification (BITE) increases maintenance efficiency and the modular construction minimises repair time by fast access to all circuit modules.

Modes: usb, data (usb), cw, Link 11
Frequency stability: $\pm 1 \times 10^6$
Power output: 400 W pep; 200 W average (ssb)
Power supply: 115 V 400 Hz, 3-phase, 100 VA transmit
Operational specification: STANAG OTAN 5035, AIR 7304

LMT 3520 hf/ssb radio

LMT 3527 hf/ssb radio

STATUS: in production for the French Navy.

Omera-Segid

Omera-Segid, 49 rue Ferdinand Berthoud, 95101 Argenteuil
TELEPHONE: (1) 39 47 09 42
TELEX: 696797

TR-AP-21A uhf radio

The TR-AP-21A is a military uhf transmitter-receiver providing 3500 channels, spaced at 50 kHz, within the frequency range 225 to 400 MHz. It is designed for service with high performance aircraft for air-to-air and air-to-ground communication and will operate reliably while subjected to accelerations of ± 10 g along lateral, longitudinal or vertical axes. Maximum operational altitude is 66 000 feet (20 000 metres) and the limiting temperature extremes are –40 and +70°C, although the latter limitation may be overriden for short periods of operation.

The transmitter output power is from 3 to 5 watts according to the frequency selected and the system may be operated in A2 or A3 mode. The system comprises the ER T6A transmitter-receiver, a BA 220A power supply converter, an SK-63A power supply unit and a remote controller.

The TR-AP-21A first entered service with the French Air Force in the late 1950s, notably aboard Dassault Mirage aircraft, but is no longer in production.

Dimensions: (transmitter-receiver) 420 × 160 × 184 mm
(control unit) 182 × 146 × 57 mm

Weight: (transmitter-receiver) 11.9 kg
(control unit) 1 kg

STATUS: in service.

TR-AP-22A uhf radio

This system conforms largely in general technical specification to Omera-Segid's TR-AP-21A unit but has a number of additional facilities.

These facilities include data transmission and reception capability, an additional built-in guard receiver section covering the frequency band 238 to 248 MHz (which remains operable irrespective of whichever channel of the main unit's 3500 frequencies are in use), and a re-broadcast facility. Additionally, the TR-AP-22A has a pre-selection facility for up to 26 channels on the main transmitter-receiver system. Transmitter power output is from 10 to 15 watts according to the frequency selected, and the range slightly exceeds that of the TR-AP-21A. A data-receive rate of up to 5000 baud is possible but the guard receiver is inhibited when the data link is in operation.

Like the TR-AP-21A, the TR-AP-22A first entered French Air Force service during the late 1950s, again principally in Dassault Mirage aircraft, and is no longer in production.

Dimensions: (transmitter-receiver) 550 × 264 × 230 mm
(control unit) 179 × 146 × 76 mm
Weight: (transmitter-receiver) 25.3 kg
(control unit) 1.2 kg

STATUS: in service.

TR-AP-28A vhf radio

Omera-Segid's TR-AP-28A is a military vhf system designed to provide air-to-air and air-to-ground communication for high-performance aircraft. It covers the frequency band from 100 to 156 MHz, in which it provides 1121 channels at increments of 50 kHz. In keeping with its high-performance role, it maintains reliable operation at accelerations of ± 10 g in lateral, longitudinal and vertical axes at altitudes of up to 66 000 feet (20 000 metres) and over a temperature range of –40 to +70°C, and at higher temperatures for short periods.

It operates in A2 and A3 modes and has a transmitter output power of 3 to 5 watts. A complete system would comprise an ER78A transmitter-receiver, controller, BA-220A power supply converter, and SK-63A power supply unit.

Like other Omera-Segid TR-AP-20 equipment, the TR-AP-28A first entered service in the late 1950s, again notably in Dassault Mirage aircraft, and production has ceased.

Dimensions: (transmitter-receiver) 420 × 160 × 184 mm
(control unit) 186 × 146 × 123 mm
Weight: (transmitter-receiver) 11.9 kg
(control unit) 1.7 kg

STATUS: in service.

Sintra

Sintra, 26 rue Malakoff, 92602 Asnières
TELEPHONE: (1) 790 65 72
TELEX: 610718

This company is a subsidiary of the Thomson-CSF Group

Saram 7-82 uhf radio

The Saram 7-82 is a uhf radio communications system for high-performance aircraft which covers 7000 channels in the band 225 to 400 MHz. It provides in the AM mode A2, radio-telephony and radio-telegraphy A3, A9 NRZ and diphase, ADF homing and beacon control transmission. In the FM mode it provides F9 NRZ and diphase, F1, L4 and L11. The system incorporates a guard receiver which can monitor any fixed frequency selected between 238 and 248 MHz.

A remotely-controlled system, the 7-82 is of solid-state, modular construction and employs digital synthesis frequency-generation tech-niques. It is a member of a family of airborne military radio systems produced by Sintra, modules of which are interchangeable within the family. For example, the synthesiser section of the 7-82 also generates vhf frequencies and may be used directly in either uhf or vhf systems.

The 7-82 is said to be particularly suited to co-located operation. It produces a low level of wide-band noise in transmission mode and, in receive mode, has no channel inhibition, irrespective of whatever frequency combination may be selected among the receivers.

In-flight, self-test facilities for the transmitter and receiver sections are incorporated.

STATUS: in production and service.

BDT 100 personal locator beacon

The Sintra BDT 100 is a personal locator beacon for use in emergency situations, enabling survivors to signal their position to a rescue aircraft and also acting as a transceiver to provide two-way communications between survivors and a rescue team.

As a beacon, the equipment transmits auto-matic swept tone radio signals alternately on the two international distress frequencies of 121.5 and 243 MHz. The continuous trans-mission cycle lasts 1.8 seconds and consists of a 600-millisecond signal on 121.5 MHz, the same on 243 MHz and a 600-millisecond pause. As a transceiver it provides communications on the same channels, or on an auxiliary channel at 282.8 MHz. The range of the equipment is a minimum of 110 km in the beacon mode and 18 km when used in the communications role.

The equipment comprises a transmitter-receiver as the main element and a power supply box containing long-life lithium bat-teries. Four different power supply boxes are available with endurances of 24, 48, 72 or 96 hours. The BDT 100 is waterproof down to a depth of 10 metres.

STATUS: in production.

Socrat (Nouvelle Société)

Société de Constructions Radiotéléphoniques (Socrat), 2 and 4 rue Kuss, Paris Cedex 13
TELEPHONE: (1) 45 89 89 58

TR-AP-138A (4600-E) vhf/AM radio

Socrat's 4600-E (designated TR-AP-138A for military use) is a vhf transmitter-receiver de-signed primarily for light military aircraft and helicopters for which it provides both air-to-air and air-to-ground communication facilities. Secondary applications include vehicular and fixed-base stations.

The 4600-E covers the vhf band from 118 to 143.975 MHz in which it provides either 520 or 1040 channels, at intervals of either 25 or 50 kHz, generated by means of frequency synthesis techniques. It offers either A1 or A3 transmission modes and has a maximum transmitted carrier wave power output of more than 20 watts.

Frequency stability is guaranteed, over a temperature range of –40 to +70° C although the equipment will operate over a rather wider temperature range.

This equipment is widely used by the French Armed Services which have installed it in a great range of fixed-wing aircraft and heli-copters. It has also proved a successful export and is in service with overseas armed forces, notably those of Argentina and Spain.

TR-AP 138(4600-E) transceiver and controller

TR-AP-138E (4652-B) vhf/FM radio

The Socrat 4652-B vhf transmitter-receiver has an almost identical specification to the com-pany's 4600-E equipment. Its unmodulated continuous wave (cw) transmitted power out-put, however, is somewhat higher, at more than 25 watts, but there are few other differences in the performance of the two systems.

The 4652-B is designed mainly with retrofit application in mind and is dimensionally identical to the earlier TR-AP-19 and TR-AP-26 systems which used conventional glass-en-velope thermionic valves. It is interchangeable with the earlier equipment without alteration to existing wiring, and electrical connections are made to match the existing interconnectors. Special protection against possible antenna mismatching has also been incorporated and automatic test circuits permit easy checking of transmitter and receiver sections when the 4652-B is matched with older antenna systems.

12000 Series vhf radio

The Socrat 12000 vhf system is designed to provide air-to-air and air-to-ground communi-cation and navigation facilities for all types of aircraft. It is available in either AM or AM/FM versions and operates in simplex mode only. Three versions of the transceiver are available and different versions of the controller can also be provided to work with the transceiver versions. Three levels of transmitter power output are available: power-down 3 watts, discrete 6 watts and normal 15 watts.

Frequency coverage is wider than most airborne vhf equipments, being from 100 to 156.975 MHz for transmission, and 100 to 173.5 MHz for reception. This allows coverage of the marine band of 100 to 108 MHz and the VOR navigation band 108 to 118 MHz. Channel spacing through the entire band is at 12.5 kHz increments. Transmission modes are A3, F3, A9, F9, DF (standby) and the synthesiser-controlled frequency-generation system possesses high stability over the operating temperature range –40 to +70° C. The system is

Socrat 6857 transceiver and controller

remote controlled and is understood to be able to operate in hostile ECM conditions.

Format: ¼ ATR (transceiver)
Weight: 6 kg (transceiver)

STATUS: in production.

6857-01 uhf radio

Socrat's 6857-01 system is a military uhf communications transmitter-receiver covering the band 225 to 399.975 MHz, in which it provides 7000 channels at increments of 25 kHz. A total of 20 preset frequencies are provided

The system operates in A1, A2 and A3 modes and has a transmitter power output of 5 watts. It employs digital synthesis techniques for fre-quency generation and can operate over a temperature range from –40 to +55° C, even in hostile ECM environments.

Two versions are available a ¼ ATR size for securing directly to the aircraft structure and a ¼ ATR Short format for rack mounting.

Weight: 4.2 kg

STATUS: in production.

TRT

Télécommunications Radioélectriques et Télé-phoniques (TRT), Defence and Avionics Com-mercial Division, 88 rue Brillat Savarin, 75640 Paris Cedex 13
TELEPHONE: (1) 581 11 12
TELEX: 250838

A leading French producer of airborne elec-tronic systems, TRT, with its subsidiary Omera, manufactures avionic equipment for military and civil markets. Products include radio communications equipment, radio-altimeters, DME, Tacan, air traffic control transponders, thermal imaging systems, missile equipment, airborne tactical radars and reconnaissance systems.

ERA-7000 radio-com family

The ERA-7000 radio-com family comprises the ERA-7000 vhf/uhf transceiver; ERA-7200 uhf transceiver; ERA-7400 vhf/uhf amplifier; ERA-8250 25-watt uhf transceiver; and the TDP-201 ECCM data link unit. These transceivers are designed for use by high performance military aircraft in air-to-air and air-to-ground tactical communications. They operate between –55 and +90° C at pressure altitudes up to 70 000 feet (21 330 metres).

ERA-7000 vhf/uhf transceiver

Designated TRAP-136 by the French Army, this equipment is a combined vhf/uhf transceiver which covers the 118 to 144 MHz vhf band and the 225 to 400 MHz uhf band. The ERA-7000 provides voice transmission (A3 mode) and data reception (F1 mode). Output power is over 5 watts.

Format: ¼ ATR Long
Weight: 8 kg

STATUS: in production and service.

ERA-7200 uhf transceiver

The ERA-7200, designated TR-AP-137 by the French Army, is derived directly from the ERA-7000, and is limited to the uhf band (225 to 400 MHz). Another version, designated TR-AP-144, can be coupled to a cipher voice unit (A9 mode).

Format: ¼ ATR Medium
Weight: 7 kg

STATUS: in production and service.

ERA-7200 uhf transceiver

ERA-7400 vhf/uhf amplifier

An optional broadband amplifier, the ERA-7400 may be used to uprate vhf and uhf transmitted power to 25 watts.

Format: ¼ ATR Long
Weight: 8 kg

STATUS: in production and service.

ERA-7400 vhf/uhf power amplifier

ERA-8250 uhf transceiver

The ERA-8250 transceiver is a 25 watts output vhf transceiver. It possesses a high degree of commonality with ERA-7000 and ERA-7200 equipment.

Format: ½ ATR Short
Weight: 12 kg

STATUS: in production and service.

ERA-7000 vhf/uhf transceiver

ERA-9000 airborne transceiver

ERA-9000 multimode radio-com family
ERA-9000 uhf/vhf radio

The ERA-9000 is a versatile all-solid-state system with a wide frequency coverage of vhf and uhf bands and a comprehensive range of facilities. Described by TRT as a 'modular growth multi-mode transceiver', the ERA-9000 uses interchangeable 'slice' modules.

Frequency coverage and operating mode are provided by the ERA-9000, 225 to 400 MHz, AM/FM; ERA-9200, 100 to 400 MHz, AM/FM; and ERA-9600, 26 to 400 MHz, AM/FM.

The transmission power output is 20 watts in AM mode or 30 watts in FM. The range of facilities includes data link and secure voice in ECCM frequency-hopping modes. Radio-relay and homing/ADF are also available. A microprocessor is used for mode and frequency selection and management, continuous self-test and automatic fault identification by software. It is also used to aid the manual test procedure. The system meets a number of military standard specifications and can operate between –40 and +80° C at pressure altitudes up to 70 000 feet (21 330 metres).

Format: 1/2 ATR Short or Long (according to version)
Weight: 8–14 kg (according to version)

STATUS: in production and service.

ERA-8500 vhf/uhf transceiver

This transceiver covers the vhf/uhf band (100 to 400 MHz) in AM modes (A3, A9) and in FM modes (F3, F9, F1). Its tuning time is compatible with very fast frequency-hopping modes. Power output is 15 watts in AM and 20 watts in FM.

Format: 1/4 ATR Long
Weight: 9 kg

STATUS: in production.

ERA-8700 uhf transceiver

This transceiver is derived from ERA-8500 radio set and is limited to the uhf band (225 to 400 MHz).

Format: 1/4 ATR Medium
Weight: 8 kg

STATUS: in production.

SICOP-500 integrated radio-com system

SICOP-500 is the ECCM radio communication integrated system proposed by TRT for air force applications, operating in the vhf/uhf band. Its capabilities include: plain data and voice communications; cipher voice communications; ECCM radio communication; jam-resistant data link; ground-to-air, air-to-air and air-to-ground; jam-resistant voice communications.

The jam-resistant operation is achieved by using frequency-hopping techniques for voice and data transmissions and error correction is provided for data transmissions. For voice transmissions the ciphering can operate both in jam-resistant and fixed-frequency modes.

The SICOP-500 comprises an ERA-8500 vhf/uhf transceiver, ERA-8700 uhf transceiver and TDP-500 ECCM radio processor unit. It is designed for use by high performance military aircraft in a jamming environment. It operates between –55 and +90°C at pressure altitude up to 96 000 feet (30 000 metres).

TDP-201 ECCM data link unit

The TDP-201, coupled to ERA-7000 or ERA-7200 transceivers, forms a jam-protected system, providing data transmission in frequency-hopping mode.

Format: 1/4 ATR Short
Weight: 6 kg

STATUS: in production.

TDP-500 ECCM radio processor unit

The TDP-500, coupled to the ERA-8500 or ERA-8700 transceivers, provides the cipher voice and jam-resistant functions in frequency-hopping modes.

Format: 1/4 ATR Short
Weight: 6 kg

STATUS: in production.

CCS 2500 cockpit control system

TRT's CCS 2500 cockpit control system is

TRT CCS 2500 cockpit control system

designed to provide centralised management for a variety of avionics equipment including communications, navigation and identification systems. It also displays information derived from the omega navigation system.

A basic CCS 2500 installation comprises three cockpit control units for captain, first officer and navigator, and two processing and coupler units installed in the aircraft's avionics compartment.

Dimensions: (cockpit control unit) 172 × 146 × 165 mm
(processing/coupler unit) 193 × 127 × 407 mm

STATUS: in production and service.

GERMANY, FEDERAL REPUBLIC

Becker

Becker Flugfunkwerk GmbH, Niederwaldstrasse 20, D-7550 Rastatt
TELEPHONE: (7222) 120
TELEX: 781271
FAX: 12217

AR 2009/25 vhf radio

One of Becker's Comm 2000 series, the AR 2009/25 is a vhf transmitter-receiver covering the 118 to 135.975 MHz band in which it provides 720 channels at 25 kHz increments. This general aviation system is suitable for light aircraft or IFR-equipped twins. It normally operates from a 14-volt dc supply but can also work from 28 volts in conjunction with a VR 2011 converter.

The AR 2009/25 is a lightweight, single-block radio designed for panel mounting. Controls comprise a single on/off switch, volume and squelch knobs and rotary frequency selectors. Transmitter power output is 6 to 10 watts.

As well as radio communication, the system also doubles as an intercom between crew members without the need for additional equipment such as amplifiers. The AR 2009/25 is operable over a temperature range from –15 to +55°C at altitudes up to 15 000 feet (4570 metres).

Dimensions: (AR 2009/25) 47.5 × 146 × 191 mm
(VR 2011 voltage regulator) 169 × 120 × 34 mm
Weight: (AR 2009/25) 1.3 kg
(VR 2011 voltage regulator) 0.4 kg

STATUS: in production and service.

AR 2010/25 vhf radio

Another member of the Becker Comm 2000 series, the Becker AR 2010/25 is virtually identical in size, appearance and general specification to the AR 2009/25. It differs

externally only in having a test button on the front facia, indicating that this model has built-in test and diagnostic circuitry. The AR 2010/25 was developed specifically for IFR operation and is suitable for commercial aircraft.

Apart from integrated self-test facilities, the AR 2010/25 has an audio output of 150 mW compared to the AR 2009/25's 70 mW output. It is also operable over a wider temperature range of –40 to +55°C and at altitudes up to 45 000 feet (13 700 metres).

Dimensions: 47.5 × 146 × 191 mm
Weight: 1.3 kg

STATUS: in production and service.

AR 2011/25 vhf radio

The AR 2011/25 is the premium model in Becker's Comm 2000 range of radio equipment. Its external appearance and dimensions are identical to the AR 2010/25 model but the AR 2011/25 meets C37b, C38b Class 1 requirements and has a power output of more than 10 watts. It has a wider operational environment envelope, remaining operable at temperatures down to –46°C. It is suitable for service in all commercial aircraft, including helicopters.

Dimensions: 47.5 × 146 × 191 mm
Weight: 1.3 kg

STATUS: in production and service.

ZG 2 and ZG 3 homing systems

Becker ZG 2 and ZG 3 systems are lightweight, compact beacon-homing receiver-indicators designed primarily for search and rescue applications in conjunction with surface-situated emergency location transmitters (ELTs). Each is built to the same general specification with the exception that the ZG 2 covers vhf frequencies while the ZG 3 receives

on uhf. The systems provide left/right indications received on preset frequencies independently of the aircraft's normal communications radio system.

Each can receive on two channels. The emergency channel is normally set to 121.5 MHz in the vhf band and to 243 MHz in uhf operation. The auxiliary channel can be set to ∓2.5 MHz of the emergency channel in vhf and within ∓5 MHz in uhf. Operational modes are A1, A2 and A3.

The units are housed in single instrument cases suitable for panel mounting in standard-sized cut-outs. Receivers are of the triple conversion superheterodyne type, with high sensitivity. The front panel contains an on/off switch with a combined sensitivity level control, a channel selector, a signal received lamp and the left/right indicator. Audio output is provided for the aircraft communication system in order that aural indication of carrier wave or modulated carrier wave signals may be detected in addition to visual indications. Bearing accuracy of the visual left/right pointer is 10° at full deflection. Normal power requirement is 28 volts dc but alternative versions working at 12 volts dc are also available. These are suitable for surface vessels or vehicles.

The systems are operable throughout a temperature range from –45 to +55°C but the upper limitation may be extended to 71°C for short duration. Maximum operational altitude is 30 000 feet (9140 metres).

Becker claims that ground-to-air ranges of 60 to 80 miles can be achieved at search aircraft altitudes of 10 000 feet (3050 metres) and from 150 to 200 miles at 30 000 feet (9140 metres) altitude when used in conjunction with the company's MR 506 ELT unit.

Dimensions: 80 × 80 × 166 mm
Weight: 1 kg

STATUS: in production and service.

ZG 360 homing system
Becker's ZG 360 homing system is a multi-channel device covering the vhf band and designed primarily to pick up signals from the automatically-activated emergency location transmitters of crashed aircraft. It has secondary functions in the forward liaison and supply role and as an emergency navigation aid. It may also be used as an additional guard receiver.

In standard form, the ZG 360 provides coverage of the vhf band from 118 to 136 MHz with 360 channels at 50 kHz increments. An optional version offering 720 channels at 25 kHz spacing is also available. The system operates independently of the normal communications radio system on the aircraft.

The ZG 360 system comprises two separate units, a receiver section and a homing indicator. Each can be panel mounted, either together or independently if cockpit layout so dictates. The indicator's full scale deflection is adjustable over a 10 to 30° range.

The antenna system consists of a pair of 50-ohm elements, which may be of rod, blade or dipole type as appropriate to aircraft performance.

The operational temperature range of the system is from –15 to +55° C with an upper limit extension to 71° C for short durations. Maximum operable altitude is 45 000 feet (13 700 metres).

Dimensions: (receiver) 48 × 146 × 235 mm
(indicator) 83 × 83 × 180 mm
Weight: (receiver) 1.3 kg
(indicator) 0.3 kg

STATUS: in production and service.

AR 3201 vhf radio
The AR 3201 is a small lightweight vhf transceiver covering the frequency range 118 to 136.975 MHz, which meets the ICAO standards for the 1990s. Of the 760 channels provided, four are pre-selectable, although the number of channels can be restricted to 720 if required. The equipment is designed primarily for light aircraft and twins for use up to 20 000 feet (6100 metres), but can also be installed in heavy twin and business jet aircraft as a COMM-3 emergency transceiver, together with the emergency power unit Becker EPU-400.

The system is designed for console mounting with a liquid-crystal frequency display and an output power of 5 to 7 watts. The microprocessor technology used allows four non-volatile channel memories and permanent storage of the distress frequency 121.5 MHz. The standard version includes intercom, panel lighting, AF auxiliary input and automatic self-test of the display. Optionally, outside air temperature and bus bar voltage can be displayed. Power supply is 10 to 15.2 volts dc.

AR 3201 vhf transceiver

Dimensions: 60.6 × 60.6 × 212.5 mm
Weight: 0.9 kg

STATUS: in production.

GK 310 portable vhf transceiver
The GK 310 is a portable version of the 760-channel AR 3201 vhf transceiver which operates over the frequency range 118 to 136.975 MHz with a power output of 5 to 7 watts. It is intended primarily for use in a number of applications ranging from balloons, ultra-light aircraft, ferry flights of non-equipped aircraft and for mobile use on airports or in the field. The equipment is designed for use with either a loudspeaker-microphone, a headset or a protective helmet. Power supply is by means of a 12-volt 2.1 Ah leads-acid battery.

Dimensions: 281 × 115 × 79 mm
Weight: 3 kg

STATUS: in production.

GK 320 portable vhf transceiver
The GK 320 is another portable version of the AR 3201 vhf transceiver for use in a multitude of applications ranging from use on ferry flights of non-equipped aircraft, mobile use on airports or in balloons, and retrieving crews or survey teams. The equipment is contained in a carrying case of high thermic resistant and shockproof plastic. Frequency cover is from 118 to 136.975 MHz, with 760 channels and a power output of 5 to 7 watts. The power supply is by a lead-acid battery with a NiCd battery pack as an available option. An integrated 110/220-volt charging device is supplied as well as storage compartments for the steeltape antenna and hand microphone. Operating time varies between 15 and 28 hours according to the receive/transmit ratio. A 12-volt dc external connector is provided and a cable for this can be supplied, as can also a charging adapter for 28-volt dc systems.

Dimensions: 325 × 247 × 85 mm
Weight: 6 kg

STATUS: in production.

AR 3202 vhf radio
The AR 3202 vhf transmitter-receiver is a member of Becker's 3000 series avionic systems which feature microprocessor control. It provides 760 channels in the vhf band which extends from 118 to 137 MHz for civil aircraft. This range can be extended to 144 or 152 MHz for military aircraft. The system features a non-volatile memory and solid-state switches which eliminate mechanical contacts giving increased reliability. Opto-electronic switching is employed for frequency selection. Frequency-generation and display are microprocessor controlled, the latter comprising two liquid-crystal presentations, one of which indicates the active channel while the other shows a pre-selected frequency.

The transmitter output power is 20 watts and the AR 3202 is suitable for both fixed-wing aircraft and helicopters.

Dimensions: 47.5 × 146 × 245 mm
Weight: 1.3 kg

STATUS: in production and service.

ARC-114(G) vhf radio
The Becker ARC-114(G) vhf/FM transmitter-receiver is a compact lightweight, single-block device for tactical military aircraft. It provides selectable communication in FM voice, digital data transmission and homing facilities on any one of 920 discrete channels in the band 30 to 79.95 MHz. The system can also provide real-time re-broadcast functions when used in conjunction with another transmitter-receiver. The system includes one fixed-tuned guard receiver channel. Tunable channel spacing is at 50 kHz increments.

The ARC-114(G)'s transmitter power output is 10 watts minimum and the equipment can operate over a temperature range of –54 to +55° C, although full performance is not attainable at temperatures below –32° C. Maximum operating altitude is 50 000 feet (15 240 metres).

Dimensions: 108 × 146 × 203 mm
Weight: 3.2 kg

STATUS: in production and service.

AR 2010/25N vhf/AM radio
Becker's AR 2010/25N vhf/AM transmitter-receiver is the premium member of the company's Comm 2000 range of equipment, and is designed for a full range of applications aboard all types of fixed-wing aircraft and helicopters.

The system offers up to 720 channels, at 25 kHz increments, over the band 118 to 135.975 MHz. It provides full intercommunication facilities without the need for additional amplifiers or accessories. The AR 2010/25N is of modular plug-in construction and contains built-in test facilities providing rapid functional analysis. Transmitter output power is from 6 to 10 watts.

Normally operated from a 28-volt dc supply, the system is compatible with a 14-volt dc supply in conjunction with a Becker VR 2011 regulator unit. Operating temperature range is from –40 to +55° C at altitudes up to 45 000 feet (13 700 metres).

Dimensions: 48 × 146 × 195 mm
Weight: 1.3 kg

STATUS: in production and service.

ZVG 2002 vhf broadband homing adapter
The Becker ZVG 2002 is a homing adapter, operating in conjunction with any airborne vhf/AM transceiver, and enabling the pilot to home onto a ground or airborne station which is transmitting in the selected channel. A panel-mounted device functioning as a normal command instrument with a range approximately the same as the associated transceiver, it covers the frequency range 118 to 135.975 MHz with a bearing accuracy of 20 degrees. The ZVG 2002 may also be adapted to operate in the uhf range of 200 to 300 MHz in sections of 20 MHz.

Dimensions: 82.55 × 82.55 × 96 mm
Weight: 0.45 kg

ASI 3100 audio selector and intercommunication system
A member of the Becker 3000 series avionic systems, the ASI 3100 controls four transmitters and up to six receivers. By addition of an auxiliary unit, a further six receiver units can be added to the audio chain. In different versions the ASI 3100 is capable of either voice-operated switch or 'push-to-talk' operation with all other stations, voice filter for ADF and navigation systems, connection of various types of microphone and emergency operation, and providing redundancy of the transmitter-receiver operation. The intercom amplifier has a common bus connecting up to six cabin and three cockpit stations. A cockpit voice recorder output is incorporated.

The system is of modular construction and may be tailored to precise customer requirements. It is suitable for fixed-wing aircraft and helicopters. To achieve maximum adaptability, a full range of sub-units has been developed to extend the function of the main AS 3100 controller to a full cabin communication and passenger entertainment system. An intercom amplifier permits communication between passengers and crew in noisy aircraft such as helicopters, and a service station allows communication between the crew and flight attendants as well as a public address facility. A tape player, used in conjunction with the public address amplifier, is the basis of passenger entertainment through headsets or loudspeakers. The public address amplifier includes one mono or stereo amplifier and a double-tone 'gong' which operates when activated by the 'fasten seat belts' or 'no smoking' signs.

An external jack-box, which is normally installed in the wheel well or any other readily accessible location, permits communication between cockpit or flight-deck and ground crew during starting and departure checks.

All units operate from a 28-volt dc supply and most can function up to 50 000 feet (15 240 metres). Most of the sub-units are also operable over a temperature range from –40 to +70° C.

Dimensions: (main control unit) 38 × 146 × 35 mm
(auxiliary unit) 29 × 146 × 26 mm
(cassette player) 57 × 146 × 170 mm
(service station) 210 × 66 × 115 mm
(public address amplifier) 129 × 45 × 245 mm
(external jack box) 117 × 80 × 80 mm

Weight: (main control unit) 0.8 kg
(auxiliary unit) 0.2 kg
(cassette player) 1 kg
(service station) 1 kg
(public address amplifier) 0.8 kg
(external jack box) 0.6 kg

STATUS: in production and service.

VCS 220 voice communication system

Becker has produced, under the generic designation of VCS 220, a series of audio selector and intercommunication units for military fixed-wing aircraft and helicopters. All, broadly, provide intercom amplification, radio and intercom functions and navaid selection. This includes emergency system and mode selection, volume control and selection choice for either 'hot microphone' or 'push-to-talk' type microphones.

Features include full redundancy of active circuitry and built-in self-test facilities. Special attention has been given to maintaining reliability under severe environmental conditions. The systems function up to 71 000 feet (21 640 metres) over temperatures ranging from –54 to +71° C at 95 per cent humidity. They withstand vibration levels of 5 g from 2 to 6 Hz. Operating voltages are either 28 volts dc or from 4 to 12 volts dc.

The range of units includes the VCS 221 A and VCS 222 A communications and navaid controllers, VCS 221-1 main controller (designed specifically for the Alpha Jet), VCS 224-1 junction box, and the VCS 225 ground crew jack-box.

Dimensions: (VCS 222 navaid auxiliary input panel) 29 × 146 × 62 mm
(VCS 221-1 main control unit) 67 × 146 × 128 mm
(VCS 225 ground crew jack-box) 54 × 82 × 56 mm
(VCS 224-1 junction box) 111 × 177 × 65 mm
Weight: (VCS 222 navaid auxiliary input panel) 0.2 kg
(VCS 221-1 main control unit) 1.3 kg
(VCS 225 ground crew jack box) 0.4 kg
(VCS 224-1 junction box) 0.5 kg

STATUS: in production and service.

Dittel

Avionic Dittel, Postfach 348, Rudolf-Diesel-Strasse 4, 8910 Landsberg/Lech
TELEPHONE: (8191) 4077
TELEX: 527202

ATR 720A vhf transceiver

The ATR 720A is a low-cost vhf transceiver suitable for light aircraft. It covers the frequency band 118 to 135.975 MHz in 25 kHz spacing (720 channels) and frequency selection is by push-button for each digit.

Power output is greater than 4.5 watts and the unit can be easily expanded to 136.975 MHz (760 channels).

Dimensions: 77 × 56 × 200 mm
Weight: 0.8 kg

ATR 720B vhf transceiver

The ATR 720B is similar to the A model but with a liquid-crystal frequency display rather than mechanical digits. It has four pre-selected frequencies for rapid selection.

Dimensions: (circular front face) 80 mm dia; 210 mm depth
Weight: 0.9 kg

ATR 720C vhf transceiver

The ATR 720C increases the Avionic Dittel range, offering 25 kHz spacing from 118 to 135.975 MHz (136.975 MHz optional) and a liquid-crystal display. Ten frequencies can be stored electronically for rapid recall. As with the A and B models, power output is greater than 4.5 watts.

Dimensions: 77 × 56 × 200 mm
Weight: 0.9 kg

ATR 720M vhf transceiver

The ATR 720M is a panel-mounted light aircraft vhf transceiver, giving more than 5.5 watts power output. Frequency coverage is as the other models in the ATR 720 series.

Dimensions: 32 × 160 × 255 mm
Weight: 1 kg

GS-1 portable vhf transceiver

The GS-1 is a portable vhf transceiver, suitable for standby use in light aircraft, gliders and other vehicles, or as a ground station. It is self-contained, including a 30-hour receive-endurance battery, or it can be powered from aircraft or domestic supplies. It covers the vhf band from 118 to 135.975 MHz in 25 kHz spacing by manual selection.

Dimensions: 245 × 260 × 85 mm
Weight: 5.5 kg

Rohde and Schwarz

Rohde and Schwarz GmbH, Muhldorfstrasse 15, D-8000 Munich 80
TELEPHONE: (89) 41291
TELEX: 523703

Rohde and Schwarz is a leading West German research, development and manufacturing company in the field of radio communications and electronic measuring equipment. As well as providing equipment for general surface and airborne communications and air-traffic control, the company produces radio monitoring equipment and is noted for precision direction-finding equipment. It is a major supplier to the West German armed forces and has provided equipment for Europe's tri-national Panavia Tornado, the Franco-German Alpha Jet strike/trainer, McDonnell Douglas Phantom F-4, and the MBB BO-105 VBH and PAH helicopters.

Vhf/uhf radio

In order to meet a common requirement for all branches of the West German armed forces, Rohde and Schwarz developed a family of vhf/uhf equipment for airborne and surface land mobile and shipboard communications purposes. It is based on a range of functional modules which may be combined on a 'building-block' basis to form a system for any particular application. Advantages are gained not only in terms of operational flexibility and logistics but also in the ability to interchange modules for servicing and repair. All members of this family of equipment employ high-density integration techniques and, except the XT 2000 uhf radio, use digital synthesis for frequency generation.

XT 3000 vhf/uhf radio

The XT 3000 is a combined vhf/uhf air-to-air and air-to-ground transmitter-receiver covering the frequency ranges 100 to 162 MHz and 225 to 400 MHz, in both FM and AM modes, for voice and data. Frequency increments are spaced at 25 kHz intervals but the channel spacing is switchable by increments of 25, 50 or 100 kHz as required. Transmitter output is 10 watts. Up to 28 operational channels, together with the international distress frequencies of 121.5 and 243 MHz, may be pre-selected on remote control units. The system incorporates built-in test equipment and inputs for special-to-type and automatic test equipment.

The XT3000 comprises a single vhf/uhf transmitter-receiver, vhf and uhf amplifiers, two

Members of Rohde and Schwarz vhf/uhf family; left, uhf power amplifier; centre, transmitter-receiver; right, vhf power amplifier. Two control units and power supply are also shown

control units for remote operation, and a channel/frequency indicator. It meets MIL-E-5400 and MIL-STDs-810B, 461, 462, 463, 781B and VG 95211.

Format: ½ ATR Short plus two ¼ ATR Short units
Weight: 30 kg

STATUS: in production and service.

XT 3011 uhf radio

The XT 3011 represents what may be regarded as the uhf section of the foregoing XT 3000 system, with which it is almost identical. Exceptions are the lack of a vhf transmitter-receiver and appropriate amplifier module, and the inclusion of an integrated but independent guard receiver.

Frequency range is from 225 to 400 MHz with a frequency spacing of 25 kHz. Channel spacing is normally 50 kHz but this may be modified to give 25 kHz increments. Modulation mode is AM only.

The system is operable from either one or two remote control positions and there are three different types of controller available. Provision is also made for a remote frequency/channel indicator.

Format: (transmitter-receiver) ½ ATR Short controllers in accordance with MIL-STD-25212
Weight: 7.5 kg

STATUS: in production and service.

XT 3013 vhf radio

This system is the vhf counterpart of the XT 3011 uhf equipment, the modules for uhf operation having been replaced by vhf equivalent units.

The XT 3013 covers the vhf band from 100 to 156 MHz at 25 kHz frequency setting and a channel spacing also of 25 kHz or, optionally, 50 kHz increments.

Format: (transmitter-receiver) ½ ATR Short controllers to MIL-STD-25212
Weight: 7.5 kg

STATUS: in production and service.

XT 3010 uhf radio

This system is essentially a cockpit-mounted variant of the XT 3011 radio, in which a standard control panel is fitted to the main electronic module stack for direct user control. Provision is also made for remote operation from another flight crew position and a remote frequency/channel indicator may also be used with this system.

Format: in accordance with MIL-STD-25212
Weight: 6.2 kg

STATUS: in production and service.

Rohde and Schwarz XT 3011 uhf transceiver with control units

Rohde and Schwarz XT 3012 vhf radio

XT 3012 vhf radio

As the XT 3010 uhf equipment relates to the XT 3011, so does its vhf equivalent, the XT 3012, relate to the XT 3013. Again, it is a cockpit-mounted vhf equivalent of the XT 3010 system and conforms to the broad vhf specification of this entire range of equipment.

Format: in accordance with MIL-STD-25212
Weight: 6.2 kg

STATUS: in production and service.

XT 2000 uhf radio

The XT 2000 system is a cockpit-mounted uhf AM emergency transmitter-receiver operating on the international distress frequency of 243 MHz. It is the only member of the Rohde and Schwarz vhf/uhf family which is non-synthesised in terms of frequency generation. Although two other frequencies in the range 242 to 244 MHz may be used, they must be selected by replacement of crystals. The distress channel may be externally switched,

with priority, irrespective of which channel is selected.

Power output of the XT 2000 transmitter is 3 watts. The unit conforms with the standard specifications of the other equipment in the Rohde and Schwarz vhf/uhf family.

Format: in accordance with MIL-STD-25212
Weight: 2.3 kg

STATUS: in production and service.

XK 401 hf/ssb radio

The XK 401 has been developed jointly by Rohde and Schwarz with Siemens AG for the tri-national Panavia Tornado, with the aim of providing reliable air-to-air and air-to-ground communication over long ranges.

The system covers the hf band from 2 to 29.999 MHz in 100 Hz increments, providing more than 280 000 channels, of which any 11

Rohde and Schwarz XK 401 transmitter-receiver

Rohde and Schwarz XT 2000 uhf emergency transmitter-receiver with XT 3011 uhf transceiver and two controllers

Rohde and Schwarz XK 401 ssb transceiver

Control unit for Rohde and Schwarz XK 401 hf system

Rohde and Schwarz uhf antenna switch unit

Rohde and Schwarz 610 vhf/uhf radio

are pre-selectable. All channels are individually selectable by means of decade switches. Operational modes are upper sideband, A3J (duplex), and continuous wave. Transmitter power output is 400 watts peak power for modulated transmission and 100 watts in carrier wave mode transmission.

The XK 401 comprises three basic units: a transmitter-receiver XK 041, a remote controller GB 041 and a power amplifier VK 241. Additionally, two optional units, an antenna tuner FK 241 and a control frequency selector, are available and are recommended for optimum operation. The system is of modular construction and solid-state components are used.

The transmitter-receiver section uses digital synthesis frequency-generation techniques and the synthesiser itself is said to possess outstandingly good noise characteristics. Two intermediate frequencies, 72.03 MHz and 30 kHz, are used for both transmission and reception paths. The receiver section has automatic squelch which operates if the hf/ssb level exceeds an adjustable threshold. In the transmitter section, the lower sideband is suppressed by mechanical filters. Built-in test equipment is incorporated.

The remote controller, which may be up to 50 metres from the other major units, contains all necessary system controls as well as storage facilities for the 11 pre-selectable channels.

Two identical amplifier modules, with a common output matching circuit, form the power amplifier section, this duplication being applied in the interests of reliability. In normal operation both modules are in use; and in the event of one module failing, the only consequence is reduction of output power rather than total power output breakdown. Thermal and mismatch overload protection circuits, which include open and short circuit protection, are provided. Heat dissipated by the amplifier section is extracted by an external ventilator.

Use of the optional antenna tuner and the control frequency selector considerably enhances system performance. The former unit matches the base impedance of the antenna to transmitter-receiver output impedance and during reception periods acts as a pre-selector. The control frequency selector is used to control the digital tuning information. Average tuning time for the system is less than half a second but use of the control frequency selector unit enables tuning information to be stored for all pre-selected channels and the tuning time is thus further reduced. No power is radiated during tuning, resulting in radio silence being maintained during such operation.

Dimensions: (transmitter-receiver) 124 × 194 × 319 mm
(power amplifier) 257 × 194 × 319 mm
(controller) 146 × 86 × 165 mm
Weight: (transmitter-receiver) 11.2 kg
(power amplifier) 17.2 kg
(controller) 1.8 kg

STATUS: in service.

GB 606 central control unit
The GB 606 series is a family of control units designed for radio and flight- and performance-management systems in larger aircraft. The control units can be designed to suit customer requirements and their detailed operation can be easily modified by changing the software. The keyboard has varying functions which are labelled on the screen and controlled by the software.

The screen has six lines each of 24 dimmable characters, displayed in red or green according to choice. There is a 48 Kbyte program memory and 4 Kbyte working memory. The interface is to MIL-STD-1553B and RS-422 standards.

Dimensions: 146 × 171 × 165 mm (18 units per MS 25212)
Weight: 3.5 kg
Power: 25 W at 28 V dc

610 series vhf/uhf radios
The Rohde and Schwarz 610 family of vhf/uhf radio communication systems comes in two basic versions, a single panel-mounted unit or a remotely-located transmitter-receiver with a panel-mounted control unit. The controllers for the vhf and uhf variants are electrically and mechanically identical. They permit parallel operation of both a vhf and a uhf transmitter-receiver from a single controller and/or the operation of a single transmitter-receiver from two control units. All systems in the 610 series provide not only radio-telephonic communication but also incorporate a 16 Kbit baseband data transmission and ADF facilities.

Technical specifications of the vhf and uhf variants are virtually identical. Each type has a transmitter power output of 105 watts peak power and carrier power of 10 watts at normal supply voltage of 28 volts, or 1 watt carrier wave with a reduced emergency power supply of 16 volts. Frequency range of the vhf systems is from 100 to 155.975 MHz and that of the uhf equipments from 225 to 399.975 MHz. Guard receivers cover the emergency channels of 121.5 and 242 MHz respectively. Frequency-

setting increments are 25 kHz in each case and the channel spacing is also 25 kHz, although in the case of uhf systems the spacing is optionally adaptable to 50 kHz. Up to 30 channels, plus the guard channel, may be pre-selected and a remote frequency/channel indicator is an optional accessory.

All 610 series transmitter-receivers are compatible with one another and the various modules have exactly-designed interfaces to permit easy replacements to be made without need for adjustment. The systems are suitable for retrofitting and Rohde and Schwarz have produced replacement kits, with tailored adaptor trays, for such aircraft as the McDonnell Douglas F-4 and RF-4E and the Lockheed F-104G Starfighter. A range of special-to-type test equipment is available, with first-line test sets which can isolate faults down to module level, and full-scale automatic test equipments for base repair facilities.

Dimensions: (XU-610 vhf panel-mounted) 146 × 124 × 165 mm
(XU-611 vhf remote controlled) 127 × 124 × 165 mm
(XD-610 uhf panel-mounted) 146 × 124 × 165 mm
(XD-611 uhf remote controlled) 127 × 124 × 165 mm
(GB-600 remote control unit) 146 × 76 × 110 mm
Weight: (XU-610 vhf panel-mounted) 4.2 kg
(XU-611 vhf remote controlled) 3.7 kg
(XD-610 uhf panel-mounted) 4.2 kg
(XD-611 uhf remote controlled) 3.7 kg
(GB-600 remote control unit) 1.9 kg

STATUS: in production.

Uhf and vhf filters
Rohde and Schwarz manufactures a range of uhf and vhf filters, which allow several transmitters and receivers to be operated simultaneously under adverse conditions with closely spaced antennas.

The uhf filter FD 220 covers the frequency range 225 to 400 MHz, and the FU 220 covers the vhf band from 100 to 163 MHz. Both units are manually tuned and are available with shockmounts for mobile use.

The FD 221 and FU 221 filters are automatically tuned versions. All the units can handle 200 watts AM and 300 watts FM.

Dimensions: (FD 220 and 221) 220 × 483 × 560 mm
(FU 220 and 221) 220 × 483 × 500 mm
Weight: (all versions) 30 kg

ISRAEL

Elta

Elta Electronic Industries Limited, PO Box 330, Ashdod 77102
TELEPHONE: (55) 30333
TELEX: 31807

Elta is a subsidiary of Israel Aircraft Industries Ltd.

EL/K-1711 uhf radio

The Elta EL/K-1711 is an airborne protected (ECCM) uhf radio equipment which offers AM or FM, normal or protected, voice or data methods of communication. The system consists of a transceiver, cockpit control unit and an optional 100-watt power amplifier. It is suitable for both new installations and retrofits. In the latter case it is particularly applicable as a replacement for the earlier uhf equipment AN/ARC-51. No technical details have been made available but it is assumed that it will cover the normal military uhf frequency band from 225 to 399.975 MHz, probably with 25 kHz channel spacing and a guard frequency at 243 MHz.

STATUS: in operational service.

EL/K-1711 radio (by pilot's left hand)

Tadiran

Tadiran Limited, 11 Ben Gurion Streèt, Givat-Shmuel, PO Box 648, Tel Aviv 61006
TELEPHONE: (3) 713 111
TELEX: 341692

RT-1200 and RT-1210 (ARC-240) uhf radios

The RT-1200 and RT-1210 radios, referred to generally as ARC-240 series, are military uhf systems designed for service in all types of aircraft. Each covers the uhf band from 225 to 400 MHz, providing 7000 channels at increments of 25 kHz. Both types operate in AM voice and modulated continuous wave (mcw) signalling modes and each has an X-mode and ADF capability in conjunction with appropriate equipment.

The RT-1200 is designed for panel or console mounting for small lightweight aircraft, while the RT-1210 is a remotely controlled version appropriate to larger aircraft or those in which cockpit space is too limited for the convenient accommodation of the full transmitter-receiver system. Normal transmitter power output of both versions is in excess of 10 watts. Up to 30 channels are pre-selectable in either variant, the channel memory for the RT-1210 being contained within the remote controller rather than in the main unit itself. Each type has a built-in independent guard receiver covering a single pre-tuned guard frequency.

The RT-1210 is of the same general specification as the RT-1200 and uses identical modules and components. A half-sized remote control unit is available and a single controller may be employed to manage a dual transmitter-receiver installation. Both the RT-1200 and the RT-1210 drive a remote frequency display and both use oven-crystal controlled digital synthesis techniques for frequency-generation.

The ARC-240 systems are designed to replace and interchange with a wide range of other military aircraft radios and are compatible with them and their ancillary equipment such as antennas and mountings.

For extended power output and range performance, the systems may be coupled with the AM-100 (ARC-240T) wide-band uhf amplifier system.

Dimensions: (RT-1200) 172 × 146 × 125 mm (RT-1210) 165 × 125 × 125 mm
Weight: (RT-1200) 4.3 kg (RT-1210) 4.1 kg

STATUS: in production and service.

AM-100 (ARC-240T) uhf amplifier

Tadiran's AM-100 uhf amplifier with an output of 100 watts is designed, primarily, to increase the transmitted power output and range of the ARC-240 series radios, but may also be applied to other military radios operating in the uhf band.

The AM-100 amplifier accepts an RF input of from 5 to 30 watts within the 225 to 400 MHz range. If the power supply voltage drops below the nominal rating of 27.5 volts, the AM-100 will continue to produce an output of 60 watts at a minimum supply of 22 volts.

Tadiran ARC-240 radio

The system's major features are low-distortion and flat-response amplification achieved by automatic level control circuitry. The AM-100 contains built-in protection against transient high-voltage surges and spikes and against overheating. The amplifier is nominally cleared for operation over a temperature range of –55 to +71°C but in fact continues to operate satisfactorily beyond these limits, particularly at the upper end of the scale. Normal cooling is by convection but forced air cooling, provided by a built-in centrifugal fan, is used in the AM-100 and is automatically operated if the amplifier temperature exceeds 60°C. The system is claimed to be exceptionally reliable and is of particularly strong construction, meeting a number of stringent military specifications.

Dimensions: 425 × 225 × 170 mm
Weight: 15 kg

STATUS: in production and service.

ITALY

Elmer

Elmer, Viale dell'Industria 4, PO Box 189, 00040 Pomezia
TELEPHONE: (6) 912 971
TELEX: 610112

SRT-170/E hf/ssb transceiver

Following research and development into hf propagation, Elmer has designed and produced an advanced hf/ssb system for aircraft and helicopters. The standard system comprises a 100-watt transceiver and a tuner directly connected to an advanced design hf loop antenna.

The transceiver is a scaled-down version of the SRT-470F 400-watt transceiver developed by Elmer for the Panavia Tornado programme, with a slightly modified receiver/exciter and a newly designed 100-watt rf power amplifier, which are contained within a 3/8 ATR Short enclosure. The tuner and loop antenna were jointly developed with British Aerospace Dynamics Group.

Extensive ground and flight tests over the last three years have used ground mock-ups and helicopters such as the Aérospatiale/Westland Gazelle, Sikorsky SH-3D, Agusta-Bell AB-212, and Agusta A-109.

The loop antenna is designed to achieve a high angle of incidence in the skywave mode to improve communication during helicopter operation near the ground and over unfavourable terrain. Attention has been given to defining the best geometric configuration and installation so as to optimise ground wave and

Elmer SRT-170/E hf/ssb control panel

NVIS (near vertical incidence skywave) propagation. Extensive field-strength measurements conducted over the hf range, have demonstrated a significant improvement over the current wire antennas.

The SRT-170/E comprises a SP-648/SM control panel; SP-649/SM receiver/exciter; SP-480/E power amplifier; SP-1036 mounting tray and AN16C tuner and loop antenna.

Its main features are as follows. All solid-stage technology with extensive use of miniaturisation techniques; frequency synthesis; fully automatic tuning including the tuner; free air convection cooling; low power consumption; remote control of operating functions by means of a serial bit stream; incorporated built-in test facilities for self-diagnosis (continuous and interruptive); compatibility with MIL-STD-1553B data-bus (optional); channel presetting by associated control panel (optional); NATO Link-11 compatibility (optional); SIMOP capability (simultaneous operation of collocated rf sets) with the addition of ancillary equipments (pre/post-selectors); tuner mounted externally to the helicopter/aircraft surface (reducing electromagnetic compatibility problems) and directly connected to the antenna; and no limitation in distance between power amplifier and tuner.

Frequency range: 2–30 MHz
Antenna: loop
Modes: A2J (cw), A3H (ame), A3J (usb, lsb)
Tuning: automatic and digital at 100 Hz minimum step
Tuning time: 2 s typical (including tuner)
Number of channels: 280 000
Frequency stability: ± 5 to 10^{-7} per day over entire temperature range
Rf power output: 100 W pep
Power consumption: (transmitter) 350 W (receiver) 50 W at 28 V dc
Dimensions: (receiver/exciter) 3/8 ATR Short
(power amplifier) 3/8 ATR Short
(control panel) 146 × 85.7 × 125 mm
Weight: (receiver/exciter) 5.1 kg
(power amplifier) 6.5 kg
(control panel) 0.9 kg
(tuner) 4.1 kg

STATUS: in current production and in service with Italian and other armed forces, with more than 300 units delivered.

AN/ARC-150 (V) transceiver

The AN/ARC-150 (V) designation represents a family of small, high performance, lightweight, airborne uhf transceivers manufactured by Elmer and based on a Magnavox design as well as Elmer's own research and development. It has produced many versions, including the 10-watt panel mounted ARC-150 (V) 10, the remote controlled 10-watt ARC-150 (V) 2, and the remote controlled 30-watt ARC-150 (V) 8.

Elmer has also developed a series of control panels and frequency/channel repeaters to meet specific installation requirements on different aircraft and helicopters. A feature of this family is 'slice' assembly, which simplifies maintenance and facilitates growth.

A series of mounting/adapters has been developed and produced to allow the basic ARC-150 transceivers to replace older uhf radios such as the AN/ARC-51BX, AN/ARC-109, AN/ARC-52, AN/ARC-552, without any mech-

Elmer SRT-170/E hf/ssb tuner and loop antenna installed in SH-3D helicopter

anical and electrical modification to the aircraft. A version of the ARC-150 (V) with ECCM capabilities has also been developed using the frequency-hopping technique 'Have Quick'. This capability is easily implemented by the substitution of the synthesiser slice and by minor changes on the control panel unit. No mechanical or electrical modifications are required on the aircraft. The modified radio retains also the normal non-hopping mode.

Frequency range: 225–400 MHz
Guard receiver: 243 MHz
Channel spacing: 25 kHz
Preset channels: 20, using electronic memory (mnos)
Frequency accuracy: 2 kHz
Power output: 10 W (30 W for the AN/ARC-150(V)8 model)
Power: 28 V dc

Operating modes: AM voice, ADF, homer, secure voice/data, ECCM
Dimensions: (10 W panel mounted (RT-1136)) 146 × 124 × 193 mm
(10 W remote (RT-1051)) 127 × 120 × 183 mm
(30 W remote (RT-1073)) 127 × 120 × 291 mm
Weight: (10 W panel mounted (RT-1136)) 4.3 kg
(10 W remote (RT-1051)) 3.7 kg
(30 W remote (RT-1073)) 6 kg

STATUS: in production and in service for Italian and other armed forces. Over 1000 units have been delivered. Installed on wide range of aircraft and helicopters including Panavia Tornado, Aermacchi MB-339, Aeritalia G91Y and F-104S fixed-wing aircraft, and Agusta A109, Agusta-Bell 212, Agusta-Sikorsky SH-3D and HH-3F and other helicopters.

Elmer ARC-150 (V) uhf AM/FM transceiver

Elmer 10-watt panel-mounted ARC-150 (V) transceiver

Elmer SRT-651 vhf/uhf transceiver

SRT-194 vhf/AM radio

Elmer's SRT-194 is a vhf/AM transmitter-receiver covering the 108 to 156 MHz band with channel spacings of 25 kHz. Up to 20 channels, plus an additional guard channel, are preselectable. The system provides air-to-air and air-to-ground communication in AM mode, together with modulated carrier wave tone and retransmit facilities; ADF and homer functions are also optionally available by the addition of appropriate ancillary equipment. The system comprises a transmitter-receiver unit with a remote controller and an additional channel/frequency selector can also be provided.

A modular 'slice' style of construction is employed and new functions can also be incorporated by simply adding appropriate slices. Construction is of all-solid-state and miniaturisation techniques are extensively employed. Power consumption is low and the thermal design is claimed to be extremely efficient, so that no forced air cooling supply is required. The system is compatible with a wide range of interphone systems, and identical pin-to-pin connections with Elmer uhf systems render the SRT-194 readily interchangeable with these without change to aircraft wiring or mountings. Transmitter output power is 15 watts minimum. The equipment conforms to MIL-5400 Class II modified specification.

Dimensions: (transmitter-receiver) 228 × 137 × 137 mm
(control panel) 83 × 146 × 181 mm
(channel/frequency indicator) 83 × 33 × 154 mm
Weight: (transmitter-receiver) 4.3 kg
(control panel) 1.4 kg
(channel/frequency indicator) 0.5 kg

STATUS: in production and service.

ARC-150 (V)8 uhf/FM AM/FM transceiver

This is a development of the basic 10-watt ARC-150 (V) transceiver incorporating the FM modulation capability by means of an additional slice. This facility makes the unit particularly suitable for use with data, frequency-shift keying and secure voice modems.

It has been demonstrated to be fully compatible with the Vinson KY-48 system both in AM and FM and in the diphase and baseband modes. It has also been successfully used as a main component of an airborne system for uhf satellite communication.

STATUS: in current production for Italian and other armed forces.

SRT-470F hf/ssb transceiver

The SRT-470F was designed to meet the hf/ssb requirements of the tri-national Tornado aircraft for air-to-air and air-to-ground communications, and consists of a transmitter-receiver, a power amplifier/antenna coupler and a panel mounted control unit. The transceiver is fully solid-state with advanced circuit techniques such as wide-band rf amplification. The antenna coupler is completely automatic and matches the power amplifier output to a notch antenna which is an integral part of the aircraft structure. The equipment operates over the frequency range 2 to 30 MHz giving 280 000 channels at 100 Hz spacing, and has a power output of 400 watts pep.

Dimensions: (transceiver) 199 × 261 × 319 mm
(power amplifier/antenna coupler) 199 × 94 × 319 mm
(controller) 86 × 146 × 125

STATUS: in production for the Italian Air Force.

SRT-651 vhf/uhf AM/FM transceiver

The SRT-651 is an all-solid-state, compact, lightweight airborne transceiver covering the 30 to 400 MHz frequency range, and using the most advanced techniques in the area of large scale integration components and microprocessors. The system comprises a CP-1200 control panel, RT-651 receiver-transmitter, and ID-1151 channel/frequency repeater (optional).

Conceived and designed for airborne use, the SRT-651 is also suitable for a wide range of applications where space and weight are limited. The flexible modular 'slice' design permits easy assembly and disassembly and allows expansion of the operating functions. The rf power rating can be easily upgraded to 30 watts by changing the transmitter slice. Different interface standards are available, including a serial data interface (Elmer type, using Control Panel CP-1200), MIL-STD-1553B or ARINC-429.

Built-in test facilites can be implemented by using different plug-in cards for the interface slice. The test result is automatically monitored on the control panel display, which gives a direct identification of the faulty slice.

The SRT-651 has facilities for ECCM, including frequency-hopping (the 'Have Quick' technique has been implemented), and spread spectrum pseudo-noise modulation. It can be operated with a wide variety of ancillaries, including homer indicators; vhf/uhf direction finders; uhf emergency beacons; KY-58 secure voice modems in diphase and baseband modes; Elmer SP-1212 ECCM spread spectrum voice and data modem; NATO multitone Link-11 data modem; and FSK modems. Various mountings can be provided to retrofit the SRT-651 directly into existing vhf and uhf installations.

Frequency range: 30–88 MHz vhf/FM, 108–156 MHz vhf/AM, 156–174 MHz vhf/FM, and 225–400 MHz uhf AM/FM
Channel spacing: 25 kHz in all bands
Guard receiver: 40.5, 121.5, 156.8 and 243 MHz automatically selected
Pre-selected channels: 20, using electronic memory
Power output: 10 W in AM, 15 W in FM
Power consumption: 120 W max on transmit, 25 W max on receive; 28 V dc
Dimensions: (RT-651) 126.7 × 120.6 × 224 mm
(CP-1200) 146 × 57.1 × 150 mm
Weight: (RT-651) 3.5 kg
(CP-1200) 1.4 kg

STATUS: in production for Italian, Brazilian and other armed forces. Future installation on new Aeritalia/Aermacchi/Embraer AM-X attack aircraft.

SRT-653 vhf/uhf transceiver

Designed to meet a requirement of the Italian Air Force for its version of the Panavia Tornado multi-role aircraft, the SRT-653 vhf/uhf transceiver comprises the RT-1051 uhf transceiver, SP-1047 vhf transceiver, SP-1203 adapter/power supply, SP-1204 mounting tray, CP-1001 control panel, and ID-1150 frequency repeater.

Elmer SRT-653 vhf/uhf transceiver

The RT-1051 uhf transceiver belongs to the basic ARC-150 (V) family. The SP-1047 vhf transceiver has been derived by Elmer from this system, and maintains its general characteristics.

Dimensions: SP-1047 and RT-1051 (installed on adapter and mounting tray) 265 × 155 × 408 mm
(CP-1001) 146 × 95.2 × 165.5 mm
(ID-1150) 86 × 50 × 159 mm
Weight: (SP-1047 and RT-1051 on tray) 13.5 kg
(CP-1001) 1.6 kg
(ID-1150) 0.35 kg

Frequency range: 108–156 and 225–400 MHz
Channel spacing: 25 KHz
Preset channels: 17 + 2 guards
Power output: 10 W
Power: 320 VA max on transmit, 140 VA max on receive, at 115 V ac, 400 Hz, 3-phase

STATUS: in current production and installed on Italian Tornado.

RV-4/213/A vhf/FM transceiver
The RV-4/213/A vhf/FM equipment is an airborne transceiver suitable for light aircraft

and helicopters. It is in effect an airborne version of the Italian Army RV-3/213/V which has been repackaged for installation in small aircraft. It is compatible with, and equivalent to, the Thomson-CSF TR-AP-113 and is intended to replace equipment of the ARC-131 type. Tuning is fully automatic and digital over the frequency range 26 to 72 MHz in 50 kHz steps. Modes of operation are F3E FM voice and homing, and rf power output is 20 watts.

Dimensions: 320 × 92 × 194 mm
Weight: 6 kg

STATUS: in production and in service.

Marconi Italiana
Marconi Italiana SpA, Via A Negrone 1A, PO Box 60, 16153 Genoa-Cornigliano
TELEPHONE: (10) 6002
TELEX: 270386

ART-312 vhf/FM transceiver
The Marconi ART-312 is a vhf/FM tactical communications system covering the frequency range 26 to 75.975 MHz at 25 kHz

channel spacing. FM voice communication and homing facilities are available on any of the 2000 channels provided, with two power outputs, a 10-watt high output and a 1-watt low. The system uses synthesiser tuning and comprises two units, a transmitter-receiver and a control unit.
The ART-312 was developed for use in small military aircraft and helicopters, and is in widespread use with the Italian armed forces.

Dimensions: (transceiver) 102 × 124 × 319 mm

ART-151 vhf/AM transceiver
The ART-151 vhf/AM transceiver provides 20 watts output power on 1599 channels in the 116 to 155.975 range. It is a solid-state remotely controlled transceiver and has been designed to operate in both voice and data modes. It is in service with the Italian Air Force.

Dimensions: 87 × 124 × 31 mm
Weight: Less than 6 kg

SWEDEN

Bofors
Bofors Aerotronics AB, S-181 81 Lidingo
TELEPHONE: (8) 765 2960
TELEX: 19188

AMR 342 vhf/uhf radio
The Bofors Aerotronics AMR 342 is a military airborne communications radio designed principally for high-performance fighters, such as the Saab Draken and Viggen aircraft. It has, however, been applied to a wider range of aircraft types.
The system is available in two versions, a combined vhf/uhf equipment or in uhf-only form, both operating in either AM or FM mode. The combined vhf/uhf version accommodates a total of 9280 channels, the vhf-only variant 7000 channels. Channel spacing is at increments of 25 kHz. According to the type of controller in use, up to 1000 channels may be pre-selected. Likewise, in some cases, the controller in use may provide pre-selected automatic selection of AM or FM mode of modulation but a manual override is also available with most controllers.
The AMR 342 in combined vhf/uhf form has both in one transceiver using the same digital synthesiser for frequency-generation. Two receivers, one main unit and an independent guard receiver are incorporated. Transmitter output power varies according to frequency band or modulation in use. In FM mode, output is 50 watts in both vhf and uhf, but in AM vhf output is 25 watts and in uhf, 20 watts. The frequency ranges covered are from 103 to 160 MHz in vhf and from 225 to 400 MHz in uhf. The guard receiver monitors the appropriate

Bofors Aerotronics AMR 345 radio

international distress frequencies on 121.5 or 243 MHz. In some applications two AMR 342 radios are installed, for example to use different transmitters for speech and data, or for ac/dc power supply.
The control unit normally used with the combined vhf/uhf version of the AMR 342 is the AMR 346. This has facilities for pre-selector tuning of up to 64 channels in addition to providing manual tuning for the entire frequency range covered by the system.
Solid-state circuitry is used throughout the AMR 342 radio, and it has comprehensive built-in self-test facilities. A mean time between failures in excess of 1300 hours has been achieved and the system operates over a temperature range of –40 to +70°C.

Dimensions: 319 × 124 × 194 mm
Weight: 12 kg

STATUS: in production and service.

AMR 345 vhf/uhf transceiver
The AMR 345 is a small vhf/uhf AM/FM panel-mounted radio for speech and data communi-

cations, covering the frequency range 104 to 162 and/or 223 to 408 MHz, with 25 kHz channel spacing. It has an ECCM capability and is designed to operate in a secure voice system. Transmitter output power is 10 watts in the AM mode and 15 watts in FM. The equipment is designed for use in air, sea and ground applications.
The transceiver memory has a flexible storage capacity for up to 500 preset channels. These channels can be used for immediate access to all frequencies at all air-bases. The front panel of the transceiver has a digital display which can be read even in bright sunlight. The keyboard is used for selecting the preset frequencies, as well as for manual setting. A built-in test facility performs a go/no go test in the aircraft.
An autonomous system is formed by the transceiver, a 28-volt power supply, a headset and an antenna. It is also operable with other systems in the aircraft by means of a serial data control link. The microprocessor control allows easy adaption of the transceiver (including the front panel function) to any system configuration. Preset channels and data back-up

Bofors Aerotronics AMR 342 transmitter-receiver and AMR 346 control unit

are stored in an electrically reprogrammable non-volatile memory.

The transceiver is EMP-hardened which ensures full operation after EMP exposure. ECCM capability is provided by use of an external ECCM control unit which can be designed in accordance with customer requirements.

The transceiver is designed to meet applicable MIL standards.

Dimensions: 230 × 146 × 76 mm
Weight: 4 kg
Frequency accuracy: 5 ppm (–40 to +70°C)

STATUS: in production. The AMR 345 is being included in the communication system for the new JAS 39 Gripen multi-role aircraft for the Royal Swedish Air Force, and has also been selected for re-equipping the Saab 105 (SK60). The radio is expected to find additional applications in maritime and ground communication systems.

AMR 347 channel selector
The Bofors Aerotronics AMR 347 is a microprocessor-based system designed to provide simplified, rapid radio channel selection in military aircraft with minimum attention from

Bofors Aerotronics AMR 347 channel selector

the pilot. The system, which was developed in co-operation with the Swedish Air Force, is installed in Swedish Air Force/Saab JA 37 fighters.

The selector is a single-box unit with a push-button keyboard, a set of sunlight-readable displays and a reprogrammable memory which can contain up to 1000 preset frequencies. These are stored in log-sequence and access to the next required channel is said to be easy and rapid. Three operating modes are available.

The first, designated MHz, is an open frequency and modulation selector. The two other modes control the memory array and in these direct access to the last digit or letter enables consecutive functions on different frequencies to be changed by using only one push-button. The second and third modes are used in Swedish Air Force service for air traffic control purposes with pre-programmed ATC frequencies and for combat control with pre-programmed combat frequencies.

A microprocessor is used to control the memory, the capacity of which permits the encoding of more than 1000 codes of 16 bits. These codes can be used for radio frequency, type of modulation and parity checking. The non-volatile memory has a retention time in excess of ten years. Both series and parallel interfacing permit the AMR 347 to be used to control any type of radio. A built-in test facility, controlled by push-buttons and the display, is incorporated. Reprogramming in the aircraft is carried out by means of a 'fill gun' with a data memory.

Dimensions: 146 × 76 × 200 mm
Weight: 2.1 kg

STATUS: in production and service.

UNITED KINGDOM

Burndept
Burndept Electronics Limited, St Fidelis Road, Erith, Kent DA8 1AU
TELEPHONE: (03224) 41155
TELEX: 896299

Sarbe search and rescue beacons
Burndept Electronics designs and manufactures a range of Sarbe search and rescue beacons designed to be stowed in flight clothing, life rafts or ejection-seats.

Sarbe 5 exists in five versions, operating on various distress frequencies of 121.5, 123.1, 243 and 282.8 MHz, providing a distress beacon

transmission which can be activated either automatically or manually, with additional facilities for two-way communication on a distress frequency channel. One version provides a second voice channel.

Sarbe 6 is another version which meets the latest NATO STANAG. It provides a distress beacon facility simultaneously on both civil and military distress frequencies of 121.5 and 243 MHz, and can be activated automatically or manually. Two-way voice can be used on both channels simultaneously, overriding the beacon facilities.

The BE 499 Tacbe is a tactical communications/beacon equipment which has been developed from the Sarbe 5 to provide a simple

two-channel ground-to-air two-way AM communications radio, with a retained selectable automatic beacon tone facility on one channel. It has a line-of-sight range and provides up to 30 hours endurance from a primary battery. Operating frequencies are preset over the vhf range 118 to 136 MHz, and uhf from 238 to 248 and 276 to 293 MHz

Dimensions: (Sarbe 6) 140 × 88 × 38 mm (BE 499 Tacbe) 114 × 76 × 30 mm
Weight: (Sarbe 6) 0.64 kg (BE 499 Tacbe) 0.57 kg

STATUS: In production and in service worldwide with military and civil operators.

Chelton
Chelton (Electrostatics) Limited, Fieldhouse Land, Marlow, Buckinghamshire SL7 1LR
TELEPHONE: (06284) 72072
TELEX: 849363

Chelton, established in 1947, specialises in the design, development and manufacture of vhf and uhf guard receivers, homing systems, antennas and a wide range of avionic auxiliary equipment and components. A recent development is the Series 700 frequency-synthesised, dual-bandwidth receivers for homing systems, with channel spacing of 25 kHz (narrow-band) and 375 kHz (broadband). The company is a supplier to more than 150 armed forces and major airlines in over 60 countries. Customers include British, European and US defence departments. The company is particularly recognised for work on static dischargers and special-to-type antenna units. Aircraft for which Chelton has supplied equipment include Tornado, Sea Harrier, Jaguar, Concorde and BAe 125, General Dynamics F-16, Boeing 737 and 757, Boeing-Vertol Chinook, Bell 214ST, Westland Lynx and Sea King and Aérospatiale Ecureuil.

System 7 homing system
Chelton's System 7 is a 'building block' homing system with broadband and emergency guard channel homing facilities. In its basic form it comprises a homing indicator unit, an antenna feed unit, an antenna system and an aircraft receiver. It provides broadband homing over the frequency spectrum of the aircraft receiver.

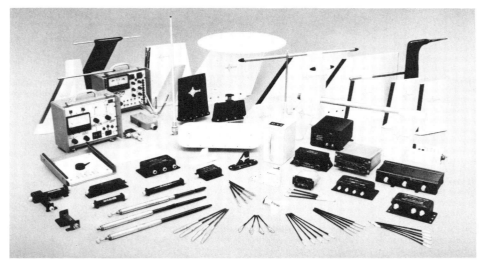
Chelton communication, navigation and special-to-type antennas, homing equipment, static dischargers and rf filters

If an independent emergency guard channel homing system is required, Chelton's Series 7-28 two-channel receiver can be incorporated.

The System 7 can be interfaced with the main aircraft receiver and the 7-28 unit, providing both broadband and guard channel homing. The system interfaces with all AM receivers such as ARC-116, ARC-159, ARC-164, ARC-182, ARC-186, VHF-20 and PTR-1751. It

will also interface with FM equipment possessing an AM facility.

The Series 7-28 emergency guard receiver is available in three variants: the 7-28-1, operating at 121.5 MHz, the 7-28-3 (243 MHz) and the 7-28-7 (156.8 MHz). Each type has a test frequency.

Two versions of the 'steer left/right' indicator unit are available, one of which enables the homing signal to be heard on the aircraft's audio system, together with four types of indicator feed unit which conduct the signal directly to existing aircraft navigation indicators or flight-directors. Whichever configuration is used, the System 7 automatically reverts to communications mode if the transmit key is depressed.

STATUS: in production and service.

Chelton Series 700 receivers

Cossor

Cossor Electronics Limited, The Pinnacles, Elizabeth Way, Harlow, Essex CM19 5BB
TELEPHONE: (0279) 26862
TELEX: 81228

CTM1080 hf data modem

The CTM1080 is a ruggedised high-frequency data modem designed for use in military or similarly hostile environments to provide reliable encoded or plain data transmission and reception in severe hf and battlefield situations. It is a multi-tone unit that can operate at data rates of between 75 and 2400 bits a second, with full forward error correction up to 1200 bits a second.

The CTM1080 can be used in both in-band and out-band diversity modes so that any

synchronous data may be sent over ground-wave or skywave hf systems within a very robust modulation format. At data rates of 1200 bits a second and below, in-band diversity can be used to provide enhanced bit-error-rate performance. The addition of a second receiver allows the out-band diversity facility to be used. The equipment is based on a Cossor-developed fast processor unit, with the addition of specially developed software functions and forward error correction facilities. Two processing modules can be accommodated in the CTM1080, each being programmed to carry out different signal processing tasks.

Through program control the equipment may be employed in a wide range of equipment applications with the hf system such as parallel data modem, serial modem, vocoder or channel evaluator.

The entire unit is housed in a ruggedised

Cossor CTM1080 hf data modem

case, with all control functions and connectors being brought to the front panel. The modem tone library is compliant with MIL-STD-188C and power consumption is only 55 watts maximum.

Dowty

Dowty Electronics Limited, Communications Division, 419 Bridport Road, Greenford Industrial Estate, Greenford, Middlesex UB6 8UA
TELEPHONE: (01) 578 0081
TELEX: 934512

D403 emergency transmitter/receiver

The D403 is intended to provide emergency communications in the event of failure of the aircraft main uhf transceiver. It has been designed for remote mounting with space in the cockpit required only for the radio's combined on/channel select switch.

In the emergency role, two-channel operation allows the initial distress call to be made on 243 MHz. Once communication on this frequency has been established, a second channel

is available for approach and landing instructions. For alternative uses, the radio can be supplied as a two-channel transmitter/receiver operating within the 238 to 248 MHz frequency range. Other models covering alternative frequency ranges are available.

Dimensions: 121 × 84 × 254 mm
Weight: 2.2 kg

STATUS: in service with a number of countries as the standard uhf standby radio.

Ferranti

Ferranti Computer Systems Limited, Western Road, Bracknell, Berkshire RG12 1RA
TELEPHONE: (0344) 483232
TELEX: 848117

Data Link G

The Ferranti Data Link G has been designed to meet a civil and military requirement for a system to pass location information from a number of aircraft to a central ground control station. In civil use Link G passes the aircraft's own position, height and status. In military applications the equipment can also be used to pass targetting information. Link G operates in a half-duplex mode at 1200 baud over vhf/uhf and hf voice networks.

The aircraft terminal incorporates a control panel and includes interfaces with the aircraft's radio navigation equipment, and encoding altimeter if available. The ground station continuously calls the aircraft terminals on a cyclic basis. Each aircraft replies sending its address and the required information. Although the system normally uses only one frequency, two may be used for faster responses, one by the ground station and one by the aircraft. Transmission over the data link is protected by a code so that any corruption of data is detected at the receiver. The message can then be retrans-

mitted if required. Comprehensive alarm facilities are provided against failure, including an emergency squawk facility. The complete system consists of a number of aircraft terminals and a ground station.

Dimensions: 165 × 57.2 × 146 mm
Weight: 1.5 kg

Data Link Y TDMA terminal

The airborne TDMA (time division multiple access) terminal enables an aircraft to participate in Ferranti Link Y networks as either net control or a dependent unit. The equipment can operate over vhf/uhf or hf radio on a single-voice frequency channel.

The terminal incorporates a microprocessor which performs the message-handling function. Other facilities of the terminal are:
(a) generation and response to control codes
(b) generation and test of error detection and correction codes
(c) modulation and demodulation of signals
(d) control of transmit/receiver switching
(e) test for signal presence and acquisition of incoming signals.

The terminal is available in two configurations, in an aircraft with a central tactical system where it would be integrated via a number of interfaces, or as an independent system where no tactical facility is carried.

Dimensions: 368 × 196.5 × 124 mm
Weight: 9.5 kg

Data Link 11 pre-processor

The Link 11 airborne data link pre-processor (DLPP) is designed to enable an aircraft to exchange tactical data with other Link 11 units. Its function is to minimise loads on the central tactical computer, to protect the central tactical system (CTS) from change when link protocols change, to offer the potential for incorporation of other forms of data link, and to simplify interoperability trials.

The equipment is available in three configurations depending on the type of CTS fitted:
(a) as an independent unit with its own display and keyboard where no CTS facilities are fitted
(b) as an independent unit, with its own display and keyboard, where a CTS with a limited track capacity is fitted. In this case it would interface with the CTS so that a limited number of designated tracks could be exchanged
(c) without the display and keyboard where a full CTS facility is incorporated, unless required to provide back-up facilities.

In cases (a) and (b) where the operator communicates with the data link through the DLPP, the following facilities are available:
(a) injection of data

(b) readout of track data, Sitrep data and tactical calculations
(c) selection of data to be displayed, adjustment of range scale and off-centering of picture
(d) solution of simple navigation problems
(e) setting-up and monitoring of the data link.

The system receives over the link data which is stored and available for immediate display. A constant update of own position is maintained. Track data from on-board sensors is also received and the operator can transmit this over the link if required. The displayed picture consists of computer-generated symbols showing position of tracks and track labels. Any size of area may be displayed, up to the limit of a million square miles operational area and the picture may be centred anywhere within this area. Two data readouts are provided on the display, one for command messages and one as the operator's interface display.

Dimensions: (pre-processor) 371 × 196.5 × 190 mm
Weight: (pre-processor) 16.5 kg

GEC Avionics

GEC Avionics Limited, Airadio Products Division and Airadio Systems Division, Christopher Martin Road, Basildon, Essex SS14 3EL
TELEPHONE: (0268) 22822
TELEX: 99225

ARC-340 (AD 190) vhf/FM radio

The GEC Avionics ARC-340 radio, marketed by the company under the designation AD 190, is a tactical vhf/FM equipment designed specifically for air-to-ground communication between military aircraft and ground forces. It is in service with the British Armed Forces, in conjunction with Clansman series vhf equipment, and is in operation in a wide range of fixed-wing aircraft and helicopters in many parts of the world.

The system provides clear and secure speech communication, data transmission, automatic re-broadcast for extending the range of tactical communications, and homing facilities. The homing mode may be used simultaneously with a communications channel without mutual interference. The range capability of the homing facility is equivalent to communications range.

The AD 190 covers the vhf band from 30 to 75.975 MHz at selectable channel spacing of either 25 or 50 kHz increments. Tuning is silent and instantaneous and transmitter output power is selectable at either 1 or 20 watts.

Multi-station, co-located operation is possible with a maximum of three systems in the same aircraft. This is subject to the proviso that antennas must be sited at least 3 feet apart and frequency separation of 3.5 per cent for three systems or 3 per cent for two systems must be maintained. The AD 190 incorporates diagnostic built-in test equipment.

Dimensions: (transmitter-receiver) 15 × 7.6 × 4.9 inches (385 × 197 × 125 mm)
(controller) 5.7 × 2.6 × 3.3 inches (146 × 67 × 85 mm)
Weight: (transmitter-receiver) 20 lb (9.2 kg)
(controller) 2 lb (0.9 kg)

STATUS: in production and service. Over 800 systems have been ordered.

AD 120 vhf/AM radio

Originally designed for civil aviation applications by the King Radio Corporation in the

GEC Avionics ARC 340 (AD 190) vhf radio system with antenna and switching unit

USA, GEC Avionics has requalified this vhf/AM system for military roles and, manufacturing under licence from King, markets the system under the designation AD 120. It serves aboard most United Kingdom military aircraft, both fixed-wing and helicopter, and has also been supplied to a number of overseas customers.

Covering the frequency band 118 to 135.975 MHz, the AD 120 provides 720 channels at a channel spacing of 25 kHz. Services provided are double sideband AM voice communication. Power output can be varied between 10 and 20 watts. The system is an all-solid-state equipment of modular construction and is designed for easy installation and maintenance. It is also exceptionally simple to operate. The system's standard remote controller possesses only five controls: an on-off switch, volume control, a test button and two rotary switches for frequency selection. Tuning is instantaneous. Automatic squelch and gain control eliminate the need for manual adjustment.

The self-test facility may be used during operation as a crew member's confidence check.

Dimensions: (transmitter-receiver) 2.3 × 4.95 × 12 inches (60 × 127 × 315 mm)
(controller) 5.7 × 1.83 × 3.5 inches (146 × 47 × 90 mm)
Weight: (transmitter-receiver) 5.06 lb (2.3 kg)
(controller) 1.32 lb (0.6 kg)

STATUS: in production and service. More than 3000 systems have been ordered to date.

AD 1550 communications control system

The AD 1550 communications control system is a single-box unit which governs an aircraft's entire intercommunication system and all incoming and outgoing audio services. This function permits all crew control station boxes to become 'passive' and also enables an aircraft manufacturer to tailor the station-boxes to match the design of the crew stations, while the electronic system concerned may be accommodated in the avionics bay.

The system's main feature is the incorporation of special techniques to improve speech intelligibility against the background noise invariably encountered on aircraft flight-decks. Although the system was designed initially for civil aircraft, these speech-intelligibility techniques have a possible read-across to military aircraft, in which the background noise environment can be appreciably more severe.

Format: 3/8 ATR

STATUS: AD 1550 system was introduced by GEC in mid-1981 and is in service on the British Aerospace 146 airliner and 125-700B business jet.

AD 3400 multimode secure radio system

The GEC Avionics AD 3400 provides, in a single transmitter-receiver, coverage of the entire airborne line-of-sight frequency band from 30 to 400 MHz with channel increments of 25 kHz. Up to 20 channels may be pre-selected in this range.

In effect, the system provides the equivalent of four radios in a single unit since it covers the following bands and modes: vhf/FM for tactical close-support, vhf/AM, the civil ATC band, vhf/FM civil and maritime bands, uhf/AM/FM military band. Its other major feature is that it provides secure communications for both speech and data transmission. For speech transmission in a secure mode, an encryption

GEC Avionics AD 120 vhf radio

GEC Avionics AD 1550 clear speech central audio unit and headphones

GEC Avionics AD 3400 transmitter-receiver unit

unit is used and for data security the appropriate modem is connected to the radio equipment.

Separate, continuously operating guard receivers are incorporated and homing and ADF facilities are also available with suitable keying and antenna installations.

Construction of the system features 'slice' techniques and large-scale integrated (lsi) circuitry. The system is based on Intel 8085 microprocessors and uses distributed processing. One of these microprocessors is in the control unit for the formatting of displayed data, to read, key-in and erase data from the non-volatile channel memory and to communicate with the main transmitter-receiver. Another is located within the transmitter-receiver to recover instructions from the controller and to apply them to the radio circuitry for control of the sub-modules and the system's bite programme. Proprietary lsi is also used in the synthesiser loops and the multi-sourced amplifiers in the receiver. Varactor diodes are employed in the power voltage control oscillators to provide a high output level with low noise. This technique reduces the amplification required in the power amplifier stage. The extensive use of lsi and proven components should provide a high measure of reliability. Serviceability is further enhanced, however, by the comprehensive BITE facility which can provide rapid fault diagnosis, particularly since intermittent fault data is stored in a non-volatile memory for maintenance purposes.

The AD 3400 has been designed for ease of installation with particular attention to retrofit requirements in older aircraft in which, says GEC, it is often possible to fit two of these systems in the space formerly occupied by a single radio. Attention has also been paid to growth potential and it is significant that although serial data between the controller and the transmitter-receiver is presently carried on an ARINC 429 highway, provision has also been made for a MIL-STD-1553B data-bus if required.

The secure speech facility is provided by an AA 34601 encryption unit. In this digital sequence, length, key setting and synchron-isation are the prominent factors which, according to GEC Avionics, 'render the system millions of times more secure than systems in current use'. Settings are programmed into the unit by a key-management equipment, the AA 34602, which is normally located in a secure area. These settings are transferred to the aircraft by means of a 'fill gun', designated AA 34604. Four independent key settings for pilot selection are stored in the aircraft encryption unit.

The system is convection cooled and is protected by a sensing device which progressively reduces power output at high temperatures to prevent transmitter damage. The transmitter is also protected from damage caused by short or open circuits at the transmitter output.

Dimensions: (AA 34001 transmitter-receiver) 7.6 × 4.9 × 10 inches (194 × 125 × 256 mm) (AA 34024 controller) 2.25 × 5.75 × 6 inches (57 × 146 × 152 mm) (AA 34601-1 encryption unit) 7.6 × 2.3 × 12.6 inches (194 × 58 × 320 mm)
Weight: (AA 34001 transmitter-receiver) 14.3 lb (6.5 kg) (AA 34024 control unit) 2.86 lb (1.3 kg) (AA 34601-1 encryption unit) 9.9 lb (4.5 kg)

STATUS: in production. In 1983 British Ministry of Defence placed £6 million order for this radio for Royal Air Force. Export orders exceed £10 million.

Direct-voice command technology

GEC Avionics has been actively involved in the development of speech technology to control avionic equipment since the mid-1970s. In early 1978 a two-year feasibility study began on the use of direct voice input (DVI) commands for use in combat aircraft. The project was funded jointly by the company and the UK Ministry of Defence, and a cockpit rig was constructed for testing and technique evaluation. From this work the feasibility of DVI for the cockpit was confirmed and development continued.

The DVI concept is intended to be an efficient multi-function command input system which permits the pilot to fly head-up in 'eyes-out-of-the-cockpit' and 'hands-on-stick-and-throttle' modes. It is seen as leading to the total exploitation of advanced avionics during high workload periods.

GEC Avionics is well advanced towards production of a flexible DVI system, tailored specifically to demanding cockpit environments, which is capable of high performance, real-time, continuous connected-word recognition. Key features will include a high noise-immunity to permit operation in harsh environments and sufficient flexibility to allow a user-program vocabulary and syntax for up to 1000 words. The system will be capable of dual-user operation to permit its employment in dual-seat cockpits, and for civil application in both fixed-wing aircraft and helicopters. It is envisaged that the system, in conjunction with a MIL-STD-1553B data-bus, will provide control functions for communications, navigation and stores management systems, and will also integrate with helmet-mounted sights and head-up displays.

STATUS: joint development of a speech recognition system is being conducted in co-operation with another GEC company, Marconi Secure Radio Systems Limited. The programme is being supported by the UK Ministry of Defence.

Secure speech communication system

The GEC Avionics secure speech communication system has been developed to enable strategic and intelligence information to be passed between aircraft and ships without risk of interception. Messages can be transmitted in the complete confidence that only the recipient can understand the contents. The system has been designed to allow integration with existing communications control equipment in naval and air force aircraft with minimum aircraft modification.

System development has involved the design and manufacture of 'A' and 'B' model line-replaceable units, system test rigs and special-to-type test equipment, and commissioning of the aircraft rigs. The production programme has been built on a combination of modern control methods and advanced automatic test procedures to ensure that the total set of units for each aircraft comes together as a proven system. All aspects of overall system cost, including those of ownership, installation, crew training, maintenance and support have been reduced by employing a modular concept. This concept allows system configuration to be cost-effective, giving maximum flexibility to meet customer requirements, with a built-in provision for expansion.

STATUS: project definition was commenced in 1978, development in 1979, and delivery of the first production system began in October 1985.

Communications/navigation interface unit (CNIU)

The communications/navigation interface unit (CNIU) developed by the Airadio Systems Division of GEC Avionics has been designed for the integration of the control functions for communications and navigation systems on modern aircraft. The system has been developed for use in a number of civil and military projects, and has found applications in a BAe

GEC Avionics AD 3400 control unit

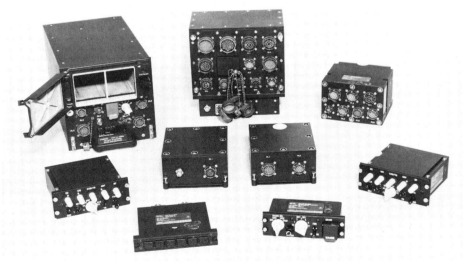

Units of GEC Avionics secure speech communication system

GEC Avionics communications navigation interface unit

(BAC) One-Eleven experimental aircraft, military helicopters, direct voice input experiments, and input/output processors for host computers.

Interface requirements are configured using a standard series of printed circuit boards which are designed to meet the needs of many different applications, and will also allow evolutionary changes to parts of the communication and navigation system without the necessity of a complete redesign. It is a microprocessor-based system which comprises a power supply, cpu card, and up to 10 input/output (I/O) cards including RS32, ARINC 429 and MIL-STD 1553B serial digital interfaces. User definable I/O connections are available and a user-programmable memory capability of 64 K bits is provided, expansible to 1 M bytes. Power consumption of the system is 15 or 45 watts depending on the configuration.

Dimensions: ½ Short ATR (195.4 × 127.4 × 315 mm)
Weight: 6 kg

STATUS: installed in the BAC One-Eleven experimental and research aircraft at the UK Royal Aircraft Establishment, Bedford.

ASD 60 data terminal system

The ASD 60 data terminal set (DTS) enables vital operational information to be transferred between military aircraft, ships, shore stations and battlefield units, and reduces congestion on the voice channels. Data generated by aircraft sensors can be processed and formed into appropriate messages for transmission over the data link. The ability of the ASD 60 to pass large amounts of information automatically, and with less risk of interception than voice transmissions, enhances the operational capability of units in the network. The system operates on hf or uhf, either as a net control station or as a picket, using a data rate of 2.4 K bits a second.

The DTS transmit function accepts digital data originating in the system digital computer and converts it to a form suitable for radio transmission. This conversion process uses digital phase shift keying techniques that encode and impress the digital data on to 15 different tones. Two data bits are impressed on each audio tone. The 15 tones, plus a Doppler tone, are then combined into a single multitone audio composite signal and used to modulate a carrier for transmission. Modulation is 40 dpsk.

ASD 60 data terminal set

The receive function accepts the audio composite signal from the receiver and, using a differentially coherent phase shift detection process, reproduces from this signal the digital data as originally impressed on the audio tones prior to transmission.

The system is remotely controlled by an integrated control and display unit and is capable of continuous, unattended operation. There are no operator controls or adjustment on the equipment that have to be set to maintain equipment operation. Built-in test and monitoring facilities provide a continuous check of equipment performance and ensure that some 90 per cent of all equipment failures will be detected. The system also incorporates automatic off-line test features which enable isolation of at least 90 per cent of equipment failures to two modules or less.

Dimensions: ½ Short ATR (195.4 × 127.4 × 315 mm)
Weight: 6 kg

Communications system for Nimrod AEW3

The mission communications sub-system for the Royal Air Force Nimrod AEW3 airborne early warning aircraft comprises hf and uhf equipment for transmission and reception of voice and data. It is controlled by a communications management computer although reversionary manual control facilities are provided.

The sub-system enables the aircraft to operate in conjunction with the UK Air Defence Ground Environment (UKADGE), Royal Navy ships and fighters such as the Panavia Tornado F2.

The definitive system comprises 127 line-replaceable units and achieves maximum

GEC Avionics Nimrod systems test-rig

GEC Avionics tactical control panel for Nimrod AEW.3

utilisation of a variety of uhf, vhf, hf and vlf radios on clear and secure voice, data and radio teletype communications, by a very flexible control system.

System development began in 1975 with a requirements study. A significant achievement was the completion of ground testing of the entire system which constituted the aircraft's tactical communication system. In addition, a special transportable ground station, which operates with the airborne system, was built for use in the flight trials and successfully proved the fully developed system.

STATUS: delivery of production equipment has been completed and assigned to production aircraft for further flight trials.

AD 150 voice enhancement systems

The AD 150 voice enhancement systems are intended to improve speech intelligibility and to reduce fatigue in noisy environments. They are designed to operate with all modern helmets and mic/tel systems for maximum flexibility of system application.

Many operational roles require the use of a 'hot' microphone facility which results in continuous noise on the crew intercom channels with a consequent degradation of speech quality and increased crew fatigue. The AA 1502 is a voice-operated switch (VOS) which is

Radio, control and integration units produced by GEC Avionics for Nimrod AEW3

GEC Avionics AA 1501/AA 1503 voice conditioning units

triggered only by a person's voice and mutes the background noise since it is off the rest of the time. A very fast acting switch is required to avoid cutting off the first syllable, and in the AA 1502 activation occurs in less than 1 millisecond. To prevent ambient noise operating the VOS, automatic noise tracking circuits ensure voice-only activation.

The AA 1501 voice conditioning unit employs all the features of the AA 1502 together with a fast acting voice-operated gain adjustment device (VOGAD) which smooths out differences in voice level to provide a constant modulation level for communication transmitters. To ensure optimum signal control and maximum speech intelligibility, the VOGAD is coupled to noise tracking speech recognition circuitry. Intelligibility is further enhanced by frequency tailoring techniques which ensure that a rising frequency response is used to reduce the effects of low frequency masking on high frequency components of speech. A separate input from a noise sensor, Type AA 1503, monitors ambient noise levels and automatically controls the output of receiver and crew interphones to eliminate the need for constant readjustment of volume controls.

Only very low power is required and the AA 1502 and AA 1501/1503 units are small enough to be mounted within the cockpit and simply plugged into the communication control system. As well as applications in all types of aircraft, they are equally suitable for use in any high ambient noise-level environment, eg battle tanks, armoured personnel carriers, self-propelled guns, ships' engine rooms. Primary settings are provided to cover the noise prints of helicopters, jet aircraft, armoured/tracked land vehicles and US standard inputs. There are six other independently settable adjustments to tailor units specifically to the customer's system and environment.

Dimensions: (AA 1501 and AA 1502) 35 × 85 × 110 mm
(AA 1503) 16 mm high × 37 mm diameter
Weights: (AA 1501 and AA 1502) 0.5 kg
(AA 1503) 50 g
STATUS: in production for British Army armoured fighting vehicles.

AD980 central suppression unit

The AD980 central suppression unit is designed for co-ordination of all transmitter and receiver equipment in military aircraft in order to prevent mutual interference between high-power systems such as radar, IFF and Tacan and communications systems. It can deal with up to 30 input and 30 output channels and can combine up to eight outputs into a single one.

Each transmitter when in operation outputs a 'suppression' pulse and can accept a blanking pulse from other systems. The central suppression unit receives and transmits these pulses and controls the blanking pulses with a hard-wired logic matrix. The matrix is programmed for particular aircraft configurations and may be changed to meet new requirements such as equipment retrofits.

Dimensions: 1.7 × 11.7 × 2.8 inches (43 × 288 × 72 mm)
Weight: 3.3 lb (1.5 kg)
STATUS: in production and service.

Digital speech communication system

The digital speech communication system can be integrated with existing communications-control equipment in Royal Navy and Royal Air Force aircraft with the minimum of aircraft modification. Specific variants of the system have been configured for Royal Navy helicopters and fixed-wing aircraft and for Royal Air Force maritime reconnaissance aircraft. Successful flight and ground trials of the systems have been conducted at the Royal Aircraft Establishment.

STATUS: in production.

Type AD 950 anti-jam communications system

The AD 950 is a low-cost anti-jam system designed to work with existing clear-speech radios. It is intended primarily for use with interceptor type fighters and receives encrypted messages from a ground controller in the presence of jamming. Typical information provided is height, bearing, distance and speed of a target, and commands to break left or right or return to base. The information is displayed on a panel-mounted receive/display unit in alphanumeric format, and is protected during transmission by the use of encryption. The encryption code is 10^{15} bits long with eight variable keys.

The complete AD 950 system consists of an airborne receiver, a portable keyboard for both ground and airborne use, a fill management and a fill-gun. Encryption codes are produced by the fill management unit and optically transferred to the airborne receiver and keyboard by the fill-gun which carries all eight codes simultaneously. Information can be transferred between aircraft using a portable keyboard. Transmit and receive units are completely independent and information can be a two-way flow if required. The airborne keyboard only requires to be connected to the

AD 950 airborne receiver

28-volt supply, microphone input, and PTT lines.

The design includes error-detection codes which ensure that the message received is correct. If the codes do not match the display is blanked. Frequency shift coding is used to transmit the message since this enhances the signal in a high-noise environment. With only 71 bits in the message, allowing for it to be repeated three times, the total transmission time is less than 0.5 second. A bit rate of up to 600 bits a second is used to avoid any problems with phase errors in narrow bandwidth receivers. Interfacing to existing installations is simple and the airborne receiver unit is a standard control type fitting. It can be used in conjunction with hf, vhf and uhf radios having normal clear speech bandwidths.

Dimensions: (receiver) 67 × 145 × 103 mm
(keyboard) 138 × 76 × 25 mm

Type AA27811 1553B interface unit

The Type AA27811 is designed to provide existing equipment with the MIL-STD-1553B facilities necessary for compatibility with other digital equipment in military aircraft wired to that standard without expensive redesign. An additional advantage is that in systems where the number of radio/telephones on the bus is reaching the limit, the unit can be used as a single radio/telephone but in turn can interface with up to three other boxes via dedicated ARINC-format highways. The unit design also allows other dedicated highway formats to be catered for, for example ARINC 568 and Panavia 64 Kbit. The AA27811 contains two printed circuit boards and for ease of access can be split down the middle to expose the two cards.

Dimensions: 296 × 43 × 127 mm
Weight: 1.32 kg
Power: 0.5 A at 28 V dc

Graseby Dynamics

Graseby Dynamics Limited, Park Avenue, Bushey, Watford, Hertfordshire WD2 2BW
TELEPHONE: (0923) 28566
TELEX: 923010

Sarsat radio beacon

This radio beacon has been developed by Graseby, part of the Cambridge Electronic Industries group, as part of a world-wide search and rescue system. This system known as Sarsat (search and rescue satellite aided tracking) is operated jointly by the USSR and Western countries using three satellites to monitor the world surface continually for distress signals. A fourth satellite is expected to come into service in 1986. The distress beacons are normally stowed aboard lifeboats and

liferafts and taken in survival packs by aircrew members. In emergency situations signals are emitted automatically, are relayed by a repeater/processor on the satellite to the nearest ground terminal and onwards to a mission control centre. The system is capable of handling multiple distress calls should there be more than one survival craft or if there are several simultaneous incidents to service. The beacons are used by civil airlines, military aircraft and shipping companies.

The Graseby beacon operates on a frequency of 406 MHz and transmits a 5-watt, 400-millisecond burst every 50 seconds. The signal data format consists of the class of user, country of origin, identity and type of emergency. It is a compact unit which includes batteries to give an operating life of 48 hours.

Weight: 2.5 kg (including batteries)

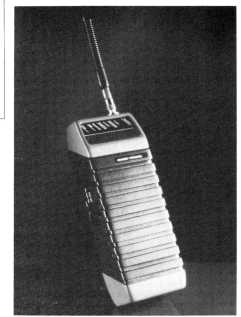

Graseby Sarsat distress beacon

Marconi Defence Systems

Marconi Defence Systems Limited, The Grove, Warren Lane, Stanmore, Middlesex HA7 4LY
TELEPHONE: (01) 954 2311
TELEX: 22616

Master airborne satcoms terminal

Marconi Defence Systems is currently involved in a joint collaborative programme with the UK Ministry of Defence in the development of an airborne satellite communications terminal. The programme is intended to develop a terminal operating in the 7 to 8 Ghz frequency range to provide beyond-line-of-sight voice and data link facilities for the long-range aircraft of the Royal Air Force, operating through the Skynet 4 military communications satellite.

Marconi has developed a compact high performance dish antenna, mounted on a microprocessor-controlled mechanical pointing gimbal with either two or three axes. The equipment uses a design known as the hybrid splashplate feed to obtain the required antenna performance. Pointing accuracy will be of the order of 1°, and selection of either a two or three-axis system would depend on cost/accuracy considerations.

STATUS: Master airborne terminal was due to be mounted in the tail fin radome of a Nimrod aircraft in early 1986 for tests with the Skynet 4 satellite later in the year.

Marconi Secure Radio Systems

Marconi Secure Radio Systems Limited, Browns Lane, The Airport, Portsmouth, Hampshire PO3 5PH
TELEPHONE: (0705) 664 966
TELEX: 86666

Voice encoder/decoder

Marconi Secure Radio Systems (MSRS) has developed a channel vocoder equipment designed to operate in environments of high acoustic noise and high bit error rates, such as those encountered in tactical military roles. The 2400 bits a second vocoder incorporates a robust pitch extractor and regenerator, automatic speech level control and forward error correction. According to MSRS the system provides improved resistance to acoustic noise and transmission errors over other systems such as linear predictive vocoders. The pitch extraction technique is less sensitive to acoustic noise and forward error correction, and median smoothing/majority vote methods applied to the pitch information result in reduced sensitivity to transmission errors.

Good communication has been achieved between operators in high acoustic noise and transmission conditions. Subjective tests, using standard helmets and face-mask microphones, have been carried out in a noise chamber where operators were subjected to levels of 120 dB (typical fighter noise) and 115 dB (fighter-bomber level), said to be representative of high-speed, low-level flight. Communication could be maintained at these noise levels for random transmission error levels of up to 2.5 per cent.

The system is almost completely digital in character and is based upon an Intel 8085 microprocessor chip combined with a special purpose digital filter bank. It is now being manufactured as a four-board pack.

Dimensions: 51 × 176 × 181 mm
Weight: 1 kg

STATUS: in production as a four-board pack.

LPC-10 JTIDS vocoder

The LPC-10 vocoder board was developed by Marconi to convert analogue speech to narrow-band digital data at 2.4 K bits a second, for incorporation in the UK Joint Tactical Information System (JTIDS) based on the Singer-Kearfott US JTIDS Class 2 Terminal. The vocoder is engineered on a standard Class 2 Terminal frame and is to the agreed US/UK/NATO standard for LPC-10. Operation is half-duplex and audio bandwidth is 100 Hz to 3.6 kHz. Power consumption is 20 watts maximum.

Dimensions: 12.4 × 6.5 × 1 inches (315 × 165 × 25 mm)

VSC 330-A modem

The VSC 330-A is a modem designed for airborne satcom terminals operating with geostationary satellites to provide highly jam-resistant communications. The equipment provides the facilities of a modem, a multiplexer and an anti-jamming device. Anti-jam, ECCM is achieved by a combination of frequency-hopping spread-spectrum and a reduction of the link data-rate giving a better power-to-bandwidth ratio. Selection of hopping rates, widths and link data-rate are set up manually by means of a front panel touch-pad, or may be preset and stored within the modem for simple selection.

The power supplies and timing clocks are provided centrally within the equipment. The operation of both transmit and receive sections of the modem are controlled by a common front panel. In other respects the transmit and receive paths of the equipment are functionally independent. In the transmit path a data terminal controller module accepts the data received from the terminals, and in some cases performs source encoding to enable a more efficient transmission. The transmit controller buffers, encodes, bit-stuffs and interleaves the data. In the receive direction the input signal from the radio receiver is filtered by a tracking filter, amplified and frequency converted. The signal processor demodulates, decodes, de-interleaves and de-stuffs the received signal. The data terminal controller then provides the interface to the terminal equipment. Traffic capability is up to five duplex channels. When jamming is less severe it is able to transmit/receive data or speech at data-rates in the 2.4 to 16 K bits a second.

The front panel of the unit contains all the operating switches and controls. Major parameters such as power on, signal acquired, buffer store full, microprocessor program running correctly are indicated as functioning correctly by light-emitting diodes. Other information is displayed numerically on the display panel and is selected by means of the touch pad. This information includes frequency, hop code, hop bandwidth, terminal data-rate, link data-rate, satellite delay and receiver noise level.

The VSC 330 is available in two other versions, the VSC 330-L for use in vehicles and the VSC 330-S for naval applications. It has been purchased by the UK Ministry of Defence for Army and NATO use.

Dimensions: (rack-mounted) 483 mm (height) 178 mm
Weight: 27 kg (60 lb)

MarCrypDix Air secure speech system

MarCrypDix Air is a secure speech encryption equipment developed specifically for installation in civil and military aircraft. It is designed to interface with any airborne digital communication system and provides high-grade speech and data encryption. There is storage for eight different key variables (code settings), fully automatic synchronisation with zero error extension and negligible range degradation.

The key variables are generated by a key management unit and are inserted either manually, using a hexadecimal keyboard, or may be generated automatically within the key management unit in a random manner. Up to 16 variables can be stored within the management unit awaiting transfer to the airborne units. Transfer of the key variables from the management unit to the encryption unit is accomplished by a device known as an optical fill-gun. This is a battery operated, pocket-sized device with an internal memory system, data interfacing being performed by light pulses. Input information is received from the management unit by a photo-diode and the stored information is transferred optically to the encryption unit via a light-emitting diode. The fill-gun mates mechanically with the management and encryption units to ensure optical alignment. Key variables stored within the gun can be transmitted to any number of encryption units. Provision is also made for the instant erasure of all information stored within the gun's memory. The number of key variables is 2^{128}, with a key variable length of 128 bits, plus 16 further bits set uniquely for each customer. An alternative version is available which restricts code settings to a sub-set of

Four-board vocoder pack

Marconi MarCrypDix Air

10^{10}, selectable manually using ten thumbwheels which avoids the introduction of a fill management system.

The equipment is self-monitoring and will stop encrypting should any fault occur in the equipment, thus ensuring that sensitive information cannot be transmitted with less than complete protection.

Although MarCrypDix Air has been designed for the airborne environment, the same unit, suitable mounted, can be employed in a ground role for ground-to-air or ship-to-air communication.

Dimensions: 193.5 × 320.5 × 58 mm
Weight: 4 kg

STATUS: in production.

MarCrypFlex

MarCrypFlex is designed for the highest grade digital crypto-protection of voice transmission over narrow-band (3 kHz) radio channels. It is contained in a single unit and consists of a 2.4 K bit a second vocoder, a standard MSRS cryptograph, a multi-tone modem and a power supply. It is available in two versions, MarCrypFlex 113 which is ruggedised for use in military environments, and a 213 version which is more suitable for use in an office place. Code setting to select any one of 10^{12} settings is carried out by the operation of thumbwheels.

The equipment will fully protect voice transmissions over any link which can carry the plain voice equivalent, including long-range hf radio circuits and satellite links.

STATUS: in production.

Marconi MarCrypFlex

SR-128 speech recogniser

The Marconi SR-128 speech recogniser system is a voice-command equipment for use in both military and civil applications. The system is

designed so that the pilot is able, in an airborne role, to control equipment non-manually. The 'personalised' voice recognition element depends on matching word commands voiced by the pilot to identical words recorded by the same pilot on templates which are transferred to a cassette and loaded into the system. In this way the likelihood of error is considerably reduced.

The SR-128 uses the mathematical technique of dynamic programming to carry out acoustic pattern matching between the input utterance and the pre-stored reference templates. This operates on the principle that, for a single speaker, the acoustic signal produced for a particular word will be similar to a pre-recorded version, or template, of that word. The template preparation procedure need only be performed once, after which the templates are transferred to a mini-cassette. Operation then becomes only a matter of loading the cassette into the aircraft's voice recognition system.

The SR-128 can apply syntax to isolated keywords. The syntax for each airborne application is prepared on the ground by the pilot/operator, using a visual display unit and keyboard before being transferred to the mini-cassette. The equipment recognises normal connected speech in real-time and outputs the recognised texts to the front panel display at the end of a phrase or sentence.

The airborne version of the SR-128 has been undergoing trials in a BAe (BAC) One-Eleven aircraft at the Royal Aircraft Establishment, Bedford. Trials are now continuing in a Buccaneer Mk2B. The primary aim of these airborne tests is to evaluate recognition performance in the presence of cockpit noise, and were conducted at 550 knots at a height of 250 feet (76 metres) with a cockpit noise level of some 115 dB. Under these conditions a 98 per cent success rate was achieved using isolated words and 95 per cent using connected speech input of three words. The trials flying is scheduled to continue throughout 1986 with in-flight application of speech recognition.

STATUS: in production.

MEL

MEL, Manor Royal, Crawley, West Sussex RH10 2PZ
TELEPHONE: (0293) 28787
TELEX: 87267

Mandarin data link

Mandarin is a secure, multi-channel microwave link with civil and military capabilities. It

provides secure transmission of digital data with simultaneous real-time video. Multiple frequency band options allow optimum choice of frequency for each application, while the ability to use multi-channels on the chosen frequency enables maximum use to be made of that frequency. This capability is particularly important where frequency availability is limited.

Data is transmitted by secure means via a data link system using burst transmission and

spread modulation. The use of spread spectrum, pseudo-random noise modulation achieves a very high degree of security whereby the transmission becomes hidden within the normal transmission noise, making it almost impossible to locate. In addition, this method gives a high resistance to jamming.

Plessey Avionics

Plessey Avionics, Martin Road, West Leigh, Havant, Hampshire PO9 5DH
TELEPHONE: (0705) 486391
TELEX: 86227

PTR 1751 uhf/AM radio

The PTR 1751 is a lightweight uhf/AM radio designed for all types of military fixed-wing aircraft and helicopters. It provides 7000 channels in the frequency band 225 to 399.975 MHz at a channel spacing of 25 kHz, and is available with either 10- or 20-watt transmitter outputs. Channel spacing of 50 kHz is available as an option.

Both versions comprise a single transmitter-receiver unit with either a manual controller or, optionally, a manual/preset controller. Each controller provides full selection of the range of 7000 channels together with control of the built-in test functions; the manual/preset unit additionally permits pre-selection of up to 30 channels, plus guard frequency, through in-

PTR 1751 radio installed in British Army Lynx helicopter

corporation of a non-volatile memory store. A remote frequency and channel indicator is also available.

Options include continuous AM monitoring on 243 MHz (the international distress frequency) via a separate guard receiver module

which plugs directly into the main transmitter-receiver chassis, an external homing unit and a wide-band secure-speech facility which requires no additional interface equipment.

The PTR 1751 conforms generally to DEF-STAN-07-55 and operates satisfactorily over a temperature range from –25 to +70° C. It is of all-solid-state modular construction with high reliability as a principal design aim, and Plessey claims a current mean time between failures of 800 hours.

Dimensions: (transmitter-receiver, 10 W version) 1/2 ATR Short × 160 mm high
(20 W version) 1/2 ATR Medium × 160 mm high
(manual controller) 146 × 48 × 108 mm
(preset controller) 146 × 95 × 108 mm
Weight: (transmitter-receiver, 10 W version) 5 kg
(20 W version) 6.7 kg
(preset controller) 1.4 kg
(manual controller) 0.7 kg
(guard receiver module) 0.3 kg

STATUS: in production and service.

PTR 1741 vhf/AM radio

The PTR 1741 equipment may be regarded as the vhf counterpart of the Plessey PTR 1751 system, both units being members of the same family of Plessey military airborne radios. Covering the frequency range from 100 to 155.975 MHz, the PTR 1741 provides 2240 channels at 25 kHz separation. If desired, however, channel spacing can be set at steps of 50 kHz to ease interface with older equipment which does not have the same frequency stability as new-generation systems. As in the case of the PTR 1751, the PTR 1741 is available with a choice of manual or manual/preset control unit; however, only a single power output level (10 watts) is available.

The PTR 1741 shares a number of common features with the PTR 1751 in regard to build standard, operating range, options etc. Both are suitable for dual uhf/vhf installation, their electrical interfaces and mechanical dimensions being identical. A single controller may be used to operate the two transceivers in such a dual installation.

Dimensions: (transmitter-receiver) ½ ATR Short × 160 mm
(manual controller) 146 × 48 × 108 mm
(preset controller) 146 × 95 × 108 mm
Weight: (transmitter-receiver) 5 kg
(manual controller) 0.7 kg
(preset controller) 1.4 kg
(guard receiver module) 0.3 kg

STATUS: in production and service.

PTR 1721 uhf/vhf radio

The PTR 1721, a combined uhf/vhf radio covering the uhf band 225 to 400 MHz and the vhf band 100 to 156 MHz, is designed for all types of military aircraft. Plessey says that the ability to communicate on vhf in addition to uhf offers increased flexibility to air forces which, on occasion, may need to operate from civil airfields equipped with vhf facilities only. The PTR 1721 has been chosen for the British version of the tri-national Panavia Tornado attack aircraft

The PTR 1721 offers up to 9240 channels, 2240 on vhf and 7000 on uhf. Channel spacing

Plessey PTR 1721 combined vhf/uhf system

in the vhf band is at 25 kHz and standard spacing in uhf is at 50 kHz although an optional interval of 25 kHz spacing in uhf is available. Frequencies are selectable directly from the remote control unit which also provides pre-selection of up to 17 channels with the addition of the uhf and vhf guards at the international distress frequencies of 243 and 121.5 MHz respectively. Frequency synthesis techniques ensure good frequency stability and channel selection characteristics.

The system is entirely solid-state in construction and, wherever possible, employs conventional technology in the interests of reliability enhancement. The receiver is varactor-tuned and the frequency synthesiser is compared against a single reference oscillator.

A sealed case houses the transmitter-receiver unit and access to the modules is gained by removal of the sides of the case, to which the modules themselves are attached. Heat is conducted from the modules via the chassis, which acts as a heat sink, to the sides of the case and hence to external air. Forced cooling is provided by the mounting tray installation which also acts as a vibration insulator. Forced-air cooling may be dispensed with, in less demanding aircraft environments.

Switching for dual control operation is external to the system and in two-seat aircraft, identical control units are installed at each crew position. Options include an antenna lobe switch for azimuth homing requirements and a uhf antenna switch for automatic direction finder operation.

The environmental temperature range extends from −40 to +70° C ambient and the normal operational altitude is to a maximum of 50 000 feet (15 240 metres), although operation is possible for short periods up to altitudes of 70 000 feet (21 330 metres).

Dimensions: (transmitter-receiver) 125 × 194 × 339 mm
(controller) 94 × 145 × 175 mm
Weight: (transmitter-receiver) 11.25 kg
(controller) 1.8 kg

STATUS: in production and service.

PVS 4750 vhf/uhf communications system

The PVS 4750 is a vhf/uhf ECM-resistant communications system and is part of a family of such equipment developed by Plessey and known under the name of Shelter (slow hop electronic threat evasion radio). The transceivers are modified versions of the PTR 1741 and PTR 1751 radios, with the use of frequency-hopping techniques on uhf to provide reliable communications in an ECM environment. They cover the frequency range from 100 to 155.975 MHz and 225 to 399.75 MHz giving 2240 channels on vhf and 7000 on uhf. Output power is 10 watts on vhf and 20 watts on uhf. The PTR 4750 can be used on all types of aircraft and helicopters, either as new installations or as a retrofit into existing aircraft.

The system comprises four principal units: an optional vhf transceiver type PTR 4741, a uhf transceiver type PTR 4751, an applique unit and a control unit. The applique unit provides the interface between the vhf and uhf transceivers, a preset channel storage, and circuitry for the ECCM operation consisting of frequency definition, control and synchronisation. The transceivers and applique unit are designed as a combined centre. A range of control units is available.

Dimensions: (complete centre) 312 × 346 × 300 mm
Weight: (complete centre) 18 kg

Racal Acoustics

Racal Acoustics Limited, Beresford Avenue, Wembley, Middlesex HA0 1RU
TELEPHONE: (01) 903 1444
TELEX: 926288

Six-Ninety series station boxes

Racal's Six-Ninety series station boxes are designed to meet a diverse range of intercommunication and audio integration requirements

Racal Six-Ninety series station boxes and amplifiers

in light-aviation, civil and military roles. The communication control systems provided range from simple-to-operate single-station boxes for light aircraft to multi-station box fits for military aircraft fulfilling a variety of roles including AEW, ASW, MR and search and rescue tasks. The Six-Ninety series boxes are versatile, interchangeable units which may be adapted to control large numbers of transmitters, receivers, intercommunication facilities and mic/tel types.

Standard or adaptively-engineered units are used to assemble Six-Ninety communication control systems which can be used irrespective of transceiver, navaid or headset type, and regardless of impedance or input/output levels. Facia panel engraving, lighting, colour, switch use and electrical characteristics are configured to meet either new or retrofitted systems in both fixed-wing aircraft and helicopters.

STATUS: in production.

Six-Ninety series audio amplifiers

Racal's Six-Ninety series of audio amplifiers has been developed to meet the requirements of aircraft intercommunication, microphone conversion and amplification, audio isolation, telephone and loudspeaker drive. Units in the range are said to be able to satisfy the on-board requirements of any type of aircraft. Inputs at telephone, electromagnetic, dynamic or carbon levels can be accommodated to provide operational flexibility with corresponding outputs to suit impedances of most current telephones, headsets or loudspeakers.

A special version, the A697 cockpit voice recorder (cvr) summing amplifier is designed to meet all possible needs in aircraft having to satisfy the mandatory requirement for cvr installations. All audio signals monitored by the pilot, co-pilot and a third crew-member, together with 'hot-mike' signals in all control column switch modes, can be recorded on channels of a cvr in association with a separate 'area' microphone.

The A697 can also provide the interface between aircraft audio systems from different manufacturers. Electromagnetic, dynamic or carbon microphones may be employed and input may be made to any one of several approved cvr systems.

In the A6912, a development of A697, the third crew member audio summing circuitry has been replaced by circuitry enabling helicopter rotor speed to be recorded. Signals relating to the rotor speed are taken from a tachometer, encoded in the A6912, and recorded on one channel of the cvr.

STATUS: in production.

RA800 IDACS integrated digital audio control system

Designed to satisfy the present and envisaged future requirements for audio management, the RA800 system can be installed in a variety of military and civil fixed-wing aircraft and helicopters with a minimum of custom engineering to integrate and control all the audio signals.

Two control units from Racal Acoustics RA800 series

Control, interfacing and amplification is provided for all audio equipment on the aircraft, together with a sophisticated and flexible intercom network for the crew. Crosstalk and interference are reduced to a minimum, and control is exercised over MIL-STD-1553B or ARINC data-links. A single communications audio management unit can support up to eight major crew stations or intercom nets, providing interfacing and control for up to eight transmitter-receivers, eight receivers, eight unswitched audios and four recorder outputs. Options include voice and tone-alerting, eight-second audio recall, speech security and active noise reduction. Built-in test routines and dual, separated power supplies ensure increased reliability and failure damage protection.

Jaguar-U (BCC 72) uhf radio

The Racal Jaguar-U system is a range of radio communications equipment, operating in the uhf band, available for airborne, shipborne and land use. The airborne version, BCC 72, is designed to operate over the frequency range 225 to 400 MHz and provides fixed-frequency or frequency-hopping FM with selectable encryption, and fixed-frequency AM compatible with current operational systems. It provides 7000 channels at 25 kHz spacing, with 30 programmable channels (including a guard channel on the distress frequency of 243 MHz) with flexibility to select clear, secure, fixed and hopping modes.

Design of Jaguar-U is based on experience gained in the Jaguar-V systems and employs the same method of medium-speed frequency-hopping to protect against interception, direc-

Racal Acoustics BCC 72 transceiver and BCC 584B control

tion-finding and jamming. To simplify frequency management the 225 to 400 MHz band has been divided into 13 sub-bands, allowing the co-sited operation of multiple radio systems without interference. However, the radio can also be programmed to use any one of three larger hop-bands. In either mode, orthoganol hop-sets are available to assist frequency management. Large numbers of nets can operate in the same frequency bands and individual bands can be 'barred' to avoid jamming or to protect other fixed-frequency stations. Selective communication can be performed within an individual net, or conversely a radio can be selectively barred from a net should it be captured. Once the radios have been programmed with hop codes and frequencies they synchronise automatically without the need for 'time of day' input. Communications security is also enhanced by the use of either a built-in encryption unit using a second keystream generator or any external 16 K-bits a second system in fixed or hopping frequency modes.

The airborne BCC 72 system consists of a rack-mounted transceiver and a panel-mounted control/display unit. The transceiver provides two output powers, either 10 mW or 15 watts on FM, and 40 mW or 40 watts pep on AM, output level being selected on the controller. Two control/display units are available, depending on the aircraft requirements. The BCC 584B gives full manual selection of any of the 7000 channels, plus selection of the 30 programmable channels, mode selection (hopping or fixed), low or high power selection and frequency/mode display by light-emitting diodes. The alternative controller provides for selection of the 30 programmable channels and the operating mode.

Dimensions: (transceiver) 230 × 90 × 350 mm (controller BCC 584B) 145 × 66 × 104 mm (controller BCC 584C) 62 × 168 × 110 mm
Weight: (transceiver) 6.5 kg (controller BCC 584B/C) 0.8 kg

STATUS: in production.

Racal-Tacticom

Racal-Tacticom Limited, PO Box 112, 472 Basingstoke Road, Reading, Berkshire RG2 0QF
TELEPHONE: (0734) 875 181
TELEX: 848011

BCC306 helicopter vhf/FM transmitter/receiver

The BCC306 is a development of the Clansman RT351 vhf receiver/transmitter which is in service with the British Army for vehicle and manpack use, and is intended for tactical use between airborne and ground-based units. The

BCC306 provides air-to-ground and air-to-air voice communications over the 30 to 76 MHz frequency range, giving 1841 channels at 25 kHz spacing. The equipment functions in the single-frequency simplex mode and uses F3E (narrow-band) modulation. It is simple to operate with channel selection and all other control functions via a remote control unit.

The system consists of the transmitter/receiver, remote control unit, power supply unit, antenna tuning unit and an aircraft interface unit, with an overall weight of 10.37 kg. The transmitter/receiver is an RT351 equipment from which all the controls have been removed and resited on a remote control unit near the pilot.

The tuning unit is capable of tuning the 1.7-metre whip antenna over the complete frequency range. The unit does not require frequency setting or antenna impedance information, antenna matching and tuning being carried out automatically following operation of the appropriate switch on the control unit.

STATUS: in operational service with the British and other armed forces. Although developed originally for installation in helicopters, especially the Westland Wessex, the equipment has also been fitted in various types of light aircraft.

UNITED STATES OF AMERICA

Allied Technology

Allied Technology Inc, 6104 Poe Avenue, Dayton, Ohio 45414

AN/ARC-84 vhf nav/com radio

The Allied Technology AN/ARC-84 is a military navigation-communications radio system operating in the vhf band. For communications purposes it provides 360 channels in the frequency range 118 to 136 MHz and, additionally, covers a further 200 channels in the range 108 to 118 MHz for reception of VOR and ILS navigation aids. Channel separation throughout the entire frequency coverage is at 50 kHz. Transmitter power output is 25 watts and the system operates in FM mode. Duplex transmit/receive mode is available over the frequency ranges 124 to 127 MHz and 133 to 136 MHz.

The system comprises three major units, separate transmitter and receiver sections and a remote controller. Transmitter and receiver both employ a mixture of conventional glass-envelope thermionic valve and solid-state technology. Crystal-controlled synthesis techniques are used for frequency-generation. The ARC-84 is operable over a temperature range of –40 to +50°C at pressure altitudes of up to 30 000 feet (9000 metres).

Dimensions: (transmitter) 7.875 × 3.625 × 17.625 inches (200 × 91 × 444 mm)
(receiver) 7.75 × 2.5 × 14.5 inches (197 × 63 × 368 mm)
(controller) 1.875 × 5.75 × 5.625 inches (48 × 146 × 143 mm)

Weight: (transmitter) 15.1 lb (6.8 kg)
(receiver) 10.7 lb (4.8 kg)
(controllers) 1.6 lb (0.72 kg)

STATUS: in service.

AN/ARC-101 vhf nav/com radio

Allied Technology's AN/ARC-101 is a military combined vhf navigation-communications radio system which has many similarities with the company's ARC-84 system. The ARC-101, however, provides over 680 communication channels in the frequency range 116 to 150 MHz and 880 channels over the range 108 to 152 MHz for communication and VOR/ILS reception respectively. Channel spacing throughout is in increments of 50 kHz. Transmitter power output is 20 watts.

Like the ARC-84, the ARC-101 employs a mix of thermionic valve and solid-state technology and has separate, remote-controlled transmitter and receiver sections with crystal-controlled, synthesised frequency-generation. The ARC-101, however, can operate to pressure altitudes up to 55 000 feet (16 700 metres), although over the same temperature range as the ARC-84, namely –40 to +70° C. It serves aboard a number of US Navy maritime-patrol aircraft.

STATUS: in service.

C-8616/ARC and C-7307/ARC-51A controllers

The C-8616/ARC is used for the manual control of radio sets AN/ARC-51B, AN/ARC-51BX, RT-767/ARC-51B, RT-742/ARC-51BX, RT-767B/ARC-51B and RT-742/ARC-51BX. The C-7307/ARC-51A serves the same purpose for the AN/ARC-51A, AN/ARC-51AX and similar variants of the AN/ARC-51A. Information on the AN/ARC-51 uhf transmitter/receiver series is given under the Lapointe Industries entry.

Both control sets enable a manual selection of any of 3500 frequencies over the range 225 to 400 MHz in 25 kHz steps. Up to 20 frequencies can be preset into a memory drum for immediate access by the operator. A hinged cover plate at the top of the panel is raised to display a legend, a slot containing eight movable pins, and an adjustment tool. Using the legend on the underside of the cover, the operator can programme each of the 20 channels with any frequency within the range.

Dimensions: (both controllers) 125 × 136 × 106 mm
Weight: (both controllers) 1.72 kg

ID-1752/ARC and ID-1472/ARC-51A frequency channel indicators

The ID-1752/ARC and ID-1472/ARC-51A are frequency channel indicators designed for use in conjunction with the AN/ARC-51A controllers described in the previous entry. They provide a remote display of the frequency selected and are normally used when the mounting position of the associated controller is advantageous for manual adjustment, but awkward for visual observation. They can be mounted at any reasonable distance from the controller.

Dimensions: 101 mm long × 51 mm dia, with a mounting flange of 63 × 63 mm.

Bendix

Bendix Avionics Division, Allied-Signal Inc, 2100 NW 62nd Street, Fort Lauderdale, Florida 33310
TELEPHONE: (305) 776 4100
TELEX: 51447
TWX: 510 955 9884

An established leader in the avionics field, the company was founded in 1914 by an engineer, Vincent Bendix, to meet the needs of the newly emerging automotive industry. From that strong financial base Bendix began acquiring, in 1928, small aviation orientated companies, which were later consolidated into a new organisation known as the Radio Division of Bendix Aviation Corporation. By 1936 the company had developed the first automatic direction finder, and during World War II built up its electronics business, notably in radar. By the late 1940s and early 1950s the product range had expanded to include airborne communication and navigation equipment, marker beacon receivers and passenger address systems. During the 1950s the company also introduced the first commercial weather avoidance radar and the first commercial self-contained Doppler navigation system.

With a view to securing a place in the growing business aircraft field, Bendix acquired in 1963 the aviation product lines of Motorola Aviation Electronics (itself a hand-down from Lear Siegler) and Transco Inc. By 1965 Bendix Avionics was offering one of the widest ranges of aviation electronics anywhere in the world. In 1966 the company moved from its original site in Maryland to Fort Lauderdale in Florida.

In response to continuing growth Bendix Avionics was re-organised in 1980 into the General Aviation Avionics Division and the Air Transport Avionics Division, the latter covering both airline and government business. In September 1982 the digital Series III family, ARINC 429 compatible, was introduced. In 1984, in a further move to consolidate its position in general aviation, the company

bought up King Radio after several years of negotiations, the price paid was about $110 million. The rationale was that the new Series III digital avionics and the King Gold Crown line, mainly analogue, would challenge other manufacturers at the medium and top end of the business aircraft market, while the King Silver Crown range would fill the gap left by dropping the Bendix BX-2000 series, which did not meet with wide acceptance.

Apart from the 'firsts' mentioned earlier, Bendix claims to be the first company to incorporate transistors into avionics equipment, to use VOR check self-test techniques, to develop super squelch, to produce flush-mounted ADF antennas, to use solid-state synchros, to use lsi technology in avionics, and to provide a digital display and multi-function capability for weather radars.

Series III integrated digital avionics system

Bendix announced a family of integrated digital avionic equipment at the 1982 convention of the National Business Aircraft Association, and the project began to crystallise the following year. Originally and informally known as the X-line, the system has now been designated Series III, indicating it to be the third range of digital electronic equipment developed by Bendix. Based on ARINC 429 data-buses, the system was also designed for compatibility with a new line of EFIS displays being produced by the company.

The current Series III family comprises the following equipment:
VCS-40 vhf communication transceiver
VNS-40 vhf navigation receiver
DFS-43 automatic direction finder
DME-44 distance measuring equipment
TRS-42 transponder
More detailed descriptions of some of these systems can be found under the appropriate section headings in this book.

STATUS: in production. An initial installation on board a Cessna Citation III was completed in December 1984, having been FAA-certified earlier in that year. The five-system suite was chosen by Fokker in May 1985 as the standard communication, navigation and identification package for its new 50-seat Fokker 50 twin-engined regional airliner.

ARINC 700 series digital integrated avionics

The ARINC 700 series CNI (communications, navigation and identification) equipment began as part of the US industry-wide momentum in the 1970s to develop new technology systems for the transport aircraft in prospect to replace the Boeing 707 and Douglas DC-8. Bendix was successful in winning the competition to provide the CNI suite for the Boeing 767, launched by United Airlines in July 1978, and the slightly later Boeing 757. The current ARINC 700 series consists of the following equipment:
RTA-44A vhf communication transceiver
RVA-36A vhf navigation receiver
DFA-75A automatic direction finder
SMA-37A distance measuring equipment
RIA-35A ILS receiver
ALA-52A radio altimeter
TRA-65A transponder

STATUS: in production and service.

MUA-46 ACARS data link

The MUA-46 data link system is an ACARS (ARINC communications addressing and reporting system) digital link used for data management between aircraft and the airline's base. Messages can be 'down-linked' through the vhf communications transceiver to a central processing system which determines the registration and flight numbers of the aircraft, the nature and destination of the message, and then processes the message. Information can

Bendix MUA-46 data link being set prior to Delta Air Lines flight

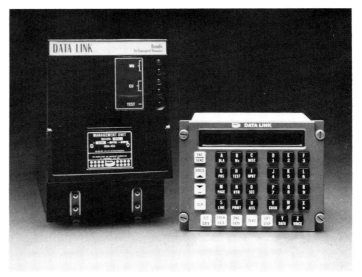

Units of Bendix MUA-46 data link set

PTA-45A data printer

also be 'up-linked' to the aircraft in the same manner. Messages can be entered manually using the control unit keyboard. The data link system can also interface with appropriate aircraft systems for automatic reporting of data, including the weather radar indicator for extended display of data information.

Optional items include a dedicated hard-copy printer, a management terminal for use by the cabin crew, and a 20-page display memory. Provisions have been made for additional ARINC 429 input/output ports, checklist and multi-function display interface, and hf modem. The system has been installed by Delta and Piedmont Airlines.

STATUS: in production. In March 1986 Bendix reported that the systems operational with

Delta Air Lines were giving an mtbf in excess of 5000 hours.

PTA-45A airborne data printer
The Bendix PTA-45A data printer is the first unit of its kind designed specifically for commercial aircraft. The equipment uses a high efficiency switching power supply which provides sufficient power to continuously print 'all-black' (all 600+ dots printing at the same time). Since normal operation runs at only 15 per cent of the machine capacity, the printer is claimed to be capable of providing up to 10 times the reliability of previously available models.

The printer permits flight crew access to hard copies of uplinked flight plans, messages, and

graphics, and can also print a copy of data link information displayed on the Bendix weather radar display. The printer may be driven by a data link system or a maintenance computer. User features include paper guides to help prevent paper jamming, a hinged spool to ease paper changing, and a paper level indicator to tell how much paper is left.

STATUS: in production and ordered for the British Aerospace 146 aircraft of Pacific Southwest Airlines.

Brelonix
Brelonix Corporation, 106 North 36th Street, Seattle, Washington 98103
TELEPHONE: (206) 282 7352

Brelonix Corporation manufactures radio communications equipment and ancillaries for aircraft, marine and land applications. It specialises in hf/ssb equipment and has supplied systems to the Royal Norwegian Air Force, the South African Air Force, and to many scheduled or charter airlines in Canada and Alaska.

SAM 100 hf/ssb radio
The Brelonix SAM 100 basic equipment provides 10-channel hf communication within the frequency band 2 to 14 MHz. Operating modes are ssb-A3J suppressed carrier and AM-A3H with 6 dB carrier signal and in each case output power is 100 watts peak. Two channels are available for semi-duplex operation. The system is remotely controlled and has voice detecting

squelch control. Frequency stability remains within ± 20 Hz over the temperature range –30 to +60° C.

Three differing antenna couplers are available, the BD model which is manually operated and which uses power peaking for fine tuning, and the BR and WBR models which are both remotely operated by Brelonix for wider frequency response. The latter is recommended by Brelonix for wider frequency response. Options include a 5-watt audio insulation amplifier and a choice of fixed, trapped dipole or whip antennas.

Dimensions: (transmitter-receiver) 6.25 × 7.5 × 17 inches (159 × 191 × 432 mm)
(controller) 2.5 × 6.25 × 3.5 inches (64 × 159 × 89 mm)
Weight: (transmitter-receiver) 12.5 lb (5.7 kg)
(controller) 1 lb (0.45 kg)

STATUS: in production and service.

SAM 70 hf/ssb radio
This is a manually operated hf unit which offers broadly the same facilities as the SAM 100

system but with only five channels, two of which may be semi-duplex, and a lower power output. Modes are the same as those for the SAM 100 equipment but power output has been increased to 40 watts peak power for both ssb and AM modes, with a 20-watt carrier signal in the latter case.

As with the SAM 100 system, three antenna couplers are available for the SAM 70 but in this instance are designed for five channel operation. The same range of antenna options is also available.

A new version of the SAM 70, the SAM 70R, was introduced by Brelonix in May 1981. This is a remotely controlled version with a miniature controller which provides the same operational facilities as the direct manual control model.

Dimensions: (transmitter-receiver) 4.25 × 6.25 × 11 inches (108 × 159 × 279 mm)
(controller) 1 × 6.25 × 2.5 inches (25 × 159 × 64 mm)
Weight: (transmitter-receiver) 5 lb (2.3 kg)
(controller) 0.5 lb (0.2 kg)

STATUS: in production and service.

Cincinnati Electronics

Cincinnati Electronics, 2630 Glendale-Milford Road, Cincinnati, Ohio 45241
TELEPHONE: (513) 733 6500
TELEX: 214452
TWX: 810 464 8151

F-102 notch filter

The F-102 is a versatile hf notch filter providing 50 dBN attenuation between a receive antenna and a receiver input, thereby reducing interference caused by close proximity high-power transmitters. Four types of filter cover the frequencies between 2 and 24 MHz.

The F-102 is particularly suited for use in swept frequency and monitoring/receiving applications, and is designed to withstand high shock and vibration levels in a military environment. Units can be connected to provide multiple notches in one or a combination of bands.

STATUS: in production.

Cincinnati Electronics F-102 notch filters in series for multiple emitters

RT-2000 radio relay for RPVs

RT-2000 radio relay for RPVs

The RT-2000 is designed to provide a retransmitting function via an RPV to relay traffic between users who are unable to intercommunicate directly. The equipment relays plain and cipher text voice and digital data and automatically selects the correct mode. From an altitude of 10 000 feet (3000 metres), the range is more than 300 km for ground-to-air-to-ground operation. An additional RPV can more than double this range by acting as a second relay.

The relay consists of two receivers and one transmitter, working in the 30 to 75.975 MHz range, and operates by retransmitting a received signal on a different frequency to that of reception. Ten different frequencies can be stored and selected remotely by an aircraft command link. Transmitter power output is 2 watts.

RT-601 millimetre-wave communicator

The RT-601 is a millimetre-wave communicator in a helmet-housed transceiver. It is designed to provide a short range, hands-off communication link for tactical operations between, for example, parachutists while descending, artillery batteries, aircraft-to-aircraft voice links, aircraft carrier flight-deck communications and command post communication. It is a voice-operated narrow beam/directional system, with an antenna beamwidth of approximately 20 degrees (depending on the antenna size), with a range of about 1 mile (1.6 km). Frequency of operation is in the Ka band and radiated power is less than 100 mW.

Weight: 14 oz (0.4kg)
plus battery 5.5 oz (0.15 kg)

RT-601 millimetre-wave communicator units

Collins

Collins Avionics Divisions, Rockwell International, 400 Collins Road NE, Cedar Rapids, Iowa 52493
TELEPHONE: (319) 395 1000
TELEX: 464421
TWX: 910 525 1321

Now a world leader in communications and electronics, the Cedar Rapids firm was founded in 1931 as the Collins Radio Company by radio enthusiast and builder Arthur A Collins, its early products being amateur radios. The company quickly expanded to meet the growing demand for broadcasting and police communications equipment. Soon it began to build radios for the emerging US airline industry, and recognition of its technical standing in this field came with the selection of a Collins hf system by Admiral Richard Byrd for his 1933 expedition to the Antarctic. Radio as an aviation product line began to assume importance when in 1937 Braniff Airways became a customer, to be followed shortly by American Airlines. During the Second World War, the company produced the important AN/ART-13 radio, with 10 pre-tuned frequencies to reduce the workload in fighter aircraft. As the company grew in the post-war era its product range swelled to include uhf, single sideband hf, radar, auto-matic flight control, and microwave and satellite communication equipment.

The company was bought by Rockwell Corporation in 1971 when it was re-organised under the title of Collins Division. Shortly afterwards it introduced Microline, a family of compatible nav/com units for light aircraft, and followed it with equivalent Pro Line and Pro Line II series for business aircraft.

Collins continues to be internationally prominent in all branches of aviation, though aeronautics accounts for only part of its business. In the avionics field the company is heavily involved in producing equipment for both the military and civil markets, details of the product lines being given under the appropriate section headings in this book.

Micro Line nav/com family

In the early 1970s, shortly after its acquisition by Rockwell, Collins introduced the first members of a family of panel-mounted navigation and communication units, called Micro Line, for light single and twin-engined aircraft. The rationale for this step was the lack of any agreed standard or specification for avionics systems appropriate to this type of aircraft, which gave rise to considerable compatibility or interchangeability problems between equipment from different manufacturers. By contrast the airlines already had the very comprehensive ARINC characteristics to ensure compatibility. Intended for low-budget operation and for aircraft with limited internal space and weight allowance, the various Micro Line units each contained the relevant processing and indication circuits and mechanisms within single boxes, so that they could be mounted as self-contained items on the pilot's panel without the need for any associated remote rack-mounted equipment.

The current Micro Line family comprises the following equipment:

VHF-250 communication transceiver
VHF-251 communication transceiver
VIR-350 navigation receiver
VIR-351 navigation receiver
IND-350/351 course deviation indicators
AUD-250 audio control unit
AMR-350 audio control unit/marker beacon receiver
MKR-350 marker beacon receiver
PWC-150 28/14-volt power converter
GLS-350 ILS glideslope receiver
ADF-650A automatic direction finder
TDR-950 transponder
WXR-150 digital weather radar
WXR-200A digital weather radar
ANS-351 area navigation system

DME-451 DME
DCE-400 distance computing equipment
APS-841 flight control system (for light helicopters).
More detailed descriptions of some of these systems can be found under the appropriate section headings in this book.

STATUS: in production and service.

Pro Line nav/com family

Introduced in 1970, the Pro Line series avionics were intended for medium and large general aviation piston- and turbine-engined twins. The family was originally designated Low Profile to emphasise the compact size and form factor of individual units, but in 1975 was re-named Pro Line in recognition of its acceptance by professional pilots in regional airlines, and by the defence forces. Unlike the self-contained members of the Micro Line, Pro Line systems comprise panel-mounted indicators and controls driven by or controlling separate rack-mounted processing and computing boxes. The size and form factor was laid down by Collins, there being no industry-wide agreement on packaging for general aviation electronics, in contrast to the highly defined ATR standards governing the characteristics of equipment for airlines. In the military field Pro Line is found on a wide range of aircraft, not only those equivalent in size and performance to general aviation types, but also on attack and surveillance aircraft such as the Douglas A-4, Northrop F-5E, Lockheed C-130 and Boeing E-3 AWACS.

The current Pro Line family comprises the following equipment:

VHF-20A/B communication transceivers
VIR-30A/31A navigation receivers
HF-200/220 hf transceivers
PN-101 pictorial navigation system
BDI-36 bearing/distance indicator
ADF-60 automatic direction finder
DME-40 DME
ALT-50/55 radio altimeters
TDR-90 transponder
ANS-31A area navigation system
NCS-31A navigation and control system
LRN-70/80/85 Omega/VLF navigation system
FPA-80 flight profile advisory system
WXR-250A digital weather radar
WXR-300 colour weather radar
346B audio control/isolation and speaker amplifiers
APS-80 autopilot
AP-105 autopilot
AP-106A autopilot
FDS-84 flight director system (comprising FD-108 flight director and FIS-70 flight instrument system)
FDS-85 flight director system (comprising FD-109 flight director and associated horizontal situation indicator)
FD-112V flight director (incorporating in one box both attitude and horizontal situation indication and commands)
ADS-80 air data system
CTL series control heads

STATUS: production is being phased out in favour of Pro Line II avionics (see below), but equipment remains in wide-scale service. Well over 100 000 Pro Line boxes and controls have been manufactured and sold.

Pro Line II digital nav/com family

Acknowledging the advances in technology, notably in signal processing, since the appearance of Pro Line in 1970, Collins decided in the late 1970s to develop a replacement. The result was Pro Line II, the first members of which (vhf communications transceiver, vhf navigation transceiver and DME) appeared in January 1983. Other units were added and there now exists a complete range of Pro Line II equipment.

Collins foresaw the financial and down-time

Pro Line II nav/com family

burden facing operators if they were obliged to replace entire avionics suites at a single stroke because of the incompatibility of analogue and digital systems. Accordingly, Pro Line II boxes contain analogue/digital and digital/analogue circuits so that individual units of the earlier family can be exchanged on a one-for-one basis as convenient, and without change to the aircraft wiring or racks. Microprocessors within the new units are programmed to accept either analogue or digital frequency tuning arrangements; the digital tuning format is designed by Collins, using the US Electronic Industry Association's RS-422A electrical characteristics. Thus there is complete functional and mechanical interchangeability between equivalent rack-mounted members of the two series. The control heads are, however, smaller, though retaining the same gas-discharge displays, but frequency storage capacity is increased. Again, the new series is generally less expensive, system for system. Finally Pro Line II includes a series of crt-based electronic flight instrument systems, an option not available with the earlier series.

A Pro Line II suite equips the world's first all-digital business aircraft, a British Aerospace 125 Series 800, which was certificated in December 1984. Collins has devised software-based bench-test checkout routines that can be run on an Apple II Plus micro-computer, the only additional equipment required being an interface card. The computer is a standard commercial item but Collins supplies the software and interface card.

The current Pro Line II family consists of the following equipment:
CTL-22 communication control unit
VHF-21/22 vhf communication transceiver
CTL-32 navigation control unit
VIR-32 navigation receiver
IND-42 DME control unit
DME-41/42 DME receiver
CTL-62 automatic direction finder control unit
CTL-92 transponder control unit
ADS-82 air data system
AHS-85 attitude/heading reference system
APS-84/95 autopilot
EFIS-74 electronic flight instrument system
EFIS-85/86 electronic flight instrument systems
More detailed descriptions of some of these units can be found under the appropriate section headings in this book.

STATUS: in production and service.

Pro Line Concept 4 integrated avionics system

The heart of the new Pro Line Concept 4 avionics line is Control Central, an integrated avionics processing system.

Pro Line Concept 4 integrated display

Functions performed by Control Central include processing of inputs from sensors located throughout the aircraft, and computation of commands and displays for autopilot functions, flight management system and electronic displays. Instead of a number of units performing flight control functions, light modules in the control work together to serve as the aircraft central processing system. The small size of the modules allows Control Central to provide fail/safe integrity, with quadruple redundancy of critical functions. Internal communications between critical functions are protected, eliminating exposed external wires and buses. If one module fails, the system automatically reconfigures to carry on operation without interruption.

Advances in component and manufacturing technology allow circuit boards in Control Central to perform the functions of up to eight boards normally used. Surface mounted devices, automatic insertion of components and manufacture control have allowed a considerable reduction in circuit board size.

As well as Control Central, Concept 4 includes EFIS and EICAS displays, radio tuning units and a WXR-850 Doppler weather radar.

STATUS: in development.

Series 500 avionics family

Developed by Collins Air Transport Division for commercial aircraft, the Series 500 avionics family is based on ARINC 500 characteristics, and is the final step in the company's range of analogue equipment. Though still analogue in nature, more advanced solid-state circuits together with a small number of components per functions (a lower parts count) make for a substantial weight saving and greater reliability over previous equipments. Series 500 boxes are available on a one-for-one replacement basis for earlier systems.

The current Series 500 family comprises the following equipment:

628T-1 hf transceiver
618M-3 vhf transceiver
490S-1 hf antenna coupler
51Y-7 (DF-206) automatic direction finder
51RV-4 VOR/ILS receiver
860E-4/860E-5 DME systems
ILS-70 instrument landing system receiver
860F-4 digital radio altimeter
621A-6A air traffic control transponder
54W-1 comparator warning monitor
346D-2/2B passenger address amplifier
FD-110 flight director system (comprising FMC-28 flight mode controller, 562A-5F5 flight computer, 329B-8J attitude director indicator, and 331-8K horizontal situation indicator)
Data link system (comprising DLC-700 control unit and 597A-1 management unit)
More detailed descriptions of some of these systems can be found under the appropriate section headings in this book.

STATUS: in production and service.

Series 700 digital avionics

Design of the Collins Series 700 digital avionics for commercial aircraft dates back to the early 1970s when US industry in general began planning for the new generation of transports then in prospect. The new aircraft emerged as the Airbus A 300 series, and the Boeing 767 and 757; the digital systems for them and later projects being defined by ARINC characteristics of the 700 series. They were introduced in 1978, launch year of the first of these new transports, the Boeing 767.

Together with the usual range of navaids and other electronic devices, the Series 700 introduces an electronic flight instrument system (EFIS) based on television-style cathode-ray tubes, and a similar display suite for engine indication and warning.

The current Series 700 family comprises the following equipment:
VHF-700 vhf radio transceiver
HF-700 hf radio transceiver
ADF-700 automatic direction finder
DME-700 DME
VOR-700 vhf navigation receiver
ILS-700 instrument landing receiver
LRA-700 low-range radio altimeter
TPR-700 transponder
PAU-700 passenger address amplifier
More detailed descriptions of some of these systems can be found under the appropriate section headings in this book.

STATUS: in production and service.

AN/ARC-94 hf radio

Consisting of a 618T-2 transmitter-receiver and a 714E-2 controller with a self-contained power supply, this configuration operates from a 27.5-volt, 4 A dc supply, or a 208-volt, three-phase, 400 Hz, 1000-watt source.

AN/ARC-102 hf radio

This system comprises a 618T-3 transmitter-receiver and 714E-2 controller with a self-

contained power supply for operation from a direct current of 27.5 volts, 35 A and 115 volts, one-phase, 400 Hz, 1 A source.

AN/ARC-153 hf radio

Collins AN/ARC-153 airborne radio provides 280 000 channels in the 2 to 30 MHz HF band. Operating in AM mode, the system has been specifically developed for service aboard US Navy/Lockheed S-3A ASW aircraft, for which it provides voice and sonobuoy data communication. Normal transmitter power output is 1 kW but this may be reduced, if desired, to 400 watts. The system is of all-solid-state construction.

STATUS: in service.

AN/ARC-156 uhf radio

The Collins AN/ARC-156 is an airborne communication system which, operating in the uhf band, provides 7000 channels within the frequency range 225 to 400 MHz. It was designed for service aboard US Navy/Lockheed S-3A and Grumman E-2C aircraft and operates in AM, FM and fsk modes. Transmitter power output is 30 watts in AM and more than 100 watts in FM and fsk. A secure speech mode is also available.

STATUS: in service.

AN/ARC-157 hf radio

This Collins military airborne long-range communication system covers the hf band from 2 to 30 MHz, providing 280 000 channels at intervals of 100 kHz. Operating modes include usb and lsb in voice and data transmission and, additionally, lsb in data transmission only. Maximum transmitter output power is 1 kW.

The system operates over a temperature range from –54 to +55°C with operation for up to 30 minutes at temperatures up to 71°C. Maximum operational pressure altitude is 15 000 feet (4570 metres).

Dimensions: 10.05 × 22.1 × 23 inches (255 × 562 × 584 m)
Weight: 103.2 lb (46.75 kg)

STATUS: in service.

AN/ARC-158 uhf transceiver

This is an all solid-state uhf transceiver, known as the RT-1017, which is the main component of the AN/ARC-158 communications system for the US Navy/Grumman E-2C aircraft. It operates over the frequency range 225 to 399.75 MHz giving 7000 channels at 25 kHz spacing with a power output of 30 watts AM or 100 watts FM or fsk. Operational modes are normal and secure AM, AM ADF, TADIL-A FM data, TADIL-C fsk data, and tactical satellite communications.

Dimensions: 222 × 152 × 483 mm
Weight: 14.52 kg

AN/ARC-159(V) uhf radio

The Collins AN/ARC-159(V) is a military airborne communications system designed principally for high-performance fixed-wing aircraft and helicopters but also applicable to the broadest range of aircraft types. It covers the uhf band from 225 to 399.975 MHz, in which range it provides 7000 channels at increments of 25 kHz. Up to 20 channels can be pre-selected. The system has a transmitter power output of 40 watts. Operating mode is AM. A built-in guard receiver monitoring the 243 MHz uhf distress frequency is incorporated and the radio also provides an azimuth homing facility.

The ARC-159(V) is available either in remotely controlled form or as a panel- or console-mounted unit in which the control facia is combined with the transmitter-receiver unit. Dual control configuration is also available. The system is particularly suited to retrofit applications and various adaptive mounting trays are available to make possible the easy replacement of earlier, larger and heavier radio equipment in a wide range of aircraft with the minimum need for modification to cable harnesses. A number of frequency channel indicator units, such as the OD-122 group, ID-1972 and ID-1984, are available as duplicate channel indicators to provide clear readouts in various parts of the aircraft.

The ARC-159(V) employs solid-state components, integrated circuits, mos-devices and thin-film techniques. These, claims Collins, result in high reliability and a mean time between failures in excess of 1000 hours. The system contains a number of protective measures to safeguard it against power surging and the effects of operation at excessive temperature. In the latter case, transmitter output power is temporarily reduced if overheating occurs but is automatically restored to normal level when temperature is reduced to within standard limits.

The ARC-159(V) is currently in service with the military arms of some 30 customer countries including the United Kingdom, France, Spain, Australia, Singapore and Israel. In the United States, it serves with the Navy's F-14, A-7 and other aircraft and the interim Lamps helicopter. The US Navy has ordered more than 4000 of these systems out of nearly 8000 produced to date.

Dimensions: 4.875 × 5.75 × 6.5 inches (124 × 146 × 165 mm)
Weight: 9.5 lb (4.35 kg)

STATUS: in production and service.

AN/ARC-171(V) uhf radio

The Collins AN/ARC-171(V) system comprises a family of uhf communications radios, each of which covers the band 225 to 399.975 MHz in which range 7000 channels are provided at increments of 25 kHz. In each case, an independent guard channel, covering the emergency frequency of 243 MHz, is also provided. Up to 20 of the normal operational channels may be pre-selected.

Members of the family between them provide a very wide range of transmission modes which include AM voice, AM secure voice, FM voice and data, fsk data and automatic re-broadcast. Certain versions also include an ADF navigation facility. With one exception, a full duplex transmission-reception type, all models use the same basic chassis configuration and the modification required to convert one version to another is accomplished by replacing circuit cards or complete modules. All equipment in the ARC-171(V) series may be remotely controlled and differing control units are employed according to the variant in use or the role in which it is employed. For example, in AM-only versions a simplified controller, covering functions common only to AM modes, is used but a more comprehensive unit is available for use with those variants which provide AM/FM/fsk facilities. An alternative type of

Collins Series 700 avionics. From left to right the units are: HFS-700, VHF-700, PAU-700, VOR-700, ILS-700, ADF-700, DME-700, LRA-700 and TPR-700

controller is provided for use with a satellite communications version.

The ARC-171(V), together with the Magnavox ARC-164, is a standard system selected for US Air Force service and is installed in a variety of aircraft, particularly those with a satellite communication requirement. Certain configurations are intended primarily for use aboard US Air Force/Boeing E-3A Sentry AWACS aircraft. The systems are in quantity production at a rate of more than 2000 units a year.

Dimensions: 16 × 9.5 × 7 inches (406 × 241 × 170 mm)
Weight: 35 lb (15.88 kg)

STATUS: in production and service.

AN/ARC-174(V) (718V-5) hf radio

The AN/ARC-174(V) is a lightweight solid-state hf communications radio. Its construction, low power consumption and the ability to tune various antennas without the need of an external antenna coupler make it suitable for light aircraft and helicopters. The system operates over the frequency range from 2 to 29.9999 MHz, providing 280 000 channels with 100 Hz channel spacing. Transmission modes available are usb, lsb, ame, cw, and secure voice upper and lower sidebands. Operation with fsk tone keyers, narrow-band secure voice systems and other similar data modems is possible. Output power is 100 watts.

The system is packaged into three units: a receiver/exciter, a power amplifier/antenna coupler, and a remote control unit. The transceiver incorporates an audio processing technique that provides an increase in the average power output of the system in voice modes with no change in peak output power level. A serial data link control system is used between the controller and the radio.

Dimensions: (receiver-transmitter) 199 × 130 × 374 mm
(amplifier-coupler) 196 × 127 × 409 mm
(control) 69 × 147 × 140 mm
Weight: (receiver-transmitter) 6.5 kg
(amplifier-coupler) 7.5 kg
(controller) 1.1 kg

AN/ARC-178(V) uhf radio

This is a family of uhf communication systems having a number of operational features in common with the Collins AN/ARC-171(V) uhf systems but which has been developed mainly for the US Navy service. The ARC-178(V) systems provide 7000 channels over the band 225 to 400 MHz and offer, according to the variant, a choice of voice, data and satcom facilities in AM, FM, fsk and spread-spectrum ECCM modes. Transmitted power output is 30 watts am or 100 watts FM or fsk.

Up to 20 channels may be pre-selected and built-in self-test facilities are provided. Changes from one configuration to another are accomplished by module or card substitution.

Dimensions: 17 × 13.125 × 7.5 inches (430 × 333 × 200 mm)
Weight: 43 lb (19.5 kg)

STATUS: in production and service.

AN/ARC-182(V) vhf/uhf radio

Collins ARC-182(V) is a combined vhf/uhf military communication system designed for all types of fixed-wing aircraft and helicopters but is small and light enough to be especially attractive for installation in the lighter aircraft classes. It covers the frequency bands from 30 to 88 MHz in FM, 116 to 156 MHz in AM, 156 to 174 MHz in FM and the uhf band 225 to 400 MHz in both AM and FM modes. Additionally, a receive-only facility, covering the band 108 to 116 MHz is also provided for navigation purposes. Channel spacing throughout the range is at 25 kHz intervals. A total of 11 960 channels,

up to 28 of which may be pre-selected, are available and guard channel coverage on the 243 MHz emergency frequency is also provided. A guard precedence mode is activated by single-switch selection.

Developed for a US Navy requirement, the ARC-182 is now also entering service with the US Air Force. The provision of a comprehensive communication system, which also covers the marine band, renders the equipment useful in a number of tactical roles especially those associated with extended economic zone maritime patrol. Later additions include a facility to scan up to five channels, allowing monitoring of multiple communication nets, and an added satellite communications ability which allows trans-oceanic flights with line-of-sight radio. For tactical operations the AN/ARC-182 can be linked to the PRC-124 manpack radio to provide a ground-to-air frequency-hopping link.

Dimensions: 6.5 × 5.75 × 4.875 inches (165 × 146 × 124 mm)
Weight: 10 lb (4.54 kg)

STATUS: in production and service. An $18.4 million development contract has been awarded to Collins to develop an anti-jam capability.

AM-7177A/ARC-182(V) transceiver system

The AM-7177A/ARC-182(V) is a variant of the AN/ARC-182(V) described above, and is a frequency agile system which is designed for simultaneous operation of several transceivers in airborne applications. The system permits operation of transmitters and receivers frequency-spaced as close as 5 per cent, with antenna isolation as small as 20 dB in the 225 to 400 MHz band in both single-channel and frequency-hopping modes.

The equipment consists of the RT-1250A/ARC-182 transceiver, the F-1556/ARC uhf high frequency agile filter, the AM-7177A/ARC uhf high power amplifier, the MT-6330/ARC mounting tray and the C-10319A/XN-3/4 controller.

The system provides enhanced AN/ARC-182 performance in uhf at 30 watts while permitting standard ARC-182 operation in the vhf band. The high power amplifier is wide-band while the electronically tuned filter, used in both transmit and receive, allows medium speed frequency hopping. In addition, a synchronous filter capable of providing high energy pulsed emitter protection is provided.

Dimensions: 188.9 × 241.3 × 444.5 mm
Weight: 14.78 kg

RT-1379( )/ASW transmitter/receiver/processor

The RT1379( )/ASW is an AN/ARC-182 derivative design providing a 5 K bits a second half-duplex or simplex rf data link using the TADIL-C message protocol and modulation. As currently configured, the radio covers the 225 to 400 MHz uhf band with 25 kHz channel spacing, and is compatible for alignment with a number of US Navy data systems. These include the Naval

Tactical Data System, Airborne Tactical Data System, Automatic Carrier Landing System AN/SPN-10/42, Precise Course Direction System AN/TPQ-10/27, and the Inertial Navigation System.

The RT-1379 interfaces with the mission computer on either of two (redundant) MIL-STD-1553B multiplex buses. Jumpers in the aircraft wing harness determine the unique multiplex address assigned to the radio.

TADIL-C address assignment is via five octal-encoded switches under a front protective cover. The radio's address is normally selected on the flight line before a mission. The last three (least significant) octal address digits can be changed by commands from the mission computer at any time, causing the radio to assume a new TADIL-C address.

Among types of information which can be handled are two-way transfer of target information, aircraft vectoring data, INS update data, landing system data, and general data reporting of aircraft status.

In general any data which is available on the aircraft multiplex bus can be transmitted by the radio. It can be modified to communicate using a message protocol other than TADIL-C format, and can accommodate other data-rates up to 16 K bits a second. The radio can also be made to operate on any channelised frequency between 30 and 400 MHz, and voice communications capability can be added.

Dimensions: 135.9 × 127 × 270.5 mm
Weight: 4.9 kg

STATUS: RT-1379( )/ASW radio is in operational service with US Navy F-18 Hornet aircraft. Among other mission duties, it has supported hands-off automatic carrier landings.

AN/ARC-186(V)/VHF-186 vhf/AM/FM radio

The Collins AN/ARC-186(V)/VHF-186 is a tactical vhf AM/FM radio communications system designed for all types of military aircraft. It has been selected by the US Air Force as the standard equipment for all aircraft requiring vhf AM/FM capability and the first aircraft to be so equipped include the General Dynamics F-16 Fighting Falcon, the Fairchild A-10 Thunderbolt II, and the Lockheed C-130 Hercules. The system is made by Collins' Government Avionics Division.

The basic ARC-186(V) is a solid-state 10-watt system of modular construction which provides 4080 channels at 25 kHz spacing. The 1760 FM channels are contained in the 30 to 88 MHz band and 2320 AM channels within the range 108 to 152 MHz. A secure speech facility can be used in both AM and FM modes, and the equipment is compatible with either 16 or 18 K bit secure systems in diphase and base-band operation.

Up to 20 channels may be programmed for pre-selection on the ground or in the air. Pre-selection is accomplished through incorporation of a non-volatile mnos memory which continues to retain data in the event of a loss of power supply. Two dedicated channel-selector switch positions cover the FM and AM emer-

Collins ARC-186 military vhf system in both direct and remote control configurations

gency channel frequencies of 40.5 and 121.5 MHz respectively.

Either panel-mounting, with direct control through an integral controller, or remote-mounting, with an identical control-panel presentation, is possible. A half-size remote controller which contains the same control functions is also available and a typical configuration in a two-seat aircraft would comprise a full-panel mount in the pilot's cockpit with a half-size controller at the co-pilot's position. In these dual-control configurations, a manual take-control switch provides full communications control for either crew member.

Conversion from panel- to remote-mount control is made by removal of the panel-type controller and replacing it with a plug-in serial control receiver module. A typical conversion is said to require less than 5 minutes. Frequency displays on both types of controller are immune to fade-out during periods of low voltage.

Circuitry of the ARC-186(V) is of modular design. Seven module cards are held in place by the body chassis or card cage and are electrically interconnected by a planar card in which all hard wiring has been virtually eliminated. All radio frequency lines in the interconnecting planar card have been buried to minimise electromagnetic interference. Individual module cards are readily removable and may be replaced in the field to reduce fault-finding and repair time.

Current options for the ARC-186(V) include AM/FM homing facility, but growth capability has been designed into the equipment from the outset and possible future developments could include selcal, burst data, and target hand-off. One present simple modification, carried out by replacement of the decoder module in the remote transceiver, permits the radio to be directly connected to a MIL-STD-1553 digital data-bus and it is claimed that the system will be equally compatible with suites of future-generation equipment. An additional possibility is the uprating of transmitter output power.

US Air Force testing has demonstrated a mean time between failures in excess of 9000 hours.

A principal design objective for the ARC-186 series equipment was that it should be capable of easy retrofit in existing installations. Since the system is considerably smaller than the equipment it is designed to replace, this is accomplished by use of plug-in adaptor trays which permit rapid replacement without disturbance to existing aircraft wiring harnesses.

Dimensions: (remote-mounted transmitter-receiver) 5 × 6.5 × 4.8 inches (127 × 165 × 123 mm)
(half-size remote control) 5.75 × 3.75 × 2.25 inches (146 × 95 × 57 mm)
(panel-mounted transmitter-receiver) 5.75 × 6.5 × 4.8 inches (146 × 165 × 123 mm)
Weight: (transmitter-receiver, panel- or remotely-mounted) 6.5 lb (2.95 kg)
(remote control) 1.75 lb (0.79 kg)
(FM homing module) 1 lb (0.45 kg)

STATUS: in production and service. More than 15 000 sets procured by US Air Force and US Army since 1979.

AN/ARC-190 (V) hf radio
The Collins AN/ARC-190(V) is a military hf transmitter-receiver designed as a replacement for a number of earlier hf systems in a US Air Force modernisation programme. The system is therefore particularly suited to retrofit applications as well as for installation as original equipment in a wide range of aircraft, such as the Rockwell B-1B, McDonnell Douglas F-15, Boeing B-52 and KC-135, and Lockheed C-130.

It covers the 2 to 30 MHz band in which it provides 280 000 channels, any 30 of which are pre-selectable, in incremental steps of 100 kHz. Operational modes include usb, lsb, ame and cw. Data transmission facilities are also available in usb and lsb modes, and in these modes

the system is able to operate with audio frequency shift keying or multitone modems.

The system is remotely controlled and dual control of the radio is possible from two crew-stations. Serial data control is applied between each of the major units in the system and this is said to render it adaptable to future facility requirements such as selcal and remote frequency management. Transmitter power output is 400 watts.

Construction of the ARC-190(V) is all-solid-state. The full system includes an antenna coupler which ensures compatibility with military-type cap or probe antenna systems. A bandpass filter unit, F-135, is available to provide added selectivity and overload protection for improved receiver performance in a strong signal environment. In the transmit mode it provides additional filtering to the exciter rf output. The system is operable to pressure altitudes up to 70 000 feet (21 300 metres) over temperatures ranging from –55 to +71° C.

Collins is developing an automatic communications processor for the ARC-190(V) which will automatically select the optimum hf frequency after the operator has selected the station to be called. The processor will also feature anti-jam modes. After test and evaluation, production of the processor is likely to commence in 1988.

Dimensions: (transmitter-receiver) 18.9 × 10.1 × 7.6 inches (480 × 257 × 194 mm)
(controller) 4.5 × 5.75 × 2.6 inches (114 × 146 × 66 mm)
(antenna coupler) 21.4 × 8.3 × 7.4 inches (545 × 211 × 189 mm)
Weight: (transmitter-receiver) 50 lb (22.68 kg)
(controller) 1.5 lb (0.68 kg)
(antenna coupler) 24 lb (10.89 kg)

STATUS: in production and service.

618T hf/ssb transmitter-receiver
The Collins 618T and its variants form a family of radio systems appropriate to a wide range of airborne applications, both civil and military. Fundamentally, they are all of similar design and construction; some units, however, are specifically tailored for new installations, some for retrofit and others for specialised applications. A description of the basic version is given below and details of the variants follow.

The 618T is a high frequency single sideband (hf/ssb) transmitter-receiver which provides voice, compatible AM, continuous wave or data communications in the 2 to 29.9999 MHz band. It is automatically tunable to any one of 28 000 channels in this range in increments of 1 kHz. Nominal transmitter output power is 400 watts pep in single sideband or 125 watts in compatible AM. The system is remotely controlled.

A notable aspect of this particular radio is that although designed for airborne use, it has been produced in large quantities for transportable, mobile, shipborne and semi-fixed stations in many parts of the world.

Accurate frequency control and stability is attained by the use of phase-locking circuits in which, by use of phase-comparison techniques, all injections to transmitter and receiver sections are related to a single, accurate frequency standard which is maintained on-frequency by temperature compensation.

The equipment is cooled by a filtered air supply which delivers a metered quantity of air to all parts requiring it from a blower installed in the front panel. However, an exhaust port is provided for use with central cooling systems, in accordance with ARINC 404. Electronic construction is of the plug-in module type, and transistor circuitry is used wherever possible.

Features include selcal and data transmission facilities. Use of selcal on AM is by way of a special audio output which enables the signals to be monitored irrespective of the mode-selection switch setting. With regard to data transmission, the 618T is suitable for frequency-shift keying or other signalling modes at rates of up to 100 words per minute. A connection is provided to an external frequency standard if so required. Selection of the data mode on the control unit automatically adjusts the receive sensitivity for maximum gain and causes the system to operate in upper side band mode.

A choice of optional accessory antenna tuning units is available to ensure maximum performance over the frequency spectrum employed, and a Collins 49T adaptor is also available to facilitate retrofit installation aboard aircraft wired for the earlier Collins 618S high frequency system.

The different variants of the basic 618T equipment are as follows:

618T-1 hf radio
This model includes a power-supply module for 400 volts, 1500 Hz primary source and is used only for certain retrofit installations. More usually, power is furnished by a Collins 516H-1 power-supply unit which provides 115 volts, 400 Hz and the direct current of 27.5 volts required to operate the equipment. This two-package system may also be used with the 49T-4 retrofit adaptor to effect direct retrofitting into a 614C-2 harness without alterations to existing cable installations.

618T-2 hf radio
Intended primarily for new installations, the 618T-2 has a high-voltage power-supply module using three-phase, 400 Hz power, together with a direct current of 27.5 volts. No external power source is required.

618T-3 hf radio
This self-contained system is intended for new installations although retrofitting is possible in some cases through employment of the 49T-3 adaptor. It is equipped with a 27.5-volt

Collins 618T equipment in mounting tray

electronic inverter-type high-voltage power supply module and requires a direct current of 27.5 volts, 35 A and 115 volts, single-phase, 400 Hz, 1 A supply.

628T-1 hf/ssb radio

The 628T-1 is an hf/ssb transceiver for long-range transport aircraft operating on overwater routes or other areas over which reliable extended-range communications are required. It provides upper side-band or AM voice or data communications on any of 24 200 channels, at 1 kHz increments, in the 2.8 to 26.999 MHz band. Tuning is automatically controlled through a remote control unit. Nominal transmission power is 200 watts peak power in ssb or 100 watts average in compatible AM.

This all-solid-state system uses digital synthesis techniques for frequency generation. High stability is maintained through temperature compensation of the frequency standard.

Mechanical design is aimed at maximising maintainability. The transmitter-receiver is housed in a case with hinged tray and fold-out doors for easy accessibility. Plug-board circuitry is used extensively and bench-testing is simplified through the provision of a built-in test connector at the rear of the casing. Special provision is made for the replacement of the power amplifier transistors without the necessity of removing the entire power amplifier board.

Transmitter cooling is achieved through a heat sink and filtered, forced-airflow, while the receiver section relies on conventional convection cooling and is not dependent on a cool air supply.

An AM selcal facility is provided by means of a special audio output through which selcal signals are monitored irrespective of the selected operating mode. Options include automatic antenna tuning couplers, to permit antenna performance optimisation over the frequency spectrum covered by the 628T-1, and a 999W-1/A1 adaptor unit which permits interchangeability with the Collins 618T-2/5 transceivers without disturbance to aircraft wiring, racks, connectors, antenna couplers or frequency selector.

Dimensions: ¾ ATR Short
Weight: 30 lb (13.6 kg)

STATUS: in production and service.

628T-2 hf radio

The 628T-2 is based on the 628T-1 and to some extent as well on the earlier 618T-2/5 equipment which is still to be found on many transport aircraft. It uses much of the operational and design experience derived from these systems. The 628T-2, however, was designed to offer certain advantages over the earlier equipment, principally higher power output, extended coverage (and hence a greater number of channels) of the hf band, and more options in terms of operating modes.

Transmitter output power is 400 watts peak power and full coverage of the hf band, from 2

to 30 MHz, permits use of up to 280 000 channels at 100 Hz increments or 28 000 channels at a separation of 1000 Hz.

Operational modes include usb, lsb, AM, carrier wave and data. Full 400 watts peak output is available in the sideband modes with 125 watts average in compatible AM and 125 nominal in continuous wave mode.

In mechanical and electronic design, the 628T-2 is very similar to the 628T-1 although Collins has incorporated a number of improvements with regard to sensitivity, cross-modulation elimination, intermediate frequency translation, and heat dissipation in the power amplifier stage. All are aimed at extension of performance and enhancement of reliability.

Dimensions: 6 MCU in accordance with ARINC 600
Weight: 28 lb (12.72 kg)

STATUS: in production and service.

628T-3 and 628T-3/A hf radios

Like the 628T-2, the 628T-3 and 628T-3/A are based on the 628T-1 design and represent the latest standard in this series. The principal difference between the 628T-3 and the 628T-3/A variant is the narrower bandwidth intermediate frequency filtration of the latter, resulting in heightened selectivity over part of the operating spectrum and slight differences in audio response.

As in the case of the 628T-2, each system provides full coverage of the 2 to 30 MHz hf band and offers 280 000 channels at 100 Hz increments. There is, however, no option for a smaller number of channels at 1000 Hz separation, and transmitter output is 200 watts peak power.

Operating modes are voice, and voice and data in both upper and lower sidebands, with compatible AM and carrier wave in usb only. Selcal facilities are as those of the 628T-1.

Format: ¾ ATR Short
Weight: 25 lb (11.36 kg)

STATUS: in production and service.

718U-4A hf transceiver

The Collins 718U-4A is a 400-watt transceiver for use in aircraft that require a high power output while limited by weight restrictions. It provides reception and transmission over the frequency range 2 to 30 MHz in 100 Hz steps, giving 280 000 channels in usb, lsb, ame, continuous wave and data modes. It can replace existing 618T transceiver installations in most aircraft without modification to existing wiring or aircraft structure. The system comprises four units: a transceiver, control unit, power amplifier/antenna coupler, and power supply. A pressurised case can be supplied for the latter two units for use in high performance aircraft at altitudes up to 69 000 feet (21 000 metres).

A panel-mounted control unit provides frequency selection with digital read-out of the frequency. Serial-digital data link control is

used between units to minimise cabling and reduce weight.

Dimensions: (transceiver) 194 × 124 × 318 mm (power amplifier) 194 × 72 × 441 mm (power supply) 194 × 75 × 441 mm (control) 67x 146 × 84 mm
Weight: (transceiver) 6.8 kg (power amplifier) 5.12 kg (power supply) 7.7 kg (controller) 1.03 kg

STATUS: in production and service with US, Canadian, French and Italian armed forces.

718U-5M (AN/ARC-174(V)5) hf transceiver

The 718U-5M is an improved version of the 718U-5 (AN/ARC-174(V)) hf transceiver, for use in small aircraft and helicopters, providing operation in usb, lsb, ame, cw and secure voice modes. It is a half-duplex communications system, designed to communicate with ARINC and maritime radio channels, having 280 000 channels in the frequency band 2 to 30 MHz. The system consists of a transceiver, power amplifier/antenna coupler and a panel-mounted controller, and has an output of 100 watts maximum.

The controller incorporates microprocessor techniques to simplify operation, and adds half-duplex capability to enable the pilot to transmit on one frequency and receive on another, as required in maritime stations. The controller non-volatile memory enables channels to be preset with mode of operation and the transmit and receive frequencies. When in half-duplex mode the control automatically selects the frequency for transmit, receive and mode, and provides frequency and channel information on the incandescent display.

STATUS: in production and service.

HF-101 hf radio

This comprises a 618T-1 transmitter-receiver and a 714E-2 control unit (see subsequent entries). Operation is from a direct current of 27.5 volts, 35 A and 115 volts, one-phase, 400 Hz, 2 A power supply.

HF-102 hf radio

The HF-102 configuration uses a 618T-2 transmitter-receiver and a 714E-2 controller with a self-contained power supply for operation from a 27.5-volt, 4 A dc supply, or a 208-volt, three-phase, 400 Hz, 1 kW source.

HF-103 hf radio

This system consists of a 618T-3 transmitter-receiver and a 714E-2 controller with a self-contained power supply for operation from a direct current of a 27.5-volt, 35 A and 115-volt, one-phase, 400 Hz, 1 A source.

HF 121 (AN/ARC-512, AN/ARC-191) radio

Collins HF-121, known under JETDS nomenclature as the AN/ARC-512 and AN/ARC-191, is a high-reliability airborne communication system which covers the full hf band from 2 to 30 MHz, in which range it provides 280 000 channels. The system comprises elements from two other Collins airborne hf systems, the AN/ARC-153 and the AN/ARC-157, both produced for the US Navy (see appropriate entries). The resultant configuration complies with MIL-E-5400.

The HF-121 can transmit and receive both data and voice signals and operates in usb, lsb, isb and ame modes. Transmitted power output level is selectable at either 100, 500 or 1000 watts. The system operates over temperatures from –54 to +55° C with short-term operation up

Collins 628T-1 hf transmitter-receiver

Collins 628T-3 hf transmitter-receiver in mounting tray

Collins HF-220 hf installation

Collins HF-230 hf radio and ITU radio telephony transceiver

to 70° C. Maximum operational pressure altitude is 15 000 feet (4600 metres).

STATUS: in production and service.

HF-200 hf radio
First introduced in 1978, the Collins HF-200 radio is designed primarily for the light aircraft segment of the general aviation market.

It is an all-solid-state radio of 100 watts peak power transmitter output and provides 18 pre-programmed channels plus two manually tunable channels, all in the 2 to 22.9999 MHz hf band. Frequency increment between any two adjacent channels is 100 Hz. Synthesis techniques are employed for frequency generation. Channel pre-programming is undertaken by Collins' dealers to user requirements by programming of the synthesiser card, while manual tuning may be carried out by approved individuals and no special equipment is required for the latter operation.

Normal operating mode is upper sideband but of the 18 pre-programmed channels, 12 are capable of half-duplex operation, which provides access to the maritime radio-telephone network. Operation in AM mode is also incorporated, in which case output is 25 watts average.

The system comprises a transmitter-receiver, power amplifier, antenna coupler unit, and a controller, of which three optional types are available to cater for differing panel-mounting requirements.

Dimensions: (transmitter-receiver) 5.04 × 4 × 11.75 inches (128 × 102 × 298 mm)
(power amplifier) 5.04 × 5 × 11.7 inches (128 × 127 × 297 mm)
(antenna coupler) 5.06 × 7.54 × 10.96 inches (129 × 192 × 278 mm)
(controller) 1.6 × 6.6 × 4.4 inches (41 × 162 × 118 mm)
Weight: (transmitter-receiver) 6.5 lb (2.94 kg)
(power amplifier) 7.25 lb (3.3 kg)
(antenna coupler) 10 lb (4.55 kg)
(controller) 1 lb (0.45 kg)

STATUS: in production and service.

HF-220 hf radio
The Collins HF-220 is an hf set designed for general aviation fixed-wing aircraft and helicopters. It covers the hf spectrum from 2 to 22.999 MHz, providing 210 000 channels at 100 Hz frequency separation. Up to 16 channels may be preset with the aid of a special programmer and all preset channels may be further programmed for half-duplex operation for radio telephone patch-through on either the US A3A mode or in the International A3J format.

Normal operating mode is upper sideband but AM operation is also possible. Transmitter

Control unit for Collins HF-220 radio

outputs are 100 watts peak power and 25 watts average respectively.

The HF-220 is a fully solid-state, synthesised equipment comprising a transmitter-receiver, power amplifier, antenna coupler and control unit. Emphasis is on simplicity, with simplified controls and a light-emitting diode readout of channel or frequency selected. Antenna tuning is fully automatic and the antenna coupler is compatible with antenna lengths of 10 to 30 feet (3 to 9 metres). An automatic probe antenna coupler, the PAC-200, has been developed by Collins and is especially designed for helicopter use with the HF-200/220 radios.

Dimensions: (transmitter-receiver) 5.04 × 4 × 11.75 inches (128 × 102 × 298 mm)
(power amplifier) 5.04 × 5 × 11.7 inches (128 × 127 × 297 mm)
(antenna coupler) 5.06 × 7.54 × 10.96 inches (129 × 192 × 278 mm)
(controller) 2.625 × 5.75 × 4.3 inches (67 × 146 × 109 mm)
Weight: (transmitter-receiver) 6.5 lb (2.94 kg)
(power amplifier) 7 lb (3.17 kg)
(antenna coupler) 10 lb (4.55 kg)
(controller) 1.5 lb (0.68 kg)

STATUS: in production and service.

HF-230 hf radio
Introduced in 1983, the Collins HF-230 radio is for use in fixed-wing aircraft and helicopters. It provides 280 000 channels at 100 Hz channel spacing between 2 and 29.999 MHz. All 176 ITU radio telephony channels are pre-programmed, giving phone-patch capability over very long ranges wherever this facility is available. Lower sideband operation is possible for international or maritime communications.

The system comprises a TCR-230 transceiver, PWR-230 power amplifier, and a range of antenna couplers. A DSA-220 adapter permits two such systems to be operated in the same aircraft. With a pressurised antenna coupler the HF-230 is approved for operation up to 55 000 feet (16 800 metres) and from –55 to +70°C. Power output is 100 watts pep.

The radio features pilot programmable channels: 40 can be simply programmed and, when

selected, channel number and frequency are displayed.

The new display unit, the CTL-230, forms part of the HF-230, and features gas-discharge symbology.

Weight: 23.2 lb (10.52 kg)

STATUS: in production.

HFS-700 hf radio
The Collins HFS-700 is an hf transmitter-receiver intended for service aboard transport aircraft which have a long-range communication requirement. Designed and developed in accordance with ARINC 719, the HFS-700 draws largely upon operational experience gained with the company's earlier 628T-1 and 618T-2/5 systems.

Covering the full hf band from 2 to 30 MHz, the HFS-700 provides 28 000 channels at increments of 1 kHz and uses ARINC 429 serial channel selection. Operating modes are usb, lsb, AM equivalent, data and cw. Nominal transmitted power outputs are 400 watts pep in ssb modes, 125 watts in compatible AM and 125 watts in cw. A special audio output permits selcal monitoring to continue in all mode settings.

The system is of all-solid-state construction and extensive use is made of cmos and linear integrated circuits. The digital frequency-generation synthesiser is locked to a highly accurate reference standard.

Features include a digital information transfer system (DITS) tuning interface and new installation concept connectors and cooling with a blower. Particular attention has been paid to control of excessive temperature during abnormal operating conditions and a special design of heat sink has been employed to maintain output transistors within derated temperature limits. Additionally, a dissipation detector reduces drive to lower levels under high dissipation conditions.

Mechanical design features include plug-in boards with hinged tray and foldout doors to allow easy access to components during servicing without use of card extenders. The system's normal operating range is from –55 to +70°C at 50 per cent duty cycle. Maximum operational pressure altitude is 40 000 feet (12 200 metres).

Format: 6 MCU per ARINC 719600
Weight: 28 lb (12.7 kg)

STATUS: in production and service.

HF-9000 series hf radios
The HF-9000 series is a new generation of lightweight hf radios developed by the Collins Defense Communications Division of Rockwell. The series is designed to meet the communications requirements of military aircraft, ranging from helicopters to high-performance fighters, with the initial emphasis on the development of a system, designated the HF-9100, for light fixed- and rotary-wing tactical aircraft. The radios employ modular design combined with fibre-optics, micropro-

Collins AN/ARC-190 hf radio

cessor technology and digital synthesisers and couplers.

The basic communications equipment includes an hf receiver-transmitter with a 175-watt hf power amplifier/antenna coupler. The transceiver design is such that it can be configured with a control unit for panel mounting in the cockpit. All control and status information between the transceiver and the power amplifier/coupler is transferred through a small fibre-optic cable, permitting fast exchange of large amounts of data between the two units. The system can be operated in simplex or half-duplex modes over the frequency range from 2 to 29.9999 MHz in 100 Hz increments. Some 20 programmable preset channels are stored in a non-volatile memory and each memory channel can store separate receive and transmit modes and frequencies. In addition 178 pre-programmed channels are available to the user. The transceiver uses a direct digital frequency synthesiser for rapid frequency changes with microprocessor control to improve stability. The power amplifier/coupler is designed to permit rapid tuning of a wide range of antennas in a variety of aircraft.

Dimensions: (transmitter-receiver) 5 × 4.75 × 6 inches (127 × 120 × 153 mm)
(power amplifier/coupler) 4.9 × 7.6 × 14 inches (123 × 193 × 356 mm)
Weight: (transmitter-receiver) 7 lb (3.2 kg)
(power amplifier/coupler) 13 lb (5.9 kg)

PAC-230 hf antenna coupler
The Collins PAC-230 is a self-contained hf antenna/antenna coupler system designed principally for helicopters in order to eliminate problems associated with long-wire antenna installations on rotary wing aircraft.

The unit, which is compatible with the Collins HF-200 and HF-220 hf radio equipments (see preceding entries), is suitable for operation in the 2 to 30 MHz part of the hf band, in which 280 000 channels are covered.

The PAC-230's principal feature is simplicity of operation, which reduces the amount of 'hands-off' flying time required to retune the set. The system is activated by the desired frequency being selected by the pilot who then keys the microphone switch in order to complete the antenna tuning cycle.

VHF-253 vhf radio
The VHF-253, which was introduced in late 1981, is a vhf transmitter-receiver covering the band 118 to 135.975 MHz at selectable 25 or 50 kHz increments. It is designed for general aviation fixed-wing aircraft and helicopters.

Collins VHF-253 system

The system is microprocessor-controlled and is claimed by Collins to be the first panel-mounted radio to offer direct and remote control access to six vhf frequencies while the active and a preset frequency are simultaneously and continuously displayed. Values of the active and preset frequencies are presented on a lcd which is back-lit for night-time operation.

As frequencies are selected for storage, on pilot demand, they appear in the preset (right-hand) window of the display and can be stored in any of four locations by pressing the appropriate storage-location button. Changes between active and preset frequencies are accomplished by depressing a transfer button at which the frequencies briefly alternate on the display until an aural tone confirms completion of the transfer. Recall into the preset window is made by depressing the appropriately numbered button.

Alternatively, recall can be accomplished from a remote switch mounted on the pilot's control yoke or, in the case of a helicopter, the cyclic-pitch control. This permits the pilot to run through the stored frequencies and transfer them to the active frequency position.

Direct tuning of the desired active frequency is also possible, if preferred. A non-volatile memory which retains all six frequencies (active, preset and four stored) guards against temporary power-loss or momentary power-supply interruption.

Large scale integration circuitry is used extensively in the VHF-253 and is expected to result in markedly improved reliability. The use of lcd as opposed to led presentation of frequency selection is also claimed to result in lower power consumption, reduced heat generation and improved reliability. Other system features include audio-gain control, 'stuck-microphone' protection and adjustable microphone gain.

An alternative version, designated VHF-253S, is available for operations in locations where ground transmitters may exhibit frequency instability. This unit can accommodate variations of up to 13 kHz from the nominal transmitted frequency.

STATUS: in production and service.

C-11029(V)/ARC-186 nav/com controller
The C-11029 is used in US Air Force aircraft as a dual controller for the AN/ARC-186 vhf radio transceiver and a VOR/ILS receiver.

Collins C-11029(V) vhf nav/com controller

Collins VHF-253 equipment in dual configuration aboard Beechcraft turboprop twin

The C-11029 display shows the communications and vhf navigation frequencies in seven-element bright red electronic digits. Frequency is changed by depressing a switch in the direction of the required change. Either incremental single digit or the more rapid slew technique may be used to select frequency digits.

In November 1983 Collins received a $4.6 million order for 800 controllers from the US Air Force, as a follow-on to a previous $2.7 million 400 unit order. Deliveries of all 1200 controllers were scheduled for completion by October 1985.

VHF-22 series vhf radio

Designed primarily for general aviation aircraft of all types, the Collins VHF-22 transmitter-receiver is a remotely-controlled, rack-mounted equipment of 20-watt transmitter output. It is available in two versions, the A equipment covering the vhf band from 118 to 135.975 MHz and the B variant from 118 to 151.975 MHz. Channel spacing is a 25 kHz. Units with broad receiver bandwidths are available.

The VHF-22 uses digital synthesis frequency-generation techniques and is of all-solid-state construction. It provides automatic carrier and phase noise squelch and automatic gain control and is designed to drive cabin audio systems of all types. Principal attractions are low weight, compactness and the low power consumption of 6.5 A during transmission. Consequently it requires no forced air supply and electronic section cooling is carried out by a combination of heat sink and convective air flow.

Either hard- or soft-mounting may be used and all connections are made through a single connector on the rear of the casing.

Dimensions: 3.3 × 3.4 × 14 inches (84 × 86 × 355 mm)
Weight: 4.6 lb (2.1 kg)

STATUS: in production and service.

VHF-700 vhf radio

The Collins VHF-700 is a vhf AM transmitter-receiver designed principally for air transport use. It covers the band 118 to 135.975 MHz with channel spacings of 25 kHz. Typical transmitted output power is 30 watts.

The system, which has been designed and developed to ARINC 716 specifications, has a number of advanced features including a new receiver design which eliminates mutual interference between equipment even when up to three transmitter-receivers are used in the same aircraft. A complete end-to-end self-test facility, which checks 99 per cent of all critical components, is built-in. The ARINC 429 serial-data-bus is also tested to ensure that the system is receiving valid information from the remote tuning selector.

The VHF-700 is microprocessor controlled, of all-solid-state construction and employs a temperature-compensated crystal oscillator as the reference standard for its digital frequency-generation synthesiser.

The system operates over a temperature range of –55 to +71°C with an ARINC-standard cooling-air supply on a continuous transmit mode or over the same temperature range with no forced air cooling on a 20 per cent transmit duty cycle. Maximum operating pressure altitude is 50 000 feet (15 200 metres).

Format: 3 MCU Short
Weight: 8.9 lb (4 kg)

STATUS: in production and service.

618M-3 and 618M-3A vhf radio

The 618M-3 radio and its variant, the 618M-3A, are air-transport category vhf transceivers of 25-watt nominal transmitter output. They were developed by Collins as retrofit replacements

Collins 618M-3 hf radio

Collins 618M-1 vhf transmitter-receiver

for the earlier 618M-1 radio and for similar ARINC 546 and 566 systems as a response to the introduction of 25 kHz spacing between channels for vhf air traffic control communication.

Coverage of the vhf band by the 618M-3 is from 118 to 135.975 MHz, while the 618M-3A version coverage extends from 116 to 151.975 MHz. The former provides 720 channels, the latter 1440 channels, in each case in 25 kHz incremental steps. Besides basic voice communications, each system also posesses data and selcal facilities.

Frequency generation is supplied from a digitally-controlled frequency standard. Low component density and use of solid-state circuitry is a feature and heat-sinking with cooling vanes assists in maintaining low transmitter temperatures. All of these factors result in improved reliability and the calculated mean time between failures is greater than 4000 hours.

Speech compression ensures good intelligibility irrespective of the user's voice characteristics or microphone technique and overmodulation is avoided by input signal amplitude limiting. Carrier-to-noise, carrier-override squelch control and automatic gain control are also incorporated.

With a cooling air supply, the continuous operating temperature range is from –54° C to +55° C; without cooling air, the continuous upper maximum limit is 30° C. On a 1- to 2-minute transmit-to-receive ratio duty cycle, operation can continue to a maximum tempera-

ture of 55° C. Maximum operational pressure altitude is 55 000 feet (16 800 metres).

Format: 1/2 ATR Short
Weight: 10 lb (4.53 kg)

STATUS: in production and service.

TACAMO communications system

The TACAMO system provides around-the-clock airborne very low frequency (vlf) communications to ensure that links with the US Navy strategic submarine fleet are available at all times. The system is a manned communication relay link to strategic forces, normally passing messages only one-way from the national command to the submarines and other strategic forces. Lockheed EC-130Q type TACAMO aircraft, with communication links to ground-based, airborne and satellite-based command centres, are always airborne, one each over the Atlantic and Pacific oceans. At present, a complete communications centre in the TACAMO aircraft allows simultaneous receive and transmit throughout the frequency range vlf to uhf. The system receives multiple-frequency low-level signals while simultaneously transmitting at highpower in a stressed environment. The vlf power amplifier provides amplification of the signal to 200 kW power and automatic tuning of the signal to the dual trailing wire antenna system. This latter consists of two antennas, one nearly 5000 feet (1500 metres) long and the other more than 28 000 feet (8500 metres). Only the short wire is charged, the energy re-radiating off the longer wire, the length of which varies with the frequency in use. The transmitted signal to the submarine is vertically polarised, with the EC-130 aircraft flying in a continuous tight turn. This allows most of the antenna system to hang vertically from the aircraft.

In view of the increased operational requirements, and the age of the basic airframe, the US Navy is to replace the EC-130 with Boeing E-6A aircraft, over the next few years. This uses the same basic airframe as the Boeing E-3A AWACS aircraft, without the rotodome and with different engines, and will be EMP-hardened in a layered approach. Protection will be applied to the hull, doors, hatches, windows, equipment consoles and some of the avionics equipment.

STATUS: Lockheed EC-130Q TACAMO aircraft have been in operational service for many years. First flight of the prototype Boeing E-6A aircraft is expected in early 1987. When deliveries begin, each EC-130Q will be flown to the Boeing factory and the avionics suite removed and installed in the E-6As.

346B-3 audio control centre

The 346B-3 audio control centre is designed to provide selection and control functions for aircraft audio equipment. The system, which is rack-mounted and remotely controlled, permits remote selection, isolation and amplification of 12 receiver audio inputs plus microphone and interphone inputs and has provision for two further inputs. Sidetone inputs for three transmitters are included together with provision for three more. Inputs may be combined in different combinations to meet various installation requirements, such as remote receiver selection, isolation, cockpit speaker amplification, passenger address amplification and a five-station interphone.

Volumes of speaker and interphone channels are separately controlled and compression amplifiers are employed to maintain desired voice outputs irrespective of input level. Outputs are 10 watts nominal into speaker loads and 100 mW into 600-ohm headset loads. Public address functions may be used for cabin, flight-deck, wheelwell, or other external speakers. The interphone for pilot, co-pilot and up to three other stations can be implemented by normal keying or by 'hot' voice-activated microphones. The 346B-3 equipment is of all-solid-state construction.

Format: ¾ ATR Short
Weight: 3.5 lb (1.59 kg)

STATUS: in production and service.

387C-4 audio control unit

This system is designed for direct panel mounting on a wide range of general aviation aircraft, from light twins to high altitude jets in which it provides flexible audio and transmitter-receiver control. With the exception of the ADF, all receiver, interphone and passenger address signals are processed by dual-channel limiter amplifiers to a standard 100 mW, 600-ohm level. When the desired speaker and interphone audio level is selected, the system maintains the level. ADF inputs continue to provide the normal build and fade type of signal. A built-in range-voice-both switch is incorporated for the ADF filter. Marker-beacon 'hi-lo' sensitivity and six-position microphone selection switches are included.

Dimensions: 5.75 × 1.875 × 4.7 inches (146 × 48 × 119 mm)
Weight: 1.6 lb (0.73 kg)

STATUS: in production and service.

714E hf control units

The 714E control units used in conjunction with the 618T series transmitter-receivers provide remote selection of the available 28 000 channels. Frequencies are indicated in a direct-reading digital display and can be selected in 1 kHz increments throughout the 2 to 29.9999 MHz range. Frequency selection is accomplished by rotating four knobs until the desired frequency appears in the window. A function selector and radio frequency sensitivity adjustment are included.

The 714E-2 control unit may be used either in new installations or as a replacement for the 614C-2 control unit in retrofit applications.

The 714E-3 is used with equipment operated in the continuous wave or data mode.

Dimensions: 5.75 × 2.5 × 6.25 inches (146 × 65 × 159 mm)
Weight: 2 lb (0.9 kg)

STATUS: in service.

490S-1 hf antenna tuning unit

The Collins 490S-1 antenna tuning unit is an hf antenna coupler which automatically matches the impedance of aircraft shunt or notch antennas to a 50-ohm transmission line. It operates in single and dual ARINC configurations throughout the entire 2 to 30 MHz hf band. Operating efficiency is claimed to range from 70 to 95 per cent, depending upon antenna impedance.

The system can automatically tune to a voltage standing-wave ratio limit of 1.3:1 across the entire hf band within an average time of 4 seconds and a maximum time of 7 seconds. It operates satisfactorily on all types of modulation at 400 watts maximum average rf input or 1000 watts pep. Primary power requirement is 115 volts ac at 400 Hz.

The 490S-1 is designed to operate ideally in an ARINC dual system installation to eliminate the possibility that a single fault would disable both systems. In order that the system may be located as closely as possible to the antenna itself, it has been designed and constructed to withstand the severe temperature and vibration environment associated with the tail/stabiliser area of the aircraft. The mean time between failures rating is in excess of 5000 hours, and the operation being constantly self-monitored.

The system is contained in a pressurised casing which is internally divided into the control and rf sections. These are separated by

a thermal shield. No cooling air supply, either blown or external, is required. The electronic elements feature the use of planar cards fabricated from multi-layer copper and glass-epoxy materials which provide high rigidity. The control circuits mounted on the cards consist mainly of integrated circuits in standard 14-lead flatpacks.

The system is operable over a temperature range from –54 to +71°C up to a pressure altitude of 50 000 feet (15 200 metres).

According to Collins, design of the 490S-1 is based on development of shunt and notch antenna systems for such aircraft as the McDonnell Douglas RF-4C, F-4M and RF-101, Dassault Mirage 1V-A, General Dynamics F-111, Grumman EA-6B military types and Gulfstream II corporate jet.

Dimensions: 7.5 × 5 × 15.75 inches (190 × 127 × 400 mm)
Weight: 16.5 lb (7.5 kg)

STATUS: in production and service.

490T-1 hf antenna coupler

The 490T-1 is a general purpose high frequency automatic antenna coupling unit for 25-foot (7.62-metre) or longer whip and wire antennas in the 2 to 30 MHz frequency range. Shorter antennas may be accommodated by the use of suitable loading coils. It features a short tuning cycle of 5 seconds maximum and 3 seconds average tuning time. Such rapid tuning capability reduces the overall rechannel time and keeps transmission to a minimum in the interests of radio silence.

The system contains an antenna transfer relay and an antenna grounding relay which can be used as needed in dual installations. The 490T-1 will operate with the 437R-1 helical monopole antenna and optional applications include exchange with either the 180L-3 or 180L-3A antenna coupler.

The unit's rapid tuning time enhances reliability since the operating elements are only energised for brief periods. The servo system is controlled by a demand surveillance technique which causes the coupler to retune if the antenna impedance changes appreciably but does not require the servo system to remain in constant operation.

The 490T-1 comprises four radio frequency assemblies, three modules, a chassis, front panel and dust cover. Solid-state logic circuits, capable of fast decisions with high speed switched and variable elements, are used to ensure reliable high speed tuning. All assemblies are removable for maintenance and repair purposes.

Dimensions: 7.7 × 10.4 × 14.4 inches (196 × 264 × 366 mm)
Weight: 19.7 lb (8.93 kg)

STATUS: in service.

AT-101 hf antenna tuning system

The AT-101 antenna tuning system is for use in aircraft employing a tailcap antenna. It consists of a 452A-1 pressurised lightning arrester and relay assembly, a 180R-4 pressurised antenna coupler assembly and a 309A-1 control unit. The lightning arrester and relay assembly serves as a mounting for the antenna coupler assembly. Provision is made for a second coupler for installations in which two transmitter-receiver units are required, permitting operation of the two receivers simultaneously on a common tailcap although allowing only one transmitter to be operated at any one time. Two optional type 156G-1 receiver modules plug into the 309A-1, permitting additional receivers to be used for monitoring purposes.

The antenna coupler assembly contains servo-controlled loading and phasing elements for resonating the antenna and matching its impedance at various operating frequencies. Maximum tuning time required is 10 seconds.

Protective circuits are incorporated to safeguard the equipment against loss of pressure or an excessive rise in temperature. There are no conventional glass envelope valves, transistors or diodes within the lightning arrester assembly.

Dimensions: (lightning arrester and relay unit) 7.5 × 10.625 × 16.4 inches (190 × 270 × 417 mm) (antenna coupler) 7.4 × 5.5 × 11.125 inches (189 × 141 × 282 mm)
(coupler control unit) ⅜ ATR Short
Weight: (lightning arrester and relay unit) 10 lb (4.54 kg)
(antenna coupler) 13 lb (6.15 kg)
(coupler control unit) 12.75 lb (5.78 kg)

STATUS: in service.

180R-6 hf antenna coupler

Used in conjunction with the 309A-2D antenna coupler control, the 180R-6 will automatically tune 45 to 100 feet (13.7 to 30.5 metres) antennas over the 2 to 30 MHz frequency range. The addition of optional plug-in 156G-1 modules in the coupler permits the use of up to three additional receivers for monitoring other frequencies.

Dimensions: 7 × 9.5 × 17.25 inches (177 × 241 × 437 mm)
(coupler controller) 3.625 × 7.625 × 14.5 inches (93 × 193 × 368 mm)
Weight: 21.51 lb (9.75 kg)
(coupler controller) 12.25 lb (5.56 kg)

STATUS: in service.

180R-12 hf antenna coupler

The 180R-12, together with the 309A-9 coupler control unit, automatically matches a probe antenna in the 2 to 30 MHz frequency range. The system is automatically tuned in a maximum of 16 seconds, but the typical tuning time is 5 seconds. The 180R-12 antenna coupler was designed for the Boeing 727 commercial transport but can be retrofitted into the earlier Boeing 707, using the 309A-9A unit.

The servo loop is activated only during tuning or when the voltage standing-wave ratio exceeds preset limits. This contributes to increased component life, but all components are tested to provide a calculated mean time between failures of 2000 hours.

High-voltage protection is provided by a ball gap which fires at a voltage lower than that which causes either external or internal arcing. This activates a circuit which cuts off transmitter power within 50 milliseconds. If the protective circuit functions due to gap-firing, the transmitter can be channelled to a new frequency and, if the excessive voltage does not exist at that frequency, the coupler will tune correctly. A sensor is incorporated to cut radio frequency power if internal air temperature rises to exceed 100° C.

Dimensions: 8.25 × 7.5 × 18.75 inches (211 × 190 × 476 mm)
Weight: 21 lb (9.53 kg)

STATUS: in service.

51X-2 vhf receiver

The Collins 51X-2 is a vhf navigation/communications receiver designed for air transport, general aviation and military aircraft applications. It covers the frequency range 108 to 151.95 MHz at increments of 50 kHz. In the communications mode, the equipment provides full coverage of the band stated unless coverage of only the 108 to 135.95 MHz band is required, in which case only a partial crystal complement is supplied.

In navigation modes, localiser coverage is over the band 108.1 to 111.9 MHz at odd-tenth MHz intervals and TVOR coverage is in the same band at even-tenth MHz intervals. VOR reception is at 50 kHz steps over the band 108 to 117.95 MHz. In all navigation modes, an output is available for either Collins 344A-1, B-1 or D-2 VOR/ILS instruments. Voice communication reception continues during navigation mode operation.

The system is remotely controlled by a choice of three types of controller, the 614U-1, 614U-3 or 614U-7. The 614U-1 provides remote frequency selection of both the 51X-2 and its associated 17L-7 vhf transmitter in single- and double-channel simplex and double-channel duplex communications system. The controller covers the range 118 to 151.95 MHz in 50 kHz steps. The 614U-3 and the 614U-7 controllers channel the 51X-2 through both the aircraft navigation and communication frequencies. In addition, both units provide for automatic selection of glideslope frequencies whenever an ils channel is selected. The 614U-7, in addition to the above, also contains circuits for automatic selection of DMET frequencies whenever a VOR channel is selected. Both controls channel the 51X-2 in 50 kHz steps over the entire vhf frequency range from 108 to 151.95 MHz.

The 51X-2 is of modular construction and is based on a bridge-type chassis, stressed at the sides. The three modules (radio frequency amplifier, variable intermediate frequency and frequency selector and fixed intermediate frequency and audio, and power supply) plug into the chassis with their interconnecting wiring between the two removable decks. Hold-down screws keep the modules in place. The electronic sections use conventional glass-envelope valves.

Format: 3/8 ATR Short
Weight: 10.5 lb (4.77 kg)

STATUS: in service.

37R-2 vhf communications antenna
This unit is designed for use on aircraft cruising up to 600 mph (965 km an hour) and is compatible with the Collins 51X-2 receiver. It is a vertically polarised antenna and provides a standing wave ratio of 2:1 or less over the range 116 to 152 MHz. It is used for both receiving and transmitting and can handle a maximum input power of 125 watts. Drag is approximately 0.518 lb (0.23 kg) at 250 mph (400 km an hour) at sea level, zero angle of attack. Under the same conditions, but at 400 mph (640 km an hour), drag rises to 1.32 lb (0.6 kg). Printed circuitry and foamed-in-place plastic are used in construction of this antenna.

Dimensions: 12.25 × 3.75 × 11.5 inches (312 × 94 × 292 mm)
Weight: 2 lb (0.9 kg)

STATUS: in service.

VP-110 voice-encryption system
The VP-110 is a voice-encryption device designed for use with airborne communications systems. A companion unit, the VP-100, performs the same function for fixed station and vehicular radios. Although aimed primarily at hf radios, the system will work equally well on both vhf and uhf equipment and ordinary telephone lines. The equipment is intended for a range of uses such as law-enforcement, business, diplomatic, government agency and selected military voice transmission applications. It has been designed for the encryption of 'sensitive' rather than restricted transmissions.

The system is cryptographically secure and eliminates all syllabic content in the encrypted mode while retaining clear voice quality and recognition. For transmission, the voice is converted into analogue signals and then divided into low and high band frequency ranges. It is then encoded and transmitted in a random mode with regard to time and frequency.

Public keying is provided, enabling private conversations between two stations without prior manual exchange of a recognition code. In this method of communication the operator selects the mode and the two units exchange a set of numbers, using a complex mathematical algorithm, which in effect establishes a signature for establishing private conversations. Eight codes, or key variables, can be entered in the unit microprocessors, providing 10 × 7[19] code possibilities.

Collins produces two versions of the VP-100, one for the US market and one for export. The former uses a data encryption standard (DES) algorithm whilst the latter is provided with a Rockwell developed algorithm.

Selscan communications control processor
The Selscan automatic communications control processor is a microprocessor-based subsystem that combines the application of selective calling (selcal), multi-channel frequency scanning and link quality analysis (LQA) techniques. It has been designed for use with hf radio systems, both ground-based and airborne, such as the ARC-190 and ARC-174. Selscan is a registered trade-mark.

The Selscan system automatically scans up to 30 preset frequencies and continuously builds and updates its own propagation or LQA data base. It also listens for incoming calls, verifies their reception and informs the operator that a call has been received and from which station. Frequency selection for preset channels is automatic and is based on stored LQA data that represents virtual real-time propagation conditions between the calling and receiving stations. Once the link has been established the hf system operation is totally independent of Selscan. The processor, however, continues to monitor system activity and returns automatically to the scanning mode following completion of a communication. This new system has the advantage that an hf radio operator no longer needs to spend time researching signal propagation data, establishing schedules, monitoring frequencies for incoming calls or making repeated voice calls to establish contact.

STATUS: in production.

IFM-101/AM-7189 vhf/fm power amplifier
This sub-system has been designed to enable standard vhf/FM radios in military tactical aircraft to communicate with the ground over extended ranges while flying at very low altitudes. It is intended primarily for helicopters and provides an effective radiated power of 40 watts nominal over the tactical communication band of 30 to 88 MHz. The complete system consists of a rack-mounted amplifier and a panel-mounted controller.

Dimensions: (amplifier) 127 × 102 × 422 mm (controller) 38 × 146 × 45 mm
Weight: (amplifier) 4.7 kg (controller) 0.41 kg

Data link systems
The Collins data link systems are designed for commercial and military transports and conform to ARINC 724 and 597. Their associated control panels have full alphanumeric keyboards with a two-line, 32-character light-emitting diode display. Information can also be displayed on a compatible weather-radar screen.

These data links can control hf and vhf radios, act as a check-list memory, and provide the interface between flight and performance-management systems and ground-based data transmitters. The operator can change the software in flight and the units have comprehensive built-in test facilities and a continuous GMT clock facility.

Collins ARINC 724/597 data link cockpit control panel

Control unit
Dimensions: 4.5 × 5.75 × 4.75 inches (114 × 146 × 120 mm)
Weight: 2.65 lb (1.2 kg)
Temperature: –15 to +70°C
Altitude: 55 000 ft (16 800 m)

Management Unit DLM-700 (to ARINC 724, 597 and 600)
Format: 4 MCU (ARINC 600)
Weight: 8.8 lb (4 kg)
Temperature: –20 to +70°C
Altitude: 55 000 ft (16 800 m)
Power: 60 W at 115 volts, 400 Hz single-phase 1 W at 28 volts dc

Management Unit 597A-1 (to ARINC 404A and 597)
As for DLM-700 except:
Format: 3/8 ATR Short (ARINC 404A)
Weight: 8 lb (3.6 kg)

STATUS: in production.

JTIDS joint tactical information distribution system
The first AN/URC-107(V) JTIDS Class 2 terminal manufactured by the Collins Government Avionics Division of Rockwell International was delivered to the US Air Force in September 1984 for testing aboard a McDonnell Douglas F-15 fighter. This terminal, built under the US Air Force 'leader-follower' concept, was the culmination of three-and-a-half years' full-scale development by Singer-Kearfott and Collins. Under this leader-follower arrangement, Singer-Kearfott led the design of the data processor group and Collins developed the receiver-transmitter. Both companies have now manufactured complete JTIDS Class 2 pre-production terminals for US Air Force, Army and NATO applications.

By the middle of 1986 Collins had shipped five complete JTIDS class 2 terminals and all 63 Collins-designed receiver-transmitters to Singer-Kearfott under the leader-follower arrangement.

346D-2/2B passenger address amplifier
The Collins 346-2 passenger address amplifier is a single-box system designed to replace earlier Collins products such as the 346D-1B equipment. Use of FET transistors in place of photo-resistors and relays, combined with integrated circuitry instead of discrete transistorised circuits, has led to substantial weight reduction and a doubled mean time between failures.

The 346D-2 provides 120 watts rms output (30 watts continuous) into a load of 41.5 ohm and 60 watts rms (15 watts continuous) into a standard 83-ohm load. Five inputs are provided with priorities arranged so that a flight crew member (number 1) input takes priority over attendants' number 2/2A inputs.

The 346D-2B variant can accept taped announcements or taped music through inputs numbers 3 and 4 respectively. Pilot and attendant priority, in that order of precedence, can override these two taped inputs. Programmable dual chime circuits permit high, low or combined chime tones to be used to call passengers, attendants or attract attention to 'seat belt/no smoking' signs.

Gain is automatically increased by 6 dB when engines are running and a further 3 dB increase is available for use during a cabin decompression emergency. The attendant's station is provided with a 20-watt output from an auxiliary amplifier.

Self-test facilities are incorporated for both ground and in-flight testing. A function switch and a light-emitting diode display are provided on the front panel to facilitate tests.

Collins states that one unit will suffice for narrow-bodied aircraft. Wide-bodied aircraft may be wired to accept one or two systems as required.

Format: 1/4 ATR Short
Weight: (346D-2) 6.1 lb (2.75 kg)
(346D-2B) 6..6 lb (3 kg)

STATUS: in production and service.

PAU-700 passenger address amplifier

The PAU-700 is the new-generation Collins passenger address amplifier designed in conformance with ARINC characteristic 715 for ARINC 600 installations. It is based largely on experience gained with the earlier Collins 346D-2/2B systems.

The PAU-700 allows pilot, crew and flight attendants to address the passengers. It provides for amplification of recorded announcements and taped music; tone annunciation of flight-attendant call, passenger call, 'no smoking' and 'fasten seat belt' signs and emergency tone.

Audio inputs to the amplifier are remotely activated and are provided by the pilot's or flight attendant's microphones, recorded announcements and by recorded music. Chime signalling is remotely selectable to provide passenger call (high chime), attendant call (high/low chime), 'fasten seat belt' or 'no smoking' (low chime) and attendant alert (three high/low chimes).

Features include two-tone oscillators which operate continuously, thus eliminating frequency swings or 'chirps' when the oscillator starts. This also provides immunity to control-line and supply-voltage transients which have caused spurious chime outputs in previous-generation systems. Solid-state circuits have also been incorporated to sense transients and to inhibit the audio when one is detected, thereby eliminating 'key-click' sounds. The system is all-solid-state and uses fet and cmos circuitry extensively. A front panel led digital display with a four-position switch and a test tone facility is provided for testing and calibration.

Operating temperature range is from –55 to +71° C and the system will continue to function at pressure altitudes over 50 000 feet (15 240 metres).

Dimensions: 2 MCU
Weight: 7.2 lb (3.27 kg)

STATUS: in production and service.

Communications Specialists

Communications Specialists Inc, 426 West Taft Avenue, Orange, California 92665
TELEPHONE: (714) 998 3021

TR-720 airband transceiver

The TR-720 is a solid-state, fully synthesised portable airband transceiver, covering both the communications band from 118 to 136 MHz and the navigation band of 108 to 118 MHz. It provides 720 communications channels at 25 kHz spacing and 200 navigation channels at 50 kHz spacing. As a hand-held device, the TR-720 is intended primarily for use in ultra-lights, gliders and balloons, in ferry flights of non-equipped aircraft, as an emergency back-up equipment in any aircraft and for channel-monitoring purposes.

The transceiver frequencies are selected by thumbwheels and a memory storage facility is incorporated whereby the operator can store three preset frequencies and select them as required. Power is supplied by a battery pack containing eight rechargeable NiCd batteries, giving a power output of 3 watts pep and 1 watt nominal. The pack is easily exchanged when the batteries lose their ability to be recharged (usually after about 1000 hours of operation). A charging unit is incorporated which will plug into the cigarette lighter of a car or aircraft. A power amplifier, CS-10, is now available to boost the power output to 10 watts.

A holder which allows the radio to be held upright in the aircraft, enabling 'hands free operation', is available.

Dimensions: 6.6 × 2.6 × 1.5 inches (169 × 64 × 38 mm)
Weight: 1.2 lb (0.55 kg) including battery pack

STATUS: in production.

Conrac

Conrac Corporation, Systems East-Division, 32 Fairfield Place, West Caldwell, New Jersey 07006
TELEPHONE: (201) 575 8000
TELEX: 138452

Conrad ACNIP communications panel

The Conrad auxiliary communication, navigation and identification panel (ACNIP) is under development for the US Navy/McDonnell Douglas AV-8B V/Stol aircraft. The panel is mounted in the cockpit and performs audio amplification, control and distribution functions and generates aural warning messages in response to a variety of discrete, serial and/or analogue inputs. It provides the code, mode and control functions for interfacing two cipher units in either a manual or remote mode. Logic-controlled 'push-to-talk' signals for telephone, ground-crew and 'hot-microphone' communication and back-up radio panel program control are provided. ACNIP has both secure and non-secure wiring, each portion being controlled by a microcomputer and designed to prevent inadvertent cross-coupling between the two modes.

The ACNIP system includes a new-generation fibre-optic receiver for on-board communication. It terminates a 40-foot (12-metre) aircraft fibre-optic link carrying 125 K baud universal asynchronous receiver-transmitter (UART) protocol optical signals. The receiver is part of the first production application of fibre-optics communication in an airborne tactical weapon system, and operates with received signal power ranging from 1 to 158 μW. The link connects the ACNIP system to a Sperry communication, navigation and identification data converter (CDNIC). Data is received and transmitted in a bit-aerial manner organised according to a UART protocol. Between the ACNIP and the CDNIC a twisted-pair shielded wire acts as a back-up for the optical link.

The receiver system is based on a National Semiconductor LH0082 hybrid transimpedance amplifier. This device best matches the performance requirements of the receiver but requires a well-regulated power supply, necessitating extra filtering circuits. The receiver circuit is contained in an aluminium case which also acts as a shield against electromagnetic interference.

Dimensions: 5.25 × 5.75 × 6.25 inches (133 × 146 × 158 mm)
Weight: 2.9 kg (6.4 lb)

C-RAN Corporation

C-RAN Corporation, 699 4th Street NW, Largo, Florida 33540
TELEPHONE: (813) 585 3850

AN/PRC-103 rescue radio

The AN/PRC-103 is a lightweight, portable, dual-channel, two-way, voice communication rescue radio equipment. It is designed primarily for paramedic use and consists of receiver/transmitter, antenna, headset, cable, microphone, earphones and battery. The equipment provides hands-free, two way communication between a para-rescue medical airman and a rescue aircraft through voice transmission on 243 MHz. An additional channel on 282.8 MHz allows rescue operations to continue after initial contact has been established. The radio is enclosed in a watertight case and is designed to operate satisfactorily after immersion in water to depths of 15.2 metres. Output power is 100 mW average.

STATUS: in operational service with the US Air Force.

AN/PRC-106 survival radio transceiver

The AN/PRC-106 is a lightweight emergency vhf/uhf beacon/transceiver which is used as a radio beacon on which search and rescue forces can locate downed aircrew, and to provide voice communications between the two parties.

The equipment operates in both beacon and voice communication modes on 121.5 and 243 MHz, with an output power of 125 mW average for the beacon transmission and 100 mW average for voice. The radio is compact and ruggedised and will operate satisfactorily after immersion in water up to depths of 15.2 metres.

Dimensions: 153 × 79 × 36 mm
Weight: 680 g (including batteries)

STATUS: in operational service with the US Air Force.

Cubic

Cubic Corporation, Defense Systems Division, 9333 Balboa Avenue, San Diego, California 92123.
TELEPHONE: (619) 277 6780
TWX: 910 335 2010

ELF V electronic location finder

The electronic location finder, ELF V, is a lightweight, high reliability tracking system that provides helicopters with high accuracy air rescue location. It is a fifth-generation system, representing 16 years of continued production and worldwide usage.

The principal components of the ELF V are a receiver, a control unit, and an antenna assembly which is normally located under the nose of the aircraft. Installation kits for specific aircraft are available.

The system permits a pilot accurately to locate downed aircrew from as far away as 100 miles (160 km). It allows the pilot to acquire a ground beacon signal, steer to the beacon, and hover over it completely blind with a positional accuracy of 3 to 6 feet (0.9 to 1.8 metres). The left/right and fore/aft steering and hovering accuracy is ±2°. The ELF V can also be used for resupplying, precise air-dropping and bad weather homing. With a minor variation it can also be used for precise azimuth and glide slope landings where the ground equipment is no more than a simple beacon.

In operation the beacon signal is received by two pairs of aircraft-mounted antennas. Phase sensing and angle(s) resolution are functions of the receiver-processor. Outputs are provided to the CDI (course deviation indicator) and VSI (vertical situation indicator) in the cockpit. Three controls are provided, an on-off switch, a 'frequency' select switch and a 'long-med-short' range select switch.

As a result of production over many years, and a cycle of continuing development, the ELF V is considerably smaller, lighter and more sensitive than earlier models, all the active circuits being located on eight removable modules.

Dimensions: (receiver) 7.8 × 5.13 × 9.63 inches (199 × 130 × 245 mm)
(control unit) 1.875 × 5.75 × 2.75 inches (47.6 × 146 × 70 mm)
(antenna assembly) 33 inches (835 mm) dia × 4 inches (100 mm) deep
Weight: (receiver) 9.5 lb (4.43 kg)
(control unit) 1 lb (0.45 kg)
(antenna assembly) 30 lb (13.6 kg)

STATUS: although primarily for helicopter use, the ELF V has been successfully tested on the fixed-wing LTV A-7 bomber. The system has recently undergone MIL-Spec qualification tests enabling it to be installed on the Sikorsky UH-60A Blackhawk helicopter.

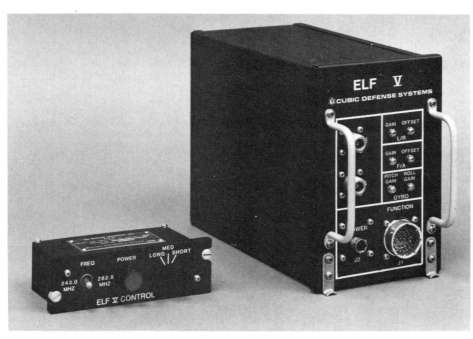

Cubic ELF V electronic location finder (receiver and control panel)

EECO

EECO Inc, 1601 East Chestnut Avenue, PO Box 659, Santa Ana, California 92702-0659
TELEPHONE: (714) 835 6000
TELEX: 678420
TWX: 910 595 1550

Radio management system

EECO's radio management system is designed to replace the separate radio controllers by combining all their control and display functions on a single panel. The system, which is suitable for all classes of civil aircraft large enough to possess a multi-radio installation, provides sufficient flexibility to interface with both digital and analogue tuning systems. It complies with ARINC 736, ARINC 429 and ARINC 720.

The all-solid-state design of the radio management system reduces cockpit space requirements. The system is microprocessor-controlled and can be used in conjunction with head-up displays or panel-mounted cathode ray tube presentations. Built-in self-test is operated periodically without interruption to operational functions and directly aids the maintenance function.

Operating temperature range is –15 to +70°C for the display and from –55 to +70°C for the processor. Maximum operating pressure altitudes for each are 15 000 feet (4570 metres) and 55 000 feet (16 760 metres) respectively.

STATUS: entering production.

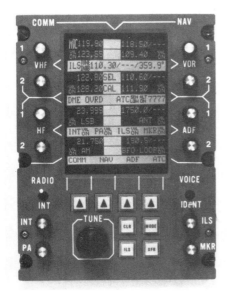

EECO ARINC 736 radio management system

EECO ARINC 720 radio frequency controller

EECO communications controller is compatible with ARINC 720 and 736 units

E-Systems

E-Systems Inc, Memcor Division, PO Box 23,500, Tampa, Florida 33630
TELEPHONE: (813) 885 7000
TELEX: 523455
TWX: 810 876 9174

AN/ARC-114A vhf radio

The E-Systems AN/ARC-114A is a military airborne radio which covers the vhf band from 30 to 75.95 MHz and provides 920 channels at a spacing of 50 kHz. The system also incorporates an independent guard receiver to monitor a single, pre-tuned guard channel in the frequency range 40 to 42 MHz.

A compact, lightweight, single-container unit, the ARC-114A is designed primarily for helicopters although its high-altitude performance, up to 50 000 feet (15 240 metres), means it is suitable for higher-flying fixed-wing aircraft.

The ARC-114A operates in FM mode and can transmit either voice or data signals. Transmitter power output is 10 watts. Facilities include azimuth homing, and a re-broadcast capability is also available when the system is used in conjunction with a suitable transmitter-receiver.

The system is of all-solid-state construction and employs crystal-controlled digital synthesis techniques for frequency-generation. In receive mode, varactor and band-selection tuning methods are employed and for transmit, the frequency selection is accomplished by indirect synthesis and digital tuning techniques. Normal operating temperature range is from –32 to +55°C.

The ARC-114A has been produced in quantity for the US Army, which has procured over 5000 systems for service in its Bell UH-1H and Sikorsky UH-60 utility helicopters.

Dimensions: 4.125 × 5.75 × 8 inches (105 × 146 × 215 mm)

Weight: 6.8 lb (3.1 kg)

STATUS: in production and service.

AN/ARC-115A vhf radio

The E-Systems AN/ARC-115A broadly provides the same range of facilities as the company's ARC-114A system but provides, within the frequency band 116 to 149.975 MHz, 1360 channels at a spacing of 25 kHz. Like the ARC-114A, it has a re-broadcast facility when operated in conjunction with suitable equipment. It also has an azimuth homing capability.

Dimensions: 4.875 × 5.75 × 9.25 inches (124 × 146 × 235 mm)

Weight: 7.2 lb (3.26 kg)

STATUS: in production and service.

E-Systems Inc, Garland Division, PO Box 660023, Dallas, Texas 75266
TELEPHONE: (214) 272 0515
TELEX: 732306

CV-3670/A digital speech processor

The CV-3670/A is an airborne digital, linear predictive data converter which provides digitised speech output at a data rate of 2.4 K bits a second, permitting narrow-band system operation on standard voice quality circuits. The output can be multiplexed with other data bit streams to allow simultaneous voice and data transmissions. Independent data clocks provided by the data terminal to the transmitter (analyser) and receiver (synthesiser) of the speech processor are used to synchronise the equipment to the receiver and transmitter of the data terminal.

The CV-3670/A is a remotely operated unit, packaged in a ½ ATR box, with voice input/output via the aircraft intercommunication system. It is operated by use of the C-10085/A controller/indicator.

Genave

Genave Inc, 802 East Lord Street, Indianapolis, Indiana 46202
TELEPHONE: (317) 546 1111

Genave manufactures a wide range of mobile, hand-held and aircraft transmitter-receiver radio-communications equipment. The aviation products are aimed at the general aviation and light aircraft markets.

AirCom vhf radio

The AirCom is a hand-held, battery-powered vhf/AM transmitter-receiver covering the frequency range 118 to 128 MHz, or optionally to 139.975 MHz, in which range it provides four channels. Channel separation is 25 kHz and the frequency spread is 10 MHz. Peak transmitter output power is 3 watts. Manual squelch adjustment is provided.

Power is supplied from eight NiCd batteries which may be recharged by a Genave charging unit. Electronic circuitry is designed for simplicity and consists mainly of a two-sided printed circuit board layout. The AirCom is a low-cost radio particularly suited for general aviation and related environments and would provide a useful emergency or standby communications set for those aircraft which are not fitted with dual communications equipment.

Dimensions: 1.8 × 2.9 × 9.1 inches (46 × 29 × 80 mm)
Weight: 1 lb (0.45 kg)

STATUS: in production and service.

Aircom 720 vhf radio

The Aircom 720 is a vhf/AM transceiver which provides all 720 channels in the 118 to 136 MHz aeronautical band. It is battery-powered by an internal NiCd pack and is fitted with a battery charger and a flexible antenna. Options include a remote speaker/microphone/headphone and leather carrying case.

Dimensions: 6.6 × 2.5 × 1.6 inches (168 × 64 × 41 mm)
Weight: 1.3 lb (0.57 kg)

STATUS: in production.

Alpha 6 vhf radio

The Alpha 6 vhf/AM transmitter-receiver provides six channels in the 118 to 136 MHz band with a channel spacing of 25 kHz at a frequency spread of 18 MHz. This hand-held equipment is battery powered, of similar construction to the Genave AirCom system, and aimed at the same market segment. Power output, however, is 2 watts carrier and 10 NiCd batteries are used as the power source.

A variant, the Alpha 6U, covers the uhf band from 350 to 400 MHz and is also available for export markets, though not for use within the USA.

Dimensions: 7.9 × 2.6 × 1.3 inches (203 × 66 × 33 mm)
Weight: 2.2 lb (1 kg)

STATUS: in production and service.

Alpha 12 vhf radio

The Genave Alpha 12 is a panel-mounted vhf/AM transmitter-receiver for light aircraft. It has a low power consumption particularly suiting it to aircraft with a limited electrical generation capacity, such as gliders, certain agricultural aircraft, or home-builts. It provides

Genave AirCom 720 hand-held radio

Genave Alpha 6 hand-held radio

12 channels in the band 118 to 135.975 MHz at a channel spacing of 25 kHz. Transmitter power output is a nominal 4 watts carrier, with 3.3 watts minimum.

Features include a mosfet, track-tuned front-end and crystal intermediate-frequency filter-

Genave Alpha 12 aircraft hand-held transmitter-receiver with microphone

Genave Alpha 120/Alpha 1200 remote transmitter-receiver

ing. A light-emitting diode is incorporated to act as a 'transmit' indicator.

Dimensions: 2.5 × 6.5 × 10 inches (63 × 165 × 254 mm)
Weight: 4 lb (1.81 kg)

STATUS: in production and service.

Alpha 100 vhf radio
The Genave Alpha 100 is a panel-mounted vhf/AM transmitter-receiver providing 100 channels in the band 118 to 127.9 MHz at a channel spacing of 100 kHz. Transmitter power output is 8 watts peak power, 2 to 3 watts carrier. Construction is fully solid-state and the receiver section is of the double conversion, super-heterodyne type and is crystal controlled. Facilities include a manually adjustable squelch disable and automatic gain control. Frequencies are selected by means of a dual knob selector with digital readout and a light-emitting diode is employed as a 'transmit' indicator.

The system has a low power requirement, in common with other Genave equipment, making it suitable for aircraft with little or no electrical generation capacity.

Dimensions: 6.5 × 2.5 × 9 inches (165 × 63 × 228 mm)
Weight: 4 lb (1.82 kg)

STATUS: in production and service.

Alpha 720 vhf radio
The Genave Alpha 720 is a panel-mounted vhf/AM transmitter-receiver providing 720 channels in the 118 to 135.975 MHz band at a channel separation of 25 kHz. Transmitter output power is 4 watts nominal. It is designed for the general aviation and light aircraft market. The system is a single-crystal unit using digital phase-locked synthesis techniques for frequency generation. Construction is fully solid-state, with extensive employment of integrated circuitry. Features include a transformerless series modulator in the transmitter section, a single conversion receiver and field-effect transistor front-end and mixer circuitry. Facilities include automatic squelch

Genave Alpha 720 synthesised transmitter-receiver

Genave Alpha 120/Alpha 1200 control head

disable and active impulse noise limitation to reduce external interference effects. Channel selection is performed by use of a dual control frequency-selector knob on the equipment's front casing, and the selection is confirmed by a dimmable incandescent readout display.

Like many Genave products, the Alpha 720 is a low power consumption system suitable for aircraft with limited electrical power. A variant, the Man-Pack, designed for portable use, is produced for gliders, home-builts and agricultural aircraft without electrical systems.

Dimensions: 2.5 × 6.5 × 10 inches (63 × 165 × 254 mm)
Weight: 4 lb (1.81 kg)

STATUS: in production and service.

GA/1000 vhf nav/com system
The Genave GA/1000 is a vhf/AM nav/com system for light aircraft. It can be panel-mounted as a single unit or, alternatively, be installed so that the VOR/LOC indicator section can be retained as a panel instrument with the control head mounted elsewhere in the cabin.

The system's communications section covers 720 channels in the vhf band 118 to 135.975 MHz at a 25 kHz channel separation. The independent navigation receiver covers the band 108 to 117.95 MHz, with 50 kHz separation and covers 200 navigation channels (160 VOR and 40 localisers). The use of separate receivers for communications and navigation functions permits radio operation without disrupting reception of navigation signals. Both sections employ hot filament digital readout displays for

Genave Alpha 720 transmitter-receiver with microphone

Genave GA/1000 nav/com radio panel

confirmation of the frequency selection. This is carried out in each case through use of dual frequency selector knobs on the front case of the control unit.

Transmitter output power is a nominal 4-watt carrier signal and a radio frequency actuated light-emitting diode is incorporated as a transmit indicator. Communications receiver gain is automatically controlled and automatic squelch with manual disable is also provided. Audio output may be either through a 4-ohm speaker

for which a 3-watt output is available, or into 600-ohm earphones from a 100 mW output. The same values are applicable to the audio outputs from the navigation receiver section. In VOR mode, transmissions from the system's transmitter section cause no visible deflection of the course-deviation indicator needle. Both VOR and localiser have ARINC standard autopilot outputs.

The GA/1000 uses solid-state integrated circuitry and a single-crystal digital synthesiser

is employed for frequency generation. Receiver demodulation circuitry is of the single conversion type. A range of antenna systems and associated antenna coupler units is available to match differing aircraft installation requirements and a speaker muting relay is also obtainable.

STATUS: in production and service.

Gould

Gould Inc, Navcom Systems Division, 4323 Arden Drive, El Monte, California 91731-1997
TELEPHONE: (213) 442 0123
TELEX: 667487
TWX: 910 587 3428

AN/ARC-98 hf radio

Gould's AN/ARC-98 is an hf/ssb airborne communication system designed to provide

extended-range communication for military aircraft, particularly for Army light fixed-wing aircraft and helicopters operating in low-altitude, nap-of-the-earth roles. The system covers the 2 to 30 MHz band in which 280 000 channels, at increments of 100 kHz, are provided. It has both voice and data transmission/reception capabilities and operating modes include usb, lsb, and ame for voice and usb or lsb for data. Transmitter power output is 400 watts.

Particular attention has been paid to ease of servicing and repair under field conditions. The system was developed to a US Army requirement and is in service in a wide range of aircraft. A variant of the ARC-98, using modules common to those in the airborne equipment, is serving in military vehicles. The system is also believed to be operated by the Iranian Armed Forces.

STATUS: in production and service.

GTE

GTE Government Systems, 100 First Avenue, Waltham, Massachusetts 02254
TELEPHONE: (617) 890 9200

GTE laser transmitter

GTE Government Systems is developing an improved airborne laser transmitter for aircraft-to-submarine communications, under contract to the US Navy. Experiments conducted during the past few years with a blue-green laser system have demonstrated the feasibility of communicating through cloud with a submerged submarine at significant depths, and the existing high performance equipment is

being upgraded to operate with a high area coverage rate in various weather conditions and water types. Laser communications systems offer a variety of benefits for use with submerged submarines compared with conventional radio communications, including less distortion as the beam travels through water, higher data-rates, greater security and less susceptibility to jamming.

It is intended that the airborne transmitter will provide nearly instantaneous communication between carrier-based aircraft and submerged submarines in tactical situations. In addition, advanced versions of the transmitter would enable a space platform to communicate both tactical and strategic messages over large areas.

Lasers are also being developed for air-to-air applications under the Have Lace laser airborne communications experimental programme. GTE is producing two terminals, mainly from off-the-shelf equipment, which include laser transmitters and receivers, as well as acquisition and tracking equipment.

STATUS: in development. The blue-green laser transmitter was due to be mounted in a Lockheed P-3C Orion aircraft for experimental flights during 1986. The Have Lace terminals will be evaluated on board a Boeing KC-135 test bed aircraft. GTE is also developing submarine optical receivers, based on atomic resonance filter techniques.

Harris

Harris Corporation, PO Box 94000, Melbourne, Florida 32902

AN/USQ-86(V) data link

The AN/USQ-86(V) is a modular integrated communications and navigation system which provides a nuclear- and ballistic-hardened anti-jam data link for air-to-ground applications. Modules of the system, which has been developed under the auspices of the US Army's

Electronics Research and Development Command, can be used with a variety of manned aircraft and remotely-piloted vehicles.

In the RPV role the system provides a secure data-link for transmitting video sensor information and telemetry from an airborne data terminal to a ground control station. This could service a number of facilities, such as capability for long-range target acquisition, artillery fire adjustment, target designation and reconnaissance.

In the manned aircraft role the system can be

used as part of a stand-off target radar system, or part of a wide-band sensor configuration for remote control airborne intercept, to compromise enemy communications, and to locate hostile command and control facilities.

The key element in all these facilities is the secure air-to-ground data link. This is achieved by a combination of spread spectrum modulation techniques, message coding, signal processing and advanced antenna techniques.

STATUS: in late development.

Hazeltine

Hazeltine Corporation, Building 10, Greenlawn, New York 11740
TELEPHONE: (516) 261 7000
TELEX: 967800

EJS (enhanced JTIDS system)

The EJS project is in advanced development to provide secure, anti-jam communications in the uhf range for air-to-air and air-to-ground communications. The programme is a successor to the Seek Talk and Have Clear projects and is stated to be the cornerstone of US

Department of Defense plans for US Air Force and US Navy interoperability in voice communications against severe jamming. EJS will use JTIDS technology, a derivative of the USN JTIDS waveform, and a new frequency to make interoperability as simple as possible. In addition, EJS will include provisions to add a data link capability under subsequent pre-planned product improvement plans.

EJS uses a combination of techniques, including spread spectrum modulation, fast frequency-hopping and voice signal processing to provide a number of capabilities. These include high jam resistance to allow dependable communications in the presence of heavy

jamming, voice conferencing with the ability to listen simultaneously to multiple audio signals transmitted from other platforms, high voice intelligibility and quality, transmission and message security, operational flexibility, and ease of use with minimal network management procedures.

STATUS: in advanced development. Hazeltine is the prime contractor for the development phase of the programme, and has been funded by over $200 million to design, develop, manufacture and test 50 prototypes. Subsequent production is expected to result in operational deployment beginning in 1989.

Hughes

Hughes Aircraft Company, Ground Systems
Group, PO Box 3310, Fullerton, California
92634
TELEPHONE: (714) 732 3232
TELEX: 685504

Hughes Aircraft Company, Microelectronic
Systems Division, 2601 Campus Drive, Irvine,
California 92715
TELEPHONE: (714) 752 3800
TELEX: 910 595 2435

AN/AXQ-14 weapon control data link for GBU-15 glide bomb

The Hughes AN/AXQ-14 is a two-way communication data link to guide the GBU-15 glide
bomb. It provides a video and command link
between the command aircraft and the weapon,
enabling the systems operator to remain in the
control loop while the weapon is being directed
to its target. In effect, the data link permits a
command authority similar to a 'fly-by-wire'
situation, whereby the operator can transmit
guidance instruction from launch to impact.
Alternatively, he may select any one of a
number of autonomous weapon-control modes,
including an override mode which permits
target updating or re-designation as required.

The extended weapon control capability
conferred by the data link contributes to
weapon system performance in terms of stand-
off range and operational utility. Target acquisition is deferred until the weapon, rather than
the command aircraft, is closer to the target.
Tactically, the aircraft can leave the target zone
immediately after launch.

The AN/AXQ-14 system comprises three
major elements: a data link pod mounted on the
command aircraft, a data link control panel,
used in conjunction with an existing display
within the aircraft, and a weapon data link
module, mounted on the rear of the weapon
itself.

The pod is an aerodynamically-shaped container mounted on a standard stores-carriage
strong-point on the fuselage centreline or on an
underwing station, according to aircraft type. It
contains four line-replaceable units comprising
an electronics section incorporating all radio
frequency generating and receiving equipment,
a de-multiplexer to decode all aircraft command
and pod control signals, an encoder, and
antenna controls; a phase-scanned array for
weapon tracking in normal operation; a forward
horn antenna which provides additional coverage; and a mission tape-recorder which maintains a permanent record of weapon video data.
The pod is suitable for high performance
aircraft, is certificated for operation at speeds in
excess of Mach 1 and is also compatible with
high/low-altitude operations. There is said to
be no compromise of aircraft performance
attributable to carriage of the pod.

Used in conjunction with an existing display
system, the aircraft control-panel acts as the
interface between the weapon system operator
and the weapon guidance system. The panel
accepts signal inputs from the aircraft as well as
from its own controls, and formats these into
discrete commands as required by the data link.
Although the panels are tailored to the individual requirements of the aircraft type and
intended customer usage, each unit accepts
the standard configurations of the GBU-15 data
link and the pod.

Attached to the aft-end of the GBU-15
weapon is the ultimate component in the data
link chain, the weapon data link module. This
simultaneously transmits video from the
weapon's seeker-head and processes incoming
command signals from the aircraft to the
weapon. Heading changes during the weapon's
flight are effected through discrete command
signals. Dual analogue voltage channels enable
the operator to slew the weapon in pitch and
yaw during approach to the target.

Digital techniques are employed in the
AN/AXQ-14 system and the transmitter is of
all-solid-state construction. The system's electronically phase-scanned antenna array provides the data link with high rate tactical
manoeuvring capability. A comprehensive
range of test equipment is provided, including a
flight checkout unit for testing aircraft cables
from the pod connection point, an aircraft
simulator unit which permits functional checks
of the control panel, and a weapon simulator
unit for test of the aircraft-mounted pod and
isolating faults down to line-replaceable unit
level. Used together, these two latter units
permit full system functional checkout.

Two primary launch modes are envisaged for
operation of the GBU-15 weapon and AN/AXQ-
14 control combination: low-altitude penetration and high-altitude stand-off. Hughes
claims that use of the data link has improved
weapon delivery accuracy over non-link

weaponry in various profiles from airborne
platforms such as the US Air Force's McDonnell
Douglas F-4 Phantom, General Dynamics F-111
and Boeing B-52 aircraft. The system is also
said to be compatible with the McDonnell
Douglas F-15, F/A-18, A-4; General Dynamics
F-111, F-16 and LTV A-7 aircraft. Potential
weapon applications include Harpoon,
Maverick, MRASM (medium-range air-to-surface missile) and cruise missiles.

At the end of December 1982, Hughes
announced a $13.4 million order from the US
Air Force for the production of additional data
link systems, bringing the total post-develop-
ment awards to nearly $54 million. Production
began in 1981 and at the end of 1983 Hughes
was negotiating with the US Air Force and
foreign air forces for further sales with a
potential value of over $46 million.

Hughes AN/AXQ-14 data link pod mounted inboard of GBU-15 glidebomb, under fuselage of US Air Force/McDonnell Douglas F-4 Phantom

Hughes Class 1 JTIDS terminal on board US Air Force E-3A AWACS aircraft

In April 1986 Hughes joined forces with Cubic Corp to bid for the full scale engineering, development and initial production of a new anti-jamming data link for the GBU-15/AGM-130. The total programme is estimated to be worth $300 million. The joint venture company is known as the Hughes-Cubic Communications Company.

STATUS: in production and service.

AN/ARC-181 TDMA radio terminal

The AN/ARC-181 time-division multiple-access (TDMA) terminal is a secure, jam-resistant radio terminal for airborne surveillance, command and control centres, developed by Hughes for US and NATO E-3A AWACS aircraft, Hawk missile batteries and NATO Air Defence Ground Environment (NADGE) centres under the Airborne Early Warning/Ground Environment Integration Segment (AEGIS) programme. Each terminal comprises a communications processor, transmitter, receiver, high-power amplifier, and control and display panel.

Part of the JTIDS programme, the terminals provide the channel for continuous communications exchange, resulting in a constantly-updated information pool which is available to all network members. Spread spectrum, data interleaving and frequency hopping techniques, give enhanced data and jam-resistance capabilities. The contracting agency is the Joint Service Program Office in the Electronic Systems Division of the US Air Force Systems Command. Sub-contractors include Siemens of West Germany and Italtel of Italy.

STATUS: first production terminals went into service with NATO/Boeing E-3A aircraft and ground stations in 1983. Hughes is to build more than 80 terminals for the USA and NATO.

Model 1150 ACIS advanced cabin interphone system

The Hughes Model 1150 ACIS is designed primarily for Boeing 747 wide-body transport aircraft. It provides up to 20 handset stations, one pilot station and 19 attendant stations which are normally located at strategic positions such as each door, the upper deck and in the galleys. This -410 version of the system has capability for three separate passenger address announcements as well as total aircraft passenger address.

The system is composed of a central switching unit, a pilot's (flight-crew) control unit, a chime light sensor and handsets.

The flight-crew's handset, used in conjunction with the control unit, provides immediate access to the flight interphone, the passenger address system, and any or all of the attendants' stations. The flight-crew control unit contains the dialling keyboard and push-button controls for other functions. The system is programmed to give the flight-crew immediate priority for any station, whether busy or otherwise.

Attendants' handsets contain the dialling keyboard, reset and push-to-talk switches. The keyboard is used to dial any other station.

The number of simultaneous calls which may be made is limited only by the number of stations installed. Operation is similar to public telephone systems with dial, ringing and engaged tones. A chime sounds at the called station and a call light is illuminated to indicate that the station is being called.

Every attendant's station can initiate an attendants' all call which places all attendants' stations on a common party line, and an all call which signals all stations, including the crew's. For direct immediate calls to the pilot, any station can place a pilot's alert call giving direct communication to the pilot's station on the flight deck. In addition, each station can have access to the passenger-address system either in first class section only, coach sections only,

upper deck only, or all together. If the pilot is using the passenger address, an attendant's station cannot interrupt but the pilot has priority to interrupt the passenger address when it is in use by an attendant. A pilot's conversation with an attendant using the passenger address is not however broadcast over the public address system.

Howler-alert is provided to signal that a handset is 'off hook' to nearby personnel but any number of 'off hook' sets will not disrupt normal operation of the system. The central switching unit is a fully-solid-state system under microprocessor control. To meet reliability standards, MIL qualified components are used wherever possible throughout the system. Telephone functions other than stated may be programmed in by modification of the software. Built-in self-test facilities for the system are incorporated into the central switching units.

Weight: (central switching unit) 8 lb (3.63 kg) (flight-crew control unit) 1.25 lb (0.59 kg) (chime light sensor) 0.25 lb (0.13 kg) (handset) 0.5 lb (0.36 kg)

STATUS: in production and service.

AN/AIC-28(V)1 audio distribution system

Hughes AN/AIC-28(V)1 audio distribution system provides Comsec-compatible internal communications among aircraft crew-members and direct access communications between them and personnel in support of aircraft mission and maintenance activities. The Audio Distribution System (ADS) consists of up to 28 subscriber control panels located at the various crew-stations, two central switching units to route audio traffic and a programming, display and test panel to control, monitor and test the ADS.

The key to achieving ADS secure communication performance lies in the microminiature lsi design which provides extremely low levels of crosstalk (greater than 100 dB between communication channels). Interlock circuitry prevents any subscriber from transmitting simultaneously on a clear and a secure or classified channel. The interlock also inhibits all direct access transmissions from a subscriber station initiating a public-address announcement.

Any subscriber control panel or station can signal any other subscriber station for a private conversation via a four-channel selective intercom. Other subscribers can be added to form progressive conferences.

Access to radios by subscribers is accomplished remotely from the programming, display and test panel. Subscribers are provided with panel controls for selection of monitor and transmit functions, and controls for adjusting the receive volume level for up to four radios at a time. The ADS retains all selective interphone and radio access operations in the event of an aircraft power failure.

The programming display and test panel enables a control operator to program selectively the direct access radio and intercom nets to the various subscribers. It also displays the current program status by visual display of the channels assigned to each subscriber. The programming display and test system also isolates faults down to a replaceable primary unit level.

The central switching unit interfaces directly with the system's subscriber stations and serves an an audio distribution matrix. It consists of an array of audio multiplexer devices with appropriate station and channel input/output audio buffers and station data buffers. All audio switching and mixing is performed within the central switching unit so that, irrespective of the total number of channels a subscriber chooses to monitor simultaneously, composite earphone audio is conveyed from the central switching unit to the subscriber station via a single twisted cable

pair. Conversely, when a subscriber transmits, the microphone audio is conveyed on a single twisted-pair to the central switching unit which distributes it to other subscriber stations and to radios on the channels on which he is transmitting. The two central switching units operate synchronously in a master-slave relationship.

Four separate types of subscriber station panels can operate with the system: a flight deck audio panel, a mission audio panel, a special audio channel, and maintenance audio panels.

All flight deck and mission audio-panels can initiate public address announcements which are broadcast over all headsets and loud-speakers in the aircraft. When in public address mode, all direct-access radio transmission from the originating subscriber station is inhibited, and when originating from the flight deck the speaker nearest the originating station is muted. All-weather 5-inch (127 mm) and 8-inch (203 mm) speakers are provided. The public address amplifier is packaged for standard air transport racking. The package contains controls and an output-power meter for testing and adjusting the 120-watt main amplifier, a 16-watt auxiliary amplifier and the speakers. Provision is made for the incorporation of recorders to log all two-way external communications.

The system has a built-in self-test capability at both system level and at subscriber station level. Replacement of primary units is possible during mission operations. A semi-automatic test station is available to facilitate ground maintenance and repair of primary units.

System features include the automatic assignment of nets on the selective intercom together with logic to decode call digits and determine which nets are available to the respective subscribers; capability of subscribers to transmit on more than one communication net; direct access to as many as 22 radios with additional access from flight-crew stations to seven navigation monitors; and capability for distributing audible alarm signals to subscribers in response to sensors and switches external to the ADS.

STATUS: in production and service.

AN/AIC-28(V)2 and AN/AIC-28(V)3 audio distribution systems

The AN/AIC-28(V)2 and AN/AIC-28(V)3 audio distribution systems are identical to the AN/AIC-28(V)1 system except for an increased number of subscriber audio stations and direct access radios available for mission operations (see previous entry). These changes required expansion of the programming display and test panel and the central switching unit.

STATUS: in production and service.

AN/AIC-28(V)4 audio distribution system for tanker aircraft

The AN/AIC-28(V)4 tanker audio distribution system provides secure communications between crew members and radio equipment. The system consists primarily of: sunlight-readable refuelling audio panels located stations for the boom operator, observers, and loadmaster; flight deck audio panels, and their associated clear-secure panels, located at the aircraft commander's, pilot's, flight engineer's and navigator's station; central switching units to route the audio traffic; a radio select panel to program access to radios; and an ADS test panel to test and fault monitor operation of the system.

The system audio traffic is divided into clear and secure communications. The central switching unit interlock circuitry prevents compromise of secure communications, such as transmitting simultaneously on a clear and secure channel. Any subscriber panel can be used to signal other subscribers for private or conference conversations, or can be used to make public address announcements, which are broadcast over the headsets and loudspeakers.

STATUS: in producton.

AN/AIC-29(V)1 intercommunicaton system

The AN/AIC-29(V)1 intercommunicaton system provides secure internal communications between helicopter crew members and direct access between crew members and radios and/or security equipment for external communications. The system can accommodate up to 15 transmit and 18 receive radio channels and provides greater than 100 dB crosstalk isolation between any transmit and any other transmit or receive channel. The system consists of six crew station units, a single maintenance station unit, and a communications switching unit. The crew station units are night-vision goggle compatible. The communications switching unit performs switching and mixing of audio channels in accordance with digital data multiplexed from each crew station unit. The switching unit's response to the multiplexed data depends on the communication plan programmed into the system. The system provides emergency back-up intercom and radio communication selection which bypasses the communications switching unit for audio transmit and receive functions.

System features include the capability for distributing audible alarms to crew stations in response to sensors/switches external to the intercommunicaton system. The system also provides interface with the MIL-STD-1553 databus, two-way chime call capability, and built-in test circuits.

STATUS: in production.

AN/AIC-30(V)1 intercommunication system

The AN/AIC-30(V)1 intercommunication system provides communications among heli-

copter crew members and direct access to radio communications in support of mission activities. The system includes a microprocessor-controlled communications switching unit which is controlled by multiplexed data from six crew station units during normal system operation. Interlock circuitry prevents a crew member from transmitting on a clear channel while receiving on a secure channel. Crosstalk isolation between any two channels is greater than 126 dB at 1 kHz.

An intercom (backup) CALL channel, which bypasses the switching unit matrix, may be activated at the crew station units. This back-up channel may be constantly monitored using the crew station unit master volume control. In addition, two cockpit crew station units are provided with a direct back-up interface to two radio transceivers. The intercom backup CALL channel is also used for extended ground communicaton operatons with the switching unit power removed. Two audio frequency amplifiers provide volume control for communications at the maintenance stations.

STATUS: in production.

AN/AIC-30(V)2 intercommunicaton system

The AN/AIC-30(V)2 intercommunication system is an enhancement to the AN/AIC-30(V)1 intercommunication system for use in larger aircraft (see previous entry). The AN/AIC-30(V)2 system increases the number of crew stations and audio frequency amplifiers, modifies the communications switching unit, expands the radio channels and hard-wired back-up channels, and provides dual headset and microphone capability at each crew station unit.

The communications switching unit modification includes added circuit cards and wiring for multiplexing control and audio access transmit and receive. A fourth connector is included on the front panel of the switching unit to accommodate the additional audio traffic.

The crew station unit front panel changes reflect the additional radios and navaids utilised for the system. Each crew station unit provides dual headset capability. The crosstalk isolation between transmit and receive channels is greater than 126 dB at 1 kHz. The crew station unit also provides intercom back-up circuits and is night-vision goggle compatible. The three audio frequency amplifiers provide headset volume control at the maintenance or remote crew stations.

STATUS: in production development.

AN/AIC-32(V)1 intercommunication system

The AN/AIC-32(V)1 intercommunication system provides secure internal communications between crew members and direct access to mission radios and security equipment for external communications. The system primary units consist of a communication control unit, four flight deck crew station units, five mission area crew station units, eight maintenance station units and one maintenance control unit.

Audio traffic from all station units is controlled by the communication control unit during normal system operations. A back-up operating mode, initiated from selected crew station units, bypasses the communication control unit to provide hard-wired access to a set of predetermined radios. A back-up intercom network, integrated with the CALL function on all station units, is also provided for emergency intercommunication between crew members.

A test switch on each crew station unit permits pre-flight verification of audio and lamp indicator circuitry and digital interface with the communication control unit. A public address system, accessed from the flight crew station, provides announcements over the loudspeakers and headsets. Auxiliary control units, located at selected mission area crew station units, expands the total system direct access capability to 30 transmit and 36 receive channels.

STATUS: in pre-production development.

ITT

ITT Aerospace/Optical Division, 3700 E. Pontiac Street, PO Box 3700, Fort Wayne, Indiana 46801
TELEPHQNE: (219) 423 9636

AN/ARC-201 vhf/fm transceiver (Sincgars-V)

The AN/ARC-201 is the first airborne vhf/FM frequency-hopping radio, and is an all solid-state equipment for use in helicopters, light observation aircraft and fighters.

The equipment operates in the 30 to 87.975 MHz band using 25 kHz channel spacing to provide 2320 channel capability in single-channel and frequency-hopping modes. A six-channel, non-volatile preset memory is incorporated for single-channel and ECCM modes. Power output is 10 watts, and an

ITT AN/ARC-201 vhf/FM transceiver

interface and controls for the AM-7189A/ARC 50-watt amplifier are incorporated.

The AN/ARC-201 is available in panel-mounted, dedicated remote, and 1553B multiplex bus remote configurations. It is interoperable with the current vhf/FM radios in the single channel mode, and with the ground-

based Sincgars radios (VRC-87 to VRC-92, and manpack PRC-119) in the frequency-hopping mode. Electro-luminescent lighting is provided on the front panel, compatible with the use of night vision goggles.

The radio has 80% commonality with the ground Sincgars communication equipment, with extensive use of lsi circuitry and microprocessors being made for high reliability. A built-in test function isolates faults to the module level with 90 per cent confidence. A data-rate adaptor interfaces the radio with data devices for data communication. An automatic single-channel cueing capability in the ECCM modes allows a single-channel user to alert members of an ECCM net. Existing Comsec devices can be used to provide secure communications in voice and data.

Dimensions: 102 × 127 × 203 mm
Weight: 3.52 kg

STATUS: in production.

King Radio

King Radio Corporation, 400 North Rogers Road, Olathe, Kansas 66062
TELEPHONE: (913) 782 0400
TELEX: 42299

King Radio is pre-eminently a manufacturer of avionics for computer and top-of-the-line business aircraft. The firm was founded in 1959 as King Radio Corporation by Edward J King, through acquisition of the Wright Airborne Electronics Company which at that time was building communications transceivers. Edward King had already built from scratch an elec-

tronics company producing magnetic amplifiers and missile guidance systems, which he sold to Collins Radio Company, becoming for three years subsequently a Collins employee. By the early 1960s King Radio had launched its first product line, the Silver Crown series, and was becoming a recognised supplier for light aircraft equipment. Silver Crown was followed by Gold Crown for the more sophisticated business aircraft, and a limited ARINC 500 line for the air transport industry. The company was producing not only ranges of panel-mounted and remote-mounted nav/com equipment, but also autopilots (initially acquired from Honey-

well, which then became a shareholder with 10 per cent holding), and weather radar. It also has military interests via DME and Tacan, and autopilots for light communication aircraft.

By 1980 Edward King was considering disposal of his company, in which he had a 54 per cent holding, and during the following year began discussions with Bendix. These reached a conclusion in 1984, with an agreement by Bendix to buy for about $110 million the majority of King's business. Its weather radar line, comprising two models, was sold to Narco, Bendix having its own more comprehensive weather radar family. Bendix plans to operate

King Radio as a wholly-owned autonomous facility at its existing Kansas location, but the radar business has been moved to Narco's plant in Pennsylvania.

Silver Crown nav/com avionics family

The Silver Crown designation applies to a family of CNI (communiciations, navigation and identification) and autopilot equipment for singles and light twins up to turbo-prop size, and for light helicopters. The various units are self-contained, with processing and presentation or indication mechanisms accommodated together in single, panel-mounted boxes without the need for remote units. Initial members of the range appeared in the 1960s, and subsequent additions have made Silver Crown into a very comprehensive suite. In 1980 the series was upgraded by the introduction of some units incorporating digital technology.

The current Silver Crown family consists of a large number of alternative units, the complete list of available units being as follows:

KRA 10A radar altimeter
KMA 24/24H audio control system with three-light marker beacon receiver or five-station intercom
KWX 50 digital weather radar
KN 53 navigation receiver
KGS 55A compass system
KN 62A digital DME
KN 63 digital DME
KT 76A transponder
KT 79 solid-state transponder
KNS 80 integrated navigation system (VOR/LOC receiver, digital DME, digital RNav computer, glideslope receiver)
KNS 81 integrated navigation system (VOR/LOC/glideslope receiver, RNav computer)
KR 86 automatic direction finder
KR 87 digital automatic direction finder
KY 92 vhf communication transceiver
KT 96 radio telephone
KA 134 audio control console/KR 22 marker beacon receiver
KX 145 nav/com transceiver
KX 155/165 nav/com transceivers
KX 170B/KX 175B nav/com transceivers
KY 196/KY 197 communication transceivers
KFC 150 2/3-axis digital flight control system
KAP 150 2-axis autopilot
KAP 100 1-axis autopilot
KFC 200 2/3-axis flight control system
KAP 200 2-axis autopilot
More detailed descriptions of some of these units can be found under the appropriate section headings in this book.

STATUS: in production for, and in service with, a wide range of light single and twin-engined aircraft.

Gold Crown nav/com avionics family

In the late 1960s the Gold Crown range of remote-mounted (ie panel-mounted flight-deck instruments driven by rack-mounted processing amplifiers) equipment was introduced to meet the larger piston and turbine twins then being developed, and to complement the Silver Crown range. In 1972 a more advanced family of digital equipment appeared and began to take a large share of this market. The most recent range is Gold Crown III, unveiled in 1981. Designed for top-of-the-line corporate and commuter twins, turbo-props and turbine helicopters, this completely redesigned family incorporates custom lsi and microprocessor technology; the remote-mounted units are some 40 per cent smaller and 30 per cent lighter than their predecessors. Being of solid-state design (claimed by King to be the first completely solid-state avionics for general aviation), the systems are substantially more reliable.

The current Gold Crown III family comprises the following equipment:

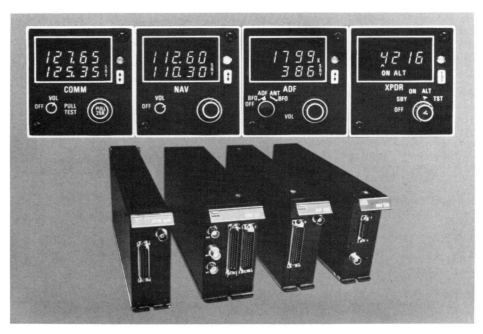

King Gold Crown III avionics family

KWX 56 colour weather radar
KFC 200 flight control system
KFC 250 autopilot/flight director system
KNR 634 digital navigation receiver
KXP 756 transponder
KDF digital automatic direction finder
KTR 908 vhf communication transceiver
KHF 950 hf single-sideband transceiver
KAA 955 audio amplifier/control
KNS 81 integrated nav/RNav system
KNI 582 indicator
KQI 553A pictorial navigation indicator
More detailed descriptions of some of these systems can be found under the appropriate section headings in this book.

STATUS: in production and service.

ARINC avionics family for air transport

Although King has only a very limited line of air transport avionics, well over 100 airlines are customers for one system or another.

The current ARINC family consists of the following equipment:
KNR 6030 VOR/ILS receiver
KDM 700B digital DME
KXP 7500 transponder
KTR 9100A vhf communication transceiver
More detailed descriptions of these systems can be found under the appropriate section headings in this book.

STATUS: in production and service.

KY 92 vhf radio

The King KY 92 is a vhf transmitter-receiver which provides two-way communication on 720 channels, at 25 kHz increments, in the 118 to 135.975 MHz band. Bandwidth is wider in receive mode, the receiver section extending coverage down to 116 MHz. Transmitter output power is 7 watts.

The system is of all-solid-state construction and employs digital synthesis techniques for frequency generation. It features a field-effect transistor rf amplifier and mixer stage to enhance interference-free reception and a crystal filter for added protection against interference from adjacent channels.

Squelch control is automatic but manual override is provided for optimisation of reception from distant stations. Volume control is sufficiently adjustable to permit the use of either headset or cabin speaker audio output.

The entire system is contained in a single unit designed for panel mounting. It is operable over a temperature range from –20 to +55° C or up to 70° C for short periods.

Dimensions: 6.25 × 1.5 × 10.5 inches (159 × 38 × 267 mm)

Weight: 2.8 lb (1.27 kg)

STATUS: in production and service.

KMA 24 and KMA 24H audio control systems

King's KMA 24 and KMA 24H are compact, lightweight systems for the integrated control of a number of radio communication and navigation systems. Designed principally for the general aviation sector, they are single-box units, for panel mounting.

The KMA 24 can control up to three transmitter-receivers and six receivers, including an internal marker-beacon receiver for which it contains an automatically-dimmed three-light presentation. In the 24H version, the internal marker-beacon facility is replaced by a five-station 'hot microphone' intercom and associated volume control. The 24H's marker switches control audio from an external marker beacon receiver. This system also features switching which gives the pilot's microphone priority over that of the co-pilot.

Both units provide transmitter-receiver and receiver outputs to speakers, headphones or both. A separate headphone-isolation amplifier maintains constant, noise-free volume levels even when several receivers are keyed simultaneously. When a microphone is keyed, all receivers are automatically muted to eliminate feedback. Both systems are offered in a choice of eight configurations with a number of different optional facilities.

The KMA 24 has two rows of alternate-action push-buttons which control all receiver audio distribution functions. The top row selects receivers for the cockpit speaker and the lower row for headphones. Each row is independent of the other and allows selection of speaker, headphones or both for all receiver combinations. A rotary selector switch connects the microphone to either COMM 1 or COMM 2 and an additional switch position permits selection of either hf or radio-telephone. International models of the KMA 24 are available with an hf switch position. Other switch positions allow selection of cabin-address and ramp-hailer facilities.

When the microphone is switched off, power to the speaker amplifier and the marker beacon is cut but the headphone amplifier remains operative. An additional option is the choice of an 'auto' feature or an additional receiver, such as an additional ADF. When engaged, this auto feature automatically matches the corresponding receiver audio with the selected transmitter. The unit operates from either 13.75- or 27.5-volt supplies without any need for adaptors or converters.

The KMA 24 has a built-in crystal-controlled superheterodyne marker-beacon receiver with a three-light display in which, claims King, good selectivity eliminates interference from FM and television broadcast stations. Automatic dimming circuits adjust the brightness of the lamps to a level appropriate to ambient cockpit lighting. Push-button controls are provided for sensitivity selection and lamp test. The unit can also drive remote marker-beacon lights.

The marker-beacon receiver system in the KMA 24H, is replaced by an intercom unit which handles up to five microphone inputs and six parallel-connected headsets. A public address position replaces the 'INT' position for the microphone switch. This arrangement allows either pilot or co-pilot to make public address announcements in aircraft equipped with a cabin speaker. If a marker-beacon system is required in an aircraft equipped with the 24H model, the King KR 21 unit can be used to provide audio beacon signals to the KMR 24H system. The KR 21 is self-contained and can be mounted on the panel. It has a similar light display to that of the KMA 24 unit.

Both the KMA 24 and the KMA 24H systems have two unswitched inputs which can be used to connect additional facilities such as a radio altimeter audio alert or a ringing signal from a radio-telephone. The systems are operable over a temperature range from –20 to +55°C or up to 70°C for brief periods.

Dimensions: (length behind panel [KMA 24]) 6.75 inches (172 mm)
(KMA 24H) 6.8 × 1.3 × 6.25 inches (173 × 33 × 159 mm)

Weight: 1.7 lb (0.77 kg)

STATUS: in production and service.

KHF-950 hf radio

Introduced in February 1981, King's first hf system, the KHF-950, has coverage up to 29.999 MHz and can consequently offer a full choice of 280 000 frequencies at 100 Hz spacing. The system operates in usb, lsb and AM modes. Transmitter output in each ssb mode is 150 watts pep and 35 watts average over the full frequency range. Since its introduction, the system has received wide support,

King KHF-950 hf radio system with standard control unit

King KHF-950 radio: left to right, KAC-952 power amplifier, KTR-953 transmitter-receiver, and KCU-951 control/display unit

and by January 1984 it was to be found on the flight-decks of more than 1000 aircraft, including general aviation types such as Canadair Challenger, Gulfstream G III, Lear 55, Citation III and Falcon 50.

The KHF-950 employs synthesised frequency-generation techniques and uses microprocessor control for easy in-flight-operation. It possesses a non-volatile memory which allows pre-selection storage for up to 99 channels and their appropriate modes but, in addition, allows the user to tune manually to any other frequency within the covered bandwidth without disturbance to the stored pre-sets.

Operation may be either simplex, for normal air traffic or similar communication, or semi-duplex which permits patch-through into public utility telephone circuits. Provision is also made for a selcal facility, and the dedicated circuits enable continuous selcal monitoring to be maintained without having to select the AM mode.

A feature of the KHF-950 is its automatic antenna tuning capability, an operation carried out simply by keying the microphone. King claims that the system will operate satisfactorily on antennas only 10 feet long. It will also tune to fixed-rod aerials and towel-rail antennas used on helicopters. King has also developed equipment to facilitate operation from shunt and notch-type antenna systems.

The KHF-950 system comprises the KCU-951 all-digital remote controller, the KAC-952 power amplifier/antenna coupler, and the KTR-953 transmitter-receiver. The remote controller (or frequency selector) is designed for panel mounting and presents channel, frequency and mode-selection data on a self-dimming gas discharge numeric display.

An optional controller is the KFS-954 which is also panel-mounted, and measures only 2.25 square inches. This unit contains storage for all 176 ITU maritime radio-telephone channel selection plus an additional pre-selected simplex air traffic or conventional airborne communication channels. By pre-programming the ITU channels, it is possible for the operator to call any radio-telephone station without having to manually select the separate transmit and receive channels required. He merely selects the radio-telephone mode and the required channel. The KHF-950 also interfaces with teletype and facsimile systems and a recently-introduced dual installation equipment allows dual-frequency reception from a single antenna.

Weight: 20.2 lb (9.16 kg)

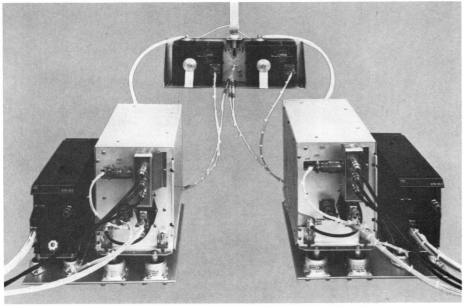

King KHF-950 radio in dual configuration

STATUS: in production and service. Version called KHF-950 has also been selected by US Army for helicopters (low-level combat role presents severe challenge to other forms of communication). Version of KHF950 system is to equip McDonnell Douglas F/A-18 Hornets being built for Royal Australian Air Force.

KHF-970 hf radio

The King KHF-970 is a derivative of the KHF-950 hf/ssb airborne radio which has been adapted specifically for military use. Selection of the system for nap-of-the-earth helicopter communications, an application for which hf is now recognised as being superior to vhf, was announced by the US Army in 1982. A surface vehicle version for military use has been designated KVR-980.

The KHF-970 covers the hf band from 2 to 29.999 MHz, providing 280 000 channels at 100 kHz increments. Transmitted output power is 150 watts. Preset channel selection from a non-volatile memory is also provided but, unlike the KHF-950, all channels and frequencies are selected on a keyboard and displayed on a cathode ray tube.

Other features include scanning of preset channels and provision for multiple selective addressing, frequency link analysis and for automated-communications operating instructions.

The US Army contract covers initial design and provides four one-year production options. The total contract could cover more than 3000 systems and is valued at approximately $40 million over a five-year period.

STATUS: in production.

KHF-990 hf radio

The King KHF-990 hf radio is a helicopter-dedicated system which draws on technology used in the company's airborne KHF-950 and marine KMC-95 systems. It provides 280 000 channels in the 2 to 30 MHz hf band at frequency increments between channels of 100 kHz. Modulation is in ssb mode and transmitted output power is 150 watts pep.

The KHF-99 has been optimised for helicopter operation. It uses the miniature KFS-594 controller, a KAC-992 combined antenna coupler/probe antenna and a remotely-located KTR-993 receiver/exciter/power amplifier. This combination, says King, provides a fully capable yet lightweight system.

The KFS-594 controller provides access to 176 permanently programmed ITU marine radiotelephone channels and to 19 programmable channels which may be selected or re-tuned by the pilot. The KAC-992 is an automatic, digital antenna coupler which is self-contained in the end of a probe antenna system. It may be mounted externally or internally with only the probe portion of the antenna protruding from the aircraft.

The KTR-993 receiver/exciter/power amplifier can be mounted in any convenient location within the helicopter with no restrictions on proximity to the other two units. It meets the TSO requirements for explosion-proof, drip-proof and salt-spray categories.

Weight: 21.8 lb (9.9 kg)

STATUS: in production.

King KX-170B and KX-175B nav/com radios

The King KX-170B is a navigation/communications transmitter-receiver providing 720 communication channels in the 118 to 135.975 MHz band and 200 navigation channels in the 108 to 117.95 MHz band. Channel separation is at 25 kHz increments in the communication section and 50 kHz increments in the navigation section. Communication transmitted power output is 7 watts unmodulated. Both sections are manually tuned with the selected frequencies being indicated digitally in windows on the front panel.

Separate receivers with independent crystal oscillators are used for the communication and navigation functions so that a fault in one has no effect on the other. Construction of the system is all-solid-state and crystal-controlled digital synthesisers are used for frequency generation. Integrated circuitry is extensively employed.

The KX-175B differs principally from the KX-170B in that it meets additional FAA TSO compliance standards and other European Civil Aviation Conference standards also. A variant, the KX-175BE has wider receiver selectivity for operation in areas where ground-station standards of frequency stability do not meet ICAO 25 kHz channel separation requirements.

Navigation indicators available for the KX-170B are the KI-208 VOR/LOC type and the KI-209 which also provides glideslope indications when used in conjunction with a suitable glideslope receiver. A choice of three different navigation indicators is available for use with the KX-175B; the KI-204 with VOR, LOC and glideslope, the KI-203 with VOR/LOC indication only and the KI-525A Pictorial Navigation Indicator for which a separate VOR/LOC converter and indicator are required. Automatically-channelled output for a 200-channel DME and 40-channel glideslope

receiver is provided together with an area navigation output.

Dimensions: 6.25 × 2.5 × 13 inches (159 × 64 × 331 mm)
Weight: 7 lb (3.8 kg)

STATUS: in production and service.

KTR 908 vhf transceiver

The KTR 908 is a remote-mounted airborne vhf communications transceiver with a standard frequency range of 118 to 135.975 MHz and channel spacing of 25 kHz. An optional extension to 151.975 MHz is available. Power output is 20 watts from a 28-volt dc power supply, and storage of two frequencies (active and standby) is provided. It is operated from a cockpit-mounted controller, KFS 598, which requires 57 mm square panel space.

Dimensions: (transceiver) 147 × 45 × 299 mm
Weight: (transceiver) 1.6 kg

STATUS: in operational service with the US Army and National Guard.

KY-196 and KY-197 vhf radios

The KY-196 is a compact, lightweight, panel-mounted transmitter-receiver particularly suitable for light aircraft. It covers the vhf band from 118 to 135.975 MHz in which range 720 channels are provided; channels are selectable at increments of 25 or 50 kHz.

A principal feature of the system is that a second frequency in addition to the one in use may be stored for immediate selection into operating mode. Both the operating and standby frequencies are presented on a self-dimming gas discharge display in a window on the front panel. When the standby channel is selected the former operating channel is entered into the standby store. Non-volatile storage, provided by an electrically alterable read only memory (EAROM) chip, ensures that both frequencies remain stored when the power supply is off or is disconnected. No separate memory power supply is required.

The KY-197 differs from the KY-196 only in that the former has a transmitted output power of 10 watts whereas the latter has a 16-watt output.

Solid-state construction is employed throughout. The system is microprocessor-controlled and digital synthesis techniques are used for frequency generation. A mosfet rf amplifier and mixer stage is used to provide clear signal reception.

Operating temperature range is from –20 to +55°C but short term operation is possible at temperatures up to 70°C.

Dimensions: 6.25 × 1.3 × 10.5 inches (159 × 33 × 267 mm)
Weight: 3.2 lb (1.45 kg)

STATUS: in production and service.

KT-96 uhf radio telephone

The King KT-96 uhf radio telephone provides air-to-ground communication facilities for interconnection with public telephone networks. Operating in duplex mode, the system provides 12 transmission channels in the 459.7 to 459.975 MHz band and 13 reception channels in the 454.675 to 454.975 MHz band. The additional reception channel is for ringing-tone signalling purposes. Transmitted output power is 10 watts minimum.

Typical operating ranges are quoted as 110 miles (177 km) at 10 000 feet (3050 metres) and 220 miles (354 km) at 40 000 feet (12 200 metres). The system may be connected to a KMA -24 audio console for use with a cabin speaker and a microphone or, optionally, to a conventional-style telephone handset and hook-switch.

The system is of all-solid-state construction, employing both silicon transistors and inte-

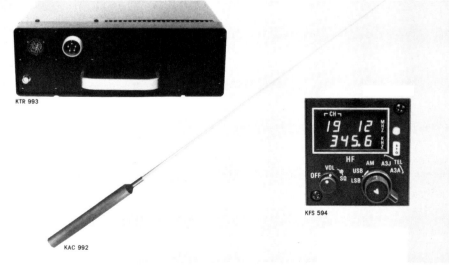

KTR 993

KAC 992

KFS 594

King KHF-990 hf radio for helicopters

grated circuitry. Operating temperature range is from –15 to +55° C.

Dimensions: 6.25 × 1.625 × 11.8 inches (159 × 41 × 230 mm)
Weight: 3.5 lb (1.59 kg)

STATUS: in production and service.

AN/ARC-199 hf radio
The King Radio AN/ARC-199 is a solid-state hf communications system providing 280 000 channels in the frequency band 2 to 30 MHz. It incorporates power management and low probability of intercept with selectable power output of 4, 40 or 150 watts pep. Twenty channels can be pre-selected for instantaneous recall and use. Automatic recognition of incoming messages by a selective addressing system is used.

The radio is microprocessor-controlled and contains a dual MIL-STD 1553B data-bus interface. Remote control can be provided through this interface, as well as through a dedicated keyboard/cathode ray tube display controller. The equipment can operate with both voice and data formats, and can be used in clear or with encryption devices. The cathode ray tube display is compatible with night-vision goggles. The antenna coupler is able to tune a variety of antennas throughout the frequency range, including long open wires, grounded wires, whips and shunt antennas.

Weight: 14.5 kg

STATUS: in service with the US Army.

AN/ARC-200 hf radio
The AN/ARC-200 is an hf/ssb communications system providing 280 000 channels in the 2 to 30 MHz frequency band. It is intended primarily for fighter aircraft and is similar in many respects to the AN/ARC-199. Power management, with a low probability of intercept, is incorporated and a selectable output power of 4, 40 or 150 watts pep is provided. Twenty channels can be pre-selected for instantaneous recall and use. Automatic recognition of incoming messages by a selective addressing system is used. Channel scanning, speech processing and selective squelch are provided.

An MDC-A3818 McDonnell Aircraft Avionics multiplex control bus is incorporated and remote control of the system is provided through this interface. Operation in clear or with encryption devices is available.

Weight: 18.2 kg

Lapointe Industries
Lapointe Industries, 155 West Main Street, Rockville, Connecticut 06066
TELEPHONE: (203) 872 8581

AN/ARC-51 uhf radio
The Lapointe AN/ARC-51 is a military airborne radio covering the uhf band from 225 to 400 MHz in which it provides 3500 channels. The system, intended mainly for high-performance aircraft, is available in a number of versions designated ARC-51A, ARC-51AX, ARC-51B and ARC-51BX. These, together with a wide range of controllers, permit users the flexibility of assembling a number of configurations to suit particular installations; many ARC-51 combinations are used in surface vehicles and in base-station applications.

All variants operate in AM/dsb mode and have a transmitted power output of 20 watts. All provide azimuth homing facilities, when used in conjunction with suitable indicator equipment, and can be used for automatic re-broadcast purposes.

Although many differing types of controller are available, a typical ARC-51 installation would provide pre-selection of up to 20 channels together with manual selection of any of the total of 3500 frequencies covered. An independent guard receiver provides simultaneous continuous monitoring of the international uhf distress frequency at 243 MHz. Other installations permit the dual control of one or more transmitter-receivers from more

than one flight-crew position, and miniature frequency indicator displays are available for use in situations where space is limited.

An interesting feature of this system is the hermetically-sealed pressurised container which allows full operational performance in unpressurised avionics bays at aircraft altitudes up to 70 000 feet (21 330 metres) when used in conjunction with a forced-air cooling supply. This sealed-case assembly also renders the system particularly suitable for operation in situations where dust or water contamination could otherwise be expected, such as in desert vehicles or high-humidity tropical environments. The system's normal operating temperature range is from –54 to +71° C.

Variants of the ARC-51 are in service with the US, Canadian, Belgian, West German, Italian, Thai and Indian armed forces.

Dimensions: 16.875 × 8.75 × 6.75 inches (429 × 222 × 171 mm)
Weight: 32 lb (14.5 kg)

STATUS: in service.

AN/ARC-73A vhf nav/com radio
Lapointe's AN/ARC-73A is an AM vhf combined navigation and communications transmitter-receiver system for light military aircraft.

The transmitter section covers the frequency band 116 to 149.95 MHz in which it provides 680 channels. The receiver, which can drive ILS and VOR indicators, covers a wider band to receive

ground-based navigation aid signals. This band extends from 108 to 151.95 MHz and provides 880 channels. Channel spacing is at 50 kHz increments in both transmitter and receiver sections. Transmitter power output is 20 watts. The system is remotely controlled.

Dimensions: (transmitter) 15.8 × 3.5 × 7.6 inches (401 × 89 × 193 mm)
(receiver) 12.5 × 3.5 × 7.5 inches (318 × 89 × 191 mm)
(controller) 6.3 × 5.7 × 2.2 inches (160 × 145 × 56 mm)
Weight: (transmitter) 14.8 lb (6.7 kg)
(receiver) 10.5 lb (4.7 kg)
(controller) 1.7 lb (0.76 kg)

STATUS: in service.

AN/PRC-90 emergency transceiver
The AN/PRC-90 is an emergency transceiver for use by downed aircrew for location and voice communication purposes. It operates on 282.8 MHz (voice-receive and transmit) and 243 MHz (voice, mcw and beacon-receive). Power output is 400 mW on voice and 500 mW pep on beacon.

Dimensions: 152 × 79 × 38 mm
Weight: 0.6 kg

STATUS: in service but being replaced by the AN/PRC-112(V). Over 35 000 sets have been produced.

M/A-COM
M/A-COM, Government Systems Division, 3033 Science Park Road, San Diego, California 92121
TELEPHONE: (619) 457 2340
TWX: 910 337 1277

MD-1035/A uhf dual modem
The MD-1035/A uhf dual modem was designed by M/A-COM to meet a US Air Force Electronic Systems Division requirement for communication over both current and anticipated satellite systems with widely varying characteristics. It provides a variety of modulation/demodulation, error-control (convolutional or block coding) and multiple access options in a single package. Communication and network control functions are performed by a flexible, multistack microcomputer which permits demodulation of virtually any digital signalling scheme through software changes. It can interface with a number of rf systems including those operating in frequency bands above the nominal uhf range.

Features include AFSAT 1 and AFSAT 2 (US Air Force Satellite) modulation schemes, dedicated interleaving, error control coding/

decoding, input/output and network control and extensive built-in test facilities which permit fault identification down to card level. Incorporated firmware changes provide an additional channel for use with the single channel transponder (SCT), 2400 bits a second for secure voice, coding and interleaving to reduce scintillation mitigation, probing for SCT report-back, 1200 bits a second and de-multiplexing for fleet broadcast and 2400 bits a second data for tactical operations.

The MD1035/A is installed in strategic force elements of the US Air Force with over 900 units having been delivered for installation in airborne terminals. The system is the basis for M/A-COM airborne command post modem/processor.

Dimensions: (control-indicator) 6.8 × 5.8 × 6.9 inches (173 × 147 × 175 mm)
(telegraph modem) 7.8 × 5 × 14.27 inches (198 × 127 × 362 mm)
(electrical equipment) 3.85 × 5.33 × 17.14 inches (98 × 135 × 435 mm)
Weight: (control-indicator) 5.6 lb (2.54 kg)
(telegraph modem) 19 lb (8.63 kg)
(electrical equipment) 3.5 lb (1.59 kg)

STATUS: in production and service.

MD-1093/ASC 30 command post modem/processor
The MD-1039/ASC 30 command post modem/processor (CPM/P) was designed primarily for terminals in airborne command-post aircraft, notably the US Air Force/Boeing E-4B, in which it allows emergency message dissemination and communications among the national command authorities, the Joint Chiefs of Staff, the commanders-in-chief, and strategic force elements such as bombers and missile launch control centres. It supports a variety of auxiliary functions including antenna pointing, plasma display control, communications network management and navigational computations. The system is based on the M/A-COM MD1035/A dual uhf modem and, likewise, operates in the uhf regime from 225 to 400 MHz. It may, however, be interfaced with equipment operating in the shf and ehf bands.

System capabilities include anti-jam and control functions, establishment of communications networks and protocols (random, TDM 1 and TDM 2), Comsec/Transec interfaces, AFSAT 1 communications, and SCT communications. It has a 40-satellite database and a software expansion capability enabling the system to accommodate new signal structures as they are introduced.

The requirements have been met through use of full duplex modulation, fast synthesisers, programmable read-only memories and computer-initiated control by a fast multi-stack microprocessor.

The MD-1093/ASC 30 is deployed with US Air Force airborne force elements.

Dimensions: (command/post processor) 19.5 × 5 × 8 inches (495 × 127 × 203 mm) (command/post modem) 19.5 × 5 × 8 inches (495 × 127 × 203 mm)
Weight: (command/post processor) 25 lb (11.36 kg)
(command/post modem) 25 lb (11.36 kg)

STATUS: in production and service.

SHF/AJ modem

The SHF/AJ is a super high frequency/anti-jam tri-modem based frequency-hopper modem for use in reliable communications and network acquisition during critical command control situations. It provides anti-jamming protection and scintillation mitigation for digital communications over military satellite links. The equipment shares the design features of the MD-1035/A uhf dual modem through the use of identical components, and similar hardware and software design.

The SHF/AJ interfaces digital baseband equipment to shf terminals via a 70 MHz intermediate frequency carrier and provides jam-resistant and secure communications. Features include four baseband input/output channels allowing TDMA or T/CDMA, data rates up to 9.6 K bits a second in stress mode, single mode of operation for both faded and non-faded operation and secure conferencing program interface. A 700 MHz if carrier is available as an alternative.

Dimensions: (standard 19-inch rack) 8.7 inches (221 mm) high × 20 inches (508 mm) deep
Weight: 40 lb (18.2 kg)

Magnavox

The Magnavox Government and Industrial Electronics Company, 1313 Production Road, Fort Wayne, Indiana 46808

TELEPHONE: (219) 429 6000
TELEX: 228472
TWX: 810 332 1610

Military radio manufacture within Magnavox, a subsidiary of North American Philips, is vested in Magnavox Electronics Systems Company at Fort Wayne. Since 1957 the company has developed a wide range of military communication systems and has produced them in large quantities, mainly for the US armed forces.

AN/ARC-131 vhf radio

The Magnavox ARC-131 is a military vhf airborne radio which provides 920 channels over the frequency range 30 to 75.95 MHz. The system has a robust construction for aircraft operating air-to-ground missions in forward combat zones. It was based largely on the US Army's AN/VRC-12 ground-vehicle radio, with which it shares a high proportion of common components.

The ARC-131 provides FM voice, homing, re-transmission and secure speech facilities. Transmitter output power is selectable at either 1 or 10 watts. Though no longer in production, the equipment was manufactured in large quantities for the US Army.

Dimensions: 7.9 × 6 × 15.6 inches (200 × 154 × 390 mm)

Weight: 30 lb (13.6 kg)

STATUS: in service.

Magnavox ARC-164 uhf radio

AN/ARC-164 uhf radio

The ARC-164 is the basic member of a family of radio communications equipment and sub-variants, each designed for particular applications, yet all with a high degree of commonality.

The basic ARC-164 covers the uhf band, providing 7000 channels over the range 225 to 400 MHz in 25 kHz increments. Any 20 channels may be pre-selected. Standard power output is 10 watts, although this may be readily uprated to 30 watts.

A fully solid-state system, the ARC-164 is distinguished by its unique 'slice' module construction, in which a series of modules, connected by a flexible harness, are simply bolted together to form the desired electronic configuration. A typical, simple system would comprise transmitter, receiver, guard-receiver, and synthesiser. The control unit may either form part of this consolidated package or be remotely located. The modular approach adopted implies 'growth' capability, extra modules being added as required. A range of optional facilities, such as data transmission, secure speech and ECCM capability are available by the addition of the appropriate slices.

A number of directly-connected or remote control units are produced for the ARC-164. These include a simple frequency selection controller; a 32-channel preset control with light-emitting diode readout of the selected channel; a 20-channel preset unit with provision for two-cockpit take-control; and a microprocessor controller with 400 preset channels (uhf, vhf, AM or FM), liquid-crystal display channel and frequency readout, and the capability of controlling up to four systems simultaneously. Additional remote frequency/channel indicators are available. Magnavox produces a variety of mounting trays to suit differing aircraft installations for new types of aircraft and for the updating of older aircraft equipment.

Perhaps the most notable feature of the ARC-164 is its high mean time between failures. In a 100 000-hour life-cycle cost verification programme, conducted by the US Air Force on Cessna T-37, Northrop T-38, Supersabre F-100 and Lockheed C-130 aircraft, the ARC-164 has demonstrated a mean time between failures of 2000 hours.

The system is fitted to a wide range of US Air Force aircraft, including the General Dynamic F-16 fighter, and that service has placed orders for over 18 000 sets. Current production total is in excess of 20 000 sets.

The ARC-164 equips the Royal Navy's Sea King helicopters and Sea Harrier V/Stol combat aircraft, and the Hawk, Jaguar and other Royal Air Force aircraft. Further Hawk aircraft produced by British Aerospace for overseas customers, notably those delivered to Kenya, are also fitted with the ARC-164, as are Strikemaster strike/trainers which have been refurbished by that manufacturer. The latter includes Strikemasters operated by the Royal Saudi Arabian Air Force.

Dimensions: (transmitter-receiver, 10 W version) ½ ATR × 7 inches (178 mm)
(30 W version) ½ ATR × 14.75 inches (374 mm)
(controller) ½ ATR × 3.3 inches (83 mm)
Weight: (transmitter-receiver, 10 W version) 8.1 lb (3.7 kg)
(30 W version) 15 lb (6.8 kg)
(controller) 4.3 lb (2 kg)

STATUS: in production and service.

AN/ARC-187 uhf radio

The Magnavox ARC-187 is a further development of the company's ARC-164 US Air Force standard system which has been adapted to meet a US Navy requirement for a low-cost terminal designed to 'talk' with communication satellites, principally the US Navy's Fltsatcom. It covers the uhf band from 225 to 400 MHz in which range it provides 7000 channels at increments of 25 kHz. Up to 20 channels may be pre-selected.

Magnavox ARC-164 installation (by pilot's right hand) in British Aerospace Hawk

Operating modes include AM with a secure speech facility and FM and fsk data transmission, in both analogue and digital form. Transmitter power output is 30 watts in AM and 100 watts in FM/fsk mode.

The system, which is remotely controlled, uses standard ARC-164 'slice' modules including a modified synthesiser section designed for compatibility with comsat data-rate requirements.

The ARC-187 has been selected for US Navy P-3 Orion aircraft.

Dimensions: (transmitter-receiver) 17.3 × 6 × 5.6 inches (440 × 153 × 143 mm) (controller) 5.2 × 5.8 × 4.8 inches (132 × 147 × 124 mm)

Weight: (transmitter-receiver) 16.28 lb (7.4 kg) (controller) 4.4 lb (2 kg)

STATUS: in production and service.

Motorola

Motorola Inc, Government Electronics Group, 8201E McDowell Road, Scottsdale, Arizona 85252
TELEPHONE: (602) 949 4176
TELEX: 667490

AN/ARC-188 aircraft/ramp intercommunication radio

The Motorola ARC-188 is designed to provide intercommunication between aircrew and attendant ground crew while the aircraft is on the ramp during start-up, run-down or ground running checks. It provides an intercommunication facility without cable systems.

The system covers the band 410 to 419.975MHz in which it provides three separate

Narco

Narco Avionics Inc, 270 Commerce Drive, Fort Washington, Pennsylvania 19034
TELEPHONE: (215) 643 2900
TELEX: 846395

HT 800 hand-held transceiver

Designed for use in microlights, ultra lights or as back-up communication in light aircraft, the HT 800 covers the vhf frequency band between 118 and 135.975 MHz with 25 kHz spacing. In addition, 10 channels can be pre-programmed. Frequency selection is by means of a 16-key keyboard and a rechargeable NiCd battery gives 2 watts output power.

Dimensions: 50 × 70 × 171 mm
Weight: 0.57 kg

STATUS: in production.

Narco HT 800 hand-held vhf transceiver

ARC-195 vhf radio

This radio may be considered as a variant of the uhf ARC-163, with which it has a component commonality of 93 per cent. It is also available in 10- and 30-watt output versions. The major differences between the two systems consist of some component value changes and the substitution, in the ARC-195, of a synthesiser with a frequency standard appropriate to the vhf section of the radio frequency spectrum.

The ARC-195 covers the vhf band from 116 to 156 MHz, providing 1750 channels at a frequency separation of 25 kHz. In other respects it is almost identical to its uhf counterpart. Control units, again, are almost identical and certain units from the ARC-164 range of controllers may be used for combined uhf/vhf operation.

STATUS: in production and service.

channels at intervals of 25 kHz. Both the aircraft transmitter-receiver and the portable ground crew unit are similar in construction but the aircraft equipment is mounted in a shock tray and is interfaced with the aircraft's internal intercommunication system. An aircraft-mounted power supply is available for recharging the batteries in the ground crew portable transmitter-receiver.

The ARC-188 was developed to meet a US Air Force requirement.

Dimensions: (aircraft radio and shockmount) 3.5 × 5.6 × 7.6 inches (89 × 142 × 193 mm)

Weight: (aircraft radio and shockmount) 2.6 lb (1.18 kg)

STATUS: in production and service.

Com 810/811 vhf radios

Narco's Com 810 and Com 811 models are solid-state, microprocessor-controlled communication systems designed principally for light and general aviation aircraft. Each covers the vhf band from 118 to 135.975 MHz in which they provide 720 channels, two of which are pre-selectable, one for 'active' the other for 'standby' use. The 810 and 811 systems are essentially similar except that the former is designed for operation from a 13.75-volt dc supply and the latter from a 27.5-volt dc supply. Both have a nominal transmitter power output of 8 watts.

The active and standby frequencies are presented on light-emitting diode displays which are automatically dimmed during darkness by a built-in photocell circuit. A transmit legend, 'XMT', is illuminated when the microphone is keyed for transmission.

New frequencies may be entered in either the active or standby positions when desired, an illuminated arrow indicating which section has been selected for new frequency entry. Frequency selection is completed by use of a concentric tuning control, the outer part of which makes frequency readout changes at the rate of one MHz per detent and the inner part providing kHz changes at 25 kHz per detent. Clockwise rotation increases the numerical value of the frequency selection and counterclockwise rotation decreases it. A transfer switch is used to exchange selected frequencies between active and standby modes.

An optional feature is a connection which enables the last entered frequencies to be retained in the system memory when the radio is inactive. This requires a trickle current of 0.1 mA from the aircraft's battery. If this circuit is not connected, then the radio automatically retunes to the 121.5 MHz (internationally designated emergency) frequency in the active mode and to 121.9 MHz (ground control) frequency in standby mode, the next time it is switched on. In the event of a display failure, the

CA-657 vhf/AM radio

The Magnavox CA-657 is another vhf variant of the ARC-164 uhf system and its derivative, the ARC-195. Its frequency band coverage, however, is somewhat broader than the ARC-195 since it covers from 100 to 159.975 MHz in which range it provides 2400 channels. Microprocessor memory-control systems used with the CA-657 permit alternative pre-selection of up to either 20 or 30 channels. Channel separation is at intervals of 25 kHz, and the transmitter output power is 10 watts.

Construction of the system is based upon that of the ARC-164, and a similarly high degree of component commonality exists.

STATUS: in production and service.

AN/PRC-112(V) survival transceiver

The AN/PRC-112(V) is a portable survival radio designed for the use of downed aircrew both as a beacon for homing purposes by the search aircraft, and as a voice communication equipment. It has a multi-mode operation, AM and swept-tone beacon, and acts as a transponder by supplying ranging and personnel identification information. It is the tri-service replacement for the AN/PRC-90, with a number of advantages such as broadband uhf coverage, channelised 25 kHz frequency operation, a transponder for ranging, individual personnel identification codes and an increased transmitter power output.

Weight: 0.7 kg

STATUS: in production.

Narco 810 vhf radio

radio automatically reverts to these frequencies and may be re-tuned to the desired channel by counting of the detent clicks of the tuning control.

Built-in automatic squelch control, de-activated by use of a pull/test switch, maintains audio silence until a signal is received. Automatic audio-levelling in both transmitter and receiver allows all signals to be heard at the same level, regardless of modulation. A built-in 10-watt amplifier, provision for multiple audio inputs and intercommunication facilities are also included.

Dimensions: (transmitter-receiver) 6.25 × 1.5 × 11 inches (159 × 38 × 279 mm)
Weight: (transmitter-receiver) 2.9 lb (1.3 kg) (mounting tray) 0.7 lb (0.34 kg)

STATUS: in production and service.

Mk 12D nav/com radio

The Narco Mk 12D is a vhf communications/navigation radio system designed principally for light and general-aviation aircraft. The communications section is a transmitter-receiver covering the vhf band from 118 to 135.975 MHz and encompassing 720 channels. Two channels are pre-selectable, one for active, the other for standby use. Transmitter power output is nominally 8 watts.

The navigation section consists of a VOR/LOC receiver covering the vhf band from 108 to 117.95 MHz, providing 200 channels. Again,

Narco Mk 12D nav/com radio

two of these are pre-selectable for active and standby use.

In each section, new frequencies may be entered into the standby mode at any time and transfer buttons are activated to exchange the selected frequency between active and standby positions.

Concentric controls are used for frequency selection. The outer part of each control makes frequency readout changes at the rate of one MHz per detent and the inner part of the control is used for selection of the kHz portions. Clockwise rotation increases the numerical value of the frequency and counter-clockwise rotation decreases it. Active frequencies and displayed data are not affected by use of these controls.

The frequencies are presented on light-emitting diode displays which are automatically self-dimming in darkness through a built-in photocell light-sensing circuit. An optional memory circuit may be connected if desired and this will retain the last selected frequency in the active-channel modes of both navigation and communication sections, the use of this facility imposing a 1 mA drain on the aircraft battery. If the circuit is not so connected then random, though valid, frequencies are displayed when the equipment is switched on.

In the event of display unserviceability, then in the navigation section case (if the indicator appears to be unaffected) signal reception can be verified by operating a volume/ident control for audible check. In the case of blackout of the communication section display, frequency verification can be made by a check transmission and response. Both sections may be re-tuned under display loss conditions by using the preset frequencies as references and counting the detent click-stops of the frequency-selection controls.

The communication and navigation sections have independent volume controls. By pulling the volume control out to the 'test' position on the communication section, the system's automatic squelch may be de-activated in order to improve weak signal reception.

The system includes a built-in 10-watt amplifier, provision for intercommunication and an optional remote active/standby frequency transfer facility.

Navigation section output will drive compatible horizontal situation indicators, area navigation systems, VOR radio magnetic indicators or the system's companion ID 824 VOR/LOC indicator. It will also drive the ID 825 VOR/ILS indicator and for full ILS capability, a combined 40-channel glideslope receiver, covering the band 329.15 to 335.0 MHz, is available.

The Mk 12D uses digital electronic techniques, is microprocessor-controlled and is of all-solid-state construction. It is available in either 28- or 14-volt dc power supply configuration.

Dimensions: 6.25 × 2.5 × 11 inches (159 × 64 × 279 mm)

Weight: (nav/com with glideslope receiver) 4.4 lb (2.0 kg)
(without glideslope receiver) 4.1 lb (1.9 kg)
(mounting tray) 0.7 lb (0.34 kg)

STATUS: in production and service.

TR1000B portable vhf radio

The Narco TR1000B is a portable vhf transmitter-receiver designed for a range of aviation-related activities, including use as a fixed or mobile ground station. It is, however, particularly suitable for airborne use, either as a primary means of communication aboard non-radio equipped aircraft, or as an emergency back-up unit for aircraft with limited communications facilities. It covers the frequency bands 118 to 135.95 MHz or 118 to 139.975 MHz, depending upon whether a Narco 360-channel or 720-channel communications transmitter-receiver is also installed in the aircraft.

For airborne use, power supply may be obtained either from two internal 14-volt 7.5 ampere-hour, gelled-electrolyte, rechargeable batteries or from any 12- to 14-volt dc source, using a cigarette-lighter adapter plug which is supplied. On the ground, the system can be operated from a 120/230 volt ac supply. The TR1000B can operate either through its attached 28-inch (711 mm) telescopic antenna or through an aircraft antenna installation if available.

A built-in, 3-inch (76 mm) speaker is normally used for audio output although a socket which accepts a standard headset jack is included. A standard microphone jack socket is also incorporated.

The TR1000B is manually tuned, the selected frequency being indicated digitally through a viewing window in the facia. The numerals may be illuminated during darkness. A battery-level meter, for checking cell charge condition, is also incorporated into the facia panel.

Dimensions: 4.75 × 16 × 11 inches (121 × 406 × 279 mm)

Weight: 19.7 lb (8.94 kg)

STATUS: in production and service.

ELT 10 emergency location transmitter

Narco's ELT 10 radio is an emergency location transmitter which provides a beacon-homing signal to assist in search and rescue operations for the occupants of a crashed or otherwise missing aircraft. The unit, when activated, transmits simultaneously on the 121.5 and 243 MHz international distress frequencies.

Narco TR1000B portable vhf radio

Initial transmitted power output, at an ambient temperature of 25°C, is 300 mW on each frequency. Transmissions are modulated by a downward swept tone between 1600 and 300 Hz at a sweep repetition rate of 2 to 4 Hz and the transmit duty cycle is continuous.

The ELT 10 is designed for permanent installation, although it is easily removed for use as a personal locator beacon. A built-in folding antenna is included, primarily for use in the personal role. An external aircraft antenna, tuned for optimum performance on the two transmission frequencies, is optionally available.

The radio may be activated by a positive-acting, fail-safe 'Rolamite' inertia switch which is said to prevent inadvertent operation. Alternatively, it may be manually activated by a built-in ON/OFF/ARM switch or remotely from a panel-mounted ON/ARM test switch.

Power is supplied from an internal sealed, nine alkaline D-cell, 13.5-volt battery pack which has a usable life of 14 months.

Dimensions: 8.8 × 2.56 × 3.125 inches (224 × 65 × 79 mm)

Weight: 3.5 lb (1.6 kg)

STATUS: in production; more than 120 000 units in service.

MB-700 microphone

The Narco MB-700 is a lightweight dynamic microphone employing a patented variable-reluctance principle for high-quality voice transmission. To achieve maximum noise-cancellation, the microphone cartridge is suspended in a special acoustic chamber which is claimed to virtually eliminate wind, engine and propeller noise. A shaped frequency response, optimised for the voice frequency range, also helps to reduce unwanted background noise. Frequency response covers the range 200 to 4000 Hz. Power level in 200 ohms is –49 dB.

The microphone is fully shielded and is shockmounted in a high-impact casing. It is supplied with a 6-foot (1.83 m) cord terminating in a right-angle plug.

Weight: 0.5 lb (198 g)

STATUS: in production and service.

CP 136 and CP 136M audio control panels

Belonging to the Centerline range, the CP 136 and the slightly larger 'M' variant provide fully solid-state control of all radios, headsets and loudspeakers using push-button control and led selection display. The units provide 10 watts across 4 ohms to speakers or 50 mW to 600-ohm headphones.

Dimensions: 28 × 159 × 213 mm
Weight: (CP 136) 0.82 kg
(CP 136M) 0.91 kg

Narco CP-136 radio system controller

Radio Systems Technology

Radio Systems Technology Inc, 13281 Grass Valley Avenue, Grass Valley, California 95945.
TELEPHONE: (916) 272 2203

Radio Systems Technology (RST) is a small organisation founded during the early 1970s to produce low-cost avionic equipment for the light aircraft market. Most of RST's systems are sold in self-build kit form. The company will provide fully assembled and tested RST-542 radios to special order.

RST-542 vhf radio

The RST-542 radio is designed for air-to-ground communications in all categories of light aircraft, including gliders. It provides crystal-controlled, six-channel communication in the vhf band from 118 to 136 MHz, using two crystals per channel. Transmitted power output is 6 watts peak power.

Comprising a single-box panel-mounted unit, the solid-state RST-542 features extensive use of integrated circuit technology. It incorporates automatic gain control and adjustable squelch facilities and provides a receiver audio output power of 4 watts minimum to drive either a 4-ohm cabin speaker or a 600-ohm headset. A principal feature, rendering it particularly suitable for gliders, is low power consumption, 60 mA in receive and 1 A in transmit mode.

Dimensions: 9 × 3 × 3 inches (229 × 76 × 76 mm)
Weight: 1.5 lb (0.68 kg)

STATUS: in production and service.

RST-571/572 vhf nav/com radios

The RST-571 and -572 are light aircraft vhf nav/com radio systems which vary from each other only in the number of channels provided by their communications sections. Each covers the band 118 to 135.95 MHz; the RST-571 provides 360 channels at 50 kHz increments, and the RST-572 provides 720 channels at increments of 25 kHz. The common navigation receiver section covers 200 channels in the band 108 to 117.95 MHz, in increments of

Radio Systems Technology RST-542 six-channel transmitter-receiver

Radio Systems Technology RST-571 360-channel nav/com system

50 kHz. In transmit mode, the transmitter-receiver section has a transmitted power output of 10 watts peak power or 2.5 watts in continuous wave.

The RST-571 and -572 are panel-mounted units using thumbwheel switches for selection

and display of the channel and for a chosen VOR radial. Both the navigation and communication sections of an equipment may be used simultaneously. To prevent the display of spurious signals, the navigation section is automatically disabled when the transmitter-receiver section is in transmit mode.

Construction is solid-state and digital synthesis techniques are used for frequency generation in both the communication and navigation-receiver sections. Employment of digital techniques permits the phase-locking of the VOR/LOC circuits to the navigation receiver signal, with consequently enhanced accuracy. Automatic gain control and adjustable squelch facilities are incorporated in the communications section.

Dimensions: 6.5 × 3.5 × 11 inches (165 × 89 × 279 mm)
Weight: 4.5 lb (2 kg)

STATUS: in production and service.

RCA

RCA Government Communications Systems, Front and Cooper Streets, Camden, New Jersey 08102
TELEPHONE: (609) 338 2105
TWX: 710 891 0101

AN/ARC-161 hf radio

RCA's AN/ARC-161 is a lightweight, compact radio designed to provide extended range communication for all types of aircraft. Covering the hf band from 2 to 30 MHz, it provides 280 000 channels at uniform spacings of 100 kHz. The system was developed primarily for the US Navy which uses it extensively aboard Lockheed P-3C Orion long-range maritime patrol aircraft. Operating modes are A3J, A3B, A9B and A7S providing voice, teletype and data transmission facilities. Secure cipher and cryptographic transmission reception is also possible. Transmitter power output is selectable at either 400 or 1000 watts.

The system's principal features include high reliability with a mean time between failures in excess of 1000 hours, and good co-location performance. Design for co-sited operation was taken into account from the outset with the requisite circuitry being built into each equipment. Consequently, an ARC-161 can be installed in an aircraft with confidence that its full co-location performance can be attained without the need to change aircraft cable harnesses.

The ARC-161 is of all-solid-state construction and uses synthesis technology for frequency generation. It may be remotely controlled over distances up to 200 feet (61 metres) and is thus particularly suited to larger aircraft.

Dimensions: 23.75 × 15.6 × 11.7 inches (603 × 396 × 277 mm)

Weight: 82 lb (37.21 kg)

STATUS: in service.

AN/ARC-170 hf radio

The RCA AN/ARC-170 is a general-purpose hf airborne radio for long-range communication in all types of military aircraft, and is also suitable for surface vehicles or vessels and base-station operation. It covers the full hf band from 2 to 30 MHz, in which range it provides 280 000 channels at spacings of 100 Hz. The system operates in A1, A3H, A3J, A3B, A9B and A7J modes. It is thus sufficiently adaptable to provide voice, data or radio-teletype, and is Link 11 compatible. Encrypted narrow-band facilities are available for secure communication purposes. Transmitted power output is normally 1 watt but an optionally available amplifier permits outputs selectable at 100, 400 and 1000 watts.

The system is of all-solid-state construction and uses a synthesiser for frequency generation. It is designed for both convection or forced-air cooling in installations where a suitable supply is available. It is particularly

suited to simultaneous, co-sited operation with one or more other AN/ARC-170 systems and is supplied permanently wired so that this mode of operation can be used without modification.

The AN/ARC-170 may be remotely controlled from distances up to 200 feet (61 metres) and is particularly suited to installation in larger, long-range aircraft. It is available with either segregated transmitter and receiver sections or in a combined transmitter-receiver form.

The system supersedes the earlier RCA AN/ARC-161 model (currently in service) over which it provides a number of additional facilities and improved performance in a more compact assembly.

Dimensions: (transmitter and receiver) 17.5 × 7.6 × 5 inches (446 × 194 × 127 mm) (combined transmitter-receiver) 22 × 9.25 × 15.6 inches (558 × 235 × 396 mm) (amplifier) 22 × 7.6 × 10.1 inches (559 × 194 × 257 mm) (controller) 3.6 × 5.75 × 2.6 inches (99 × 146 × 67 mm)

Weight: (transmitter and receiver) 20 lb (9.07 kg) (combined transmitter-receiver) 70.5 lb (31.97 kg) (amplifier) 43 lb (19.5 kg) (controller) 1.5 lb (0.68 kg)

STATUS: in service.

RF Products

RF Products Inc, Davis and Copewood Streets, Camden, New Jersey 08103
TELEPHONE: (609) 365 5500
TWX: 710 891 7087

RF Products (formerly TRW RF Filter Products) specialises in the development and production of tunable bandpass filters and multi-couplers. These are produced mainly for the military market, in which the proliferation of communications systems aboard aircraft often creates problems caused by the need to site radios and antennas in close proximity. Major programmes to which RF has contributed include the US Air Force/NATO/Boeing E-3A Sentry (AWACS), US Air Force/Boeing E-4B Advanced Airborne Command Post, US Army Guardrail, EH-60B 'Quick Fix' IIB countermeasures helicopter, the US Navy AEGIS and the Royal Air Force Nimrod AEW aircraft.

Automatically tuned uhf multi-couplers

RF's airborne antenna multi-couplers use a building-block, modular approach design in which plug-in tunable filters are used interchangeably as single-channel devices or in two-, three-, four- or five-port multi-couplers, covering the uhf band from 225 to 400 MHz. The radio frequency capability of these systems is 120 watts peak or 50 watts carrier wave. Good selectivity characteristics are claimed for these units, with the following performance:

	399 MHz	300 MHz	225 MHz
3 dB	± 2.9	± 2.7	± 2.5
45 dB	± 12.5	± 9.5	± 8.0
80 dB	± 38.0	± 33.0	± 30.0

For 50 dB attenuation, channel separations are 15 MHz at 399 MHz, 12 MHz at 300 MHz and 10 MHz at 225 MHz. The tuning times are a maximum of 10 seconds but 6 seconds is said to be typical.

RF's tunable bandpass filters use digital techniques for remote tuning. This is accomplished by converting digital information from the radio equipment's digital synthesiser, or another source, to rotational command information. The only tuning element in the filter is an air-variable capacitor which can be stepped to a position corresponding to a pre-calibrated frequency by a digitally-controlled stepping motor. Digital logic circuitry permits fast and accurate tuning by precise positioning of the filter tuning shaft to within ±0.04°. Angular shaft position information is supplied to the digital control circuitry by a position-sensing circuit comprising a light-emitting diode and a photocell.

On receipt of a 'tune-initiate' command, the stepping motor rotates the filter tuning shaft to the upper end of the tuning range which is used as a reference. The control circuit then counts a number of pre-determined steps corresponding to the frequency code input, stopping the motor at the desired shaft position for the frequency desired. On completion, a 'tune-cycle complete' signal is generated. An optional self-initiated tuning module can be made available.

Dimensions: (5-port multi-coupler) 14 × 12 × 15 inches (350 × 300 × 380 mm)
(4-port multi-coupler) 14 × 12 × 15 inches (350 × 300 × 380 mm)
(3-port multi-coupler) 9 × 14 × 15 inches (230 × 350 × 380 mm)
(2-port multi-coupler) 7 × 10 × 15 inches (180 × 250 × 380 mm)
(single filter) 7 × 7 × 11 inches (180 × 180 × 280 mm)
Weights: (5-port multi-coupler) 62 lb (28 kg)
(4-port multi-coupler) 57 lb (26 kg)
(3-port multi-coupler) 42 lb (19 kg)
(2-port multi-coupler) 28 lb (13 kg)
(single filter) 15 lb (7 kg)

STATUS: in production and service.

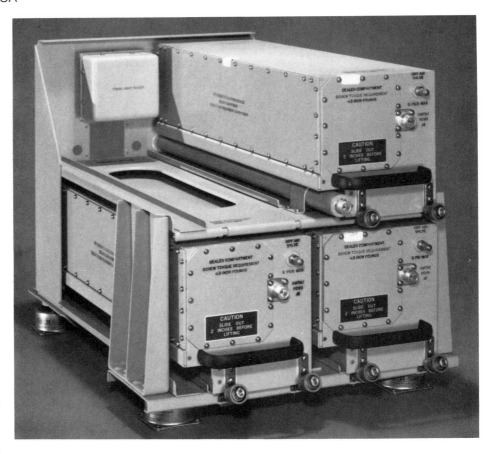

RF Products Type 7 three-port multi-coupler

RF Products Type 7 two-port multi-coupler

RF Products Type 7 uhf lightweight tunable filter

RF Products Type 6 uhf filter

Type 6 tunable uhf filter

The RF uhf Type 6 tunable filter covers the band 225 to 399.9 MHz and is provided with a 243 MHz guard channel by-pass filter to ensure reception of emergency transmissions regardless of the programmed operating frequency. Its radio frequency power handling capability is 100 watts peak or continuous wave.

RF Products five-port uhf multi-coupler

RF Products Type 7 four-port multi-coupler

The unit comprises a pressurised four-pole cavity band-pass filter which is compatible with modern uhf transmitter-receivers both dimensionally and electrically. It is thus particularly suitable for retrofit applications in which a compact transmitter-receiver/filter packaging is required. The system has a mean time between failures of 5000 hours.

Dimensions: 7 × 3.5 × 14.8 inches (177 × 88 × 375 mm)
Weight: 13 lb (5.9 kg)

STATUS: in production and service.

Type 7 two-, three- and four-port uhf/vhf multi-couplers

RF Type 7 multi-couplers are available in versions covering the military vhf band from 110 to 160 MHz and the uhf band from 225 to 400 MHz. They are designed to ensure reliable operation of multiple transmitter and receiver installations in severe co-location environments and may be used interchangeably as single-channel, two-, three- or four-port units. This is accomplished by use of modular building blocks.

The systems are remotely tuned by use of digital control circuitry combined with stepper motor-driven filter tuning shafts (see entry for RF automatically tuned uhf multi-couplers). Power handling capability is 150 watts maximum and the average tuning time is 5 seconds with a maximum of 10 seconds. Adjacent channel operation is 3 MHz minimum. Two-port multi-couplers from the Type 7 series are used aboard the US Air Force/Boeing E-3A

Sentry AWACS and the Royal Air Force Nimrod AEW3 aircraft.

Dimensions: 14 × 15 × 26 inches (355 × 381 × 660 mm)
Weight: (3-port) 105 lb (47.72 kg)
(4-port) 140 lb (63.63 kg)

STATUS: in production and service.

Type 7 digitally-tuned uhf filter

RF's Type 7 digitally-tuned uhf filter covers the frequency range 225 to 399.99 MHz. It is designed to permit multiple transmitter-receiver operation in severe co-location situations. The system is available in two versions, a 0.7 per cent bandwidth version with a radio frequency input power limit of 150 watts continuous wave, or a 2 per cent bandwidth unit with 200 watts peak power. The filter is of the four-pole cavity type and is capacitively loaded and tuned. Tuning is digital parallel in 100 kHz steps.

Dimensions: 21.5 × 6.75 × 6.5 inches (546 × 171 × 165 mm)
Weight: 28 lb (12.72 kg)

STATUS: in production and service.

SCI Systems

SCI Systems Inc, 8600 South Memorial Highway, PO Box 4000, Huntsville, Alabama 35802
TELEPHONE: (205) 882 4800
TELEX: 782421

F-15 integrated communications/navigation identification control panels (INICP)

The INICP is a group of 6 cockpit-mounted interconnected control panels used in the McDonnell Douglas F-15 aircraft. The group is the centre command and control point for the complete communication, navigation and identification system.

An integrated communications control panel is the heart of the system. Using dual microprocessor control, this unit switches audio paths, controls and displays frequencies for the

aircraft radios (including 40 channels of preset non-volatile memory), generates and formats synthesised 'voice alerting' messages, generates audio warning tones, and has provision for antenna selection. Built-in test circuitry is claimed to detect 95% of all electronic component failures.

A second panel, for main communications control, is mounted on a head-up display to provide the pilot with selection of frequency, preset channels and volume control for one of the aircraft radio sets. It also has code selection controls for IFF mode 3/A operation, a master caution annunciator, and two electronic warfare warning displays.

The remaining panels are the IFF control, the air-to-air interrogator control, and a take-control panel which is only used in the trainer configuration of the F-15.

Dimensions: 181 × 146 × 165 mm
Weight: 4.4 kg

Integrated radio control panel

This panel is designed to provide the pilot and/or co-pilot with the means to manage all communications equipment on board a commercial or military aircraft. Two interface units are available, MIL-STD-1553 for military applications and ARINC 429 for commercial aircraft.

The primary display medium is a cathode ray tube (crt) which displays mode and frequency information. In addition to the crt, several dedicated controls are provided for those functions which require immediate access. The unit has a non-volatile memory for power-off data retention.

Dimensions: 203 × 146 × 200 mm
Weight: 4.5 kg approx

Singer Kearfott

The Singer Company, Kearfott Division, 1150 McBride Avenue, Little Falls, New Jersey 07424
TELEPHONE: (201) 785 6000
TELEX: 133440
TWX: 710 988 5700

JTIDS joint tactical information distribution system

JTIDS is a US joint-service jam-resistant, secure communications system. Using time-division multiple access (TDMA) technology, it provides a flexible, multiple-user tactical information exchange service between surface and airborne military units and their command. It consists of digital data and voice links operating in the 960 to 1215 MHz band and employs spread-spectrum techniques for increased resistance to jamming.

Each member of a JTIDS network is allocated time slots to accommodate his input messages into the system. Such inputs would automatically include identification and navigation data in addition to specialised tactical information such as target acquisition details. All users can continuously monitor and sample the database for such information as they require.

Participants routinely inject information into the net through their regular broadcast slots without necessarily knowing the identity or location of the ultimate recipients who, likewise, do not necessarily know the whereabouts or

identity of the provider of the data. Typical JTIDS messages from an airborne data source would cover identity, location, altitude, speed and heading; tactical information regarding the mission such as target acquisition, weapon or stores availability, fuel remaining and equipment status would follow.

JTIDS communication security is maintained by coding and frequency hopping techniques. Each transmission consists of a pulse stream, the leading pulses of which are synchronisation pulses for locking transmissions to the hopped frequency sequence. Identification pulses and the message itself then follow.

JTIDS is one of the most ambitious communications projects, civil or military, ever undertaken. It permits the interchange of essential tactical information between aircraft, surface vessels, mobile or fixed-base land stations and, most importantly, the command, on a scale not previously envisaged. The technical problems which have had to be resolved in order to implement the system have required an exceptional research and development effort on the part of the manufacturers producing the equipment for the system.

With regard to the airborne elements, the principal contractors are The Singer Company's Kearfott Division, prime contractor and systems integrator, working with Rockwell Collins which provides engineering support and development as an alternative production source.

Responsibilities for the surface elements are divided among the companies concerned in

various ways, according to differing terminal classifications and individual service requirements. A principal contractor, with regard to ground-based air-defence radars within NATO, is IBM.

Initial emphasis in development of JTIDS hardware was on the Class 1 terminal system, which is intended for land-based application as well as airborne use. Its most notable application to date is aboard the earlier US Air Force/Boeing E-3A Sentry AWACS aircraft.

Developed by Hughes Aircraft, the Class 1 terminals underwent considerable development refinement during the earliest history of the programme. Work on the initial Class 1 terminal, designated AN/ARC-181, commenced under a contract placed in 1974 and first deliveries of the resultant hardware took place in January 1977. While this proved successful in over 7500 hours of operation, which included flight tests aboard an E-3A, work was already proceeding on the improved B-waveform model, airborne development of which was completed by May 1978. Tests included live jamming and were completed ahead of schedule with all requirements and specifications being successfully satisfied, an impressive performance in view of the considerable technical challenge presented by the task.

The ARC-181 terminal comprises three electronic units, together occupying approximately five ATR rack spaces, together with a controller. Hughes, however, was already working on a privately-funded JTIDS Class 1 terminal project, in parallel with the main programme, with the aim of size and weight reduction. This resulted in the Hughes Improved Terminal (HIT) with a reduction of some 40 per cent in volume, a corresponding weight decrease and no compromise in terms of performance, while maintaining full TDMA compatibility. These HIT terminals are now standard equipment for US Air Force and NATO E-3A AWACS aircraft.

Airborne Class 2 terminals for the US Air Force and Army are being developed by the team of Singer Kearfott and Rockwell Collins under a contract placed in June 1981. Under this contract, 20 terminals were delivered for flight test aboard US Air Force F-15 air-superiority fighters and US Army ground elements. A proportion will subsequently be passed to the US Army for further tests as part of the hybrid position locating and reporting system (PLRS)/JTIDS system, known as PJH. Development of both PLRS and PJH is a Hughes Aircraft responsibility.

Under the JTIDS Class 2 contract, Collins will supply transmitter-receiver sub-systems to Kearfott, which is developing processor and input-output units. Kearfott is responsible for systems integration as well as overall programme management, and has received some $30 million worth of the total contract value.

The Singer Class 2 terminal has applications in both surface and airborne roles. It comprises a two-box package occupying about 1.6 cubic feet (0.045 cubic metres) and weighing approximately 95 lb (43 kg). The Collins-built transmitter-receiver section contains all TDMA radio frequency functions. An independent receiver-processor provides a Tacan function which operates simultaneously with JTIDS communications. The Singer-constructed equipment is divisible into two line-replaceable units, the data processor and input/output adapters. The former carries out signal and digital message processing; the latter provides a unique interface with the host platform. A software-controlled relative navigation function is incorporated and automatically provides position information with respect to other co-operating tactical entities, during both active and passive operation.

The key output of the system is the display of information received from the JTIDS network, and the US Air Force is evaluating a number of possibilities. One of these is a multi-function display which will present JTIDS information as selected by the pilot. The important factor remains the ability of JTIDS to permit target assignment by a command and control authority, by self-assignment or by co-ordination between members of a flight. One

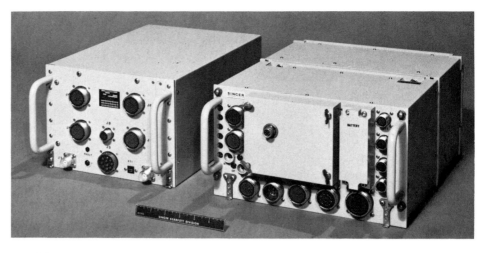

JTIDS Class 2 airborne transmitter-receiver terminal manufactured by Kearfott in association with Rockwell Collins

JTIDS information shown on colour display of McDonnell Douglas F-15 aircraft. Aircraft is shown at centre of display, concentric circles indicate distance in nautical miles, round symbols indicate friendly aircraft, rectangles are unknowns and triangles are enemy aircraft.

particular area of evaluation, according to Singer, is the intra-cockpit correlation of radar tracks and JTIDS data display together with inter-cockpit exchange of radar and JTIDS data between members of a flight.

The Kearfott Division of Singer has been awarded a follow-on contract for 20 additional Army JTIDS Class 2 Terminals. These will be used to support the Intra-Army and Joint Army-Air Force testing at Eglin Air Force Base scheduled for 1986. During these tests the terminals will interface with the Army's air defence, field artillery and manoeuvre control systems.

JTIDS has also been specified for use as part of the British air defence network. Development and production of the UK system, including application to improved UK ADGE sites and to Nimrod and Tornado aircraft, is expected to cost $225 million. It is planned to enter service during the late 1980s. Development contracts have been placed with GEC Avionics and Plessey in the UK and Singer in the US. Smaller contracts will also be placed with other UK companies. Co-operation between the UK and the US will minimise duplication and take advantage of completed development work.

STATUS: in full-scale development. The US Air Force in mid-1984 took delivery of the first of 40 JTIDS Class 2 terminals being bought for operational assessment. They are being installed on US Air Force/McDonnell Douglas F-15 Eagle fighters, and US Army equipment. Production decision is anticipated in early 1987 after the test programme has been completed. The first flight test was completed successfully in December 1985.

NB See also EJS (Enhanced JTIDS System) under Hazeltine Corporation.

Sperry

Sperry Corporation, Defence Products Group, 13133 N 34th Street, PO Box 4648, Clearwater, Florida 33518
TELEPHONE: (813) 577 1900
TELEX: 523453

Sperry millimetre wave radio

A millimetre wave-based armour/air covert net communications system is being developed by Sperry under a contract from the US Army Communications and Electronics Command. It is a modular radio, intended for use in helicopters or armoured fighting vehicles to provide secure jam- and intercept-resistant communications in the battlefield environment.

The basic design is that of a lightweight, binocular-equipped directional radio to enable communications over an extended range between a single helicopter or vehicle and ground personnel. The development contract was awarded in mid-1984 and the programme was expected to be completed in 18 months.

Sunair

Sunair Electronics Inc, 3101 SW 3rd Avenue, Fort Lauderdale, Florida 33315
TELEPHONE: (305) 525 1505
TELEX: 514443

ASB-500 hf/ssb radio

The Sunair ASB-500 has been specifically designed for aircraft and helicopters requiring a large number of operational frequencies but where space and weight are limiting factors. It covers the hf band from 2 to 18 MHz with 12 000 synthesised channels in usb, lsb and ame modes with 500 kHz spacing.

The radio, comprising a receiver, exciter and power amplifier, is combined in a ½ ATR case. The controller is panel-mounted and includes a six-digit light-emitting diode frequency display and illuminated status and antenna coupler tuning monitors.

The ASB-500 is certificated to the relevant FCC and FAA TSOs and can operate at 30 000 feet (9140 metres), between –46 and +55° C and in 95 per cent humidity.

Format: ½ ATR Short case

ASB-850 hf/ssb radio

Sunair's ASB-850 is a military hf transceiver

covering the band 2 to 30 MHz in which it provides 280 000 channels at a spacing of 100 kHz. It is a particularly light and compact system designed for light fixed-wing aircraft and helicopters operating in tactical roles. The system is remotely controlled from a miniature panel-mounted unit which contains a light-emitting diode frequency-selection display. It operates in usb, lsb and ame modes and has transmitter power outputs of 100 watts in single sideband and 40 watts nominal in cw.

The ASB-850 is of all-solid-state construction and is claimed to be of exceptionally robust design and manufacture. It can operate over the temperature range –46 to +55° C at pressure altitudes up to 30 000 feet (9140 metres) and in conditions of high relative humidity.

An automatic antenna coupler is incorporated within the system and this unit can tune the antenna to the frequency selected in 3 seconds or less.

Dimensions: (transmitter-receiver) ½ ATR (amplifier/antenna coupler) ½ ATR (controller) 5.75 × 5 × 2.625 inches (146 × 127 × 67 mm)
Weight: (transmitter-receiver) 15.8 lb (7.11 kg) (amplifier/antenna coupler) 22 lb (9.68 kg) (controller) 1.8 lb (0.81 kg)

STATUS: in service.

ACU-150D hf antenna coupler

The Sunair ACU-150D is a 100-watt digital automatic hf antenna coupler designed for many applications including aircraft. It is compact, of low weight, and will tune many types of hf antenna over the 2 to 18 MHz frequency band.

Operation can be automatic and non-automatic; the ACU-150D will remember the last 10 tuned frequencies and can retune to any one of them in 0.3 second. Non-memory tuning takes about 2 seconds.

Dimensions: 178 × 152 × 305 mm
Weight: 4.05 kg

Sunair ASB-500 hf/ssb transceiver

Sunair ASB-850 hf/ssb airborne transceiver

Telex

Telex Communications Inc, 9600 Aldrich Avenue South, Minneapolis, Minnesota 55420
TELEPHONE: (612) 884 4051
TELEX: 297053

Telex Communications manufactures a diverse range of electronic, electromechanical and acoustic products. Its aviation products include pilot communication headsets, microphones, aircraft intercom systems and hand-held transceivers. Telex supplies a number of leading airframe manufacturers and international airlines with such equipment.

TC-200 intercommunication system

The Telex Model TC-200 intercommunication system is designed for a range of airborne applications. As an intercom, it can provide a voice activated link between the pilot and co-pilot, and four stations can be added to the system for passengers. An audio entertainment system can also be incorporated. The intercom allows for normal radio use by the pilot without intercom interference. The TC-200 is lightweight 1.18 lb (0.53 kg) and can be installed on or below the instrument panel. The unit can also be carried on board and used as a portable system with power supplied through the cigarette lighter jack.

Telex TC-200 aircraft intercommunication system

For expansion, the TC-400 gives two and the TC-600 four extensions.

STATUS: in production and service.

Terra

Terra Corporation, 3520 Pan American Freeway North East, Albuquerque, New Mexico 87107
TELEPHONE: (505) 884 2321

Terra manufactures a range of compact vhf transceivers, and other avionic equipment for use in light or ultra-light aircraft.

TX 10 vhf transceiver

The TX 10 is a 10-channel crystal-controlled vhf transceiver which can be panel-mounted and powered from aircraft supplies or carried separately and driven from an attached battery pack. Crystals for the TX 10 are available for the complete 118 to 135.975 MHz band at 25 kHz spacing.

Dimensions: 41 × 81 × 229 mm
Weight: 0.43 kg

TPX-720 vhf radio

Terra's TPX-720 transmitter-receiver is a hand-held unit with built-in antenna and speaker. Covering the vhf band from 118 to 135.975 MHz, the TPX-720 provides AM communications on 720 channels in that band. It is designed principally for light aircraft with limited or no electrical or radio facilities and as a back-up system for aircraft with radio. In addition to the 720-channel two-way communications facility, the receiver section also covers 200 VOR channels. Channel separation is 25 kHz over the communications range, and in the navigation range, a 50 kHz separation applies.

Two transmitter power-output levels are available. In the high range, output is 8 watts peak power with a carrier signal of 2.5 watts. Alternatively, the low range may be used with the lower 2-watt peak power output but with an increased carrier signal power of 5 watts. The unit is powered from 10 rechargeable NiCd batteries.

All controls are in the head of the case and a miniature digital read-out thumbwheel switch is used for channel selection. Digital synthesis techniques are employed for frequency-generation and electronic construction is of cmos integrated circuits. The receiver is of double conversion superheterodyne design and uses crystal monolithic filtering to achieve high selectivity. A wide range of optional accessory equipment is available. It includes adaptors for operating directly from an aircraft electrical power supply, headsets with or without microphone attachment, separate microphone, charging units, and a push-to-talk switch for use on control columns.

Dimensions: 41 × 81 × 229 mm
(antenna) 235 mm
Weight: 0.57 kg

STATUS: in production and service.

Texas Instruments

Texas Instruments Inc, PO Box 226015, M/S 3127, Dallas, Texas 75266
TELEPHONE: (214) 995 1188
TELEX: 470900

Speech recognition technology

Texas Instruments has been engaged in the development of speech technology for military purposes over a number of years. This includes the use of speech recognition for purposes such as voice-verification systems for security and voice interactive systems for control purposes in aircraft. The company employs a concept of a common programmable speech module, using very large scale integration techniques, which allows software configuration, system updates and algorithm modifications to hardware. Development is being carried out in operator prompting, status and warning messages, command recognition, syntax and grammar driven tasks, and artificial intelligence applications.

STATUS: in development.

US Army

Avionics Laboratory, US Army Avionics Research and Development Activity, Fort Monmouth, New Jersey

ICNIA integrated communications/navigation identification avionics

Sponsored by the US Air Force and Army, a major, long-term development programme is under way to integrate the mechanics and functions of many of the nav/com systems on tactical military aircraft. The aims are to reduce the number of equipment boxes and the high level of duplication in such areas as information processing, power supplies and displays, to save on weight, volume and cost, and to reappraise the pilot/aircraft interfaces. A reduction in pilot workload is seen to be of particular importance, especially in combat helicopters, where the pilot is involved in nap-of-the-earth navigation and weapons-management.

The first application of ICNIA will be the US Army's LHX scout/observation helicopter. The Army plans to acquire as many as 5000 such helicopters following their expected introduction at the end of the decade.

Initial estimates of the potential savings to be achieved with ICNIA have been drawn up for the LHX project. ICNIA could be applied to the following systems on this helicopter: the radio-frequency system comprising vhf AM/FM radio, uhf/AM radio, IFF transponder, ILS receiver, ADF, hf voice/data system, Sincgars (single-channel ground and airborne radio system), microwave landing receiver, PLSS (position location reporting system), JTIDS Class 2 secure data system, and Navstar GPS satellite navigation receiver. The total weight of this equipment based on conventional current technology and packaging would be 294 lb (133.4 kg) and it would occupy about 4.45 cu feet (0.13 cu metre). A comparable ICNIA system would weigh about 165 lb (74.8 kg) and occupy 2.65 cu ft (0.08 cu metre).

Two teams are in competition to develop prototype equipment. They are ITT with Texas Instruments, which have received a letter contract for $11.19 million, and TRW Electronic Systems Group with Rockwell's Collins Government Avionics Division, Singer-Kearfott, and General Dynamics/Fort Worth, which have received $11.20 million. The programme will incorporate advanced technology electronics; TRW plans to use five types of vhsic chip, representing several hundred in its vhsic-based architecture now under development. The company claims that vhsic chips will undertake multiplexing, signal-processing and arithmetic logic and will assist in fault isolation down to chip level.

The ICNIA programme will be largely influenced by the results from another test programme, ARTI (advanced rotorcraft technology integration). The objective of this programme is to test under realistic conditions a number of control, display and instrumentation ideas intended to facilitate low-level single-seat helicopter operations. Five companies have been chosen by the US Army for initial analysis of ARTI requirements. They are Bell, Boeing Vertol, Hughes Helicopters, IBM and Sikorsky. The Army will choose two or more teams to go forward to flight demonstration. Sikorsky is to modify an S-76 helicopter by adding a single-seat cabin to the nose.

STATUS: ICNIA is under assessment.

Wulfsberg Electronics

Wulfsberg Electronics Inc, 11300 West 89th Street, Overland Park, Kansas 66214
TELEPHONE: (913) 492 3000
TELEX: 424242

Wulfsberg Electronics manufactures vhf and uhf radio communication systems for the civil aviation market. As well as producing equipment for the air transport sector, the company produces equipment for the general aviation market, particularly for heavier, longer-range corporate aircraft, and for helicopters.

It was announced in March 1984 that the company had been acquired by the Sundstrand Corporation and would become part of the Sundstrand Data Control division.

WT-200B vhf radio

The Wulfsberg WT-200B is the newest version of the basic WT-200 vhf/AM transceiver. It uses a digital frequency synthesiser to provide communications over the frequency range 118 to 151.975 MHz at 25 kHz spacing. A single wire change allows operation down to 116 MHz. Transmitter power output is 20 watts. A voice-compressor offering up to 20 dB of compression is available, in addition to an audio clipper which provides 6 to 10 dB of clipping. Shop adjustable (S+N)/N and carrier squelch are provided. Frequency control is through a remote standard two-out-of-five control. All transmitter spurious outputs are more than 90 dB down. The transmitter is certified to 50 000 feet (15 240 metres) and for operation between –55 and +70°C. The frequency-synthesised unit is of all-solid-state construction.

The WT-200B is designed for use in helicopters, general aviation and feeder airlines and is especially useful for aircraft in an environment that permits operation above 136 MHz.

Dimensions: 2.88 × 5.22 × 13.2 inches (73 × 133 × 337 mm)
Weight: 5.6 lb (2.54 kg)

STATUS: in production and service.

Flexcomm communications system

Flexcomm, as the name implies, is a flexible communications system providing very wide frequency coverage capable of operating from a single control, the C-1000. This provides for thumbwheel control of frequency as well as 30 programmable preset channels. A complete Flexcomm system consists of the C-1000 control, the RT-30 (30 to 50 MHz mf), RT-118 (118 to 138 MHz AM), RT-138 (138-174 MHz FM) and the RT-450 (450-470 MHz FM). Each unit may have its own control or all may share a common bus. Selection of frequency at the control chooses the appropriate radio telephone unit. A single mount may be installed and the FM radio telephone units may be interchanged. It is only necessary to connect the proper antenna to the radio telephone unit.

The system has found wide acceptance in the helicopter field, particularly through involved agency work. A total of 18 830 channels are available with all FM radio telephone units.

Each FM set may carry an optional guard receiver. In addition to straight simplex operation, semi-duplex (talk on one frequency, listen on another) may be used for operation with repeaters. All uhf international marine frequencies are covered.

STATUS: in production and service.

C-1000 controller

The C-1000 is a fully frequency-agile control unit which provides thumbwheel control of all available Flexcomm channels and provides for the storage of up to 30 preset channels for simplex or semi-duplex operation. Any of the channels may be changed by the operator. However, this capability may be disabled if the operator so desires. It also provides control of CTCSS tones in both receive or transmit. The C-1000 has edge-lighting, using either 5 or 28 volts. When used with a complete Flexcomm system it automatically selects the appropriate radio telephone unit depending on the desired frequency.

The system also contains full discrete switches for use with external devices such as antennas and DTMF coders.

The earom memory chips remember their programmed channel information indefinitely without external power.

Dimensions: 5.75 × 3 × 7.5 inches (146 × 76 × 191 mm)
Weight: 2.6 lb (1.2 kg)

STATUS: in production and service.

Wulfsberg RT-30 vhf transmitter/receiver

RT-30 vhf/FM radio

The RT-30 FM transceiver uses a digital frequency synthesiser to provide FM communications over the frequency range 29.7 to 49.99 MHz. There are no band spread limitations; the receiver can operate at one frequency extreme with the transmitter at the other.

Fully solid-state, it also provides 32 crystal-controlled sub-audible CTCSS tones. An available system includes a single-channel guard-receiver operating anywhere in the band. Separate audio inputs and outputs are provided for use with external CTCSS tones, tone bursts, DTMF encoders, voice scramblers and data.

Nominal power output is 10 watts continuous between –40 and +60°C. The receiver provides a usable sensitivity of 0.6 μV, 12 dB SINAD.

Dimensions: 4.38 × 10.5 × 5 inches (111 × 266 × 127 mm)
Weight: 7.5 lb (3.4 kg)

STATUS: in production and service.

RT-118 vhf/AM transceiver

The RT-118 vhf/AM transceiver employs a digital synthesiser to provide AM communications over the frequency range 118 and 138 MHz. Using a C-1000 control, all simplex frequencies are available and up to 32 channels may be stored in memory. This solid-state transceiver is of plug-in, modular construction using bifurcated gold-card edge connectors. The RT-118 operates well at high or low ambient temperatures and is ideal for fixed- or rotary-wing aircraft, particularly when used with other Flexcomm radio telephone units. The transmitter provides 20 watts rf output between –55 and +70°C and employs up to 20 dB of squelch compression as well as 6 to 10 dB of clipping. Spurious signals are at least 90 dB down.

Dimensions: 3.14 × 13.27 × 5.22 inches (80 × 337 × 133 mm)
Weight: 5.6 lb (2.54 kg)

STATUS: in production and service.

RT-138 vhf/FM transceiver

The RT-138 uses a digital frequency synthesiser to provide FM communications over the frequency range 138 to 173.9975 MHz. There are no band spread limitations; the receiver can operate at one frequency extreme with the transmitter at the other. Of fully solid-state construction, it also provides 32 crystal-controlled sub-audible CTCSS tones. A single-channel guard-receiver is available which can operate anywhere in the band.

Separate audio inputs and outputs are provided for use with external CTCSS tones, tone bursts, DTMF encoders, voice scramblers, data etc.

Nominal power output is 10 watts continuous between –40 and +50°C. The receiver provides a usable sensitivity of 0.35 μV, 12 dB SINAD.

Dimensions: 4.38 × 10.5 × 5 inches (111 × 266 × 127 mm)
Weight: 7.5 lb (3.4 kg)

STATUS: in production and service.

RT-450 uhf/FM transceiver

The RT-450 uses a digital frequency synthesiser to provide FM communications over the frequency range 450 to 470 MHz. There are no band-spread limitations; the receiver can operate at one frequency extreme and the transmitter at the other.

Fully solid-state, the system also provides 32 crystal-controlled sub-audible CTCSS tones, and a single-channel guard-receiver is also available operating anywhere in the band.

Separate audio inputs and outputs are provided for use with external CTCSS tones, tone bursts, DTMF encoders, voice-scramblers, data etc.

Nominal power output is 10 watts continuous between –40 to +60°C. The receiver provides a useful sensitivity of 0.4 μV, 12 db SINAD.

Dimensions: 4.38 × 10.5 × 5 inches (111 × 266 × 127 mm)
Weight: 7.5 lb (3.4 kg)

STATUS: in production and service.

RT-7200 vhf/FM radio

The Wulfsberg RT-7200 transmitter-receiver is an air-to-ground communications system suitable for a wide range of fixed-wing aircraft or helicopters. It provides FM operation over a choice of 7200 channels within the band 138 to 174 MHz at either 25 or 30 kHz increments but can be tuned to the much lower incremental value of 5 kHz. Like other equipment in the Wulfsberg range, the RT-7200 provides a semi-duplex facility with automatic push-to-talk tuning.

In simplex operation, frequency selection may be made on an individual dial-up basis using conventional thumbwheel controls; a digital readout on the C-722 remote controller confirms the selection. Alternatively, up to 15 channels are available on a programmable pre-selection basis. Power output is operator selectable at either 1 or 10 watts.

The system has an optional two-channel built-in guard receiver which conforms to the same general specification as the main equipment's receiver section. There is no frequency separation restriction if the optional guard receiver is fitted.

Automatic signal-to-noise squelch, with manual override, is provided and a separate input is included for encoder, voice-scrambler, data or other systems. Construction is all-solid-state and digital synthesis is employed for frequency generation.

Dimensions: 5 × 12.6 × 5 inches (127 × 320 × 127 mm)
Weight: 9.3 lb (4.21 kg)

STATUS: in production and service.

RT-9600 vhf/FM radio

This equipment is virtually identical to the RT-7200 system (see separate entry) in its general technical specification. The major difference is in its frequency coverage and channel capacity; ie it covers the frequency range 150 to 174 MHz but provides a total of 9600 channels in this band. Normal frequency spacing is 25 or 50 kHz but the RT-9600 (nomenclatured AN/ARC-513) is in fact capable of tuning to a finer 2.5 kHz incremental spacing.

The C-722A and the C-962A controls are edge lit, using 5 or 28 volts dc supply for lighting. Unlike the Flexcomm units, the RT-7200 and the RT-9600 have an additional AM/if detector to provide an output for a DF or ADF system.

Dimensions: 5 × 12.6 × 5.4 inches (127 × 320 × 132 mm)
Weight: 9.3 lb (4.21 kg)

STATUS: in production and service.

Wulfsberg RT-7200 vhf radio with C722 controller

Components of Wulfsberg Flitefone 40 system

Flitefone III, IV, V and VI uhf/FM radio telephones

Wulfsberg Flitefone III and Flitefone IV are full-duplex uhf/FM airborne radio telephone systems which operate in conjunction with ground stations, providing almost total coverage of the USA and part of the more heavily populated areas of Canada. The systems, which provide air-to-ground and ground-to-air communication via the public telephone networks, are installed, according to Wulfsberg, in the large majority of corporate aircraft built in the USA and Canada. Also, many are fitted to European-manufactured corporate aircraft destined for US users. The equipment is also used in other countries.

The latest system is Flitefone VI, the sixth generation of this equipment. It is compatible with the current manual ground stations and the new air-ground-radio telephone automated service (AGRAS) stations. It consists basically of an RT-18D transceiver of 10 watts output, an MT-28A rigid mounting and either an AT-270 rod or an AT-461 blade antenna. The rod antenna is suitable for aircraft up to 250 knots (463 km an hour) while the blade type will

function at all subsonic speeds. Cockpit control is exercised by a C-118D remote controller which provides pushbutton channel selection, intercomm and transmit control and handset volume variation. Two types of cabin control are available, the WH-6 where the keyboard is in the base of the unit and the WH-10 with a lighted keyboard built into the handset.

Flitefone VI is microprocessor-based so that selection of a ground station channel is automatic when the handset is removed. After the number is dialled the equipment will select the best usable AGRAS ground station first. If they are all busy, or none are within range, the best manual ground station will be selected. Manual override to a selected channel is also available.

The radio telephone also includes an hf option that expands its utility by interconnecting the handset to the aircraft hf transceiver. A selector switch on the handset is simply moved to the hf position. A noise-cancelling microphone in the mouthpiece is fitted to improve the quality of calls made from an aircraft.

The new automated ground station system,

Wulfsberg Flitefone IV system with radio telephone handset

AGRAS, has also been developed by Wulfsberg and provides for direct dial telephone calls from air-to-ground and vice versa thereby eliminating the need for a ground station operator or telephone operator assistance on every call.

STATUS: in production and service.

Training systems

Boeing has been awarded a $1.83 million contract from the US Air Force to re-develop a military simulator using the Ada software language. The 19-month programme will be conducted by the Boeing Military Airplane Company's Simulation and Training Systems organization in Huntsville, Alabama. The Ada software language implemented under the contract will be tested and validated on a Boeing-built E-3A AWACS simulator located at Tinker Air Force base, shown above

BELGIUM

ACEC

ACEC Ateliers de Constructions Electriques de Charleroi SA, Group Systèmes Electroniques, Division EAT, BP 8, 6000 Charleroi
TELEPHONE: (71) 44 21 11
TELEX: 51227

Series 320 simulators

ACEC has built a number of flight simulators for third-level airlines and the general aviation community. Aircraft so modelled include Pilatus Britten-Norman, SIAI SM1019, Embraer EMB121 Xingu, and SNIAS Caravelle VI transport. Four models are offered: Model 320 is a representative, non-specific cockpit; Model 321 covers piston twins; Model 322 turboprop twins; and Model 323 jet twins. Simulators can be mounted on platforms having three, four or six degrees of freedom. The computing system that drives the simulator can store simultaneously details of up to 512 navigation beacons throughout the world. Visual systems can include two four or six windows. Performance levels meet the requirements of the FAA's FAR Part 21.

ACEC simulator for general aviation

BRAZIL

Centro Tecnico Aeroespacial

Centro Tecnico Aeroespacial (CTA), CP 6001, 12220 São José dos Campos (SP)
TELEPHONE: (123) 21 13 11
TELEX: 1133393

Flight simulator for Tucano

CTA has developed a simulator which represents the Tucano turboprop trainer built by Embraer for the Brazilian Air Force Academy. The two-axis simulator is controlled by a minicomputer of Brazilian design and took 30 months to develop.

STATUS: in production. Ten simulators will, reportedly, be required, and cost to other countries is put at approximately US $1 million.

CANADA

CAE

CAE Electronics Limited, Box 1800, Saint-Laurent, Quebec H4L 4X4
TELEPHONE: (514) 341 6780
TELEX: 05-824856
TWX: 610 422 3063

CAE Electronics designs and produces flight simulators for commercial and military aircraft. The company has introduced many innovations in simulation technology, such as digital control loading, digital motion and an advanced rotor model. Commercial aircraft simulators have been made for the Airbus A300 and A310, the Boeing 727, 737, 747, 757 and 767, the McDonnell Douglas DC-8, -9, -10 and MD-80 series, the Lockheed L-1011, the Fokker F28, Fokker 50 and 100, and the Canadair CL-600. CAE has manufactured simulators for military flight, tactical, mission, combat, patrol and transport aircraft and helicopters. CAE has supplied simulators for the Boeing E-3A AWACS, Lockheed C-130 Hercules and P-3C Orion, Panavia Tornado, McDonnell Douglas A-4S1 Skyhawk and the Northrop F-5E Tiger II. Helicopter simulators have been built for the Agusta AB-205 and AB-212, Bell UH-1D, Boeing-Vertol CH-47, Sikorsky CH-53, the Westland Sea Lynx and the Sea King Mk 41. CAE is currently producing six Lockheed C-5B Galaxy weapon system trainers for the US Air Force.
    CAE has provided five full flight simulators to United Airlines (two for B-727s, one for DC-10,

One of three CF-18 flight, tactics and mission simulators developed by CAE for the Canadian Armed Forces

and two for B-767s) and has a technology-sharing agreement with the airline for access to extended aircraft performance data. The Canadian company anticipates that the inclusion of such data in future designs, and for restrospective improvements to existing simulators, will enhance their value as training devices.

The latest and most sophisticated simulator from CAE is that for the Canadian Armed Force's CF-18 Hornet. Cockpit controls and systems are fully represented and the *g*-system simulates load factors up to 7 *g*.

A CAE simulator of the MD-80 and incorporating a McDonnell Douglas Vital 4 visual system was installed at FlightSafety's facility in St Louis in 1985 for training TWA crews.

CAE claims to be the only manufacturer to have built simulators for all three 'glass cockpit' transports: Airbus A310 and Boeing 757 and 767.

Helicopter simulation

CAE claims to be the first company to successfully develop an accurate simulation of the complex dynamics of a helicopter's rotor blades, greatly enhancing the realism of helicopter flight training on a simulator during all phases of flight.

The generic real-time blade element software treats the helicopter, the main rotor hub and hinges, and the rotor blades themselves as separate elements, the blade flapping and lagging being faithfully replicated. Velocities are calculated at five points along each blade to allow for the varying Mach number effects.

The use of a very high-speed computer, some six to ten times faster than a conventional computer used in simulation, enables a typical

A310 simulator manufactured by CAE for Lufthansa

rotor blade to be simulated at every 7° during its rotation, enabling realistic effects to be portrayed through the simulator to the pilot.

STATUS: in development.

FRANCE

CGA Alcatel

CGA Alcatel Thomson, 33 rue Emeriau, 75725 Paris Cedex 15
TELEPHONE: (1) 45 71 11 55

Mirage 2000 maintenance trainer

The CGA Alcatel maintenance and armament trainer can simulate over 200 different systems failures selected by the instructor for the trainees to rectify. The instructors have a series of interactive displays, each controlling an aircraft system.

The first model of this trainer was delivered to the French Army at Mont de Marsan in July 1984 and subsequently orders have been received from India, Abu Dhabi and Egypt.

STATUS: in production.

Sogitec

Sogitec SA, 27 rue de Vanves, 92100 Boulogne
TELEPHONE: (1) 609 91 01
TELEX: 260922

Air-to-ground mission simulator

In late 1983 Sogitec delivered what it claims to be the first European-built air-to-ground mission simulator. It permits combat pilots to 'fly' a complete mission and uses the GI 1000 computer-generated image visual system. To improve realism, a Sogitec 4X computer-generated noise system can be added, which can include simulation of air-to-ground radar returns. Whereas the GI 1000 can generate 1000 polygonal surfaces, improvements are now in progress as preparation for the GI 10 000, for which Sogitec has a French Army contract. The system is aimed principally at aircraft and tank simulation.

Air-to-ground radar simulators

Sogitec mission simulators include an air-to-ground radar simulator. The terrain database used for the image visual system is also used to generate the information to simulate the radar map.

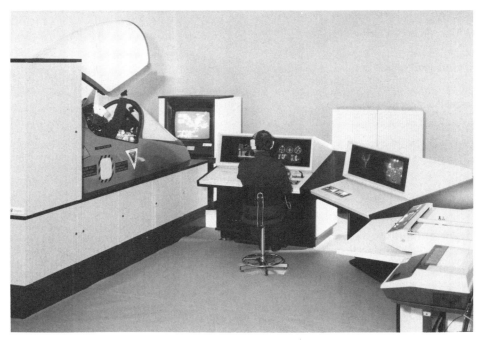

Sogitec air-to-ground mission simulator with computer-generated image

MBB 339A maintenance trainer

Sogitec is developing a maintenance trainer for the Italian Air Force's Aermacchi MB-339A aircraft; delivery to the Lecce Air Force Base was scheduled for 1986.

The trainer enables maintenance personnel to train on the aircraft's systems without resort to a real aircraft, thus reducing costly aircraft down time spent purely on training, and also eliminating any possible damage to an aircraft caused by mistakes in rectifications.

The simulator consists of a main cockpit unit, including the avionics, and five peripheral units simulating the fuel, air conditioning, hydraulics, electrical generation and electro-hydraulic ancilliary systems, all coupled to an instructor's station.

STATUS: in production for Italian Air Force.

GI 1000 day-mode computer-generated image visual system

The GI 1000 day visual system is used in conjunction with Sogitec's mission simulators, permitting pilots to be trained in continuous, full-task missions including navigation and multi-target strike. The system provides realistic detail, weather effects, and intensity and direction of lighting. It can handle more than 1000 polygon surfaces in real time, and display a continuous and inter-related terrain database equivalent to an area of 400 by 400 km. This database can be supplemented with air-to-ground radar echoes, simulating a typical radar-mapping mode.

The special-purpose computer developed by Sogitec has 1800 K bytes of memory, three dedicated microprogrammed processors, a machine cycle time of 200 μs, 32-bit data and 128-bit instructions, and Fast/Schottky technology.

STATUS: in production and service.

GI 10 000 CGI visual system

The first model of Sogitec's GI 10 000 daylight CGI simulator visual system was delivered in spring 1986 to the Centre Electronique de l'Armement (CELAR). The GI 10 000 is designed mainly for helicopter and tank simulation applications and can create a 10 000 polygon scene, with moving objection driven without restriction from the digital database.

CELAR has ordered a second system.

STATUS: in development.

Sogitec Alpha Jet flight procedure trainer

Projection systems

Sogitec has developed a multi-target and ground/sky projection system for use in air combat training domes.

Lockheed placed the first order for such a system in 1985 for development work on the Advanced Tactical Fighter.

STATUS: in production.

Thomson-CSF

Thomson-CSF, Simulator Division, 3 avenue Albert Einstein, BP 116, 78192 Trappes Cedex
TELEPHONE: (1) 30 50 61 01
TELEX: 204780

Flight simulation

In 1966 Thomson-CSF built the first digital simulator in Europe, not for training but to assist in the Concorde development programme. Its activities now embrace all types of systems, from cockpit procedures trainers up to full flight and mission simulators, for military and commercial aircraft and helicopters, armoured vehicles, ships, and nuclear and other power stations.

Responding to the standards for new simulation equipment set by the FAA with the aim of reducing training time on actual aircraft, Thomson-CSF's commercial-aircraft simulators embody a number of standard facilities. These are: six-degrees-of-freedom motion systems

mounted on hydrostatic bearings; g-seats; advanced digital-based control loading systems; computer-generated image displays; high-resolution alphanumerics and graphics displays, together with pre-programmed lessons, enabling greater consistency of teaching at the instructor's console; real-time computers specially adapted to simulator requirements, permitting optimisation of performance and flexibility; programming in high-level or assembler language; and built-in test equipment that continuously verifies the fidelity of simulation.

In 1971 Thomson-CSF built the first simulator for the Airbus A300 airliner, and has since been awarded 16 out of 22 contracts let for simulators by operators of the A300/A300-600/A310 family. Among these is the first flight simulator to represent the two-crew configuration A300 (first adopted by the Indonesian airline, Garuda), the first A300-600 simulator, and the first A310 simulator. The latter was initially employed in 1983, to assist with the certification of the A310 itself, and was later used to qualify

flight crews from Swissair, Lufthansa, KLM, British Caledonian and other A310 operators.

In late 1983 the company delivered a flight simulator to the Soviet airline Aeroflot to support the Ilyushin Il-86 airliner. This system has a six-degrees-of-freedom motion base and a night visual system. However, a Soviet request for a daytime visual system was reportedly turned down by the French Government on the grounds of an unacceptable technology transfer, and the possibility that it could be used for military applications.

For military users Thomson-CSF offers two simulator configurations: a trailer installation for customers needing mobility, and a fixed version where the training site is permanently located. These military systems embody motion platforms with four or six degrees of freedom, g-trousers to provide initial acceleration cues, g-seat to reproduce sustained acceleration, and computer-generated image displays representing airfield or aircraft carrier layouts, air-to-ground attack situations, air-to-air combat, formation flight, and air-refuelling.

For simulators appropriate to the newest combat aircraft, such as the Dassault Mirage 2000, with their advanced radar and weapon-delivery systems, enhanced facilities include devices to represent the operational environment more realistically. These include digital simulation of the radar landmass, air-to-air radar, electronic countermeasures, digital displays for ground-attack, reconnaissance and penetration, and the provision of special systems to develop proficiency in air combat.

Three such mission simulators have been built: one for the Mirage F1CR training centre in Strasbourg, a second for the Mirage 2000 base at Dijon (where the first Mirage 2000 unit became operational in the summer of 1984), and the last is a Mirage 2000 simulator for export. The system for Dijon was delivered in April 1985.

Air-combat simulators

1985 saw the entry into service of an air-combat simulator developed and built by Thomson-CSF at the French Air Force Mont de Marsan experimental base in south-west France. It was a first worldwide for any air force and was specially designed for the training of pilots.

This multi-dome system consists of a hall containing three 8-metre diameter domes with the appropriate instructor installation alongside. The system is today in its first phase, with two spheres operational, and provides combat training for two Mirage F1C interceptor cockpits, and one-on-one fights. Images of the sky, ground, enemy aircraft and missiles fired are projected onto the inner surface of the domes.

The horizon projector, which gives a view of the ground varying with altitude, provides the pilots with roll and pitch angle references. The target projector image is generated by a VISA CGI visual system which takes into account the dimensions and altitude of the target with respect to the fighter for ranges from 60 to 6000 metres. The cockpits do not contain all the aircraft operational systems but only those needed during the combat phase (flight controls, air-to-air weapons system functions). Acceleration cues are reproduced by a g-seat and an anti-g system.

The flight director's station has been specifically designed to facilitate training and the analysis of exercise results. Three colour graphic screens and four monochrome screens provide repetitions of each pilot's radar screen and head-up display and give continuous information and views of the combat sequences (perspective, cockpit and combined views, utilisation and initialisation pages).

Artist's impression of two Thomson-CSF air-combat simulators and control console

It is possible to replace an aircraft by a target manually flown from the flight director's console with a joy-stick and a throttle. Particular attention has been paid to replay and exercise debriefing procedures which makes the simulator a powerful tool for tactical situation analysis. The exercise in progress is systematically recorded to enable the flight director to safeguard any part of it. Recorded exercise sequences can be replayed immediately after the exercise or at any later time and can be completely analysed to provide statistical data or to define combat tactics. The third dome is scheduled to be commissioned in 1987, enabling three aircraft (four if the instructor is included) to fly against each other.

Mirage 2000C simulator

1985 saw the inauguration of the air defence Mirage 2000C flight and mission simulator at Dijon. This is the first simulator to bring together flight simulator functions and those of air combat missions. The simulator is conventional with the exception that its visual system reflects the mission of the aircraft.

The complete cockpit, in which all controls and indicators are operational, is equipped with a sound system and an anti-g device and is representative of the first Mirage 2000 air defence aircraft fitted with the RDM radar delivered to the French Air Force. Training in all

normal and emergency procedures involving all systems is possible on this simulator.

In the centre of the sphere is the cockpit and the horizon projector displays an image of the sky and the ground. The images are obtained from slides which are changed to suit the differences in lighting and scale of the ground when the aircraft changes altitude. The horizon line seen from the cockpit changes with the aircraft pitch and roll angles and with altitude.

Above the horizon projector, and a little behind it to reduce masking effects for the pilot, are the two target projectors. The images of the two aircraft, generated by a special purpose computer, appear exactly as they would be seen in real flight, taking into account the target's dimensions and their distance from the aircraft. The targets fly pre-programmed paths but can also be steered by the instructor. The target generators and projectors can also be used to display landmarks or runways. The missile projector provides a realistic display of the trajectory of missiles fired by the pilot or of missiles fired against him by the enemy aircraft.

The pilot is placed in an environment including four hostile aircraft. Having located them on his radar screen and analysed the situation, he can carry out an interception, make visual contact with two targets, switch on his weapons and countermeasures systems and fire his missiles. In this way he becomes accustomed to handling complex hostile situations.

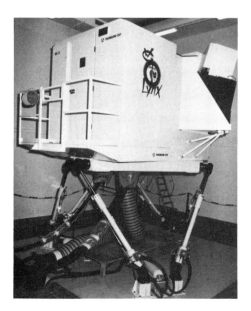

One of two WG13 Lynx simulators manufactured by Thomson-CSF for the French Fleet Air Arm

Mirage F1CR simulator with VISA 3 visual system installed at French Air Force Base at Strasbourg

Thomson-CSF A300-600 simulator operated by Aeroformation at Toulouse

Mirage F1CR simulator

Operational since early 1985 at the Strasbourg Air Base, the Mirage F1CR reconnaissance fighter simulator is the first in France to be equipped with a high performance CGI visual system (VISA 3). This system includes a synthetic image generator using the same terrain database as the ground-mapping radar. The image of the ground seen by the pilots is perfectly correlated with the image of the same ground seen on the radar screen.

All missions can take place at night or at dusk under various meteorological conditions and in a hostile environment. From his console, the instructor can bring radar or missile batteries into action, fire ground-to-air missiles and fly an interceptor to perturb the pilot in his mission. All the air-to-air and air-to-ground functions of the Cyrano IVMR radar are simulated using a specific simulator taking its information from a digital database for a real terrain. 250 000 km² of digitised terrain is available for each mission.

The pilot can thus be trained efficiently in normal and emergency procedures and also in reconnaissance missions: position fixes can be obtained from the real Parameter Insertion Module data, the corresponding fixes appearing in the external visual system. Low altitude flying is also possible, by making use of the visual system of the altitude references obtained from the terrain database and by the reproduction of a hostile environment. Similarly by using the correlation between the radar image and the visual image, air combat and ground attack missions can be accomplished.

Typical VISA 4 visual system presentation as selected for Mirage 2000N simulator

Mirage 2000N simulator

Early in 1986 Thomson-CSF announced that it was to supply the French Air Force with a simulator of the Mirage 2000N. This will permit full training for the aircraft's low-level nuclear strike role and will have a five-window VISA 4 visual system, giving a field of view of 150° laterally and up to 100° vertically. In May 1986 Thomson-CSF announced that a VISA CGI visual system had been chosen to equip this simulator.

Helicopter simulators

Helicopter activities included the manufacture in 1978 of a simulator for the French Air Force/Aérospatiale SA 330 and two simulators for the French Navy/Westland/Aérospatiale WG-13; the latter are equipped with night/dusk visual systems to permit landing practice aboard a helicopter carrier.

Civilian aircraft simulators

After an international call for proposals, Thomson-CSF was selected in November 1985 by Airbus Industrie to supply six simulators for the new A320 aircraft. This was the largest single order ever placed for Airbus simulators.

This new generation of simulators meets the most recent FAA standards and is equipped with MAGIC (Multiple Action Global Interactive Control) touch-activated instructor's station using computer assisted techniques. The rapid development of enhanced reliability software and its updating by users is facilitated by a methodology meeting the most stringent international standards and based on the use of a

software workshop called FIRST (Fast Industrialisation and Reliability Software Tools). Moreover, many on-board computers will be simulated, thus enabling the simulators to be adapted to the different versions of the A320 by a simple software change. This large contract makes Thomson-CSF world leader for Airbus simulators with nearly 70 per cent of the world market. In May 1986 a seventh A320 simulator was ordered, bringing to 24 the total number of Airbus simulators ordered from Thomson-CSF.

Other simulators

In late 1985 the company delivered two WG13 Lynx simulators to the French Navy's bases at Laneveoc-Poulmic and Saint-Mandrier. Thomson-CSF night/dusk visual systems are fitted. A Puma helicopter simulator, with six degrees of freedom motion system and full day/night visual system was delivered to the French Army at Francazal Air Base at the end of 1985. This simulator permits full mission training as well as being able to reproduce search and rescue tasks and operations from oil rigs and urban areas.

In addition to simulators representing specific aircraft types, Thomson-CSF provides families of procedures trainers under the designations 350 (fixed-wing transport and executive aircraft), 531 (combat aircraft and trainers) and 150H (helicopters) for training in IFR flight and radio navigation. Aircraft whose equipment is simulated for this purpose include Aérospatiale N262, Dassault Falcon 20, Beech C90 King Air, Fokker F27, Dornier/Breguet Br. 1150, MS 760 Paris, Caravelle III, Airbus A300, Boeing 727, Embraer Xingu and Aérospatiale SA-341 Gazelle.

Military aircraft simulators include Dassault Mirage, Dassault-Breguet/Dornier Alpha Jet, Sepecat Jaguar, Aérospatiale/MBB Transall,

Dassault-Breguet Etendard IV and Super Etendard and Aérospatiale SA330 Puma and Westland/Aérospatiale WG-13 helicopters.

Computer-generated imagery

Thomson-CSF's Simulator Division has developed a wide range of visual systems based on video disks, model boards and computer-generated images. Except for several specific cases, the general trend has been to use computer-generated imagery. Thomson-CSF has developed a whole range of these generators, known as VISA, well adapted to the operational requirements of the following applications:

VISA 10: target images for air combat and gunnery simulators,

VISA 2: landscape and target images for training in approach and landing, ground attack, navigation and interception,

VISA 3: combines the features of VISA 2 with simulation of real ground elevation and slope for nap-of-the-earth flying,

VISA 4: uses general texturing to give a very realistic representation of the visual environment at near, middle and far distances and is Thomson-CSF's prestige product which is to equip the tank driving simulators for West Germany. VISA 4 is suited for training in operations in hilly and wooded country, particularly suited therefore, for flight and gunnery simulators for helicopters and tanks.

Thomson-CSF has also developed a companion range of projection devices to exploit fully the quality of the computer-generated images including: various systems of projection in domes for air combat; high brightness tube projectors using a mixture of raster scanning and calligraphic drawing; and shadow mask tube monitors using a mixture of raster scanning and calligraphic drawing.

GERMANY, FEDERAL REPUBLIC

MBB

Messerschmitt-Bölkow-Blohm GmbH, Defence Systems Group Division, Postfach 80 11 49, D-8000 Munich 80
TELEPHONE: (89) 600 02127
TELEX: 52870

Computer-generated image visual and digital landmass simulation systems

Designed for operation in conjunction with flight simulators to be used by the West German Navy for training crews on its Panavia Tornados, MBB's first production-standard CGI visual system was commissioned in mid-1983. Matching the capability of the aircraft itself, it will permit simulation of ground attack and terrain-following missions, exercises for which previous visual systems were said to be inadequate.

MBB had earlier produced a prototype to a German military specification written in 1975. The specification proved to be inadequate and was rewritten in 1978 to form the basis of the MBB visual system. The key to the improved realism is a special-purpose computer, together with software changes, taking account of factors such as field of view foreshortening, allowing correlation between pilot's view of the outside world and the radar map display. The system also allows automatic data processing from a General Electric digital landmass simulator.

By combining the proven day-dusk-night capabilities of an advanced computer image generation system with the realism of full colour, high-resolution display modules, an advanced design, modular family of computer-generated visual systems suited to a variety of vehicle simulators can be offered including:
Flight simulators for both fixed-winged aircraft and helicopters;
Ground vehicles, particularly conduct of fire trainers for armoured vehicles;
Ships and amphibious vehicles such as air-cushion craft.

MBB's system design features meet flight simulator requirements for basic and advanced flight training, air combat manoeuvres, air-to-

MBB visual system, with control-configured version of McDonnell Douglas F-15 fighter, and cloud obscuration of the ground

ground weapon delivery and ground-to-air evasion and low altitude navigation missions.

The system's all-electronic approach to visual simulation offers significant advantages in training flexibility, system reliability, cost of operations and maintenance. In addition, the full colour day-dusk-night and variable visibility capability, supplemented with a broad choice of image display module configuration, permits an unlimited range of simulated conditions and manoeuvres to be fully supported by realistic visuals for current and future training needs.

Eight visual systems and the same number of digital radar landmass simulators are on order for the West German Air Force and Navy for training crews on Panavia Tornados (seven of the systems have been delivered). Visual systems have been ordered for research simulators in Germany and Italy and two digital

radar landmass simulators were delivered to the Italian Air Force.

MBB's CGI visual system and the digital radar landmass simulation systems were developed in co-operation with General Electric Inc of the USA.

In addition, MBB offers simulation systems for special sensor elements and part-task-trainers.

	Prototype	Production
Edges	10 000	40 000
Edges/scene	2000	8000
Light points	1000	4000
Colours/scene	64	256
Television lines	525	875
Raster elements/ lines	512	1000

STATUS: in production.

INDIA

Aeronautical Development Establishment

Aeronautical Development Establishment, Chinnya Mission Hospital Road, Indiranagar, Bangalore
TELEPHONE: (812) 557 404

During 1985 India announced that simulators of the HJT-16 Kiran trainer and the Ajeet fighter had been developed by the Aeronautical Development Establishment and had been evaluated by Indian Air Force test pilots.

STATUS: in development.

ISRAEL

Elisra

Elisra Electronic Systems Limited, 48 Mivtza Kadesh Street, Bene Beraq 51203
TELEPHONE: (3) 787 141
TELEX: 33553

EW Crewtrainer

The Crewtrainer is an aircrew training system capable of simulating the electronic warfare environment during a mission. The crew member sits in front of displays and controls identical to those in the real aircraft and is presented with displays which are either record-

ings of actual missions or simulated. He can then evade the threats, for example, using the control column to simulate manoeuvres, or act against them in other ways.

STATUS: in development.

SPAIN

Ceselsa

Cecsa Sistemas Electrónicos SA (Ceselsa), Division de Sistemas de Simulación, Plaza de la Castellana 143, Piso 6, Madrid 28046
TELEPHONE: (1) 450 7002
TELEX: 43646

SVC-101 operational flight trainer for Casa C-101

This simulator is designed to familiarise student pilots with the equipment, and handling and performance characteristics of the Casa C-101 two-seat trainer. It can simulate normal and emergency procedures, take-off, approach and landing, instrument flight and navigation, and tactical missions. Emphasis has been given to the accurate simulation of control forces, and the feel system is provided by a high-perfor-mance hydrostatic control-loading system. The central processing system is based on a 32-bit VAX 11/780 computer, with 1.5 M bytes of main memory and 56 M bytes of disk storage. The simulator does not incorporate a visual system, but is said to be compatible with any of the current visual equipment.

STATUS: in production.

Ceselsa SVC-101 simulator with instructor station

SWEDEN

Saab-Scania

Saab-Scania AB, Aircraft Division, S-581 88 Linköping
TELEPHONE: (13) 18 13 65
TELEX: 50040

JAS 39 simulator

Saab-Scania has developed a flight simulator of the JAS 39 Gripen, using Ericsson computers and cockpit displays.

STATUS: in development as part of JAS 39 programme.

UNITED KINGDOM

British Aerospace

British Aerospace plc, Warton Division, Warton Aerodrome, Preston PR4 1AX
TELEPHONE: (0772) 633 333
TELEX: 67627

Twin-dome air combat simulator

Following an order announced in September 1984, the Royal Air Force installed its first twin-dome combat simulator at RAF Coningsby in 1986. It will be used to train Panavia Tornado F2 and McDonnell Phantom FGR2 crews in the techniques for interception and air fighting. Several times since 1977 the RAF had evaluated the single-dome simulator at Warton and, beginning in 1982, regular training exercises were conducted on the then new twin-dome system. The benefits of this were highlighted when crews of Phantoms scheduled for service in the Falklands after the conflict there were able to 'fly' combat against simulated Dassault

Mirage IIIs, the standard Argentine Air Force fighter.

The British Aerospace system uses inflatable domes, 9.1 metres in diameter, containing target and sky/ground projectors. The target in each dome is generated by a television camera viewing a servo driven model. Images of the sky and ground, and also missiles in flight, are also simulated.

The RAF air combat simulator uses a Gould SEL 32/97 computer; other computer options are available.

The system accommodates many modes of operation, for example pilot versus pilot, pilot versus instructor, pilot versus computer, pilot plus pilot versus instructor, and substantial de-briefings are held after every session. The purpose of the simulator is to demonstrate the art of combat, and not necessarily to reduce the actual flying time.

STATUS: in production.

View inside one dome as pilot chases image of Dassault Mirage III 'flown' by instructor in another dome

Easams

Easams Limited, Lyon Way, Frimley Road, Camberley, Surrey GU18 5EX
TELEPHONE: (0276) 63377
TELEX: 858118

Tornado interceptor trainer

The first of two Tornado air interception trainers (TAIT) was handed over to the Royal Air Force in October 1985. The trainer is used by pilots at No 229 Operational Conversion Unit, RAF Coningsby, who are converting onto the Tornado F2/3 for the real-time practising of air interception manoeuvres.

The TAIT includes mathematical modelling of the Tornado's flight characteristics, its avionics and weapons systems; the modelling of the aircraft's radar is said to be the first of its kind in such a system.

The aircrew sit in full-size replicas of the Tornado's two-man cockpit and carry out tactics and attacks as in real life, the exercises being controlled from an instructor's station, which has repeaters of the crew's displays.

The TAIT software is compiled in Fortran 77 and runs on a DEC VAX 11/782 computer. A PDP11/23 computer is used for the radar simulation.

STATUS: one system delivered October 1985, second system delivered mid-1986.

Ferranti

Ferranti Computer Systems Limited, Cheadle Heath Division, Bird Hall Lane, Stockport, Cheshire SK3 0XQ
TELEPHONE: (061) 442 0771
TELEX: 666803

Cockpit emergency procedures trainers

Ferranti has delivered two cockpit emergency and procedure trainers (CEPT) for the Royal Air Force's main versions of the F-4 Phantom: the F-4M and the F-4J.

The F-4M CEPT contract was awarded to Ferranti in November 1983. The existing Phantom mission simulators, which are located at all the RAF's main Phantom operating bases in the UK and West Germany, have been in service for many years, and were about to be successively updated by their manufacturer, so the new CEPT was to be trailer-mounted so that it could be based wherever the main simulator was being updated and then easily moved to the next when required.

Ferranti's previous work with simulation had not involved cockpits, but the company had experience of providing fixed-base operators' procedures trainers (of the Lynx, Sea King and Jetstream) to the Royal Navy, and having won the F-4M CEPT contract, the first job was to acquire a cockpit: for the F-4M this was brought from Germany, arriving early in 1984.

The bare cockpit then had to be refurbished, and mounted in a standard trailer that could be towed on the roads by an RAF tractor unit. The inside of the trailer is arranged so that the cockpit is at one end, and behind it is the computer – a Ferranti Argus 700 GX main processor and a 700 GL secondary processor –

Instructor's station of Ferranti Lynx observer procedure trainer

and a two-station instructor's console. There is also a small briefing/debriefing area, and the floor level of the instruction area is arranged so that the instructors can walk on the same level to be alongside the cockpit, making it easy to watch the trainee in action.

The cockpit is fully authentic, apart from displays such as the radar which are not required in this training role. As its name implies, the CEPT is for training pilots in both emergency and standard procedures and the trainer is therefore used both for basic instruction when a pilot joins the squadron, and to keep crews fully trained in the correct procedures to follow should an emergency arise.

The instructor's station has two consoles, on which the entire systems of the Phantom can be displayed in schematic or tabular form, and the pilot's actions also monitored. The flight instruments are also relayed to the consoles, and displayed on the TV screens in pictorial form.

The CEPT has no motion system or visual display, these being considered unnecessary complications in what is a basic simulator, but the cockpit is vibrated during flight to give the pilot some sense of motion. The canopies are opaque, so that the flight being conducted is, in effect, taking place at night. The F-4M CEPT was delivered in May 1985, and is now in service with the RAF.

In the meantime, a contract was placed to supply an F-4J CEPT, using the same principles; different computer software had to be written for the -J and a set of different instruments acquired and fitted to the skeletal cockpit, which also had to be extensively refurbished.

The F-4J CEPT was delivered to the RAF early in 1986, and, being the RAF's only F-4J simulator, will be used just as much as a conversion trainer as a CEPT.

Sea King Mk 5 rear crew trainer

The Sea King Mk 5 rear crew trainer delivered to RNAS Culdrose in 1986 is used to train aircrew in anti-submarine warfare operations.

The system comprises three rear crew trainer cabins (RCTCs). Each RCTC is initially housed

Ferranti cockpit emergency procedures trainer for Royal Air Force Phantoms is housed in mobile trailer for easy transport between bases

in two containers: one houses a fully-equipped replica of a Sea King Mk 5 rear cabin, and the other contains the computer, instructor's console and simulation equipment.

The trainee teams sit at operational positions, seeing and hearing realistic responses on their sonar, sonobuoy receiver, radar, navigation, ESM, and communications equipment as they carry out the procedures for detecting, tracking and attacking the target submarine.

Environmental factors such as sea state, tidal and bathythermal conditions and bottom topography are all carefully modelled. Similarly, flight profiles, weapon trajectories and ship and submarine behavioural characteristics are all accurately generated according to type.

Training sessions are controlled by the instructors either from VDU-based consoles or from a portable local control panel. Each exercise is set-up from the instructor's console which has facilities for monitoring the complete tactical situation and injecting faults into the system. The local control panel may be used within the rear cabin for more direct control and closer monitoring of trainee performance.

Each RCTC is capable of linking up to a dynamic cockpit simulator to form an integrated training system. This will enable three teams to operate independently, allowing individual levels of training, or jointly to permit full mission training.

STATUS: in service.

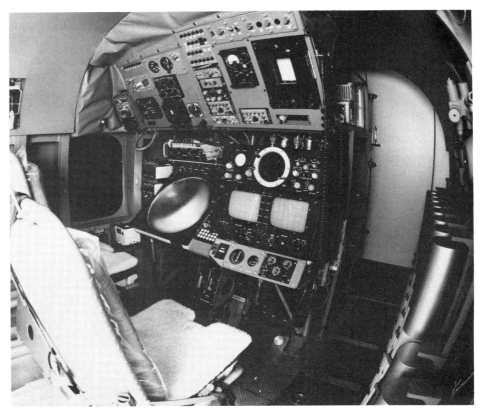

Ferranti Westland Sea King rear crew trainer, used for anti-submarine warfare training

Lynx observer procedure trainer

The Lynx helicopter observer procedure trainer (LOPT) which went operational in 1984 at HMS Osprey, Portland, comprises two independent but identically equipped cubicles, each fully simulating an observer's cockpit area. The system is linked up to a full mission Lynx simulator to form an integrated training system.

The trainer is based around Ferranti Argus 700 computers linked in a multiprocessor configuration. Full and sophisticated simulation of the Lynx's Ferranti Seaspray target acquisition radar is accomplished by a comprehensive radar effects simulator trainer system (CREST). Complex software simulates an electronic warfare environment and a conventional Doppler simulation technique is used for the tactical air navigation system (TANS).

STATUS: operational at HMS Osprey, Portland, Dorset.

Flytsim

Flytsim Limited, Unit 9, Valley Centre, Slater Street, High Wycombe, Buckinghamshire HP13 6EQ
TELEPHONE: (0494) 459545
TELEX: 849462

Flight simulation

Flytsim designs and builds high-quality simulators and training aids tailored to customer requirements. Target markets comprise flying schools, air-taxi and charter operators, aeronautical colleges or universities with appropriate departments, research establishments, feeder and international airlines, and military forces.

The company specialises in single- and twin-engined simulators, procedures trainers, and other instructional systems for piston-engined and jet aircraft, and most of them are built around five standard units, tailored as necessary to particular needs. These are:

Digital computer; designed and built by Flytsim for radio aids computation, permitting any combination of ground stations from 10 to 140 to be incorporated into the equipment. These stations can represent any standard navigation or communications facility, for example, ADF, VOR, ILS, DME, VOR/DME, or Vortac. Other facilities include digital QDM selection, digital station idents, and automatic landing markers.

Solid-state flight computer; this can be programmed to represent the full flight envelope of any type of aircraft from light singles to wide-bodies.

Two-axis hydraulic motion system; this permits excursions up to ±10° in pitch and roll. The system also includes effects to simulate take-off and landing roll, application of brakes, and runway irregularities.

Simulator; a glass-fibre cockpit, cabin, or flight-deck section.

Plotters; an on-board plotter records the approach pattern of the simulator, and a second 36 by 36 inch (914 by 914 mm) unit

Flytsim representation of Boeing 707-369 for Kuwait Airways

external to the simulator displays en route or approach patterns.

A Viscount 806 simulator is installed at the Aylesbury facility: it is used primarily for training British Air Ferries' crews.

In July 1986 Flytsim delivered a tilt-rotor research simulator to Boeing Vertol.

GEC Avionics

GEC Avionics Limited, Maritime Aircraft Systems Division, Airport Works, Rochester, Kent ME1 2XX
TELEPHONE: (0634) 44400
TELEX: 96333

ACT-1 airborne crew trainer

The airborne crew trainer (ACT-1) is used by crews of Royal Air Force Nimrod MR2s to simulate sonobuoy returns without real submarines or sonobuoys and independent of other air traffic.

ACT-1 comprises a portable instructor station, which is plugged into the aircraft's AQS-901 sonobuoy processing system (described below), with a reel of magnetic tape containing the computer program. The crew member acting as instructor keys into the exercise control unit the position, course and speed of target submarines and sonobuoy data, such as the type of buoy, its active frequency and last known location. In this way a simulated threat situation is set up. Submarine manoeuvres can be injected at any time to evade location and trainees will see corresponding sonobuoy reactions on the five screens which make up the AQS-901 installation.

ACT-1 is used mainly when the aircraft is in transit or on training missions. However, with the ability to switch instantaneously to 'live' data processing, the aircraft is always operational when required. It is important to note that ACT-1, the first airborne training aid in squadron service, complements rather than replaces the ground simulator.

STATUS: in service with the Royal Air Force.

ACT-2 airborne crew trainer

ACT-2 is under development for use with the AQS-902 and AQS-903 and other advanced acoustic processors. It is designed for operation in both fixed-wing aircraft and helicopters, providing the latter with a dipping sonar trainer in addition to sonobuoy simulation.

STATUS: under development.

Marconi Instruments

Marconi Instruments Limited, Longacres, St Albans, Hertfordshire AL4 0JN
TELEPHONE: (0727) 59292
TELEX: 23350

Rediffusion Simulation

Rediffusion Simulation Limited, Gatwick Road, Crawley, Sussex RH10 2RL
TELEPHONE: (0293) 28811
TELEX: 87327

With its background of navigation-trainer production during the Second World War, in 1951 this company built the world's first electronic flight simulator, for British Overseas Airways Corporation. It represented the Boeing 377 Stratocruiser. Since then the company has built more than 300 full flight simulators for the world's airlines and military forces.

Rediffusion produces a comprehensive range of pilot and aircrew simulators and training aids. These include fixed-base, flight-deck procedures trainers, and full flight simulators with six degrees of motion, monitor or projection based visual systems and a touch activated simulator control instructor station.

Flight simulator for 737-300

In 1983 Rediffusion was contracted by Orion Airways to build the world's first flight simulator for the new Boeing 737-300 airliner. The £4.5 million simulator was installed in 1985 at Orion's flight training centre at the East Midlands Airport. Crew training began in January 1985 and is being conducted under British certification procedures, although the simulator is designed to meet the technical requirements of Phase 2 of the FAA's Advanced Simulation Plan. It is also designed for quick reconfiguration to Orion's -200s fitted with the SP77 autopilot, and can be converted to represent the same aircraft with the SP177 system.

Visual simulation of actual airports under night or dusk conditions will be provided by a Novoview SP1 computer-generated image system linked to a WIDE display.

In May 1986 British Caledonian ordered a 737-300 simulator, valued at £5 million for installation in its training centre.

Flight simulator for 747

Rediffusion has begun work on a flight simulator for the Boeing 747 wide-body airliner that will bring together some of the company's most recent technical advances. Ordered in April 1985 for Air France for late 1986 delivery, it will in particular be the first device to incorporate TASC (touch activated simulator control), Rediffusion's new innovative instructor station. The company says that TASC is the first flight simulator instruction facility to combine touch-screen technology with microprocessor control. The principal benefit is to reduce the time spent by the instructor in planning and monitoring exercises so that more attention can be given to crew training. The system also incorporates a six-window Novoview SP1 computer-generated image visual system, and will be driven by a 32-bit Gould Concept 32 host computer. Software is being developed by Rediffusion from Boeing flight-test data, and the simulator is designed to meet the French equivalent of the FAA's Phase 2 Advanced Simulation Plan training requirements.

STATUS: in production.

Sea King simulator

The Rediffusion simulator of the Royal Navy's Sea King helicopter was delivered in mid-1986. Claimed to be the most advanced of its kind, it features the Novoview SP3T visual system and

Sea King AEW crew trainer

In May 1986 Marconi Instruments announced that it was to supply the UK MoD with a mission simulator of the Royal Navy's Westland Sea King airborne early warning variant. The simulator, valued at £1.75 million, will be a representation of the Sea King's rear cabin and will be installed at RNAS Culdrose. Marconi Instruments is also currently supplying crew trainers for the RAF's Nimrod MR2 aircraft.

Computer-generated imagery view presented on Suprawide to flight deck of Rediffusion British Aerospace 146 simulator

WIDE II display and interfaces with any of three rear crew trainers produced by Ferranti (which see). Pilot cues are enhanced by sound and vibration effects.

BAe 146 simulator facility

A Rediffusion simulator equips the British Aerospace BAe 146 simulator facility at Hatfield, Hertfordshire: it features the Novoview SP3T visual system and WIDE display. The simulator cost £5 million and a similar system is installed at the San Diego, California headquarters of Pacific Southwest Airlines.

B-1B training system

Rediffusion is supplying simulation equipment to Boeing Military, under a contract signed in 1985, for the B-1B operational training programme. If all the options in the contract are exercised, Rediffusion will supply five weapon systems trainers, two mission trainers, one software support centre and six cockpit procedures trainers.

Deliveries began in March 1985 and will continue until August 1987.

Full mission simulation

In the military arena Rediffusion provides equipment for transport, combat types and helicopters, to allow all levels of training right up to full-mission tactical operation. It has built the flight station for a US Air Force/Boeing E-3A full mission simulator and a B-52 in-flight refuelling simulator; current contracts incude supplying major simulation components for the Boeing training programmes associated with the US Air Force/Rockwell B-1B bomber and E-3A Sentry.

As well as the Royal Navy/Westland Sea King simulator, British Ministry of Defence orders have included the UK's first full mission

helicopter simulator for the Royal Navy/Westland Lynx, the flight stations for the Royal Air Force's GR1 and F2 Tornados, and operational simulators for both VC10 transport and tanker variants, which feature in-flight refuelling capability.

Interactive training systems

Completing a comprehensive range of training equipment is Rediffusion's new computer-based open learning concept, interactive training systems (ITS). Operationally ITS translates conventional training programmes into dynamic, self-teach lessons, stored in a computer and accessed through a keyboard and touch sensitive screen. This technique has been found to reduce learning time by 30 per cent or more. From a single student terminal, systems can be extended to produce group networks and multi-media teaching packages which interface with pre-programmed video, audio or slides presentations. At its most sophisticated, real-time systems familiarisation trainers can be created by adding a second monitor and an extra processor incorporating simulation software. The student operates the system from a graphic control panel on a touch-sensitive screen and sees the reaction in a dynamic systems schematic on the second screen. Such software could represent the precise characteristics of a specific aircraft so that students can experience normal and emergency procedures without risk or the cost of using operational equipment or full-scale flight simulators.

Novoview SP visual systems

More than 300 Novoview computer-generated image visual systems have been built by Rediffusion for its own and other manufacturers simulators. In the last four years the company claims to have taken 70 per cent of the visual simulation commercial market. The series covers the full range of training needs, from night/dusk to daylight with added two-dimensional texture. All weather, cloud and visibility conditions can be selected by the instructor. Novoview SPI provides night and dusk scenes and has 6000 lights, with five light colours, 200 to 400 surfaces, 64 grey shades, 256 edges, a refresh rate of 30 Hz, and, with SPI/T, the option of texture. Novoview SP3 produces night, dusk and bright daylight conditions, while SP3T adds texture to surface detail. The most recent applications for SP3T include the first two simulators for the British Aerospace 146 and also for the Royal Navy's Westland HAS Mk 5 Sea King helicopter.

In mid-1985 the US Air Force ordered the first four SP3T visual systems for the Undergraduate Pilot Training (UPT) programme being managed by the Air Training Command. A further eight SP3Ts, and a visual system support station were ordered (total value $7 million) in early 1986. Over the following two years the US Air Force plans to purchase a further 12 SP3Ts.

A US Navy CH-46D simulator is also to have a SP3T visual system: the $5 million order was announced in October 1985. At the end of 1985 a Boeing 757 flight simulator, equipped with a SP3T visual system and operated by Northwest Orient Airlines was approved to FAA Phase 3 standards, claimed to be the first such approval given for a visual system with computer-generated imagery.

Rediffusion's most sophisticated computer-generated image visual system – the CT6 series – now allows military flight training to enter the full tactical mission area with a new degree of realism.

It extends the capability of the CT5A which was selected by US Marine Corps for its Boeing Vertol CH-46 and Sikorsky CH-53 helicopter crew training, McDonnell Douglas AV-8B tactical mission simulators and for a highly successful US Air Force F-15 simulator evaluation programme.

The CT6 reaches new levels of realism with features such as two-dimensional modulation and contour texture, including photobased texture images. With the ability to apply texture to any surface, an inherently high surface capacity and advanced hardware database management, the CT6 has the highest scene density capability of any comparable CGI visual in the world. And in 1986 the firm will launch SP-X, a brand new family of CGI visual systems that is upwardly expansible from basic night/dusk to full daylight – all with the option of texture on any polygon regardless of its orientation in the database and with options on picture resolution for design-to-cost flexibility.

SP-X visual system

Rediffusion announced this new, sophisticated visual system in November 1985, claiming it to be a major technical breakthrough.

The family of SP-X visual systems covers the full range from night/dusk to full daylight and all models have the option of full texture. Specifically SP-X offers high scene quality and stability, transparency, luminous polygons, pixel fog, fade level of detail management and polygon format for ease of modelling.

There are also a wide range of configurations and feature options which allow operators design-to-cost flexibility. For example, because the number of pixels produced by an image generator impacts on both picture quality and system price, varying pixel count levels will be offered.

Operationally SP-X is said to be unique in that a combination of calligraphic light points and raster lights provide crisp night/dusk scenes and new levels of realism. Hardware texture, too, is available to enhance scene quality providing detailed height and speed cues and peripheral information expected in the real world.

Texture technology has been expanded in SP-X so that it can be applied to any polygon

Simulated scene photographed directly from Rediffusion's CT6 display system

Rediffusion SP-X visual display introduced in late 1985, showing canopy transparency and all aspect texture

regardless of its orientation in the database. SP-X texture also modulates from colour to colour which is far more realistic and efficient than previous techniques. Other major advantages include special motion effects and a powerful anti-aliasing clamping feature.

WIDE wide-angle infinity display equipment

In November 1982 the first of the company's simulators to be equipped with the WIDE visual display system was put into service by British Airways Helicopters at Aberdeen. This displays the computer-generated image 'outside world' scenes to the flight-deck, in this case of a Boeing Vertol 234 transport helicopter simulator. It is used both for training in hazardous North Sea oil related operations and by the Royal Air Force. In November 1983 the first WIDE system in the USA went into service on a Rediffusion Boeing 737 simulator bought by Delta Air Lines.

WIDE is claimed to represent a major breakthrough in flight simulation, and is the

Rediffusion new-generation Touch-Activated Simulator Control, launched in March 1985

culmination of a 10-year research and development programme by Rediffusion. By replacing conventional monitor-based displays with a triple-projector system, WIDE for the first

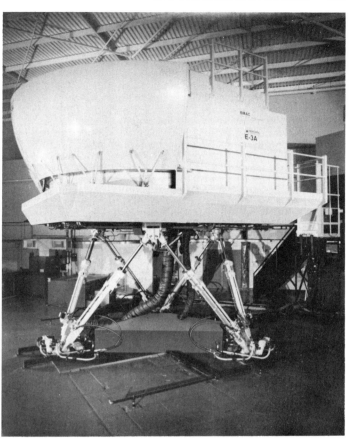

Rediffusion simulator for Boeing 767 is jointly owned by Britannia Airways and Norway's Braathens SAFE

Rediffusion simulator, with WIDE, representing Boeing E-3A AWACS built for Boeing to train US Air Force aircrews in air-refuelling techniques

time allows cross-cockpit views without optical limitations. The exceptional degree of realism is conferred by the use of a technique whereby the outside world is displayed on a 20-foot curved screen surrounding the simulator flight-deck, replacing the conventional monitor displays at each window, with their attendant discontinuities. The screen is vacuum-formed and the back-projected image is collimated (that is, seen at infinity, as the outside world would be viewed) and spans 150° horizontally and 40° vertically. The system is certificated by the UK's Civil Aviation Authority for all mandatory six-month proficiency tests. Flight crews are scheduled to fly only about four hours in the real aircraft during conversion to type, thereafter relying solely on the simulator. The British Airways Helicopters simulator and its WIDE system is making a particular contribution in North Sea operations because of its suitability for teaching pilots the difficulties of perception in poor weather. Other WIDE customers include

Singapore Airlines, British Aerospace, Alitalia, Helikopter Service, Orion Airways, Britannia Airways and Braathens SAFE (which jointly operate a system) and Boeing's B-1B bomber training complex.

During May 1984 the company announced a new system, WIDE II, with a 200° field of view and having applications for rotary-wing aircraft and specialist fixed-wing training. WIDE II retains the earlier configuration by using projectors mounted above the simulator cab and throwing images onto a back-projection screen. This system, however, has five projectors instead of three, and a specially extended back-production screen and mirror.

In January 1986 Rediffusion extended the range with Suprawide which uses new crt technology with better picture quality and brightness.

STATUS: in production and service. By January 1986 40 WIDE systems had been ordered, were

under construction or had been built for military and commercial operators. The first Suprawide system was incorporated in a BAe 146 simulator for Pacific Southwest Airlines. This was operational in 1986. Recent orders for WIDE systems include ten for FlightSafety International, for installation on Gulfstream III, Boeing 737, helicopter and business jet simulators.

In June 1986 Rediffusion announced that the WIDE II had been selected for the US Navy's EA-6B Prowler operational flight trainer programme. Two systems are to be supplied to prime contractor Reflectone. Novoview SP3T image generators are included in the systems.

Technology

In 1985 Rediffusion also introduced a new simulator concept designed specifically for commuter aircraft such as the ATR 42 and

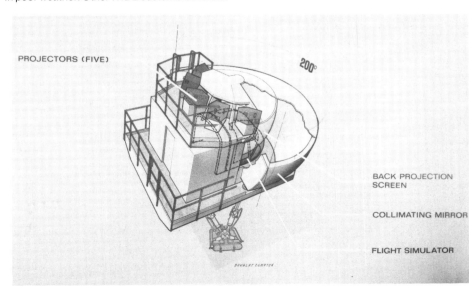

Artist's impression of Rediffusion WIDE II display associated with helicopter simulator

PROJECTORS (FIVE)

200°

BACK PROJECTION SCREEN

COLLIMATING MIRROR

FLIGHT SIMULATOR

Rediffusion WIDE visual system in British Airways Helicopters' flight simulator at Aberdeen

Shorts 330. The new concept is designed to offer the commuter operator a wide range of cost/capability options, from a generic simulation to full FAA Phase 2 capability. Using modular construction techniques and proportionally smaller cockpits and motion systems, the new approach makes full flight simulation commercially viable for commuter operators.

Other proven state-of-the-art technology typically incorporated in Rediffusion's advanced flight simulators includes hydrostatic motion systems with digital control loading and six degrees of freedom; the touch-activated simulator control (TASC) together with an instructor station which allows exercises to be set up and monitored in spontaneous, semi- or fully-automatic modes; and precise cockpit/flight deck replicas with what the company calls Total Sound Environment.

Realism in the flight-deck configuration is matched by 32-bit computing resolution packages. In 1985 Gould launched SCI-Clone/32, a brand new distributed computing system which looks set to become the industry standard. The first systems will be installed on Rediffusion simulators, a Boeing 727-2S2F for Federal Express and a Boeing 747 for Brazilian national airline Varig. (The Federal Express order included DC-10-30 cockpit procedures trainers and was valued at nearly $14 million in mid-1985.)

SCI-Clone was designed from the outset around the real-time simulation task and, unlike a micro-based system, any one of its distributed processors has the proven ability to handle the critical path in the most complex simulation activity, the flight dynamics.

Operationally, the processors, which are grouped together to form a 'node', are interconnected in the computer by both a high and low-speed link. The high-speed link, or reflective memory bus, is absolutely unique in allowing immediate access throughout the complex, to common information needed to perform the simulation task. By isolating this from the 'housekeeping' activities carried on the low-speed link, the computer's entire high-speed capability is concentrated exactly where it is needed operating the simulator.

SCI-Clone's major advantage over mainframes is that computing power can be tailored to precise training demands from the smallest aircraft types to the most complex glass cockpit aircraft of the future. As more data becomes available, or as training needs extend, that computing power may be expanded simply by adding processors or a complete node.

In 1985 the company captured around 55 per cent of the commercial flight simulator market and with the range of procedural and flight trainers, built at Rediffusion's Aylesbury plant in Buckinghamshire, has now designed equipment for all the latest aircraft types including Concorde, Boeing 737, 747, 757 and 767, McDonnell Douglas MD-80, BAe 146 and the Airbus.

Rediffusion Simulation and Link-Miles

Rediffusion Simulation Limited, Gatwick Road, Crawley, Sussex RH10 2RL
TELEPHONE: (0293) 28811
TELEX: 87327

Singer Link-Miles Limited, 27 Churchill Industrial Estate, Lancing, West Sussex BN15 8UE
TELEPHONE: (0903) 755 882
TELEX: 87165

Flight simulator for Tornado F2

Rediffusion and Link-Miles are jointly developing and building four flight simulators to support the Royal Air Force's 165 Tornado F2 interceptors. The Operational Conversion Unit for the type will be set up at RAF Coningsby in Lincolnshire, where the simulators will be located. They will be used to provide conversion training from other types of aircraft (notably the Royal Air Force interceptor, the British Aerospace Lightning) and proficiency training.

This collaboration is an extension of an earlier programme, begun in 1977, to provide six simulators for the GR1 version of the Tornado.

STATUS: in production.

Singer Link-Miles

Singer Link-Miles Limited, 27 Churchill Industrial Estate, Lancing, West Sussex BN15 8UE
TELEPHONE: (0903) 755 881
TELEX: 87165

More details of Singer Link-Miles' work can be found listed under the parent company, Singer-Link of the USA.

Micro-processor simulation technology

In parallel with the revolutionary changes taking place in the aircrew's work-station, where conventional instruments are giving way to high resolution, multi-function crt displays and where airborne computing has made such significant advances, major improvements in performance and fidelity of today's training simulators have been achieved by Singer Link-Miles through the introduction of new concept computing.

The sophistication of flight-deck displays as seen in fighter cockpits, contemporary airliners and general aviation aircraft has another parallel in the increasing standard of proficiency required of the pilots who fly them.

Singer Link-Miles has refined and developed the ultimate training tool – the flight simulator – using one major determinator: how is the device best configured to achieve cost-effective training?

In response to the requirements of operators, pilots and simulator instructors, Singer Link-Miles has designed and developed MST (microprocessor simulation technology) based on a functionally distributed, parallel computing system and introducing a new concept instructor's station, TMS (training management system).

MST benefits are applied across the full range of simulators and training aids from the company. The same technology is also applied to the IMAGE range of CGI visual systems.

MST functionally distributed computing (FDS) employs vlsi circuitry based upon the INTEL range of 8086, 80286 and 80287 micro-

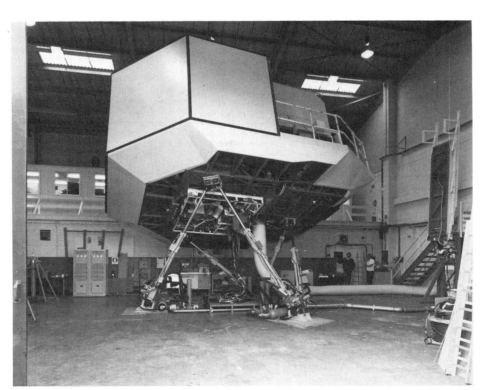

Singer Link-Miles Sea Harrier flight simulator

Singer Link-Miles KC-10 flight simulator

processors. Multibus II and double-extended Eurocards ensure standardisation and integrity backed up by long-term future commitment from suppliers and second-source insurance.

The superseding of the general-purpose minicomputer, conventionally adapted for simulator application, with a parallel, modular computing system (FDS) has come about as a result of Singer Link-Miles' search for a technology tailor-made for the requirement. Adaptation of a general-purpose device for the very special computing task represents an awkward compromise for the company. The benefits of customised computing are immediate and compelling. Those benefits are realised in terms of performance, reliability, fidelity and cost effectiveness. The system is tailor-made for the application.

Essentially the computing requirement (dealt with by the mini-computer sequentially) is split into major functions to each of which is dedicated a microprocessor-based cpu. Thus computing power is shared out precisely in accordance with demand.

Currently the FDS system will handle up to 16 major functions (eg Flight, Systems, TMS) - typically the Boeing B737-300 has six. Each major function can have up to six microprocessors. The computing power of one such major function can amount to the total of a Gould SEL 32/77.

With intelligent I/O (Intel 80188 based) as much as half of the total cpu computing power is available and dedicated to diagnostics. In-built debug facilities enable instant identification of faults down to printed circuit board and, in some cases, component level. Rapid rectification is ensured by a high level of interchangeability and low component count. Modularity means that a requirement for

rectification in one location has no adverse effect on the rest of the system.

The ease of maintenance aspect of MST has introduced new standards of user independence and flexibility.

That flexibility is also evident in the new MST instructor's facility, TMS. Here the company emphasises user-friendliness and has achieved a concept which accelerates the transfer of knowledge. It provides the instructor with full control of the exercise, total flexibility of programme, whilst enabling him to concentrate more on teaching the crew up front.

Civil aviation
In December 1985, Scandinavian Airlines System ordered a Singer Link-Miles MD-82 flight simulator.

Earlier in the year Thai Airways International chose MST for its Airbus A300-600 simulator. In the same year Singer Link-Miles was selected by Airbus Industrie to supply an A310-300 simulator for its own training establishment in the USA. The simulator can be readily converted to a A300-600 system and, while initially operating to FAA Phase 2 standard, will later be converted to Phase 3.

These contracts are the latest in a series which include BAe Jetstream and Concorde, Boeing 747, 737 and 727, Lockheed L1011, Douglas DC-10, KC-10 and MD-82 and Airbus A300-600 and A310.

Taking advantage of the flexibility inherent in the modular architecture of MST, Singer Link-Miles is extending into total training systems. The total training package sees the simulator installed interactively with its associated part task trainers. cockpit procedures trainers and computer-based training equipment.

In May 1986 Orion Airways ordered a Singer

Link-Miles simulator of the Boeing 737-300 valued at £6 million. Orion has formed a joint venture flight training company with Singer Link-Miles.

Military aviation
The advantages of the parallel microprocessor computing system, FDS – functionally distributed simulation – were pioneered and first realised on a flight simulator for the British Aerospace Hawk aircraft flown by the Finnish Air Force. The success of this application has led to further Hawk simulator contracts for other air forces (see separate entry).

In addition to Hawk, Singer Link-Miles has, to date, supplied BAe Nimrod AEW 3, Lockheed C-130 (seven alone to the Royal Air Force), Sepecat Jaguar GR1 (six models), the high technology aspects of Tornado (six of the GR1 and four of the F2) and are exclusive suppliers of BAe Harrier and Sea Harrier simulators (three GR1, two FRS1 and one AV-8A).

The Westland Lynx helicopter simulator supplied to the British Army Air Corps is housed in a relocatable building. Westland's development programme for the EH 101 includes a Singer Link-Miles' research and development simulator with IMAGE II T dusk/night CGI visual system with texture. Both Sea King and Commando flight simulators have been supplied by Singer Link-Miles (four variants of the Sea King).

Hawk simulators
Singer Link-Miles has provided full mission simulators of the British Aerospace Hawk to the Finnish, Kuwaiti and United Arab Emirates Air Forces and in March 1986 announced a contract to supply a similar system to the Royal Saudi Air Force. This simulator will incorporate microprocessor simulation technology (see separate entry), the company's IMAGE IIIT computer-generated visual system, with three windows, and a six-degrees-of-freedom motion system: it is scheduled for delivery in the summer of 1987.

IMAGE visual systems
Derived from the pioneering application of microprocessor technology to the specialised computing requirements of simulation, FDS, the IMAGE range of CGI visual systems share both the technology and the benefits.

IMAGE II provides dusk/night visual scenes in raster and calligraphic light point displays to conform with FAA Phase 2 standards. IMAGE III is a day/dusk/night visual system to comply with FAA Phase 3 requirements.

MST's attribute of upgradability is particularly relevant to the IMAGE microprocessor computing system where upgrade from dusk/night IMAGE II to day/dusk/night IMAGE III, from a computing point of view, involves nothing more than the addition of the required circuit boards.

Similarly texture can be added to either II or III by adding the necessary printed circuit boards.

Parallel processing allows the addition of further displays without the need to purchase the next more powerful mini-computer or consider initial purchase of superfluous computing power.

Singer Link-Miles Nimrod AEW 3 simulator in final stages of production has six-window, six-channel IMAGE II visual system

Tector
Tector Limited, Operational Displays Division, Woodhill Road, Collingham, Newark, Nottingham NG23 7NR
TELEPHONE: (0636) 892 246
TELEX: 377119

Opdis computer-generated image visual displays
This company manufactures the Opdis (Operational Displays Division) range of computer-generated visual image systems, primarily for use with flight simulators and trainers. These low-cost solid-state visual systems provide daylight images of the outside world in full colour. Representation is highly stylised, and includes a runway with vasis and papis (pre-

cision approach path indicators), or a runway with full approach and runway lighting, and surface markings. Ground images can also be generated to serve as navigation fixpoints, targets for weapon-aiming, or for other reference purposes.

Two versions are available: Opdis IVM, which generates a daytime image of a flat terrain in which various fixed or moving targets can be set, and Opdis IVC, which provides day/dusk/

night lighting conditions. This equipment has six degrees of freedom, and can be used for basic or advanced training by military and commercial operators. The multi-channel, multi-window configuration permits the use of the wide-angle displays essential for some aspects of pilot training. Tector also manufactures a range of projection and infinity-collimated display equipment for use with Opdis.

STATUS: in production for wide range of civil and military applications, including British Aerospace VC10 and Merchantman (Vickers Vanguard) transports, Dassault-Breguet/Dornier Alpha Jet and Saab Draken. A number have been ordered by European and American customers for research purposes.

Tecstar computer-generated image system

Formerly known as Atigs (aerial target image-generation system), the Tecstar system computes three-dimensional, fully manoeuvrable images representing targets or objects of interest on ground, sea or in the air. These images can be keyed in to the Opdis ground/sky background, or used in isolation as targets projected on to the domes of air combat simulators.

STATUS: in production. Supplied to British Aerospace at Warton and the Royal Aircraft Establishment at Farnborough for air combat domes. On order by the Austrian Army for an anti-aircraft gunnery trainer. The BAe system was integrated into a twin-dome simulator in late 1985.

Single-channel (one window) Tector system for Cessna 421 twin. Instructor's console nearest camera

Tecstar computer-generated image of Northrop F-5 fighter

Weston Simfire

Weston Simfire, Great Cambridge Road, Enfield, Middlesex EN1 3RX
TELEPHONE: (01) 367 5500
TELEX: 24724

Simstrike weapons training aid

Since the late 1960s the British Army and a number of other armies have used a laser-based training aid called Simfire to represent the impact points of projectiles fired from the guns of main battle tanks. A laser projector strapped to the gun barrel, and with automatic correction for target range (allowing for trajectory and other factors), fires an eye-safe laser beam at the target. If the gunner's aim was sufficiently accurate, as sensed by special detection equipment on the target, then smoke-pots on it ignite to indicate a 'kill'. The advantages over using real ammunition are cost savings ($250 a round, and $25 000 for gun barrels that have to be renewed every 300 to 500 firings), and the ability to get away from 'live' ranges as the system can be used anywhere. Equally important is the reality of engaging a target equipped with appropriate audio and visual effects to indicate a 'hit', and which, if similarly equipped with a laser projector, can 'fire' back. There can be also no dispute about whether an exchange resulted in a 'kill' or not.

The system has been very successful, some 4000 systems being used by about 35 countries.

Mindful of the losses to US helicopters in Viet-Nam resulting from ground-fire, the British Army in the mid-1970s began a programme called Simstrike to adapt Simfire technology to the armed attack version of its new Westland Lynx utility helicopter. This aircraft is equipped with the Hughes Aircraft TOW anti-tank missile and sighting system. The laser beam represents the flight-path of the missile, and after an appropriate 'time of flight' a 'kill' is indicated by ignition of smoke-pots on the target. The helicopter is also equipped with laser detectors and smoke-pots so making it vulnerable to anti-aircraft 'fire' from a ground-based Simflak laser-based missile emulator, and the system provides an effective way of improving the gunner's guidance and aiming proficiency and at the same time giving the pilot practice in using terrain and appropriate manoeuvres to avoid being 'shot down'. The Simstrike laser weapon projector is built into the Hughes sight system (so that the gunner becomes familiar with the actual operational equipment), and its characteristics are matched to those of the TOW missile for the greatest realism; for example the laser measures range to target, and if the gunner 'fires' at a distance greater than the effective range of the missile (about 4 km) no 'kill' is recorded.

Since Viet-Nam, any engagement by tech-nologically advanced forces has involved the use of armed helicopters. With their speed and mobility, they may be the first line of defence against a massive armoured advance. However, they are expensive and the estimation of numbers needed to counter a particular threat is largely based on the results of exercises, in which 'losses' are determined by arbitrary means since there is usually no conclusive way of deciding whether a tank or aircraft has been 'destroyed' or merely 'damaged'.

The Simstrike system permits not only the effective training of aircrews in aircraft and weapons management, but also the accurate estimation of attrition in any particular exercise. Furthermore, new tactical variations can be tried out to see in practice if they demonstrate the theoretical advantages claimed for them.

STATUS: system was originally tested in ground/air form with 'proof of concept' equipment aboard Westland/Aérospatiale Gazelle helicopter at Army Air Corps centre at Middle Wallop in Wiltshire during 1979/80. Trials are under way with air-to-ground equipment aboard Lynx helicopter.

Simgat weapons-evasion training aid

An airborne detection and indication system designed to train helicopter and fixed-wing pilots in evading ground fire (especially in nap-of-the-earth flying), Simgat registers the effects of anti-aircraft 'strikes' from Simflak or from an umpire's gun. Laser detectors fitted to the aircraft respond to 'ground-fire' and indicate to the pilot, via a control/display unit, the approximate direction of the 'anti-aircraft' site, and whether a 'kill' has been registered. A 'kill' is also indicated to ground observers by two flashing beacons on the aircraft, and, optionally, by the activation of smoke pots.

STATUS: in service with British Army Westland Lynx, Aérospatiale/Westland, Gazelle and Hughes Scout helicopters.

UNITED STATES OF AMERICA

Advanced Technology Systems

Advanced Technology Systems, PO Box 950, 17-01 Pollitt Drive, Fair Lawn, New Jersey 07410
TELEPHONE: (201) 794 0200
TWX: 710 988 2262

Advanced Technology Systems is a division of The Austin Company

Computrol computer-generated image visual system

Computrol is a day/dusk/night computer-generated image system for airline training. It provides full-colour scenes, including blue lights, and can display up to 30 000 edges and 10 000 light points, or any combination of these, to create images in gradual transition from day to night. Dynamic simulation shows moving objects, with no restriction on numbers of such objects.

STATUS: in production.

ATC

ATC Flight Simulator Company, 1650 19th Street, Santa Monica, California 90404
TELEPHONE: (213) 453 3557
TELEX: 284687

ATC-610 flight simulator/ procedures trainer

This is a 'personal', desk-top flight simulator representing the general configuration of light, single-engined aircraft, for private ownership, flying schools and clubs. The principal unit is a portable panel with an array of typical flight and navigation instruments, panel-mounted yoke providing pitch and roll inputs, and power quadrant with propeller, power, and mixture controls. A separate rudder pedal assembly permits co-ordination of all control axes.

Optional equipment includes a monitor communications module, marker-beacon audio unit, flight-plotter, and more extensive navigation facilities. The monitor unit permits instructor briefings and pupil/instructor conversations to be recorded for analysis. Use of the approved plotter permits the ATC-610 to be used for the entire period of simulation training time allowed by FAR Part 141 to count towards a pupil's total training time.

Dimensions:
(simulator) 29.1 × 21.3 × 17.1 inches (739 × 541 × 434 mm)
(rudder pedal unit) 18 × 7.3 × 19.8 inches (457 × 185 × 503 mm)
(flight plotter) 19 × 6 × 14 inches (483 × 152 × 356 mm)

STATUS: in production and available.

ATC-710 flight simulator/ procedures trainer

A development of the ATC-610 for professional pilots, the ATC-710 has an enclosure that, in conjunction with an approved flight-plotter, meets FAA requirements for private, commercial, and instrument ratings. Customers include flight schools, fixed-based operators, and corporate training departments. A model aircraft linked to the simulator and mounted above the glare shield moves against a land/sky background to provide realistic responses to pilot inputs for those needing an FAA-approved, moving visual reference model.

An instructor's console is an option where ATC-710 simulators are used singly, but is standard where a number of units are to be operated simultaneously. It provides two-way communication with any or all stations, master tape player, intercom, and headsets.

ATC Model 610 desk-top simulator

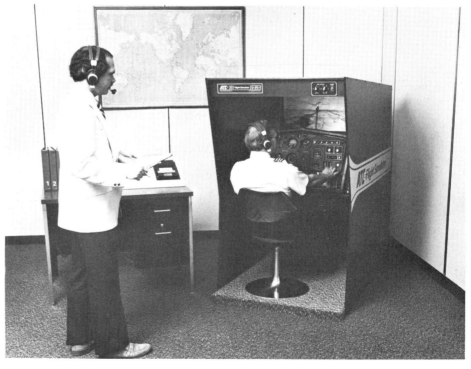

ATC-710 flight simulator/procedures trainer

Dimensions:
(simulator) 41 × 60 × 66 inches (1041 × 1524 × 1676 mm)
(instructor's console) 24 × 36 × 40 inches (610 × 914 × 1016 mm)
(flight plotter) 19 × 6 × 14 inches (483 × 152 × 356 mm)

STATUS: in production and available.

ATC-810 flight simulator/procedures trainer

The ATC-810 is a microprocessor-based procedures and IFR simulator that represents the dashboard layout and performance of 6500 to 8000 lb (2950 to 3630 kg) turbo-charged cabin-class twin-engined aircraft such as Piper Navajo/Chieftain, Beech Baron/Duke, or the Cessna 400 series. It is designed for air-taxi and Part 135 commuter operators, flight schools, airlines, and corporate and other multi-engined users.

The simulator has a comprehensive set of flight systems and navigation instruments and controls, and failures can be commanded by means of a fault console, operated by an instructor, that plugs into the simulator. The system's prom creates a 150 × 150 nautical mile navigational region representing an ATC region, with 65 airports, 36 VORS, 18 ADFS, 38 ILS approaches, four Category II approaches, and 153 instrument approaches. Options include a visual display system, with take-off and landing capability, flight director, and area navigation.

Dimensions: (simulator) 44 × 67.5 × 60 inches (1118 × 1714 × 1524 mm)

STATUS: in production.

ATC-910 flight simulator for turboprops

The ATC-910 represents top-of-the-range general aviation turboprop aircraft of up to 14 000 lb (6350 kg) all-up weight. The principal difference between this and simulators designed for piston-engined types lies in the representation of the propeller turbine characteristics. The simulation of emergency procedures in flight involving a powerplant shutdown is impracticable in this class of aircraft, but the ATC-910 permits realistic, repetitive practice in both normal and emergency engine-handling procedure.

The microprocessor-based system has full cockpit systems and procedures facilities for visual and IFR flight, and it can create a navigation situation to duplicate exactly a typical air traffic control environment. Realism is improved with the sound of propeller, boost pumps, engine start, landing gear and stall warning.

ATC-212H IFR flight simulator for helicopters

Representing the general characteristics of light to medium turbine-powered helicopters, the ATC-212H is an IFR flight simulator for flying schools, corporate or charter fleet organisations, fixed-base operators, educational institutions, military establishments, and private owners. Performance and handling very closely follow actual aircraft characteristics, and rotor torque behaviour can be switched to represent clockwise or anti-clockwise rotation. Realistic response to control inputs are provided and correlated by hybrid computer circuitry. Dynamically balanced cyclic and collective-pitch controls can be adjusted to provide the desired control stabilisation. The computer is programmed for a speed range of 40 to 160 knots. Winds up to 50 knots can be introduced from 12 directions, and five degrees of turbulence set up to affect pitch, roll, and altitude.

Accessories include an x-y flight plotter, monitor communications system, headset, and an enclosure for IFR training in accordance with FAA FAR Part 141. The flight plotter can be

Interior of ATC-212H simulator

used with standard low-altitude charts to provide records for post-exercise study.

The 212H offers *ab initio* helicopter training, the FAA allowing five hours of the time a student has spent on the ATC-212H towards a rotary-wing flight training course.

By manipulating the simulator controls the student actually flies a 4-foot (1.22-metre) model, enhancing the value of training.

Dimensions: (simulator) 34 × 46 × 64 inches (864 × 1168 × 1626 mm)
(flight plotter) 19 × 6 × 14 inches (483 × 152 × 356 mm)

STATUS: in production.

Aviation Simulation Technology

Aviation Simulation Technology Inc, Hanscom Field East, Bedford, Massachusetts 01730
TELEPHONE: (617) 274 6600
TELEX: 469479

Flight simulation

The company builds simulators representing single- and multi-engined aircraft in the general aviation sector. The principal market is for instrument and procedures training, though it also provides an introduction to the airborne environment and aircraft handling.

All units employ four independent computers to familiarise the student with the four separate aspects of flying: handling, instruments and systems, avionics and navigation. They are supplied with computer-generated visual displays, claimed to be unique among general aviation simulators, that present sky, horizon, ground, and runway symbology. Proms, either standard or to special order, and containing the appropriate data, permit students to fly routes and approaches into any airport in the world. Flights can be started at any desired location, and instantly repositioned, for example, to return to an initial approach position so that consecutive approaches can be flown in quick succession. Digital design and modular construction allow new facilities or technology to be rapidly and economically incorporated as it becomes available.

A number of accessories are offered, including x-y plotters, digitally adjustable from 0.1 to 99.9 nautical miles an inch, that can be used with any standard aeronautical chart or blank paper. A portable console permits the instructor to fail instruments and engines from outside the cockpit. Conversion kits permit customers with multi-engine simulators to convert them to single-engine units.

Simulators and displays have full FAA approval, and time on them can be counted towards the total flight and ground time needed for a student to qualify for his licence. Four models are available.

Model 201: basic representative single-engine simulator
Model 300: basic representative cabin-class twin

As supplied with the full range of options, they become, respectively:
Mooney 200X: fully comprehensive simulator, with layout resembling that of the Mooney light single
Model 300X: fully comprehensive simulator representing cabin-class twin.

STATUS: all in production.

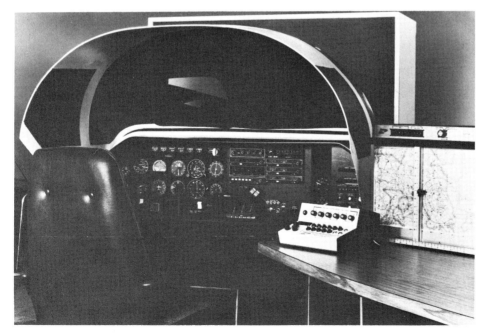

Aviation Simulation Technology multi-engine flight simulator

Colour visual system

A derivative of earlier systems designed for AST's 500 series simulators, this new colour visual system was first demonstrated publicly in May 1984, and is intended as an option for these simulators. The system meets most of the requirements of the FAA criteria for training recurrency appropriate to operators coming under FAR Parts 121, 125 and 135. Airport scenes include runways with numbers and markings, taxiways with defined surfaces, and ramps and terminal buildings. Full lighting systems, including approach, visual approach slope indicator, runway edge and taxi lights, are also presented. The system is basically non-specific, but can be programmed to represent particular airport scenes in accordance with definable geographic and terrain characteristics.

A new low-cost visual system was launched in September 1985, selling for less than $20 000. It will create a scene consisting of horizon, ground and runway, related to the aircraft's attitude and speed. Day/night and visibility conditions can be selected, and a choice made between a number of actual runway simulations. Better detail, smoothness and update rate are claimed for the new system, compared with its predecessors.

Boeing

Boeing Military Airplane Company, PO Box 7730, Wichita, Kansas 67277
TELEPHONE: (316) 526 3153
TELEX: 417484
TWX: 910 741 6900

Flight/mission simulation for B-1B

In August 1983 Boeing's Military Airplane Company was awarded a $5.5 million, 14-month contract by the US Air Force Aeronautical Systems Division at Wright-Patterson Air Force Base to conduct preliminary designs and develop specifications for a flight and mission simulator for the Rockwell B-1B bomber. A similar contract was also given to a competing team comprising Singer-Link, Rockwell and AAI Corporation.

In October 1984 Boeing was judged the winner and awarded an $89 million contract to proceed with development, production and support activities. If all the contract options are exercised, Boeing will build five weapons system trainers, two mission trainers, one software support centre, and up to six cockpit procedures trainers. Each weapons system trainer will consist of a flight station for pilot and co-pilot, and a compartment for the operators of the offensive and defensive systems. The mission trainer will be identical to the offensive/defensive station, and will train operators before they are integrated with flight crews in the weapons system trainer. The cockpit procedures trainer will provide early experience for all four crew-members until the weapons system trainers become available.

In December 1984 UK company Rediffusion was named as a major sub-contractor on Boeing's B-1B simulation programme. Under an initial sub-contract it will supply two fully integrated flight-decks, two motion systems, two Total Sound Environment systems, instructor-station equipment and a WIDE (wide-angle infinity display equipment) visual display system, as well as linkage and control loading.

In early 1986 Boeing Military awarded a $2.6 million contract to IBM for the supply of eleven avionic computer controls and three memory storage units for the B-1B simulator.

STATUS: first simulator was delivered in summer 1986, by which time the total value of contracts placed by the US Air Force was $158 million.

Simulator for E-3A

In January 1986 Boeing Military received a $14 million contract from the US Air Force's Electronic Systems Division for the development of a simulator for certification testing of hardware and software for the Boeing E-3 airborne warning and control system. The facility for inter-operability testing will be operational in April 1987 and will consist of duplicates of E-3 mission avionics, mission simulations and communications interfaces.

Burtek

Burtek Inc, PO Box 1677, 7041 East 15th Street, Tulsa, Oklahoma 74101
TELEPHONE: (918) 836 4621
TELEX: 492438

Founded in 1939, Burtek Inc was purchased in 1979 by the Thomson Corporation of America, a US holding company of the Thomson-CSF Group of France.

Flight simulation

More than 50 military services, commercial airlines and aircraft manufacturers operate Burtek flight and maintenance simulators and operator and evacuation trainers.

At the lower end of the flight simulation product line, cockpit procedures trainers facilitate the conversion of crews to new aircraft and feature the dynamic operation of all aircraft systems and controls under normal and emergency conditions. Limited flight capability may also be incorporated. In the range of cockpit systems simulation equipment, the level of simulation is increased and the flight and navigation capability expanded; some cockpit-simulation systems are virtually fixed-based simulators.

A significant activity is the manufacture of operational flight trainers for the Northrop F-5. Digital software controls all aircraft responses and provides such special dynamic effects as pre-stall buffet and the vibration resulting from weapons firing. The equipment fully duplicates the fire control system and lead-computing optical sight, seat-shaker and g-suit, and a visual simulation system permits tactical training throughout the F-5's operational envelope.

Burtek is currently refurbishing the US Air Force's fleet of eight Lockheed C-141 StarLifter flight simulators. This programme includes replacing the existing digital computers and software with state-of-the-art digital technology, replacing crew-members' and instructors' stations with colour graphic cathode ray tubes, incorporating automatic fault-insertion, and updating the motion system with new digital control loading.

Simulated aircraft maintenance trainers will provide dynamic 'hands-on' training for diagnostic, repair and function test of the integrated

Burtek cockpit systems simulator for Boeing 747 in 'swung-open' position

Burtek cockpit procedures trainer for US Air Force/Lockheed C-141 StarLifter transport

avionics and flight control system in the new McDonnell Douglas F/A-18 and CF-18 Hornet fighters. The avionics maintenance simulator includes a separate cockpit with simulation of cathode ray tube and head-up displays. Cur- rently under development for F/A-18 and Canadian CF-18 pilots is a Hotas (hands on throttle and stick) training aid to familiarise them with the fighter's sophisticated avionics, instrumentation and controls.

STATUS: Bell TH-57, OH-58, Sikorsky CH-53, UH-60 helicopters and LTV A-7, McDonnell Douglas F-4, F-15, F/A-18, DC-8, DC-9, DC-10, Northrop F-5, General Dynamics F-111, Boeing 707, 727, 737, 747, Lockheed L-1011, and Airbus A300 and A310 aircraft.

Collins

Collins Avionics Division, Rockwell Internat- ional Corporation, 400 Collins Road NE, Cedar Rapids, Iowa 52498
TELEPHONE: (319) 395 1000
TELEX: 464421
TWX: 910 525 1321

Electronic flight instrument procedures trainers

To prepare pilots for the new generation of electronic flight-deck instrument displays, Col- lins began in the early 1980s designing a family of procedures trainers to familiarise them with these computer-based information systems. In December 1978 Collins won the Boeing con- tract for the EFIS electronic flight instrument systems on the new 767 and 757 families of 'digital' airliners, and saw the demand for EFIS simulators as a further penetration of the market for these devices.

These procedures trainers incorporate the EFIS and EICAS displays (see separate entries) and their control panels, autopilot mode panel, automatic landing status annunciator, and maintenance control/display panel. They are employed in three ways. First, to demonstrate the major features of these displays, and their failure indications. Secondly, they can be 'flown' to simulate flight management planning take-off, engine failure, flight-plan route changes, and automatic landing. Thirdly, the simulated aircraft can be 'flown' on autopilot, providing hands-on training and familiarisation with this system. Weather radar information fed into the EFIS from a video recorder is an optional extra. One of two systems ordered by Delta in March 1982 is used for EFIS and autopilot training, the other for EICAS familiar- isation.

STATUS: in production.

Approach demonstrator

A joint venture by avionics company Collins and simulation services organisation Flight- Safety International, the approach demon- strator is designed to teach student pilots how to execute safe approaches, particularly in bad weather or where severe wind-shear conditions are encountered.

The simulator employs the new Collins APS-85 digital autopilot system, designed for general aviation, and includes an EFIS-85 electronic flight instrument system and an ADS-82 air data system and instruments. The simulator is driven by a FlightSafety computer, which employs three different FAA wind-shear models, with lateral, longitudinal and vertical shear. In effect, the computer sets the approach conditions, and the autopilot 'flies' the simulator through them. The EFIS display provides a profile page for each approach situation, detailing wind changes and deviations from the ILS beam resulting from these changes.

The dual-channel, digital, fail-passive auto- pilot uses inputs from the Collins AHS-85 strapdown inertial attitude and heading refer- ence system. Since the autopilot receives acceleration signals (for example an increased sink-rate) in advance of significant displace- ments, it can initiate corrective action much sooner, so reducing deviation from the com- mand glide-path.

STATUS: demonstrator. The system was first shown at the 1984 Convention of the National Business Aircraft Association.

One of two Collins EFIS procedures ordered by Delta, sharing two EICAS displays and captain's electronic ADIs and HSIs, and autopilot annunciator and controls

Collins approach demonstrator, built in association with FlightSafety

Cubic

Cubic Corporation, 9333 Balboa Avenue, San Diego, California 92123
TELEPHONE: (619) 277 6780
TWX: 910 335 2010

Training system for B-1B

Cubic is an established manufacturer of com- puter-driven training systems, its largest current contract being for the US Air Force/Rockwell B-1B cockpit procedures trainer which repli- cates the four crew stations in the aircraft.

Six units are being built, for delivery to Dyess Air Force Base, Texas, under a $15.8 million contract from Boeing and deliveries started in 1986.

STATUS: in production.

Cubic cockpit procedures trainer under development for B-1B

Electronic Workshop

Electronic Workshop Inc, Display Workshop Division, 150 Huyshope Avenue, Hartford, Connecticut 06106
TELEPHONE: (203) 246 8557

Flight simulation

This company, formed in 1951, specialises in cockpit and flight-deck procedures trainers built to customer specifications. A major line is the production of Boeing 727 simulators, but other cockpit procedures trainers have been built for operators of corporate and commuter aircraft. The B-727 units are claimed to be an inexpensive alternative to full-scale simulators, and are FAA-approved.

Cockpit procedures trainer by Electronic Workshop representing principal flight-deck control and instrument panels of Boeing's Advanced 727-200 transport. Instructor's console is on left

Frasca

Frasca International Inc, 606 South Neil Street, Champaign, Illinois 61820
TELEPHONE: (217) 359 3951
TELEX: 206549

Frasca has ceased manufacture of the Models 121, 122 and 210T (see the 1985–86 edition of *Jane's Avionics*). The model numbering system indicates the number of seats (first digit), level of technology (second digit) and finally the number of engines in the aircraft being simulated.

Model 125H flight simulator

The Model 125H is a low-cost simulator duplicating the features of typical light turbine helicopters, but can be programmed to characterise particular types of aircraft in that category. It can represent all manoeuvres encountered in normal instrument flight, and special additional capabilities include translation and autorotation.

The system comprises two principal units: an instrument panel and surround that represents the cockpit environment, and an instructor's console. An x-y plotter with initial position-setting facility and instructor's controls tracks the course followed during the exercise over an area of 60 by 85 nautical miles at a scale of 1 inch to 2.5 nautical miles.

STATUS: in production.

Frasca Model 141 instrument panel showing standard equipment

Model 141 single-engined aircraft simulator

The Model 141 simulates a single-seat, single-engined aircraft and is designed to enable training to be carried out to FAA Part 141 requirements.

Using a microprocessor-based computer generation simulation (CGS) navigating system, up to 300 navigation beacons spread over an area of 1000 nautical miles square can be programmed, including all the characteristics of real life beacons, such as frequency, location and type, etc.

An optional plotter is available, with variable scaling and which can trace the simulated flight path on actual approach plates. The instructor can start the exercise anywhere in the training area, with any flight parameters, at will. King Silver Crown avionics are fitted as standard. All flight characteristics are faithfully simulated, including noise, compass effects, brakes, the effects of ice and the depletion of fuel contents.

STATUS: in production.

Frasca Model 210 simulator built for Oxford Air Training School

Model 142 twin-engined aircraft simulator

The Model 142 is very similar to the Model 141, with the addition of a second throttle, magneto switch and propeller pitch and mixture controls. Fuel tank crossfeed is also installed.

STATUS: in production.

Model 242 twin turboprop flight simulators

The Model 242 has replaced the Model 210T in the Frasca range. It uses the same technology as the Model 141, but has two seats and controls for two engines. Among several additional features on the Model 242 are that the effects of outside air temperature and altitude on performance are computed, and propeller feathering and windmilling are realistic. Situations such as 'low oil pressure' can be simulated, with all the subsequent real life effects.

STATUS: in production.

Models 300H/205 and 300H/206 flight simulators

These two simulators represent, respectively, the Bell 205/UH-1H/212/Agusta A109 helicopters, and the Bell 206 and equivalent helicopters. Each comprises two units: the instrument panel and surround, which together reproduce the cockpit environment, and an instructor's console. A special feature of the 300H/205 is the portable instructor's panel that allows him to control most functions while seated next to the student. Plotters based on x-y co-ordinates permit courses to be tracked during exercises over areas of 97.5 × 135 nautical miles (Model 300H/205) and 60 × 85 nautical miles (Model 300H/206), both at scales of 1 inch to 2.5 nautical miles.

STATUS: in production.

Frasca Model 242 simulator

Frasca Model 300H simulator of Bell UH-1H helicopter

Frasca Model 300H simulator of Bell 206 helicopter

General Electric

General Electric Company, Defense Systems
Division, Simulation and Control Systems
Department, PO Box 2500, Daytona Beach,
Florida 32015
TELEPHONE: (904) 258 2511
TELEX: 566556
TWX: 810 832 6223

Center for advanced airmanship (CAA)

Located in the Phoenix area, the CAA which cost $12 million and was inaugurated in February 1986 focuses on providing a sound fundamental baseline for pilots of Northrop F-5 user nations. Because the CAA training system is designed to complement user nations' existing aircraft flight training programmes, the baseline that the CAA provides allows pilots to achieve higher levels of proficiency at reduced cost and/or training time.

The training objectives of the CAA programme are threefold:
(1) to accelerate and enhance the process of safe transition to the F-5 aircraft
(2) to confront the pilot with emergency and other task saturated situations
(3) to embed the fundamentals of flight discipline and both air and surface attack in a hostile environment.

To accomplish these objectives, General Electric has developed a completely integrated training system. An F-5E simulator (cockpit and advanced CGI visual system) is incorporated with an instructional sub-system comprised of computer-based academic training and simulation-based exercises. These two basic components, academic and simulation, are presented by an international cadre of highly experienced, professional, ex-military fighter instructor pilots.

Student flow through the CAA training cycle is controlled by the student's individualised training syllabus tailored to sequence the rate and flow based on individual pilot skills.

The academic element is the first step in the CAA training system. It is here that all new material is introduced to the student prior to its use in the simulator. A GE developed computer-based training (CBT) system is the primary medium for the academic instruction. Seminar/classroom instruction conducted by

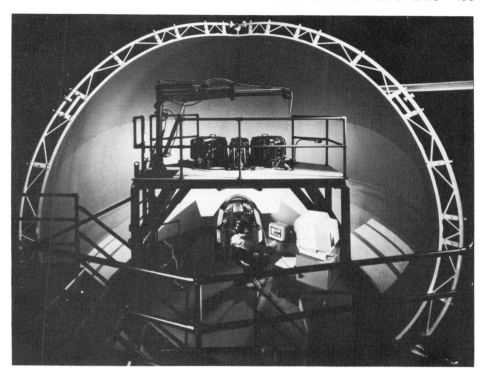

F-5E mission simulator at General Electric's Center for Advanced Airmanship, Tempe, Arizona. Shown is wide-angle dome housing F-5E cockpit and instructor station. Visual system is three-channel Compu-Scene III visual simulation system by General Electric

CAA instructors is used to supplement the CBT.

The CBT brings strength to the academic element, both in the flexibility provided by the modular design of the courseware and the control provided by the computer managed instruction (CMI). All bits of information about a particular aircraft system, procedure, or other area of interest are contained in one location (module). Within each information module are sub-modules that accommodate the multiple aircraft avionic and flight control configurations. All of this information is then sequenced in a logical format determined by the level of difficulty. The CAA then uses the CMI system to sort the modular hierarchy of configuration and level of difficulty so that automatically a course of study tailored to the needs of the individual

student can be designed and built. This courseware structure gives the CAA extreme flexibility in designing courses that match the student's F-5E experience level and his country's aircraft configuration. The modular structure of the courseware precludes duplication of information allowing updates and changes to be incorporated through a user level authoring system as soon as they occur. This enables CAA instructor pilots (IPs) to keep the courseware current and accurate at all times.

The students receive their computer-based instruction in a student workstation that has been specifically designed for the CAA. The objective is to provide the student, who may never have used a computer, with a comfortable, functional, non-threatening, user-friendly environment. Each student CBT station is housed in an attractive enclosure for privacy. The only hardware that the student sees are the two screens that are used to present the course material. Noticeably absent from the station is a computer keyboard. Interaction is accomplished with either the touch sensitive screen or a simple six-button keypad. The student does not have to remember cryptic computer commands to progress through the course. His only decision about using the system is whether to sit up in attentive position to use the handheld keypad. The workstation also has desktop space, bookshelves, individual lighting, and sufficient room for an IP to join the student for one on one tutoring, if required.

Simulation, the second element of the CAA training system, is provided by combining an F-5E cockpit, GE's Compu-Scene III image generator (IG), and a newly-developed instructor operation station (IOS). As in the academic training, the simulator training will be made up of those tasks required of a qualified, fully trained F-5E pilot. The simulator training standards will reflect the overall course objectives and will parallel the training objectives currently set by the US Air Force for F-5E aircraft flight training. Providing the means for practical, task intensive training, the simulator is used to reinforce those tasks previously introduced in the academic element by allowing the student to practise them. At the same time, the simulator training allows the student to begin developing the judgment and psychomotor skills necessary to fly a fighter aircraft in a task saturated, near combat environment.

F-5E mission simulator at General Electric's Center for Advanced Airmanship overflies Williams Air Force Base

Overseeing this development is the third, and most important, element of the CAA training system: the instructor pilots. All former fighter instructors, with most having combat experience, this group is able to draw on their recent training and operational experiences to guide the student from his first engine start, through offensive aerial combat manoeuvring, to his final Lo-Lo-Lo surface attack profile.

Instructional briefings/debriefing by the IPs before and after each simulated sortie provide the continuity that closes the loop between the academic element and the simulator. Here the instructor reinforces the theory of the academic presentations and explains their practical applications in the simulator and aircraft.

For the instructor to be effective in this role, the training system with which he is working must be efficient. The system must free him from the role of hardware technician. The CAA has reduced and automated the administrative and technical functions by developing an 'integrated training system' that reduces the IP's non-instructional workload and makes the instructor's teaching time more effective. This integrated training system ensures that the CAA is providing a standardised, objective, quality, cost-effective training programme.

Integration of the training cycle is accomplished by connecting three primary hardware components: the graphics and parametric recording computers at the IOS, the CBTs at the student workstations, and the computers in the briefing/debriefing rooms to one local area network (LAN). The networking of these components brings several major benefits to the CAA that are not enjoyed in other training systems. The main areas that benefit from integration are:

1. Instructor operation station
Networking the simulator instructor operation station produces several significant improvements that increase the IPs capabilities and further reduces his workload. The integrated workstation provides the IP with a user friendly interface through a touch-sensitive screen. The IP can quickly program the simulator configuration, mission position and initial conditions. Emergencies and other inputs can also be made in real-time with a quick touch of the screen. Selection of a head-up display (HUD) video camera or an over-the-shoulder camera that will record and present flight instruments and avionics displays is also made with the touch screen. The entire simulator control process can be automated, if desired, so that the IP can devote his complete attention to monitoring the progress of the student.

The touch screen is also used by the IP to control the graphics computer that is displaying and recording the positions and parametric data of the simulator and any opposition fighters. This system also displays and captures all weapons scores (bombs, missiles, strafe, etc) and passes them to a central record keeping file via the LAN. These scores are then available to the IP during the debriefing and completion of the student gradesheet.

Using a GE-developed authoring system, modification to the IOS is just as easy as its operation. This authoring system allows changes to be made quickly and without the aid of a computer programmer, thus ensuring that the CAA system is always current.

2. Automated scheduling and syllabus implementation
The addition of IBM computers to the LAN allows CAA personnel to perform automated scheduling of students, IPs, and facilities. The scheduling program incorporates the academic and simulator syllabus flows into a curriculum, automatically monitoring and scheduling the academic and simulator training. Automated record keeping of planned versus actual schedules are maintained by the system for auditing and analysis.

3. Enhanced briefings and debriefings
After sortie completion, all graphic and parametric data is passed over the network to the brief/debrief room. Here, aided by the student's HUD camera video and the graphic data, the IP is able to recreate any situation from the just completed sortie.

Gradesheets are prepared using CBT software accessible from the debriefing computer system. To help in student trend analysis, the IP can review all previous gradesheets from the brief/debrief system. If he desires to recommend remedial training for the student, he can then make an immediate input to the automated scheduler.

4. Performance monitoring
Possibly the most powerful aspect of the integrated system is its performance monitoring capabilities. As noted, all of the students weapons scores are recorded and archived. All gradesheet inputs are also archived into the central network storage files. In addition, the CBT system automatically records and saves performance information as the student progresses through each academic subject. This data is compiled and reviewed by the CAA Chief Pilot and the instructors. This ensures that the student's progress is well monitored and that any problems are rectified while the student is still in the CAA. This data will be used by the CAA to analyse their own effectiveness and find those areas of the course that possibly need more, or a different type of emphasis. As the CAA database grows, through an analysis of student experience levels versus historical performance data collected from students with similar experience, the CAA will be able to provide a syllabus even more tailored to specific student needs.

The CAA instructional system enhances the fundamental approach to aviation training. Implementation of training through the use of state-of-the-art equipment and techniques provides current, high quality, standardised training. Integration of the training system to reduce the IP's non-instructional workload and thus enable him to focus his attention on the student's training needs is a major improvement in fighter training. The CAA has the first truly automated, integrated training system capable of producing a high quality, personalised training package. With its Center for Advanced Airmanship, GE has evolved from a manufacturer of state-of-the-art training components, into a state-of-the-art provider of fighter training services.

Advanced computer-generated visual systems
The Simulation and Control Systems Departments embraces three areas; advanced computer-generated visual systems for flight simulators, digital radar landmass simulation, and mission simulators, in particular for the Northrop F-5 light fighter and training services such as the newly established Center For Advanced Airmanship.

Compu-Scene visual simulation systems
Although more than 50 per cent of the international simulation business has been military, with an appreciable percentage taken by the offshore operators, there has been a swing towards the airlines. The impetus for this development has come from rapidly increasing fuel costs, which have led regulatory bodies (predominantly the FAA) to sanction the use of flight simulators for greater proportions of regular proficiency training. The goal is 100 per cent, and there is a demand for considerable refinements in visual simulation to assist this progression.

Many airlines have current-generation simulators on six-degrees-of-freedom motion bases that cost around $6 million. These systems are typically amortized over 15 years, and such investments cannot be discarded prematurely.

As a result, General Electric policy has been to anticipate sales of advanced visual systems to bring simulators up to the standards required by the new training schemes.

GE's visual system is called Compu-Scene and is a development of the class of equipment largely pioneered by the company itself. GE built the world's first computer-image generation system in 1958, and supplied it to the US Army and Navy for a joint programme called JANIP (Joint Army Navy Instrumentation Program). In 1962 it delivered to NASA the first computer-image generation system to train astronauts in space rendezvous and docking manoeuvres. This was followed up in 1972 and 1974 when the first such systems were delivered respectively to the US Navy and Air Force. In 1975 GE provided the first full day/night computer-image generation system for commercial flying training.

Compu-Scene is an all-electronic, full-colour, day/dusk/night, variable-visibility 'outside world' display with a broad choice of image display configurations. The system is completely modular and based on two types of equipment, an image generator and an image display. Integration with existing simulators is straightforward, the image displays being typically 'wrap-around' boxed units mounted in front of the simulator enclosure. Among many major capabilities are texture generation, circular feature generation, and curved-surface shading. The first produces texture patterns of selectable colour to provide enhanced motion and distance cues on bounded terrain areas and model face surfaces. Texture algorithms are modified to produce modulated patterns to represent sea, cloud, crops, or various types of terrain. Circular-feature generation software creates circles, ellipses, spheres and ellipsoids of selectable colours to represent clusters of circular objects such as trees, clouds, storage tanks, silos and similar cultural objects. Such features can be generated with curves instead of straight lines, resulting in great economy in computer storage space; a single curve can be generated that might require approximately 50 edges. The last-named capability permits the use of continuously variable shade or tone of colour across the face of an object to make it appear curved, even though the object is modelled with flat faces.

High resolution is achieved by a suitable choice of raster density between 625, 763, 875 and 1023 lines, and by the capacity of the image-storage system, which can provide up to 8000 edges and the same number of light points.

In 1982 GE announced a new family of modular computer-generated image visual systems called Compu-Scene III. It can produce full colour, day/dusk and night scenes at high resolution for one to three channels with 4000 edges. Its modular configuration permits both capacity and the number of channels to be increased as necessary. In addition to the high scene content for conventional training, the system can provide weapons effects, six-degrees-of-freedom moving models, reticles and alphanumerics for optical displays, and face-blending to improve dynamic scene changes. Available options are texturing to improve scene realism, the addition of essential cues, unattainable with edges alone, for the satisfactory simulation of low-level flight, and comprehensive correction distortion for use with the dome displays employed in combat simulation. As a result of the wide range of options available in the Compu-Scene III the requirements of individual users can be economically met with customised equipment.

In 1984 the company announced Compu-Scene IV, in which the flexibility afforded by real-time computer-image generation is combined with a quality of image virtually equivalent to photographic standard. The system was designed to provide effective visual simulation of the environment as seen by pilots flying terrain-following or nap-of-the-earth missions.

Compu-Scene IV depiction of section of Northrop F-5 fighters

At altitude the ground is seen through cloud cover, with sun-illumination effects. Weapon-delivery effects such as impact and fragmentation are also included. Compu-Scene IV has been chosen to train pilots of attack helicopters, being able to represent realistically the fast-moving environment as it is seen at low level in such aircraft, providing close-up ground detail as viewed from below tree-top level. The system is also appropriate to other situations calling for high degrees of operator skill, for example management of air-cushion vehicles and tanks.

Of customers announced to date, Sikorsky will integrate a Compu-Scene IV with a Singer-Link System for research into different helicopter configurations, (notably the proposed LHX light battlefield scout helicopter), McDonnell Douglas has ordered a fourth Compu-Scene IV to be used by its new subsidiary, Hughes Helicopters will use another (again for research into advanced helicopters such as LHX), and Lockheed-California will employ its system for research purposes to test new aircraft designs and modifications, including the ATF advanced tactical fighter planned by the US Air Force as its McDonnell Douglas F-15 successor.

In December 1985 GE announced its biggest-ever order for Compu-Scene visuals. The US Air Force ordered four Compu-Scene IIs as part of Lockheed C-130 simulators used by both the US Air Force and Marines. The order was valued at $18 million and deliveries start in late 1986, with all four operational by mid-1987.

A Compu-Scene IV is being delivered to McDonnell Douglas for helicopter work as part of the LHX programme.

STATUS: production of Compu-Scene IV under way, with deliveries beginning in 1986. Compu-Scene III in production and service. Other versions of earlier Compu-Scene visuals are in use in military and commercial aircraft simulators.

Mission simulator system for F-5

General Electric's mission simulator for the Northrop F-5 closely reproduces the capability and environment of the tactical fighter to provide training in visual and instrument meteorological conditions. The system is based on real-time digital simulation and includes a full-colour GE computer-image generation visual system.

The mission simulator system accurately simulates the F-5 instruments, navigation, communications, and fire-control systems, and provides aural cues, realistic stick-force feel, and outside-world scenes. The system can be supplied in either single or dual cockpit configurations. The basic single-cockpit system comprises representative pilot enclosure, instructor station, and three-window continuous-scene visual display, all controlled by a general-purpose digital computer. The advantage of the dual-cockpit configuration is that it permits simultaneous use of a single display by two pilots, in order to practise formation flying, one-versus-one combat, or even two-versus-one if a computer-controlled model is programmed into the scene.

The instructor is seated directly behind the student, permitting full communication and also allowing them to share the same display. The visual system depicts ground scenes and sky equivalent to an area of 340 × 340 nautical miles and extending up to 60 000 feet. The data base is three-dimensional and can include hills as well as fields, man-made objects and bodies of water. The density and distribution of data base detail is great enough to support speed, altitude, and distance estimation.

Operating costs of the mission simulator system, according to GE, are conservatively estimated at between 10 and 15 per cent of the cost involved in operating an F-5.

STATUS: in production.

Digital radar landmass simulation

The training of radar operators in conjunction with pilots is now recognised as a critical factor in ensuring the combat-readiness of aircrews. Advanced flight or mission simulators in conjunction with a digital radar landmass simulation (DRLMS) system can now represent every aspect of a mission, particularly the navigation procedures and workload in the approach to the target, weapon-delivery phase, and subsequent escape.

General Electric's DRLMS programme began in 1971 with an experimental radar prediction device, and the company now produces several families of radar-image simulation equipment for all three US armed services and overseas customers.

The DRLMS generates a real-time simulated radar image of the ground area around the aircraft from a digital database, with data retrieval under computer control. The data covers the viewing area within range of the radar system and is correlated with the visual system scenes. The computation by special-purpose hardware of the radar image point, sweep by sweep and scan by scan, is based on the geometry of the radar imaging situation. As with the actual aircraft system, the radar picture is created in real-time.

The inclusion of a special console in most GE DRLMS systems facilitates editing and updating digital databases. The device also allows operators to monitor the radar image while the simulator is running a mission, and to control the DRLMS for maintenance and test purposes. The data displays are 360° plan-position indicator-style representations of the information held in the system's memory about the region being displayed.

GE DRLMS equipment has been delivered to the US Navy for use with Grumman/A-6E Prowler simulators and to the US Air Force for Boeing B-52 simulators, and nine systems are being supplied to MBB in West Germany for Panavia Tornado simulators. Other systems have been built to train operators on naval ships, as radar prediction systems representing multiple aircraft radars, and for research into advanced radar simulation and database development. GE is also the contractor for the DRLMS to be used in conjunction with the simulator for the General Dynamics F-16 fighter. It will be the first system to represent advanced, on-board digital signal processing.

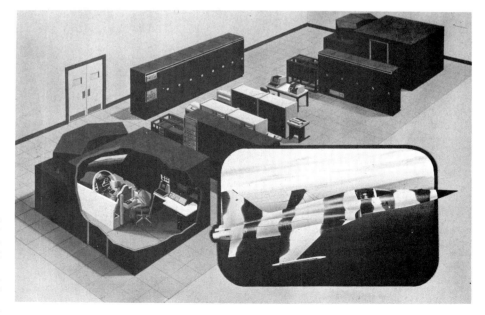

General Electric mission simulation system for Northrop F-5. Inset, F-5E fighter

The F-16 DRLMS and associated weapons-system trainer will provide realistic simulation in all the radar modes available with this aircraft, including Doppler beam sharpening. The General Dynamics F-16 DRLMS can also represent terrain-following and terrain-avoidance, as well as special effects such as chaff dispensing and jamming, weather, beacons, and radar faults.

STATUS: earlier DRLMS systems are in service, and General Dynamics F-16 system entered production in June 1983.

Mission simulator visual system for C-130

As the Viet-Nam war demonstrated, assault transport aircraft need the same skills in low-flying and caution in the battle area as interdiction and strike aircraft. Formation flying in large aircraft, often necessary for accurate para-drops, also calls for precision handling. For these reasons the US Air Force has established at its Little Rock, Arkansas Military Airlift Command training base two Lockheed C-130 mission simulators equipped with a General Electric visual system that can provide crew introduction and proficiency training 20 hours a day, 7 days a week.

Each visual system can generate up to 8000 edges and 4000-point light sources simultaneously, with eight levels of detail, and can provide up to seven simultaneous models, including formation aircraft and surface-to-air missiles. To accomplish this the image generator uses one of the most powerful special-purpose computers ever built. An innovative technique known as scene texturing provides the essential visual cues for manoeuvres such as low-altitude parachute extraction, assault landings and contour flying.

The displays are formed from high-resolution, full-colour cathode ray tube images and are presented in six units in front of the aircraft windows on the six-degrees-of-freedom moving platform. Each scene is an independent 'snapshot' determined by instantaneous computations of aircraft position and attitude. A unique feature is that two of the side-mounted displays (one each side) can be lowered to give the navigator the view he needs to identify drop references, or raised to show airfield references while banking during a turn. Instructors can call up weather effects such as fog, storm and cloud and can trigger hostile action such as anti-aircraft tracer and shell bursts and the flight of missiles. The instructor can also operate the system in a crash override mode, in which a training mission that would otherwise have been abandoned due to a fatal error on the student's part can be continued. A freeze mode can also be selected so that a particular situation can be discussed, and the previous 7.5 minutes of the mission can be replayed for the same purpose.

STATUS: in service.

Advanced visual technology system

In 1984 General Electric delivered a device called the advanced technology system to the US Air Force. It will provide 'outside world' imagery for the full spectrum of tactical aircraft missions, including take-off and landing, low-level flight, air-combat, weapon delivery, and air-refuelling. Image quality is significantly improved by the addition of colour, circular features, improved edge smoothing and two-dimensional texturing. A new feature is three-dimensional texturing, which provides high-density ground clutter to represent small-scale detail such as cacti, shrubs and rocks. Such texturing gives improved terrain relief in judging altitude during low-level high-speed flight.

STATUS: in service.

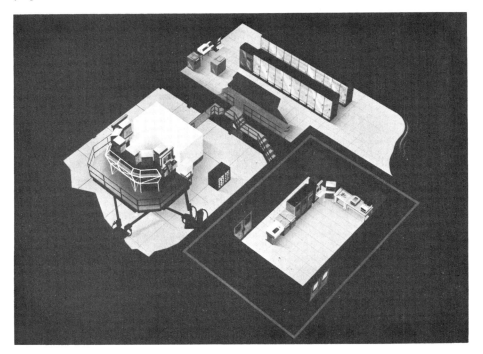

Lockheed C-130 mission simulator with General Electric visual system display, image generator and database facility

Visual system for helicopters

General Electric, under a $10 million contract awarded by the US Army (though administered by the US Air Force) in April 1983, has been developing a computer-based visual system that will provide better realism for training Hughes AM-64A Apache battlefield helicopter crews. The system uses a process called cell-texturing, in which the scenery is generated by photographs, suitably digitised and with details blurred or blended to remove the 'staircase' effect characteristic of computer-generated images.

The US Army requirement is to provide much more surface detail than is required with fixed-wing aircraft, and the problem in accomplishing this by conventional computer-generation techniques is that of providing the number of 'faces' needed to prevent the typically solid, cartoon-like presentation and achieve smooth texturing. The lack of adequate texturing is much more important in nap-of-the-earth helicopter operation than with fixed-wing missions. For example, a typical computer-generated tree can consume 50 to 60 faces, and so generating a realistic woodland is not possible if the computer can store only 2000 faces.

Video for the simulator is generated by a computer, as well as colour for each picture element and other details such as fog, haze or shading.

STATUS: competitive development programme; other contractor is Honeywell. US Army is conducting evaluation during 1985.

Daylight touchdown scene in General Electric C-130 mission simulator

Goodyear

Defense Systems Division, Goodyear Aerospace Corporation, 1210 Massillon Road, Akron, Ohio 44315-0001
TELEPHONE: (216) 796 4929
TELEX: 986439
TWX: 810 431 2080

Goodyear's association with flight simulation began about 1951, when the company was contracted by the US Navy to provide suitable simulation facilities for the series of 'blimps', or non-rigid airships, being built by Goodyear for maritime patrol.

Simulator for F-15

Goodyear is responsible for the flight simulators representing the US Air Force/McDonnell Douglas F-15 Eagle fighter, and has been building them since 1973, the year the F-15 was ordered into production. Since then the company has built 10 simulators for installation at Tactical Air Command bases. The F-15 was optimised as an air-superiority fighter, and the simulators reflect this single-mode intention. The first seven simulators were built with six-degrees-of-freedom motion system, and *g*-cueing (combination of *g*-seat and *g*-suit) was built into the seventh and subsequent simulators. The first ten simulators were driven by Harris 6024 computers, but Gould 32/8780 processors are employed from simulator number 11 onwards, and that processor will be retrofitted to the first 10 simulators, more than doubling their speed and capacity. Early systems were programmed in assembly language, but Fortran is being used from the 11th onwards.

Simulators 11, 12 and 13 have been ordered and will be delivered in the three years up to the end of 1988. The 13th is to be enhanced so as to match the performance of the MSIP multistage improvement programme aircraft, and will use an actual F-15 central processor.

STATUS: in production. The standard of simulation appropriate to the MSIP aircraft, together with enhancement in visual displays (see below) will probably form the basis of Goodyear's submission for a simulation system to represent the new F-15E dual-role (air-to-air/air-to-ground) fighter.

F-15E weapon system trainer development

Goodyear is currently under contract for the development and production of both hardware and software of the high resolution sensor simulator, which will include simulation of the synthetic aperture radar and infra-red and electro-optic sensors of the F-15E.

STATUS: in development.

UK's Rediffusion CT5A visual system is part of Goodyear's enhanced simulation; here a warship is taking 'hits' fore and aft

Computer-generated image simulation system

Goodyear Aerospace, Rediffusion Inc and Evans & Sutherland have jointly produced an image visual system with more effective cues to provide better training for tactical-combat pilots. It was evaluated during a three-month programme in 1984 by some 48 US Air Force Pilots currently flying McDonnell Douglas F-15, General Dynamics F-16 and Fairchild Republic A-10 aircraft. The system originated in a 1982 request from US Air Force Tactical Air Command for visual systems to accompany its simulators for these aircraft types. They were to be limited field (less than 360°), state-of-the-art types, and suitable for representing air combat, air-to-ground missions, and airfield procedures. Goodyear (which has been building F-15 simulators for the US Air Force) judged Rediffusion's CT5A visual system to be the most suitable, with its large database for instantaneous display and on-line storage, and advanced animation techniques. The database effectively contains a 300 by 300 nautical mile region of representative terrain, plus a 100 by 20 nautical mile corridor of 'real world' terrain representing beginning at a specific US Air Force base. The CT5A is an improved version of the Rediffusion CT5, and was developed by that company in conjunction with Evans & Sutherland. It uses four calligraphic projectors throwing images on to the specially smoothed interior of a 20-foot diameter dome.

STATUS: in assessment for possible future combat simulation systems.

Gould

Gould Defense Systems Group, Systems and Simulation Division, 5902 Breckenridge Parkway, Tampa, Florida 33610-4233
TELEPHONE: (813) 628 6100
TWX: 810 876 0809

Gould's Systems and Simulation Division, with facilities located in Tampa, Florida, and Melville, New York, has a wide array of proven experience in design, production, integration and support of aviation, electronic, marine and land-based simulation and training systems for military and commercial customers worldwide.

The Division has recently expanded its capabilities with the development of an innovative modular design approach to visual integration. Employing advanced Fresnel Technology, Gould/KFO's visual image display system is a unique integration of video technology and optical design. The rear projected image provides a high performance, low-cost display three times the brightness of conventional mirror/beam splitter systems.

Gould's aviation training experience encompasses full flight simulators, operational flight trainers, cockpit procedures trainers and part-task trainers for tactical, combat and transport aircraft including the McDonnell Douglas F-4 and F/A-18, Fairchild A-10, Grumman E-2C, Lockheed C-130 and C-5, and Beechcraft T-34 and T-44. Current programmes include the following.

F-18 hands-on-throttle-and-stick part-task trainers for the US Navy

In August 1985 the Division was awarded a contract by NTSC to upgrade two part-task trainers currently in operation at NAS Lemore, California, and NAS Cecil Field, Jacksonville, Florida. The systems delivered by Gould are designed to train radar intercept geometry and provide an introduction to the basic capabilities of the combined use of the Hotas, up-front control, master monitor display, multi-functional display and master armament panel. The contract initially awarded to Gould in August 1979 is now valued at over $9 million.

C-130H aircrew training system for Nigeria

Awarded in August 1985 the Division's Nigerian C-130H ATS programme effort includes all aspects of the training system, courseware, training devices, training system management, training support, programme evaluation and operational equipment. The existing Nigerian

Gould C-130 simulator

Cockpit of the Gould simulator for McDonnell Douglas F/A-18

C-130H simulator with six-degrees-of-freedom motion system, delivered by Gould, will be upgraded with the integration of a McDonnell Douglas Vital IV three-channel four-window dusk/night computer-generated image visual system. Contract value from initiation in December 1980 to delivery of the total training system in August 1986 is estimated at over $14 million.

T-34 cockpit procedures trainers

Initiated June 1978, Gould is under contract to provide the US Navy with a total of four T-34 cockpit procedures trainers (CPTs) and 25 T-34 instrument flight trainers (IFTs). The CPTs, located at NAS Milton, Florida, provide effective

basic training and testings of pilots in normal and emergency aircraft procedures. The T-34C IFTs employ digital simulation techniques to provide equivalent aircraft flight training for ground, take-off, climb, cruise, descent, landing, navigation and communications, as well as procedures training. The IFTs are located at NAS Pensicola, Whiting and Corpus Christi. The contract will be completed with the delivery of the final two IFTs in July 1986 at NAS Whiting. Total value is approximately $32.2 million.

C-5B aircrew training system

The contract for the C-5B aircrew training system for the US Air Force was awarded in October 1984. Teamed with United Airlines and CAE, Gould is responsible for the development of one new C-5B cockpit procedures trainer

and refurbishment of two existing systems previously delivered to the US Air Force by Gould.

Seagull computer system

Under a 1984 agreement, Gould and the UK company Rediffusion are jointly developing a functionally distributed computer system called Seagull. Gould is developing the equipment, Rediffusion the software. The first system will be supplied to Rediffusion, but Gould will make it available later to other simulator manufacturers. The system is based on Gould's Concept/32 host processor. Prototypes were to be supplied to Rediffusion in early 1985, and the first of the UK company's simulators to employ Seagull is scheduled for delivery in 1987.

STATUS: in development.

Honeywell

Honeywell Inc, Systems and Research Center, Honeywell Plaza, Minneapolis, Minnesota 55408
TELEPHONE: (612) 870 2920
TWX: 910 576 2692

CGSI computer-generated synthesised imagery

Computer-generated synthesised imagery (CGSI) is an innovative approach to the storage, processing and display of realistic graphic and photographic images that can be applied to all forms of simulation training and decision aids.

In the seven years that this technology has been under development at Honeywell, it has risen from a research idea to a state-of-the-art, real-time image generation system.

CGSI is a composite of the best of two earlier technologies, computer-generated imagery (CGI) with its full freedom of view point, object motion and ability to generate representations of objects which do not yet exist; and computer synthesised imagery (CSI) which digitises information from photographs for realistic detail.

The flexibility of mathematical calculation of view point is combined with the realism of actual photographs in one process. Stored in digital form and displayed on a video screen, the photographic quality images can be derived from any region of the electromagnetic spec-

trum: radar, infra-red, visual, or X-ray. Digital storage permits rapid computer processing, and the high fidelity of the original image provides a realism unique to this technology. Scenes can be composed, objects inserted, and view points changed in real-time so that the display correctly simulates the view by a trainee or a sensor.

The main tasks of CGSI can be subdivided into information storage, processing, and display. Two types of information are stored in two related sub-systems: the object library and the gaming area database. The object database consists of stored photos. The gaming area database contains maps of different terrains. These maps are grids on which the locations of various objects are plotted. Each coordinate in the grid is assigned an elevation, so that objects moving within the image can be made to follow the surface of the terrain.

Object sizing, positioning, intensity modulation, and the calculation of moving objects are accomplished by the CGSI processing system. Mathematical models which determine how fast and in which directions objects (vehicles) in the scene can move, and operator control inputs, which choose courses within these limitations, compute the locations and the viewing direction of the sensor system in each moving vehicle.

Control data for vehicle simulation are distributed to the individual processing channel controllers, which transform the gaming area coordinates to screen coordinates, change intensities of the images, and command the

object library for retrieval of proper image data. The library supplies the proper image to the processing channels. Each processing channel changes a stored image to scene conditions by changing image position, size, rotation, and perspective.

The scene construction module inserts objects (targets) into the background. The objects occlude the more distant background. Translucent special effects are added after the scene assembly. Transmissivity masks are stored in the system's object library and are processed like objects in their own special processing channel, which also controls their transmissivity.

Honeywell's warp algorithm allows the system to display on-screen a different object view than the one stored in its library. Size and position parameters for each pixel are determined from the input image corner coordinates, and the output corner coordinates are determined by incremental interpolation.

The resulting scene is displayed on video screen to the operator, who makes his decisions on the basis of the visual cues. When there is a sensor and computer in the loop instead of a human operator, the output of the computer is fed directly to the appropriate vehicle controls.

Honeywell is currently working on a number of application programs which capitalise on the CGSI system's abilities: CGSI/real-time feasibility demonstration jointly funded by the Navy (NTEC), Army (PMTRADE), and Air Force (HRL); the real-time image scenario processing system, funded by the Army (WSMR) and visual system component development programme, Army (PMTRADE); and the infra-red digital injection system funded by the Air Force (Eglin Air Force Base).

All these system applications use the same major hardware and software components. Varying the number of channel processors, the structure of the gaming area, and the contents of the library, a wide range of real-time simulation, sensor stimulation, and training exercises can be conducted in any region of the spectrum.

STATUS: while expanded technical development continues, CGSI is currently in use in engineering simulators and under proposal for various trainer applications.

Typical Honeywell CGSI presentation

Hughes Aircraft
Hughes Aircraft Company, Support Systems, 1501 Hughes Way, PO Box 9399, Building A1/4A301, Long Beach, California 90810
TELEPHONE: (213) 513 3000
TWX: 910 346 6332

Combat simulator for F/A-18
Under contract to the US Navy, Hughes is producing four combat simulators to train Navy and Marine Corps pilots for the McDonnell F/A-18 Hornet. Each simulator comprises two 40-foot (12.2-metre) diameter spheres, on the inside surfaces of which are projected high-resolution targets set against a background of earth and sky. A computer system provides the visual, aural and other perceptions necessary to provide realism in this type of training; a notable refinement is the provision of 'greyout' and 'blackout' cues at high g.

The simulator can be operated in two modes. In the first, one or two pilots, each in their own domes, can fly against 'targets', flown by a computer program called Adaptive Manoeuvring Logic. This can be selected to drive the targets in either offensive or defensive manoeuvres. In the second mode, two pilots can engage in combat against one another or against instructor-flown targets.

The earth-sky background is produced on the inner surface of the sphere by fixed light valve projectors, while the target imagery is produced by four projectors mounted near centre.

A large ground area, over 600 × 600 nautical miles enables a variety of terrain and cultural features to be modelled. At altitude, a perspective grid of lines on the earth gives ground reference to the pilot. At lower altitudes more detail is produced by the computer, including power poles and towers to give depth perception. Ground objects appear in three-dimensional perspective. These include fuel and industrial complexes.

The computer image generation system produces the high fidelity surface scenes required for the full range of air-to-surface weapons tasks.

Air-to-air targets have high resolution and are produced with details such as afterburner

Hughes Aircraft combat simulator for US Navy/McDonnell Douglas F/A-18 with manoeuvring targets

visual cues and control surface positions. During an air-to-air attack, missile smoke trails, explosions and gunfire are realistically portrayed.

The simulator is also used for night carrier landing training, when the aircraft can be landed on board, slewed to the catapult and launched. The pilot can then complete a visual pattern before landing back on the carrier.

For normal instrument flying when operating from land bases, an instrument approach can be flown down to cloudbreak. An airfield will be added at a future date.

The central control room has an instructor's station, with displays which mirror the cockpit indications. Additionally, he can introduce a range of malfunctions of increasing severity which are realistically reproduced in the cockpit, with caution lights, audio alerts and voice alert. There is a freeze and playback facility so that the trainee can see the effects of his actions.

The instructor's position has a stick and throttle to enable him to manoeuvre the target aircraft projected onto the dome. His display at the console can be selected to give an ACMR presentation.

Multiple targets can be presented on his screen and can be viewed from any angle with the chosen viewpoint selected by rotating a rolling ball inset into the console. The target can be an instructor-flown adversary target (IFT) or the instructor can select one of five levels of difficulty with the computer flying the target.

The computer program driving the target aircraft, known as adaptive manoeuvring logic (AML), has total knowledge of the combat situation and selects the optimum manoeuvre relative to the trainee's actions. Level 1 is a non-manoeuvring target, with the levels increasing in difficulty until level 5, the most difficult, being virtually unbeatable.

Targets programmed into the El Toro simulator are the Northrop F-5, US Navy/Grumman F-14 Tomcat, McDonnell Douglas F-15 and F/A-18, General Dynamics F-16 and MiG-21 and -23. Each has the appropriate flight dynamics for that particular adversary aircraft.

The two domes can operate in an independent mode, where each of the two pilots can be separately trained in air combat missions against a target flown by the instructor or by the AML. Another option is to train two pilot-trainees simultaneously and they can engage in air combat against each other, or both can engage the IFT or AML target.

Each cockpit is an exact replica of a McDonnell Douglas F/A-18 cockpit. Canopy, windscreen and leading edge extensions are reproduced to give the same exterior field of view as that of the actual F/A-18. A head-up display (HUD) is fitted, together with the full complement of aircraft controls, displays and instruments.

In addition to giving pilots experience in combat tactics and aircraft management, the system also familiarises them with the variety of modes available with the Hughes Aircraft AN/APG-65 radar on the F/A-18, and the

display of its functions and commands on the head-up display.

Especially important is the fidelity of the radar and weapons control system simulation in order to support appropriate HUD weapons solutions for simulated guns, missiles, bombs and rockets, as well as to ensure exact correlation with target imagery.

The comprehensive flight dynamics simulation models the engines, flight controls, weight and balance together with aerodynamic data. Force feel cues, buffet and wing rock are provided.

Numerous motion cues are incorporated. An aural system generates the sounds of engines, gunfire and missile launches, as well as the headset-produced aural warning tones.

The pilot normally flies the simulator wearing his torso harness and g-suit. Inflatable bladders in the seat simulate g with realism further enhanced by variations in shoulder harness tension in response to g-forces. Simultaneously a pneumatic system varies the pressure in the pilot's g-suit with greyout cues simulating high-g conditions. A buffet system simulates aircraft stalls, gunfire and speedbrake extension.

STATUS: in production. Simulators have been delivered to the US Naval Air Stations at Lemoore and Cecil Field and to the US Marine Corps at El Toro. The fourth simulator is destined for Lemoore NAS.

Intersim

Intersim Inc, 2700 North Hemlock Circle, Broken Arrow, Oklahoma 74012
TELEPHONE: (918) 258 8585
TELEX: 497457

Flight simulators

Intersim, which changed its name from GMI in early-1986, was established in April 1978 by George Moody, a former Rediffusion Simulation engineer, and from a nucleus of flight simulation engineers and designers now has 200 employees in manufacturing and engineering facilities in Tulsa, Oklahoma, and in Sussex in the UK.

Intersim's simulation activities now encompass all aspects of aircraft crew training, from full flight simulators, cockpit procedures trainers and cockpit systems simulators down to trainers for cabin procedures and evacuation, emergency evacuation, ground maintenance, classroom instruction and self-teaching, and simulator modification and engineering.

The company uses Computer Automation, DEC PDP 11/34, PDP 11/55 and VAX 11-780, and SEL 32 (with structured Fortran) computers in its flight simulators. Intersim's preferred system is the SEL 32 with structured Fortran. Intersim's first programme was to complete the majority of the software and systems engineering for a Boeing 737 full-flight simulator for American Airlines Training Corporation at Fort Worth, Texas. This involved the major conversion of a Singer-Link system with a new flight-deck and visual system, but using a Singer three-axis motion system and GP-4 computer. Intersim currently has in production simulators for the US Air Force (Boeing RC-135), Finnair (Aérospatiale/Aeritalia ATR42, to FAA Phase 2 standard), Flight International (Boeing 727-100/200), Airborne Express (Douglas DC-9-30), and Embraer (Embraer EMB-120 Brasilia). It has already built simulators for Continental (DC-9-30 Phase 2), USAir

Intersim full flight simulator for Scandinavian Airlines System's Saab SF-340

(Boeing 737-200) and SAS (Saab 340, claimed to be the first simulator for this feederliner). The company has installed and integrated in its simulators most of the currently available computer-generated image visual systems.

These simulators incorporate the Intersim-designed Smart (simulation management and reporting terminal) system, which permits instructor control of the equipment using a discrete control panel and cathode ray tube,

Intersim's digital sound system, digital interface module, and fully hydrostatic control loading and motion systems.

Late in 1985 Intersim received an order to co-produce five Embraer Tucano simulators in conjunction with ABC Sistemas of Brazil. The simulators will use Gould series 32 computers and deliveries will be completed in 1987.

Loral

Loral Corporation, 600 Third Avenue, New York 10016
TELEPHONE: (212) 697 1105
TELEX: 644018

AGES air-to-ground engagement system

Following a laser-based weapons-effects system called Miles (multiple integrated laser engagement system) to represent the fall of shot of main battle-tank guns, designed to train weapons operators, Loral has developed equipment using similar technology for tactical and support helicopters called AGES (air-to-ground engagement system).

AGES comprises two types of equipment: a laser detection system to record laser radiation directed at the aircraft, simulating the firing of anti-aircraft guns or missiles, and an airborne laser designator representing the flight of anti-armour missiles from the helicopter to the target. The system was designed for the Bell AH-1 Cobra attack helicopter and the Bell UH-1 Huey and OH-58 Kiowa transport and observation helicopters. Utility or transport helicopters such as the latter two would have just the receiving system, the purpose of which is to teach the pilot or crew in the clearest way

possible how to traverse heavily defended areas with the least risk. Attack helicopters would in addition have the laser transmitter, representing the flight path of the missile, to give the gunner realistic practice in handling his weapons while the pilot is manoeuvring the aircraft behind trees or other natural cover in an effort to avoid return fire.

The receiving equipment comprises a number of laser detectors mounted on the airframe to afford full 360° azimuth coverage. With each signal received, a built-in probability criterion is applied to assess whether a 'near miss', 'hit', or 'kill' has been achieved. When a laser signal has been received, an audio tone is inserted into the aircraft's intercom system, a lamp in the cockpit warns the crew, and a strobe light mounted on a skid flashes to alert observers on the ground. If the 'hit' is sufficiently substantial, a smoke canister (also mounted on a skid) is activated as well to indicate the event to ground observers and gun crews.

The attack system represents the effect of the Hughes TOW anti-tank missile, 20 mm cannon, and 2.75-inch rocket. The laser is eye-safe and so there are no restrictions on its use during exercises.

STATUS: following evaluation by US Army at Fort Bliss, Texas, beginning in September 1981, system is in production. It was initially deployed in 1984.

Bell UH-1 helicopter with Loral laser detectors attached to fabric belt passing around nose. Strobe light and smoke canister are mounted on left skid at rear

LTV

LTV Aerospace and Defense Company, Vought Missiles and Advanced Programs Division, PO Box 225907, Dallas, Texas 75265
TELEPHONE: (214) 266 9014

Photography/computer-generated image visual system

In an attempt to provide a new level of realism, LTV Aerospace has developed, under a US Navy contract, a system that combines new photographic techniques with computer image generation. High fidelity moving scenes are then projected on to six contiguous screens that surround the simulator cockpit.

The key to realism is a database of aerial photographic images generated at precise positions throughout a geographic region of interest. Each image, in video disk storage,

provides 360° viewing for the observer. A typical database gaming area for tactical missions comprises some 6000 such images. When processed by computer, this database provides unlimited, six-degrees-of-freedom movement throughout the gaming area.

The system in use with the US Navy uses 16 on-line video disks to store the data for a typical mission. Images are retrieved from the disk storage as the observer's location changes within the gaming area. When retrieved, each image is digitised, stored in a buffer memory and then, under computer control, transformed or manipulated to provide the appropriate perspective at a refresh rate of 60 Hz. As the observer moves from the region covered by one image into that of another, the system retrieves that image and processes it in the same way. The process of successive image retrieval, processing and presentation is continuous throughout the gaming area.

Developments with the system are directed towards the simulation of low-level, high-speed missions. Under US Air Force contract, the system is being extended to represent the environment as it appears at 200 feet altitude and 600 knots. During 1984 simulations in non-real time were conducted at 80 feet and 600 knots. They were based on algorithms to provide three-dimensional natural and cultural terrain imagery, using Defense Mapping Agency digital terrain elevation data and digital feature analysis. Incorporation of three-dimensional features in concert with high-speed capabilities is expected by 1987, in an operational system.

STATUS: one system is operational with the US Navy, but requirements for a second have been deleted.

McDonnell Douglas Electronics

McDonnell Douglas Electronics Company, Box 426, 2600 North Third Street, St Charles, Missouri 63301
TELEPHONE: (314) 925 4000
TELEX: 447369

Vital visual simulation systems

The first visual simulation system to be marketed by McDonnell Douglas was Vital, which appeared in 1969 and was a fore-runner in low-cost, calligraphic, computer-generated image visual simulation.

Vital II and III visual simulation systems

In 1972 the Vital II computer-generated image night-only visual system became the first such display to be approved by the FAA. Vital III, introduced in 1975, offered significant improvements in the representations of runway surfaces, together with up to 600 light-points.

Captain's-eye view of runway as seen by McDonnell Douglas Vital IV simulation system representing Douglas DC-10 of French airline UTA

La Guardia Airport as represented by McDonnell Douglas Vital IV computer-generated image system

Vital IV visual simulation system

Vital IV, introduced in 1977, has up to 6700 light-points, 250 multi-coloured surfaces, a wider range of low-light conditions, realistic landing-light lobes and moving air and ground hazards. It was granted FAA approval to Phase 2 standards in 1982. It builds on the successful base of computer-generated image technology to provide finer and sharper images with improved detail with the option of panoramic views across all flight-deck windows, and presents images to the flight crew in realistic perspective and virtually without distortion.

The Vital IV system comprises three standard equipment modules packaged as required to match any simulator flight-deck:

Image generator: single cabinet containing a general-purpose 64 K computer that shares the job of equation-solving with special processing circuitry to generate the real-time scenes. Programs and extra scenes are stored on-line in a high-speed disk unit.

Display unit: combination of electronic cathode ray tube and reflective virtual-image optics mounted on the simulator flight-deck.

Instructor's control panel: a small facility that may be incorporated into the simulator's instructor system, giving him control over such characteristics of the visual scene as runway light visibility and weather.

Types of scene: airports, surrounding districts, aircraft carriers, general terrain
Scene conditions: IFR, VFR, twilight, night, dull day, above, in, or below cloud
Database coverage: normally 170 × 170 n miles or, with combined database techniques, unlimited
Optical resolution: <3 minutes
Lightpoints: 400 000 + carried on-line, with 6700 displayed at any one time. Used for airport lighting, roads, moving traffic, beacons, and other illuminated representations
Surfaces: 15 000 + carried on-line, with 250 displayed at any one time. Used to represent terrain, buildings, moving vehicles, runways, and other hard surfaces
Scene access: on-line access to 50 separately stored scenes, each loaded and brought into view in a few seconds. Alternatively, large contiguous areas represented by automatically accessed sub-scenes
Weather: variable visibility, cloud, fog, haze, and thunderstorms
Colours: 10 + separate colours, ranging from red through green

McDonnell Douglas air combat simulator for two-seat F-15

Dynamics: smooth, flicker-free bright image with motion in direct response to simulated aircraft manoeuvres, without smearing, comet-tailing, or break-up. 30 new pictures presented each second. All practicable manoeuvres and speeds
Occultation: complex buildings, mountains, and vehicles programmed to have a solid appearance, occulting lights and solid objects behind them from the pilot's point of view

STATUS: in production for commercial and military customers.

Vital V visual simulation system

Introduced during 1982, Vital V extends the range of colours to include cyan to red, using a high-resolution beam penetration display. It uses parallel distributive processing and incorporates individual surface texturing for a realistic representation of blowing sand and snow, cloud-tops and water. The image generation equipment is a high-speed general-purpose computer with memory options between 64 and 256 K. It is pre-wired to accommodate plug-in processing elements during later expansion of hardware.

Vital VI visual simulation system

This is a full-colour daylight system, designed to comply with FAA Phase 3 requirements and launched in 1983. It features a shadow-mask cathode ray tube display, at the same time incorporating all the characteristics of Vital IV and V.

Multiview

Multiview is the latest addition to the Vital family, allowing the same, continuous view out of the cockpit for all crew members, regardless of their position in the cockpit. A wraparound pneumatically formed mirror surface presents this picture and four colour projectors produce the computer-generated imagery on the screen, over a 40° vertical field of view, with either 65°, 180° or 210° horizontally.

Dimensions: 25 × 13.5 × 18.5 ft (7.7 × 4 × 5.6 m)
Weight: 2200 lb (1000 kg)
Writing speed: 1 in/μs (2.5 cm/μs)
Lightpoint resolution: 3 arc min

STATUS: in development.

Combat simulator for F-15

McDonnell Douglas has been expanding its air-combat facility at St Louis with the addition of a 40-foot diameter projection dome housing a representative cockpit for the McDonnell Douglas F-15E Strike Eagle version of its top-selling combat aircraft, chosen in early 1984 by the US Air Force as its new multi-mission fighter. The new manned air combat simulator being used to test these capabilities is the fifth domed simulator to be installed at the plant since the facility was established in 1969. It went 'operational' with US Air Force pilots in March 1983. The principal aim of the trials is to assess crew workload in the new role, using an improved Hughes APG-63 radar, terrain-following system, forward-looking infra-red sensors, and navigation equipment.

Air and ground combat scenes can be projected on to the inner surface of the dome while one or two pilots 'fly' a choice of missions using the cockpit's fully integrated flight controls and displays. The detailed ground scenes are provided by a 105-foot long terrain board simulating an area of more than 200 square miles. A camera slaved to aircraft manoeuvres and scaled for speed and altitude projects the ground views. For air-to-air scenes, television

McDonnell Douglas Multiview simulator visual system

images of small-scale aircraft models are projected.

Heart of the new simulator is a Control Data Cyber 760 computer which simulates the F-15's performance and handling characteristics and drives the cockpit instruments and displays.

The simulation facility, now with three 40-foot and two 20-foot domes, was originally set up to assist in the early stages of the F-15's design and development. It has since grown to include simulations of the US Navy/McDonnell Douglas F/A-18 fighter and AV-8B V/Stol close-support aircraft, and for more general development and investigation of advanced fighter tactics.

Simulator for US Marine Corps AV-8B

McDonnell Douglas has supplied the flight dynamics software, and Evans and Sutherland two CT-5 visual systems for the US Marines' AV-8B Harrier 2 training facility at Cherry Point Air Base.

The operational flight simulator uses computer-generated imagery depicting actual ships and permits the practice of V/Stol operations in all weathers.

The weapons tactics trainer permits exercises to be practised using the aircraft's electronic warfare equipment, and a severe EW environment can be created to act against the pilot. This latter simulator was accepted in July 1985.

Pacer

Pacer Systems Inc, 87 Second Avenue, Northwest Industrial Park, Burlington, Massachusetts 01803
TELEPHONE: (617) 272 5995
TWX: 710 332 6400

Pacer engages in diverse activities including the development of air data systems, training and simulation, computer-based ship-stabilising system, programme management support, and systems engineering and integration.

CPIFT cockpit procedures and instrument flight trainer

The company has developed a device called CPIFT (cockpit procedures and instrument flight trainer) in which the capabilities of a conventional cpt are expanded so that it can be 'flown' and navigated. The microprocessor-based system combines all the functions of a cockpit procedures trainer with most of those of a full simulator at a cost claimed to be lower than that of any other system with this capability.

Simulators are built to customers' specifi-

cations around six sub-systems: cockpit or flight-deck simulation and enclosure, power plant, aircraft systems, radio/navigation, flight dynamics, and instructor station. Moving-base capability and visual displays are only provided by special order. Costs are between 25 and 50 per cent of those appropriate to the aircraft they represent.

Instructor's station for Pacer's de Havilland Canada DHC-6 Stol transport simulator

Pacer CPIFT simulator system for Short 330 30-seat commuter airliner

---

Rediflite

Rediflite Inc, 2201 Arlington Downs Road, Arlington, Texas 76011
TELEPHONE: (817) 469 8411
TELEX: 758308

Triad flight simulator for helicopters

The drawback with flight simulators for general aviation aircraft and helicopters is their very high cost in proportion to the aircraft they represent. For example, a full flight simulator for a typical wide-body airliner costs around one-third the value of the transport, but a comparable device for an up-market business twin or helicopter such as a Bell 206 might equate to five times the value. This is because the nature and quantity of electronic guidance, control and systems monitoring equipment required to operate an aircraft is largely independent of its size, as is the computing power needed to duplicate the characteristics of the aircraft and drive the simulator. Thus the general aviation community is barred by cost from realising the advantages that simulators are increasingly bringing to the airlines.

A possible solution to this problem is the Triad helicopter flight simulator being tested and evaluated in the USA. Triad is a collabor-

ative initiative by Rediffusion Inc of Arlington, Texas (a subsidiary of the British firm Rediffusion) and Offshore Logistics of Lafayette, Louisiana, which in early 1982 launched a joint company, Rediflite Inc, to pursue the venture. The motiviation behind this activity comes from the Louisiana company's subsidiary, Air Logistics Inc, which operates a sizeable fleet of helicopters supporting oil-drilling activities in the Gulf of Mexico.

The Triad system employs three elements: an actual aircraft, a Gould SEL 32 host computer, and a visual and sound system. The aircraft forms effectively a fixed-base simulator, signals from its flight control instrument and other systems being brought out to the computer via suitable connections. The visual system surrounds the cockpit in the same way as it would a conventional simulator. The intention would be that one example in an operator's fleet of aircraft would be appropriately modified during manufacture and would be withdrawn from day-to-day flying as required for conversion or proficiency checks. The system would be suitable for a mixed fleet, the computer being programmed to represent the handling and other characteristics of each type.

A Bell 206 helicopter has been modified to represent the simulator, and by March 1983 had been 'flown' by some 270 pilots. FAA certi-

fication of the device as a fixed-base simulator appears likely, though a Rediflite scheme to increase the realism by superimposing typical vibration spectra on the airframe was turned down as possibly affecting the subsequent airworthiness of the aircraft. Some simulator time will also be credited towards airborne training, though no details have been worked out.

Pilot comments from the evaluation emphasised several aspects of helicopter simulation, reflecting largely the manoeuvrability of rotary-wing aircraft and their very large angles of movement relative to the fore and aft axis. They showed up a need for: good visual cues within the region of peripheral vision (this region was found to be significantly more important than originally expected, calling for visual systems with a much wider field of view); more vertical objects to serve as visual references during hovering, approach and landing than for simulators representing fixed-wing aircraft; and good visual and sound cues, which are more meaningful for helicopter pilots than for those flying conventional aircraft.

Modifications to make the helicopter compatible with the computer system account for about 10 lb (4.5 kg) extra weight in wiring and connectors and the circuit panel. The cost of conversion and software development is estimated at $500 000 to $700 000.

Early in 1984 it was announced that Sikorsky had bought a Rediflite system to support the research and development activities associated with its ARTI programme. The system chosen, Triad 3, incorporates a Rediffusion Novoview SP3T computer-generated image visual system, providing a daylight VFR environment (Triad 1 is a basic training system with no visual facility, while Triad 2 provides a computer-generated image night/dusk environment). ARTI is the technology demonstration programme based on a UH-60A helicopter, that will provide the avionics technology foundation for the US Army's proposed LHX light observation helicopter. This aircraft, a replacement for the Bell OH-58 Kiowa, will represent a major procurement with the production of 5000 aircraft due to begin towards the end of the decade.

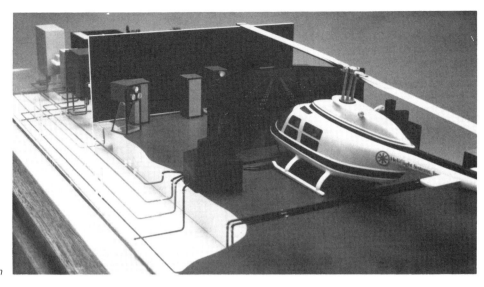

Demonstration model of Rediflite's Triad system

Reflectone

Reflectone Inc, 5125 Tampa West Boulevard, Tampa, Florida 33614
TELEPHONE: (813) 885 7481
TELEX: 5267101
TWX: 810 876 0840

Operational flight simulator for 747-300

Developed for Swissair, this simulator uses a six-degrees-of-freedom hydrostatic motion system with extended 2.5-metre stroke to represent the new Boeing 747-300. The Reflectone simulator (claimed to be the first of its kind) was shipped to Zurich in November 1983. It is capable of crew-training to FAA Phase 3 standards, which permit air carriers to conduct all pilot training on suitable simulators.

A particular feature of the simulator is the inclusion of two Color Graphic 7 cathode ray tubes, permitting the pilot instructor and engineer instructor to operate in parallel and providing better utilisation and more realism. The simulator employs four SEL 32/77 computers, three of which drive the main simulator enclosure (using 1.3 M words of 32-bit memory) and the fourth to control the Hitachi Denshi day/night computer-generated image visual system.

Operational flight simulator for A310

This Reflectone system has a six-degrees-of-freedom hydrostatic motion base with 2.5 metres of movement, and is designed to represent Swissair's Airbus A310 airliner with a two-man flight crew. The system was commissioned in 1984.

The simulator will be capable of FAA Phase 3 operation, so that all pilot training can be done on the ground. It uses four SEL 32/77 computers with 5.25 M bytes of mos memory and another of the same type to drive the Reflectone-developed digital control loading system. An accompanying day/night visual system by Hitachi-Denshi uses one of the SEL 32/77 computers, using 1.25 M bytes of its memory.

Operational flight trainer for A-10

Built for the US Air Force to represent the Fairchild A-10, the system permits proficiency development in all phases of instrument flight from pre-flight and start-up through navigation and combat, visual and GCA approach, to landing and post-flight de-briefing. The weapon system can also simulate the electronic warfare environment and, while the simulator is mounted on a fixed base, a g-seat with inflatable air bladders provides cues represent-

ing the onset of acceleration and deceleration. A microprocessor-driven digital control loading system simulates the control forces.

Night/dusk capability is enhanced by a McDonnell Douglas Vital V visual display mounted on the windscreen. The trainer uses three SEL 32/35 computers each with 128 K words of core memory and three 40 M byte discs. One computer is largely dedicated to electronic warfare simulation and has an associated instructor's station to provide separate ew training. Extensive weapons simulation includes trajectories and guidance of 'smart' bombs. Scoring algorithms for the 30 mm cannon on this aircraft, and its bombs and rockets, are also included.

STATUS: production run of 14 systems was completed in 1981, and devices are deployed world-wide.

Operational flight trainer for CH-46 helicopter

Developed for the US Navy and Marine Corps, these systems (one for Boeing Vertol CH-46D and two for CH-46E) represent the tandem-rotor helicopter and are intended to build up and maintain pilot proficiency in all aspects of operation. The system is mounted on a six-degrees-of-freedom motion base and is controlled by a Harris Slash-4 computer with 96 K words of 24-bit memory and more than 40 M bytes of disk storage. The full-daylight computer-generated image system on the USMC simulators is provided by Rediffusion Simu-

lation in collaboration with Evans and Sutherland Computer Corporation. The visual display, added in late 1980, permits a number of new training missions to be flown: landing and take-offs in confined areas, operations aboard LST-class ships, IFR and VFR in poor weather in daylight, dusk, or darkness, flight with a slung load, and formation flying.

The system employs a remote trainer control panel at the centre console of the flight-deck, permitting the instructor to fly as a pilot or co-pilot and so exercise limited control of the training session. In addition it allows the student to conduct self-training, without instructor assistance, and enables playback of the most recent five minutes of flight.

The third system was completed in 1982.

Weapons system trainer for SH-3H helicopter

Developed for the US Navy, this device represents the most widely used twin-engined helicopter with emergency amphibious capability, and is intended to develop and maintain pilot proficiency in all aspects of flight, including the detection, classification, tracking and attacking of hostile submarines.

The system is mounted on a six-degrees-of-motion base and controlled by two Harris Slash-5 computers, each with 64 K words of memory and 40 M bytes of disk storage. A McDonnell Douglas Vital night/dusk visual system, with seven windows and five channels, augments realism for IFR and VFR missions

Reflectone operational flight trainer for Fairchild A-10

from aircraft carriers, frigates, land bases and confined areas in all weather conditions.

A separate station is provided for the sensor operator and his sonar and MAD equipment. It can be operated in conjunction with, or independently of, the trainer. An acoustic generator was being incorporated during 1984 to provide realistic target acoustic signatures and active target echoes for the sonar data computer.

STATUS: two trainers have been built (last in January 1983) for US Navy in California and Florida.

Instrument flight simulator for S-76 helicopter

This six-degrees-of-freedom motion simulator was designed for the American Airlines Training Corporation to duplicate the characteristics of the Sikorsky S-76 helicopter. It develops crew proficiency in the operation of controls, interpretation of instruments, management of navigation and communication systems, and in coping with emergency situations through instructor-induced faults.

The simulator comprises a fully representative cockpit, motion and hydraulic system, and instructor's station mounted on the motion base behind the cockpit, and a Harris Corporation Slash-6 computer with 48 K words of memory. The system characteristics are based on flight test data obtained by American

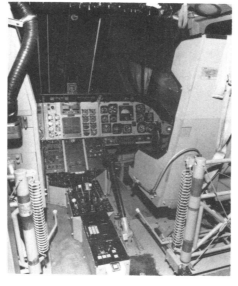

Reflectone instrument flight-deck of Sikorsky S-76 helicopter

Airlines specifically for this application, and include duplication of the vibration spectrum generated by the main rotor and transmission. A Rediffusion SP1 visual system has also been incorporated to permit VFR training.

Enclosure of Reflectone S-76 simulator

Operational flight trainer for T-46A

In late 1984 Reflectone was chosen to design and build the operational flight trainer for the US Air Force's new Fairchild T-46A primary trainer. An initial contract valued at $42 million calls for the manufacture of six T-46A facilities, each with four representative two-seat cockpits.

Singer-Link

Link Flight Simulation Division, The Singer Company, Corporate Drive, Binghamton, New York 13902
TELEPHONE: (607) 772 3011
TELEX: 932423
TWX: 510 252 0195

This entry should be read in conjunction with that of the British company Singer Link-Miles. Projects below denoted as 'Link-Miles' are primarily the responsibility of the British company.

Advanced simulator technology (AST)

The '100 per cent simulation' goal, around which AST technology was developed, was defined in the USA during 1980 by FAR 121-14C requirements, which specify simulator characteristics appropriate to three phases of training capability concerned with crew proficiency, conversion, and upgrading.

A new instructor's station within the simulated flight compartment can seat the two instructors serving the two pilots and flight engineer, and an observer. Twin, interchangeable cathode ray tube systems provide alphanumeric or graphics information, and the control panels are identical. Control panels and the new interface equipment operate independently of the displays, connected only by software. The programmable nature of the displays permits changes in the type or arrangement of information without electrical or mechanical changes.

The most significant development in the AST motion system is the new actuator assembly, using much smoother, hydrostatic bearings. A new ultrasonic linear displacement transducer eliminates all mechanical connection between actuator and position sensor, resulting in a very clean feedback signal. More than two-thirds of the plumbing of previous designs has been eliminated and platform, joint and electronic assemblies have been simplified, reducing the likelihood of oil leaks, simplifying maintenance and improving accessibility and reliability. The interconnection between the simulator's flight

compartment and the computer complex has been greatly simplified by the substitution of a digital bus, using serialised data transmission. This permits simpler cable runs, less documentation, and easier trouble-shooting.

AST simulators are packaged into an octagonal enclosure that houses both simulator flight-deck and electronics cabinets, reducing by as much as 80 per cent the space previously needed for these systems. The proximity of simulator flight-deck and electronics reduces the lengths of interconnecting cables, so cutting down the risk of electrical interference.

Total training systems

In recent years there has been a growing trend to place complete responsibility for the entire training process — mission and task analysis,

training objectives definition, media development, training device design and production, device maintenance and overall administration — under the responsibility of a single management or contractor team. This total training system (TTS) approach is being endorsed by many US Air Force, Army, Navy, Marine Corps and commercial airline training experts as a very practical option for aircrew and mission training.

Details on some of the military and commercial flight TTS programmes in which Singer-Link is a key team member follow.

Hercules Flight Training Center

This joint venture operated by Lockheed-Georgia and Singer-Link started business in March 1985. It provides complete training for

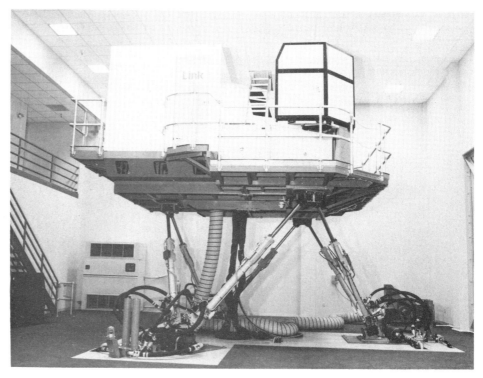

C-130/L-100 simulator at Lockheed/Singer-Link Hercules Flight Training Center

pilots, co-pilots, flight engineers, and mainten-ance technicians for the Lockheed C-130/L-100 Hercules cargo/troop transport. The C-130 and its L-100 commercial derivative are in use in 57 countries worldwide.

The Hercules Training Center includes class-rooms with modern audio-visual and electronic teaching aids, briefing rooms enabling students to review and measure their skills against optimum performance in critical situations, instructors from the nearby Lockheed manufac-turing plant, and state-of-the-art Singer-Link simulation.

The simulator is mounted on a six-degrees-of-freedom motion system and features an IMAGE II night/dusk computer-generated visual system. It is visually certified by the FAA. The simulator can be configured for the C-130H model, the C-130-30 stretched version, or the L-100-30.

By 1987 it is estimated that the Center could be training over 200 crews per year. The facility was built with a second simulator bay and there is scope for future expansion.

KC-10 Extender training

Turnkey aircrew training is the US Air Force approach at Barksdale Air Force Base, Louisiana, for pilots, co-pilots, flight engineers, and boom operators of the McDonnell Douglas KC-10A Extender tanker/transport. The pro-gramme — one of the first military total contractor-managed operations — en-compasses initial, refresher, upgrade, and instructor training.

The programme, managed by American Airlines Training Corporation, covers topics such as flight instruments, fuel system, hy-draulics, communications, cargo loading, iner-tial navigation, mission planning, and receiver/ tanker air refuelling. The instruction is self-paced, progressing from basic texts to training in a Singer-Link six-degrees-of-freedom motion simulator with IMAGE III full colour daylight/ dusk/night visual system.

The latest KC-10A simulator, the pro-gramme's third, features the new AWARD (aviation wide-angle reflected display) system with IMAGE III, developed by Singer Link-Miles. The simulators are operated 16 hours a day, seven days per week.

Aircrew training for the KC-10A at Barksdale has demonstrated a 99.4 per cent success rate — defined as the ability of the trainee to pass a SACR 60-4 standard evaluation check on the first attempt — which virtually matches the US Air Force's 99.5 per cent mission completion success rate for 45 000 flight hours completed

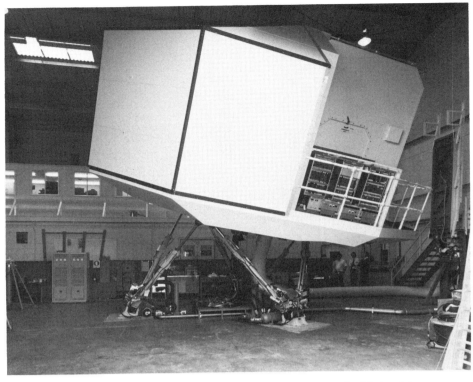

AWARD wide-angle visual display system on Singer-Link's KC-10A simulator

up to spring 1985. More than 1800 crew members have 'graduated' from the training programme.

Undergraduate Naval Flight Officer programme

The US Navy's initial TTS venture is the UNFO programme at Pensacola Naval Air Station, Florida, which aims to teach radar intercept and ground mapping to Naval flight officers. The Navy owns no flight training equipment, but simply purchases the services of a team of training contractors.

The prime contractor, Cessna Aircraft, fur-nishes the training aircraft, designated T-47A, a derivative of the Citation S/II. Modifications include simulated combat features such as shorter wingspan, a flat plate antenna in the larger nose, and two Emerson-supplied on-board student radar scopes. Northrop World-wide Aircraft Services is responsible for provid-ing pilots and plane maintenance.

Singer-Link produced the six part-task trainers (PTTs), which account for nearly 75 per cent of the instruction time. Four of the PTTs are devoted to air-to-air intercept radar training (AIRT), the other two to ground-mapping radar training (GMRT).

The AIRT simulation equipment generates radar imagery with as many as six independent airborne targets during simulated air-to-air engagement. In addition, cloud and storm activity can be added by the instructor; the resulting backscatter effect of such weather conditions on the radar is realistically portrayed.

The GMRT uses a digital radar landmass simulator (DRLMS) to provide real-world dis-plays on the radar indicator. The DRLMS developed by Singer-Link's Advanced Products Operation in Sunnyvale, California, for UNFO has significant features not available on pre-vious radar simulation programmes, and yet is half the size. The 200-foot resolution includes shadows, range attenuation, earth curvature, aspect angle, far-shore brightening, low-level effects, antenna patterns, and pulse width effects.

After completing the 160-hour programme at Pensacola, the student pilot officers will be ready for radar operation on fighters such as the Grumman F-14 and McDonnell Douglas F/A-18.

UC-12B training

US Navy and Marine Corps pilots are training on another modified corporate aircraft, the UC-12B, the commercial version of the Beech-craft King Air 200.

The US Navy/Marines have contracted Simu-Flite Training International, which is a division of The Singer Company, to provide the UC-12B flight training programme utilising SimuFlite's state-of-the-art personal learning system, the FasTrak system. FasTrak includes a large computer database, digital touch-sensitive video terminals, and videodisc technology, plus classroom instruction by highly skilled instruc-tor pilots.

Training analysis and media selection tech-niques optimise the mix of computer-based training and instructor-led classes in the devel-opment of the ground school training for 13 aircraft types: Learjet 24/25, 35/36 (the first business jet Phase II simulator), and 55, Cessna Citation II and III, Dassault Falcon 10/100, 20, and 50, Gulfstream II and III (with EFIS), British

Part-task radar trainer supplied by Singer-Link to US Navy's UNFO programme

Aerospace 125, IAI Westwind 2, and the King Air 200/C-12/UC-12B (the first Phase II turboprop simulator). All of the simulators are equipped with IMAGE III daylight/dusk/night visual systems.

Ground school training courses range from recurrent training to abnormal and emergency procedures training to transition training. The transition training results in a pilot receiving a zero flight hour type rating entirely by virtue of the ground school and simulator training. Current operations include training for maintenance technicians, flight engineers, chief pilots, and aviation managers.

SimuFlite, the first to apply computer-based training to the business jet training community, uses flight deck management (FDM) as an integral part of all courses. FDM uses a cockpit recording system; the monitored results of the recording taken during each mission are used in the debriefing sessions.

The UC-12B programme, over a three-year period, will train some 400 US Navy, Marine Corps, and Naval Reserve pilots. SimuFlite, which opened its doors in 1984, expects to add other military training curricula in the future, as well as keeping up to date with new corporate jet technology.

People Express Training Center

In mid-1985, People Express Airlines announced an agreement with Singer-Link for a newly integrated training facility at Newark International Airport. The operation includes classrooms, briefing rooms, interactive audiovisual carrels, part-task trainers, and a quartet of simulators.

The simulators were relocated from previous training sites. All four simulators (two Boeing-727's, a 737, and a 747) are fitted with IMAGE II dusk/night visual systems. The B-747, mounted on a six-degrees-of-freedom motion system, is Phase II certified; the other simulators are visually certified.

The seven-year contract calls for training of nearly 4000 People Express pilots/co-pilots (called flight managers) and customer service personnel. Singer-Link manages the entire training programme, including organising, scheduling, and monitoring.

Singer-Link is also providing training to other operators of the three Boeing commercial aircraft. Customers include Flying Tigers, Viewtop Corporation, Tower Air, and the Federal Aviation Administration.

MST micro simulation technology

Singer-Link in 1985 introduced a new generation of simulators employing microprocessor technology to increase the flexibility, reliability and cost-effectiveness of any training system.

The new design, called Micro Simulation Technology (MST), is suited to the full range of simulator applications.

MST utilises vlsi circuit technology in a functionally and physically distributed modular system. This architecture provides each major function of the simulator: motion system, visual system, sensor systems and control loading, with its own dedicated computational power.

Each major element of an MST simulator also has its own dedicated microprocessor for independent and comprehensive diagnostics. This ensures rapid fault identification and rectification to the board and component levels. This diagnostic facility, coupled with the inherent reliability of vlsi technology, simplifies maintenance, increases simulator availability, and enhances cost-effectiveness.

The modular design of the MST concept means the simulation systems can be upgraded as training requirements expand. Substantial amounts of additional computing power can be added in small packages at modest cost. In addition, future advances in microprocessor technology can be incorporated into an existing MST-based system.

The use of vlsi technology also significantly reduces the component count, power consumption, and spare parts inventory required in simulator operations.

The MST concept is suitable to military and commercial flight simulators, general aviation, ground and marine simulators, and computer-generated visual systems.

The first simulator employing the new MST concept is a combined flight and maintenance simulator for Thai Airways Airbus A300-600 aircraft. The simulator, now in production at the Singer Link-Miles facility, will meet FAA Phase III standards.

The second MST commercial simulator is for the Scandanavian Airlines System's McDonnell Douglas MD-82 aircraft.

Microflite flight simulator

A new family of flight simulators, the Microflite series of cost-effective fundamental aviation trainers, was introduced in 1985. Microflite is a microprocessor-based, cost-effective modular design which can be configured to simulate a wide range of aircraft types including civil and military trainers, mid-to-intermediate range transport and utility aircraft, fighter planes and helicopters. Microflite technology is extremely adaptable and easy to operate and maintain.

The incorporation of microprocessor technology has made it possible to configure Microflite simulators with modular elements for all computations, instrument drives and control effects.

Microflite computational systems offer several specific advantages:

1. vlsi circuits provide complex functions with a minimum of external connections and system components to give higher reliability and greatly reduced requirements for space and power

2. monoboard computers increase computational speeds, thus reducing the impediments to real-time response

3. modular expansibility permits the addition of substantial amounts of computing power at modest cost in very small packages

4. distributed processing permits the use of dedicated systems for specific simulator functions such as flight, engines, and other aircraft systems, avoiding the need for the large, monolithic central computer systems used in traditional simulator design.

In addition, optional system enhancements are available. These include a full daylight/dusk/night visual system which provides all the cues required for a fundamental visual training curriculum as well as motion systems in a range of complexity to meet the specific needs of each user.

IMAGE III-T visual system (Link-Miles)

Link-Miles has introduced the industry's first daylight microprocessor-based visual system with texture: IMAGE III-T. This new visual system with texture provides important height, speed, and positional cues such as required for helicopter and ground vehicle training exercises. The new hardware and associated software can produce a wide range of different texture patterns including concrete, grass, forests, clouds, water, and desert. Additional patterns may be introduced to accommodate specific training requirements.

The IMAGE product line is the first visual system package to employ microprocessor instead of traditional mainframe or minicomputer systems architecture. The result is a system that is easier to operate, maintain, and update, and therefore offers a lower overall life-cycle cost. Furthermore, because of its modular architecture, IMAGE can be easily expanded to meet new training needs or modified to incorporate future microelectronic advances.

Using this latest digital image generation technology, true-to-life scenes are presented to the crew through the simulator cockpit windows, providing the necessary visual cues needed for various aspects of aircraft flight

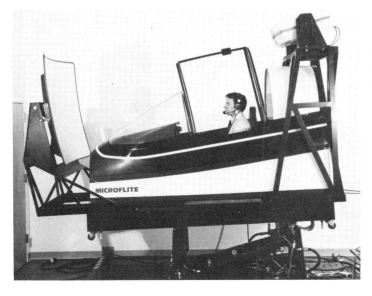

Singer-Link Microflite flight simulator

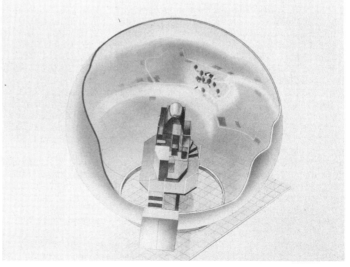

Artist's conception of Singer-Link's Esprit visual display system for tactical helicopter and fighter aircraft training

training such as takeoff, landing, in-flight refuelling, systems performance such as navigation, and weapons firing.

Esprit visual display system

Singer-Link has successfully completed the testing of the Esprit proof-of-concept test-bed, including human-in-the-loop evaluations.

The Esprit (eye-slaved projected raster inset) test-bed demonstrates the feasibility of an eye-slaved, area-of-interest (AOI) display where the high-resolution AOI is always centred around the trainee's eye line-of-sight, moving as the eye moves. The background is filled in with reduced-resolution images, requiring fewer channels and thereby reducing cost and complexity. The combination provides an effective, high-resolution picture over a very large field of view at an affordable cost.

A full-scale Esprit prototype display, including a 24-foot (7.3 metres) diameter dome, is scheduled to be completed by mid-1986. High-quality visual simulation required in tactical combat training will be provided by the dome display, offering a 3-arcminute/OLP (optical line pass) resolution over a 270° horizontal × 130° vertical field of view. The high-resolution picture will be projected onto a high-gain, high-brightness spherical screen. Other features previously demonstrated on the proof-of-concept test-bed will be incorporated. These include: smooth electronic blending of AOI/background images, true-perspective (parallax) simulation, and special effects such as collapsing field of view when pulling high 'gs' (g-dimming). Distortion correction for dome projection will also be provided.

When combined with a state-of-the-art digital image generation system, the Esprit eye-slaved display promises to provide the type of sophisticated, realistic simulation necessary for tactical fighters and attack helicopters. Initial transition training, refinement of tactical skills, and development of more effective air-to-air and air-to-ground tactics, including operations at the extremes of the flight envelope, will all be possible and cost-effective with this technology.

MOD DIG visual system

MOD DIG is a new day/dusk/night full colour digital image generator which simulates out-the-window, visionics, and sensor imagery and will operate in concert with a full mission simulator.

MOD DIG is designed to support effectively nap-of-the-earth flight training for helicopters, terrain following flight training for fighter and bomber aircraft, space flight training, and armoured vehicle training.

A wide variety of enemy threats is possible, ranging from small arms ground fire to missiles to fixed wing and rotary wing aircraft. Special visual effects are tracers, missile launching, trajectory and weapon impact effects, flares with ground illumination, and slewable light lobes.

Sensors supported include forward-looking infra-red, night-vision goggles, low-light level television, telescopic sight unit and laser range finder (LRF).

Singer-Link has incorporated a host of design innovations, yielding the most realistic, high fidelity simulation picture ever: techniques such as 2 × 2 spatial filtering (anti-aliasing), 128 increments of translucency, multiple types of texture including blending factor texture and concurrent texture, smooth colour shading, geographic photomapping, atmospheric effects, and library objects.

Even more significant is the fact that these various MOD DIG image features can be simulated *simultaneously* without 'strange' effects or overloading. It is claimed that no other system on the market has been able to achieve this interaction quality.

Extensive diagnostic software, including built-in fault isolation to the single-card level

and a maintenance test stand for chip-level detection, means MOD DIG will be highly reliable and easily repairable.

MOD DIG, as the name implies, will be able to grow over its lifetime via modular additions if the requirements of a particular application grow. New features can be added as they are developed by Singer-Link (without hardware change) permitting expansion of processing capacity by the parallel addition of identical modules.

MOD DIG is compatible with various image display systems, including Link's Esprit foveal/peripheral full field of view dome projection system.

Weather radar simulation

Singer-Link's weather radar display system simulates the weather radar transmitter/receiver of an actual aircraft, producing real-time weather pattern data displays in colour and monochrome.

Key features include unlimited geographic coverage of the gaming area, altitude coverage from sea level to 60 000 feet, capacity for 10 instructor-selectable scenarios with up to 10 storms per scenario, up to 20 radar cells per storm, and up to 16 horizontal slices (nested contours) per cell (also, the capability of simulating light, moderate, or severe turbulence levels to further define each slice).

Displays include simulated range attenuation, precipitation attenuation, ground clutter, and beam smearing due to finite beam widths. Storm position information is available to the simulator's visual system for weather correlation.

The weather radar display system, designed for FAA Phase II or III simulators, is capable of being upgraded with aircraft hardware which may be developed in the future for windshear detection and avoidance.

TEM thunderstorm environmental model

Singer-Link has developed and refined a thunderstorm environmental model (TEM) which yields high-fidelity simulation of real-world weather, including windshear and microburst effects commonly associated with aircraft crashes.

The four-dimensional (the three physical dimensions plus time) modelling, correlated with colour radar representations, will enable pilots to train in turbulent weather conditions in the complete safety of a flight simulator. An instructor can select mild or severe thunderstorm conditions and, by positioning the storm with respect to the runway in time and space, present dynamic ever-changing thunderstorm conditions.

Simulated meterological parameters include:
1. three wind components (longitudinal, lateral, vertical)
2. turbulence intensities for each wind component
3. horizontal and vertical turbulence scale length for horizontal and vertical components
4. temperature and pressure changes resulting from cold downdrafts and their outflow
5. precipitation rate as a function of location and time
6. RVR inputs to the visual system as a function of precipitation rate

The model is a terminal area thunderstorm simulation suitable for FAA Phase II, III, and LOFT applications. A storm size of 20 by 20 nautical miles by 3200 feet is provided. The model contains approximately 1 M byte of data and can, in most cases, be contained in a satellite computer/microprocessor.

The TEM can be integrated into new and existing simulators with or without Singer-Link's weather radar display system.

DRLMS digital radar landmass simulation

In the late 1960s and early 1970s, digital technology reached a state where digital radar landmass simulation became a reality. Techniques developed by Singer-Link made it possible to digitise transparencies and provide the first digital databases for use in radar simulation. For the first time these databases became easily modifiable and changes in gaming areas became a simple task of changing disc packs.

These techniques were directly transferable to systems developed later in the 1970s which used digital databases developed by the Defense Mapping Agency (DMA).

Today, the digital radar landmass simulator (DRLMS) has evolved to the state where the system, now one-quarter the size of the first DRLMS, has better resolution, a more realistic image, a more detailed database, and more capable fault isolation.

Radar simulation is an integral part of any weapon system trainer. In addition to training crew members for navigation and weapons delivery, it is also a valuable tool for learning aerial refuelling techniques, target identification, low-level terrain-following and terrain-avoidance, and weather detection and avoidance. Simulated missions allow complete crew co-ordination training. The radar presentations viewed during the simulation are exactly as they would be on an actual mission. A digital radar landmass simulator allows training in both air-to-air and air-to-ground roles.

Current DRLMS characteristics include:
1. resolution down to less than 10 feet
2. solid state district memory data storage
3. DMA digital databases
4. weather simulation
5. special features such as terrain-following/terrain-avoidance, station-keeping equipment, IFF, beacon simulation
6. ability to simulate multiple radars, such as the EF-111 DRLMS where three separate radars are simulated simultaneously
7. ability to provide multiple independent radar simulation stations using one DRLMS
8. ground texture to add detail to the DMA data for enhanced realism
9. automatic fault isolation to one replaceable unit
10. simulation of synthetic aperture radar/Doppler beam sharpening systems
11. microprocessor target generator for target skinpaint, beacon and jamming simulation

The SAR/DBS DRLMS developed by Singer-Link has the ability to provide both real-beam and synthetic aperture radar simulation as well as terrain-following simulation. All the major effects of SAR/DBS processing are simulated. These include geometric distortions (keystone effect, elevation layover, motion compensation, moving objects) and resolution improvement/degradation due to weather, antenna pointing angle, multiple look processing, motion compensation, Doppler filter sidelobes, rf overload, surface wind, and pulse compression.

F-111/FB-111 simulator

To achieve and maintain concurrency with modifications to the aircraft, major simulator improvements include DRLMS (digital radar landmass system), new computational systems, rehosting the computer assembly language, built-in diagnostics, and modern instructor/operator stations. Thirteen devices in seven different configurations are being upgraded for the US Air Force Tactical Air Command, Strategic Air Command, and US Air Force - Europe training locations. The programme, which will run until 1990, also provides for two software support centres to manage and modify software, effectively extending the useful life of the simulators (built in the 1960s and 1970s) into the 21st century.

Cockpit of Singer-Link's F-16 tactical flight simulator

Refuelling training in B-52 weapon systems trainer built for the US Air Force by Singer-Link

F-16 Falcon simulator

Eleven new trainer flight simulators plus modification upgrades are covered by new contracts. Each simulator consists of a cockpit, instructor station, digital computer, and interface equipment. Full simulation of the single-seat tactical aircraft's highly complex on-board avionics system is provided, plus g-seat/g-suit motion cues.

The new simulators will be delivered to the US Air Force, Belgium, and the Netherlands.

To date, 25 F-16 simulators are in use by nine countries, another 11 are in production, and there are contract options on 14 more — a total of 50, which would be the second largest military simulator production (trailing only the F-4's 71 units, also by Singer-Link).

F-4 Phantom simulator

Singer-Link will modify several of the US Air Force F-4 simulators to reflect changes in the aircraft. F-4G modifications include replacement of navigation and weapons release computer systems simulation with stimulation of actual aircraft components, display system replacement, replacement of the existing radar landmass analog antenna system with a digital antenna system (DAS), incorporating simulation of the LRU-1 operational flight programme (OFP) and digital automatic acquisition, and update the existing OFP for the AN/APR-38 simulation.

On the RF-4C, F-4C, F-4D, F-4E and F-4G, simulation of the existing centreline fuel tank will be replaced with simulation of the high performance centreline tank. On the RF-4C and F-4E, a computer-based diagnostic test of the AN/APR-46 radar warning system will be incorporated.

The F-4 Flight Simulator Update Program (FSUP) will replace existing computation systems and peripherals with new state-of-the-art equipment and software; incorporate the new aircraft modification airborne video tape recorder in the F-4E and F-4G; add existing F-4G modifications to the F-4E; improve existing performance and operational capabilities; provide new interfaces for unmodified systems; update technical and software documentation; and provide a software support centre to maintain configuration control.

B-52 strategic bomber simulator

Prototype modifications for the US Air Force B-52 weapon systems trainers include an upgraded offensive avionics suite, enhanced radar system, simulation of a new aircraft monitoring and control system, and a new centre of gravity/fuel level advisory system. These simulator modifications parallel changes in the technology of the actual aircraft.

The B-52 WST, designed and built by Singer-Link, is the most sophisticated fully integrated strategic mission simulator in use in the world today. It allows the six-man crew to carry out anticipated combat missions under realistic conditions, including hostile environments.

UH-60A Black Hawk simulator

The US Army's utility helicopter, the Black Hawk, is used for transporting troops, equipment and supplies, and medical relief. The 18 simulators in production at Link include four-window ATACDIG (Army tactical digital image generator) visual systems (with texture), six-degrees-of-freedom motion systems, seat shakers, and realistic sound cues. Aircrews can practise basic flight, transition, advanced flight such as nap-of-the-earth low-level tactics, pinnacle and slope landing, and slingload training.

CH-47D Chinook simulator

Two new D model CH-47 Chinook Helicopter simulators are under construction for the US Army, while work progresses on the conversion of four existing C models to the D configuration so as to make all simulators in the field consistent and representative of the current Chinook helicopter.

The conversion emphasis is on the visual system, which is to be converted from a closed-circuit television image generation system coupled with a three-dimensional terrain model. The new system is ATACDIG, which will be provided for both the converted and newly-produced Chinook simulators.

Additional elements in the C-to-D conversion include a full range of new avionics, changes to the engine simulation, different rotor characteristics, and Doppler navigation.

AH-1S Cobra simulator

Fort Lewis, Washington, became the sixth US Army installation to begin AH-1S helicopter training in a simulator when it dedicated a Singer-Link Cobra flight weapons simulator (FWS) in October 1985.

Other AH-1S Cobra simulators are also in service at Fort Hood, Texas; Fort Campbell, Kentucky; Hanau, West Germany; and Illesheim, West Germany. Fort Rucker, Alabama, utilises the prototype AH-1Q simulator, which was modified and updated to the AH-1S configuration.

The AH-1S Cobra FWS is the world's first simulator to feature a full colour laser image generation (LIG) visual system. The LIG visual system provides total nap-of-the-earth flight and tactical training capabilities across a terrain model board representing a 140 km^2 gaming area.

The laser system, which grew out of an Army R&D award to Link, uses a three-rotational degree of freedom optical probe mounted on a gantry which provides horizontal and vertical cues relative to the model board and simulated altitude. The full colour light produced in the laser beam is reflected from the model board to banks of photomultipliers. The result is a very high fidelity visual scene.

LIG images are accurate even during the type of rapid scene changes which are typical of low-altitude flight, and are particularly important for training in air-to-ground weapons delivery, nap-of-the-earth, and unmasking manoeuvres. The visual system provides highly realistic targets, threats, and weapons firing effects, including display of tracers and ordnance impacts.

Each Cobra simulator has a pair of cockpits, one for the pilot and one for the gunner. Each cockpit also has its own instructor station. Pilots and gunners may train individually, or as a team.

The pilot can simulate firing all weapons except the TOW missile. The gunner, who also serves as navigator, can fire all weapons, including the wire-guided TOW, 20/30 mm cannons, and 2.75-inch FFARs – all through a telescopic sight unit (TSU). The TSU symbology provides the gunner with a visual scene, including targets, weapons effects, and the horizon — all correlated with the visual system. There are 26 'scorable' targets on the 24 × 64 foot model board.

The Cobra gunship simulators are mounted on six-degrees-of-freedom motion systems. The Cobra FWS also utilises Singer-Link advanced technology to simulate systems such as Doppler navigation, fire control computer, laser range-finder, and head-up display.

In production are three additional AH-1 flight weapons simulators. These devices will feature Singer-Link's ATACDIG computer-generated visual system with realistic moving targets, three-window scenes for both pilot and gunner, and improved TSU.

Cobra simulators 7, 8, and 9 will be installed at Fort Rucker and at National Guard sites in Pennsylvania and Arizona, with delivery dates scheduled for 1987 and 1988.

AH-64 Apache simulator

The prototype and six production units are currently being built for the Apache CMS (combat mission simulator), the most sophisticated helicopter simulator ever designed. It requires 29 computers to simulate realistically the complex flight and weapons systems of the US Army's most advanced gunship.

The simulator integrates tactics, terrain, weather effects, aircraft sensors, crew communications, enemy threats, and weapons delivery into an absolutely safe, cost-effective, positive training scenario. The two cockpits – one for the pilot, one for the co-pilot/gunner – are mounted on six-degrees-of-freedom motion systems and use Singer-Link's ATACDIG visual system with texture. Each cockpit is capable of independent or fully integrated operation.

AH-1T Sea Cobra simulator

US Marine Corps Air Stations at Camp Pendleton, California, and New River, North Carolina, are the sites for the first-ever Sea Cobra AH-1T weapon system trainers (WST).

The TOW/Sidewinder missile-carrying attack helicopter simulators will feature dual cockpits for independent or integrated mission training. Both the pilot station and gunner station, with onboard instructor stations, are mounted on 60-inch-stroke, six-degrees-of-freedom motion platforms.

Flight modes simulated include hover, transition, cruise, and autorotation. Atmospheric effects are also present: pressure variation, non-standard temperatures, wind velocity and direction, and icing conditions.

A unique feature is simulation of the tactical systems, such as the helmet-mounted sight system, the radar warning receiver display system, the telescopic sight unit controls and eyepiece, and the Navy armament control and delivery systems panel. The use of actual aircraft control heads and displays connected to modular simulator software facilitates updating the simulator to match future aircraft changes.

AH-1T tactical systems simulation encompasses all weapons and stores that the Sea Cobra can use, including the TOW air-to-ground and AIM-9 Sidewinder air-to-air missiles.

Instructor features include autofly, store/reset, freeze, hardcopy for record keeping and syllabus development, and record/playback — enabling a trainee to observe a just-completed sequence, evaluate it with his instructor, and then repeat the mission correctly.

The AH-1T simulators are also designed for future upgrade to the AH-1W Supercobra configuration.

F-14 Tomcat simulator

The US Navy FY 84 Update calls for 30 modifications, including safety, computer rehost, cockpit configuration changes, maintenance, documentation, and instructor station changes. The major upgrade is the computer rehost, replacing three Gould 3275 central processing units with two Gould 9750 pro-

Co-pilot/gunner station of the dual-cockpit AH-64 Apache combat mission simulator by Singer-Link

cessors. The new concept 9750s will greatly enhance speed and capacity, allowing flexibility for future modifications and additional simulation features.

P-3C Orion simulator

This flight improvement programme to six operational flight trainers (OFTs), originally built for the US Navy by Singer-Link more than a decade ago based on limited flight data about the then new anti-submarine warfare aircraft, will incorporate design modifications reflecting recent test data. Improvements include revision of the basic software model, a new digital control loading system, an automatic test guide, and revision of the simulated nosewheel hardware/software. The result will be a more realistic simulated airframe performance.

The P-3C OFTs will be linked to tactics trainers, produced by Singer-Link's facility in Silver Springs, Maryland, plus visual systems enabling independent or integrated training.

SH-60B LAMPS Mk III Sea Hawk simulator

LAMPS (light airborne multi-purpose system) simulators numbers 3 and 4 will be configured as weapon systems trainers (WSTs). The units will be mounted on six-degrees-of-freedom motion systems and will feature six-window dusk/night visual systems. Delivery will be to the US Navy's North Island, California, and Mayport, Florida, Naval Air Stations in 1986.

RAF Tornado F Mk 2 simulator (Link-Miles)

The Royal Air Force has commissioned production of four Link-Miles' flight simulators for the Panavia Tornado. These are in addition to six Tornado simulators previously delivered.

One of the most advanced simulators ever built for the Royal Air Force, the Tornado units will enable aircrews to train under realistic combat conditions, including a wide range of malfunctions and emergencies. Every aspect of the aircraft's mission capability will be simulated — radar, navigation, communications, weapons selection and delivery.

UAE Hawk simulator (Link-Miles)

The United Arab Emirates have ordered a microprocessor-based light jet trainer for the

British Aerospace Hawk. The Link-Miles simulator will have a three-window IMAGE III-T full colour daylight/dusk/night visual system with texture and will be mounted on a six-degrees-of-freedom motion system. It will also feature two instructor stations – one on-board, the other off-board – from which the instructor can control training scenarios. Delivery is scheduled for 1987.

Link-Miles' simulation of the Hawk, initially developed in 1981, was the first to replace mainframe computers with a fully distributed, multi-processing computing system using microprocessors — significantly lowering the life-cycle costs of flight, submarine, and ground vehicle simulation.

Airbus A310-300 simulator (Link-Miles)

Airbus Industrie selected Link-Miles to build a flight simulator for its A310-300 model. The simulator will be rapidly convertible to an A300-600 model, incorporating several engine options and two flight management systems.

The simulator will be built to FAA Phase II standard, with eventual upgrade to Phase III. Installation will be at the new Airbus Training Center in the USA.

Thai Airways International A300-600 simulator (Link-Miles)

The first Link-Miles MST (microprocessor simulation technology) concept simulator will replicate the Airbus A300-600 aircraft for Thai Airways International. The simulator will feature a six-window IMAGE III-T full colour daylight/dusk/night visual system with texture plus a six-degrees-of-freedom motion system.

The combination flight and maintenance simulator is due to be installed in 1987 at Thai Airways' Bangkok training facility.

American Airlines MD-80 simulator (Link-Miles)

This award is the third of three Link-Miles IMAGE III full colour daylight visual systems for MD-80 flight simulators ordered by American Airlines.

The simulators, installed in 1985 following a 1984 contract award, provide training for airline pilots who are making a transition from other types of aircraft, as well as recurrent pilot proficiency training, first officer upgrade to captain's level, and maintenance training for ground crews and engineers.

USAir cockpit procedures trainers
USAir has contracted with Singer-Link for two cockpit procedures trainers (CPTs) — one for the three crew member B-727-200 aircraft, the other for the two crew member BAC-111 airliner. The integration package for both CPTs will feature Singer-Link's advanced digital microprocessor-based approach, which is far more flexible than the analogue computer system traditionally used for CPTs. Cockpit procedures trainers provide aircrew members with basic familiarisation with the configured aircraft's instrument displays, including operable lights and switches, as well as several emergency situations.

SAS MD-82 simulator (Link-Miles)
Scandinavian Airline System (SAS) selected Link-Miles to provide a flight simulator for their MD-82 aircraft. The simulator will be built to FAA Phase II standards, including a six-degrees-of-freedom motion system and state-of-the-art instructor station.

The SAS MD-82 simulator will be the second simulator to incorporate fully Link-Miles' new microprocessor simulation technology (MST). It will also feature the new Link-Miles training management system (TMS), an ergonomically designed system incorporating high resolution graphic displays on touch sensitive cathode ray tube screens.

SMS Shuttle mission simulator
The SMS is actually two simulators: the motion base simulator sits on hydraulic lifts and rotates on three axes giving the crew a sensation of space flight, and the fixed base simulator is used for exercises from launch to landing. Both contain visual and sound effects that accurately depict all phases of the entire flight from launch to landing.

The SMS is the most complex simulator ever built. The Shuttle cockpit contains 2000 separate displays and controls, the SMS has the exact same configuration; the Shuttle's four main computers hold 16 million bytes of program and data, the SMS uses more than 413 million bytes of program and data; counting the four computers and backup computer, the Shuttle is capable of approximately 2 million operations per second, the SMS can perform approximately 4 million operations per second.

The SMS, originally built to integrate with Mission Control Center (MCC), is also being used to connect training sessions and test data links between the Johnson Space Center mission control, Marshall Space Flight Center, West Germany's Spacelab Mission Control Center at Oberpfaffenhoffen, and Sunnyvale, California.

The SMS has been regularly upgraded to remain concurrent with the Shuttle programme's expanding technology. Major enhancements are expected in the near future, both because the present computers are crammed to capacity and to accommodate the requirements of the planned Space Station programme.

Visulink laser image generator
Singer-Link's laser image generator system provides high resolution and scene detail at all altitudes, but is particularly suitable for simulating low-altitude flight. It offers marked improvement in performance and lower maintenance costs while retaining many of the components of the high-resolution systems based on television cameras. Thus, says Singer-Link, it is a natural step in the evolution of advanced visual display technology.

The system employs a multi-coloured laser beam to scan a high-detail model board within the pilot's field of view. The reflected light from the board is detected by a bank of photomultipliers and processed to generate full-colour signals to the simulator display, which may be a crt or projector. The principal differences from the television-based system lie in the replacement of the TV camera by a bank of light-sensitive photomultipliers and the substitution of the lights illuminating the model by a scanning laser beam. Benefits claimed for the system are: no degradation of resolution due to image lag; simpler and more stable colour registration and alignment; improved signal/noise ratio; improved resolution due to more efficient use of projection aperture; elimination of power demands to drive the lighting system; better simulation of night and dusk as well as daylight; greater depth of field; and wider field of view.

The wide field of view and picture quality associated with the system are particularly appropriate to nap-of-earth attack helicopter missions and V/Stol operations.

STATUS: in production.

Smiths Industries Inc
Smiths Industries Aerospace and Defense Systems Inc, PO Box 5389, Clearwater, Florida 33518
TELEPHONE: (813) 531 7781
TELEX: 6815304
TWX: 810 866 4106

Programmable weather radar simulation system
Coincident with the appearance of the Boeing 767 in commercial service is an upgrading of flight simulator technology to include a real-time programmable weather radar presentation. This augmentation, introduced on the Singer-Link flight simulation equipment used by American Airlines during late 1982, was developed by the US subsidiary of the UK company, Smiths Industries, to enhance the realism of training schedules.

Simulated weather patterns in colour are generated by a dedicated Digital Equipment Corporation PDP 11/24 computer that communicates with an SEL Computer Systems 32/77 host processor, the latter providing dynamic instrument readings, motion, sound and visual cues that reflect the weather situation.

The simulated weather displayed on the EFIS-700 electronic flight instrument system multi-purpose cathode ray tubes of the 767 can be moved across the screen or rotated, expanded or contracted, grown or decayed, according to the commands of the instructor at the console control. Turbulence experienced through the motion system, and the effects of windshear (a dangerous situation, reflected in rapidly decaying airspeed) can be keyed to the imminence of storms and poor weather. For aircraft without the EFIS-700 system the simulation facility can interface with a straightforward ARINC 708 weather radar. The modular high-order language used for the weather simulation facilitates the updating or customising of data and imposes minimal penalties of processing time and memory on the host computer. Interface with the simulator is a 16-bit direct memory channel between the weather system's PDP 11/24 and the high-speed data interface of the SEL 32/77 computer. With the facility offered by the direct link between the weather simulation and the motion system, improved training realism can accommodate such behaviour as the degradation of aircraft performance caused by heavy rainfall.

The system permits resolution of weather in terminal areas at 1000-foot intervals up to 5000 feet, and en route resolution at 5000-foot intervals up to 60 000 feet. Storm information is stored on disc memory in the PDP 11/24, and one of the main challenges has been the correlation of weather information with the simulator's visual database in the host computer. Despite these difficulties, several notable storms and windshear situations have been reproduced, including the environment at Kennedy Airport, New York, on 25 June 1975, when an Eastern Airlines Boeing 727 airliner crashed during final approach as the result of windshear.

STATUS: in production and service.

Sperry
Sperry Corporation, Systems Management Group, 12010 Sunrise Valley Drive, Reston, Virginia 22091
TELEPHONE: (703) 620 7000
TELEX: 620 7800
TWX: 710 833 9032

Sperry's simulation activities complement its large-scale and historical involvement in autopilots, flight control systems, and navigation and flight instruments, utilising in particular its extensive experience in mathematical modelling and real-time data processing. Apart from aircraft, Sperry builds simulators and training aids for military ships and vehicles.

Weapons systems trainer for EA-6B
Sperry's trainer for the Grumman EA-6B Prowler is the US Navy's largest and most technically advanced flight simulation device. It accurately reproduces the EA-6B's performance and operational characteristics, the environment within which it operates, and the inter-relationship of performance and environment. High fidelity is achieved by extensive simulation of radars, visual scenes, and radio communications. The system incorporates the most recent features of digital flight simulation technology, permitting it to cover the entire flight envelope and every variation of flying qualities, including stalls and spins.

For realism in its displays, the system includes a digital radar landmass simulation of search-radar returns, with a storage capacity of

Sperry weapons-system trainer for US Navy/ Grumman EA-6B

Instructor's station for Sperry's Grumman EA-6B simulator

reproduction of performance and flight characteristics throughout entire flight envelopes; they include buffet and rotor stall effects, autorotation, power settling, ground effect, ground resonance, and variable turbulence.

The simulators for the US Marine Corps CH-53D and -E assault helicopters comprise the complete cockpit mounted on a six-degrees-of-freedom motion platform, instructor station, and digital computation system. The cockpit module has a six-window, six-channel CGI visual system with 196° horizontal and 60° vertical field of view. Sperry has built three CH-53 simulators, the most recent being delivered in 1984.

1.6 million square miles. Terrain and cultural features that can be stored for display include shadows, refraction and earth curvature, far-shore brightening, range attenuation, moving targets, and emitter occulting. Electronic countermeasures simulation covers tactical and defensive jamming, communications jamming, and the effects of chaff dispensations.

The four-seat cockpit, fully representative of the EA-6B, is mounted on a six-degrees-of-freedom motion platform, and a computer-generated visual-image system displays outside-world scenes for both shore- and carrier-based operations. As many as three instructors can co-ordinate and control a training mission, integrating the operation of the aircraft with the military environment. The instructor station has five cathode ray tube graphics displays with interactive terminals, and repeat displays there permit the instructors to monitor the out-of-window scenes, radar and other displays.

Canberra-based software development and maintenance company, and Thorn-EMI, a leading Australian defence electronics supplier.

The systems being built for the Royal Australian Air Force are similar to those for the US Navy, of which the first was delivered in July 1983 and the second in February 1985. Six more are to follow.

Operational flight trainer for CH-53 helicopter

Sperry's experience in helicopter flight simulation includes the design, development and construction of training equipment for the Sikorsky CH-53, HH-3F, HH-53C and HH-52A, Bell UH-1E and TH-1L, and Boeing Vertol CH-47 helicopters. This work has resulted in the development of new mathematical modelling techniques for helicopter aerodynamics and engine operation. These mathematical models now permit complete and accurate

Operational flight trainer for T-45

To provide familiarisation in the units that will operate the 300-strong fleet of British Aerospace/McDonnell Douglas T-45 trainers, Sperry will supply 20 operational flight trainers and 10 instrument flight trainers. Instructor stations for both are being designed to maximise effective dialogue and transfer of skills between instructor and student. Computers will be used, as far as possible, to control the trainers, permitting instructors to concentrate on teaching. Both types of trainer will be linked to the central computerised training integration system that will evaluate the training scheme, recommend special routines, and compare the performance of individuals to that of a hypothetical 'average' student, for simulator post-flight assessment.

Simulator for HU-25A

Sperry has supplied the US Coast Guard with a flight simulator for the Dassault HU-25A Falcon Jets used by that service. It incorporates a high-fidelity flight-deck in which the instructor can sit either at a console, or between the two pilots, controlling the exercise via a hand-held keyboard. The system employs six-degrees-of-freedom motion base, and a computer-generated image visual system. Close attention was given to refining the control loading system to enhance the capability and realism of the

Operational flight trainer for F/A-18

The Sperry simulator of the McDonnell Douglas F/A-18 represents the systems and performance of the aircraft, the environment in which it operates, and the relationship between aircraft and carrier. Specifically, it represents the environment of the USS *Nimitz*, including approach, arrested landing, wave-off, bolter, touch-and-go, barricade arrestment, deck taxi, and catapult tensioning and launch.

The simulator comprises a representative F/A-18 cockpit, a three-window, three-channel visual system, and a facility for transmitting seat and buffet loads and shocks to the cockpit floor. A *g*-seat and *g*-suit system combined with seat-buffet generator provide acceleration cues. The instructor station can be operated individually or jointly by an instructor and a device operator, and can also accommodate an observer. Faults can be inserted or programmed from the crt keyboard, and the design can be modified and expanded in accordance with aircraft development and new training requirements.

In January 1983 the Australian Government contracted Sperry to produce two F/A-18 operational flight simulators at a cost of $22.5 million, along with supporting equipment and services, to support the fleet of 75 aircraft ordered by the Royal Australian Air Force. The first simulator was delivered in 1985 and the second will follow in 1987. A key factor in the choice of Sperry as contractor was the company's proposal for a significant work-sharing programme with the Australian aerospace industry. Major partners are C-3 Limited, a

Full flight simulator for US Marine Corps/Sikorsky CH-53 helicopter, with full colour visual system

primary and secondary flight controls, protect the integrity of the schedule, ease software development, and improve reliability.

The company has also built for the same customer a simulator for the Aérospatiale HH-65A Dauphin 2 helicopter.

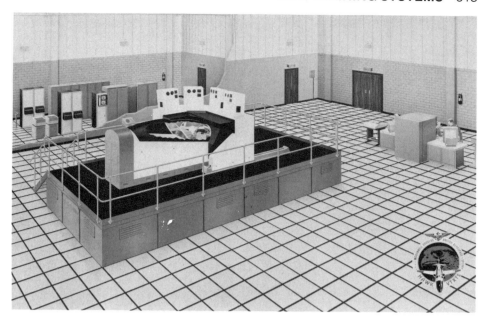

Impression of Sperry operational flight trainer for US Navy/British Aerospace T-45A (deck-landing version of UK's Hawk trainer)

US Coast Guard/Sperry simulators for Dassault HU-25A (foreground) and Aérospatiale HH-65A

ADDENDA

Radar

CANADA

MacDonald Dettwiler

MacDonald Dettwiler Technologies Limited, Airborne Radar Division, 3751 Shell Road, Richmond, British Columbia V6X 2Z9
TELEPHONE: (604) 278 3411
TELEX: 04-355599

IRIS integrated radar imaging system (see page 19)

This entry is supplemental to the information given in the main Radar section. MacDonald Dettwiler manufactures an airborne imaging radar reconnaissance system, IRIS. A major advantage afforded by the IRIS is real-time, tactical operation. The full resolution, full swath, on-board processing unique to the IRIS, enables the simultaneous viewing of reconnaissance images in the air and, via downlink, on the ground. In a battlefield or other active military situation, this capability allows the integration of the airborne reconnaissance system with a ground-based, tactical command and communications system to coordinate and control ground forces and air strike support. Intelligence derived from the IRIS reconnaissance images on the position of opposing forces, can be used by field commanders to direct ground operations and airborne support, within seconds of acquisition of the reconnaissance data.

The IRIS consists of three segments: an airborne segment, a transportable ground segment, and a precision analysis facility. The standard equipment for these segments is augmented by options for image display, storage and interpretation. Drawing on these options, the IRIS can be configured to suit precisely a variety of user requirements.

The airborne segment consists of an airborne imaging synthetic aperture radar with on-board digital processing, downlink transmission, high density magnetic tape data recording and image production capabilities. Images are processed in real-time and displayed on a video monitor and hardcopy paper strip. Two operational modes are provided with the capability of switching, in-flight, from a wide swath to a high resolution mode in seconds, providing resolutions of 18 × 18 metres and 3 × 3 metres at swath widths of 60 km and 10 km respectively. Images are produced from distances of up to 100 km and altitudes of up to 15 km, providing the maximum in stand-off range. The airborne segment has been packaged in compact, rugged modules easily configured on high performance executive-size turboprop and turbojet aircraft. In addition, independent processors are included for simultaneous fixed and moving target imaging. Moving and fixed targets can be displayed independently or superimposed in colour in a single image.

The transportable ground segment consists of a downlink receiver, digital data processor and tactical workstation. The tactical workstation is rugged, transportable and inexpensive – ideal for rapid deployment in critical reconnaissance areas. It features high density magnetic tape data recording, continuous paper strip printing, continuous image and frame image video display capabilities.

The precision analysis facility provides radar interpreters with the ability to analyse reconnaissance data received via downlink or recorded on board the aircraft and the transportable ground segment. The facility can archive and retrieve multiple data sets from the airborne and transportable segments, as well as maps and other interpretation aids entered at the facility. The precision analysis facility allows for in-depth interpretation of imagery, change detection, map mosaic-ing and preparation of precision hardcopy film products.

Four IRIS (and IRIS prototypes) are currently operational and the prototype IRIS has demonstrated very high reliability in over 4000 hours of operation in a difficult environment (the Canadian Arctic).

STATUS: in production and service.

NETHERLANDS

Hollandse Signaalapparaten

Hollandse Signaalapparaten BV, PO Box 42, Zuidelijke Havenweg 40, 7550 GD Hengelo
TELEPHONE: (74) 483 094
TELEX: 44310

Vesta airborne transponder

Vesta is a lightweight, compact transponder unit for use aboard helicopters, light aircraft or patrol boats. It enables accurate tracking of friendly helicopters or patrol boats on the radar display, even in heavy clutter environments. The transponder principle is based on a radar-triggered vhf reply. Operation is possible with any synchronised surveillance radar (either shipborne or shore-based) in the 1 to 10 GHz band.

For every radar pulse the transponder receives, a vhf reply pulse is transmitted, followed by a code pulse. Up to five helicopters can be identified by means of pre-selected codes (extension up to 32 is possible). The return signal is received and processed by the base installed Vesta receiver, which identifies and decodes the transponder reply. Unwanted vhf reply pulses are rejected by digital filters controlled by the allocated radar at the basis.

The transponder unit consists of a fully solid-state transponder, a control unit, two radar

Hollandse Signaalapparaten Vesta transponder system

pick-up antennas and a vhf transmitting antenna.

Two radar pick-up antennas are used to guarantee a combined sensitivity pattern which is virtually omni-directional. The control unit has only two switches, one for sensitivity selection and power on/off, and the other for code selection.

Transmitter frequency: (vhf) A band
Radar reception frequency: 1-10 GHz
Transmitter peakpower: 10W
Operational range: 0-32 n miles
Altitude: 18 000 ft
Pulse spacing: 2.2 μs nominal
Number of helicopter codes: 5 standard (optional extension up to 32)

UNITED STATES OF AMERICA

General Electric

General Electric Company, Aerospace Electronic Systems Department, French Road, Utica, New York 13503
TELEPHONE: (315) 793 7000

AN/APS-145 AEW radar

General Electric is developing an advanced version of the AN/APS-138 radar which is the prime sensor aboard the Grumman E-2C Hawkeye AEW aircraft. The new radar, designated AN/APS-145 was offered in spring 1986 by Lockheed as part of its proposals to sell the British Royal Air Force P-3 AEW aircraft to replace the troubled Nimrod AEW3. Grumman has also proposed using the APS-145 radar suite fitted into both E-2Cs and reconverted Nimrods for the Royal Air Force. In this latter guise the aircraft would have the above fuselage rotordome characteristic of the E-2 and P-3.

The APS-145 incorporates the modifications developed for the APS-138 and -139 radars (the latter enters service in 1988) and will be produced, and retrofitted to all US Navy E-2Cs, from 1990. The APS-145 development specifically addresses the problem of overland clutter – the radar is said to perform very well over sea and desert, but rapidly degrades as the terrain becomes more rugged. To reduce the false alarm rate a feature known as 'environmental processing' is being developed, which adjusts the sensitivity of the radar cell by cell, according to the clutter and traffic in each cell.

To enable large aircraft to be detected at ranges out to 350 nautical miles, a new lower prf is being introduced, and to match this development the E-2C's rotordome will be slowed from 6 to 5 rpm. A third prf is also being introduced with the APS-145, allowing the radar to operate with different prfs during scanning, eliminating problems with target 'blind speed' caused by single prf operation.

STATUS: in flight and ground test. Scheduled to enter service in 1990.

Hughes Aircraft

Hughes Aircraft Company, Radar Systems Group, PO Box 92426, Los Angeles, California 90009
TELEPHONE: (213) 648 2345
TWX: 910 348 6681

ASARS-2 advanced synthetic aperture system
(see Page 61)

ASARS-2, flown on the US Air Force's TR-1 aircraft, is a high-altitude, side-looking reconnaissance system that provides real-time, high-resolution radar ground maps in all weather and at ranges well in excess of electro-optical devices. Long-range radar mapping imagery is superior to photographic or optical techniques because the atmosphere absorbs more radiation in the visible and infra-red spectrums than in the frequencies used by radars. The Hughes Radar Systems Group developed the system under contract from the US Air Force's Aeronautical Systems Division which is located at Wright-Patterson Air Force Base in Ohio.

First production Hughes Aircraft ASARS-2 undergoes its final test

Electro-optics

GERMANY, FEDERAL REPUBLIC

MBB

Messerschmitt-Bölkow-Blohm GmbH, Postfach
80 11 60, D-8000 Munich 80
TELEPHONE: (89) 6000 0
TELEX: 528 7975

Reconnaissance pod

In late summer 1985 General Dynamics and MBB conducted a joint flight test evaluation of the MBB-developed Tornado reconnaissance pod on the US F-16. The programme consisted of three complete missions which were said to be a success.

Primarily designed and produced for the Tornado aircraft of the German Navy and the Italian Air Force, MBB's reconnaissance pod has a mission-effective sensor package, integral structure, good maintainability thanks to modular design, and a high growth-potential. The sensors, housed in the pod's centre module, are two Zeiss camera systems (LHOV and LLDC) and a Texas Instruments infra-red line-scanner (IRLS). The LHOV camera can be either focused from 300 metres to infinity during flight or be operated with fixed focus and with the different horizontal, oblique or vertical modes relative to the direction of flight. In addition, the camera also supplies stereoscopic pictures for direct interpretation.

The LLDC system produces a distortion-free imagery of the terrain from horizon to horizon at low-level during day/night missions. The Texas Instruments IRLS (RS-710) scans thermal sources under all light conditions and records them on film material. This sensor has a terrain coverage very similar to that of the optical LLDC system, which makes intelligence interpretation of targets easier by combining the imageries of the two sensors.

Apart from the image-generating sensors, the rear module of the MBB reconnaissance pod is equipped with Litef's reconnaissance interface system (RIS). The RIS controls the overall system fully automatically and co-ordinates the sensor and aircraft data. Geographical latitude and longitude of targets overflown, altitude, velocity, heading, time, etc are recorded by RIS as digital information on the film's 'data block'. The hybrid electronic package with an integrated microprocessor has built-in test equipment (BITE). In the pod's format module, the environmental control unit (ECU) is installed, which stabilises the sensors to an operating temperature of 35°C. With Tornado, the ECU runs with bleed air; in the case of the F-16 demonstrations, this could not be realised. Peripheric equipment of the Tornado recce pod consists of an operator's console in the rear cockpit, plus two camera visors for the pilots.

The MBB reconnaissance pod was tested on a F-16B, and General Dynamics made the necessary modifications, in particular wiring, and prepared the aircraft for the three planned reconnaissance missions (two day, one night). Main objective of the joint flight demonstration was to prove the compatibility of the F-16 with the pod, especially under the operationally important low-level high-speed conditions (approximately 200 feet AGL, 600 knots IAS).

General Dynamics F-16 equipped with MBB reconnaissance pod takes off for night-sortie

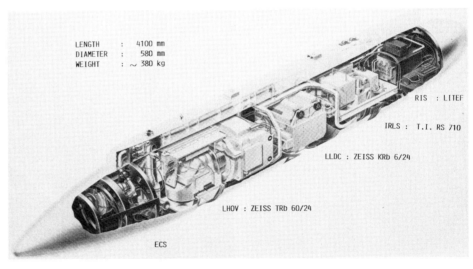

LENGTH	:	4100 mm
DIAMETER	:	580 mm
WEIGHT	:	~ 380 kg

RIS : LITEF
IRLS : T.I. RS 710
LLDC : ZEISS KRb 6/24
LHOV : ZEISS TRb 60/24
ECS

MBB reconnaissance pod

The F-16 flew the missions with a mission-typical stores configuration consisting of MBB's pod (centre line station 5), two Sidewinder AIM9P on wing-tip stations 1 and 9 plus two 370-gallon fuel tanks on inboard wing stations 4 and 6 during the second and third flights.

MBB claims that the camera systems and the IRLS worked extremely well and the imagery proved to be of excellent quality. The two day missions (second flight: up to 14 200 feet AGL, 400 to 600 knots IAS) focused primarily on military targets such as built-up areas, bridges, crossroads, oil tanks, and dams, which were recorded with all three imaging sensors. During the night flight (between 500 and 1000 feet AGL, 350 knots IAS), the cameras were used in addition to the IRLS which had priority for naval targets in the Gulf of Mexico area.

GTE Sylvania

GTE Sylvania Systems Group, Western Division, 100 Ferguson Drive, Mountain View, California 94042
TELEPHONE: (415) 966 9111

Have Lace programme (see page 92)

The Have Lace flight programme began in August 1985 and lasted until February 1986. Funded by the US Air Force's Satellite Communications Group in the Avionics Laboratory of the Wright-Patterson Laboratory, the experiment successfully demonstrated that air-to-air communications could be successfully made between two aircraft using a laser beam. Simple, low-powered infra-red laser diodes, an off the shelf gimballed head and inexpensive optics were used in this trial, which also showed the feasibility of using laser communications for high-speed, secure, jam-resistant data links. A total of 50 hours of flight-testing was undertaken, using two Boeing C-135 aircraft. Tests were carried out in various weather conditions, both overland and over water and at distances of up to 100 miles (160 km) between the two co-operating aircraft. Three months of

Boeing C-135 aircraft carrying Have Lace system

ground-to-air tests followed the air-to-air trials.

Results from the Have Lace programme will be carried forward into a follow-on project, due to start in 1988; this will develop prototype laser

communications equipment for specific applications such as data transfer between reconnaissance aircraft and airborne command posts.

Electronic warfare

UNITED KINGDOM

Plessey Avionics

Plessey Avionics, Martin Road, West Leigh, Havant, Hampshire PO9 5D
TELEPHONE: (0706) 486 391
TELEX: 86227

MAW missile approach warner

Plessey has developed a radar-based electronic warfare suite designed to equip the Royal Air Force's Tornado and Harrier aircraft. Reportedly using a small pulse-Doppler radar and weighing only 11 kg, the system is due to go on trial at the US Navy Weapons Center at China

Lake in late 1986 as part of a US evaluation. The radar is mounted in the tail of the aircraft and can be linked to a chaff dispenser for the automatic deployment of countermeasures.

STATUS: in development.

UNITED STATES OF AMERICA

Dalmo Victor

Dalmo Victor Operations, Bell Aerospace, Division of Textron Inc, 1515 Industrial Way, Belmont, California 94002
TELEPHONE: (415) 595 1414
TWX: 910 376 4400

Radar warning receivers for Korea
(see page 126)

In early 1986 Dalmo Victor won a 'multi-million' dollar contract to supply radar warning receivers to a military aircraft programme being undertaken in the Republic of Korea. No details of type, cost or quantity were available.

Data recording

UNITED KINGDOM

Racal Avionics

Burlington House, 118 Burlington Road, New Malden, Surrey KT3 4NR
TELEPHONE: (01) 942-2488
TELEX: 22891

Health Monitoring Enhancement for RAMS

In mid 1986 Racal Avionics and Stewart Hughes Limited signed an agreement to provide for the addition of a health and useage monitoring facility to the Racal Avionics Management System (RAMS).

Navigation

UNITED KINGDOM

British Aerospace

British Aerospace, Electronic Systems and Equipment Division, Downshire Way, Bracknell, Berkshire RG12 1QL
TELEPHONE: (0344) 483 222
TELEX: 848129

Beacon navigation system for RPVs

The British Aerospace Beacon navigation system has been developed under a Ministry of Defence contract to enable RPVs to fly designated routes over enemy territory while on surveillance missions. An airborne transmitter-receiver operates in conjunction with a series of ground beacons which have individual codes

and remain passive until interrogated by the airborne transmitter. The coded signals are then transmitted to the RPV for a time just long enough for it to establish the 'fix', or receive an instruction, minimising the chance of jamming or detection by the enemy.

STATUS: in development.

Radio communications

NETHERLANDS

Hollandse Signaalapparaten

Hollandse Signaalappareten BV, PO Box 42, Zuidelijke Havenweg 40, 7550 GD Hengelo
TELEPHONE: (74) 483 094
TELEX: 44310

Vesta-VC

The Vesta transponder receiver system (see separate entry) can easily be extended with a voice channel data link. This makes it suitable for data transfer and over the horizon targeting by using the helicopter.

The data link is established via an existing communications voice channel. The Vesta transponder switches the available radio in the helicopter from voice to data after which the data is transmitted via the voice channel in a frequency shift keyed signal.

An interface and extractor unit on the base station then converts this data to an 'own position' reference and interfaces with the radar displays, the data handling system (the weapon control system). Vesta is compatible with all airborne and base system interfaces.

STATUS: over 50 data airborne transponders have been delivered to several navies. They are fitted into Westland Lynx and Agusta AB 212 helicopters.

UNITED STATES OF AMERICA

Cubic

Cubic Corporation, Defense Systems Division, 9333 Balboa Avenue, San Diego, California 92123
TELEPHONE: (619) 277 6780
TWX: 910 335 2010

AN/URQ-34 anti-jam tactical data link

The AN/URQ-34 is an anti-jam tactical data link transmission system, operating in the Ku frequency band, and built to support US Army real-time combat sensors. It was originally developed for the stand-off target acquisition system (SOTAS), and was delivered and deployed for special projects after SOTAS was discontinued. A further development of the AN/URQ-34 was selected for production to late 1985 as the surveillance data link for the US Air Force managed Joint-Stars programme. The AN/URQ-34 uses fast frequency-hopping technology to defeat hostile ECM, and has a demonstrated range of 240 km. Input of digital data can be clear or encrypted and the data-rate is 25 to 100 K bytes nominal. It meets all applicable MIL and SOTAS standards for environment, nuclear hardening, ballistic hardening, mtbf and mttr. The mtbf is given as 480 hours (lower limit) and 1680 hours (upper limit). The AN/URQ-34 makes extensive use of built-in-test equipment giving fault location to module level.

Lear Siegler

Lear Siegler Inc, Avionic Systems Division, 7-11 Vreeland Road, Florham Park, New Jersey 07932
TELEPHONE: (201) 822 1300
TELEX: 136521
TWX: 710 986 8504

C-10382/A communication system control set

The primary function of the communication system control set (CSC) is to provide the pilot with integrated and centralised control, data transferring capability over the power switching, mode selection, operating frequencies, interconnections and signal-flow routes of the aircraft's CNI equipment. The CSC provides for highly efficient operations of communications by integrating these primary CNI systems controls into a single, convenient, easy-to-operate pilot's control panel.

On the US Navy/McDonnell-Douglas F/A-18A aircraft the controls and displays of the cockpit control panel are human-engineered to optimise pilot control of the CNI equipment. The control panel is positioned to allow the pilot to keep his eyes focused straight ahead, with only the pertinent information and controls he needs presented in his field of view.

A redundant MIL-STD-1553 multiplex bus provides connection between the CSC and the AN/AYK-14(V) mission computer for the flow of information and control. The mission computer provides CNI control signals, BIT commands and information for the control panel's alphanumeric display. In return, the CSC transmits equipment status, received CNI data, operating options, and BIT response to the mission computer. Dedicated serial digital lines interface the control panel with the CSC.

To process data to and from the CSC, mission computer, control panel and CNI

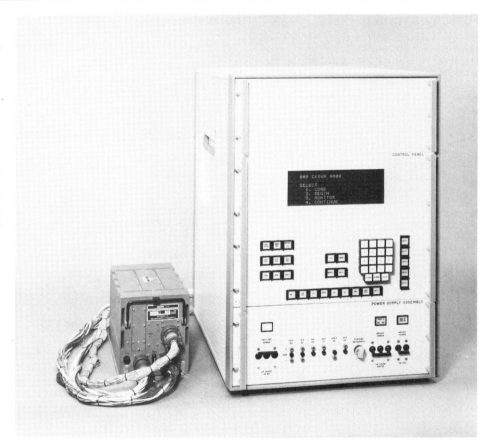

Lear Siegler C-10382/A communications system control set

equipment, the CSC interfaces serial digital signals, discrete signals, analogue signals, synchro signals, avionic multiplex bus signals and audio signals. Because the CSC micro-computer processes at least 1300 parameters per second, it controls and processes the data required, leaving 40 per cent of real-time available for growth.

M/A COM

M/A COM, Government Systems Division, 3033 Science Park Road, San Diego, California 92121
TELEPHONE: (619) 457 2340
TWX: 910 337 1277

OM-73 QPSK/BPSK modem set

The OM-73 modem set provides offset quadra-phase-shift keyed (QPSK) and bi-phase-shift keyed (BPSK) modulation and demodulation capability. The modem set is capable of providing digital communications at all data-rates from 16 bytes a second to 20 Mbps in a variety of configurations. Functions include operation of up to eight duplex or sixteen simplex links, all under control of a single master controller; forward error correction coding/decoding, tight spectral occupancy; and excellent bit error rate performance in the presence of tightly packed channels. The OM-73 represents state-of-the-art digital tech-nology. It is compatible with existing equipment for end-to-end operation over the DSCS II or III satellites, and it is operable with the MD-1002/G, KY801B, Spectrum Efficient Network (SENU), and many other systems.

The modem set consists of a mainframe tray and three modules; controller, transmitter and receiver. Each module is a stand-alone unit with integral power supply, backplane and display. The application of widespread vlsi technology to the design has yielded a much more compact and efficient modem than the suite it replaces.

STATUS: M/A COM is under contract with the US Army to deliver 236 modem sets.

Md-1035/A uhf dual modem
(see page 51)

Information has recently been released on upgrade work to the Md-1035/A, described in the main Radio Communications section, as follows.

The dual modem upgrade programme is extending the equipment capability further, to operate in a MILSTAR uhf network and to receive the SCT AFSAT I and II type downlinks. The upgrade will eliminate LES 8/9 functions and add such improvements as extension of TDM operations to additional channels on SDS satellites, and allowing EAM message termin-ation via either ETX or MRK termination characters. MILSTAR operation will be provided by the DMU's ability to process the uhf-AFSAT II Robust signalling modulation/demodulation and networking algorithms. The modem will be capable of transmitting and receiving on any one of the four TDM channels of the MILSTAR uhf communications sub-systems. The up-graded modem will receive modified versions of the AFSAT II signalling from the SDS and DSCS III SCTs and AFSATCOM I channel 1.5 from the DSCS III SCT. This programme includes development of a MILSTAR payload and command post simulator to allow DMU testing prior to the availability of the MILSTAT system.

Index

Elta Electronics
The Innovative Response

In our history of defense, the art of the unconventional, innovative response dates back a long way. Today, in defense electronics, Elta continues this tradition. Elta specializes in applying superior technology to specific operational problems — producing original solutions that help to overcome the enemy's quantitative advantages. Elta's professional performance record is based on more than two decades of combat-proven experience. A technological lead in electronics, backed by the aerospace expertise of Israel Aircraft Industries.

Elta provides a diversified range of cost-effective answers to multi-threat combat scenarios. Products. Systems. Technology transfers. Joint ventures. A solid professional reputation for battlefield reliability.

- ■ RADAR ■ EW ■ SIGINT ■ COMPUTERS
- ■ COMMUNICATIONS ■ SIGNAL PROCESSING

IAI **ELTA Electronics Industries Ltd**
a subsidiary of ISRAEL AIRCRAFT INDUSTRIES LTD / Electronics Division

Ashdod, Israel: P.O.Box 330 Ashdod 77102
Tel: (55)30333. Telex: 371874 ELTA IL. Cables: ELTASHDOD.
New York: Israel Aircraft Industries Inc.
Tel: (212)620-4410. Telex: 230-125180 ISRAIR.
Brussels: IAI European Marketing Office.
Tel: (2)5131455. Telex: 62718 ISRAVI B.

Printed and bound in Great Britain by Biddles Ltd, Guildford and King's Lynn